VOLUME **1**

Physics
for Scientists and Engineers

TENTH EDITION

Raymond A. Serway
Emeritus, James Madison University

John W. Jewett, Jr.
Emeritus, California State
Polytechnic University, Pomona

With contributions from Vahé Peroomian
University of Southern California

About the Cover
The cover shows a six-propeller drone carrying a pilot
cable almost 5 kilometers across the deep canyon through
which the Dadu River flows during the Xingkang Bridge
construction project in the Sichuan Province in China. This
method avoids the requirement to use boats on the fast-
flowing river or other methods such as manned helicopters
and small rockets. It also cuts the costs for laying the
cable to about 20% of that of traditional methods. Once
the pilot cable is laid, it can be used to pull heavier cables
across the gorge.

 CENGAGE

Australia • Brazil • Mexico • Singapore • United Kingdom • United States

Physics for Scientists and Engineers
Volume 1, **Tenth Edition**

Raymond A. Serway, John W. Jewett, Jr

Product Director: Dawn Giovanniello

Product Manager: Rebecca Berardy Schwartz

Content Developer: Ed Dodd

Product Assistant: Caitlyn Ghegan

Media Developer: Sheila Moran

Marketing Manager: Tom Ziolkowski

Content Project Manager: Tanya Nigh

Production Service: MPS Limited

Photo/Text Researcher: LDI

Art Director: Cate Barr

Cover/Text Designer: Shawn Girsberger

Cover and Title Page Image: Zhang Jian/Chengdu Economic Daily/VCG/Getty Images

Compositor: MPS Limited

For product information and technology assistance, contact us at
Cengage Customer & Sales Support, 1-800-354-9706.

For permission to use material from this text or product, submit all requests online at **www.cengage.com/permissions**.
Further permissions questions can be e-mailed to
permissionrequest@cengage.com.

Library of Congress Control Number: 2017953590

Student Edition:
ISBN: 978-1-337-55357-5

Loose-leaf Edition:
ISBN: 978-1-337-55359-9

Cengage
20 Channel Center Street
Boston, MA 02210
USA

Cengage is a leading provider of customized learning solutions with employees residing in nearly 40 different countries and sales in more than 125 countries around the world. Find your local representative at **www.cengage.com**.

Cengage products are represented in Canada by Nelson Education, Ltd.

To learn more about Cengage platforms and services, visit **www.cengage.com**. To register or access your online learning solution or purchase materials for your course, visit **www.cengagebrain.com**.

Printed in the United States of America
Print Number: 01   Print Year: 2017

We dedicate this book to our wives,
Elizabeth and Lisa,
and all our children and grandchildren
for their loving understanding
when we spent time on writing instead of being with them.

Brief Contents

Contents

Appendices

About the Authors

Raymond A. Serway received his doctorate at Illinois Institute of Technology and is Professor Emeritus at James Madison University. In 2011, he was awarded with an honorary doctorate degree from his alma mater, Utica College. He received the 1990 Madison Scholar Award at James Madison University, where he taught for 17 years. Dr. Serway began his teaching career at Clarkson University, where he conducted research and taught from 1967 to 1980. He was the recipient of the Distinguished Teaching Award at Clarkson University in 1977 and the Alumni Achievement Award from Utica College in 1985. As Guest Scientist at the IBM Research Laboratory in Zurich, Switzerland, he worked with K. Alex Müller, 1987 Nobel Prize recipient. Dr. Serway also was a visiting scientist at Argonne National Laboratory, where he collaborated with his mentor and friend, the late Dr. Sam Marshall. Dr. Serway is the coauthor of *College Physics,* Eleventh Edition; *Principles of Physics,* Fifth Edition; *Essentials of College Physics; Modern Physics,* Third Edition; and the high school textbook *Physics,* published by Holt McDougal. In addition, Dr. Serway has published more than 40 research papers in the field of condensed matter physics and has given more than 60 presentations at professional meetings. Dr. Serway and his wife, Elizabeth, enjoy traveling, playing golf, fishing, gardening, singing in the church choir, and especially spending quality time with their four children, ten grandchildren, and a recent great grandson.

John W. Jewett, Jr. earned his undergraduate degree in physics at Drexel University and his doctorate at Ohio State University, specializing in optical and magnetic properties of condensed matter. Dr. Jewett began his academic career at Stockton University, where he taught from 1974 to 1984. He is currently Emeritus Professor of Physics at California State Polytechnic University, Pomona. Through his teaching career, Dr. Jewett has been active in promoting effective physics education. In addition to receiving four National Science Foundation grants in physics education, he helped found and direct the Southern California Area Modern Physics Institute (SCAMPI) and Science IMPACT (Institute for Modern Pedagogy and Creative Teaching). Dr. Jewett's honors include the Stockton Merit Award at Stockton University in 1980, selection as Outstanding Professor at California State Polytechnic University for 1991–1992, and the Excellence in Undergraduate Physics Teaching Award from the American Association of Physics Teachers (AAPT) in 1998. In 2010, he received an Alumni Lifetime Achievement Award from Drexel University in recognition of his contributions in physics education. He has given more than 100 presentations both domestically and abroad, including multiple presentations at national meetings of the AAPT. He has also published 25 research papers in condensed matter physics and physics education research. Dr. Jewett is the author of *The World of Physics: Mysteries, Magic, and Myth,* which provides many connections between physics and everyday experiences. In addition to his work as the coauthor for *Physics for Scientists and Engineers,* he is also the coauthor on *Principles of Physics,* Fifth Edition, as well as *Global Issues,* a four-volume set of instruction manuals in integrated science for high school. Dr. Jewett enjoys playing keyboard with his all-physicist band, traveling, underwater photography, learning foreign languages, and collecting antique quack medical devices that can be used as demonstration apparatus in physics lectures. Most importantly, he relishes spending time with his wife, Lisa, and their children and grandchildren.

Preface

I n writing this Tenth Edition of *Physics for Scientists and Engineers,* **we continue** our ongoing efforts to improve the clarity of presentation and include new pedagogical features that help support the learning and teaching processes. Drawing on positive feedback from users of the Ninth Edition, data gathered from both professors and students who use WebAssign, as well as reviewers' suggestions, we have refined the text to better meet the needs of students and teachers.

This textbook is intended for a course in introductory physics for students majoring in science or engineering. The entire contents of the book in its extended version could be covered in a three-semester course, but it is possible to use the material in shorter sequences with the omission of selected chapters and sections. The mathematical background of the student taking this course should ideally include one semester of calculus. If that is not possible, the student should be enrolled in a concurrent course in introductory calculus.

Content

The material in this book covers fundamental topics in classical physics and provides an introduction to modern physics. The book is divided into six parts. Part 1 (Chapters 1 to 14) deals with the fundamentals of Newtonian mechanics and the physics of fluids; Part 2 (Chapters 15 to 17) covers oscillations, mechanical waves, and sound; Part 3 (Chapters 18 to 21) addresses heat and thermodynamics; Part 4 (Chapters 22 to 33) treats electricity and magnetism; Part 5 (Chapters 34 to 37) covers light and optics; and Part 6 (Chapters 38 to 44) deals with relativity and modern physics.

Objectives

This introductory physics textbook has three main objectives: to provide the student with a clear and logical presentation of the basic concepts and principles of physics, to strengthen an understanding of the concepts and principles through a broad range of interesting real-world applications, and to develop strong problem-solving skills through an effectively organized approach. To meet these objectives, we emphasize well-organized physical arguments and a focused problem-solving strategy. At the same time, we attempt to motivate the student through practical examples that demonstrate the role of physics in other disciplines, including engineering, chemistry, and medicine.

An Integrative Approach to Course Materials

This new edition takes an *integrative approach* to course material with an optimized, protected, online-only problem experience combined with rich textbook content designed to support an active classroom experience. This new optimized online homework set is built on contextual randomizations and answer-dependent student remediation for every problem. With this edition, you'll have an integrative approach that seamlessly matches curated content to the learning environment for which it was intended—from in-class group problem solving to online homework that utilizes targeted feedback. This approach engages and guides students where they are at—whether they are studying online or with the textbook.

Students often approach an online homework problem by googling to find the right equation or explanation of the relevant concept; however, this approach has eroded the value attributed to online homework as students leave the support of the program for unrelated help elsewhere and encounter imprecise information.

Students don't need to leave WebAssign to get help when they are stuck—each problem has feedback that addresses the misconception or error a student made to reach the wrong answer. Each optimized problem also features comprehensive written solutions, and many have supporting video solutions that go through one contextual variant of the problem one step at a time. Since the optimized problem set is not in print, the content is protected from "solution providers" and will be augmented every year with updates to the targeted feedback based on actual student answers.

Working in tandem with the optimized online homework, the printed textbook has been designed for an active learning experience that supports activities in the classroom as well as after-class practice and review. New content includes *Think–Pair–Share* activities, context-rich problems, and a greater emphasis on symbolic and conceptual problems. *All* of the printed textbook's problems will also be available to assign in WebAssign.

Changes in the Tenth Edition

A large number of changes and improvements were made for the Tenth Edition of this text. Some of the new features are based on our experiences and on current trends in science education. Other changes were incorporated in response to comments and suggestions offered by users of the Ninth Edition and by reviewers of the manuscript. The features listed here represent the major changes in the Tenth Edition.

WebAssign for *Physics for Scientists and Engineers*

WebAssign is a flexible and fully customizable online instructional solution that puts powerful tools in the hands of instructors, enabling you deploy assignments, instantly assess individual student and class performance, and help your students master the course concepts. With WebAssign's powerful digital platform and content specific to *Physics for Scientists and Engineers,* you can tailor your course with a wide range of assignment settings, add your own questions and content, and access student and course analytics and communication tools. WebAssign for *Physics for Scientists and Engineers* includes the following new features for this edition.

Optimized Problems. Only available online via WebAssign, this problem set combines new assessments with classic problems from *Physics for Scientists and Engineers* that have been optimized with just-in-time targeted feedback tailored to student responses and full student-focused solutions. Moving these problems so that they are only available online allows instructors to make full use of the capability of WebAssign to provide their students with dynamic assessment content, and reduces the opportunity for students to find online solutions through anti-search-engine optimizations. These problems reduce these opportunities both by making the text of the problem less searchable and by providing immediate assistance to students within the homework platform.

Interactive Video Vignettes (IVV) encourage students to address their alternate conceptions outside of the classroom and can be used for pre-lecture activities in traditional or even workshop physics classrooms. Interactive Video Vignettes include online video analysis and interactive individual tutorials to address learning difficulties identified by PER (Physics Education Research). Within the WebAssign platform there are additional conceptual questions immediately following each IVV in order to evaluate student engagement with the material and reinforce the message around these classic misconceptions. A screen shot from one of the Interactive Video Vignettes appears on the next page:

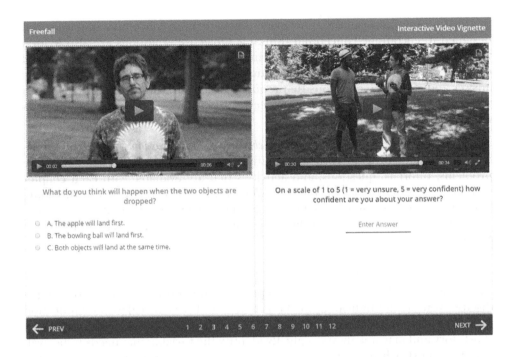

New MCAT-Style Passage Problem Modules. Available only in WebAssign, these 30 brand-new modules are modeled after the new MCAT exam's "passage problems." Each module starts with a text passage (often with accompanying photos/figures) followed by 5–6 multiple-choice questions. The passage and the questions are usually not confined to a single chapter, and feedback is available with each question.

New Life Science Problems. The online-only problems set for each chapter in WebAssign features two new life science problems that highlight the relevance of physics principles to those students taking the course who are majoring in one of the life sciences.

New What If? Problem Extensions. The online-only problems set for each chapter in WebAssign contains 6 new **What If? extensions** to existing problems. What If? extensions extend students' understanding of physics concepts beyond the simple act of arriving at a numerical result.

Pre-Lecture Explorations combine interactive simulations with conceptual and analytical questions that guide students to a deeper understanding and help promote a robust physical intuition.

An Expanded Offering of All-New Integrated Tutorials. These Integrated Tutorials strengthen students' problem-solving skills by guiding them through the steps in the book's problem-solving process, and include meaningful feedback at each step so students can practice the problem-solving process and improve their skills. The feedback also addresses student preconceptions and helps them to catch algebraic and other mathematical errors. Solutions are carried out symbolically as long as possible, with numerical values substituted at the end. This feature promotes conceptual understanding above memorization, helps students understand the effects of changing the values of each variable in the problem, avoids unnecessary repetitive substitution of the same numbers, and eliminates round-off errors.

Increased Number of Fully Worked-Out Problem Solutions. Hundreds of solutions have been newly added to online end-of-chapter problems. Solutions step through problem-solving strategies as they are applied to specific problems.

Objective and Conceptual Questions Now Exclusively Available in WebAssign. **Objective Questions** are multiple-choice, true/false, ranking, or other multiple-guess-type questions. Some require calculations designed to facilitate students' familiarity with the equations, the variables used, the concepts the variables represent, and the relationships between the concepts. Others are more conceptual in nature and designed to encourage conceptual thinking. Objective Questions are also written with the personal response system user in mind, and most of the questions could easily be used in these systems. **Conceptual Questions** are more traditional short-answer and essay-type questions that require students to think conceptually about a physical situation. More than 900 Objective and Conceptual Questions are available in WebAssign.

New Physics for Scientists and Engineers WebAssign Implementation Guide. The Implementation Guide provides instructors with occurrences of the different assignable problems, tutorials, questions, and activities that are available with each chapter of *Physics for Scientists and Engineers* in WebAssign. Instructors can use this manual when making decisions about which and how many assessment items to assign. To facilitate this, an overview of how the assignable items are integrated into the course is included.

New Assessment Items

New Context-Rich Problems. Context-rich problems (identified with a **CR** icon) always discuss "you" as the individual in the problem and have a real-world connection instead of discussing blocks on planes or balls on strings. They are structured like a short story and may not always explicitly identify the variable that needs to be evaluated. Context-rich problems may relate to the opening storyline of the chapter, might involve "expert witness" scenarios, which allow students to go beyond mathematical manipulation by designing an argument based on mathematical results, or ask for decisions to be made in real situations. Selected new context-rich problems will only appear online in WebAssign. An example of a new context-rich problem appears below:

> **20.** **CR** There is a 5K event coming up in your town. While talking to your grandmother, who uses an electric scooter for mobility, she says that she would like to accompany you on her scooter while you walk the 5.00-km distance. The manual that came with her scooter claims that the fully charged battery is capable of providing 120 Wh of energy before being depleted. In preparation for the race, you go for a "test drive": beginning with a fully charged battery, your grandmother rides beside you as you walk 5.00 km on flat ground. At the end of the walk, the battery usage indicator shows that 40.0% of the original energy in the battery remains. You also know that the combined weight of the scooter and your grandmother is 890 N. A few days later, filled with confidence that the battery has sufficient energy, you and your grandmother drive to the 5K event. Unbeknownst to you, the 5K route is not on flat ground, but is all uphill, ending at a point higher than the starting line. A race official tells you that the total amount of vertical displacement on the route is 150 m. Should your grandmother accompany you on the walk, or will she be stranded when her battery runs out of energy? Assume that the only difference between your test drive and the actual event is the vertical displacement.

New Think–Pair–Share Problems and Activities. Think–Pair–Share problems and activities are similar to context-rich problems, but tend to benefit more from group discussion because the solution is not as straightforward as for a single-concept problem. Some Think–Pair–Share problems require the group to discuss and make decisions; others are made more challenging by the fact that some information is not and cannot be known. All chapters in the text have at least one Think–Pair–Share problem or activity; several more per chapter will be available only in WebAssign. Examples of a Think–Pair–Share Problem and a Think–Pair–Share Activity appear on the next page:

1. You are working as a delivery person for a dairy store. In the back of your pickup truck is a crate of eggs. The dairy company has run out of bungee cords, so the crate is not tied down. You have been told to drive carefully because the coefficient of static friction between the crate and the bed of the truck is 0.600. You are not worried, because you are traveling on a road that appears perfectly straight. Due to your confidence and inattention, your speed has crept upward to 45.0 mi/h. Suddenly, you see a curve ahead with a warning sign saying, "Danger: unbanked curve with radius of curvature 35.0 m." You are 15.0 m from the beginning of the curve. What can you do to save the eggs: (i) take the curve at 45.0 mi/h, (ii) brake to a stop before entering the curve to think about it, or (iii) slow down to take the curve at a slower speed? Discuss these options in your group and determine if there is a best course of action.

3. **ACTIVITY** (a) Place ten pennies on a horizontal meterstick, with a penny at 10 cm, 20 cm, 30 cm, etc., out to 100 cm. Carefully pick up the meterstick, keeping it horizontal, and have a member of the group make a video recording of the following event, using a smartphone or other device. While the video recording is underway, release the 100-cm end of the meterstick while the 0-cm end rests on someone's finger or the edge of the desk. By stepping through the video images or watching the video in slow motion, determine which pennies first lose contact with the meterstick as it falls. (b) Make a theoretical determination of which pennies should first lose contact and compare to your experimental result.

Content Changes

Reorganized Chapter 16 (Wave Motion). This combination of Chapters 16 and 17 from the last edition brings all of the fundamental material on traveling mechanical waves on strings and sound waves through materials together in one chapter. This allows for more close comparisons between the features of the two types of waves that are similar, such as derivations of the speed of the wave. The section on reflection and transmission of waves, details of which are not necessary in a chapter on traveling waves, was moved into Chapter 17 (Superposition and Standing Waves) for this edition, where it fits more naturally in a discussion of the effects of boundary conditions on waves.

Reorganization of Chapters 22–24. Movement of the material on continuous distribution of charge out of Chapter 22 (Electric Fields) to Chapter 23 (Continuous Charge Distributions and Gauss's Law) results in a chapter that is a more gradual introduction for students into the new and challenging topic of electricity. The chapter now involves only electric fields due to point charges and uniform electric fields due to parallel plates.

Chapter 23 previously involved only the analysis of electric fields due to continuous charge distributions using Gauss's law. Movement of the material on continuous distribution of charge into Chapter 23 results in an entire chapter based on the analysis of fields from continuous charge distributions, using two techniques: integration and Gauss's law.

Chapter 23 previously contained a discussion of four properties of isolated charged conductors. Three of the properties were discussed and argued from basic principles, while the student was referred to necessary material in the next chapter (on Electric Potential) for a discussion of the fourth property. With the movement of this discussion into Chapter 24 for this edition, the student has learned all of the necessary basic material *before* the discussion of properties of isolated charged conductors, and all four properties can be argued from basic principles together.

Reorganized Chapter 43 (Nuclear Physics). Chapters 44 (Nuclear Structure) and 45 (Applications of Nuclear Physics) in the last edition have been combined in this edition. This new Chapter 43 allows all of the material on nuclear physics to be studied together. As a consequence, we now have a series of the final five chapters of the text that each cover in one chapter focused applications of the fundamental principles studied before: Chapter 40 (Quantum Mechanics), Chapter 41 (Atomic Physics), Chapter 42 (Molecules and Solids), Chapter 43 (Nuclear Physics), and Chapter 44 (Particle Physics).

New Storyline Approach to Chapter-Opening Text. Each chapter opens with a *Storyline* section. This feature provides a continuous storyline through the whole book of "you" as an inquisitive physics student observing and analyzing phenomena seen in everyday life. Many chapters' Storyline involves measurements made with a smartphone, observations of YouTube videos, or investigations on the Internet.

New Chapter-Opening **Connections.** The start of each chapter also features a *Connections* section that shows how the material in the chapter connects to previously studied material and to future material. The Connections section provides a "big picture" of the concepts, explains why this chapter is placed in this particular location relative to the other chapters, and shows how the structure of physics builds on previous material.

Text Features

Most instructors believe that the textbook selected for a course should be the student's primary guide for understanding and learning the subject matter. Furthermore, the textbook should be easily accessible and should be styled and written to facilitate instruction and learning. With these points in mind, we have included many pedagogical features, listed below, that are intended to enhance its usefulness to both students and instructors.

Problem Solving and Conceptual Understanding

Analysis Model Approach to Problem Solving. Students are faced with hundreds of problems during their physics courses. A relatively small number of fundamental principles form the basis of these problems. When faced with a new problem, a physicist forms a model of the problem that can be solved in a simple way by identifying the fundamental principle that is applicable in the problem. For example, many problems involve conservation of energy, Newton's second law, or kinematic equations. Because the physicist has studied these principles and their applications extensively, he or she can apply this knowledge as a model for solving a new problem. Although it would be ideal for students to follow this same process, most students have difficulty becoming familiar with the entire palette of fundamental principles that are available. It is easier for students to identify a situation rather than a fundamental principle.

The *Analysis Model approach* lays out a standard set of situations that appear in most physics problems. These situations are based on an entity in one of four simplification models: *particle, system, rigid object,* and *wave*. Once the simplification model is identified, the student thinks about what the entity is doing or how it interacts with its environment. This leads the student to identify a particular Analysis Model for the problem. For example, if an object is falling, the object is recognized as a particle experiencing an acceleration due to gravity that is constant. The student has learned that the Analysis Model of a *particle under constant acceleration* describes this situation. Furthermore, this model has a small number of equations associated with it for use in starting problems, the kinematic equations presented in Chapter 2. Therefore, an understanding of the situation has led to an Analysis Model, which then identifies a very small number of equations to start the problem, rather than the myriad equations that students see in the text. In this way, the use of Analysis Models leads the student to identify the fundamental principle. As the student gains more experience, he or she will lean less on the Analysis Model approach and begin to identify fundamental principles directly.

The Analysis Model Approach to Problem Solving is presented in full in Chapter 2 (Section 2.4, pages 30–32), and provides students with a structured process for solving problems. In all remaining chapters, the strategy is employed explicitly in every example so that students learn how it is applied. Students are encouraged to follow this strategy when working end-of-chapter problems.

Analysis Model descriptive boxes appear at the end of any section that introduces a new Analysis Model. This feature recaps the Analysis Model introduced in the section and provides examples of the types of problems that a student could solve using the Analysis Model. These boxes function as a "refresher" before students see the Analysis Models in use in the worked examples for a given section. The approach is further reinforced in the end-of-chapter summary under the heading *Analysis Models for Problem Solving,* and through the **Analysis Model Tutorials** that are based on selected end-of-chapter problems and appear in WebAssign.

Analysis Model Tutorials. John Jewett developed 165 tutorials (ones that appear in the printed text's problem sets are indicated by an AMT icon) that strengthen students' problem-solving skills by guiding them through the steps in the problem-solving process. Important first steps include making predictions and focusing on physics concepts before solving the problem quantitatively. A critical component of these tutorials is the selection of an appropriate Analysis Model to describe what is going on in the problem. This step allows students to make the important link between the situation in the problem and the mathematical representation of the situation. Analysis Model tutorials include meaningful feedback at each step to help students practice the problem-solving process and improve their skills. In addition, the feedback addresses student misconceptions and helps them to catch algebraic and other mathematical errors. Solutions are carried out symbolically as long as possible, with numerical values substituted at the end. This feature helps students understand the effects of changing the values of each variable in the problem, avoids unnecessary repetitive substitution of the same numbers, and eliminates round-off errors. Feedback at the end of the tutorial encourages students to compare the final answer with their original predictions.

Worked Examples. All in-text worked examples are presented in a two-column format to better reinforce physical concepts. The left column shows textual information that describes the steps for solving the problem. The right column shows the mathematical manipulations and results of taking these steps. This layout facilitates matching the concept with its mathematical execution and helps students organize their work. The examples closely follow the Analysis Model Approach to Problem Solving introduced in Section 2.4 to reinforce effective problem-solving habits. All worked examples in the text may be assigned for homework in WebAssign. A sample of a worked example can be found on the next page.

Examples consist of two types. The first (and most common) example type presents a problem and numerical answer. The second type of example is conceptual in nature. To accommodate increased emphasis on understanding physical concepts, the many conceptual examples are labeled as such and are designed to help students focus on the physical situation in the problem. Solutions in worked examples are presented symbolically as far as possible, with numerical values substituted at the end. This approach will help students think symbolically when they solve problems instead of unnecessarily inserting numbers into intermediate equations.

What If? Approximately one-third of the worked examples in the text contain a What If? feature. At the completion of the example solution, a What If? question offers a variation on the situation posed in the text of the example. This feature encourages students to think about the results of the example, and it also assists in conceptual understanding of the principles. What If? questions also prepare students to encounter novel problems that may be included on exams. Selected end-of-chapter problems also include this feature.

Quick Quizzes. Students are provided an opportunity to test their understanding of the physical concepts presented through Quick Quizzes. The questions require students to make decisions on the basis of sound reasoning, and some of the questions have been written to help students overcome common misconceptions. Quick Quizzes have been cast in an objective format, including multiple-choice, true–false, and ranking. Answers to all Quick Quiz questions are found at the end of the text. Many instructors choose to use such questions in a "peer instruction" teaching style or with the use of personal response system "clickers," but they can be used in standard quiz format as well. An example of a Quick Quiz follows below.

QUICK QUIZ 7.5 A dart is inserted into a spring-loaded dart gun by pushing the spring in by a distance x. For the next loading, the spring is compressed a distance $2x$. How much faster does the second dart leave the gun compared with the first? **(a)** four times as fast **(b)** two times as fast **(c)** the same **(d)** half as fast **(e)** one-fourth as fast

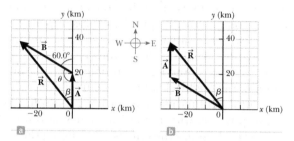

All worked examples are also available to be assigned as interactive examples in WebAssign.

Example 3.2 | A Vacation Trip

A car travels 20.0 km due north and then 35.0 km in a direction 60.0° west of north as shown in Figure 3.11a. Find the magnitude and direction of the car's resultant displacement.

> *Each solution has been written to closely follow the Analysis Model Approach to Problem Solving as outlined in Section 2.4 (pages 30–32), so as to reinforce good problem-solving habits.*

SOLUTION

Conceptualize The two vectors $\vec{A}$ and $\vec{B}$ that appear in Figure 3.11a help us conceptualize the problem. The resultant vector $\vec{R}$ has also been drawn. We expect its magnitude to be a few tens of kilometers. The angle β that the resultant vector makes with the y axis is expected to be less than 60°, the angle that vector $\vec{B}$ makes with the y axis.

Figure 3.11 (Example 3.2) (a) Graphical method for finding the resultant displacement vector $\vec{R} = \vec{A} + \vec{B}$. (b) Adding the vectors in reverse order $(\vec{B} + \vec{A})$ gives the same result for $\vec{R}$.

Categorize We can categorize this example as a simple analysis problem in vector addition. The displacement $\vec{R}$ is the resultant when the two individual displacements $\vec{A}$ and $\vec{B}$ are added. We can further categorize it as a problem about the analysis of triangles, so we appeal to our expertise in geometry and trigonometry.

Analyze In this example, we show two ways to analyze the problem of finding the resultant of two vectors. The first way is to solve the problem geometrically, using graph paper and a protractor to measure the magnitude of $\vec{R}$ and its direction in Figure 3.11a. (In fact, even when you know you are going to be carrying out a calculation, you should sketch the vectors to check your results.) With an ordinary ruler and protractor, a large diagram typically gives answers to two-digit but not to three-digit precision. Try using these tools on $\vec{R}$ in Figure 3.11a and compare to the trigonometric analysis below!

The second way to solve the problem is to analyze it using algebra and trigonometry. The magnitude of $\vec{R}$ can be obtained from the law of cosines as applied to the triangle in Figure 3.11a (see Appendix B.4).

> *Each step of the solution is detailed in a two-column format. The left column provides an explanation for each mathematical step in the right column, to better reinforce the physical concepts.*

Use $R^2 = A^2 + B^2 - 2AB \cos \theta$ from the law of cosines to find R:

$$R = \sqrt{A^2 + B^2 - 2AB \cos \theta}$$

Substitute numerical values, noting that $\theta = 180° - 60° = 120°$:

$$R = \sqrt{(20.0 \text{ km})^2 + (35.0 \text{ km})^2 - 2(20.0 \text{ km})(35.0 \text{ km}) \cos 120°}$$
$$= 48.2 \text{ km}$$

Use the law of sines (Appendix B.4) to find the direction of $\vec{R}$ measured from the northerly direction:

$$\frac{\sin \beta}{B} = \frac{\sin \theta}{R}$$

$$\sin \beta = \frac{B}{R} \sin \theta = \frac{35.0 \text{ km}}{48.2 \text{ km}} \sin 120° = 0.629$$

$$\beta = 38.9°$$

The resultant displacement of the car is 48.2 km in a direction 38.9° west of north.

Finalize Does the angle β that we calculated agree with an estimate made by looking at Figure 3.11a or with an actual angle measured from the diagram using the graphical method? Is it reasonable that the magnitude of $\vec{R}$ is larger than that of both $\vec{A}$ and $\vec{B}$? Are the units of $\vec{R}$ correct?

Although the head to tail method of adding vectors works well, it suffers from two disadvantages. First, some people find using the laws of cosines and sines to be awkward. Second, a triangle only results if you are adding two vectors. If you are adding three or more vectors, the resulting geometric shape is usually not a triangle. In Section 3.4, we explore a new method of adding vectors that will address both of these disadvantages.

WHAT IF? Suppose the trip were taken with the two vectors in reverse order: 35.0 km at 60.0° west of north first and then 20.0 km due north. How would the magnitude and the direction of the resultant vector change?

Answer They would not change. The commutative law for vector addition tells us that the order of vectors in an addition is irrelevant. Graphically, Figure 3.11b shows that the vectors added in the reverse order give us the same resultant vector.

> *What If? statements appear in about one-third of the worked examples and offer a variation on the situation posed in the text of the example. For instance, this feature might explore the effects of changing the conditions of the situation, determine what happens when a quantity is taken to a particular limiting value, or question whether additional information can be determined about the problem situation. This feature encourages students to think about the results of the example and assists in conceptual understanding of the principles.*

Pitfall Preventions. More than two hundred Pitfall Preventions (such as the one to the right) are provided to help students avoid common mistakes and misunderstandings. These features, which are placed in the margins of the text, address both common student misconceptions and situations in which students often follow unproductive paths.

Summaries. Each chapter contains a summary that reviews the important concepts and equations discussed in that chapter. The summary is divided into three sections: Definitions, Concepts and Principles, and Analysis Models for Problem Solving. In each section, flash card–type boxes focus on each separate definition, concept, principle, or analysis model.

Problems Sets. For the Tenth Edition, the authors reviewed each question and problem and incorporated revisions designed to improve both readability and assignability.

Problems. An extensive set of problems is included at the end of each chapter; in all, the printed textbook contains more than 2 000 problems, while another 1 500 optimized problems are available only in WebAssign. Answers for odd-numbered problems in the printed text are provided at the end of the book, and solutions for all printed text problems are found in the *Instructor's Solutions Manual.*

The end-of-chapter problems are organized by the sections in each chapter (about two-thirds of the problems are keyed to specific sections of the chapter). Within each section, the problems now "platform" students to higher-order thinking by presenting all the straightforward problems in the section first, followed by the intermediate problems. (The problem numbers for straightforward problems are printed in **black;** intermediate-level problems are in blue.) The *Additional Problems* section contains problems that are not keyed to specific sections. At the end of each chapter is the *Challenge Problems* section, which gathers the most difficult problems for a given chapter in one place. (Challenge Problems have problem numbers marked in red.)

There are several kinds of problems featured in this text:

V *Watch It* video solutions available in WebAssign explain fundamental problem-solving strategies to help students step through selected problems.

Q|C *Quantitative/Conceptual problems* contain parts that ask students to think both quantitatively and conceptually. An example of a Quantitative/Conceptual problem appears here:

> The problem is identified with a **Q|C** icon.

> Parts (a)–(c) of the problem ask for quantitative calculations.

35. **Q|C** A horizontal spring attached to a wall has a force constant of $k = 850$ N/m. A block of mass $m = 1.00$ kg is attached to the spring and rests on a frictionless, horizontal surface as in Figure P8.35. (a) The block is pulled to a position $x_i = 6.00$ cm from equilibrium and released. Find the elastic potential energy stored in the spring when the block is 6.00 cm from equilibrium and when the block passes through equilibrium. (b) Find the speed of the block as it passes through the equilibrium point. (c) What is the speed of the block when it is at a position $x_i/2 = 3.00$ cm? (d) Why isn't the answer to part (c) half the answer to part (b)?

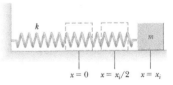

$x = 0$ $x = x_i/2$ $x = x_i$

Figure P8.35

> Part (d) asks a conceptual question about the situation.

S *Symbolic problems* ask students to solve a problem using only symbolic manipulation. Reviewers of the Ninth Edition (as well as the majority of respondents to a large survey) asked specifically for an increase in the number of symbolic problems found in the text because it better reflects the way instructors want their students to think when solving physics problems. An example of a Symbolic problem appears on the next page:

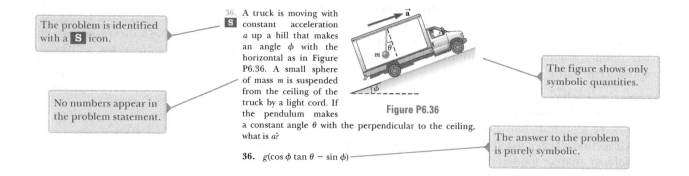

The problem is identified with a **S** icon.

No numbers appear in the problem statement.

36. A truck is moving with constant acceleration a up a hill that makes an angle ϕ with the horizontal as in Figure P6.36. A small sphere of mass m is suspended from the ceiling of the truck by a light cord. If the pendulum makes a constant angle θ with the perpendicular to the ceiling, what is a?

Figure P6.36

The figure shows only symbolic quantities.

36. $g(\cos \phi \tan \theta - \sin \phi)$

The answer to the problem is purely symbolic.

GP *Guided Problems* help students break problems into steps. A physics problem typically asks for one physical quantity in a given context. Often, however, several concepts must be used and a number of calculations are required to obtain that final answer. Many students are not accustomed to this level of complexity and often don't know where to start. A Guided Problem breaks a standard problem into smaller steps, enabling students to grasp all the concepts and strategies required to arrive at a correct solution. Unlike standard physics problems, guidance is often built into the problem statement. Guided Problems are reminiscent of how a student might interact with a professor in an office visit. These problems (there is one in every chapter of the text) help train students to break down complex problems into a series of simpler problems, an essential problem-solving skill. An example of a Guided Problem appears here:

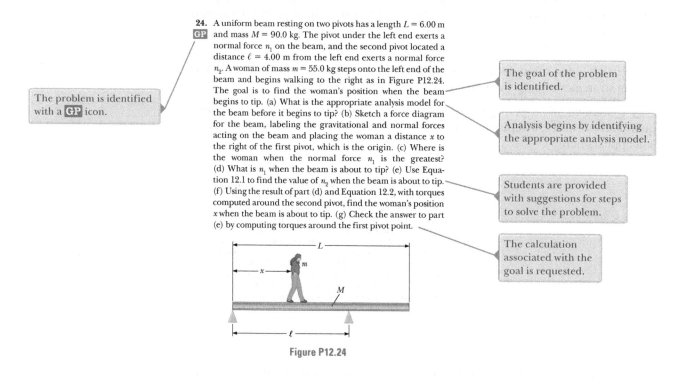

The problem is identified with a **GP** icon.

24. A uniform beam resting on two pivots has a length $L = 6.00$ m and mass $M = 90.0$ kg. The pivot under the left end exerts a normal force n_1 on the beam, and the second pivot located a distance $\ell = 4.00$ m from the left end exerts a normal force n_2. A woman of mass $m = 55.0$ kg steps onto the left end of the beam and begins walking to the right as in Figure P12.24. The goal is to find the woman's position when the beam begins to tip. (a) What is the appropriate analysis model for the beam before it begins to tip? (b) Sketch a force diagram for the beam, labeling the gravitational and normal forces acting on the beam and placing the woman a distance x to the right of the first pivot, which is the origin. (c) Where is the woman when the normal force n_1 is the greatest? (d) What is n_1 when the beam is about to tip? (e) Use Equation 12.1 to find the value of n_2 when the beam is about to tip. (f) Using the result of part (d) and Equation 12.2, with torques computed around the second pivot, find the woman's position x when the beam is about to tip. (g) Check the answer to part (e) by computing torques around the first pivot point.

The goal of the problem is identified.

Analysis begins by identifying the appropriate analysis model.

Students are provided with suggestions for steps to solve the problem.

The calculation associated with the goal is requested.

Figure P12.24

Biomedical problems. These problems (indicated with a **BIO** icon) highlight the relevance of physics principles to those students taking this course who are majoring in one of the life sciences.

T *Master It Tutorials* available in WebAssign help students solve problems by having them work through a stepped-out solution.

Impossibility problems. Physics education research has focused heavily on the problem-solving skills of students. Although most problems in this text are structured in the form of providing data and asking for a result of computation, two problems in each chapter, on average, are structured as impossibility problems. They begin with the phrase *Why is the following situation impossible?* That is followed by the description of a situation. The striking aspect of these problems is that no question is asked of the students, other than that in the initial italics. The student must determine what questions need to be asked and what calculations need to be performed. Based on the results of these calculations, the student must determine why the situation described is not possible. This determination may require information from personal experience, common sense, Internet or print research, measurement, mathematical skills, knowledge of human norms, or scientific thinking.

These problems can be assigned to build critical thinking skills in students. They are also fun, having the aspect of physics "mysteries" to be solved by students individually or in groups. An example of an impossibility problem appears here:

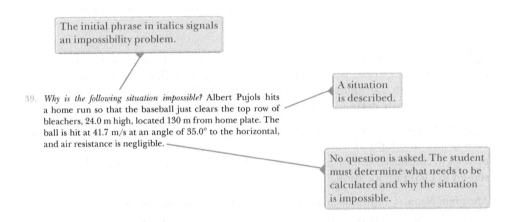

The initial phrase in italics signals an impossibility problem.

39. *Why is the following situation impossible?* Albert Pujols hits a home run so that the baseball just clears the top row of bleachers, 24.0 m high, located 130 m from home plate. The ball is hit at 41.7 m/s at an angle of 35.0° to the horizontal, and air resistance is negligible.

A situation is described.

No question is asked. The student must determine what needs to be calculated and why the situation is impossible.

Paired problems. These problems are otherwise identical, one asking for a numerical solution and one asking for a symbolic derivation. There is at least one pair of these problems in most chapters, indicated by cyan shading in the end-of-chapter problems set.

Review problems. Many chapters include review problems requiring the student to combine concepts covered in the chapter with those discussed in previous chapters. These problems (marked **Review**) reflect the cohesive nature of the principles in the text and verify that physics is not a scattered set of ideas. When facing a real-world issue such as global warming or nuclear weapons, it may be necessary to call on ideas in physics from several parts of a textbook such as this one.

"Fermi problems." One or more problems in most chapters ask the student to reason in order-of-magnitude terms.

Design problems. Several chapters contain problems that ask the student to determine design parameters for a practical device so that it can function as required.

Calculus-based problems. Every chapter contains at least one problem applying ideas and methods from differential calculus and one problem using integral calculus.

Artwork. Every piece of artwork in the Tenth Edition is in a modern style that helps express the physics principles at work in a clear and precise fashion. *Focus pointers* are included with many figures in the text; these either point out important aspects of a figure or guide students through a process illustrated by the artwork or photo. This format helps those students who are more visual learners. An example of a figure with a focus pointer appears on the next page.

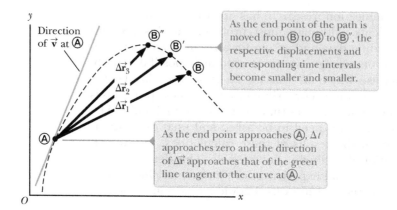

As the end point of the path is moved from ⒷⓉto Ⓑ' to Ⓑ", the respective displacements and corresponding time intervals become smaller and smaller.

As the end point approaches Ⓐ, Δt approaches zero and the direction of $\Delta\vec{r}$ approaches that of the green line tangent to the curve at Ⓐ.

Figure 4.2 As a particle moves between two points, its average velocity is in the direction of the displacement vector $\Delta\vec{r}$. By definition, the instantaneous velocity at Ⓐ is directed along the line tangent to the curve at Ⓐ.

Math Appendix. The math appendix (Appendix B), a valuable tool for students, shows the math tools in a physics context. This resource is ideal for students who need a quick review on topics such as algebra, trigonometry, and calculus.

Helpful Features

Style. To facilitate rapid comprehension, we have written the book in a clear, logical, and engaging style. We have chosen a writing style that is somewhat informal and relaxed so that students will find the text appealing and enjoyable to read. New terms are carefully defined, and we have avoided the use of jargon.

Important Definitions and Equations. Most important definitions are set in **boldface** or are highlighted with a background screen for added emphasis and ease of review. Similarly, important equations are also highlighted with a background screen to facilitate location.

Marginal Notes. Comments and notes appearing in the margin with a ▶ icon can be used to locate important statements, equations, and concepts in the text.

Pedagogical Use of Color. Readers should consult the **pedagogical color chart** (inside the front cover) for a listing of the color-coded symbols used in the text diagrams. This system is followed consistently throughout the text.

Mathematical Level. We have introduced calculus gradually, keeping in mind that students often take introductory courses in calculus and physics concurrently. Most steps are shown when basic equations are developed, and reference is often made to mathematical appendices near the end of the textbook. Although vectors are discussed in detail in Chapter 3, vector products are introduced later in the text, where they are needed in physical applications. The dot product is introduced in Chapter 7, which addresses energy of a system; the cross product is introduced in Chapter 11, which deals with angular momentum.

Significant Figures. In both worked examples and end-of-chapter problems, significant figures have been handled with care. Most numerical examples are worked to either two or three significant figures, depending on the precision of the data provided. End-of-chapter problems regularly state data and answers to three-digit precision. When carrying out estimation calculations, we shall typically work with a single significant figure. (More discussion of significant figures can be found in Chapter 1, pages 13–15.)

Units. The international system of units (SI) is used throughout the text. The U.S. customary system of units is used only to a limited extent in the chapters on mechanics and thermodynamics.

Appendices and Endpapers. Several appendices are provided near the end of the textbook. Most of the appendix material represents a review of mathematical concepts and techniques used in the text, including scientific notation, algebra, geometry, trigonometry, differential calculus, and integral calculus. Reference to these appendices is made throughout the text. Most mathematical review sections in the appendices include worked examples and exercises with answers. In addition to the mathematical reviews, the appendices contain tables of physical data, conversion factors, and the SI units of physical quantities as well as a periodic table of the elements. Other useful information—fundamental constants and physical data, planetary data, a list of standard prefixes, mathematical symbols, the Greek alphabet, and standard abbreviations of units of measure—appears on the endpapers.

Course Solutions That Fit Your Teaching Goals and Your Students' Learning Needs

Recent advances in educational technology have made homework management systems and audience response systems powerful and affordable tools to enhance the way you teach your course. Whether you offer a more traditional text-based course, are interested in using or are currently using an online homework management system such as WebAssign, or are ready to turn your lecture into an interactive learning environment, you can be confident that the text's proven content provides the foundation for each and every component of our technology and ancillary package.

Lecture Presentation Resources

Cengage Learning Testing Powered by Cognero is a flexible, online system that allows you to author, edit, and manage test bank content from multiple Cengage Learning solutions, create multiple test versions in an instant, and deliver tests from your LMS, your classroom, or wherever you want.

Instructor Resource Website for **Serway/Jewett** *Physics for Scientists and Engineers, Tenth Edition.* The Instructor Resource Website contains a variety of resources to aid you in preparing and presenting text material in a manner that meets your personal preferences and course needs. The posted *Instructor's Solutions Manual* presents complete worked solutions for all of the printed textbook's end-of-chapter problems and answers for all even-numbered problems. Robust PowerPoint lecture outlines that have been designed for an active classroom are available, with reading check questions and Think–Pair–Share questions as well as the traditional section-by-section outline. Images from the textbook can be used to customize your own presentations. Available online via www.cengage.com/login.

CengageBrain.com

To register or access your online learning solution or purchase materials for your course, visit **www.cengagebrain.com**.

CENGAGE **brain**

Student Resources

Physics Laboratory Manual, Fourth Edition by David Loyd (Angelo State University) Ideal for use with any introductory physics text, Loyd's *Physics Laboratory Manual* is suitable for either calculus- or algebra/trigonometry-based physics courses. Designed to help students demonstrate a physical principle and teach techniques of careful measurement, Loyd's *Physics Laboratory Manual* also emphasizes conceptual understanding and includes a thorough discussion of physical theory to help students see the connection between the lab and the lecture. Many labs give students hands-on experience with statistical analysis, and now five computer-assisted data-entry labs are included in the printed manual. The fourth edition maintains the minimum equipment requirements to allow for maximum flexibility and to make the most of preexisting lab equipment. For instructors interested in using some of Loyd's experiments, a customized lab manual is another option available through the Cengage Learning Custom Solutions program. Now, you can select specific experiments from Loyd's *Physics Laboratory Manual*, include your own original lab experiments, and create one affordable bound book. Contact your Cengage Learning representative for more information on our Custom Solutions program. Available with InfoTrac® Student Collections http://gocengage.com/infotrac.

Physics Laboratory Experiments, Eighth Edition by Jerry D. Wilson (Lander College) and Cecilia A. Hernández (American River College). This market-leading manual for the first-year physics laboratory course offers a wide range of class-tested experiments designed specifically for use in small to midsize lab programs. A series of integrated experiments emphasizes the use of computerized instrumentation and includes a set of "computer-assisted experiments" to allow students and instructors to gain experience with modern equipment. It also lets instructors determine the appropriate balance of traditional versus computer-based experiments for their courses. By analyzing data through two different methods, students gain a greater understanding of the concepts behind the experiments. The Eighth Edition is updated with 4 new economical labs to accommodate shrinking department budgets and 30 new Pre-Lab Demonstrations, designed to capture students' interest prior to the lab and requiring only widely available materials and items.

Teaching Options

The topics in this textbook are presented in the following sequence: classical mechanics, oscillations and mechanical waves, and heat and thermodynamics, followed by electricity and magnetism, electromagnetic waves, optics, relativity, and modern physics. This presentation represents a traditional sequence, with the subject of mechanical waves being presented before electricity and magnetism. Some instructors may prefer to discuss both mechanical and electromagnetic waves together after completing electricity and magnetism. In this case, Chapters 16 and 17 could be covered along with Chapter 33. The chapter on relativity is placed near the end of the text because this topic often is treated as an introduction to the era of "modern physics." If time permits, instructors may choose to cover Chapter 38 after completing Chapter 13 as a conclusion to the material on Newtonian mechanics. For those instructors teaching a two-semester sequence, some sections and chapters could be deleted without any loss of continuity. The following sections can be considered optional for this purpose:

2.9	Kinematic Equations Derived from Calculus	**28.6**	The Hall Effect
4.6	Relative Velocity and Relative Acceleration	**29.6**	Magnetism in Matter
6.3	Motion in Accelerated Frames	**30.6**	Eddy Currents
6.4	Motion in the Presence of Resistive Forces	**33.6**	Production of Electromagnetic Waves by an Antenna
7.9	Energy Diagrams and Equilibrium of a System	**35.5**	Lens Aberrations
9.9	Rocket Propulsion	**35.6**	Optical Instruments
11.5	The Motion of Gyroscopes and Tops	**37.5**	Diffraction of X-Rays by Crystals
14.8	Other Applications of Fluid Dynamics	**38.9**	The General Theory of Relativity
15.6	Damped Oscillations	**40.6**	Applications of Tunneling
15.7	Forced Oscillations	**41.9**	Spontaneous and Stimulated Transitions
17.8	Nonsinusoidal Waveforms	**41.10**	Lasers
25.7	An Atomic Description of Dielectrics	**42.7**	Semiconductor Devices
26.5	Superconductors	**43.11**	Radiation Damage
27.5	Household Wiring and Electrical Safety	**43.12**	Uses of Radiation from the Nucleus
28.3	Applications Involving Charged Particles Moving in a Magnetic Field	**43.13**	Nuclear Magnetic Resonance and Magnetic Resonance Imaging

Acknowledgments

This Tenth Edition of *Physics for Scientists and Engineers* was prepared with the guidance and assistance of many professors who reviewed selections of the manuscript, the prerevision text, or both. We wish to acknowledge the following scholars and express our sincere appreciation for their suggestions, criticisms, and encouragement:

John F. DiTusa, *Louisiana State University;* Hani Dulli, *Texas Tech University;* Eric Hudson, *Pennsylvania State University;* David Joffe, *Kennesaw State University;* Yibin Pan, *University of Wisconsin-Madison;* Mark Rzchowski, *University of Wisconsin-Madison;* Joseph Scanio, *University of Cincinnati;* Brian Utter, *Bucknell University*

During our work on this revision, we worked with development partners to help us determine the course of the revision; we would like to thank them:

Tom Barrett, *The Ohio State University;* Ken Bolland, *The Ohio State University;* Colleen Countryman, *North Carolina State University;* Dawn Hollenbeck, *Rochester Institute of Technology;* Kathleen Koenig, *University of Cincinnati;* David Lamp, *Texas Tech University;* Rafael Lopez-Mobilia, *The University of Texas at San Antonio;* Yibin Pan, *University of Wisconsin-Madison;* Chandralekha Singh, *University of Pittsburgh;* Michael Thackston, *Kennesaw State University;* Michael Ziegler, *The Ohio State University*

Prior to our work on this revision, we conducted a survey of professors; their feedback and suggestions helped shape this revision, and so we would like to thank the survey participants:

Steve Alexander, *Southwestern University;* Sanjeev Arora, *Fort Valley State University;* Erik Aver, *Gonzaga University;* David Berube, *Loyola Marymount University;* Muhammad Bhatti, *The University of Texas Rio Grande Valley;* Jeffrey Bierman, *Gonzaga University;* Ken Bolland, *The Ohio State University;* John Bulman, *Loyola Marymount University;* Hani Dulli, *Texas Tech University;* Eric Hudson, *Pennsylvania State University;* Satyanaraya Kachiraju, *The University of Texas Rio Grande Valley;* Brent McDaniel, *Kennesaw State University;* Lisa Paulius, *Western Michigan University;* Linh Pham, *University of San Diego;* Charles Ruggiero, *The Ohio State University at Marion;* Mackay Salley, *Wofford College;* Jeff Sanny, *Loyola Marymount University;* Joseph Scanio, *University of Cincinnati;* Jeffrey Schwartz, *University of California at Los Angeles;* Amit Sharma, *Wright State University;* Mark Spraker, *University of North Georgia;* Anthony Teate, *James Madison University;* Lih-Sin The, *Clemson University*

This title was carefully checked for accuracy by Michael Faux, *SUNY College at Oneonta.* We thank him for his diligent efforts under schedule pressure.

Belal Abas, Zinoviy Akkerman, Eric Boyd, Hal Falk, Melanie Martin, Steve McCauley, and Glenn Stracher made corrections to problems taken from previous editions. Harvey Leff provided invaluable guidance on the restructuring of the discussion of entropy in Chapter 21. We are grateful to Vahé Peroomian for preparing an excellent *Instructor's Solutions Manual.* Matt Kohlmyer (Senior Instructional Content Developer at Cengage), Marllin L. Simon (Professor Emeritus, *Auburn University*), and Susan English all worked hard on the optimization of the online-only homework problems, and we thank them. Linnea Cookson provided an excellent accuracy check of the Analysis Model tutorials.

Special thanks and recognition go to the professional staff at Cengage—in particular, Rebecca Berardy Schwartz, Michael Jacobs, Ed Dodd, Tanya Nigh, Teresa Trego, Lorreen Towle, Tom Ziolkowski, Cate Barr, and Caitlin Ghegan—for their fine work during the development, production, and promotion of this textbook. We recognize the skilled production service and excellent artwork provided by Ed Dionne and the staff at MPS Limited and the dedicated image and text research of Cheryl Du Bois and Ragu Veeraragavan, respectively.

Finally, we are deeply indebted to our wives, children, and grandchildren for their love, support, and long-term sacrifices.

Raymond A. Serway
St. Petersburg, Florida

John W. Jewett, Jr.
Anaheim, California

To the Student

t is appropriate to offer some words of advice that should be of benefit to you, the student. Before doing so, we assume you have read the Preface, which describes the various features of the text and support materials that will help you through the course.

How to Study

Instructors are often asked, "How should I study physics and prepare for examinations?" There is no simple answer to this question, but we can offer some suggestions based on our own experiences in learning and teaching over the years.

First and foremost, maintain a positive attitude toward the subject matter, keeping in mind that physics is the most fundamental of all natural sciences. Other science courses that follow will use the same physical principles, so it is important that you understand and are able to apply the various concepts and theories discussed in the text.

Concepts and Principles

It is essential that you understand the basic concepts and principles before attempting to solve assigned problems. You can best accomplish this goal by carefully reading the textbook before you attend your lecture on the covered material. When reading the text, you should jot down those points that are not clear to you. Also be sure to make a diligent attempt at answering the questions in the Quick Quizzes as you come to them in your reading. We have worked hard to prepare questions that help you judge for yourself how well you understand the material. Study the **What If?** features that appear in many of the worked examples carefully. They will help you extend your understanding beyond the simple act of arriving at a numerical result. The Pitfall Preventions will also help guide you away from common misunderstandings about physics. During class, take careful notes and ask questions about those ideas that are unclear to you. Keep in mind that few people are able to absorb the full meaning of scientific material after only one reading; several readings of the text and your notes may be necessary. Your lectures and laboratory work supplement the textbook and should clarify some of the more difficult material. You should minimize your memorization of material. Successful memorization of passages from the text, equations, and derivations does not necessarily indicate that you understand the material. Your understanding of the material will be enhanced through a combination of efficient study habits, discussions with other students and with instructors, and your ability to solve the problems presented in the textbook. Ask questions whenever you believe that clarification of a concept is necessary.

Study Schedule

It is important that you set up a regular study schedule, preferably a daily one. Make sure that you read the syllabus for the course and adhere to the schedule set by your instructor. The lectures will make much more sense if you read the corresponding text material *before* attending them. As a general rule, you should devote about two hours of study time for each hour you are in class. If you are having trouble with the course, seek the advice of the instructor or other students who have taken the course. You may find it necessary to seek further instruction from experienced students. Very often, instructors offer review sessions in addition to regular class periods. Avoid the practice of delaying study until a day or two before an exam.

More often than not, this approach has disastrous results. Rather than undertake an all-night study session before a test, briefly review the basic concepts and equations, and then get a good night's rest.

You can purchase any Cengage Learning product at your local college store or at our preferred online store **CengageBrain.com.**

Use the Features

You should make full use of the various features of the text discussed in the Preface. For example, marginal notes are useful for locating and describing important equations and concepts, and **boldface** indicates important definitions. Many useful tables are contained in the appendices, but most are incorporated in the text where they are most often referenced. Appendix B is a convenient review of mathematical tools used in the text.

Answers to Quick Quizzes and odd-numbered problems are given at the end of the textbook. The table of contents provides an overview of the entire text, and the index enables you to locate specific material quickly. Footnotes are sometimes used to supplement the text or to cite other references on the subject discussed.

After reading a chapter, you should be able to define any new quantities introduced in that chapter and discuss the principles and assumptions that were used to arrive at certain key relations. In some cases, you may find it necessary to refer to the textbook's index to locate certain topics. You should be able to associate with each physical quantity the correct symbol used to represent that quantity and the unit in which the quantity is specified. Furthermore, you should be able to express each important equation in concise and accurate prose.

Problem Solving

R. P. Feynman, Nobel laureate in physics, once said, "You do not know anything until you have practiced." In keeping with this statement, we strongly advise you to develop the skills necessary to solve a wide range of problems. Your ability to solve problems will be one of the main tests of your knowledge of physics; therefore, you should try to solve as many problems as possible. It is essential that you understand basic concepts and principles before attempting to solve problems. It is good practice to try to find alternate solutions to the same problem. For example, you can solve problems in mechanics using Newton's laws, but very often an alternative method that draws on energy considerations is more direct. You should not deceive yourself into thinking that you understand a problem merely because you have seen it solved in class. You must be able to solve the problem and similar problems on your own.

The approach to solving problems should be carefully planned. A systematic plan is especially important when a problem involves several concepts. First, read the problem several times until you are confident you understand what is being asked. Look for any key words that will help you interpret the problem and perhaps allow you to make certain assumptions. Your ability to interpret a question properly is an integral part of problem solving. Second, you should acquire the habit of writing down the information given in a problem and those quantities that need to be found; for example, you might construct a table listing both the quantities given and the quantities to be found. This procedure is sometimes used in the worked examples of the textbook. Finally, after you have decided on the method you believe is appropriate for a given problem, proceed with your solution. The Analysis Model Approach to Problem Solving will guide you through complex problems. If you follow the steps of this procedure *(Conceptualize, Categorize, Analyze, Finalize),* you will find it easier to come up with a solution and gain more from your efforts. This strategy, located in Section 2.4 (pages 30–32), is used in all worked examples in the remaining chapters so that you can learn how to apply it. Specific problem-solving strategies for certain types of situations are included in the text and appear with a special heading. These specific strategies follow the outline of the Analysis Model Approach to Problem Solving.

Often, students fail to recognize the limitations of certain equations or physical laws in a particular situation. It is very important that you understand and remember the assumptions that underlie a particular theory or formalism. For example, certain equations in kinematics apply only to a particle moving with constant acceleration. These equations are not valid for describing motion whose acceleration is not constant, such as the motion of an object connected to a spring or the motion of an object through a fluid. Study the Analysis Models for Problem Solving in the chapter summaries carefully so that you know how each model can be applied to a specific situation. The analysis models provide you with a logical structure for solving problems and help you develop your thinking skills to become more like those of a physicist. Use the analysis model approach to save you hours of looking for the correct equation and to make you a faster and more efficient problem solver.

Experiments

Physics is a science based on experimental observations. Therefore, we recommend that you try to supplement the text by performing various types of "hands-on" experiments either at home or in the laboratory. These experiments can be used to test ideas and models discussed in class or in the textbook. For example, the common Slinky toy is excellent for studying traveling waves, a ball swinging on the end of a long string can be used to investigate pendulum motion, various masses attached to the end of a vertical spring or rubber band can be used to determine its elastic nature, an old pair of polarized sunglasses and some discarded lenses and a magnifying glass are the components of various experiments in optics, and an approximate measure of the free-fall acceleration can be determined simply by measuring with a stopwatch the time interval required for a ball to drop from a known height. The list of such experiments is endless. When physical models are not available, be imaginative and try to develop models of your own.

New Media

If available, we strongly encourage you to use the **WebAssign** product that is available with this textbook. It is far easier to understand physics if you see it in action, and the materials available in WebAssign will enable you to become a part of that action.

It is our sincere hope that you will find physics an exciting and enjoyable experience and that you will benefit from this experience, regardless of your chosen profession. Welcome to the exciting world of physics!

The scientist does not study nature because it is useful; he studies it because he delights in it, and he delights in it because it is beautiful. If nature were not beautiful, it would not be worth knowing, and if nature were not worth knowing, life would not be worth living.

—Henri Poincaré

Physics

for Scientists and Engineers

TENTH EDITION

PART 1

Mechanics

Physics, the most fundamental physical science, is concerned with the fundamental principles of the Universe. It is the foundation upon which the other sciences—astronomy, biology, chemistry, and geology—are based. It is also the basis of a large number of engineering applications. The beauty of physics lies in the simplicity of its fundamental principles and in the manner in which just a small number of concepts and models can alter and expand our view of the world around us.

The study of physics can be divided into six main areas:

1. *classical mechanics,* concerning the motion of objects that are large relative to atoms and move at speeds much slower than the speed of light
2. *relativity,* a theory describing objects moving at any speed, even speeds approaching the speed of light
3. *thermodynamics,* dealing with heat, temperature, and the statistical behavior of systems with large numbers of particles
4. *electromagnetism,* concerning electricity, magnetism, and electro-magnetic fields
5. *optics,* the study of the behavior of light and its interaction with materials
6. *quantum mechanics,* a collection of theories connecting the behavior of matter at the submicroscopic level to macroscopic observations

The disciplines of mechanics and electromagnetism are basic to all other branches of classical physics (developed before 1900) and modern physics (c. 1900–present). The first part of this textbook deals with classical mechanics, sometimes referred to as *Newtonian mechanics* or simply *mechanics.* Many principles and models used to understand mechanical systems retain their importance in the theories of other areas of physics and can later be used to describe many natural phenomena. Therefore, classical mechanics is of vital importance to students from all disciplines. ■

The Toyota Mirai, a fuel-cell-powered automobile available to the public, albeit in limited quantities. A fuel cell converts hydrogen fuel into electricity to drive the motor attached to the wheels of the car. Automobiles, whether powered by fuel cells, gasoline engines, or batteries, use many of the concepts and principles of mechanics that we will study in this first part of the book. Quantities that we can use to describe the operation of vehicles include position, velocity, acceleration, force, energy, and momentum. (*Chris Graythen/Getty Images Sport/Getty Images*)

1

Physics and Measurement

Stonehenge, in southern England, was built thousands of years ago. Various hypotheses have been proposed about its function, including a burial ground, a healing site, and a place for ancestor worship. One of the more intriguing ideas suggests that Stonehenge was an observatory, allowing measurements of some of the quantities discussed in this chapter, such as position of objects in space and time intervals between repeating celestial events. (*Image copyright Stephen Inglis. Used under license from Shutterstock.com*)

STORYLINE **Each chapter in this textbook will begin with a paragraph** related to a storyline that runs throughout the text. The storyline centers on *you*: an inquisitive physics student. You could live anywhere in the world, but let's say you live in southern California, where one of the authors lives. Most of your observations will occur there, although you will take trips to other locations. As you go through your everyday activities, you see physics in action all around you. In fact, you can't get away from physics! As you observe phenomena at the beginning of each chapter, you will ask yourself, "Why does that happen?" You might take measurements with your smartphone. You might look for related videos on YouTube or photographs on an image search site. You are lucky indeed because, in addition to those resources, you have this textbook and the expertise of your instructor to help you understand the exciting physics surrounding you. Let's look at your first observations as we begin your storyline. You have just bought this textbook and have flipped through some of its pages. You notice a page of conversions on the inside back cover. You notice in the entries under "Length" the unit of a *light-year*. You say, "Wait a minute! (You will say this often in the upcoming chapters.) How can a unit based on a *year* be a unit of *length*?" As you look farther down the page, you see 1 kg ≈ 2.2 lb (lb is the abbreviation for *pound*; lb is from Latin *libra pondo*) under the heading "Some Approximations Useful for Estimation Problems." Noticing the "approximately equal" sign (≈), you wonder what the *exact* conversion is and look upward on the page to the heading "Mass," since a kilogram is a unit of mass. The relation between kilograms and pounds is not there! Why not? Your physics adventure has begun!

CONNECTIONS The second paragraph in each chapter will explain how the material in the chapter connects to that in the previous chapter and/or future chapters. This feature will help you see that the textbook is not a collection of unrelated chapters, but rather is a structure of understanding that we are building, step by step. These paragraphs will provide a roadmap through the concepts and principles as they are introduced in the text. They will justify why the material in that chapter is presented at that time and help you to see the "big picture" of the study of physics. In this first chapter, of course, we cannot connect to a previous chapter. We will simply look ahead to the present chapter, in which we discuss some preliminary concepts of measurement, units, modeling, and estimation that we will need throughout *all* the chapters of the text.

1.1 Standards of Length, Mass, and Time

To describe natural phenomena, we must make measurements of various aspects of nature. Each measurement is associated with a physical quantity, such as the length of an object. The laws of physics are expressed as mathematical relationships among physical quantities that we will introduce and discuss throughout the book. In mechanics, the three fundamental quantities are *length, mass,* and *time.* All other quantities in mechanics can be expressed in terms of these three.

If we are to report the results of a measurement to someone who wishes to reproduce this measurement, a *standard* must be defined. For example, if someone reports that a wall is 2 meters high and our standard unit of length is defined to be 1 meter, we know that the height of the wall is twice our basic length unit. Whatever is chosen as a standard must be readily accessible and must possess some property that can be measured reliably. Measurement standards used by different people in different places—throughout the Universe—must yield the same result. In addition, standards used for measurements must not change with time.

In 1960, an international committee established a set of standards for the fundamental quantities of science. It is called the **SI** (Système International), and its fundamental units of length, mass, and time are the *meter, kilogram,* and *second,* respectively. Other standards for SI fundamental units established by the committee are those for temperature (the *kelvin*), electric current (the *ampere*), luminous intensity (the *candela*), and the amount of substance (the *mole*).

Length

We can identify **length** as the distance between two points in space. In 1120, the king of England decreed that the standard of length in his country would be named the *yard* and would be precisely equal to the distance from the tip of his nose to the end of his outstretched arm. Similarly, the original standard for the foot adopted by the French was the length of the royal foot of King Louis XIV. Neither of these standards is constant in time; when a new king took the throne, length measurements changed! The French standard prevailed until 1799, when the legal standard of length in France became the **meter** (m), defined as one ten-millionth of the distance from the equator to the North Pole along one particular longitudinal line that passes through Paris. Notice that this value is an Earth-based standard that does not satisfy the requirement that it can be used throughout the Universe.

Table 1.1 (page 4) lists approximate values of some measured lengths. You should study this table as well as the next two tables and begin to generate an intuition for what is meant by, for example, a length of 20 centimeters, a mass of 100 kilograms, or a time interval of 3.2×10^7 seconds.

As recently as 1960, the length of the meter was defined as the distance between two lines on a specific platinum–iridium bar stored under controlled conditions in France. Current requirements of science and technology, however, necessitate more accuracy than that with which the separation between the lines on the bar can be determined. In the 1960s and 1970s, the meter was defined to be equal to

PITFALL PREVENTION 1.1
Reasonable Values Generating intuition about typical values of quantities when solving problems is important because you must think about your end result and determine if it seems reasonable. For example, if you are calculating the mass of a housefly and arrive at a value of 100 kg, this answer is *unreasonable* and there is an error somewhere.

TABLE 1.1   Approximate Values of Some Measured Lengths

	Length (m)
Distance from the Earth to the most remote known quasar	2.7×10^{26}
Distance from the Earth to the most remote normal galaxies	3×10^{26}
Distance from the Earth to the nearest large galaxy (Andromeda)	2×10^{22}
Distance from the Sun to the nearest star (Proxima Centauri)	4×10^{16}
One light-year	9.46×10^{15}
Mean orbit radius of the Earth about the Sun	1.50×10^{11}
Mean distance from the Earth to the Moon	3.84×10^{8}
Distance from the equator to the North Pole	1.00×10^{7}
Mean radius of the Earth	6.37×10^{6}
Typical altitude (above the surface) of a satellite orbiting the Earth	2×10^{5}
Length of a football field	9.1×10^{1}
Length of a housefly	5×10^{-3}
Size of smallest dust particles	$\sim 10^{-4}$
Size of cells of most living organisms	$\sim 10^{-5}$
Diameter of a hydrogen atom	$\sim 10^{-10}$
Diameter of an atomic nucleus	$\sim 10^{-14}$
Diameter of a proton	$\sim 10^{-15}$

Figure 1.1 (a) International Prototype of the Kilogram, an accurate copy of the International Standard Kilogram kept at Sèvres, France, is housed under a double bell jar in a vault at the National Institute of Standards and Technology. (b) A cesium fountain atomic clock. The clock will neither gain nor lose a second in 20 million years.

1 650 763.73 wavelengths[1] of orange-red light emitted from a krypton-86 lamp. In October 1983, however, the meter was redefined as **the distance traveled by light in vacuum during a time interval of 1/299 792 458 second.** In effect, this latest definition establishes that the speed of light in vacuum is precisely 299 792 458 meters per second. This definition of the meter is valid throughout the Universe based on our assumption that light is the same everywhere. The speed of light also allows us to define the **light-year**, as mentioned in the introductory storyline: the distance that light travels through empty space in one year. Use this definition and the speed of light to verify the length of a light-year in meters as given in Table 1.1.

Mass

We will find that the **mass** of an object is related to the amount of material that is present in the object, or to how much that object resists changes in its motion. Mass is an inherent property of an object and is independent of the object's surroundings and of the method used to measure it. The SI fundamental unit of mass, the **kilogram** (kg), is defined as **the mass of a specific platinum–iridium alloy cylinder kept at the International Bureau of Weights and Measures at Sèvres, France.** This mass standard was established in 1887 and has not been changed since that time because platinum–iridium is an unusually stable alloy. A duplicate of the Sèvres cylinder is kept at the National Institute of Standards and Technology (NIST) in Gaithersburg, Maryland (Fig. 1.1a). Table 1.2 lists approximate values of the masses of various objects.

In Chapter 5, we will discuss the difference between mass and weight. In anticipation of that discussion, let's look again at the approximate equivalence mentioned in the introductory storyline: 1 kg ≈ 2.2 lb. It would never be correct to claim that a number of kilograms *equals* a number of pounds, because these units represent different variables. A kilogram is a unit of *mass*, while a pound is a unit of *weight*. That's why an equality between kilograms and pounds is not given in the section of conversions for mass on the inside back cover of the textbook.

---

[1] We will use the standard international notation for numbers with more than three digits, in which groups of three digits are separated by spaces rather than commas. Therefore, 10 000 is the same as the common American notation of 10,000. Similarly, $\pi = 3.14159265$ is written as 3.141 592 65.

TABLE 1.2 Approximate Masses of Various Objects

	Mass (kg)
Observable Universe	$\sim 10^{52}$
Milky Way galaxy	$\sim 10^{42}$
Sun	1.99×10^{30}
Earth	5.98×10^{24}
Moon	7.36×10^{22}
Shark	$\sim 10^{3}$
Human	$\sim 10^{2}$
Frog	$\sim 10^{-1}$
Mosquito	$\sim 10^{-5}$
Bacterium	$\sim 1 \times 10^{-15}$
Hydrogen atom	1.67×10^{-27}
Electron	9.11×10^{-31}

TABLE 1.3 Approximate Values of Some Time Intervals

	Time Interval (s)
Age of the Universe	4×10^{17}
Age of the Earth	1.3×10^{17}
Average age of a college student	6.3×10^{8}
One year	3.2×10^{7}
One day	8.6×10^{4}
One class period	3.0×10^{3}
Time interval between normal heartbeats	8×10^{-1}
Period of audible sound waves	$\sim 10^{-3}$
Period of typical radio waves	$\sim 10^{-6}$
Period of vibration of an atom in a solid	$\sim 10^{-13}$
Period of visible light waves	$\sim 10^{-15}$
Duration of a nuclear collision	$\sim 10^{-22}$
Time interval for light to cross a proton	$\sim 10^{-24}$

Time

Before 1967, the standard of **time** was defined in terms of the *mean solar day.* (A solar day is the time interval between successive appearances of the Sun at the highest point it reaches in the sky each day.) The fundamental unit of a **second** (s) was defined as $(\frac{1}{60})(\frac{1}{60})(\frac{1}{24})$ of a mean solar day. This definition is based on the rotation of one planet, the Earth. Therefore, this motion does not provide a time standard that is universal.

In 1967, the second was redefined to take advantage of the high precision attainable in a device known as an *atomic clock* (Fig. 1.1b), which measures vibrations of cesium atoms. One second is now defined as **9 192 631 770 times the period of vibration of radiation from the cesium-133 atom.**[2] Approximate values of time intervals are presented in Table 1.3.

You should note that we will use the notations *time* and *time interval* differently. A **time** is a description of an instant relative to a reference time. For example, $t = 10.0$ s refers to an instant 10.0 s after the instant we have identified as $t = 0$. As another example, a *time* of 11:30 a.m. means an instant 11.5 hours after our reference time of midnight. On the other hand, a **time interval** refers to *duration*: he required 30.0 minutes to finish the task. It is common to hear a "time of 30.0 minutes" in this latter example, but we will be careful to refer to measurements of duration as time intervals.

Units and Quantities In addition to SI, another system of units, the *U.S. customary system,* is still used in the United States despite acceptance of SI by the rest of the world. In this system, the units of length, mass, and time are the foot (ft), slug, and second, respectively. In this book, we shall use SI units because they are almost universally accepted in science and industry. We shall make some limited use of U.S. customary units in the study of classical mechanics.

In addition to the fundamental SI units of meter, kilogram, and second, we can also use other units, such as millimeters and nanoseconds, where the prefixes *milli-* and *nano-* denote multipliers of the basic units based on various powers of ten. Prefixes for the various powers of ten and their abbreviations are listed in Table 1.4 (page 6). For example, 10^{-3} m is equivalent to 1 millimeter (mm), and 10^{3} m corresponds to 1 kilometer (km). Likewise, 1 kilogram (kg) is 10^{3} grams (g), and 1 mega volt (MV) is 10^{6} volts (V).

---

[2]Period is defined as the time interval needed for one complete vibration.

TABLE 1.4	Prefixes for Powers of Ten					
Power	**Prefix**	**Abbreviation**		**Power**	**Prefix**	**Abbreviation**
10^{-24}	yocto	y		10^{3}	kilo	k
10^{-21}	zepto	z		10^{6}	mega	M
10^{-18}	atto	a		10^{9}	giga	G
10^{-15}	femto	f		10^{12}	tera	T
10^{-12}	pico	p		10^{15}	peta	P
10^{-9}	nano	n		10^{18}	exa	E
10^{-6}	micro	μ		10^{21}	zetta	Z
10^{-3}	milli	m		10^{24}	yotta	Y
10^{-2}	centi	c				
10^{-1}	deci	d				

The variables length, mass, and time are examples of *fundamental quantities*. Most other variables are *derived quantities,* those that can be expressed as a mathematical combination of fundamental quantities. Common examples are *area* (a product of two lengths) and *speed* (a ratio of a length to a time interval).

▷ A table of the letters in the Greek alphabet is provided on the back endpaper of this book.

Another example of a derived quantity is **density.** The density ρ (Greek letter rho) of any substance is defined as its *mass per unit volume:*

$$\rho \equiv \frac{m}{V} \tag{1.1}$$

In terms of fundamental quantities, density is a ratio of a mass to a product of three lengths. Aluminum, for example, has a density of 2.70×10^{3} kg/m^3, and iron has a density of 7.86×10^{3} kg/m^3. An extreme difference in density can be imagined by thinking about holding a 10-centimeter (cm) cube of Styrofoam in one hand and a 10-cm cube of lead in the other. See Table 14.1 in Chapter 14 for densities of several materials.

QUICK QUIZ 1.1 In a machine shop, two cams are produced, one of aluminum and one of iron. Both cams have the same mass. Which cam is larger? **(a)** The aluminum cam is larger. **(b)** The iron cam is larger. **(c)** Both cams have the same size.

1.2 Modeling and Alternative Representations

Most courses in general physics require the student to learn the skills of problem solving, and examinations usually include problems that test such skills. This section describes some useful ideas that will enable you to enhance your understanding of physical concepts, increase your accuracy in solving problems, eliminate initial panic or lack of direction in approaching a problem, and organize your work.

One of the primary problem-solving methods in physics is to form an appropriate **model** of the problem. **A model is a simplified substitute for the real problem that allows us to solve the problem in a relatively simple way.** As long as the predictions of the model agree to our satisfaction with the actual behavior of the real system, the model is valid. If the predictions do not agree, the model must be refined or replaced with another model. The power of modeling is in its ability to reduce a wide variety of very complex problems to a limited number of classes of problems that can be approached in similar ways.

In science, a model is very different from, for example, an architect's scale model of a proposed building, which appears as a smaller version of what it represents.

A scientific model is a theoretical construct and may have no visual similarity to the physical problem. A simple application of modeling is presented in Example 1.1, and we shall encounter many more examples of models as the text progresses.

Models are needed because the actual operation of the Universe is extremely complicated. Suppose, for example, we are asked to solve a problem about the Earth's motion around the Sun. The Earth is very complicated, with many processes occurring simultaneously. These processes include weather, seismic activity, and ocean movements as well as the multitude of processes involving human activity. Trying to maintain knowledge and understanding of all these processes is an impossible task.

The modeling approach recognizes that none of these processes affects the motion of the Earth around the Sun to a measurable degree. Therefore, these details are all ignored. In addition, as we shall find in Chapter 13, the size of the Earth does not affect the gravitational force between the Earth and the Sun; only the masses of the Earth and Sun and the distance between their centers determine this force. In a simplified model, the Earth is imagined to be a particle, an object with mass but zero size. This replacement of an extended object by a particle is called the **particle model,** which is used extensively in physics. By analyzing the motion of a particle with the mass of the Earth in orbit around the Sun, we find that the predictions of the particle's motion are in excellent agreement with the actual motion of the Earth.

The two primary conditions for using the particle model are as follows:

- The size of the actual object is of no consequence in the analysis of its motion.
- Any internal processes occurring in the object are of no consequence in the analysis of its motion.

Both of these conditions are in action in modeling the Earth as a particle. Its radius is not a factor in determining its motion, and internal processes such as thunderstorms, earthquakes, and manufacturing processes can be ignored.

Four categories of models used in this book will help us understand and solve physics problems. The first category is the **geometric model.** In this model, we form a geometric construction that represents the real situation. We then set aside the real problem and perform an analysis of the geometric construction. Consider a popular problem in elementary trigonometry, as in the following example.

Example 1.1 Finding the Height of a Tree

You wish to find the height of a tree but cannot measure it directly. You stand 50.0 m from the tree and determine that a line of sight from the ground to the top of the tree makes an angle of 25.0° with the ground. How tall is the tree?

SOLUTION

Figure 1.2 shows the tree and a right triangle corresponding to the information in the problem superimposed over it. (We assume that the tree is exactly perpendicular to a perfectly flat ground.) In the triangle, we know the length of the horizontal leg and the angle between the hypotenuse and the horizontal leg. We can find the height of the tree by calculating the length of the vertical leg. We do so with the tangent function:

$$\tan \theta = \frac{\text{opposite side}}{\text{adjacent side}} = \frac{h}{50.0 \text{ m}}$$

$$h = (50.0 \text{ m}) \tan \theta = (50.0 \text{ m}) \tan 25.0° = 23.3 \text{ m}$$

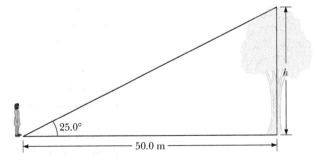

Figure 1.2 (Example 1.1) The height of a tree can be found by measuring the distance from the tree and the angle of sight to the top above the ground. This problem is a simple example of geometrically *modeling* the actual problem.

You may have solved a problem very similar to Example 1.1 but never thought about the notion of modeling. From the modeling approach, however, once we draw the triangle in Figure 1.2, the triangle is a geometric model of the real problem; it is a *substitute*. Until we reach the end of the problem, we do not imagine the problem to be about a *tree* but to be about a *triangle*. We use trigonometry to find the vertical leg of the triangle, leading to a value of 23.3 m. Because this leg *represents* the height of the tree, we can now return to the original problem and claim that the height of the tree is 23.3 m.

Other examples of geometric models include modeling the Earth as a perfect sphere, a pizza as a perfect disk, a meter stick as a long rod with no thickness, and an electric wire as a long, straight cylinder.

The particle model is an example of the second category of models, which we will call **simplification models.** In a simplification model, details that are not significant in determining the outcome of the problem are ignored. When we study rotation in Chapter 10, objects will be modeled as *rigid objects*. All the molecules in a rigid object maintain their exact positions with respect to one another. We adopt this simplification model because a spinning rock is much easier to analyze than a spinning block of gelatin, which is *not* a rigid object. Other simplification models will assume that quantities such as friction forces are negligible, remain constant, or are proportional to some power of the object's speed. We will assume *uniform* metal beams in Chapter 12, *laminar* flow of fluids in Chapter 14, *massless* springs in Chapter 15, *symmetric* distributions of electric charge in Chapter 23, *resistance-free* wires in Chapter 27, *thin* lenses in Chapter 34. These, and many more, are simplification models.

The third category is that of **analysis models,** which are general types of problems that we have solved before. An important technique in problem solving is to cast a new problem into a form similar to one we have already solved and which can be used as a model. As we shall see, there are about two dozen analysis models that can be used to solve most of the problems you will encounter. All of the analysis models in classical physics will be based on four simplification models: *particle, system, rigid object,* and *wave.* We will see our first analysis models in Chapter 2, where we will discuss them in more detail.

The fourth category of models is **structural models.** These models are generally used to understand the behavior of a system that is far different in scale from our macroscopic world—either much smaller or much larger—so that we cannot interact with it directly. As an example, the notion of a hydrogen atom as an electron in a circular orbit around a proton is a structural model of the atom. The ancient *geocentric* model of the Universe, in which the Earth is theorized to be at the center of the Universe, is an example of a structural model for something larger in scale than our macroscopic world.

Intimately related to the notion of modeling is that of forming **alternative representations** of the problem that you are solving. **A representation is a method of viewing or presenting the information related to the problem.** Scientists must be able to communicate complex ideas to individuals without scientific backgrounds. The best representation to use in conveying the information successfully will vary from one individual to the next. Some will be convinced by a well-drawn graph, and others will require a picture. Physicists are often persuaded to agree with a point of view by examining an equation, but non-physicists may not be convinced by this mathematical representation of the information.

A word problem, such as those at the ends of the chapters in this book, is one representation of a problem. In the "real world" that you will enter after graduation, the initial representation of a problem may be just an existing situation, such as the effects of climate change or a patient in danger of dying. You may have to identify the important data and information, and then cast the situation yourself into an equivalent word problem!

Considering alternative representations can help you think about the information in the problem in several different ways to help you understand and solve it. Several types of representations can be of assistance in this endeavor:

- **Mental representation.** From the description of the problem, imagine a scene that describes what is happening in the word problem, then let time progress so that you understand the situation and can predict what changes will occur in the situation. This step is critical in approaching *every* problem.
- **Pictorial representation.** Drawing a picture of the situation described in the word problem can be of great assistance in understanding the problem. In Example 1.1, the pictorial representation in Figure 1.2 allows us to identify the triangle as a geometric model of the problem. In architecture, a blueprint is a pictorial representation of a proposed building.

  Generally, a pictorial representation describes *what you would see* if you were observing the situation in the problem. For example, Figure 1.3 shows a pictorial representation of a baseball player hitting a short pop foul. Any coordinate axes included in your pictorial representation will be in two dimensions: *x* and *y* axes.

- **Simplified pictorial representation.** It is often useful to redraw the pictorial representation without complicating details by applying a simplification model. This process is similar to the discussion of the particle model described earlier. In a pictorial representation of the Earth in orbit around the Sun, you might draw the Earth and the Sun as spheres, with possibly some attempt to draw continents to identify which sphere is the Earth. In the simplified pictorial representation, the Earth and the Sun would be drawn simply as dots, representing particles, with appropriate labels. Figure 1.4 shows a simplified pictorial representation corresponding to the pictorial representation of the baseball trajectory in Figure 1.3. The notations v_x and v_y refer to the components of the velocity vector for the baseball. We will study vector components in Chapter 3. We shall use such simplified pictorial representations throughout the book.

- **Graphical representation.** In some problems, drawing a graph that describes the situation can be very helpful. In mechanics, for example, position–time graphs can be of great assistance. Similarly, in thermodynamics, pressure–volume graphs are essential to understanding. Figure 1.5 shows a graphical representation of the position as a function of time of a block on the end of a vertical spring as it oscillates up and down. Such a graph is helpful for understanding simple harmonic motion, which we study in Chapter 15.

  A graphical representation is different from a pictorial representation, which is also a two-dimensional display of information but whose axes, if any, represent *length* coordinates. In a graphical representation, the axes may represent *any* two related variables. For example, a graphical representation may have axes for temperature and time. The graph in Figure 1.5 has axes of vertical position *y* and time *t*. Therefore, in comparison to a pictorial representation, a graphical representation is generally *not* something you would see when observing the situation in the problem with your eyes.

- **Tabular representation.** It is sometimes helpful to organize the information in tabular form to help make it clearer. For example, some students find that making tables of known quantities and unknown quantities is helpful. The periodic table of the elements is an extremely useful tabular representation of information in chemistry and physics.

- **Mathematical representation.** The ultimate goal in solving a problem is often the mathematical representation. You want to move from the information contained in the word problem, through various representations of the problem that allow you to understand what is happening, to one or more equations that represent the situation in the problem and that can be solved mathematically for the desired result.

Figure 1.3 A pictorial representation of a pop foul being hit by a baseball player.

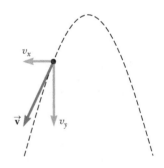

Figure 1.4 A simplified pictorial representation for the situation shown in Figure 1.3.

Figure 1.5 A graphical representation of the position as a function of time of a block hanging from a spring and oscillating.

1.3 Dimensional Analysis

In physics, the word *dimension* denotes the physical nature of a quantity. The distance between two points, for example, can be measured in feet, meters, or furlongs, which are all different units for expressing the dimension of length.

The symbols we use in this book to specify the dimensions of length, mass, and time are L, M, and T, respectively.[3] We shall often use brackets [ ] to denote the dimensions of a physical quantity. For example, the symbol we use for speed in this book is v, and in our notation, the dimensions of speed are written $[v] = L/T$. As another example, the dimensions of area A are $[A] = L^2$. The dimensions and units of area, volume, speed, and acceleration are listed in Table 1.5. The dimensions of other quantities, such as force and energy, will be described as they are introduced in the text.

In many situations, you may have to check a specific equation to see if it matches your expectations. A useful procedure for doing that, called **dimensional analysis,** can be used because dimensions can be treated as algebraic quantities. For example, quantities can be added or subtracted only if they have the same dimensions. Furthermore, the terms on both sides of an equation must have the same dimensions. By following these simple rules, you can use dimensional analysis to determine whether an expression has the correct form. Any relationship can be correct only if the dimensions on both sides of the equation are the same.

To illustrate this procedure, suppose you are interested in an equation for the position x of a car at a time t if the car starts from rest at $x = 0$ and moves with constant acceleration a. The correct expression for this situation is $x = \frac{1}{2}at^2$ as we show in Chapter 2. The quantity x on the left side has the dimension of length. For the equation to be dimensionally correct, the quantity on the right side must also have the dimension of length. We can perform a dimensional check by substituting the dimensions for acceleration, L/T^2 (Table 1.5), and time, T, into the equation. That is, the dimensional form of the equation $x = \frac{1}{2}at^2$ is

$$L = \frac{L}{T^2} \cdot T^2 = L$$

The dimensions of time cancel as shown, leaving the dimension of length on the right-hand side to match that on the left.

A more general procedure using dimensional analysis is to set up an expression of the form

$$x \propto a^n t^m$$

where n and m are exponents that must be determined and the symbol $\propto$ indicates a proportionality. This relationship is correct only if the dimensions of both sides are the same. Because the dimension of the left side is length, the dimension of the right side must also be length. That is,

$$[a^n t^m] = L = L^1 T^0$$

PITFALL PREVENTION 1.2

Symbols for Quantities Some quantities have a small number of symbols that represent them. For example, the symbol for time is almost always t. Other quantities might have various symbols depending on the usage. Length may be described with symbols such as x, y, and z (for position); r (for radius); a, b, and c (for the legs of a right triangle); ℓ (for the length of an object); d (for a distance); h (for a height); and so forth.

TABLE 1.5 Dimensions and Units of Four Derived Quantities

Quantity	Area (A)	Volume (V)	Speed (v)	Acceleration (a)
Dimensions	L^2	L^3	L/T	L/T^2
SI units	m^2	m^3	m/s	m/s^2
U.S. customary units	ft^2	ft^3	ft/s	ft/s^2

---

[3]The *dimensions* of a quantity will be symbolized by a capitalized, nonitalic letter such as L or T. The *algebraic symbol* for the quantity itself will be an italicized letter such as L for the length of an object or t for time.

Because the dimensions of acceleration are L/T^2 and the dimension of time is T, we have

$$(L/T^2)^n T^m = L^1 T^0 \quad \rightarrow \quad (L^n T^{m-2n}) = L^1 T^0$$

The exponents of L and T must be the same on both sides of the equation. From the exponents of L, we see immediately that $n = 1$. From the exponents of T, we see that $m - 2n = 0$, which, once we substitute for n, gives us $m = 2$. Returning to our original expression $x \propto a^n t^m$, we conclude that $x \propto at^2$.

QUICK QUIZ 1.2 True or False: Dimensional analysis can give you the numerical value of constants of proportionality that may appear in an algebraic expression.

Example 1.2 Analysis of an Equation

Show that the expression $v = at$, where v represents speed, a acceleration, and t an instant of time, is dimensionally correct.

SOLUTION

Identify the dimensions of v from Table 1.5:

$$[v] = \frac{L}{T}$$

Identify the dimensions of a from Table 1.5 and multiply by the dimensions of t:

$$[at] = \frac{L}{T^2}\, T = \frac{L}{T}$$

Therefore, $v = at$ is dimensionally correct because we have the same dimensions on both sides. (If the expression were given as $v = at^2$, it would be dimensionally *incorrect*. Try it and see!)

Example 1.3 Analysis of a Power Law

Suppose we are told that the acceleration a of a particle moving with uniform speed v in a circle of radius r is proportional to some power of r, say r^n, and some power of v, say v^m. Determine the values of n and m and write the simplest form of an equation for the acceleration.

SOLUTION

Write an expression for a with a dimensionless constant of proportionality k:

$$a = kr^n v^m$$

Substitute the dimensions of a, r, and v:

$$\frac{L}{T^2} = L^n \left(\frac{L}{T}\right)^m = \frac{L^{n+m}}{T^m}$$

Equate the exponents of L and T so that the dimensional equation is balanced:

$$n + m = 1 \text{ and } m = 2$$

Solve the two equations for n:

$$n = -1$$

Write the acceleration expression:

$$a = kr^{-1} v^2 = k\frac{v^2}{r}$$

In Section 4.4 on uniform circular motion, we show that $k = 1$ if a consistent set of units is used. The constant k would not equal 1 if, for example, v were in km/h and you wanted a in m/s².

1.4 Conversion of Units

Sometimes it is necessary to convert units from one measurement system to another or convert within a system (for example, from kilometers to meters). Conversion factors between SI and U.S. customary units of length are as follows:

$$1 \text{ mile} = 1\,609 \text{ m} = 1.609 \text{ km} \qquad 1 \text{ ft} = 0.304\,8 \text{ m} = 30.48 \text{ cm}$$
$$1 \text{ m} = 39.37 \text{ in.} = 3.281 \text{ ft} \qquad 1 \text{ in.} = 0.025\,4 \text{ m} = 2.54 \text{ cm (exactly)}$$

A more complete list of conversion factors can be found in Appendix A.

Like dimensions, units can be treated as algebraic quantities that can cancel each other. For example, suppose we wish to convert 15.0 in. to centimeters. Because 1 in. is defined as exactly 2.54 cm, we find that

$$15.0 \text{ in.} = (15.0 \text{ in.})\left(\frac{2.54 \text{ cm}}{1 \text{ in.}}\right) = 38.1 \text{ cm}$$

where the ratio in parentheses is equal to 1. We express 1 as 2.54 cm/1 in. (rather than 1 in./2.54 cm) so that the unit "inch" in the denominator cancels with the unit in the original quantity. The remaining unit is the centimeter, our desired result.

QUICK QUIZ 1.3 The distance between two cities is 100 mi. What is the number of kilometers between the two cities? **(a)** smaller than 100 **(b)** larger than 100 **(c)** equal to 100

Example 1.4 **Is He Speeding?**

On an interstate highway in a rural region of Wyoming, a car is traveling at a speed of 38.0 m/s. Is the driver exceeding the speed limit of 75.0 mi/h?

SOLUTION

Convert meters to miles and seconds to hours:
$$(38.0 \text{ m/s})\left(\frac{1 \text{ mi}}{1\,609 \text{ m}}\right)\left(\frac{60 \text{ s}}{1 \text{ min}}\right)\left(\frac{60 \text{ min}}{1 \text{ h}}\right) = 85.0 \text{ mi/h}$$

The driver is indeed exceeding the speed limit and should slow down.

WHAT IF? What if the driver were from outside the United States and is familiar with speeds measured in kilometers per hour? What is the speed of the car in km/h?

Answer We can convert our final answer to the appropriate units:

$$(85.0 \text{ mi/h})\left(\frac{1.609 \text{ km}}{1 \text{ mi}}\right) = 137 \text{ km/h}$$

Figure 1.6 shows an automobile speedometer displaying speeds in both mi/h and km/h. Can you check the conversion we just performed using this photograph?

Figure 1.6 The speedometer of a vehicle that shows speeds in both miles per hour and kilometers per hour.

1.5 Estimates and Order-of-Magnitude Calculations

Suppose someone asks you the number of bits of data on a typical Blu-ray Disc. In response, it is not generally expected that you would provide the exact number but rather an estimate, which may be expressed in scientific notation. The estimate may be made even more approximate by expressing it as an **order of magnitude**, which is a power of 10 determined as follows:

1. Express the number in scientific notation, with the multiplier of the power of 10 between 1 and 10 and a unit.
2. If the multiplier is less than 3.162 (the square root of 10), the order of magnitude of the number is the power of 10 in the scientific notation. If the multiplier is greater than 3.162, the order of magnitude is one larger than the power of 10 in the scientific notation.

We use the symbol $\sim$ for "is on the order of." Use the procedure above to verify the orders of magnitude for the following lengths:

$$0.008\ 6\ \text{m} \sim 10^{-2}\ \text{m} \qquad 0.002\ 1\ \text{m} \sim 10^{-3}\ \text{m} \qquad 720\ \text{m} \sim 10^{3}\ \text{m}$$

Usually, when an order-of-magnitude estimate is made, the results are reliable to within about a factor of 10.

Inaccuracies caused by guessing too low for one number are often canceled by other guesses that are too high. You will find that with practice your guesstimates become better and better. Estimation problems can be fun to work because you freely drop digits, venture reasonable approximations for unknown numbers, make simplifying assumptions, and turn the question around into something you can answer in your head or with minimal mathematical manipulation on paper. Because of the simplicity of these types of calculations, they can be performed on a small scrap of paper and are often called *back-of-the-envelope calculations*.

Example 1.5 **Breaths in a Lifetime**

Estimate the number of breaths taken during an average human lifetime.

SOLUTION

We start by guessing that the typical human lifetime is about 70 years. Think about the average number of breaths that a person takes in 1 min. This number varies depending on whether the person is exercising, sleeping, angry, serene, and so forth. To the nearest order of magnitude, we shall choose 10 breaths per minute as our estimate. (This estimate is certainly closer to the true average value than an estimate of 1 breath per minute or 100 breaths per minute.)

Find the approximate number of minutes in a year:

$$1\ \text{yr} \left(\frac{400\ \text{days}}{1\ \text{yr}} \right) \left(\frac{25\ \text{h}}{1\ \text{day}} \right) \left(\frac{60\ \text{min}}{1\ \text{h}} \right) = 6 \times 10^{5}\ \text{min}$$

Find the approximate number of minutes in a 70-year lifetime:

$$\text{number of minutes} = (70\ \text{yr})(6 \times 10^{5}\ \text{min/yr})$$
$$= 4 \times 10^{7}\ \text{min}$$

Find the approximate number of breaths in a lifetime:

$$\text{number of breaths} = (10\ \text{breaths/min})(4 \times 10^{7}\ \text{min})$$
$$= 4 \times 10^{8}\ \text{breaths}$$

Therefore, a person takes on the order of 10^{9} breaths in a lifetime. Notice how much simpler it is in the first calculation above to multiply 400×25 than it is to work with the more accurate 365×24.

WHAT IF? What if the average lifetime were estimated as 80 years instead of 70? Would that change our final estimate?

Answer We could claim that $(80\ \text{yr})(6 \times 10^{5}\ \text{min/yr}) = 5 \times 10^{7}\ \text{min}$, so our final estimate should be 5×10^{8} breaths. This answer is still on the order of 10^{9} breaths, so an order-of-magnitude estimate would be unchanged.

1.6 Significant Figures

When certain quantities are measured, the measured values are known only to within the limits of the experimental uncertainty. The value of this uncertainty can depend on various factors, such as the quality of the apparatus, the skill of the experimenter, and the number of measurements performed. The number of

significant figures in a measurement can be used to express something about the uncertainty. The number of significant figures is related to the number of numerical digits used to express the measurement, as we discuss below.

As an example of significant figures, suppose we are asked to measure the radius of a Blu-ray Disc using a meterstick as a measuring instrument. Let us assume the accuracy to which we can measure the radius of the disc is ±0.1 cm. Because of the uncertainty of ±0.1 cm, if the radius is measured to be 6.0 cm, we can claim only that its radius lies somewhere between 5.9 cm and 6.1 cm. In this case, we say that the measured value of 6.0 cm has two significant figures. Note that *the significant figures include the first estimated digit*. Therefore, we could write the radius as (6.0 ± 0.1) cm.

Zeros may or may not be significant figures. Those used to position the decimal point in such numbers as 0.03 and 0.007 5 are not significant. Therefore, there are one and two significant figures, respectively, in these two values. When the zeros come after other digits, however, there is the possibility of misinterpretation. For example, suppose the mass of an object is given as 1 500 g. This value is ambiguous because we do not know whether the last two zeros are being used to locate the decimal point or whether they represent significant figures in the measurement. To remove this ambiguity, it is common to use scientific notation to indicate the number of significant figures. In this case, we would express the mass as 1.5×10^3 g if there are two significant figures in the measured value, 1.50×10^3 g if there are three significant figures, and 1.500×10^3 g if there are four. The same rule holds for numbers less than 1, so 2.3×10^{-4} has two significant figures (and therefore could be written 0.000 23) and 2.30×10^{-4} has three significant figures (and therefore written as 0.000 230).

In problem solving, we often combine quantities mathematically through multiplication, division, addition, subtraction, and so forth. When doing so, you must make sure that the result has the appropriate number of significant figures. A good rule of thumb to use in determining the number of significant figures that can be claimed in a multiplication or a division is as follows:

> When multiplying several quantities, the number of significant figures in the final answer is the same as the number of significant figures in the quantity having the smallest number of significant figures. The same rule applies to division.

Let's apply this rule to find the area of the Blu-ray Disc whose radius we measured above. Using the equation for the area of a circle,

$$A = \pi r^2 = \pi (6.0 \text{ cm})^2 = 1.1 \times 10^2 \text{ cm}^2$$

If you perform this calculation on your calculator, you will likely see 113.097 335 5. It should be clear that you don't want to keep all of these digits, but you might be tempted to report the result as 113 cm². This result is not justified because it has three significant figures, whereas the radius only has two. Therefore, we must report the result with only two significant figures as shown above.

For addition and subtraction, you must consider the number of decimal places when you are determining how many significant figures to report:

> When numbers are added or subtracted, the number of decimal places in the result should equal the smallest number of decimal places of any term in the sum or difference.

As an example of this rule, consider the sum

$$23.2 + 5.174 = 28.4$$

Notice that we do not report the answer as 28.374 because the lowest number of decimal places is one, for 23.2. Therefore, our answer must have only one decimal place.

PITFALL PREVENTION 1.4
Read Carefully Notice that the rule for addition and subtraction is different from that for multiplication and division. For addition and subtraction, the important consideration is the number of *decimal places*, not the number of *significant figures*.

The rule for addition and subtraction can often result in answers that have a different number of significant figures than the quantities with which you start. For example, consider these operations that satisfy the rule:

$$1.000\ 1 + 0.000\ 3 = 1.000\ 4$$
$$1.002 - 0.998 = 0.004$$

In the first example, the result has five significant figures even though one of the terms, 0.000 3, has only one significant figure. Similarly, in the second calculation, the result has only one significant figure even though the numbers being subtracted have four and three, respectively.

In this book, most of the numerical examples and end-of-chapter problems will yield answers having three significant figures. When carrying out estimation calculations, we shall typically work with a single significant figure.

◀ Significant figure guidelines used in this book

If the number of significant figures in the result of a calculation must be reduced, there is a general rule for rounding numbers: the last digit retained is increased by 1 if the last digit dropped is greater than 5. (For example, 1.346 becomes 1.35.) If the last digit dropped is less than 5, the last digit retained remains as it is. (For example, 1.343 becomes 1.34.) If the last digit dropped is equal to 5, the remaining digit should be rounded to the nearest even number. (This rule helps avoid accumulation of errors in long arithmetic processes.)

In a long calculation involving multiple steps, it is very important to delay the rounding of numbers until you have the final result, in order to avoid error accumulation. Wait until you are ready to copy the final answer from your calculator before rounding to the correct number of significant figures. In this book, we display numerical values rounded off to two or three significant figures. This occasionally makes some mathematical manipulations look odd or incorrect. For instance, looking ahead to Example 3.5 on page 62, you will see the operation −17.7 km + 34.6 km = 17.0 km. This looks like an incorrect subtraction, but that is only because we have rounded the numbers 17.7 km and 34.6 km for display. If all digits in these two intermediate numbers are retained and the rounding is only performed on the final number, the correct three-digit result of 17.0 km is obtained.

PITFALL PREVENTION 1.5
Symbolic Solutions When solving problems, it is very useful to perform the solution completely in algebraic form and wait until the very end to enter numerical values into the final symbolic expression. This method will save many calculator keystrokes, especially if some quantities cancel so that you never have to enter their values into your calculator! In addition, you will only need to round once, on the final result.

Example 1.6   Installing a Carpet

A carpet is to be installed in a rectangular room whose length is measured to be 12.71 m and whose width is measured to be 3.46 m. Find the area of the room.

SOLUTION

If you multiply 12.71 m by 3.46 m on your calculator, you will see an answer of 43.976 6 m². How many of these numbers should you claim? Our rule of thumb for multiplication tells us that you can claim only the number of significant figures in your answer as are present in the measured quantity having the lowest number of significant figures. In this example, the lowest number of significant figures is three in 3.46 m, so we should express our final answer as 44.0 m².

Summary

▷ Definitions

The three fundamental physical quantities of mechanics are **length, mass,** and **time,** which in the SI system have the units **meter** (m), **kilogram** (kg), and **second** (s), respectively. These fundamental quantities cannot be defined in terms of more basic quantities.

The **density** of a substance is defined as its *mass per unit volume:*

$$\rho \equiv \frac{m}{V} \tag{1.1}$$

continued

Concepts and Principles

The method of **dimensional analysis** is very powerful in solving physics problems. Dimensions can be treated as algebraic quantities. By making estimates and performing order-of-magnitude calculations, you should be able to approximate the answer to a problem when there is not enough information available to specify an exact solution completely.

Problem-solving skills and physical understanding can be improved by **modeling** the problem and by constructing **alternative representations** of the problem. Models helpful in solving problems include **geometric, simplification, analysis, and structural models.** Helpful representations include the **mental, pictorial, simplified pictorial, graphical, tabular,** and **mathematical representations.**

When you compute a result from several measured numbers, each of which has a certain accuracy, you should give the result with the correct number of **significant figures.**

When **multiplying** several quantities, the number of significant figures in the final answer is the same as the number of significant figures in the quantity having the smallest number of significant figures. The same rule applies to **division.**

When numbers are **added** or **subtracted,** the number of decimal places in the result should equal the smallest number of decimal places of any term in the sum or difference.

Think–Pair–Share

See the Preface for an explanation of the icons used in this problems set. For additional assessment items for this section, go to **WEBASSIGN** *From Cengage*

1. A student is supplied with a stack of copy paper, ruler, compass, scissors, and a sensitive balance. He cuts out various shapes in various sizes, calculates their areas, measures their masses, and prepares the graph of Figure TP1.1. (a) Consider the fourth experimental point from the top. How far is it vertically from the best-fit straight line? Express your answer as a difference in vertical-axis coordinate. (b) Express your answer as a percentage. (c) Calculate the slope of the line. (d) State what the graph demonstrates, referring to the shape of the graph and the results of parts (b) and (c). (e) Describe whether this result should be expected theoretically. (f) Describe the physical meaning of the slope.

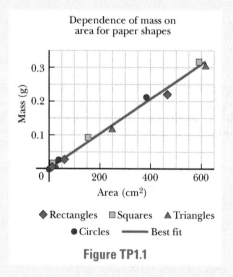

Figure TP1.1

2. **ACTIVITY** Have each person in the group measure the height of another person using a meter stick with metric distances on one side and U.S. customary distances, such as inches, on the other side. Record the height to the nearest centimeter and to the nearest half-inch. For each person, divide his or her height in centimeters by the height in inches. Compare the results of this division for everyone in your group. What can you say about the results?

3. **ACTIVITY** Gather together a number of U.S. pennies, either from your instructor or from the members of your group. Divide up the pennies into two samples: (1) those with dates of 1981 or earlier, and (2) those with dates of 1983 and later (exclude 1982 pennies from your sample). Find the total mass of all the pennies in each sample. Then divide each of these total masses by the number of pennies in its corresponding sample, to find the average penny mass in each sample. Discuss why the results are different for the two samples.

4. **ACTIVITY** Discuss in your group the process by which you can obtain the best measurement of the thickness of a single sheet of paper in Chapters 1–5 of this book. Perform that measurement and express it with an appropriate number of significant figures and uncertainty. From that measurement, predict the total thickness of the pages in Volume 1 of this book (Chapters 1–21). After making your prediction, measure the thickness of Volume 1. Is your measurement within the range of your prediction and its associated uncertainty?

Problems

See the Preface for an explanation of the icons used in this problems set. For additional assessment items for this section, go to ⚡ **WEBASSIGN** From Cengage

Note: Consult the endpapers, appendices, and tables in the text whenever necessary in solving problems. For this chapter, Table 14.1 and Appendix B.3 may be particularly useful. Answers to odd-numbered problems appear in the back of the book.

SECTION 1.1 Standards of Length, Mass, and Time

1. (a) Use information on the endpapers of this book to calculate the average density of the Earth. (b) Where does the value fit among those listed in Table 14.1 in Chapter 14? Look up the density of a typical surface rock like granite in another source and compare it with the density of the Earth.

2. A proton, which is the nucleus of a hydrogen atom, can be modeled as a sphere with a diameter of 2.4 fm and a mass of 1.67×10^{-27} kg. (a) Determine the density of the proton. (b) State how your answer to part (a) compares with the density of osmium, given in Table 14.1 in Chapter 14.

3. Two spheres are cut from a certain uniform rock. One has radius 4.50 cm. The mass of the other is five times greater. Find its radius.

4. What mass of a material with density ρ is required to make a hollow spherical shell having inner radius r_1 and outer radius r_2?

5. You have been hired by the defense attorney as an expert witness in a lawsuit. The plaintiff is someone who just returned from being a passenger on the first orbital space tourist flight. Based on a travel brochure offered by the space travel company, the plaintiff expected to be able to see the Great Wall of China from his orbital height of 200 km above the Earth's surface. He was unable to do so, and is now demanding that his fare be refunded and to receive additional financial compensation to cover his great disappointment. Construct the basis for an argument for the defense that shows that his expectation of seeing the Great Wall from orbit was unreasonable. The Wall is 7 m wide at its widest point and the normal visual acuity of the human eye is 3×10^{-4} rad. (Visual acuity is the smallest subtended angle that an object can make at the eye and still be recognized; the subtended angle in radians is the ratio of the width of an object to the distance of the object from your eyes.)

SECTION 1.2 Modeling and Alternative Representations

6. A surveyor measures the distance across a straight river by the following method (Fig. P1.6). Starting directly across from a tree on the opposite bank, she walks $d = 100$ m along the riverbank to establish a baseline. Then she sights across to the tree. The angle from her baseline to the tree is $\theta = 35.0°$. How wide is the river?

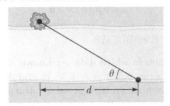

Figure P1.6

7. A crystalline solid consists of atoms stacked up in a repeating lattice structure. Consider a crystal as shown in Figure P1.7a. The atoms reside at the corners of cubes of side $L = 0.200$ nm. One piece of evidence for the regular arrangement of atoms comes from the flat surfaces along which a crystal separates, or cleaves, when it is broken. Suppose this crystal cleaves along a face diagonal as shown in Figure P1.7b. Calculate the spacing d between two adjacent atomic planes that separate when the crystal cleaves.

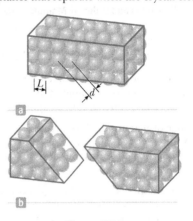

Figure P1.7

SECTION 1.3 Dimensional Analysis

8. The position of a particle moving under uniform acceleration is some function of time and the acceleration. Suppose we write this position as $x = ka^m t^n$, where k is a dimensionless constant. Show by dimensional analysis that this expression is satisfied if $m = 1$ and $n = 2$. Can this analysis give the value of k?

9. Which of the following equations are dimensionally correct? (a) $v_f = v_i + ax$ (b) $y = (2 \text{ m}) \cos(kx)$, where $k = 2 \text{ m}^{-1}$

10. (a) Assume the equation $x = At^3 + Bt$ describes the motion of a particular object, with x having the dimension of length and t having the dimension of time. Determine the dimensions of the constants A and B. (b) Determine the dimensions of the derivative $dx/dt = 3At^2 + B$.

SECTION 1.4 Conversion of Units

11. A solid piece of lead has a mass of 23.94 g and a volume of 2.10 cm³. From these data, calculate the density of lead in SI units (kilograms per cubic meter).

12. *Why is the following situation impossible?* A student's dormitory room measures 3.8 m by 3.6 m, and its ceiling is 2.5 m high. After the student completes his physics course, he displays his dedication by completely wallpapering the walls of the room with the pages from his copy of volume 1 (Chapters 1–21) of this textbook. He even covers the door and window.

13. One cubic meter (1.00 m³) of aluminum has a mass of 2.70×10^3 kg, and the same volume of iron has a mass of 7.86×10^3 kg. Find the radius of a solid aluminum sphere that will balance a solid iron sphere of radius 2.00 cm on an equal-arm balance.

14. Let ρ_{Al} represent the density of aluminum and ρ_{Fe} that of iron.
[S] Find the radius of a solid aluminum sphere that balances a solid iron sphere of radius r_{Fe} on an equal-arm balance.

15. One gallon of paint (volume $= 3.78 \times 10^{-3}$ m^3) covers an
[T] area of 25.0 m^2. What is the thickness of the fresh paint on the wall?

16. An auditorium measures 40.0 m $\times$ 20.0 m $\times$ 12.0 m. The
[V] density of air is 1.20 kg/m^3. What are (a) the volume of the room in cubic feet and (b) the weight of air in the room in pounds?

SECTION 1.5 Estimates and Order-of-Magnitude Calculations

Note: In your solutions to Problems 17 and 18, state the quantities you measure or estimate and the values you take for them.

17. (a) Compute the order of magnitude of the mass of a bathtub half full of water. (b) Compute the order of magnitude of the mass of a bathtub half full of copper coins.

18. To an order of magnitude, how many piano tuners reside in New York City? The physicist Enrico Fermi was famous for asking questions like this one on oral Ph.D. qualifying examinations.

19. Your roommate is playing a video game from the latest
[CR] *Star Wars* movie while you are studying physics. Distracted by the noise, you go to see what is on the screen. The game involves trying to fly a spacecraft through a crowded field of asteroids in the asteroid belt around the Sun. You say to him, "Do you know that the game you are playing is very unrealistic? The asteroid belt is not that crowded and you don't have to maneuver through it like that!" Distracted by your statement, he accidentally allows his spacecraft to strike an asteroid, just missing the high score. He turns to you in disgust and says, "Yeah, prove it." You say, "Okay, I've learned recently that the highest concentration of asteroids is in a doughnut-shaped region between the Kirkwood gaps at radii of 2.06 AU and 3.27 AU from the Sun. There are an estimated 10^9 asteroids of radius 100 m or larger, like those in your video game, in this region . . ." Finish your argument with a calculation to show that the number of asteroids in the space near a spacecraft is tiny. (An astronomical unit—AU—is the mean distance of the Earth from the Sun: 1 AU $= 1.496 \times 10^{11}$ m.)

SECTION 1.6 Significant Figures

Note: Appendix B.8 on propagation of uncertainty may be useful in solving some problems in this section.

20. How many significant figures are in the following numbers?
[V] (a) 78.9 $\pm$ 0.2 (b) 3.788 $\times$ 10^9 (c) 2.46 $\times$ 10^{-6} (d) 0.005 3

21. The *tropical year,* the time interval from one vernal equinox to the next vernal equinox, is the basis for our calendar. It contains 365.242 199 days. Find the number of seconds in a tropical year.

Note: The next seven problems call on mathematical skills from your prior education that will be useful throughout this course.

22. **Review.** The average density of the planet Uranus is 1.27 $\times$ 10^3 kg/m^3. The ratio of the mass of Neptune to that of Uranus is 1.19. The ratio of the radius of Neptune to that of Uranus is 0.969. Find the average density of Neptune.

23. **Review.** In a community college parking lot, the number of ordinary cars is larger than the number of sport utility vehicles by 94.7%. The difference between the number of cars and the number of SUVs is 18. Find the number of SUVs in the lot.

24. **Review.** Find every angle θ between 0 and 360° for which the ratio of sin θ to cos θ is -3.00.

25. **Review.** The ratio of the number of sparrows visiting a bird feeder to the number of more interesting birds is 2.25. On a morning when altogether 91 birds visit the feeder, what is the number of sparrows?

26. **Review.** Prove that one solution of the equation

$$2.00x^4 - 3.00x^3 + 5.00x = 70.0$$

is $x = -2.22$.

27. **Review.** From the set of equations
[S]

$$p = 3q$$

$$pr = qs$$

$$\tfrac{1}{2}pr^2 + \tfrac{1}{2}qs^2 = \tfrac{1}{2}qt^2$$

involving the unknowns p, q, r, s, and t, find the value of the ratio of t to r.

28. **Review.** Figure P1.28 shows students studying the ther-
[QC] mal conduction of energy into cylindrical blocks of ice. As
[S] we will see in Chapter 19, this process is described by the equation

$$\frac{Q}{\Delta t} = \frac{k\pi d^2(T_h - T_c)}{4L}$$

For experimental control, in one set of trials all quantities except d and Δt are constant. (a) If d is made three times larger, does the equation predict that Δt will get larger or get smaller? By what factor? (b) What pattern of proportionality of Δt to d does the equation predict? (c) To display this proportionality as a straight line on a graph, what quantities should you plot on the horizontal and vertical axes? (d) What expression represents the theoretical slope of this graph?

Figure P1.28

ADDITIONAL PROBLEMS

29. In a situation in which data are known to three significant digits, we write 6.379 m = 6.38 m and 6.374 m = 6.37 m. When a number ends in 5, we arbitrarily choose to write 6.375 m = 6.38 m. We could equally well write 6.375 m = 6.37 m, "rounding down" instead of "rounding up," because

we would change the number 6.375 by equal increments in both cases. Now consider an order-of-magnitude estimate, in which factors of change rather than increments are important. We write 500 m ~ 10^3 m because 500 differs from 100 by a factor of 5 while it differs from 1 000 by only a factor of 2. We write 437 m ~ 10^3 m and 305 m ~ 10^2 m. What distance differs from 100 m and from 1 000 m by equal factors so that we could equally well choose to represent its order of magnitude as ~ 10^2 m or as ~ 10^3 m?

30. (a) What is the order of magnitude of the number of micro organisms in the human intestinal tract? A typical bacterial length scale is 10^{-6} m. Estimate the intestinal volume and assume 1% of it is occupied by bacteria. (b) Does the number of bacteria suggest whether the bacteria are beneficial, dangerous, or neutral for the human body? What functions could they serve?

31. The distance from the Sun to the nearest star is about 4×10^{16} m. The Milky Way galaxy (Fig. P1.31) is roughly a disk of diameter 10^{21} m and thickness ~ 10^{19} m. Find the order of magnitude of the number of stars in the Milky Way. Assume the distance between the Sun and our nearest neighbor is typical.

Figure P1.31 The Milky Way galaxy.

32. *Why is the following situation impossible?* In an effort to boost interest in a television game show, each weekly winner is offered an additional $1 million bonus prize if he or she can personally count out that exact amount from a supply of one-dollar bills. The winner must do this task under supervision by television show executives and within one 40-hour work week. To the dismay of the show's producers, most contestants succeed at the challenge.

33. Bacteria and other prokaryotes are found deep underground, in water, and in the air. One micron (10^{-6} m) is a typical length scale associated with these microbes. (a) Estimate the total number of bacteria and other prokaryotes on the Earth. (b) Estimate the total mass of all such microbes.

34. A spherical shell has an outside radius of 2.60 cm and an inside radius of a. The shell wall has uniform thickness and is made of a material with density 4.70 g/cm^3. The space inside the shell is filled with a liquid having a density of 1.23 g/cm^3. (a) Find the mass m of the sphere, including its contents, as a function of a. (b) For what value of the variable a does m have its maximum possible value? (c) What is this maximum mass? (d) Explain whether the value from part (c) agrees with the result of a direct calculation of the mass of a solid sphere of uniform density made of the same material as the shell. (e) **What If?** Would the answer to part (a) change if the inner wall were not concentric with the outer wall?

35. Air is blown into a spherical balloon so that, when its radius is 6.50 cm, its radius is increasing at the rate 0.900 cm/s. (a) Find the rate at which the volume of the balloon is increasing. (b) If this volume flow rate of air entering the balloon is constant, at what rate will the radius be increasing when the radius is 13.0 cm? (c) Explain physically why the answer to part (b) is larger or smaller than 0.9 cm/s, if it is different.

36. In physics, it is important to use mathematical approximations. (a) Demonstrate that for small angles (< 20°)

$$\tan \alpha \approx \sin \alpha \approx \alpha = \frac{\pi \alpha'}{180°}$$

where α is in radians and α' is in degrees. (b) Use a calculator to find the largest angle for which $\tan \alpha$ may be approximated by α with an error less than 10.0%.

37. The consumption of natural gas by a company satisfies the empirical equation $V = 1.50t + 0.008\ 00t^2$, where V is the volume of gas in millions of cubic feet and t is the time in months. Express this equation in units of cubic feet and seconds. Assume a month is 30.0 days.

38. A woman wishing to know the height of a mountain measures the angle of elevation of the mountaintop as 12.0°. After walking 1.00 km closer to the mountain on level ground, she finds the angle to be 14.0°. (a) Draw a picture of the problem, neglecting the height of the woman's eyes above the ground. *Hint:* Use two triangles. (b) Using the symbol y to represent the mountain height and the symbol x to represent the woman's original distance from the mountain, label the picture. (c) Using the labeled picture, write two trigonometric equations relating the two selected variables. (d) Find the height y.

CHALLENGE PROBLEM

39. A woman stands at a horizontal distance x from a mountain and measures the angle of elevation of the mountaintop above the horizontal as θ. After walking a distance d closer to the mountain on level ground, she finds the angle to be ϕ. Find a general equation for the height y of the mountain in terms of d, ϕ, and θ, neglecting the height of her eyes above the ground.

2 Motion in One Dimension

The introductory storyline involves a long, straight road like this one, where the power poles are equally spaced. (*John Arehart/Shutterstock.com*)

STORYLINE **You are a passenger in a car being driven by a friend** down a straight road. You notice that the telephone poles, streetlight poles, or electric power poles on the side of the road are located at equal distances from each other. You pull out your smartphone and use it as a stopwatch to measure the time intervals required for you to pass between adjacent pairs of poles.[1] When your friend tells you that the car is moving at a fixed speed, you notice that all of these time intervals are the same. Now, the driver begins to slow down for a traffic light. You again measure the time intervals and find that each one is longer than the one before. After the car pulls away from the traffic light and speeds up, the time intervals between poles become shorter. Does this behavior make sense? When the car is moving at a constant speed again, you use the time interval between poles and the driving speed reported by your friend to calculate the distance between the poles. You excitedly tell your friend to pull over so you can pace out the distance between the poles. How accurate was your calculation?

CONNECTIONS We begin our study of physics with the topic of *kinematics*. In this broad topic, we generally investigate *motion*: the motion of objects without regard for interactions with the environment that influence the motion. Motion is what many of the early scientists studied. Early astronomers in Greece, China, the Middle East, and Central America observed the motion of objects in the night sky. Galileo Galilei studied the motion of objects rolling down inclined planes. Isaac Newton pondered the nature of falling objects. From everyday experience, we recognize that motion of an object represents a continuous change in the object's

---

[1]A number of specialized smartphone apps can be downloaded and used to make numerical measurements, such as speed and acceleration. In our storylines, however, we will restrict our smartphone use mostly to apps that are standard on the phone as purchased.

position. In this chapter, we will analyze the motion of an object along a straight line, like the car in the storyline. We will use measurements of length and time as described in Chapter 1 to quantify the motion. An object moving vertically and subject to gravity is an important application of one-dimensional motion, and will also be studied in this chapter. Remember our discussion of making models for physical situations in Section 1.2. In our study, we use the simplification model mentioned in that section and called the particle model, and describe the moving object as a particle regardless of its size. In general, a particle is a point-like object, that is, an object that has mass but is of infinitesimal size. In Section 1.2, we discussed the fact that the motion of the Earth around the Sun can be treated as if the Earth were a particle. We will return to this model for the Earth when we study planetary orbits in Chapter 13. As an example on a much smaller scale, it is possible to explain the pressure exerted by a gas on the walls of a container by treating the gas molecules as particles, without regard for the internal structure of the molecules; we will see this analysis in Chapter 20. For now, let us apply the particle model to a wide variety of moving objects in this chapter. An understanding of motion will be essential throughout the rest of this book: the motion of planets in Chapter 13 on gravity, the motion of electrons in electric circuits in Chapter 26, the motion of light waves in Chapter 34 on optics, the motion of quantum particles tunneling through barriers in Chapter 40.

2.1 Position, Velocity, and Speed of a Particle

A particle's **position** x is the location of the particle with respect to a chosen reference point that we can consider to be the origin of a coordinate system. The motion of a particle is completely known if the particle's position in space is known at all times.

◄ Position

Consider a car moving back and forth along the x axis as in Figure 2.1a (page 22). The numbers under the horizontal line are position markers for the car, similar to the equally spaced poles in the introductory storyline. When we begin collecting position data, the car is 30 m to the right of the reference position $x = 0$. We will use the particle model by identifying some point on the car, perhaps the front door handle, as a particle representing the entire car.

We start our clock, and once every 10 s we note the car's position. As you can see from Table 2.1, the car moves to the right (which we have defined as the positive direction) during the first 10 s of motion, from position Ⓐ to position Ⓑ. After Ⓑ, the position values begin to decrease, suggesting the car is backing up from position Ⓑ through position Ⓕ. In fact, at Ⓓ, 30 s after we start measuring, the car is at the origin of coordinates (see Fig. 2.1a). It continues moving to the left and is more than 50 m to the left of $x = 0$ when we stop recording information after our sixth data point. A graphical representation of this information is presented in Figure 2.1b. Such a plot is called a *position–time graph*.

Notice the alternative representations of information, as discussed in Section 1.2, that we have used for the motion of the car. Figure 2.1a is a pictorial representation, whereas Figure 2.1b is a graphical representation. Table 2.1 is a tabular representation of the same information. The ultimate goal, as mentioned in Section 1.2, is a mathematical representation, which can be analyzed to solve for some requested piece of information.

In the introductory storyline, you observed the change in the position of your car relative to the power poles. The **displacement** Δx of a particle is defined as its change in position in some time interval. As the particle moves from an initial position x_i to a final position x_f, its displacement is given by

TABLE 2.1 Position of the Car at Various Times

Position	t (s)	x (m)
Ⓐ	0	30
Ⓑ	10	52
Ⓒ	20	38
Ⓓ	30	0
Ⓔ	40	−37
Ⓕ	50	−53

$$\Delta x \equiv x_f - x_i \qquad (2.1)$$

◄ Displacement

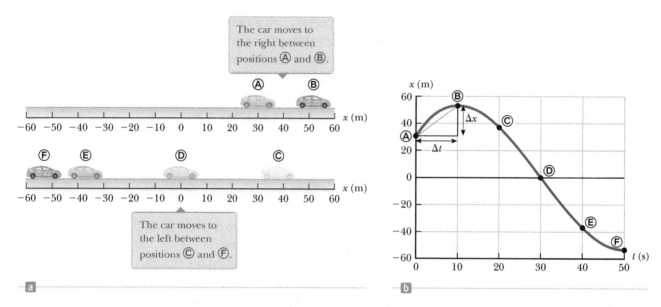

Figure 2.1 A car moves back and forth along a straight line. Because we are interested only in the car's translational motion, we can model it as a particle. Several representations of the information about the motion of the car can be used. Table 2.1 is a tabular representation of the information. (a) A pictorial representation of the motion of the car. (b) A graphical representation (position–time graph) of the motion of the car.

Figure 2.2 On this basketball court, players run back and forth for the entire game. The distance that the players run over the duration of the game is nonzero. The displacement of the players over the duration of the game is approximately zero because they keep returning to the same point over and over again.

We use the capital Greek letter delta (Δ) to denote the *change* in a quantity. From this definition, we see that Δx is positive if x_f is greater than x_i and negative if x_f is less than x_i. Given the data in Table 2.1, we can easily determine the displacement of the car for various time intervals.

It is very important to recognize the difference between displacement and distance traveled. **Distance** is the length of a path followed by a particle. Consider, for example, the basketball players in Figure 2.2. If a player runs from his own team's basket down the court to the other team's basket and then returns to his own basket, the *displacement* of the player during this time interval is zero because he ended up at the same point as he started: $x_f = x_i$, so $\Delta x = 0$. During this time interval, however, he moved through a *distance* of twice the length of the basketball court. Distance is always represented as a positive number, whereas displacement can be either positive or negative.

Displacement is an example of a vector quantity. Many other physical quantities, including position, velocity, and acceleration, also are vectors. In general, a **vector quantity** requires the specification of both direction and magnitude. For example, in the case of the car in Figure 2.1, by how much did the position of the car change (*magnitude*) and in what *direction*—forward or backward? By contrast, a **scalar quantity** has a numerical value and no direction. Distance is a scalar: how far did the car move, as measured by its odometer, in a certain time interval? In this chapter, we use positive (+) and negative (−) signs to indicate vector direction. For example, for horizontal motion let us arbitrarily specify to the right as being the positive direction. It follows that any object always moving to the right undergoes a positive displacement $\Delta x > 0$, and any object moving to the left undergoes a negative displacement so that $\Delta x < 0$. We shall treat vector quantities in greater detail in Chapter 3.

One very important point has not yet been mentioned. Notice that the data in Table 2.1 result only in the six data points in the graph in Figure 2.1b. Therefore, the motion of the particle is not completely known because we don't know its position at *all* times. The smooth curve drawn through the six points in the graph is only a *possibility* of the actual motion of the car. We only have information about six instants of time; we have no idea what happened between the data points. The smooth curve is a *guess* as to what happened, but keep in mind that it is *only* a guess. If the smooth curve does represent the actual motion of the car, the graph contains complete information about the entire 50-s interval during which we watch the car move.

QUICK QUIZ 2.1 Which of the following choices best describes what can be determined exactly from Table 2.1 and Figure 2.1 for the entire 50-s interval? (a) The distance the car moved. (b) The displacement of the car. (c) Both (a) and (b). (d) Neither (a) nor (b).

It is much easier to see changes in position from the graph than from a verbal description or even a table of numbers. For example, it is clear that the car covers more ground during the middle of the 50-s interval than at the end. Between positions Ⓒ and Ⓓ, the car changes position by almost 40 m, but during the last 10 s, between positions Ⓔ and Ⓕ, it changes position by less than half that much. A common way of comparing these different motions is to divide the displacement Δx that occurs between two clock readings by the value of that particular time interval Δt. The result turns out to be a very useful ratio, one that we shall use many times. This ratio has been given a special name: the *average velocity*. The **average velocity** $v_{x,\mathrm{avg}}$ of a particle is defined as the particle's displacement Δx divided by the time interval Δt during which that displacement occurs:

$$v_{x,\mathrm{avg}} \equiv \frac{\Delta x}{\Delta t} \qquad\qquad (2.2)$$

◀ Average velocity

where the subscript x indicates motion along the x axis. From this definition we see that average velocity has dimensions of length divided by time (L/T), or meters per second in SI units.

The average velocity of a particle moving in one dimension can be positive or negative, depending on the sign of the displacement. (The time interval Δt is always positive.) If the coordinate of the particle increases in time (that is, if $x_f > x_i$), Δx is positive and $v_{x,\mathrm{avg}} = \Delta x / \Delta t$ is positive. This case corresponds to a particle moving in the positive x direction, that is, toward larger values of x. If the coordinate decreases in time (that is, if $x_f < x_i$), Δx is negative and hence $v_{x,\mathrm{avg}}$ is negative. This case corresponds to a particle moving in the negative x direction.

We can interpret average velocity geometrically by drawing a straight line between any two points on the position–time graph in Figure 2.1b. This line forms the hypotenuse of a right triangle of height Δx and base Δt. The slope of this line is the ratio $\Delta x / \Delta t$, which is what we have defined as average velocity in Equation 2.2. For example, the line between positions Ⓐ and Ⓑ in Figure 2.1b has a slope equal to the average velocity of the car between those two times, $(52\ \mathrm{m} - 30\ \mathrm{m})/(10\ \mathrm{s} - 0) = 2.2\ \mathrm{m/s}$.

In everyday usage, the terms *speed* and *velocity* are interchangeable. In physics, however, there is a clear distinction between these two quantities. Consider a marathon runner who runs a distance d of more than 40 km and yet ends up at her starting point. Her total displacement is zero, so her average velocity is zero! Nonetheless, we need to be able to quantify how fast she was running. A slightly different ratio accomplishes that for us. The **average speed** v_{avg} of a particle, a scalar quantity, is defined as the total distance d traveled divided by the total time interval required to travel that distance:

$$v_{\mathrm{avg}} \equiv \frac{d}{\Delta t} \qquad\qquad (2.3)$$

◀ Average speed

The SI unit of average speed is the same as the unit of average velocity: meters per second. Unlike average velocity, however, average speed has no direction and is always expressed as a positive number. Notice the clear distinction between the definitions of average velocity and average speed: average velocity (Eq. 2.2) is the *displacement* divided by the time interval, whereas average speed (Eq. 2.3) is the *distance* divided by the time interval.

Knowledge of the average velocity or average speed of a particle does not provide information about the details of the trip. For example, suppose it takes you 45.0 s to travel 100 m down a long, straight hallway toward your departure gate at an airport. At the 100-m mark, you realize you missed the restroom, and you return back 25.0 m along the same hallway, taking 10.0 s to make the return trip. The magnitude of your average *velocity* is +75.0 m/55.0 s = +1.36 m/s. The average *speed* for your trip is 125 m/55.0 s = 2.27 m/s. You may have traveled at various speeds during the walk and, of course, you changed direction. Neither average velocity nor average speed provides information about these details.

QUICK QUIZ 2.2 Under which of the following conditions is the magnitude of the average velocity of a particle moving in one dimension smaller than the average speed over some time interval? **(a)** A particle moves in the $+x$ direction without reversing. **(b)** A particle moves in the $-x$ direction without reversing. **(c)** A particle moves in the $+x$ direction and then reverses the direction of its motion. **(d)** There are no conditions for which this is true.

Example 2.1   **Calculating the Average Velocity and Speed**

Find the displacement, average velocity, and average speed of the car in Figure 2.1a between positions Ⓐ and Ⓕ.

SOLUTION

Consult Figure 2.1 to form a mental image of the car and its motion. We model the car as a particle. From the position–time graph given in Figure 2.1b, notice that $x_Ⓐ = 30$ m at $t_Ⓐ = 0$ s and that $x_Ⓕ = -53$ m at $t_Ⓕ = 50$ s.

Use Equation 2.1 to find the displacement of the car:

$$\Delta x = x_Ⓕ - x_Ⓐ = -53 \text{ m} - 30 \text{ m} = -83 \text{ m}$$

This result means that the car ends up 83 m in the negative direction (to the left, in this case) from where it started. This number has the correct units and is of the same order of magnitude as the supplied data. A quick look at Figure 2.1a indicates that it is the correct answer.

Use Equation 2.2 to find the car's average velocity:

$$v_{x,\text{avg}} = \frac{x_Ⓕ - x_Ⓐ}{t_Ⓕ - t_Ⓐ}$$

$$= \frac{-53 \text{ m} - 30 \text{ m}}{50 \text{ s} - 0 \text{ s}} = \frac{-83 \text{ m}}{50 \text{ s}} = -1.7 \text{ m/s}$$

We cannot unambiguously find the average speed of the car from the data in Table 2.1 because we do not have information about the positions of the car between the data points. If we adopt the assumption that the details of the car's position are described by the curve in Figure 2.1b, the distance traveled is 22 m (from Ⓐ to Ⓑ) plus 105 m (from Ⓑ to Ⓕ), for a total of 127 m.

Use Equation 2.3 to find the car's average speed:

$$v_{\text{avg}} = \frac{127 \text{ m}}{50 \text{ s}} = 2.5 \text{ m/s}$$

Notice that the average speed is positive, as it must be. Suppose the red-brown curve in Figure 2.1b were different so that between 0 s and 10 s it went from Ⓐ up to 100 m and then came back down to Ⓑ. The average speed of the car would change because the distance is different, but the average velocity would not change.

2.2 Instantaneous Velocity and Speed

Often we need to know the velocity of a particle at a particular instant in time t rather than the average velocity over a finite time interval Δt. In other words, you would like to be able to specify your velocity just as precisely as you can specify your position by noting what is happening at a specific clock reading, that is, at some

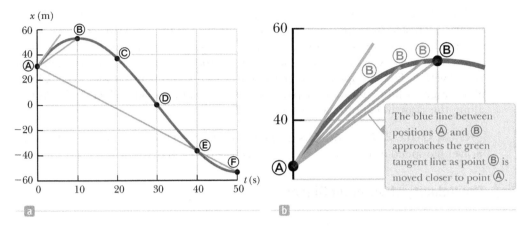

Figure 2.3 (a) Graph representing the motion of the car in Figure 2.1. (b) An enlargement of the upper-left-hand corner of the graph.

specific instant. What does it mean to talk about how quickly something is moving if we "freeze time" and talk only about an individual instant? If the time interval has a value of zero, the displacement of the object is also zero, so the average velocity from Equation 2.2 would seem to be 0/0. How do we evaluate that ratio? In the late 1600s, with the invention of calculus, scientists began to understand how to answer that question and describe an object's motion at any moment in time.

To see how that is done, consider Figure 2.3a, which is a reproduction of the graph in Figure 2.1b. What is the particle's velocity at $t = 0$? We have already discussed the average velocity for the interval during which the car moved from position Ⓐ to position Ⓑ (given by the slope of the blue line) and for the interval during which it moved from Ⓐ to Ⓕ (represented by the slope of the longer blue line and calculated in Example 2.1). The car starts out by moving to the right, which we defined to be the positive direction. Therefore, being positive, the value of the average velocity during the interval from Ⓐ to Ⓑ is more representative of the initial velocity than is the value of the average velocity during the interval from Ⓐ to Ⓕ, which we determined to be negative in Example 2.1. Now let us focus on the short blue line and imagine sliding point Ⓑ to the left along the curve, toward point Ⓐ, as in Figure 2.3b. The line between the points becomes steeper and steeper, and as the two points become extremely close together, the line becomes a tangent line to the curve, indicated by the green line in Figure 2.3b. The slope of this tangent line represents the velocity of the car at point Ⓐ. What we have done is determine the *instantaneous velocity* at that moment. In other words, the **instantaneous velocity** v_x equals the limiting value of the ratio $\Delta x/\Delta t$ as Δt approaches zero:[2]

$$v_x \equiv \lim_{\Delta t \to 0} \frac{\Delta x}{\Delta t} \tag{2.4}$$

In calculus notation, this limit is called the *derivative* of x with respect to t, written dx/dt:

$$v_x \equiv \lim_{\Delta t \to 0} \frac{\Delta x}{\Delta t} = \frac{dx}{dt} \tag{2.5}$$

◀ Instantaneous velocity

The instantaneous velocity can be positive, negative, or zero. When the slope of the position–time graph is positive, such as at any time during the first 10 s in Figure 2.3, v_x is positive and the car is moving toward larger values of x. After point Ⓑ, v_x is negative because the slope is negative and the car is moving toward smaller values of x. At point Ⓑ, the slope and the instantaneous velocity are zero and the car is momentarily at rest.

---

[2]As mentioned previously, the displacement Δx also approaches zero as Δt approaches zero, so the ratio $\Delta x/\Delta t$ looks like 0/0. The ratio can be evaluated in the limit in this situation, however. As Δx and Δt become smaller and smaller, the ratio $\Delta x/\Delta t$ approaches a value equal to the slope of the line tangent to the x-versus-t curve.

From here on, we use the word *velocity* to designate instantaneous velocity. When we are interested in *average velocity*, we shall always use the adjective *average*.

The **instantaneous speed** of a particle is defined as the magnitude of its instantaneous velocity. As with average speed, instantaneous speed has no direction associated with it. For example, if one particle has an instantaneous velocity of +25 m/s along a given line and another particle has an instantaneous velocity of −25 m/s along the same line, both have a speed[3] of 25 m/s.

QUICK QUIZ 2.3 Are officers in the highway patrol more interested in **(a)** your average speed or **(b)** your instantaneous speed as you drive?

Conceptual Example 2.2 **The Velocity of Different Objects**

Consider the following one-dimensional motions: **(A)** a ball thrown directly upward rises to a highest point and falls back into the thrower's hand; **(B)** a race car starts from rest and speeds up to 100 m/s; and **(C)** a spacecraft drifts through space at constant velocity. Are there any points in the motion of these objects at which the instantaneous velocity has the same value as the average velocity over the entire motion? If so, identify the point(s).

SOLUTION

(A) The average velocity for the thrown ball is zero because the ball returns to the starting point; therefore, its displacement is zero. There is one point at which the instantaneous velocity is zero: at the top of the motion.

(B) The car's average velocity cannot be evaluated unambiguously with the information given, but it must have some value between 0 and 100 m/s. Because the car will have every instantaneous velocity between 0 and 100 m/s at some time during the interval, there must be some instant at which the instantaneous velocity is equal to the average velocity over the entire motion.

(C) Because the spacecraft's instantaneous velocity is constant, its instantaneous velocity at *any* time and its average velocity over *any* time interval are the same.

Example 2.3 **Average and Instantaneous Velocity**

A particle moves along the x axis. Its position varies with time according to the expression $x = -4t + 2t^2$, where x is in meters and t is in seconds.[4] The position–time graph for this motion is shown in Figure 2.4a. Because the position of the particle is given by a mathematical function, the motion of the particle is known at all times, *unlike* that of the car in Figure 2.1, where data is only provided at six instants of time. Notice that the particle moves in the negative x direction for the first second of motion, is momentarily at rest at the moment $t = 1$ s, and moves in the positive x direction at times $t > 1$ s.

(A) Determine the displacement of the particle in the time intervals $t = 0$ to $t = 1$ s and $t = 1$ s to $t = 3$ s.

SOLUTION

From the graph in Figure 2.4a, form a mental representation of the particle's motion. Keep in mind that the particle does not move in a curved path in space such as that shown by the red-brown curve in the graphical representation. The particle moves only along the x axis in one dimension as shown in Figure 2.4b. At $t = 0$, is it moving to the right or to the left?

During the first time interval, the slope is negative and hence the average velocity is negative. Therefore, we know that the displacement between Ⓐ and Ⓑ must be a negative number having units of meters. Similarly, we expect the displacement between Ⓑ and Ⓓ to be positive.

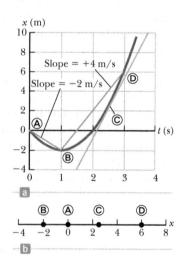

Figure 2.4 (Example 2.3) (a) Position–time graph for a particle having an x coordinate that varies in time according to the expression $x = -4t + 2t^2$. (b) The particle moves in one dimension along the x axis.

continued

---

[3]As with velocity, we drop the adjective for instantaneous speed. *Speed* means "instantaneous speed."

[4]Simply to make it easier to read, we write the expression as $x = -4t + 2t^2$ rather than as $x = (-4.00 \text{ m/s})t + (2.00 \text{ m/s}^2)t^{2.00}$. When an equation summarizes measurements, consider its coefficients and exponents to have as many significant figures as other data quoted in a problem. Consider its coefficients to have the units required for dimensional consistency. When we start our clocks at $t = 0$, we usually do not mean to limit the precision to a single digit. Consider any zero value in this book to have as many significant figures as you need.

2.3 continued

In the first time interval, set $t_i = t_{\text{Ⓐ}} = 0$ and $t_f = t_{\text{Ⓑ}} = 1$ s. Substitute these values into $x = -4t + 2t^2$ and use Equation 2.1 to find the displacement:

$$\Delta x_{\text{Ⓐ}\rightarrow\text{Ⓑ}} = x_f - x_i = x_{\text{Ⓑ}} - x_{\text{Ⓐ}}$$
$$= [-4(1) + 2(1)^2] - [-4(0) + 2(0)^2] = -2 \text{ m}$$

For the second time interval ($t = 1$ s to $t = 3$ s), set $t_i = t_{\text{Ⓑ}} = 1$ s and $t_f = t_{\text{Ⓓ}} = 3$ s:

$$\Delta x_{\text{Ⓑ}\rightarrow\text{Ⓓ}} = x_f - x_i = x_{\text{Ⓓ}} - x_{\text{Ⓑ}}$$
$$= [-4(3) + 2(3)^2] - [-4(1) + 2(1)^2] = +8 \text{ m}$$

These displacements can also be read directly from the position–time graph.

(B) Calculate the average velocity during these two time intervals.

SOLUTION

In the first time interval, use Equation 2.2 with $\Delta t = t_f - t_i = t_{\text{Ⓑ}} - t_{\text{Ⓐ}} = 1$ s:

$$v_{x,\text{avg (Ⓐ}\rightarrow\text{Ⓑ)}} = \frac{\Delta x_{\text{Ⓐ}\rightarrow\text{Ⓑ}}}{\Delta t} = \frac{-2 \text{ m}}{1 \text{ s}} = -2 \text{ m/s}$$

In the second time interval, $\Delta t = 2$ s:

$$v_{x,\text{avg (Ⓑ}\rightarrow\text{Ⓓ)}} = \frac{\Delta x_{\text{Ⓑ}\rightarrow\text{Ⓓ}}}{\Delta t} = \frac{8 \text{ m}}{2 \text{ s}} = +4 \text{ m/s}$$

These values are the same as the slopes of the blue lines joining these points in Figure 2.4a.

(C) Find the instantaneous velocity of the particle at $t = 2.5$ s.

SOLUTION

Calculate the slope of the green line at $t = 2.5$ s (point Ⓒ) in Figure 2.4a by reading position and time values for the ends of the green line from the graph:

$$v_x = \frac{10 \text{ m} - (-4 \text{ m})}{3.8 \text{ s} - 1.5 \text{ s}} = +6 \text{ m/s}$$

Notice that this instantaneous velocity is on the same order of magnitude as our previous results, that is, a few meters per second. Is that what you would have expected?

2.3 Analysis Model: Particle Under Constant Velocity

In Section 1.2 we discussed the importance of making models. As mentioned there, a particularly important model used in the solution to physics problems is an *analysis model*. **An analysis model is a common situation that occurs time and again when solving physics problems.** Because it represents a common situation, it also represents a common type of problem that we have solved before. When you identify an analysis model in a new problem, the solution to the new problem can be modeled after that of the previously solved problem. Analysis models help us to recognize those common situations and guide us toward a solution to the problem. The form that an analysis model takes is a description of either (1) the behavior of some physical entity or (2) the interaction between that entity and the environment. When you encounter a new problem, you should identify the fundamental details of the problem, ignore details that are not important, and attempt to recognize which of the situations you have already seen that might be used as a model for the new problem. For example, suppose an automobile is moving along a straight freeway at a constant speed. Is it important that it is an automobile? Is it important that it is a freeway? If the answers to both questions are no, but the car moves in a straight line at constant speed, we model the automobile as a *particle under constant velocity*, which we will discuss in this section. Once the problem has been modeled, it is no longer about an automobile. It is about a particle undergoing a certain type of motion, a motion that we have studied before.

◀ Analysis model

This method is somewhat similar to the common practice in the legal profession of finding "legal precedents." If a previously resolved case can be found that is very similar legally to the current one, it is used as a model and an argument is made in court to link them logically. The finding in the previous case can then be used to sway the finding in the current case. We will do something similar in physics. For a given problem, we search for a "physics precedent," a model with which we are already familiar and that can be applied to the current problem.

All of the analysis models that we will develop are based on four fundamental simplification models. The first of the four is the particle model discussed in the introduction to this chapter. We will look at a particle under various behaviors and environmental interactions. Further analysis models are introduced in later chapters based on simplification models of a *system*, a *rigid object*, and a *wave*. Once we have introduced these analysis models, we shall see that they appear again and again in different problem situations.

When solving a problem, you should avoid browsing through the chapter looking for an equation that contains the unknown variable that is requested in the problem. In many cases, the equation you find may have nothing to do with the problem you are attempting to solve. It is *much* better to take this first step: **Identify the analysis model that is appropriate for the problem.** To do so, think carefully about what is going on in the problem and match it to a situation you have seen before. Once the analysis model is identified, there are a small number of equations from which to choose that are appropriate for that model, sometimes only one equation. Therefore, **the model tells you which equation(s) to use for the mathematical representation.**

Let us use Equation 2.2 to build our first analysis model for solving problems. We imagine a particle moving with a constant velocity. The model of a **particle under constant velocity** can be applied in *any* situation in which an entity that can be modeled as a particle is moving with constant velocity. This situation occurs frequently, so this model is important.

If the velocity of a particle is constant, its instantaneous velocity at any instant during a time interval is the same as the average velocity over the interval. That is, $v_x = v_{x,\text{avg}}$. Therefore, substituting v_x for $v_{x,\text{avg}}$ in Equation 2.2 gives us an equation to be used in the mathematical representation of this situation:

$$v_x = \frac{\Delta x}{\Delta t} \tag{2.6}$$

Remembering that $\Delta x = x_f - x_i$, we see that $v_x = (x_f - x_i)/\Delta t$, or

$$x_f = x_i + v_x \Delta t$$

This equation tells us that the position of the particle is given by the sum of its original position x_i at time $t = 0$ plus the displacement $v_x \Delta t$ that occurs during the time interval Δt. In practice, we usually choose the time at the beginning of the interval to be $t_i = 0$ and the time at the end of the interval to be $t_f = t$, so our equation becomes

$$x_f = x_i + v_x t \quad \text{(for constant } v_x\text{)} \tag{2.7}$$

Equations 2.6 and 2.7 are the primary equations used in the model of a particle under constant velocity. Whenever you have identified the analysis model in a problem to be the particle under constant velocity, you can immediately turn to these equations.

Figure 2.5 is a graphical representation of the particle under constant velocity. On this position–time graph, the slope of the line representing the motion is constant and equal to the magnitude of the velocity. Equation 2.7, which is the equation of a straight line, is the mathematical representation of the particle under constant velocity model. The slope of the straight line is v_x and the y intercept is x_i in both representations.

In the opening storyline, the particle under constant velocity model was represented by the part of the motion taking place at "fixed speed." You found in the

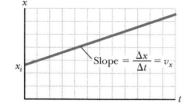

Figure 2.5 Position–time graph for a particle under constant velocity. The value of the constant velocity is the slope of the line.

Position as a function of ▶ time for the particle under constant velocity model storyline that the time intervals between poles were always the same in this case. Is this result consistent with Equation 2.7? Example 2.4 below shows a numerical application of the particle under constant velocity model.

Example 2.4 Modeling a Runner as a Particle

A kinesiologist is studying the biomechanics of the human body. (*Kinesiology* is the study of the movement of the human body. Notice the connection to the word *kinematics*.) She determines the velocity of an experimental subject while he runs along a straight line at a constant rate. The kinesiologist starts the stopwatch at the moment the runner passes a given point and stops it after the runner has passed another point 20 m away. The time interval indicated on the stopwatch is 4.0 s.

(A) What is the runner's velocity?

SOLUTION

We model the moving runner as a particle because the size of the runner and the movement of arms and legs are unnecessary details. Because the problem states that the subject runs "at a constant rate," we can model him as a *particle under constant velocity*.

Having identified the model, we can use Equation 2.6 to find the constant velocity of the runner:

$$v_x = \frac{\Delta x}{\Delta t} = \frac{x_f - x_i}{\Delta t} = \frac{20 \text{ m} - 0}{4.0 \text{ s}} = 5.0 \text{ m/s}$$

(B) If the runner continues his motion after the stopwatch is stopped, what is his position after 10 s have passed?

SOLUTION

Use Equation 2.7 and the velocity found in part (A) to find the position of the particle at time $t = 10$ s:

$$x_f = x_i + v_x t = 0 + (5.0 \text{ m/s})(10 \text{ s}) = 50 \text{ m}$$

Is the result for part (A) a reasonable speed for a human? How does it compare to world-record speeds in 100-m and 200-m sprints? Notice the value in part (B) is more than twice that of the 20-m position at which the stopwatch was stopped. Is this value consistent with the time of 10 s being more than twice the time of 4.0 s?

The mathematical manipulations for the particle under constant velocity stem from Equation 2.6 and its descendent, Equation 2.7. These equations can be used to solve for any variable in the equations that happens to be unknown if the other variables are known. For example, in part (B) of Example 2.4, we find the position when the velocity and the time are known. Similarly, if we know the velocity and the final position, we could use Equation 2.7 to find the time at which the runner is at this position.

A particle under constant velocity moves with a constant speed along a straight line. Now consider a particle moving with a constant speed through a distance d along a *curved* path. As we will see in Section 2.5 below, a change in the direction of motion of a particle signifies a change in the velocity of a particle even though its speed is constant; there is a change in the speed *vector*. Therefore, our particle moving along a curved path is not represented by the particle under constant velocity model. However, it can be represented with the model of a **particle under constant speed.** The primary equation for this model is Equation 2.3, with the average speed v_{avg} replaced by the constant speed v:

$$v = \frac{d}{\Delta t} \tag{2.8}$$

As an example, imagine a particle moving at a constant speed in a circular path. If the speed is 5.00 m/s and the radius of the path is 10.0 m, we can calculate the time interval required to complete one trip around the circle:

$$v = \frac{d}{\Delta t} \rightarrow \Delta t = \frac{d}{v} = \frac{2\pi r}{v} = \frac{2\pi (10.0 \text{ m})}{5.00 \text{ m/s}} = 12.6 \text{ s}$$

Particle Under Constant Velocity

Imagine a moving object that can be modeled as a particle. If it moves at a constant speed through a displacement Δx in a straight line in a time interval Δt, its constant velocity is

$$v_x = \frac{\Delta x}{\Delta t} \qquad \text{(2.6)}$$

The position of the particle as a function of time is given by

$$x_f = x_i + v_x t \qquad \text{(2.7)}$$

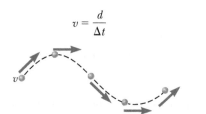

Examples:

- a meteoroid traveling through gravity-free space
- a car traveling at a constant speed on a straight highway
- a runner traveling at constant speed on a perfectly straight path
- an object moving at terminal speed through a viscous medium (Chapter 6)

Particle Under Constant Speed

Imagine a moving object that can be modeled as a particle. If it moves at a constant speed through a distance d along a straight line or a curved path in a time interval Δt, its constant speed is

$$v = \frac{d}{\Delta t} \qquad \text{(2.8)}$$

Examples:

- a planet traveling around a perfectly circular orbit
- a car traveling at a constant speed on a curved racetrack
- a runner traveling at constant speed on a curved path
- a charged particle moving through a uniform magnetic field (Chapter 28)

2.4 The Analysis Model Approach to Problem Solving

We have just seen our first analysis models: the particle under constant velocity and the particle under constant speed. Now, what do we do with these models? The analysis models fit into a general method of solving problems that we describe below. In particular, pay attention to the "Categorize" step in the discussion below. That is where you identify the analysis model to be applied to the problem. After that, the problem is solved using the equation or equations that you have already learned to be associated with that model. This is the way physicists approach complex situations and complicated problems, and break them into manageable pieces. It is an extremely useful skill for you to learn. It may look complicated at first, but it will become easier and of second nature as you practice it!

Conceptualize

- The first things to do when approaching a problem are to *think about* and *understand* the situation. Study carefully any representations of the information (for example, diagrams, graphs, tables, or photographs) that accompany the problem. Imagine a movie, running in your mind, of what happens in the problem: the mental representation.
- If a pictorial representation is not provided, you should almost always make a quick drawing of the situation. Indicate any known values, perhaps in a table or directly on your sketch.
- Now focus on what algebraic or numerical information is given in the problem. Carefully read the problem statement, looking for key phrases such as "starts from rest" ($v_i = 0$) or "stops" ($v_f = 0$).

- Now focus on the expected result of solving the problem. Exactly what is the question asking? Will the final result be numerical, algebraic, or verbal? Do you know what units to expect?
- Don't forget to incorporate information from your own experiences and common sense. What should a reasonable answer look like? For example, you wouldn't expect to calculate the speed of an automobile to be 5×10^6 m/s.

Categorize

- Once you have a good idea of what the problem is about, you need to *simplify* the problem. Use a simplification model to remove the details that are not important to the solution. For example, model a moving object as a particle. If appropriate, ignore air resistance or friction between a sliding object and a surface.
- Once the problem is simplified, it is important to *categorize* the problem in one of two ways. Is it a simple *substitution problem* such that numbers can be substituted into a simple equation or a definition? If so, the problem is likely to be finished when this substitution is done. If not, you face what we call an *analysis problem:* the situation must be analyzed more deeply to generate an appropriate equation and reach a solution.
- If it is an analysis problem, it needs to be categorized further. Have you seen this type of problem before? Does it fall into the growing list of types of problems that you have solved previously? If so, identify any *analysis model(s)* appropriate for the problem to prepare for the Analyze step below. Being able to classify a problem with an analysis model can make it much easier to lay out a plan to solve it.

Analyze

- Now you must analyze the problem and strive for a mathematical solution. Because you have already categorized the problem and identified an analysis model, it should not be too difficult to select relevant equations that apply to the type of situation in the problem. For example, if the problem involves a particle under constant velocity, Equation 2.7 is relevant.
- Use algebra (and calculus, if necessary) to solve symbolically for the unknown variable in terms of what is given. Finally, substitute in the appropriate numbers, calculate the result, and round it to the proper number of significant figures.

Finalize

- Examine your numerical answer. Does it have the correct units? Does it meet your expectations from your conceptualization of the problem? What about the algebraic form of the result? Does it make sense? Examine the variables in the problem to see whether the answer would change in a physically meaningful way if the variables were drastically increased or decreased or even became zero. Looking at limiting cases to see whether they yield expected values is a very useful way to make sure that you are obtaining reasonable results.
- Think about how this problem compared with others you have solved. How was it similar? In what critical ways did it differ? Why was this problem assigned? Can you figure out what you have learned by doing it? If it is a new category of problem, be sure you understand it so that you can use it as a model for solving similar problems in the future.

When solving complex problems, you may need to identify a series of subproblems and apply the Analysis Model Approach to each. For simple problems, you probably don't need this approach. When you are trying to solve a problem and you don't know what to do next, however, remember the steps in the approach and use them as a guide.

In the rest of this book, we will label the *Conceptualize, Categorize, Analyze,* and *Finalize* steps in the worked examples. If a worked example is identified as a substitution problem in the *Categorize* step, there will generally not be *Analyze* and *Finalize* sections labeled in the solution.

To show how to apply this approach, we reproduce Example 2.4 below, with the steps of the approach labeled.

Example 2.4 | **Modeling a Runner as a Particle**

A kinesiologist is studying the biomechanics of the human body. (*Kinesiology* is the study of the movement of the human body. Notice the connection to the word *kinematics*.) She determines the velocity of an experimental subject while he runs along a straight line at a constant rate. The kinesiologist starts the stopwatch at the moment the runner passes a given point and stops it after the runner has passed another point 20 m away. The time interval indicated on the stopwatch is 4.0 s.

(A) What is the runner's velocity?

SOLUTION

Conceptualize We model the moving runner as a particle because the size of the runner and the movement of arms and legs are unnecessary details.

Categorize Because the problem states that the subject runs "at a constant rate," we can model him as a *particle under constant velocity*.

Analyze Having identified the model, we can use Equation 2.6 to find the constant velocity of the runner:

$$v_x = \frac{\Delta x}{\Delta t} = \frac{x_f - x_i}{\Delta t} = \frac{20 \text{ m} - 0}{4.0 \text{ s}} = 5.0 \text{ m/s}$$

(B) If the runner continues his motion after the stopwatch is stopped, what is his position after 10 s have passed?

SOLUTION

Use Equation 2.7 and the velocity found in part (A) to find the position of the particle at time $t = 10$ s:

$$x_f = x_i + v_x t = 0 + (5.0 \text{ m/s})(10 \text{ s}) = 50 \text{ m}$$

Finalize Is the result for part (A) a reasonable speed for a human? How does it compare to world-record speeds in 100-m and 200-m sprints? Notice the value in part (B) is more than twice that of the 20-m position at which the stopwatch was stopped. Is this value consistent with the time of 10 s being more than twice the time of 4.0 s?

2.5 Acceleration

In Example 2.3, we worked with a common situation in which the velocity of a particle changes while the particle is moving. When the velocity of a particle changes with time, the particle is said to be *accelerating*. For example, the magnitude of a car's velocity increases when you step on the gas and decreases when you apply the brakes. Both of these actions result in an acceleration of the car. Let us see how to quantify acceleration.

Suppose an object that can be modeled as a particle moving along the *x* axis has an initial velocity v_{xi} at time t_i at position Ⓐ and a final velocity v_{xf} at time t_f at position Ⓑ as in Figure 2.6a. The red-brown curve in Figure 2.6b shows how the velocity varies with time. The **average acceleration** $a_{x,\text{avg}}$ of the particle is defined as the *change* in velocity Δv_x divided by the time interval Δt during which that change occurs:

Average acceleration ▶

$$a_{x,\text{avg}} \equiv \frac{\Delta v_x}{\Delta t} = \frac{v_{xf} - v_{xi}}{t_f - t_i} \qquad (2.9)$$

As with velocity, when the motion being analyzed is one dimensional, we can use positive and negative signs to indicate the direction of the acceleration. Because the dimensions of velocity are L/T and the dimension of time is T, acceleration has

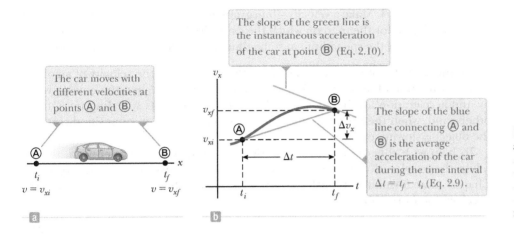

Figure 2.6 (a) A car, modeled as a particle, moving along the x axis from Ⓐ to Ⓑ, has velocity v_{xi} at $t = t_i$ and velocity v_{xf} at $t = t_f$. (b) Velocity–time graph (red-brown) for the particle moving in a straight line.

dimensions of length divided by time squared, or L/T². The SI unit of acceleration is meters per second squared (m/s²). It might be easier to interpret these units if you think of them as meters per second per second. For example, suppose an object has an acceleration of +2 m/s². You can interpret this value by forming a mental image of the object having a velocity that is along a straight line and is increasing by 2 m/s during every time interval of 1 s. If the object starts from rest, you should be able to picture it moving at a velocity of +2 m/s after 1 s, at +4 m/s after 2 s, and so on.

When your friend sped up from the traffic light in the opening storyline, you found that the time intervals between poles on the side of the road decreased. Is that result consistent with your expectations? Each new displacement between poles is undertaken at a higher speed, so the time intervals between poles become smaller.

In some situations, the value of the average acceleration may be different over different time intervals. It is therefore useful to define the **instantaneous acceleration** as the limit of the average acceleration as Δt approaches zero. This concept is analogous to the definition of instantaneous velocity discussed in Section 2.2. If we imagine that point Ⓐ is brought closer and closer to point Ⓑ in Figure 2.6a and we take the limit of $\Delta v_x/\Delta t$ as Δt approaches zero, we obtain the instantaneous acceleration at point Ⓑ:

$$a_x \equiv \lim_{\Delta t \to 0} \frac{\Delta v_x}{\Delta t} = \frac{dv_x}{dt} \tag{2.10}$$

That is, the instantaneous acceleration equals the derivative of the velocity with respect to time, which by definition is the slope of the velocity–time graph. The slope of the green line in Figure 2.6b is equal to the instantaneous acceleration at point Ⓑ. Notice that Figure 2.6b is a *velocity–time* graph, not a *position–time* graph like Figures 2.1b, 2.3, 2.4, and 2.5. Therefore, we see that just as the velocity of a moving particle is the slope at a point on the particle's x–t graph, the acceleration of a particle is the slope at a point on the particle's v_x–t graph. One can interpret the derivative of the velocity with respect to time as the time rate of change of velocity. If a_x is positive, the acceleration is in the positive x direction; if a_x is negative, the acceleration is in the negative x direction.

Figure 2.7 illustrates how an acceleration–time graph is related to a velocity-time graph. The acceleration at any time is the slope of the velocity–time graph at that time. Positive values of acceleration correspond to those points in Figure 2.7a where the velocity is increasing in the positive x direction. The acceleration reaches a maximum at time $t_Ⓐ$, when the slope of the velocity–time graph is a maximum. The acceleration then goes to zero at time $t_Ⓑ$, when the velocity is a maximum (that is, when the slope of the v_x–t graph is zero). The acceleration is negative when the velocity is decreasing in the positive x direction, and it reaches its most negative value at time $t_Ⓒ$.

◀ Instantaneous acceleration

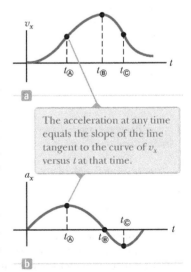

Figure 2.7 (a) The velocity–time graph for a particle moving along the x axis. (b) The instantaneous acceleration can be obtained from the velocity–time graph.

Make a velocity–time graph for the car in Figure 2.1a. Suppose the speed limit for the road on which the car is driving is 30 km/h. True or False? The car exceeds the speed limit at some time within the time interval 0–50 s.

For the case of motion in a straight line, the direction of the velocity of an object and the direction of its acceleration are related as follows. When the object's velocity and acceleration are in the same direction, the object is speeding up. On the other hand, when the object's velocity and acceleration are in opposite directions, the object is slowing down.

To help with this discussion of the signs of velocity and acceleration, we can relate the acceleration of an object to the total *force* exerted on the object. In Chapter 5, we formally establish that **the force on an object is proportional to the acceleration of the object:**

$$F_x \propto a_x \qquad (2.11)$$

This proportionality indicates that acceleration is caused by force. Furthermore, force and acceleration are both vectors, and the vectors are in the same direction. Therefore, let us think about the signs of velocity and acceleration by imagining a force applied to an object and causing it to accelerate. Let us assume the velocity and acceleration are in the same direction. This situation corresponds to an object that experiences a force acting in the same direction as its velocity. In this case, the object speeds up! Now suppose the velocity and acceleration are in opposite directions. In this situation, the object moves in some direction and experiences a force acting in the opposite direction. Therefore, the object slows down! It is very useful to equate the direction of the acceleration to the direction of a force because it is easier from our everyday experience to think about what effect a force will have on an object than to think only in terms of the direction of the acceleration.

PITFALL PREVENTION 2.4
Negative Acceleration Keep in mind that *negative acceleration does not necessarily mean that an object is slowing down.* If the acceleration is negative and the velocity is negative, the object is speeding up!

PITFALL PREVENTION 2.5
Deceleration The word *deceleration* has the common popular connotation of *slowing down.* We will not use this word in this book because it confuses the definition we have given for negative acceleration.

QUICK QUIZ 2.5 If a car is traveling eastward and slowing down, what is the direction of the force on the car that causes it to slow down? **(a)** eastward **(b)** westward **(c)** neither eastward nor westward

From now on, we shall use the term *acceleration* to mean instantaneous acceleration. When we mean average acceleration, we shall always use the adjective *average*. Because $v_x = dx/dt$, the acceleration can also be written as

$$a_x = \frac{dv_x}{dt} = \frac{d}{dt}\left(\frac{dx}{dt}\right) = \frac{d^2x}{dt^2} \qquad (2.12)$$

That is, in one-dimensional motion, the acceleration of a particle equals the *second derivative* of the particle's position x with respect to time.

Conceptual Example **2.5** | **Graphical Relationships Between *x*, *v_x*, and *a_x***

The position of an object moving along the *x* axis varies with time as in Figure 2.8a. Graph the velocity versus time and the acceleration versus time for the object.

SOLUTION

The velocity at any instant is the slope of the tangent to the *x–t* graph at that instant. Between $t = 0$ and $t = t_{\circledA}$, the slope of the *x–t* graph increases uniformly, so the velocity increases linearly as shown in Figure 2.8b. Between $t_{\circledA}$ and $t_{\circledB}$, the slope of the *x–t* graph is constant, so the velocity remains constant. Between $t_{\circledB}$ and $t_{\circledC}$, the slope of the *x–t* graph decreases, so the value of the velocity in the v_x–*t* graph decreases. At $t_{\circledC}$,

the slope of the *x–t* graph is zero, so the velocity is zero at that instant. Between $t_{\circledC}$ and $t_{\circledD}$, the slope of the *x–t* graph and therefore the velocity are negative and decrease uniformly in this interval. In the interval $t_{\circledD}$ to $t_{\circledE}$, the slope of the *x–t* graph is still negative, and at $t_{\circledE}$ it goes to zero. Finally, after $t_{\circledE}$, the slope of the *x–t* graph is zero, meaning that the object is at rest for $t > t_{\circledE}$.

continued

2.5 continued

The acceleration at any instant is the slope of the tangent to the v_x–t graph at that instant. The graph of acceleration versus time for this object is shown in Figure 2.8c. The acceleration is constant and positive between 0 and $t_Ⓐ$, where the slope of the v_x–t graph is positive. It is zero between $t_Ⓐ$ and $t_Ⓑ$ and for $t > t_Ⓕ$ because the slope of the v_x–t graph is zero at these times. It is negative between $t_Ⓑ$ and $t_Ⓔ$ because the slope of the v_x–t graph is negative during this interval. Between $t_Ⓔ$ and $t_Ⓕ$, the acceleration is positive like it is between 0 and $t_Ⓐ$, but higher in value because the slope of the v_x–t graph is steeper.

Notice that the sudden changes in acceleration shown in Figure 2.8c are unphysical. Such instantaneous changes cannot occur in reality.

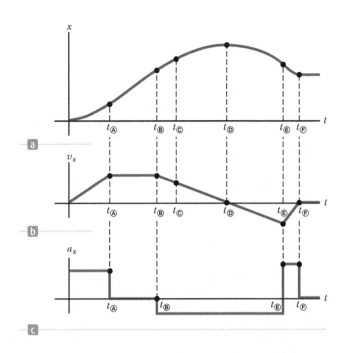

Figure 2.8 (Conceptual Example 2.5) (a) Position–time graph for an object moving along the x axis. (b) The velocity–time graph for the object is obtained by measuring the slope of the position–time graph at each instant. (c) The acceleration–time graph for the object is obtained by measuring the slope of the velocity–time graph at each instant.

Example 2.6 | Average and Instantaneous Acceleration

The velocity of a particle moving along the x axis varies according to the expression $v_x = 40 - 5t^2$, where v_x is in meters per second and t is in seconds.

(A) Find the average acceleration in the time interval $t = 0$ to $t = 2.0$ s.

SOLUTION

Conceptualize Think about what the particle is doing from the mathematical representation. Is it moving at $t = 0$? In which direction? Does it speed up or slow down? Figure 2.9 is a v_x–t graph that was created from the velocity versus time expression given in the problem statement. Because the slope of the entire v_x–t curve is negative, we expect the acceleration to be negative.

Categorize The solution to this problem does not require either of the analysis models we have developed so far, and can be solved with simple mathematics. Therefore, we categorize the problem as a substitution problem.

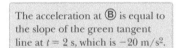
The acceleration at Ⓑ is equal to the slope of the green tangent line at $t = 2$ s, which is -20 m/s².

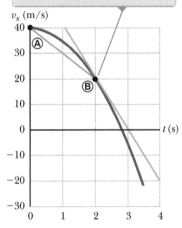

Figure 2.9 (Example 2.6) The velocity–time graph for a particle moving along the x axis according to the expression $v_x = 40 - 5t^2$.

Find the velocities at $t_i = t_Ⓐ = 0$ and $t_f = t_Ⓑ = 2.0$ s by substituting these values of t into the expression for the velocity:

$$v_{xⒶ} = 40 - 5t_Ⓐ^2 = 40 - 5(0)^2 = +40 \text{ m/s}$$

$$v_{xⒷ} = 40 - 5t_Ⓑ^2 = 40 - 5(2.0)^2 = +20 \text{ m/s}$$

Use Equation 2.9 to find the average acceleration in the specified time interval $\Delta t = t_Ⓑ - t_Ⓐ = 2.0$ s:

$$a_{x,\text{avg}} = \frac{v_{xf} - v_{xi}}{t_f - t_i} = \frac{v_{xⒷ} - v_{xⒶ}}{t_Ⓑ - t_Ⓐ} = \frac{20 \text{ m/s} - 40 \text{ m/s}}{2.0 \text{ s} - 0 \text{ s}}$$

$$= -10 \text{ m/s}^2$$

The negative sign is consistent with our expectations: the average acceleration, represented by the slope of the blue line joining the initial and final points on the velocity–time graph, is negative.

continued

2.6 continued

(B) Determine the acceleration at $t = 2.0$ s.

Knowing that the initial velocity at any time t is $v_{xi} = 40 - 5t^2$, find the velocity at any later time $t + \Delta t$:

$$v_{xf} = 40 - 5(t + \Delta t)^2 = 40 - 5t^2 - 10t\,\Delta t - 5(\Delta t)^2$$

Find the change in velocity over the time interval Δt:

$$\Delta v_x = v_{xf} - v_{xi} = -10t\,\Delta t - 5(\Delta t)^2$$

To find the acceleration at any time t, divide this expression by Δt and take the limit of the result as Δt approaches zero:

$$a_x = \lim_{\Delta t \to 0} \frac{\Delta v_x}{\Delta t} = \lim_{\Delta t \to 0}(-10t - 5\,\Delta t) = -10t$$

Substitute $t = 2.0$ s:

$$a_x = (-10)(2.0)\ \text{m/s}^2 = -20\ \text{m/s}^2$$

Because the velocity of the particle is positive and the acceleration is negative at this instant, the particle is slowing down.

Finalize Notice that the answers to parts (A) and (B) are different. The average acceleration in part (A) is the slope of the blue line in Figure 2.9 connecting points Ⓐ and Ⓑ. The instantaneous acceleration in part (B) is the slope of the green line tangent to the curve at point Ⓑ. Notice also that the acceleration is *not* constant in this example. Situations involving constant acceleration are treated in Section 2.7.

So far, we have evaluated the derivatives of a function by starting with the definition of the function and then taking the limit of a specific ratio. If you are familiar with calculus, you should recognize that there are specific rules for taking derivatives. These rules, which are listed in Appendix B.6, enable us to evaluate derivatives quickly. For instance, one rule tells us that the derivative of any constant is zero. As another example, suppose x is proportional to some power of t such as in the expression

$$x = At^n$$

where A and n are constants. (This expression is a very common functional form.) The derivative of x with respect to t is

$$\frac{dx}{dt} = nAt^{n-1}$$

Applying these rules to Example 2.6, in which $v_x = 40 - 5t^2$, we quickly find that the acceleration is $a_x = dv_x/dt = -10t$, as we found in part (B) of the example.

2.6 Motion Diagrams

The concepts of velocity and acceleration are often confused with each other, but in fact they are quite different quantities. In forming a mental representation of a moving object, a pictorial representation called a *motion diagram* is sometimes useful to describe the velocity and acceleration while an object is in motion.

A motion diagram can be formed by imagining a *stroboscopic* photograph of a moving object, which shows several images of the object taken as the strobe light flashes at a constant rate. Figure 2.1a is a motion diagram for the car studied in Section 2.1. Figure 2.10 represents three sets of strobe photographs of cars moving along a straight roadway in a single direction, from left to right. The time intervals between flashes of the stroboscope are equal in each part of the diagram. So as to not confuse the two vector quantities, we use red arrows for velocity and purple arrows for acceleration in Figure 2.10. The arrows are shown at several instants during the motion of the object. Let us describe the motion of the car in each diagram.

In Figure 2.10a, the images of the car are equally spaced, showing us that the car moves through the same displacement in each time interval. This equal spacing is consistent with the car moving with *constant positive velocity* and *zero acceleration*. We

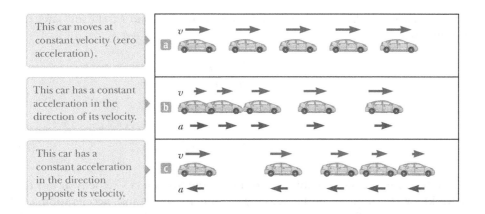

Figure 2.10 Motion diagrams of a car moving along a straight roadway in a single direction. The velocity at each instant is indicated by a red arrow, and the constant acceleration is indicated by a purple arrow.

could model the car as a particle and describe it with the particle under constant velocity model. The red velocity arrows are all of equal length, and there is no purple acceleration arrow shown because it is of length zero.

In Figure 2.10b, the images become farther apart as time progresses. In this case, the red velocity arrows increase in length with time because the car's displacement between adjacent positions increases in time. These features suggest the car is moving with a *positive velocity* and a *positive acceleration.* The velocity and acceleration are in the same direction. In terms of our earlier force discussion, imagine a force pulling on the car in the same direction it is moving: it speeds up.

In Figure 2.10c, we can tell that the car slows as it moves to the right because its displacement between adjacent images decreases with time. This case suggests the car moves to the right with a negative acceleration. The lengths of the velocity arrows decrease in time and eventually reach zero. From this diagram, we see that the acceleration and velocity arrows are *not* in the same direction. The car is moving with a *positive velocity,* but with a *negative acceleration.* (This type of motion is exhibited by a car that skids to a stop after its brakes are applied.) The velocity and acceleration are in opposite directions. In terms of our earlier force discussion, imagine a force pulling on the car opposite to the direction it is moving: it slows down.

Each purple acceleration arrow in parts (b) and (c) of Figure 2.10 is the same length. Therefore, these diagrams represent motion of a *particle under constant acceleration.* This important analysis model will be discussed in the next section.

QUICK QUIZ 2.6 Which one of the following statements is true? **(a)** If a car is traveling eastward, its acceleration must be eastward. **(b)** If a car is slowing down, its acceleration must be negative. **(c)** A particle with constant acceleration can never stop and stay stopped.

2.7 Analysis Model: Particle Under Constant Acceleration

If the acceleration of a particle varies in time, its motion can be complex and difficult to analyze. A very common and simple type of one-dimensional motion, however, is that in which the acceleration is constant. In such a case, the average acceleration $a_{x,\text{avg}}$ over any time interval is numerically equal to the instantaneous acceleration a_x at any instant within the interval, and the velocity changes at the same rate throughout the motion. This situation occurs often enough that we identify it as an analysis model: the **particle under constant acceleration.** In the discussion that follows, we generate several equations that describe the motion of a particle for this model.

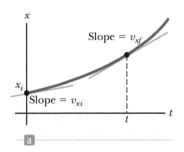

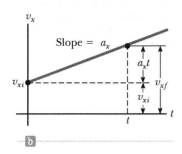

Figure 2.11 A particle under constant acceleration a_x moving along the *x* axis: (a) the position–time graph, (b) the velocity–time graph, and (c) the acceleration–time graph.

▷ Position as a function of velocity and time for the particle under constant acceleration model

▷ Position as a function of time for the particle under constant acceleration model

If we replace $a_{x,\text{avg}}$ by a_x in Equation 2.9 and take $t_i = 0$ and t_f to be any later time t, we find that

$$a_x = \frac{v_{xf} - v_{xi}}{t - 0}$$

or

$$v_{xf} = v_{xi} + a_x t \quad \text{(for constant } a_x\text{)} \tag{2.13}$$

This powerful expression enables us to determine an object's velocity at *any* time t if we know the object's initial velocity v_{xi} and its (constant) acceleration a_x. A velocity–time graph for this constant-acceleration motion is shown in Figure 2.11b. The graph is a straight line, the slope of which is the acceleration a_x; the (constant) slope is consistent with $a_x = dv_x/dt$ being a constant. Notice that the slope is positive, which indicates a positive acceleration. If the acceleration were negative, the slope of the line in Figure 2.11b would be negative. When the acceleration is constant, the graph of acceleration versus time (Fig. 2.11c) is a straight line having a slope of zero.

Because velocity at constant acceleration varies linearly in time according to Equation 2.13, we can express the average velocity in any time interval as the arithmetic mean of the initial velocity v_{xi} and the final velocity v_{xf}:

$$v_{x,\text{avg}} = \frac{v_{xi} + v_{xf}}{2} \quad \text{(for constant } a_x\text{)} \tag{2.14}$$

Notice that this expression for average velocity applies *only* in situations in which the acceleration is constant.

We can now use Equations 2.1, 2.2, and 2.14 to obtain the position of an object as a function of time. Recalling that Δx in Equation 2.2 represents $x_f - x_i$ and recognizing that $\Delta t = t_f - t_i = t - 0 = t$, we find that

$$x_f - x_i = v_{x,\text{avg}} t = \tfrac{1}{2}(v_{xi} + v_{xf})t$$

$$x_f = x_i + \tfrac{1}{2}(v_{xi} + v_{xf})t \quad \text{(for constant } a_x\text{)} \tag{2.15}$$

This equation provides the final position of the particle at time t in terms of the initial and final velocities.

We can obtain another useful expression for the position of a particle under constant acceleration by substituting Equation 2.13 into Equation 2.15:

$$x_f = x_i + \tfrac{1}{2}[v_{xi} + (v_{xi} + a_x t)]t$$

$$x_f = x_i + v_{xi}t + \tfrac{1}{2}a_x t^2 \quad \text{(for constant } a_x\text{)} \tag{2.16}$$

This equation provides the final position of the particle at time t in terms of the initial position, the initial velocity, and the constant acceleration.

The position–time graph for motion at constant (positive) acceleration shown in Figure 2.11a is obtained from Equation 2.16. Notice that the curve is a parabola. The slope of the tangent line to this curve at $t = 0$ equals the initial velocity v_{xi}, and the slope of the tangent line at any later time t equals the velocity v_{xf} at that time.

Finally, we can obtain an expression for the final velocity that does not contain time as a variable by substituting the value of t from Equation 2.13 into Equation 2.15:

$$x_f = x_i + \tfrac{1}{2}(v_{xi} + v_{xf})\left(\frac{v_{xf} - v_{xi}}{a_x}\right) = x_i + \frac{v_{xf}^2 - v_{xi}^2}{2a_x}$$

$$v_{xf}^2 = v_{xi}^2 + 2a_x(x_f - x_i) \quad \text{(for constant } a_x\text{)} \tag{2.17}$$

◀ Velocity as a function of position for the particle under constant acceleration model

This equation provides the final velocity in terms of the initial velocity, the constant acceleration, and the position of the particle.

For motion at *zero* acceleration, we see from Equations 2.13 and 2.16 that

$$\left.\begin{array}{l} v_{xf} = v_{xi} = v_x \\ x_f = x_i + v_x t \end{array}\right\} \quad \text{when } a_x = 0$$

That is, when the acceleration of a particle is zero, its velocity is constant and its position changes linearly with time. In terms of models, when the acceleration of a particle is zero, the particle under constant acceleration model reduces to the particle under constant velocity model (Section 2.3).

Equations 2.13 through 2.17 are **kinematic equations** that may be used to solve any problem involving a particle under constant acceleration in one dimension. These equations are listed together below for convenience. The choice of which equation you use in a given situation depends on what you know beforehand. Sometimes it is necessary to use two of these equations to solve for two unknowns. You should recognize that the quantities that vary during the motion are position x_f, velocity v_{xf}, and time t.

You will gain a great deal of experience in the use of these equations by solving a number of exercises and problems. Many times you will discover that more than one method can be used to obtain a solution. Remember that these equations of kinematics *cannot* be used in a situation in which the acceleration varies with time. They can be used only when the acceleration is constant.

ⓠUICK QUIZ 2.7 In Figure 2.12, match each v_x–t graph on the top with the a_x–t graph on the bottom that best describes the motion.

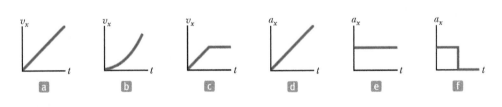

Figure 2.12 (Quick Quiz 2.7) Parts (a), (b), and (c) are v_x–t graphs of objects in one-dimensional motion. The possible accelerations of each object as a function of time are shown in scrambled order in (d), (e), and (f).

ANALYSIS MODEL **Particle Under Constant Acceleration**

Imagine a moving object that can be modeled as a particle. If it begins from position x_i and initial velocity v_{xi} and moves in a straight line with a constant acceleration a_x, its subsequent position and velocity are described by the following kinematic equations:

$$v_{xf} = v_{xi} + a_x t \tag{2.13}$$

$$v_{x,\text{avg}} = \frac{v_{xi} + v_{xf}}{2} \tag{2.14}$$

Examples

- a car accelerating at a constant rate along a straight freeway

$$x_f = x_i + \tfrac{1}{2}(v_{xi} + v_{xf})t \tag{2.15}$$

- a dropped object in the absence of air resistance (Section 2.8)

$$x_f = x_i + v_{xi}t + \tfrac{1}{2}a_x t^2 \tag{2.16}$$

- an object on which a constant net force acts (Chapter 5)

$$v_{xf}^2 = v_{xi}^2 + 2a_x(x_f - x_i) \tag{2.17}$$

- a charged particle in a uniform electric field (Chapter 22)

Example **2.7**   **Carrier Landing**

A jet lands on an aircraft carrier at a speed of 140 mi/h ($\approx$ 63 m/s).

(A) What is its acceleration (assumed constant) if it stops in 2.0 s due to an arresting cable that snags the jet and brings it to a stop?

SOLUTION

Conceptualize   You might have seen movies or television shows in which a jet lands on an aircraft carrier and is brought to rest surprisingly fast by an arresting cable. A careful reading of the problem reveals that in addition to being given the initial speed of 63 m/s, we also know that the final speed is zero. We define our x axis as the direction of motion of the jet. Notice that we have no information about the change in position of the jet while it is slowing down.

Categorize   Because the acceleration of the jet is assumed constant, we model it as a *particle under constant acceleration.*

Analyze   Equation 2.13 is the only equation in the particle under constant acceleration model that does not involve position, so we use it to find the acceleration of the jet, modeled as a particle:

$$a_x = \frac{v_{xf} - v_{xi}}{t} \approx \frac{0 - 63 \text{ m/s}}{2.0 \text{ s}}$$
$$= -32 \text{ m/s}^2$$

(B) If the jet touches down at position $x_i = 0$, what is its final position?

SOLUTION

Use Equation 2.15 to solve for the final position: $x_f = x_i + \frac{1}{2}(v_{xi} + v_{xf})t = 0 + \frac{1}{2}(63 \text{ m/s} + 0)(2.0 \text{ s}) = $ 63 m

Finalize   Given the size of aircraft carriers, a length of 63 m seems reasonable for stopping the jet. The idea of using arresting cables to slow down landing aircraft and enable them to land safely on ships originated at about the time of World War I. The cables are still a vital part of the operation of modern aircraft carriers.

WHAT IF?   Suppose the jet lands on the deck of the aircraft carrier with a speed faster than 63 m/s but has the same acceleration due to the cable as that calculated in part (A). How will that change the answer to part (B)?

Answer   If the jet is traveling faster at the beginning, it will stop farther away from its starting point, so the answer to part (B) should be larger. Mathematically, we see in Equation 2.15 that if v_{xi} is larger, then x_f will be larger.

Example **2.8**   **Watch Out for the Speed Limit!**

You are driving at a constant speed of 45.0 m/s when you pass a trooper on a motorcycle hidden behind a billboard. One second after your car passes the billboard, the trooper sets out from the billboard to catch you, accelerating at a constant rate of 3.00 m/s². How long does it take the trooper to overtake your car?

SOLUTION

Conceptualize   This example represents a class of problems called *context-rich* problems. These problems involve real-world situations that one might encounter in one's daily life. These problems also involve "you" as opposed to an unspecified particle or object. With you as the character in the problem, *you* can make the connection between physics and everyday life!

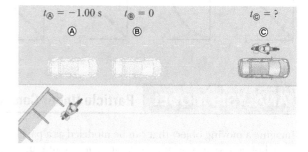

Figure 2.13 (Example 2.8) You are in a speeding car that passes a hidden trooper.

Categorize   A pictorial representation (Fig. 2.13) helps clarify the sequence of events. Your car is modeled as a *particle under constant velocity*, and the trooper is modeled as a *particle under constant acceleration.*

Analyze   First, we write expressions for the position of each vehicle as a function of time. It is convenient to choose the position of the billboard as the origin and to set $t_\text{B} = 0$ as the time the trooper begins moving. At that instant, your car has already

continued

2.8 continued

traveled a distance of 45.0 m from the billboard because it has traveled at a constant speed of $v_x = 45.0$ m/s for 1 s. Therefore, the initial position of your car is $x_{\circledB} = 45.0$ m.

Using the particle under constant velocity model, apply Equation 2.7 to give your car's position at any time t:

$$x_{car} = x_{\circledB} + v_{x\,car}t$$

A quick check shows that at $t = 0$, this expression gives your car's correct initial position when the trooper begins to move: $x_{car} = x_{\circledB} = 45.0$ m.

The trooper starts from rest at $t_{\circledB} = 0$ and accelerates at $a_x = 3.00$ m/s² away from the origin. Use Equation 2.16 to give her position at any time t:

$$x_f = x_i + v_{xi}t + \tfrac{1}{2}a_x t^2$$
$$x_{trooper} = 0 + (0)t + \tfrac{1}{2}a_x t^2 = \tfrac{1}{2}a_x t^2$$

Set the positions of your car and the trooper equal to represent the trooper overtaking your car at position ©:

$$x_{trooper} = x_{car}$$
$$\tfrac{1}{2}a_x t^2 = x_{\circledB} + v_{x\,car}t$$

Rearrange to give a quadratic equation:

$$\tfrac{1}{2}a_x t^2 - v_{x\,car}t - x_{\circledB} = 0$$

Solve the quadratic equation for the time at which the trooper catches your car (for help in solving quadratic equations, see Appendix B.2):

$$t = \frac{v_{x\,car} \pm \sqrt{v_{x\,car}^2 + 2a_x x_{\circledB}}}{a_x}$$

(1) $$t = \frac{v_{x\,car}}{a_x} \pm \sqrt{\frac{v_{x\,car}^2}{a_x^2} + \frac{2x_{\circledB}}{a_x}}$$

Evaluate the solution, choosing the positive root because that is the only choice consistent with a time $t > 0$:

$$t = \frac{45.0 \text{ m/s}}{3.00 \text{ m/s}^2} + \sqrt{\frac{(45.0 \text{ m/s})^2}{(3.00 \text{ m/s}^2)^2} + \frac{2(45.0 \text{ m})}{3.00 \text{ m/s}^2}} = 31.0 \text{ s}$$

Finalize Why didn't we choose $t = 0$ as the time at which your car passes the trooper? If we did so, we would not be able to use the particle under constant acceleration model for the trooper. Her acceleration would be zero for the first second and then 3.00 m/s² for the remaining time. By defining the time $t = 0$ as when the trooper begins moving, we can use the particle under constant acceleration model for her movement for all positive times.

WHAT IF? What if the trooper had a more powerful motorcycle with a larger acceleration? How would that change the time at which the trooper catches your car?

Answer If the motorcycle has a larger acceleration, the trooper should catch up to your car sooner, so the answer for the time should be less than 31 s. Because all terms on the right side of Equation (1) have the acceleration a_x in the denominator, we see symbolically that increasing the acceleration will decrease the time at which the trooper catches your car.

2.8 Freely Falling Objects

It is well known that, in the absence of air resistance, all objects dropped near the Earth's surface fall toward the Earth with the same constant acceleration under the influence of the Earth's gravity, regardless of their mass. It was not until about 1600 that this conclusion was accepted. Before that time, the teachings of the Greek philosopher Aristotle (384–322 BC) had held that heavier objects fall faster than lighter ones.

The Italian Galileo Galilei (1564–1642) originated our present-day ideas concerning falling objects. There is a legend that he demonstrated the behavior of falling objects by observing that two different weights dropped simultaneously from the Leaning Tower of Pisa hit the ground at approximately the same time. Although there is some doubt that he carried out this particular experiment, it is well established that Galileo performed many experiments on objects moving on inclined planes. In his experiments, he rolled balls down a slight incline and measured the distances they covered in successive time intervals. The purpose of the incline was to reduce the acceleration, which made it possible for him to make accurate measurements of the time intervals. By gradually increasing the slope of the incline,

Galileo Galilei
Italian physicist and astronomer
(1564–1642)
Galileo formulated the laws that govern the motion of objects in free fall and made many other significant discoveries in physics and astronomy. Galileo publicly defended Nicolaus Copernicus's assertion that the Sun is at the center of the Universe (the heliocentric system).

PITFALL PREVENTION 2.6

g and g Be sure not to confuse the italic symbol *g* for free-fall acceleration with the nonitalic symbol g used as the abbreviation for the unit gram.

PITFALL PREVENTION 2.7

The Sign of *g* Keep in mind that *g* is a *positive number*. It is tempting to substitute -9.80 m/s² for *g*, but resist the temptation. Downward gravitational acceleration is indicated explicitly by stating the acceleration as $a_y = -g$.

PITFALL PREVENTION 2.8

Acceleration at the Top of the Motion A common misconception is that the acceleration of a projectile at the top of its trajectory is zero. Although the velocity at the top of the motion of an object thrown upward momentarily goes to zero, *the acceleration is still that due to gravity* at this point. If the velocity and acceleration were both zero, the projectile would stay at the top.

he was finally able to draw conclusions about freely falling objects because a freely falling ball is equivalent to a ball moving down a vertical incline.

You might want to try the following experiment. Simultaneously drop a coin and a piece of paper from the same height. The coin will always reach the ground faster. Now, crumple the paper into a tight ball and repeat the experiment. Since you've minimized the effects of air resistance, the coin and the paper will have the same motion and will hit the floor at the same time. In the idealized case, in which air resistance is absent, such motion is referred to as *free-fall* motion. If this same experiment could be conducted in a vacuum, in which air resistance is truly negligible, the paper and the coin would fall with the same acceleration even when the paper is not crumpled. On August 2, 1971, astronaut David Scott conducted such a demonstration on the Moon. He simultaneously released a hammer and a feather, and the two objects fell together to the lunar surface. This simple demonstration surely would have pleased Galileo!

When we use the expression *freely falling object,* we do not necessarily refer to an object dropped from rest. A freely falling object is any object moving freely under the influence of gravity alone, regardless of its initial motion. Objects thrown upward or downward and those released from rest are all falling freely once they are released. Any freely falling object experiences an acceleration directed *downward*, regardless of its initial motion.

We shall denote the magnitude of the *free-fall acceleration,* also called the *acceleration due to gravity,* by the symbol *g*. The value of *g* decreases with increasing altitude above the Earth's surface. Furthermore, slight variations in *g* occur with changes in latitude. At the Earth's surface, the value of *g* is approximately 9.80 m/s². Unless stated otherwise, we shall use this value for *g* when performing calculations. For making quick estimates, use $g \sim 10$ m/s².

If we neglect air resistance and assume the free-fall acceleration does not vary with altitude over short vertical distances, the motion of a freely falling object moving vertically is equivalent to the motion of a particle under constant acceleration in one dimension. Therefore, the equations developed in Section 2.7 for the particle under constant acceleration model can be applied. The only modification for freely falling objects that we need to make in these equations is to note that the motion is in the vertical direction (the *y* direction) rather than in the horizontal direction (*x*) and that the acceleration is downward and has a magnitude of 9.80 m/s². Therefore, we choose $a_y = -g = -9.80$ m/s², where the negative sign means that the acceleration of a freely falling object is downward. In Chapter 13, we shall study how to deal with variations in *g* with altitude.

QUICK QUIZ 2.8 Consider the following choices: (a) increases, (b) decreases, (c) increases and then decreases, (d) decreases and then increases, (e) remains the same. From these choices, select what happens to **(i)** the acceleration and **(ii)** the speed of a ball after it is thrown upward into the air.

Conceptual Example 2.9 **The Daring Skydivers**

A skydiver jumps out of a hovering helicopter. A few seconds later, another skydiver jumps out, and they both fall along the same vertical line. Ignore air resistance so that both skydivers fall with the same acceleration. Does the difference in their speeds stay the same throughout the fall? Does the vertical distance between them stay the same throughout the fall?

SOLUTION

At any given instant, the speeds of the skydivers are different because one had a head start. In any time interval Δt after this instant, however, the two skydivers increase their speeds by the same amount because they have the same acceleration. Therefore, the difference in their speeds remains the same throughout the fall.

The first jumper always has a greater speed than the second. Therefore, in a given time interval, the first skydiver covers a greater distance than the second. Consequently, the separation distance between them increases.

Example 2.10 | Not a Bad Throw for a Rookie!

A stone thrown from the top of a building is given an initial velocity of 20.0 m/s straight upward. The stone is launched 50.0 m above the ground, and the stone just misses the edge of the roof on its way down as shown in Figure 2.14.

(A) Using $t_{Ⓐ} = 0$ as the time the stone leaves the thrower's hand at position Ⓐ, determine the time at which the stone reaches its maximum height.

SOLUTION

Conceptualize You most likely have experience with dropping objects or throwing them upward and watching them fall, so this problem should describe a familiar experience. To simulate this situation, toss a small object upward and notice the time interval required for it to fall to the floor. Now imagine throwing that object upward from the roof of a building.

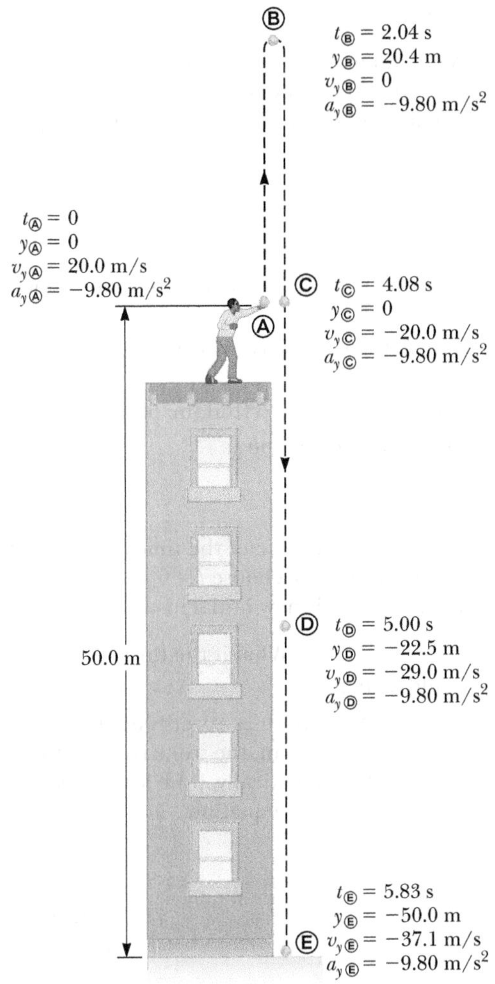

Figure 2.14 (Example 2.10) Position, velocity, and acceleration values at various times for a freely falling stone thrown initially upward with a velocity $v_{yi} = 20.0$ m/s. Many of the quantities in the labels for points in the motion of the stone are calculated in the example. Can you verify the other values that are not?

Ⓑ $t_{Ⓑ} = 2.04$ s
$y_{Ⓑ} = 20.4$ m
$v_{yⒷ} = 0$
$a_{yⒷ} = -9.80$ m/s^2

$t_{Ⓐ} = 0$
$y_{Ⓐ} = 0$
$v_{yⒶ} = 20.0$ m/s
$a_{yⒶ} = -9.80$ m/s^2

Ⓒ $t_{Ⓒ} = 4.08$ s
$y_{Ⓒ} = 0$
$v_{yⒸ} = -20.0$ m/s
$a_{yⒸ} = -9.80$ m/s^2

Ⓐ

50.0 m

Ⓓ $t_{Ⓓ} = 5.00$ s
$y_{Ⓓ} = -22.5$ m
$v_{yⒹ} = -29.0$ m/s
$a_{yⒹ} = -9.80$ m/s^2

$t_{Ⓔ} = 5.83$ s
$y_{Ⓔ} = -50.0$ m
Ⓔ $v_{yⒺ} = -37.1$ m/s
$a_{yⒺ} = -9.80$ m/s^2

Categorize Because the stone is in free fall, it is modeled as a *particle under constant acceleration* due to gravity.

Analyze Recognize that the initial velocity is positive because the stone is launched upward. The velocity will change sign after the stone reaches its highest point, but the acceleration of the stone will *always* be downward so that it will always have a negative value. Choose an initial point just after the stone leaves the person's hand and a final point at the top of its flight.

Use Equation 2.13 to calculate the time at which the stone reaches its maximum height:

$$v_{yf} = v_{yi} + a_y t \rightarrow t = \frac{v_{yf} - v_{yi}}{a_y} = \frac{v_{yⒷ} - v_{yⒶ}}{-g}$$

Substitute numerical values, recognizing that $v = 0$ at point Ⓑ:

$$t = t_{Ⓑ} = \frac{0 - 20.0 \text{ m/s}}{-9.80 \text{ m/s}^2} = 2.04 \text{ s}$$

(B) Find the maximum height of the stone.

SOLUTION

As in part (A), choose the initial and final points at the beginning and the end of the upward flight.

Set $y_{Ⓐ} = 0$ and substitute the time from part (A) into Equation 2.16 to find the maximum height:

$$y_{max} = y_{Ⓑ} = y_{Ⓐ} + v_{xⒶ} t + \tfrac{1}{2} a_y t^2$$

$$y_{Ⓑ} = 0 + (20.0 \text{ m/s})(2.04 \text{ s}) + \tfrac{1}{2}(-9.80 \text{ m/s}^2)(2.04 \text{ s})^2 = 20.4 \text{ m}$$

(C) Determine the velocity of the stone when it returns to the height from which it was thrown.

SOLUTION

Choose the initial point where the stone is launched and the final point when it passes this position coming down.

Substitute known values into Equation 2.17:

$$v_{yⒸ}^2 = v_{yⒶ}^2 + 2a_y(y_{Ⓒ} - y_{Ⓐ})$$

$$v_{yⒸ}^2 = (20.0 \text{ m/s})^2 + 2(-9.80 \text{ m/s}^2)(0 - 0) = 400 \text{ m}^2/\text{s}^2$$

$$v_{yⒸ} = -20.0 \text{ m/s}$$

continued

2.10 continued

When taking the square root, we could choose either a positive or a negative root. We choose the negative root because we know that the stone is moving downward at point ©. The velocity of the stone when it arrives back at its original height is equal in magnitude to its initial velocity but is opposite in direction.

(D) Find the velocity and position of the stone at $t = 5.00$ s.

SOLUTION

Choose the initial point just after the throw and the final point 5.00 s later.

Calculate the velocity at Ⓓ from Equation 2.13:

$$v_{y\text{Ⓓ}} = v_{y\text{Ⓐ}} + a_y t = 20.0 \text{ m/s} + (-9.80 \text{ m/s}^2)(5.00 \text{ s}) = -29.0 \text{ m/s}$$

Use Equation 2.16 to find the position of the stone at $t_{\text{Ⓓ}} = 5.00$ s:

$$y_{\text{Ⓓ}} = y_{\text{Ⓐ}} + v_{y\text{Ⓐ}}t + \tfrac{1}{2}a_y t^2$$
$$= 0 + (20.0 \text{ m/s})(5.00 \text{ s}) + \tfrac{1}{2}(-9.80 \text{ m/s}^2)(5.00 \text{ s})^2$$
$$= -22.5 \text{ m}$$

Finalize   The choice of the time defined as $t = 0$ is arbitrary and up to you to select as the problem solver. As an example of this arbitrariness, choose $t = 0$ as the time at which the stone is at the highest point in its motion. Then solve parts (C) and (D) again using this new initial instant and notice that your answers are the same as those above.

WHAT IF?   What if the throw were from 30.0 m above the ground instead of 50.0 m? Which answers in parts (A) to (D) would change?

Answer   None of the answers would change. All the motion takes place in the air during the first 5.00 s. (Notice that even for a throw from 30.0 m, the stone is above the ground at $t = 5.00$ s.) Therefore, the height from which the stone is thrown is not an issue. Mathematically, if we look back over our calculations, we see that we never entered the height from which the stone is thrown into any equation.

2.9 Kinematic Equations Derived from Calculus

PITFALL PREVENTION 2.9
Previous Experience with Integration This section assumes the reader is familiar with the techniques of integral calculus. If you have not yet studied integration in your calculus course, you should skip this section or cover it after you become familiar with integration.

The velocity of a particle moving in a straight line can be determined as the derivative of the position with respect to time. It is also possible to find the position of a particle if its velocity is known as a function of time. In calculus, the procedure used to perform this task is referred to either as *integration* or as finding the *antiderivative*.

Suppose the v_x–t graph for a particle moving along the x axis is as shown in Figure 2.15. Let us divide the time interval $t_f - t_i$ into many small intervals, each of duration Δt_n. From the definition of average velocity, we see that the displacement of the particle during any small interval, such as the one shaded in Figure 2.15, is given by $\Delta x_n = v_{xn,\text{avg}} \Delta t_n$, where $v_{xn,\text{avg}}$ is the average velocity in that interval. Therefore, the displacement during this small interval is simply the area of the shaded rectangle in Figure 2.15. The total displacement for the interval $t_f - t_i$ is the sum of the areas of all the rectangles from t_i to t_f:

$$\Delta x = \sum_n v_{xn,\text{avg}} \Delta t_n$$

where the symbol Σ (uppercase Greek sigma) signifies a sum over all terms, that is, over all values of n. Now, as the intervals are made smaller and smaller, the number of terms in the sum increases and the sum approaches a value equal to the area under the curve in the velocity–time graph. Therefore, in the limit $n \to \infty$, or $\Delta t_n \to 0$, the displacement is

$$\Delta x = \lim_{\Delta t_n \to 0} \sum_n v_{xn,\text{avg}} \Delta t_n \qquad (2.18)$$

The limit of the sum shown in Equation 2.18 is called a **definite integral** and so the displacement of the particle can be written as

◀ Definite integral

$$\Delta x = \int_{t_i}^{t_f} v_x(t)\, dt \qquad (2.19)$$

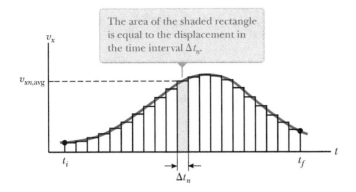

The area of the shaded rectangle is equal to the displacement in the time interval Δt_n.

Figure 2.15 Velocity versus time for a particle moving along the x axis. The total area under the curve is the total displacement of the particle.

where $v_x(t)$ denotes the velocity at any time t. If the explicit functional form of $v_x(t)$ is known and the limits are given, the integral can be evaluated.

Kinematic Equations

We now use the defining equations for acceleration and velocity to derive two of our kinematic equations, Equations 2.13 and 2.16.

The defining equation for acceleration (Eq. 2.10),

$$a_x = \frac{dv_x}{dt}$$

may be written as $dv_x = a_x\, dt$ or, in terms of an integral (or antiderivative), as

$$v_{xf} - v_{xi} = \int_0^t a_x\, dt$$

For the special case in which the acceleration is constant, a_x can be removed from the integral to give

$$v_{xf} - v_{xi} = a_x \int_0^t dt = a_x(t - 0) = a_x t \qquad (2.20)$$

which is Equation 2.13 in the particle under constant acceleration model.

Now let us consider the defining equation for velocity (Eq. 2.5):

$$v_x = \frac{dx}{dt}$$

We can write this equation as $dx = v_x\, dt$ or in integral form as

$$x_f - x_i = \int_0^t v_x\, dt$$

Because $v_x = v_{xf} = v_{xi} + a_x t$, this expression becomes

$$x_f - x_i = \int_0^t (v_{xi} + a_x t)\, dt = \int_0^t v_{xi}\, dt + a_x \int_0^t t\, dt = v_{xi}(t - 0) + a_x\left(\frac{t^2}{2} - 0\right)$$

$$x_f - x_i = v_{xi}t + \tfrac{1}{2}a_x t^2$$

which is Equation 2.16 in the particle under constant acceleration model.

PITFALL PREVENTION 2.10

Integration is an Area If this discussion of integration is confusing to you, just remember that the integral of a function is simply the area between the function and the x axis between the limits of integration. If the function has a simple shape, the area can be easily calculated without integration. For example, if the function is a constant, so that its graph is a horizontal line, the area is just that of the rectangle between the line and the x axis!

Summary

▶ Definitions

When a particle moves along the x axis from some initial position x_i to some final position x_f, its **displacement** is

$$\Delta x \equiv x_f - x_i \qquad (2.1)$$

The **average velocity** of a particle during some time interval is the displacement Δx divided by the time interval Δt during which that displacement occurs:

$$v_{x,\text{avg}} \equiv \frac{\Delta x}{\Delta t} \qquad (2.2)$$

The **average speed** of a particle is equal to the ratio of the total distance it travels to the total time interval during which it travels that distance:

$$v_{\text{avg}} \equiv \frac{d}{\Delta t} \qquad (2.3)$$

The **instantaneous velocity** of a particle is defined as the limit of the ratio $\Delta x/\Delta t$ as Δt approaches zero. By definition, this limit equals the derivative of x with respect to t, or the time rate of change of the position:

$$v_x \equiv \lim_{\Delta t \to 0} \frac{\Delta x}{\Delta t} = \frac{dx}{dt} \qquad (2.5)$$

The **instantaneous speed** of a particle is equal to the magnitude of its instantaneous velocity.

The **average acceleration** of a particle is defined as the ratio of the change in its velocity Δv_x divided by the time interval Δt during which that change occurs:

$$a_{x,\text{avg}} \equiv \frac{\Delta v_x}{\Delta t} = \frac{v_{xf} - v_{xi}}{t_f - t_i} \qquad (2.9)$$

The **instantaneous acceleration** is equal to the limit of the ratio $\Delta v_x/\Delta t$ as Δt approaches 0. By definition, this limit equals the derivative of v_x with respect to t, or the time rate of change of the velocity:

$$a_x \equiv \lim_{\Delta t \to 0} \frac{\Delta v_x}{\Delta t} = \frac{dv_x}{dt} \qquad (2.10)$$

▶ Concepts and Principles

When an object's velocity and acceleration are in the same direction, the object is speeding up. On the other hand, when the object's velocity and acceleration are in opposite directions, the object is slowing down. Remembering that $F_x \propto a_x$ is a useful way to identify the direction of the acceleration by associating it with a force.

An object falling freely in the presence of the Earth's gravity experiences free-fall acceleration directed toward the center of the Earth. If air resistance is neglected, if the motion occurs near the surface of the Earth, and if the range of the motion is small compared with the Earth's radius, the free-fall acceleration $a_y = -g$ is constant over the range of motion, where g is equal to 9.80 m/s².

Complicated problems are best approached in an organized manner. Recall and apply the *Conceptualize, Categorize, Analyze,* and *Finalize* steps of the **Analysis Model Approach to Problem Solving** when you need them.

An important aid to problem solving is the use of **analysis models.** Analysis models are situations that we have seen in previous problems. Each analysis model has one or more equations associated with it. When solving a new problem, identify the analysis model that corresponds to the problem. The model will tell you which equations to use. The first three analysis models introduced in this chapter are summarized below.

▶ Analysis Models for Problem Solving

Particle Under Constant Velocity. If a particle moves in a straight line with a constant speed v_x, its constant velocity is given by

$$v_x = \frac{\Delta x}{\Delta t} \qquad (2.6)$$

and its position is given by

$$x_f = x_i + v_x t \qquad (2.7)$$

Particle Under Constant Speed. If a particle moves a distance d along a curved or straight path with a constant speed, its constant speed is given by

$$v = \frac{d}{\Delta t} \qquad (2.8)$$

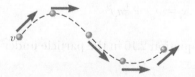

Particle Under Constant Acceleration. If a particle moves in a straight line with a constant acceleration a_x, its motion is described by the kinematic equations:

$$v_{xf} = v_{xi} + a_x t \qquad (2.13)$$

$$v_{x,avg} = \frac{v_{xi} + v_{xf}}{2} \qquad (2.14)$$

$$x_f = x_i + \tfrac{1}{2}(v_{xi} + v_{xf})t \qquad (2.15)$$

$$x_f = x_i + v_{xi}t + \tfrac{1}{2}a_x t^2 \qquad (2.16)$$

$$v_{xf}^2 = v_{xi}^2 + 2a_x(x_f - x_i) \qquad (2.17)$$

Think–Pair–Share

See the Preface for an explanation of the icons used in this problems set. For additional assessment items for this section, go to **WEBASSIGN** From Cengage

1. You are at a carnival playing the "Strike-the-Bell" game, as shown in Figure TP2.1. The goal is to hit the end of the lever with a hammer, sending a hard object upward along the frictionless vertical track so as to strike a bell at the top. Showing off your control for the crowd, you hit the lever several times in a row in such a way that the hard object rises to a height $h = 4.50$ m and just touches the bell, which makes a gentle ringing sound. Now, to really impress the crowd, you swing the hammer with a mighty motion, hit the lever, and project the object upward with twice the initial speed of your previous demonstrations. Unbeknownst to you, on the previous demonstration, the bell came loose and slipped off to the side, so that, on this demonstration, the object bypasses the bell and is projected straight up into the air. What is the total time interval between when the object begins its upward motion and then later lands on the ground beside the apparatus?

Figure TP2.1

Stephen Bjorck/Getty Images

2. Your group is at the top of a cliff of height $h = 75.0$ m. At the bottom of the cliff is a pool of water. You split the group in two. The first half of the group volunteers a member to drop a rock from rest so that it falls straight downward and makes a splash in the water. The second half of the group volunteers a member to, after some time interval has passed since the first rock was dropped, throw a second rock straight downward so that both rocks arrive at the water at the same time. You test the performance by listening for a single splash made by the rocks simultaneously hitting the water. (a) If the second rock is thrown 1.00 s after the first rock is released, with what speed must the second rock be thrown? (b) If the fastest anyone in your group can throw the rock is 40.0 m/s, what is the longest time interval that can pass between the release of the rocks so that a single splash is heard? (c) If there is no limit as to how fast the rock can be thrown, what is the longest time interval that can pass between the release of the rocks so that a single splash is heard?

3. **ACTIVITY** Have your partner hold a ruler vertically with the zero end at the bottom. Place your open finger and thumb at the zero position. Without warning, your partner should release the ruler and you should catch it as soon as you see it moving. From the position of your finger on the ruler, determine your reaction time. Repeat the experiment a number of times to estimate the uncertainty in your reaction time. Have each member of your group catch the ruler and compare your reaction times.

4. **ACTIVITY** The Acela is an electric train on the Washington–New York–Boston run, carrying passengers at speeds as high as 170 mi/h. A velocity–time graph for the Acela is shown in Figure TP2.4. (a) Describe the train's motion in each successive time interval. (b) Find the train's peak positive acceleration in the motion graphed. (c) Find the train's displacement in miles between $t = 0$ and $t = 200$ s.

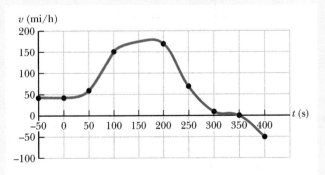

Figure TP2.4 Velocity–time graph for the Acela.

Problems

See the Preface for an explanation of the icons used in this problems set.
For additional assessment items for this section, go to ✦**WEBASSIGN**
From Cengage

SECTION 2.1 Position, Velocity, and Speed

1. The speed of a nerve impulse in the human body is about
BIO 100 m/s. If you accidentally stub your toe in the dark, estimate
the time it takes the nerve impulse to travel to your brain.

2. A particle moves according to the equation $x = 10t^2$, where x
V is in meters and t is in seconds. (a) Find the average velocity
for the time interval from 2.00 s to 3.00 s. (b) Find the average velocity for the time interval from 2.00 to 2.10 s.

3. The position of a pinewood derby car was observed at various times; the results are summarized in the following table.
Find the average velocity of the car for (a) the first second,
(b) the last 3 s, and (c) the entire period of observation.

t (s)	0	1.0	2.0	3.0	4.0	5.0
x (m)	0	2.3	9.2	20.7	36.8	57.5

SECTION 2.2 Instantaneous Velocity and Speed

4. An athlete leaves one end of a pool of length L at $t = 0$
S and arrives at the other end at time t_1. She swims back and
arrives at the starting position at time t_2. If she is swimming
initially in the positive x direction, determine her average velocities symbolically in (a) the first half of the swim,
(b) the second half of the swim, and (c) the round trip.
(d) What is her average speed for the round trip?

5. A position–time graph for a particle moving along the
x axis is shown in Figure P2.5. (a) Find the average velocity
in the time interval $t = 1.50$ s to $t = 4.00$ s. (b) Determine
the instantaneous velocity at $t = 2.00$ s by measuring the
slope of the tangent line shown in the graph. (c) At what
value of t is the velocity zero?

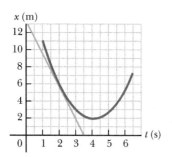

Figure P2.5

SECTION 2.3 Analysis Model: Particle Under Constant Velocity

6. A car travels along a straight line at a constant speed of
AMT 60.0 mi/h for a distance d and then another distance d in
the same direction at another constant speed. The average
velocity for the entire trip is 30.0 mi/h. (a) What is the constant speed with which the car moved during the second distance d? (b) **What If?** Suppose the second distance d were
traveled in the opposite direction; you forgot something and
had to return home at the same constant speed as found in
part (a). What is the average velocity for this trip? (c) What is
the average speed for this new trip?

7. A person takes a trip, driving with a constant speed of 89.5
T km/h, except for a 22.0-min rest stop. If the person's average speed is 77.8 km/h, (a) how much time is spent on the
trip and (b) how far does the person travel?

SECTION 2.5 Acceleration

8. A child rolls a marble on a bent track that is 100 cm long as
shown in Figure P2.8. We use x to represent the position of
the marble along the track. On the horizontal sections from
$x = 0$ to $x = 20$ cm and from $x = 40$ cm to $x = 60$ cm, the
marble rolls with constant speed. On the sloping sections,
the marble's speed changes steadily. At the places where
the slope changes, the marble stays on the track and does
not undergo any sudden changes in speed. The child gives
the marble some initial speed at $x = 0$ and $t = 0$ and then
watches it roll to $x = 90$ cm, where it turns around, eventually
returning to $x = 0$ with the same speed with which the child
released it. Prepare graphs of x versus t, v_x versus t, and a_x versus t, vertically aligned with their time axes identical, to show
the motion of the marble. You will not be able to place numbers other than zero on the horizontal axis or on the velocity
or acceleration axes, but show the correct graph shapes.

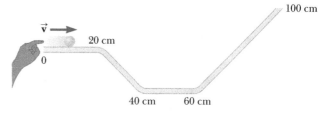

Figure P2.8

9. Figure P2.9 shows a graph of v_x versus t for the motion of
a motorcyclist as he starts from rest and moves along the
road in a straight line. (a) Find the average acceleration for
the time interval $t = 0$ to $t = 6.00$ s. (b) Estimate the time
at which the acceleration has its greatest positive value and
the value of the acceleration at that instant. (c) When is
the acceleration zero? (d) Estimate the maximum negative
value of the acceleration and the time at which it occurs.

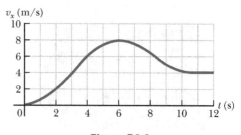

Figure P2.9

10. (a) Use the data in Problem 3 to construct a smooth graph
of position versus time. (b) By constructing tangents to
the $x(t)$ curve, find the instantaneous velocity of the car at
several instants. (c) Plot the instantaneous velocity versus
time and, from this information, determine the average
acceleration of the car. (d) What was the initial velocity of
the car?

11. A particle starts from rest and accelerates as shown in Figure P2.11. Determine (a) the particle's speed at $t = 10.0$ s and at $t = 20.0$ s, and (b) the distance traveled in the first 20.0 s.

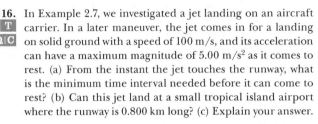

Figure P2.11

SECTION 2.6 Motion Diagrams

12. Draw motion diagrams for (a) an object moving to the right at constant speed, (b) an object moving to the right and speeding up at a constant rate, (c) an object moving to the right and slowing down at a constant rate, (d) an object moving to the left and speeding up at a constant rate, and (e) an object moving to the left and slowing down at a constant rate. (f) How would your drawings change if the changes in speed were not uniform, that is, if the speed were not changing at a constant rate?

13. Each of the strobe photographs (a), (b), and (c) in Figure P2.13 was taken of a single disk moving toward the right, which we take as the positive direction. Within each photograph the time interval between images is constant. For each photograph, prepare graphs of x versus t, v_x versus t, and a_x versus t, vertically aligned with their time axes identical, to show the motion of the disk. You will not be able to place numbers other than zero on the axes, but show the correct shapes for the graph lines.

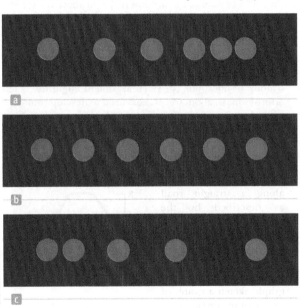

Charles D. Winters

Figure P2.13

SECTION 2.7 Analysis Model: Particle Under Constant Acceleration

14. An electron in a cathode-ray tube accelerates uniformly from 2.00×10^4 m/s to 6.00×10^6 m/s over 1.50 cm. (a) In what time interval does the electron travel this 1.50 cm? (b) What is its acceleration?

15. A parcel of air moving in a straight tube with a constant acceleration of -4.00 m/s^2 has a velocity of 13.0 m/s at 10:05:00 a.m. (a) What is its velocity at 10:05:01 a.m.? (b) At 10:05:04 a.m.? (c) At 10:04:59 a.m.? (d) Describe the shape of a graph of velocity versus time for this parcel of air. (e) Argue for or against the following statement: "Knowing the single value of an object's constant acceleration is like knowing a whole list of values for its velocity."

16. In Example 2.7, we investigated a jet landing on an aircraft carrier. In a later maneuver, the jet comes in for a landing on solid ground with a speed of 100 m/s, and its acceleration can have a maximum magnitude of 5.00 m/s^2 as it comes to rest. (a) From the instant the jet touches the runway, what is the minimum time interval needed before it can come to rest? (b) Can this jet land at a small tropical island airport where the runway is 0.800 km long? (c) Explain your answer.

17. An object moving with uniform acceleration has a velocity of 12.0 cm/s in the positive x direction when its x coordinate is 3.00 cm. If its x coordinate 2.00 s later is -5.00 cm, what is its acceleration?

18. Solve Example 2.8 by a graphical method. On the same graph, plot position versus time for the car and the trooper. From the intersection of the two curves, read the time at which the trooper overtakes the car.

19. A glider of length ℓ moves through a stationary photogate on an air track. A photogate (Fig. P2.19) is a device that measures the time interval Δt_d during which the glider blocks a beam of infrared light passing across the photogate. The ratio $v_d = \ell/\Delta t_d$ is the average velocity of the glider over this part of its motion. Suppose the glider moves with constant acceleration. (a) Argue for or against the idea that v_d is equal to the instantaneous velocity of the glider when it is halfway through the photogate in space. (b) Argue for or against the idea that v_d is equal to the instantaneous velocity of the glider when it is halfway through the photogate in time.

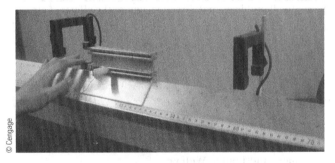

© Cengage

Figure P2.19 Problems 19 and 21.

20. *Why is the following situation impossible?* Starting from rest, a charging rhinoceros moves 50.0 m in a straight line in 10.0 s. Her acceleration is constant during the entire motion, and her final speed is 8.00 m/s.

21. A glider of length 12.4 cm moves on an air track with constant acceleration (Fig P2.19). A time interval of 0.628 s elapses between the moment when its front end passes a fixed point Ⓐ along the track and the moment when its back end passes this point. Next, a time interval of 1.39 s elapses between the moment when the back end of the glider passes the point Ⓐ and the moment when the front end of the glider passes a second point Ⓑ farther down the track. After that, an additional 0.431 s elapses until the back end of the glider passes point Ⓑ. (a) Find the average speed of the glider as it passes point Ⓐ. (b) Find the acceleration of the glider. (c) Explain how you can compute the acceleration without knowing the distance between points Ⓐ and Ⓑ.

22. In the particle under constant acceleration model, we identify the variables and parameters v_{xi}, v_{xf}, a_x, t, and

$x_f - x_i$. Of the equations in the model, Equations 2.13–2.17, the first does not involve $x_f - x_i$, the second and third do not contain a_x, the fourth omits v_{xf}, and the last leaves out t. So, to complete the set, there should be an equation *not* involving v_{xi}. Derive it from the others.

23. At $t = 0$, one toy car is set rolling on a straight track with initial position 15.0 cm, initial velocity −3.50 cm/s, and constant acceleration 2.40 cm/s². At the same moment, another toy car is set rolling on an adjacent track with initial position 10.0 cm, initial velocity +5.50 cm/s, and constant acceleration zero. (a) At what time, if any, do the two cars have equal speeds? (b) What are their speeds at that time? (c) At what time(s), if any, do the cars pass each other? (d) What are their locations at that time? (e) Explain the difference between question (a) and question (c) as clearly as possible.

24. You are observing the poles along the side of the road as described in the opening storyline of the chapter. You have already stopped and measured the distance between adjacent poles as 40.0 m. You are now driving again and have activated your smartphone stopwatch. You start the stopwatch at $t = 0$ as you pass pole #1. At pole #2, the stopwatch reads 10.0 s. At pole #3, the stopwatch reads 25.0 s. Your friend tells you that he was pressing the brake and slowing down the car uniformly during the entire time interval from pole #1 to pole #3. (a) What was the acceleration of the car between poles #1 and #3? (b) What was the velocity of the car at pole #1? (c) If the motion of the car continues as described, what is the number of the *last* pole passed before the car comes to rest?

SECTION 2.8 Freely Falling Objects

Note: In all problems in this section, ignore the effects of air resistance.

25. *Why is the following situation impossible?* Emily challenges David to catch a $1 bill as follows. She holds the bill vertically as shown in Figure P2.25, with the center of the bill between but not touching David's index finger and thumb. Without warning, Emily releases the bill. David catches the bill without moving his hand downward. David's reaction time is equal to the average human reaction time.

© Cengage

Figure P2.25

26. An attacker at the base of a castle wall 3.65 m high throws a rock straight up with speed 7.40 m/s from a height of 1.55 m above the ground. (a) Will the rock reach the top of the wall? (b) If so, what is its speed at the top? If not, what initial speed must it have to reach the top? (c) Find the change in speed of a rock thrown straight down from the top of the wall at an initial speed of 7.40 m/s and moving between the same two points. (d) Does the change in speed of the downward-moving rock agree with the magnitude of the speed change of the rock moving upward between the same elevations? (e) Explain physically why it does or does not agree.

27. The height of a helicopter above the ground is given by $h = 3.00t^3$, where h is in meters and t is in seconds. At $t = 2.00$ s, the helicopter releases a small mailbag. How long after its release does the mailbag reach the ground?

28. A ball is thrown upward from the ground with an initial speed of 25 m/s; at the same instant, another ball is dropped from a building 15 m high. After how long will the balls be at the same height above the ground?

29. A student throws a set of keys vertically upward to her sorority sister, who is in a window 4.00 m above. The second student catches the keys 1.50 s later. (a) With what initial velocity were the keys thrown? (b) What was the velocity of the keys just before they were caught?

30. At time $t = 0$, a student throws a set of keys vertically upward to her sorority sister, who is in a window at distance h above. The second student catches the keys at time t. (a) With what initial velocity were the keys thrown? (b) What was the velocity of the keys just before they were caught?

31. You have been hired by the prosecuting attorney as an expert witness in a robbery case. The defendant is accused of stealing an expensive and massive diamond ring in its box from a jewelry store. A witness to the alleged crime testified that she saw the defendant run from the store, stop next to an apartment building, and throw the box straight upward to an accomplice leaning out of a fourth-floor window. When captured, the defendant did not have the stolen box with him and claimed innocence. When the witness testified in court about the defendant's throwing of the box to an accomplice, the defending attorney argued that it would be impossible to throw the box upward that high to reach the window in question. The bottom of the window is 19.0 m above the sidewalk. You have set up a demonstration in which the defendant was asked by the judge to throw a baseball horizontally as fast as he could and a radar device was used to determine that he can throw the ball at 20 m/s. (a) What testimony can you provide about the ability of the defendant to throw the box to the window in question? (b) What argument might the defense attorney make about the process used to develop your expert testimony? What might be your counter argument? Ignore any effects of air resistance on the box.

SECTION 2.9 Kinematic Equations Derived from Calculus

32. A student drives a moped along a straight road as described by the velocity–time graph in Figure P2.32. Sketch this graph in the middle of a sheet of graph paper. (a) Directly above your graph, sketch a graph of the position versus time, aligning the time coordinates of the two graphs. (b) Sketch a graph of the acceleration versus time directly below the velocity–time graph, again aligning the time coordinates. On each graph, show the numerical values of x and a_x for all points of inflection. (c) What is the acceleration at $t = 6.00$ s? (d) Find the position (relative to the starting point) at $t = 6.00$ s. (e) What is the moped's final position at $t = 9.00$ s?

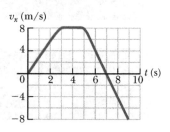

Figure P2.32

33. Automotive engineers refer to the time rate of change of acceleration as the "jerk." Assume an object moves in one dimension such that its jerk J is constant. (a) Determine expressions for its acceleration $a_x(t)$, velocity $v_x(t)$, and position $x(t)$, given that its initial acceleration, velocity, and position are a_{xi}, v_{xi}, and x_i, respectively. (b) Show that $a_x^2 = a_{xi}^2 + 2J(v_x - v_{xi})$.

ADDITIONAL PROBLEMS

34. In Figure 2.11b, the area under the velocity–time graph
`Q|C` and between the vertical axis and time t (vertical dashed
`S` line) represents the displacement. As shown, this area consists of a rectangle and a triangle. (a) Compute their areas.
(b) Explain how the sum of the two areas compares with the
expression on the right-hand side of Equation 2.16.

35. The froghopper *Philaenus spumarius* is supposedly the best
`BIO` jumper in the animal kingdom. To start a jump, this insect
can accelerate at 4.00 km/s^2 over a distance of 2.00 mm as it
straightens its specially adapted "jumping legs." Assume the
acceleration is constant. (a) Find the upward velocity with
which the insect takes off. (b) In what time interval does it
reach this velocity? (c) How high would the insect jump if
air resistance were negligible? The actual height it reaches is
about 70 cm, so air resistance must be a noticeable force on
the leaping froghopper.

36. A woman is reported to have fallen 144 ft from the 17th
floor of a building, landing on a metal ventilator box that
she crushed to a depth of 18.0 in. She suffered only minor
injuries. Ignoring air resistance, calculate (a) the speed of
the woman just before she collided with the ventilator and
(b) her average acceleration while in contact with the box.
(c) Modeling her acceleration as constant, calculate the
time interval it took to crush the box.

37. At $t = 0$, one athlete in a race running on a long, straight
`Q|C` track with a constant speed v_1 is a distance d_1 behind a sec-
`S` ond athlete running with a constant speed v_2. (a) Under
what circumstances is the first athlete able to overtake the
second athlete? (b) Find the time t at which the first athlete
overtakes the second athlete, in terms of d_1, v_1, and v_2. (c)
At what minimum distance d_2 from the leading athlete must
the finish line be located so that the trailing athlete can at
least tie for first place? Express d_2 in terms of d_1, v_1, and v_2 by
using the result of part (b).

38. *Why is the following situation impossible?* A freight train is lum-
bering along at a constant speed of 16.0 m/s. Behind the
freight train on the same track is a passenger train traveling
in the same direction at 40.0 m/s. When the front of the passenger train is 58.5 m from the back of the freight train, the
engineer on the passenger train recognizes the danger and
hits the brakes of his train, causing the train to move with
acceleration -3.00 m/s^2. Because of the engineer's action,
the trains do not collide.

39. Hannah tests her new sports car by racing with Sam, an
`AMT` experienced racer. Both start from rest, but Hannah
`T` leaves the starting line 1.00 s after Sam does. Sam moves
with a constant acceleration of 3.50 m/s^2, while Hannah
maintains an acceleration of 4.90 m/s^2. Find (a) the time at
which Hannah overtakes Sam, (b) the distance she travels
before she catches him, and (c) the speeds of both cars at the
instant Hannah overtakes Sam.

40. Two objects, A and B, are connected by hinges to a rigid
`Q|C` rod that has a length L. The objects slide along perpendic-
`S` ular guide rails as shown in Figure P2.40. Assume object A
slides to the left with a constant speed v. (a) Find the velocity v_B of object B as a function of the angle θ. (b) Describe v_B
relative to v. Is v_B always smaller
than v, larger than v, or the same
as v, or does it have some other
relationship?

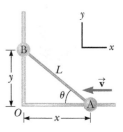

Figure P2.40

41. Lisa rushes down onto a subway
platform to find her train already
departing. She stops and watches
the cars go by. Each car is 8.60 m
long. The first moves past her in
1.50 s and the second in 1.10 s.
Find the constant acceleration of
the train.

CHALLENGE PROBLEMS

42. Two thin rods are fastened to
the inside of a circular ring as
shown in Figure P2.42. One
rod of length D is vertical, and
the other of length L makes an
angle θ with the horizontal. The
two rods and the ring lie in a
vertical plane. Two small beads
are free to slide without friction
along the rods. (a) If the two
beads are released from rest
simultaneously from the positions shown, use your intuition

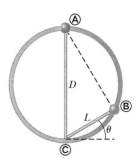

Figure P2.42

and guess which bead reaches the bottom first. (b) Find an
expression for the time interval required for the red bead to
fall from point Ⓐ to point Ⓒ in terms of g and D. (c) Find an
expression for the time interval required for the blue bead
to slide from point Ⓑ to point Ⓒ in terms of g, L, and θ.
(d) Show that the two time intervals found in parts (b) and
(c) are equal. *Hint:* What is the angle between the chords of
the circle Ⓐ Ⓑ and Ⓑ Ⓒ? (e) Do these results surprise you?
Was your intuitive guess in part (a) correct? This problem
was inspired by an article by Thomas B. Greenslade, Jr.,
"Galileo's Paradox," *Phys. Teach.* **46**, 294 (May 2008).

43. In a women's 100-m race, accelerating uniformly, Laura
takes 2.00 s and Healan 3.00 s to attain their maximum
speeds, which they each maintain for the rest of the race.
They cross the finish line simultaneously, both setting a
world record of 10.4 s. (a) What is the acceleration of each
sprinter? (b) What are their respective maximum speeds?
(c) Which sprinter is ahead at the 6.00-s mark, and by how
much? (d) What is the maximum distance by which Healan
is behind Laura, and at what time does that occur?

44. **Review.** You are sitting in your car at rest at a traffic light
with a bicyclist at rest next to you in the adjoining bicycle lane. As soon as the traffic light turns green, your car
speeds up from rest to 50.0 mi/h with constant acceleration
9.00 mi/h/s and thereafter moves with a constant speed of
50.0 mi/h. At the same time, the cyclist speeds up from
rest to 20.0 mi/h with constant acceleration 13.0 mi/h/s
and thereafter moves with a constant speed of 20.0 mi/h.
(a) For what time interval after the light turned green is
the bicycle ahead of your car? (b) What is the maximum
distance by which the bicycle leads your car during this
time interval?

3 Vectors

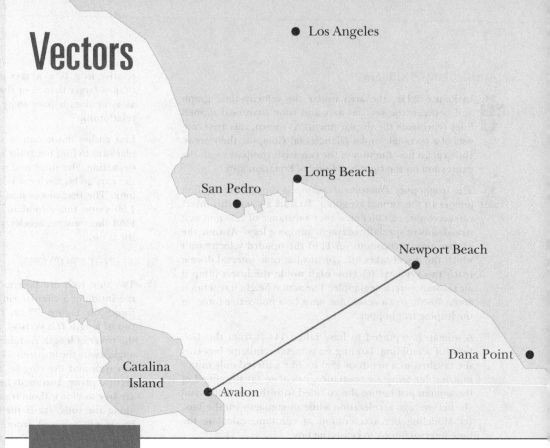

Los Angeles

Long Beach

San Pedro

Newport Beach

Catalina Island

Dana Point

Avalon

Catalina Island can be reached from different starting points along the Los Angeles–Orange County coast. The opening storyline refers to a trip to Avalon beginning in Newport Beach.

STORYLINE **Your road trip in Chapter 2 takes you toward the ocean.** You end up in Newport Beach, California. Your friend who was driving the car in Chapter 2 owns a sailboat and asks you to pilot the boat from Newport Beach to Catalina Island, which is 26 miles off the coast. Your friend challenges you to pilot the boat along a perfectly straight line. Always up for a challenge, you agree, settle into the captain's chair, and then panic. You know you have to travel 26 miles in a straight line, but what should you set as the heading for the boat? The distance of 26 miles is not sufficient information to allow you to travel to Catalina Island in a straight line. You realize that your trip will require both the distance to Catalina Island and the *direction* in which you must travel. You ask your friend the appropriate direction to Catalina Island and he gives you a heading as an angle south of due west. You open the compass app on your smartphone, find the appropriate direction, and set sail!

CONNECTIONS If you move only along a straight line, as in the previous chapter, then a single number (with a positive or negative sign) can be used to specify your position with respect to the origin. In this chapter, we will study the positions of objects or points in two- or three-dimensional space that require two types of information: distance from a reference point and direction relative to a reference axis. Quantities that require these two types of information are called *vectors*. We will learn various properties of vectors and will see how to add and subtract vectors. Vector quantities are used throughout this text. In addition to the position vectors studied in this chapter, we will see other vector quantities in subsequent chapters, such as velocity, acceleration, force, and electric field. Therefore, it is imperative that you master the techniques discussed in this chapter.

3.1 Coordinate Systems

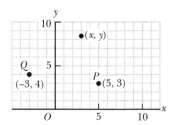

Many aspects of physics involve a description of a location in space. In Chapter 2, for example, we saw that the mathematical description of an object's motion requires a method for describing the object's position at various times. In two dimensions, this description is accomplished with the use of the Cartesian coordinate system, in which perpendicular axes intersect at a point defined as the origin O (Fig. 3.1). Cartesian coordinates of a point in space, representing the x and y values of the point, and expressed as (x, y), are also called *rectangular coordinates*.

Figure 3.1 Designation of points in a Cartesian coordinate system. Every point is labeled with coordinates (x, y).

Sometimes it is more convenient to represent a point in a plane by its *plane polar coordinates* (r, θ) as shown in Figure 3.2a. In this *polar coordinate system,* r is the distance from the origin to the point having Cartesian coordinates (x, y) and θ is the angle between a fixed axis and a line drawn from the origin to the point. The fixed axis is often the positive x axis, and θ is usually measured counterclockwise from it. From the right triangle in Figure 3.2b, we find that $\sin \theta = y/r$ and that $\cos \theta = x/r$. (A review of trigonometric functions is given in Appendix B.4.) Therefore, starting with the plane polar coordinates of any point, we can obtain the Cartesian coordinates by using the equations

$$x = r \cos \theta \qquad (3.1)$$

$$y = r \sin \theta \qquad (3.2)$$

◀ Cartesian coordinates in terms of polar coordinates

Conversely, if we know the Cartesian coordinates, the definitions of trigonometry tell us that the polar coordinates are given by

$$\tan \theta = \frac{y}{x} \qquad (3.3)$$

◀ Polar coordinates in terms of Cartesian coordinates

$$r = \sqrt{x^2 + y^2} \qquad (3.4)$$

Equation 3.4 is the familiar Pythagorean theorem.

These four expressions relating the coordinates (x, y) to the coordinates (r, θ) apply only when θ is defined as shown in Figure 3.2a—in other words, when positive θ is an angle measured counterclockwise from the positive x axis. (Some scientific calculators perform conversions between Cartesian and polar coordinates based on these standard conventions.) If the reference axis for the polar angle θ is chosen to be one other than the positive x axis or if the sense of increasing θ is chosen differently, the expressions relating the two sets of coordinates will be different from those above.

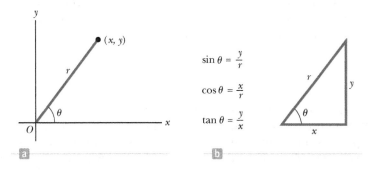

Figure 3.2 (a) The plane polar coordinates of a point are represented by the distance r and the angle θ, where θ is measured counterclockwise from the positive x axis. (b) The right triangle used to relate (x, y) to (r, θ).

The Cartesian coordinates of a point in the *xy* plane are $(x, y) = (-3.50, -2.50)$ m as shown in Figure 3.3. Find the polar coordinates of this point.

SOLUTION

Conceptualize The drawing in Figure 3.3 helps us conceptualize the problem. We wish to find *r* and *θ*. Based on the figure and the data given in the problem statement, we expect *r* to be a few meters and *θ* to be between 180° and 270°.

Categorize Based on the statement of the problem and the Conceptualize step, we recognize that we are simply converting from Cartesian coordinates to polar coordinates. We therefore categorize this example as a substitution problem.
As mentioned in Section 2.4, substitution problems generally

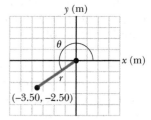

Figure 3.3 (Example 3.1) Finding polar coordinates when Cartesian coordinates are given.

do not have an extensive Analyze step other than the substitution of numbers into a given equation. Similarly, the Finalize step consists primarily of checking the units and making sure that the answer is reasonable and consistent with our expectations. Therefore, for substitution problems, we will not label Analyze or Finalize steps.

Use Equation 3.4 to find *r*:

$$r = \sqrt{x^2 + y^2} = \sqrt{(-3.50 \text{ m})^2 + (-2.50 \text{ m})^2} = 4.30 \text{ m}$$

Use Equation 3.3 to find *θ*:

$$\tan \theta = \frac{y}{x} = \frac{-2.50 \text{ m}}{-3.50 \text{ m}} = 0.714$$

$$\theta = 216°$$

Notice that you must use the signs of *x* and *y* to find that the point lies in the third quadrant of the coordinate system. That is, *θ* = 216°, not 35.5°, whose tangent is also 0.714. Answers to both *r* and *θ* agree with our expectations in the Conceptualize step.

3.2 Vector and Scalar Quantities

We now formally describe the difference between scalar quantities and vector quantities. When you want to know the temperature outside so that you will know how to dress, the only information you need is a number and the unit "degrees C" or "degrees F." Temperature is therefore an example of a *scalar quantity:*

> A **scalar quantity** is completely specified by a single value with an appropriate unit and has no direction.

Other examples of scalar quantities are volume, mass, speed, time, and time intervals. Some scalars are always positive, such as mass and speed. Others, such as temperature, can have either positive or negative values. The rules of ordinary arithmetic are used to manipulate scalar quantities.

If you are preparing to pilot a small plane and need to know the wind velocity, you must know both the speed of the wind and its direction. Because direction is important for its complete specification, velocity is a *vector quantity:*

> A **vector quantity** is completely specified by a number with an appropriate unit (the *magnitude* of the vector) plus a direction.

Another example of a vector quantity is displacement, as you know from Chapter 2. Suppose a particle moves from some point Ⓐ to some point Ⓑ along a straight path as shown in Figure 3.4. We represent this displacement by drawing an arrow from Ⓐ to Ⓑ, with the tip of the arrow pointing away from the starting point. The direction of the arrowhead represents the direction of the displacement, and the length of the arrow represents the magnitude of the displacement. If the particle travels along some other path from Ⓐ to Ⓑ such as

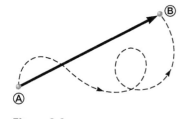

Figure 3.4 As a particle moves from Ⓐ to Ⓑ either along the straight line *or* along an arbitrary path represented by the broken line, its displacement is a vector quantity shown by the arrow drawn from Ⓐ to Ⓑ.

shown by the broken line in Figure 3.4, its displacement is still the arrow drawn from Ⓐ to Ⓑ. Displacement depends only on the initial and final positions, so the displacement vector is independent of the path taken by the particle between these two points.

In this text, we use a boldface letter with an arrow over the letter, such as $\vec{A}$, to represent a vector. Another common notation for vectors with which you should be familiar is a simple boldface character: **A**. The magnitude of the vector $\vec{A}$ is written either A or $|\vec{A}|$. The magnitude of a vector has physical units, such as meters for displacement or meters per second for velocity. The magnitude of a vector is *always* a positive number.

What about the vector to follow in our opening storyline? What heading did your friend give you to Catalina Island? You can use a latitude and longitude finder online to find the coordinates for the opening of Newport Harbor and for Avalon Harbor. Then, putting these coordinates into a distance and azimuth calculator online, you find that the distance is 30.7 mi, with a heading of 236.2° relative to due east. (Note that Catalina is described as "26 miles across the sea" in a popular song from the 1950s, but we need to travel a bit farther to make this trip. An online calculation shows the distance between San Pedro and Avalon to be 27 miles, which might be the origin of the song.)

Ⓠ**UICK QUIZ 3.1** Which of the following are vector quantities and which are scalar quantities? **(a)** your age **(b)** acceleration **(c)** velocity **(d)** speed **(e)** mass

3.3 Basic Vector Arithmetic

For many purposes, two vectors $\vec{A}$ and $\vec{B}$ may be defined to be *equal* if they have the same magnitude and if they point in the same direction. That is, $\vec{A} = \vec{B}$ only if $A = B$ and if $\vec{A}$ and $\vec{B}$ point in the same direction along parallel lines. For example, all the vectors in Figure 3.5 are equal even though they have different starting points. This property allows us to move a vector to a position parallel to itself in a diagram without affecting the vector.

The rules for **vector addition** are conveniently described by a graphical method. To add vector $\vec{B}$ to vector $\vec{A}$, first draw vector $\vec{A}$ on graph paper, with its magnitude represented by a convenient length scale, and then draw vector $\vec{B}$ to the same scale, with its tail starting from the tip of $\vec{A}$, as shown in Figure 3.6. The **resultant vector** $\vec{R} = \vec{A} + \vec{B}$ is the vector drawn from the tail of $\vec{A}$ to the tip of $\vec{B}$.

A geometric construction can also be used to add more than two vectors as shown in Figure 3.7 for the case of three vectors. The resultant vector $\vec{R} = \vec{A} + \vec{B} + \vec{C}$ is the vector that completes the polygon. In other words, $\vec{R}$ is the vector drawn from the tail of the first vector to the tip of the last vector. This technique for adding vectors is often called the "head to tail method."

PITFALL PREVENTION 3.1
Vector Addition Versus Scalar Addition Notice that $\vec{A} + \vec{B} = \vec{C}$ is very different from $A + B = C$. The first equation is a vector sum, which must be handled carefully, such as with the graphical method. The second equation is a simple algebraic addition of numbers that is handled with the normal rules of arithmetic.

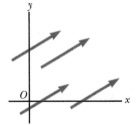

Figure 3.5 These four vectors are equal because they have equal lengths and point in the same direction.

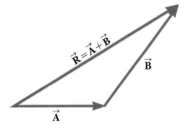

Figure 3.6 When vector $\vec{B}$ is added to vector $\vec{A}$ the resultant $\vec{R}$ is the vector that runs from the tail of $\vec{A}$ to the tip of $\vec{B}$.

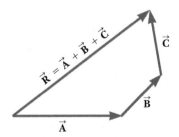

Figure 3.7 Geometric construction for summing three vectors. The resultant vector $\vec{R}$ is by definition the one that completes the polygon.

When two vectors are added, the sum is independent of the order of the addition. (This fact may seem trivial, but as you will see in Chapter 11, the order is important when vectors are multiplied. Procedures for multiplying vectors are discussed in Chapters 7 and 11.) This property, which can be seen from the geometric construction in Figure 3.8, is known as the **commutative law of addition:**

$$\vec{\mathbf{A}} + \vec{\mathbf{B}} = \vec{\mathbf{B}} + \vec{\mathbf{A}} \tag{3.5}$$

When three or more vectors are added, their sum is independent of the way in which the individual vectors are grouped together. A geometric proof of this rule for three vectors is given in Figure 3.9, where two ways of adding the same three vectors are shown. This property is called the **associative law of addition:**

$$\vec{\mathbf{A}} + (\vec{\mathbf{B}} + \vec{\mathbf{C}}) = (\vec{\mathbf{A}} + \vec{\mathbf{B}}) + \vec{\mathbf{C}} \tag{3.6}$$

We have described adding displacement vectors in this section because these types of vectors are easy to visualize. We can also add other types of vectors, such as velocity, force, and electric field vectors, which we will do in later chapters. When two or more vectors are added together, they must all have the same units and they must all be the same type of quantity. It would be meaningless to add a velocity vector (for example, 60 km/h to the east) to a displacement vector (for example, 200 km to the north) because these vectors represent different physical quantities. The same rule also applies to scalars. For example, it would be meaningless to add time intervals to temperatures.

The operation of **vector subtraction** makes use of the definition of the negative of a vector. The negative of the vector $\vec{\mathbf{A}}$ is defined as the vector that when added to $\vec{\mathbf{A}}$ gives zero for the vector sum. That is, $\vec{\mathbf{A}} + (-\vec{\mathbf{A}}) = 0$. The vectors $\vec{\mathbf{A}}$ and $-\vec{\mathbf{A}}$ have the same magnitude but point in opposite directions. We define the operation $\vec{\mathbf{A}} - \vec{\mathbf{B}}$ as vector $-\vec{\mathbf{B}}$ added to vector $\vec{\mathbf{A}}$:

$$\vec{\mathbf{A}} - \vec{\mathbf{B}} = \vec{\mathbf{A}} + (-\vec{\mathbf{B}}) \tag{3.7}$$

The geometric construction for subtracting two vectors in this way is illustrated in Figure 3.10a.

Another way of looking at vector subtraction is to notice that the difference $\vec{\mathbf{A}} - \vec{\mathbf{B}}$ between two vectors $\vec{\mathbf{A}}$ and $\vec{\mathbf{B}}$ is what you have to add to the second vector to obtain the first. In this case, as Figure 3.10b shows, the vector $\vec{\mathbf{A}} - \vec{\mathbf{B}}$ points from the tip of the second vector to the tip of the first.

Scalar multiplication of vectors is straightforward. If vector $\vec{\mathbf{A}}$ is multiplied by a positive scalar quantity m, the product $m\vec{\mathbf{A}}$ is a vector that has the same direction as $\vec{\mathbf{A}}$ and magnitude mA. If vector $\vec{\mathbf{A}}$ is multiplied by a negative scalar quantity $-m$,

Commutative law of addition ▶

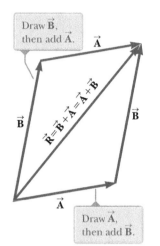

Figure 3.8 This construction shows that $\vec{\mathbf{A}} + \vec{\mathbf{B}} = \vec{\mathbf{B}} + \vec{\mathbf{A}}$ or, in other words, that vector addition is commutative.

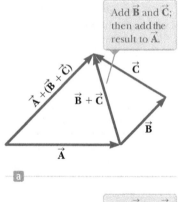

a

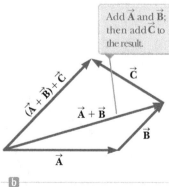

b

Figure 3.9 Geometric constructions for verifying the associative law of addition. (a) Vectors $\vec{\mathbf{B}}$ and $\vec{\mathbf{C}}$ are added first and added to $\vec{\mathbf{A}}$. (b) Vectors $\vec{\mathbf{A}}$ and $\vec{\mathbf{B}}$ are added first, and then $\vec{\mathbf{C}}$ is added.

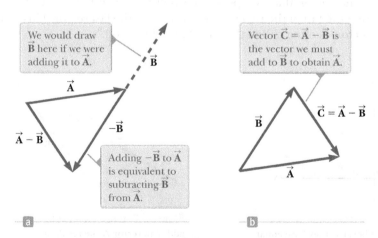

a   b

Figure 3.10 (a) Subtracting vector $\vec{\mathbf{B}}$ from vector $\vec{\mathbf{A}}$. The vector $-\vec{\mathbf{B}}$ is equal in magnitude to vector $\vec{\mathbf{B}}$ and points in the opposite direction. (b) A second way of looking at vector subtraction.

the product $-m\vec{\mathbf{A}}$ is directed opposite $\vec{\mathbf{A}}$. For example, the vector $5\vec{\mathbf{A}}$ is five times as long as $\vec{\mathbf{A}}$ and points in the same direction as $\vec{\mathbf{A}}$; the vector $-\frac{1}{3}\vec{\mathbf{A}}$ is one-third the length of $\vec{\mathbf{A}}$ and points in the direction opposite $\vec{\mathbf{A}}$.

QUICK QUIZ 3.2 The magnitudes of two vectors $\vec{\mathbf{A}}$ and $\vec{\mathbf{B}}$ are $A = 12$ units and $B = 8$ units. Which pair of numbers represents the *largest* and *smallest* possible values for the magnitude of the resultant vector $\vec{\mathbf{R}} = \vec{\mathbf{A}} + \vec{\mathbf{B}}$? **(a)** 14.4 units, 4 units **(b)** 12 units, 8 units **(c)** 20 units, 4 units **(d)** none of these answers

QUICK QUIZ 3.3 If vector $\vec{\mathbf{B}}$ is added to vector $\vec{\mathbf{A}}$, which *two* of the following choices must be true for the resultant vector to be equal to zero? **(a)** $\vec{\mathbf{A}}$ and $\vec{\mathbf{B}}$ are parallel and in the same direction. **(b)** $\vec{\mathbf{A}}$ and $\vec{\mathbf{B}}$ are parallel and in opposite directions. **(c)** $\vec{\mathbf{A}}$ and $\vec{\mathbf{B}}$ have the same magnitude. **(d)** $\vec{\mathbf{A}}$ and $\vec{\mathbf{B}}$ are perpendicular.

Example 3.2 **A Vacation Trip**

A car travels 20.0 km due north and then 35.0 km in a direction 60.0° west of north as shown in Figure 3.11a. Find the magnitude and direction of the car's resultant displacement.

SOLUTION

Conceptualize The two vectors $\vec{\mathbf{A}}$ and $\vec{\mathbf{B}}$ that appear in Figure 3.11a help us conceptualize the problem. The resultant vector $\vec{\mathbf{R}}$ has also been drawn. We expect its magnitude to be a few tens of kilometers. The angle β that the resultant vector makes with the y axis is expected to be less than 60°, the angle that vector $\vec{\mathbf{B}}$ makes with the y axis.

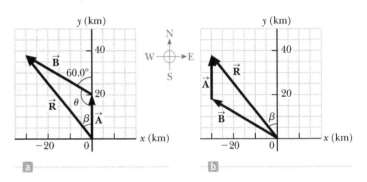

Figure 3.11 (Example 3.2) (a) Graphical method for finding the resultant displacement vector $\vec{\mathbf{R}} = \vec{\mathbf{A}} + \vec{\mathbf{B}}$. (b) Adding the vectors in reverse order ($\vec{\mathbf{B}} + \vec{\mathbf{A}}$) gives the same result for $\vec{\mathbf{R}}$.

Categorize We can categorize this example as a simple analysis problem in vector addition. The displacement $\vec{\mathbf{R}}$ is the resultant when the two individual displacements $\vec{\mathbf{A}}$ and $\vec{\mathbf{B}}$ are added. We can further categorize it as a problem about the analysis of triangles, so we appeal to our expertise in geometry and trigonometry.

Analyze In this example, we show two ways to analyze the problem of finding the resultant of two vectors. The first way is to solve the problem geometrically, using graph paper and a protractor to measure the magnitude of $\vec{\mathbf{R}}$ and its direction in Figure 3.11a. (In fact, even when you know you are going to be carrying out a calculation, you should sketch the vectors to check your results.) With an ordinary ruler and protractor, a large diagram typically gives answers to two-digit but not to three-digit precision. Try using these tools on $\vec{\mathbf{R}}$ in Figure 3.11a and compare to the trigonometric analysis below!

The second way to solve the problem is to analyze it using algebra and trigonometry. The magnitude of $\vec{\mathbf{R}}$ can be obtained from the law of cosines as applied to the triangle in Figure 3.11a (see Appendix B.4).

Use $R^2 = A^2 + B^2 - 2AB \cos \theta$ from the law of cosines to find R:

$$R = \sqrt{A^2 + B^2 - 2AB \cos \theta}$$

Substitute numerical values, noting that $\theta = 180° - 60° = 120°$:

$$R = \sqrt{(20.0 \text{ km})^2 + (35.0 \text{ km})^2 - 2(20.0 \text{ km})(35.0 \text{ km}) \cos 120°}$$
$$= 48.2 \text{ km}$$

Use the law of sines (Appendix B.4) to find the direction of $\vec{\mathbf{R}}$ measured from the northerly direction:

$$\frac{\sin \beta}{B} = \frac{\sin \theta}{R}$$

$$\sin \beta = \frac{B}{R} \sin \theta = \frac{35.0 \text{ km}}{48.2 \text{ km}} \sin 120° = 0.629$$

$$\beta = 38.9°$$

The resultant displacement of the car is 48.2 km in a direction 38.9° west of north.

continued

3.2 continued

Finalize   Does the angle β that we calculated agree with an estimate made by looking at Figure 3.11a or with an actual angle measured from the diagram using the graphical method? Is it reasonable that the magnitude of $\vec{\mathbf{R}}$ is larger than that of both $\vec{\mathbf{A}}$ and $\vec{\mathbf{B}}$? Are the units of $\vec{\mathbf{R}}$ correct?

Although the head to tail method of adding vectors works well, it suffers from two disadvantages. First, some people

find using the laws of cosines and sines to be awkward. Second, a triangle only results if you are adding two vectors. If you are adding three or more vectors, the resulting geometric shape is usually not a triangle. In Section 3.4, we explore a new method of adding vectors that will address both of these disadvantages.

WHAT IF?   Suppose the trip were taken with the two vectors in reverse order: 35.0 km at 60.0° west of north first and then 20.0 km due north. How would the magnitude and the direction of the resultant vector change?

Answer   They would not change. The commutative law for vector addition tells us that the order of vectors in an addition is irrelevant. Graphically, Figure 3.11b shows that the vectors added in the reverse order give us the same resultant vector.

3.4 Components of a Vector and Unit Vectors

The graphical method of adding vectors is not recommended whenever high accuracy is required or in three-dimensional problems. In this section, we describe a method of adding vectors that makes use of the projections of vectors along coordinate axes. These projections are called the **components** of the vector or its **rectangular components.** Any vector can be completely described by its components.

Consider a vector $\vec{\mathbf{A}}$ lying in the xy plane and making an arbitrary angle θ with the positive x axis as shown in Figure 3.12a. This vector can be expressed as the sum of two other *component vectors* $\vec{\mathbf{A}}_x$, which is parallel to the x axis, and $\vec{\mathbf{A}}_y$, which is parallel to the y axis. From the figure, we see that the three vectors form a right triangle and that $\vec{\mathbf{A}} = \vec{\mathbf{A}}_x + \vec{\mathbf{A}}_y$. We shall often refer to the "components of a vector $\vec{\mathbf{A}}$," written A_x and A_y (without the boldface notation). Figure 3.12b shows the component vector $\vec{\mathbf{A}}_y$ moved to the left so that it lies along the y axis. We see that the component A_x represents the projection of $\vec{\mathbf{A}}$ along the x axis, and the component A_y represents the projection of $\vec{\mathbf{A}}$ along the y axis. These components can be positive or negative. The component A_x is positive if the component vector $\vec{\mathbf{A}}_x$ points in the positive x direction and is negative if $\vec{\mathbf{A}}_x$ points in the negative x direction. A similar statement is made for the component A_y.

From Figure 3.12 and the definition of sine and cosine, we see that $\cos \theta = A_x/A$ and that $\sin \theta = A_y/A$. Hence, the components of $\vec{\mathbf{A}}$ are

$$A_x = A \cos \theta \tag{3.8}$$

$$A_y = A \sin \theta \tag{3.9}$$

PITFALL PREVENTION 3.2

x and *y* **Components**   Equations 3.8 and 3.9 associate the cosine of the angle with the *x* component and the sine of the angle with the *y* component. This association is true *only* because we measured the angle θ with respect to the *x* axis, so do not memorize these equations. If θ is measured with respect to the *y* axis (as in some problems), these equations will be incorrect. Think about which side of the triangle containing the components is adjacent to the angle and which side is opposite and then assign the cosine and sine accordingly.

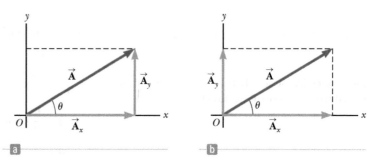

Figure 3.12 (a) A vector $\vec{\mathbf{A}}$ lying in the xy plane can be represented as a vector sum of its component vectors $\vec{\mathbf{A}}_x$ and $\vec{\mathbf{A}}_y$. These three vectors form a right triangle. (b) The y component vector $\vec{\mathbf{A}}_y$ can be moved to the left so that it lies along the y axis.

The magnitudes of these components are the lengths of the two sides of a right triangle with a hypotenuse of length A. Therefore, the magnitude and direction of $\vec{\mathbf{A}}$ are related to its components through the expressions

$$A = \sqrt{A_x^2 + A_y^2} \tag{3.10}$$

$$\theta = \tan^{-1}\left(\frac{A_y}{A_x}\right) \tag{3.11}$$

Notice that the signs of the components A_x and A_y depend on the angle θ. For example, if $\theta = 120°$, A_x is negative and A_y is positive. If $\theta = 225°$, both A_x and A_y are negative. Figure 3.13 summarizes the directions of the component vectors and signs of the components when $\vec{\mathbf{A}}$ lies in the various quadrants.

When solving problems in two dimensions, you can specify a vector $\vec{\mathbf{A}}$ either with its components A_x and A_y or with its magnitude and direction A and θ.

In many applications, it is convenient to express the components in a coordinate system having axes that are not horizontal and vertical but that are still perpendicular to each other. For example, we will consider the motion of objects sliding down inclined planes. For these examples, it is often convenient to orient the x axis parallel to the plane and the y axis perpendicular to the plane.

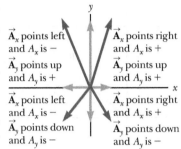

Figure 3.13 The signs of the components of a vector $\vec{\mathbf{A}}$ depend on the quadrant in which the vector is located.

QUICK QUIZ 3.4 Choose the correct response to make the sentence true: A component of a vector is **(a)** always, **(b)** never, or **(c)** sometimes larger than the magnitude of the vector.

Vector quantities often are expressed in terms of unit vectors. A **unit vector** is a dimensionless vector having a magnitude of exactly 1. Unit vectors are used to specify a given direction and have no other physical significance. They are used solely as a bookkeeping convenience in describing a direction in space. We shall use the symbols $\hat{\mathbf{i}}$, $\hat{\mathbf{j}}$, and $\hat{\mathbf{k}}$ to represent unit vectors pointing in the positive x, y, and z directions, respectively. (The "hats," or circumflexes, on the symbols are a standard notation for unit vectors.) The unit vectors $\hat{\mathbf{i}}$, $\hat{\mathbf{j}}$, and $\hat{\mathbf{k}}$ form a set of mutually perpendicular vectors in a right-handed coordinate system as shown in Figure 3.14a. The magnitude of each unit vector equals 1; that is, $|\hat{\mathbf{i}}| = |\hat{\mathbf{j}}| = |\hat{\mathbf{k}}| = 1$.

Consider a vector $\vec{\mathbf{A}}$ lying in the xy plane as shown in Figure 3.14b. The product of the component A_x and the unit vector $\hat{\mathbf{i}}$ is the component vector $\vec{\mathbf{A}}_x = A_x\hat{\mathbf{i}}$, which lies on the x axis and has magnitude $|A_x|$. Likewise, $\vec{\mathbf{A}}_y = A_y\hat{\mathbf{j}}$ is the component vector of magnitude $|A_y|$ lying on the y axis. Therefore, the unit-vector notation for the vector $\vec{\mathbf{A}}$ is

$$\vec{\mathbf{A}} = A_x\hat{\mathbf{i}} + A_y\hat{\mathbf{j}} \tag{3.12}$$

Consider now the polar coordinates shown for the point in Figure 3.2. The point in the first quadrant in that figure is reproduced in Figure 3.15. Notice that we can

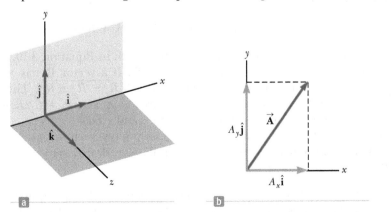

Figure 3.14 (a) The unit vectors $\hat{\mathbf{i}}$, $\hat{\mathbf{j}}$, and $\hat{\mathbf{k}}$ are directed along the x, y, and z axes, respectively. (b) Vector $\vec{\mathbf{A}} = A_x\hat{\mathbf{i}} + A_y\hat{\mathbf{j}}$ lying in the xy plane has components A_x and A_y.

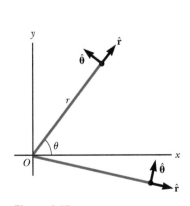

Figure 3.15 Unit vectors for a point described by polar coordinates.

identify radial and angular unit vectors $\hat{\mathbf{r}}$ and $\hat{\boldsymbol{\theta}}$. Just like for rectangular coordinates, these vectors are of unit length. Unlike rectangular coordinates, however, the directions of radial and angular unit vectors depend on the point, as shown by the point in the fourth quadrant in Figure 3.15.

Now let us see how to use components to add vectors when the graphical method is not sufficiently accurate. Suppose we wish to add vector $\vec{\mathbf{B}}$ to vector $\vec{\mathbf{A}}$ in Equation 3.12, where vector $\vec{\mathbf{B}}$ has components B_x and B_y. Because of the bookkeeping convenience of the unit vectors, all we do is add the x and y components separately. The resultant vector $\vec{\mathbf{R}}$ is

$$\vec{\mathbf{R}} = \vec{\mathbf{A}} + \vec{\mathbf{B}} = (A_x\hat{\mathbf{i}} + A_y\hat{\mathbf{j}}) + (B_x\hat{\mathbf{i}} + B_y\hat{\mathbf{j}})$$

or, rearranging terms,

$$\vec{\mathbf{R}} = (A_x + B_x)\hat{\mathbf{i}} + (A_y + B_y)\hat{\mathbf{j}} \tag{3.13}$$

Because $\vec{\mathbf{R}} = R_x\hat{\mathbf{i}} + R_y\hat{\mathbf{j}}$, we see that the components of the resultant vector are

$$R_x = A_x + B_x$$
$$R_y = A_y + B_y \tag{3.14}$$

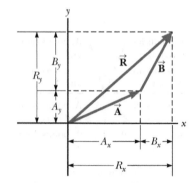

Figure 3.16 This geometric construction for the sum of two vectors shows the relationship between the components of the resultant $\vec{\mathbf{R}}$ and the components of the individual vectors.

Therefore, we see that in the component method of adding vectors, we add all the x components together to find the x component of the resultant vector and use the same process for the y components. We can check this addition by components with a geometric construction as shown in Figure 3.16.

The magnitude of $\vec{\mathbf{R}}$ and the angle it makes with the x axis are obtained from its components using the relationships

$$R = \sqrt{R_x^2 + R_y^2} = \sqrt{(A_x + B_x)^2 + (A_y + B_y)^2} \tag{3.15}$$

$$\tan\theta = \frac{R_y}{R_x} = \frac{A_y + B_y}{A_x + B_x} \tag{3.16}$$

PITFALL PREVENTION 3.3
Tangents on Calculators
Equation 3.16 involves the calculation of an angle by means of a tangent function. Generally, the inverse tangent function on calculators provides an angle between $-90°$ and $+90°$. As a consequence, if the vector you are studying lies in the second or third quadrant, the angle measured from the positive x axis will be the angle your calculator returns plus 180°.

At times, we need to consider situations involving motion in three component directions. The extension of our methods to three-dimensional vectors is straightforward. If $\vec{\mathbf{A}}$ and $\vec{\mathbf{B}}$ both have x, y, and z components, they can be expressed in the form

$$\vec{\mathbf{A}} = A_x\hat{\mathbf{i}} + A_y\hat{\mathbf{j}} + A_z\hat{\mathbf{k}} \tag{3.17}$$

$$\vec{\mathbf{B}} = B_x\hat{\mathbf{i}} + B_y\hat{\mathbf{j}} + B_z\hat{\mathbf{k}} \tag{3.18}$$

The sum of $\vec{\mathbf{A}}$ and $\vec{\mathbf{B}}$ is

$$\vec{\mathbf{R}} = (A_x + B_x)\hat{\mathbf{i}} + (A_y + B_y)\hat{\mathbf{j}} + (A_z + B_z)\hat{\mathbf{k}} \tag{3.19}$$

Notice that Equation 3.19 differs from Equation 3.13: in Equation 3.19, the resultant vector also has a z component $R_z = A_z + B_z$. If a vector $\vec{\mathbf{R}}$ has x, y, and z components, the magnitude of the vector is $R = \sqrt{R_x^2 + R_y^2 + R_z^2}$. The angle θ_x that $\vec{\mathbf{R}}$ makes with the x axis is found from the expression $\cos\theta_x = R_x/R$, with similar expressions for the angles with respect to the y and z axes.

The extension of our method to adding more than two vectors is also straightforward using the component method. For example, $\vec{\mathbf{A}} + \vec{\mathbf{B}} + \vec{\mathbf{C}} = (A_x + B_x + C_x)\hat{\mathbf{i}} + (A_y + B_y + C_y)\hat{\mathbf{j}} + (A_z + B_z + C_z)\hat{\mathbf{k}}$.

QUICK QUIZ 3.5 For which of the following vectors is the magnitude of the vector equal to one of the components of the vector? **(a)** $\vec{\mathbf{A}} = 2\hat{\mathbf{i}} + 5\hat{\mathbf{j}}$ **(b)** $\vec{\mathbf{B}} = -3\hat{\mathbf{j}}$ **(c)** $\vec{\mathbf{C}} = +5\hat{\mathbf{k}}$

Example 3.3   The Sum of Two Vectors

Find the sum of two vectors $\vec{A}$ and $\vec{B}$ lying in the xy plane and given by

$$\vec{A} = (2.0\,\hat{i} + 2.0\,\hat{j}) \quad \text{and} \quad \vec{B} = (2.0\,\hat{i} - 4.0\,\hat{j})$$

SOLUTION

Conceptualize   You can conceptualize the situation by drawing the vectors on graph paper. Do this and then draw an approximation of the expected resultant vector.

Categorize   We categorize this example as a simple substitution problem. Comparing this expression for $\vec{A}$ with the general expression $\vec{A} = A_x\hat{i} + A_y\,\hat{j} + A_z\hat{k}$, we see that $A_x = 2.0$, $A_y = 2.0$, and $A_z = 0$. Likewise, $B_x = 2.0$, $B_y = -4.0$, and $B_z = 0$. We can use a two-dimensional approach because there are no z components.

Use Equation 3.13 to obtain the resultant vector $\vec{R}$:

$$\vec{R} = (A_x + B_x)\hat{i} + (A_y + B_y)\hat{j} = (2.0 + 2.0)\hat{i} + (2.0 - 4.0)\hat{j}$$

$$= 4.0\,\hat{i} - 2.0\,\hat{j}$$

Use Equation 3.15 to find the magnitude of $\vec{R}$:

$$R = \sqrt{R_x^2 + R_y^2} = \sqrt{(4.0)^2 + (-2.0)^2} = \sqrt{20} = 4.5$$

Find the direction of $\vec{R}$ from Equation 3.16:

$$\tan\theta = \frac{R_y}{R_x} = \frac{-2.0}{4.0} = -0.50$$

Your calculator likely gives the answer $-27°$ for $\theta = \tan^{-1}(-0.50)$. This answer is correct if we interpret it to mean $27°$ *clockwise* from the x axis. Our standard form has been to quote the angles measured *counterclockwise* from the $+x$ axis, and that angle for this vector is $\theta = 333°$.

Example 3.4   The Resultant Displacement

A particle undergoes three consecutive displacements: $\Delta\vec{r}_1 = (15\,\hat{i} + 30\,\hat{j} + 12\,\hat{k})$ cm, $\Delta\vec{r}_2 = (23\,\hat{i} - 14\,\hat{j} - 5.0\,\hat{k})$ cm, and $\Delta\vec{r}_3 = (-13\,\hat{i} + 15\,\hat{j})$ cm. Find unit-vector notation for the resultant displacement and its magnitude.

SOLUTION

Conceptualize   Although x is sufficient to locate a point in one dimension, we need a vector $\vec{r}$ to locate a point in two or three dimensions. The notation $\Delta\vec{r}$ is a generalization of the one-dimensional displacement Δx in Equation 2.1. Three-dimensional displacements are more difficult to conceptualize than those in two dimensions because they cannot be drawn on paper like the latter.

For this problem, let us imagine that you start with your pencil at the origin of a piece of graph paper on which you have drawn x and y axes. Move your pencil 15 cm to the right along the x axis, then 30 cm upward along the y axis, and then 12 cm *perpendicularly toward you away* from the graph paper. This procedure provides the displacement described by $\Delta\vec{r}_1$. From this point, move your pencil 23 cm to the right parallel to the x axis, then 14 cm parallel to the graph paper in the $-y$ direction, and then 5.0 cm perpendicularly away from you toward the graph paper. You are now at the displacement from the origin described by $\Delta\vec{r}_1 + \Delta\vec{r}_2$. From this point, move your pencil 13 cm to the left in the $-x$ direction, and (finally!) 15 cm parallel to the graph paper along the y axis. Your final position is at a displacement $\Delta\vec{r}_1 + \Delta\vec{r}_2 + \Delta\vec{r}_3$ from the origin.

Categorize   Despite the difficulty in conceptualizing in three dimensions, we can categorize this problem as a substitution problem because of the careful bookkeeping methods that we have developed for vectors. The mathematical manipulation keeps track of this motion along the three perpendicular axes in an organized, compact way, as we see below.

To find the resultant displacement, add the three vectors:

$$\Delta\vec{r} = \Delta\vec{r}_1 + \Delta\vec{r}_2 + \Delta\vec{r}_3$$

$$= (15 + 23 - 13)\hat{i}\,\text{cm} + (30 - 14 + 15)\hat{j}\,\text{cm} + (12 - 5.0 + 0)\hat{k}\,\text{cm}$$

$$= (25\,\hat{i} + 31\,\hat{j} + 7.0\,\hat{k})\,\text{cm}$$

Find the magnitude of the resultant vector:

$$R = \sqrt{R_x^2 + R_y^2 + R_z^2}$$

$$= \sqrt{(25\,\text{cm})^2 + (31\,\text{cm})^2 + (7.0\,\text{cm})^2} = 40\,\text{cm}$$

Example **3.5** Taking a Hike

A hiker begins a trip by first walking 25.0 km southeast from her car. She stops and sets up her tent for the night. On the second day, she walks 40.0 km in a direction 60.0° north of east, at which point she discovers a forest ranger's tower.

(A) Determine the components of the hiker's displacement for each day.

SOLUTION

Conceptualize We conceptualize the problem by drawing a sketch as in Figure 3.17. If we denote the displacement vectors on the first and second days by $\vec{\mathbf{A}}$ and $\vec{\mathbf{B}}$, respectively, and use the car as the origin of coordinates, we obtain the vectors shown in Figure 3.17. The sketch allows us to estimate the resultant vector as shown.

Categorize Having drawn the resultant $\vec{\mathbf{R}}$, we can now categorize this problem as one we've solved before: an addition of two vectors. You should now have a hint of the power of categorization in that many new problems are very similar to problems we have already solved if we are careful to conceptualize them. Once we have drawn the displacement vectors and categorized the problem, this problem is no longer about a hiker, a walk, a car, a tent, or a tower. It is a problem about vector addition, one that we have already solved.

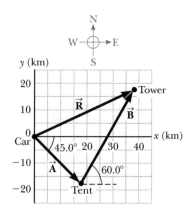

Figure 3.17 (Example 3.5) The total displacement of the hiker is the vector $\vec{\mathbf{R}} = \vec{\mathbf{A}} + \vec{\mathbf{B}}$.

Analyze Displacement $\vec{\mathbf{A}}$ has a magnitude of 25.0 km and is directed 45.0° below the positive x axis.

Find the components of $\vec{\mathbf{A}}$ using Equations 3.8 and 3.9:

$$A_x = A \cos(-45.0°) = (25.0 \text{ km})(0.707) = \boxed{17.7 \text{ km}}$$

$$A_y = A \sin(-45.0°) = (25.0 \text{ km})(-0.707) = \boxed{-17.7 \text{ km}}$$

The negative value of A_y indicates that the hiker ends up below the x axis on the first day. The signs of A_x and A_y also are evident from Figure 3.17.

Find the components of $\vec{\mathbf{B}}$ using Equations 3.8 and 3.9:

$$B_x = B \cos 60.0° = (40.0 \text{ km})(0.500) = \boxed{20.0 \text{ km}}$$

$$B_y = B \sin 60.0° = (40.0 \text{ km})(0.866) = \boxed{34.6 \text{ km}}$$

(B) Determine the components of the hiker's resultant displacement $\vec{\mathbf{R}}$ for the trip. Find an expression for $\vec{\mathbf{R}}$ in terms of unit vectors.

SOLUTION

Use Equation 3.14 to find the components of the resultant displacement $\vec{\mathbf{R}} = \vec{\mathbf{A}} + \vec{\mathbf{B}}$:

$$R_x = A_x + B_x = 17.7 \text{ km} + 20.0 \text{ km} = \boxed{37.7 \text{ km}}$$

$$R_y = A_y + B_y = -17.7 \text{ km} + 34.6 \text{ km} = \boxed{17.0 \text{ km}}$$

Write the total displacement in unit-vector form:

$$\vec{\mathbf{R}} = (37.7\hat{\mathbf{i}} + 17.0\hat{\mathbf{j}}) \text{ km}$$

Finalize Looking at the graphical representation in Figure 3.17, we estimate the position of the tower to be about (38 km, 17 km), which is consistent with the components of $\vec{\mathbf{R}}$ in our result for the final position of the hiker. Also, both components of $\vec{\mathbf{R}}$ are positive, putting the final position in the first quadrant of the coordinate system, which is also consistent with Figure 3.17.

WHAT IF?   After reaching the tower, the hiker wishes to return to her car along a single straight line. What are the components of the vector representing this hike? What should the direction of the hike be?

Answer The desired vector $\vec{\mathbf{R}}_{car}$ is the negative of vector $\vec{\mathbf{R}}$:

$$\vec{\mathbf{R}}_{car} = -\vec{\mathbf{R}} = (-37.7\hat{\mathbf{i}} - 17.0\hat{\mathbf{j}}) \text{ km}$$

The direction is found by calculating the angle that the vector makes with the x axis:

$$\tan\theta = \frac{R_{car,y}}{R_{car,x}} = \frac{-17.0 \text{ km}}{-37.7 \text{ km}} = 0.450$$

which gives an angle of $\theta = 204.2°$, or 24.2° south of west.

Summary

▶ Definitions

Scalar quantities are those that have only a numerical value and no associated direction.

Vector quantities have both magnitude and direction and obey the laws of vector addition. The magnitude of a vector is *always* a positive number.

▶ Concepts and Principles

When two or more vectors are added together, they must all have the same units and they all must be the same type of quantity. We can add two vectors $\vec{A}$ and $\vec{B}$ graphically. In this method (Fig. 3.6), the resultant vector $\vec{R} = \vec{A} + \vec{B}$ runs from the tail of $\vec{A}$ to the tip of $\vec{B}$.

If a vector $\vec{A}$ has an x component A_x and a y component A_y, the vector can be expressed in unit-vector form as $\vec{A} = A_x\hat{i} + A_y\hat{j}$. In this notation, $\hat{i}$ is a unit vector pointing in the positive x direction and $\hat{j}$ is a unit vector pointing in the positive y direction. Because $\hat{i}$ and $\hat{j}$ are unit vectors, $|\hat{i}| = |\hat{j}| = 1$.

A second method of adding vectors involves **components** of the vectors. The x component A_x of the vector $\vec{A}$ is equal to the projection of $\vec{A}$ along the x axis of a coordinate system, where $A_x = A \cos \theta$. The y component A_y of $\vec{A}$ is the projection of $\vec{A}$ along the y axis, where $A_y = A \sin \theta$.

We can find the resultant of two or more vectors by resolving all vectors into their x and y components, adding their resultant x and y components, and then using the Pythagorean theorem to find the magnitude of the resultant vector. We can find the angle that the resultant vector makes with respect to the x axis by using a suitable trigonometric function.

Think–Pair–Share

See the Preface for an explanation of the icons used in this problems set. For additional assessment items for this section, go to ✦ **WEBASSIGN** From Cengage

1. You are working at a radar station for the Coast Guard. While everyone else is out to lunch, you hear a distress call from a sinking ship. The ship is located at a distance of 51.2 km from the station, at a bearing of 36° west of north. On your radar screen, you see the locations of four other ships as follows:

Ship #	Distance from Station (km)	Bearing	Maximum Speed (km/h)
1	36.1	42° W of N	30.0
2	37.3	61° W of N	38.0
3	10.2	36° W of N	32.0
4	51.2	79° W of N	45.0

Quick! Which ship do you contact to help the sinking ship? Which ship will get there in the shortest time interval? Assume that each ship would accelerate quickly to its maximum speed and then maintain that constant speed in a straight line for the entire trip to the sinking ship.

2. **ACTIVITY** On a paper map of the United States, locate Memphis, Albuquerque, and Chicago. Draw a vector from Albuquerque to Memphis and another vector from Memphis to Chicago. Using the scale on the map, determine the straight-line distances between Albuquerque and Memphis, and between Memphis and Chicago. Use a protractor to measure the angles of your two vectors with respect to latitude and longitude lines. From this information, determine the straight-line distance in miles between Albuquerque and Chicago.

Problems

See the Preface for an explanation of the icons used in this problems set. For additional assessment items for this section, go to ✦ **WEBASSIGN** From Cengage

SECTION 3.1 Coordinate Systems

1. Two points in the xy plane have Cartesian coordinates (2.00, −4.00) m and (−3.00, 3.00) m. Determine (a) the distance between these points and (b) their polar coordinates.

2. Two points in a plane have polar coordinates (2.50 m, 30.0°) and (3.80 m, 120.0°). Determine (a) the Cartesian coordinates of these points and (b) the distance between them.

3. The polar coordinates of a certain point are ($r = 4.30$ cm, $\theta = 214°$). (a) Find its Cartesian coordinates x and y. Find the polar coordinates of the points with Cartesian coordinates (b) $(-x, y)$, (c) $(-2x, -2y)$, and (d) $(3x, -3y)$.

4. Let the polar coordinates of the point (x, y) be (r, θ). Determine the polar coordinates for the points (a) $(-x, y)$, (b) $(-2x, -2y)$, and (c) $(3x, -3y)$.

SECTION 3.2 Vector and Scalar Quantities

5. *Why is the following situation impossible?* A skater glides along a circular path. She defines a certain point on the circle as

her origin. Later on, she passes through a point at which the distance she has traveled along the path from the origin is smaller than the magnitude of her displacement vector from the origin.

SECTION 3.3 Basic Vector Arithmetic

6. Vector $\vec{A}$ has a magnitude of 29 units and points in the positive y direction. When vector $\vec{B}$ is added to $\vec{A}$, the resultant vector $\vec{A} + \vec{B}$ points in the negative y direction with a magnitude of 14 units. Find the magnitude and direction of $\vec{B}$.

7. A force $\vec{F}_1$ of magnitude 6.00 units acts on an object at the origin in a direction $\theta = 30.0°$ above the positive x axis (Fig. P3.7). A second force $\vec{F}_2$ of magnitude 5.00 units acts on the object in the direction of the positive y axis. Find graphically the magnitude and direction of the resultant force $\vec{F}_1 + \vec{F}_2$.

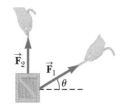

Figure P3.7

8. Three displacements are $\vec{A} = 200$ m due south, $\vec{B} = 250$ m due west, and $\vec{C} = 150$ m at $30.0°$ east of north. (a) Construct a separate diagram for each of the following possible ways of adding these vectors: $\vec{R}_1 = \vec{A} + \vec{B} + \vec{C}$; $\vec{R}_2 = \vec{B} + \vec{C} + \vec{A}$; $\vec{R}_3 = \vec{C} + \vec{B} + \vec{A}$. (b) Explain what you can conclude from comparing the diagrams.

9. The displacement vectors $\vec{A}$ and $\vec{B}$ shown in Figure P3.9 both have magnitudes of 3.00 m. The direction of vector $\vec{A}$ is $\theta = 30.0°$. Find graphically (a) $\vec{A} + \vec{B}$, (b) $\vec{A} - \vec{B}$, (c) $\vec{B} - \vec{A}$, and (d) $\vec{A} - 2\vec{B}$. (Report all angles counterclockwise from the positive x axis.)

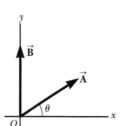

Figure P3.9
Problems 9 and 25.

10. A roller-coaster car moves 200 ft horizontally and then rises 135 ft at an angle of $30.0°$ above the horizontal. It next travels 135 ft at an angle of $40.0°$ downward. What is its displacement from its starting point? Use graphical techniques.

SECTION 3.4 Components of a Vector and Unit Vectors

11. A minivan travels straight north in the right lane of a divided highway at 28.0 m/s. A camper passes the minivan and then changes from the left lane into the right lane. As it does so, the camper's path on the road is a straight displacement at 8.50° east of north. To avoid cutting off the minivan, the north–south distance between the camper's back bumper and the minivan's front bumper should not decrease. (a) Can the camper be driven to satisfy this requirement? (b) Explain your answer.

12. A person walks 25.0° north of east for 3.10 km. How far would she have to walk due north and due east to arrive at the same location?

13. Your dog is running around the grass in your back yard. He undergoes successive displacements 3.50 m south, 8.20 m northeast, and 15.0 m west. What is the resultant displacement?

14. Given the vectors $\vec{A} = 2.00\hat{i} + 6.00\hat{j}$ and $\vec{B} = 3.00\hat{i} - 2.00\hat{j}$, (a) draw the vector sum $\vec{C} = \vec{A} + \vec{B}$ and the vector difference $\vec{D} = \vec{A} - \vec{B}$. (b) Calculate

$\vec{C}$ and $\vec{D}$, in terms of unit vectors. (c) Calculate $\vec{C}$ and $\vec{D}$ in terms of polar coordinates, with angles measured with respect to the positive x axis.

15. The helicopter view in Fig. P3.15 shows two people pulling on a stubborn mule. The person on the right pulls with a force $\vec{F}_1$ of magnitude 120 N and direction of $\theta_1 = 60.0°$. The person on the left pulls with a force $\vec{F}_2$ of magnitude 80.0 N and direction of $\theta_2 = 75.0°$. Find (a) the single force that is equivalent to the two forces shown and (b) the force that a third person would have to exert on the mule to make the resultant force equal to zero. The forces are measured in units of newtons (symbolized N).

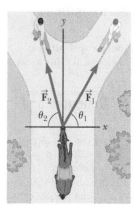

Figure P3.15

16. A snow-covered ski slope makes an angle of $35.0°$ with the horizontal. When a ski jumper plummets onto the hill, a parcel of splashed snow is thrown up to a maximum displacement of 1.50 m at $16.0°$ from the vertical in the uphill direction as shown in Figure P3.16. Find the components of its maximum displacement (a) parallel to the surface and (b) perpendicular to the surface.

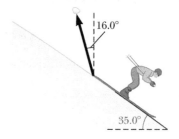

Figure P3.16

17. Consider the three displacement vectors $\vec{A} = (3\hat{i} - 3\hat{j})$ m, $\vec{B} = (\hat{i} - 4\hat{j})$ m, and $\vec{C} = (-2\hat{i} + 5\hat{j})$ m. Use the component method to determine (a) the magnitude and direction of $\vec{D} = \vec{A} + \vec{B} + \vec{C}$ and (b) the magnitude and direction of $\vec{E} = -\vec{A} - \vec{B} + \vec{C}$.

18. Vector $\vec{A}$ has x and y components of -8.70 cm and 15.0 cm, respectively; vector $\vec{B}$ has x and y components of 13.2 cm and -6.60 cm, respectively. If $\vec{A} - \vec{B} + 3\vec{C} = 0$, what are the components of $\vec{C}$?

19. The vector $\vec{A}$ has x, y, and z components of 8.00, 12.0, and -4.00 units, respectively. (a) Write a vector expression for $\vec{A}$ in unit-vector notation. (b) Obtain a unit-vector expression for a vector $\vec{B}$ one-fourth the length of $\vec{A}$ pointing in the same direction as $\vec{A}$. (c) Obtain a unit-vector expression for a vector $\vec{C}$ three times the length of $\vec{A}$ pointing in the direction opposite the direction of $\vec{A}$.

20. Given the displacement vectors $\vec{A} = (3\hat{i} - 4\hat{j} + 4\hat{k})$ m and $\vec{B} = (2\hat{i} + 3\hat{j} - 7\hat{k})$ m, find the magnitudes of the following vectors and express each in terms of

its rectangular components. (a) $\vec{C} = \vec{A} + \vec{B}$ (b) $\vec{D} = 2\vec{A} - \vec{B}$

21. Vector $\vec{A}$ has a negative x component 3.00 units in
[T] length and a positive y component 2.00 units in length.
(a) Determine an expression for $\vec{A}$ in unit-vector notation. (b) Determine the magnitude and direction of $\vec{A}$.
(c) What vector $\vec{B}$ when added to $\vec{A}$ gives a resultant vector
with no x component and a negative y component 4.00 units
in length?

22. Three displacement vectors of a croquet ball are shown in Figure P3.22,
where $|\vec{A}| = 20.0$ units, $|\vec{B}| =$
40.0 units, and $|\vec{C}| = 30.0$ units. Find
(a) the resultant in unit-vector notation and (b) the magnitude and direction of the resultant displacement.

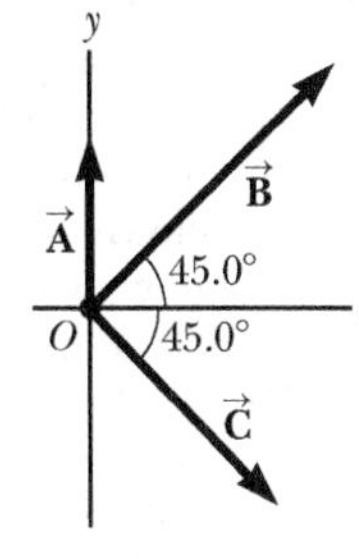

23. (a) Taking $\vec{A} = (6.00\,\hat{i} - 8.00\,\hat{j})$ units,
[QC] $\vec{B} = (-8.00\,\hat{i} + 3.00\,\hat{j})$ units, and
$\vec{C} = (26.0\,\hat{i} + 19.0\,\hat{j})$ units, determine a
and b such that $a\vec{A} + b\vec{B} + \vec{C} = 0$.
(b) A student has learned that a single equation cannot be
solved to determine values for more than one unknown in
it. How would you explain to him that both a and b can be
determined from the single equation used in part (a)?

Figure P3.22

24. Vector $\vec{B}$ has x, y, and z components of 4.00, 6.00, and
3.00 units, respectively. Calculate (a) the magnitude of $\vec{B}$
and (b) the angle that $\vec{B}$ makes with each coordinate axis.

25. Use the component method to add the vectors $\vec{A}$
and $\vec{B}$ shown in Figure P3.9. Both vectors have magnitudes of 3.00 m and vector $\vec{A}$ makes an angle of
$\theta = 30.0°$ with the x axis. Express the resultant $\vec{A} + \vec{B}$ in
unit-vector notation.

26. A girl delivering newspapers covers her route by traveling 3.00 blocks west, 4.00 blocks north, and then 6.00
blocks east. (a) What is her resultant displacement?
(b) What is the total distance she travels?

27. A man pushing a mop across a floor causes it to undergo
[T] two displacements. The first has a magnitude of 150 cm and
makes an angle of 120° with the positive x axis. The resultant
displacement has a magnitude of 140 cm and is directed at
an angle of 35.0° to the positive x axis. Find the magnitude
and direction of the second displacement.

28. Figure P3.28 illustrates typical proportions of male (m)
[BIO] and female (f) anatomies. The displacements $\vec{d}_{1m}$ and $\vec{d}_{1f}$
from the soles of the feet to the navel have magnitudes of

104 cm and 84.0 cm, respectively. The displacements $\vec{d}_{2m}$
and $\vec{d}_{2f}$ from the navel to outstretched fingertips have magnitudes of 100 cm and 86.0 cm, respectively. Find the vector
sum of these displacements $\vec{d}_3 = \vec{d}_1 + \vec{d}_2$ for both people.

29. **Review.** As it passes over Grand Bahama Island, the
[AMT] eye of a hurricane is moving in a direction 60.0° north
[GP] of west with a speed of 41.0 km/h. (a) What is the
unit-vector expression for the velocity of the hurricane?
It maintains this velocity for 3.00 h, at which time the
course of the hurricane suddenly shifts due north, and
its speed slows to a constant 25.0 km/h. This new velocity is maintained for 1.50 h. (b) What is the unit-vector
expression for the new velocity of the hurricane?
(c) What is the unit-vector expression for the displacement of the hurricane during the first 3.00 h?
(d) What is the unit-vector expression for the displacement of the hurricane during the latter 1.50 h?
(e) How far from Grand Bahama is the eye 4.50 h after it
passes over the island?

30. In an assembly operation illustrated in Figure P3.30, a
robot moves an object first straight upward and then also
to the east, around an arc forming one-quarter of a circle
of radius 4.80 cm that lies in
an east–west vertical plane.
The robot then moves the
object upward and to the
north, through one-quarter
of a circle of radius 3.70 cm
that lies in a north–south
vertical plane. Find (a) the
magnitude of the total displacement of the object
and (b) the angle the total
displacement makes with
the vertical.

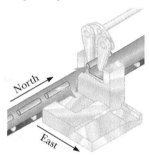

Figure P3.30

31. **Review.** You are standing on the ground at the origin of a
[AMT] coordinate system. An airplane flies over you with constant
velocity parallel to the x axis and at a fixed height of 7.60×10^3 m. At time $t = 0$, the airplane is directly above you so
that the vector leading from you to it is $\vec{P}_0 = 7.60 \times 10^3\,\hat{j}$ m.
At $t = 30.0$ s, the position vector leading from you to the air-
plane is $\vec{P}_{30} = (8.04 \times 10^3\,\hat{i} + 7.60 \times 10^3\,\hat{j})$ m as suggested in
Figure P3.31. Determine the magnitude and orientation of
the airplane's position vector at $t = 45.0$ s.

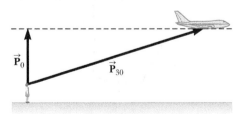

Figure P3.31

32. *Why is the following situation impossible?* A shopper pushing
a cart through a market follows directions to the canned
goods and moves through a displacement $8.00\,\hat{i}$ m down one
aisle. He then makes a 90.0° turn and moves 3.00 m along
the y axis. He then makes another 90.0° turn and moves
4.00 m along the x axis. *Every* shopper who follows these
directions correctly ends up 5.00 m from the starting point.

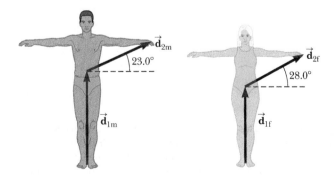

Figure P3.28

33. In Figure P3.33, the line segment represents a path from the point with position vector $(5\hat{\mathbf{i}} + 3\hat{\mathbf{j}})$ m to the point with location $(16\hat{\mathbf{i}} + 12\hat{\mathbf{j}})$ m. Point Ⓐ is along this path, a fraction f of the way to the destination. (a) Find the position vector of point Ⓐ in terms of f. (b) Evaluate the expression from part (a) for $f = 0$. (c) Explain whether the result in part (b) is reasonable. (d) Evaluate the expression for $f = 1$. (e) Explain whether the result in part (d) is reasonable.

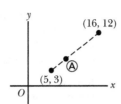

Figure P3.33 Point Ⓐ is a fraction f of the distance from the initial point (5, 3) to the final point (16, 12).

ADDITIONAL PROBLEMS

34. You are spending the summer as an assistant learning how to navigate on a large ship carrying freight across Lake Erie. One day, you and your ship are to travel across the lake a distance of 200 km traveling due north from your origin port to your destination port. Just as you leave your origin port, the navigation electronics go down. The captain continues sailing, claiming he can depend on his years of experience on the water as a guide. The engineers work on the navigation system while the ship continues to sail, and winds and waves push it off course. Eventually, enough of the navigation system comes back up to tell you your location. The system tells you that your current position is 50.0 km north of the origin port and 25.0 km east of the port. The captain is a little embarrassed that his ship is so far off course and barks an order to you to tell him immediately what heading he should set from your current position to the destination port. Give him an appropriate heading angle.

35. A person going for a walk follows the path shown in Figure P3.35. The total trip consists of four straight-line paths. At the end of the walk, what is the person's resultant displacement measured from the starting point?

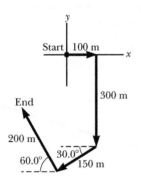

Figure P3.35

36. A ferry transports tourists between three islands. It sails from the first island to the second island, 4.76 km away, in a direction 37.0° north of east. It then sails from the second island to the third island in a direction 69.0° west of north. Finally it returns to the first island, sailing in a direction 28.0° east of south. Calculate the distance between (a) the second and third islands and (b) the first and third islands.

37. Two vectors $\vec{\mathbf{A}}$ and $\vec{\mathbf{B}}$ have precisely equal magnitudes. For the magnitude of $\vec{\mathbf{A}} + \vec{\mathbf{B}}$ to be 100 times larger than the magnitude of $\vec{\mathbf{A}} - \vec{\mathbf{B}}$, what must be the angle between them?

38. Two vectors $\vec{\mathbf{A}}$ and $\vec{\mathbf{B}}$ have precisely equal magnitudes. For the magnitude of $\vec{\mathbf{A}} + \vec{\mathbf{B}}$ to be larger than the magnitude of $\vec{\mathbf{A}} - \vec{\mathbf{B}}$ by the factor n, what must be the angle between them?

39. Review. The biggest stuffed animal in the world is a snake 420 m long, constructed by Norwegian children. Suppose the snake is laid out in a park as shown in Figure P3.39, forming two straight sides of a 105° angle, with one side 240 m long. Olaf and Inge run a race they invent. Inge runs directly from the tail of the snake to its head, and Olaf starts from the same place at the same moment but runs along the snake. (a) If both children run steadily at 12.0 km/h, Inge reaches the head of the snake how much earlier than Olaf? (b) If Inge runs the race again at a constant speed of 12.0 km/h, at what constant speed must Olaf run to reach the end of the snake at the same time as Inge?

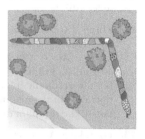

Figure P3.39

40. Ecotourists use their global positioning system indicator to determine their location inside a botanical garden as latitude 0.002 43 degree south of the equator, longitude 75.642 38 degrees west. They wish to visit a tree at latitude 0.001 62 degree north, longitude 75.644 26 degrees west. (a) Determine the straight-line distance and the direction in which they can walk to reach the tree as follows. First model the Earth as a sphere of radius 6.37×10^6 m to determine the westward and northward displacement components required, in meters. Then model the Earth as a flat surface to complete the calculation. (b) Explain why it is possible to use these two geometrical models together to solve the problem.

41. A vector is given by $\vec{\mathbf{R}} = 2\hat{\mathbf{i}} + \hat{\mathbf{j}} + 3\hat{\mathbf{k}}$. Find (a) the magnitudes of the x, y, and z components; (b) the magnitude of $\vec{\mathbf{R}}$; and (c) the angles between $\vec{\mathbf{R}}$ and the x, y, and z axes.

42. You are working as an assistant to an air-traffic controller at the local airport, from which small airplanes take off and land. Your job is to make sure that airplanes are not closer to each other than a minimum safe separation distance of 2.00 km. You observe two small aircraft on your radar screen, out over the ocean surface. The first is at altitude 800 m above the surface, horizontal distance 19.2 km, and 25.0° south of west. The second aircraft is at altitude 1 100 m, horizontal distance 17.6 km, and 20.0° south of west. Your supervisor is concerned that the two aircraft are too close together and asks for a separation distance for the two airplanes. (Place the x axis west, the y axis south, and the z axis vertical.)

43. Review. The instantaneous position of an object is specified by its position vector leading from a fixed origin to the location of the object, modeled as a particle. Suppose for a certain object the position vector is a function of time given by $\vec{\mathbf{r}} = 4\hat{\mathbf{i}} + 3\hat{\mathbf{j}} - 2t\,\hat{\mathbf{k}}$, where

$\vec{r}$ is in meters and t is in seconds. (a) Evaluate $d\vec{r}/dt$. (b) What physical quantity does $d\vec{r}/dt$ represent about the object?

44. Vectors $\vec{A}$ and $\vec{B}$ have equal magnitudes of 5.00. The sum of $\vec{A}$ and $\vec{B}$ is the vector $6.00\hat{j}$. Determine the angle between $\vec{A}$ and $\vec{B}$.

45. A rectangular parallelepiped has dimensions a, b, and c as shown in Figure P3.45. (a) Obtain a vector expression for the face diagonal vector $\vec{R}_1$. (b) What is the magnitude of this vector? (c) Notice that $\vec{R}_1$, $c\hat{k}$, and $\vec{R}_2$ make a right triangle. Obtain a vector expression for the body diagonal vector $\vec{R}_2$.

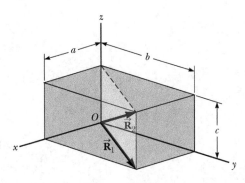

Figure P3.45

CHALLENGE PROBLEM

46. A pirate has buried his treasure on an island with five trees located at the points (30.0 m, −20.0 m), (60.0 m, 80.0 m), (−10.0 m, −10.0 m), (40.0 m, −30.0 m), and (−70.0 m,

60.0 m), all measured relative to some origin, as shown in Figure P3.46. His ship's log instructs you to start at tree A and move toward tree B, but to cover only one-half the distance between A and B. Then move toward tree C, covering one-third the distance between your current location and C. Next move toward tree D, covering one-fourth the distance between where you are and D. Finally move toward tree E, covering one-fifth the distance between you and E, stop, and dig. (a) Assume you have correctly determined the order in which the pirate labeled the trees as A, B, C, D, and E as shown in the figure. What are the coordinates of the point where his treasure is buried? (b) **What If?** What if you do not really know the way the pirate labeled the trees? What would happen to the answer if you rearranged the order of the trees, for instance, to B (30 m, −20 m), A (60 m, 80 m), E (−10 m, −10 m), C (40 m, −30 m), and D (−70 m, 60 m)? State reasoning to show that the answer does not depend on the order in which the trees are labeled.

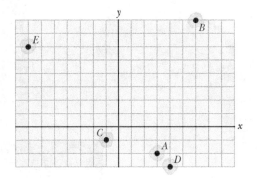

Figure P3.46

Motion in Two Dimensions

Compare the shapes of the paths of: a teenager jumping off a cliff; sparks generated by a welder at work; water projected into a park fountain; the stream from a water fountain. (*Top Left: André Berg/ EyeEm/Getty Images; Top Right: wi6995/Shutterstock.com; Bottom Right: Kristina Postnikova/ Shutterstock.com; Bottom Left: Flashon Studio/Shutterstock.com*)

68

STORYLINE **In the preceding chapter, you are sailing for Catalina** Island in a sailboat. As you approach the island, you see teenagers doing flips off a cliff. You take out your smartphone, open the camera, and take photographs of a teenager as he travels along his trajectory. Using a special app on your phone, you combine several pictures into one so that you can see several images of the falling teenager. He seems to be following a path of a particular shape as he falls. What do you think that shape is? As you enter the harbor on Catalina Island, you see a welder performing repairs on a metal boat. A shower of sparks occurs. You look at the paths of the individual sparks and notice their shape. On shore, you see a fountain in a park in which streams of water are projected at an angle into the air and follow a certain shape as they come back down. You get a drink from a water fountain and notice the shape of the water projected from the fountain. What *is* that shape that you are seeing over and over again?

CONNECTIONS In Chapter 2, we studied motion in one dimension. In Chapter 3, we learned about vector quantities in general, addition of vectors, and vector components. We focused there on position vectors. In this chapter, we will see how to use vectors from Chapter 3 to modify our mathematical expressions for position, velocity, and acceleration from Chapter 2 to account for motion in two dimensions. We will study two important types of two-dimensional motion: projectile motion, such as that of a thrown baseball or the diving teenager in the previous paragraph, and circular motion, such as the idealized motion of a planet around a star. We also discuss the concept of relative motion, which shows why observers in different frames of reference may measure different positions and velocities for a given particle. This chapter will complete our discussion of ways to describe the motion of a particle, and will set us up for Chapter 5, in which we study the cause of changes in the motion of a particle.

4.1 The Position, Velocity, and Acceleration Vectors

In one dimension, a single numerical value describes a particle's position, but in two dimensions, we indicate its position by its **position vector** $\vec{\mathbf{r}}$, drawn from the origin of some coordinate system to the location of the particle in the xy plane as in Figure 4.1. At time t_i, the particle is at point Ⓐ, described by position vector $\vec{\mathbf{r}}_i$. At some later time t_f, it is at point Ⓑ, described by position vector $\vec{\mathbf{r}}_f$. The path followed by the particle from Ⓐ to Ⓑ is not necessarily a straight line. As the particle moves from Ⓐ to Ⓑ in the time interval $\Delta t = t_f - t_i$, its position vector changes from $\vec{\mathbf{r}}_i$ to $\vec{\mathbf{r}}_f$. As we learned in Chapter 2, displacement is a vector, and the displacement of the particle is the difference between its final position and its initial position. We now define the **displacement vector** $\Delta\vec{\mathbf{r}}$ for a particle such as the one in Figure 4.1 as the difference between its final position vector and its initial position vector:

$$\Delta\vec{\mathbf{r}} \equiv \vec{\mathbf{r}}_f - \vec{\mathbf{r}}_i \tag{4.1}$$

◀ Displacement vector (Compare to Equation 2.1)

The direction of $\Delta\vec{\mathbf{r}}$ is indicated in Figure 4.1. As we see from the figure, the magnitude of $\Delta\vec{\mathbf{r}}$ is *less* than the distance traveled along the curved path followed by the particle.

As we saw in Chapter 2, it is often useful to quantify motion by looking at the displacement divided by the time interval during which that displacement occurs, which gives the rate of change of position. Two-dimensional (or three-dimensional) kinematics is similar to one-dimensional kinematics, but we must now use full vector notation rather than positive and negative signs to indicate the direction of motion.

We define the **average velocity** $\vec{\mathbf{v}}_{avg}$ of a particle during the time interval Δt as the displacement of the particle divided by the time interval:

$$\vec{\mathbf{v}}_{avg} \equiv \frac{\Delta\vec{\mathbf{r}}}{\Delta t} \tag{4.2}$$

◀ Average velocity (Compare to Equation 2.2)

Multiplying or dividing a vector quantity by a positive scalar quantity such as Δt changes only the magnitude of the vector, not its direction. Because displacement is a vector quantity and the time interval is a positive scalar quantity, we conclude that the average velocity is a vector quantity directed along $\Delta\vec{\mathbf{r}}$.

The average velocity between points is *independent of the path* taken. That is because average velocity is proportional to displacement, which depends only on the initial and final position vectors and not on the path taken. As with one-dimensional motion, we conclude that if a particle starts its motion at some point and returns to this point via any path, its average velocity is zero for this trip because its displacement is zero. Consider again our basketball players on the court in Figure 2.2 (page 22). We previously considered only their one-dimensional motion back and forth between the baskets. In reality, however, they move over a two-dimensional surface, running back and forth between the baskets as well as left and right across the width of the court. Starting from one basket, a given player may follow a very complicated two-dimensional path. Upon returning to the original basket, however, a player's average velocity is zero because the player's displacement for the whole trip is zero.

Consider again the motion of a particle between two points in the xy plane as shown in Figure 4.2 (page 70). The dashed curve shows the path of the particle from point Ⓐ to point Ⓑ. As the time interval over which we observe the motion becomes smaller and smaller—that is, as Ⓑ is moved to Ⓑ′ and then to Ⓑ″ and so on—the direction of the displacement approaches that of the green line tangent to the path at Ⓐ. The **instantaneous velocity** $\vec{\mathbf{v}}$ is defined as the limit of the average velocity $\Delta\vec{\mathbf{r}}/\Delta t$ as Δt approaches zero:

$$\vec{\mathbf{v}} \equiv \lim_{\Delta t \to 0} \frac{\Delta\vec{\mathbf{r}}}{\Delta t} = \frac{d\vec{\mathbf{r}}}{dt} \tag{4.3}$$

◀ Instantaneous velocity (Compare to Equation 2.5)

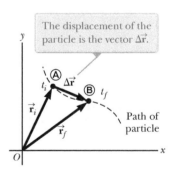

Figure 4.1 A particle moving in the xy plane is located with the position vector $\vec{\mathbf{r}}$ drawn from the origin to the particle. The displacement of the particle as it moves from Ⓐ to Ⓑ in the time interval $\Delta t = t_f - t_i$ is equal to the vector $\Delta\vec{\mathbf{r}} = \vec{\mathbf{r}}_f - \vec{\mathbf{r}}_i$.

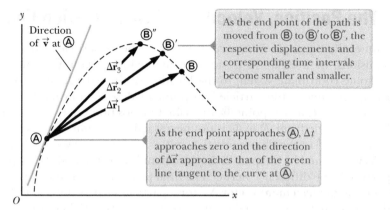

Figure 4.2 As a particle moves between two points, its average velocity is in the direction of the displacement vector $\Delta\vec{\mathbf{r}}$. By definition, the instantaneous velocity at Ⓐ is directed along the line tangent to the curve at Ⓐ.

That is, the instantaneous velocity at point Ⓐ equals the derivative of the position vector with respect to time, evaluated at point Ⓐ. The direction of the instantaneous velocity vector at any point in a particle's path is along a line tangent to the path at that point and in the direction of motion.

The magnitude of the instantaneous velocity vector $v = |\vec{\mathbf{v}}|$ of a particle is called the *speed* of the particle, which is a scalar quantity.

As a particle moves from one point to another along some path, its instantaneous velocity vector changes from $\vec{\mathbf{v}}_i$ at time t_i to $\vec{\mathbf{v}}_f$ at time t_f. Knowing the velocity at these points allows us to determine the average acceleration of the particle. The **average acceleration** $\vec{\mathbf{a}}_{avg}$ of a particle is defined as the change in its instantaneous velocity vector $\Delta\vec{\mathbf{v}}$ divided by the time interval Δt during which that change occurs:

◀ Average acceleration
(Compare to Equation 2.9)

$$\vec{\mathbf{a}}_{avg} \equiv \frac{\Delta\vec{\mathbf{v}}}{\Delta t} = \frac{\vec{\mathbf{v}}_f - \vec{\mathbf{v}}_i}{t_f - t_i} \tag{4.4}$$

Because $\vec{\mathbf{a}}_{avg}$ is the ratio of a vector quantity $\Delta\vec{\mathbf{v}}$ and a positive scalar quantity Δt, we conclude that average acceleration is a vector quantity directed along $\Delta\vec{\mathbf{v}}$. As indicated in Figure 4.3, the vector $\Delta\vec{\mathbf{v}}$ is the difference between vectors $\vec{\mathbf{v}}_f$ and $\vec{\mathbf{v}}_i$: $\Delta\vec{\mathbf{v}} = \vec{\mathbf{v}}_f - \vec{\mathbf{v}}_i$.

When the average acceleration of a particle changes during different time intervals, it is useful to define its instantaneous acceleration. The **instantaneous acceleration** $\vec{\mathbf{a}}$ is defined as the limiting value of the ratio $\Delta\vec{\mathbf{v}}/\Delta t$ as Δt approaches zero:

◀ Instantaneous acceleration
(Compare to Equation 2.10)

$$\vec{\mathbf{a}} \equiv \lim_{\Delta t \to 0} \frac{\Delta\vec{\mathbf{v}}}{\Delta t} = \frac{d\vec{\mathbf{v}}}{dt} \tag{4.5}$$

In other words, the instantaneous acceleration equals the derivative of the velocity vector with respect to time.

Various changes can occur when a particle accelerates in two dimensions. First, the magnitude of the velocity vector (the speed) may change with time as in one-dimensional motion. Second, the direction of the velocity vector may change with time even if its magnitude (speed) remains constant. Finally, both the magnitude and the direction of the velocity vector may change simultaneously.

PITFALL PREVENTION 4.1

Vector Addition As mentioned in Chapter 3, vector addition can be applied to *any* type of vector quantity. Figure 4.3, for example, shows the addition of *velocity* vectors using the graphical approach.

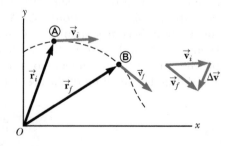

Figure 4.3 A particle moves from position Ⓐ to position Ⓑ. Its velocity vector changes from $\vec{\mathbf{v}}_i$ to $\vec{\mathbf{v}}_f$. The vector diagram at the right of the figure shows how to determine the vector $\Delta\vec{\mathbf{v}}$ from the initial and final velocities.

QUICK QUIZ 4.1 Consider the following controls in an automobile in motion: gas pedal, brake, steering wheel. What are the controls in this list that cause an acceleration of the car? **(a)** all three controls **(b)** the gas pedal and the brake **(c)** only the brake **(d)** only the gas pedal **(e)** only the steering wheel

4.2 Two-Dimensional Motion with Constant Acceleration

In Section 2.7, we investigated one-dimensional motion of a particle under constant acceleration and developed the particle under constant acceleration model. Let us now consider two-dimensional motion during which the acceleration of a particle remains constant in both magnitude and direction. As we shall see, this approach is useful for analyzing some common types of motion.

In section 4.1, we considered position vectors for a particle and represented them as arrows. Now, let us recall our discussion of vector components in Section 3.4. Consider a particle located in the xy plane at a position having Cartesian coordinates (x, y) as in Figure 4.4. The point can be specified by the position vector $\vec{\mathbf{r}}$, which in unit-vector form is given by

$$\vec{\mathbf{r}} = x\hat{\mathbf{i}} + y\hat{\mathbf{j}} \tag{4.6}$$

where x, y, and $\vec{\mathbf{r}}$ change with time as the particle moves while the unit vectors $\hat{\mathbf{i}}$ and $\hat{\mathbf{j}}$ remain constant.

We need to emphasize an important point regarding two-dimensional motion. Imagine an air hockey puck moving in a straight line along a perfectly level, friction-free surface of an air hockey table. Figure 4.5a shows a motion diagram from an overhead point of view of this puck. Recall that in Section 2.4 we related the acceleration of an object to a force on the object. Because there are no forces on the puck in the horizontal plane, it moves with constant velocity in the x direction. Now suppose you blow a quick puff of air on the puck as it passes your position, with the force from your puff of air *exactly* in the y direction. Because the force from this puff of air has no component in the x direction, it causes no acceleration in the x direction. It only causes a momentary acceleration in the y direction, causing the puck to have a constant y component of velocity once the force from the puff of air is removed. After your puff of air on the puck, its velocity component in the x direction is unchanged as shown in Figure 4.5b. The y component of the puck in Equation 4.6 remained constant before the puff of air, but is increasing afterward. The generalization of this simple experiment is that **motion in two dimensions can be modeled as two *independent* motions in each of the two perpendicular directions associated with the x and y axes. That is, any influence in the y direction does not affect the motion in the x direction and vice versa.**

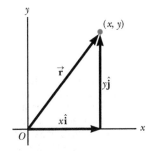

Figure 4.4 The point whose Cartesian coordinates are (x, y) can be represented by the position vector $\vec{\mathbf{r}} = x\hat{\mathbf{i}} + y\hat{\mathbf{j}}$.

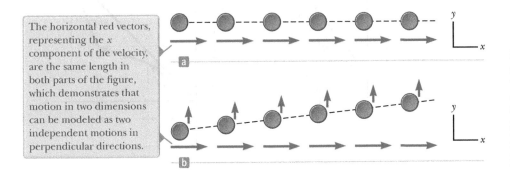

The horizontal red vectors, representing the x component of the velocity, are the same length in both parts of the figure, which demonstrates that motion in two dimensions can be modeled as two independent motions in perpendicular directions.

Figure 4.5 (a) A puck moves across a horizontal air hockey table at constant velocity in the x direction. (b) After a puff of air in the y direction is applied to the puck, the puck has gained a y component of velocity, but the x component is unaffected by the force in the perpendicular direction.

If the position vector of a particle is known, the velocity of the particle can be obtained from Equations 4.3 and 4.6, which give

$$\vec{v} = \frac{d\vec{r}}{dt} = \frac{dx}{dt}\hat{i} + \frac{dy}{dt}\hat{j} = v_x\hat{i} + v_y\hat{j} \qquad (4.7)$$

Because the acceleration $\vec{a}$ of the particle is assumed constant in this discussion, its components a_x and a_y also are constants. Therefore, we can model the particle as a particle under constant acceleration independently in each of the two directions and apply the equations of kinematics separately to the x and y components of the velocity vector. Substituting, from Equation 2.13, $v_{xf} = v_{xi} + a_x t$ and $v_{yf} = v_{yi} + a_y t$ into Equation 4.7 to determine the final velocity at any time t, we obtain

$$\vec{v}_f = (v_{xi} + a_x t)\hat{i} + (v_{yi} + a_y t)\hat{j} = (v_{xi}\hat{i} + v_{yi}\hat{j}) + (a_x\hat{i} + a_y\hat{j})t$$

◀ Velocity vector as a function of time for a particle under constant acceleration in two dimensions (Compare to Equation 2.13)

$$\vec{v}_f = \vec{v}_i + \vec{a}t \quad \text{(for constant } \vec{a}\text{)} \qquad (4.8)$$

This result states that the velocity of a particle at some time t equals the vector sum of its initial velocity $\vec{v}_i$ at time $t = 0$ and the additional velocity $\vec{a}t$ acquired at time t as a result of constant acceleration.

Similarly, from Equation 2.16 we know that the x and y coordinates of a particle under constant acceleration are

$$x_f = x_i + v_{xi}t + \tfrac{1}{2}a_x t^2 \qquad y_f = y_i + v_{yi}t + \tfrac{1}{2}a_y t^2$$

Substituting these expressions into Equation 4.6 (and labeling the final position vector $\vec{r}_f$) gives

$$\vec{r}_f = (x_i + v_{xi}t + \tfrac{1}{2}a_x t^2)\hat{i} + (y_i + v_{yi}t + \tfrac{1}{2}a_y t^2)\hat{j}$$

$$= (x_i\hat{i} + y_i\hat{j}) + (v_{xi}\hat{i} + v_{yi}\hat{j})t + \tfrac{1}{2}(a_x\hat{i} + a_y\hat{j})t^2$$

◀ Position vector as a function of time for a particle under constant acceleration in two dimensions (Compare to Equation 2.16)

$$\vec{r}_f = \vec{r}_i + \vec{v}_i t + \tfrac{1}{2}\vec{a}t^2 \quad \text{(for constant } \vec{a}\text{)} \qquad (4.9)$$

Equation 4.9 tells us that the position vector $\vec{r}_f$ of a particle is the vector sum of the original position $\vec{r}_i$, a displacement $\vec{v}_i t$ arising from the initial velocity of the particle, and a displacement $\tfrac{1}{2}\vec{a}t^2$ resulting from the constant acceleration of the particle.

We can consider Equations 4.8 and 4.9 to be the mathematical representation of a two-dimensional version of the particle under constant acceleration model. Graphical representations of Equations 4.8 and 4.9 are shown in Figure 4.6. The components of the position and velocity vectors are also illustrated in the figure.

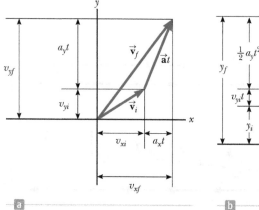
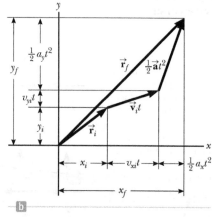

Figure 4.6 Vector representations and components of (a) the velocity and (b) the position of a particle under constant acceleration in two dimensions.

Example 4.1 | Motion in a Plane

A particle moves in the xy plane, starting from the origin at $t = 0$ with an initial velocity having an x component of 20 m/s and a y component of -15 m/s. The particle experiences an acceleration in the x direction, given by $a_x = 4.0$ m/s^2.

(A) Determine the total velocity vector at any later time.

SOLUTION

Conceptualize The components of the initial velocity tell us that the particle starts by moving toward the right and downward. The x component of velocity starts at 20 m/s and increases by 4.0 m/s every second. The y component of velocity never changes from its initial value of -15 m/s. We sketch a motion diagram of the situation in Figure 4.7. Because the particle is accelerating in the $+x$ direction, its velocity component in this direction increases and the path curves as shown in the diagram. Notice that the spacing between successive images increases as time goes on because the speed is increasing. The placement of the acceleration (purple) and velocity (red) vectors in Figure 4.7 helps us further conceptualize the situation.

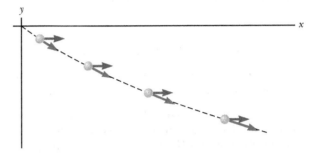

Figure 4.7 (Example 4.1) Motion diagram for the particle. Velocity vectors are shown in red and acceleration vectors in purple.

Categorize Because the initial velocity has components in both the x and y directions, we categorize this problem as one involving a particle moving in two dimensions. Because the particle only has an x component of acceleration, we model it as a *particle under constant acceleration* in the x direction and a *particle under constant velocity* in the y direction.

Analyze To begin the mathematical analysis, we set $v_{xi} = 20$ m/s, $v_{yi} = -15$ m/s, $a_x = 4.0$ m/s^2, and $a_y = 0$.

Use Equation 4.8 for the velocity vector:

$$\vec{v}_f = \vec{v}_i + \vec{a}t = (v_{xi} + a_x t)\hat{i} + (v_{yi} + a_y t)\hat{j}$$

Substitute numerical values in metric units:

$$\vec{v}_f = [20 + (4.0)t]\hat{i} + [-15 + (0)t]\hat{j}$$

$$(1) \quad \vec{v}_f = [(20 + 4.0t)\hat{i} - 15\hat{j}]$$

Finalize Notice from this expression that the x component of velocity increases in time while the y component remains constant; this result is consistent with our prediction.

(B) Calculate the velocity and speed of the particle at $t = 5.0$ s and the angle the velocity vector makes with the x axis.

SOLUTION

Analyze
Evaluate the result from Equation (1) at $t = 5.0$ s:

$$\vec{v}_f = \{[20 + 4.0(5.0)]\hat{i} - 15\hat{j}\} = (40\hat{i} - 15\hat{j}) \text{ m/s}$$

Determine the angle θ that $\vec{v}_f$ makes with the x axis at $t = 5.0$ s:

$$\theta = \tan^{-1}\left(\frac{v_{yf}}{v_{xf}}\right) = \tan^{-1}\left(\frac{-15 \text{ m/s}}{40 \text{ m/s}}\right) = -21°$$

Evaluate the speed of the particle as the magnitude of $\vec{v}_f$:

$$v_f = |\vec{v}_f| = \sqrt{v_{xf}^2 + v_{yf}^2} = \sqrt{(40)^2 + (-15)^2} \text{ m/s} = 43 \text{ m/s}$$

Finalize The negative sign for the angle θ indicates that the velocity vector is directed at an angle of 21° below the positive x axis. Notice that if we calculate v_i from the x and y components of $\vec{v}_i$, we find that $v_f > v_i$. Is that consistent with our prediction?

continued

4.1 continued

(C) Determine the x and y coordinates of the particle at any time t and its position vector at this time.

SOLUTION

Analyze

Use Equation 4.9 for the position vector:

$$\vec{\mathbf{r}}_f = \vec{\mathbf{r}}_i + \vec{\mathbf{v}}_i t + \tfrac{1}{2}\vec{\mathbf{a}}t^2 = \left(x_i + v_{xi}t + \tfrac{1}{2}a_x t^2\right)\hat{\mathbf{i}} + \left(y_i + v_{yi}t + \tfrac{1}{2}a_y t^2\right)\hat{\mathbf{j}}$$

Substitute numerical values in metric units:

$$\vec{\mathbf{r}}_f = \left[0 + (20)t + \tfrac{1}{2}(4.0)t^2\right]\hat{\mathbf{i}} + \left[0 + (-15)t + \tfrac{1}{2}(0)t^2\right]\hat{\mathbf{j}}$$

$$\vec{\mathbf{r}}_f = (20t + 2.0t^2)\hat{\mathbf{i}} - 15t\,\hat{\mathbf{j}}$$

Finalize   Let us now consider a limiting case for very large values of t.

WHAT IF?   What if we wait a very long time and then observe the motion of the particle? How would we describe the motion of the particle for large values of the time?

Answer   Looking at Figure 4.7, we see the path of the particle curving toward the x axis. There is no reason to assume this tendency will change, which suggests that the path will become more and more parallel to the x axis as time grows large. Mathematically, Equation (1) shows that the y component of the velocity remains constant while the x component grows linearly with t. Therefore, when t is very large, the x component of the velocity will be much larger than the y component, suggesting that the velocity vector becomes more and more parallel to the x axis. The magnitudes of both x_f and y_f continue to grow with time, although x_f grows much faster.

PITFALL PREVENTION 4.2
Acceleration at the Highest Point
As discussed in Pitfall Prevention 2.8, many people claim that the acceleration of a projectile at the topmost point of its trajectory is zero. This mistake arises from confusion between zero vertical velocity and zero acceleration. If the projectile were to experience zero acceleration at the highest point, its velocity at that point would not change; rather, the projectile would move horizontally at constant speed from then on! That does not happen, however, because the acceleration is *not* zero anywhere along the trajectory.

4.3 Projectile Motion

Anyone who has observed a baseball in motion has observed projectile motion. The ball moves in a curved path and returns to the ground. **Projectile motion** of an object is simple to analyze if we make two assumptions: (1) the free-fall acceleration is constant over the range of motion and is directed downward (i.e., $a_x = 0$, $a_y = -g$),[1] and (2) the effect of air resistance is negligible.[2] With these assumptions, we find that the path of a projectile, which we call its *trajectory,* is *always* a parabola as shown in Figure 4.8. **We use these assumptions throughout this chapter.** The parabola is the shape for *all* the trajectories described in the opening storyline for this chapter: the diving teenager, the sparks caused by the welder, the water in the park fountain, and the water in the drinking fountain.

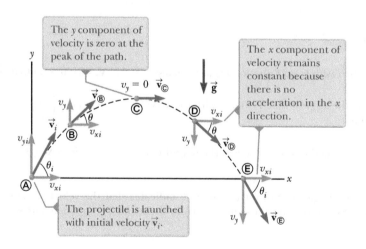

Figure 4.8 The parabolic path of a projectile that leaves the origin with a velocity $\vec{\mathbf{v}}_i$. The velocity vector $\vec{\mathbf{v}}$ changes with time in both magnitude and direction. This change is the result of acceleration $\vec{\mathbf{a}} = \vec{\mathbf{g}}$ in the negative y direction.

---

[1] This assumption is reasonable as long as the range of motion is small compared with the radius of the Earth (6.4×10^6 m). In effect, this assumption is equivalent to assuming the Earth is flat over the range of motion considered.

[2] This assumption is often *not* justified, especially at high velocities. In addition, any spin imparted to a projectile, such as that applied when a pitcher throws a curve ball, can give rise to some very interesting effects associated with aerodynamic forces, which will be discussed in Chapter 14.

The expression for the position vector of the projectile as a function of time follows directly from Equation 4.9, with its acceleration being that due to gravity, $\vec{\mathbf{a}} = \vec{\mathbf{g}}$:

$$\vec{\mathbf{r}}_f = \vec{\mathbf{r}}_i + \vec{\mathbf{v}}_i t + \tfrac{1}{2}\vec{\mathbf{g}} t^2 \qquad (4.10)$$

where the initial x and y components of the velocity of the projectile are

$$v_{xi} = v_i \cos \theta_i \qquad v_{yi} = v_i \sin \theta_i \qquad (4.11)$$

A pictorial representation of the path of a particle described by the position function in Equation 4.10 is shown in Figure 4.9 for a projectile launched from the origin, so that $\vec{\mathbf{r}}_i = 0$. The final position of a particle can be considered to be the superposition of its initial position $\vec{\mathbf{r}}_i$; the term $\vec{\mathbf{v}}_i t$, which is its displacement if no acceleration were present; and the term $\tfrac{1}{2}\vec{\mathbf{g}} t^2$ that arises from its acceleration due to gravity. In other words, if there were no gravitational acceleration, the particle would continue to move along a straight path in the direction of $\vec{\mathbf{v}}_i$. Therefore, the vertical distance $\tfrac{1}{2}\vec{\mathbf{g}} t^2$ through which the particle "falls" off the straight-line path is the same distance that an object dropped from rest would fall during the same time interval.

In Section 4.2, we stated that two-dimensional motion with constant acceleration can be analyzed as a combination of two independent motions in the x and y directions, with accelerations a_x and a_y. Projectile motion can also be handled in this way, with acceleration $a_x = 0$ in the x direction and a constant acceleration $a_y = -g$ in the y direction. Therefore, when solving projectile motion problems, use two analysis models: (1) the particle under constant velocity in the horizontal direction (Eq. 2.7),

$$x_f = x_i + v_{xi}t \qquad (4.12)$$

and (2) the particle under constant acceleration in the vertical direction (Eqs. 2.13–2.17 with x changed to y and $a_y = -g$),

$$v_{yf} = v_{yi} - gt \qquad (4.13)$$

$$v_{y,\text{avg}} = \frac{v_{yi} + v_{yf}}{2} \qquad (4.14)$$

$$y_f = y_i + \tfrac{1}{2}(v_{yi} + v_{yf})t \qquad (4.15)$$

$$y_f = y_i + v_{yi}t - \tfrac{1}{2}gt^2 \qquad (4.16)$$

$$v_{yf}^2 = v_{yi}^2 - 2g(y_f - y_i) \qquad (4.17)$$

The horizontal and vertical components of a projectile's motion are completely independent of each other and can be handled separately, with time t as the common variable for both components.

QUICK QUIZ 4.2 **(i)** As a projectile thrown at an upward angle moves in its parabolic path (such as in Fig. 4.9), at what point along its path are the velocity and acceleration vectors for the projectile perpendicular to each other? (a) nowhere (b) the highest point (c) the launch point **(ii)** From the same choices, at what point are the velocity and acceleration vectors for the projectile parallel to each other?

Horizontal Range and Maximum Height of a Projectile

Before embarking on some examples, let us consider a special case of projectile motion that occurs often. Assume a projectile is launched from the origin at $t_i = 0$ with a positive v_{yi} component as shown in Figure 4.10 and returns to *the same horizontal level*. This situation is common in sports, where baseballs, footballs, and golf balls often land at the same level from which they were launched.

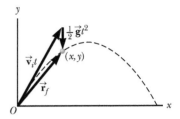

Figure 4.9 The position vector $\vec{\mathbf{r}}_f$ of a projectile launched from the origin whose initial velocity at the origin is $\vec{\mathbf{v}}_i$. The vector $\vec{\mathbf{v}}_i t$ would be the displacement of the projectile if gravity were absent, and the vector $\tfrac{1}{2}\vec{\mathbf{g}} t^2$ is its vertical displacement from a straight-line path due to its downward gravitational acceleration.

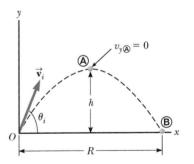

Figure 4.10 A projectile launched over a flat surface from the origin at $t_i = 0$ with an initial velocity $\vec{\mathbf{v}}_i$. The maximum height of the projectile is h, and the horizontal range is R. At Ⓐ, the peak of the trajectory, the particle has coordinates $(R/2, h)$.

Two points in this motion are especially interesting to analyze: the peak point Ⓐ, which has Cartesian coordinates $(R/2, h)$, and the point Ⓑ, which has coordinates $(R, 0)$. The distance R is called the *horizontal range* of the projectile, and the distance h is its *maximum height*. Let us find h and R mathematically in terms of v_i, θ_i, and g.

We can determine h by noting that at the peak $v_{y\text{Ⓐ}} = 0$. Therefore, from the particle under constant acceleration model, we can use Equation 4.13 to determine the time $t_\text{Ⓐ}$ at which the projectile reaches the peak:

$$v_{yf} = v_{yi} - gt \quad \rightarrow \quad 0 = v_i \sin \theta_i - gt_\text{Ⓐ}$$

$$t_\text{Ⓐ} = \frac{v_i \sin \theta_i}{g} \qquad (4.18)$$

Substituting this expression for $t_\text{Ⓐ}$ into Equation 4.16 and replacing $y_f = y_\text{Ⓐ}$ with h, we obtain an expression for h in terms of the magnitude and direction of the initial velocity vector:

$$y_f = y_i + v_{yi}t - \tfrac{1}{2}gt^2 \quad \rightarrow \quad h = (v_i \sin \theta_i)\frac{v_i \sin \theta_i}{g} - \tfrac{1}{2}g\left(\frac{v_i \sin \theta_i}{g}\right)^2$$

$$h = \frac{v_i^2 \sin^2 \theta_i}{2g} \qquad (4.19)$$

Because of the symmetry of the trajectory, the projectile covers the upward part of the trajectory to the top in exactly the same time interval as it requires to come back to the ground from the topmost point. Therefore, the range R is the horizontal position of the projectile at a time that is twice the time at which it reaches its peak, that is, at time $t_\text{Ⓑ} = 2t_\text{Ⓐ}$. Using the particle under constant velocity model, noting that $v_{xi} = v_{x\text{Ⓑ}} = v_i \cos \theta_i$, and setting $x_\text{Ⓑ} = R$ at $t = 2t_\text{Ⓐ}$, we find from Equation 4.12 that

$$x_f = x_i + v_{xi}t \quad \rightarrow \quad R = v_{xi}t_\text{Ⓑ} = (v_i \cos \theta_i)2t_\text{Ⓐ}$$

$$= (v_i \cos \theta_i)\frac{2v_i \sin \theta_i}{g} = \frac{2v_i^2 \sin \theta_i \cos \theta_i}{g}$$

PITFALL PREVENTION 4.3
The Range Equation Equation 4.20 is useful for calculating R only for a symmetric path as shown in Figure 4.11. If the path is not symmetric, *do not use this equation*. The particle under constant velocity and particle under constant acceleration models are the important starting points because they give the position and velocity components of *any* projectile moving with constant acceleration in two dimensions at *any* time t, symmetric path or not.

Using the identity $\sin 2\theta = 2 \sin \theta \cos \theta$ (see Appendix B.4), we can write R in the more compact form

$$R = \frac{v_i^2 \sin 2\theta_i}{g} \qquad (4.20)$$

The maximum value of R from Equation 4.20 is $R_{\max} = v_i^2/g$. This result makes sense because the maximum value of $\sin 2\theta_i$ is 1, which occurs when $2\theta_i = 90°$. Therefore, R is a maximum when $\theta_i = 45°$.

Figure 4.11 illustrates various trajectories for a projectile having a given initial speed but launched at different angles. As you can see, the range is a maximum for $\theta_i = 45°$. In addition, for any θ_i other than 45°, a point having Cartesian coordinates $(R, 0)$ can be reached by using either one of two complementary values of θ_i for

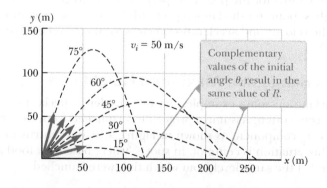

Figure 4.11 A projectile launched over a flat surface from the origin with an initial speed of 50 m/s at various angles of projection.

which sin $2\theta_i$ gives the same result, such as 75° and 15°. Of course, the maximum height and time of flight for one of these values of θ_i are different from the maximum height and time of flight for the complementary value. The time of flight depends only on v_{yi} and is independent of v_{xi}.

QUICK QUIZ 4.3 Rank the launch angles for the five paths in Figure 4.11 with respect to time of flight from the shortest time of flight to the longest.

PROBLEM-SOLVING STRATEGY **Projectile Motion**

We suggest you use the following approach when solving projectile motion problems.

1. Conceptualize. Think about what is going on physically in the problem. Establish the mental representation by imagining the projectile moving along its trajectory.

2. Categorize. Confirm that the problem involves a particle in free fall and that air resistance is neglected. Select a coordinate system with x in the horizontal direction and y in the vertical direction. Use the particle under constant velocity model for the x component of the motion. Use the particle under constant acceleration model for the y direction. In the special case of the projectile returning to the same level from which it was launched, use Equations 4.19 and 4.20.

3. Analyze. If the initial velocity vector is given, resolve it into x and y components. Select the appropriate equation(s) from the particle under constant acceleration model (4.13 through 4.17) for the vertical motion and use these along with Equation 4.12 for the horizontal motion to solve for the unknown(s).

4. Finalize. Once you have determined your result, check to see if your answers are consistent with the mental and pictorial representations and your results are realistic.

Example 4.2 **The Long Jump**

A long jumper (Fig. 4.12) leaves the ground at an angle of 20.0° above the horizontal and at a speed of 11.0 m/s.

(A) How far does he jump in the horizontal direction?

SOLUTION

Conceptualize The arms and legs of a long jumper move in a complicated way, but we will ignore this motion. We model the long jumper as a particle and conceptualize his motion as equivalent to that of a simple projectile.

Figure 4.12 (Example 4.2) Ashton Eaton of the United States competes in the men's decathlon long jump at the 2016 Rio de Janeiro Olympic Games.

Categorize We categorize this example as a projectile motion problem. Because the initial speed and launch angle are given and because the final height is the same as the initial height, we further categorize this problem as satisfying the conditions for which Equations 4.19 and 4.20 can be used. This approach is the most direct way to analyze this problem, although the general methods that have been described will always give the correct answer.

Analyze
Use Equation 4.20 to find the range of the jumper:

$$R = \frac{v_i^2 \sin 2\theta_i}{g} = \frac{(11.0 \text{ m/s})^2 \sin 2(20.0°)}{9.80 \text{ m/s}^2} = 7.94 \text{ m}$$

(B) What is the maximum height reached?

SOLUTION

Analyze
Find the maximum height reached by using Equation 4.19:

$$h = \frac{v_i^2 \sin^2 \theta_i}{2g} = \frac{(11.0 \text{ m/s})^2 (\sin 20.0°)^2}{2(9.80 \text{ m/s}^2)} = 0.722 \text{ m}$$

Finalize Find the answers to parts (A) and (B) using the general method. The results should agree. Treating the long jumper as a particle is an oversimplification. Nevertheless, the values obtained are consistent with experience in sports. We can model a complicated system such as a long jumper as a particle and still obtain reasonable results.

Example 4.3　**A Bull's-Eye Every Time**

In a popular lecture demonstration, a projectile is aimed directly at a target and fired in such a way that the projectile leaves the gun at the same time the target is dropped from rest. Show that the projectile hits the falling target.

SOLUTION

Conceptualize　We conceptualize the problem by studying Figure 4.13a. Notice that the problem does not ask for numerical values. The expected result must involve an algebraic argument.

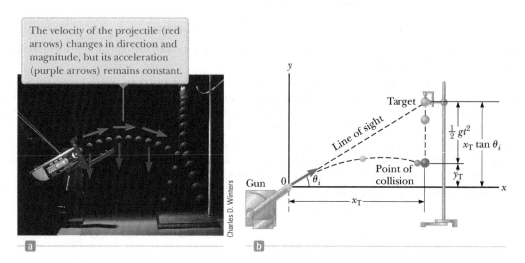

The velocity of the projectile (red arrows) changes in direction and magnitude, but its acceleration (purple arrows) remains constant.

Charles D. Winters

Figure 4.13　(Example 4.3) (a) Multiflash photograph of the projectile–target demonstration. If the gun is aimed directly at the target and is fired at the same instant the target begins to fall, the projectile will hit the target. (b) Schematic diagram of the projectile–target demonstration.

Categorize　Because both objects are subject only to gravity, we categorize this problem as one involving two objects in free fall, the target moving in one dimension and the projectile moving in two. The target T is modeled as a *particle under constant acceleration* in one dimension. The projectile P is modeled as a *particle under constant acceleration* in the y direction and a *particle under constant velocity* in the x direction.

Analyze　Figure 4.13b shows that the initial y coordinate y_{iT} of the target is $x_T \tan \theta_i$ and its initial velocity is zero. It falls with acceleration $a_y = -g$.

Write an expression for the y coordinate of the target at any moment after release, noting that its initial velocity is zero:

(1)　$y_T = y_{iT} + (0)t - \tfrac{1}{2}gt^2 = x_T \tan \theta_i - \tfrac{1}{2}gt^2$

Write an expression for the y coordinate of the projectile at any moment:

(2)　$y_P = y_{iP} + v_{yiP}t - \tfrac{1}{2}gt^2 = 0 + (v_{iP} \sin \theta_i)t - \tfrac{1}{2}gt^2 = (v_{iP} \sin \theta_i)t - \tfrac{1}{2}gt^2$

Write an expression for the x coordinate of the projectile at any moment:

$x_P = x_{iP} + v_{xiP}t = 0 + (v_{iP} \cos \theta_i)t = (v_{iP} \cos \theta_i)t$

Solve this expression for time as a function of the horizontal position of the projectile:

$t = \dfrac{x_P}{v_{iP} \cos \theta_i}$

Substitute this expression into Equation (2):

(3)　$y_P = (v_{iP} \sin \theta_i)\left(\dfrac{x_P}{v_{iP} \cos \theta_i}\right) - \tfrac{1}{2}gt^2 = x_P \tan \theta_i - \tfrac{1}{2}gt^2$

Finalize　Compare Equations (1) and (3). We see that when the x coordinates of the projectile and target are the same—that is, when $x_T = x_P$—their y coordinates given by Equations (1) and (3) are the same and a collision results.

Example 4.4 That's Quite an Arm!

A stone is thrown from the top of a building upward at an angle of 30.0° to the horizontal with an initial speed of 20.0 m/s as shown in Figure 4.14. The height from which the stone is thrown is 45.0 m above the ground.

(A) How long does it take the stone to reach the ground?

SOLUTION

Conceptualize Study Figure 4.14, in which we have indicated the trajectory and various parameters of the motion of the stone.

Categorize We categorize this problem as a projectile motion problem. The stone is modeled as a *particle under constant acceleration* in the *y* direction and a *particle under constant velocity* in the *x* direction.

Analyze We have the information $x_i = y_i = 0$, $y_f = -45.0$ m, $a_y = -g$, and $v_i = 20.0$ m/s (the numerical value of y_f is negative because we have chosen the point of the throw as the origin).

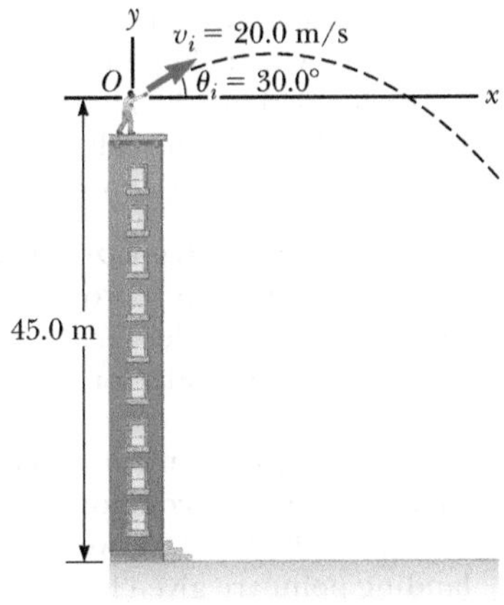

Figure 4.14 (Example 4.4) A stone is thrown from the top of a building.

Find the initial *x* and *y* components of the stone's velocity:

$$v_{xi} = v_i \cos \theta_i = (20.0 \text{ m/s}) \cos 30.0° = 17.3 \text{ m/s}$$

$$v_{yi} = v_i \sin \theta_i = (20.0 \text{ m/s}) \sin 30.0° = 10.0 \text{ m/s}$$

Express the vertical position of the stone from the particle under constant acceleration model:

$$y_f = y_i + v_{yi}t - \tfrac{1}{2}gt^2$$

Substitute numerical values:

$$-45.0 \text{ m} = 0 + (10.0 \text{ m/s})t + \tfrac{1}{2}(-9.80 \text{ m/s}^2)t^2$$

Solve the quadratic equation for *t*:

$$t = 4.22 \text{ s}$$

(B) What is the speed of the stone just before it strikes the ground?

SOLUTION

Analyze Use the velocity equation in the particle under constant acceleration model to obtain the *y* component of the velocity of the stone just before it strikes the ground:

$$v_{yf} = v_{yi} - gt$$

Substitute numerical values, using $t = 4.22$ s:

$$v_{yf} = 10.0 \text{ m/s} + (-9.80 \text{ m/s}^2)(4.22 \text{ s}) = -31.3 \text{ m/s}$$

Use this component with the horizontal component $v_{xf} = v_{xi} = 17.3$ m/s to find the speed of the stone at $t = 4.22$ s:

$$v_f = \sqrt{v_{xf}^2 + v_{yf}^2} = \sqrt{(17.3 \text{ m/s})^2 + (-31.3 \text{ m/s})^2} = 35.8 \text{ m/s}$$

Finalize Is it reasonable that the *y* component of the final velocity is negative? Is it reasonable that the final speed is larger than the initial speed of 20.0 m/s?

WHAT IF? What if a horizontal wind is blowing in the same direction as the stone is thrown and it causes the stone to have a horizontal acceleration component $a_x = 0.500$ m/s²? Which part of this example, (A) or (B), will have a different answer?

Answer Recall that the motions in the *x* and *y* directions are independent. Therefore, the horizontal wind cannot affect the vertical motion. The vertical motion determines the time of the projectile in the air, so the answer to part (A) does not change. The wind causes the horizontal velocity component to increase with time, so the final speed will be larger in part (B). Taking $a_x = 0.500$ m/s², we find $v_{xf} = 19.4$ m/s and $v_f = 36.9$ m/s.

Example 4.5 The End of the Ski Jump

A ski jumper leaves the ski track moving in the horizontal direction with a speed of 25.0 m/s as shown in Figure 4.15. The landing incline below her falls off with a slope of 35.0°. Where does she land on the incline?

SOLUTION

Conceptualize We can conceptualize this problem based on memories of observing winter ski jumping competitions. We estimate the skier to be airborne for perhaps 4 s and to travel a distance of about 100 m horizontally. We should expect the value of d, the distance traveled along the incline, to be of the same order of magnitude.

Categorize We categorize the problem as one of a particle in projectile motion. As with other projectile motion problems, we use the *particle under constant velocity* model for the horizontal motion and the *particle under constant acceleration* model for the vertical motion.

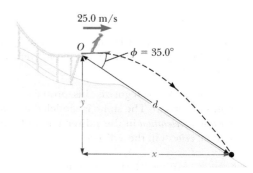

Analyze It is convenient to select the beginning of the jump as the origin. The initial velocity components are $v_{xi} = 25.0$ m/s and $v_{yi} = 0$. From the right triangle in Figure 4.15, we see that the jumper's x and y coordinates at the landing point are given by $x_f = d \cos \phi$ and $y_f = -d \sin \phi$.

Figure 4.15 (Example 4.5) A ski jumper leaves the track moving in a horizontal direction.

Express the coordinates of the jumper as a function of time, using the particle under constant velocity model for x and the position equation from the particle under constant acceleration model for y:

$$(1) \quad x_f = v_{xi}t \quad \rightarrow \quad (2) \quad d \cos \phi = v_{xi}t$$

$$(3) \quad y_f = v_{yi}t - \tfrac{1}{2}gt^2 \quad \rightarrow \quad (4) \quad -d \sin \phi = -\tfrac{1}{2}gt^2$$

Solve Equation (2) for t and substitute the result into Equation (4):

$$-d \sin \phi = -\tfrac{1}{2}g\left(\frac{d \cos \phi}{v_{xi}}\right)^2$$

Solve for d and substitute numerical values:

$$d = \frac{2v_{xi}^2 \sin \phi}{g \cos^2 \phi} = \frac{2(25.0 \text{ m/s})^2 \sin 35.0°}{(9.80 \text{ m/s}^2) \cos^2 35.0°} = 109 \text{ m}$$

Evaluate the x and y coordinates of the point at which the skier lands:

$$x_f = d \cos \phi = (109 \text{ m}) \cos 35.0° = 89.3 \text{ m}$$

$$y_f = -d \sin \phi = -(109 \text{ m}) \sin 35.0° = -62.5 \text{ m}$$

Finalize Let us compare these results with our expectations. We expected the horizontal distance to be on the order of 100 m, and our result of 89.3 m is indeed on this order of magnitude. It might be useful to calculate the time interval that the jumper is in the air and compare it with our estimate of about 4 s.

WHAT IF? Suppose everything in this example is the same except the ski jump is curved so that the jumper is projected upward at an angle from the end of the track. Is this design better in terms of maximizing the length of the jump?

Answer If the initial velocity has an upward component, the skier will be in the air longer and should therefore travel farther. Tilting the initial velocity vector upward, however, will reduce the horizontal component of the initial velocity. Therefore, angling the end of the ski track upward at a *large* angle may actually *reduce* the distance. Consider the extreme case: the skier is projected at 90° to the horizontal and simply goes up and comes back down at the end of the ski track! This argument suggests that there must be an optimal angle between 0° and 90° that represents a balance between making the flight time longer and the horizontal velocity component smaller.

Let us find this optimal angle mathematically. We modify Equations (1) through (4) in the following way, assuming the skier is projected at an angle θ with respect to the horizontal over a landing incline sloped with an arbitrary angle ϕ:

$$(1) \text{ and } (2) \quad \rightarrow \quad x_f = (v_i \cos \theta)t = d \cos \phi$$

$$(3) \text{ and } (4) \quad \rightarrow \quad y_f = (v_i \sin \theta)t - \tfrac{1}{2}gt^2 = -d \sin \phi$$

By eliminating the time t between these equations and using differentiation to maximize d in terms of θ, we arrive (after several steps; see Problem 52) at the following equation for the angle θ that gives the maximum value of d:

$$\theta = 45° - \frac{\phi}{2}$$

For the slope angle in Figure 4.15, $\phi = 35.0°$; this equation results in an optimal launch angle of $\theta = 27.5°$. For a slope angle of $\phi = 0°$, which represents a horizontal plane (no slope), this equation gives an optimal launch angle of $\theta = 45°$, as we would expect (see Figure 4.11).

4.4 Analysis Model: Particle in Uniform Circular Motion

Figure 4.16a shows a car moving in a circular path; we describe this motion by calling it **circular motion.** If the car is moving on this path with *constant speed v,* we call it **uniform circular motion.** Because it occurs so often, this type of motion is recognized as an analysis model called the **particle in uniform circular motion.** We discuss this model in this section.

It is often surprising to students to find that even though an object moves at a constant speed in a circular path, *it still has an acceleration.* To see why, consider the defining equation for acceleration, $\vec{\mathbf{a}} = d\vec{\mathbf{v}}/dt$ (Eq. 4.5). Notice that the acceleration depends on the change in the *velocity.* Because velocity is a vector quantity, an acceleration can occur in two ways as mentioned in Section 4.1: by a change in the *magnitude* of the velocity and by a change in the *direction* of the velocity. The latter situation occurs for an object moving with constant speed in a circular path. The constant-magnitude velocity vector is always tangent to the path of the object and perpendicular to the radius of the circular path. Therefore, the direction of the velocity vector is always changing.

Let us first argue that the acceleration vector in uniform circular motion is always perpendicular to the path and, therefore, always points toward the center of the circle. If that were not true, there would be a component of the acceleration parallel to the path and therefore parallel to the velocity vector. Such an acceleration component would lead to a change in the speed of the particle along the path. This situation, however, is inconsistent with our setup of the situation: the particle moves with constant speed along the path. Therefore, for *uniform* circular motion, the acceleration vector can only have a component perpendicular to the path, which is toward the center of the circle.

Let us now find the magnitude of the acceleration of the particle. Consider the diagram of the position and velocity vectors in Figure 4.16b. The figure also shows the vector representing the change in position $\Delta\vec{\mathbf{r}}$ for an arbitrary time interval. The particle follows a circular path of radius *r,* part of which is shown by the dashed curve. The particle is at Ⓐ at time t_i, and its velocity at that time is $\vec{\mathbf{v}}_i$; it is at Ⓑ at some later time t_f, and its velocity at that time is $\vec{\mathbf{v}}_f$. Let us also assume $\vec{\mathbf{v}}_i$ and $\vec{\mathbf{v}}_f$ differ only in direction; their magnitudes are the same (that is, $v_i = v_f = v$ because it is *uniform* circular motion).

In Figure 4.16c, the velocity vectors in Figure 4.16b have been redrawn tail to tail. The vector $\Delta\vec{\mathbf{v}}$ connects the tips of the vectors, representing the vector addition $\vec{\mathbf{v}}_f = \vec{\mathbf{v}}_i + \Delta\vec{\mathbf{v}}$. In both Figures 4.16b and 4.16c, we can identify triangles that help us analyze the motion. The angle $\Delta\theta$ between the two position vectors in Figure 4.16b is the same as the angle between the velocity vectors in Figure 4.16c because the velocity vector $\vec{\mathbf{v}}$ is always perpendicular to the position vector $\vec{\mathbf{r}}$. Therefore, the two triangles are *similar.* (Two triangles are similar if the angle between any two sides is the same for both triangles and if the ratio of the lengths of these sides is the same.) We can now write a relationship between the lengths of the sides for the two triangles in Figures 4.16b and 4.16c:

$$\frac{|\Delta\vec{\mathbf{v}}|}{v} = \frac{|\Delta\vec{\mathbf{r}}|}{r}$$

where $v = v_i = v_f$ and $r = r_i = r_f$. This equation can be solved for $|\Delta\vec{\mathbf{v}}|$, and the expression obtained can be substituted into Equation 4.4, $\vec{\mathbf{a}}_{avg} = \Delta\vec{\mathbf{v}}/\Delta t$, to give

PITFALL PREVENTION 4.4

Acceleration of a Particle in Uniform Circular Motion
Remember that acceleration in physics is defined as a change in the *velocity,* not a change in the *speed* (contrary to the everyday interpretation). In circular motion, the velocity vector is always changing in direction, so there is indeed an acceleration.

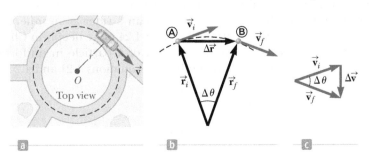

Figure 4.16 (a) A car moving along a circular path at constant speed experiences uniform circular motion. (b) As a particle moves along a portion of a circular path from Ⓐ to Ⓑ, its velocity vector changes from $\vec{\mathbf{v}}_i$ to $\vec{\mathbf{v}}_f$. (c) The construction for determining the direction of the change in velocity $\Delta\vec{\mathbf{v}}$, which is toward the center of the circle for small $\Delta\vec{\mathbf{r}}$.

the magnitude of the average acceleration over the time interval for the particle to move from Ⓐ to Ⓑ:

$$|\vec{\mathbf{a}}_{avg}| = \frac{|\Delta\vec{\mathbf{v}}|}{\Delta t} = \frac{v|\Delta\vec{\mathbf{r}}|}{r\Delta t}$$

Now imagine that points Ⓐ and Ⓑ in Figure 4.16b become extremely close together. As Ⓐ and Ⓑ approach each other, Δt approaches zero, $|\Delta\vec{\mathbf{r}}|$ approaches the distance traveled by the particle along the circular path, and the ratio $|\Delta\vec{\mathbf{r}}|/\Delta t$ approaches the speed v. In addition, the average acceleration becomes the instantaneous acceleration at point Ⓐ. Hence, in the limit $\Delta t \to 0$, the magnitude of the acceleration is

◀ Centripetal acceleration for a particle in uniform circular motion

$$a_c = \frac{v^2}{r} \tag{4.21}$$

An acceleration of this nature is called a **centripetal acceleration** (*centripetal* means *center-seeking*). The subscript on the acceleration symbol reminds us that the acceleration is centripetal.

In many situations, it is convenient to describe the motion of a particle moving with constant speed in a circle of radius r in terms of the **period** T, which is defined as the time interval required for one complete revolution of the particle. In the time interval T, the particle moves a distance of $2\pi r$, which is equal to the circumference of the particle's circular path. Therefore, because its speed is equal to the circumference of the circular path divided by the period, or $v = 2\pi r/T$, it follows that

◀ Period of circular motion for a particle in uniform circular motion

$$T = \frac{2\pi r}{v} \tag{4.22}$$

The period of a particle in uniform circular motion is a measure of the number of seconds for one revolution of the particle around the circle. The inverse of the period is the *rotation rate* and is measured in revolutions per second. Because one full revolution of the particle around the circle corresponds to an angle of 2π radians, the product of 2π and the rotation rate gives the **angular speed** ω of the particle, measured in radians/s or s^{-1}:

$$\omega = \frac{2\pi}{T} \tag{4.23}$$

Combining this equation with Equation 4.22, we find a relationship between angular speed and the translational speed with which the particle travels in the circular path:

$$\omega = 2\pi\left(\frac{v}{2\pi r}\right) = \frac{v}{r} \quad \to \quad v = r\omega \tag{4.24}$$

PITFALL PREVENTION 4.5

Centripetal Acceleration Is Not Constant We derived the magnitude of the centripetal acceleration vector and found it to be constant for uniform circular motion, but *the centripetal acceleration vector is not constant*. It always points toward the center of the circle, but it continuously changes direction as the object moves around the circular path.

Equation 4.24 demonstrates that, for a fixed angular speed, the translational speed becomes larger as the radial position becomes larger. Therefore, for example, if a merry-go-round rotates at a fixed angular speed ω, a rider at an outer position at large r will be traveling through space faster than a rider at an inner position at smaller r. We will investigate Equations 4.23 and 4.24 more deeply in Chapter 10.

We can express the centripetal acceleration of a particle in uniform circular motion in terms of angular speed by combining Equations 4.21 and 4.24:

$$a_c = \frac{(r\omega)^2}{r}$$

$$a_c = r\omega^2 \tag{4.25}$$

Equations 4.21–4.25 are to be used when the particle in uniform circular motion model is identified as appropriate for a given situation.

QUICK QUIZ 4.4 A particle moves in a circular path of radius r with speed v. It then increases its speed to $2v$ while traveling along the same circular path. (i) The centripetal acceleration of the particle has changed by what factor? Choose one: (a) 0.25 (b) 0.5 (c) 2 (d) 4 (e) impossible to determine (ii) From the same choices, by what factor has the period of the particle changed?

ANALYSIS MODEL | **Particle in Uniform Circular Motion**

Imagine a moving object that can be modeled as a particle. If it moves in a circular path of radius r at a constant speed v, the magnitude of its centripetal acceleration is

$$a_c = \frac{v^2}{r} \qquad \text{(4.21)}$$

and the **period** of the particle's motion is given by

$$T = \frac{2\pi r}{v} \qquad \text{(4.22)}$$

The **angular speed** of the particle is

$$\omega = \frac{2\pi}{T} \qquad \text{(4.23)}$$

Examples:

- a rock twirled in a circle on a string of constant length
- a planet traveling around a perfectly circular orbit (Chapter 13)
- a charged particle moving in a uniform magnetic field (Chapter 28)
- an electron in orbit around a nucleus in the Bohr model of the hydrogen atom (Chapter 41)

Example 4.6 | **The Centripetal Acceleration of the Earth**

(A) What is the centripetal acceleration of the Earth as it moves in its orbit around the Sun?

SOLUTION

Conceptualize We will model the Earth as a particle and approximate the Earth's orbit as circular (it's actually slightly elliptical, as we discuss in Chapter 13).

Categorize The Conceptualize step allows us to categorize this problem as one of a *particle in uniform circular motion.*

Analyze We do not know the orbital speed of the Earth to substitute into Equation 4.21. With the help of Equation 4.22, however, we can recast Equation 4.21 in terms of the period of the Earth's orbit, which we know is one year, and the radius of the Earth's orbit around the Sun, which is 1.496×10^{11} m.

Combine Equations 4.21 and 4.22:

$$a_c = \frac{v^2}{r} = \frac{\left(\frac{2\pi r}{T}\right)^2}{r} = \frac{4\pi^2 r}{T^2}$$

Substitute numerical values:

$$a_c = \frac{4\pi^2 (1.496 \times 10^{11} \text{ m})}{(1 \text{ yr})^2} \left(\frac{1 \text{ yr}}{3.156 \times 10^7 \text{ s}}\right)^2 = 5.93 \times 10^{-3} \text{ m/s}^2$$

(B) What is the angular speed of the Earth in its orbit around the Sun?

SOLUTION

Analyze

Substitute numerical values into Equation 4.23:

$$\omega = \frac{2\pi}{1 \text{ yr}} \left(\frac{1 \text{ yr}}{3.156 \times 10^7 \text{ s}}\right) = 1.99 \times 10^{-7} \text{ s}^{-1}$$

Finalize The acceleration in part (A) is much smaller than the free-fall acceleration on the surface of the Earth. An important technique we learned here is replacing the speed v in Equation 4.21 in terms of the period T of the motion. In many problems, it is more likely that T is known rather than v. In part (B), we see that the angular speed of the Earth is very small, which is to be expected because the Earth takes an entire year to go around the circular path once.

4.5 Tangential and Radial Acceleration

Let us consider a more general motion than that presented in Section 4.4. A particle moves to the right along a curved path, and its velocity changes *both* in direction and in magnitude as described in Figure 4.17. In this situation, the velocity vector is always tangent to the path; the acceleration vector $\vec{\mathbf{a}}$, however, is at some angle to the path. At each of three points Ⓐ, Ⓑ, and Ⓒ in Figure 4.17, the dashed blue circles represent the curvature of the actual path at each point. The radius of each circle is equal to the path's radius of curvature at each point.

As the particle moves along the curved path in Figure 4.17, the direction of the total acceleration vector $\vec{\mathbf{a}}$ changes from point to point. At any instant, this vector can be resolved into two components based on an origin at the center of the dashed circle corresponding to that instant: a radial component a_r along the radius of the circle and a tangential component a_t perpendicular to this radius. The *total* acceleration vector $\vec{\mathbf{a}}$ can be written as the vector sum of the component vectors:

Total acceleration ▶
$$\vec{\mathbf{a}} = \vec{\mathbf{a}}_r + \vec{\mathbf{a}}_t \tag{4.26}$$

The tangential acceleration component causes a change in the speed v of the particle. This component is parallel to the instantaneous velocity, and its magnitude is given by

Tangential acceleration ▶
$$a_t = \left| \frac{dv}{dt} \right| \tag{4.27}$$

The radial acceleration component arises from a change in direction of the velocity vector and is given by

Radial acceleration ▶
$$a_r = -a_c = -\frac{v^2}{r} \tag{4.28}$$

where r is the radius of curvature of the path at the point in question. We recognize the magnitude of the radial component of the acceleration as the centripetal acceleration discussed in Section 4.4 with regard to the particle in uniform circular motion model. Even in situations in which a particle moves along a curved path with a varying speed, however, Equation 4.21 can be used for the centripetal acceleration. In this situation, the equation gives the *instantaneous* centripetal acceleration at any time. The negative sign in Equation 4.28 indicates that the direction of the centripetal acceleration is toward the center of the circle representing the radius of curvature. The direction is opposite that of the radial unit vector $\hat{\mathbf{r}}$, which always points away from the origin at the center of the circle. (See Fig. 3.15.)

Because $\vec{\mathbf{a}}_r$ and $\vec{\mathbf{a}}_t$ are perpendicular component vectors of $\vec{\mathbf{a}}$, it follows that the magnitude of $\vec{\mathbf{a}}$ is $a = \sqrt{a_r^2 + a_t^2}$. At a given speed, a_r is large when the radius of curvature is small (as at points Ⓐ and Ⓑ in Fig. 4.17) and small when r is large (as at point Ⓒ). The direction of $\vec{\mathbf{a}}_t$ is either in the same direction as $\vec{\mathbf{v}}$ (if v is increasing) or opposite $\vec{\mathbf{v}}$ (if v is decreasing, as it must be at point Ⓑ).

In uniform circular motion, where v is constant, $a_t = 0$ and the acceleration is always completely radial as described in Section 4.4. In other words, uniform circular motion is a special case of motion along a general curved path. Furthermore, if the direction of $\vec{\mathbf{v}}$ does not change, there is no radial acceleration and the motion is one dimensional (in this case, $a_r = 0$, but a_t may not be zero).

Figure 4.17 The motion of a particle along an arbitrary curved path lying in the *xy* plane. If the velocity vector $\vec{\mathbf{v}}$ (always tangent to the path) changes in direction and magnitude, the components of the acceleration $\vec{\mathbf{a}}$ are a tangential component a_t and a radial component a_r.

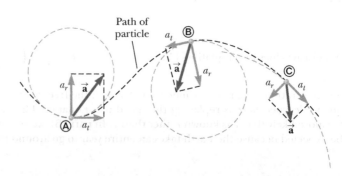

UICK QUIZ 4.5 A particle moves along a path, and its speed increases with time. **(i)** In which of the following cases are its acceleration and velocity vectors parallel? (a) when the path is circular (b) when the path is straight (c) when the path is a parabola (d) never **(ii)** From the same choices, in which case are its acceleration and velocity vectors perpendicular everywhere along the path?

Example 4.7 | Over the Rise

A car leaves a stop sign and exhibits a constant acceleration of 0.300 m/s² parallel to the roadway. The car passes over a rise in the roadway such that the top of the rise is shaped like an arc of a circle of radius 500 m. At the moment the car is at the top of the rise, its velocity vector is horizontal and has a magnitude of 6.00 m/s. What are the magnitude and direction of the total acceleration vector for the car at this instant?

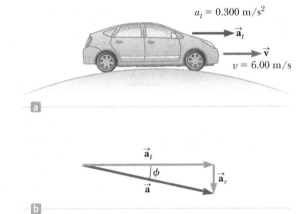

SOLUTION

Conceptualize Conceptualize the situation using Figure 4.18a and any experiences you have had in driving over rises on a roadway.

Categorize Because the accelerating car is moving along a curved path, we categorize this problem as one involving a particle experiencing both tangential and radial acceleration. We recognize that it is a relatively simple substitution problem.

Figure 4.18 (Example 4.7) (a) A car passes over a rise that is shaped like an arc of a circle. (b) The total acceleration vector $\vec{a}$ is the sum of the tangential and radial acceleration vectors $\vec{a}_t$ and $\vec{a}_r$.

The tangential acceleration vector has magnitude 0.300 m/s² and is horizontal. The radial acceleration is given by Equation 4.28, with $v = 6.00$ m/s and $r = 500$ m. The radial acceleration vector is directed straight downward.

Evaluate the radial acceleration:

$$a_r = -\frac{v^2}{r} = -\frac{(6.00 \text{ m/s})^2}{500 \text{ m}} = -0.072\,0 \text{ m/s}^2$$

Find the magnitude of $\vec{a}$:

$$a = \sqrt{a_r^2 + a_t^2} = \sqrt{(-0.072\,0 \text{ m/s}^2)^2 + (0.300 \text{ m/s}^2)^2}$$

$$= 0.309 \text{ m/s}^2$$

Find the angle ϕ (see Fig. 4.18b) between $\vec{a}$ and the horizontal:

$$\phi = \tan^{-1}\frac{a_r}{a_t} = \tan^{-1}\left(\frac{-0.072\,0 \text{ m/s}^2}{0.300 \text{ m/s}^2}\right) = -13.5°$$

4.6 Relative Velocity and Relative Acceleration

In this section, we describe how observations made by different observers in different frames of reference are related to one another. A frame of reference can be described by a Cartesian coordinate system for which an observer is at rest with respect to the origin.

Let us conceptualize a sample situation in which there will be different observations for different observers. Consider the two observers A and B along the number line in Figure 4.19a. Observer A is located 5 units to the right of observer B. Both observers measure the position of point P, which is located 5 units to the right of observer A. Suppose each observer decides that he is located at the origin of an x axis as in Figure 4.19b. Notice that the two observers disagree on the value of the position of point P. Observer A claims point P is located at a position with a value of $x_A = +5$, whereas observer B claims it is located at a position with a value of $x_B = +10$. Both observers are correct, even though they make different measurements. Their measurements differ because they are making the measurement from different frames of reference.

Imagine now that observer B in Figure 4.19b is moving to the right along the x_B axis. Now the two measurements are even more different. Observer A claims point P

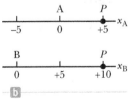

Figure 4.19 Different observers make different measurements. (a) Observer A is located 5 units to the right of Observer B. Both observers measure the position of a particle at P. (b) If both observers see themselves at the origin of their own coordinate system, they disagree on the value of the position of the particle at P.

The woman standing on the beltway sees the man moving with a slower speed than does the woman observing the man from the stationary floor.

Figure 4.20 Two observers measure the speed of a man walking on a moving beltway.

remains at rest at a position with a value of +5, whereas observer B claims the position of P continuously changes with time, even passing him and moving behind him! Again, both observers are correct, with the difference in their measurements arising from their different frames of reference.

We explore this phenomenon further by considering two observers watching a man walking on a moving beltway at an airport in Figure 4.20. The woman standing on the moving beltway sees the man moving at a normal walking speed. The woman observing from the stationary floor sees the man moving with a higher speed because the beltway speed combines with his walking speed. Both observers look at the same man and arrive at different values for his speed. Both are correct; the difference in their measurements results from the relative velocity of their frames of reference.

In a more general situation, consider a particle located at point P in Figure 4.21. Imagine that the motion of this particle is being described by two observers, observer A in a reference frame S_A fixed relative to the Earth and a second observer B in a reference frame S_B moving to the right relative to S_A (and therefore relative to the Earth) with a constant velocity $\vec{v}_{BA}$. In this discussion of relative velocity, we use a double-subscript notation; the first subscript represents what is being observed, and the second represents who is doing the observing. Therefore, the notation $\vec{v}_{BA}$ means the velocity of observer B (and the attached frame S_B) as measured by observer A. With this notation, observer B measures A to be moving to the left with a velocity $\vec{v}_{AB} = -\vec{v}_{BA}$. For purposes of this discussion, let us place each observer at her or his respective origin.

We define the time $t = 0$ as the instant at which the origins of the two reference frames coincide in space. Therefore, at time t, the origins of the reference frames will be separated by a distance $v_{BA}t$. We label the position P of the particle relative to observer A with the position vector $\vec{r}_{PA}$ and that relative to observer B with the position vector $\vec{r}_{PB}$, both at time t. From Figure 4.21, we see that the vectors $\vec{r}_{PA}$ and $\vec{r}_{PB}$ are related to each other through the expression

$$\vec{r}_{PA} = \vec{r}_{PB} + \vec{v}_{BA}t \tag{4.29}$$

By differentiating Equation 4.29 with respect to time, noting that $\vec{v}_{BA}$ is constant, we obtain

$$\frac{d\vec{r}_{PA}}{dt} = \frac{d\vec{r}_{PB}}{dt} + \vec{v}_{BA}$$

$$\vec{u}_{PA} = \vec{u}_{PB} + \vec{v}_{BA} \tag{4.30}$$

where $\vec{u}_{PA}$ is the velocity of the particle at P measured by observer A and $\vec{u}_{PB}$ is its velocity measured by B. (We use the symbol $\vec{u}$ for particle velocity rather than $\vec{v}$, which we have already used for the relative velocity of two reference frames.) Equation 4.30 is demonstrated by the red vectors at the top of Figure 4.21. Vector $\vec{u}_{PB}$ is the velocity of the particle at time t as seen by observer B. When you add the relative velocity $\vec{v}_{BA}$ of the frames, the sum is the velocity of the particle as measured by observer A.

Equations 4.29 and 4.30 are known as **Galilean transformation equations.** They relate the position and velocity of a particle as measured by observers in relative motion.

Although observers in two frames measure different velocities for the particle, they measure the *same acceleration* when $\vec{v}_{BA}$ is constant. We can verify that by taking the time derivative of Equation 4.30:

$$\frac{d\vec{u}_{PA}}{dt} = \frac{d\vec{u}_{PB}}{dt} + \frac{d\vec{v}_{BA}}{dt}$$

Because $\vec{v}_{BA}$ is constant, $d\vec{v}_{BA}/dt = 0$. Therefore, we conclude that $\vec{a}_{PA} = \vec{a}_{PB}$ because $\vec{a}_{PA} = d\vec{u}_{PA}/dt$ and $\vec{a}_{PB} = d\vec{u}_{PB}/dt$. That is, the acceleration of the particle measured by an observer in one frame of reference is the same as that measured by any other observer moving with constant velocity relative to the first frame.

Figure 4.21 A particle located at P is described by two observers, one in the fixed frame of reference S_A and the other in the frame S_B, which moves to the right with a constant velocity $\vec{v}_{BA}$. The vector $\vec{r}_{PA}$ is the particle's position vector relative to S_A, and $\vec{r}_{PB}$ is its position vector relative to S_B. The red vectors at the top of the figure show a vector addition for the velocities of the particle at time t, representing Equation 4.30.

Example 4.8 A Boat Crossing a River

A boat crossing a wide river moves with a speed of 10.0 km/h relative to the water. The water in the river has a uniform speed of 5.00 km/h due east relative to the Earth.

(A) If the boat heads due north, determine the velocity of the boat relative to an observer standing on either bank.

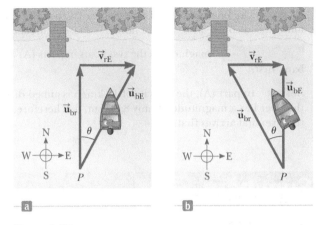

Figure 4.22 (Example 4.8) (a) A boat aims directly across a river and ends up downstream. (b) To move directly across the river, the boat must aim upstream.

SOLUTION

Conceptualize Imagine moving in a boat across a river while the current pushes you down the river. You will not be able to move directly across the river, but will end up downstream as suggested in Figure 4.22a. Imagine observer A on the shore, so that she is on the Earth, represented by letter E. Observer B is represented by letter r in the figure; this observer is on a cork floating in the river, at rest with respect to the water and carried along with the current. When the boat begins from point P and is aimed straight across the river, the velocities $\vec{\mathbf{u}}_{br}$, the boat relative to the river, and $\vec{\mathbf{v}}_{rE}$, the river relative to the Earth, add to give the velocity $\vec{\mathbf{v}}_{bE}$, the velocity of the boat relative to observer A on the Earth. Compare the vector addition in Figure 4.22a to that in Figure 4.21. As the boat moves, it will follow along vector $\vec{\mathbf{v}}_{bE}$, as suggested by its position after some time in Figure 4.22a.

Categorize Because of the combined velocities of you relative to the river and the river relative to the Earth, we can categorize this problem as one involving relative velocities.

Analyze We know $\vec{\mathbf{u}}_{br}$, the velocity of the *boat* relative to the *river,* and $\vec{\mathbf{v}}_{rE}$, the velocity of the *river* relative to the *Earth.* What we must find is $\vec{\mathbf{u}}_{bE}$, the velocity of the *boat* relative to the *Earth.* The relationship between these three quantities is $\vec{\mathbf{u}}_{bE} = \vec{\mathbf{u}}_{br} + \vec{\mathbf{v}}_{rE}$. The terms in the equation must be manipulated as vector quantities; the vectors are shown in Figure 4.22a. The quantity $\vec{\mathbf{u}}_{br}$ is due north; $\vec{\mathbf{v}}_{rE}$ is due east; and the vector sum of the two, $\vec{\mathbf{u}}_{bE}$, is at an angle θ as defined in Figure 4.22a.

Find the speed u_{bE} of the boat relative to the Earth using the Pythagorean theorem:

$$u_{bE} = \sqrt{u_{br}^{2} + v_{rE}^{2}} = \sqrt{(10.0 \text{ km/h})^2 + (5.00 \text{ km/h})^2}$$

$$= 11.2 \text{ km/h}$$

Find the direction of $\vec{\mathbf{u}}_{bE}$:

$$\theta = \tan^{-1}\left(\frac{v_{rE}}{u_{br}}\right) = \tan^{-1}\left(\frac{5.00}{10.0}\right) = 26.6°$$

Finalize The boat is moving at a speed of 11.2 km/h in the direction 26.6° east of north relative to the Earth. Notice that the speed of 11.2 km/h is faster than your boat speed of 10.0 km/h. The current velocity adds to yours to give you a higher speed. Notice in Figure 4.22a that your resultant velocity is at an angle to the direction straight across the river, so you will end up downstream, as we predicted.

(B) If the boat travels with the same speed of 10.0 km/h relative to the river and is to travel due north as shown in Figure 4.22b, what should its heading be?

SOLUTION

Conceptualize/Categorize This question is an extension of part (A), so we have already conceptualized and categorized the problem. In this case, however, we must aim the boat upstream so as to go straight across the river.

Analyze The analysis now involves the new triangle shown in Figure 4.22b. As in part (A), we know $\vec{\mathbf{v}}_{rE}$ and the magnitude of the vector $\vec{\mathbf{u}}_{br}$, and we want $\vec{\mathbf{u}}_{bE}$ to be directed across the river. Notice the difference between the triangle in Figure 4.22a and the one in Figure 4.22b: the hypotenuse in Figure 4.22b is no longer $\vec{\mathbf{u}}_{bE}$.

Use the Pythagorean theorem to find u_{bE}:

$$u_{bE} = \sqrt{u_{br}^{2} - v_{rE}^{2}} = \sqrt{(10.0 \text{ km/h})^2 - (5.00 \text{ km/h})^2} = 8.66 \text{ km/h}$$

Find the direction in which the boat is heading:

$$\theta = \tan^{-1}\left(\frac{v_{rE}}{u_{bE}}\right) = \tan^{-1}\left(\frac{5.00}{8.66}\right) = 30.0°$$

continued

4.8 continued

Finalize   The boat must head upstream so as to travel directly northward across the river. For the given situation, the boat must steer a course 30.0° west of north. For faster currents, the boat must be aimed upstream at larger angles.

WHAT IF?   Imagine that the two boats in parts (A) and (B) are racing across the river. Which boat arrives at the opposite bank first?

Answer   In part (A), the velocity of 10 km/h is aimed directly across the river. In part (B), the velocity that is directed across the river has a magnitude of only 8.66 km/h. Therefore, the boat in part (A) has a larger velocity component directly across the river and arrives first.

Summary

▶ Definitions

The **displacement vector** $\Delta \vec{r}$ for a particle is the difference between its final position vector and its initial position vector:

$$\Delta \vec{r} \equiv \vec{r}_f - \vec{r}_i \tag{4.1}$$

The **average velocity** of a particle during the time interval Δt is defined as the displacement of the particle divided by the time interval:

$$\vec{v}_{avg} \equiv \frac{\Delta \vec{r}}{\Delta t} \tag{4.2}$$

The **instantaneous velocity** of a particle is defined as the limit of the average velocity as Δt approaches zero:

$$\vec{v} \equiv \lim_{\Delta t \to 0} \frac{\Delta \vec{r}}{\Delta t} = \frac{d\vec{r}}{dt} \tag{4.3}$$

The **average acceleration** of a particle is defined as the change in its instantaneous velocity vector divided by the time interval Δt during which that change occurs:

$$\vec{a}_{avg} \equiv \frac{\Delta \vec{v}}{\Delta t} = \frac{\vec{v}_f - \vec{v}_i}{t_f - t_i} \tag{4.4}$$

The **instantaneous acceleration** of a particle is defined as the limiting value of the average acceleration as Δt approaches zero:

$$\vec{a} \equiv \lim_{\Delta t \to 0} \frac{\Delta \vec{v}}{\Delta t} = \frac{d\vec{v}}{dt} \tag{4.5}$$

Projectile motion is one type of two-dimensional motion, exhibited by an object launched into the air near the Earth's surface and experiencing free fall. If the projectile is launched at an upward angle from the horizontal, it will follow a path described mathematically as a parabola.

A particle moving in a circular path with constant speed is exhibiting **uniform circular motion**.

▶ Concepts and Principles

If a particle moves with *constant* acceleration $\vec{a}$ and has velocity $\vec{v}_i$ and position $\vec{r}_i$ at $t = 0$, its velocity and position vectors at some later time t are

$$\vec{v}_f = \vec{v}_i + \vec{a}t \tag{4.8}$$

$$\vec{r}_f = \vec{r}_i + \vec{v}_i t + \tfrac{1}{2}\vec{a}t^2 \tag{4.9}$$

For two-dimensional motion in the xy plane under constant acceleration, each of these vector expressions is equivalent to two component expressions: one for the motion in the x direction and one for the motion in the y direction.

It is useful to think of projectile motion in terms of a combination of two analysis models: (1) the particle under constant velocity model in the x direction and (2) the particle under constant acceleration model in the vertical direction with a constant downward acceleration of magnitude $g = 9.80$ m/s².

If a particle moves along a curved path in such a way that both the magnitude and the direction of $\vec{v}$ change in time, the particle has an acceleration vector that can be described by two component vectors: (1) a radial component vector $\vec{a}_r$ that causes the change in direction of $\vec{v}$ and (2) a tangential component vector $\vec{a}_t$ that causes the change in magnitude of $\vec{v}$. The magnitude of $\vec{a}_r$ is v^2/r, and the magnitude of $\vec{a}_t$ is $|dv/dt|$.

A particle in uniform circular motion undergoes a radial acceleration $\vec{a}_r$ because the direction of $\vec{v}$ changes in time. This acceleration is called **centripetal acceleration**, and its direction is always toward the center of the circle.

The velocity $\vec{u}_{PA}$ of a particle measured in a fixed frame of reference S_A can be related to the velocity $\vec{u}_{PB}$ of the same particle measured in a moving frame of reference S_B by

$$\vec{u}_{PA} = \vec{u}_{PB} + \vec{v}_{BA} \tag{4.30}$$

where $\vec{v}_{BA}$ is the velocity of S_B relative to S_A.

➤ Analysis Model for Problem Solving

Particle in Uniform Circular Motion If a particle moves in a circular path of radius r with a constant speed v, the magnitude of its centripetal acceleration is given by

$$a_c = \frac{v^2}{r} \qquad (4.21)$$

and the **period** of the particle's motion is given by

$$T = \frac{2\pi r}{v} \qquad (4.22)$$

The **angular speed** of the particle is

$$\omega = \frac{2\pi}{T} \qquad (4.23)$$

Think–Pair–Share

See the Preface for an explanation of the icons used in this problems set. For additional assessment items for this section, go to **WEBASSIGN** *From Cengage*

1. You watch your toddler nephew rolling marbles toward the top of a staircase. There are 12 steps, each 30.0 cm deep horizontally, and separated by 20.0 cm vertically. The marbles leave the upper landing horizontally and are projected into the air, bouncing down the steps until they arrive at the lower floor. This gets you wondering the following: (a) How fast must the marble be rolled so that it misses bouncing off the *first* step below the upper landing? (b) How fast must the marble be rolled so that it misses bouncing off the *second* step below the upper landing? (c) Is it possible for your toddler nephew to roll the marble fast enough to miss *all* the steps? (d) Suppose the marble is projected with a speed such that it lands on the sixth step and bounces upward at the same angle at which it struck the step, with the same speed. Argue that the marble will not hit another step before striking the floor of the lower landing.

2. **ACTIVITY** Place a penny at the corner of a table as shown in the overhead view in Figure TP4.2. Place a ruler next to the penny and another penny on the top of the part of the ruler

that hangs off the edge of the table. Hold the end of the ruler on the table with one hand and use your other hand to flick the end of the ruler with the penny parallel to the table surface. This will project the penny sitting on the corner of the table in a horizontal direction. At the same time, the ruler will slide out from under the second penny, which will fall straight down. Using your smartphone audio recorder, make an audio recording of the two falling pennies. From the recording, determine the time interval between the landing of the two pennies on the floor. What should the time interval be theoretically?

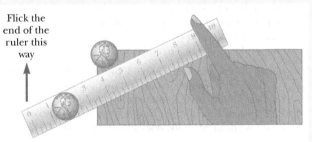

Flick the end of the ruler this way

Figure TP4.2

Problems

See the Preface for an explanation of the icons used in this problems set. For additional assessment items for this section, go to **WEBASSIGN** *From Cengage*

SECTION 4.1 The Position, Velocity, and Acceleration Vectors

1. Suppose the position vector for a particle is given as a function of time by $\vec{\mathbf{r}}(t) = x(t)\hat{\mathbf{i}} + y(t)\hat{\mathbf{j}}$, with $x(t) = at + b$ and $y(t) = ct^2 + d$, where $a = 1.00$ m/s, $b = 1.00$ m, $c = 0.125$ m/s^2, and $d = 1.00$ m. (a) Calculate the average velocity during the time interval from $t = 2.00$ s to $t = 4.00$ s. (b) Determine the velocity and the speed at $t = 2.00$ s.

2. The coordinates of an object moving in the xy plane vary with time according to the equations $x = -5.00 \sin \omega t$ and $y = 4.00 - 5.00 \cos \omega t$, where ω is a constant, x and y are in meters, and t is in seconds. (a) Determine the components

of velocity of the object at $t = 0$. (b) Determine the components of acceleration of the object at $t = 0$. (c) Write expressions for the position vector, the velocity vector, and the acceleration vector of the object at any time $t > 0$. (d) Describe the path of the object in an xy plot.

SECTION 4.2 Two-Dimensional Motion with Constant Acceleration

3. The vector position of a particle varies in time according to the expression $\vec{\mathbf{r}} = 3.00\hat{\mathbf{i}} - 6.00t^2\hat{\mathbf{j}}$, where $\vec{\mathbf{r}}$ is in meters and t is in seconds. (a) Find an expression for the velocity of the particle as a function of time. (b) Determine the acceleration of the particle as a function of time. (c) Calculate the particle's position and velocity at $t = 1.00$ s.

4. It is not possible to see very small objects, such as viruses,
BIO using an ordinary light microscope. An electron microscope, however, can view such objects using an electron
beam instead of a light beam. Electron microscopy has
proved invaluable for investigations of viruses, cell mem-
branes and subcellular structures, bacterial surfaces, visual
receptors, chloroplasts, and the contractile properties of
muscles. The "lenses" of an electron microscope consist of
electric and magnetic fields that control the electron beam.
As an example of the manipulation of an electron beam, consider an electron traveling away from the origin along the
x axis in the xy plane with initial velocity $\vec{\mathbf{v}}_i = v_i\hat{\mathbf{i}}$. As it passes
through the region $x = 0$ to $x = d$, the electron experiences
acceleration $\vec{\mathbf{a}} = a_x\hat{\mathbf{i}} + a_y\hat{\mathbf{j}}$, where a_x and a_y are constants. For
the case $v_i = 1.80 \times 10^7$ m/s, $a_x = 8.00 \times 10^{14}$ m/s², and $a_y = 1.60 \times 10^{15}$ m/s², determine at $x = d = 0.010\ 0$ m (a) the position of the electron, (b) the velocity of the electron, (c) the
speed of the electron, and (d) the direction of travel of the
electron (i.e., the angle between its velocity and the x axis).

5. **Review.** A snowmobile is originally at the point with position vector 29.0 m at 95.0° counterclockwise from the x
axis, moving with velocity 4.50 m/s at 40.0°. It moves with
constant acceleration 1.90 m/s² at 200°. After 5.00 s have
elapsed, find (a) its velocity and (b) its position vector.

SECTION 4.3 Projectile Motion

Note: Ignore air resistance in all problems and take $g = 9.80$ m/s²
at the Earth's surface.

6. In a local bar, a customer slides an empty beer mug down
S the counter for a refill. The height of the counter is h. The
mug slides off the counter and strikes the floor at distance d
from the base of the counter. (a) With what velocity did the
mug leave the counter? (b) What was the direction of the
mug's velocity just before it hit the floor?

7. Mayan kings and many school sports teams are named
BIO for the puma, cougar, or mountain lion—*Felis concolor*—
the best jumper among animals. It can jump to a height of
12.0 ft when leaving the ground at an angle of 45.0°. With
what speed, in SI units, does it leave the ground to make
this leap?

8. A projectile is fired in such a way that its horizontal range is
equal to three times its maximum height. What is the angle
of projection?

9. The speed of a projectile when it reaches its maximum
height is one-half its speed when it is at half its maximum
height. What is the initial projection angle of the projectile?

10. A rock is thrown upward from level ground in such a way
QC that the maximum height of its flight is equal to its hori-
S zontal range R. (a) At what angle θ is the rock thrown? (b)
In terms of its original range R, what is the range $R_{\max}$ the
rock can attain if it is launched at the same speed but at the
optimal angle for maximum range? (c) **What If?** Would your
answer to part (a) be different if the rock is thrown with the
same speed on a different planet? Explain.

11. A firefighter, a distance d from a burning building, directs
S a stream of water from a fire hose at angle θ_i above the
horizontal as shown in Figure P4.11. If the initial speed of
the stream is v_i, at what height h does the water strike the
building?

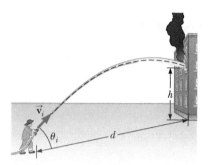

Figure P4.11

12. A basketball star covers 2.80 m horizontally in a jump to
dunk the ball (Fig. P4.12a). His motion through space
can be modeled precisely as that of a particle at his *center of mass,* which we will define in Chapter 9. His center of mass is at elevation 1.02 m when he leaves the
floor. It reaches a maximum height of 1.85 m above the
floor and is at elevation 0.900 m when he touches down
again. Determine (a) his time of flight (his "hang time"),
(b) his horizontal and (c) vertical velocity components
at the instant of takeoff, and (d) his takeoff angle. (e) For
comparison, determine the hang time of a whitetail deer
making a jump (Fig. P4.12b) with center-of-mass elevations
$y_i = 1.20$ m, $y_{\max} = 2.50$ m, and $y_f = 0.700$ m.

a   b

Figure P4.12

13. A student stands at the edge of a
GP cliff and throws a stone horizontally over the edge with a speed
of $v_i = 18.0$ m/s. The cliff is $h = 50.0$ m above a body of water as
shown in Figure P4.13. (a) What
are the coordinates of the initial
position of the stone? (b) What
are the components of the initial
velocity of the stone? (c) What is
the appropriate analysis model
for the vertical motion of the
stone? (d) What is the appropriate analysis model for the
horizontal motion of the stone?
(e) Write symbolic equations for
the x and y components of the
velocity of the stone as a function of time. (f) Write symbolic
equations for the position of the stone as a function of time.
(g) How long after being released does the stone strike the
water below the cliff? (h) With what speed and angle of
impact does the stone land?

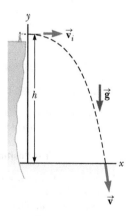

Figure P4.13

14. The record distance in the sport of throwing cowpats is 81.1 m. This record toss was set by Steve Urner of the United States in 1981. Assuming the initial launch angle was 45° and neglecting air resistance, determine (a) the initial speed of the projectile and (b) the total time interval the projectile was in flight. (c) How would the answers change if the range were the same but the launch angle were greater than 45°? Explain.

15. A home run is hit in such a way that the baseball just clears a wall 21.0 m high, located 130 m from home plate. The ball is hit at an angle of 35.0° to the horizontal, and air resistance is negligible. Find (a) the initial speed of the ball, (b) the time it takes the ball to reach the wall, and (c) the velocity components and the speed of the ball when it reaches the wall. (Assume the ball is hit at a height of 1.00 m above the ground.)

16. A projectile is fired from the top of a cliff of height h above the ocean below. The projectile is fired at an angle θ above the horizontal and with an initial speed v_i. (a) Find a symbolic expression in terms of the variables v_i, g, and θ for the time at which the projectile reaches its maximum height. (b) Using the result of part (a), find an expression for the maximum height h_{max} above the ocean attained by the projectile in terms of h, v_i, g, and θ.

17. A boy stands on a diving board and tosses a stone into a swimming pool. The stone is thrown from a height of 2.50 m above the water surface with a velocity of 4.00 m/s at an angle of 60.0° above the horizontal. As the stone strikes the water surface, it immediately slows down to exactly half the speed it had when it struck the water and maintains that speed while in the water. After the stone enters the water, it moves in a straight line in the direction of the velocity it had when it struck the water. If the pool is 3.00 m deep, how much time elapses between when the stone is thrown and when it strikes the bottom of the pool?

SECTION 4.4 Analysis Model: Particle in Uniform Circular Motion

Note: Problems 3 and 9 in Chapter 6 can also be assigned with this section.

18. In Example 4.6, we found the centripetal acceleration of the Earth as it revolves around the Sun. From information on the endpapers of this book, compute the centripetal acceleration of a point on the surface of the Earth at the equator caused by the rotation of the Earth about its axis.

19. The astronaut orbiting the Earth in Figure P4.19 is preparing to dock with a Westar VI satellite. The satellite is in a circular orbit 600 km above the Earth's surface, where the free-fall acceleration is 8.21 m/s². Take the radius of the Earth as 6 400 km. Determine the speed of the satellite and the time interval required to complete one orbit around the Earth, which is the period of the satellite.

Figure P4.19

20. An athlete swings a ball, connected to the end of a chain, in a horizontal circle. The athlete is able to rotate the ball at the rate of 8.00 rev/s when the length of the chain is 0.600 m. When he increases the length to 0.900 m, he is able to rotate the ball only 6.00 rev/s. (a) Which rate of rotation gives the greater speed for the ball? (b) What is the centripetal acceleration of the ball at 8.00 rev/s? (c) What is the centripetal acceleration at 6.00 rev/s?

21. The athlete shown in Figure P4.21 rotates a 1.00-kg discus along a circular path of radius 1.06 m. The maximum speed of the discus is 20.0 m/s. Determine the magnitude of the maximum radial acceleration of the discus.

Figure P4.21

22. A tire 0.500 m in radius rotates at a constant rate of 200 rev/min. Find the speed and acceleration of a small stone lodged in the tread of the tire (on its outer edge).

SECTION 4.5 Tangential and Radial Acceleration

23. (a) Can a particle moving with instantaneous speed 3.00 m/s on a path with radius of curvature 2.00 m have an acceleration of magnitude 6.00 m/s²? (b) Can it have an acceleration of magnitude 4.00 m/s²? In each case, if the answer is yes, explain how it can happen; if the answer is no, explain why not.

24. A ball swings counterclockwise in a vertical circle at the end of a rope 1.50 m long. When the ball is 36.9° past the lowest point on its way up, its total acceleration is $(-22.5\,\hat{\mathbf{i}} + 20.2\,\hat{\mathbf{j}})$ m/s². For that instant, (a) sketch a vector diagram showing the components of its acceleration, (b) determine the magnitude of its radial acceleration, and (c) determine the speed and velocity of the ball.

SECTION 4.6 Relative Velocity and Relative Acceleration

25. A bolt drops from the ceiling of a moving train car that is accelerating northward at a rate of 2.50 m/s². (a) What is the acceleration of the bolt relative to the train car? (b) What is the acceleration of the bolt relative to the Earth? (c) Describe the trajectory of the bolt as seen by an observer inside the train car. (d) Describe the trajectory of the bolt as seen by an observer fixed on the Earth.

26. The pilot of an airplane notes that the compass indicates a heading due west. The airplane's speed relative to the air is 150 km/h. The air is moving in a wind at 30.0 km/h toward the north. Find the velocity of the airplane relative to the ground.

27. You are taking flying lessons from an experienced pilot. You and the pilot are up in the plane, with you in the pilot seat. The control tower radios the plane, saying that, while you have been airborne, a 25-mi/h crosswind has arisen, with the direction of the wind perpendicular to the runway on which you plan to land. The pilot tells you that your normal airspeed as you land will be 80 mi/h relative to the

ground. This speed is relative to the air, in the direction in which the nose of the airplane points. He asks you to determine the angle at which the aircraft must be "crabbed," that is, the angle between the centerline of the aircraft and the centerline of the runway that will allow the airplane's velocity relative to the ground to be parallel to the runway.

28. A car travels due east with a speed of 50.0 km/h. Raindrops are falling at a constant speed vertically with respect to the Earth. The traces of the rain on the side windows of the car make an angle of 60.0° with the vertical. Find the velocity of the rain with respect to (a) the car and (b) the Earth.

29. A science student is riding on a flatcar of a train traveling along a straight, horizontal track at a constant speed of 10.0 m/s. The student throws a ball into the air along a path that he judges to make an initial angle of 60.0° with the horizontal and to be in line with the track. The student's professor, who is standing on the ground nearby, observes the ball to rise vertically. How high does she see the ball rise?

30. A river has a steady speed of 0.500 m/s. A student swims upstream a distance of 1.00 km and swims back to the starting point. (a) If the student can swim at a speed of 1.20 m/s in still water, how long does the trip take? (b) How much time is required in still water for the same length swim? (c) Intuitively, why does the swim take longer when there is a current?

31. A river flows with a steady speed v. A student swims upstream a distance d and then back to the starting point. The student can swim at speed c in still water. (a) In terms of d, v, and c, what time interval is required for the round trip? (b) What time interval would be required if the water were still? (c) Which time interval is larger? Explain whether it is always larger.

32. You are participating in a summer internship with the Coast Guard. You have been assigned the duty of determining the direction in which a Coast Guard speedboat should travel to intercept unidentified vessels. One day, the radar operator detects an unidentified vessel at a distance of 20.0 km from the radar installation in the direction 15.0° east of north. The vessel is traveling at 26.0 km/h on a course at 40.0° east of north. The Coast Guard wishes to send a speedboat, which travels at 50.0 km/h, to travel in a straight line from the radar installation to intercept and investigate the vessel, and asks you for the heading for the speedboat to take. Express the direction as a compass bearing with respect to due north.

33. A farm truck moves due east with a constant velocity of 9.50 m/s on a limitless, horizontal stretch of road. A boy riding on the back of the truck throws a can of soda upward (Fig. P4.33) and

Figure P4.33

catches the projectile at the same location on the truck bed, but 16.0 m farther down the road. (a) In the frame of reference of the truck, at what angle to the vertical does the boy throw the can? (b) What is the initial speed of the can relative to the truck? (c) What is the shape of the can's trajectory as seen by the boy?

An observer on the ground watches the boy throw the can and catch it. In this observer's frame of reference, (d) describe the shape of the can's path and (e) determine the initial velocity of the can.

ADDITIONAL PROBLEMS

34. A ball on the end of a string is whirled around in a horizontal circle of radius 0.300 m. The plane of the circle is 1.20 m above the ground. The string breaks and the ball lands 2.00 m (horizontally) away from the point on the ground directly beneath the ball's location when the string breaks. Find the radial acceleration of the ball during its circular motion.

35. *Why is the following situation impossible?* A normally proportioned adult walks briskly along a straight line in the $+x$ direction, standing straight up and holding his right arm vertical and next to his body so that the arm does not swing. His right hand holds a ball at his side a distance h above the floor. When the ball passes above a point marked as $x = 0$ on the horizontal floor, he opens his fingers to release the ball from rest relative to his hand. The ball strikes the ground for the first time at position $x = 7.00h$.

36. A particle starts from the origin with velocity $5\hat{\mathbf{i}}$ m/s at $t = 0$ and moves in the xy plane with a varying acceleration given by $\vec{\mathbf{a}} = (6\sqrt{t}\,\hat{\mathbf{j}})$, where $\vec{\mathbf{a}}$ is in meters per second squared and t is in seconds. (a) Determine the velocity of the particle as a function of time. (b) Determine the position of the particle as a function of time.

37. Lisa in her Lamborghini accelerates at $(3.00\hat{\mathbf{i}} - 2.00\hat{\mathbf{j}})$ m/s², while Jill in her Jaguar accelerates at $(1.00\hat{\mathbf{i}} + 3.00\hat{\mathbf{j}})$ m/s². They both start from rest at the origin. After 5.00 s, (a) what is Lisa's speed with respect to Jill, (b) how far apart are they, and (c) what is Lisa's acceleration relative to Jill?

38. A boy throws a stone horizontally from the top of a cliff of height h toward the ocean below. The stone strikes the ocean at distance d from the base of the cliff. In terms of h, d, and g, find expressions for (a) the time t at which the stone lands in the ocean, (b) the initial speed of the stone, (c) the speed of the stone immediately before it reaches the ocean, and (d) the direction of the stone's velocity immediately before it reaches the ocean.

39. *Why is the following situation impossible?* Albert Pujols hits a home run so that the baseball just clears the top row of bleachers, 24.0 m high, located 130 m from home plate. The ball is hit at 41.7 m/s at an angle of 35.0° to the horizontal, and air resistance is negligible.

40. As some molten metal splashes, one droplet flies off to the east with initial velocity v_i at angle θ_i above the horizontal, and another droplet flies off to the west with the same speed at the same angle above the horizontal as shown in Figure P4.40. In terms of v_i and θ_i, find the distance between the two droplets as a function of time.

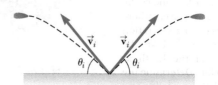

Figure P4.40

41. An astronaut on the surface of the Moon fires a cannon to launch an experiment package, which leaves the barrel moving horizontally. Assume the free-fall acceleration on the Moon is one-sixth of that on the Earth. (a) What must the muzzle speed of the package be so that it travels completely around the Moon and returns to its original location? (b) What time interval does this trip around the Moon require?

42. A pendulum with a cord of length $r = 1.00$ m swings in a vertical plane (Fig. P4.42). When the pendulum is in the two horizontal positions $\theta = 90.0°$ and $\theta = 270°$, its speed is 5.00 m/s. Find the magnitude of (a) the radial acceleration and (b) the tangential acceleration for these positions. (c) Draw vector diagrams to determine the direction of the total acceleration for these two positions. (d) Calculate the magnitude and direction of the total acceleration at these two positions.

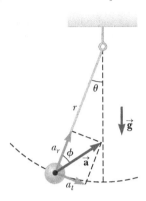

Figure P4.42

43. A spring cannon is located at the edge of a table that is 1.20 m above the floor. A steel ball is launched from the cannon with speed v_i at 35.0° above the horizontal. (a) Find the horizontal position of the ball as a function of v_i at the instant it lands on the floor. We write this function as $x(v_i)$. Evaluate x for (b) $v_i = 0.100$ m/s and for (c) $v_i = 100$ m/s. (d) Assume v_i is close to but not equal to zero. Show that one term in the answer to part (a) dominates so that the function $x(v_i)$ reduces to a simpler form. (e) If v_i is very large, what is the approximate form of $x(v_i)$? (f) Describe the overall shape of the graph of the function $x(v_i)$.

44. A projectile is launched from the point $(x = 0, y = 0)$, with velocity $(12.0\hat{\mathbf{i}} + 49.0\hat{\mathbf{j}})$ m/s, at $t = 0$. (a) Make a table listing the projectile's distance $|\vec{\mathbf{r}}|$ from the origin at the end of each second thereafter, for $0 \le t \le 10$ s. Tabulating the x and y coordinates and the components of velocity v_x and v_y will also be useful. (b) Notice that the projectile's distance from its starting point increases with time, goes through a maximum, and starts to decrease. Prove that the distance is a maximum when the position vector is perpendicular to the velocity. *Suggestion:* Argue that if $\vec{\mathbf{v}}$ is not perpendicular to $\vec{\mathbf{r}}$, then $|\vec{\mathbf{r}}|$ must be increasing or decreasing. (c) Determine the magnitude of the maximum displacement. (d) Explain your method for solving part (c).

45. A fisherman sets out upstream on a river. His small boat, powered by an outboard motor, travels at a constant speed v in still water. The water flows at a lower constant speed v_w. The fisherman has traveled upstream for 2.00 km when his ice chest falls out of the boat. He notices that the chest is missing only after he has gone upstream for another 15.0 min. At that point, he turns around and heads back downstream, all the time traveling at the same speed relative to the water. He catches up with the floating ice chest just as he returns to his starting point. How fast is the river flowing? Solve this problem in two ways. (a) First, use the Earth as a reference frame. With respect to the Earth, the boat travels upstream at speed $v - v_w$ and downstream at $v + v_w$. (b) A second much simpler and more elegant solution is obtained by using the water as the reference frame. This approach has important applications in many more complicated problems; examples are calculating the motion of rockets and satellites and analyzing the scattering of subatomic particles from massive targets.

46. An outfielder throws a baseball to his catcher in an attempt to throw out a runner at home plate. The ball bounces once before reaching the catcher. Assume the angle at which the bounced ball leaves the ground is the same as the angle at which the outfielder threw it as shown in Figure P4.46, but that the ball's speed after the bounce is one-half of what it was before the bounce. (a) Assume the ball is always thrown with the same initial speed and ignore air resistance. At what angle θ should the fielder throw the ball to make it go the same distance D with one bounce (blue path) as a ball thrown upward at 45.0° with no bounce (green path)? (b) Determine the ratio of the time interval for the one-bounce throw to the flight time for the no-bounce throw.

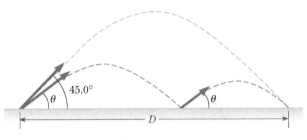

Figure P4.46

47. Do not hurt yourself; do not strike your hand against anything. Within these limitations, describe what you do to give your hand a large acceleration. Compute an order-of-magnitude estimate of this acceleration, stating the quantities you measure or estimate and their values.

48. You are on the Pirates of the Caribbean attraction in the Magic Kingdom at Disney World. Your boat rides through a pirate battle, in which cannons on a ship and in a fort are firing at each other. While you are aware that the splashes in the water do not represent actual cannonballs, you begin to wonder about such battles in the days of the pirates. Suppose the fort and the ship are separated by 75.0 m. You see that the cannons in the fort are aimed so that their cannonballs would be fired horizontally from a height of 7.00 m above the water. (a) You wonder at what speed they must be fired in order to hit the ship before falling in the water. (b) Then, you think about the sludge that must build up inside the barrel of a cannon. This sludge should slow down the cannonballs. A question occurs in your mind: if the cannonballs can be fired at only 50.0% of the speed found earlier, is it possible to fire them upward at some angle to the horizontal so that they would reach the ship?

CHALLENGE PROBLEMS

49. A skier leaves the ramp of a ski jump with a velocity of $v = 10.0$ m/s at $\theta = 15.0°$ above the horizontal as shown in Figure P4.49 (page 94). The slope where she will land is inclined downward at $\phi = 50.0°$, and air resistance is negligible. Find (a) the distance from the end of the ramp to where the jumper lands and (b) her velocity components just before the landing. (c) Explain how you think the results might be affected if air resistance were included.

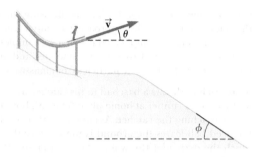

Figure P4.49

50. A projectile is fired up an incline (incline angle ϕ) with an initial speed v_i at an angle θ_i with respect to the horizontal ($\theta_i > \phi$) as shown in Figure P4.50. (a) Show that the projectile travels a distance d up the incline, where

$$d = \frac{2v_i^2 \cos\theta_i \sin(\theta_i - \phi)}{g \cos^2 \phi}$$

(b) For what value of θ_i is d a maximum, and what is that maximum value?

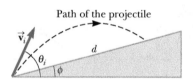

Path of the projectile

Figure P4.50

51. Two swimmers, Chris and Sarah, start together at the same point on the bank of a wide stream that flows with a speed v. Both move at the same speed c (where $c > v$) relative to the water. Chris swims downstream a distance L and then upstream the same distance. Sarah swims so that her motion relative to the Earth is perpendicular to the banks of the stream. She swims the distance L and then back the same distance, with both swimmers returning to the starting point. In terms of L, c, and v, find the time intervals required (a) for Chris's round trip and (b) for Sarah's round trip. (c) Explain which swimmer returns first.

52. In the What If? section of Example 4.5, it was claimed that the maximum range of a ski jumper occurs for a launch angle θ given by

$$\theta = 45° - \frac{\phi}{2}$$

where ϕ is the angle the hill makes with the horizontal in Figure 4.15. Prove this claim by deriving the equation above.

53. A fireworks rocket explodes at height h, the peak of its vertical trajectory. It throws out burning fragments in all directions, but all at the same speed v. Pellets of solidified metal fall to the ground without air resistance. Find the smallest angle that the final velocity of an impacting fragment makes with the horizontal.

The Laws of Motion

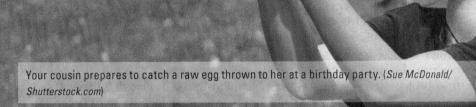

Your cousin prepares to catch a raw egg thrown to her at a birthday party. (*Sue McDonald/ Shutterstock.com*)

STORYLINE **You have returned home from your trip to Catalina** Island in the previous two chapters. Your family is having a picnic to celebrate a birthday, so there are many people in your backyard on a beautiful day. Someone suggests an egg toss contest. You decide to offer some advice to your cousin and instruct her to move her hands backward just as she catches the egg. Your cousin looks you in the eye and says, "Why?" You are tempted to say, "Because that's just how you do it," but then consider the deeper implications of your cousin's question. Why is it that you move your hands backward? What happens if you hold your hands in a fixed position and catch the egg? Should you have your cousin try this? You take your cousin to the computer and have her search for YouTube videos involving catching an egg, and then you notice some videos showing the results of throwing an egg into a vertical sheet. As you and your cousin watch these videos, both of you begin to understand the physics of throwing and catching eggs.

CONNECTIONS In the previous chapters, we learned how to describe the motion of particles and objects that can be modeled as particles. We saw motion changing in various ways. The acceleration of a car is a change in its velocity. The direction of the velocity of a thrown baseball changes as it flies through the air. We can *describe* these changes with the material in the previous chapters, but what *causes* these changes? Such a question represents a transition from kinematics, the description of motion, to *dynamics*, the study of causes of changes in motion. We will see that *force* is the cause of changes in motion, and will study the effects of force through the laws of motion as handed down to us by Isaac Newton. The notion of force will be used again and again in future chapters: gravitational forces in Chapter 13, electric forces in Chapter 22, magnetic forces in Chapter 28, nuclear forces in Chapter 43, and more.

5.1 The Concept of Force

Everyone has a basic understanding of the concept of force from everyday experience. When you push your empty dinner plate away, you exert a force on it. Similarly, you exert a force on a ball when you throw or kick it. In these examples, the word *force* refers to an interaction with an object by means of muscular activity and some change in the object's velocity. Forces do not always cause motion, however. For example, when you are sitting, a gravitational force acts on your body and yet you remain stationary. As a second example, you can push (in other words, exert a force) on a large boulder and not be able to move it.

What force (if any) causes the Moon to orbit the Earth? Newton answered this and related questions by stating that forces are what cause any change in the velocity of an object. The Moon's velocity changes in direction as it moves in a nearly circular orbit around the Earth. This change in velocity is caused by the gravitational force exerted by the Earth on the Moon.

When a coiled spring is pulled, as in Figure 5.1a, the spring stretches. When a stationary cart is pulled, as in Figure 5.1b, the cart moves. When a football is kicked, as in Figure 5.1c, it is both deformed and set in motion. These situations are all examples of a class of forces called *contact forces*. That is, they involve physical contact between two objects. Other examples of contact forces are the force exerted by gas molecules on the walls of a container and the force exerted by your feet on the floor.

Another class of forces, known as *field forces*, does not involve physical contact between two objects. These forces act through empty space. The gravitational force of attraction between two objects with mass, illustrated in Figure 5.1d, is an example of this class of force. The gravitational force keeps objects bound to the Earth and the planets in orbit around the Sun. Another common field force is the electric force that one electric charge exerts on another (Fig. 5.1e), such as the attractive electric force between an electron and a proton that form a hydrogen atom. A third example of a field force is the force a bar magnet exerts on a piece of iron (Fig. 5.1f).

The distinction between contact forces and field forces is not as sharp as you may have been led to believe by the previous discussion. When examined at the atomic level, all the forces we classify as contact forces turn out to be caused by electric (field) forces of the type illustrated in Figure 5.1e. Nevertheless, in developing models for macroscopic phenomena, it is convenient to use both classifications of forces. The only known *fundamental* forces in nature are all field forces: (1) *gravitational forces* between objects, (2) *electromagnetic forces* between electric charges, (3) *strong forces* between subatomic particles, and (4) *weak forces* that arise in certain radioactive decay processes. In classical physics, we are concerned only with gravitational and electromagnetic forces. We will discuss strong and weak forces in Chapter 44.

Isaac Newton
English physicist and mathematician
(1642–1727)
Isaac Newton was one of the most brilliant scientists in history. Before the age of 30, he formulated the basic concepts and laws of mechanics, discovered the law of universal gravitation, and invented the mathematical methods of calculus. As a consequence of his theories, Newton was able to explain the motions of the planets, the ebb and flow of the tides, and many special features of the motions of the Moon and the Earth. He also interpreted many fundamental observations concerning the nature of light. His contributions to physical theories dominated scientific thought for two centuries and remain important today.

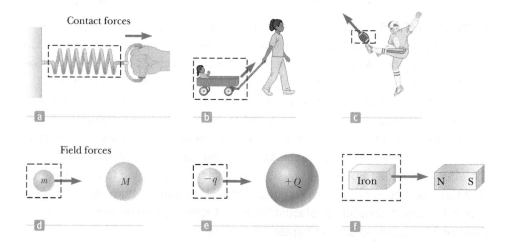

Figure 5.1 Some examples of applied forces. In each case, a force is exerted on the object within the boxed area. Some agent in the environment external to the boxed area exerts a force on the object.

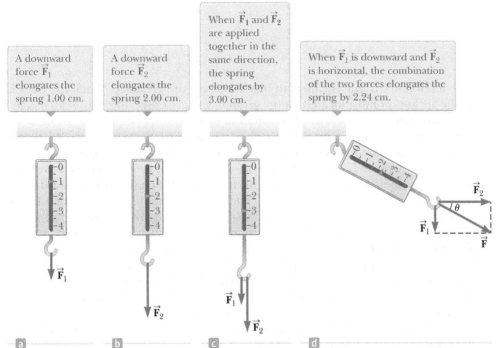

When $\vec{F}_1$ and $\vec{F}_2$ are applied together in the same direction, the spring elongates by 3.00 cm.

A downward force $\vec{F}_1$ elongates the spring 1.00 cm.

A downward force $\vec{F}_2$ elongates the spring 2.00 cm.

When $\vec{F}_1$ is downward and $\vec{F}_2$ is horizontal, the combination of the two forces elongates the spring by 2.24 cm.

Figure 5.2 The vector nature of a force is tested with a spring scale.

The Vector Nature of Force

It is possible to use the deformation of a spring to measure force. Suppose a vertical force is applied to a spring scale that has a fixed upper end as shown in Figure 5.2a. The spring elongates when the force is applied, and a pointer on the scale reads the extension of the spring. We can calibrate the spring by defining a reference force $\vec{F}_1$ as the force that produces a pointer reading of 1.00 cm. If we now apply a different downward force $\vec{F}_2$ whose magnitude is twice that of the reference force $\vec{F}_1$ as seen in Figure 5.2b, the pointer moves to 2.00 cm. Figure 5.2c shows that the combined effect of the two collinear forces is the sum of the effects of the individual forces.

Now suppose the two forces are applied simultaneously with $\vec{F}_1$ downward and $\vec{F}_2$ horizontal as illustrated in Figure 5.2d. In this case, the pointer reads 2.24 cm. The single force $\vec{F}$ that would produce this same reading is the sum of the two vectors $\vec{F}_1$ and $\vec{F}_2$ as described in Figure 5.2d. That is, $|\vec{F}_1| = \sqrt{F_1^2 + F_2^2} = 2.24$ units, and its direction is $\theta = \tan^{-1}(-0.500) = -26.6°$. Because forces have been experimentally verified to behave as vectors, you *must* use the rules of vector addition to obtain the net force on an object.

5.2 Newton's First Law and Inertial Frames

We begin our study of forces by imagining some physical situations involving a puck on a perfectly level air hockey table (Fig. 5.3). You expect that the puck will remain stationary when it is placed gently at rest on the table. Now imagine your air hockey table is located on a train moving with constant velocity along a perfectly smooth track. If the puck is placed on the table, the puck again remains where it is placed. If the train were to accelerate, however, the puck would start moving along the table opposite the direction of the train's acceleration, just as a set of papers on your dashboard falls onto the floor of your car when you step on the accelerator.

As we saw in Section 4.6, a moving object can be observed from any number of reference frames. **Newton's first law of motion,** sometimes called the *law of inertia,* defines a special set of reference frames called *inertial frames.* This law can be stated

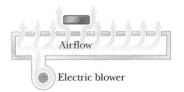

Airflow

Electric blower

Figure 5.3 On an air hockey table, air blown through holes in the surface supports the puck and allows it to move almost without friction across the table. If the table is not accelerating, a puck placed on the table will remain at rest.

in a theoretical manner as follows:

A theoretical statement ▶
of Newton's first law

If an object does not interact with other objects, it is possible to identify a reference frame in which the object has zero acceleration.

Inertial frame of reference ▶

Such a reference frame is called an **inertial frame of reference.** When the puck is on the air hockey table located on the ground, you are observing it from an inertial reference frame; there are no horizontal interactions of the puck with any other objects, and you observe it to have zero acceleration in that direction. When you are on the train moving at constant velocity, you are also observing the puck from an inertial reference frame. Any reference frame that moves with constant velocity relative to an inertial frame is itself an inertial frame. When you and the train accelerate, however, you are observing the puck from a **noninertial reference frame** because you and the train are accelerating relative to the inertial reference frame of the Earth's surface. While the puck appears to be accelerating according to your observations, a reference frame can be identified in which the puck has zero acceleration. For example, an observer standing outside the train on the ground sees the puck sliding relative to the table but always moving with the same velocity with respect to the ground as the train had before it started to accelerate (because there is almost no friction to "tie" the puck and the train together). Therefore, Newton's first law is still satisfied even though your observations as a rider on the train show an apparent acceleration relative to you.

A reference frame that moves with constant velocity relative to the distant stars is the best approximation of an inertial frame, and for our purposes we can consider the Earth as being such a frame. The Earth is not really an inertial frame because of its orbital motion around the Sun and its rotational motion about its own axis, both of which involve centripetal accelerations. These accelerations are small compared with g, however, and can often be neglected. For this reason, we model the Earth as an inertial frame, along with any other frame attached to it.

Let us assume we are observing an object from an inertial reference frame. (We will return to observations made in noninertial reference frames in Section 6.3.) Before about 1600, scientists believed that the natural state of matter was the state of rest. Observations showed that moving objects eventually stopped moving. Galileo was the first to take a different approach to motion and the natural state of matter. He devised thought experiments and concluded that it is not the nature of an object to stop once set in motion: rather, it is its nature to *resist changes in its motion*. In his words, "Any velocity once imparted to a moving body will be rigidly maintained as long as the external causes of retardation are removed." For example, a spacecraft drifting through empty space with its engine turned off will keep moving forever. It would *not* seek a "natural state" of rest.

Given our discussion of observations made from inertial reference frames, we can pose a more practical statement of Newton's first law of motion than that in the previous screened statement:

A more practical statement ▶
of Newton's first law

In the absence of external forces and when viewed from an inertial reference frame, an object at rest remains at rest and an object in motion continues in motion with a constant velocity (that is, with a constant speed in a straight line).

In other words, **when no force acts on an object, the acceleration of the object is zero.** From the first law, we conclude that any *isolated object* (one that does not interact with its environment) is either at rest and stays at rest, or is moving with constant velocity. The tendency of an object to resist any attempt to change its velocity is called **inertia.** Given the statement of the first law above, we can conclude that an object that is accelerating must be experiencing a force. In turn, from the first law, Definition of force ▶ we can define **force** as **that which causes a change in motion of an object.**

PITFALL PREVENTION 5.1
Newton's First Law Newton's first law does *not* say what happens for an object with *zero net force*, that is, multiple forces that cancel; it says what happens *in the absence of external forces*. This subtle but important difference allows us to define force as that which causes a change in the motion. The description of an object under the effect of forces that balance is covered by Newton's second law.

QUICK QUIZ 5.1 Which of the following statements is correct? **(a)** It is possible for an object to have motion in the absence of forces on the object. **(b)** It is possible to have forces on an object in the absence of motion of the object. **(c)** Neither statement **(a)** nor statement **(b)** is correct. **(d)** Both statements **(a)** and **(b)** are correct.

5.3 Mass

Imagine playing catch with either a basketball or a bowling ball. Which ball is more likely to keep moving when you try to catch it? Which ball requires more effort to throw it? The bowling ball requires more effort. In the language of physics, we say that the bowling ball is more resistant to changes in its velocity than the basketball. How can we quantify this concept?

Mass is that property of an object that specifies how much resistance an object exhibits to changes in its velocity, and as we learned in Section 1.1, the SI unit of mass is the kilogram. Experiments show that the greater the mass of an object, the less that object accelerates under the action of a given applied force.

◀ Definition of mass

To describe mass quantitatively, we conduct experiments in which we compare the accelerations a given force produces on different objects. Suppose a force acting on an object of mass m_1 produces a change in motion of the object that we can quantify with the object's acceleration $\vec{a}_1$, and the *same force* acting on an object of mass m_2 produces an acceleration $\vec{a}_2$. The ratio of the two masses is defined as the *inverse* ratio of the magnitudes of the accelerations produced by the force:

$$\frac{m_1}{m_2} \equiv \frac{a_2}{a_1} \tag{5.1}$$

For example, if a given force acting on a 3-kg object produces an acceleration of 4 m/s^2, the same force applied to an object with twice the mass, 6 kg, produces an acceleration with half the magnitude, 2 m/s^2. According to a huge number of similar observations, we conclude that the magnitude of the acceleration of an object is inversely proportional to its mass when acted on by a given force. If one object has a known mass, the mass of the other object can be obtained from acceleration measurements.

As mentioned in Chapter 1, mass is an inherent property of an object and is independent of the object's surroundings and of the method used to measure it. Also, mass is a scalar quantity and thus obeys the rules of ordinary arithmetic. For example, if you combine a 3-kg mass with a 5-kg mass, the total mass is 8 kg. This result can be verified experimentally by comparing the acceleration that a known force gives to several objects separately with the acceleration that the same force gives to the same objects combined as one unit.

Mass should not be confused with weight. Mass and weight are two different quantities. The weight of an object is equal to the magnitude of the gravitational force exerted on the object and varies with location (see Section 5.5). For example, a person weighing 180 lb on the Earth weighs only about 30 lb on the Moon. On the other hand, the mass of an object is the same everywhere: an object having a mass of 2 kg on the Earth also has a mass of 2 kg on the Moon.

◀ Mass and weight are different quantities

5.4 Newton's Second Law

Newton's first law explains what happens to an object when *no* forces act on it: it maintains its original motion; it either remains at rest or moves in a straight line with constant speed. Newton's second law answers the question of what happens to an object when one or more forces act on it.

Imagine performing an experiment in which you push a block of mass m across a frictionless, horizontal surface. When you exert some horizontal force $\vec{F}$ on the block, it moves with some acceleration $\vec{a}$. If you apply a force twice as great on the same block, experimental results show that the acceleration of the block doubles;

PITFALL PREVENTION 5.2
Force Is the Cause of Changes in Motion An object can have motion in the absence of forces as described in Newton's first law. Therefore, don't interpret force as the cause of *motion*. Force is the cause of *changes in motion*.

if you increase the applied force to $3\vec{\mathbf{F}}$, the acceleration triples; and so on. From such observations, we conclude that the acceleration of an object is directly proportional to the force acting on it: $\vec{\mathbf{F}} \propto \vec{\mathbf{a}}$. This idea was first introduced in Section 2.4 when we discussed the direction of the acceleration of an object. We also know from Equation 5.1 that the magnitude of the acceleration of an object is inversely proportional to its mass: $|\vec{\mathbf{a}}| \propto 1/m$.

These experimental observations are summarized in **Newton's second law:**

> When viewed from an inertial reference frame, the acceleration of an object is directly proportional to the net force acting on it and inversely proportional to its mass:

$$\vec{\mathbf{a}} \propto \frac{\sum \vec{\mathbf{F}}}{m}$$

PITFALL PREVENTION 5.3

$m\vec{\mathbf{a}}$ **Is Not a Force** Equation 5.2 does *not* say that the product $m\vec{\mathbf{a}}$ is a force. All forces on an object are added vectorially to generate the net force on the left side of the equation. This net force is then equated to the product of the mass of the object and the acceleration that results from the net force. Do *not* include an "$m\vec{\mathbf{a}}$ force" in your analysis of the forces on an object.

If we choose a proportionality constant of 1, we can relate mass, acceleration, and force through the following mathematical statement of Newton's second law:[1]

◀ Newton's second law

$$\sum \vec{\mathbf{F}} = m\vec{\mathbf{a}} \tag{5.2}$$

In both the textual and mathematical statements of Newton's second law, we have indicated that the acceleration is due to the *net force* $\sum \vec{\mathbf{F}}$ acting on an object. The **net force** on an object is the vector sum of all forces acting on the object. (Other names used for the net force include the *total force*, the *resultant force*, and the *unbalanced force*.) In solving a problem using Newton's second law, it is imperative to determine the correct net force on an object. Many forces may be acting on an object, but there is only one acceleration of the object.

Equation 5.2 is a vector expression and hence is equivalent to three component equations:

◀ Newton's second law: component form

$$\sum F_x = ma_x \qquad \sum F_y = ma_y \qquad \sum F_z = ma_z \tag{5.3}$$

QUICK QUIZ 5.2 An object experiences no acceleration. Which of the following *cannot* be true for the object? **(a)** A single force acts on the object. **(b)** No forces act on the object. **(c)** Forces act on the object, but the forces cancel.

QUICK QUIZ 5.3 You push an object, initially at rest, across a frictionless floor with a constant force for a time interval Δt, resulting in a final speed of v for the object. You then repeat the experiment, but with a force that is twice as large. What time interval is now required to reach the same final speed v? **(a)** $4\,\Delta t$ **(b)** $2\,\Delta t$ **(c)** Δt **(d)** $\Delta t/2$ **(e)** $\Delta t/4$

The SI unit of force is the **newton** (N). A force of 1 N is the force that, when acting on an object of mass 1 kg, produces an acceleration of 1 m/s². From this definition and Newton's second law, we see that the newton can be expressed in terms of the following fundamental units of mass, length, and time:

◀ Definition of the newton

$$1\ \text{N} \equiv 1\ \text{kg} \cdot \text{m/s}^2 \tag{5.4}$$

In the U.S. customary system, the unit of force is the **pound** (lb). A force of 1 lb is the force that, when acting on a 1-slug mass,[2] produces an acceleration of 1 ft/s²:

$$1\ \text{lb} \equiv 1\ \text{slug} \cdot \text{ft/s}^2$$

A convenient approximation is $1\ \text{N} \approx \frac{1}{4}\ \text{lb}$.

---

[1]Equation 5.2 is valid only when the speed of the object is much less than the speed of light. We treat the relativistic situation in Chapter 38.

[2]The *slug* is the unit of mass in the U.S. customary system and is that system's counterpart of the SI unit the *kilogram*. Because most of the calculations in our study of classical mechanics are in SI units, the slug is seldom used in this text.

Why do you move your hands backward when you catch the egg in the opening storyline? Imagine holding your hands stiffly and not moving them as you catch the egg. Then the egg will hit your hand and be brought to rest in a very short time interval. As a result, the magnitude of the acceleration of the egg will be large. According to Equation 5.2, this will require a large force from your hands. This large force is sufficient to break the shell of the egg. If you move your hands backward, however, and slowly bring the egg to rest, the acceleration is of a much smaller magnitude. This, in turn, requires a much smaller force, which can keep the shell of the egg intact.

Throwing the egg into the sheet is similar: when the egg strikes the sheet, the sheet moves in the same direction in response, bringing the egg to a lower velocity over a relatively long distance.

Example **5.1** **An Accelerating Hockey Puck**

A hockey puck having a mass of 0.30 kg slides on the frictionless, horizontal surface of an ice rink. Two hockey sticks strike the puck simultaneously, exerting the forces on the puck shown in Figure 5.4. The force $\vec{F}_1$ has a magnitude of 5.0 N, and is directed at $\theta = 20°$ below the x axis. The force $\vec{F}_2$ has a magnitude of 8.0 N and its direction is $\phi = 60°$ above the x axis. Determine both the magnitude and the direction of the puck's acceleration.

S O L U T I O N

Conceptualize Study Figure 5.4. Using your expertise in vector addition from Chapter 3, predict the approximate direction of the net force vector on the puck. The acceleration of the puck will be in the same direction.

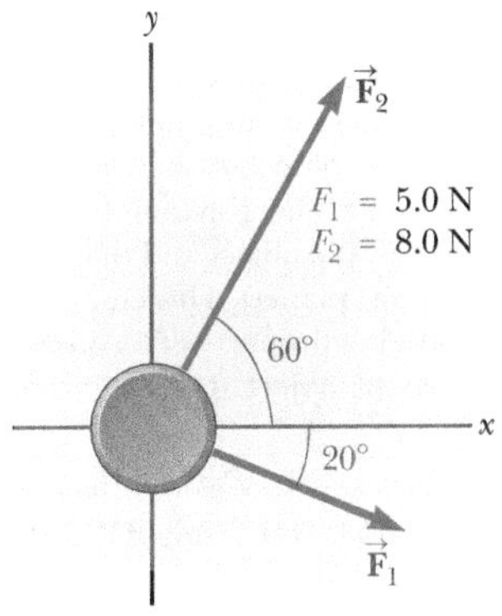

Figure 5.4 (Example 5.1) A hockey puck moving on a frictionless surface is subject to two forces $\vec{F}_1$ and $\vec{F}_2$.

Categorize Because we can determine a net force and we want an acceleration, this problem is categorized as one that may be solved using Newton's second law. In Section 5.7, we will formally introduce the *particle under a net force* analysis model to describe a situation such as this one.

Analyze Find the component of the net force acting on the puck in the x direction:	$\sum F_x = F_{1x} + F_{2x} = F_1 \cos \theta + F_2 \cos \phi$
Find the component of the net force acting on the puck in the y direction:	$\sum F_y = F_{1y} + F_{2y} = F_1 \sin \theta + F_2 \sin \phi$
Use Newton's second law in component form (Eq. 5.3) to find the x and y components of the puck's acceleration:	$a_x = \dfrac{\sum F_x}{m} = \dfrac{F_1 \cos \theta + F_2 \cos \phi}{m}$ $a_y = \dfrac{\sum F_y}{m} = \dfrac{F_1 \sin \theta + F_2 \sin \phi}{m}$
Substitute numerical values:	$a_x = \dfrac{(5.0 \text{ N}) \cos(-20°) + (8.0 \text{ N}) \cos(60°)}{0.30 \text{ kg}} = 29 \text{ m/s}^2$ $a_y = \dfrac{(5.0 \text{ N}) \sin(-20°) + (8.0 \text{ N}) \sin(60°)}{0.30 \text{ kg}} = 17 \text{ m/s}^2$
Find the magnitude of the acceleration:	$a = \sqrt{(29 \text{ m/s}^2)^2 + (17 \text{ m/s}^2)^2} = 34 \text{ m/s}^2$
Find the direction of the acceleration relative to the positive x axis:	$\theta = \tan^{-1}\left(\dfrac{a_y}{a_x}\right) = \tan^{-1}\left(\dfrac{17}{29}\right) = 31°$

Finalize The vectors in Figure 5.4 can be added graphically to check the reasonableness of our answer. Because the acceleration vector is along the direction of the resultant force, a drawing showing the resultant force vector helps us check the validity of the answer. (Try it!)

continued

5.1 continued

WHAT IF? Suppose three hockey sticks strike the puck simultaneously, with two of them exerting the forces shown in Figure 5.4. The result of the three forces is that the hockey puck shows *no* acceleration. What must be the components of the third force?

Answer If there is zero acceleration, the net force acting on the puck must be zero. Therefore, the three forces must cancel. The components of the third force must be of equal magnitude and opposite sign compared to the components of the net force applied by the first two forces so that all the components add to zero. Therefore, $F_{3x} = -\sum F_x = -(0.30 \text{ kg})(29 \text{ m/s}^2) = -8.7 \text{ N}$ and $F_{3y} = -\sum F_y = -(0.30 \text{ kg})(17 \text{ m/s}^2) = -5.2 \text{ N}$.

5.5 The Gravitational Force and Weight

All objects are attracted to the Earth. The attractive force exerted by the Earth on an object is called the **gravitational force** $\vec{F}_g$. This force is directed toward the center of the Earth,[3] and its magnitude is called the **weight** of the object.

We saw in Section 2.6 that a freely falling object experiences an acceleration $\vec{g}$ acting toward the center of the Earth. Applying Newton's second law $\sum \vec{F} = m\vec{a}$ to a freely falling object of mass m, with $\vec{a} = \vec{g}$ and $\sum \vec{F} = \vec{F}_g$, gives

$$\vec{F}_g = m\vec{g} \tag{5.5}$$

Therefore, the weight of an object, being defined as the magnitude of $\vec{F}_g$, is given by

$$F_g = mg \tag{5.6}$$

Because it depends on g, weight varies with geographic location. Because g decreases with increasing distance from the center of the Earth, objects weigh less at higher altitudes than at sea level. For example, a 1 000-kg pallet of bricks used in the construction of the Empire State Building in New York City weighed 9 800 N at street level, but weighed about 1 N less by the time it was lifted from sidewalk level to the top of the building. As another example, suppose a student has a mass of 70.0 kg. The student's weight in a location where $g = 9.80$ m/s² is 686 N (about 150 lb). At the top of a mountain, however, where $g = 9.77$ m/s², the student's weight is only 684 N. Therefore, if you want to lose weight without going on a diet, climb a mountain or weigh yourself at 30 000 ft during an airplane flight!

Equation 5.6 indicates that there is a clear difference between mass and weight. The life-support unit strapped to the back of astronaut Harrison Schmitt in Figure 5.5 weighed 300 lb on the Earth and had a mass of 136 kg. During his training, a 50-lb mockup with a mass of 23 kg was used. Although this strategy effectively simulated the reduced *weight* the unit would have on the Moon, it did not correctly mimic the unchanging *mass*. It was more difficult to accelerate the 136-kg unit (perhaps by jumping or twisting suddenly) on the Moon than it was to accelerate the 23-kg unit on the Earth.

Equation 5.6 quantifies the gravitational force on the object, but notice that this equation does not require the object to be moving. Even for a stationary object or for an object on which several forces act, Equation 5.6 can be used to calculate the magnitude of the gravitational force. The result is a subtle shift in the interpretation of m in the equation. The mass m in Equation 5.6 determines the strength of the gravitational attraction between the object and the Earth. This role is completely different from that previously described for mass, that of measuring the resistance to changes in motion in response to an external force. In that role, mass is also called **inertial mass.** We call m in Equation 5.6 the **gravitational mass.** Even though this quantity is different in behavior from inertial mass, it is one of the

PITFALL PREVENTION 5.4

"Weight of an Object" We are familiar with the everyday phrase, the "weight of an object." Weight, however, is not an inherent property of an object; rather, it is a measure of the gravitational force between the object and the Earth (or other planet). Therefore, weight is a property of a *system* of items: the object and the Earth.

PITFALL PREVENTION 5.5

Kilogram Is Not a Unit of Weight You may have seen the "conversion" 1 kg = 2.2 lb. Despite popular statements of weights expressed in kilograms, the kilogram is not a unit of *weight*, it is a unit of *mass*. The conversion statement is not an equality; it is an *equivalence* that is valid only on the Earth's surface. We first raised this issue in the Chapter 1 storyline.

Figure 5.5 Astronaut Harrison Schmitt carries a backpack on the Moon.

---
[3]This statement ignores that the mass distribution of the Earth is not perfectly spherical.

experimental conclusions in Newtonian dynamics that gravitational mass and inertial mass have the same value.

Although this discussion has focused on the gravitational force on an object due to the Earth, the concept is generally valid on any planet. The value of g will vary from one planet to the next, but the magnitude of the gravitational force will always be given by the value of mg.

QUICK QUIZ 5.4 Suppose you are talking by interplanetary telephone to a friend who lives on the Moon. He tells you that he has just won a newton of gold in a contest. Excitedly, you tell him that you entered the Earth version of the same contest and also won a newton of gold! Who is richer? **(a)** You are. **(b)** Your friend is. **(c)** You are equally rich.

Conceptual Example 5.2 How Much Do You Weigh in an Elevator?

You have most likely been in an elevator that accelerates upward as it moves toward a higher floor. In this case, you feel heavier. In fact, if you are standing on a bathroom scale at the time, the scale measures a force having a magnitude that is greater than your weight. Therefore, you have tactile and measured evidence that leads you to believe you are heavier in this situation. *Are* you heavier?

SOLUTION

No; your weight is unchanged. Your experiences are due to your being in a noninertial reference frame. To provide the acceleration upward, the floor or scale must exert on your feet an upward force that is greater in magnitude than your weight. It is this greater force you feel, which you interpret as feeling heavier. The scale reads the force with which it pushes up on you, not your weight (unless you are at rest), and so its reading increases. We will examine the effect of the acceleration of an elevator on apparent weight in Example 5.8.

5.6 Newton's Third Law

If you press against a corner of this textbook with your fingertip, the book pushes back and makes a small dent in your skin. If you push harder, the book does the same and the dent in your skin is a little larger. This simple activity illustrates that forces are *interactions* between two objects: when your finger pushes on the book, the book pushes back on your finger. This important principle is known as **Newton's third law:**

If two objects interact, the force $\vec{\mathbf{F}}_{12}$ exerted by object 1 on object 2 is equal in magnitude and opposite in direction to the force $\vec{\mathbf{F}}_{21}$ exerted by object 2 on object 1:

◀ Newton's third law

$$\vec{\mathbf{F}}_{12} = -\vec{\mathbf{F}}_{21} \qquad (5.7)$$

When it is important to designate forces as interactions between two objects, we will use this subscript notation, where $\vec{\mathbf{F}}_{ab}$ means "the force exerted *by* a *on* b." The third law is illustrated in Figure 5.6. The force that object 1 exerts on object 2 is popularly called the *action force*, and the force of object 2 on object 1 is called the *reaction force*. These italicized terms are not scientific terms; furthermore, either force can be labeled the action or reaction force. We will use these terms for convenience. In all cases, the action and reaction forces act on *different* objects and must be of the same type (gravitational, electrical, etc.). For example, the force acting on a freely falling projectile is the gravitational force exerted by the Earth on the projectile $\vec{\mathbf{F}}_g = \vec{\mathbf{F}}_{Ep}$ (E = Earth, p = projectile), and the magnitude of this force is mg. The reaction to this force is the gravitational force exerted by the projectile on the Earth $\vec{\mathbf{F}}_{pE} = -\vec{\mathbf{F}}_{Ep}$. The reaction force $\vec{\mathbf{F}}_{pE}$ must accelerate the Earth toward the projectile just as the action force $\vec{\mathbf{F}}_{Ep}$ accelerates the projectile toward

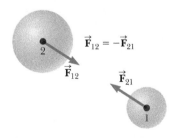

Figure 5.6 Newton's third law. The force $\vec{\mathbf{F}}_{12}$ exerted by object 1 on object 2 is equal in magnitude and opposite in direction to the force $\vec{\mathbf{F}}_{21}$ exerted by object 2 on object 1.

Figure 5.7 (a) When a computer monitor is at rest on a table, the forces acting on the monitor are the normal force $\vec{\mathbf{n}}$ and the gravitational force $\vec{\mathbf{F}}_g$. The reaction to $\vec{\mathbf{n}}$ is the force $\vec{\mathbf{F}}_{mt}$ exerted by the monitor on the table. The reaction to $\vec{\mathbf{F}}_g$ is the force $\vec{\mathbf{F}}_{mE}$ exerted by the monitor on the Earth. (b) A *force diagram* shows the forces on the monitor. (c) A *free-body diagram* shows the monitor as a black dot with the forces acting on it.

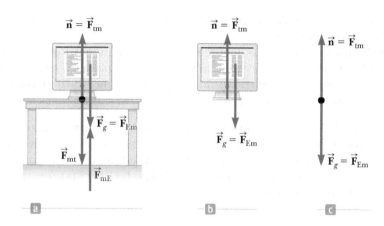

the Earth. Because the Earth has such a large mass, however, its acceleration due to this reaction force is negligibly small.

Consider a computer monitor at rest on a table as in Figure 5.7a. The gravitational force on the monitor is $\vec{\mathbf{F}}_g = \vec{\mathbf{F}}_{Em}$. The reaction to this force is the force $\vec{\mathbf{F}}_{mE} = -\vec{\mathbf{F}}_{Em}$ exerted by the monitor on the Earth. The monitor does not accelerate because it is held up by the table. The table exerts on the monitor an upward force $\vec{\mathbf{n}} = \vec{\mathbf{F}}_{tm}$, called the **normal force.** (*Normal* in this context means *perpendicular*.) In general, whenever an object is in contact with a surface, the surface exerts a normal force on the object. The normal force on the monitor can have any value needed, up to the point of breaking the table. Because the monitor has zero acceleration, Newton's second law applied to the monitor gives us $\sum \vec{\mathbf{F}} = \vec{\mathbf{n}} + m\vec{\mathbf{g}} = 0$, so $n\hat{\mathbf{j}} - mg\hat{\mathbf{j}} = 0$, or $n = mg$. The normal force balances the gravitational force on the monitor, so the net force on the monitor is zero. The reaction force to $\vec{\mathbf{n}}$ is the force exerted by the monitor downward on the table, $\vec{\mathbf{F}}_{mt} = -\vec{\mathbf{F}}_{tm} = -\vec{\mathbf{n}}$.

Notice that the forces acting on the monitor are $\vec{\mathbf{F}}_g$ and $\vec{\mathbf{n}}$ as shown in Figure 5.7b. The two forces $\vec{\mathbf{F}}_{mE}$ and $\vec{\mathbf{F}}_{mt}$ are exerted on objects other than the monitor.

Figure 5.7 illustrates an extremely important step in solving problems involving forces. Figure 5.7a shows many of the forces in the situation: those acting on the monitor, one acting on the table, and one acting on the Earth. Figure 5.7b, by contrast, shows only the forces acting on *one object,* the monitor, and is called a **force diagram** or a *diagram showing the forces on the object.* The important pictorial representation in Figure 5.7c is called a **free-body diagram.** In a free-body diagram, the particle model is used by representing the object as a dot and showing the forces that act on the object as being applied to the dot. When analyzing an object subject to forces, we are interested in the net force acting on one object, which we will model as a particle. Therefore, a free-body diagram helps us isolate only those forces on the object and eliminate the other forces from our analysis.

QUICK QUIZ 5.5 **(i)** If a fly collides with the windshield of a fast-moving bus, which experiences an impact force with a larger magnitude? (a) The fly. (b) The bus. (c) The same force is experienced by both. **(ii)** Which experiences the greater acceleration? (a) The fly. (b) The bus. (c) The same acceleration is experienced by both.

Conceptual Example 5.3   **You Push Me and I'll Push You**

A large man and a small boy stand facing each other on frictionless ice. They put their hands together and push against each other so that they move apart.

(A) Who moves away with the higher speed?

5.3 continued

This situation is similar to what we saw in Quick Quiz 5.5. According to Newton's third law, the force exerted by the man on the boy and the force exerted by the boy on the man are a third-law pair of forces, so they must be equal in magnitude. (A bathroom scale placed between their hands would read the same, regardless of which way it faced.) Therefore, the boy, having the smaller mass, experiences the greater acceleration. Both individuals accelerate for the same amount of time, but the greater acceleration of the boy over this time interval results in his moving away from the interaction with the higher speed.

(B) Who moves farther while their hands are in contact?

Because the boy has the greater acceleration and therefore the greater average velocity, he moves farther than the man during the time interval during which their hands are in contact.

5.7 Analysis Models Using Newton's Second Law

In this section, we discuss two analysis models for solving problems in which objects are either in equilibrium ($\vec{\mathbf{a}} = 0$) or accelerating under the action of constant external forces. Remember that when Newton's laws are applied to an object, we are interested only in external forces that act on the object. If the objects are modeled as particles, we need not worry about rotational motion such as spinning. For now, we also neglect the effects of friction in those problems involving motion, which is equivalent to stating that the surfaces are *frictionless*. (The friction force is discussed in Section 5.8.)

We usually neglect the mass of any ropes, strings, or cables involved. In this approximation, the magnitude of the force exerted by any element of the rope on the adjacent element is the same for all elements along the rope. In problem statements, the synonymous terms *light* and *of negligible mass* are used to indicate that a mass is to be ignored when you work the problems. When a rope attached to an object is pulling on the object, the rope exerts a force on the object in a direction away from the object, parallel to the rope. The magnitude T of that force is called the **tension** in the rope. Because it is the magnitude of a vector quantity, tension is a scalar quantity.

Analysis Model: The Particle in Equilibrium

If the acceleration of an object modeled as a particle is zero, the object is treated with the **particle in equilibrium** model. In this model, the net force on the object is zero:

$$\sum \vec{\mathbf{F}} = 0 \qquad (5.8)$$

Consider a lamp suspended from a light chain fastened to the ceiling as in Figure 5.8a. The force diagram for the lamp (Fig. 5.8b) shows that the forces acting on the lamp are the downward gravitational force $\vec{\mathbf{F}}_g$ and the upward force $\vec{\mathbf{T}}$ exerted by the chain. Because there are no forces in the x direction, $\sum F_x = 0$ provides no helpful information. The condition $\sum F_y = 0$ gives

$$\sum F_y = T - F_g = 0 \text{ or } T = F_g$$

Again, notice that $\vec{\mathbf{T}}$ and $\vec{\mathbf{F}}_g$ are *not* an action–reaction pair because they act on the same object, the lamp. The reaction force to $\vec{\mathbf{T}}$ is a downward force exerted by the lamp on the chain.

Example 5.4 (page 107) shows an application of the particle in equilibrium model.

Figure 5.8 (a) A lamp suspended from a ceiling by a chain of negligible mass. (b) The forces acting on the lamp are the gravitational force $\vec{\mathbf{F}}_g$ and the force $\vec{\mathbf{T}}$ exerted by the chain.

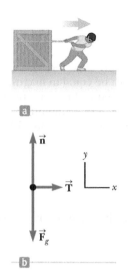

Figure 5.9 (a) A crate being pulled to the right on a frictionless floor. (b) The free-body diagram representing the external forces acting on the crate.

Analysis Model: The Particle Under a Net Force

If an object experiences an acceleration, its motion can be analyzed with the **particle under a net force** model. The appropriate equation for this model is Newton's second law, Equation 5.2:

$$\sum \vec{F} = m\vec{a} \tag{5.2}$$

Consider a crate being pulled to the right on a frictionless, horizontal floor as in Figure 5.9a. Of course, the floor directly under the boy must have friction; otherwise, his feet would simply slip when he tries to pull on the crate! Suppose you wish to find the acceleration of the crate and the force the floor exerts on it. The forces acting on the crate are illustrated in the free-body diagram in Figure 5.9b. Notice that the horizontal force $\vec{T}$ being applied to the crate acts through the rope. The magnitude of $\vec{T}$ is equal to the tension in the rope. In addition to the force $\vec{T}$, the free-body diagram for the crate includes the gravitational force $\vec{F}_g$ and the normal force $\vec{n}$ exerted by the floor on the crate.

We can now apply Newton's second law in component form to the crate. The only force acting in the x direction is $\vec{T}$. Applying $\sum F_x = ma_x$ to the horizontal motion gives

$$\sum F_x = T = ma_x \quad \text{or} \quad a_x = \frac{T}{m}$$

No acceleration occurs in the y direction because the crate moves only horizontally. Therefore, we use the particle in equilibrium model in the y direction. Applying the y component of Equation 5.8 yields

$$\sum F_y = n - F_g = 0 \quad \text{or} \quad n = F_g$$

That is, the normal force has the same magnitude as the gravitational force but acts in the opposite direction.

If $\vec{T}$ is a constant force, the acceleration $a_x = T/m$ also is constant. Hence, the crate is also modeled as a particle under constant acceleration in the x direction, and the equations of kinematics from Chapter 2 can be used to obtain the crate's position x and velocity v_x as functions of time.

Notice from this discussion two concepts that will be important in future problem solving: (1) *In a given problem, it is possible to have different analysis models applied in different directions.* The crate in Figure 5.9 is a particle in equilibrium in the vertical direction and a particle under a net force in the horizontal direction. (2) *It is possible to describe an object by multiple analysis models.* The crate is a particle under a net force in the horizontal direction and is also a particle under constant acceleration in the same direction.

In the situation just described, the magnitude of the normal force $\vec{n}$ is equal to the magnitude of $\vec{F}_g$, but that is not always the case, as noted in Pitfall Prevention 5.6. For example, suppose a book is lying on a table and you push down on the book with a force $\vec{F}$ as in Figure 5.10. Because the book is at rest and therefore not accelerating, $\sum F_y = 0$, which gives $n - F_g - F = 0$, or $n = F_g + F = mg + F$. In this situation, the normal force is *greater* than the gravitational force. Other examples in which $n \neq F_g$ are presented later.

Several examples below demonstrate the use of the particle in equilibrium model and the particle under a net force model.

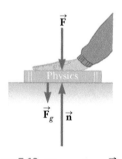

Figure 5.10 When a force $\vec{F}$ pushes vertically downward on another object, the normal force $\vec{n}$ on the object is greater than the gravitational force: $n = F_g + F$.

ANALYSIS MODEL   **Particle in Equilibrium**

Imagine an object that can be modeled as a particle. If it has several forces acting on it so that the forces all cancel, giving a net force of zero, the object will have an acceleration of zero. This condition is mathematically described as

$$\sum \vec{F} = 0 \tag{5.8}$$

$$\vec{a} = 0$$

$$\overset{m}{\underset{\sum \vec{F} = 0}{\longleftarrow \circ \longrightarrow}}$$

ANALYSIS MODEL   **Particle in Equilibrium** *continued*

Examples

- a chandelier hanging over a dining room table
- an object moving at terminal speed through a viscous medium (Chapter 6)
- a steel beam in the frame of a building (Chapter 12)
- a boat floating on a body of water (Chapter 14)

ANALYSIS MODEL   **Particle Under a Net Force**

Imagine an object that can be modeled as a particle. If it has one or more forces acting on it so that there is a net force on the object, it will accelerate in the direction of the net force. The relationship between the net force and the acceleration is

$$\sum \vec{F} = m\vec{a} \qquad \text{(5.2)}$$

Examples:

- a crate pushed across a factory floor
- a falling object acted upon by a gravitational force
- a piston in an automobile engine pushed by hot gases (Chapter 21)
- a charged particle in an electric field (Chapter 22)

Example 5.4   A Traffic Light at Rest

A traffic light weighing 122 N hangs from a cable tied to two other cables fastened to a support as in Figure 5.11a. The upper cables make angles of $\theta_1 = 37.0°$ and $\theta_2 = 53.0°$ with the horizontal. These upper cables are not as strong as the vertical cable and will break if the tension in them exceeds 100 N. Does the traffic light remain hanging in this situation, or will one of the cables break?

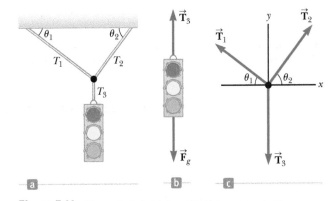

Figure 5.11 (Example 5.4) (a) A traffic light suspended by cables. (b) The forces acting on the traffic light. (c) The free-body diagram for the knot where the three cables are joined.

SOLUTION

Conceptualize Inspect the drawing in Figure 5.11a. Let us assume the cables do not break and nothing is moving.

Categorize If nothing is moving, no part of the system is accelerating. We can now model the light as a *particle in equilibrium* on which the net force is zero. Similarly, the net force on the knot (Fig. 5.11c) is zero, so it is also modeled as a *particle in equilibrium.*

Analyze We construct a diagram of the forces acting on the traffic light, shown in Figure 5.11b, and a free-body diagram for the knot that holds the three cables together, shown in Figure 5.11c. This knot is a convenient object to choose because all the forces of interest act along lines passing through the knot.

From the particle in equilibrium model, apply Equation 5.8 for the traffic light in the y direction:

$$\sum F_y = 0 \quad \rightarrow \quad T_3 - F_g = 0$$
$$T_3 = F_g$$

Choose the coordinate axes as shown in Figure 5.11c and resolve the forces acting on the knot into their components:

Force	x Component	y Component
$\vec{T}_1$	$-T_1 \cos \theta_1$	$T_1 \sin \theta_1$
$\vec{T}_2$	$T_2 \cos \theta_2$	$T_2 \sin \theta_2$
$\vec{T}_3$	0	$-F_g$

Apply the particle in equilibrium model to the knot:

(1) $\sum F_x = -T_1 \cos \theta_1 + T_2 \cos \theta_2 = 0$

(2) $\sum F_y = T_1 \sin \theta_1 + T_2 \sin \theta_2 + (-F_g) = 0$

continued

5.4 continued

Equation (1) shows that the horizontal components of $\vec{T}_1$ and $\vec{T}_2$ must be equal in magnitude, and Equation (2) shows that the sum of the vertical components of $\vec{T}_1$ and $\vec{T}_2$ must balance the downward force $\vec{T}_3$, which is equal in magnitude to the weight of the light.

Solve Equation (1) for T_2 in terms of T_1:

$$(3) \quad T_2 = T_1\left(\frac{\cos\theta_1}{\cos\theta_2}\right)$$

Substitute this value for T_2 into Equation (2):

$$T_1\sin\theta_1 + T_1\left(\frac{\cos\theta_1}{\cos\theta_2}\right)(\sin\theta_2) - F_g = 0$$

Solve for T_1:

$$T_1 = \frac{F_g}{\sin\theta_1 + \cos\theta_1\tan\theta_2}$$

Substitute numerical values:

$$T_1 = \frac{122\text{ N}}{\sin 37.0° + \cos 37.0°\tan 53.0°} = 73.4\text{ N}$$

Using Equation (3), evaluate T_2:

$$T_2 = (73.4\text{ N})\left(\frac{\cos 37.0°}{\cos 53.0°}\right) = 97.4\text{ N}$$

Both values are less than 100 N (just barely for T_2), so the cables will not break.

Finalize   Let us finalize this problem by imagining a change in the system, as in the following What If?

WHAT IF?   Suppose the two angles in Figure 5.11a are equal. What would be the relationship between T_1 and T_2?

Answer   We can argue from the symmetry of the problem that the two tensions T_1 and T_2 would be equal to each other. Mathematically, if the equal angles are called θ, Equation (3) becomes

$$T_2 = T_1\left(\frac{\cos\theta}{\cos\theta}\right) = T_1$$

which also tells us that the tensions are equal. Without knowing the specific value of θ, we cannot find the values of T_1 and T_2. The tensions will be equal to each other, however, regardless of the value of θ.

Conceptual Example 5.5   **Forces Between Cars in a Train**

Train cars are connected by *couplers*, which are under tension as the locomotive pulls the train. Imagine you are on a train speeding up with a constant acceleration. As you move through the train from the locomotive to the last car, measuring the tension in each set of couplers, does the tension increase, decrease, or stay the same? When the engineer applies the brakes, the couplers are under compression. How does this compression force vary from the locomotive to the last car? (Assume only the brakes on the wheels of the engine are applied.)

SOLUTION

While the train is speeding up, tension decreases from the front of the train to the back. The coupler between the locomotive and the first car must apply enough force to accelerate the rest of the cars. As you move back along the train, each coupler is accelerating less mass behind it. The last coupler has to accelerate only the last car, and so it is under the least tension.

When the brakes are applied, the force again decreases from front to back. The coupler connecting the locomotive to the first car must apply a large force to slow down the rest of the cars, but the final coupler must apply a force large enough to slow down only the last car.

Example 5.6   **The Runaway Car**

A car of mass m is on an icy driveway inclined at an angle θ as in Figure 5.12a.

(A) Find the acceleration of the car, assuming the driveway is frictionless.

SOLUTION

Conceptualize   Use Figure 5.12a to conceptualize the situation. From everyday experience, we know that a car on an icy incline will accelerate down the incline. (The same thing happens to a car on a hill with its brakes not set.)

5.6 continued

Categorize We categorize the car as a *particle under a net force* because it accelerates. Furthermore, this example belongs to a very common category of problems in which an object moves under the influence of gravity on an inclined plane.

Analyze Figure 5.12b shows the free-body diagram for the car. The only forces acting on the car are the normal force $\vec{\mathbf{n}}$ exerted by the inclined plane, which acts perpendicular to the plane, and the gravitational force $\vec{\mathbf{F}}_g = m\vec{\mathbf{g}}$, which acts vertically downward. For problems involving inclined planes, it is convenient to choose the coordinate axes with x along the incline and y perpendicular to it as in Figure 5.12b. Using similar triangles, we can show that the angle between the gravitational force $\vec{\mathbf{F}}_g$ and the negative y axis in part b of Figure 5.12 is

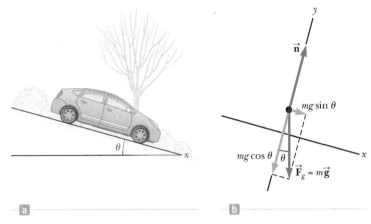

Figure 5.12 (Example 5.6) (a) A car on a frictionless incline. (b) The free-body diagram for the car. The black dot represents the position of the center of mass of the car. We will learn about center of mass in Chapter 9.

equal to the angle θ that the incline makes with the horizontal in part a. With these axes, we represent the gravitational force by a component of magnitude $mg\sin\theta$ along the positive x axis and one of magnitude $mg\cos\theta$ along the negative y axis. Our choice of axes results in the car being modeled as a particle under a net force in the x direction and a particle in equilibrium in the y direction.

Apply these models to the car:

(1) $\sum F_x = mg\sin\theta = ma_x$

(2) $\sum F_y = n - mg\cos\theta = 0$

Solve Equation (1) for a_x:

(3) $a_x = g\sin\theta$

Finalize Note that the acceleration component a_x is independent of the mass of the car. It depends only on the angle of inclination and on g.

From Equation (2), we conclude that the component of $\vec{\mathbf{F}}_g$ perpendicular to the incline is balanced by the normal force; that is, $n = mg\cos\theta$. This situation is a case in which the normal force is *not* equal in magnitude to the weight of the object (as discussed in Pitfall Prevention 5.6 on page 104).

It is possible, although inconvenient, to solve the problem with "standard" horizontal and vertical axes. You may want to try it, just for practice.

(B) Suppose the car is released from rest at the top of the incline and the distance from the car's front bumper to the bottom of the incline is d. How long does it take the front bumper to reach the bottom of the hill, and what is the car's speed as it arrives there?

SOLUTION

Conceptualize Imagine the car is sliding down the hill and you use a stopwatch to measure the entire time interval until it reaches the bottom.

Categorize This part of the problem belongs to kinematics rather than to dynamics, and Equation (3) shows that the acceleration a_x is constant. Therefore, you should categorize the car in this part of the problem as a *particle under constant acceleration.*

Analyze Defining the initial position of the front bumper as $x_i = 0$ and its final position as $x_f = d$, and recognizing that $v_{xi} = 0$, choose Equation 2.16 from the particle under constant acceleration model:

$x_f = x_i + v_{xi}t + \frac{1}{2}a_x t^2 \quad \rightarrow \quad d = \frac{1}{2}a_x t^2$

Solve for t:

(4) $t = \sqrt{\dfrac{2d}{a_x}} = \sqrt{\dfrac{2d}{g\sin\theta}}$

Use Equation 2.17, with $v_{xi} = 0$, to find the final velocity of the car:

$v_{xf}^2 = 2a_x d$

(5) $v_{xf} = \sqrt{2a_x d} = \sqrt{2gd\sin\theta}$

continued

5.6 c o n t i n u e d

Finalize   We see from Equations (4) and (5) that the time t at which the car reaches the bottom and its final speed v_{xf} are independent of the car's mass, as was its acceleration. Notice that we have combined techniques from Chapter 2 with new techniques from this chapter in this example. As we learn more techniques in later chapters, this process of combining analysis models and information from several parts of the book will occur more often. In these cases, use the Analysis Model Approach to Problem Solving discussed in Chapter 2 to help you work your way through new problems.

WHAT IF?   What previously solved problem does this situation become if $\theta = 90°$?

Answer   Imagine θ going to 90° in Figure 5.12. The inclined plane becomes vertical, and the car is an object in free fall! Equation (3) becomes

$$a_x = g \sin \theta = g \sin 90° = g$$

which is indeed the free-fall acceleration. (We find $a_x = g$ rather than $a_x = -g$ because we have chosen positive x to be downward in Fig. 5.12.) Notice also that the condition $n = mg \cos \theta$ gives us $n = mg \cos 90° = 0$. That is consistent with the car falling downward *next to* the vertical plane, in which case there is no contact force between the car and the plane.

Example 5.7   One Block Pushes Another

Two blocks of masses m_1 and m_2, with $m_1 > m_2$, are placed in contact with each other on a frictionless, horizontal surface as in Figure 5.13a. A constant horizontal force $\vec{F}$ is applied to m_1 as shown.

(A) Find the magnitude of the acceleration of the system.

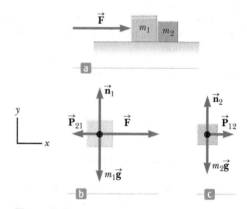

Figure 5.13   (Example 5.7) (a) A force is applied to a block of mass m_1, which pushes on a second block of mass m_2. (b) The forces acting on m_1. (c) The forces acting on m_2.

SOLUTION

Conceptualize   Conceptualize the situation by using Figure 5.13a and realize that both blocks must experience the *same* acceleration because they are in contact with each other and remain in contact throughout the motion.

Categorize   We categorize this problem as one involving a *particle under a net force* because a force is applied to a system of blocks and we are looking for the acceleration of the system.

Analyze   First model the combination of two blocks as a single particle under a net force. Apply Newton's second law to the combination in the x direction to find the acceleration:

$$\sum F_x = F = (m_1 + m_2)a_x$$

$$(1) \quad a_x = \frac{F}{m_1 + m_2}$$

Finalize   The acceleration given by Equation (1) is the same as that of a single object of mass $m_1 + m_2$ and subject to the same force.

(B) Determine the magnitude of the contact force between the two blocks.

SOLUTION

Conceptualize   The contact force is internal to the system of two blocks. Therefore, we cannot find this force by modeling the whole system (the two blocks) as a single particle.

Categorize   Now consider each of the two blocks individually by categorizing each as a *particle under a net force*.

Analyze   We construct a diagram of forces acting on the object for each block as shown in Figures 5.13b and 5.13c, where the contact force is denoted by $\vec{P}$. From Figure 5.13c, we see that the only horizontal force acting on m_2 is the contact force $\vec{P}_{12}$ (the force exerted by m_1 on m_2), which is directed to the right.

Apply Newton's second law to m_2:

$$(2) \quad \sum F_x = P_{12} = m_2 a_x$$

Substitute the value of the acceleration a_x given by Equation (1) into Equation (2):

$$(3) \quad P_{12} = m_2 a_x = \left(\frac{m_2}{m_1 + m_2}\right)F$$

5.7 continued

Finalize This result shows that the contact force P_{12} is *less* than the applied force *F*. The force required to accelerate block 2 alone must be less than the force required to produce the same acceleration for the two-block system.

To finalize further, let us check this expression for P_{12} by considering the forces acting on m_1, shown in Figure 5.13b. The horizontal forces acting on m_1 are the applied force $\vec{\mathbf{F}}$ to the right and the contact force $\vec{\mathbf{P}}_{21}$ to the left (the force exerted by m_2 on m_1). From Newton's third law, $\vec{\mathbf{P}}_{21}$ is the reaction force to $\vec{\mathbf{P}}_{12}$, so $P_{21} = P_{12}$.

Apply Newton's second law to m_1:

$$(4) \quad \sum F_x = F - P_{21} = F - P_{12} = m_1 a_x$$

Solve for P_{12} and substitute the value of a_x from Equation (1):

$$P_{12} = F - m_1 a_x = F - m_1 \left(\frac{F}{m_1 + m_2} \right) = \left(\frac{m_2}{m_1 + m_2} \right) F$$

This result agrees with Equation (3), as it must.

WHAT IF? Imagine that the force $\vec{\mathbf{F}}$ in Figure 5.13 is applied toward the left on the right-hand block of mass m_2. Is the magnitude of the force $\vec{\mathbf{P}}_{12}$ the same as it was when the force was applied toward the right on m_1?

Answer When the force is applied toward the left on m_2, the contact force must accelerate m_1. In the original situation, the contact force accelerates m_2. Because $m_1 > m_2$, more force is required, so the magnitude of $\vec{\mathbf{P}}_{12}$ is greater than in the original situation. To see this mathematically, modify Equation (4) appropriately and solve for $\vec{\mathbf{P}}_{12}$.

Example 5.8 **Weighing a Fish in an Elevator**

A person weighs a fish of mass *m* on a spring scale attached to the ceiling of an elevator as illustrated in Figure 5.14.

(A) Show that if the elevator accelerates either upward or downward, the spring scale gives a reading that is different from the weight of the fish.

SOLUTION

Conceptualize The reading on the scale is related to the extension of the spring in the scale, which is related to the force on the end of the spring as in Figure 5.2. Imagine that the fish is hanging on a string attached to the end of the spring. In this case, the magnitude of the force exerted on the spring is equal to the tension *T* in the string. Therefore, we are looking for *T*. The force $\vec{\mathbf{T}}$ pulls down on the spring and pulls up on the fish.

Categorize We can categorize this problem by identifying the fish as a *particle in equilibrium* if the elevator is not accelerating or as a *particle under a net force* if the elevator is accelerating.

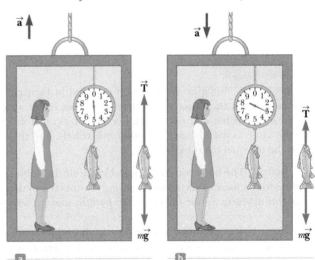

When the elevator accelerates upward, the spring scale reads a value greater than the weight of the fish.

When the elevator accelerates downward, the spring scale reads a value less than the weight of the fish.

Figure 5.14 (Example 5.8) A fish is weighed on a spring scale in an accelerating elevator car.

Analyze Inspect the diagrams of the forces acting on the fish in Figure 5.14 and notice that the external forces acting on the fish are the downward gravitational force $\vec{\mathbf{F}}_g = m\vec{\mathbf{g}}$ and the force $\vec{\mathbf{T}}$ exerted by the string. If the elevator is either at rest or moving at constant velocity, the fish is a particle in equilibrium, so $\sum F_y = T - F_g = 0$ or $T = F_g = mg$. (Remember that the scalar *mg* is the weight of the fish.)

Now suppose the elevator is moving with an acceleration $\vec{\mathbf{a}}$ relative to an observer standing outside the elevator in an inertial frame. The fish is now a particle under a net force.

Apply Newton's second law to the fish:

$$\sum F_y = T - mg = ma_y$$

Solve for *T*:

$$(1) \quad T = ma_y + mg = mg \left(\frac{a_y}{g} + 1 \right) = F_g \left(\frac{a_y}{g} + 1 \right)$$

where we have chosen upward as the positive *y* direction. We conclude from Equation (1) that the scale reading *T* is greater than the fish's weight *mg* if $\vec{\mathbf{a}}$ is upward, so a_y is positive (Fig. 5.14a), and that the reading is less than *mg* if $\vec{\mathbf{a}}$ is downward, so a_y is negative (Fig. 5.14b).

continued

5.8 continued

(B) Evaluate the scale readings for a 40.0-N fish if the elevator moves with an acceleration $a_y = \pm 2.00$ m/s².

SOLUTION

Evaluate the scale reading from Equation (1) if $\vec{\mathbf{a}}$ is upward: $T = (40.0 \text{ N})\left(\dfrac{2.00 \text{ m/s}^2}{9.80 \text{ m/s}^2} + 1\right) = 48.2 \text{ N}$

Evaluate the scale reading from Equation (1) if $\vec{\mathbf{a}}$ is downward: $T = (40.0 \text{ N})\left(\dfrac{-2.00 \text{ m/s}^2}{9.80 \text{ m/s}^2} + 1\right) = 31.8 \text{ N}$

Finalize   Take this advice: if you buy a fish by weight in an elevator, make sure the fish is weighed while the elevator is either at rest or accelerating downward! Furthermore, notice that from the information given here, one cannot determine the direction of the velocity of the elevator.

WHAT IF?   Suppose the woman in Figure 5.14 tires of watching the scale and exits the elevator. Then the elevator cable breaks and the elevator and its remaining contents are in free fall. What happens to the reading on the scale?

Answer   If the elevator falls freely, the fish's acceleration is $a_y = -g$. We see from Equation (1) that the scale reading T is zero in this case; that is, the fish *appears* to be weightless.

Example **5.9**   **The Atwood Machine**

When two objects of unequal mass are hung vertically over a frictionless pulley of negligible mass as in Figure 5.15a, the arrangement is called an *Atwood machine*. The device is sometimes used in the laboratory to determine the value of g by measuring the acceleration of the objects. Determine the magnitude of the acceleration of the two objects and the tension in the lightweight string.

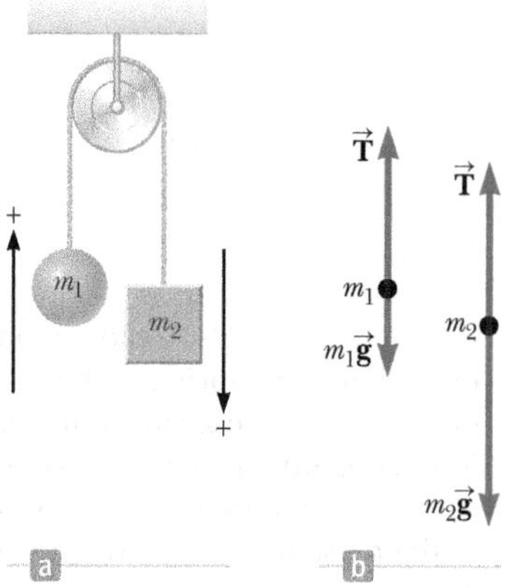

SOLUTION

Conceptualize   Imagine the situation pictured in Figure 5.15a in action: as one object moves upward, the other object moves downward. Because the objects are connected by an inextensible string, the distance one object travels in a given time interval must be the same as the distance the other one travels, and their velocities and accelerations must be of equal magnitude.

Categorize   The objects in the Atwood machine are subject to the gravitational force as well as to the forces exerted by the strings connected to them. Therefore, we can categorize this problem as one involving two *particles under a net force*.

Figure 5.15   (Example 5.9) The Atwood machine. (a) Two objects connected by a massless inextensible string over a frictionless pulley. (b) The free-body diagrams for the two objects.

Analyze   The free-body diagrams for the two objects are shown in Figure 5.15b. Two forces act on each object: the upward force $\vec{\mathbf{T}}$ exerted by the string and the downward gravitational force. In problems such as this one in which the pulley is modeled as massless and frictionless, the tension in the string on both sides of the pulley is the same. If the pulley has mass or is subject to friction, the tensions on either side are not the same and the situation requires techniques we will learn in Chapter 10.

We must be very careful with signs in problems such as this one. In Figure 5.15a, notice that if object 1 accelerates upward, object 2 accelerates downward. Therefore, for consistency with signs, if we define the upward direction as positive for object 1, we must define the downward direction as positive for object 2. With this sign convention, both objects accelerate in the same direction as defined by the choice of sign. Furthermore, according to this sign convention, the y component of the net force exerted on object 1 is $T - m_1 g$, and the y component of the net force exerted on object 2 is $m_2 g - T$.

From the particle under a net force model, apply Newton's second law to object 1:   (1) $\sum F_y = T - m_1 g = m_1 a_y$

Apply Newton's second law to object 2:   (2) $\sum F_y = m_2 g - T = m_2 a_y$

Add Equation (2) to Equation (1), noticing that T cancels: $- m_1 g + m_2 g = m_1 a_y + m_2 a_y$

5.9 continued

Solve for the acceleration:

$$(3) \quad a_y = \left(\frac{m_2 - m_1}{m_1 + m_2} \right) g$$

Substitute Equation (3) into Equation (1) to find T:

$$(4) \quad T = m_1(g + a_y) = \left(\frac{2m_1 m_2}{m_1 + m_2} \right) g$$

Finalize The acceleration given by Equation (3) can be interpreted as the ratio of the magnitude of the unbalanced force on the system $(m_2 - m_1)g$ to the total mass of the system $(m_1 + m_2)$, as expected from Newton's second law. Notice that the sign of the acceleration depends on the relative masses of the two objects; if $m_2 > m_1$, the acceleration is positive, corresponding to downward motion for m_2 and upward for m_1. However, if $m_1 > m_2$, Equation (3) gives a negative acceleration, indicating that m_1 moves downward and m_2 moves upward.

WHAT IF? Describe the motion of the system if the objects have equal masses, that is, $m_1 = m_2$.

Answer If we have the same mass on both sides, the system is balanced and should not accelerate. Mathematically, we see that if $m_1 = m_2$, Equation (3) gives us $a_y = 0$.

WHAT IF? What if one of the masses is much larger than the other: $m_1 \gg m_2$?

Answer In the case in which one mass is infinitely larger than the other, we can ignore the effect of the smaller mass. Therefore, the larger mass should simply fall as if the smaller mass were not there. We see that if $m_1 \gg m_2$, Equation (3) gives us $a_y = -g$.

Example 5.10 **Acceleration of Two Objects Connected by a Cord**

A ball of mass m_1 and a block of mass m_2 are attached by a lightweight cord that passes over a frictionless pulley of negligible mass as in Figure 5.16a. The block lies on a frictionless incline of angle θ. Find the magnitude of the acceleration of the two objects and the tension in the cord.

SOLUTION

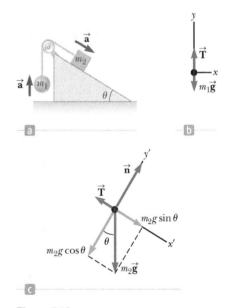

Conceptualize Imagine the objects in Figure 5.16 in motion. If m_2 moves down the incline, then m_1 moves upward. Because the objects are connected by a cord (which we assume does not stretch), their accelerations have the same magnitude. Notice the normal coordinate axes in Figure 5.16b for the ball and the "tilted" axes for the block in Figure 5.16c. Just as we chose the positive direction to be different for each of the objects in Example 5.9, we are free to choose entirely different coordinate axes for the two objects here.

Categorize We can identify forces on each of the two objects and we are looking for an acceleration, so we categorize the objects as *particles under a net force*. For the block, this model is only valid for the x' direction. In the y' direction, we apply the *particle in equilibrium* model because the block does not accelerate in that direction.

Figure 5.16 (Example 5.10) (a) Two objects connected by a lightweight cord strung over a frictionless pulley. (b) The free-body diagram for the ball. (c) The free-body diagram for the block. (The incline is frictionless.)

Analyze Consider the free-body diagrams shown in Figures 5.16b and 5.16c.

Apply Newton's second law in the y direction to the ball, choosing the upward direction as positive:

$$(1) \quad \sum F_y = T - m_1 g = m_1 a_y = m_1 a$$

For the ball to accelerate upward, it is necessary that $T > m_1 g$. In Equation (1), we replaced a_y with a because the acceleration has only a y component.

For the block, we have chosen the x' axis along the incline as in Figure 5.15c. For consistency with our choice for the ball, we choose the positive x' direction to be down the incline.

continued

5.10 continued

Apply the particle under a net force model to the block in the x' direction and the particle in equilibrium model in the y' direction:

(2) $\sum F_{x'} = m_2 g \sin \theta - T = m_2 a_{x'} = m_2 a$

(3) $\sum F_{y'} = n - m_2 g \cos \theta = 0$

In Equation (2), we replaced $a_{x'}$ with a because the two objects have accelerations of equal magnitude a.

Solve Equation (1) for T:

(4) $T = m_1(g + a)$

Substitute this expression for T into Equation (2):

$m_2 g \sin \theta - m_1(g + a) = m_2 a$

Solve for a:

(5) $a = \left(\dfrac{m_2 \sin \theta - m_1}{m_1 + m_2} \right) g$

Substitute this expression for a into Equation (4) to find T:

(6) $T = \left[\dfrac{m_1 m_2 (\sin \theta + 1)}{m_1 + m_2} \right] g$

Finalize   The block accelerates down the incline only if $m_2 \sin \theta > m_1$. If $m_1 > m_2 \sin \theta$, the acceleration is up the incline for the block and downward for the ball. Also notice that the result for the acceleration, Equation (5), can be interpreted as the magnitude of the net external force acting on the ball–block system divided by the total mass of the system; this result is consistent with Newton's second law.

WHAT IF?   What happens in this situation if $\theta = 90°$?

Answer   If $\theta = 90°$, the inclined plane becomes vertical and there is no interaction between its surface and m_2. Therefore, this problem becomes the Atwood machine of Example 5.9. Letting $\theta \rightarrow 90°$ in Equations (5) and (6) causes them to reduce to Equations (3) and (4) of Example 5.9!

WHAT IF?   What if $m_1 = 0$?

Answer   If $m_1 = 0$, then m_2 is simply sliding down an inclined plane without interacting with m_1 through the string. Therefore, this problem becomes the sliding car problem in Example 5.6. Letting $m_1 \rightarrow 0$ in Equation (5) causes it to reduce to Equation (3) of Example 5.6!

5.8   Forces of Friction

When an object is in motion either on a surface or in a viscous medium such as air or water, there is resistance to the motion because the object interacts with its surroundings. We call such resistance a **force of friction.** Forces of friction are very important in our everyday lives. They allow us to walk or run and are necessary for the motion of wheeled vehicles.

Imagine that you are working in your garden and have filled a trash can with yard clippings. You then try to drag the trash can across the surface of your concrete patio as in Figure 5.17a. This surface is *real*, not an idealized, frictionless surface. If we apply an external horizontal force $\vec{\mathbf{F}}$ to the trash can, acting to the right, the trash can remains stationary when $\vec{\mathbf{F}}$ is small. The force on the trash can that counteracts $\vec{\mathbf{F}}$ and keeps it from moving acts toward the left and is called the

◀ Force of static friction

force of static friction $\vec{\mathbf{f}}_s$. As long as the trash can is not moving, $f_s = F$. Therefore, if $\vec{\mathbf{F}}$ is increased, $\vec{\mathbf{f}}_s$ also increases. Likewise, if $\vec{\mathbf{F}}$ decreases, $\vec{\mathbf{f}}_s$ also decreases.

Experiments show that the friction force arises from the nature of the two surfaces: because of their roughness, contact is made only at a few locations where peaks of the material touch. At these locations, the friction force arises in part because one peak physically blocks the motion of a peak from the opposing surface and in part from chemical bonding ("spot welds") of opposing peaks as they come into contact. Although the details of friction are quite complex at the atomic level, this force ultimately involves an electrical interaction between atoms or molecules.

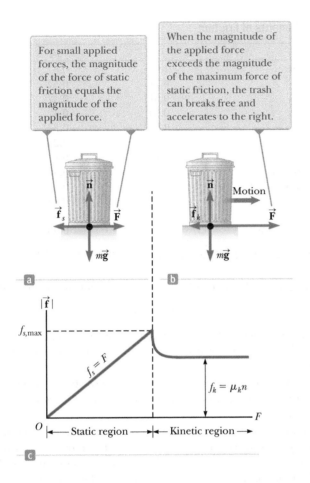

For small applied forces, the magnitude of the force of static friction equals the magnitude of the applied force.

When the magnitude of the applied force exceeds the magnitude of the maximum force of static friction, the trash can breaks free and accelerates to the right.

Figure 5.17 (a) and (b) When pulling on a trash can, the direction of the force of friction $\vec{\mathbf{f}}$ between the can and a rough surface is opposite the direction of the applied force $\vec{\mathbf{F}}$. (c) A graph of friction force versus applied force. Notice that $f_{s,\text{max}} > f_k$.

If we increase the magnitude of $\vec{\mathbf{F}}$ as in Figure 5.17b, the trash can eventually slips. When the trash can is on the verge of slipping, f_s has its maximum value $f_{s,\text{max}}$ as shown in Figure 5.17c. When F exceeds $f_{s,\text{max}}$, the trash can moves and accelerates to the right. We call the friction force for an object in motion the **force of kinetic friction** $\vec{\mathbf{f}}_k$. When the trash can is in motion, the force of kinetic friction on the can is less than $f_{s,\text{max}}$ (Fig. 5.17c). The net force $F - f_k$ in the x direction produces an acceleration to the right, according to Newton's second law. If $F = f_k$, the acceleration is zero and the trash can moves to the right with constant speed. If the applied force $\vec{\mathbf{F}}$ is removed from the moving can, the friction force $\vec{\mathbf{f}}_k$ acting to the left provides an acceleration of the trash can in the $-x$ direction and eventually brings it to rest, again consistent with Newton's second law.

◀ Force of kinetic friction

Experimentally, we find that, to a good approximation, both $f_{s,\text{max}}$ and f_k are proportional to the magnitude of the normal force exerted on an object by the surface. The following descriptions of the force of friction are based on experimental observations and serve as the simplification model we shall use for forces of friction in problem solving:

- The magnitude of the force of static friction between any two surfaces in contact can have the values

$$f_s \leq \mu_s n \qquad (5.9)$$

where the dimensionless constant μ_s is called the **coefficient of static friction** and n is the magnitude of the normal force exerted by one surface on the other. The equality in Equation 5.9 holds when the surfaces are on the verge of slipping, that is, when $f_s = f_{s,\text{max}} = \mu_s n$. This situation is called *impending motion*. The inequality holds when the surfaces are not on the verge of slipping.

PITFALL PREVENTION 5.9
The Equal Sign Is Used in Limited Situations In Equation 5.9, the equal sign is used *only* in the case in which the surfaces are just about to break free and begin sliding. Do not fall into the common trap of using $f_s = \mu_s n$ in *any* static situation.

TABLE 5.1   Coefficients of Friction

	μ_s	μ_k
Rubber on concrete	1.0	0.8
Steel on steel	0.74	0.57
Aluminum on steel	0.61	0.47
Glass on glass	0.94	0.4
Copper on steel	0.53	0.36
Wood on wood	0.25–0.5	0.2
Waxed wood on wet snow	0.14	0.1
Waxed wood on dry snow	—	0.04
Metal on metal (lubricated)	0.15	0.06
Teflon on Teflon	0.04	0.04
Ice on ice	0.1	0.03
Synovial joints in humans	0.01	0.003

Note: All values are approximate. In some cases, the coefficient of friction can exceed 1.0.

- The magnitude of the force of kinetic friction acting between two surfaces is

$$f_k = \mu_k n \tag{5.10}$$

where μ_k is the **coefficient of kinetic friction.** Although the coefficient of kinetic friction can vary with speed, we shall usually neglect any such variations in this text.
- The values of μ_k and μ_s depend on the nature of the surfaces, but μ_k is generally less than μ_s. Typical values range from around 0.03 to 1.0. Table 5.1 lists some reported values.
- The direction of the friction force on an object is parallel to the surface with which the object is in contact and opposite to the actual motion (kinetic friction) or the impending motion (static friction) of the object relative to the surface.
- The coefficients of friction are nearly independent of the area of contact between the surfaces. We might expect that placing an object on the side having the most area might increase the friction force. Although this method provides more points in contact, the weight of the object is spread out over a larger area and the individual points are not pressed together as tightly. Because these effects approximately compensate for each other, the friction force is independent of the area.

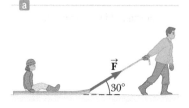

Figure 5.18 (Quick Quiz 5.7) A father slides his daughter on a sled either by (a) pushing down on her shoulders or (b) pulling up on a rope.

QUICK QUIZ 5.6 You press your physics textbook flat against a vertical wall with your hand. What is the direction of the friction force exerted by the wall on the book? **(a)** downward **(b)** upward **(c)** out from the wall **(d)** into the wall

QUICK QUIZ 5.7 Charlie is playing with his daughter Torrey in the snow. She sits on a sled and asks him to slide her across a flat, horizontal field. Charlie has a choice of **(a)** pushing her from behind by applying a force downward on her shoulders at 30° below the horizontal (Fig. 5.18a) or **(b)** attaching a rope to the front of the sled and pulling with a force at 30° above the horizontal (Fig. 5.18b). Which would be easier for him and why?

Example 5.11   **Experimental Determination of μ_s and μ_k**

The following is a simple method of measuring coefficients of friction. Suppose a block is placed on a rough surface inclined relative to the horizontal as shown in Figure 5.19. The incline angle is increased until the block starts to move. Show that you can obtain μ_s by measuring the critical angle θ_c at which this slipping just occurs.

SOLUTION

Conceptualize   Consider Figure 5.19 and imagine that the block tends to slide down the incline due to the gravitational force. To simulate the situation, place a coin on this book's cover and tilt the book until the coin begins to slide. Notice how this

5.11 continued

example differs from Example 5.6. When there is no friction on an incline, *any* angle of the incline will cause a stationary object to begin moving. When there is friction, however, there is no movement of the object for angles less than the critical angle.

Categorize The block is subject to various forces. Because we are raising the plane to the angle at which the block is just ready to begin to move but is not moving, we categorize the block as a *particle in equilibrium*.

Analyze The diagram in Figure 5.19 shows the forces on the block: the gravitational force $m\vec{g}$, the normal force $\vec{n}$, and the force of static friction $\vec{f}_s$. We choose x to be parallel to the plane and y perpendicular to it.

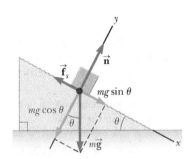

Figure 5.19 (Example 5.11) The external forces exerted on a block lying on a rough incline are the gravitational force $m\vec{g}$, the normal force $\vec{n}$, and the force of friction $\vec{f}_s$. For convenience, the gravitational force is resolved into a component $mg\sin\theta$ along the incline and a component $mg\cos\theta$ perpendicular to the incline.

From the particle in equilibrium model, apply Equation 5.8 to the block in both the x and y directions:

$$(1) \quad \sum F_x = mg\sin\theta - f_s = 0$$

$$(2) \quad \sum F_y = n - mg\cos\theta = 0$$

Substitute $mg = n/\cos\theta$ from Equation (2) into Equation (1):

$$(3) \quad f_s = mg\sin\theta = \left(\frac{n}{\cos\theta}\right)\sin\theta = n\tan\theta$$

When the incline angle is increased until the block is on the verge of slipping, the force of static friction has reached its maximum value $\mu_s n$. The angle θ in this situation is the critical angle θ_c. Make these substitutions in Equation (3):

$$\mu_s n = n\tan\theta_c$$
$$\mu_s = \tan\theta_c$$

We have shown, as requested, that the coefficient of static friction is related only to the critical angle. For example, if the block just slips at $\theta_c = 20.0°$, we find that $\mu_s = \tan 20.0° = 0.364$.

Finalize Once the block starts to move at $\theta \geq \theta_c$, it accelerates down the incline and the force of friction is $f_k = \mu_k n$.

WHAT IF? How could you determine μ_k for the block and incline?

Answer If θ is reduced to a value less than θ_c, it may be possible to find an angle θ_c' such that the block moves down the incline with constant speed as a particle in equilibrium again ($a_x = 0$). In this case, use Equations (1) and (2) with f_s replaced by f_k to find μ_k: $\mu_k = \tan\theta_c'$, where $\theta_c' < \theta_c$.

Example 5.12 **The Sliding Hockey Puck**

A hockey puck on a frozen pond is given an initial speed of 20.0 m/s. If the puck always remains on the ice and slides 115 m before coming to rest, determine the coefficient of kinetic friction between the puck and ice.

SOLUTION

Conceptualize Imagine that the puck in Figure 5.20 slides to the right. The kinetic friction force acts to the left and slows the puck, which eventually comes to rest due to that force.

Categorize The forces acting on the puck are identified in Figure 5.20, but the text of the problem provides kinematic variables. Therefore, we categorize the problem in several ways. First, it involves modeling the puck as a *particle under a net force* in the horizontal direction: kinetic friction causes the puck to accelerate. There is no acceleration of the puck in the vertical direction, so we use the *particle in equilibrium* model for that direction. Furthermore, because we model the force of kinetic friction as independent of speed, the acceleration of the puck is constant. So, we can also categorize this problem by modeling the puck as a *particle under constant acceleration*.

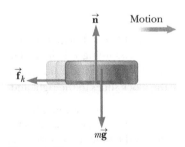

Figure 5.20 (Example 5.12) After the puck is given an initial velocity to the right, the only external forces acting on it are the gravitational force $m\vec{g}$, the normal force $\vec{n}$, and the force of kinetic friction $\vec{f}_k$.

continued

5.12 continued

Analyze   First, let's find the acceleration algebraically in terms of the coefficient of kinetic friction, using Newton's second law. Once we know the acceleration of the puck and the distance it travels, the equations of kinematics can be used to find the numerical value of the coefficient of kinetic friction.

Apply the particle under a net force model in the x direction to the puck:

(1) $\sum F_x = -f_k = ma_x$

Apply the particle in equilibrium model in the y direction to the puck:

(2) $\sum F_y = n - mg = 0$

Substitute $n = mg$ from Equation (2) and $f_k = \mu_k n$ into Equation (1):

$$-\mu_k n = -\mu_k mg = ma_x$$
$$a_x = -\mu_k g$$

The negative sign means the acceleration is to the left in Figure 5.20. Because the velocity of the puck is to the right, the puck is slowing down. The acceleration is independent of the mass of the puck and is constant because we assume μ_k remains constant.

Apply the particle under constant acceleration model to the puck, choosing Equation 2.17 from the model, $v_{xf}^2 = v_{xi}^2 + 2a_x(x_f - x_i)$, with $x_i = 0$ and $v_{xf} = 0$:

$$0 = v_{xi}^2 + 2a_x x_f = v_{xi}^2 - 2\mu_k g x_f$$

Solve for the coefficient of kinetic friction:

$$\mu_k = \frac{v_{xi}^2}{2gx_f}$$

Substitute the numerical values:

$$\mu_k = \frac{(20.0 \text{ m/s})^2}{2(9.80 \text{ m/s}^2)(115 \text{ m})} = 0.177$$

Finalize   Notice that μ_k is dimensionless, as it should be, and that it has a low value, consistent with an object sliding on ice.

Example **5.13**   Acceleration of Two Connected Objects When Friction Is Present

A block of mass m_2 on a rough, horizontal surface is connected to a ball of mass m_1 by a lightweight cord over a lightweight, frictionless pulley as shown in Figure 5.21a. A force of magnitude F at an angle θ with the horizontal is applied to the block as shown, and the block slides to the right. The coefficient of kinetic friction between the block and surface is μ_k. Determine the magnitude of the acceleration of the two objects.

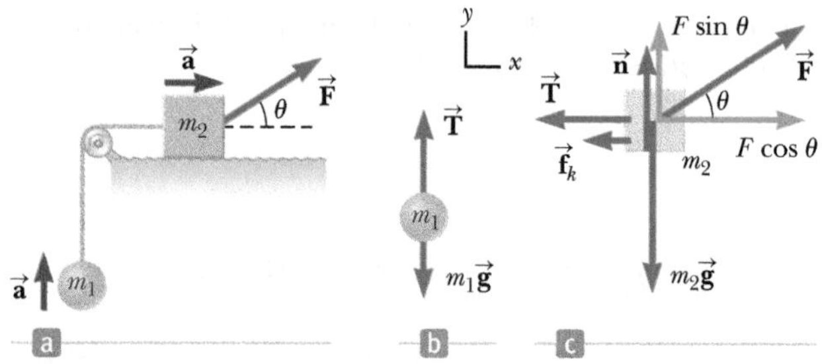

Figure 5.21   (Example 5.13) (a) The external force $\vec{F}$ applied as shown can cause the block to accelerate to the right. (b, c) Diagrams showing the forces on the two objects, assuming the block accelerates to the right and the ball accelerates upward.

SOLUTION

Conceptualize   Imagine what happens as $\vec{F}$ is applied to the block. Assuming $\vec{F}$ is large enough to break the block free from static friction but not large enough to lift the block, the block slides to the right and the ball rises.

Categorize   We can identify forces and we want an acceleration, so we categorize this problem as one involving two *particles under a net force*, the ball and the block. Because we assume that the block does not rise into the air due to the applied force, we model the block as a *particle in equilibrium* in the vertical direction.

Analyze   First draw force diagrams for the two objects as shown in Figures 5.21b and 5.21c. Notice that the string exerts a force of magnitude T on both objects. The applied force $\vec{F}$ has x and y components $F\cos\theta$ and $F\sin\theta$, respectively. Because the two objects are connected, we can equate the magnitudes of the x component of the acceleration of the block and the y component of the acceleration of the ball and call them both a. Let us assume the motion of the block is to the right.

Apply the particle under a net force model to the block in the horizontal direction:

(1) $\sum F_x = F\cos\theta - f_k - T = m_2 a_x = m_2 a$

5.13 continued

Because the block moves only horizontally, apply the particle in equilibrium model to the block in the vertical direction:

(2) $\sum F_y = n + F\sin\theta - m_2 g = 0$

Apply the particle under a net force model to the ball in the vertical direction:

(3) $\sum F_y = T - m_1 g = m_1 a_y = m_1 a$

Solve Equation (2) for n:

$n = m_2 g - F\sin\theta$

Substitute n into $f_k = \mu_k n$ from Equation 5.10:

(4) $f_k = \mu_k(m_2 g - F\sin\theta)$

Substitute Equation (4) and the value of T from Equation (3) into Equation (1):

$F\cos\theta - \mu_k(m_2 g - F\sin\theta) - m_1(a + g) = m_2 a$

Solve for a:

(5) $a = \dfrac{F(\cos\theta + \mu_k \sin\theta) - (m_1 + \mu_k m_2)g}{m_1 + m_2}$

Finalize   The acceleration of the block can be either to the right or to the left depending on the sign of the numerator in Equation (5). If the velocity is to the left, we must reverse the sign of f_k in Equation (1) because the force of kinetic friction must oppose the motion of the block relative to the surface. In this case, the value of a is the same as in Equation (5), with the two plus signs in the numerator changed to minus signs.

What does Equation (5) reduce to if the force $\vec{\mathbf{F}}$ is removed and the surface becomes frictionless? Call this expression Equation (6). Does this algebraic expression match your intuition about the physical situation in this case? Now go back to Example 5.10 and let angle θ go to zero in Equation (5) of that example. How does the resulting equation compare with your Equation (6) here in Example 5.13? Should the algebraic expressions compare in this way based on the physical situations?

Summary

▶ Definitions

An **inertial frame of reference** is a frame in which an object that does not interact with other objects experiences zero acceleration. Any frame moving with constant velocity relative to an inertial frame is also an inertial frame.

We define **force** as **that which causes a change in motion of an object.**

▶ Concepts and Principles

Newton's first law states that it is possible to find an inertial frame in which an object that does not interact with other objects experiences zero acceleration, or, equivalently, in the absence of an external force, when viewed from an inertial frame, an object at rest remains at rest and an object in uniform motion in a straight line maintains that motion.

Newton's second law states that the acceleration of an object is directly proportional to the net force acting on it and inversely proportional to its mass.

Newton's third law states that if two objects interact, the force exerted by object 1 on object 2 is equal in magnitude and opposite in direction to the force exerted by object 2 on object 1.

The maximum **force of static friction** $\vec{\mathbf{f}}_{s,max}$ between an object and a surface is proportional to the normal force acting on the object. In general, $f_s \le \mu_s n$, where μ_s is the **coefficient of static friction** and n is the magnitude of the normal force.

The **gravitational force** exerted on an object is equal to the product of its mass (a scalar quantity) and the free-fall acceleration:

$$\vec{\mathbf{F}}_g = m\vec{\mathbf{g}} \qquad (5.5)$$

The **weight** of an object is the magnitude of the gravitational force acting on the object:

$$F_g = mg \qquad (5.6)$$

When an object slides over a surface, the magnitude of the **force of kinetic friction** $\vec{\mathbf{f}}_k$ is given by $f_k = \mu_k n$, where μ_k is the **coefficient of kinetic friction.**

continued

▶ Analysis Models for Problem Solving

Particle Under a Net Force   If a particle of mass m experiences a nonzero net force, its acceleration is related to the net force by Newton's second law:

$$\sum \vec{F} = m \vec{a} \qquad (5.2)$$

Particle in Equilibrium   If a particle maintains a constant velocity (so that $\vec{a} = 0$), which could include a velocity of zero, the forces on the particle balance and Newton's second law reduces to

$$\sum \vec{F} = 0 \qquad (5.8)$$

Think–Pair–Share

See the Preface for an explanation of the icons used in this problems set. For additional assessment items for this section, go to ▚ **WEBASSIGN** From Cengage

1. You are a member of an expert witness group that provides scientific services to the legal community. Your group has been asked by a defense attorney to argue at trial that a driver was not exceeding the speed limit. You are provided with the following data: The mass of the car is 1.50×10^3 kg. The mass of the driver is 95.0 kg. The coefficient of kinetic friction between the car's tires and the roadway is 0.580. The coefficient of static friction between the car's tires and the roadway is 0.820. The posted speed limit on the road is 25 mi/h. The roadway was dry and the weather was sunny at the time of the incident.

   You are also provided with the following description of the incident: The driver was driving up a hill that makes an angle of 17.5° with the horizontal. The driver saw a dog run into the street, slammed on the brakes and left a skid mark 17.0 m long. The car came to rest at the end of the skid mark. The driver did not hit the dog, but the sound of the screeching tires drew the attention of a nearby policeman, who ticketed the driver for speeding. (a) Should your group agree to offer testimony for the defense in this case? (b) Why or why not?

2. Consider the egg-catching activity discussed in the chapter-opening storyline. Discuss in your group and make estimates of the following. Identify a separation distance between the thrower and the catcher of the egg, and determine a typical speed with which the egg must be thrown so that it covers the distance without hitting the ground or passing over the head of the catcher. Estimate the mass and diameter of the egg (the shorter diameter, perpendicular to the longest dimension). From these data, estimate the force on the egg exerted by your hand if your hand is held stiffly and doesn't move when the egg hits it. Now, simulate moving your hands backward while you catch the egg. Have a group member estimate the distance over which your hands move in this process. From these data, estimate the force on the egg exerted by your hand in this catching process. Compare the forces of your hand on the egg between these two methods of catching the egg.

3. **ACTIVITY** A simple procedure can be followed to measure the coefficient of friction using the technique discussed in Example 5.11. Lay your book down on a table and place a coin on the cover far from the spine. Slowly open the book cover so that it forms an inclined plane down which the coin will slide. Watch carefully and stop opening the cover at the instant the coin begins to slide. (a) Measure this critical angle that the book cover makes with the horizontal with a protractor. From this angle, determine the coefficient of static friction between the coin and the cover. (b) Place a loop of tape between two coins and repeat the procedure above for the two-coin stack. How does the coefficient of static friction for the stack compare to that for the single coin?

Problems

See the Preface for an explanation of the icons used in this problems set. For additional assessment items for this section, go to ▚ **WEBASSIGN** From Cengage

SECTION 5.1 The Concept of Force

1. **BIO** A certain orthodontist uses a wire brace to align a patient's crooked tooth as in Figure P5.1. The tension in the wire is adjusted to have a magnitude of 18.0 N. Find the magnitude of the net force exerted by the wire on the crooked tooth.

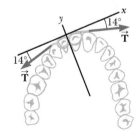

Figure P5.1

2. One or more external forces, large enough to be easily measured, are exerted on each object enclosed in a dashed box shown in Figure 5.1. Identify the reaction to each of these forces.

SECTION 5.4 Newton's Second Law

3. A 3.00-kg object undergoes an acceleration given by **V** $\vec{a} = (2.00\hat{i} + 5.00\hat{j})$ m/s². Find (a) the resultant force acting on the object and (b) the magnitude of the resultant force.

4. The average speed of a nitrogen molecule in air is about 6.70×10^2 m/s, and its mass is 4.68×10^{-26} kg. (a) If it takes 3.00×10^{-13} s for a nitrogen molecule to hit a wall and

rebound with the same speed but moving in the opposite direction, what is the average acceleration of the molecule during this time interval? (b) What average force does the molecule exert on the wall?

5. Two forces, $\vec{F}_1 = (-6.00\hat{i} - 4.00\hat{j})$ N and $\vec{F}_2 = (-3.00\hat{i} + 7.00\hat{j})$ N, act on a particle of mass 2.00 kg that is initially at rest at coordinates (−2.00 m, +4.00 m). (a) What are the components of the particle's velocity at $t = 10.0$ s? (b) In what direction is the particle moving at $t = 10.0$ s? (c) What displacement does the particle undergo during the first 10.0 s? (d) What are the coordinates of the particle at $t = 10.0$ s?

6. The force exerted by the wind on the sails of a sailboat is
[T] 390 N north. The water exerts a force of 180 N east. If the boat (including its crew) has a mass of 270 kg, what are the magnitude and direction of its acceleration?

7. Review. Three forces acting on an object are given by $\vec{F}_1 =$
[V] $(-2.00\hat{i} + 2.00\hat{j})$ N, and $\vec{F}_2 = (5.00\hat{i} - 3.00\hat{j})$ N, and $\vec{F}_3 = (-45.0\hat{i})$ N. The object experiences an acceleration of magnitude 3.75 m/s². (a) What is the direction of the acceleration? (b) What is the mass of the object? (c) If the object is initially at rest, what is its speed after 10.0 s? (d) What are the velocity components of the object after 10.0 s?

8. If a single constant force acts on an object that moves on a
[S] straight line, the object's velocity is a linear function of time. The equation $v = v_i + at$ gives its velocity v as a function of time, where a is its constant acceleration. What if velocity is instead a linear function of position? Assume that as a particular object moves through a resistive medium, its speed decreases as described by the equation $v = v_i - kx$, where k is a constant coefficient and x is the position of the object. Find the law describing the total force acting on this object.

SECTION 5.5 The Gravitational Force and Weight

9. Review. The gravitational force exerted on a baseball is 2.21 N down. A pitcher throws the ball horizontally with velocity 18.0 m/s by uniformly accelerating it along a straight horizontal line for a time interval of 170 ms. The ball starts from rest. (a) Through what distance does it move before its release? (b) What are the magnitude and direction of the force the pitcher exerts on the ball?

10. Review. The gravitational force exerted on a baseball
[S] is $-F_g\hat{j}$. A pitcher throws the ball with velocity $v\hat{i}$ by uniformly accelerating it along a straight horizontal line for a time interval of $\Delta t = t - 0 = t$. (a) Starting from rest, through what distance does the ball move before its release? (b) What force does the pitcher exert on the ball?

11. Review. An electron of mass 9.11×10^{-31} kg has an initial
[T] speed of 3.00×10^5 m/s. It travels in a straight line, and its speed increases to 7.00×10^5 m/s in a distance of 5.00 cm. Assuming its acceleration is constant, (a) determine the magnitude of the force exerted on the electron and (b) compare this force with the weight of the electron, which we ignored.

12. If a man weighs 900 N on the Earth, what would he weigh on Jupiter, where the free-fall acceleration is 25.9 m/s²?

13. You stand on the seat of a chair and then hop off. (a) During the time interval you are in flight down to the floor, the Earth moves toward you with an acceleration of what order of magnitude? In your solution, explain your logic. Model the Earth as a perfectly solid object. (b) The Earth moves toward you through a distance of what order of magnitude?

SECTION 5.6 Newton's Third Law

14. A brick of mass M has been placed on a rubber cushion of mass m. Together they are sliding to the right at constant velocity on an ice-covered parking lot. (a) Draw a free-body diagram of the brick and identify each force acting on it. (b) Draw a free-body diagram of the cushion and identify each force acting on it. (c) Identify all of the action–reaction pairs of forces in the brick–cushion–planet system.

SECTION 5.7 Analysis Models Using Newton's Second Law

15. Review. Figure P5.15 shows a worker poling a boat—a very efficient mode of transportation—across a shallow lake. He pushes parallel to the length of the light pole, exerting a force of magnitude 240 N on the bottom of the lake. Assume the pole lies in the vertical plane containing the keel of the boat. At one moment, the pole makes an angle of 35.0° with the vertical and the water exerts a horizontal drag force of 47.5 N on the boat, opposite to its forward velocity of magnitude 0.857 m/s. The

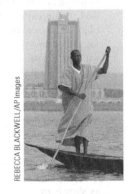

REBECCA BLACKWELL/AP Images

Figure P5.15

mass of the boat including its cargo and the worker is 370 kg. (a) The water exerts a buoyant force vertically upward on the boat. Find the magnitude of this force. (b) Model the forces as constant over a short interval of time to find the velocity of the boat 0.450 s after the moment described.

16. An iron bolt of mass 65.0 g hangs from a string 35.7 cm long. The top end of the string is fixed. Without touching it, a magnet attracts the bolt so that it remains stationary, but is displaced horizontally 28.0 cm to the right from the previously vertical line of the string. The magnet is located to the right of the bolt and on the same vertical level as the bolt in the final configuration. (a) Draw a free-body diagram of the bolt. (b) Find the tension in the string. (c) Find the magnetic force on the bolt.

17. A block slides down a frictionless plane having an inclina-
[V] tion of $\theta = 15.0°$. The block starts from rest at the top, and the length of the incline is 2.00 m. (a) Draw a free-body diagram of the block. Find (b) the acceleration of the block and (c) its speed when it reaches the bottom of the incline.

18. A bag of cement whose weight is F_g
[S] hangs in equilibrium from three wires as shown in Figure P5.18. Two of the wires make angles θ_1 and θ_2 with the horizontal. Assuming the system is in equilibrium, show that the tension in the left-hand wire is

$$T_1 = \frac{F_g \cos \theta_2}{\sin (\theta_1 + \theta_2)}$$

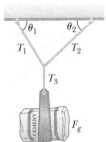

Figure P5.18

19. The distance between two telephone poles is 50.0 m. When a 1.00-kg bird lands on the telephone wire midway between the poles, the wire sags 0.200 m. (a) Draw a free-body diagram of the bird. (b) How much tension does the bird produce in the wire? Ignore the weight of the wire.

20. An object of mass $m = 1.00$ kg is observed to have an acceleration $\vec{a}$ with a magnitude of 10.0 m/s² in a direction 60.0°

east of north. Figure P5.20 shows a view of the object from above. The force $\vec{\mathbf{F}}_2$ acting on the object has a magnitude of 5.00 N and is directed north. Determine the magnitude and direction of the one other horizontal force $\vec{\mathbf{F}}_1$ acting on the object.

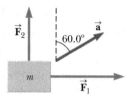

Figure P5.20

21. A simple accelerometer is constructed inside a car by suspending an object of mass m from a string of length L that is tied to the car's ceiling. As the car accelerates the string–object system makes a constant angle of θ with the vertical. (a) Assuming that the string mass is negligible compared with m, derive an expression for the car's acceleration in terms of θ and show that it is independent of the mass m and the length L. (b) Determine the acceleration of the car when $\theta = 23.0°$.

22. An object of mass $m_1 = 5.00$ kg placed on a frictionless, horizontal table is connected to a string that passes over a pulley and then is fastened to a hanging object of mass $m_2 = 9.00$ kg as shown in Figure P5.22. (a) Draw free-body diagrams of both objects. Find (b) the magnitude of the acceleration of the objects and (c) the tension in the string.

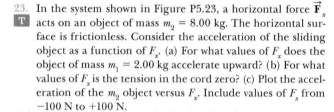

Figure P5.22
Problems 22 and 29.

23. In the system shown in Figure P5.23, a horizontal force $\vec{\mathbf{F}}_x$ acts on an object of mass $m_2 = 8.00$ kg. The horizontal surface is frictionless. Consider the acceleration of the sliding object as a function of F_x. (a) For what values of F_x does the object of mass $m_1 = 2.00$ kg accelerate upward? (b) For what values of F_x is the tension in the cord zero? (c) Plot the acceleration of the m_2 object versus F_x. Include values of F_x from -100 N to $+100$ N.

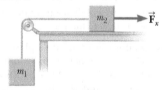

Figure P5.23

24. A car is stuck in the mud. A tow truck pulls on the car with the arrangement shown in Figure P5.24. The tow cable is under a tension of 2 500 N and pulls downward and to the left on the pin at its upper end. The light pin is held in equilibrium by forces exerted by the two bars A and B. Each bar is a *strut*; that is, each is a bar whose weight is small compared to the forces it exerts and which exerts forces only through hinge pins at its ends. Each strut exerts a force directed parallel to its length. Determine the force of tension or compression in each strut. Proceed as follows. Make a guess as to which way (pushing or pulling) each force acts on the top pin. Draw a free-body diagram of the pin. Use the condition for equilibrium of the pin to translate the free-body diagram into equations. From the equations calculate the forces exerted by struts A and B. If you obtain a positive answer, you correctly guessed the direction of the force. A negative answer means that the direction should be reversed, but the absolute value correctly gives the

magnitude of the force. If a strut pulls on a pin, it is in tension. If it pushes, the strut is in compression. Identify whether each strut is in tension or in compression.

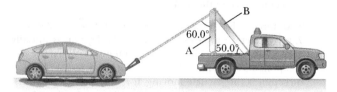

Figure P5.24

25. An object of mass m_1 hangs from a string that passes over a very light fixed pulley P_1 as shown in Figure P5.25. The string connects to a second very light pulley P_2. A second string passes around this pulley with one end attached to a wall and the other to an object of mass m_2 on a frictionless, horizontal table. (a) If a_1 and a_2 are the accelerations of m_1 and m_2, respectively, what is the relation between these accelerations? Find expressions for (b) the tensions in the strings and (c) the accelerations a_1 and a_2 in terms of the masses m_1 and m_2, and g.

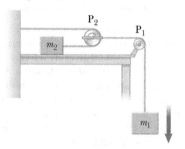

Figure P5.25

SECTION 5.8 Forces of Friction

26. *Why is the following situation impossible?* Your 3.80-kg physics book is placed next to you on the horizontal seat of your car. The coefficient of static friction between the book and the seat is 0.650, and the coefficient of kinetic friction is 0.550. You are traveling forward at 72.0 km/h and brake to a stop with constant acceleration over a distance of 30.0 m. Your physics book remains on the seat rather than sliding forward onto the floor.

27. Consider a large truck carrying a heavy load, such as steel beams. A significant hazard for the driver is that the load may slide forward, crushing the cab, if the truck stops suddenly in an accident or even in braking. Assume, for example, that a 10 000-kg load sits on the flatbed of a 20 000-kg truck moving at 12.0 m/s. Assume that the load is not tied down to the truck, but has a coefficient of friction of 0.500 with the flatbed of the truck. (a) Calculate the minimum stopping distance for which the load will not slide forward relative to the truck. (b) Is any piece of data unnecessary for the solution?

28. Before 1960, people believed that the maximum attainable coefficient of static friction for an automobile tire on a roadway was $\mu_s = 1$. Around 1962, three companies independently developed racing tires with coefficients of 1.6. This problem shows that tires have improved further since then. The shortest time interval in which a piston-engine car initially at rest has covered a distance of one-quarter mile is

about 4.43 s. (a) Assume the car's rear wheels lift the front wheels off the pavement as shown in Figure P5.28. What minimum value of μ_s is necessary to achieve the record time? (b) Suppose the driver were able to increase his or her engine power, keeping other things equal. How would this change affect the elapsed time?

Figure P5.28

29. A 9.00-kg hanging object is connected by a light, inextensible cord over a light, frictionless pulley to a 5.00-kg block that is sliding on a flat table (Fig. P5.22). Taking the coefficient of kinetic friction as 0.200, find the tension in the string.

30. The person in Figure P5.30 weighs 170 lb. As seen from the front, each light crutch makes an angle of 22.0° with the vertical. Half of the person's weight is supported by the crutches. The other half is supported by the vertical forces of the ground on the person's feet. Assuming that the person is moving with constant velocity and the force exerted by the ground on the crutches acts along the crutches, determine (a) the smallest possible coefficient of friction between crutches and ground and (b) the magnitude of the compression force in each crutch.

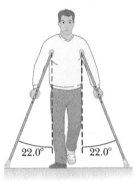

Figure P5.30

31. Three objects are connected on a table as shown in Figure P5.31. The coefficient of kinetic friction between the block of mass m_2 and the table is 0.350. The objects have masses of $m_1 = 4.00$ kg, $m_2 = 1.00$ kg, and $m_3 = 2.00$ kg, and the pulleys are frictionless. (a) Draw a free-body diagram of each object. (b) Determine the acceleration of each object, including its direction. (c) Determine the tensions in the two cords. **What If?** (d) If the tabletop were smooth, would the tensions increase, decrease, or remain the same? Explain.

Figure P5.31

32. You are working as a letter sorter in a U.S Post Office. Postal regulations require that employees' footwear must have a minimum coefficient of static friction of 0.5 on a specified tile surface. You are wearing athletic shoes for which you do not know the coefficient of static friction. In order to determine the coefficient, you imagine that there is an emergency and start running across the room. You have a coworker time you, and find that you can begin at rest and move 4.23 m in 1.20 s. If you try to move faster than this, your feet slip. Assuming your acceleration is constant, does your footwear qualify for the postal regulation?

33. You have been called as an expert witness for a trial in which a driver has been charged with speeding but is claiming innocence. He claims to have slammed on his brakes to avoid rear-ending another car, but tapped the back of the other car just as he came to rest. You have been hired by the prosecution to prove that the driver was indeed speeding. You have received data as follows from the police: Skid marks left by the driver are 56.0 m long and the roadway is level. Tires matching those on the car of the driver have been dragged over the same roadway to determine that the coefficient of kinetic friction between the tires and the roadway is 0.82 at all points along the skid mark. The speed limit on the road is 35 mi/h. Construct an argument to be used in court to show that the driver was indeed speeding.

34. A block of mass 3.00 kg is pushed up against a wall by a force $\vec{\mathbf{P}}$ that makes an angle of $\theta = 50.0°$ with the horizontal as shown in Figure P5.34. The coefficient of static friction between the block and the wall is 0.250. (a) Determine the possible values for the magnitude of $\vec{\mathbf{P}}$ that allow the block to remain stationary. (b) Describe what happens if $|\vec{\mathbf{P}}|$ has a larger value and what happens if it is smaller. (c) Repeat parts (a) and (b), assuming the force makes an angle of $\theta = 13.0°$ with the horizontal.

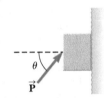

Figure P5.34

35. **Review.** A Chinook salmon can swim underwater at 3.58 m/s, and it can also jump vertically upward, leaving the water with a speed of 6.26 m/s. A record salmon has length 1.50 m and mass 61.0 kg. Consider the fish swimming straight upward in the water below the surface of a lake. The gravitational force exerted on it is very nearly canceled out by a buoyant force exerted by the water as we will study in Chapter 14. The fish experiences an upward force P exerted by the water on its threshing tail fin and a downward fluid friction force that we model as acting on its front end. Assume the fluid friction force disappears as soon as the fish's head breaks the water surface and assume the force on its tail is constant. Model the gravitational force as suddenly switching full on when half the length of the fish is out of the water. Find the value of P.

36. A 5.00-kg block is placed on top of a 10.0-kg block (Fig. P5.36). A horizontal force of 45.0 N is applied to the 10-kg block, and the 5.00-kg block is tied to the wall. The coefficient of kinetic friction between all moving surfaces is 0.200. (a) Draw a free-body diagram for each block and identify the action–reaction forces between the blocks. (b) Determine the tension in the string and the magnitude of the acceleration of the 10.0-kg block.

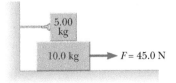

Figure P5.36

ADDITIONAL PROBLEMS

37. A black aluminum glider floats on a film of air above a level aluminum air track. Aluminum feels essentially no force in a magnetic field, and air resistance is negligible. A strong magnet is attached to the top of the glider, forming a total mass of 240 g. A piece of scrap iron attached to one end stop on the track attracts the magnet with a force of 0.823 N when the iron and the magnet are separated by 2.50 cm. (a) Find the acceleration of the glider at this instant. (b) The scrap iron is now attached to another green glider, forming total mass 120 g. Find the acceleration of each glider when the gliders are simultaneously released at 2.50-cm separation.

38. *Why is the following situation impossible?* A book sits on an inclined plane on the surface of the Earth. The angle of the plane with the horizontal is 60.0°. The coefficient of kinetic friction between the book and the plane is 0.300. At time $t = 0$, the book is released from rest. The book then slides through a distance of 1.00 m, measured along the plane, in a time interval of 0.483 s.

39. Two blocks of masses m_1 and m_2 are placed on a table in contact with each other as discussed in Example 5.7 and shown in Figure 5.13a. The coefficient of kinetic friction between the block of mass m_1 and the table is μ_1, and that between the block of mass m_2 and the table is μ_2. A horizontal force of magnitude F is applied to the block of mass m_1. We wish to find P, the magnitude of the contact force between the blocks. (a) Draw diagrams showing the forces for each block. (b) What is the net force on the system of two blocks? (c) What is the net force acting on m_1? (d) What is the net force acting on m_2? (e) Write Newton's second law in the x direction for each block. (f) Solve the two equations in two unknowns for the acceleration a of the blocks in terms of the masses, the applied force F, the coefficients of friction, and g. (g) Find the magnitude P of the contact force between the blocks in terms of the same quantities.

40. A 1.00-kg glider on a horizontal air track is pulled by a string at an angle θ. The taut string runs over a pulley and is attached to a hanging object of mass 0.500 kg as shown in Figure P5.40. (a) Show that the speed v_x of the glider and the speed v_y of the hanging object are related by $v_x = uv_y$, where $u = z(z^2 - h_0^2)^{-1/2}$. (b) The glider is released from rest. Show that at that instant the acceleration a_x of the glider and the acceleration a_y of the hanging object are related by $a_x = ua_y$. (c) Find the tension in the string at the instant the glider is released for $h_0 = 80.0$ cm and $\theta = 30.0°$.

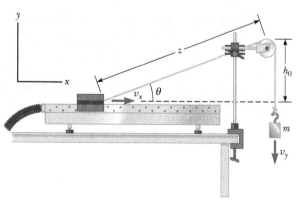

Figure P5.40

41. An inventive child named Nick wants to reach an apple in a tree without climbing the tree. Sitting in a chair connected to a rope that passes over a frictionless pulley (Fig. P5.41), Nick pulls on the loose end of the rope with such a force that the spring scale reads 250 N. Nick's true weight is 320 N, and the chair weighs 160 N. Nick's feet are not touching the ground. (a) Draw one pair of diagrams showing the forces for Nick and the chair considered as separate systems and another diagram for Nick and the chair considered as one system. (b) Show that the acceleration of the system is *upward* and find its magnitude. (c) Find the force Nick exerts on the chair.

Figure P5.41 Problems 41 and 44.

42. A rope with mass m_r is attached to a block with mass m_b as in Figure P5.42. The block rests on a frictionless, horizontal surface. The rope does not stretch. The free end of the rope is pulled to the right with a horizontal force $\vec{F}$. (a) Draw force diagrams for the rope and the block, noting that the tension in the rope is not uniform. (b) Find the acceleration of the system in terms of m_b, m_r, and F. (c) Find the magnitude of the force the rope exerts on the block. (d) What happens to the force on the block as the rope's mass approaches zero? What can you state about the tension in a *light* cord joining a pair of moving objects?

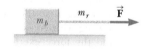

Figure P5.42

43. In Example 5.7, we pushed on two blocks on a table. Suppose three blocks are in contact with one another on a frictionless, horizontal surface as shown in Figure P5.43. A horizontal force $\vec{F}$ is applied to m_1. Take $m_1 = 2.00$ kg, $m_2 = 3.00$ kg, $m_3 = 4.00$ kg, and $F = 18.0$ N. (a) Draw a separate free-body diagram for each block. (b) Determine the acceleration of the blocks. (c) Find the *resultant* force on each block. (d) Find the magnitudes of the contact forces between the blocks. (e) You are working on a construction project. A coworker is nailing up plasterboard on one side of a light partition, and you are on the opposite side, providing "backing" by leaning against the wall with your back pushing on it. Every hammer blow makes your back sting. The supervisor helps you put a heavy block of wood between the wall and your back. Using the situation analyzed in parts (a) through (d) as a model, explain how this change works to make your job more comfortable.

Figure P5.43

44. In the situation described in Problem 41 and Figure P5.41, the masses of the rope, spring balance, and pulley are negligible. Nick's feet are not touching the ground. (a) Assume Nick is momentarily at rest when he stops pulling down on the rope and passes the end of the rope to another child, of weight 440 N, who is standing on the ground next to him. The rope does not break. Describe the ensuing motion. (b) Instead, assume Nick is momentarily at rest when he ties the end of the rope to a strong hook projecting from the tree trunk. Explain why this action can make the rope break.

45. A crate of weight F_g is pushed by a force $\vec{P}$ on a horizontal floor as shown in Figure P5.45. The coefficient of static friction is μ_s, and $\vec{P}$ is directed at angle θ below the horizontal. (a) Show that the minimum value of P that will move the crate is given by

Figure P5.45

$$P = \frac{\mu_s F_g \sec \theta}{1 - \mu_s \tan \theta}$$

(b) Find the condition on θ in terms of μ_s for which motion of the crate is impossible for any value of P.

46. In Figure P5.46, the pulleys and the cord are light, all surfaces are frictionless, and the cord does not stretch. (a) How does the acceleration of block 1 compare with the acceleration of block 2? Explain your reasoning. (b) The mass of block 2 is 1.30 kg. Find its acceleration as it depends on the mass m_1 of block 1. (c) **What If?** What does the result of part (b) predict if m_1 is very much less than 1.30 kg? (d) What does the result of part (b) predict if m_1 approaches infinity? (e) In this last case, what is the tension in the cord? (f) Could you anticipate the answers to parts (c), (d), and (e) without first doing part (b)? Explain.

Figure P5.46

47. You are working as an expert witness for the defense of a container ship captain whose ship ran into a reef surrounding a Caribbean island. The captain is being charged with intentionally running the ship into the reef. In discovery, the following information has been presented, and attorneys on both sides have stipulated that the information is correct: The ship was traveling at 2.50 m/s toward the reef when a mechanical failure caused the rudder to jam in the straight-ahead position. At that point in time, the ship was 900 m from the reef. The wind was blowing directly toward the reef, and exerting a constant force of 9.00×10^3 N on the boat in a direction toward the reef. The mass of the ship and its cargo was 5.50×10^7 kg. During the preparation for the trial, the captain claims that without control of the direction of travel, the only choice he had was to put the engines in reverse at maximum power, such that the total force exerted by the frictional drag force of the water and the force of the water on the propellers was 1.25×10^5 N in a direction away from the reef. From this information, construct a convincing argument that nothing the captain could do in this situation could have prevented the ship from striking the reef.

48. A flat cushion of mass m is released from rest at the corner of the roof of a building, at height h. A wind blowing along the side of the building exerts a constant horizontal force of magnitude F on the cushion as it drops as shown in Figure P5.48. The air exerts no vertical force. (a) Show that the path of the cushion is a straight line. (b) Does the cushion fall with constant velocity? Explain. (c) If $m = 1.20$ kg, $h = 8.00$ m, and $F = 2.40$ N, how far from the building will the cushion hit the level ground? **What If?** (d) If the cushion is thrown downward with a nonzero speed at the top of the building, what will be the shape of its trajectory? Explain.

Figure P5.48

49. What horizontal force must be applied to a large block of mass M shown in Figure P5.49 so that the tan blocks remain stationary relative to M? Assume all surfaces and the pulley are frictionless. Notice that the force exerted by the string accelerates m_2.

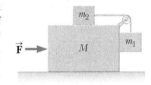

Figure P5.49 Problems 49 and 53

50. An 8.40-kg object slides down a fixed, frictionless, inclined plane. Use a computer to determine and tabulate (a) the normal force exerted on the object and (b) its acceleration for a series of incline angles (measured from the horizontal) ranging from 0° to 90° in 5° increments. (c) Plot a graph of the normal force and the acceleration as functions of the incline angle. (d) In the limiting cases of 0° and 90°, are your results consistent with the known behavior?

CHALLENGE PROBLEMS

51. A block of mass 2.20 kg is accelerated across a rough surface by a light cord passing over a small pulley as shown in Figure P5.51. The tension T in the cord is maintained at 10.0 N, and the pulley is 0.100 m above the top of the block. The coefficient of kinetic friction is 0.400. (a) Determine the acceleration of the block when $x = 0.400$ m. (b) Describe the general behavior of the acceleration as the block slides from a location where x is large to $x = 0$. (c) Find the maximum value of the acceleration and the position x for which it occurs. (d) Find the value of x for which the acceleration is zero.

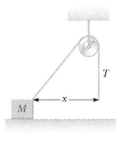

Figure P5.51

52. *Why is the following situation impossible?* A 1.30-kg toaster is not plugged in. The coefficient of static friction between the toaster and a horizontal countertop is 0.350. To make the toaster start moving, you carelessly pull on its electric cord. Unfortunately, the cord has become frayed from your previous similar actions and will break if the tension in the cord exceeds 4.00 N. By pulling on the cord at a particular angle, you successfully start the toaster moving without breaking the cord.

53. Initially, the system of objects shown in Figure P5.49 is held motionless. The pulley and all surfaces and wheels are frictionless. Let the force $\vec{F}$ be zero and assume that m_1 can move only vertically. At the instant after the system of objects is released, find (a) the tension T in the string, (b) the acceleration of m_2, (c) the acceleration of M, and (d) the acceleration of m_1. (*Note:* The pulley accelerates along with the cart.)

54. A mobile is formed by supporting four metal butterflies of equal mass m from a string of length L. The points of support are evenly spaced a distance ℓ apart as shown in Figure P5.54. The string forms an angle θ_1 with the ceiling at each endpoint. The center section of string is horizontal. (a) Find the tension in each section of string in terms of θ_1, m, and g. (b) In terms of θ_1, find the angle θ_2 that the sections of string between the outside butterflies and the inside butterflies form with the horizontal. (c) Show that the distance D between the endpoints of the string is

$$D = \frac{L}{5}\left\{2\cos\theta_1 + 2\cos\left[\tan^{-1}\left(\tfrac{1}{2}\tan\theta_1\right)\right] + 1\right\}$$

55. In Figure P5.55, the incline has mass M and is fastened to the stationary horizontal tabletop. The block of mass m is placed near the bottom of the incline and is released with a quick push that sets it sliding upward. The block stops near the top of the incline as shown in the figure and then slides down again, always without friction. Find the force that the tabletop exerts on the incline throughout this motion in terms of m, M, g, and θ.

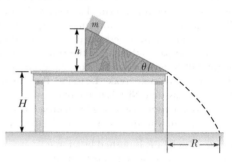

Figure P5.55

Figure P5.54

Circular Motion and Other Applications of Newton's Laws

The Mad Tea Party at Disneyland is a ride with lots of circular motion. Each cup rotates around a central axis. In addition, six cups are mounted on a rotating turntable. Furthermore, three such turntables are mounted on a large turntable rotating in the opposite direction to the smaller turntables. (*Pascal Le Segretain/Getty Images News/Getty Images*)

STORYLINE **You have no classes today and decide to spend the day** at Disneyland with a friend. It's a weekday, so the lines are relatively short. In fact, even the line for the Mad Tea Party is short! You have read online that this ride is available at all of Disney's parks around the world, even at the newest Disneyland in Shanghai, China. For this ride, you and your friend sit in a large tea cup that spins rapidly. While your friend pulls on the wheel in the center of the teacup to make it spin quickly, you hang your smartphone from a string to form a pendulum. You dangle the pendulum from your hand at the rim of the tea cup. You notice that the pendulum does not hang straight down! You open a special app on the smartphone that gives you a readout of the angle of the phone with respect to the vertical and hang the phone again as a pendulum. Why does the pendulum deviate from the vertical? In what direction does the pendulum deviate from the vertical? What happens to the angle reading on the phone as you move your hand holding the pendulum toward the center of the tea cup? Why does the reading change in this way?

CONNECTIONS In this chapter, we expand on the circular motion we studied in Chapter 4, combining it with our new knowledge about force from Chapter 5. What forces act on an object when it is in circular motion? In addition, we consider some other cases in which Newton's laws help us to understand the motion. We will consider how the laws of physics appear when one is in an accelerated frame of reference, such as the spinning teacup in the storyline. We will also extend our discussions of friction from Chapter 5 by looking at resistive forces on an object, such as air resistance. Unlike in our model of kinetic friction, these forces vary in magnitude according to the speed of the object relative to the surrounding medium. In future chapters, we will see several examples of circular motion, such as planets in orbit in Chapter 13, charged particles moving in circular paths in magnetic fields in Chapter 28, and electrons in circular orbits in the Bohr theory of the hydrogen atom in Chapter 41. The action of particles undergoing resistive forces will appear in future chapters, and we will see electrical analogs to resistive forces in various kinds of electric circuits in Chapters 27 and 31. The material in this chapter on accelerated reference frames will be followed up with the discussion of general relativity in Chapter 38.

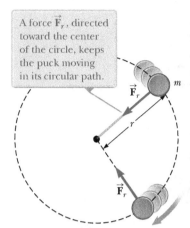

A force $\vec{\mathbf{F}}_r$, directed toward the center of the circle, keeps the puck moving in its circular path.

Figure 6.1 An overhead view of a puck moving in a circular path in a horizontal plane.

6.1 Extending the Particle in Uniform Circular Motion Model

In Section 4.4, we discussed the analysis model of a particle in uniform circular motion, in which a particle moves with constant speed v in a circular path having a radius r. The particle experiences an acceleration that has a magnitude

$$a_c = \frac{v^2}{r}$$

The acceleration is called *centripetal acceleration* because $\vec{\mathbf{a}}_c$ is directed toward the center of the circle. Furthermore, $\vec{\mathbf{a}}_c$ is *always* perpendicular to $\vec{\mathbf{v}}$. (If there were a component of acceleration parallel to $\vec{\mathbf{v}}$, the particle's speed would be changing.)

Let us now extend the particle in uniform circular motion model from Section 4.4 by incorporating the concept of force. Consider a puck of mass m that is tied to a string of length r and moves at constant speed in a horizontal, circular path as illustrated in the overhead view in Figure 6.1. Its weight is supported by the normal force from a frictionless table, and the string is anchored to a peg at the center of the circular path of the puck. Why does the puck move in a circle? According to Newton's first law, the puck would move in a straight line if there were no force on it; the string, however, prevents motion along a straight line by exerting on the puck a radial force $\vec{\mathbf{F}}_r$ that makes it follow the circular path. This force is directed along the string toward the center of the circle as shown in Figure 6.1.

If Newton's second law is applied along the radial direction, the net force causing the centripetal acceleration can be related to the acceleration as follows:

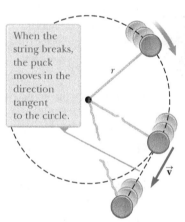

When the string breaks, the puck moves in the direction tangent to the circle.

Figure 6.2 The string holding the puck in its circular path breaks.

Force causing centripetal ▶ acceleration

$$\sum F = ma_c = m\frac{v^2}{r} \tag{6.1}$$

A force causing a centripetal acceleration acts toward the center of the circular path and causes a change in the direction of the velocity vector. If that force should vanish, the object would no longer move in its circular path; instead, it would move along a straight-line path tangent to the circle. This idea is illustrated in Figure 6.2 for the puck moving in a circular path at the end of a string in a horizontal plane. If the string breaks at some instant, the puck moves along the straight-line path that is tangent to the circle at the position of the puck at this instant.

PITFALL PREVENTION 6.1

Direction of Travel When the String Is Cut Study Figure 6.2 very carefully. Many students (wrongly) think that the puck will move *radially* away from the center of the circle when the string is cut. The velocity of the puck is *tangent* to the circle. By Newton's first law, the puck continues to move in the same direction in which it is moving just as the force from the string disappears.

QUICK QUIZ 6.1 You are riding on a Ferris wheel that is rotating with constant speed. The car in which you are riding always maintains its correct upward orientation; it does not invert. **(i)** What is the direction of the normal force on you from the seat when you are at the top of the wheel? (a) upward (b) downward (c) impossible to determine **(ii)** From the same choices, what is the direction of the net force on you when you are at the top of the wheel?

ANALYSIS MODEL Particle in Uniform Circular Motion (Extension)

Imagine a moving object that can be modeled as a particle. If it moves in a circular path of radius r at a constant speed v, it experiences a centripetal acceleration. Because the particle is accelerating, there must be a net force acting on the particle. That force is directed toward the center of the circular path and is given by

$$\sum F = ma_c = m\frac{v^2}{r} \qquad \text{(6.1)}$$

Examples

- the tension in a string of constant length acting on a rock twirled in a circle

- the gravitational force acting on a planet traveling around the Sun in a perfectly circular orbit (Chapter 13)
- the magnetic force acting on a charged particle moving in a uniform magnetic field (Chapter 28)
- the electric force acting on an electron in orbit around a nucleus in the Bohr model of the hydrogen atom (Chapter 41)

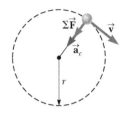

Example 6.1   The Conical Pendulum

A small ball of mass m is suspended from a string of length L. The ball revolves with constant speed v in a horizontal circle of radius r as shown in Figure 6.3. (Because the string sweeps out the surface of a cone, the system is known as a *conical pendulum*.) Find an expression for v in terms of the length of the string and the angle it makes with the vertical in Figure 6.3.

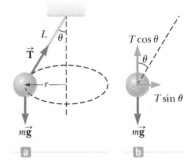

SOLUTION

Conceptualize Imagine the motion of the ball in Figure 6.3a and convince yourself that the string sweeps out a cone and that the ball moves in a horizontal circle. What happens if the ball moves with a higher speed?

Categorize The ball in Figure 6.3 does not accelerate vertically. Therefore, we model it as a *particle in equilibrium* in the vertical direction. It experiences a centripetal acceleration in the horizontal direction, so it is modeled as a *particle in uniform circular motion* in this direction.

Figure 6.3 (Example 6.1) (a) A conical pendulum. The path of the ball is a horizontal circle. (b) The forces acting on the ball.

Analyze Let θ represent the angle between the string and the vertical. In the diagram of forces acting on the ball in Figure 6.3b, the force $\vec{T}$ exerted by the string on the ball is resolved into a vertical component $T\cos\theta$ and a horizontal component $T\sin\theta$ acting toward the center of the circular path.

Apply the particle in equilibrium model in the vertical direction:

$$\sum F_y = T\cos\theta - mg = 0$$

$$(1) \quad T\cos\theta = mg$$

Use Equation 6.1 from the particle in uniform circular motion model in the horizontal direction:

$$(2) \quad \sum F_x = T\sin\theta = ma_c = \frac{mv^2}{r}$$

Divide Equation (2) by Equation (1) and use $\sin\theta/\cos\theta = \tan\theta$:

$$\tan\theta = \frac{v^2}{rg}$$

Solve for v:

$$v = \sqrt{rg\tan\theta}$$

Incorporate $r = L\sin\theta$ from the geometry in Figure 6.3a:

$$(3) \quad v = \sqrt{Lg\sin\theta\tan\theta}$$

Finalize Notice that the speed is independent of the mass of the ball. Consider what happens when θ goes to $90°$ so that the string is horizontal. Because the tangent of $90°$ is infinite, the speed v is infinite, which tells us the string cannot possibly be horizontal. If it were, there would be no vertical component of the force $\vec{T}$ to balance the gravitational force on the ball. The situation in this problem is similar in some ways to your experience on the Mad Tea Party ride. As you move your hand holding the hanging smartphone toward the center of your spinning cup, the speed v of the phone changes, resulting in a different angle θ, as suggested by Equation (3).

Example 6.2 How Fast Can It Spin?

A puck of mass 0.500 kg is attached to the end of a cord 1.50 m long. The puck moves in a horizontal circle as shown in Figure 6.1. If the cord can withstand a maximum tension of 50.0 N, what is the maximum speed at which the puck can move before the cord breaks? Assume the string remains horizontal during the motion.

SOLUTION

Conceptualize It makes sense that the stronger the cord, the faster the puck can move before the cord breaks. Also, we expect a more massive puck to break the cord at a lower speed. (Imagine whirling a bowling ball on the cord!)

Categorize Because the puck moves in a circular path, we model it as a *particle in uniform circular motion*.

Analyze Incorporate the tension and the centripetal acceleration into Newton's second law as described by Equation 6.1:

$$T = m\frac{v^2}{r}$$

Solve for v:

$$(1) \quad v = \sqrt{\frac{Tr}{m}}$$

Find the maximum speed the puck can have, which corresponds to the maximum tension the string can withstand:

$$v_{max} = \sqrt{\frac{T_{max}r}{m}} = \sqrt{\frac{(50.0\ \text{N})(1.50\ \text{m})}{0.500\ \text{kg}}} = 12.2\ \text{m/s}$$

Finalize Equation (1) shows that v increases with T and decreases with larger m, as we expected from our conceptualization of the problem.

WHAT IF? Suppose the puck moves in a circle of larger radius at the same speed v. Is the cord more likely or less likely to break?

Answer The larger radius means that the change in the direction of the velocity vector will be smaller in a given time interval. Therefore, the acceleration is smaller and the required tension in the string is smaller. As a result, the string is less likely to break when the puck travels in a circle of larger radius.

Example 6.3 What Is the Maximum Speed of the Car?

A 1 500-kg car moving on a flat, horizontal road negotiates a curve as shown in the overhead view in Figure 6.4a. If the radius of the curve is 35.0 m and the coefficient of static friction between the tires and dry pavement is 0.523, find the maximum speed the car can have and still make the turn successfully.

SOLUTION

Conceptualize Imagine that the curved roadway is part of a large circle so that the car is moving in a circular path.

Categorize Based on the Conceptualize step of the problem, we model the car as a *particle in uniform circular motion* in the horizontal direction. The car is not accelerating vertically, so it is modeled as a *particle in equilibrium* in the vertical direction.

Analyze The back view in Figure 6.4b shows the forces on the car. The force that enables the car to remain in its circular path is the force of static friction. (It is *static* because no slipping occurs at the point of contact between road and tires. If this force of static friction were zero—for example, if the car were on an icy road—the car would continue in a straight line and slide off the curved road.) The maximum speed v_{max} the car can have around the curve is the speed at which it is on the verge of skidding outward. At this point, the friction force has its maximum value $f_{s,max} = \mu_s n$.

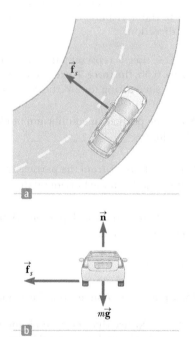

Figure 6.4 (Example 6.3) (a) The force of static friction directed toward the center of the curve keeps the car moving in a circular path. (b) The forces acting on the car.

6.3 c o n t i n u e d

Apply Equation 6.1 from the particle in uniform circular motion model in the radial direction for the maximum speed condition:

$$(1) \quad f_{s,\text{max}} = \mu_s n = m\frac{v_{\text{max}}^2}{r}$$

Apply the particle in equilibrium model to the car in the vertical direction:

$$\sum F_y = 0 \quad \rightarrow \quad n - mg = 0 \quad \rightarrow \quad n = mg$$

Solve Equation (1) for the maximum speed and substitute for n:

$$(2) \quad v_{\text{max}} = \sqrt{\frac{\mu_s n r}{m}} = \sqrt{\frac{\mu_s mgr}{m}} = \sqrt{\mu_s gr}$$

Substitute numerical values:

$$v_{\text{max}} = \sqrt{(0.523)(9.80 \text{ m/s}^2)(35.0 \text{ m})} = 13.4 \text{ m/s}$$

Finalize This speed is equivalent to 30.0 mi/h. Therefore, if the speed limit on this roadway is higher than 30 mi/h, this roadway could benefit greatly from some banking, as in the next example! Notice that the maximum speed does not depend on the mass of the car, which is why curved highways do not need multiple speed limits to cover the various masses of vehicles using the road.

W H A T   I F ? Suppose a car travels this curve on a wet day and begins to skid on the curve when its speed reaches only 8.00 m/s. What can we say about the coefficient of static friction in this case?

Answer The coefficient of static friction between the tires and a wet road should be smaller than that between the tires and a dry road. This expectation is consistent with experience with driving because a skid is more likely on a wet road than a dry road.

To check our suspicion, we can solve Equation (2) for the coefficient of static friction:

$$\mu_s = \frac{v_{\text{max}}^2}{gr}$$

Substituting the numerical values gives

$$\mu_s = \frac{v_{\text{max}}^2}{gr} = \frac{(8.00 \text{ m/s})^2}{(9.80 \text{ m/s}^2)(35.0 \text{ m})} = 0.187$$

which is indeed smaller than the coefficient of 0.523 for the dry road.

Example **6.4** The Banked Roadway

You are a civil engineer who has been given the assignment to redesign the curved roadway in Example 6.3 in such a way that a car will not have to rely on friction to round the curve without skidding. In other words, a car moving at the designated speed can negotiate the curve even when the road is covered with ice. Such a road is usually *banked*, which means that the roadway is tilted toward the inside of the curve. Suppose the designated speed for the road is to be 13.4 m/s (30.0 mi/h) and the radius of the curve is 35.0 m. You need to determine the angle at which the roadway on the curve should be banked.

S O L U T I O N

Conceptualize The difference between this example and Example 6.3 is that the car is no longer moving on a flat roadway. Figure 6.5 shows the banked roadway, with the center of the circular path of the car far to the left of the figure. Notice that the horizontal component of the normal force participates in causing the car's centripetal acceleration. Note in Figure 6.5 that, unlike the motion on an inclined plane in Figure 5.12, we use a coordinate system with x horizontal and y upward. This choice is made so that the centripetal acceleration of the car is purely in the x direction.

Categorize As in Example 6.3, the car is modeled as a *particle in equilibrium* in the vertical direction and a *particle in uniform circular motion* in the horizontal direction.

Analyze On a level (unbanked) road, the force that causes the centripetal acceleration is the force of static friction between tires and the road as we saw in the preceding example.

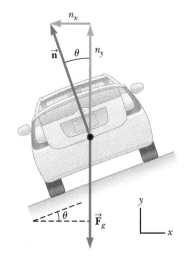

Figure 6.5 (Example 6.4) A car moves into the page and is rounding a curve on a road banked at an angle θ to the horizontal. When friction is neglected, the force that causes the centripetal acceleration and keeps the car moving in its circular path is the horizontal component of the normal force.

continued

6.4 continued

If the road is banked at an angle θ as in Figure 6.5, however, the normal force $\vec{\mathbf{n}}$ has a horizontal component toward the center of the curve. Because the road is to be designed so that the force of static friction is zero, the component $n_x = n \sin \theta$ is the only force that causes the centripetal acceleration.

Write Newton's second law for the car in the radial direction, which is the $-x$ direction:

$$(1) \quad \sum F_r = n \sin \theta = \frac{mv^2}{r}$$

Apply the particle in equilibrium model to the car in the vertical direction:

$$\sum F_y = n \cos \theta - mg = 0$$

$$(2) \quad n \cos \theta = mg$$

Divide Equation (1) by Equation (2):

$$(3) \quad \tan \theta = \frac{v^2}{rg}$$

Solve for the angle θ and substitute numerical values:

$$\theta = \tan^{-1} \left[\frac{(13.4 \text{ m/s})^2}{(35.0 \text{ m})(9.80 \text{ m/s}^2)} \right] = 27.6°$$

Finalize   Equation (3) shows that the banking angle is independent of the mass of the vehicle negotiating the curve. If a car rounds the curve at a speed less than 13.4 m/s, the centripetal acceleration decreases. Therefore, the normal force, which is unchanged, is sufficient to cause *two* accelerations: the lower centripetal acceleration and an acceleration of the car down the inclined roadway. Consequently, an additional friction force parallel to the roadway and upward is needed to keep the car from sliding down the bank (to the left in Fig. 6.5). Similarly, a driver attempting to negotiate the curve at a speed greater than 13.4 m/s has to depend on friction to keep from sliding up the bank (to the right in Fig. 6.5). See problem 41 for an analysis of this situation.

 WHAT IF?   Imagine that this same roadway were built on Mars in the future to connect different colony centers. Could it be traveled at the same speed?

Answer   The reduced gravitational force on Mars would mean that the car is not pressed as tightly to the roadway. The reduced normal force results in a smaller component of the normal force toward the center of the circle. This smaller component would not be sufficient to provide the centripetal acceleration associated with the original speed. The centripetal acceleration must be reduced, which can be done by reducing the speed v.

Mathematically, notice that Equation (3) shows that the speed v is proportional to the square root of g for a roadway of fixed radius r banked at a fixed angle θ. Therefore, if g is smaller, as it is on Mars, the speed v with which the roadway can be safely traveled is also smaller.

Example **6.5**   Riding the Ferris Wheel

A child of mass m rides on a Ferris wheel as shown in Figure 6.6a. The child moves in a vertical circle of radius 10.0 m at a constant speed of 3.00 m/s.

(A) Determine the force exerted by the seat on the child at the bottom of the ride. Express your answer in terms of the weight of the child, mg.

SOLUTION

Conceptualize   Look carefully at Figure 6.6a. Based on experiences you may have had on a Ferris wheel or driving over small hills on a roadway, you would expect to feel lighter at the top of the path. Similarly, you would expect to feel heavier at the bottom of the path. At both the bottom of the path and the top, the normal and gravitational forces on the child act in *opposite* directions. The vector sum of these two forces gives a force of constant magnitude that keeps the child moving in a circular path at a constant speed. To yield net force vectors with the same magnitude, the normal force at the bottom must be greater than that at the top.

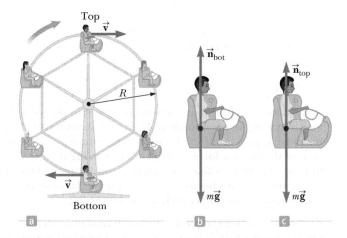

Figure 6.6   (Example 6.5) (a) A child rides on a Ferris wheel. (b) The forces acting on the child at the bottom of the path. (c) The forces acting on the child at the top of the path.

6.5 continued

Categorize Because the speed of the child is constant, we can categorize this problem as one involving a *particle* (the child) *in uniform circular motion*, complicated by the gravitational force acting at all times on the child.

Analyze We draw a diagram of forces acting on the child at the bottom of the ride as shown in Figure 6.6b. The only forces acting on him are the downward gravitational force $\vec{\mathbf{F}}_g = m\vec{\mathbf{g}}$ and the upward force $\vec{\mathbf{n}}_{bot}$ exerted by the seat. The centripetal acceleration of the child at this point is upward and the net upward force on the child has a magnitude $n_{bot} - mg$.

Using the particle in uniform circular motion model, apply Newton's second law to the child in the radial direction when he is at the bottom of the ride:

$$\sum F = n_{bot} - mg = m\frac{v^2}{r}$$

Solve for the force exerted by the seat on the child:

$$n_{bot} = mg + m\frac{v^2}{r} = mg\left(1 + \frac{v^2}{rg}\right)$$

Substitute numerical values given for the speed and radius:

$$n_{bot} = mg\left[1 + \frac{(3.00 \text{ m/s})^2}{(10.0 \text{ m})(9.80 \text{ m/s}^2)}\right]$$

$$= 1.09 \, mg$$

Hence, the magnitude of the force $\vec{\mathbf{n}}_{bot}$ exerted by the seat on the child is *greater* than the weight of the child by a factor of 1.09. So, the child experiences an apparent weight that is greater than his true weight by a factor of 1.09.

(B) Determine the force exerted by the seat on the child at the top of the ride.

SOLUTION

Analyze The diagram of forces acting on the child at the top of the ride is shown in Figure 6.6c. The centripetal acceleration of the child at this point is downward and the net downward force has a magnitude $mg - n_{top}$.

Apply Newton's second law to the child at this position:

$$\sum F = mg - n_{top} = m\frac{v^2}{r}$$

Solve for the force exerted by the seat on the child:

$$n_{top} = mg - m\frac{v^2}{r} = mg\left(1 - \frac{v^2}{rg}\right)$$

Substitute numerical values:

$$n_{top} = mg\left[1 - \frac{(3.00 \text{ m/s})^2}{(10.0 \text{ m})(9.80 \text{ m/s}^2)}\right]$$

$$= 0.908 \, mg$$

In this case, the magnitude of the force exerted by the seat on the child is *less* than his true weight by a factor of 0.908, and the child feels lighter.

Finalize The variations in the normal force are consistent with our prediction in the Conceptualize step of the problem.

WHAT IF? Suppose a defect in the Ferris wheel mechanism causes the speed of the child to increase to 10.0 m/s. What does the child experience at the top of the ride in this case?

Answer If the calculation above is performed with $v = 10.0$ m/s, the magnitude of the normal force at the top of the ride is negative, which is impossible. We interpret it to mean that the required downward centripetal acceleration of the child is larger than that due to gravity. As a result, the child will lose contact with the seat and will only stay in his circular path if there is a safety bar or a seat belt that provides a downward force on him to keep him in his seat. At the bottom of the ride, the normal force is 2.02 *mg*, which would be uncomfortable.

6.2 **Nonuniform Circular Motion**

In Chapter 4, we found that if a particle moves with varying speed in a circular path, there is, in addition to the radial component of acceleration, a tangential component having magnitude $|dv/dt|$. Therefore, the force acting on the particle must also have a tangential and a radial component. Because the total acceleration is $\vec{\mathbf{a}} = \vec{\mathbf{a}}_r + \vec{\mathbf{a}}_t$, the total force exerted on the particle is $\sum \vec{\mathbf{F}} = \sum \vec{\mathbf{F}}_r + \sum \vec{\mathbf{F}}_t$

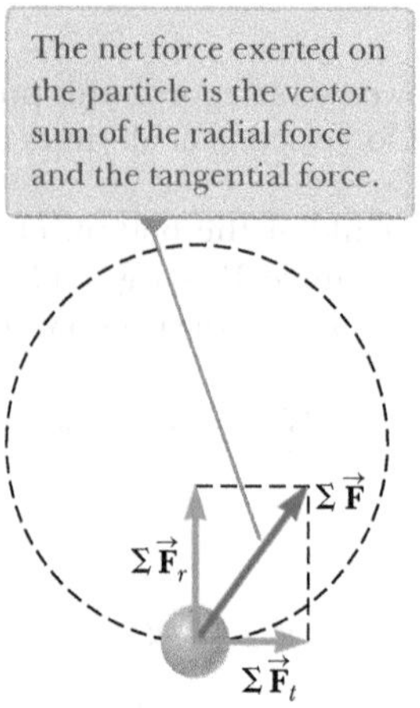

The net force exerted on the particle is the vector sum of the radial force and the tangential force.

Figure 6.7 When the net force acting on a particle moving in a circular path has a tangential component ΣF_t, the particle's speed changes.

$$\Sigma \vec{\mathbf{F}}$$

$$\Sigma \vec{\mathbf{F}}_r$$

$$\Sigma \vec{\mathbf{F}}_t$$

as shown in Figure 6.7. (We express the radial and tangential forces as net forces with the summation notation because each force could consist of multiple forces that combine.) The vector $\Sigma \vec{\mathbf{F}}_r$ is directed toward the center of the circle and is responsible for the centripetal acceleration. The vector $\Sigma \vec{\mathbf{F}}_t$ tangent to the circle is responsible for the tangential acceleration, which represents a change in the particle's speed with time.

QUICK QUIZ 6.2 A bead slides at constant speed along a curved wire lying on a horizontal surface as shown in Figure 6.8. **(a)** Draw the vectors representing the force exerted by the wire on the bead at points Ⓐ, Ⓑ, and Ⓒ. **(b)** Suppose the bead in Figure 6.8 speeds up with constant tangential acceleration as it moves toward the right. Draw the vectors representing the forces on the bead at points Ⓐ, Ⓑ, and Ⓒ.

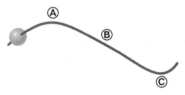

Figure 6.8 (Quick Quiz 6.2) A bead slides along a curved wire.

Example **6.6** Keep Your Eye on the Ball

A small sphere of mass m is attached to the end of a cord of length R and set into motion in a *vertical* circle about a fixed point O as illustrated in Figure 6.9. Determine the tangential acceleration of the sphere and the tension in the cord at any instant when the speed of the sphere is v and the cord makes an angle θ with the vertical.

SOLUTION

Conceptualize Compare the motion of the sphere in Figure 6.9 with that of the child in Figure 6.6a associated with Example 6.5. Both objects travel in a circular path. Unlike the child in Example 6.5, however, the speed of the sphere is *not* uniform in this example because, at most points along the path, a tangential component of acceleration arises from the gravitational force exerted on the sphere.

Categorize We model the sphere as a *particle under a net force* and moving in a circular path, but it is not a particle in *uniform* circular motion. We need to use the techniques discussed in this section on nonuniform circular motion.

Analyze From the force diagram in Figure 6.9, we see that the only forces acting on the sphere are the gravitational force $\vec{\mathbf{F}}_g = m\vec{\mathbf{g}}$ exerted by the Earth and the force $\vec{\mathbf{T}}$ exerted by the cord. We resolve $\vec{\mathbf{F}}_g$ into a tangential component $mg \sin \theta$ and a radial component $mg \cos \theta$.

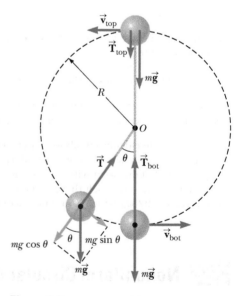

Figure 6.9 (Example 6.6) The forces acting on a sphere of mass m connected to a cord of length R and rotating in a vertical circle centered at O. Forces acting on the sphere are shown when the sphere is at the top and bottom of the circle and at an arbitrary location.

6.6 continued

From the particle under a net force model, apply Newton's second law to the sphere in the tangential direction:

$$\sum F_t = mg \sin \theta = ma_t$$

$$a_t = g \sin \theta$$

Apply Newton's second law to the forces acting on the sphere in the radial direction, noting that both $\vec{\mathbf{T}}$ and $\vec{\mathbf{a}}_r$ are directed toward O. As noted in Section 4.5, we can use Equation 4.21 for the instantaneous centripetal acceleration of a particle even when it moves in nonuniform circular motion:

$$\sum F_r = T - mg \cos \theta = \frac{mv^2}{R}$$

$$T = mg \left(\frac{v^2}{Rg} + \cos \theta \right)$$

Finalize Let us evaluate this result at the top and bottom of the circular path (Fig. 6.9):

$$T_{top} = mg \left(\frac{v_{top}^2}{Rg} - 1 \right) \qquad T_{bot} = mg \left(\frac{v_{bot}^2}{Rg} + 1 \right)$$

These results have similar mathematical forms as those for the normal forces n_{top} and n_{bot} on the child in Example 6.5, which is consistent with the normal force on the child playing a similar physical role in Example 6.5 as the tension in the string plays in this example. Keep in mind, however, that the normal force $\vec{\mathbf{n}}$ on the child in Example 6.5 is always upward, whereas the force $\vec{\mathbf{T}}$ in this example changes direction because it must always point inward along the string. Also note that v in the expressions above varies for different positions of the sphere, as indicated by the subscripts, whereas v in Example 6.5 is constant.

WHAT IF? What if the sphere is set in motion with a slower speed?

(A) What speed would the sphere have as it passes over the top of the circle if the tension in the cord goes to zero instantaneously at this point?

Answer Let us set the tension equal to zero in the expression for T_{top}:

$$0 = mg \left(\frac{v_{top}^2}{Rg} - 1 \right) \quad \rightarrow \quad v_{top} = \sqrt{gR}$$

(B) What if the sphere is set in motion such that the speed at the top is less than this value? What happens?

Answer In this case, the sphere never reaches the top of the circle. At some point on the way up, the tension in the string goes to zero and the sphere becomes a projectile. It follows a segment of a parabolic path, with its peak below the topmost position of the sphere shown in Figure 6.9, rejoining the circular path on the other side when the tension becomes nonzero again.

6.3 Motion in Accelerated Frames

Newton's laws of motion, which we introduced in Chapter 5, describe observations that are made in an inertial frame of reference. In this section, we analyze how Newton's laws are applied by an observer in a noninertial frame of reference, that is, one that is accelerating. For example, recall the discussion of the air hockey table on a train in Section 5.2. The train moving at constant velocity represents an inertial frame. An observer on the train sees the puck at rest remain at rest, and Newton's first law appears to be obeyed. The accelerating train is not an inertial frame. According to you as the observer on this train, there appears to be no force on the puck, yet it accelerates from rest toward the back of the train, appearing to violate Newton's first law. This property is a general property of observations made in noninertial frames: there appear to be unexplained accelerations of objects that are not "fastened" to the frame. Newton's first law is not violated, of course. It only appears to be violated because of observations made from a noninertial frame.

On the accelerating train, as you watch the puck accelerating toward the back of the train, you might conclude based on your belief in Newton's second law that a force has acted on the puck to cause it to accelerate. We call an apparent force such as this one a **fictitious force** because it is not a real force and is due only to

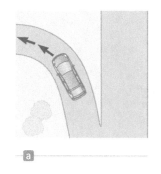

From the passenger's frame of reference, a force appears to push her toward the right door, but it is a fictitious force.

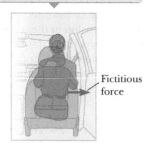

Fictitious force

b

Relative to the reference frame of the Earth, the car seat applies a real force (friction) toward the left on the passenger, causing her to change direction along with the rest of the car.

Real force

c

Figure 6.10 (a) A car approaching a curved exit ramp. What causes a passenger in the front seat to move toward the right-hand door? (b) Passenger's frame of reference. (c) Reference frame of the Earth.

observations made in an accelerated reference frame. A fictitious force appears to act on an object in the same way as a real force. Real forces are always interactions between two objects, however, and you cannot identify a second object for a fictitious force. (What second object is interacting with the puck to cause it to accelerate?) In general, simple fictitious forces appear to act in the direction *opposite* that of the acceleration of the noninertial frame. For example, the train accelerates forward and there appears to be a fictitious force causing the puck to slide toward the back of the train.

The train example describes a fictitious force due to a change in the train's speed. Another fictitious force is due to the change in the *direction* of the velocity vector. To understand the motion of a system that is noninertial because of a change in direction, consider a car traveling along a highway at a high speed and approaching a curved exit ramp on the left as shown in Figure 6.10a. As the car takes the sharp left turn on the ramp, a person sitting in the passenger seat leans or slides to the right and hits the door. At that point the force exerted by the door on the passenger keeps her from being ejected from the car. What causes her to move toward the door? A popular but incorrect explanation is that a force acting toward the right in Figure 6.10b pushes the passenger outward from the center of the circular path. Although often called the "centrifugal force," it is a fictitious force. The car represents a noninertial reference frame that has a centripetal acceleration toward the center of its circular path. As a result, the passenger feels an apparent force which is outward from the center of the circular path, or to the right in Figure 6.10b, in the direction opposite that of the acceleration.

Let us address this phenomenon in terms of Newton's laws. Before the car enters the ramp, the passenger is moving in a straight-line path. As the car enters the ramp and travels a curved path, the passenger tends to move along the original straight-line path, which is in accordance with Newton's first law: the natural tendency of an object is to continue moving in a straight line. If a sufficiently large force (toward the center of curvature) acts on the passenger as in Figure 6.10c, however, she moves in a curved path along with the car. This force is the force of friction between her and the car seat. If this friction force is not large enough, the seat follows a curved path while the passenger tends to continue in the straight-line path of the car before the car began the turn. Therefore, from the point of view of an observer in the car, the passenger leans or slides to the right relative to the seat. Eventually, she encounters the door, which provides a force large enough to enable her to follow the same curved path as the car.

Another interesting fictitious force is the "Coriolis force." It is an apparent force caused by changing the radial position of an object in a rotating coordinate system.

For example, suppose you and a friend are on opposite sides of a rotating circular platform and you decide to throw a baseball to your friend. Figure 6.11a represents what an observer would see if the ball is viewed while the observer is hovering at rest above the rotating platform. According to this observer, who is in an inertial frame, the ball follows a straight line as it must according to Newton's first law. At $t = 0$ you throw the ball toward your friend, but by the time t_f when the ball has crossed the platform, your friend has moved to a new position and can't catch the ball. Now, however, consider the situation from your friend's viewpoint. Your friend is in a noninertial reference frame because he is undergoing a centripetal acceleration relative to the inertial frame of the Earth's surface. He starts off seeing the baseball coming toward him, but as it crosses the platform, it veers to one side as shown in Figure 6.11b. Therefore, your friend on the rotating platform states that the ball does not obey Newton's first law and claims that a sideways force is causing the ball to follow a curved path. This fictitious force is called the Coriolis force.

Fictitious forces may not be real forces, but they can have real effects. An object on your dashboard *really* slides off if you press the accelerator of your car. As you ride on a merry-go-round, you feel pushed toward the outside as if due to the fictitious "centrifugal force." You are likely to fall over and injure yourself due to the

By the time t_f that the ball arrives at the other side
of the platform, your friend is no longer there to
catch it. According to this observer, the ball follows
a straight-line path, consistent with Newton's laws.

From your friend's point of view, the ball veers to
one side during its flight. Your friend introduces a
fictitious force to explain this deviation from the
expected path.

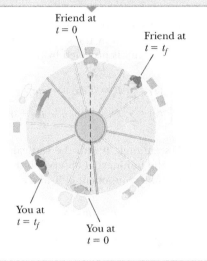

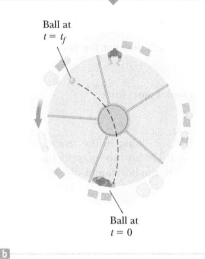

a   **b**

Figure 6.11 You and your friend stand at the edge of a rotating circular platform. You throw the
ball at $t = 0$ in the direction of your friend. (a) Overhead view observed by someone in an inertial
reference frame attached to the Earth. The ground appears stationary, and the platform rotates
clockwise. (b) Overhead view observed by someone in an inertial reference frame attached to the
platform. The platform appears stationary, and the ground rotates counterclockwise.

Coriolis force if you walk along a radial line while a merry-go-round rotates. (One
of the authors did so and suffered a separation of the ligaments from his ribs when
he fell over.) The Coriolis force due to the rotation of the Earth is responsible for
rotations of hurricanes and for large-scale ocean currents.

QUICK QUIZ 6.3 Consider the passenger in the car making a left turn in
Figure 6.10. Which of the following is correct about forces in the horizontal
direction if she is making contact with the right-hand door? **(a)** The passenger
is in equilibrium between real forces acting to the right and real forces acting
to the left. **(b)** The passenger is subject only to real forces acting to the right.
(c) The passenger is subject only to real forces acting to the left. **(d)** None of
those statements is true.

Example 6.7   Fictitious Forces in Circular Motion

Consider the experiment described in the opening storyline: you are riding on the Mad Tea Party ride and holding your
smartphone hanging from a string. Now suppose your friend stands on solid ground beside the ride watching you. You
hold the upper end of the string above a point near the outer rim of the spinning tea cup. Both the inertial observer (your
friend) and the noninertial observer (you) agree that the string makes an angle θ with respect to the vertical. You claim
that a force, which we know to be fictitious, causes the observed deviation of the string from the vertical. How is the magnitude of this force related to the smartphone's centripetal acceleration measured by the inertial observer?

SOLUTION

Conceptualize Place yourself in the role of each of the two observers. The inertial observer on the ground knows that the
smartphone has a centripetal acceleration and that the deviation of the string is related to this acceleration. As the noninertial
observer on the teacup, imagine that you ignore any effects of the spinning of the teacup, so you have no knowledge of any
centripetal acceleration. Because you are unaware of this acceleration, you claim that a force is pushing sideways on the smartphone to cause the deviation of the string from the vertical. To make the conceptualization more real, try running from rest
while holding a hanging object on a string and notice that the string is at an angle to the vertical while you are accelerating, as
if a force is pushing the object backward.

continued

6.7 c o n t i n u e d

Categorize   For the inertial observer, we model the smartphone as a *particle under a net force* in the horizontal direction and a *particle in equilibrium* in the vertical direction. For the noninertial observer, the smartphone is modeled as a *particle in equilibrium* in both directions.

Analyze   The geometry for the spinning and hanging smartphone will be similar to that shown for the ball in Figure 6.3b. According to the inertial observer at rest, the forces on the smartphone are the force $\vec{T}$ exerted by the string and the gravitational force. The inertial observer concludes that the smartphone's centripetal acceleration is provided by the horizontal component of $\vec{T}$.

For this observer, apply the particle under a net force and particle in equilibrium models:

Inertial observer
$$\begin{cases} (1)\ \sum F_x = T\sin\theta = ma_c \\ (2)\ \sum F_y = T\cos\theta - mg = 0 \end{cases}$$

According to the noninertial observer riding in the teacup, the string also makes an angle θ with the vertical; to that observer, however, the smartphone is at rest and so its acceleration is zero. Therefore, the noninertial observer introduces a force (which we know to be fictitious) in the horizontal direction to balance the horizontal component of $\vec{T}$ and claims that the net force on the smartphone is zero.

Apply the particle in equilibrium model for this observer in both directions:

Noninertial observer
$$\begin{cases} \sum F'_x = T\sin\theta - F_{\text{fictitious}} = 0 \\ \sum F'_y = T\cos\theta - mg = 0 \end{cases}$$

These expressions are equivalent to Equations (1) and (2) if $F_{\text{fictitious}} = ma_c$, where a_c is the centripetal acceleration of the smartphone according to the inertial observer.

Finalize   The angle of the string will depend on where the upper end of the string is held relative to the center of the teacup. If the string is held directly over the center, for example, the smartphone is not moving in a circular path, it has no centripetal acceleration due to the motion of the teacup, and the string will not deviate from the vertical. (In practice, it may deviate slightly due to the rotation of the turntables on which the teacup is mounted.)

WHAT IF?   Suppose you wish to measure the centripetal acceleration of the smartphone from your observations. How could you do so?

Answer   Our intuition tells us that the angle θ the string makes with the vertical should increase as the acceleration increases. By solving Equations (1) and (2) simultaneously for a_c, we find that $a_c = g\tan\theta$. Therefore, you can determine the magnitude of the centripetal acceleration of the smartphone by measuring the angle θ and using that relationship. Because the deflection of the string from the vertical serves as a measure of acceleration, *a simple pendulum can be used as an accelerometer.*

6.4 Motion in the Presence of Resistive Forces

In Chapter 5, we described the force of kinetic friction exerted on an object moving on some surface. We completely ignored any interaction between the object and the medium through which it moves. Now consider the effect of that medium, which can be either a liquid or a gas. The medium exerts a **resistive force $\vec{R}$** on the object moving through it. Some examples are the air resistance associated with moving vehicles (sometimes called *air drag*) and the viscous forces that act on objects moving through a liquid. The magnitude of $\vec{R}$ depends on factors such as the speed of the object, and the direction of $\vec{R}$ is always opposite the direction of the object's motion relative to the medium. This direction may or may not be in the direction opposite the object's velocity according to the observer. For example, if a marble is dropped into a bottle of shampoo, the marble moves downward and the resistive force is upward, resisting the falling of the marble. In contrast, imagine the moment at which there is no wind and you are looking at a flag hanging limply on a flagpole. When a breeze begins to blow toward the right, the flag moves toward the right. In this case, the drag force on the flag from the moving air is to the right and the motion of the flag in response is also to the right, the *same* direction as the drag force. Because the air moves toward the right with respect to the flag, the flag moves to the left relative to the air. Therefore, the direction of the drag force is indeed opposite to the direction of the motion of the flag with respect to the air!

The magnitude of the resistive force can depend on speed in a complex way, and here we consider only two simplified models. In the first model, we assume the resistive force is proportional to the velocity of the moving object; this model is valid for objects falling slowly through a liquid and for very small objects, such as dust particles, moving through air. In the second model, we assume a resistive force that is proportional to the square of the speed of the moving object; large objects, such as skydivers moving through air in free fall, experience such a force.

Model 1: Resistive Force Proportional to Object Velocity

If we model the resistive force acting on an object moving through a liquid or gas as proportional to the object's velocity, the resistive force can be expressed as

$$\vec{\mathbf{R}} = -b\vec{\mathbf{v}} \tag{6.2}$$

where b is a constant whose value depends on the properties of the medium and on the shape and dimensions of the object and $\vec{\mathbf{v}}$ is the velocity of the object relative to the medium. The negative sign indicates that $\vec{\mathbf{R}}$ is in the opposite direction to $\vec{\mathbf{v}}$.

Consider a small sphere of mass m released from rest in a liquid as in Figure 6.12a. Assuming the only forces acting on the sphere are the resistive force $\vec{\mathbf{R}} = -b\vec{\mathbf{v}}$ and the gravitational force $\vec{\mathbf{F}}_g$, let us describe its motion.[1] We model the sphere as a particle under a net force. Applying Newton's second law to the vertical motion of the sphere and choosing the downward direction to be positive, we obtain

$$\sum F_y = ma \;\;\rightarrow\;\; mg - bv = ma \tag{6.3}$$

where the acceleration of the sphere is downward. Solving Equation 6.3 for a and noting that a is equal to dv/dt gives

$$\frac{dv}{dt} = g - \frac{b}{m}v \tag{6.4}$$

This equation is called a *differential equation*; it contains both v and the derivative of v. The methods of solving such an equation may not be familiar to you as yet. Notice, however, that initially when $v = 0$, the magnitude of the resistive force is

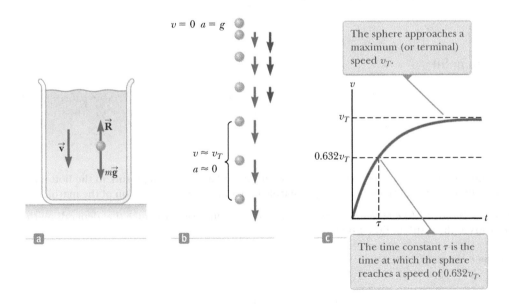

Figure 6.12 (a) A small sphere falling through a liquid. (b) A motion diagram of the sphere as it falls. Velocity vectors (red) and acceleration vectors (purple) are shown for each image after the first one. (c) A speed–time graph for the sphere.

---

[1] A *buoyant force* is also acting on the submerged object. This force is constant, and its magnitude is equal to the weight of the displaced liquid. This force can be modeled by changing the apparent weight of the sphere by a constant factor, so we will ignore the force here. We will discuss buoyant forces in Chapter 14.

also zero and the acceleration of the sphere is simply g. As t increases, the magnitude of the resistive force increases and the acceleration decreases. The acceleration approaches zero when the magnitude of the resistive force approaches the sphere's weight so that the net force on the sphere is zero. In this situation, the speed of the sphere approaches its **terminal speed** v_T.

Terminal speed ▶

The terminal speed is obtained from Equation 6.4 by setting $dv/dt = 0$, which gives

$$0 = g - \frac{b}{m}v_T \quad \text{or} \quad v_T = \frac{mg}{b} \tag{6.5}$$

Because you may not be familiar with differential equations yet, we won't show the explicit details of the process that gives the expression for v for all times t. If $v = 0$ at $t = 0$, this expression is

$$v = \frac{mg}{b}(1 - e^{-bt/m}) = v_T(1 - e^{-t/\tau}) \tag{6.6}$$

This function is plotted in Figure 6.12c. The symbol e represents the base of the natural logarithm and is also called *Euler's number:* $e = 2.718\,28$. The **time constant** $\tau = m/b$ (Greek letter tau) is the time at which the sphere released from rest at $t = 0$ reaches 63.2% of its terminal speed; when $t = \tau$, Equation 6.6 yields $v = 0.632v_T$. (The number 0.632 is $1 - e^{-1}$.)

We can check that Equation 6.6 is a solution to Equation 6.4 by direct differentiation:

$$\frac{dv}{dt} = \frac{d}{dt}\left[\frac{mg}{b}(1 - e^{-bt/m})\right] = \frac{mg}{b}\left(0 + \frac{b}{m}e^{-bt/m}\right) = ge^{-bt/m}$$

(See Appendix Table B.4 for the derivative of e raised to some power.) This is the left side of Equation 6.4. The right side is

$$g - \frac{b}{m}v = g - \frac{b}{m}\left[\frac{mg}{b}\left(1 - e^{-bt/m}\right)\right]$$

$$= ge^{-bt/m}$$

Because the results for both sides of Equation 6.4 are the same, Equation 6.6 represents a solution to Equation 6.4.

Example **6.8** Sphere Falling in Oil

A small sphere of mass 2.00 g is released from rest in a large vessel filled with oil, where it experiences a resistive force proportional to its speed. The sphere reaches a terminal speed of 5.00 cm/s. Determine the time constant τ and the time at which the sphere reaches 90.0% of its terminal speed.

SOLUTION

Conceptualize With the help of Figure 6.12, imagine dropping the sphere into the oil and watching it sink to the bottom of the vessel. If you have some thick shampoo in a clear container, drop a marble in it and observe the motion of the marble.

Categorize We model the sphere as a *particle under a net force*, with one of the forces being a resistive force that depends on the speed of the sphere. This model leads to the result in Equation 6.5.

Analyze From Equation 6.5, evaluate the coefficient b:

$$b = \frac{mg}{v_T}$$

Evaluate the time constant τ:

$$\tau = \frac{m}{b} = m\left(\frac{v_T}{mg}\right) = \frac{v_T}{g}$$

Substitute numerical values:

$$\tau = \frac{5.00 \text{ cm/s}}{980 \text{ cm/s}^2} = 5.10 \times 10^{-3} \text{ s}$$

6.8 continued

Find the time t at which the sphere reaches a speed of $0.900v_T$ by setting $v = 0.900v_T$ in Equation 6.6 and solving for t:

$$0.900v_T = v_T(1 - e^{-t/\tau})$$

$$1 - e^{-t/\tau} = 0.900$$

$$e^{-t/\tau} = 0.100$$

$$-\frac{t}{\tau} = \ln (0.100) = -2.30$$

$$t = 2.30\tau = 2.30(5.10 \times 10^{-3} \text{ s}) = 11.7 \times 10^{-3} \text{ s}$$

$$= 11.7 \text{ ms}$$

Finalize   The sphere reaches 90.0% of its terminal speed in a very short time interval. You should have also seen this behavior if you performed the activity with the marble and the shampoo. Because of the short time interval required to reach terminal velocity, you may not have noticed the time interval at all. The marble may have appeared to immediately begin moving through the shampoo at a constant velocity.

Model 2: Resistive Force Proportional to Object Speed Squared

For objects moving at high speeds through air, such as airplanes, skydivers, cars, and baseballs, the resistive force is reasonably well modeled as proportional to the square of the speed. In these situations, the magnitude of the resistive force can be expressed as

$$R = \tfrac{1}{2}D\rho Av^2 \tag{6.7}$$

where D is a dimensionless empirical quantity called the *drag coefficient*, ρ is the density of air, and A is the cross-sectional area of the moving object measured in a plane perpendicular to its velocity. The drag coefficient has a value of about 0.5 for spherical objects but can have a value as great as 2 for irregularly shaped objects.

Let us analyze the motion of a falling object subject to an upward air resistive force of magnitude $R = \tfrac{1}{2}D\rho Av^2$. Suppose an object of mass m is released from rest. As Figure 6.13 shows, the object experiences two external forces:[2] the downward gravitational force $\vec{\mathbf{F}}_g = m\vec{\mathbf{g}}$ and the upward resistive force $\vec{\mathbf{R}}$. Hence, the magnitude of the net force is

$$\sum F = mg - \tfrac{1}{2}D\rho Av^2 \tag{6.8}$$

where we have taken downward to be the positive vertical direction. Modeling the object as a particle under a net force, with the net force given by Equation 6.8, we find that the object has a downward acceleration of magnitude

$$a = g - \left(\frac{D\rho A}{2m}\right)v^2 \tag{6.9}$$

We can calculate the terminal speed v_T by noticing that when the gravitational force is balanced by the resistive force, the net force on the object is zero and therefore its acceleration is zero. Setting $a = 0$ in Equation 6.9 gives

$$0 = g - \left(\frac{D\rho A}{2m}\right)v_T^{\,2}$$

so, solving for v_T,

$$v_T = \sqrt{\frac{2mg}{D\rho A}} \tag{6.10}$$

Table 6.1 (page 142) lists the terminal speeds for several objects falling through air.

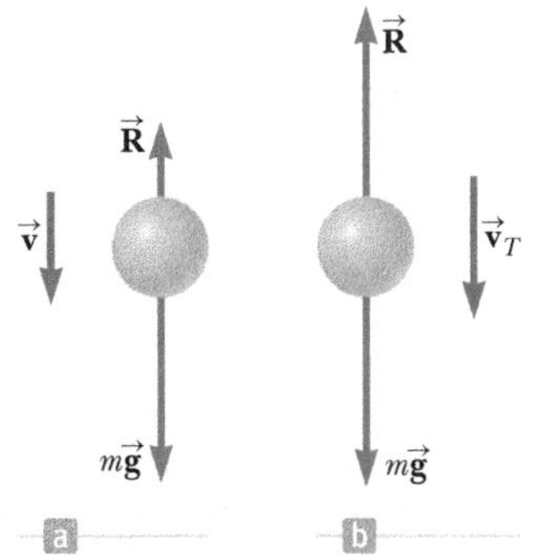

Figure 6.13 (a) An object falling through air experiences a resistive force $\vec{\mathbf{R}}$ and a gravitational force $\vec{\mathbf{F}}_g = m\vec{\mathbf{g}}$. (b) The object reaches terminal speed when the net force acting on it is zero, that is, when $\vec{\mathbf{R}} = -\vec{\mathbf{F}}_g$ or $R = mg$.

---

[2]As with Model 1, there is also an upward buoyant force that we neglect.

TABLE 6.1 Terminal Speed for Various Objects Falling Through Air

Object	Mass (kg)	Cross-Sectional Area (m²)	v_T (m/s)
Skydiver	75	0.70	60
Baseball (radius 3.7 cm)	0.145	4.2×10^{-3}	43
Golf ball (radius 2.1 cm)	0.046	1.4×10^{-3}	44
Hailstone (radius 0.50 cm)	4.8×10^{-4}	7.9×10^{-5}	14
Raindrop (radius 0.20 cm)	3.4×10^{-5}	1.3×10^{-5}	9.0

QUICK QUIZ 6.4 A basketball and a 2-inch-diameter steel ball, having the same mass, are dropped through air from rest such that their bottoms are initially at the same height above the ground, on the order of 1 m or more. Which one strikes the ground first? **(a)** The steel ball strikes the ground first. **(b)** The basketball strikes the ground first. **(c)** Both strike the ground at the same time.

Conceptual Example 6.9   The Skysurfer

Consider a skysurfer (Fig. 6.14) who jumps from a plane with his feet attached firmly to his surfboard, does some tricks, and then opens his parachute. Describe the forces acting on him during these maneuvers.

SOLUTION

When the surfer first steps out of the plane, he has no vertical velocity. The downward gravitational force causes him and the board to accelerate toward the ground. As their downward speed increases, so does the upward resistive force exerted by the air on the surfer and the board. This upward force reduces their acceleration, and so their speed increases more slowly. Eventually, they are going so fast that the upward resistive force matches the downward gravitational force. Now the net force is zero and they no longer accelerate, but instead reach their terminal speed. At some point after reaching terminal speed, he opens his parachute, resulting in a drastic increase in the upward resistive force. The net force (and therefore the acceleration) is now upward, in the direction opposite the direction of the velocity. The downward velocity therefore decreases rapidly, and the resistive force on the parachute also decreases. Eventually, the upward resistive force and the downward gravitational force balance each other again and a much smaller terminal speed is reached, permitting a safe landing.

Figure 6.14 (Conceptual Example 6.9) A skysurfer.

(Contrary to popular belief, the velocity vector of a skydiver never points upward. You may have seen a video in which a skydiver appears to "rocket" upward once the parachute opens. In fact, what happens is that the skydiver slows down but the person holding the camera continues falling at high speed.)

Example 6.10   Resistive Force Exerted on a Baseball

A pitcher hurls a 0.145-kg baseball past a batter at 40.2 m/s (= 90 mi/h). Find the resistive force acting on the ball at this speed.

SOLUTION

Conceptualize   This example is different from the previous ones in that the object is now moving horizontally through the air instead of moving vertically under the influence of gravity and the resistive force. The resistive force causes the ball to slow down, and gravity causes its trajectory to curve downward. We simplify the situation by assuming the velocity vector is exactly horizontal at the instant it is traveling at 40.2 m/s.

Categorize   In general, the ball is a *particle under a net force*. Because we are considering only one instant of time, however, we are not concerned about acceleration, so the problem involves only finding the value of one of the forces.

6.10 continued

Analyze To determine the drag coefficient D, imagine that we drop the baseball and allow it to reach terminal speed. Solve Equation 6.10 for D:

$$D = \frac{2mg}{v_T^2 \rho A}$$

Use this expression for D in Equation 6.7 to find an expression for the magnitude of the resistive force:

$$R = \tfrac{1}{2}D\rho A v^2 = \frac{1}{2}\left(\frac{2mg}{v_T^2 \rho A}\right)\rho A v^2 = mg\left(\frac{v}{v_T}\right)^2$$

Substitute numerical values, using the terminal speed from Table 6.1:

$$R = (0.145 \text{ kg})(9.80 \text{ m/s}^2)\left(\frac{40.2 \text{ m/s}}{43 \text{ m/s}}\right)^2 = 1.2 \text{ N}$$

Finalize The magnitude of the resistive force is similar in magnitude to the weight of the baseball, which is about 1.4 N. Therefore, air resistance plays a major role in the motion of the ball, as evidenced by the variety of curve balls, floaters, sinkers, and the like thrown by baseball pitchers.

Summary

▶ Concepts and Principles

A particle moving in uniform circular motion has a centripetal acceleration; this acceleration must be provided by a net force directed toward the center of the circular path.

An observer in a noninertial (accelerating) frame of reference introduces **fictitious forces** when applying Newton's second law in that frame.

An object moving through a liquid or gas experiences a speed-dependent **resistive force.** This resistive force is in a direction opposite that of the velocity of the object relative to the medium and generally increases with speed. The magnitude of the resistive force depends on the object's size and shape and on the properties of the medium through which the object is moving. In the limiting case for a falling object, when the magnitude of the resistive force equals the object's weight, the object reaches its **terminal speed.**

▶ Analysis Model for Problem Solving

Particle in Uniform Circular Motion (Extension) With our new knowledge of forces, we can extend the model of a particle in uniform circular motion, first introduced in Chapter 4. Newton's second law applied to a particle moving in uniform circular motion states that the net force causing the particle to undergo a centripetal acceleration (Eq. 4.21) is related to the acceleration according to

$$\sum F = ma_c = m\frac{v^2}{r} \tag{6.1}$$

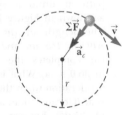

Think–Pair–Share

1. You are working as a delivery person for a dairy store. In the back of your pickup truck is a crate of eggs. The dairy company has run out of bungee cords, so the crate is not tied down. You have been told to drive carefully because the coefficient of static friction between the crate and the bed of the truck is 0.600. You are not worried, because you are traveling on a road that appears perfectly straight. Due to your confidence and inattention, your speed has crept

upward to 45.0 mi/h. Suddenly, you see a curve ahead with a warning sign saying, "Danger: unbanked curve with radius of curvature 35.0 m." You are 15.0 m from the beginning of the curve. What can you do to save the eggs: (i) take the curve at 45.0 mi/h, (ii) brake to a stop before entering the curve to think about it, or (iii) slow down to take the curve at a slower speed? Discuss these options in your group and determine if there is a best course of action.

2. **ACTIVITY** Find a YouTube video that shows the complete cycle for an amusement park ride called the "Roundup." In this ride, a rider stands against a wall at the edge of a disk

rotating around a vertical axis. When the disk reaches its operating speed, an arm raises the disk through an angle so that the disk rotates around an axis that is almost horizontal. As a result, the rider moves over the top of a vertical circle, seemingly unsupported, but does not fall downward. By using the height of a typical person on the ride, estimate the radius of the disk, using a stopped image of the disk at its highest angle. Begin the video again and use your smartphone stopwatch to measure the period of rotation of the disk. (a) From this information, calculate the centripetal acceleration of a rider at the top of the ride. (b) How does this acceleration compare to that due to gravity? (c) Why doesn't a rider at the top fall downward?

3. **ACTIVITY** Find a YouTube video that shows the complete cycle for an amusement park ride called the "Rotor." In this ride, a rider stands against a wall in a cylinder rotating around a vertical axis. When the cylinder reaches its operating speed, the floor drops away and riders remain suspended on the wall. By using the height of a typical person on the ride, estimate the radius of the cylinder, using a stopped image of the ride. Begin the video again and use your smartphone stopwatch to measure the period of rotation of the cylinder. From this information, determine the minimum coefficient of static friction necessary between the rider and the wall to keep the rider suspended.

Problems

See the Preface for an explanation of the icons used in this problems set.
For additional assessment items for this section, go to ⚡ **WEBASSIGN**
From Cengage

SECTION 6.1 Extending the Particle in Uniform Circular Motion Model

1. In the Bohr model of the hydrogen atom, an electron moves in a circular path around a proton. The speed of the electron is approximately 2.20×10^6 m/s. Find (a) the force acting on the electron as it revolves in a circular orbit of radius 0.529×10^{-10} m and (b) the centripetal acceleration of the electron.

2. Whenever two *Apollo* astronauts were on the surface of the Moon, a third astronaut orbited the Moon. Assume the orbit to be circular and 100 km above the surface of the Moon, where the acceleration due to gravity is 1.52 m/s². The radius of the Moon is 1.70×10^6 m. Determine (a) the astronaut's orbital speed and (b) the period of the orbit.

3. A car initially traveling eastward turns north by traveling in a circular path at uniform speed as shown in Figure P6.3. The length of the arc *ABC* is 235 m, and the car completes the turn in 36.0 s. (a) What is the acceleration when the car is at *B* located at an angle of 35.0°? Express your answer in terms of the unit vectors $\hat{\mathbf{i}}$ and $\hat{\mathbf{j}}$. Determine (b) the car's average speed and (c) its average acceleration during the 36.0-s interval.

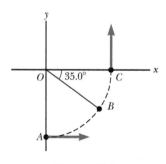

Figure P6.3

4. A curve in a road forms part of a horizontal circle. As a car goes around it at constant speed 14.0 m/s, the total horizontal force on the driver has magnitude 130 N. What is the total horizontal force on the driver if the speed on the same curve is 18.0 m/s instead?

5. In a cyclotron (one type of particle accelerator), a deuteron (of mass 2.00 u) reaches a final speed of 10.0% of the speed of light while moving in a circular path of radius 0.480 m. What magnitude of magnetic force is required to maintain the deuteron in a circular path?

6. *Why is the following situation impossible?* The object of mass $m = 4.00$ kg in Figure P6.6 is attached to a vertical rod by

two strings of length $\ell = 2.00$ m. The strings are attached to the rod at points a distance $d = 3.00$ m apart. The object rotates in a horizontal circle at a constant speed of $v = 3.00$ m/s, and the strings remain taut. The rod rotates along with the object so that the strings do not wrap onto the rod. **What If?** Could this situation be possible on another planet?

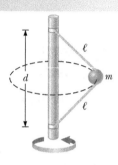

Figure P6.6

7. **CR** You are working during your summer break as an amusement park ride operator. The ride you are controlling consists of a large vertical cylinder that spins about its axis fast enough that any person inside is held up against the wall when the floor drops away (Fig. P6.7). The coefficient of static friction between a person of mass m and the wall is μ_s, and the radius of the cylinder is R. You are rotating the ride with an angular speed ω suggested by your supervisor. (a) Suppose a very heavy person enters the ride. Do you need to increase the angular speed so that this person will not slide down the wall? (b) Suppose someone enters the ride wearing a very slippery satin workout outfit. In this case, do you need to increase the angular speed so that this person will not slide down the wall?

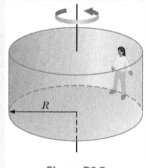

Figure P6.7

8. **CR** A driver is suing the state highway department after an accident on a curved freeway. The driver lost control and crashed into a tree located a short distance from the outside edge of the curved roadway. The driver is claiming that the radius of curvature of the unbanked roadway was too small for the speed limit, causing him to slide outward on the curve and hit the tree. You have been hired as an expert witness for the defense, and have been requested to use your knowledge of physics to testify that the radius of curvature of the roadway is appropriate for the speed limit. State regulations show that the radius of curvature of an unbanked roadway on which the speed limit is 65 mi/h must be at least 150 m. You build an accelerometer, which is a plumb bob

with a protractor that you attach to the roof of your car. An associate riding in your car with you observes that the plumb bob hangs at an angle of 15.0° from the vertical when the car is driven at a safer speed of 23.0 m/s on the curve in question. What is your testimony regarding the radius of the curve?

SECTION 6.2 Nonuniform Circular Motion

9. A hawk flies in a horizontal arc of radius 12.0 m at constant speed 4.00 m/s. (a) Find its centripetal acceleration. (b) It continues to fly along the same horizontal arc, but increases its speed at the rate of 1.20 m/s². Find the acceleration (magnitude and direction) in this situation at the moment the hawk's speed is 4.00 m/s.

10. A 40.0-kg child swings in a swing supported by two chains, each 3.00 m long. The tension in each chain at the lowest point is 350 N. Find (a) the child's speed at the lowest point and (b) the force exerted by the seat on the child at the lowest point. (Ignore the mass of the seat.)

11. A child of mass m swings in a swing supported by two chains, each of length R. If the tension in each chain at the lowest point is T, find (a) the child's speed at the lowest point and (b) the force exerted by the seat on the child at the lowest point. (Ignore the mass of the seat.)

12. One end of a cord is fixed and a small 0.500-kg object is attached to the other end, where it swings in a section of a vertical circle of radius 2.00 m as shown in Figure P6.12. When $\theta = 20.0°$, the speed of the object is 8.00 m/s. At this instant, find (a) the tension in the string, (b) the tangential and radial components of acceleration, and (c) the total acceleration. (d) Is your answer changed if the object is swinging down toward its lowest point instead of swinging up? (e) Explain your answer to part (d).

Figure P6.12

13. A roller coaster at the Six Flags Great America amusement park in Gurnee, Illinois, incorporates some clever design technology and some basic physics. Each vertical loop, instead of being circular, is shaped like a teardrop (Fig. P6.13). The cars ride on the inside of the loop at the top, and the speeds are fast enough to ensure the cars remain on the track. The biggest loop is 40.0 m high. Suppose the speed at the top of the loop is 13.0 m/s and the corresponding centripetal acceleration of the riders is $2g$. (a) What is the radius of the arc of the teardrop at the top? (b) If the total mass of a car plus the riders is M, what force does the rail exert on the car at the top? (c) Suppose the roller coaster had a circular loop of radius 20.0 m. If the cars have the same speed, 13.0 m/s at the top, what is the centripetal acceleration of the riders at the top? (d) Comment on the normal force at the top in the situation described in part (c) and on the advantages of having teardrop-shaped loops.

Figure P6.13

Frank Cezus/Photographer's Choice/Getty Images

SECTION 6.3 Motion in Accelerated Frames

14. An object of mass $m = 5.00$ kg, attached to a spring scale, rests on a frictionless, horizontal surface as shown in Figure P6.14. The spring scale, attached to the front end of a boxcar, reads zero when the car is at rest. (a) Determine the acceleration of the car if the spring scale has a constant reading of 18.0 N when the car is in motion. (b) What constant reading will the spring scale show if the car moves with constant velocity? Describe the forces on the object as observed (c) by someone in the car and (d) by someone at rest outside the car.

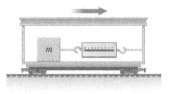

Figure P6.14

15. A person stands on a scale in an elevator. As the elevator starts, the scale has a constant reading of 591 N. As the elevator later stops, the scale reading is 391 N. Assuming the magnitude of the acceleration is the same during starting and stopping, determine (a) the weight of the person, (b) the person's mass, and (c) the acceleration of the elevator.

16. **Review.** A student, along with her backpack on the floor next to her, is in an elevator that is accelerating upward with acceleration a. The student gives her backpack a quick kick at $t = 0$, imparting to it speed v and causing it to slide across the elevator floor. At time t, the backpack hits the opposite wall a distance L away from the student. Find the coefficient of kinetic friction μ_k between the backpack and the elevator floor.

17. A small container of water is placed on a turntable inside a microwave oven, at a radius of 12.0 cm from the center. The turntable rotates steadily, turning one revolution in each 7.25 s. What angle does the water surface make with the horizontal?

SECTION 6.4 Motion in the Presence of Resistive Forces

18. The mass of a sports car is 1 200 kg. The shape of the body is such that the aerodynamic drag coefficient is 0.250 and the frontal area is 2.20 m². Ignoring all other sources of friction, calculate the initial acceleration the car has if it has been traveling at 100 km/h and is now shifted into neutral and allowed to coast.

19. **Review.** A window washer pulls a rubber squeegee down a very tall vertical window. The squeegee has mass 160 g and is mounted on the end of a light rod. The coefficient of kinetic friction between the squeegee and the dry glass is 0.900. The window washer presses it against the window with a force having a horizontal component of 4.00 N. (a) If she pulls the squeegee down the window at constant velocity, what vertical force component must she exert? (b) The window washer increases the downward force component by 25.0%, while all other forces remain the same. Find the squeegee's acceleration in this situation. (c) The squeegee is moved into a wet portion of the window, where its motion is resisted by a fluid drag force R proportional to its velocity according to $R = -20.0v$, where R is in newtons and v is in meters per second. Find the terminal velocity that the squeegee approaches, assuming the window washer exerts the same force described in part (b).

20. A small piece of Styrofoam packing material is dropped from a height of 2.00 m above the ground. Until it reaches terminal speed, the magnitude of its acceleration is given by $a = g - Bv$. After falling 0.500 m, the Styrofoam effectively reaches terminal speed and then takes 5.00 s more to reach the ground. (a) What is the value of the constant B? (b) What is the acceleration at $t = 0$? (c) What is the acceleration when the speed is 0.150 m/s?

21. A small, spherical bead of mass 3.00 g is released from rest at $t = 0$ from a point under the surface of a viscous liquid. The terminal speed is observed to be $v_T = 2.00$ cm/s. Find (a) the value of the constant b that appears in Equation 6.2, (b) the time t at which the bead reaches $0.632v_T$, and (c) the value of the resistive force when the bead reaches terminal speed.

22. Assume the resistive force acting on a speed skater is proportional to the square of the skater's speed v and is given by $f = -kmv^2$, where k is a constant and m is the skater's mass. The skater crosses the finish line of a straight-line race with speed v_i and then slows down by coasting on his skates. Show that the skater's speed at any time t after crossing the finish line is $v(t) = v_i/(1 + ktv_i)$.

23. You can feel a force of air drag on your hand if you stretch your arm out of the open window of a speeding car. *Note:* Do not endanger yourself. What is the order of magnitude of this force? In your solution, state the quantities you measure or estimate and their values.

ADDITIONAL PROBLEMS

24. A car travels clockwise at constant speed around a circular section of a horizontal road as shown in the aerial view of Figure P6.24. Find the directions of its velocity and acceleration at (a) position Ⓐ and (b) position Ⓑ.

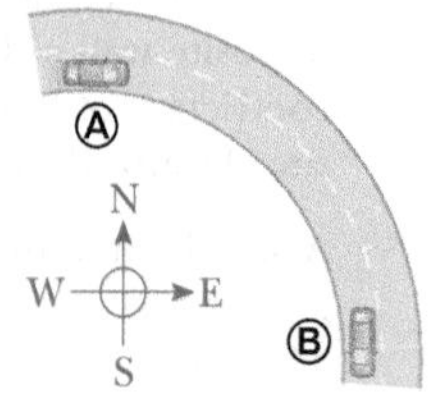

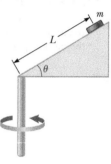

Figure P6.24

25. A string under a tension of 50.0 N is used to whirl a rock in a horizontal circle of radius 2.50 m at a speed of 20.4 m/s on a frictionless surface as shown in Figure P6.25. As the string is pulled in, the speed of the rock increases. When the string on the table is 1.00 m long and the speed of the rock is 51.0 m/s, the string breaks. What is the breaking strength, in newtons, of the string?

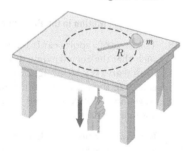

Figure P6.25

26. Disturbed by speeding cars outside his workplace, Nobel laureate Arthur Holly Compton designed a speed bump (called the "Holly hump") and had it installed. Suppose a 1 800-kg car passes over a hump in a roadway that follows the arc of a circle of radius 20.4 m as shown in Figure P6.26. (a) If the car travels at

Figure P6.26
Problems 26 and 27.

30.0 km/h, what force does the road exert on the car as the car passes the highest point of the hump? (b) **What If?** What is the maximum speed the car can have without losing contact with the road as it passes this highest point?

27. A car of mass m passes over a hump in a road that follows the arc of a circle of radius R as shown in Figure P6.26. (a) If the car travels at a speed v, what force does the road exert on the car as the car passes the highest point of the hump? (b) **What If?** What is the maximum speed the car can have without losing contact with the road as it passes this highest point?

28. A child's toy consists of a small wedge that has an acute angle θ (Fig. P6.28). The sloping side of the wedge is frictionless, and an object of mass m on it remains at constant height if the wedge is spun at a certain constant speed. The wedge is spun by rotating, as an axis, a vertical rod that is firmly attached to the wedge at the bottom end. Show that, when the object sits at rest at a point at distance L up along the wedge, the speed of the object must be $v = (gL \sin \theta)^{1/2}$.

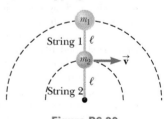

Figure P6.28

29. A seaplane of total mass m lands on a lake with initial speed $v_i\hat{\mathbf{i}}$. The only horizontal force on it is a resistive force on its pontoons from the water. The resistive force is proportional to the velocity of the seaplane: $\vec{\mathbf{R}} = -b\vec{\mathbf{v}}$. Newton's second law applied to the plane is $-bv\hat{\mathbf{i}} = m(dv/dt)\hat{\mathbf{i}}$. From the fundamental theorem of calculus, this differential equation implies that the speed changes according to

$$\int_{v_i}^{v} \frac{dv}{v} = -\frac{b}{m} \int_0^t dt$$

(a) Carry out the integration to determine the speed of the seaplane as a function of time. (b) Sketch a graph of the speed as a function of time. (c) Does the seaplane come to a complete stop after a finite interval of time? (d) Does the seaplane travel a finite distance in stopping?

30. An object of mass $m_1 = 4.00$ kg is tied to an object of mass $m_2 = 3.00$ kg with String 1 of length $\ell = 0.500$ m. The combination is swung in a vertical circular path on a second string, String 2, of length $\ell = 0.500$ m. During the motion, the two strings are collinear at all times as shown in Figure P6.30. At the top of its motion, m_2 is traveling at $v = 4.00$ m/s. (a) What is the tension in String 1 at this instant? (b) What is the tension in String 2 at this instant? (c) Which string will break first if the combination is rotated faster and faster?

String 1 ℓ
String 2 ℓ

Figure P6.30

31. A ball of mass $m = 0.275$ kg swings in a vertical circular path on a string $L = 0.850$ m long as in Figure P6.31. (a) What are the forces acting on the ball at any point on the path? (b) Draw force diagrams for the ball when it is at the bottom of the circle and when it is at the top. (c) If

its speed is 5.20 m/s at the top of the circle, what is the tension in the string there? (d) If the string breaks when its tension exceeds 22.5 N, what is the maximum speed the ball can have at the bottom before that happens?

Figure P6.31

32. *Why is the following situation impossible?* A mischievous child goes to an amusement park with his family. On one ride, after a severe scolding from his mother, he slips out of his seat and climbs to the top of the ride's structure, which is shaped like a cone with its axis vertical and its sloped sides making an angle of $\theta = 20.0°$ with the horizontal as shown in Figure P6.32. This part of the structure rotates about the vertical central axis when the ride operates. The child sits on the sloped surface at a point $d = 5.32$ m down the sloped side from the center of the cone and pouts. The coefficient of static friction between the boy and the cone is 0.700. The ride operator does not notice that the child has slipped away from his seat and so continues to operate the ride. As a result, the sitting, pouting boy rotates in a circular path at a speed of 3.75 m/s.

Figure P6.32

33. The pilot of an airplane executes a loop-the-loop maneuver in a vertical circle. The speed of the airplane is 300 mi/h at the top of the loop and 450 mi/h at the bottom, and the radius of the circle is 1 200 ft. (a) What is the pilot's apparent weight at the lowest point if his true weight is 160 lb? (b) What is his apparent weight at the highest point? (c) **What If?** Describe how the pilot could experience weightlessness if both the radius and the speed can be varied. *Note:* His apparent weight is equal to the magnitude of the force exerted by the seat on his body.

34. A basin surrounding a drain has the shape of a circular cone opening upward, having everywhere an angle of 35.0° with the horizontal. A 25.0-g ice cube is set sliding around the cone without friction in a horizontal circle of radius R. (a) Find the speed the ice cube must have as a function of R. (b) Is any piece of data unnecessary for the solution? Suppose R is made two times larger. (c) Will the required speed increase, decrease, or stay constant? If it changes, by what factor? (d) Will the time interval required for each revolution increase, decrease, or stay constant? If it changes, by what factor? (e) Do the answers to parts (c) and (d) seem contradictory? Explain.

35. **Review.** While learning to drive, you are in a 1 200-kg car moving at 20.0 m/s across a large, vacant, level parking lot. Suddenly you realize you are heading straight toward the brick sidewall of a large supermarket and are in danger of running into it. The pavement can exert a maximum horizontal force of 7 000 N on the car. (a) Explain why you should expect the force to have a well-defined maximum value. (b) Suppose you apply the brakes and do not turn the steering wheel. Find the minimum distance you must be from the wall to avoid a collision. (c) If you do not brake but instead maintain constant speed and turn the steering wheel, what is the minimum distance you must be from the wall to avoid a collision? (d) Of the two methods in parts (b) and (c), which is better for avoiding a collision? Or should you use both the brakes and the steering wheel, or neither? Explain. (e) Does the conclusion in part (d) depend on the numerical values given in this problem, or is it true in general? Explain.

36. A truck is moving with constant acceleration a up a hill that makes an angle ϕ with the horizontal as in Figure P6.36. A small sphere of mass m is suspended from the ceiling of the truck by a light cord. If the pendulum makes a constant angle θ with the perpendicular to the ceiling, what is a?

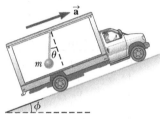

Figure P6.36

37. Because the Earth rotates about its axis, a point on the equator experiences a centripetal acceleration of 0.033 7 m/s², whereas a point at the poles experiences no centripetal acceleration. If a person at the equator has a mass of 75.0 kg, calculate (a) the gravitational force (true weight) on the person and (b) the normal force (apparent weight) on the person. (c) Which force is greater? Assume the Earth is a uniform sphere and take $g = 9.800$ m/s².

38. A puck of mass m_1 is tied to a string and allowed to revolve in a circle of radius R on a frictionless, horizontal table. The other end of the string passes through a small hole in the center of the table, and an object of mass m_2 is tied to it (Fig. P6.38). The suspended object remains in equilibrium while the puck on the tabletop revolves. Find symbolic expressions for (a) the tension in the string, (b) the radial force acting on the puck, and (c) the speed of the puck. (d) Qualitatively describe what will happen in the motion of the puck if the value of m_2 is increased by placing a small additional load on the puck. (e) Qualitatively describe what will happen in the motion of the puck if the value of m_2 is instead decreased by removing a part of the hanging load.

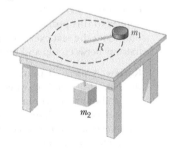

Figure P6.38

39. Galileo thought about whether acceleration should be defined as the rate of change of velocity over time or as the rate of change in velocity over distance. He chose the former, so let's use the name "vroomosity" for the rate of change of velocity over distance. For motion of a particle on a straight line with constant acceleration, the equation

$v = v_i + at$ gives its velocity v as a function of time. Similarly, for a particle's linear motion with constant vroomosity k, the equation $v = v_i + kx$ gives the velocity as a function of the position x if the particle's speed is v_i at $x = 0$. (a) Find the law describing the total force acting on this object of mass m. Describe an example of such a motion or explain why it is unrealistic for (b) the possibility of k positive and (c) the possibility of k negative.

40. Members of a skydiving club were given the following data to use in planning their jumps. In the table, d is the distance fallen from rest by a skydiver in a "free-fall stable spread position" versus the time of fall t. (a) Convert the distances in feet into meters. (b) Graph d (in meters) versus t. (c) Determine the value of the terminal speed v_T by finding the slope of the straight portion of the curve. Use a least-squares fit to determine this slope.

t (s)	d (ft)	t (s)	d (ft)	t (s)	d (ft)
0	0	7	652	14	1 831
1	16	8	808	15	2 005
2	62	9	971	16	2 179
3	138	10	1 138	17	2 353
4	242	11	1 309	18	2 527
5	366	12	1 483	19	2 701
6	504	13	1 657	20	2 875

41. A car rounds a banked curve as discussed in Example 6.4 and shown in Figure 6.5. The radius of curvature of the road is R, the banking angle is θ, and the coefficient of static friction is μ_s. (a) Determine the range of speeds the car can have without slipping up or down the road. (b) Find the minimum value for μ_s such that the minimum speed is zero.

42. In Example 6.5, we investigated the forces a child experiences on a Ferris wheel. Assume the data in that example applies to this problem. What force (magnitude and direction) does the seat exert on a 40.0-kg child when the child is halfway between top and bottom?

43. **Review.** A piece of putty is initially located at point A on the rim of a grinding wheel rotating at constant angular speed about a horizontal axis. The putty is dislodged from point A when the diameter through A is horizontal. It then rises vertically and returns to A at the instant the wheel completes one revolution. From this information, we wish to find the speed v of the putty when it leaves the wheel and the force holding it to the wheel. (a) What analysis model is appropriate for the motion of the putty as it rises and falls? (b) Use this model to find a symbolic expression for the time interval between when the putty leaves point A and when it arrives back at A, in terms of v and g. (c) What is the appropriate analysis model to describe point A on the wheel? (d) Find the period of the motion of point A in terms of the tangential speed v and the radius R of the wheel. (e) Set the time interval from part (b) equal to the period from part (d) and solve for the speed v of the putty as it leaves the wheel. (f) If the mass of the putty is m, what is the magnitude of the force that held it to the wheel before it was released?

44. A model airplane of mass 0.750 kg flies with a speed of 35.0 m/s in a horizontal circle at the end of a 60.0-m-long control wire as shown in Figure P6.44a. The forces exerted

on the airplane are shown in Figure P6.44b: the tension in the control wire, the gravitational force, and aerodynamic lift that acts at $\theta = 20.0°$ inward from the vertical. Compute the tension in the wire, assuming it makes a constant angle of $\theta = 20.0°$ with the horizontal.

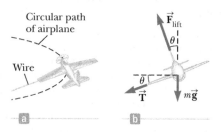

Figure P6.44

CHALLENGE PROBLEMS

45. A 9.00-kg object starting from rest falls through a viscous medium and experiences a resistive force given by Equation 6.2. The object reaches one half its terminal speed in 5.54 s. (a) Determine the terminal speed. (b) At what time is the speed of the object three-fourths the terminal speed? (c) How far has the object traveled in the first 5.54 s of motion?

46. For $t < 0$, an object of mass m experiences no force and moves in the positive x direction with a constant speed v_i. Beginning at $t = 0$, when the object passes position $x = 0$, it experiences a net resistive force proportional to the square of its speed: $\vec{F}_{net} = -mkv^2\,\hat{i}$, where k is a constant. The speed of the object after $t = 0$ is given by $v = v_i/(1 + kv_i t)$. (a) Find the position x of the object as a function of time. (b) Find the object's velocity as a function of position.

47. A golfer tees off from a location precisely at $\phi_i = 35.0°$ north latitude. He hits the ball due south, with range 285 m. The ball's initial velocity is at 48.0° above the horizontal. Suppose air resistance is negligible for the golf ball. (a) For how long is the ball in flight? The cup is due south of the golfer's location, and the golfer would have a hole-in-one if the Earth were not rotating. The Earth's rotation makes the tee move in a circle of radius $R_E \cos \phi_i = (6.37 \times 10^6 \text{ m}) \cos 35.0°$ as shown in Figure P6.47. The tee completes one revolution each day. (b) Find the eastward speed of the tee relative to the stars. The hole is also moving east, but it is 285 m farther south and thus at a slightly lower latitude ϕ_f. Because the hole moves in a slightly larger circle, its speed must be greater than that of the tee. (c) By how much does the hole's speed exceed that of the tee? During the time interval the ball is in flight, it moves upward and downward as well as southward with the projectile motion you studied in Chapter 4, but it also moves eastward with the speed you found in part (b). The hole moves to the east at a faster speed, however, pulling ahead of the ball with the relative speed you found in

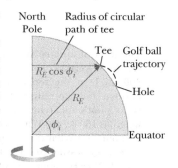

Figure P6.47

part (c). (d) How far to the west of the hole does the ball land?

48. A single bead can slide with negligible friction on a stiff wire that has been bent into a circular loop of radius 15.0 cm as shown in Figure P6.48. The circle is always in a vertical plane and rotates steadily about its vertical diameter with a period of 0.450 s. The position of the bead is described by the angle θ that the radial line, from the center of the loop to the bead, makes with the vertical. (a) At what angle up from the bottom of the circle can the bead stay motionless relative to the turning circle? (b) **What If?** Repeat the problem, this time taking the period of the circle's rotation as 0.850 s. (c) Describe how the solution to part (b) is different from the solution to part (a). (d) For any period or loop size, is there always an angle at which the bead can stand still relative to the loop? (e) Are there ever more than two angles? Arnold Arons suggested the idea for this problem.

Figure P6.48

49. Because of the Earth's rotation, a plumb bob does not hang exactly along a line directed to the center of the Earth. How much does the plumb bob deviate from a radial line at 35.0° north latitude? Assume the Earth is spherical.

50. You have a great job working at a major league baseball stadium for the summer! At this stadium, the speed of every pitch is measured using a radar gun aimed at the pitcher by an operator behind home plate. The operator has so much experience with this job that he has perfected a technique by which he can make each measurement at the exact instant at which the ball leaves the pitcher's hand. Your supervisor asks you to construct an algorithm that will provide the speed of the ball as it crosses home plate, 18.3 m from the pitcher, based on the measured speed v_i of the ball as it leaves the pitcher's hand. The speed at home plate will be lower due to the resistive force of the air on the baseball. The vertical motion of the ball is small, so, to a good approximation, we can consider only the horizontal motion of the ball. You begin to develop your algorithm by applying the particle under a net force to the baseball in the horizontal direction. A pitch is measured to have a speed of 40.2 m/s as it leaves the pitcher's hand. You need to tell your supervisor how fast it was traveling as it crossed home plate. (*Hint:* Use the chain rule to express acceleration in terms of a derivative with respect to x, and then solve a differential equation for v to find an expression for the speed of the baseball as a function of its position. The function will involve an exponential. Also make use of Table 6.1.)

7 Energy of a System

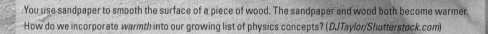

You use sandpaper to smooth the surface of a piece of wood. The sandpaper and wood both become warmer. How do we incorporate *warmth* into our growing list of physics concepts? (*DJTaylor/Shutterstock.com*)

STORYLINE **Your observations as an inquisitive physics student** have worn you out and you decide to spend a quiet day at home. You go into your garage to do further work on a carpentry project you started a while ago. You are thinking about how much you have learned about mechanics in your investigations as you find your sandpaper and a piece of wood that needs to be smoothed. You begin to sand the wood, still thinking that your studies of mechanics make a very complete description of nature and the Universe. Then you notice the sandpaper and the wood, along with your fingers, becoming *warmer* as you sand. "Wait. This is new!" you think. You are applying forces to the sandpaper, and it accelerates, therefore changing its velocity. There is friction between the sandpaper and the wood. This is all *mechanics*; you have thought about all of these concepts and have learned about them in previous chapters. But *warmth?* What's that all about? Maybe you have more thinking to do!

CONNECTIONS In this chapter, we are going to investigate a quantity that is very different from those studied in the previous chapters. Chapters 2 through 6 dealt with *change*. Velocity is a *change* in position, and acceleration is a *change* in velocity (Chapters 2 and 4). Force is the cause of *changes* in motion (Chapter 5). In this chapter and the next, we will study a quantity, energy, that is *conserved*. That is, the total energy in an isolated system *does not change* during any process that occurs in the system. Or if the total energy in a system does change, for example, if it increases, we find that the energy of the surroundings of the system decreases by the same amount! Therefore, the energy of the entire Universe is fixed; it has the same value at all times! Our analysis models presented in earlier chapters were based on the motion of a *particle*, or an object that could be modeled as a particle. We begin our new approach by focusing our attention on a new simplification model, a *system*, and analysis models based on the model of a system. These analysis models will be formally introduced in Chapter 8. In this chapter, we introduce systems and three ways to store energy in a system. We begin by making a connection between a familiar concept, force, and our new topic, *energy*. We will identify several forms in which energy can exist in a system. Even though this new quantity has a different nature from our previously studied quantities, it is very important and allows us to solve an entirely new class of problems. Furthermore, you might be happy to find out that energy is a scalar, so we don't have to perform complicated vector calculations! As we continue to study physics in the rest of the chapters in this book, we will see very often that we can take a force approach to a new area of study and we can also take an energy approach. The two approaches are complementary.

7.1 Systems and Environments

In the system model, we focus our attention on a small portion of the Universe—the **system**—and ignore details of the rest of the Universe outside of the system. A critical skill in applying the system model to problems is *identifying the system.*

A valid system

- may be a single object or particle
- may be a collection of objects or particles
- may be a region of space (such as the interior of an automobile engine combustion cylinder)
- may vary with time in size and shape (such as a rubber ball, which deforms upon striking a wall)

Identifying the need for a system approach to solving a problem (as opposed to a particle approach) is part of the Categorize step in the Analysis Model Approach to Problem Solving outlined in Chapter 2. Identifying the particular system is a second part of this step.

No matter what the particular system is in a given problem, we identify a **system boundary**, an imaginary surface (not necessarily coinciding with a physical surface) that divides the Universe into the system and the **environment** surrounding the system.

As an example, imagine a force applied to an object in empty space. We can define the object as the system and its outer surface as the system boundary. The force applied to it is an influence on the system from the environment that acts across the system boundary. We will see how to analyze this situation from a system approach in a subsequent section of this chapter.

Another example was seen in Example 5.10, where the system can be defined as the combination of the ball, the block, and the cord. The influence from the environment includes the gravitational forces on the ball and the block, the normal and friction forces on the block, and the force exerted by the pulley on the cord. The forces exerted by the cord on the ball and the block are internal to the system and therefore are not included as an influence from the environment.

There are a number of mechanisms by which a system can be influenced by its environment. The first one we shall investigate is *work*.

> **PITFALL PREVENTION 7.1**
> **Identify the System** The most important *first* step to take in solving a problem using the energy approach is to identify the appropriate system of interest.

7.2 Work Done by a Constant Force

Almost all the terms we have used thus far—velocity, acceleration, force, and so on—convey a similar meaning in physics as they do in everyday life. Now, however, we encounter a term whose meaning in physics is distinctly different from its everyday meaning: work.

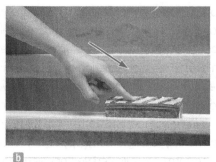

Charles D. Winters

Figure 7.1 An eraser being pushed along a chalkboard tray by a force acting at different angles with respect to the horizontal direction.

To understand what work as an influence on a system means to the physicist, consider the situation illustrated in Figure 7.1. A force $\vec{\mathbf{F}}$ is applied to a chalkboard eraser, which we identify as the system, and the eraser slides along the tray. If we want to know how effective the force is in moving the eraser, we must consider not only the magnitude of the force but also its direction. Notice that the finger in Figure 7.1 applies forces in three different directions on the eraser. Assuming the magnitude of the applied force is the same in all three photographs, the push applied in Figure 7.1b is more effective in moving the eraser than the push in Figure 7.1a. On the other hand, Figure 7.1c shows a situation in which the applied force does not move the eraser at all, regardless of how hard it is pushed (unless, of course, we apply a force so great that we break the chalkboard tray!). These results suggest that when analyzing forces to determine the influence they have on the system, we must consider the vector nature of forces. We must also consider the magnitude of the force. Moving a force with a magnitude of $|\vec{\mathbf{F}}| = 2$ N through a displacement represents a greater influence on the system than moving a force of magnitude 1 N through the same displacement. The magnitude of the displacement is also important. Moving the eraser 3 m along the tray represents a greater influence than moving it 2 cm if the same force is used in both cases.

Let us examine the situation in Figure 7.2, where the object (the system) undergoes a displacement along a straight line while acted on by a constant force of magnitude F that makes an angle θ with the direction of the displacement. We formally define the work done by the force on the system as follows:

> The **work** W done on a system by an agent exerting a constant force on the system is the product of the magnitude F of the force, the magnitude Δr of the displacement of the point of application of the force, and $\cos \theta$, where θ is the angle between the force and displacement vectors:

▶ **Work done by a constant force**

$$W \equiv F \Delta r \cos \theta \qquad (7.1)$$

Notice in Equation 7.1 that work is a scalar, even though it is defined in terms of two vectors in Figure 7.2, a force $\vec{\mathbf{F}}$ and a displacement $\Delta \vec{\mathbf{r}}$. In Section 7.3, we explore how to combine two vectors to generate a scalar quantity.

Notice also that the displacement in Equation 7.1 is that of *the point of application of the force.* If the force is applied to a particle or a rigid object that can be modeled as a particle, this displacement is the same as that of the particle. For a deformable system, however, these displacements are not the same. For example, imagine pressing in on the sides of a balloon with both hands. The center of the balloon moves through zero displacement. The points of application of the forces from your hands on the sides of the balloon, however, do indeed move through a displacement as the balloon is compressed, and that is the displacement to be used in Equation 7.1. We will see other examples of deformable systems, such as springs and samples of gas contained in a vessel.

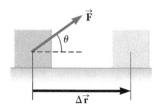

Figure 7.2 An object on a table undergoes a displacement $\Delta \vec{\mathbf{r}}$ under the action of a constant force $\vec{\mathbf{F}}$.

As an example of the distinction between the definition of work and our everyday understanding of the word, consider holding a heavy chair at arm's length for 3 min. At the end of this time interval, your tired arms may lead you to think you have done a considerable amount of work on the chair. According to our definition, however, you have done no work on it whatsoever. You exert a force to support the chair, but you do not move it. A force does no work on an object if the force does not move through a displacement. If $\Delta r = 0$, Equation 7.1 gives $W = 0$, which is the situation depicted in Figure 7.1c.

Also notice from Equation 7.1 that the work done by a force on a moving object is zero when the force applied is perpendicular to the displacement of its point of application. That is, if $\theta = 90°$, then $W = 0$ because $\cos 90° = 0$. For example, in Figure 7.3, the work done by the normal force on the object and the work done by the gravitational force on the object are both zero because both forces are perpendicular to the displacement and have zero components along an axis in the direction of $\Delta \vec{r}$.

The sign of the work also depends on the direction of $\vec{F}$ relative to $\Delta \vec{r}$. The work done by the applied force on a system is positive when the projection of $\vec{F}$ onto $\Delta \vec{r}$ is in the same direction as the displacement. For example, when an object is lifted, the work done by the applied force on the object is positive because the direction of that force is upward, in the same direction as the displacement of its point of application. When the projection of $\vec{F}$ onto $\Delta \vec{r}$ is in the direction opposite the displacement, W is negative. For example, as an object is lifted, the work done by the gravitational force on the object is negative. The factor $\cos \theta$ in the definition of W (Eq. 7.1) automatically takes care of the sign.

If an applied force $\vec{F}$ is in the same direction as the displacement $\Delta \vec{r}$, then $\theta = 0$ and $\cos 0 = 1$. In this case, Equation 7.1 gives

$$W = F \Delta r$$

The units of work are those of force multiplied by those of length. Therefore, the SI unit of work is the **newton · meter** ($N \cdot m = kg \cdot m^2/s^2$). This combination of units is used so frequently that it has been given a name of its own, the **joule** (J).

An important consideration for a system approach to problems is that **work is an energy transfer.** For now, **energy** sounds mysterious, because we have not studied it yet. It is difficult to define energy, other than to say that it is a physical quantity that is conserved. In that behavior, it is similar to *money*. When a financial transaction occurs in your checking account, money is transferred across the boundary of your account: for example, inward by deposits and outward by withdrawals. When a physical process occurs, energy is transferred across the boundary of a system. Our understanding of energy will improve as we investigate various examples in this chapter.

If W is the work done on a system and W is positive, energy is transferred *to* the system; if W is negative, energy is transferred *from* the system. Therefore, if a system interacts with its environment, this interaction can be described as a transfer of energy across the system boundary. The result is a change in the energy stored in the system. We will learn about the first type of energy storage in Section 7.5, after we investigate more aspects of work.

QUICK QUIZ 7.1 The gravitational force exerted by the Sun on the Earth holds the Earth in an orbit around the Sun. Let us assume that the orbit is perfectly circular. The work done by this gravitational force during a short time interval in which the Earth moves through a displacement in its orbital path is **(a)** zero **(b)** positive **(c)** negative **(d)** impossible to determine

QUICK QUIZ 7.2 Figure 7.4 shows four situations in which a force is applied to an object. In all four cases, the force has the same magnitude, and the displacement of the object is to the right and of the same magnitude. Rank the situations in order of the work done by the force on the object, from most positive to most negative.

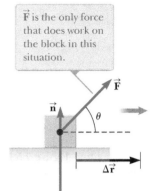

$\vec{F}$ is the only force that does work on the block in this situation.

Figure 7.3 An object is displaced on a frictionless, horizontal surface. The normal force $\vec{n}$ and the gravitational force $m\vec{g}$ do no work on the object.

PITFALL PREVENTION 7.3
Cause of the Displacement We can calculate the work done by a force on an object, but that force is *not* necessarily the cause of the object's displacement. For example, if you lift an object, (negative) work is done on the object by the gravitational force, although gravity is not the cause of the object moving upward!

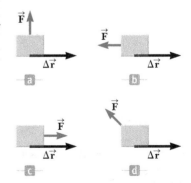

Figure 7.4 (Quick Quiz 7.2) A block is pulled by a force in four different directions. In each case, the displacement of the block is to the right and of the same magnitude.

Example 7.1 Mr. Clean

A man cleaning a floor pulls a vacuum cleaner with a force of magnitude $F = 50.0$ N at an angle of 30.0° with the horizontal (Fig. 7.5). Calculate the work done by the force on the vacuum cleaner as the vacuum cleaner is displaced 3.00 m to the right.

50.0 N

30.0°

Figure 7.5 (Example 7.1) A vacuum cleaner being pulled at an angle of 30.0° from the horizontal.

SOLUTION

Conceptualize Figure 7.5 helps conceptualize the situation. Think about an experience in your life in which you pulled an object across the floor with a rope or cord.

Categorize We are asked for the work done on an object by a force and are given the force on the object, the displacement of the object, and the angle between the two vectors, so we categorize this example as a substitution problem. We identify the vacuum cleaner as the system.

Use the definition of work (Eq. 7.1):

$$W = F \Delta r \cos \theta = (50.0 \text{ N})(3.00 \text{ m})(\cos 30.0°)$$
$$= 130 \text{ J}$$

Notice in this situation that the normal force $\vec{\mathbf{n}}$ and the gravitational $\vec{\mathbf{F}}_g = m\vec{\mathbf{g}}$ do no work on the vacuum cleaner because these forces are perpendicular to the displacements of their points of application. Furthermore, there was no mention of whether there was friction between the vacuum cleaner and the floor. The presence or absence of friction is not important when calculating the work done by the applied force. In addition, this work does not depend on whether the vacuum moved at constant velocity or if it accelerated.

PITFALL PREVENTION 7.4

Work Is a Scalar Although Equation 7.3 defines the work in terms of two vectors, *work is a scalar;* there is no direction associated with it. *All* types of energy and energy transfer are scalars. This fact is a major advantage of the energy approach because we don't need vector calculations!

7.3 The Scalar Product of Two Vectors

Because of the way the force and displacement vectors are combined in Equation 7.1, it is helpful to use a convenient mathematical tool called the **scalar product** of two vectors. We write this **scalar product** of vectors $\vec{\mathbf{A}}$ and $\vec{\mathbf{B}}$ as $\vec{\mathbf{A}} \cdot \vec{\mathbf{B}}$. (Because of the dot symbol, the scalar product is often called the **dot product**.)

The scalar product of any two vectors $\vec{\mathbf{A}}$ and $\vec{\mathbf{B}}$ is defined as a scalar quantity equal to the product of the magnitudes of the two vectors and the cosine of the angle θ between them:

◀ Scalar product of any two vectors $\vec{\mathbf{A}}$ and $\vec{\mathbf{B}}$

$$\vec{\mathbf{A}} \cdot \vec{\mathbf{B}} \equiv AB \cos \theta \qquad (7.2)$$

As is the case with any multiplication, $\vec{\mathbf{A}}$ and $\vec{\mathbf{B}}$ need not have the same units.

By comparing this definition with Equation 7.1, we can express Equation 7.1 as a scalar product:

$$W = F \Delta r \cos \theta = \vec{\mathbf{F}} \cdot \Delta \vec{\mathbf{r}} \qquad (7.3)$$

In other words, $\vec{\mathbf{F}} \cdot \Delta \vec{\mathbf{r}}$ is a shorthand notation for $F \Delta r \cos \theta$.

Before continuing with our discussion of work, let us investigate some properties of the dot product. Figure 7.6 shows two vectors $\vec{\mathbf{A}}$ and $\vec{\mathbf{B}}$ and the angle θ between them used in the definition of the dot product. In Figure 7.6, $B \cos \theta$ is the projection of $\vec{\mathbf{B}}$ onto $\vec{\mathbf{A}}$. Therefore, Equation 7.2 means that $\vec{\mathbf{A}} \cdot \vec{\mathbf{B}}$ is the product of the magnitude of $\vec{\mathbf{A}}$ and the projection of $\vec{\mathbf{B}}$ onto $\vec{\mathbf{A}}$.

From the right-hand side of Equation 7.2, we also see that the scalar product is **commutative.**[1] That is,

$$\vec{\mathbf{A}} \cdot \vec{\mathbf{B}} = \vec{\mathbf{B}} \cdot \vec{\mathbf{A}}$$

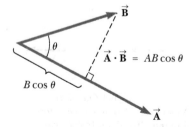

$\vec{\mathbf{B}}$

θ

$\vec{\mathbf{A}} \cdot \vec{\mathbf{B}} = AB \cos \theta$

$B \cos \theta$

$\vec{\mathbf{A}}$

Figure 7.6 The scalar product $\vec{\mathbf{A}} \cdot \vec{\mathbf{B}}$ equals the magnitude of $\vec{\mathbf{A}}$ multiplied by $B \cos \theta$, which is the projection of $\vec{\mathbf{B}}$ onto $\vec{\mathbf{A}}$.

---

[1]The commutativity of the dot product means that $\vec{\mathbf{A}} \cdot \vec{\mathbf{B}}$ also equals the product of the magnitude of $\vec{\mathbf{B}}$ and the projection of $\vec{\mathbf{A}}$ onto $\vec{\mathbf{B}}$. In Chapter 11, you will see another way of combining vectors that proves useful in physics and is not commutative.

Finally, the scalar product obeys the distributive **law of multiplication,** so

$$\vec{A} \cdot (\vec{B} + \vec{C}) = \vec{A} \cdot \vec{B} + \vec{A} \cdot \vec{C}$$

The scalar product is simple to evaluate from Equation 7.2 when $\vec{A}$ is either perpendicular or parallel to $\vec{B}$. If $\vec{A}$ is perpendicular to $\vec{B}$ ($\theta = 90°$), then $\vec{A} \cdot \vec{B} = 0$. (The equality $\vec{A} \cdot \vec{B} = 0$ also holds in the more trivial case in which either $\vec{A}$ or $\vec{B}$ is zero.) If vector $\vec{A}$ is parallel to vector $\vec{B}$ and the two point in the same direction ($\theta = 0$), then $\vec{A} \cdot \vec{B} = AB$. If vector $\vec{A}$ is parallel to vector $\vec{B}$ but the two point in opposite directions ($\theta = 180°$), then $\vec{A} \cdot \vec{B} = -AB$. The scalar product is negative when $90° < \theta \le 180°$.

The unit vectors $\hat{\mathbf{i}}$, $\hat{\mathbf{j}}$, and $\hat{\mathbf{k}}$, which were defined in Chapter 3, lie in the positive x, y, and z directions, respectively, of a right-handed coordinate system. Therefore, it follows from the definition of $\vec{A} \cdot \vec{B}$ that the scalar products of these unit vectors are

$$\hat{\mathbf{i}} \cdot \hat{\mathbf{i}} = \hat{\mathbf{j}} \cdot \hat{\mathbf{j}} = \hat{\mathbf{k}} \cdot \hat{\mathbf{k}} = 1 \tag{7.4}$$

$$\hat{\mathbf{i}} \cdot \hat{\mathbf{j}} = \hat{\mathbf{i}} \cdot \hat{\mathbf{k}} = \hat{\mathbf{j}} \cdot \hat{\mathbf{k}} = 0 \tag{7.5}$$

◀ Scalar products of unit vectors

Equations 3.17 and 3.18 state that two vectors $\vec{A}$ and $\vec{B}$ can be expressed in unit-vector form as

$$\vec{A} = A_x\hat{\mathbf{i}} + A_y\hat{\mathbf{j}} + A_z\hat{\mathbf{k}}$$

$$\vec{B} = B_x\hat{\mathbf{i}} + B_y\hat{\mathbf{j}} + B_z\hat{\mathbf{k}}$$

Using these expressions for the vectors and the information given in Equations 7.4 and 7.5 shows that the scalar product of $\vec{A}$ and $\vec{B}$ reduces to

$$\vec{A} \cdot \vec{B} = A_x B_x + A_y B_y + A_z B_z \tag{7.6}$$

(Details of the derivation are left for you in Problem 5 at the end of the chapter.) In the special case in which $\vec{A} = \vec{B}$, we see that

$$\vec{A} \cdot \vec{A} = A_x^2 + A_y^2 + A_z^2 = A^2$$

QUICK QUIZ 7.3 Which of the following statements is true about the relationship between the dot product of two vectors and the product of the magnitudes of the vectors? (a) $\vec{A} \cdot \vec{B}$ is larger than AB. (b) $\vec{A} \cdot \vec{B}$ is smaller than AB. (c) $\vec{A} \cdot \vec{B}$ could be larger or smaller than AB, depending on the angle between the vectors. (d) $\vec{A} \cdot \vec{B}$ could be equal to AB.

Example 7.2   The Scalar Product

The vectors $\vec{A}$ and $\vec{B}$ are given by $\vec{A} = 2\hat{\mathbf{i}} + 3\hat{\mathbf{j}}$ and $\vec{B} = -\hat{\mathbf{i}} + 2\hat{\mathbf{j}}$.

(A) Determine the scalar product $\vec{A} \cdot \vec{B}$.

SOLUTION

Conceptualize   There is no physical system to imagine here. Rather, it is purely a mathematical exercise involving two vectors.

Categorize   Because we have a definition for the scalar product, we categorize this example as a substitution problem.

Substitute the specific vector expressions for $\vec{A}$ and $\vec{B}$:

$$\vec{A} \cdot \vec{B} = (2\hat{\mathbf{i}} + 3\hat{\mathbf{j}}) \cdot (-\hat{\mathbf{i}} + 2\hat{\mathbf{j}})$$

$$= -2\hat{\mathbf{i}} \cdot \hat{\mathbf{i}} + 2\hat{\mathbf{i}} \cdot 2\hat{\mathbf{j}} - 3\hat{\mathbf{j}} \cdot \hat{\mathbf{i}} + 3\hat{\mathbf{j}} \cdot 2\hat{\mathbf{j}}$$

$$= -2(1) + 4(0) - 3(0) + 6(1) = -2 + 6 = 4$$

The same result is obtained when we use Equation 7.6 directly, where $A_x = 2$, $A_y = 3$, $B_x = -1$, and $B_y = 2$.

continued

7.2 continued

(B) Find the angle θ between $\vec{\mathbf{A}}$ and $\vec{\mathbf{B}}$.

SOLUTION

Evaluate the magnitudes of $\vec{\mathbf{A}}$ and $\vec{\mathbf{B}}$ using the Pythagorean theorem:

$$A = \sqrt{A_x^2 + A_y^2} = \sqrt{(2)^2 + (3)^2} = \sqrt{13}$$

$$B = \sqrt{B_x^2 + B_y^2} = \sqrt{(-1)^2 + (2)^2} = \sqrt{5}$$

Use Equation 7.2 and the result from part (A) to find the angle:

$$\cos\theta = \frac{\vec{\mathbf{A}} \cdot \vec{\mathbf{B}}}{AB} = \frac{4}{\sqrt{13}\sqrt{5}} = \frac{4}{\sqrt{65}}$$

$$\theta = \cos^{-1}\frac{4}{\sqrt{65}} = 60.3°$$

Example **7.3** **Work Done by a Constant Force**

A particle moving in the xy plane undergoes a displacement given by $\Delta\vec{\mathbf{r}} = (2.0\hat{\mathbf{i}} + 3.0\hat{\mathbf{j}})$ m as a constant force $\vec{\mathbf{F}} = (5.0\hat{\mathbf{i}} + 2.0\hat{\mathbf{j}})$ N acts on the particle. Calculate the work done by $\vec{\mathbf{F}}$ on the particle.

SOLUTION

Conceptualize   Although this example is a little more physical than the previous one in that it identifies a force and a displacement, it is similar in terms of its mathematical structure.

Categorize   Because we are given force and displacement vectors and asked to find the work done by this force on the particle, we categorize this example as a substitution problem.

Substitute the expressions for $\vec{\mathbf{F}}$ and $\Delta\vec{\mathbf{r}}$ into Equation 7.3 and use Equations 7.4 and 7.5:

$$W = \vec{\mathbf{F}} \cdot \Delta\vec{\mathbf{r}} = [(5.0\hat{\mathbf{i}} + 2.0\hat{\mathbf{j}})\,\text{N}] \cdot [(2.0\hat{\mathbf{i}} + 3.0\hat{\mathbf{j}})\,\text{m}]$$

$$= (5.0\hat{\mathbf{i}} \cdot 2.0\hat{\mathbf{i}} + 5.0\hat{\mathbf{i}} \cdot 3.0\hat{\mathbf{j}} + 2.0\hat{\mathbf{j}} \cdot 2.0\hat{\mathbf{i}} + 2.0\hat{\mathbf{j}} \cdot 3.0\hat{\mathbf{j}})\,\text{N} \cdot \text{m}$$

$$= [10 + 0 + 0 + 6]\,\text{N} \cdot \text{m} = 16\,\text{J}$$

7.4 **Work Done by a Varying Force**

Now consider a particle being displaced along the x axis under the action of a force that *varies* with position. In such a situation, we cannot use Equation 7.1 to calculate the work done by the force because this relationship applies only when $\vec{\mathbf{F}}$ is constant in magnitude and direction. The red-brown curve in Figure 7.7a shows a varying force applied on a particle that moves from initial position x_i to final position x_f. Imagine a particle undergoing a very small displacement Δx, shown in the figure. The x component F_x of the force is approximately constant over this small interval; for this small displacement, we can approximate the work done on the particle by the force using Equation 7.1 as

$$W \approx F_x \Delta x$$

which is the area of the shaded rectangle in Figure 7.7a. If the F_x versus x curve is divided into a large number of such intervals, the total work done for the displacement from x_i to x_f is approximately equal to the sum of a large number of such terms:

$$W \approx \sum_{x_i}^{x_f} F_x \Delta x$$

If the size of the small displacements is allowed to approach zero, the number of terms in the sum increases without limit but the value of the sum approaches a definite value equal to the area bounded by the F_x curve and the x axis, expressed as an integral:

$$\lim_{\Delta x \to 0} \sum_{x_i}^{x_f} F_x \Delta x = \int_{x_i}^{x_f} F_x \, dx$$

Therefore, we can express the work done by F_x on the system of the particle as it moves from x_i to x_f as

$$W = \int_{x_i}^{x_f} F_x \, dx \tag{7.7}$$

This equation reduces to Equation 7.1 when the component $F_x = F \cos \theta$ remains constant.

If more than one force acts on a system *and the system can be modeled as a particle,* the points of application of all forces move through the same displacement, and the total work done on the system is just the work done by the net force. If we express the net force in the x direction as $\sum F_x$, the total work, or *net work*, done as the particle moves from x_i to x_f is

$$\sum W = W_{\text{ext}} = \int_{x_i}^{x_f} \left(\sum F_x \right) dx \quad \text{(particle)}$$

For the general case of a net force $\sum \vec{F}$ whose magnitude and direction may both vary, we use the scalar product,

$$\sum W = W_{\text{ext}} = \int \left(\sum \vec{F} \right) \cdot d\vec{r} \quad \text{(particle)} \tag{7.8}$$

where the integral is calculated over the path that the particle takes through space. The subscript "ext" on work reminds us that the net work is done by an *external* agent on the system. We will use this notation in this chapter as a reminder and to differentiate this work from an *internal* work to be described shortly.

If the system cannot be modeled as a particle (for example, if the system is deformable), we cannot use Equation 7.8 because different forces on the system may move through different displacements. In this case, we must evaluate the work done by each force separately and then add the works algebraically to find the net work done on the system:

$$\sum W = W_{\text{ext}} = \sum_{\text{forces}} \left(\int \vec{F} \cdot d\vec{r} \right) \quad \text{(deformable system)}$$

Example 7.4 Calculating Total Work Done from a Graph

A force acting on a particle varies with x as shown in Figure 7.8. Calculate the work done by the force on the particle as it moves from $x = 0$ to $x = 6.0$ m.

SOLUTION

Conceptualize Imagine a particle subject to the force in Figure 7.8. The force remains constant as the particle moves through the first 4.0 m and then decreases linearly to zero at 6.0 m. In terms of earlier discussions of motion, the particle could be modeled as a particle under constant acceleration for the first 4.0 m because the force is constant. Between 4.0 m and 6.0 m, however, the motion does not fit into one of our earlier analysis models because the acceleration of the particle is changing. If the particle starts from rest, its speed increases throughout the motion, and the particle is always moving in the positive x direction. These details about its speed and direction are not necessary for the calculation of the work done, however.

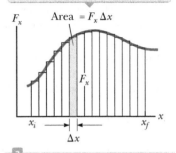
The total work done for the displacement from x_i to x_f is approximately equal to the sum of the areas of all the rectangles.

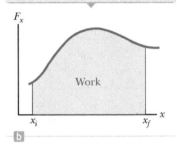
The work done by the component F_x of the varying force as the particle moves from x_i to x_f is *exactly* equal to the area under the curve.

Figure 7.7 (a) The work done on a particle by the force component F_x for the small displacement Δx is $F_x \Delta x$, which equals the area of the shaded rectangle. (b) The width Δx of each rectangle is shrunk to zero.

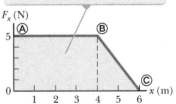
The net work done by this force is the area under the curve.

Figure 7.8 (Example 7.4) The force acting on a particle is constant for the first 4.0 m of motion and then decreases linearly with x from $x_{\circledB} = 4.0$ m to $x_{\circledC} = 6.0$ m.

continued

7.4 continued

Categorize Because the force varies during the motion of the particle, we must use the techniques for work done by varying forces. In this case, the graphical representation in Figure 7.8 can be used to evaluate the work done.

Analyze The work done by the force is equal to the area under the curve from $x_\text{Ⓐ} = 0$ to $x_\text{Ⓒ} = 6.0$ m. This area is equal to the area of the rectangular section from Ⓐ to Ⓑ plus the area of the triangular section from Ⓑ to Ⓒ.

Evaluate the area of the rectangle:

$$W_\text{Ⓐ to Ⓑ} = (5.0\text{ N})(4.0\text{ m}) = 20\text{ J}$$

Evaluate the area of the triangle:

$$W_\text{Ⓑ to Ⓒ} = \tfrac{1}{2}(5.0\text{ N})(2.0\text{ m}) = 5.0\text{ J}$$

Find the total work done by the force on the particle:

$$W_\text{Ⓐ to Ⓒ} = W_\text{Ⓐ to Ⓑ} + W_\text{Ⓑ to Ⓒ} = 20\text{ J} + 5.0\text{ J} = \boxed{25\text{ J}}$$

Finalize Because the graph of the force consists of straight lines, we can use rules for finding the areas of simple geometric models to evaluate the total work done in this example. If a force does not vary linearly, as in Figure 7.7, such rules cannot be used and the force function must be integrated as in Equation 7.7 or 7.8.

Work Done by a Spring

A model of a common physical system on which the force varies with position is shown in Figure 7.9. The system is a block on a frictionless, horizontal surface and connected to a spring. For many springs, if the spring is either stretched or compressed a small distance from its unstretched (equilibrium) configuration, it exerts on the block a force component that can be mathematically modeled as

Spring force ▶

$$F_s = -kx \qquad (7.9)$$

where x is the position of the block relative to its equilibrium ($x = 0$) position and k is a positive constant called the **force constant** or the **spring constant** of the spring. In other words, the force required to stretch or compress a spring is proportional

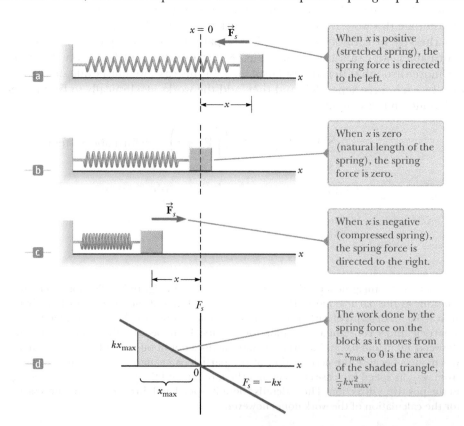

Figure 7.9 The force exerted by a spring on a block varies with the block's position x relative to the equilibrium position $x = 0$.
(a) x is positive. (b) x is zero. (c) x is negative. (d) Graph of F_s versus x for the block–spring system.

When x is positive (stretched spring), the spring force is directed to the left.

When x is zero (natural length of the spring), the spring force is zero.

When x is negative (compressed spring), the spring force is directed to the right.

The work done by the spring force on the block as it moves from $-x_{max}$ to 0 is the area of the shaded triangle, $\tfrac{1}{2}kx_{max}^2$.

to the amount of stretch or compression x. This force law for springs is known as **Hooke's law.** The value of k is a measure of the *stiffness* of the spring. Stiff springs have large k values, and soft springs have small k values. As can be seen from Equation 7.9, the units of k are N/m.

The vector form of Equation 7.9 is

$$\vec{\mathbf{F}}_s = F_s\hat{\mathbf{i}} = -kx\hat{\mathbf{i}} \tag{7.10}$$

where we have chosen the x axis to lie along the direction the spring extends or compresses.

The negative sign in Equations 7.9 and 7.10 signifies that the force exerted by the spring is always directed *opposite* the displacement from equilibrium. When $x > 0$ as in Figure 7.9a so that the block is to the right of the equilibrium position and the spring is stretched, the spring force is directed to the left, in the negative x direction. When $x < 0$ as in Figure 7.9c, the block is to the left of equilibrium, the spring is compressed, and the spring force is directed to the right, in the positive x direction. When $x = 0$ as in Figure 7.9b, the spring is unstretched and $F_s = 0$. Because the spring force always acts toward the equilibrium position ($x = 0$), it is sometimes called a *restoring force.*

If the spring is compressed until the block is at the point $-x_{max}$ and is then released, the block moves from $-x_{max}$ through zero to $+x_{max}$. It then reverses direction, returns to $-x_{max}$, and continues oscillating back and forth. We will study these oscillations in more detail in Chapter 15. For now, let's investigate the work done by the spring on the block over small portions of one oscillation.

Suppose the block has been pushed to the left to a position $-x_{max}$ and is then released as shown in Figure 7.10. We identify the block as our system and calculate the work W_s done by the spring force on the block as the block moves from $x_i = -x_{max}$ to $x_f = 0$. Applying Equation 7.8 and assuming the block may be modeled as a particle, we obtain

$$W_s = \int \vec{\mathbf{F}}_s \cdot d\vec{\mathbf{r}} = \int_{x_i}^{x_f}(-kx\hat{\mathbf{i}})\cdot(dx\hat{\mathbf{i}}) = \int_{-x_{max}}^{0}(-kx)\,dx = \tfrac{1}{2}kx_{max}^2 \tag{7.11}$$

where we have used the integral $\int x^n\,dx = x^{n+1}/(n+1)$ with $n = 1$. The work done by the spring force is positive because the force is in the same direction as its displacement (both are to the right in Figure 7.10 during the time interval considered). Because the block arrives at $x = 0$ with some speed, it will continue moving until it reaches a position $+x_{max}$. The work done by the spring force on the block as it moves from $x_i = 0$ to $x_f = x_{max}$ is $W_s = -\tfrac{1}{2}kx_{max}^2$. The work is negative because for this part of the motion the spring force is to the left and its displacement is to the right. Therefore, the *net* work done by the spring force on the block as it moves from $x_i = -x_{max}$ to $x_f = x_{max}$ is *zero.*

Figure 7.9d is a plot of F_s versus x. Equation 7.9 indicates that F_s is proportional to x, so the graph of F_s versus x is a straight line. The work calculated in Equation 7.11 is the area of the shaded triangle, corresponding to the displacement from $-x_{max}$ to 0. Because the triangle has base x_{max} and height kx_{max}, its area is $\tfrac{1}{2}kx_{max}^2$, agreeing with the work done by the spring calculated in Equation 7.11 by integration.

If the block undergoes an arbitrary displacement from $x = x_i$ to $x = x_f$, the work done by the spring force on the block is

$$W_s = \int_{x_i}^{x_f}(-kx)\,dx = \tfrac{1}{2}kx_i^2 - \tfrac{1}{2}kx_f^2 \tag{7.12}$$

◀ Work done by a spring

From Equation 7.12, we see that the work done by the spring force is zero for any motion that ends where it began ($x_i = x_f$). We shall make use of this important result in Chapter 8 when we describe the motion of this system in greater detail.

Equations 7.11 and 7.12 describe the work done by the spring on the block. Now let us consider the work done on the block by an *external agent* as the agent applies

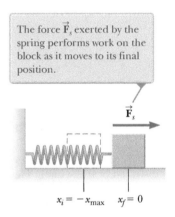

The force $\vec{\mathbf{F}}_s$ exerted by the spring performs work on the block as it moves to its final position.

$\vec{\mathbf{F}}_s$

$x_i = -x_{max}$ $x_f = 0$

Figure 7.10 A block is pushed to the initial position $x_i = -x_{max}$ and then released from rest. We identify the final position as the equilibrium position $x_f = 0$.

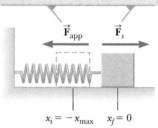

If the process of moving the block is carried out very slowly, then $\vec{\mathbf{F}}_{app}$ is equal in magnitude and opposite in direction to $\vec{\mathbf{F}}_s$ at all times.

$\vec{\mathbf{F}}_{app}$ $\vec{\mathbf{F}}_s$

$x_i = -x_{max}$ $x_f = 0$

Figure 7.11 A block moves from $x_i = -x_{max}$ to $x_f = 0$ on a frictionless surface as a force $\vec{\mathbf{F}}_{app}$ is applied to the block.

a force on the block and the block moves *very slowly* from $x_i = -x_{max}$ to $x_f = 0$ as in Figure 7.11. Compare the two figures carefully. In Figure 7.10, the spring expands freely. In Figure 7.11, however, the *applied force* $\vec{\mathbf{F}}_{app}$ pushes inward and prevents this free expansion. The magnitude of the applied force is adjusted so that the block moves to its final position very slowly. We can calculate the work done by the applied force by noting that at any value of the position, $\vec{\mathbf{F}}_{app}$ is equal in magnitude and opposite in direction to the spring force $\vec{\mathbf{F}}_s$, so $\vec{\mathbf{F}}_{app} = F_{app}\,\hat{\mathbf{i}} = -\vec{\mathbf{F}}_s = -(-kx\hat{\mathbf{i}}) = kx\hat{\mathbf{i}}$. Therefore, the work done by this applied force (the external agent) on the system of the block for the motion described is

$$W_{ext} = \int \vec{\mathbf{F}}_{app} \cdot d\vec{\mathbf{r}} = \int_{x_i}^{x_f}(kx\hat{\mathbf{i}}) \cdot (dx\hat{\mathbf{i}}) = \int_{-x_{max}}^{0} kx\,dx = -\tfrac{1}{2}kx_{max}^2$$

This work is equal to the negative of the work done by the spring force for this displacement (Eq. 7.11). The work is negative because the external agent must push inward on the spring to prevent it from expanding, and this direction is opposite the direction of the displacement of the point of application of the force as the block moves from $-x_{max}$ to 0.

For an arbitrary displacement of the block, the work done on the system by the external agent is

$$W_{ext} = \int_{x_i}^{x_f} kx\,dx = \tfrac{1}{2}kx_f^2 - \tfrac{1}{2}kx_i^2 \tag{7.13}$$

Notice that this equation is the negative of Equation 7.12.

QUICK QUIZ 7.4 A dart is inserted into a spring-loaded dart gun by pushing the spring in by a distance x. For the next loading, the spring is compressed a distance $2x$. How much work is required to load the second dart compared with that required to load the first? **(a)** four times as much **(b)** two times as much **(c)** the same **(d)** half as much **(e)** one-fourth as much

---

Example 7.5   Measuring *k* for a Spring

A common technique used to measure the force constant of a spring is demonstrated by the setup in Figure 7.12. The spring is hung vertically (Fig. 7.12a), and an object of mass m is attached to its lower end. Under the action of the "load" mg, the spring stretches a distance d from its equilibrium position (Fig. 7.12b).

(A) If a spring is stretched 2.0 cm by a suspended object having a mass of 0.55 kg, what is the force constant of the spring?

SOLUTION

Conceptualize   Figure 7.12b shows what happens to the spring when the object is attached to it. Simulate this situation by hanging an object on a rubber band.

Categorize   The object in Figure 7.12b is at rest and not accelerating, so it is modeled as a *particle in equilibrium*.

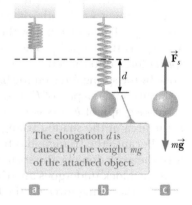

The elongation d is caused by the weight mg of the attached object.

$\vec{\mathbf{F}}_s$

d

$m\vec{\mathbf{g}}$

a   b   c

Figure 7.12 (Example 7.5) Determining the force constant k of a spring.

Analyze   Because the object is in equilibrium, the net force on it is zero and the upward spring force balances the downward gravitational force $m\vec{\mathbf{g}}$ (Fig. 7.12c).

Apply the particle in equilibrium model to the object: $\vec{\mathbf{F}}_s + m\vec{\mathbf{g}} = 0 \;\rightarrow\; F_s - mg = 0 \;\rightarrow\; F_s = mg$

Apply Hooke's law to give the magnitude $F_s = kd$ and solve for k: $k = \dfrac{mg}{d} = \dfrac{(0.55\ \text{kg})(9.80\ \text{m/s}^2)}{2.0 \times 10^{-2}\ \text{m}} = 2.7 \times 10^2\ \text{N/m}$

(B) How much work is done by the spring on the object as it stretches through this distance?

7.5 continued

SOLUTION

Use Equation 7.12 to find the work done by the spring on the object:

$$W_s = 0 - \tfrac{1}{2}kd^2 = -\tfrac{1}{2}(2.7 \times 10^2 \text{ N/m})(2.0 \times 10^{-2} \text{ m})^2$$
$$= -5.4 \times 10^{-2} \text{ J}$$

Finalize This work is negative because the spring force acts upward on the object, but its point of application (where the spring attaches to the object) moves downward. As the object moves through the 2.0-cm distance, the gravitational force also does work on it. This work is positive because the gravitational force is downward and so is the displacement of the point of application of this force. Would we expect the work done by the gravitational force, as the applied force in a direction opposite to the spring force, to be the negative of the answer above? Let's find out.

Evaluate the work done by the gravitational force on the object:

$$W = \vec{\mathbf{F}} \cdot \Delta\vec{\mathbf{r}} = (mg)(d)\cos 0 = mgd$$
$$= (0.55 \text{ kg})(9.80 \text{ m/s}^2)(2.0 \times 10^{-2} \text{ m}) = 1.1 \times 10^{-1} \text{ J}$$

If you expected the work done by gravity simply to be that done by the spring with a positive sign, you may be surprised by this result! To understand why that is not the case, we need to explore further, as we do in the next section.

7.5 Kinetic Energy and the Work–Kinetic Energy Theorem

When energy transfers across the boundary of a system, the amount of energy stored in the system changes. We have investigated work in some depth and have identified it as a mechanism for transferring energy into a system. We have stated that work is an influence on a system from the environment, but we have not yet discussed the result of this influence on the system. One possible result of doing work on a system is that the system changes its speed: a common experience is to push on an object and observe it changing its state from rest to motion. In this section, we investigate this situation and introduce our first type of energy storage in a system, called *kinetic energy*.

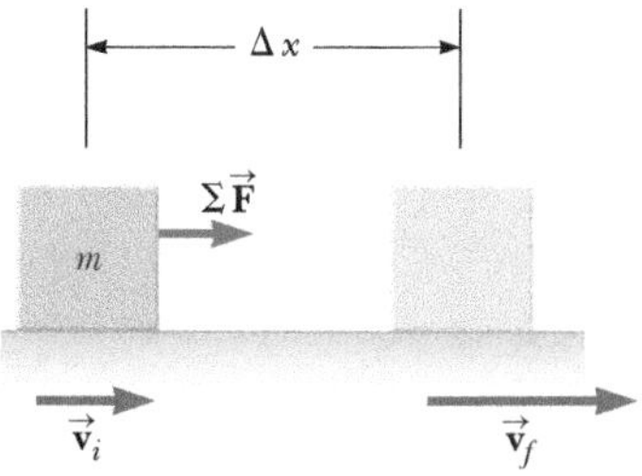

Consider a system consisting of a single object. Figure 7.13 shows a block of mass m moving through a displacement directed to the right under the action of a net force $\sum\vec{\mathbf{F}}$, also directed to the right. We know from Newton's second law that the block moves with an acceleration $\vec{\mathbf{a}}$. If the block (and therefore the force) moves through a displacement $\Delta\vec{\mathbf{r}} = \Delta x\hat{\mathbf{i}} = (x_f - x_i)\hat{\mathbf{i}}$, the net work done on the block by the external net force $\sum\vec{\mathbf{F}}$ is given by Equation 7.7:

$$W_{\text{ext}} = \int_{x_i}^{x_f} \sum F \, dx \qquad (7.14)$$

Using Newton's second law, we substitute for the magnitude of the net force $\sum F = ma$ and then perform the following chain-rule manipulations on the integrand:

$$W_{\text{ext}} = \int_{x_i}^{x_f} ma \, dx = \int_{x_i}^{x_f} m\frac{dv}{dt} \, dx = \int_{x_i}^{x_f} m\frac{dv}{dx}\frac{dx}{dt} \, dx = \int_{v_i}^{v_f} mv \, dv$$

$$W_{\text{ext}} = \tfrac{1}{2}mv_f^2 - \tfrac{1}{2}mv_i^2 \qquad (7.15)$$

where v_i is the speed of the block at $x = x_i$ and v_f is its speed at x_f.

Equation 7.15 was generated for the specific situation of one-dimensional motion, but it is a general result. It tells us that the work done by the net force on a particle of mass m is equal to the difference between the initial and final values of a quantity $\tfrac{1}{2}mv^2$. This quantity is so important that it has been given a special name, **kinetic energy:**

$$K \equiv \tfrac{1}{2}mv^2 \qquad (7.16) \qquad \blacktriangleleft \text{ Kinetic energy}$$

Figure 7.13 An object undergoing a displacement $\Delta\vec{\mathbf{r}} = \Delta x\hat{\mathbf{i}}$ and a change in velocity under the action of a net force $\sum\vec{\mathbf{F}}$.

TABLE 7.1 Kinetic Energies for Various Objects

Object	Mass (kg)	Speed (m/s)	Kinetic Energy (J)
Earth orbiting the Sun	5.97×10^{24}	2.98×10^4	2.65×10^{33}
Moon orbiting the Earth	7.35×10^{22}	1.02×10^3	3.82×10^{28}
Rocket moving at escape speed[a]	500	1.12×10^4	3.14×10^{10}
Automobile at 65 mi/h	2 000	29	8.4×10^5
Running athlete	70	10	3 500
Stone dropped from 10 m	1.0	14	98
Golf ball at terminal speed	0.046	44	45
Raindrop at terminal speed	3.5×10^{-5}	9.0	1.4×10^{-3}
Oxygen molecule in air	5.3×10^{-26}	500	6.6×10^{-21}

[a]Escape speed is the minimum speed an object must reach near the Earth's surface to move infinitely far away from the Earth.

Kinetic energy represents the energy associated with the motion of the particle. Note that kinetic energy is a scalar quantity and has the same units as work. For example, a 2.0-kg object moving with a speed of 4.0 m/s has a kinetic energy of 16 J. Table 7.1 lists the kinetic energies for various objects.

Equation 7.15 states that the work done on a particle by a net force $\sum \vec{F}$ acting on it equals the change in kinetic energy of the particle. It is often convenient to write Equation 7.15 in the form

$$W_{\text{ext}} = K_f - K_i = \Delta K \qquad (7.17)$$

Another way to write it is $K_f = K_i + W_{\text{ext}}$, which tells us that the final kinetic energy of an object is equal to its initial kinetic energy plus the change in energy due to the net work done on it.

We have generated Equation 7.17 by imagining doing work on a particle. If we identify the particle as a system, we have increased the amount of energy stored in the system by doing work on it. We have stored the energy in the particular form of kinetic energy, represented by motion of the system through space. We could also do work on a deformable system, in which members of the system move with respect to one another. In this case, we also find that Equation 7.17 is valid as long as the net work is found by adding up the works done by each force and adding, as discussed earlier with regard to Equation 7.8. The kinetic energy K of the system is the sum of the kinetic energies of all members of the system.

Equation 7.17 is an important result known as the **work–kinetic energy theorem:**

Work–kinetic energy theorem ▶

> When work is done on a system and the only change in the system is in the speeds of its members, the net work done on the system equals the change in kinetic energy of the system, as expressed by Equation 7.17: $W = \Delta K$.

The work–kinetic energy theorem indicates that the kinetic energy of a system *increases* if the net work done on it is *positive*: energy is being transferred *into* the system. The kinetic energy *decreases* if the net work is *negative*: energy is being transferred *out of* the system.

Because we have so far only investigated translational motion through space, we arrived at the work–kinetic energy theorem by analyzing situations involving translational motion. Another type of motion is *rotational motion*, in which an object spins about an axis. We will study this type of motion in Chapter 10. The work–kinetic energy theorem is also valid for systems that undergo a change in the rotational speed due to work done on the system. A windmill serves as an example of work (done by the wind) causing rotational motion.

The work–kinetic energy theorem will clarify a result seen earlier in this chapter that may have seemed odd. In Section 7.4, we arrived at a result of zero net work done when we let a spring push a block from $x_i = -x_{max}$ to $x_f = x_{max}$. Notice that because the speed of the block is continually changing, it may seem complicated to analyze this process. The quantity ΔK in the work–kinetic energy theorem, however, only refers to the initial and final configurations of the system. It does not depend on the particular path followed by any members of the system. Therefore, because the speed of the block is zero at both the initial and final points of the motion, the net work done on the block is zero. We will often see this concept of path independence in similar approaches to problems.

Let us also return to the mystery in the Finalize step at the end of Example 7.5. Why was the work done by gravity not just the value of the work done by the spring with a positive sign? Notice that the work done by gravity is larger than the magnitude of the work done by the spring. Therefore, the total work done by all forces on the object is positive. Imagine now how to create the situation in which the *only* forces on the object are the spring force and the gravitational force. You must support the object at the highest point and then remove your hand and let the object fall. If you do so, you know that when the object reaches a position 2.0 cm below your hand, it will be *moving*, which is consistent with Equation 7.17. Positive net work is done on the object, and the result is that it has a kinetic energy as it passes through the 2.0-cm point.

The only way to prevent the object from having a kinetic energy after moving through 2.0 cm is to slowly lower it with your hand. Then, however, there is a third force doing work on the object, the normal force from your hand. If this work is calculated and added to that done by the spring force and the gravitational force, the net work done on the object is zero, which is consistent because it is not moving at the 2.0-cm point.

Earlier, we indicated that work can be considered as a mechanism for transferring energy into a system. Equation 7.17 is a mathematical statement of this concept. When work W_{ext} is done on a system, the result is a transfer of energy across the boundary of the system. The result on the system, in the case of Equation 7.17, is a change ΔK in kinetic energy. In the next section, we investigate another type of energy that can be stored in a system as a result of doing work on the system.

QUICK QUIZ 7.5 A dart is inserted into a spring-loaded dart gun by pushing the spring in by a distance x. For the next loading, the spring is compressed a distance $2x$. How much faster does the second dart leave the gun compared with the first? **(a)** four times as fast **(b)** two times as fast **(c)** the same **(d)** half as fast **(e)** one-fourth as fast

---

Example 7.6 A Block Pulled on a Frictionless Surface

A 6.0-kg block initially at rest is pulled to the right along a frictionless, horizontal surface by a constant horizontal force of magnitude 12 N. Find the block's speed after it has moved through a horizontal distance of 3.0 m.

SOLUTION

Conceptualize Figure 7.14 illustrates this situation. Imagine pulling a toy car across a table with a horizontal rubber band attached to the front of the car. The force is maintained constant by ensuring that the stretched rubber band always has the same length.

Categorize We could apply the equations of kinematics to determine the answer, but let us practice the energy approach. The block is the system, and three external forces act on the system. The normal force balances the gravitational force on the block, and neither of these vertically acting forces does work on the block because their points of application are not vertically displaced.

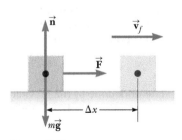

Figure 7.14 (Example 7.6) A block pulled to the right on a frictionless surface by a constant horizontal force.

continued

7.6 continued

Analyze   The net external force acting on the block is the horizontal 12-N force.

Use the work–kinetic energy theorem for the block, noting that its initial kinetic energy is zero:

$$W_{\text{ext}} = \Delta K = K_f - K_i = \tfrac{1}{2}mv_f^2 - 0 = \tfrac{1}{2}mv_f^2$$

Solve for v_f and use Equation 7.1 for the work done on the block by $\vec{\mathbf{F}}$:

$$v_f = \sqrt{\frac{2W_{\text{ext}}}{m}} = \sqrt{\frac{2F\Delta x}{m}}$$

Substitute numerical values:

$$v_f = \sqrt{\frac{2(12\text{ N})(3.0\text{ m})}{6.0\text{ kg}}} = 3.5\text{ m/s}$$

Finalize   You should solve this problem again by modeling the block as a *particle under a net force* to find its acceleration and then as a *particle under constant acceleration* to find its final velocity. In Chapter 8, we will see that the energy procedure followed above is an example of the analysis model of the *nonisolated system*.

WHAT IF?   Suppose the magnitude of the force in this example is doubled to $F' = 2F$. The 6.0-kg block accelerates to 3.5 m/s due to this applied force while moving through a displacement $\Delta x'$. How does the displacement $\Delta x'$ compare with the original displacement Δx?

Answer   If we pull harder, the block should accelerate to a given speed in a shorter distance, so we expect that $\Delta x' < \Delta x$. In both cases, the block experiences the same change in kinetic energy ΔK. Therefore, the same work is done on the block in both cases. Mathematically, from the work–kinetic energy theorem, we find that

$$W_{\text{ext}} = F'\Delta x' = \Delta K = F\Delta x$$

$$\Delta x' = \frac{F}{F'}\Delta x = \frac{F}{2F}\Delta x = \tfrac{1}{2}\Delta x$$

and the distance is shorter as suggested by our conceptual argument.

Conceptual Example 7.7   **Does the Ramp Lessen the Work Required?**

A man wishes to load a refrigerator onto a truck using a ramp at angle θ as shown in Figure 7.15. He claims that less work would be required to load the truck if the length L of the ramp were increased so that the angle θ would be smaller. Is his claim valid?

SOLUTION

No. Suppose the refrigerator is wheeled on a hand truck up the ramp at constant speed. In this case, for the system of the refrigerator and the hand truck, $\Delta K = 0$. The normal force exerted by the ramp on the system is directed at 90° to the displacement of its point of application and so does no work on the system. Because $\Delta K = 0$, the work–kinetic energy theorem applied to the refrigerator gives

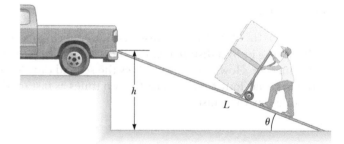

Figure 7.15 (Conceptual Example 7.7) A refrigerator attached to a frictionless, wheeled hand truck is moved up a ramp at constant speed.

$$W_{\text{ext}} = W_{\text{by man}} + W_{\text{by gravity}} = 0$$

The work done by the gravitational force equals the product of the weight mg of the system, the distance L through which the refrigerator is displaced, and $\cos(\theta + 90°)$. Therefore,

$$W_{\text{by man}} = -W_{\text{by gravity}} = -(mg)(L)[\cos(\theta + 90°)]$$
$$= mgL\sin\theta = mgh$$

where $h = L\sin\theta$ is the height of the ramp at the truck. Therefore, the man must do the same amount of work mgh on the system *regardless* of the length of the ramp. The work depends only on the height of the ramp. Although less force is required with a longer ramp, the point of application of that force moves through a greater displacement.

7.6 Potential Energy of a System

So far in this chapter, we have defined a system in general, but have focused our attention primarily on single particles or objects under the influence of external forces. Let us now consider systems of two or more particles or objects interacting via a force that is *internal* to the system. The kinetic energy of such a system is the algebraic sum of the kinetic energies of all members of the system. There may be systems, however, in which one object is so massive that it can be modeled as stationary and its kinetic energy can be neglected. For example, if we consider a ball–Earth system as the ball falls to the Earth, the kinetic energy of the system can be considered as just the kinetic energy of the ball. The Earth moves so slowly in this process that we can ignore its kinetic energy. On the other hand, the kinetic energy of a system of two electrons must include the kinetic energies of both particles.

Let us imagine a system consisting of a book and the Earth, interacting via the gravitational force. We do some work on the system by lifting the book slowly from rest through a vertical displacement $\Delta \vec{r} = (y_f - y_i)\hat{j}$ as in Figure 7.16. According to our discussion of work as an energy transfer, this work done on the system must appear as an increase in energy of the system. The book is at rest before we perform the work and is at rest after we perform the work. Therefore, there is no change in the kinetic energy of the system.

Because the energy change of the system is not in the form of kinetic energy, the work-kinetic energy theorem does not apply here and the energy change must appear as some form of energy storage other than kinetic energy. After lifting the book, we could release it and let it fall back to the position y_i. Notice that the book (and therefore, the system) now has kinetic energy and that its source is in the work that was done in lifting the book. While the book was at the highest point, the system had the *potential* to possess kinetic energy, but it did not do so until the book was allowed to fall. Therefore, we call the energy storage mechanism before the book is released **potential energy.** We will find that the potential energy of a system can only be associated with specific types of forces acting between members of a system. The amount of potential energy in the system is determined by the *configuration* of the system. Moving members of the system to different positions or rotating them may change the configuration of the system and therefore its potential energy.

Let us now derive an expression for the potential energy associated with an object at a given location above the surface of the Earth. Consider an external agent lifting an object of mass m from an initial height y_i above the ground to a final height y_f as in Figure 7.16. We assume the lifting is done slowly, with no acceleration, so the applied force from the agent is equal in magnitude to the gravitational force on the object: the object is modeled as a particle in equilibrium moving at constant velocity. The work done by the external agent on the system (object and the Earth) as the object undergoes this upward displacement is given by the product of the upward applied force $\vec{F}_{app}$ and the upward displacement of this force, $\Delta \vec{r} = \Delta y \hat{j}$:

$$W_{ext} = (\vec{F}_{app}) \cdot \Delta \vec{r} = (mg\hat{j}) \cdot [(y_f - y_i)\hat{j}] = mgy_f - mgy_i \qquad (7.18)$$

where this result is the net work done on the system because the applied force is the only force on the system from the environment. (Remember that the gravitational force is *internal* to the system.) Notice the similarity between Equation 7.18 and Equation 7.15. In each equation, the work done on a system equals a difference between the final and initial values of a quantity. In Equation 7.15, the work represents a transfer of energy into the system and the increase in energy of the system is kinetic in form. In Equation 7.18, the work represents a transfer of energy into the system and the system energy appears in a different form, which we have called potential energy.

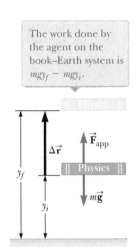

The work done by the agent on the book–Earth system is $mgy_f - mgy_i$.

Figure 7.16 An external agent lifts a book slowly from a height y_i to a height y_f.

Therefore, we can identify the quantity mgy as the **gravitational potential energy** U_g of the system of an object of mass m and the Earth:

Gravitational ▶
potential energy

$$U_g \equiv mgy \qquad (7.19)$$

The units of gravitational potential energy are joules, the same as the units of work and kinetic energy. Potential energy, like work and kinetic energy, is a scalar quantity. Notice that Equation 7.19 is valid only for objects near the surface of the Earth, where g is approximately constant.[2]

Using our definition of gravitational potential energy, Equation 7.18 can now be rewritten as

$$W_{ext} = \Delta U_g \qquad (7.20)$$

which mathematically describes that the net external work done on the system in this situation appears as a change in the gravitational potential energy of the system.

Equation 7.20 is similar in form to the work–kinetic energy theorem, Equation 7.17. In Equation 7.17, work is done on a system and energy appears in the system as kinetic energy, representing *motion* of the members of the system. In Equation 7.20, work is done on the system and energy appears in the system as potential energy, representing a change in the *configuration* of the members of the system.

Gravitational potential energy depends only on the vertical height of the object above the surface of the Earth. The same amount of work must be done on an object–Earth system whether the object is lifted vertically from the Earth or is pushed starting from the same point up a frictionless incline, ending up at the same height. We verified this statement for a specific situation of rolling a refrigerator up a ramp in Conceptual Example 7.7. This statement can be shown to be true in general by calculating the work done on an object by an agent moving the object through a displacement having both vertical and horizontal components:

$$W_{ext} = (\vec{\mathbf{F}}_{app}) \cdot \Delta\vec{\mathbf{r}} = (mg\hat{\mathbf{j}}) \cdot [(x_f - x_i)\,\hat{\mathbf{i}} + (y_f - y_i)\hat{\mathbf{j}}] = mgy_f - mgy_i$$

where there is no term involving x in the final result because $\hat{\mathbf{j}} \cdot \hat{\mathbf{i}} = 0$.

In solving problems, you must choose a reference configuration for which the gravitational potential energy of the system is set equal to some reference value, which is normally zero. The choice of reference configuration is completely arbitrary because the important quantity is the *difference* in potential energy, and this difference is independent of the choice of reference configuration.

It is often convenient to choose as the reference configuration for zero gravitational potential energy the configuration in which an object is at the surface of the Earth, but this choice is not essential. Often, the statement of the problem suggests a convenient configuration to use.

QUICK QUIZ 7.6 Choose the correct answer. The gravitational potential energy of a system **(a)** is always positive **(b)** is always negative **(c)** can be negative or positive

Example 7.8   The Proud Athlete and the Sore Toe

A trophy being shown off by a careless athlete slips from the athlete's hands and drops on his foot. Choosing floor level as the $y = 0$ point of your coordinate system, estimate the change in gravitational potential energy of the trophy–Earth system as the trophy falls. Repeat the calculation, using the top of the athlete's head as the origin of coordinates.

SOLUTION

Conceptualize The trophy changes its vertical position with respect to the surface of the Earth. Associated with this change in position is a change in the gravitational potential energy of the trophy–Earth system.

---
[2]The assumption that g is constant is valid as long as the vertical displacement of the object is small compared with the Earth's radius.

7.8 continued

Categorize We evaluate a change in gravitational potential energy defined in this section, so we categorize this example as a substitution problem. Because there are no numbers provided in the problem statement, it is also an estimation problem.

The problem statement tells us that the reference configuration of the trophy–Earth system corresponding to zero potential energy is when the bottom of the trophy is at the floor. To find the change in potential energy for the system, we need to estimate a few values. Let's say the trophy has a mass of approximately 2 kg, and the top of a person's foot is about 0.05 m above the floor. Also, let's assume the trophy falls from a height of 1.4 m.

Calculate the gravitational potential energy of the trophy–Earth system just before the trophy is released:

$$U_i = mgy_i = (2 \text{ kg})(9.80 \text{ m/s}^2)(1.4 \text{ m}) = 27.4 \text{ J}$$

Calculate the gravitational potential energy of the trophy–Earth system when the trophy reaches the athlete's foot:

$$U_f = mgy_f = (2 \text{ kg})(9.80 \text{ m/s}^2)(0.05 \text{ m}) = 0.98 \text{ J}$$

Evaluate the change in gravitational potential energy of the trophy–Earth system:

$$\Delta U_g = 0.98 \text{ J} - 27.4 \text{ J} = -26.4 \text{ J}$$

We should probably keep only two digits because of the roughness of our estimates; therefore, we estimate that the change in gravitational potential energy is -26 J. The system had about 27 J of gravitational potential energy before the trophy began its fall and approximately 1 J of potential energy as the trophy reaches the top of the foot.

The second case presented indicates that the reference configuration of the system for zero potential energy is chosen to be when the trophy is on the athlete's head (even though the trophy is never at this position in its motion). We estimate this position to be 2.0 m above the floor.

Calculate the gravitational potential energy of the trophy–Earth system just before the trophy is released from its position 0.6 m below the athlete's head:

$$U_i = mgy_i = (2 \text{ kg})(9.80 \text{ m/s}^2)(-0.6 \text{ m}) = -11.8 \text{ J}$$

Calculate the gravitational potential energy of the trophy–Earth system when the trophy reaches the athlete's foot located 1.95 m below the athlete's head:

$$U_f = mgy_f = (2 \text{ kg})(9.80 \text{ m/s}^2)(-1.95 \text{ m}) = -38.2 \text{ J}$$

Evaluate the change in gravitational potential energy of the trophy–Earth system:

$$\Delta U_g = -38.2 \text{ J} - (-11.8 \text{ J}) = -26.4 \text{ J} \approx -26 \text{ J}$$

This value is the same as before, as it must be. The change in potential energy is independent of the choice of configuration of the system representing the zero of potential energy. If we wanted to keep only one digit in our estimates, we could write the final result as $3 \times 10^1 \text{ J}$.

Elastic Potential Energy

Because members of a system can interact with one another by means of different types of forces, it is possible that there are different types of potential energy in a system. We have just become familiar with gravitational potential energy of a system in which members interact via the gravitational force. Let us explore a second type of potential energy that a system can possess.

Consider a system consisting of a block and a spring as shown in Figure 7.17 (page 168). In Section 7.4, we identified *only* the block as the system. Now we include both the block and the spring in the system and recognize that the spring force is the interaction between the two members of the system. The force that the spring exerts on the block is given by $F_s = -kx$ (Eq. 7.9). The external work done by an applied force F_{app} on the block–spring system as the block moves from x_i to x_f is given by Equation 7.13:

$$W_{\text{ext}} = \tfrac{1}{2}kx_f^2 - \tfrac{1}{2}kx_i^2 \tag{7.21}$$

In this situation, the initial and final x coordinates of the block are measured from its equilibrium position, $x = 0$. Again (as in the gravitational case, Eq. 7.18) the

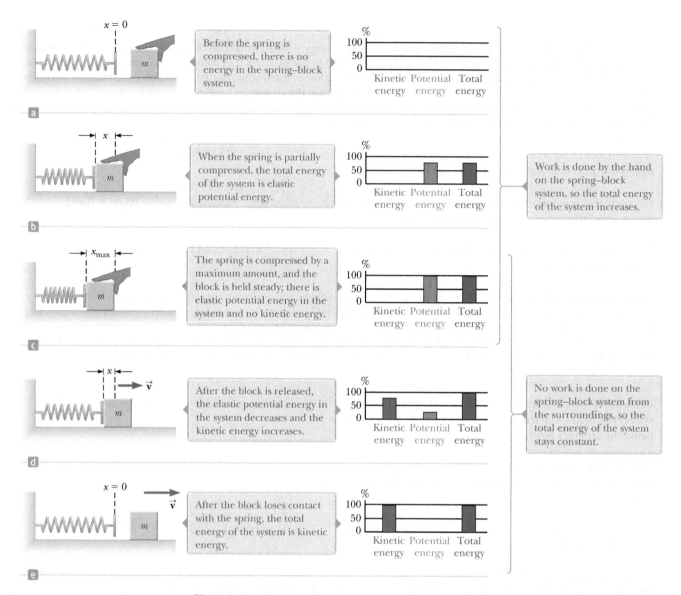

Figure 7.17 A spring on a frictionless, horizontal surface is compressed a distance x_{max} when a block of mass m is pushed against it. The block is then released and the spring pushes it to the right, where the block eventually loses contact with the spring. Parts (a) through (e) show various instants in the process. Energy bar charts on the right of each part of the figure help keep track of the energy in the system.

work done on the system is equal to the difference between the initial and final values of an expression related to the system's configuration. The **elastic potential energy** function associated with the block–spring system is defined by

Elastic potential energy ▷

$$U_s \equiv \tfrac{1}{2}kx^2 \qquad (7.22)$$

Equation 7.21 can be expressed as

$$W_{ext} = \Delta U_s \qquad (7.23)$$

Compare this equation to Equations 7.17 and 7.20. In all three situations, external work is done on a system and a form of energy storage in the system changes as a result.

The elastic potential energy of the system can be thought of as the energy stored in the deformed spring (one that is either compressed or stretched from its equilibrium position). The elastic potential energy stored in a spring is zero whenever the spring is undeformed ($x = 0$). Energy is stored in the spring only when the spring is either stretched or compressed. Because the elastic potential energy is proportional to x^2, we see that U_s is always positive in a deformed spring. Everyday examples of the storage of elastic potential energy can be found in old-style clocks or watches that operate from a wound-up spring and small wind-up toys for children.

Consider Figure 7.17 once again, which shows a spring on a frictionless, horizontal surface. When a block is pushed against the spring by an external agent, the elastic potential energy and the total energy of the system increase as indicated in Figure 7.17b. When the spring is compressed a distance x_{max} (Fig. 7.17c), the elastic potential energy stored in the spring is $\frac{1}{2}kx_{max}^2$. When the external force is removed, the only force on the block is that due to the spring, and the block moves to the right. The elastic potential energy of the system decreases, whereas the kinetic energy increases and the total energy remains fixed (Fig. 7.17d). When the spring returns to its original length, the stored elastic potential energy is completely transformed into kinetic energy of the block (Fig. 7.17e).

Figure 7.18 (Quick Quiz 7.7) A ball connected to a massless spring suspended vertically. What forms of potential energy are associated with the system when the ball is displaced downward?

QUICK QUIZ 7.7 A ball is connected to a light spring suspended vertically as shown in Figure 7.18. When pulled downward from its equilibrium position and released, the ball oscillates up and down. **(i)** In the system of *the ball, the spring, and the Earth,* what forms of energy are there during the motion? (a) kinetic and elastic potential (b) kinetic and gravitational potential (c) kinetic, elastic potential, and gravitational potential (d) elastic potential and gravitational potential **(ii)** In the system of *the ball and the spring,* what forms of energy are there during the motion? Choose from the same possibilities (a) through (d).

Energy Bar Charts

Figure 7.17 shows an important graphical representation of information related to energy of systems called an **energy bar chart.** The vertical axis represents the amount of energy of a given type in the system. The horizontal axis shows the types of energy in the system. The bar chart in Figure 7.17a shows that the system contains zero energy because the spring is relaxed and the block is not moving. Between Figure 7.17a and Figure 7.17c, the hand does work on the system, compressing the spring and storing elastic potential energy in the system. In Figure 7.17d, the block has been released and is moving to the right while still in contact with the spring. The height of the bar for the elastic potential energy of the system decreases, the kinetic energy bar increases, and the total energy bar remains fixed. In Figure 7.17e, the spring has returned to its relaxed length and the system now contains only kinetic energy associated with the moving block.

Energy bar charts can be a very useful representation for keeping track of the various types of energy in a system. For practice, try making energy bar charts for the book–Earth system in Figure 7.16 when the book is dropped from the higher position. Figure 7.18 associated with Quick Quiz 7.7 shows another system for which drawing an energy bar chart would be a good exercise. We will show energy bar charts in some figures in this chapter.

7.7 Conservative and Nonconservative Forces

We now introduce a third type of energy that a system can possess and store. Imagine that the book in Figure 7.19a (page 170) has been accelerated by your hand and is now sliding to the right on the surface of a heavy table and slowing down due to the friction force. Suppose the *surface* is the system. Then, from our discussion of work, we can argue that the friction force from the sliding book does work on the surface. The friction force on the surface is to the right and the displacement of the point of application of the force is to the right because the book has moved to the right. The work done on the surface is therefore positive, but the surface is not

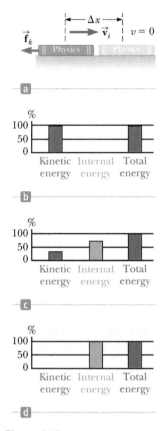

Figure 7.19 (a) A book sliding to the right on a horizontal surface slows down in the presence of a force of kinetic friction acting to the left on the book. (b) An energy bar chart showing the energy in the system of the book and the surface at the initial instant of time. The energy of the system is all kinetic energy. (c) While the book is sliding, the kinetic energy of the system decreases as it is transformed to internal energy. (d) After the book has stopped, the energy of the system is all internal energy.

moving after the book has stopped. Positive work has been done on the surface, yet there is no increase in the surface's kinetic energy. Nor is there any change in the potential energy of any system. So work has been done, but where is the energy?

From your everyday experience with sliding over surfaces with friction, you can probably guess that the surface will be *warmer* after the book slides over it. This is what you found when you sanded the wood in the opening storyline for this chapter. The work that was done on the surface has gone into warming the surface rather than increasing its speed or changing the configuration of a system. We call the energy associated with the temperature of a system its **internal energy,** symbolized E_{int}. (We will define internal energy more generally in Chapter 19.) In this case, the work done on the surface does indeed represent energy transferred into the system, but it appears in the system as internal energy rather than kinetic or potential energy.

Now consider the book and the surface in Figure 7.19a together as a system. After the book is released, and while it is slowing down, no work is done on this system. Initially, the system has kinetic energy because the book is moving. While the book is sliding, the internal energy of the system increases: the book and the surface are warmer than before. When the book stops, the kinetic energy has been completely transformed to internal energy. We can consider the friction force within the system—that is, between the book and the surface—as a *transformation mechanism* for energy. This force transforms the kinetic energy of the system into internal energy. Rub your hands together briskly to experience this effect!

Figures 7.19b through 7.19d show energy bar charts for the situation in Figure 7.19a. In Figure 7.19b, the bar chart shows that the system contains kinetic energy at the instant the book is released by your hand. We define the reference amount of internal energy in the system as zero at this instant. Figure 7.19c shows the kinetic energy transforming to internal energy as the book slows down due to the friction force. In Figure 7.19d, after the book has stopped sliding, the kinetic energy is zero, and the system now contains only internal energy E_{int}. Notice that the total energy bar in red has not changed during the process. The amount of internal energy in the system after the book has stopped is equal to the amount of kinetic energy in the system at the initial instant. This equality is described by an important principle called *conservation of energy*. We will explore this principle in Chapter 8.

Now consider in more detail an object moving downward near the surface of the Earth. The work done by the gravitational force on the object does not depend on whether it falls vertically or slides down a sloping incline with friction. All that matters is the change in the object's elevation. The energy transformation to internal energy due to friction on that incline, however, depends very much on the distance the object slides. The longer the incline, the more potential energy is transformed to internal energy. In other words, the path makes no difference when we consider the work done by the gravitational force, but it does make a difference when we consider the energy transformation due to friction forces. We can use this varying dependence on path to classify forces as either *conservative* or *nonconservative*. Of the two forces just mentioned, the gravitational force is conservative and the friction force is nonconservative.

Conservative Forces

Conservative forces have these two equivalent properties:

▶ Properties of conservative forces

1. The work done by a conservative force on a particle moving between any two points is independent of the path taken by the particle.
2. The work done by a conservative force on a particle moving through any closed path is zero. (A closed path is one for which the beginning point and the endpoint are identical.)

The gravitational force is one example of a conservative force; the force that an ideal spring exerts on any object attached to the spring is another. The work done by the gravitational force on an object moving between any two points near the

Earth's surface is $W_g = -mg\hat{\mathbf{j}} \cdot [(y_f - y_i)\hat{\mathbf{j}}] = mgy_i - mgy_f$. From this equation, notice that W_g depends only on the initial and final y coordinates of the object and hence is independent of the path. Furthermore, W_g is zero when the object moves over any closed path (where $y_i = y_f$).

For the case of the object–spring system, the work W_s done by the spring force is given by $W_s = \frac{1}{2}kx_i^2 - \frac{1}{2}kx_f^2$ (Eq. 7.12). We see that the spring force is conservative because W_s depends only on the initial and final x coordinates of the object and is zero for any closed path (where $x_i = x_f$).

Nonconservative Forces

A force is **nonconservative** if it does not satisfy properties 1 and 2 above. The work done by a nonconservative force is path-dependent. We define the sum of the kinetic and potential energies of a system as the **mechanical energy** of the system:

$$E_{\text{mech}} \equiv K + U \qquad (7.24)$$

where K includes the kinetic energy of all moving members of the system and U includes all types of potential energy in the system. For a book falling under the action of the gravitational force, the mechanical energy of the book–Earth system remains fixed; gravitational potential energy transforms to kinetic energy, and the total energy of the system remains constant. Nonconservative forces acting within a system, however, cause a *change* in the mechanical energy of the system. For example, for a book sent sliding on a horizontal surface that is not frictionless (Fig. 7.19a), the mechanical energy of the book–surface system is transformed to internal energy as we discussed earlier. Only part of the book's kinetic energy is transformed to internal energy in the book. The rest appears as internal energy in the surface. (When you trip and slide across a gymnasium floor, not only does the skin on your knees warm up, so does the floor!) Because the force of kinetic friction transforms the mechanical energy of a system into internal energy, it is a nonconservative force.

As an example of the path dependence of the work for a nonconservative force, consider Figure 7.20. Suppose you displace a book between two points on a table. If the book is displaced in a straight line along the blue path between points Ⓐ and Ⓑ in Figure 7.20, you do a certain amount of work against the kinetic friction force to keep the book moving at a constant speed. Now, imagine that you push the book along the brown semicircular path in Figure 7.20. You perform more work against friction along this curved path than along the straight path because the curved path is longer. The work done on the book depends on the path, so the friction force *cannot* be conservative.

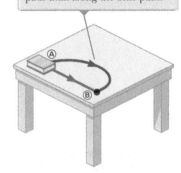

The work done in moving the book is greater along the brown path than along the blue path.

Figure 7.20 The work done against the force of kinetic friction depends on the path taken as the book is moved from Ⓐ to Ⓑ.

7.8 Relationship Between Conservative Forces and Potential Energy

We can associate a **potential energy function** U for a system with a force acting between members of the system, but *we can do so only if the force is conservative.* In general, the work W_{int} done by a conservative force on an object that is a member of a system as the system changes from one configuration to another is equal to the initial value of the potential energy of the system minus the final value:

$$W_{\text{int}} = U_i - U_f = -\Delta U \qquad (7.25)$$

The subscript "int" in Equation 7.25 reminds us that the work we are discussing is done by one member of the system on another member and is therefore *internal* to the system. It is different from the work W_{ext} done *on* the system as a whole by an external agent. As an example, compare Equation 7.25 with the equation for the work done by an external agent on a block–spring system (Eq. 7.23) as the extension of the spring changes.

PITFALL PREVENTION 7.9
Similar Equation Warning Compare Equation 7.25 with Equation 7.20. These equations are similar except for the negative sign, which is a common source of confusion. Equation 7.20 tells us that positive work done *by an outside agent* on a system causes an increase in the potential energy of the system (with no change in the kinetic or internal energy). Equation 7.25 states that positive work done on a component of a system by a conservative force *internal to the system* causes a decrease in the potential energy of the system.

Let us imagine a system of particles in which a conservative force $\vec{\mathbf{F}}$ acts between the particles. Imagine also that the configuration of the system changes due to the motion of one particle along the x axis. Then we can evaluate the internal work done by this force as the particle moves along the x axis[3] using Equations 7.7 and 7.25:

$$W_{\text{int}} = \int_{x_i}^{x_f} F_x \, dx = -\Delta U \tag{7.26}$$

where F_x is the component of $\vec{\mathbf{F}}$ in the direction of the displacement. We can also express Equation 7.26 as

$$\Delta U = U_f - U_i = -\int_{x_i}^{x_f} F_x \, dx \tag{7.27}$$

Therefore, ΔU is negative when F_x and dx are in the same direction, as when an object is lowered in a gravitational field or when a spring pushes an object toward equilibrium.

It is often convenient to establish some particular location x_i of one member of a system as representing a reference configuration and measure all potential energy differences with respect to it. We can then define the potential energy function as

$$U_f(x) = -\int_{x_i}^{x_f} F_x \, dx + U_i \tag{7.28}$$

The value of U_i is often taken to be zero for the reference configuration. It does not matter what value we assign to U_i because any nonzero value merely shifts $U_f(x)$ by a constant amount and only the *change* in potential energy is physically meaningful.

If the point of application of the force undergoes an infinitesimal displacement dx, we can express the infinitesimal change in the potential energy of the system dU as

$$dU = -F_x \, dx$$

Therefore, the conservative force is related to the potential energy function through the relationship[4]

▶ Relation of force between members of a system to the potential energy of the system

$$F_x = -\frac{dU}{dx} \tag{7.29}$$

That is, the x component of a conservative force acting on a member within a system equals the negative derivative of the potential energy of the system with respect to x.

We can easily check Equation 7.29 for the two examples already discussed. In the case of the deformed spring, $U_s = \frac{1}{2}kx^2$; therefore,

$$F_s = -\frac{dU_s}{dx} = -\frac{d}{dx}(\tfrac{1}{2}kx^2) = -kx$$

which corresponds to the restoring force in the spring (Hooke's law). Because the gravitational potential energy function is $U_g = mgy$, it follows from Equation 7.29 that $F_g = -mg$ when we differentiate U_g with respect to y instead of x.

We now see that U is an important function because a conservative force can be derived from it. Furthermore, Equation 7.29 should clarify that adding a constant to the potential energy is unimportant because the derivative of a constant is zero.

---

[3]For a general displacement, the work done in two or three dimensions also equals $-\Delta U$, where $U = U(x, y, z)$. We write this equation formally as $W_{\text{int}} = \int_i^f \vec{\mathbf{F}} \cdot d\vec{\mathbf{r}} = U_i - U_f$.

[4]In three dimensions, the expression is

$$\vec{\mathbf{F}} = -\frac{\partial U}{\partial x}\hat{\mathbf{i}} - \frac{\partial U}{\partial y}\hat{\mathbf{j}} - \frac{\partial U}{\partial z}\hat{\mathbf{k}}$$

where $(\partial U/\partial x)$ and so forth are partial derivatives. In the language of vector calculus, $\vec{\mathbf{F}}$ equals the negative of the *gradient* of the scalar quantity $U(x, y, z)$.

QUICK QUIZ 7.8 What does the slope of a graph of $U(x)$ versus x represent? **(a)** the magnitude of the force on the object **(b)** the negative of the magnitude of the force on the object **(c)** the x component of the force on the object **(d)** the negative of the x component of the force on the object

7.9 Energy Diagrams and Equilibrium of a System

The motion of a system can often be understood qualitatively through a graph of its potential energy versus the position of a member of the system. Consider the potential energy function for a block–spring system, given by $U_s = \frac{1}{2}kx^2$. This function is plotted versus x in Figure 7.21a, where x is the position of the block.

As we saw in Quick Quiz 7.8, the x component of the force is equal to the negative of the slope of the U-versus-x curve. When the block is placed at rest at the equilibrium position of the spring ($x = 0$), where $F_s = 0$, it will remain there unless some external force F_{ext} acts on it. If this external force stretches the spring from equilibrium, x is positive and the slope dU/dx is positive; therefore, the force F_s exerted by the spring is negative and the block accelerates back toward $x = 0$ when released. If the external force compresses the spring, x is negative and the slope is negative; therefore, F_s is positive and again the mass accelerates toward $x = 0$ upon release.

From this analysis, we conclude that the $x = 0$ position for a block–spring system is one of **stable equilibrium.** That is, any movement away from this position results in a force directed back toward $x = 0$. In general, configurations of a system in stable equilibrium correspond to those for which $U(x)$ for the system has a minimum.

Another simple mechanical system with a configuration of stable equilibrium is a ball rolling about in the bottom of a bowl. Anytime the ball is displaced from its lowest position, it tends to return to that position when released.

Now consider a particle moving along the x axis under the influence of a conservative force F_x, where the U-versus-x curve is as shown in Figure 7.22. Once again, $F_x = 0$ at $x = 0$, and so the particle is in equilibrium at this point. This position, however, is one of **unstable equilibrium** for the following reason. Suppose the particle is displaced to the right ($x > 0$). Because the slope is negative for $x > 0$, $F_x = -dU/dx$ is positive and the particle accelerates away from $x = 0$. If instead the particle is at $x = 0$ and is displaced to the left ($x < 0$), the force is negative because the slope is positive for $x < 0$ and the particle again accelerates away from the equilibrium position. The position $x = 0$ in this situation is one of unstable equilibrium because for any displacement from this point, the force pushes the particle farther away from equilibrium and toward a position of lower potential energy. A pencil balanced on its point is in a position of unstable equilibrium. If the pencil is displaced slightly from its absolutely vertical position and is then released, it will surely fall over. In general, configurations of a system in unstable equilibrium correspond to those for which $U(x)$ for the system has a maximum.

Finally, a configuration called **neutral equilibrium** arises when U is constant over some region. Small displacements of an object from a position in this region produce neither restoring nor disrupting forces. A ball lying on a flat, horizontal surface is an example of an object in neutral equilibrium.

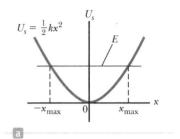

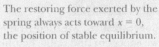
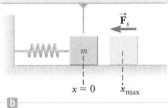

Figure 7.21 (a) Potential energy as a function of x for the frictionless block–spring system shown in (b). For a given energy E of the system, the block oscillates between the turning points, which have the coordinates $x = \pm x_{\text{max}}$.

PITFALL PREVENTION 7.10
Energy Diagrams A common mistake is to think that potential energy on the graph in an energy diagram represents the height of some object. For example, that is not the case in Figure 7.21, where the block is only moving horizontally.

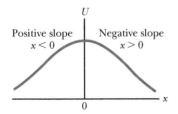

Figure 7.22 A plot of U versus x for a particle that has a position of unstable equilibrium located at $x = 0$. For any finite displacement of the particle, the force on the particle is directed away from $x = 0$.

Example **7.9**   Force and Energy on an Atomic Scale

The potential energy associated with the force between two neutral atoms in a molecule can be modeled by the Lennard–Jones potential energy function:

$$U(r) = 4\epsilon \left[\left(\frac{\sigma}{r} \right)^{12} - \left(\frac{\sigma}{r} \right)^{6} \right]$$

where r is the separation of the atoms. The function $U(r)$ contains two parameters σ and ϵ that are determined from experiments. Sample values for the interaction between two atoms in a molecule are $\sigma = 0.263$ nm and $\epsilon = 1.51 \times 10^{-22}$ J. Using a spreadsheet or similar tool, graph this function and find the most likely distance between the two atoms.

SOLUTION

Conceptualize   We identify the two atoms in the molecule as a system. Based on our understanding that stable molecules exist, we expect to find stable equilibrium when the two atoms are separated by some equilibrium distance.

Categorize   Because a potential energy function exists, we categorize the force between the atoms as conservative. For a conservative force, Equation 7.29 describes the relationship between the force and the potential energy function.

Analyze   Stable equilibrium exists for a separation distance at which the potential energy of the system of two atoms (the molecule) is a minimum.

Take the derivative of the function $U(r)$:

$$\frac{dU(r)}{dr} = 4\epsilon \frac{d}{dr}\left[\left(\frac{\sigma}{r} \right)^{12} - \left(\frac{\sigma}{r} \right)^{6} \right] = 4\epsilon \left[\frac{-12\sigma^{12}}{r^{13}} + \frac{6\sigma^{6}}{r^{7}} \right]$$

Minimize the function $U(r)$ by setting its derivative equal to zero:

$$4\epsilon \left[\frac{-12\sigma^{12}}{r_{eq}^{13}} + \frac{6\sigma^{6}}{r_{eq}^{7}} \right] = 0 \quad \rightarrow \quad r_{eq} = (2)^{1/6}\sigma$$

Evaluate r_{eq}, the equilibrium separation of the two atoms in the molecule:

$$r_{eq} = (2)^{1/6}(0.263 \text{ nm}) = 2.95 \times 10^{-10} \text{ m}$$

We graph the Lennard–Jones function on both sides of this critical value to create our energy diagram as shown in Figure 7.23.

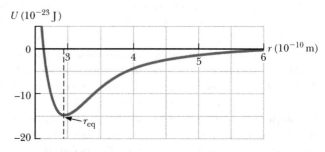

Finalize   Notice that $U(r)$ is extremely large when the atoms are very close together, is a minimum when the atoms are at their critical separation, and then increases again as the atoms move apart. When $U(r)$ is a minimum, the atoms are in stable equilibrium, indicating that the most likely separation between them occurs at this point.

Figure 7.23   (Example 7.9) Potential energy curve associated with a molecule. The distance r is the separation between the two atoms making up the molecule.

Summary

▶ Definitions

A **system** is most often a single particle, a collection of particles, or a region of space, and may vary in size and shape. A **system boundary** separates the system from the **environment.**

The **work** W done on a system by an agent exerting a constant force $\vec{F}$ on the system is the product of the magnitude Δr of the displacement of the point of application of the force and the component $F\cos\theta$ of the force along the direction of the displacement $\Delta \vec{r}$:

$$W \equiv F\Delta r \cos\theta \tag{7.1}$$

If a varying force does work on a particle as the particle moves along the x axis from x_i to x_f, the work done by the force on the particle is given by

$$W = \int_{x_i}^{x_f} F_x \, dx \qquad (7.7)$$

where F_x is the component of force in the x direction.

The **scalar product** (dot product) of two vectors $\vec{A}$ and $\vec{B}$ is defined by the relationship

$$\vec{A} \cdot \vec{B} \equiv AB \cos\theta \qquad (7.2)$$

where the result is a scalar quantity and θ is the angle between the two vectors. The scalar product obeys the commutative and distributive laws.

The **kinetic energy** of a particle of mass m moving with a speed v is

$$K \equiv \tfrac{1}{2}mv^2 \qquad (7.16)$$

If a particle of mass m is at a distance y above the Earth's surface ($y = 0$), the **gravitational potential energy** of the particle–Earth system is

$$U_g \equiv mgy \qquad (7.19)$$

The **elastic potential energy** stored in a spring of force constant k is

$$U_s \equiv \tfrac{1}{2}kx^2 \qquad (7.22)$$

A force is **conservative** if the work it does on a particle that is a member of the system as the particle moves between two points is independent of the path the particle takes between the two points. Furthermore, a force is conservative if the work it does on a particle is zero when the particle moves through an arbitrary closed path and returns to its initial position. A force that does not meet these criteria is said to be **nonconservative.**

The **total mechanical energy of a system** is defined as the sum of the kinetic energy and the potential energy:

$$E_{\text{mech}} \equiv K + U \qquad (7.24)$$

▶ Concepts and Principles

The **work–kinetic energy theorem** states that if work is done on a system by external forces and the only change in the system is in the speeds of its members,

$$W_{\text{ext}} = K_f - K_i = \Delta K = \tfrac{1}{2}mv_f^2 - \tfrac{1}{2}mv_i^2 \qquad (7.15, 7.17)$$

If the only change is in the configuration of the system,

$$W_{\text{ext}} = \Delta U \qquad (7.20, 7.23)$$

A **potential energy function** U can be associated only with a conservative force. If a conservative force $\vec{F}$ acts between members of a system while one member moves along the x axis from x_i to x_f, the change in the potential energy of the system equals the negative of the work done by that force:

$$U_f - U_i = -\int_{x_i}^{x_f} F_x \, dx \qquad (7.27)$$

Systems can be in three types of equilibrium configurations when the net force on a member of the system is zero. Configurations of **stable equilibrium** correspond to those for which $U(x)$ has a minimum.

Configurations of **unstable equilibrium** correspond to those for which $U(x)$ has a maximum.

Neutral equilibrium arises when U is constant as a member of the system moves over some region.

Think–Pair–Share

See the Preface for an explanation of the icons used in this problems set. For additional assessment items for this section, go to ⚡ WEBASSIGN From Cengage

1. You are working in a manufacturing plant. One of the machines in the plant uses a spring. For a new process, it would be desirable for springs with different force constants to be used during different portions of the process. Your boss asks you for advice on how to change out one spring for another quickly so that the entire process can take place in a reasonable amount of time. You suggest that, rather than changing out springs continuously, you can change the force constant of *one* long spring by clamping it at various locations to define a new fixed end of the spring. Then the effective spring consists only of those coils beyond the

clamp. You design a system consisting of one long spring with N coils and a force constant k. You design a clamping system that will isolate part of the spring, leaving N' coils free beyond the fixed clamp. (a) Write an expression for the force constant k' of the free end of the spring in terms of k, N, and N'. (b) The end of the unclamped, relaxed spring is grasped and pulled outward by a distance x. In the process, the hand holding the end of the spring does work W on the spring. Now, the spring is returned to its relaxed state and then clamped at its center point. The free end of the clamped, relaxed spring is grasped and pulled outward by the same distance x. How much work does the hand do on the spring in this case? W? $2W$? $4W$? Another value?

2. **ACTIVITY** In the table of data, we see minimum stopping distances d as a function of the initial speed v of a car. Work in your group to answer the following. (a) If you double the initial speed, does it take twice the distance to stop the car? (b) Assume the stopping distance is proportional to the speed of the car raised to some power: $d \propto v^n$. Use graphing techniques to determine n. (c) Why does the stopping distance depend on the particular value of n that you found in (b)?

Speed (mi/h)	Stopping Distance (ft)
20	22.5
25	35.0
30	50.4
35	68.6
40	89.6
45	113.5
50	140.0
55	169.5
60	201.7
65	236.7
70	274.5

Problems

See the Preface for an explanation of the icons used in this problems set. For additional assessment items for this section, go to **WEBASSIGN** From Cengage

SECTION 7.2 Work Done by a Constant Force

1. A shopper in a supermarket pushes a cart with a force of 35.0 N directed at an angle of 25.0° below the horizontal. The force is just sufficient to balance various friction forces, so the cart moves at constant speed. (a) Find the work done by the shopper on the cart as she moves down a 50.0-m-long aisle. (b) The shopper goes down the next aisle, pushing horizontally and maintaining the same speed as before. If the friction force doesn't change, would the shopper's applied force be larger, smaller, or the same? (c) What about the work done on the cart by the shopper?

2. The record number of boat lifts, including the boat and its ten crew members, was achieved by Sami Heinonen and Juha Räsänen of Sweden in 2000. They lifted a total mass of 653.2 kg approximately 4 in. off the ground a total of 24 times. Estimate the total work done by the two men on the boat in this record lift, ignoring the negative work done by the men when they lowered the boat back to the ground.

3. In 1990, Walter Arfeuille of Belgium lifted a 281.5-kg object through a distance of 17.1 cm using only his teeth. (a) How much work was done on the object by Arfeuille in this lift, assuming the object was lifted at constant speed? (b) What total force was exerted on Arfeuille's teeth during the lift?

4. Spiderman, whose mass is 80.0 kg, is dangling on the free end of a 12.0-m-long rope, the other end of which is fixed to a tree limb above. By repeatedly bending at the waist, he is able to get the rope in motion, eventually getting it to swing enough that he can reach a ledge when the rope makes a 60.0° angle with the vertical. How much work was done by the gravitational force on Spiderman in this maneuver?

SECTION 7.3 The Scalar Product of Two Vectors

5. For any two vectors $\vec{A}$ and $\vec{B}$, show that $\vec{A} \cdot \vec{B} = A_x B_x + A_y B_y + A_z B_z$. *Suggestions:* Write $\vec{A}$ and $\vec{B}$ in unit-vector form and use Equations 7.4 and 7.5.

6. Vector $\vec{A}$ has a magnitude of 5.00 units, and vector $\vec{B}$ has a magnitude of 9.00 units. The two vectors make an angle of 50.0° with each other. Find $\vec{A} \cdot \vec{B}$.

Note: In Problems 7 and 8, calculate numerical answers to three significant figures as usual.

7. Find the scalar product of the vectors in Figure P7.7.

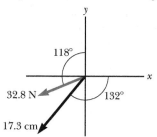

Figure P7.7

8. Using the definition of the scalar product, find the angles between (a) $\vec{A} = 3\hat{i} - 2\hat{j}$ and $\vec{B} = 4\hat{i} - 4\hat{j}$, (b) $\vec{A} = -2\hat{i} + 4\hat{j}$ and $\vec{B} = 3\hat{i} - 4\hat{j} + 2\hat{k}$, and (c) $\vec{A} = \hat{i} - 2\hat{j} + 2\hat{k}$ and $\vec{B} = 3\hat{j} + 4\hat{k}$.

SECTION 7.4 Work Done by a Varying Force

9. A particle is subject to a force F_x that varies with position as shown in Figure P7.9. Find the work done by the force on the particle as it moves (a) from $x = 0$ to $x = 5.00$ m, (b) from $x = 5.00$ m to $x = 10.0$ m, and (c) from $x = 10.0$ m to $x = 15.0$ m. (d) What is the total work done by the force over the distance $x = 0$ to $x = 15.0$ m?

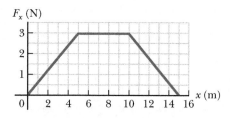

Figure P7.9 Problems 9 and 22.

10. In a control system, an accelerometer consists of a 4.70-g object sliding on a calibrated horizontal rail. A low-mass spring attaches the object to a flange at one end of the rail. Grease on the rail makes static friction negligible, but rapidly damps out vibrations of the sliding object. When subject to a steady acceleration of $0.800g$, the object should be at a location 0.500 cm away from its equilibrium position. Find the force constant of the spring required for the calibration to be correct.

11. When a 4.00-kg object is hung vertically on a certain light **AMT** spring that obeys Hooke's law, the spring stretches 2.50 cm. **T** If the 4.00-kg object is removed, (a) how far will the spring stretch if a 1.50-kg block is hung on it? (b) How much work must an external agent do to stretch the same spring 4.00 cm from its unstretched position?

12. Express the units of the force constant of a spring in SI fun-**S** damental units.

13. The tray dispenser in your cafeteria has broken and is not **CR** repairable. The custodian knows that you are good at designing things and asks you to help him build a new dispenser out of spare parts he has on his workbench. The tray dispenser supports a stack of trays on a shelf that is supported by four springs, one at each corner of the shelf. Each tray is rectangular, with dimensions 45.3 cm by 35.6 cm. Each tray is 0.450 cm thick and has a mass of 580 g. The custodian asks you to design a new four-spring dispenser such that when a tray is removed, the dispenser pushes up the remaining stack so that the top tray is at the same position as the just-removed tray was. He has a wide variety of springs that he can use to build the dispenser. Which springs should he use?

14. A light spring with force constant 3.85 N/m is compressed by 8.00 cm as it is held between a 0.250-kg block on the left and a 0.500-kg block on the right, both resting on a horizontal surface. The spring exerts a force on each block, tending to push the blocks apart. The blocks are simultaneously released from rest. Find the acceleration with which each block starts to move, given that the coefficient of kinetic friction between each block and the surface is (a) 0, (b) 0.100, and (c) 0.462.

15. A small particle of mass **S** m is pulled to the top of a frictionless half-cylinder (of radius R) by a light cord that passes over the top of the cylinder as illustrated in Figure P7.15. (a) Assuming the particle moves at a constant speed,

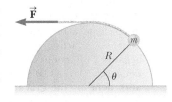

Figure P7.15

show that $F = mg\cos\theta$. *Note:* If the particle moves at constant speed, the component of its acceleration tangent to the cylinder must be zero at all times. (b) By directly integrating $W = \int \vec{\mathbf{F}} \cdot d\vec{\mathbf{r}}$, find the work done in moving the particle at constant speed from the bottom to the top of the half-cylinder.

16. The force acting on a particle is $F_x = (8x - 16)$, where F is in newtons and x is in meters. (a) Make a plot of this force versus x from $x = 0$ to $x = 3.00$ m. (b) From your graph, find the net work done by this force on the particle as it moves from $x = 0$ to $x = 3.00$ m.

17. When different loads hang on a spring, the spring **Q|C** stretches to different lengths as shown in the following

table. (a) Make a graph of the applied force versus the extension of the spring. (b) By least-squares fitting, determine the straight line that best fits the data. (c) To complete part (b), do you want to use all the data points, or should you ignore some of them? Explain. (d) From the slope of the best-fit line, find the spring constant k. (e) If the spring is extended to 105 mm, what force does it exert on the suspended object?

F (N)	2.0	4.0	6.0	8.0	10	12	14	16	18	20	22
L (mm)	15	32	49	64	79	98	112	126	149	175	190

18. A 100-g bullet is fired from a rifle having a barrel 0.600 m long. Choose the origin to be at the location where the bullet begins to move. Then the force (in newtons) exerted by the expanding gas on the bullet is $15\,000 + 10\,000x - 25\,000x^2$, where x is in meters. (a) Determine the work done by the gas on the bullet as the bullet travels the length of the barrel. (b) **What If?** If the barrel is 1.00 m long, how much work is done, and (c) how does this value compare with the work calculated in part (a)?

19. (a) A force $\vec{\mathbf{F}} = (4x\hat{\mathbf{i}} + 3y\hat{\mathbf{j}})$, where $\vec{\mathbf{F}}$ is in newtons and x **V** and y are in meters, acts on an object as the object moves in the x direction from the origin to $x = 5.00$ m. Find the work $W = \int \vec{\mathbf{F}} \cdot d\vec{\mathbf{r}}$ done by the force on the object. (b) **What If?** Find the work $W = \int \vec{\mathbf{F}} \cdot d\vec{\mathbf{r}}$ done by the force on the object if it moves from the origin to (5.00 m, 5.00 m) along a straight-line path making an angle of 45.0° with the positive x axis. Is the work done by this force dependent on the path taken between the initial and final points?

20. **Review.** The graph in Figure P7.20 specifies a functional relationship between the two variables u and v. (a) Find $\int_a^b u\,dv$. (b) Find $\int_b^a u\,dv$. (c) Find $\int_a^b v\,du$.

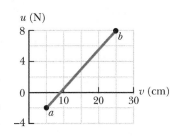

Figure P7.20

SECTION 7.5 Kinetic Energy and the Work–Kinetic Energy Theorem

21. A 0.600-kg particle has a speed of 2.00 m/s at point Ⓐ and **V** kinetic energy of 7.50 J at point Ⓑ. What is (a) its kinetic energy at Ⓐ, (b) its speed at Ⓑ, and (c) the net work done on the particle by external forces as it moves from Ⓐ to Ⓑ?

22. A 4.00-kg particle is subject to a net force that varies with **V** position as shown in Figure P7.9. The particle starts from rest at $x = 0$. What is its speed at (a) $x = 5.00$ m, (b) $x = 10.0$ m, and (c) $x = 15.0$ m?

23. A 2 100-kg pile driver is used to drive a steel I-beam into the **T** ground. The pile driver falls 5.00 m before coming into contact with the top of the beam, and it drives the beam 12.0 cm farther into the ground before coming to rest. Using energy considerations, calculate the average force the beam exerts on the pile driver while the pile driver is brought to rest.

24. **Review.** In an electron microscope, there is an electron gun **AMT** that contains two charged metallic plates 2.80 cm apart. An electric force accelerates each electron in the beam from rest to 9.60% of the speed of light over this distance.

(a) Determine the kinetic energy of the electron as it leaves the electron gun. Electrons carry this energy to a phosphorescent viewing screen where the microscope's image is formed, making it glow. For an electron passing between the plates in the electron gun, determine (b) the magnitude of the constant electric force acting on the electron, (c) the acceleration of the electron, and (d) the time interval the electron spends between the plates.

25. **Review.** You can think of the work–kinetic energy theorem as a second theory of motion, parallel to Newton's laws in describing how outside influences affect the motion of an object. In this problem, solve parts (a), (b), and (c) separately from parts (d) and (e) so you can compare the predictions of the two theories. A 15.0-g bullet is accelerated from rest to a speed of 780 m/s in a rifle barrel of length 72.0 cm. (a) Find the kinetic energy of the bullet as it leaves the barrel. (b) Use the work–kinetic energy theorem to find the net work that is done on the bullet. (c) Use your result to part (b) to find the magnitude of the average net force that acted on the bullet while it was in the barrel. (d) Now model the bullet as a particle under constant acceleration. Find the constant acceleration of a bullet that starts from rest and gains a speed of 780 m/s over a distance of 72.0 cm. (e) Modeling the bullet as a particle under a net force, find the net force that acted on it during its acceleration. (f) What conclusion can you draw from comparing your results of parts (c) and (e)?

26. You are lying in your bedroom, resting after doing your physics homework. As you stare at your ceiling, you come up with the idea for a new game. You grab a dart with a sticky nose and a mass of 19.0 g. You also grab a spring that has been lying on your desk from some previous project. You paint a target pattern on your ceiling. Your new game is to place the spring vertically on the floor, place the sticky-nose dart facing upward on the spring, and push the spring downward until the coils all press together, as on the right in Figure P7.26. You will then release the spring, firing the dart up toward the target on your ceiling, where its sticky nose will make it hang from the ceiling. The spring has an uncompressed end-to-end length of 5.00 cm, as shown on the left in Figure P7.26, and can be compressed to an end-to-end length of 1.00 cm when the coils are all pressed together. Before trying the game, you hold the upper end of the spring in one hand and hang a bundle of ten identical darts from the lower end of the spring. The spring extends by 1.00 cm due to the weight of the darts. You are so excited about the new game that, before doing a test of the game, you run out to gather your friends to show them. When your friends are in your room watching and you show them the first firing of your new game, why are you embarrassed?

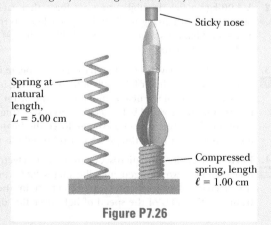

Spring at natural length, L = 5.00 cm

Sticky nose

Compressed spring, length ℓ = 1.00 cm

Figure P7.26

27. **Review.** A 5.75-kg object passes through the origin at time $t = 0$ such that its x component of velocity is 5.00 m/s and its y component of velocity is −3.00 m/s. (a) What is the kinetic energy of the object at this time? (b) At a later time $t = 2.00$ s, the particle is located at $x = 8.50$ m and $y = 5.00$ m. What constant force acted on the object during this time interval? (c) What is the speed of the particle at $t = 2.00$ s?

28. **Review.** A 7.80-g bullet moving at 575 m/s strikes the hand of a superhero, causing the hand to move 5.50 cm in the direction of the bullet's velocity before stopping. (a) Use work and energy considerations to find the average force that stops the bullet. (b) Assuming the force is constant, determine how much time elapses between the moment the bullet strikes the hand and the moment it stops moving.

SECTION 7.6 Potential Energy of a System

29. A 0.20-kg stone is held 1.3 m above the top edge of a water well and then dropped into it. The well has a depth of 5.0 m. Relative to the configuration with the stone at the top edge of the well, what is the gravitational potential energy of the stone–Earth system (a) before the stone is released and (b) when it reaches the bottom of the well? (c) What is the change in gravitational potential energy of the system from release to reaching the bottom of the well?

30. A 1 000-kg roller coaster car is initially at the top of a rise, at point Ⓐ. It then moves 135 ft, at an angle of 40.0° below the horizontal, to a lower point Ⓑ. (a) Choose the car at point Ⓑ to be the zero configuration for gravitational potential energy of the roller coaster–Earth system. Find the potential energy of the system when the car is at points Ⓐ and Ⓑ, and the change in potential energy as the car moves between these points. (b) Repeat part (a), setting the zero configuration with the car at point Ⓐ.

SECTION 7.7 Conservative and Nonconservative Forces

31. A 4.00-kg particle moves from the origin to position Ⓒ, having coordinates $x = 5.00$ m and $y = 5.00$ m (Fig. P7.31). One force on the particle is the gravitational force acting in the negative y direction. Using Equation 7.3, calculate the work done by the gravitational force on the particle as it goes from O to Ⓒ along (a) the purple path, (b) the red path, and (c) the blue path. (d) Your results should all be identical. Why?

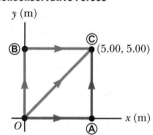

Figure P7.31
Problems 31 through 33

32. (a) Suppose a constant force acts on an object. The force does not vary with time or with the position or the velocity of the object. Start with the general definition for work done by a force

$$W = \int_i^f \vec{\mathbf{F}} \cdot d\vec{\mathbf{r}}$$

and show that the force is conservative. (b) As a special case, suppose the force $\vec{\mathbf{F}} = (3\hat{\mathbf{i}} + 4\hat{\mathbf{j}})$ N acts on a particle that moves from O to Ⓒ in Figure P7.31. Calculate the work done by $\vec{\mathbf{F}}$ on the particle as it moves along each one of the three paths shown in the figure and show that the work done along the three paths is identical. (c) **What If?** Is the work done also identical along the three paths for the force $\vec{\mathbf{F}} = (4x\hat{\mathbf{i}} + 3y\hat{\mathbf{j}})$, where $\vec{\mathbf{F}}$ is in newtons and x and y are in meters, from Problem 19? (d) **What If?** Suppose the force is given by $\vec{\mathbf{F}} = (y\hat{\mathbf{i}} - x\hat{\mathbf{j}})$, where $\vec{\mathbf{F}}$ is in newtons and x and y are in meters. Is the work done identical along the three paths for this force?

33. A force acting on a particle moving in the xy plane is given by $\vec{\mathbf{F}} = (2y\hat{\mathbf{i}} + x^2\hat{\mathbf{j}})$, where $\vec{\mathbf{F}}$ is in newtons and x and y are in meters. The particle moves from the origin to a final position having coordinates $x = 5.00$ m and $y = 5.00$ m as shown in Figure P7.31. Calculate the work done by $\vec{\mathbf{F}}$ on the particle as it moves along (a) the purple path, (b) the red path, and (c) the blue path. (d) Is $\vec{\mathbf{F}}$ conservative or nonconservative? (e) Explain your answer to part (d).

SECTION 7.8 Relationship Between Conservative Forces and Potential Energy

34. *Why is the following situation impossible?* A librarian lifts a book from the ground to a high shelf, doing 20.0 J of work in the lifting process. As he turns his back, the book falls off the shelf back to the ground. The gravitational force from the Earth on the book does 20.0 J of work on the book while it falls. Because the work done was 20.0 J + 20.0 J = 40.0 J, the book hits the ground with 40.0 J of kinetic energy.

35. A single conservative force acts on a 5.00-kg particle within a system due to its interaction with the rest of the system. The equation $F_x = 2x + 4$ describes the force, where F_x is in newtons and x is in meters. As the particle moves along the x axis from $x = 1.00$ m to $x = 5.00$ m, calculate (a) the work done by this force on the particle, (b) the change in the potential energy of the system, and (c) the kinetic energy the particle has at $x = 5.00$ m if its speed is 3.00 m/s at $x = 1.00$ m.

36. A potential energy function for a system in which a two-dimensional force acts is of the form $U = 3x^3y - 7x$. Find the force that acts at the point (x, y).

37. The potential energy of a system of two particles separated by a distance r is given by $U(r) = A/r$, where A is a constant. Find the radial force $\vec{\mathbf{F}}_r$ that each particle exerts on the other.

SECTION 7.9 Energy Diagrams and Equilibrium of a System

38. For the potential energy curve shown in Figure P7.38, (a) determine whether the force F_x is positive, negative, or zero at the five points indicated. (b) Indicate points of stable, unstable, and neutral equilibrium. (c) Sketch the curve for F_x versus x from $x = 0$ to $x = 9.5$ m.

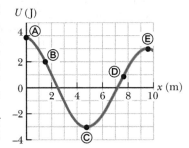

Figure P7.38

39. A right circular cone can theoretically be balanced on a horizontal surface in three different ways. Sketch these three equilibrium configurations and identify them as positions of stable, unstable, or neutral equilibrium.

ADDITIONAL PROBLEMS

40. The potential energy function for a system of particles is given by $U(x) = -x^3 + 2x^2 + 3x$, where x is the position of one particle in the system. (a) Determine the force F_x on the particle as a function of x. (b) For what values of x is the force equal to zero? (c) Plot $U(x)$ versus x and F_x versus x and indicate points of stable and unstable equilibrium.

41. You have a new internship, where you are helping to design a new freight yard for the train station in your city. There will be a number of dead-end sidings where single cars can be stored until they are needed. To keep the cars from running off the tracks at the end of the siding, you have designed a combination of two coiled springs as illustrated in Figure P7.41. When a car moves to the right in the figure and strikes the springs, they exert a force to the left on the car to slow it down.

Both springs are described by Hooke's law and have spring constants $k_1 = 1\,600$ N/m and $k_2 = 3\,400$ N/m. After the first spring compresses by a distance of $d = 30.0$ cm, the second spring acts with the first to increase the force to the left on the car in Figure P7.41. When the spring with spring constant k_2 compresses by 50.0 cm, the coils of both springs are pressed together, so that the springs can no longer compress. A typical car on the siding has a mass of 6 000 kg. When you present your design to your supervisor, he asks you for the maximum speed that a car can have and be stopped by your device.

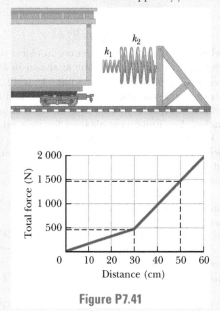

Figure P7.41

42. When an object is displaced by an amount x from stable equilibrium, a restoring force acts on it, tending to return the object to its equilibrium position. The magnitude of the restoring force can be a complicated function of x. In such cases, we can generally imagine the force function $F(x)$ to be expressed as a power series in x as $F(x) = -(k_1x + k_2x^2 + k_3x^3 + \cdots)$. The first term here is just Hooke's law, which describes the force exerted by a simple spring for small displacements. For small excursions from equilibrium, we generally ignore the higher-order terms, but in some cases it may be desirable to keep the second term as well. If we model the restoring force as $F = -(k_1x + k_2x^2)$, how much work is done on an object in displacing it from $x = 0$ to $x = x_{max}$ by an applied force $-F$?

43. A particle moves along the x axis from $x = 12.8$ m to $x = 23.7$ m under the influence of a force

$$F = \frac{375}{x^3 + 3.75x}$$

where F is in newtons and x is in meters. Using numerical integration, determine the work done by this force on the particle during this displacement. Your result should be accurate to within 2%.

44. *Why is the following situation impossible?* In a new casino, a supersized pinball machine is introduced. Casino advertising boasts that a professional basketball player can lie on top of the machine and his head and feet will not hang off the edge! The ball launcher in the machine sends metal balls up one side of the machine and then into play. The spring in the launcher (Fig. P7.44) has a force constant of 1.20 N/cm. The surface on which the ball moves is inclined $\theta = 10.0°$ with respect to the horizontal. The spring is initially compressed its maximum distance $d = 5.00$ cm. A ball of mass 100 g is projected into play by releasing the plunger. Casino visitors find the play of the giant machine quite exciting.

Figure P7.44

45. **Review.** Two constant forces act on an object of mass $m = 5.00$ kg moving in the xy plane as shown in Figure P7.45. Force $\vec{F}_1$ is 25.0 N at 35.0°, and force $\vec{F}_2$ is 42.0 N at 150°. At time $t = 0$, the object is at the origin and has velocity $(4.00\hat{i} + 2.50\hat{j})$ m/s. (a) Express the two forces in unit-vector notation. Use unit-vector notation for your other answers. (b) Find the total force exerted on the object. (c) Find the object's acceleration. Now, considering the instant $t = 3.00$ s, find (d) the object's velocity, (e) its position, (f) its kinetic energy from $\frac{1}{2}mv_f^2$, and (g) its kinetic energy from $\frac{1}{2}mv_i^2 + \sum \vec{F} \cdot \Delta\vec{r}$. (h) What conclusion can you draw by comparing the answers to parts (f) and (g)?

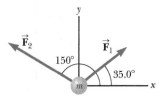

Figure P7.45

46. (a) Take $U = 5$ for a system with a particle at position $x = 0$ and calculate the potential energy of the system as a function of the particle position x. The force on the particle is given by $(8e^{-2x})\hat{i}$. (b) Explain whether the force is conservative or nonconservative and how you can tell.

47. An inclined plane of angle $\theta = 20.0°$ has a spring of force constant $k = 500$ N/m fastened securely at the bottom so that the spring is parallel to the surface as shown in Figure P7.47. A block of mass $m = 2.50$ kg is placed on the plane at a distance $d = 0.300$ m from

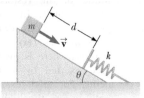

Figure P7.47
Problems 47 and 48.

the spring. From this position, the block is projected downward toward the spring with speed $v = 0.750$ m/s. By what distance is the spring compressed when the block momentarily comes to rest?

48. **S** An inclined plane of angle θ has a spring of force constant k fastened securely at the bottom so that the spring is parallel to the surface. A block of mass m is placed on the plane at a distance d from the spring. From this position, the block is projected downward toward the spring with speed v as shown in Figure P7.47. By what distance is the spring compressed when the block momentarily comes to rest?

49. **CR** Over the Christmas break, you are making some extra money for buying presents by working in a factory, helping to move crates around. At one particular time, you find that all the handtrucks, dollies, and carts are in use, so you must move a crate across the room a straight-line distance of 35.0 m without the assistance of these devices. You notice that the crate has a rope attached to the middle of one of its vertical faces. You decide to move the crate by pulling on the rope. The crate has a mass of 130 kg, and the coefficient of kinetic friction between the crate and the concrete floor is 0.350. (a) Determine the angle relative to the horizontal at which you should pull upward on the rope so that you can move the crate over the desired distance with the force of the *smallest* magnitude. (b) At this angle of pulling on the rope, how much work do you do in dragging the crate over the desired distance?

CHALLENGE PROBLEM

50. A particle of mass $m = 1.18$ kg is attached between two identical springs on a frictionless, horizontal tabletop. Both springs have spring constant k and are initially unstressed, and the particle is at $x = 0$. (a) The particle is pulled a distance x along a direction perpendicular to the initial configuration of the springs as shown in Figure P7.50. Show that the force exerted by the springs on the particle is

$$\vec{F} = -2kx\left(1 - \frac{L}{\sqrt{x^2 + L^2}}\right)\hat{i}$$

(b) Show that the potential energy of the system is

$$U(x) = kx^2 + 2kL(L - \sqrt{x^2 + L^2})$$

(c) Make a plot of $U(x)$ versus x and identify all equilibrium points. Assume $L = 1.20$ m and $k = 40.0$ N/m. (d) If the particle is pulled 0.500 m to the right and then released, what is its speed when it reaches $x = 0$?

Overhead view

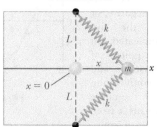

Figure P7.50

Conservation of Energy

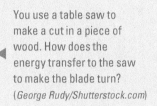

8

You use a table saw to make a cut in a piece of wood. How does the energy transfer to the saw to make the blade turn? (*George Rudy/Shutterstock.com*)

STORYLINE **In the previous chapter, you rubbed sandpaper on** wood and we associated the resulting warmth with internal energy. Now you look around the garage for more examples of energy. Your car is at rest in the garage now, but it has kinetic energy when in operation. How does it get that energy? From gasoline! But how did the gasoline get in the car? At the gasoline station! But where did the gasoline station get it? From the refinery! But where did the refinery get it? These questions go on and on! To get your mind off these questions, you start a long cut on a piece of wood with your table saw. Wait a minute! When in operation, the saw blade has rotational kinetic energy. Where does that energy come from? Ah-ha, you plugged it in, so energy is coming from the plug in the wall! But how did the energy get to the plug? It must come through the power lines from a power plant! But where does the power plant get the energy . . . ? As you continue to look around your garage, you see that energy must transfer into various devices for them to operate. And that energy must transfer out of something: a gasoline tank, a wall plug, batteries, and so on.

CONNECTIONS In the previous chapter, we found that energy can belong to a system in different forms. In this chapter, we will investigate ways that energy can *transfer* into or out of a system, or *transform* within a system. For example, in the system of the sandpaper and the wood for your carpentry project in Chapter 7, the kinetic energy of the sandpaper *transforms* to internal energy. On the other hand, for the system of your table saw in this chapter, energy *transfers* into the system by electricity to make it operate. We will see the full power of the energy approach in this chapter, embodied in the principle of *conservation of energy*. This approach will give us tools to solve problems that would be extremely difficult to

181

solve with Newton's laws. In future chapters, we will see many cases in which the conservation of energy principle is applied in a variety of situations.

8.1 Analysis Model: Nonisolated System (Energy)

As we have seen, an object, modeled as a particle, can be acted on by various forces, resulting in a change in its kinetic energy according to the work–kinetic energy theorem from Chapter 7. If we choose the object as the system, this very simple situation is the first example of a *nonisolated system,* for which energy crosses the boundary of the system during some time interval due to an interaction with the environment. This scenario is common in physics problems. If a system does not interact with its environment, it is an *isolated system,* which we will study in Section 8.2.

The work–kinetic energy theorem (Eq. 7.17) is our first example of an energy equation appropriate for a nonisolated system. In the case of that theorem, the interaction of the system with its environment is the work done by the external force, and the quantity in the system that changes is the kinetic energy.

So far, we have seen only one way to transfer energy into a system: work. We mention below a few other ways to transfer energy into or out of a system. The details of these processes will be studied in other sections of the book, but they should be familiar to you from everyday experience. We illustrate mechanisms to transfer energy in Figure 8.1 and summarize them as follows.

Work, as we have learned in Chapter 7, is a method of transferring energy to a system by applying a force to the system such that the point of application of the force undergoes a displacement (Fig. 8.1a).

Mechanical waves (Chapters 16–17) are a means of transferring energy by allowing a disturbance to propagate through air or another medium. It is the method by which energy (which you detect as *sound*) leaves the system of your clock radio through the loudspeaker and enters your ears to stimulate the hearing process (Fig. 8.1b). Other examples of mechanical waves are seismic waves and ocean waves.

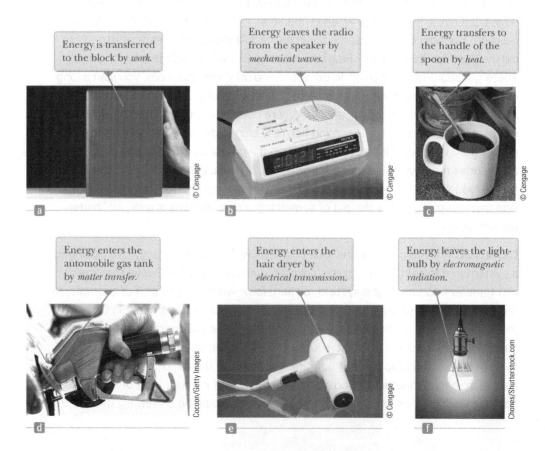

Figure 8.1 Energy transfer mechanisms. In each case, the system into which or from which energy is transferred is indicated.

Heat (Chapter 19) is a mechanism of energy transfer that is driven by a temperature difference between a system and its environment. For example, imagine dividing a metal spoon into two parts: the handle, which we identify as the system, and the portion submerged in a cup of coffee, which is part of the environment (Fig. 8.1c). The handle of the spoon becomes hot because fast-moving electrons and atoms in the submerged portion bump into slower ones in the nearby part of the handle. These particles move faster because of the collisions and bump into the next group of slow particles. Therefore, the internal energy of the spoon handle rises from energy transfer due to this collision process.

Matter transfer (Chapter 19) involves situations in which matter physically crosses the boundary of a system, carrying energy with it. Examples include filling the tank of your car in the opening storyline with gasoline (Fig. 8.1d) and carrying energy to the rooms of your home by circulating warm air from the furnace, a process called *convection*.

Electrical transmission (Chapters 26 and 27) involves energy transfer into or out of a system by means of electric currents. It is how energy transfers into your hair dryer (Fig. 8.1e), home theater system, or any other electrical device, such as the table saw in your garage in the opening storyline.

Electromagnetic radiation (Chapter 33) refers to electromagnetic waves such as visible light (Fig. 8.1f), microwaves, and radio waves crossing the boundary of a system. Examples of this method of transfer include cooking a baked potato in your microwave oven and energy traveling from the Sun to the Earth by light through space.[1]

A central feature of the energy approach is the notion that we can neither create nor destroy energy, that energy is always *conserved*. This feature has been tested in countless experiments, and no experiment has ever shown this statement to be incorrect. Therefore, **if the total amount of energy in a system changes, it can *only* be because energy has crossed the boundary of the system by a transfer mechanism such as one of the methods listed above.**

Energy is one of several quantities in physics that are conserved. We will see other conserved quantities in subsequent chapters. There are many physical quantities that do not obey a conservation principle. For example, there is no conservation of force principle or conservation of velocity principle. Similarly, in areas other than physical quantities, such as in everyday life, some quantities are conserved and some are not. For example, the money in the system of your bank account is a conserved quantity. The only way the account balance changes is if money crosses the boundary of the system by deposits or withdrawals. On the other hand, the number of people in the system of a country is not conserved. Although people indeed cross the boundary of the system, which changes the total population, the population can also change by people dying and by giving birth to new babies. Even if no people cross the system boundary, the births and deaths will change the number of people in the system. There is no equivalent in the concept of energy to dying or giving birth. The general statement of the principle of **conservation of energy** can be described mathematically with the **conservation of energy equation** as follows:

$$\Delta E_{\text{system}} = \sum T \tag{8.1}$$

◀ Conservation of energy

where E_{system} is the total energy of the system, including all methods of energy storage (kinetic, potential, and internal), T (for *transfer*) is the amount of energy transferred across the system boundary by some mechanism, and the sum is over all possible transfer mechanisms. Two of our transfer mechanisms have well-established symbolic notations. For work, $T_{\text{work}} = W$ as discussed in Chapter 7, and for heat, $T_{\text{heat}} = Q$ as defined in Chapter 19. (Now that we are familiar with work, we can simplify the appearance of equations by letting the simple symbol W represent the external work W_{ext} on a system. For internal work, we will *always* use

PITFALL PREVENTION 8.1
Heat Is Not a Form of Energy
The word *heat* is one of the most misused words in our popular language. Heat is a method of *transferring* energy, *not* a form of storing energy. Therefore, phrases such as "heat content," "the heat of the summer," and "the heat escaped" all represent uses of this word that are inconsistent with our physics definition. See Chapter 19.

---

[1]Electromagnetic radiation and work done by field forces are the only energy transfer mechanisms that do not require molecules of the environment to be available at the system boundary. Therefore, systems surrounded by a vacuum (such as planets) can only exchange energy with the environment by means of these two possibilities.

W_{int} to differentiate it from W.) The other four members of our list do not have established symbols, so we will call them T_{MW} (mechanical waves), T_{MT} (matter transfer), T_{ET} (electrical transmission), and T_{ER} (electromagnetic radiation).

The full expansion of Equation 8.1 is

The expanded conservation ▶
of energy equation

$$\Delta K + \Delta U + \Delta E_{\text{int}} = W + Q + T_{\text{MW}} + T_{\text{MT}} + T_{\text{ET}} + T_{\text{ER}} \qquad (8.2)$$

which is the primary mathematical representation of the energy version of the analysis model of the **nonisolated system.** (We will see other versions of the nonisolated system model, involving linear momentum and angular momentum, in later chapters.) In most cases, Equation 8.2 reduces to a much simpler one because some of the terms are zero for the specific situation. If, for a given system, all terms on the right side of the conservation of energy equation are zero, the system is an *isolated system,* which we study in the next section.

The conservation of energy equation is no more complicated in theory than the process of balancing your checking account statement. If your account is the system, the change in the account balance for a given month is the sum of all the transfers: deposits, withdrawals, fees, interest, and checks written. You may find it useful to think of energy as the *currency of nature!*

Equation 8.2 represents a *general* situation; it covers all possibilities for situations in classical physics that we will find throughout this book. You don't need to memorize different energy equations for different situations. Equation 8.2 is the *only* equation you need to begin an energy approach to a problem solution. When using it to solve a problem, the procedure is to analyze the situation and set terms in Equation 8.2 that don't apply to the situation equal to zero. This will reduce Equation 8.2 to a smaller equation that is appropriate to the situation. For example, suppose a force is applied to a nonisolated system and the point of application of the force moves through a displacement. Now suppose the only change in the system is in the speed of one or more components of the system. Then Equation 8.2 reduces to

$$\Delta K = W \qquad (8.3)$$

which is the work–kinetic energy theorem. This theorem is a special case of the more general principle of conservation of energy. We shall see several more special cases in future chapters.

QUICK QUIZ 8.1 Consider a block sliding over a horizontal surface with friction. Ignore any sound the sliding might make. **(i)** If the system is the *block,* this system is (a) isolated (b) nonisolated (c) impossible to determine **(ii)** If the system is the *surface,* describe the system from the same set of choices. **(iii)** If the system is the *block and the surface,* describe the system from the same set of choices.

ANALYSIS MODEL **Nonisolated System (Energy)**

Imagine you have identified a system to be analyzed and have defined a system boundary. Energy can exist in the system in three forms: kinetic, potential, and internal. The total of that energy can be changed when energy crosses the system boundary by any of six transfer methods shown in the diagram here. The total change in the energy in the system is equal to the total amount of energy that has crossed the system boundary. The mathematical statement of that concept is expressed in the **conservation of energy equation:**

$$\Delta E_{\text{system}} = \Sigma\, T \qquad (8.1)$$

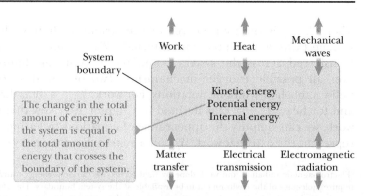

The full expansion of Equation 8.1 shows the specific types of energy storage and transfer:

$$\Delta K + \Delta U + \Delta E_{int} = W + Q + T_{MW} + T_{MT} + T_{ET} + T_{ER} \tag{8.2}$$

For a specific problem, this equation is generally reduced to a smaller number of terms by eliminating the terms that are equal to zero because they are not appropriate to the situation. See Conceptual Example 8.1, below.

Examples:

- a force does work on a system of a single object, changing its speed: the work–kinetic energy theorem, $W = \Delta K$
- a gas contained in a vessel has work done on it and experiences a transfer of energy by heat, resulting in a change in its temperature: the first law of thermodynamics, $\Delta E_{int} = W + Q$ (Chapter 19)
- an incandescent light bulb is turned on, with energy entering the filament by electricity, causing its temperature to increase, and leaving by light: $\Delta E_{int} = T_{ET} + T_{ER}$ (Chapter 26)
- a photon enters a metal, causing an electron to be ejected from the metal: the photoelectric effect, $\Delta K + \Delta U = T_{ER}$ (Chapter 39)

Conceptual Example 8.1 | **Reducing Equation 8.2 in Specific Situations**

When using Equation 8.2 to solve a problem, the following steps should be remembered: (1) define the system; (2) identify the beginning and end of the time interval of interest; (3) identify initial and final configurations of the system (positions of objects in gravitational situations, extensions of springs, etc.) and assign appropriate reference values of potential energy; (4) write Equation 8.2, eliminating or setting terms equal to zero that do not apply in the situation. Consider the following examples. For each example, the system is provided and the time interval is from before the device is turned on until it has been operating for a few moments.

(A) Your television set.

$$\Delta K + \Delta U + \Delta E_{int} = W + Q + T_{MW} + T_{MT} + T_{ET} + T_{ER} \quad \rightarrow \quad \Delta E_{int} = Q + T_{MW} + T_{ET} + T_{ER}$$

Your television set is a nonisolated system, warming up after it is turned on, taking in energy by electricity in order to operate, and emitting energy by sound from the speakers, light from the screen, and heat from warm surfaces.

(B) Your gasoline-powered lawn mower. The time interval includes the process of filling the tank with gasoline.

$$\Delta K + \Delta U + \Delta E_{int} = W + Q + T_{MW} + T_{MT} + T_{ET} + T_{ER} \quad \rightarrow \quad \Delta K + \Delta U + \Delta E_{int} = W + Q + T_{MW} + T_{MT}$$

Your lawn mower is a nonisolated system, with a moving blade, an increased potential energy in the fuel that was added, and an increasing temperature as it operates. Energy entered the lawn mower when it was filled with fuel and leaves by sound and heat from hot surfaces. The goal of the device is the work that it does on grass as the grass is cut. Notice that there may be electrical processes associated with the spark plug of the mower engine, but these processes are *internal* to the system, so they do not represent energy crossing the boundary.

(C) A cup of tea being warmed in a microwave oven.

$$\Delta K + \Delta U + \Delta E_{int} = W + Q + T_{MW} + T_{MT} + T_{ET} + T_{ER} \quad \rightarrow \quad \Delta E_{int} = Q + T_{ER}$$

Your cup of tea is a nonisolated system, with an increasing temperature. The input of energy is by electromagnetic radiation: the microwaves. The heat term represents some energy transferring out of the hot tea into the lower-temperature air surrounding the cup.

8.2 Analysis Model: Isolated System (Energy)

In this section, we study another very common scenario in physics problems: a system is chosen such that no energy crosses the system boundary by any method. We begin by considering a gravitational situation. Think about the book–Earth system in Figure 7.16 in the preceding chapter. After we have lifted the book, there is gravitational potential energy stored in the system, which can be calculated from the work done by the external agent on the system, using $W = \Delta U_g$.

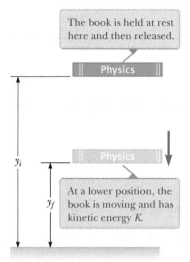

Figure 8.2 A book is released from rest and falls due to work done by the gravitational force on the book.

PITFALL PREVENTION 8.2

Conditions on Equation 8.5 Equation 8.5 is only true for a system in which conservative forces act. We will see how to handle nonconservative forces in Sections 8.3 and 8.4.

▶ The mechanical energy of an isolated system with no nonconservative forces acting is conserved.

(Check to see that this equation, which we've seen before as Eq. 7.20, is contained within Eq. 8.2 above.)

Now imagine dropping the book from the position to which you lifted it, as shown in Figure 8.2. The book–Earth system now does not interact with the environment, since your hand is no longer in contact with the book. As the book falls, the kinetic energy of the system, which is due to the motion of the book alone, increases, and the gravitational potential energy of the system decreases. From Equation 8.2, we see that

$$\Delta K + \Delta U_g = 0 \tag{8.4}$$

The left side of this equation represents a sum of changes of the energy stored in the system. There are no transfers of energy of any kind across the boundary of the system, so we set all terms on the right side of Equation 8.2 equal to zero; the book–Earth system is *isolated* from the environment. We developed this equation for a gravitational system, but it can be shown to be valid for a system with any type of potential energy. Therefore, for this isolated system,

$$\Delta K + \Delta U = 0 \tag{8.5}$$

(Check to see that this equation is contained within Eq. 8.2.) Notice what happens in this process. Energy is not *transferred* across the boundary of an isolated system. Rather, energy is *transformed* within the system, from one type to another. In the case of the falling book in Figure 8.2, the *transformation mechanism* is the internal work done on the book within the system by the gravitational force.

We defined in Chapter 7 the sum of the kinetic and potential energies of a system as its mechanical energy:

$$E_{\text{mech}} \equiv K + U \tag{8.6}$$

where U represents the total of *all* types of potential energy. Because the system under consideration is isolated, Equations 8.5 and 8.6 tell us that the mechanical energy of the system is conserved:

$$\Delta E_{\text{mech}} = 0 \tag{8.7}$$

Equation 8.7 is a statement of **conservation of mechanical energy** for an isolated system with no nonconservative forces acting. The mechanical energy in such a system is conserved: the sum of the kinetic and potential energies remains constant:

Let us now write the changes in energy in Equation 8.5 explicitly:

$$(K_f - K_i) + (U_f - U_i) = 0$$
$$K_f + U_f = K_i + U_i \tag{8.8}$$

For the gravitational situation of the falling book, Equation 8.8 can be written as

$$\tfrac{1}{2}mv_f^2 + mgy_f = \tfrac{1}{2}mv_i^2 + mgy_i \tag{8.9}$$

where $v_i = 0$ if the book in Figure 8.2 is dropped from rest. As the book falls to the Earth, the book–Earth system loses potential energy and gains kinetic energy such that the total of the two types of energy always remains constant: $E_{\text{total},i} = E_{\text{total},f}$.

If there are nonconservative forces acting within the system, mechanical energy is transformed to internal energy as discussed in Section 7.7. If nonconservative forces act in an isolated system, the total energy of the system is conserved, although the mechanical energy is not. In that case, we can express the conservation of energy of the system as

$$\Delta E_{\text{system}} = 0 \tag{8.10}$$

▶ The total energy of an isolated system is conserved.

where E_{system} includes all kinetic, potential, and internal energies. This equation is the most general statement of the energy version of the **isolated system** model. It is equivalent to Equation 8.2 with all terms on the right-hand side equal to zero. When using the isolated or nonisolated system models as discussed here, we will add the qualifier *for energy* or *(energy)*. We will find in the next few chapters that there are isolated and nonisolated system models for other quantities as well.

Associated with this most general equation for the isolated system are a variety of new transformation mechanisms. Examples include nonconservative forces (the warming of the sandpaper in the Chapter 7 storyline), chemical reactions (an exploding firecracker), and nuclear reactions (operation of a nuclear reactor).

QUICK QUIZ 8.2 A rock of mass m is dropped to the ground from a height h. A second rock, with mass $2m$, is dropped from the same height. When the second rock strikes the ground, what is its kinetic energy? **(a)** twice that of the first rock **(b)** four times that of the first rock **(c)** the same as that of the first rock **(d)** half as much as that of the first rock **(e)** impossible to determine

QUICK QUIZ 8.3 Three identical balls are thrown from the top of a building, all with the same initial speed. As shown in Figure 8.3, the first is thrown horizontally, the second at some angle above the horizontal, and the third at some angle below the horizontal. Neglecting air resistance, rank the speeds of the balls at the instant each hits the ground.

Figure 8.3 (Quick Quiz 8.3) Three identical balls are thrown with the same initial speed from the top of a building.

ANALYSIS MODEL Isolated System (Energy)

Imagine you have identified a system to be analyzed and have defined a system boundary. Energy can exist in the system in three forms: kinetic, potential, and internal. Imagine also a situation in which no energy crosses the boundary of the system by any method. Then, the system is isolated; energy transforms from one form to another and Equation 8.2 becomes

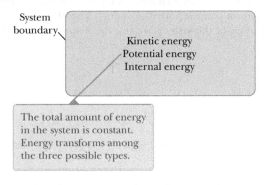

System boundary

Kinetic energy
Potential energy
Internal energy

The total amount of energy in the system is constant. Energy transforms among the three possible types.

$$\Delta E_{\text{system}} = 0 \tag{8.10}$$

If no nonconservative forces act within the isolated system, the mechanical energy of the system is conserved, so

$$\Delta E_{\text{mech}} = 0 \tag{8.7}$$

Examples:

- an object is in free-fall; gravitational potential energy transforms to kinetic energy: $\Delta K + \Delta U = 0$
- a basketball rolling across a gym floor comes to rest; kinetic energy transforms to internal energy: $\Delta K + \Delta E_{\text{int}} = 0$
- a pendulum is raised and released with an initial speed; its motion eventually stops due to air resistance; gravitational potential energy and kinetic energy transform to internal energy, $\Delta K + \Delta U + \Delta E_{\text{int}} = 0$ (Chapter 15)
- a battery is connected to a resistor; chemical potential energy in the battery transforms to internal energy in both the battery and the resistor: $\Delta U + \Delta E_{\text{int}} = 0$ (Chapter 27)

Example 8.2 Ball in Free Fall

A ball of mass m is dropped from a height h above the ground as shown in Figure 8.4 (page 188).

(A) Neglecting air resistance, determine the speed of the ball when it is at a height y above the ground. Choose the system as the ball and the Earth.

SOLUTION

Conceptualize Figure 8.4 and our everyday experience with falling objects allow us to conceptualize the situation. Although we can readily solve this problem with the techniques of Chapter 2, let us practice an energy approach.

Categorize As suggested in the problem, we identify the system as the ball and the Earth. Because there is neither air resistance nor any other interaction between the system and the environment, the system is isolated and we use the *isolated system* model for energy. The only force between members of the system is the gravitational force, which is conservative.

continued

8.2 continued

Analyze   Because the system is isolated and there are no nonconservative forces acting within the system, we apply the principle of conservation of mechanical energy to the ball–Earth system. At the instant the ball is released, its kinetic energy is $K_i = 0$ and the gravitational potential energy of the system is $U_{gi} = mgh$. When the ball is at a position y above the ground, its kinetic energy is $K_f = \frac{1}{2}mv_f^2$ and the potential energy relative to the ground is $U_{gf} = mgy$.

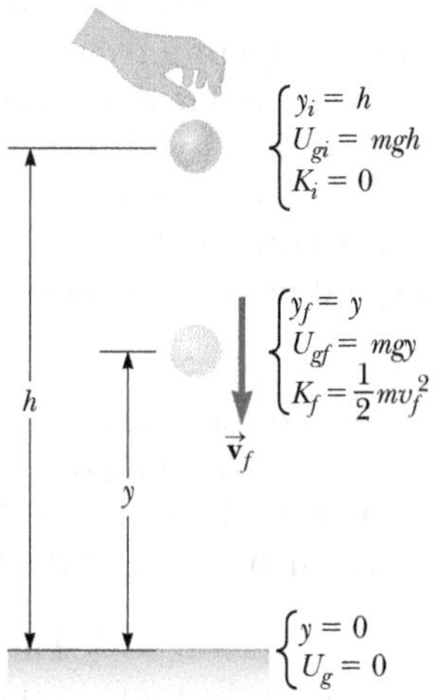

Write the appropriate reduction of Equation 8.2, noting that the only types of energy in the system that change are kinetic energy and gravitational potential energy:

$$\Delta K + \Delta U_g = 0$$

Substitute for the energies:

$$(\tfrac{1}{2}mv_f^2 - 0) + (mgy - mgh) = 0$$

Solve for v_f:

$$v_f^2 = 2g(h - y) \;\rightarrow\; v_f = \sqrt{2g(h - y)}$$

The speed is always positive. If you had been asked to find the ball's velocity, you would use the negative value of the square root as the y component to indicate the downward motion.

(B) Find the speed of the ball again at height y by choosing the ball as the system.

SOLUTION

Categorize   In this case, the only type of energy in the system that changes is kinetic energy. A single object that can be modeled as a particle cannot possess potential energy. The effect of gravity is to do work on the ball across the boundary of the system. We use the *nonisolated system* model for energy.

Figure 8.4 (Example 8.2) A ball is dropped from a height h above the ground. Initially, the total energy of the ball–Earth system is gravitational potential energy, equal to mgh relative to the ground. At the position y, the total energy is the sum of the kinetic and potential energies.

(Figure labels: $y_i = h$, $U_{gi} = mgh$, $K_i = 0$; $y_f = y$, $U_{gf} = mgy$, $K_f = \frac{1}{2}mv_f^2$; $\vec{v}_f$; h; y; $y = 0$, $U_g = 0$)

Analyze   Write the appropriate reduction of Equation 8.2:

$$\Delta K = W$$

Substitute for the initial and final kinetic energies and the work done by gravity:

$$(\tfrac{1}{2}mv_f^2 - 0) = \vec{F}_g \cdot \Delta\vec{r} = -mg\hat{j} \cdot \Delta y\hat{j}$$
$$= -mg\Delta y = -mg(y - h) = mg(h - y)$$

Solve for v_f:

$$v_f^2 = 2g(h - y) \;\rightarrow\; v_f = \sqrt{2g(h - y)}$$

Finalize   The final result is the same, regardless of the choice of system. In your future problem solving, keep in mind that the choice of system is yours to make. Sometimes the problem is much easier to solve if a judicious choice is made as to the system to analyze.

WHAT IF?   What if the ball were thrown downward from its highest position with a speed v_i? What would its speed be at height y?

Answer   If the ball is thrown downward initially, we would expect its speed at height y to be larger than if simply dropped. Make your choice of system, either the ball alone or the ball and the Earth. You should find that either choice gives you the following result:

$$v_f = \sqrt{v_i^2 + 2g(h - y)}$$

Example **8.3**   **A Grand Entrance**

You are part of the stage crew for a theatrical company and are designing an apparatus to support an actor of mass 65.0 kg who is to "fly" down to the stage during the performance of a play. You attach the actor's harness to a 130-kg sandbag by means of a lightweight steel cable running smoothly over two frictionless pulleys as in Figure 8.5a. You need 3.00 m of cable between the harness and the nearest pulley so that the pulley can be hidden behind a curtain. For the apparatus to work successfully, the sandbag must never lift above the floor as the actor swings from above the stage to the floor. Let us call the initial angle that the actor's cable makes with the vertical θ. What is the maximum value θ can have before the sandbag lifts off the floor?

8.3 continued

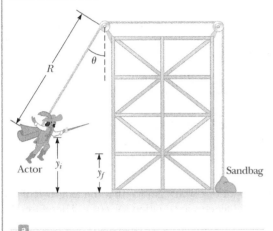

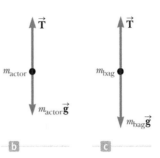

Figure 8.5 (Example 8.3) (a) An actor uses some clever staging to make his entrance. (b) The free-body diagram for the actor at the bottom of the circular path. (c) The free-body diagram for the sandbag if the normal force from the floor goes to zero.

SOLUTION

Conceptualize We must use several concepts to solve this problem. Imagine what happens as the actor approaches the bottom of the swing. At the bottom, the cable is vertical and must support his weight as well as provide centripetal acceleration of his body in the upward direction. At this point in his swing, the tension in the cable is the highest and the sandbag is most likely to lift off the floor.

Categorize Looking first at the swinging of the actor from the initial point to the lowest point, we model the actor and the Earth as an *isolated system* for energy. We ignore air resistance, so there are no nonconservative forces acting. You might initially be tempted to model the system as nonisolated because of the interaction of the system with the cable, which is in the environment. The force applied to the actor by the cable, however, is always perpendicular to each element of the displacement of the actor and hence does no work. Therefore, in terms of energy transfers across the boundary, the system is isolated.

Analyze We first find the actor's speed as he arrives at the floor as a function of the initial angle θ and the radius R of the circular path through which he swings. We use the particle model by choosing a particular point on the actor's body.

From the isolated system model, make the appropriate reduction of Equation 8.2 for the actor–Earth system:

(1) $\Delta K + \Delta U_g = 0$

Let y_i be the initial height of the actor above the floor and v_f be his speed at the instant before he lands. (Notice that $K_i = 0$ because the actor starts from rest.) Insert the energies into Equation (1) and solve for the final speed of the actor.

$(\frac{1}{2}m_{actor}v_f^2 - 0) + (m_{actor}gy_f - m_{actor}gy_i) = 0$

(2) $v_f^2 = 2g(y_f - y_i)$

From the geometry in Figure 8.5a, notice that $y_f - y_i = R - R\cos\theta = R(1 - \cos\theta)$. Use this relationship in Equation (2).

(3) $v_f^2 = 2gR(1 - \cos\theta)$

Categorize Next, focus on the instant the actor is at the lowest point. Because the tension in the cable is transferred as a force applied to the sandbag, we model the actor at this instant as a *particle under a net force*. Because the actor moves along a circular arc, he experiences at the bottom of the swing a centripetal acceleration of v_f^2/R directed upward.

Analyze Apply Newton's second law from the particle under a net force model to the actor at the bottom of his path, using the free-body diagram in Figure 8.5b as a guide, and recognizing the acceleration as centripetal:

$\sum F_y = T - m_{actor}g = m_{actor}\dfrac{v_f^2}{R}$

(4) $T = m_{actor}g + m_{actor}\dfrac{v_f^2}{R}$

Categorize Finally, notice that the sandbag lifts off the floor when the upward force exerted on it by the cable exceeds the gravitational force acting on it; the normal force from the floor is zero when that happens. We do *not*, however, want the sandbag to lift off the floor. The sandbag must remain at rest, so we model it as a *particle in equilibrium*.

continued

8.3 continued

Analyze   A force T of the magnitude given by Equation (4) is transmitted by the cable to the sandbag. If the sandbag remains at rest but is just ready to be lifted off the floor if any more force were applied by the cable, the normal force on it becomes zero and the particle in equilibrium model tells us that $T = m_{bag}g$ as in Figure 8.5c.

Substitute this condition and Equation (3) into Equation (4):

$$m_{bag}g = m_{actor}g + m_{actor}\frac{2gR(1 - \cos\theta)}{R}$$

Solve for $\cos\theta$ and substitute the given parameters:

$$\cos\theta = \frac{3m_{actor} - m_{bag}}{2m_{actor}} = \frac{3(65.0\text{ kg}) - 130\text{ kg}}{2(65.0\text{ kg})} = 0.500$$

$$\theta = 60.0°$$

Finalize   Here we had to combine several analysis models from different areas of our study. Notice that the length R of the cable from the actor's harness to the leftmost pulley did not appear in the final algebraic equation for $\cos\theta$. Therefore, the final answer is independent of R.

Example 8.4   The Spring-Loaded Popgun

The launching mechanism of a popgun consists of a trigger-released spring (Fig. 8.6a). The spring is compressed to a position $y_Ⓐ$ and the trigger is fired. The projectile of mass m rises to a maximum position $y_Ⓒ$ above the position at which it leaves the spring, indicated in Figure 8.6b as position $y_Ⓑ = 0$. Consider a firing of the gun for which $m = 35.0$ g, $y_Ⓐ = -0.120$ m, and $y_Ⓒ = 20.0$ m.

(A) Neglecting all resistive forces, determine the spring constant.

SOLUTION

Conceptualize   Imagine the process illustrated in parts (a) and (b) of Figure 8.6. The projectile starts from rest at Ⓐ, speeds up as the spring pushes upward on it, leaves the spring at Ⓑ, and then slows down as the gravitational force pulls downward on it, eventually coming to rest at point Ⓒ. Notice that there are two types of potential energy in this system: gravitational and elastic.

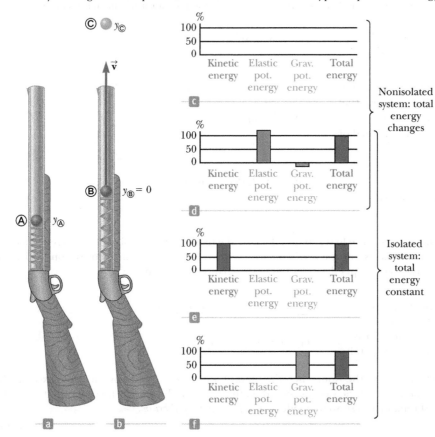

Figure 8.6 (Example 8.4) A spring-loaded popgun (a) before firing and (b) when the spring extends to its relaxed length. (c) An energy bar chart for the popgun–projectile–Earth system before the popgun is loaded. The energy in the system is zero. (d) The popgun is loaded by means of an external agent doing work on the system to push the spring downward. Therefore the system is nonisolated during this process. After the popgun is loaded, elastic potential energy is stored in the spring and the gravitational potential energy of the system is lower because the projectile is below point Ⓑ. (e) As the projectile passes through point Ⓑ, all of the energy of the isolated system is kinetic. (f) When the projectile reaches point Ⓒ, all of the energy of the isolated system is gravitational potential.

8.4 continued

Categorize We identify the system as the projectile, the spring, and the Earth. We ignore both air resistance on the projectile and friction in the gun, so we model the system as isolated for energy with no nonconservative forces acting.

Analyze Because the projectile starts from rest, its initial kinetic energy is zero. We choose the zero configuration for the gravitational potential energy of the system to be when the projectile leaves the spring at Ⓑ. For this configuration, the elastic potential energy is also zero.

After the gun is fired, the projectile rises to a maximum height $y_{©}$. The final kinetic energy of the projectile is zero.

From the isolated system model for energy, write a conservation of mechanical energy equation for the system between configurations when the projectile is at points Ⓐ and ©:

(1) $\Delta K + \Delta U_g + \Delta U_s = 0$

Substitute for the initial and final energies:

$(0 - 0) + (mgy_{©} - mgy_{Ⓐ}) + (0 - \frac{1}{2}kx^2) = 0$

Solve for k:

$$k = \frac{2mg(y_{©} - y_{Ⓐ})}{x^2}$$

Substitute numerical values:

$$k = \frac{2(0.035\,0\text{ kg})(9.80\text{ m/s}^2)[20.0\text{ m} - (-0.120\text{ m})]}{(0.120\text{ m})^2} = 958\text{ N/m}$$

(B) Find the speed of the projectile as it moves through the equilibrium position Ⓑ of the spring as shown in Figure 8.6b.

> S O L U T I O N

Analyze The energy of the system as the projectile moves through the equilibrium position of the spring includes only the kinetic energy of the projectile $\frac{1}{2}mv_{Ⓑ}^2$. Both types of potential energy are equal to zero for this configuration of the system.

Write Equation (1) again for the system between configurations for which the projectile is at points Ⓐ and Ⓑ:

$\Delta K + \Delta U_g + \Delta U_s = 0$

Substitute for the initial and final energies:

$(\frac{1}{2}mv_{Ⓑ}^2 - 0) + (0 - mgy_{Ⓐ}) + (0 - \frac{1}{2}kx^2) = 0$

Solve for $v_{Ⓑ}$:

$$v_{Ⓑ} = \sqrt{\frac{kx^2}{m} + 2gy_{Ⓐ}}$$

Substitute numerical values:

$$v_{Ⓑ} = \sqrt{\frac{(958\text{ N/m})(0.120\text{ m})^2}{(0.035\,0\text{ kg})} + 2(9.80\text{ m/s}^2)(-0.120\text{ m})} = 19.8\text{ m/s}$$

Finalize This example is the first one we have seen in which we must include two different types of potential energy. Notice in part (A) that we never needed to consider anything about the speed of the ball between points Ⓐ and ©, which is part of the power of the energy approach: changes in kinetic and potential energy depend only on the initial and final values, not on what happens between the configurations corresponding to these values.

8.3 Situations Involving Kinetic Friction

Consider again the book in Figure 7.19a sliding to the right on the surface of a heavy table and slowing down due to the friction force. Work is done by the friction force on the book because there is a force and a displacement. Keep in mind, however, that our equations for work involve the displacement *of the point of application of the force*. A simple model of the friction force between the book and the surface is shown in Figure 8.7a (page 192). We have represented the entire friction force between the book and surface as being due to two identical teeth that have been spot-welded together.[2] One tooth projects upward from the surface, the other

---

[2]Figure 8.7 and its discussion are inspired by a classic article on friction: B. A. Sherwood and W. H. Bernard, "Work and heat transfer in the presence of sliding friction," *American Journal of Physics,* **52**:1001, 1984.

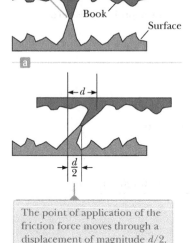

The entire friction force is modeled to be applied at the interface between two identical teeth projecting from the book and the surface.

Book

Surface

a

←d→

$\frac{d}{2}$

The point of application of the friction force moves through a displacement of magnitude $d/2$.

b

Figure 8.7 (a) A simplified model of friction between a book and a surface. (b) The book is moved to the right by a distance d.

downward from the book, and they are welded at the points where they touch. The friction force acts at the junction of the two teeth. Imagine that the book slides a small distance d to the right as in Figure 8.7b. Because the teeth are modeled as identical, the junction of the teeth moves to the right by a distance $d/2$. Therefore, the displacement of the point of application of the friction force is $d/2$, but the displacement of the book is d!

In reality, however, the friction force is spread out over the entire contact area of an object sliding on a surface, so the force is not localized at a point. In addition, because the magnitudes of the friction forces at various points are constantly changing as individual spot welds occur, the surface and the book deform locally, and so on, the displacement of the point of application of the friction force is not at all the same as the displacement of the book. In fact, the displacement of the point of application of the friction force is not calculable and so neither is the work done by the friction force.

The work–kinetic energy theorem is valid for a particle or an object that can be modeled as a particle. When a friction force acts, however, we cannot calculate the work done by friction. For such situations, Newton's second law is still valid for the system even though the work–kinetic energy theorem is not. The case of a nondeformable object like our book sliding on the surface[3] can be handled in a relatively straightforward way.

Starting from a situation in which forces, including friction, are applied to the book, we can follow a similar procedure to that done in developing Equation 7.17. Let us start by writing Equation 7.8 for all forces on an object other than friction:

$$\sum W_{\text{other forces}} = \int \left(\sum \vec{\mathbf{F}}_{\text{other forces}} \right) \cdot d\vec{\mathbf{r}} \tag{8.11}$$

The $d\vec{\mathbf{r}}$ in this equation is the displacement of the object because for forces other than friction, under the assumption that these forces do not deform the object, this displacement is the same as the displacement of the point of application of the forces. To each side of Equation 8.11 let us add the integral of the scalar product of the force of kinetic friction and $d\vec{\mathbf{r}}$. In doing so, we are not defining this quantity as work! We are simply saying that it is a quantity that can be calculated mathematically and will turn out to be useful to us in what follows.

$$\sum W_{\text{other forces}} + \int \vec{\mathbf{f}}_k \cdot d\vec{\mathbf{r}} = \int \left(\sum \vec{\mathbf{F}}_{\text{other forces}} \right) \cdot d\vec{\mathbf{r}} + \int \vec{\mathbf{f}}_k \cdot d\vec{\mathbf{r}}$$

$$= \int \left(\sum \vec{\mathbf{F}}_{\text{other forces}} + \vec{\mathbf{f}}_k \right) \cdot d\vec{\mathbf{r}}$$

The integrand on the right side of this equation is the net force $\sum \vec{\mathbf{F}}$ on the object, so

$$\sum W_{\text{other forces}} + \int \vec{\mathbf{f}}_k \cdot d\vec{\mathbf{r}} = \int \sum \vec{\mathbf{F}} \cdot d\vec{\mathbf{r}}$$

Incorporating Newton's second law $\sum \vec{\mathbf{F}} = m\vec{\mathbf{a}}$ gives

$$\sum W_{\text{other forces}} + \int \vec{\mathbf{f}}_k \cdot d\vec{\mathbf{r}} = \int m\vec{\mathbf{a}} \cdot d\vec{\mathbf{r}} = \int m\frac{d\vec{\mathbf{v}}}{dt} \cdot d\vec{\mathbf{r}} = \int_{t_i}^{t_f} m\frac{d\vec{\mathbf{v}}}{dt} \cdot \vec{\mathbf{v}}\, dt \tag{8.12}$$

where we have used Equation 4.3 to rewrite $d\vec{\mathbf{r}}$ as $\vec{\mathbf{v}}\, dt$. The scalar product obeys the product rule for differentiation (see Eq. B.30 in Appendix B.6), so the derivative of the scalar product of $\vec{\mathbf{v}}$ with itself can be written

$$\frac{d}{dt}(\vec{\mathbf{v}} \cdot \vec{\mathbf{v}}) = \frac{d\vec{\mathbf{v}}}{dt} \cdot \vec{\mathbf{v}} + \vec{\mathbf{v}} \cdot \frac{d\vec{\mathbf{v}}}{dt} = 2\frac{d\vec{\mathbf{v}}}{dt} \cdot \vec{\mathbf{v}}$$

---

[3]The overall shape of the book remains the same, which is why we say it is nondeformable. On a microscopic level, however, there is deformation of the book's face as it slides over the surface.

We used the commutative property of the scalar product to justify the final expression in this equation. Consequently,

$$\frac{d\vec{\mathbf{v}}}{dt} \cdot \vec{\mathbf{v}} = \tfrac{1}{2}\frac{d}{dt}(\vec{\mathbf{v}} \cdot \vec{\mathbf{v}}) = \tfrac{1}{2}\frac{dv^2}{dt}$$

Substituting this result into Equation 8.12 gives

$$\sum W_{\text{other forces}} + \int \vec{\mathbf{f}}_k \cdot d\vec{\mathbf{r}} = \int_{t_i}^{t_f} m\left(\tfrac{1}{2}\frac{dv^2}{dt}\right)dt = \tfrac{1}{2}m\int_{v_i}^{v_f} d(v^2) = \tfrac{1}{2}mv_f^2 - \tfrac{1}{2}mv_i^2 = \Delta K$$

Looking at the left side of this equation, notice that in the inertial frame of the surface, $\vec{\mathbf{f}}_k$ and $d\vec{\mathbf{r}}$ will be in opposite directions for every increment $d\vec{\mathbf{r}}$ of the path followed by the object. Therefore, $\vec{\mathbf{f}}_k \cdot d\vec{\mathbf{r}} = -f_k\,dr$. The previous expression now becomes

$$\sum W_{\text{other forces}} - \int f_k\,dr = \Delta K$$

In our model for friction, the magnitude of the kinetic friction force is constant, so f_k can be brought out of the integral. The remaining integral $\int dr$ is simply the sum of increments of length along the path, which is the total path length d. Therefore,

$$W - f_k d = \Delta K \tag{8.13}$$

where W represents the work done on the object by all forces other than friction. Equation 8.13 can be used when a friction force acts on an object. The change in kinetic energy is equal to the work done by all forces other than friction minus a term $f_k d$ associated with the friction force.

Considering the sliding book situation again, let's identify the larger system of the book *and* the surface as the book slows down under the influence of a friction force alone. There is no work done across the boundary of this system by other forces because the system does not interact with the environment. There are no other types of energy transfer occurring across the boundary of the system, assuming we ignore the inevitable sound the sliding book makes! In this case, Equation 8.2 becomes

$$\Delta K + \Delta E_{\text{int}} = 0$$

The change in kinetic energy of this book–surface system is the same as the change in kinetic energy of the book alone because the book is the only part of the system that is moving. Therefore, incorporating Equation 8.13 with no work done by other forces gives

$$-f_k d + \Delta E_{\text{int}} = 0$$

$$\Delta E_{\text{int}} = f_k d \tag{8.14}$$

◀ Change in internal energy due to a constant friction force within the system

Equation 8.14 tells us that the increase in internal energy of the system is equal to the product of the friction force and the path length through which the block moves. In summary, a friction force transforms kinetic energy in a system to internal energy. If work is done on the system by forces other than friction, Equation 8.13, with the help of Equation 8.14, can be written as

$$W = \Delta K + \Delta E_{\text{int}} \tag{8.15}$$

which is a reduced form of Equation 8.2 and represents the nonisolated system model for energy for a system within which a nonconservative force acts. For any system in which the force of kinetic friction acts between members of the system, we can write the full form of Equation 8.2, reduce it accordingly, and then use Equation 8.14 to substitute for the change in the internal energy.

QUICK QUIZ 8.4 You are traveling along a freeway at 65 mi/h. Your car has
kinetic energy. You suddenly skid to a stop because of congestion in traffic.
Where is the kinetic energy your car once had? **(a)** It is all in internal energy
in the road. **(b)** It is all in internal energy in the tires. **(c)** Some of it has transformed to internal energy and some of it transferred away by mechanical waves.
(d) It is all transferred away from your car by various mechanisms.

Example 8.5   A Block Pulled on a Rough Surface

A 6.0-kg block initially at rest is pulled to the right along a horizontal surface by a
constant horizontal force of magnitude 12 N.

(A) Find the speed of the block after it has moved 3.0 m if the surfaces in contact
have a coefficient of kinetic friction of 0.15.

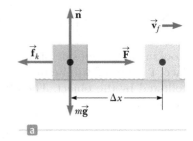

SOLUTION

Conceptualize   This example is similar to Example 7.6 (page 163), but modified so that
the surface is no longer frictionless. The rough surface applies a friction force on the
block opposite to the applied force. As a result, we expect the speed to be lower than that
found in Example 7.6.

Categorize   The block is pulled by a force and the surface is rough, so the block and the
surface are modeled as a *nonisolated system* for energy with a nonconservative force acting.

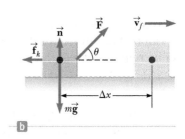

Analyze   Figure 8.8a illustrates this situation. Neither the normal force nor the gravitational force does work on the system because their points of application are displaced
horizontally.

Figure 8.8   (Example 8.5) (a) A
block pulled to the right on a rough
surface by a constant horizontal
force. (b) The applied force is at an
angle θ to the horizontal.

Write the appropriate reduction of Equation 8.2:

$$(1) \quad \Delta K + \Delta E_{int} = W$$

Find the work done on the system by the applied force
just as in Example 7.6, noting that $\Delta x = d$ because the
motion is in a straight line:

$$W = F\,\Delta x = Fd$$

Apply the *particle in equilibrium* model to the block in the
vertical direction:

$$\sum F_y = 0 \quad \rightarrow \quad n - mg = 0 \quad \rightarrow \quad n = mg$$

Find the magnitude of the friction force:

$$f_k = \mu_k n = \mu_k mg$$

Substitute the energies into Equation (1), using
Equation 8.14 for ΔE_{int}, and solve for the final
speed of the block:

$$(\tfrac{1}{2}mv_f^2 - 0) + (\mu_k mg)d = Fd$$

$$v_f = \sqrt{2d\left(\frac{F}{m} - \mu_k g\right)}$$

Substitute numerical values:

$$v_f = \sqrt{2(3.0 \text{ m})\left[\frac{12 \text{ N}}{6.0 \text{ kg}} - (0.15)(9.80 \text{ m/s}^2)\right]} = 1.8 \text{ m/s}$$

Finalize   As expected, this value is less than the 3.5 m/s found in the case of the block sliding on a frictionless surface (see
Example 7.6). The difference in kinetic energies between the block in Example 7.6 and the block in this example is equal to
the increase in internal energy of the block–surface system in this example.

(B) Suppose the force $\vec{F}$ is applied at an angle θ as shown in Figure 8.8b. At what angle should the force be applied to
achieve the largest possible speed after the block has moved 3.0 m to the right?

SOLUTION

Conceptualize   You might guess that $\theta = 0$ would give the largest speed because the force would have the largest component
possible in the direction parallel to the surface. Think about $\vec{F}$ applied at an arbitrary nonzero angle, however. Although
the horizontal component of the force would be reduced, the vertical component of the force would reduce the normal force, in
turn reducing the force of friction, which suggests that the speed could be maximized by pulling at an angle other than $\theta = 0$.

8.5 continued

Categorize As in part (A), we model the block and the surface as a *nonisolated system* for energy with a nonconservative force acting.

Analyze Write the appropriate reduction of Equation 8.2:

(1) $\Delta K + \Delta E_{int} = W$

Find the work done by the applied force:

(2) $W = F\Delta x \cos\theta = Fd\cos\theta$

Apply the particle in equilibrium model to the block in the vertical direction:

$\sum F_y = n + F\sin\theta - mg = 0$

Solve for n:

(2) $n = mg - F\sin\theta$

Substitute for the energy changes in Equation (1) and solve for the final kinetic energy of the block:

$(K_f - 0) + f_k d = W \quad \rightarrow \quad K_f = W - f_k d$

Substitute the results found in Equations (1) and (2):

$K_f = Fd\cos\theta - \mu_k nd = Fd\cos\theta - \mu_k(mg - F\sin\theta)d$

Maximizing the speed is equivalent to maximizing the final kinetic energy. Consequently, differentiate K_f with respect to θ and set the result equal to zero:

$\dfrac{dK_f}{d\theta} = -Fd\sin\theta - \mu_k(0 - F\cos\theta)d = 0$

$-\sin\theta + \mu_k\cos\theta = 0$

$\tan\theta = \mu_k$

Evaluate θ for $\mu_k = 0.15$:

$\theta = \tan^{-1}(\mu_k) = \tan^{-1}(0.15) = 8.5°$

Finalize Notice that the angle at which the speed of the block is a maximum is indeed not $\theta = 0$. When the angle exceeds 8.5°, the horizontal component of the applied force is too small to be compensated by the reduced friction force and the speed of the block begins to decrease from its maximum value.

Example 8.6 A Block–Spring System

A block of mass 1.6 kg is attached to a horizontal spring that has a force constant of 1 000 N/m as shown in Figure 8.9a. The spring is compressed 2.0 cm and is then released from rest as in Figure 8.9b.

(A) Calculate the speed of the block as it passes through the equilibrium position $x = 0$ if the surface is frictionless.

SOLUTION

Conceptualize This situation has been discussed before, and it is easy to visualize the block being pushed to the right by the spring and moving with some speed at $x = 0$.

Categorize We identify the system as the block and model the block as a *nonisolated system* for energy.

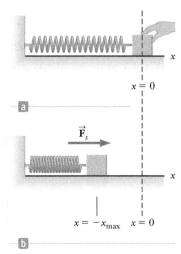

Figure 8.9 (Example 8.6) (a) A block attached to a spring is pushed inward from an initial position $x = 0$ by an external agent. (b) At position $x = -x_{max}$, the block is released from rest and the spring pushes it to the right.

Analyze Write the appropriate reduction of Equation 8.2 for the block being pushed by the spring:

(1) $\Delta K = W_s$

Use Equation 7.11 to find the work done by the spring on the system:

(2) $W_s = \frac{1}{2}kx_{max}^2$

Substitute the initial and final kinetic energies on the left of Equation (1) and the expression for the work in Equation (2) on the right:

$(\frac{1}{2}mv_f^2 - 0) = \frac{1}{2}kx_{max}^2 \quad \rightarrow \quad v_f = x_{max}\sqrt{\dfrac{k}{m}}$

Substitute numerical values:

$v_f = (0.020 \text{ m})\sqrt{\dfrac{1\,000 \text{ N/m}}{1.6 \text{ kg}}} = 0.50 \text{ m/s}$

continued

8.6 continued

Finalize Although this problem could have been solved in Chapter 7, it is presented here to provide contrast with the following part (B), which requires the techniques of this chapter.

(B) Calculate the speed of the block as it passes through the equilibrium position if a constant friction force of 4.0 N retards its motion from the moment it is released.

SOLUTION

Conceptualize The correct answer must be less than that found in part (A) because the friction force retards the motion.

Categorize We identify the system as the block and the surface, a *nonisolated system* for energy because of the work done by the spring. There is a nonconservative force acting within the system: the friction between the block and the surface.

Analyze Write the appropriate reduction of Equation 8.2 and substitute for the energy changes:

$$\Delta K + \Delta E_{int} = W_s \rightarrow (\tfrac{1}{2}mv_f^2 - 0) + f_k d = W_s$$

Solve for v_f:

$$v_f = \sqrt{\frac{2}{m}(W_s - f_k d)}$$

Substitute for the work done by the spring, found in part (A):

$$v_f = \sqrt{\frac{2}{m}(\tfrac{1}{2}kx_{max}^2 - f_k d)}$$

Substitute numerical values:

$$v_f = \sqrt{\frac{2}{1.6\text{ kg}}[\tfrac{1}{2}(1\ 000\text{ N/m})(0.020\text{ m})^2 - (4.0\text{ N})(0.020\text{ m})]} = 0.39\text{ m/s}$$

Finalize As expected, this value is less than the 0.50 m/s found in part (A).

WHAT IF? What if the friction force were increased to 10.0 N? What is the block's speed at $x = 0$?

Answer In this case, the value of $f_k d$ as the block moves to $x = 0$ is

$$f_k d = (10.0\text{ N})(0.020\text{ m}) = 0.20\text{ J}$$

which is equal in magnitude to the kinetic energy at $x = 0$ for the frictionless case. (Verify it!). Therefore, all the kinetic energy has been transformed to internal energy by friction when the block arrives at $x = 0$, and its speed at this point is $v = 0$.

In this situation as well as that in part (B), the speed of the block reaches a maximum at some position other than $x = 0$. Problem 27 asks you to locate these positions.

8.4 Changes in Mechanical Energy for Nonconservative Forces

In the discussion leading to Equation 8.14, which identifies the change in internal energy of a system due to friction, we considered nonconservative forces that affected only the *kinetic* energy of the system. Now, however, suppose the book on the surface that we were discussing there is part of a system that also exhibits a change in potential energy. In this case, $f_k d$ is the change in internal energy due to a decrease in the *mechanical* energy of the system because of the force of kinetic friction. For example, if the book moves on an incline that is not frictionless, there is a change in both the kinetic energy and the gravitational potential energy of the book–incline–Earth system. Consequently, Equation 8.2 can be written as

$$\Delta K + \Delta U + \Delta E_{int} = 0 \tag{8.16}$$

where ΔE_{int} is given by Equation 8.14.

Example 8.7 | Crate Sliding Down a Ramp

A 3.00-kg crate slides down a ramp. The ramp is 1.00 m in length and inclined at an angle of 30.0° as shown in Figure 8.10. The crate starts from rest at the top and experiences a constant friction force of magnitude 5.00 N. The crate continues to move a short distance on the horizontal floor after it leaves the ramp, and then comes to rest.

(A) Use energy methods to determine the speed of the crate at the bottom of the ramp.

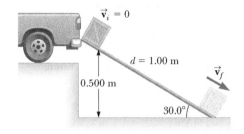

SOLUTION

Conceptualize Imagine the crate sliding down the ramp in Figure 8.10. The larger the friction force, the more slowly the crate will slide.

Figure 8.10 (Example 8.7) A crate slides down a ramp with friction under the influence of gravity. The potential energy of the system decreases, whereas the kinetic and internal energies increase.

Categorize We identify the crate, the surface, and the Earth as an *isolated system* for energy with a nonconservative force acting. We consider the time interval from when the crate leaves the top of the ramp until it reaches the bottom.

Analyze Because $v_i = 0$, the initial kinetic energy of the system when the crate is at the top of the ramp is zero. If the y coordinate is measured from the bottom of the ramp (the final position of the crate, for which we choose the gravitational potential energy of the system to be zero) with the upward direction being positive, then $y_i = 0.500$ m.

Write the conservation of energy equation (Eq. 8.2) for this system:

$$\Delta K + \Delta U + \Delta E_{\text{int}} = 0$$

Substitute for the energies:

$$(\tfrac{1}{2}mv_f^2 - 0) + (0 - mgy_i) + f_k d = 0$$

Solve for v_f:

$$(1) \quad v_f = \sqrt{\frac{2}{m}(mgy_i - f_k d)}$$

Substitute numerical values:

$$v_f = \sqrt{\frac{2}{3.00 \text{ kg}}[(3.00 \text{ kg})(9.80 \text{ m/s}^2)(0.500 \text{ m}) - (5.00 \text{ N})(1.00 \text{ m})]} = 2.54 \text{ m/s}$$

(B) How far does the crate slide on the horizontal floor if it continues to experience a friction force of magnitude 5.00 N?

SOLUTION

Analyze This part of the problem is handled in exactly the same way as part (A), but in this case we consider the time interval from the moment the crate begins to slide at the top of the ramp until it comes to rest on the floor.

Write the conservation of energy equation for this situation:

$$\Delta K + \Delta U + \Delta E_{\text{int}} = 0$$

Substitute for the energies noting that the crate slides over a total distance, ramp and floor, that we call d_{total}:

$$(0 - 0) + (0 - mgy_i) + f_k d_{\text{total}} = 0$$

Solve for the distance d_{total} and substitute numerical values:

$$d_{\text{total}} = \frac{mgy_i}{f_k} = \frac{(3.00 \text{ kg})(9.80 \text{ m/s}^2)(0.500 \text{ m})}{5.00 \text{ N}} = 2.94 \text{ m}$$

Subtracting the 1.00-m length of the ramp gives us 1.94 m that the crate slides across the floor.

Finalize For comparison, you may want to calculate the speed of the crate at the bottom of the ramp in the case in which the ramp is frictionless. Also notice that the increase in internal energy of the system for the entire motion of the crate is $f_k d_{\text{total}} = (5.00 \text{ N})(2.94 \text{ m}) = 14.7 \text{ J}$. This energy is shared between the crate and the surface, each of which is a bit warmer than before.

Also notice that the distance d the object slides on the horizontal surface is infinite if the surface is frictionless. Is that consistent with your conceptualization of the situation?

WHAT IF? A cautious worker decides that the speed of the crate when it arrives at the bottom of the ramp may be so large that its contents may be damaged. Therefore, he replaces the ramp with a longer one such that the new ramp makes an angle of 25.0° with the ground. Does this new ramp reduce the speed of the crate as it reaches the ground?

continued

8.7 continued

Answer   Because the ramp is longer, the friction force acts over a longer distance and transforms more of the mechanical energy into internal energy. The result is a reduction in the kinetic energy of the crate, and we expect a lower speed as it reaches the ground.

Find the length d of the new ramp:

$$\sin 25.0° = \frac{0.500 \text{ m}}{d} \quad \rightarrow \quad d = \frac{0.500 \text{ m}}{\sin 25.0°} = 1.18 \text{ m}$$

Find v_f from Equation (1) in part (A):

$$v_f = \sqrt{\frac{2}{3.00 \text{ kg}}[(3.00 \text{ kg})(9.80 \text{ m/s}^2)(0.500 \text{ m}) - (5.00 \text{ N})(1.18 \text{ m})]} = 2.42 \text{ m/s}$$

The final speed is indeed lower than in the higher-angle case.

Example 8.8   Block–Spring Collision

A block having a mass of 0.80 kg is given an initial velocity $v_Ⓐ = 1.2$ m/s to the right and collides with a spring whose mass is negligible and whose force constant is $k = 50$ N/m as shown in Figure 8.11.

(A) Assuming the surface to be frictionless, calculate the maximum compression of the spring after the collision.

SOLUTION

Conceptualize   The various parts of Figure 8.11 help us imagine what the block will do in this situation. All motion takes place in a horizontal plane, so we do not need to consider changes in gravitational potential energy. Before the collision, when the block is at Ⓐ, it has kinetic energy and the spring is uncompressed, so the elastic potential energy stored in the system is zero. Therefore, the total mechanical energy of the system before the collision is just $\frac{1}{2}mv_Ⓐ^2$. After the collision, when the block is at Ⓒ, the spring is fully compressed; now the block is at rest and so has zero kinetic energy. The elastic potential energy stored in the system, however, has its maximum value $\frac{1}{2}kx^2 = \frac{1}{2}kx_{max}^2$, where the

Figure 8.11   (Example 8.8) A block sliding on a frictionless, horizontal surface collides with a light spring. (a) Initially, the block slides to the right, approaching the spring. (b) The block strikes the spring and begins to compress it. (c) The block stops momentarily at the maximum compression of the spring. (d) The spring pushes the block to the left. As the spring returns to its equilibrium length, the block continues moving to the left. The energy equations at the right show the energies of the system in the frictionless case in part (A).

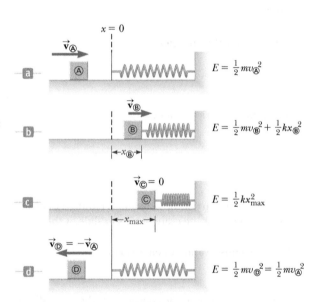

origin of coordinates $x = 0$ is chosen to be the equilibrium position of the spring and x_{max} is the maximum compression of the spring, which in this case happens to be $x_Ⓒ$. The total mechanical energy of the system is conserved because no nonconservative forces act on objects within the isolated system.

Categorize   We identify the system to be the block and the spring and model it as an *isolated system* for energy with no nonconservative forces acting.

Analyze   Write the appropriate reduction of Equation 8.2 between points Ⓐ and Ⓒ:

$$\Delta K + \Delta U = 0$$

Substitute for the energies:

$$(0 - \tfrac{1}{2}mv_Ⓐ^2) + (\tfrac{1}{2}kx_{max}^2 - 0) = 0$$

Solve for x_{max} and evaluate:

$$x_{max} = \sqrt{\frac{m}{k}}\,v_Ⓐ = \sqrt{\frac{0.80 \text{ kg}}{50 \text{ N/m}}}(1.2 \text{ m/s}) = 0.15 \text{ m}$$

8.8 continued

(B) Suppose a constant force of kinetic friction acts between the block and the surface, with $\mu_k = 0.50$. If the speed of the block at the moment it collides with the spring is $v_\text{Ⓐ} = 1.2$ m/s, what is the maximum compression $x_\text{©}$ in the spring?

SOLUTION

Conceptualize Because of the friction force, we expect the compression of the spring to be smaller than in part (A) because some of the block's kinetic energy is transformed to internal energy in the block and the surface.

Categorize We identify the system as the block, the surface, and the spring. This is an *isolated system* for energy but now involves a nonconservative force.

Analyze In this case, the mechanical energy $E_\text{mech} = K + U_s$ of the system is *not* conserved because a friction force acts on the block. From the *particle in equilibrium* model in the vertical direction, we see that $n = mg$.

Evaluate the magnitude of the friction force: $\qquad f_k = \mu_k n = \mu_k mg$

Write the appropriate reduction of Equation 8.2 $\qquad \Delta K + \Delta U + \Delta E_\text{int} = 0$
for this situation:

Substitute the initial and final energies: $\qquad (0 - \frac{1}{2}mv_\text{Ⓐ}^2) + (\frac{1}{2}kx_\text{©}^2 - 0) + \mu_k mgx_\text{©} = 0$

Rearrange the terms into a quadratic equation: $\qquad kx_\text{©}^2 + 2\mu_k mgx_\text{©} - mv_\text{Ⓐ}^2 = 0$

Solve the quadratic equation: $\qquad x_\text{©} = \dfrac{\mu_k mg}{k}\left(\pm\sqrt{1 + \dfrac{kv_\text{Ⓐ}^2}{\mu_k^2 mg^2}} - 1\right)$

Substituting numerical values gives $x_\text{©} = 0.092$ m and $x_\text{©} = -0.25$ m. The physically meaningful root is $x_\text{©} = 0.092$ m.

Finalize The negative root does not apply to this situation because the block must be to the right of the origin (positive value of x) when it comes to rest. Notice that the value of 0.092 m is less than the distance obtained in the frictionless case of part (A) as we expected.

Example 8.9 **Connected Blocks in Motion**

Two blocks are connected by a light string that passes over a frictionless pulley as shown in Figure 8.12. The block of mass m_1 lies on a horizontal surface and is connected to a spring of force constant k. The system is released from rest when the spring is unstretched. If the hanging block of mass m_2 falls a distance h before coming to rest, calculate the coefficient of kinetic friction between the block of mass m_1 and the surface.

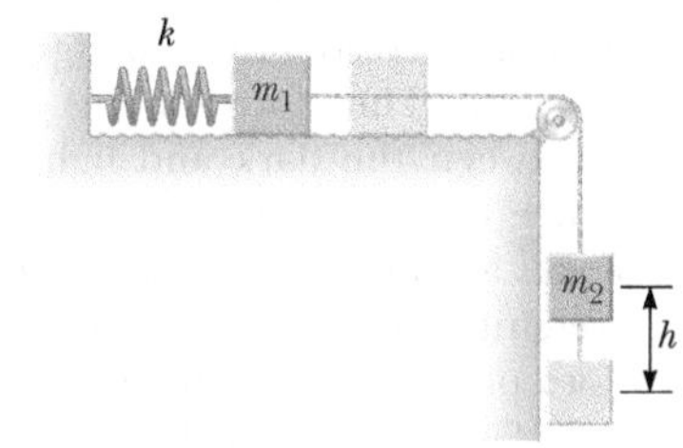

SOLUTION

Conceptualize The key word *rest* appears twice in the problem statement. This word suggests that the configurations of the system associated with rest are good candidates for the initial and final configurations because the kinetic energy of the system is zero for these configurations.

Categorize In this situation, the system consists of the two blocks, the spring, the surface, and the Earth. This is an *isolated system* with a nonconservative force acting. We also model the sliding block as a *particle in equilibrium* in the vertical direction, leading to $n = m_1 g$.

Figure 8.12 (Example 8.9) As the hanging block moves from its highest elevation to its lowest, the system loses gravitational potential energy but gains elastic potential energy in the spring. Some mechanical energy is transformed to internal energy because of friction between the sliding block and the surface.

Analyze We need to consider two forms of potential energy for the system, gravitational and elastic: $\Delta U_g = U_{gf} - U_{gi}$ is the change in the system's gravitational potential energy, and $\Delta U_s = U_{sf} - U_{si}$ is the change in the system's elastic potential energy. The change in the gravitational potential energy of the system is associated with only the falling block because the vertical coordinate of the horizontally sliding block does not change.

continued

8.9 continued

Write the appropriate reduction of Equation 8.2:

$$(1) \quad \Delta K + \Delta U_g + \Delta U_s + \Delta E_{\text{int}} = 0$$

Substitute for the energies for the time interval beginning upon release and ending when the system is again at rest, noting that as the hanging block falls a distance h, the horizontally moving block moves the same distance h to the right, and the spring stretches by a distance h:

$$(0 - 0) + (0 - m_2gh) + (\tfrac{1}{2}kh^2 - 0) + f_kh = 0$$

Substitute for the friction force:

$$-m_2gh + \tfrac{1}{2}kh^2 + \mu_k m_1 gh = 0$$

Solve for μ_k:

$$\mu_k = \frac{m_2g - \tfrac{1}{2}kh}{m_1g}$$

Finalize This setup represents a method of measuring the coefficient of kinetic friction between an object and some surface. Notice how we have solved the examples in this chapter using the energy approach. We begin with Equation 8.2 and then tailor it to the physical situation. This process may include deleting terms, such as all terms on the right-hand side of Equation 8.2 in this example. It can also include expanding terms, such as rewriting ΔU due to two types of potential energy in this example.

Conceptual Example 8.10   **Interpreting the Energy Bars**

The energy bar charts in Figure 8.13 show three instants in the motion of the system in Figure 8.12 and described in Example 8.9. For each bar chart, identify the configuration of the system that corresponds to the chart.

SOLUTION

In Figure 8.13a, there is no kinetic energy in the system. Therefore, nothing in the system is moving. The bar chart shows that the system contains only gravitational potential energy and no internal energy yet, which corresponds to the configuration with the darker blocks in Figure 8.12 and represents the instant just after the system is released.

In Figure 8.13b, the system contains four types of energy. The height of the gravitational potential energy bar is at 50%, which tells us that the hanging block has moved halfway between its position corresponding to Figure 8.13a and the position defined as $y = 0$. Therefore, in this configuration, the hanging block is between the dark and light images of the hanging block in Figure 8.12. The system has gained kinetic energy because the blocks are moving, elastic potential energy because the spring is stretching, and internal energy because of friction between the block of mass m_1 and the surface.

In Figure 8.13c, the height of the gravitational potential energy bar is zero, telling us that the hanging block is at $y = 0$. In addition, the height of the kinetic energy bar is zero, indicating that the blocks have stopped moving momentarily. Therefore, the configuration of the system is that shown by the light images of the blocks in Figure 8.12. The height of the elastic potential energy bar is high because the spring is stretched its maximum amount. The height of the internal energy bar is higher than in Figure 8.13b because the block of mass m_1 has continued to slide over the surface after the configuration shown in Figure 8.13b.

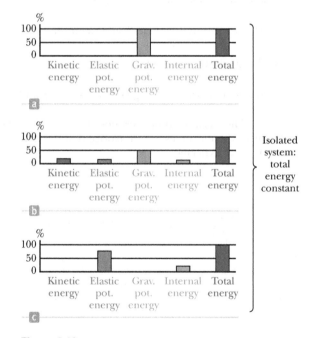

Figure 8.13 (Conceptual Example 8.10) Three energy bar charts are shown for the system in Figure 8.12.

8.5   **Power**

Consider Conceptual Example 7.7 again, which involved rolling a refrigerator up a ramp into a truck. Suppose the man is not convinced the work is the same regardless of the ramp's length and sets up a long ramp with a gentle rise. Although he does the same amount of work as someone using a shorter ramp, he takes longer to do the work because he has to move the refrigerator over a greater distance. Although the work done on both ramps is the same, there is *something* different about the tasks: the *time interval* during which the work is done.

The time rate of energy transfer is called the **instantaneous power** P and is defined as

$$P \equiv \frac{dE}{dt} \tag{8.17}$$

◀ Definition of power

We will focus on work as the energy transfer method in this discussion, but keep in mind that the notion of power is valid for *any* means of energy transfer discussed in Section 8.1. If an external force is applied to an object (which we model as a particle) and if the work done by this force on the object in the time interval Δt is W, the **average power** during this interval is

$$P_{\text{avg}} = \frac{W}{\Delta t}$$

Therefore, in Conceptual Example 7.7, although the same work is done in rolling the refrigerator up both ramps, less power is required for the longer ramp.

In a manner similar to the way we approached the definition of velocity and acceleration, the instantaneous power is the limiting value of the average power as Δt approaches zero:

$$P = \lim_{\Delta t \to 0} \frac{W}{\Delta t} = \frac{dW}{dt}$$

where we have represented the infinitesimal value of the work done by dW. We find from Equation 7.3 that $dW = \vec{\mathbf{F}} \cdot d\vec{\mathbf{r}}$. Therefore, for a constant force, the instantaneous power can be written

$$P = \frac{dW}{dt} = \vec{\mathbf{F}} \cdot \frac{d\vec{\mathbf{r}}}{dt} = \vec{\mathbf{F}} \cdot \vec{\mathbf{v}} \tag{8.18}$$

where $\vec{\mathbf{v}} = d\vec{\mathbf{r}}/dt$.

The SI unit of power is joules per second (J/s), also called the **watt** (W) after James Watt:

$$1 \text{ W} = 1 \text{ J/s} = 1 \text{ kg} \cdot \text{m}^2/\text{s}^3$$

◀ The watt

A unit of power in the U.S. customary system is the **horsepower** (hp):

$$1 \text{ hp} = 746 \text{ W}$$

A unit of energy (or work) can now be defined in terms of the unit of power. One **kilowatt-hour** (kWh) is the energy transferred in 1 h at the constant rate of 1 kW = 1 000 J/s. The amount of energy represented by 1 kWh is

$$1 \text{ kWh} = (10^3 \text{ W})(3\,600 \text{ s}) = 3.60 \times 10^6 \text{ J}$$

A kilowatt-hour is a unit of energy, not power. When you pay your electric bill, you are buying energy, and the amount of energy transferred by electrical transmission into a home during the period represented by the electric bill is usually expressed in kilowatt-hours. For example, your bill may state that you used 900 kWh of energy during a month and that you are being charged at the rate of 11¢ per kilowatt-hour. Your obligation is then $99 for this amount of energy. As another example, suppose an electric bulb is rated at 100 W. In 1.00 h of operation, it would have energy transferred to it by electrical transmission in the amount of (0.100 kW)(1.00 h) = 0.100 kWh = 3.60×10^5 J.

> **PITFALL PREVENTION 8.3**
> **W, *W*, and watts** Do not confuse the symbol W for the watt with the italic symbol *W* for work. Also, remember that the watt already represents a rate of energy transfer, so "watts per second" does not make sense. The watt is *the same as* a joule per second.

Example 8.11 | Power Delivered by an Elevator Motor

An elevator car (Fig. 8.14a) has a mass of 1 600 kg and is carrying passengers having a combined mass of 200 kg. A constant friction force of 4 000 N retards its motion.

(A) How much power must a motor deliver to lift the elevator car and its passengers at a constant speed of 3.00 m/s?

SOLUTION

Conceptualize The motor must supply the force of magnitude T that pulls the elevator car upward.

Categorize The friction force increases the power necessary to lift the elevator. The problem states that the speed of the elevator is constant, which tells us that $a = 0$. We model the elevator as a *particle in equilibrium*.

Figure 8.14 (Example 8.11) (a) The motor exerts an upward force $\vec{\mathbf{T}}$ on the elevator car. The magnitude of this force is the total tension T in the cables connecting the car and motor. The downward forces acting on the car are a friction force $\vec{\mathbf{f}}$ and the gravitational force $\vec{\mathbf{F}}_g = M\vec{\mathbf{g}}$. (b) The free-body diagram for the elevator car.

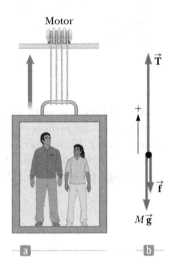

Analyze The free-body diagram in Figure 8.14b specifies the upward direction as positive. The *total* mass M of the system (car plus passengers) is equal to 1 800 kg.

Using the particle in equilibrium model, apply Newton's second law to the car:

$$\sum F_y = T - f - Mg = 0$$

Solve for T:

$$T = Mg + f$$

Use Equation 8.18 and that $\vec{\mathbf{T}}$ is in the same direction as $\vec{\mathbf{v}}$ to find the power:

$$P = \vec{\mathbf{T}} \cdot \vec{\mathbf{v}} = Tv = (Mg + f)v$$

Substitute numerical values:

$$P = [(1\,800 \text{ kg})(9.80 \text{ m/s}^2) + (4\,000 \text{ N})](3.00 \text{ m/s}) = 6.49 \times 10^4 \text{ W}$$

(B) What power must the motor deliver at the instant the speed of the elevator is v if the motor is designed to provide the elevator car with an upward acceleration of 1.00 m/s^2?

SOLUTION

Conceptualize In this case, the motor must supply the force of magnitude T that pulls the elevator car upward with an increasing speed. We expect that more power will be required to do that than in part (A) because the motor must now perform the additional task of accelerating the car.

Categorize In this case, we model the elevator car as a *particle under a net force* because it is accelerating.

Analyze Using the particle under a net force model, apply Newton's second law to the car:

$$\sum F_y = T - f - Mg = Ma$$

Solve for T:

$$T = M(a + g) + f$$

Use Equation 8.18 to obtain the required power:

$$P = Tv = [M(a + g) + f]v$$

Substitute numerical values:

$$P = [(1\,800 \text{ kg})(1.00 \text{ m/s}^2 + 9.80 \text{ m/s}^2) + 4\,000 \text{ N}]v$$

$$= (2.34 \times 10^4)v$$

where v is the instantaneous speed of the car in meters per second and P is in watts.

Finalize To compare with part (A), let $v = 3.00$ m/s, giving a power of

$$P = (2.34 \times 10^4 \text{ N})(3.00 \text{ m/s}) = 7.02 \times 10^4 \text{ W}$$

which is larger than the power found in part (A), as expected.

Summary

▶ Definitions

A **nonisolated system** is one for which energy crosses the boundary of the system. An **isolated system** is one for which no energy crosses the boundary of the system.

The **instantaneous power** P is defined as the time rate of energy transfer:

$$P \equiv \frac{dE}{dt} \qquad (8.17)$$

▶ Concepts and Principles

For a nonisolated system, we can equate the change in the total energy stored in the system to the sum of all the transfers of energy across the system boundary, which is a statement of **conservation of energy.** For an isolated system, the total energy is constant.

If a friction force of magnitude f_k acts over a distance d within a system, the change in internal energy of the system is

$$\Delta E_{\text{int}} = f_k d \qquad (8.14)$$

▶ Analysis Models for Problem Solving

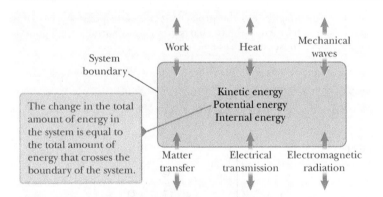

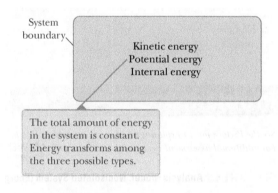

Nonisolated System (Energy). The most general statement describing the behavior of a nonisolated system is the **conservation of energy equation:**

$$\Delta E_{\text{system}} = \Sigma\, T \qquad (8.1)$$

Including the types of energy storage and energy transfer that we have discussed gives

$$\Delta K + \Delta U + \Delta E_{\text{int}} = W + Q + T_{\text{MW}} + T_{\text{MT}} + T_{\text{ET}} + T_{\text{ER}} \qquad (8.2)$$

For a specific problem, this equation is generally reduced to a smaller number of terms by eliminating the terms that are not appropriate to the situation.

Isolated System (Energy). The total energy of an isolated system is conserved, so

$$\Delta E_{\text{system}} = 0 \qquad (8.10)$$

If no nonconservative forces act within the isolated system, the mechanical energy of the system is conserved, so

$$\Delta E_{\text{mech}} = 0 \qquad (8.7)$$

which can be written as

$$\Delta K + \Delta U = 0 \qquad (8.5)$$

If a nonconservative force such as friction acts within the system, there is a change in internal energy, so

$$\Delta K + \Delta U + \Delta E_{\text{int}} = 0 \qquad (8.16)$$

Think–Pair–Share

See the Preface for an explanation of the icons used in this problems set.
For additional assessment items for this section, go to **WEBASSIGN** From Cengage

1. You are a member of an expert witness group that provides scientific services to the legal community. Your group has been asked by a defense attorney to prove at trial that a driver was not exceeding the speed limit. You are provided with the following data: The mass of the car is 1.50×10^3 kg. The mass of the driver is 95.0 kg. The coefficient of kinetic friction between the car's tires and the roadway is 0.580. The coefficient of static friction between the car's tires and the roadway is 0.820. The posted speed limit on the road is 25 mi/h. The roadway was dry and the weather was sunny at the time of the incident.

   You are also provided with the following description of the incident: The driver was driving up a hill that makes an angle of 17.5° with the horizontal. The driver saw a dog run into the street, slammed on the brakes and left a skid mark 17.0 m long. The car came to rest at the end of the skid mark. The driver did not hit the dog, but the sound of the screeching tires drew the attention of a nearby policeman, who ticketed the driver for speeding.

   Should your group agree to offer testimony for the defense in this case? (Notice that this problem is the same as Think–Pair–Share Problem 5.1 (see page 120), but we want to use an energy approach here for comparison.)

2. You are working on a team of expert witnesses for an automobile company. The company is being sued by a persistent inventor who is frustrated that the company will not adopt his idea of a car that is operated solely by solar power. Prepare an argument for your company to use at trial to show that there is simply not enough energy delivered to a normal-sized car by solar energy to operate the car on streets and highways. Use the fact that the maximum intensity of sunlight available, near the equator, is 1 000 W/m^2.

3. **ACTIVITY** (a) Draw a simple diagram of a house and indicate all major means of energy transfer between the house and the environment. (b) What means can be used to combat or take advantage of the energy transfers to keep the temperature of the house fixed at a lower monthly cost for utility bills? (c) The following are words used in architecture when discussing energy considerations for a building: Insolation, Infiltration. Assign these architectural words to specific corresponding terms in Eq. 8.2.

4. **ACTIVITY** Consider the popgun in Example 8.4. Suppose the projectile mass, compression distance, and spring constant remain the same as given or calculated in the example. Suppose, however, there is a friction force of magnitude 2.00 N acting on the projectile as it rubs against the interior of the barrel. The vertical length from point Ⓐ to the end of the barrel is 0.600 m. (a) After the spring is compressed and the popgun fired, to what height does the projectile rise above point Ⓑ? (b) Draw four energy bar charts for this situation, analogous to those in Figures 8.6c–d.

Problems

See the Preface for an explanation of the icons used in this problems set.
For additional assessment items for this section, go to **WEBASSIGN** From Cengage

SECTION 8.1 Analysis Model: Nonisolated System (Energy)

1. A ball of mass m falls from a height h to the floor. (a) Write the appropriate version of Equation 8.2 for the system of the ball and the Earth and use it to calculate the speed of the ball just before it strikes the Earth. (b) Write the appropriate version of Equation 8.2 for the system of the ball and use it to calculate the speed of the ball just before it strikes the Earth.

SECTION 8.2 Analysis Model: Isolated System (Energy)

2. A 20.0-kg cannonball is fired from a cannon with muzzle speed of 1 000 m/s at an angle of 37.0° with the horizontal. A second ball is fired at an angle of 90.0°. Use the isolated system model to find (a) the maximum height reached by each ball and (b) the total mechanical energy of the ball–Earth system at the maximum height for each ball. Let $y = 0$ at the cannon.

3. A block of mass $m = 5.00$ kg is released from point Ⓐ and slides on the frictionless track shown in Figure P8.3. Determine (a) the block's speed at points Ⓑ and Ⓒ and (b) the net work done by the gravitational force on the block as it moves from point Ⓐ to point Ⓒ.

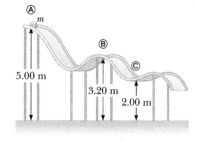

Figure P8.3

4. At 11:00 a.m. on September 7, 2001, more than one million British schoolchildren jumped up and down for one minute to simulate an earthquake. (a) Find the energy stored in the children's bodies that was converted into internal energy in the ground and their bodies and propagated into the ground by seismic waves during the experiment. Assume 1 050 000 children of average mass 36.0 kg jumped 12 times each, raising their centers of mass by 25.0 cm each time and briefly resting between one jump and the next. (b) Of the energy that propagated into the ground, most produced high-frequency "microtremor" vibrations that were rapidly damped and did not travel far. Assume 0.01% of the total energy was carried away by long-range seismic

waves. The magnitude of an earthquake on the Richter scale is given by

$$M = \frac{\log E - 4.8}{1.5}$$

where E is the seismic wave energy in joules. According to this model, what was the magnitude of the demonstration quake?

5. A light, rigid rod is 77.0 cm long. Its top end is pivoted on a frictionless, horizontal axle. The rod hangs straight down at rest with a small, massive ball attached to its bottom end. You strike the ball, suddenly giving it a horizontal velocity so that it swings around in a full circle. What minimum speed at the bottom is required to make the ball go over the top of the circle?

6. **Review.** The system shown in Figure P8.6 consists of a light, inextensible cord, light, frictionless pulleys, and blocks of equal mass. Notice that block B is attached to one of the pulleys. The system is initially held at rest so that the blocks are at the same height above the ground. The blocks are then released. Find the speed of block A at the moment the vertical separation of the blocks is h.

Figure P8.6

SECTION 8.3 Situations Involving Kinetic Friction

7. A crate of mass 10.0 kg is pulled up a rough incline with an initial speed of 1.50 m/s. The pulling force is 100 N parallel to the incline, which makes an angle of 20.0° with the horizontal. The coefficient of kinetic friction is 0.400, and the crate is pulled 5.00 m. (a) How much work is done by the gravitational force on the crate? (b) Determine the increase in internal energy of the crate–incline system owing to friction. (c) How much work is done by the 100-N force on the crate? (d) What is the change in kinetic energy of the crate? (e) What is the speed of the crate after being pulled 5.00 m?

8. A 40.0-kg box initially at rest is pushed 5.00 m along a rough, horizontal floor with a constant applied horizontal force of 130 N. The coefficient of friction between box and floor is 0.300. Find (a) the work done by the applied force, (b) the increase in internal energy in the box–floor system as a result of friction, (c) the work done by the normal force, (d) the work done by the gravitational force, (e) the change in kinetic energy of the box, and (f) the final speed of the box.

9. A smooth circular hoop with a radius of 0.500 m is placed flat on the floor. A 0.400-kg particle slides around the inside edge of the hoop. The particle is given an initial speed of 8.00 m/s. After one revolution, its speed has dropped to 6.00 m/s because of friction with the floor. (a) Find the energy transformed from mechanical to internal in the particle–hoop–floor system as a result of friction in one revolution. (b) What is the total number of revolutions the particle makes before stopping? Assume the friction force remains constant during the entire motion.

SECTION 8.4 Changes in Mechanical Energy for Nonconservative Forces

10. As shown in Figure P8.10, a green bead of mass 25 g slides along a straight wire. The length of the wire from point Ⓐ to point Ⓑ is 0.600 m, and point Ⓐ is 0.200 m higher than point Ⓑ. A constant friction force of magnitude 0.025 0 N acts on the bead. (a) If the bead is released from rest at point Ⓐ, what is its speed at point Ⓑ? (b) A red bead of mass 25 g slides along a curved wire, subject to a friction force with the same constant magnitude as that on the green bead. If the green and red beads are released simultaneously from rest at point Ⓐ, which bead reaches point Ⓑ with a higher speed? Explain.

Figure P8.10

11. At time t_i, the kinetic energy of a particle is 30.0 J and the potential energy of the system to which it belongs is 10.0 J. At some later time t_f, the kinetic energy of the particle is 18.0 J. (a) If only conservative forces act on the particle, what are the potential energy and the total energy of the system at time t_f? (b) If the potential energy of the system at time t_f is 5.00 J, are any nonconservative forces acting on the particle? (c) Explain your answer to part (b).

12. A 1.50-kg object is held 1.20 m above a relaxed massless, vertical spring with a force constant of 320 N/m. The object is dropped onto the spring. (a) How far does the object compress the spring? (b) **What If?** Repeat part (a), but this time assume a constant air-resistance force of 0.700 N acts on the object during its motion. (c) **What If?** How far does the object compress the spring if the same experiment is performed on the Moon, where $g = 1.63$ m/s² and air resistance is neglected?

13. A child of mass m starts from rest and slides without friction from a height h along a slide next to a pool (Fig. P8.13). She is launched from a height $h/5$ into the air over the pool. We wish to find the maximum height she reaches above the water in her projectile motion. (a) Is the child–Earth system isolated or nonisolated? Why? (b) Is there a nonconservative force acting within the system? (c) Define the configuration of the system when the child is at the water level as having zero gravitational potential energy. Express the total energy of the system when the child is at the top of the waterslide. (d) Express the total energy of the system when the child is at the launching point. (e) Express the total energy of the system when the child is at the highest point in her projectile motion. (f) From parts (c) and (d), determine her initial

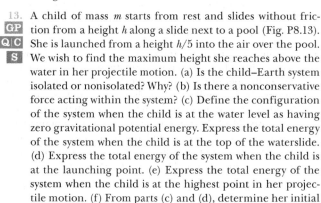

Figure P8.13

speed v_i at the launch point in terms of g and h. (g) From parts (d), (e), and (f), determine her maximum airborne height y_{max} in terms of h and the launch angle θ. (h) Would your answers be the same if the waterslide were not frictionless? Explain.

14. An 80.0-kg skydiver jumps out of a balloon at an altitude of 1 000 m and opens his parachute at an altitude of 200 m. (a) Assuming the total retarding force on the skydiver is constant at 50.0 N with the parachute closed and constant at 3 600 N with the parachute open, find the speed of the skydiver when he lands on the ground. (b) Do you think the skydiver will be injured? Explain. (c) At what height should the parachute be opened so that the final speed of the skydiver when he hits the ground is 5.00 m/s? (d) How realistic is the assumption that the total retarding force is constant? Explain.

15. You have spent a long day skiing and are tired. You are standing at the top of a hill, looking at the lodge at the bottom of the hill. You are so tired that you want to simply start from rest and coast down the slope, without pushing with your poles or doing anything else to change your motion. You want to let gravity do all the work! You have a choice of two trails to reach the lodge. Both trails have the same coefficient of friction μ_k. In addition, both trails represent the same horizontal separation between the initial and final points. Trail A has a short, steep downslope and then a long, flat coast to the lodge. Trail B has a long, gentle downslope and then a short remaining flat coast to the lodge. Which trail will result in your arriving at the lodge with the highest final speed?

SECTION 8.5 Power

16. The electric motor of a model train accelerates the train from rest to 0.620 m/s in 21.0 ms. The total mass of the train is 875 g. (a) Find the minimum power delivered to the train by electrical transmission from the metal rails during the acceleration. (b) Why is it the minimum power?

17. An energy-efficient lightbulb, taking in 28.0 W of power, can produce the same level of brightness as a conventional lightbulb operating at power 100 W. The lifetime of the energy-efficient bulb is 10 000 h and its purchase price is $4.50, whereas the conventional bulb has a lifetime of 750 h and costs $0.42. Determine the total savings obtained by using one energy-efficient bulb over its lifetime as opposed to using conventional bulbs over the same time interval. Assume an energy cost of $0.200 per kilowatt-hour.

18. An older-model car accelerates from 0 to speed v in a time interval of Δt. A newer, more powerful sports car accelerates from 0 to $2v$ in the same time period. Assuming the energy coming from the engine appears only as kinetic energy of the cars, compare the power of the two cars.

19. Make an order-of-magnitude estimate of the power a car engine contributes to speeding the car up to highway speed. In your solution, state the physical quantities you take as data and the values you measure or estimate for them. The mass of a vehicle is often given in the owner's manual.

20. There is a 5K event coming up in your town. While talking to your grandmother, who uses an electric scooter for mobility, she says that she would like to accompany you on her scooter while you walk the 5.00-km distance. The manual that came with her scooter claims that the fully charged battery is capable of providing 120 Wh of energy before being depleted. In preparation for the race, you go for a "test drive": beginning with a fully charged battery, your grandmother rides beside you as you walk 5.00 km on flat ground. At the end of the walk, the battery usage indicator shows that 40.0% of the original energy in the battery remains. You also know that the combined weight of the scooter and your grandmother is 890 N. A few days later, filled with confidence that the battery has sufficient energy, you and your grandmother drive to the 5K event. Unbeknownst to you, the 5K route is not on flat ground, but is all uphill, ending at a point higher than the starting line. A race official tells you that the total amount of vertical displacement on the route is 150 m. Should your grandmother accompany you on the walk, or will she be stranded when her battery runs out of energy? Assume that the only difference between your test drive and the actual event is the vertical displacement.

21. For saving energy, bicycling and walking are far more efficient means of transportation than is travel by automobile. For example, when riding at 10.0 mi/h, a cyclist uses food energy at a rate of about 400 kcal/h above what he would use if merely sitting still. (In exercise physiology, power is often measured in kcal/h rather than in watts. Here 1 kcal = 1 nutritionist's Calorie = 4 186 J.) Walking at 3.00 mi/h requires about 220 kcal/h. It is interesting to compare these values with the energy consumption required for travel by car. Gasoline yields about 1.30×10^8 J/gal. Find the fuel economy in equivalent miles per gallon for a person (a) walking and (b) bicycling.

22. Energy is conventionally measured in Calories as well as in joules. One Calorie in nutrition is one kilocalorie, defined as 1 kcal = 4 186 J. Metabolizing 1 g of fat can release 9.00 kcal. A student decides to try to lose weight by exercising. He plans to run up and down the stairs in a football stadium as fast as he can and as many times as necessary. To evaluate the program, suppose he runs up a flight of 80 steps, each 0.150 m high, in 65.0 s. For simplicity, ignore the energy he uses in coming down (which is small). Assume a typical efficiency for human muscles is 20.0%. This statement means that when your body converts 100 J from metabolizing fat, 20 J goes into doing mechanical work (here, climbing stairs). The remainder goes into extra internal energy. Assume the student's mass is 75.0 kg. (a) How many times must the student run the flight of stairs to lose 1.00 kg of fat? (b) What is his average power output, in watts and in horsepower, as he runs up the stairs? (c) Is this activity in itself a practical way to lose weight?

ADDITIONAL PROBLEMS

23. A block of mass $m = 200$ g is released from rest at point Ⓐ along the horizontal diameter on the inside of hemispherical bowl of radius $R = 30.0$ cm, and the surface of the bowl is rough (Fig. P8.23). The block's speed at point Ⓑ is 1.50 m/s.

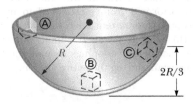

Figure P8.23

(a) What is its kinetic energy at point Ⓑ? (b) How much mechanical energy is transformed into internal energy as the block moves from point Ⓐ to point Ⓑ? (c) Is it possible to determine the coefficient of friction from these results in any simple manner? (d) Explain your answer to part (c).

24. Make an order-of-magnitude estimate of your power output as you climb stairs. In your solution, state the physical quantities you take as data and the values you measure or estimate for them. Do you consider your peak power or your sustainable power?

25. You are working with a team that is designing a new roller coaster–type amusement park ride for a major theme park. You are present for the testing of the ride, in which an empty 250-kg car is sent along the entire ride. Near the end of the ride, the car is at near rest at the top of a 110-m tall track. It then enters a final section, rolling down an undulating hill to ground level. The total length of track for this final section from the top to the ground is 250 m. For the first 230 m, a constant friction force of 50.0 N acts from computer-controlled brakes. For the last 20 m, which is horizontal at ground level, the computer increases the friction force to a value required for the speed to be reduced to zero just as the car arrives at the point on the track at which the passengers exit. (a) Determine the required constant friction force for the last 20 m for the empty test car. (b) Find the highest speed reached by the car during the final section of track length 250 m. (c) You are asked by your team supervisor to determine the answers to parts (a) and (b) for a fully loaded car with an upper limit of 450 kg of passenger mass. Find these new values. (d) The required friction force in part (c) is well within design limits. The fastest speed, however, is well below that of current leading rides, so you would like to increase the maximum speed. You can't make the tower taller above ground, so you decide to include a feature where part of the track goes *underground*. Determine the depth to which the underground part of the ride must go to increase the maximum speed to 55.0 m/s. Assume the overall length of the first part of the track remains at 230 m and the length of track from the top to the lowest underground point is 150 m. The same 50.0-N friction force acts on the entire 230-m section of track. (e) Is the construction in part (d) feasible?

26. **Review.** As shown in Figure P8.26, a light string that does not stretch changes from horizontal to vertical as it passes over the edge of a table. The string connects m_1, a 3.50-kg block originally at rest on the horizontal table at a height $h = 1.20$ m above the floor, to m_2, a hanging 1.90-kg block originally a distance $d = 0.900$ m above the floor. Neither the surface of the table nor its edge exerts a force of kinetic friction. The blocks start to move from rest. The sliding block m_1 is projected horizontally after reaching the edge of the table. The hanging block m_2 stops without bouncing when it strikes the floor. Consider the two blocks plus the Earth as the system. (a) Find the speed at which m_1 leaves the edge of the table. (b) Find the impact speed of m_1 on the floor. (c) What is the shortest length of the string so that

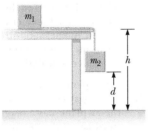

Figure P8.26

it does not go taut while m_1 is in flight? (d) Is the energy of the system when it is released from rest equal to the energy of the system just before m_1 strikes the ground? (e) Why or why not?

27. Consider the block–spring–surface system in part (B) of Example 8.6. (a) Using an energy approach, find the position x of the block at which its speed is a maximum. (b) In the **What If?** section of this example, we explored the effects of an increased friction force of 10.0 N. At what position of the block does its maximum speed occur in this situation?

28. *Why is the following situation impossible?* A softball pitcher has a strange technique: she begins with her hand at rest at the highest point she can reach and then quickly rotates her arm backward so that the ball moves through a half-circle path. She releases the ball when her hand reaches the bottom of the path. The pitcher maintains a component of force on the 0.180-kg ball of constant magnitude 12.0 N in the direction of motion around the complete path. As the ball arrives at the bottom of the path, it leaves her hand with a speed of 25.0 m/s.

29. Jonathan is riding a bicycle and encounters a hill of height 7.30 m. At the base of the hill, he is traveling at 6.00 m/s. When he reaches the top of the hill, he is traveling at 1.00 m/s. Jonathan and his bicycle together have a mass of 85.0 kg. Ignore friction in the bicycle mechanism and between the bicycle tires and the road. (a) What is the total external work done on the system of Jonathan and the bicycle between the time he starts up the hill and the time he reaches the top? (b) What is the change in potential energy stored in Jonathan's body during this process? (c) How much work does Jonathan do on the bicycle pedals within the Jonathan–bicycle–Earth system during this process?

30. Jonathan is riding a bicycle and encounters a hill of height h. At the base of the hill, he is traveling at a speed v_i. When he reaches the top of the hill, he is traveling at a speed v_f. Jonathan and his bicycle together have a mass m. Ignore friction in the bicycle mechanism and between the bicycle tires and the road. (a) What is the total external work done on the system of Jonathan and the bicycle between the time he starts up the hill and the time he reaches the top? (b) What is the change in potential energy stored in Jonathan's body during this process? (c) How much work does Jonathan do on the bicycle pedals within the Jonathan–bicycle–Earth system during this process?

31. As the driver steps on the gas pedal, a car of mass 1 160 kg accelerates from rest. During the first few seconds of motion, the car's acceleration increases with time according to the expression

$$a = 1.16t - 0.210t^2 + 0.240t^3$$

where t is in seconds and a is in m/s². (a) What is the change in kinetic energy of the car during the interval from $t = 0$ to $t = 2.50$ s? (b) What is the minimum average power output of the engine over this time interval? (c) Why is the value in part (b) described as the *minimum* value?

32. As it plows a parking lot, a snowplow pushes an ever-growing pile of snow in front of it. Suppose a car moving through the air is similarly modeled as a cylinder of area A pushing a growing disk of air in front of it. The originally stationary air is set into motion at the constant speed v of the cylinder

as shown in Figure P8.32. In a time interval Δt, a new disk of air of mass Δm must be moved a distance $v \, \Delta t$ and hence must be given a kinetic energy $\frac{1}{2}(\Delta m)v^2$. Using this model, show that the car's power loss owing to air resistance is $\frac{1}{2}\rho A v^3$

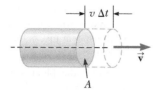

Figure P8.32

and that the resistive force acting on the car is $\frac{1}{2}\rho A v^2$, where ρ is the density of air. Compare this result with the empirical expression $\frac{1}{2}D\rho A v^2$ for the resistive force.

33. Heedless of danger, a child leaps onto a pile of old mattresses to use them as a trampoline. His motion between two particular points is described by the energy conservation equation

$$\tfrac{1}{2}(46.0 \text{ kg})(2.40 \text{ m/s})^2 + (46.0 \text{ kg})(9.80 \text{ m/s}^2)(2.80 \text{ m} + x) = \tfrac{1}{2}(1.94 \times 10^4 \text{ N/m})x^2$$

(a) Solve the equation for x. (b) Compose the statement of a problem, including data, for which this equation gives the solution. (c) Add the two values of x obtained in part (a) and divide by 2. (d) What is the significance of the resulting value in part (c)?

34. **Review.** *Why is the following situation impossible?* A new high-speed roller coaster is claimed to be so safe that the passengers do not need to wear seat belts or any other restraining device. The coaster is designed with a vertical circular section over which the coaster travels on the inside of the circle so that the passengers are upside down for a short time interval. The radius of the circular section is 12.0 m, and the coaster enters the bottom of the circular section at a speed of 22.0 m/s. Assume the coaster moves without friction on the track and model the coaster as a particle.

35. A horizontal spring attached to a wall has a force constant of $k = 850$ N/m. A block of mass $m = 1.00$ kg is attached to the spring and rests on a frictionless, horizontal surface as in Figure P8.35. (a) The block is pulled to a position $x_i = 6.00$ cm from equilibrium and released. Find the elastic potential energy stored in the spring when the block is 6.00 cm from equilibrium and when the block passes through equilibrium. (b) Find the speed of the block as it passes through the equilibrium point. (c) What is the speed of the block when it is at a position $x_i/2 = 3.00$ cm? (d) Why isn't the answer to part (c) half the answer to part (b)?

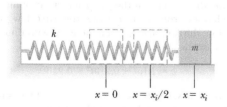

Figure P8.35

36. More than 2 300 years ago, the Greek teacher Aristotle wrote the first book called *Physics*. Put into more precise terminology, this passage is from the end of its Section Eta:

Let P be the power of an agent causing motion; w, the load moved; d, the distance covered; and Δt, the time interval required. Then (1) a power equal to P will in

an interval of time equal to Δt move $w/2$ a distance $2d$; or (2) it will move $w/2$ the given distance d in the time interval $\Delta t/2$. Also, if (3) the given power P moves the given load w a distance $d/2$ in time interval $\Delta t/2$, then (4) $P/2$ will move $w/2$ the given distance d in the given time interval Δt.

(a) Show that Aristotle's proportions are included in the equation $P\Delta t = bwd$, where b is a proportionality constant. (b) Show that our theory of motion includes this part of Aristotle's theory as one special case. In particular, describe a situation in which it is true, derive the equation representing Aristotle's proportions, and determine the proportionality constant.

37. **Review.** As a prank, someone has balanced a pumpkin at the highest point of a grain silo. The silo is topped with a hemispherical cap that is frictionless when wet. The line from the center of curvature of the cap to the pumpkin makes an angle $\theta_i = 0°$ with the vertical. While you happen to be standing nearby in the middle of a rainy night, a breath of wind makes the pumpkin start sliding downward from rest. It loses contact with the cap when the line from the center of the hemisphere to the pumpkin makes a certain angle with the vertical. What is this angle?

38. **Review.** *Why is the following situation impossible?* An athlete tests her hand strength by having an assistant hang weights from her belt as she hangs onto a horizontal bar with her hands. When the weights hanging on her belt have increased to 80% of her body weight, her hands can no longer support her and she drops to the floor. Frustrated at not meeting her hand-strength

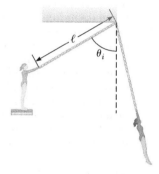

Figure P8.38

goal, she decides to swing on a trapeze. The trapeze consists of a bar suspended by two parallel ropes, each of length ℓ, allowing performers to swing in a vertical circular arc (Fig. P8.38). The athlete holds the bar and steps off an elevated platform, starting from rest with the ropes at an angle $\theta_i = 60.0°$ with respect to the vertical. As she swings several times back and forth in a circular arc, she forgets her frustration related to the hand-strength test. Assume the size of the performer's body is small compared to the length ℓ and air resistance is negligible.

39. An airplane of mass 1.50×10^4 kg is in level flight, initially moving at 60.0 m/s. The resistive force exerted by air on the airplane has a magnitude of 4.0×10^4 N. By Newton's third law, if the engines exert a force on the exhaust gases to expel them out of the back of the engine, the exhaust gases exert a force on the engines in the direction of the airplane's travel. This force is called thrust, and the value of the thrust in this situation is 7.50×10^4 N. (a) Is the work done by the exhaust gases on the airplane during some time interval equal to the change in the airplane's kinetic energy? Explain. (b) Find the speed of the airplane after it has traveled 5.0×10^2 m.

40. A pendulum, comprising a light string of length L and a small sphere, swings in the vertical plane. The string hits a peg located a distance d below the point of suspension

(Fig. P8.40). (a) Show that if the sphere is released from a height below that of the peg, it will return to this height after the string strikes the peg. (b) Show that if the pendulum is released from rest at the horizontal position ($\theta = 90°$) and is to swing in a complete circle centered on the peg, the minimum value of d must be $3L/5$.

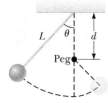

Figure P8.40

41. A ball whirls around in a *vertical* circle at the end of a
S   string. The other end of the string is fixed at the center of the circle. Assuming the total energy of the ball–Earth system remains constant, show that the tension in the string at the bottom is greater than the tension at the top by six times the ball's weight.

42. You are working in the distribution center of a large online
CR   shopping site. Efforts are being made to increase the number of packages per unit time that are being loaded onto a conveyor belt to be carried to waiting trucks. But the motor driving the conveyor belt is having difficulty keeping up with the increased demands. Your supervisor has asked you to determine the requirements for a new motor that can provide enough power to keep the conveyor belt moving smoothly under the increased loading rate. You are given the following information: The design goal is to have 50.0-kg packages loaded onto the belt at several locations at an average rate of 5.00 packages per second. The belt moves at a horizontal speed of 1.35 m/s. Humans at the various locations along the belt place the package on the belt so that it is initially at rest relative to the floor of the building just before being dropped from negligible height onto the belt. Your task is to determine the minimum power the driving motor must have to accelerate these packages and keep the belt moving at constant speed.

43. Consider the block–spring collision discussed in Example 8.8. (a) For the situation in part (B), in which the surface exerts a friction force on the block, show that the block never arrives back at $x = 0$. (b) What is the maximum value of the coefficient of friction that would allow the block to return to $x = 0$?

CHALLENGE PROBLEMS

44. Starting from rest, a 64.0-kg person bungee jumps from
QC   a tethered hot-air balloon 65.0 m above the ground. The bungee cord has negligible mass and unstretched length 25.8 m. One end is tied to the basket of the balloon and the other end to a harness around the person's body. The cord is modeled as a spring that obeys Hooke's law with a spring constant of 81.0 N/m, and the person's body is modeled as

a particle. The hot-air balloon does not move. (a) Express the gravitational potential energy of the person–Earth system as a function of the person's variable height y above the ground. (b) Express the elastic potential energy of the cord as a function of y. (c) Express the total potential energy of the person–cord–Earth system as a function of y. (d) Plot a graph of the gravitational, elastic, and total potential energies as functions of y. (e) Assume air resistance is negligible. Determine the minimum height of the person above the ground during his plunge. (f) Does the potential energy graph show any equilibrium position or positions? If so, at what elevations? Are they stable or unstable? (g) Determine the jumper's maximum speed.

45. **Review.** A uniform board of length L is sliding along a
S   smooth, frictionless, horizontal plane as shown in Figure P8.45a. The board then slides across the boundary with a rough horizontal surface. The coefficient of kinetic friction between the board and the second surface is μ_k. (a) Find the acceleration of the board at the moment its front end has traveled a distance x beyond the boundary. (b) The board stops at the moment its back end reaches the boundary as shown in Figure P8.45b. Find the initial speed v of the board.

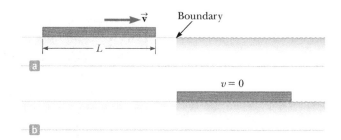

Figure P8.45

46. A uniform chain of length 8.00 m initially lies stretched out on a horizontal table. (a) Assuming the coefficient of static friction between chain and table is 0.600, show that the chain will begin to slide off the table if at least 3.00 m of it hangs over the edge of the table. (b) Determine the speed of the chain as its last link leaves the table, given that the coefficient of kinetic friction between the chain and the table is 0.400.

47. **What If?** Consider the roller coaster described in Prob-
QC   lem 34. Because of some friction between the coaster and the track, the coaster enters the circular section at a speed of 15.0 m/s rather than the 22.0 m/s in Problem 34. Is this situation *more* or *less* dangerous for the passengers than that in Problem 34? Assume the circular section is still frictionless.

9

Linear Momentum and Collisions

1

2

3

Your excitement about winning a game of pool is overcome by the physics of an interesting shot made by your friend. The purple ball in the middle never moves during the process.

STORYLINE **You decide to play pool at the student center at the** university. You and your friend are in the middle of a game when one shot made by your friend fascinates you and starts your mind thinking again. The initial situation is shown in diagram #1 above. Two balls, yellow and purple, are at rest and touching each other. You friend shoots the white cue ball along a line drawn through the centers of all three balls, and the cue ball makes a direct hit, so that the centers of all three balls are momentarily lined up, as in diagram #2. The cue ball stops and only the yellow ball moves away from the collision, as shown in diagram #3. The purple ball in the middle remains stationary during the entire interaction. You think, "Wait a minute! Why did that happen? The energy of the system of three balls must be conserved. So why couldn't *both* of the initially stationary balls move off after the collision at smaller speeds so that their kinetic energies add up to that of the cue ball?" Your friend pleads with you to continue the game, but your mind is elsewhere, analyzing this interesting situation.

CONNECTIONS While the energy approach studied in the previous chapters is powerful, there are still some problems we cannot solve in an easy way with the physics we've studied so far. In this chapter, we find that there is another conserved quantity besides energy. While this new quantity is a combination of mass and velocity, similar to kinetic energy, it is a vector, very different from energy. We find that the new conservation principle for this quantity, *momentum*, allows us to solve even more new types of problems, such as the one in the storyline. This conservation principle is particularly useful in analyzing collisions between two or more objects. As with energy, the analysis of systems is important; we will generate momentum principles for both isolated and nonisolated systems. Furthermore, our study of momentum in systems will lead to the important concept of the *center of mass* of a system of particles. The principles associated with momentum will join those associated with energy in several future chapters to allow us to understand many physical situations.

9.1 Linear Momentum

In Chapter 8, we studied situations that are difficult to analyze with Newton's laws. We were able to solve problems involving these situations by identifying a system and applying a conservation principle, conservation of energy. Let us consider another situation and see if we can solve it with the models we have developed so far:

A 60-kg archer stands at rest on frictionless ice and fires a 0.030-kg arrow horizontally at 85 m/s. With what velocity does the archer move across the ice after firing the arrow?

From Newton's third law, we know that the force that the bow exerts on the arrow is paired with a force in the opposite direction on the bow (and the archer). This force causes the archer to slide backward on the ice with the speed requested in the problem. We cannot determine this speed using motion models such as the particle under constant acceleration because we don't have any information about the acceleration of the archer. We cannot use force models such as the particle under a net force because we don't know anything about forces in this situation. Energy models are of no help because we know nothing about the work done in pulling the bowstring back or the elastic potential energy in the system related to the taut bowstring.

Despite our inability to solve the archer problem using models learned so far, this problem is very simple to solve if we introduce a new quantity that describes motion, *linear momentum*. To generate this new quantity, consider an isolated system of two particles (Fig. 9.1) with masses m_1 and m_2 moving with velocities $\vec{v}_1$ and $\vec{v}_2$ at an instant of time. Because the system is isolated, the only force on one particle is that from the other particle. If a force from particle 1 (for example, a gravitational force) acts on particle 2, there must be a second force—equal in magnitude but opposite in direction—that particle 2 exerts on particle 1. That is, the forces on the particles form a Newton's third law action–reaction pair, and $\vec{F}_{12} = -\vec{F}_{21}$. We can express this condition as

$$\vec{F}_{21} + \vec{F}_{12} = 0$$

From a system point of view, this equation says that if we add up the forces on the particles in an isolated system, the sum is zero.

Let us further analyze this situation by incorporating Newton's second law, Equation 5.2. At the instant shown in Figure 9.1, the interacting particles in the system have accelerations corresponding to the forces on them. Therefore, replacing the force on each particle in the previous equation with $m\vec{a}$ for the particle gives

$$m_1\vec{a}_1 + m_2\vec{a}_2 = 0$$

Now we replace each acceleration with its definition from Equation 4.5:

$$m_1\frac{d\vec{v}_1}{dt} + m_2\frac{d\vec{v}_2}{dt} = 0$$

If the masses m_1 and m_2 are constant, we can bring them inside the derivative operation, which gives

$$\frac{d(m_1\vec{v}_1)}{dt} + \frac{d(m_2\vec{v}_2)}{dt} = 0$$

$$\frac{d}{dt}(m_1\vec{v}_1 + m_2\vec{v}_2) = 0 \tag{9.1}$$

Notice that the derivative of the sum $m_1\vec{v}_1 + m_2\vec{v}_2$ with respect to time is zero. Consequently, this sum must be constant over an arbitrary time interval. We saw in Chapter 8 that the total energy of an isolated system is constant over a time interval, because energy is conserved. We learn from this discussion that the quantity

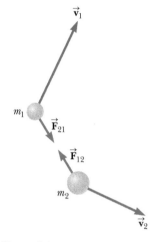

Figure 9.1 Two particles interact with each other. According to Newton's third law, we must have $\vec{F}_{12} = -\vec{F}_{21}$.

$m\vec{\mathbf{v}}$ for a particle is important in that the sum of these quantities for an isolated system of particles is also conserved. We call this quantity *linear momentum:*

Definition of linear ▶ momentum of a particle

> The **linear momentum** $\vec{\mathbf{p}}$ of a particle or an object that can be modeled as a particle of mass m moving with a velocity $\vec{\mathbf{v}}$ is defined to be the product of the mass and velocity of the particle:

$$\vec{\mathbf{p}} \equiv m\vec{\mathbf{v}} \tag{9.2}$$

Linear momentum is a vector quantity because it equals the product of a scalar quantity m and a vector quantity $\vec{\mathbf{v}}$. Its direction is along $\vec{\mathbf{v}}$, it has dimensions ML/T, and its SI unit is kg · m/s.

If a particle is moving in an arbitrary direction, $\vec{\mathbf{p}}$ has three components, and Equation 9.2 is equivalent to the component equations

$$p_x = mv_x \qquad p_y = mv_y \qquad p_z = mv_z$$

As you can see from its definition, the concept of momentum[1] provides a quantitative distinction between heavy and light particles moving at the same velocity. For example, the magnitude of the momentum of a bowling ball is much greater than that of a tennis ball moving at the same speed. Newton called the product $m\vec{\mathbf{v}}$ *quantity of motion;* this term is perhaps a more graphic description than our present-day word *momentum,* which comes from the Latin word for movement.

We have seen another quantity, kinetic energy, that is a combination of mass and speed. It would be a legitimate question to ask why we need a second quantity, momentum, based on mass and velocity. There are clear differences between kinetic energy and momentum. First, kinetic energy is a scalar, whereas momentum is a vector. Consider a system of two equal-mass particles heading toward each other along a line with equal speeds. There is kinetic energy associated with this system because members of the system are moving. Because of the vector nature of momentum, however, the momentum of this system is zero. A second major difference is that kinetic energy can transform to other types of energy, such as potential energy or internal energy. There is only one type of linear momentum, so we see no such transformations when using a momentum approach to a problem. These differences are sufficient to make models based on momentum separate from those based on energy, providing an independent tool to use in solving problems.

Using Newton's second law of motion, we can relate the linear momentum of a particle to the resultant force acting on the particle. We start with Newton's second law and substitute the definition of acceleration:

$$\sum \vec{\mathbf{F}} = m\vec{\mathbf{a}} = m\frac{d\vec{\mathbf{v}}}{dt}$$

In Newton's second law, the mass m is assumed to be constant. Therefore, we can bring m inside the derivative operation to give us

Newton's second law ▶ for a particle

$$\sum \vec{\mathbf{F}} = \frac{d(m\vec{\mathbf{v}})}{dt} = \frac{d\vec{\mathbf{p}}}{dt} \tag{9.3}$$

This equation shows that **the time rate of change of the linear momentum of a particle is equal to the net force acting on the particle.** In Chapter 5, we identified force as that which causes a change in the motion of an object (Section 5.2). In Newton's second law (Eq. 5.2), we used acceleration $\vec{\mathbf{a}}$ to represent the change in motion. We see now in Equation 9.3 that we can use the derivative of momentum $\vec{\mathbf{p}}$ with respect to time to represent the change in motion.

---

[1]In this chapter, the terms *momentum* and *linear momentum* have the same meaning. Later, in Chapter 11, we shall use the term *angular momentum* for a different quantity when dealing with rotational motion.

This alternative form of Newton's second law is the form in which Newton presented the law, and it is actually more general than the form introduced in Chapter 5. In addition to situations in which the velocity vector varies with time, we can use Equation 9.3 to study phenomena in which the mass changes. For example, the mass of a rocket changes as fuel is burned and ejected from the rocket. We cannot use $\sum \vec{\mathbf{F}} = m\vec{\mathbf{a}}$ to analyze rocket propulsion; we must use a momentum approach, as we will show in Section 9.9.

ⓠUICK QUIZ 9.1 Two objects have equal kinetic energies. How do the magnitudes of their momenta compare? **(a)** $p_1 < p_2$ **(b)** $p_1 = p_2$ **(c)** $p_1 > p_2$ **(d)** not enough information to tell

ⓠUICK QUIZ 9.2 Your physical education teacher throws a baseball to you at a certain speed and you catch it. The teacher is next going to throw you a medicine ball whose mass is ten times the mass of the baseball. You are given the following choices: You can have the medicine ball thrown with **(a)** the same speed as the baseball, **(b)** the same momentum, or **(c)** the same kinetic energy. Rank these choices from easiest to hardest to catch.

9.2 Analysis Model: Isolated System (Momentum)

Using the definition of momentum, Equation 9.1 can be written

$$\frac{d}{dt}(\vec{\mathbf{p}}_1 + \vec{\mathbf{p}}_2) = 0$$

Because the time derivative of the total momentum $\vec{\mathbf{p}}_{\text{tot}} = \vec{\mathbf{p}}_1 + \vec{\mathbf{p}}_2$ is *zero*, we conclude that the *total* momentum of the isolated system of the two particles in Figure 9.1 must remain constant:

$$\vec{\mathbf{p}}_{\text{tot}} = \text{constant} \tag{9.4}$$

or, equivalently, over some time interval,

$$\Delta\vec{\mathbf{p}}_{\text{tot}} = 0 \tag{9.5}$$

Equation 9.5 can be written for a two-particle system as

$$\vec{\mathbf{p}}_{1i} + \vec{\mathbf{p}}_{2i} = \vec{\mathbf{p}}_{1f} + \vec{\mathbf{p}}_{2f}$$

where $\vec{\mathbf{p}}_{1i}$ and $\vec{\mathbf{p}}_{2i}$ are the initial values and $\vec{\mathbf{p}}_{1f}$ and $\vec{\mathbf{p}}_{2f}$ are the final values of the momenta for the two particles for the time interval during which the particles interact. This equation in component form demonstrates that the total momenta in the x, y, and z directions are all independently conserved:

$$p_{1ix} + p_{2ix} = p_{1fx} + p_{2fx} \qquad p_{1iy} + p_{2iy} = p_{1fy} + p_{2fy} \qquad p_{1iz} + p_{2iz} = p_{1fz} + p_{2fz} \tag{9.6}$$

Equation 9.5 is the mathematical statement of a new analysis model, the **isolated system (momentum)**. It can be extended to any number of particles in an isolated system, as we show in Section 9.7. For momentum, an isolated system is one on which no external forces act. We studied the energy version of the isolated system model in Chapter 8 ($\Delta E_{\text{system}} = 0$) and now we have a momentum version. In general, Equation 9.5 can be stated in words as follows:

> Whenever two or more particles in an isolated system interact, the total momentum of the system does not change.

This statement tells us that the total momentum of an isolated system at all times equals its initial momentum.

PITFALL PREVENTION 9.1
Momentum of an Isolated *System* Is Conserved Although the momentum of an isolated *system* is conserved, the momentum of one *particle* within an isolated system is not necessarily conserved because other particles in the system may be interacting with it. Avoid applying conservation of momentum to a single particle.

◀ The momentum version of the isolated system model

Notice that we have made no statement concerning the type of forces acting on the particles of the system. Furthermore, we have not specified whether the forces are conservative or nonconservative. We have also not indicated whether or not the forces are constant. The only requirement is that the forces must be *internal* to the system. This single requirement should give you a hint about the power of this new model.

ANALYSIS MODEL   **Isolated System (Momentum)**

Imagine you have identified a system to be analyzed and have defined a system boundary. If there are no external forces on the system, the system is *isolated*. In that case, the total momentum of the system, which is the vector sum of the momenta of all members of the system, is conserved:

$$\Delta \vec{\mathbf{p}}_{\text{tot}} = 0 \qquad\qquad \textbf{(9.5)}$$

Examples:

- a cue ball strikes another ball on a pool table
- a spacecraft fires its rockets and moves faster through space (Section 9.9)
- molecules in a gas at a specific temperature move about and strike each other (Chapter 20)
- an incoming particle strikes a nucleus, creating a new nucleus and a different outgoing particle (Chapter 43)
- an electron and a positron annihilate to form two outgoing photons (Chapter 44)

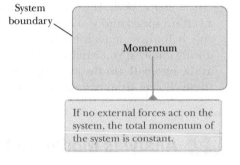

If no external forces act on the system, the total momentum of the system is constant.

Example **9.1**   **The Archer**

Let us consider the situation proposed at the beginning of Section 9.1. A 60-kg archer stands at rest on frictionless ice and fires a 0.030-kg arrow horizontally at 85 m/s (Fig. 9.2). With what velocity does the archer move across the ice after firing the arrow?

SOLUTION

Conceptualize   You may have conceptualized this problem already when it was introduced at the beginning of Section 9.1. Imagine the arrow being fired one way and the archer recoiling in the opposite direction.

Categorize   As discussed in Section 9.1, we cannot solve this problem with models based on motion, force, or energy. Nonetheless, we *can* solve this problem very easily with an approach involving momentum.

   Let us take the system to consist of the archer (including the bow) and the arrow. The system is not isolated because the gravitational force and the normal force from the ice act on the system. These forces, however, are vertical and perpendicular to the motion of the system. There are no external forces in the horizontal direction, and we can apply the *isolated system (momentum)* model in terms of momentum components in this direction.

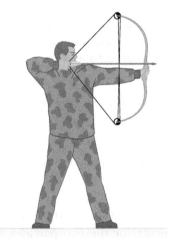

Figure 9.2   (Example 9.1) An archer fires an arrow horizontally to the right. Because he is standing on frictionless ice, he will begin to slide to the left across the ice.

Analyze   The total horizontal momentum of the system before the arrow is fired is zero because nothing in the system is moving. Therefore, the total horizontal momentum of the system after the arrow is fired must also be zero. We choose the direction of firing of the arrow as the positive *x* direction. Identifying the archer as particle 1 and the arrow as particle 2, we have $m_1 = 60$ kg, $m_2 = 0.030$ kg, and $\vec{\mathbf{v}}_{2f} = 85\hat{\mathbf{i}}$ m/s.

Using the isolated system (momentum) model, begin with Equation 9.5:

$$\Delta \vec{\mathbf{p}} = 0 \quad \rightarrow \quad \vec{\mathbf{p}}_f - \vec{\mathbf{p}}_i = 0 \quad \rightarrow \quad \vec{\mathbf{p}}_f = \vec{\mathbf{p}}_i \quad \rightarrow \quad m_1 \vec{\mathbf{v}}_{1f} + m_2 \vec{\mathbf{v}}_{2f} = 0$$

Solve this equation for $\vec{\mathbf{v}}_{1f}$ and substitute numerical values:

$$\vec{\mathbf{v}}_{1f} = -\frac{m_2}{m_1}\vec{\mathbf{v}}_{2f} = -\left(\frac{0.030 \text{ kg}}{60 \text{ kg}}\right)(85\hat{\mathbf{i}} \text{ m/s}) = -0.042\hat{\mathbf{i}} \text{ m/s}$$

9.1 continued

Finalize The negative sign for $\vec{\mathbf{v}}_{1f}$ indicates that the archer is moving to the left in Figure 9.2 after the arrow is fired, in the direction opposite the direction of motion of the arrow, in accordance with Newton's third law. Because the archer is much more massive than the arrow, his acceleration and consequent velocity are much smaller than the acceleration and velocity of the arrow. Notice that this problem sounds very simple, but we could not solve it with models based on motion, force, or energy. Our new momentum model, however, shows us that it not only *sounds* simple, it *is* simple!

WHAT IF? What if the arrow were fired in a direction that makes an angle θ with the horizontal? How will that change the recoil velocity of the archer?

Answer The recoil velocity should decrease in magnitude because only a component of the velocity of the arrow is in the x direction. Conservation of momentum in the x direction gives

$$m_1 v_{1f} + m_2 v_{2f} \cos \theta = 0$$

leading to

$$v_{1f} = -\frac{m_2}{m_1} v_{2f} \cos \theta$$

For $\theta = 0$, $\cos \theta = 1$ and the final velocity of the archer reduces to the value when the arrow is fired horizontally. For nonzero values of θ, the cosine function is less than 1 and the recoil velocity is less than the value calculated for $\theta = 0$. If $\theta = 90°$, then $\cos \theta = 0$ and $v_{1f} = 0$, so there is no recoil velocity. In this case, the arrow is fired directly upward and the archer is simply pushed downward harder against the ice as the arrow is fired.

Example **9.2**   **Can We Really Ignore the Kinetic Energy of the Earth?**

In Section 7.6, we claimed that we can ignore the kinetic energy of the Earth when considering the energy of a system consisting of the Earth and a dropped ball. Verify this claim.

SOLUTION

Conceptualize Imagine dropping a ball at the surface of the Earth. From your point of view, the ball falls while the Earth remains stationary. By Newton's third law, however, the Earth experiences an upward force and therefore an upward acceleration while the ball falls. In the calculation below, we will show that this motion is extremely small and can be ignored.

Categorize We identify the system as the ball and the Earth. We assume there are no forces on the system from outer space, so the system is isolated. Let's use the *momentum* version of the *isolated system* model.

Analyze We begin by setting up a ratio of the kinetic energy of the Earth to that of the ball. We identify v_E and v_b as the speeds of the Earth and the ball, respectively, after the ball has fallen through some distance.

Use the definition of kinetic energy to set up this ratio:

$$(1) \quad \frac{K_E}{K_b} = \frac{\frac{1}{2} m_E v_E^2}{\frac{1}{2} m_b v_b^2} = \left(\frac{m_E}{m_b}\right)\left(\frac{v_E}{v_b}\right)^2$$

Apply the isolated system (momentum) model, recognizing that the initial momentum of the system is zero:

$$\Delta \vec{\mathbf{p}} = 0 \quad \rightarrow \quad p_i = p_f \quad \rightarrow \quad 0 = m_b v_b + m_E v_E$$

Solve the equation for the ratio of velocity components:

$$\frac{v_E}{v_b} = -\frac{m_b}{m_E}$$

Take the absolute value of this ratio to make it a ratio of speeds and substitute for v_E/v_b in Equation (1):

$$\frac{K_E}{K_b} = \left(\frac{m_E}{m_b}\right)\left(\frac{m_b}{m_E}\right)^2 = \frac{m_b}{m_E}$$

Substitute order-of-magnitude numbers for the masses:

$$\frac{K_E}{K_b} = \frac{m_b}{m_E} \sim \frac{1 \text{ kg}}{10^{25} \text{ kg}} \sim 10^{-25}$$

Finalize The kinetic energy of the Earth is a very small fraction of the kinetic energy of the ball, so we are justified in ignoring it in the kinetic energy of the system.

9.3 Analysis Model: Nonisolated System (Momentum)

In the previous section, we found that the momentum of a system is conserved if there are no external forces on the system. What if there *is* an external force on the system? According to Equation 9.3, the momentum of a particle changes if a net force acts on the particle. The same can be said about a net force applied to a system as we will show explicitly in Section 9.7: the momentum of a system changes if a net force from the environment acts on the system. This may sound similar to our discussion of energy in Chapter 8: the energy of a system changes if energy crosses the boundary of the system to or from the environment. In this section, we consider a *nonisolated system*. For energy considerations, a system is nonisolated if energy transfers across the boundary of the system by any of the means listed in Section 8.1. For momentum considerations, a system is nonisolated if a net force acts on the system for a time interval. In this case, we can imagine momentum being transferred to the system from the environment by means of the net force. Knowing the change in momentum caused by a force is useful in solving some types of problems. To build a better understanding of this important concept, let us assume a net force $\sum \vec{\mathbf{F}}$ acts on a system consisting of a single particle and this force may vary with time. According to Newton's second law, in the form expressed in Equation 9.3, $\sum \vec{\mathbf{F}} = d\vec{\mathbf{p}}/dt$, we can write

$$d\vec{\mathbf{p}} = \sum \vec{\mathbf{F}}\, dt \tag{9.7}$$

We can integrate[2] this expression to find the change in the momentum of a particle when the force acts over some time interval. If the momentum of the particle changes from $\vec{\mathbf{p}}_i$ at time t_i to $\vec{\mathbf{p}}_f$ at time t_f, integrating Equation 9.7 gives

$$\Delta\vec{\mathbf{p}} = \vec{\mathbf{p}}_f - \vec{\mathbf{p}}_i = \int_{t_i}^{t_f} \sum \vec{\mathbf{F}}\, dt \tag{9.8}$$

To evaluate the integral, we need to know how the net force varies with time. The quantity on the right side of this equation is a vector called the **impulse** of the net force $\sum \vec{\mathbf{F}}$ acting on a particle over the time interval $\Delta t = t_f - t_i$:

Impulse of a force ▶

$$\vec{\mathbf{I}} \equiv \int_{t_i}^{t_f} \sum \vec{\mathbf{F}}\, dt \tag{9.9}$$

From its definition, we see that impulse $\vec{\mathbf{I}}$ is a vector quantity having a magnitude equal to the area under the force–time curve as described in Figure 9.3a. It is assumed the force varies in time in the general manner shown in the figure and is nonzero in the time interval $\Delta t = t_f - t_i$. The direction of the impulse vector is the same as the direction of the change in momentum. Impulse has the dimensions of momentum, that is, ML/T. Impulse is *not* a property of a particle; rather, it is a measure of the degree to which an external force changes the particle's momentum.

Because the net force imparting an impulse to a particle can generally vary in time, it is convenient to define a time-averaged net force:

$$\left(\sum \vec{\mathbf{F}}\right)_{avg} \equiv \frac{1}{\Delta t}\int_{t_i}^{t_f} \sum \vec{\mathbf{F}}\, dt \tag{9.10}$$

where $\Delta t = t_f - t_i$. (This equation is an application of the mean value theorem of calculus.) Therefore, we can express Equation 9.9 as

$$\vec{\mathbf{I}} = \left(\sum \vec{\mathbf{F}}\right)_{avg} \Delta t \tag{9.11}$$

This time-averaged force, shown in Figure 9.3b, can be interpreted as the constant force that would give to the particle in the time interval Δt the same impulse that the time-varying force gives over this same interval.

---

[2]Here we are integrating force with respect to time. Compare this strategy with our efforts in Chapter 7, where we integrated force with respect to position to find the work done by the force.

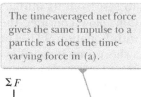

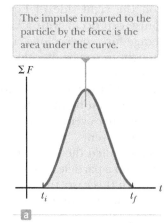

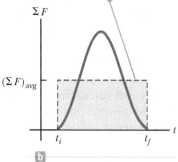

Figure 9.3 (a) A net force acting on a particle may vary in time. (b) The value of the constant force $(\Sigma F)_{avg}$ (horizontal dashed line) is chosen so that the area $(\Sigma F)_{avg} \Delta t$ of the rectangle is the same as the area under the curve in (a).

In principle, if $\Sigma \vec{F}$ is known as a function of time, the impulse can be calculated from Equation 9.9. The calculation becomes especially simple if the force acting on the particle is constant. In this case, $(\Sigma \vec{F})_{avg} = \Sigma \vec{F}$, where $\Sigma \vec{F}$ is the constant net force, and Equation 9.11 becomes

$$\vec{I} = \sum \vec{F} \, \Delta t \qquad \text{(constant net force)} \qquad (9.12)$$

Combining Equations 9.8 and 9.9 gives us an important statement known as the **impulse–momentum theorem:**

The change in the momentum of a particle is equal to the impulse of the net force acting on the particle:

$$\Delta \vec{p} = \vec{I} \qquad (9.13)$$

◄ Impulse–momentum theorem for a particle

This statement is equivalent to Newton's second law. When we say that an impulse is given to a particle, we mean that momentum is transferred from an external agent to that particle. Equation 9.13 is identical in form to the conservation of energy equation, Equation 8.1, and its full expansion, Equation 8.2. Equation 9.13 is the most general statement of the principle of **conservation of momentum** and is called the **conservation of momentum equation.** In the case of a momentum approach, isolated systems tend to appear in problems more often than nonisolated systems, so, in practice, the conservation of momentum equation is often identified as the special case of Equation 9.5.

The left side of Equation 9.13 represents the change in the momentum of the system, which in our discussion so far is a single particle. The right side is a measure of how much momentum crosses the boundary of the system due to the net force being applied to the system. Equation 9.13 is the mathematical statement of a new analysis model, the **nonisolated system (momentum)** model. Although this equation is similar in form to Equation 8.2, there are several differences in its application to problems. First, Equation 9.13 is a vector equation, whereas Equation 8.2 is a scalar equation. Therefore, directions are important for Equation 9.13. Second, there is only one type of momentum and therefore only one way to store momentum in a system. In contrast, as we see from Equation 8.2, there are three ways to store energy in a system: kinetic, potential, and internal. Third, there is only one way to transfer momentum into a system: by the application of a force on the system over a time interval. Equation 8.2 shows six ways we have identified as transferring energy into a system. Therefore, there is no expansion of Equation 9.13 analogous to Equation 8.2.

As a real-world example of Equation 9.13, consider the crash-test dummy in Figure 9.4 (page 218), representing a human driver in an accident. As the car

Figure 9.4 A crash-test dummy is brought to rest by an air bag in a test collision. The air bag increases the time interval during which the dummy is brought to rest, thereby decreasing the force on the dummy. Air bags in automobiles have saved countless human lives in accidents.

is brought to rest from its initial speed, the dummy experiences a given change in momentum. Now consider the impulse on the right side of Equation 9.13, expressed with Equation 9.11. The same impulse can occur with a large average force over a short time interval or a small average force over a long time interval. In the absence of an air bag, the dummy is brought to rest by the sudden collision of his head with the steering wheel or dashboard. This is an example of the former possibility, and the large average force could result in serious injury to a human driver. If an air bag is present, however, the dummy can be brought to rest gradually over a longer time interval, resulting in a smaller average force. As a result, there is a possibility of avoiding injury to a human driver.

In many physical situations, we shall use what is called the **impulse approximation,** in which we assume one of the forces exerted on a particle acts for a short time but is much greater than any other force present. In this case, the net force $\sum \vec{\mathbf{F}}$ in Equation 9.9 is replaced with a single force $\vec{\mathbf{F}}$ to find the impulse on the particle. This approximation is especially useful in treating collisions in which the duration of the collision is very short. When this approximation is made, the single force is referred to as an *impulsive force*. For example, when a baseball is struck with a bat, the time of the collision is about 0.01 s and the average force that the bat exerts on the ball in this time is typically several thousand newtons. Because this contact force is much greater than the magnitude of the gravitational force, the impulse approximation justifies our ignoring the gravitational forces exerted on the ball and bat during the collision. When we use this approximation, it is important to remember that $\vec{\mathbf{p}}_i$ and $\vec{\mathbf{p}}_f$ represent the momenta *immediately* before and after the collision, respectively. Therefore, in any situation in which it is proper to use the impulse approximation, the particle moves very little during the collision.

QUICK QUIZ 9.3 Two objects are at rest on a frictionless surface. Object 1 has a greater mass than object 2. **(i)** When a constant force is applied to object 1, it accelerates through a distance d in a straight line. The force is removed from object 1 and is applied to object 2. At the moment when object 2 has accelerated through the same distance d, which statements are true? (a) $p_1 < p_2$ (b) $p_1 = p_2$ (c) $p_1 > p_2$ (d) $K_1 < K_2$ (e) $K_1 = K_2$ (f) $K_1 > K_2$ **(ii)** When a force is applied to object 1, it accelerates for a time interval Δt. The force is removed from object 1 and is applied to object 2. From the same list of choices, which statements are true after object 2 has accelerated for the same time interval Δt?

QUICK QUIZ 9.4 Rank an automobile dashboard, seat belt, and air bag, each used alone in separate collisions from the same speed, in terms of (a) the impulse and (b) the average force each delivers to a front-seat passenger, from greatest to least.

ANALYSIS MODEL **Nonisolated System (Momentum)**

Imagine you have identified a system to be analyzed and have defined a system boundary. If external forces are applied on the system, the system is *nonisolated*. In that case, the change in the total momentum of the system is equal to the impulse on the system, a statement known as the **impulse–momentum theorem**:

$$\Delta \vec{\mathbf{p}} = \vec{\mathbf{I}} \qquad (9.13)$$

Examples:

- a baseball is struck by a bat
- a spool sitting on a table is pulled by a string (Example 10.14 in Chapter 10)
- a gas molecule strikes the wall of the container holding the gas (Chapter 20)
- photons strike an absorbing surface and exert pressure on the surface (Chapter 33)

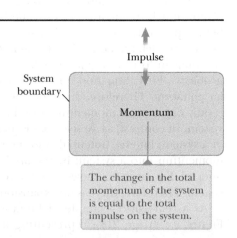

| Example **9.3** | **How Good Are the Bumpers?** |

In a particular crash test, a car of mass 1 500 kg collides with a wall as shown in Figure 9.5. The initial and final velocities of the car are $\vec{\mathbf{v}}_i = -15.0\hat{\mathbf{i}}$ m/s and $\vec{\mathbf{v}}_f = 2.60\hat{\mathbf{i}}$ m/s, respectively. If the collision lasts 0.150 s, find the impulse on the car during the collision and the average net force exerted on the car.

SOLUTION

Conceptualize The collision time is short, so we can imagine the car being brought to rest very rapidly and then moving back in the opposite direction with a reduced speed.

Categorize Let us assume the net force exerted on the car by the wall and friction from the ground is large compared with other forces on the car (such as air resistance). Furthermore, the gravitational force and the normal force exerted by the road on the car are perpendicular to the

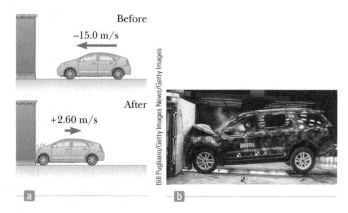

Figure 9.5 (Example 9.3) (a) This car's momentum changes as a result of its collision with the wall. (b) In a crash test, much of the car's initial kinetic energy is transformed into energy associated with the damage to the car.

motion and therefore do not affect the horizontal momentum. Therefore, we categorize the problem as one in which we can apply the impulse approximation in the horizontal direction. We also see that the car's momentum changes due to an impulse from the environment. Therefore, we can apply the *nonisolated system (momentum)* model to the system of the car.

Analyze

Use Equation 9.13 to find the impulse on the car:

$$\vec{\mathbf{I}} = \Delta\vec{\mathbf{p}} = \vec{\mathbf{p}}_f - \vec{\mathbf{p}}_i = m\vec{\mathbf{v}}_f - m\vec{\mathbf{v}}_i = m(\vec{\mathbf{v}}_f - \vec{\mathbf{v}}_i)$$

$$= (1\,500 \text{ kg})[2.60\hat{\mathbf{i}} \text{ m/s} - (-15.0\hat{\mathbf{i}} \text{ m/s})] = 2.64 \times 10^4 \hat{\mathbf{i}} \text{ kg} \cdot \text{m/s}$$

Use Equation 9.11 to evaluate the average net force exerted on the car:

$$\left(\sum\vec{\mathbf{F}}\right)_{\text{avg}} = \frac{\vec{\mathbf{I}}}{\Delta t} = \frac{2.64 \times 10^4 \hat{\mathbf{i}} \text{ kg} \cdot \text{m/s}}{0.150 \text{ s}} = 1.76 \times 10^5 \hat{\mathbf{i}} \text{ N}$$

Finalize The net force found above is a combination of the normal force on the car from the wall and any friction force between the tires and the ground as the front of the car crumples. If the brakes are not operating while the crash occurs and the crumpling metal does not interfere with the free rotation of the tires, this friction force could be relatively small due to the freely rotating wheels. Notice that the signs of the velocities in this example indicate the reversal of directions. What would the mathematics be describing if both the initial and final velocities had the same sign?

WHAT IF? What if the car did not rebound from the wall? Suppose the final velocity of the car is zero and the time interval of the collision remains at 0.150 s. Would that represent a larger or a smaller net force on the car?

Answer In the original situation in which the car rebounds, the net force on the car does two things during the time interval: (1) it stops the car, and (2) it causes the car to move away from the wall at 2.60 m/s after the collision. If the car does not rebound, the net force is only doing the first of these steps—stopping the car—which requires a *smaller* force.

Mathematically, in the case of the car that does not rebound, the impulse is

$$\vec{\mathbf{I}} = \Delta\vec{\mathbf{p}} = \vec{\mathbf{p}}_f - \vec{\mathbf{p}}_i = 0 - (1\,500 \text{ kg})(-15.0\hat{\mathbf{i}} \text{ m/s}) = 2.25 \times 10^4 \hat{\mathbf{i}} \text{ kg} \cdot \text{m/s}$$

The average net force exerted on the car is

$$\left(\sum\vec{\mathbf{F}}\right)_{\text{avg}} = \frac{\vec{\mathbf{I}}}{\Delta t} = \frac{2.25 \times 10^4 \hat{\mathbf{i}} \text{ kg} \cdot \text{m/s}}{0.150 \text{ s}} = 1.50 \times 10^5 \hat{\mathbf{i}} \text{ N}$$

which is indeed smaller than the previously calculated value, as was argued conceptually.

9.4 Collisions in One Dimension

In this section, we use the isolated system (momentum) model to describe what happens when two particles collide. The term **collision** represents an event during which two particles come close to each other and interact by means of forces. The

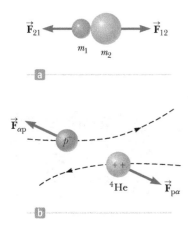

Figure 9.6 (a) The collision between two objects as the result of direct contact. (b) The "collision" between two charged particles.

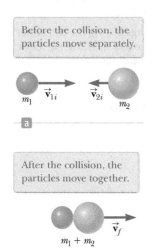

Figure 9.7 Schematic representation of a perfectly inelastic head-on collision between two particles.

interaction forces are assumed to be much greater than any external forces present, so we can use the impulse approximation.

A collision may involve physical contact between two macroscopic objects as described in Figure 9.6a, but the notion of what is meant by a collision must be generalized because "physical contact" on a submicroscopic scale is ill-defined and hence meaningless. To understand this concept, consider a collision on an atomic scale (Fig. 9.6b) such as the collision of a proton with an alpha particle (the nucleus of a helium atom). Because the particles are both positively charged, they repel each other due to the strong electrostatic force between them at close separations and never come into "physical contact."

When two particles of masses m_1 and m_2 collide as shown in Figure 9.6, the impulsive forces may vary in time in complicated ways, such as that shown in Figure 9.3. Regardless of the complexity of the time behavior of the impulsive force, however, this force is internal to the system of two particles. Therefore, the two particles form an isolated system and the momentum of the system must be conserved in *any* collision.

In contrast, the total kinetic energy of the system of particles may or may not be conserved, depending on the type of collision. In fact, collisions are categorized as being either *elastic* or *inelastic* depending on whether or not kinetic energy is conserved.

An **elastic collision** between two particles, or objects that can be modeled as particles, is one in which the total kinetic energy (as well as total momentum) of the system is the same before and after the collision. Collisions between certain objects in the macroscopic world, such as billiard balls, are only *approximately* elastic because some deformation and loss of kinetic energy take place. For example, you can hear a billiard ball collision, so you know that some of the energy is being transferred away from the system by sound. An elastic collision must be perfectly silent! *Truly* elastic collisions occur between atomic and subatomic particles. These collisions are described by the isolated system model for both energy and momentum.

An **inelastic collision** is one in which the total kinetic energy of the system is not the same before and after the collision (even though the momentum of the system is conserved). Inelastic collisions are further divided into two types. When the objects stick together after they collide, as happens when a meteorite collides with the Earth, the collision is called **perfectly inelastic.** When the colliding objects do not stick together but some kinetic energy is transformed or transferred away, the collision is called **inelastic** (with no modifying adverb). The collision of a rubber ball bouncing from a hard surface is inelastic, but not perfectly inelastic, because the ball does not stick to the surface. It is not elastic because some of the initial kinetic energy of the ball has been transformed to internal energy in the ball and the surface as the ball deformed during the time interval of contact.

In the remainder of this section, we investigate the mathematical details for collisions in one dimension and consider the two extreme cases, perfectly inelastic and elastic collisions.

Perfectly Inelastic Collisions

Consider two particles of masses m_1 and m_2 moving with initial velocities $\vec{\mathbf{v}}_{1i}$ and $\vec{\mathbf{v}}_{2i}$ along the same straight line as shown in Figure 9.7. The two particles collide head-on, stick together, and then move with some common velocity $\vec{\mathbf{v}}_f$ after the collision. For example, two carts with Velcro on their bumpers colliding on an air track will behave in this way. Because the momentum of an isolated system is conserved in *any* collision, we can say that the total momentum before the collision equals the total momentum of the composite system after the collision:

$$\Delta\vec{\mathbf{p}} = 0 \quad \rightarrow \quad \vec{\mathbf{p}}_i = \vec{\mathbf{p}}_f \quad \rightarrow \quad m_1\vec{\mathbf{v}}_{1i} + m_2\vec{\mathbf{v}}_{2i} = (m_1 + m_2)\vec{\mathbf{v}}_f \tag{9.14}$$

Solving for the final velocity gives

$$\vec{\mathbf{v}}_f = \frac{m_1\vec{\mathbf{v}}_{1i} + m_2\vec{\mathbf{v}}_{2i}}{m_1 + m_2} \tag{9.15}$$

Elastic Collisions

Consider two particles of masses m_1 and m_2 moving with initial velocities $\vec{\mathbf{v}}_{1i}$ and $\vec{\mathbf{v}}_{2i}$ along the same straight line as shown in Figure 9.8. The two particles collide head-on and then leave the collision site with different velocities, $\vec{\mathbf{v}}_{1f}$ and $\vec{\mathbf{v}}_{2f}$. In an elastic collision, both the momentum and kinetic energy of the system are conserved. Therefore, considering velocities along the horizontal direction in Figure 9.8, we have

$$p_i = p_f \quad \rightarrow \quad m_1 v_{1i} + m_2 v_{2i} = m_1 v_{1f} + m_2 v_{2f} \tag{9.16}$$

$$K_i = K_f \quad \rightarrow \quad \tfrac{1}{2} m_1 v_{1i}{}^2 + \tfrac{1}{2} m_2 v_{2i}{}^2 = \tfrac{1}{2} m_1 v_{1f}{}^2 + \tfrac{1}{2} m_2 v_{2f}{}^2 \tag{9.17}$$

Because all velocities in Figure 9.8 are either to the left or the right, they can be represented by the corresponding speeds along with algebraic signs indicating direction. We shall indicate v as positive if a particle moves to the right and negative if it moves to the left.

In a typical problem involving elastic collisions, there are two unknown quantities, and Equations 9.16 and 9.17 can be solved simultaneously to find them. An alternative approach, however—one that involves a little mathematical manipulation of Equation 9.17—often simplifies this process. To see how, let us cancel the factor $\tfrac{1}{2}$ in Equation 9.17 and rewrite it by gathering terms with subscript 1 on the left and 2 on the right:

$$m_1(v_{1i}{}^2 - v_{1f}{}^2) = m_2(v_{2f}{}^2 - v_{2i}{}^2)$$

Factoring both sides of this equation gives

$$m_1(v_{1i} - v_{1f})(v_{1i} + v_{1f}) = m_2(v_{2f} - v_{2i})(v_{2f} + v_{2i}) \tag{9.18}$$

Next, let us separate the terms containing m_1 and m_2 in Equation 9.16 in a similar way to obtain

$$m_1(v_{1i} - v_{1f}) = m_2(v_{2f} - v_{2i}) \tag{9.19}$$

To obtain our final result, we divide Equation 9.18 by Equation 9.19 and obtain

$$v_{1i} + v_{1f} = v_{2f} + v_{2i}$$

Now rearrange terms once again so as to have initial quantities on the left and final quantities on the right:

$$v_{1i} - v_{2i} = -(v_{1f} - v_{2f}) \tag{9.20}$$

This equation, in combination with Equation 9.16, can be used to solve problems dealing with elastic collisions. This pair of equations (Eqs. 9.16 and 9.20) is easier to handle than the pair of Equations 9.16 and 9.17 because there are no quadratic terms like there are in Equation 9.17. According to Equation 9.20, the *relative* velocity of the two particles before the collision, $v_{1i} - v_{2i}$, equals the negative of their relative velocity after the collision, $-(v_{1f} - v_{2f})$.

Suppose the masses and initial velocities of both particles are known. Equations 9.16 and 9.20 can be solved for the final velocities in terms of the initial velocities because there are two equations and two unknowns:

$$v_{1f} = \left(\frac{m_1 - m_2}{m_1 + m_2}\right) v_{1i} + \left(\frac{2m_2}{m_1 + m_2}\right) v_{2i} \tag{9.21}$$

$$v_{2f} = \left(\frac{2m_1}{m_1 + m_2}\right) v_{1i} + \left(\frac{m_2 - m_1}{m_1 + m_2}\right) v_{2i} \tag{9.22}$$

It is important to use the appropriate signs for v_{1i} and v_{2i} in Equations 9.21 and 9.22.

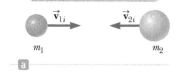

Before the collision, the particles move separately.

After the collision, the particles continue to move separately with new velocities.

Figure 9.8 Schematic representation of an elastic head-on collision between two particles.

PITFALL PREVENTION 9.3

Not a General Equation Equation 9.20 can only be used in a very *specific* situation, a one-dimensional, elastic collision between two objects. The *general* concept is conservation of momentum (and conservation of kinetic energy if the collision is elastic) for an isolated system.

Let us consider some special cases. If $m_1 = m_2$, Equations 9.21 and 9.22 show that $v_{1f} = v_{2i}$ and $v_{2f} = v_{1i}$, which means that the particles exchange velocities if they have equal masses. That is approximately what one observes in head-on billiard ball collisions: the cue ball stops and the struck ball moves away from the collision with the same velocity the cue ball had.

If particle 2 is initially at rest, then $v_{2i} = 0$, and Equations 9.21 and 9.22 become

◀ **Elastic collision: particle 2 initially at rest**

$$v_{1f} = \left(\frac{m_1 - m_2}{m_1 + m_2} \right) v_{1i} \qquad (9.23)$$

$$v_{2f} = \left(\frac{2m_1}{m_1 + m_2} \right) v_{1i} \qquad (9.24)$$

If m_1 is much greater than m_2 and $v_{2i} = 0$, we see from Equations 9.23 and 9.24 that $v_{1f} \approx v_{1i}$ and $v_{2f} \approx 2v_{1i}$. That is, when a very heavy particle collides head-on with a very light one that is initially at rest, the heavy particle continues its motion unaltered after the collision and the light particle rebounds with a speed equal to about twice the initial speed of the heavy particle. An example of such a collision is that of a moving heavy atom, such as uranium, striking a light atom, such as hydrogen.

If m_2 is much greater than m_1 and particle 2 is initially at rest, then $v_{1f} \approx -v_{1i}$ and $v_{2f} \approx 0$. That is, when a very light particle collides head-on with a very heavy particle that is initially at rest, the light particle has its velocity reversed and the heavy one remains approximately at rest. For example, imagine what happens when you throw a table tennis ball at a bowling ball as in Quick Quiz 9.6 below.

ⓠUICK QUIZ 9.5 In a perfectly inelastic one-dimensional collision between two moving objects, what condition alone is necessary so that the final kinetic energy of the system is zero after the collision? **(a)** The objects must have initial momenta with the same magnitude but opposite directions. **(b)** The objects must have the same mass. **(c)** The objects must have the same initial velocity. **(d)** The objects must have the same initial speed, with velocity vectors in opposite directions.

ⓠUICK QUIZ 9.6 A table-tennis ball is thrown at a stationary bowling ball. The table-tennis ball makes a one-dimensional elastic collision and bounces back along the same line. Compared with the bowling ball after the collision, does the table-tennis ball have **(a)** a larger magnitude of momentum and more kinetic energy, **(b)** a smaller magnitude of momentum and more kinetic energy, **(c)** a larger magnitude of momentum and less kinetic energy, **(d)** a smaller magnitude of momentum and less kinetic energy, or **(e)** the same magnitude of momentum and the same kinetic energy?

PROBLEM-SOLVING STRATEGY **One-Dimensional Collisions**

You should use the following approach when solving collision problems in one dimension:

1. Conceptualize. Imagine the collision occurring in your mind. Draw simple diagrams of the particles before and after the collision and include appropriate velocity vectors. At first, you may have to guess at the directions of the final velocity vectors.

2. Categorize. Is the system of particles isolated? If so, use the isolated system (momentum) model. Further categorize the collision as elastic, inelastic, or perfectly inelastic.

3. Analyze. Set up the appropriate mathematical representation for the problem. If the collision is perfectly inelastic, use Equation 9.15. If the collision is elastic, use Equations 9.16 and 9.20. If the collision is inelastic, use Equation 9.16. To find the final velocities in this case, you will need some additional information.

4. Finalize. Once you have determined your result, check to see if your answers are consistent with the mental and pictorial representations and that your results are realistic.

Example 9.4 The Executive Stress Reliever

An ingenious device that illustrates conservation of momentum and kinetic energy is shown in Figure 9.9a. It consists of five identical hard balls supported by strings of equal lengths. When ball 1 is pulled out and released, after the almost-elastic collision between it and ball 2, ball 1 stops and ball 5 moves out as shown in Figure 9.9b. If balls 1 and 2 are pulled out and released, they stop after the collision and balls 4 and 5 swing out, and so forth. Even if four balls (1–4) are pulled out and released, four balls (2–5) swing out after the collision! Is it ever possible that when ball 1 is released, it stops after the collision and balls 4 and 5 will swing out on the opposite side and travel with half the speed of ball 1 as in Figure 9.9c?

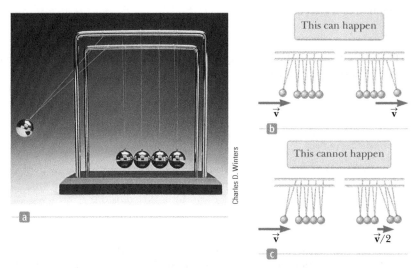

Figure 9.9 (Example 9.4) (a) An executive stress reliever. (b) If one ball swings down, we see one ball swing out at the other end. (c) Is it possible for one ball to swing down and two balls to leave the other end with half the speed of the first ball? In (b) and (c), the velocity vectors shown represent those of the balls immediately before and immediately after the collision.

SOLUTION

Conceptualize With the help of Figure 9.9c, imagine one ball coming in from the left and two balls exiting the collision on the right. That is the phenomenon we want to test to see if it could ever happen.

Categorize Because of the very short time interval between the arrival of the ball from the left and the departure of the ball(s) from the right, we can use the impulse approximation to ignore the gravitational forces on the balls and model the five balls as an *isolated system* in terms of both *momentum* and *energy*. Because the balls are hard, we can categorize the collisions between them as elastic for purposes of calculation.

Analyze Let's consider the situation shown in Figure 9.9c. The momentum of the system just before the collision is mv, where m is the mass of ball 1 and v is its speed immediately before the collision. After the collision, we imagine that ball 1 stops and balls 4 and 5 swing out, each moving with speed $v/2$. The total momentum of the system after the collision would be $m(v/2) + m(v/2) = mv$. Therefore, the momentum of the system is conserved in the situation shown in Figure 9.9c!

The kinetic energy of the system immediately before the collision is $K_i = \frac{1}{2}mv^2$ and that after the collision is $K_f = \frac{1}{2}m(v/2)^2 + \frac{1}{2}m(v/2)^2 = \frac{1}{4}mv^2$. This calculation shows that the kinetic energy of the system is *not* conserved, which is inconsistent with our assumption that the collisions are elastic.

Finalize Our analysis shows that it is *not* possible for balls 4 and 5 to swing out when only ball 1 is released. The only way to conserve both momentum and kinetic energy of the system is for one ball to move out when one ball is released, two balls to move out when two are released, and so on. A similar analysis can be applied to the billiard ball collision in the opening storyline. In that case, there are two billiard balls in contact rather than four steel balls as in Figure 9.9. When the cue ball strikes the pair of balls, the only way to conserve both momentum and kinetic energy for the system of three balls is for only one ball to leave the collision. Therefore, the purple ball remains stationary, just like balls 2 through 4 in Figure 9.9.

WHAT IF? Consider what would happen if balls 4 and 5 are glued together. Now what happens when ball 1 is pulled out and released?

Answer In this situation, balls 4 and 5 *must* move together as a single object after the collision. We have argued that both momentum and energy of the system cannot be conserved in this case. We assumed, however, ball 1 stopped after striking ball 2. What if we do not make this assumption? Consider the conservation equations with the assumption that ball 1 moves after the collision. For conservation of momentum,

$$p_i = p_f$$

$$mv_{1i} = mv_{1f} + 2mv_{4,5}$$

where $v_{4,5}$ refers to the final speed of the ball 4–ball 5 combination. Conservation of kinetic energy gives us

$$K_i = K_f$$

$$\tfrac{1}{2}mv_{1i}^2 = \tfrac{1}{2}mv_{1f}^2 + \tfrac{1}{2}(2m)v_{4,5}^2$$

continued

9.4 continued

Combining these equations gives

$$v_{4,5} = \tfrac{2}{3}v_{1i} \qquad v_{1f} = -\tfrac{1}{3}v_{1i}$$

Therefore, balls 4 and 5 move together as one object after the collision while ball 1 bounces back from the collision with one third of its original speed.

Example **9.5**   Carry Collision Insurance!

An 1 800-kg car stopped at a traffic light is struck from the rear by a 900-kg car. The two cars become entangled, moving along the same path as that of the originally moving car. If the smaller car were moving at 20.0 m/s before the collision, what is the velocity of the entangled cars after the collision?

SOLUTION

Conceptualize   This kind of collision is easily visualized, and one can predict that after the collision both cars will be moving in the same direction as that of the initially moving car. Because the initially moving car has only half the mass of the stationary car, we expect the final velocity of the cars to be relatively small.

Categorize   We identify the two cars as an *isolated system* in terms of *momentum* in the horizontal direction and apply the impulse approximation during the short time interval of the collision. The phrase "become entangled" tells us to categorize the collision as perfectly inelastic.

Analyze   The magnitude of the total momentum of the system before the collision is equal to that of the smaller car because the larger car is initially at rest.

Use the isolated system model for momentum: $\Delta \vec{p} = 0 \;\rightarrow\; p_i = p_f \;\rightarrow\; m_1 v_i = (m_1 + m_2)v_f$

Solve for v_f and substitute numerical values: $v_f = \dfrac{m_1 v_i}{m_1 + m_2} = \dfrac{(900 \text{ kg})(20.0 \text{ m/s})}{900 \text{ kg} + 1\,800 \text{ kg}} = 6.67 \text{ m/s}$

Finalize   Because the final velocity is positive, the direction of the final velocity of the combination is the same as the velocity of the initially moving car as predicted. The speed of the combination is also much lower than the initial speed of the moving car.

WHAT IF?   Suppose we reverse the masses of the cars. What if a stationary 900-kg car is struck by a moving 1 800-kg car? Is the final speed the same as before?

Answer   Intuitively, we can guess that the final speed of the combination is higher than 6.67 m/s if the initially moving car is the more massive car. Mathematically, that should be the case because the system has a larger momentum if the initially moving car is the more massive one. Solving for the new final velocity, we find

$$v_f = \dfrac{m_1 v_i}{m_1 + m_2} = \dfrac{(1\,800 \text{ kg})(20.0 \text{ m/s})}{1\,800 \text{ kg} + 900 \text{ kg}} = 13.3 \text{ m/s}$$

which is two times the previous final velocity.

Example **9.6**   The Ballistic Pendulum

The ballistic pendulum (Fig. 9.10) is an apparatus used to measure the speed of a fast-moving projectile such as a bullet. A projectile of mass m_1 is fired into a large block of wood of mass m_2 suspended from some light wires. The projectile embeds in the block, and the entire system swings through a height h. How can we determine the speed of the projectile from a measurement of h?

SOLUTION

Conceptualize   Figure 9.10a helps conceptualize the situation. Run the animation in your mind: the projectile enters the pendulum, which swings up to some height at which it momentarily comes to rest.

9.6 continued

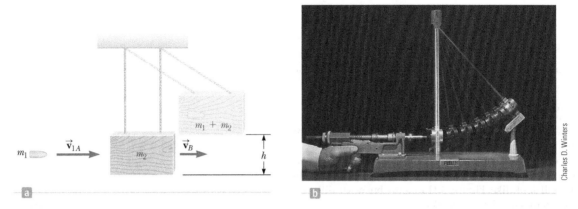

Figure 9.10 (Example 9.6) (a) Diagram of a ballistic pendulum. Notice that $\vec{v}_{1A}$ is the velocity of the projectile immediately before the collision and $\vec{v}_B$ is the velocity of the projectile–block system immediately after the perfectly inelastic collision. (b) Multiflash photograph of a ballistic pendulum used in the laboratory.

Categorize Let's focus first on the collision between the projectile and the block. The projectile and the block form an *isolated system* in terms of *momentum* in the horizontal direction if we identify configuration *A* as immediately before the collision and configuration *B* as immediately after the collision. Because the projectile imbeds in the block, we can categorize the collision between them as perfectly inelastic.

Analyze To analyze the collision, we use Equation 9.15, which gives the speed of the system immediately after the collision when we assume the impulse approximation.

Noting that $v_{2A} = 0$, write Equation 9.15 for v_B:

$$(1) \quad v_B = \frac{m_1 v_{1A}}{m_1 + m_2}$$

Categorize For the second process, during which the projectile–block combination swings upward to height *h* (ending at a configuration we'll call *C*), we focus on a *different* system, that of the projectile, the block, and the Earth. We categorize this part of the problem as one involving an *isolated system* for *energy* with no nonconservative forces acting.

Analyze Write an expression for the total kinetic energy of the system immediately after the collision:

$$(2) \quad K_B = \tfrac{1}{2}(m_1 + m_2)v_B^2$$

Substitute the value of v_B from Equation (1) into Equation (2):

$$K_B = \frac{m_1^2 v_{1A}^2}{2(m_1 + m_2)}$$

This kinetic energy of the system immediately after the collision is *less* than the initial kinetic energy of the projectile as is expected in an inelastic collision.

We define the gravitational potential energy of the system for configuration *B* to be zero. Therefore, $U_B = 0$, whereas $U_C = (m_1 + m_2)gh$.

Apply the isolated system model for energy (Eq. 8.2) to the system:

$$\Delta K + \Delta U = 0 \quad \rightarrow \quad (K_C - K_B) + (U_C - U_B) = 0$$

Substitute the energies:

$$\left[0 - \frac{m_1^2 v_{1A}^2}{2(m_1 + m_2)}\right] + [(m_1 + m_2)gh - 0] = 0$$

Solve for v_{1A}:

$$v_{1A} = \left(\frac{m_1 + m_2}{m_1}\right)\sqrt{2gh}$$

Finalize We had to solve this problem in two steps. Each step involved a different system and a different analysis model: isolated system (momentum) for the first step and isolated system (energy) for the second. Because the collision was assumed to be perfectly inelastic, some mechanical energy was transformed to internal energy during the collision. Therefore, it would have been *incorrect* to apply the isolated system (energy) model to the entire process by equating the initial kinetic energy of the incoming projectile with the final gravitational potential energy of the projectile–block–Earth combination.

Example 9.7   A Two-Body Collision with a Spring

A block of mass $m_1 = 1.60$ kg initially moving to the right with a speed of 4.00 m/s on a frictionless, horizontal track collides with a light spring attached to a second block of mass $m_2 = 2.10$ kg initially moving to the left with a speed of 2.50 m/s as shown in Figure 9.11a. The spring constant is 600 N/m.

(A) Find the velocities of the two blocks when they are again moving separately after the collision.

SOLUTION

Conceptualize With the help of Figure 9.11a, run an ani-mation of the collision in your mind. Figure 9.11b shows an instant during the collision when the spring is compressed. Eventually, block 1 and the spring will again separate, so the system will look like Figure 9.11a again but with different velocity vectors for the two blocks.

Categorize Because the spring force is conservative, kinetic energy in the system of two blocks and the spring is not transformed to internal energy during the compression of the spring. Ignoring any sound made when the block hits the spring, we can categorize the collision as being elastic and categorize the two blocks and the spring as an *isolated system* for both *energy* and *momentum*.

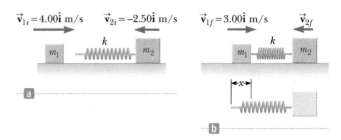

Figure 9.11 (Example 9.7) A moving block approaches a second moving block that is attached to a spring.

Analyze Because momentum of the system is conserved, apply Equation 9.16:

(1) $m_1 v_{1i} + m_2 v_{2i} = m_1 v_{1f} + m_2 v_{2f}$

Because the collision is elastic, apply Equation 9.20:

(2) $v_{1i} - v_{2i} = -(v_{1f} - v_{2f})$

Multiply Equation (2) by m_1:

(3) $m_1 v_{1i} - m_1 v_{2i} = -m_1 v_{1f} + m_1 v_{2f}$

Add Equations (1) and (3):

$2m_1 v_{1i} + (m_2 - m_1)v_{2i} = (m_1 + m_2)v_{2f}$

Solve for v_{2f}:

$$v_{2f} = \frac{2m_1 v_{1i} + (m_2 - m_1)v_{2i}}{m_1 + m_2}$$

Substitute numerical values:

$$v_{2f} = \frac{2(1.60 \text{ kg})(4.00 \text{ m/s}) + (2.10 \text{ kg} - 1.60 \text{ kg})(-2.50 \text{ m/s})}{1.60 \text{ kg} + 2.10 \text{ kg}} = 3.12 \text{ m/s}$$

Solve Equation (2) for v_{1f} and substitute numerical values:

$v_{1f} = v_{2f} - v_{1i} + v_{2i} = 3.12 \text{ m/s} - 4.00 \text{ m/s} + (-2.50 \text{ m/s}) = -3.38 \text{ m/s}$

Finalize Notice that both blocks have reversed direction due to the collision. Also notice that we did not need to know any-thing about the spring to find the answer in this part of the problem. The spring is just another mechanism for the two blocks to exert forces of equal magnitude and opposite direction on one another, just like those between the objects and particles shown in Figure 9.6.

(B) Determine the velocity of block 2 during the collision, at the instant block 1 is moving to the right with a velocity of +3.00 m/s as in Figure 9.11b.

SOLUTION

Conceptualize Focus your attention now on Figure 9.11b, which represents the final configuration of the system at the end of the time interval of interest.

Categorize Because the momentum of the *isolated system* of two blocks and the spring are conserved *throughout* the collision, the collision can be categorized as elastic for *any* final instant of time. Let us now choose the final instant to be when block 1 is moving with a velocity of +3.00 m/s.

9.7 continued

Analyze   Apply Equation 9.16:

$$m_1 v_{1i} + m_2 v_{2i} = m_1 v_{1f} + m_2 v_{2f}$$

Solve for v_{2f}:

$$v_{2f} = \frac{m_1 v_{1i} + m_2 v_{2i} - m_1 v_{1f}}{m_2}$$

Substitute numerical values:

$$v_{2f} = \frac{(1.60 \text{ kg})(4.00 \text{ m/s}) + (2.10 \text{ kg})(-2.50 \text{ m/s}) - (1.60 \text{ kg})(3.00 \text{ m/s})}{2.10 \text{ kg}}$$

$$= -1.74 \text{ m/s}$$

Finalize   The negative value for v_{2f} means that block 2 is still moving to the left at the instant we are considering.

(C) Determine the distance the spring is compressed at that instant.

SOLUTION

Conceptualize   Once again, focus on the configuration of the system shown in Figure 9.11b.

Categorize   For the system of the spring and two blocks, no friction or other nonconservative forces act within the system. Therefore, we categorize the system as an *isolated system* in terms of *energy* with no nonconservative forces acting. The system also remains an *isolated system* in terms of *momentum*.

Analyze   We choose the initial configuration of the system to be that existing immediately before block 1 strikes the spring and the final configuration to be that when block 1 is moving to the right at 3.00 m/s.

Write the appropriate reduction of Equation 8.2:

$$\Delta K + \Delta U = 0$$

Evaluate the energies, recognizing that two objects in the system have kinetic energy and that the potential energy is elastic:

$$[(\tfrac{1}{2} m_1 v_{1f}^2 + \tfrac{1}{2} m_2 v_{2f}^2) - (\tfrac{1}{2} m_1 v_{1i}^2 + \tfrac{1}{2} m_2 v_{2i}^2)] + (\tfrac{1}{2} k x^2 - 0) = 0$$

Solve for x^2:

$$x^2 = \frac{1}{k}[m_1(v_{1i}^2 - v_{1f}^2) + m_2(v_{2i}^2 - v_{2f}^2)]$$

Substitute numerical values:

$$x^2 = \left(\frac{1}{600 \text{ N/m}}\right)\{(1.60 \text{ kg})[(4.00 \text{ m/s})^2 - (3.00 \text{ m/s})^2] + (2.10 \text{ kg})[(2.50 \text{ m/s})^2 - (1.74 \text{ m/s})^2]\}$$

$$\rightarrow x = 0.173 \text{ m}$$

Finalize   This answer is not the maximum compression of the spring because the two blocks are still moving toward each other at the instant shown in Figure 9.11b. Can you determine the maximum compression of the spring?

9.5   Collisions in Two Dimensions

In Section 9.2, we showed that the momentum of a system of two particles is conserved when the system is isolated. For any collision of two particles, this result implies that the momentum in each of the directions x, y, and z is conserved. An important subset of collisions takes place in a plane. The game of billiards is a familiar example involving multiple collisions of objects moving on a two-dimensional surface. For such two-dimensional collisions between two particles, we obtain two component equations for conservation of momentum:

$$m_1 v_{1ix} + m_2 v_{2ix} = m_1 v_{1fx} + m_2 v_{2fx}$$

$$m_1 v_{1iy} + m_2 v_{2iy} = m_1 v_{1fy} + m_2 v_{2fy}$$

where the three subscripts on the velocity components in these equations represent, respectively, the identification of the object (1, 2), initial and final values (i, f), and the velocity component (x, y).

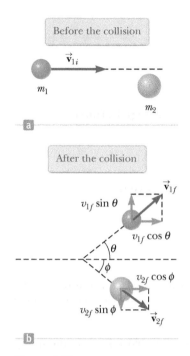

a

b

Figure 9.12 An elastic, glancing collision between two particles.

Let us consider a specific two-dimensional problem in which particle 1 of mass m_1 collides with particle 2 of mass m_2 initially at rest as in Figure 9.12a. After the collision (Fig. 9.12b), particle 1 moves at an angle θ with respect to the horizontal and particle 2 moves at an angle ϕ with respect to the horizontal. This event is called a *glancing* collision. Applying the law of conservation of momentum in component form and noting that the initial y component of the momentum of the two-particle system is zero gives

$$\Delta p_x = 0 \quad \rightarrow \quad p_{ix} = p_{fx} \quad \rightarrow \quad m_1 v_{1i} = m_1 v_{1f} \cos\theta + m_2 v_{2f} \cos\phi \qquad (9.25)$$

$$\Delta p_y = 0 \quad \rightarrow \quad p_{iy} = p_{fy} \quad \rightarrow \quad 0 = m_1 v_{1f} \sin\theta - m_2 v_{2f} \sin\phi \qquad (9.26)$$

where the minus sign in Equation 9.26 is included because after the collision particle 2 has a y component of velocity that is downward. (The symbols v in these particular equations are speeds, not velocity components. The direction of the component vector is indicated explicitly with plus or minus signs.) We now have two independent equations. As long as no more than two of the seven quantities in Equations 9.25 and 9.26 are unknown, we can solve the problem.

If the collision is elastic, we can also use Equation 9.17 (conservation of kinetic energy) with $v_{2i} = 0$:

$$K_i = K_f \quad \rightarrow \quad \tfrac{1}{2} m_1 v_{1i}{}^2 = \tfrac{1}{2} m_1 v_{1f}{}^2 + \tfrac{1}{2} m_2 v_{2f}{}^2 \qquad (9.27)$$

Knowing the initial speed of particle 1 and both masses, we are left with four unknowns (v_{1f}, v_{2f}, θ, and ϕ). Because we have only three equations, one of the four remaining quantities must be given to determine the motion after the elastic collision from conservation principles alone.

If the collision is inelastic, kinetic energy is *not* conserved and Equation 9.27 does *not* apply. Then we have four unknowns and only two equations!

PROBLEM-SOLVING STRATEGY **Two-Dimensional Collisions**

The following procedure is recommended when dealing with problems involving collisions between two particles in two dimensions.

1. Conceptualize. Imagine the collisions occurring and predict the approximate directions in which the particles will move after the collision. Set up a coordinate system and define your velocities in terms of that system. It is convenient to have the x axis coincide with one of the initial velocities. Sketch the coordinate system, draw and label all velocity vectors, and include all the given information.

2. Categorize. Is the system of particles truly isolated? If so, categorize the collision as elastic, inelastic, or perfectly inelastic.

3. Analyze. Write expressions for the x and y components of the momentum of each object before and after the collision. Remember to include the appropriate signs for the components of the velocity vectors and pay careful attention to signs throughout the calculation.

Apply the isolated system model for momentum $\Delta \vec{\mathbf{p}} = 0$. When applied in each direction, this equation will generally reduce to $p_{ix} = p_{fx}$ and $p_{iy} = p_{fy}$, where each of these terms refer to the sum of the momenta of all objects in the system. Write expressions for the *total* momentum in the x direction *before* and *after* the collision and equate the two. Repeat this procedure for the total momentum in the y direction.

Proceed to solve the momentum equations for the unknown quantities. If the collision is inelastic, kinetic energy is *not* conserved and additional information is probably required. If the collision is perfectly inelastic, the final velocities of the two objects are equal.

If the collision is elastic, kinetic energy is conserved and you can equate the total kinetic energy of the system before the collision to the total kinetic energy after the collision, providing an additional relationship between the velocity magnitudes.

4. Finalize. Once you have determined your result, check to see if your answers are consistent with the mental and pictorial representations and that your results are realistic.

Example 9.8 Collision at an Intersection

A 1 500-kg car traveling east with a speed of 25.0 m/s collides at an intersection with a 2 500-kg truck traveling north at a speed of 20.0 m/s as shown in Figure 9.13. Find the direction and magnitude of the velocity of the wreckage after the collision, assuming the vehicles stick together after the collision.

SOLUTION

Conceptualize Figure 9.13 should help you conceptualize the situation before and after the collision. Let us choose east to be along the positive x direction and north to be along the positive y direction.

Categorize Because we consider instants of time immediately before and immediately after the collision as defining our time interval, we ignore the small effect that friction would have on the wheels of the vehicles and model the two vehicles as an *isolated system* in terms of *momentum*. We also ignore the vehicles' sizes and model them as particles. The collision is perfectly inelastic because the car and the truck stick together after the collision.

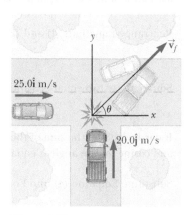

Figure 9.13 (Example 9.8) An eastbound car colliding with a northbound truck.

Analyze Before the collision, the only object having momentum in the x direction is the car. Therefore, the magnitude of the total initial momentum of the system (car plus truck) in the x direction is that of only the car. Similarly, the total initial momentum of the system in the y direction is that of the truck. Immediately after the collision, let us assume the wreckage moves at an angle θ with respect to the x axis with speed v_f.

Apply the isolated system model for momentum in the x direction:

$$\Delta p_x = 0 \quad \rightarrow \quad \sum p_{xi} = \sum p_{xf} \quad \rightarrow \quad (1) \quad m_1 v_{1i} = (m_1 + m_2)v_f \cos\theta$$

Apply the isolated system model for momentum in the y direction:

$$\Delta p_y = 0 \quad \rightarrow \quad \sum p_{yi} = \sum p_{yf} \quad \rightarrow \quad (2) \quad m_2 v_{2i} = (m_1 + m_2)v_f \sin\theta$$

Divide Equation (2) by Equation (1):

$$\frac{m_2 v_{2i}}{m_1 v_{1i}} = \frac{\sin\theta}{\cos\theta} = \tan\theta$$

Solve for θ and substitute numerical values:

$$\theta = \tan^{-1}\left(\frac{m_2 v_{2i}}{m_1 v_{1i}}\right) = \tan^{-1}\left[\frac{(2\,500\text{ kg})(20.0\text{ m/s})}{(1\,500\text{ kg})(25.0\text{ m/s})}\right] = 53.1°$$

Use Equation (2) to find the value of v_f and substitute numerical values:

$$v_f = \frac{m_2 v_{2i}}{(m_1 + m_2)\sin\theta} = \frac{(2\,500\text{ kg})(20.0\text{ m/s})}{(1\,500\text{ kg} + 2\,500\text{ kg})\sin 53.1°} = 15.6\text{ m/s}$$

Finalize Notice that the angle θ is qualitatively in agreement with Figure 9.13. Also notice that the final speed of the combination is less than the initial speeds of the two cars. This result is consistent with the kinetic energy of the system being reduced in an inelastic collision. It might help if you draw the momentum vectors of each vehicle before the collision and the two vehicles together after the collision.

Example 9.9 Proton–Proton Collision

A proton collides elastically with another proton that is initially at rest. The incoming proton has an initial speed of 3.50×10^5 m/s and makes a glancing collision with the second proton as in Figure 9.12. (At close separations, the protons exert a repulsive electrostatic force on each other.) After the collision, one proton moves off at an angle of 37.0° to the original direction of motion and the second deflects at an angle of ϕ to the same axis. Find the final speeds of the two protons and the angle ϕ.

SOLUTION

Conceptualize This collision is like that shown in Figure 9.12, which will help you conceptualize the behavior of the system. We define the x axis to be along the direction of the velocity vector of the initially moving proton.

Categorize The pair of protons form an *isolated system*. Both momentum and kinetic energy of the system are conserved in this glancing elastic collision.

continued

9.9 continued

Analyze Using the isolated system model for both momentum and energy for a two-dimensional elastic collision, set up the mathematical representation with Equations 9.25 through 9.27:

(1) $v_{1i} = v_{1f} \cos \theta + v_{2f} \cos \phi$
(2) $0 = v_{1f} \sin \theta - v_{2f} \sin \phi$
(3) $v_{1i}^2 = v_{1f}^2 + v_{2f}^2$

Rearrange Equations (1) and (2):

$v_{2f} \cos \phi = v_{1i} - v_{1f} \cos \theta$

$v_{2f} \sin \phi = v_{1f} \sin \theta$

Square these two equations and add them:

$v_{2f}^2 \cos^2 \phi + v_{2f}^2 \sin^2 \phi =$
$v_{1i}^2 - 2v_{1i}v_{1f} \cos \theta + v_{1f}^2 \cos^2 \theta + v_{1f}^2 \sin^2 \theta$

Incorporate that the sum of the squares of sine and cosine for *any* angle is equal to 1:

(4) $v_{2f}^2 = v_{1i}^2 - 2v_{1i}v_{1f} \cos \theta + v_{1f}^2$

Substitute Equation (4) into Equation (3):

$v_{1f}^2 + (v_{1i}^2 - 2v_{1i}v_{1f} \cos \theta + v_{1f}^2) = v_{1i}^2$
(5) $v_{1f}^2 - v_{1i}v_{1f} \cos \theta = 0$

One possible solution of Equation (5) is $v_{1f} = 0$, which corresponds to a head-on, one-dimensional collision in which the first proton stops and the second continues with the same speed in the same direction. That is not the solution we want.

Divide both sides of Equation (5) by v_{1f} and solve for the remaining factor of v_{1f}:

$v_{1f} = v_{1i} \cos \theta = (3.50 \times 10^5 \text{ m/s}) \cos 37.0° = 2.80 \times 10^5 \text{ m/s}$

Use Equation (3) to find v_{2f}:

$v_{2f} = \sqrt{v_{1i}^2 - v_{1f}^2} = \sqrt{(3.50 \times 10^5 \text{ m/s})^2 - (2.80 \times 10^5 \text{ m/s})^2}$
$= 2.11 \times 10^5 \text{ m/s}$

Use Equation (2) to find ϕ:

(2) $\phi = \sin^{-1}\left(\frac{v_{1f} \sin \theta}{v_{2f}}\right) = \sin^{-1}\left[\frac{(2.80 \times 10^5 \text{ m/s}) \sin 37.0°}{(2.11 \times 10^5 \text{ m/s})}\right]$
$= 53.0°$

Finalize It is interesting that $\theta + \phi = 90°$. This result is *not* accidental. Whenever two objects of equal mass collide elastically in a glancing collision and one of them is initially at rest, their final velocities are perpendicular to each other.

9.6 The Center of Mass

In this section, we describe the overall motion of a system in terms of a special point called the **center of mass** of the system. The system can be either a small number of distinct particles or an extended, continuous object, such as a gymnast leaping through the air. We shall see that the translational motion of the center of mass of the system is the same as if all the mass of the system were concentrated at that point. That is, the system moves as if the net external force were applied to a single particle located at the center of mass. This model, the *particle model,* was introduced in Chapter 2. This behavior is independent of other motion, such as rotation or vibration of the system or deformation of the system (for instance, when a gymnast folds her body).

Consider a system consisting of a pair of particles that have different masses and are connected by a light, rigid rod (Fig. 9.14). The position of the center of mass of a system can be described as being the *average position* of the system's mass. The center of mass of the system is located somewhere on the line joining the two particles and is closer to the particle having the larger mass. If a single force is applied at a point on the rod above the center of mass, the system rotates clockwise (see Fig. 9.14a). If the force is applied at a point on the rod below the center of mass, the system rotates counterclockwise (see Fig. 9.14b). If the force is applied at the center of mass, the system moves in the direction of the force without rotating (see Fig. 9.14c). The center of mass of an object can be located with this procedure.

The center of mass of the pair of particles described in Figure 9.15 is located on the x axis and lies somewhere between the particles. Its x coordinate is given by

$$x_{CM} \equiv \frac{m_1 x_1 + m_2 x_2}{m_1 + m_2} \tag{9.28}$$

For example, if $x_1 = 0$, $x_2 = d$, and $m_2 = 2m_1$, we find that $x_{CM} = \frac{2}{3}d$. That is, the center of mass lies closer to the more massive particle. If the two masses are equal, the center of mass lies midway between the particles.

We can extend this concept to a system of many particles with masses m_i in three dimensions. The x coordinate of the center of mass of n particles is defined to be

$$x_{CM} \equiv \frac{m_1 x_1 + m_2 x_2 + m_3 x_3 + \cdots + m_n x_n}{m_1 + m_2 + m_3 + \cdots + m_n} = \frac{\sum\limits_i m_i x_i}{\sum\limits_i m_i} = \frac{\sum\limits_i m_i x_i}{M} = \frac{1}{M}\sum\limits_i m_i x_i \tag{9.29}$$

where x_i is the x coordinate of the ith particle and the total mass is $M \equiv \sum_i m_i$ where the sum runs over all n particles. The y and z coordinates of the center of mass are similarly defined by the equations

$$y_{CM} \equiv \frac{1}{M}\sum\limits_i m_i y_i \quad \text{and} \quad z_{CM} \equiv \frac{1}{M}\sum\limits_i m_i z_i \tag{9.30}$$

The center of mass can be located in three dimensions by its position vector $\vec{r}_{CM}$. The components of this vector are x_{CM}, y_{CM}, and z_{CM}, defined in Equations 9.29 and 9.30. Therefore,

$$\vec{r}_{CM} = x_{CM}\hat{\mathbf{i}} + y_{CM}\hat{\mathbf{j}} + z_{CM}\hat{\mathbf{k}} = \frac{1}{M}\sum\limits_i m_i x_i \hat{\mathbf{i}} + \frac{1}{M}\sum\limits_i m_i y_i \hat{\mathbf{j}} + \frac{1}{M}\sum\limits_i m_i z_i \hat{\mathbf{k}}$$

$$\vec{r}_{CM} \equiv \frac{1}{M}\sum\limits_i m_i \vec{r}_i \tag{9.31}$$

where $\vec{r}_i$ is the position vector of the ith particle, defined by

$$\vec{r}_i \equiv x_i\hat{\mathbf{i}} + y_i\hat{\mathbf{j}} + z_i\hat{\mathbf{k}}$$

Although locating the center of mass for an extended, continuous object is somewhat more cumbersome than locating the center of mass of a small number of particles, the basic ideas we have discussed still apply. Think of an extended object as a system containing a large number of small mass elements such as the cube in Figure 9.16 (page 232). Because the separation between elements is very small, the object can be considered to have a continuous mass distribution. By dividing the object into elements of mass Δm_i with coordinates x_i, y_i, z_i, we see that the x coordinate of the center of mass is approximately

$$x_{CM} \approx \frac{1}{M}\sum\limits_i x_i \Delta m_i$$

with similar expressions for y_{CM} and z_{CM}. If we let the number of elements n approach infinity, the size of each element approaches zero and x_{CM} is given precisely. In this limit, we replace the sum by an integral and Δm_i by the differential element dm:

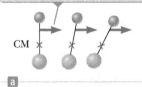

The system rotates clockwise when a force is applied above the center of mass.

CM

a

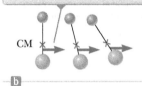

The system rotates counterclockwise when a force is applied below the center of mass.

CM

b

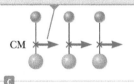

The system moves in the direction of the force without rotating when a force is applied at the center of mass.

CM

c

Figure 9.14 A force is applied to a system of two particles of unequal mass connected by a light, rigid rod.

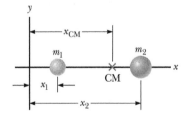

Figure 9.15 The center of mass of two particles of unequal mass on the x axis is located at x_{CM}, a point between the particles, closer to the one having the larger mass.

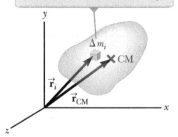

An extended object can be considered to be a distribution of small elements of mass Δm_i.

Figure 9.16 The center of mass is located at the vector position $\vec{\mathbf{r}}_{CM}$, which has coordinates x_{CM}, y_{CM}, and z_{CM}.

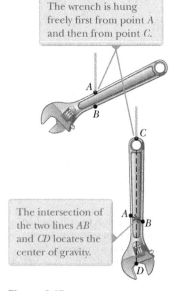

The wrench is hung freely first from point A and then from point C.

The intersection of the two lines AB and CD locates the center of gravity.

Figure 9.17 An experimental technique for determining the center of gravity of a wrench.

$$x_{CM} = \lim_{\Delta m_i \to 0} \frac{1}{M} \sum_i x_i \, \Delta m_i = \frac{1}{M} \int x \, dm \qquad (9.32)$$

Likewise, for y_{CM} and z_{CM} we obtain

$$y_{CM} = \frac{1}{M} \int y \, dm \quad \text{and} \quad z_{CM} = \frac{1}{M} \int z \, dm \qquad (9.33)$$

We can express the vector position of the center of mass of an extended object in the form

$$\vec{\mathbf{r}}_{CM} = \frac{1}{M} \int \vec{\mathbf{r}} \, dm \qquad (9.34)$$

which is equivalent to the three expressions given by Equations 9.32 and 9.33.

The center of mass of any symmetric object of uniform density lies on an axis of symmetry and on any plane of symmetry. For example, the center of mass of a uniform rod lies in the rod, midway between its ends. The center of mass of a sphere or a cube lies at its geometric center.

Because an extended object is a continuous distribution of mass, each small mass element is acted upon by the gravitational force. The net effect of all these forces is equivalent to the effect of a single force $M\vec{\mathbf{g}}$ acting through a special point, called the **center of gravity.** If $\vec{\mathbf{g}}$ is constant over the mass distribution, the center of gravity coincides with the center of mass. If an extended object is pivoted at its center of gravity, it balances in any orientation.

The center of gravity of an irregularly shaped object such as a wrench can be determined by suspending the object first from one point and then from another. In Figure 9.17, a wrench is hung from point A and a vertical line AB (which can be established with a plumb bob) is drawn when the wrench has stopped swinging. The wrench is then hung from point C, and a second vertical line CD is drawn. The center of gravity is halfway through the thickness of the wrench, under the intersection of these two lines. In general, if the wrench is hung freely from any point, the vertical line through this point must pass through the center of gravity.

QUICK QUIZ 9.7 A baseball bat of uniform density is cut at the location of its center of mass as shown in Figure 9.18. Which piece has the smaller mass? **(a)** the piece on the right **(b)** the piece on the left **(c)** both pieces have the same mass **(d)** impossible to determine

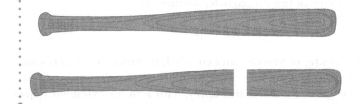

Figure 9.18 (Quick Quiz 9.7) A baseball bat cut at the location of its center of mass.

Example 9.10 **The Center of Mass of Three Particles**

A system consists of three particles located as shown in Figure 9.19. Find the center of mass of the system. The masses of the particles are $m_1 = m_2 = 1.0$ kg and $m_3 = 2.0$ kg.

SOLUTION

Conceptualize Figure 9.19 shows the three masses. Your intuition should tell you that the center of mass is located somewhere in the region between the blue particle and the pair of tan particles as shown in the figure.

Figure 9.19 (Example 9.10) Two particles are located on the x axis, and a single particle is located on the y axis as shown. The vector indicates the location of the system's center of mass.

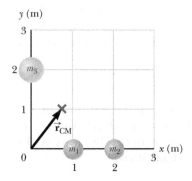

9.10 continued

Categorize We categorize this example as a substitution problem because we will be using the equations for the center of mass developed in this section.

Use the defining equations for the coordinates of the center of mass and notice that $z_{CM} = 0$:

$$x_{CM} = \frac{1}{M} \sum_i m_i x_i = \frac{m_1 x_1 + m_2 x_2 + m_3 x_3}{m_1 + m_2 + m_3}$$

$$= \frac{(1.0 \text{ kg})(1.0 \text{ m}) + (1.0 \text{ kg})(2.0 \text{ m}) + (2.0 \text{ kg})(0)}{1.0 \text{ kg} + 1.0 \text{ kg} + 2.0 \text{ kg}} = \frac{3.0 \text{ kg} \cdot \text{m}}{4.0 \text{ kg}} = 0.75 \text{ m}$$

$$y_{CM} = \frac{1}{M} \sum_i m_i y_i = \frac{m_1 y_1 + m_2 y_2 + m_3 y_3}{m_1 + m_2 + m_3}$$

$$= \frac{(1.0 \text{ kg})(0) + (1.0 \text{ kg})(0) + (2.0 \text{ kg})(2.0 \text{ m})}{4.0 \text{ kg}} = \frac{4.0 \text{ kg} \cdot \text{m}}{4.0 \text{ kg}} = 1.0 \text{ m}$$

Write the position vector of the center of mass:

$$\vec{\mathbf{r}}_{CM} \equiv x_{CM} \hat{\mathbf{i}} + y_{CM} \hat{\mathbf{j}} = (0.75 \hat{\mathbf{i}} + 1.0 \hat{\mathbf{j}}) \text{ m}$$

Example **9.11** **The Center of Mass of a Rod**

(A) Show that the center of mass of a rod of mass M and length L lies midway between its ends, assuming the rod has a uniform mass per unit length.

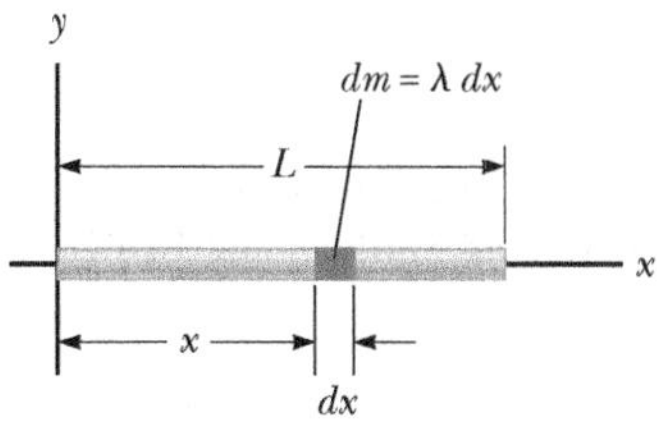

SOLUTION

Conceptualize The rod is shown aligned along the x axis in Figure 9.20, so $y_{CM} = z_{CM} = 0$. What is your prediction of the value of x_{CM}?

Categorize We categorize this example as an analysis problem because we need to divide the rod into small mass elements to perform the integration in Equation 9.32.

Figure 9.20 (Example 9.11) The geometry used to find the center of mass of a uniform rod.

Analyze The mass per unit length (this quantity is called the *linear mass density*) can be written as $\lambda = M/L$ for the uniform rod. If the rod is divided into elements of length dx, the mass of each element is $dm = \lambda \, dx$.

Use Equation 9.32 to find an expression for x_{CM}:

$$x_{CM} = \frac{1}{M} \int x \, dm = \frac{1}{M} \int_0^L x\lambda \, dx = \frac{\lambda}{M} \frac{x^2}{2} \Big|_0^L = \frac{\lambda L^2}{2M}$$

Substitute $\lambda = M/L$:

$$x_{CM} = \frac{L^2}{2M} \left(\frac{M}{L} \right) = \tfrac{1}{2} L$$

One can also use symmetry arguments to obtain the same result.

(B) Suppose a rod is *nonuniform* such that its mass per unit length varies linearly with x according to the expression $\lambda = \alpha x$, where α is a constant. Find the x coordinate of the center of mass as a fraction of L.

SOLUTION

Conceptualize Because the mass per unit length is not constant in this case but is proportional to x, elements of the rod to the right are more massive than elements near the left end of the rod.

Categorize This problem is categorized similarly to part (A), with the added twist that the linear mass density is not constant.

Analyze We replace dm in Equation 9.32 by $\lambda \, dx$, where, in this case, $\lambda = \alpha x$.

Use Equation 9.32 to find an expression for x_{CM}:

$$x_{CM} = \frac{1}{M} \int x \, dm = \frac{1}{M} \int_0^L x\lambda \, dx = \frac{1}{M} \int_0^L x\alpha x \, dx$$

$$= \frac{\alpha}{M} \int_0^L x^2 \, dx = \frac{\alpha L^3}{3M}$$

continued

9.11 continued

Find the total mass of the rod:

$$M = \int dm = \int_0^L \lambda \, dx = \int_0^L \alpha x \, dx = \frac{\alpha L^2}{2}$$

Substitute M into the expression for x_{CM}:

$$x_{CM} = \frac{\alpha L^3}{3\alpha L^2/2} = \tfrac{2}{3}L$$

Finalize   Notice that the center of mass in part (B) is farther to the right than that in part (A). That result is reasonable because the elements of the rod become more massive as one moves to the right along the rod in part (B).

9.7 Systems of Many Particles

Consider a system of two or more particles for which we have identified the center of mass. We can begin to further understand the physical significance and utility of the center of mass concept by taking the time derivative of the position vector for the center of mass given by Equation 9.31. From Section 4.1, we know that the time derivative of a position vector is by definition the velocity vector. Assuming M remains constant for a system of particles—that is, no particles enter or leave the system—we obtain the following expression for the **velocity of the center of mass** of the system:

◀ Velocity of the center of mass of a system of particles

$$\vec{\mathbf{v}}_{CM} = \frac{d\vec{\mathbf{r}}_{CM}}{dt} = \frac{1}{M}\sum_i m_i \frac{d\vec{\mathbf{r}}_i}{dt} = \frac{1}{M}\sum_i m_i \vec{\mathbf{v}}_i \tag{9.35}$$

where $\vec{\mathbf{v}}_i$ is the velocity of the ith particle. Rearranging Equation 9.35 gives

◀ Total momentum of a system of particles

$$M\vec{\mathbf{v}}_{CM} = \sum_i m_i \vec{\mathbf{v}}_i = \sum_i \vec{\mathbf{p}}_i = \vec{\mathbf{p}}_{tot} \tag{9.36}$$

Therefore, the total linear momentum of the system equals the total mass multiplied by the velocity of the center of mass. In other words, the total linear momentum of the system is equal to that of a single particle of mass M moving with a velocity $\vec{\mathbf{v}}_{CM}$.

Differentiating Equation 9.35 with respect to time, we obtain the **acceleration of the center of mass** of the system:

◀ Acceleration of the center of mass of a system of particles

$$\vec{\mathbf{a}}_{CM} = \frac{d\vec{\mathbf{v}}_{CM}}{dt} = \frac{1}{M}\sum_i m_i \frac{d\vec{\mathbf{v}}_i}{dt} = \frac{1}{M}\sum_i m_i \vec{\mathbf{a}}_i \tag{9.37}$$

Rearranging this expression and using Newton's second law gives

$$M\vec{\mathbf{a}}_{CM} = \sum_i m_i \vec{\mathbf{a}}_i = \sum_i \vec{\mathbf{F}}_i \tag{9.38}$$

where $\vec{\mathbf{F}}_i$ is the net force on particle i.

The forces on any particle in the system may include both external forces (from outside the system) and internal forces (from within the system). By Newton's third law, however, the internal force exerted by particle 1 on particle 2, for example, is equal in magnitude and opposite in direction to the internal force exerted by particle 2 on particle 1. Therefore, when we sum over all internal force vectors in Equation 9.38, they cancel in pairs and we find that the net force on the system is caused *only* by external forces. We can then write Equation 9.38 in the form

◀ Newton's second law for a system of particles

$$\sum \vec{\mathbf{F}}_{ext} = M\vec{\mathbf{a}}_{CM} \tag{9.39}$$

That is, the net external force on a system of particles equals the total mass of the system multiplied by the acceleration of the center of mass. Comparing Equation 9.39 with Newton's second law for a single particle, we see that the particle model we have used in several chapters can be described in terms of the center of mass:

> The center of mass of a system of particles having combined mass M moves like an equivalent single particle of mass M would move under the influence of the net external force on the system.

Let us integrate Equation 9.39 over a finite time interval:

$$\int \sum \vec{\mathbf{F}}_{ext}\, dt = \int M\vec{\mathbf{a}}_{CM}\, dt = \int M\frac{d\vec{\mathbf{v}}_{CM}}{dt}\, dt = M \int d\vec{\mathbf{v}}_{CM} = M\Delta\vec{\mathbf{v}}_{CM}$$

Notice that this equation can be written as

$$\Delta\vec{\mathbf{p}}_{tot} = \vec{\mathbf{I}} \tag{9.40}$$

◀ Impulse–momentum theorem for a system of particles

where $\vec{\mathbf{I}}$ is the impulse imparted to the system by external forces and $\vec{\mathbf{p}}_{tot}$ is the momentum of the system. Equation 9.40 is the generalization of the impulse–momentum theorem for a particle (Eq. 9.13) to a system of many particles. It is also the mathematical representation of the nonisolated system (momentum) model for a system of many particles.

Finally, if the net external force on a system is zero so that the system is isolated, it follows from Equation 9.39 that

$$M\vec{\mathbf{a}}_{CM} = M\frac{d\vec{\mathbf{v}}_{CM}}{dt} = 0$$

Therefore, the isolated system model for momentum for a system of many particles is described by

$$\Delta\vec{\mathbf{p}}_{tot} = 0 \tag{9.41}$$

which can be rewritten as

$$M\vec{\mathbf{v}}_{CM} = \vec{\mathbf{p}}_{tot} = \text{constant} \quad (\text{when } \sum \vec{\mathbf{F}}_{ext} = 0) \tag{9.42}$$

That is, the total linear momentum of a system of particles is conserved if no net external force is acting on the system. It follows that for an isolated system of particles, both the total momentum and the velocity of the center of mass are constant in time. This statement is a generalization of the isolated system (momentum) model for a many-particle system.

Suppose the center of mass of an isolated system consisting of two or more members is at rest. The center of mass of the system remains at rest if there is no net force on the system. For example, consider a system of a swimmer standing on a raft, with the system initially at rest. When the swimmer dives horizontally off the raft, the raft moves in the direction opposite that of the swimmer and the center of mass of the system remains at rest (if we neglect friction between raft and water). Furthermore, the linear momentum of the diver is equal in magnitude to that of the raft, but opposite in direction.

QUICK QUIZ 9.8 A cruise ship is moving at constant speed through the water. The vacationers on the ship are eager to arrive at their next destination. They decide to try to speed up the cruise ship by gathering at the bow (the front) and running together toward the stern (the back) of the ship. **(i)** While they are running toward the stern, is the speed of the ship (a) higher than it was before, (b) unchanged, (c) lower than it was before, or (d) impossible to determine? **(ii)** The vacationers stop running when they reach the stern of the ship. After they have all stopped running, is the speed of the ship (a) higher than it was before they started running, (b) unchanged from what it was before they started running, (c) lower than it was before they started running, or (d) impossible to determine?

Conceptual Example 9.12   Exploding Projectile

A projectile fired into the air suddenly explodes into several fragments (Fig. 9.21).

(A) What can be said about the motion of the center of mass of the system made up of all the fragments after the explosion?

SOLUTION

Neglecting air resistance, the only external force on the projectile is the gravitational force. Therefore, if the projectile did not explode, it would continue to move along the parabolic path indicated by the dashed line in Figure 9.21. Because the forces caused by the explosion are internal, they do not affect the motion of the center of mass of the system (the fragments). Therefore, after the explosion, the center of mass of the fragments follows the same parabolic path the projectile would have followed if no explosion had occurred.

Figure 9.21 (Conceptual Example 9.12) When a projectile explodes into several fragments, the center of mass of the system made up of all the fragments follows the same parabolic path the projectile would have taken had there been no explosion.

(B) If the projectile did not explode, it would land at a distance R from its launch point. Suppose the projectile explodes and splits into two pieces of equal mass. One piece lands at a distance $2R$ to the right of the launch point. Where does the other piece land?

SOLUTION

As discussed in part (A), the center of mass of the two-piece system lands at a distance R from the launch point. One of the pieces lands at a farther distance R from the landing point (or a distance $2R$ from the launch point), to the right in Figure 9.21. Because the two pieces have the same mass, the other piece must land a distance R to the left of the landing point in Figure 9.21, which places this piece right back at the launch point!

Example 9.13   The Exploding Rocket

A rocket is fired vertically upward. At the instant it reaches an altitude of 1 000 m and a speed of $v_i = 300$ m/s, it explodes into three fragments having equal mass. One fragment moves upward with a speed of $v_1 = 450$ m/s following the explosion. The second fragment has a speed of $v_2 = 240$ m/s and is moving east right after the explosion. What is the velocity of the third fragment immediately after the explosion?

SOLUTION

Conceptualize   Picture the explosion in your mind, with one piece going upward and a second piece moving horizontally toward the east. Do you have an intuitive feeling about the direction in which the third piece moves?

Categorize   This example is a two-dimensional problem because we have two fragments moving in perpendicular directions after the explosion as well as a third fragment moving in an unknown direction in the plane defined by the velocity vectors of the other two fragments. We assume the time interval of the explosion is very short, so we use the impulse approximation in which we ignore the gravitational force and air resistance. Because of the short time interval and the ignoring of external forces, the center of mass of the system remains fixed in space during the explosion. Therefore, the rocket is an *isolated system* in terms of *momentum*. Equation 9.41 describes the situation, and the total momentum $\vec{\mathbf{p}}_i$ of the rocket immediately before the explosion must equal the total momentum $\vec{\mathbf{p}}_f$ of the fragments immediately after the explosion.

Analyze   Because the three fragments have equal mass, the mass of each fragment is $M/3$, where M is the total mass of the rocket. We will let $\vec{\mathbf{v}}_3$ represent the unknown velocity of the third fragment.

Use the isolated system (momentum) model to equate the initial and final momenta of the system and express the momenta in terms of masses and velocities:

$$\Delta \vec{\mathbf{p}} = 0 \quad \rightarrow \quad \vec{\mathbf{p}}_i = \vec{\mathbf{p}}_f \quad \rightarrow \quad M\vec{\mathbf{v}}_i = \frac{M}{3}\vec{\mathbf{v}}_1 + \frac{M}{3}\vec{\mathbf{v}}_2 + \frac{M}{3}\vec{\mathbf{v}}_3$$

Solve for $\vec{\mathbf{v}}_3$:

$$\vec{\mathbf{v}}_3 = 3\vec{\mathbf{v}}_i - \vec{\mathbf{v}}_1 - \vec{\mathbf{v}}_2$$

Substitute the numerical values:

$$\vec{\mathbf{v}}_3 = 3(300\hat{\mathbf{j}} \text{ m/s}) - (450\hat{\mathbf{j}} \text{ m/s}) - (240\hat{\mathbf{i}} \text{ m/s}) = (-240\hat{\mathbf{i}} + 450\hat{\mathbf{j}}) \text{ m/s}$$

Finalize Notice that this event is the reverse of a perfectly inelastic collision. There is one object before the collision and three objects afterward. Imagine running a movie of the event backward: the three objects would come together and become a single object. In a perfectly inelastic collision, the kinetic energy of the system decreases. If you were to calculate the kinetic energy before and after the event in this example, you would find that the kinetic energy of the system increases. (Try it!) This increase in kinetic energy comes from the potential energy stored in whatever fuel exploded to cause the breakup of the rocket.

9.8 Deformable Systems

So far in our discussion of mechanics, we have analyzed the motion of particles or nondeformable objects that can be modeled as particles. The discussion in Section 9.7 can be applied to an analysis of the motion of deformable systems. For example, suppose you stand on a skateboard and push off a wall, setting yourself in motion away from the wall. Your body has deformed during this event: your arms were bent before the event, and they straightened out while you pushed off the wall. How would we describe this event?

The force from the wall on your hands moves through no displacement; the force is always located at the interface between the wall and your hands. Therefore, the force does no work on the system, which is you and your skateboard. Pushing off the wall, however, does indeed result in a change in the kinetic energy of the system. If you try to use the work–kinetic energy theorem, $W = \Delta K$, to describe this event, you will notice that the left side of the equation is zero but the right side is not zero. The work–kinetic energy theorem is not valid for this event and is often not valid for systems that are deformable.

To analyze the motion of deformable systems, we appeal to Equation 8.2, the conservation of energy equation, and Equation 9.40, the impulse–momentum theorem. For the example of you pushing off the wall on your skateboard, identifying the system as you and the skateboard, Equation 8.2 gives

$$\Delta K + \Delta U = 0$$

where ΔK is the change in kinetic energy, which is related to the increased speed of the system, and ΔU is the decrease in potential energy stored in your body from previous meals. This equation tells us that the system transformed potential energy in your body into kinetic energy by virtue of the muscular exertion necessary to push off the wall. Notice that the system is isolated in terms of energy but nonisolated in terms of momentum.

Applying Equation 9.40 to the system in this situation gives us

$$\Delta \vec{\mathbf{p}}_{\text{tot}} = \vec{\mathbf{I}} \quad \rightarrow \quad m \, \Delta \vec{\mathbf{v}} = \int \vec{\mathbf{F}}_{\text{wall}} \, dt$$

where $\vec{\mathbf{F}}_{\text{wall}}$ is the force exerted by the wall on your hands, m is the mass of you and the skateboard, and $\Delta \vec{\mathbf{v}}$ is the change in the velocity of the system during the event. To evaluate the right side of this equation, we would need to know how the force from the wall varies in time. In general, this process might be complicated. In the case of constant forces, or well-behaved forces, however, the integral on the right side of the equation can be evaluated.

Deformable systems occur often in common situations. Any time you run or jump, your body is a deformable system. A gymnast or a platform diver performing a routine is a deformable system. In Example 9.14 (page 238), we investigate a deformable system with two blocks and a spring. Beginning in Chapter 18, we will look at very important deformable systems: samples of gas changing in size as they undergo thermodynamic processes.

| Example **9.14** | **Pushing on a Spring**[3] |

As shown in Figure 9.22a, two blocks are at rest on a frictionless, level table. Both blocks have the same mass m, and they are connected by a spring of negligible mass. The separation distance of the blocks when the spring is relaxed is L. During a time interval Δt, a constant force of magnitude F is applied horizontally to the left block, moving it through a distance x_1 as shown in Figure 9.22b. During this time interval, the right block moves through a distance x_2. At the end of this time interval, the force F is removed.

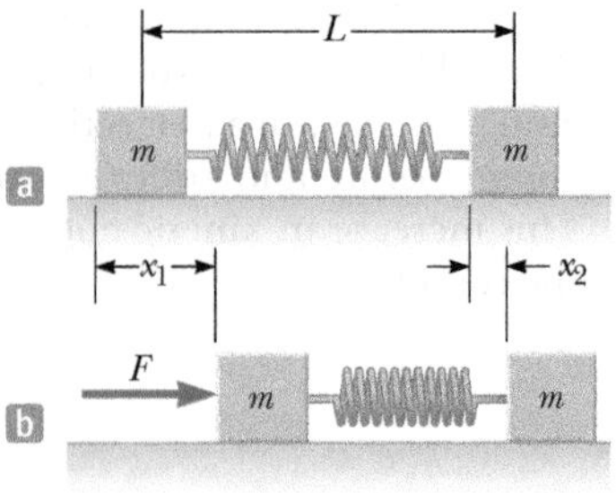

(A) Find the resulting speed $\vec{v}_{CM}$ of the center of mass of the system.

Figure 9.22 (Example 9.14) (a) Two blocks of equal mass are connected by a spring. (b) The left block is pushed with a constant force of magnitude F and moves a distance x_1 during some time interval. During this same time interval, the right block moves through a distance x_2.

SOLUTION

Conceptualize   Imagine what happens as you push on the left block. It begins to move to the right in Figure 9.22, and the spring begins to compress. As a result, the spring pushes to the right on the right block, which begins to move to the right. At any given time, the blocks are generally moving with different velocities. As the center of mass of the system moves to the right with a constant speed after the force is removed, the two blocks oscillate back and forth with respect to the center of mass.

Categorize   We apply three analysis models in this problem: the deformable system of two blocks and a spring is modeled as a *nonisolated system* in terms of *energy* because work is being done on it by the applied force. It is also modeled as a *nonisolated system* in terms of *momentum* because of the force acting on the system during a time interval. Because the applied force on the system is constant, the acceleration of its center of mass is constant and the center of mass is modeled as a *particle under constant acceleration*.

Analyze   Using the nonisolated system (momentum) model, we apply the impulse–momentum theorem to the system of two blocks, recognizing that the force F is constant during the time interval Δt while the force is applied.

Write Equation 9.40 for the system:

$$\Delta p_x = I_x \quad \rightarrow \quad (2m)(v_{CM} - 0) = F\,\Delta t$$

$$(1) \quad 2m v_{CM} = F\,\Delta t$$

During the time interval Δt, the center of mass of the system moves a distance $\frac{1}{2}(x_1 + x_2)$. Use this fact to express the time interval in terms of $v_{CM,avg}$:

$$\Delta t = \frac{\frac{1}{2}(x_1 + x_2)}{v_{CM,avg}}$$

Because the center of mass is modeled as a particle under constant acceleration, the average velocity of the center of mass is the average of the initial velocity, which is zero, and the final velocity v_{CM}:

$$\Delta t = \frac{\frac{1}{2}(x_1 + x_2)}{\frac{1}{2}(0 + v_{CM})} = \frac{(x_1 + x_2)}{v_{CM}}$$

Substitute this expression into Equation (1):

$$2m v_{CM} = F\,\frac{(x_1 + x_2)}{v_{CM}}$$

Solve for v_{CM}:

$$v_{CM} = \sqrt{F\,\frac{(x_1 + x_2)}{2m}}$$

(B) Find the total energy of the system associated with vibration relative to its center of mass after the force F is removed.

SOLUTION

Analyze   The vibrational energy is all the energy of the system other than the kinetic energy associated with translational motion of the center of mass. To find the vibrational energy, we apply the conservation of energy equation (Eq. 8.2). The kinetic energy of the system can be expressed as $K = K_{CM} + K_{vib}$, where K_{vib} is the kinetic energy of the blocks relative to the center of mass due to their vibration. The potential energy of the system is U_{vib}, which is the potential energy stored in the spring when the separation of the blocks is some value other than L.

---

[3]Example 9.14 was inspired in part by C. E. Mungan, "A primer on work–energy relationships for introductory physics," *The Physics Teacher* **43**:10, 2005.

9.14 continued

From the nonisolated system (energy) model, express Equation 8.2 for this system:	(2) $\Delta K_{CM} + \Delta K_{vib} + \Delta U_{vib} = W$
Express Equation (2) in an alternate form, noting that $K_{vib} + U_{vib} = E_{vib}$:	$\Delta K_{CM} + \Delta E_{vib} = W$
Substitute for each of the terms in this equation:	$(K_{CM} - 0) + (E_{vib} - 0) = Fx_1 \;\rightarrow\; E_{vib} = Fx_1 - K_{CM}$
Use the result from part (A):	$E_{vib} = Fx_1 - \tfrac{1}{2}(2m)v_{CM}^2 = Fx_1 - \tfrac{1}{2}(2m)\left[F\dfrac{(x_1 + x_2)}{2m}\right] = F\dfrac{(x_1 - x_2)}{2}$

Finalize Neither of the two answers in this example depends on the spring length, the spring constant, or the time interval. Notice also that the magnitude x_1 of the displacement of the point of application of the applied force is different from the magnitude $\tfrac{1}{2}(x_1 + x_2)$ of the displacement of the center of mass of the system. This difference reminds us that the displacement in the definition of work (Eq. 7.1) is that of the point of application of the force.

9.9 Rocket Propulsion

When ordinary vehicles such as cars are propelled, the driving force for the motion is friction. In the case of the car, the driving force is the force exerted by the road on the car. We can model the car as a nonisolated system in terms of momentum. An impulse is applied to the car from the roadway, and the result is a change in the momentum of the car as described by Equation 9.40.

A rocket moving in space, however, has no road to push against. The rocket is an isolated system in terms of momentum. Therefore, the source of the propulsion of a rocket must be something other than an external force. The operation of a rocket depends on the law of conservation of linear momentum as applied to an isolated system, where the system is the rocket plus its ejected fuel.

Rocket propulsion can be understood by first considering our archer standing on frictionless ice in Example 9.1. Imagine the archer fires several arrows horizontally. For each arrow fired, the archer receives a compensating momentum in the opposite direction. As more arrows are fired, the archer moves faster and faster across the ice. In addition to this analysis in terms of momentum, we can also understand this phenomenon in terms of Newton's second and third laws. Every time the bow pushes an arrow forward, the arrow pushes the bow (and the archer) backward, and these forces result in an acceleration of the archer. Figure 9.23a shows this mechanism used for maneuvering an astronaut in space. Instead of firing arrows like the archer, the astronaut fires short bursts of nitrogen gas.

In a similar manner, as a rocket moves in free space, its linear momentum changes when some of its mass is ejected in the form of exhaust gases. Because the gases are given momentum when they are ejected out of the engine, the rocket receives a compensating momentum in the opposite direction. Therefore, the rocket is accelerated as a result of the "push," or thrust, from the exhaust gases. In free space, the center of mass of the system (rocket plus expelled gases) moves uniformly, independent of the propulsion process.[4]

Suppose at some time t the magnitude of the momentum of a rocket plus its fuel is Mv, where v is the speed of the rocket relative to the Earth (Fig. 9.23b). Over a short time interval Δt, the rocket ejects fuel of mass Δm. At the end of this interval, the rocket's mass is $M - \Delta m$ and its speed is $v + \Delta v$, where Δv is the change in speed of

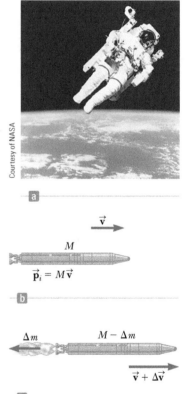

Courtesy of NASA

a

$\vec{v}$

M

$\vec{p}_i = M\vec{v}$

b

Δm $M - \Delta m$

$\vec{v} + \Delta\vec{v}$

c

Figure 9.23 Rocket propulsion. (a) The force from a nitrogen-propelled, hand-controlled device allows an astronaut to move about freely in space without restrictive tethers, using the thrust force from the expelled nitrogen. (b) The initial mass of a rocket plus all its fuel is M at a time t, and its speed is v. (c) At a time $t + \Delta t$, the rocket's mass has been reduced to $M - \Delta m$ and an amount of fuel Δm has been ejected. The rocket's speed increases by an amount Δv.

---

[4]The rocket and the archer represent cases of the reverse of a perfectly inelastic collision: momentum is conserved, but the kinetic energy of the rocket–exhaust gas system increases (at the expense of chemical potential energy in the fuel), as does the kinetic energy of the archer–arrow system (at the expense of potential energy from the archer's previous meals when he pulls back on the bowstring and stretches it).

the rocket (Fig. 9.23c). If the fuel is ejected with a speed v_e relative to the rocket (the subscript e stands for *exhaust*, and v_e is usually called the *exhaust speed*), the velocity of the fuel relative to the Earth is $v - v_e$. Because the system of the rocket and the ejected fuel is isolated, we apply the isolated system model for momentum and obtain

$$\Delta p = 0 \quad \rightarrow \quad p_i = p_f \quad \rightarrow \quad Mv = (M - \Delta m)(v + \Delta v) + \Delta m(v - v_e)$$

Simplifying this expression gives

$$M\Delta v - \Delta m\Delta v = v_e \, \Delta m \tag{9.43}$$

Solving for the change in speed, we find

$$\Delta v = \frac{v_e \Delta m}{M - \Delta m} \tag{9.44}$$

This equation is valid for a one-time ejection of mass from the rocket. It is also valid for any situation in which an object ejects mass, causing the object to move in the opposite direction. The equation can be applied to the archer problem in Example 9.1, recognizing that the initial mass of the system was that of both the archer and the arrow, $M = 60.030$ kg.

If we now take the limit as Δt goes to zero, we let $\Delta v \rightarrow dv$ and $\Delta m \rightarrow dm$ in Equation 9.43. In addition, we ignore the term $dm \, dv$ because this product of two infinitesimal quantities is much smaller than the other terms in the equation. Furthermore, the increase in the exhaust mass dm corresponds to an equal decrease in the rocket mass, so $dm = -dM$. Using this fact gives

$$M \, dv = v_e \, dm = -v_e \, dM \tag{9.45}$$

Now divide the equation by M and integrate, taking the initial mass of the rocket plus fuel to be M_i and the final mass of the rocket plus its remaining fuel to be M_f. The result is

$$\int_{v_i}^{v_f} dv = -v_e \int_{M_i}^{M_f} \frac{dM}{M}$$

◀ Expression for rocket propulsion

$$v_f - v_i = v_e \ln\left(\frac{M_i}{M_f}\right) \tag{9.46}$$

which is the basic expression for rocket propulsion. First, Equation 9.46 tells us that the increase in rocket speed is proportional to the exhaust speed v_e of the ejected gases. Therefore, the exhaust speed should be very high. Second, the increase in rocket speed is proportional to the natural logarithm of the ratio M_i/M_f. Therefore, this ratio should be as large as possible; that is, the mass of the rocket without its fuel should be as small as possible and the rocket should carry as much fuel as possible.

The **thrust** on the rocket is the force exerted on it by the ejected exhaust gases. We obtain the following expression for the thrust from Newton's second law and Equation 9.45:

$$\text{Thrust} = M \frac{dv}{dt} = \left| v_e \frac{dM}{dt} \right| \tag{9.47}$$

This expression shows that the thrust increases as the exhaust speed increases and as the rate of change of mass (called the *burn rate*) increases.

Example 9.15 A Rocket in Space

A rocket moving in space, far from all other objects, has a speed of 3.0×10^3 m/s relative to the Earth. Its engines are turned on, and fuel is ejected in a direction opposite the rocket's motion at a speed of 5.0×10^3 m/s relative to the rocket.

(A) What is the speed of the rocket relative to the Earth once the rocket's mass is reduced to half its mass before ignition?

9.15 continued

SOLUTION

Conceptualize   Figure 9.23 shows the situation in this problem. From the discussion in this section and scenes from science fiction movies, we can easily imagine the rocket accelerating to a higher speed as the engine operates.

Categorize   This problem is a substitution problem in which we use given values in the equations derived in this section.

Solve Equation 9.46 for the final velocity and substitute the known values:

$$v_f = v_i + v_e \ln\left(\frac{M_i}{M_f}\right)$$

$$= 3.0 \times 10^3 \text{ m/s} + (5.0 \times 10^3 \text{ m/s})\ln\left(\frac{M_i}{0.50M_i}\right)$$

$$= 6.5 \times 10^3 \text{ m/s}$$

(B) What is the thrust on the rocket if it burns fuel at the rate of 50 kg/s?

SOLUTION

Use Equation 9.47, noting that $dM/dt = 50$ kg/s:

$$\text{Thrust} = \left| v_e \frac{dM}{dt} \right| = (5.0 \times 10^3 \text{ m/s})(50 \text{ kg/s}) = 2.5 \times 10^5 \text{ N}$$

Example **9.16**   **Fighting a Fire**

Two firefighters must apply a total force of 600 N to steady a hose that is discharging water at the rate of 3 600 L/min. Estimate the speed of the water as it exits the nozzle.

SOLUTION

Conceptualize   As the water leaves the hose, it acts in a way similar to the gases being ejected from a rocket engine. As a result, a force (thrust) acts on the firefighters in a direction opposite the direction of motion of the water. In this case, we want the end of the hose to be modeled as a particle in equilibrium rather than to accelerate as in the case of the rocket. Consequently, the firefighters must apply a force of magnitude equal to the thrust in the opposite direction to keep the end of the hose stationary.

Categorize   This example is a substitution problem in which we use given values in an equation derived in this section. The water exits at 3 600 L/min, which is 60 L/s. Knowing that 1 L of water has a mass of 1 kg, we estimate that about 60 kg of water leaves the nozzle each second.

Use Equation 9.47 for the thrust:

$$\text{Thrust} = \left| v_e \frac{dM}{dt} \right|$$

Solve for the exhaust speed:

$$v_e = \frac{\text{Thrust}}{|dM/dt|}$$

Substitute numerical values:

$$v_e = \frac{600 \text{ N}}{60 \text{ kg/s}} = 10 \text{ m/s}$$

Summary

> **Definitions**

The **linear momentum** $\vec{p}$ of a particle of mass m moving with a velocity $\vec{v}$ is

$$\vec{p} \equiv m\vec{v} \qquad (9.2)$$

The **impulse** imparted to a particle by a net force $\sum\vec{F}$ is equal to the time integral of the force:

$$\vec{I} \equiv \int_{t_i}^{t_f} \sum \vec{F} \, dt \qquad (9.9)$$

continued

An **inelastic collision** is one for which the total kinetic energy of the system of colliding particles is not conserved. A **perfectly inelastic collision** is one in which the colliding particles stick together after the collision. An **elastic collision** is one in which the kinetic energy of the system is conserved.

The position vector of the **center of mass** of a system of particles is defined as

$$\vec{r}_{CM} \equiv \frac{1}{M} \sum_i m_i \vec{r}_i \qquad (9.31)$$

where $M = \sum_i m_i$ is the total mass of the system and $\vec{r}_i$ is the position vector of the ith particle.

▶ Concepts and Principles

The position vector of the center of mass of an extended object can be obtained from the integral expression

$$\vec{r}_{CM} = \frac{1}{M} \int \vec{r}\, dm \qquad (9.34)$$

The velocity of the center of mass for a system of particles is

$$\vec{v}_{CM} = \frac{1}{M} \sum_i m_i \vec{v}_i \qquad (9.35)$$

The total momentum of a system of particles equals the total mass multiplied by the velocity of the center of mass.

Newton's second law applied to a system of particles is

$$\sum \vec{F}_{ext} = M\vec{a}_{CM} \qquad (9.39)$$

where $\vec{a}_{CM}$ is the acceleration of the center of mass and the sum is over all external forces. The center of mass moves like an imaginary particle of mass M under the influence of the resultant external force on the system.

▶ Analysis Models for Problem Solving

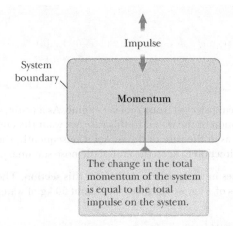

Nonisolated System (Momentum). If a system interacts with its environment in the sense that there is an external force on the system, the behavior of the system is described by the **impulse–momentum theorem:**

$$\Delta \vec{p}_{tot} = \vec{I} \qquad (9.40)$$

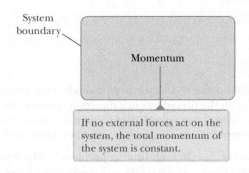

Isolated System (Momentum). The total momentum of an isolated system (no external forces) is conserved regardless of the nature of the forces between the members of the system:

$$\Delta \vec{p}_{tot} = 0 \qquad (9.41)$$

The system may be isolated in terms of momentum but nonisolated in terms of energy, as in the case of inelastic collisions.

Think–Pair–Share

See the Preface for an explanation of the icons used in this problems set. For additional assessment items for this section, go to ✹ **WEBASSIGN** From Cengage

1. You are a football player on your school's team. In practice, you kick a 0.400-kg football from the ground at the 20-yard line of a football field laid out exactly along a north–south line. The ball's initial velocity vector is directed at 30.0° above the horizontal, toward the north, and has a magnitude of 27.0 m/s. You watch the ball as it rises in the air. Just as the ball reaches its highest point in its parabolic

trajectory, a 4.18-kg eagle flying due south and along a horizontal line collides with the ball. Assume that the highly inflated ball makes an elastic collision with the hard beak of the bird, and that the ball rebounds from the collision with a velocity vector that is horizontal and due south. The ball lands back at the exact point on the ground from which it was kicked. (a) How fast was the eagle flying? (Ignore air resistance.) (b) How fast and in what direction is the eagle moving just after the collision? (Assume no flapping of wings has occurred yet!) (c) In reality, the collision will not be elastic: some kinetic energy will be converted to other forms. With the assumption of some kinetic energy lost in an inelastic collision, will the required speed of the eagle be higher or lower than in part (a)?

2. **ACTIVITY** Carefully draw a right triangle on a piece of cardboard, such that one of its non-hypotenuse legs is 30–40 cm in length and the other leg is much shorter. Measure the exact midpoint of each of the three sides of the triangle and mark these three points. Draw a line across the triangle, from a corner of the triangle to the midpoint of the opposite side. Repeat for the other two corners. The three lines happen to intersect at the center of mass of the triangle. Draw a fourth line perpendicular to the longer non-hypotenuse leg, passing through the center of mass, and ending as it crosses the hypotenuse of the triangle. Punch a hole in the cardboard just inside the edge of the triangle where the fourth line crosses the hypotenuse. Carefully cut the triangle out of the cardboard. Tie a string through the hole and hang the triangle from the string. The longer side of the triangle should be parallel to the table. Why should this be true? Now measure the distance along the longer leg from the smaller angle to the fourth line. What fraction of the entire longer leg is this distance?

Problems

See the Preface for an explanation of the icons used in this problems set. For additional assessment items for this section, go to **WEBASSIGN** From Cengage

SECTION 9.1 Linear Momentum

1. A particle of mass m moves with momentum of magnitude p.
   S (a) Show that the kinetic energy of the particle is $K = p^2/2m$. (b) Express the magnitude of the particle's momentum in terms of its kinetic energy and mass.

2. A 3.00-kg particle has a velocity of $(3.00\hat{\mathbf{i}} - 4.00\hat{\mathbf{j}})$ m/s. (a) Find its x and y components of momentum. (b) Find the magnitude and direction of its momentum.

3. A baseball approaches home plate at a speed of 45.0 m/s, moving horizontally just before being hit by a bat. The batter hits a pop-up such that after hitting the bat, the baseball is moving at 55.0 m/s straight up. The ball has a mass of 145 g and is in contact with the bat for 2.00 ms. What is the average vector force the ball exerts on the bat during their interaction?

SECTION 9.2 Analysis Model: Isolated System (Momentum)

4. A 65.0-kg boy and his 40.0-kg sister, both wearing roller
   QC blades, face each other at rest. The girl pushes the boy hard, sending him backward with velocity 2.90 m/s toward the west. Ignore friction. (a) Describe the subsequent motion of the girl. (b) How much potential energy in the girl's body is converted into mechanical energy of the boy–girl system? (c) Is the momentum of the boy–girl system conserved in the pushing-apart process? If so, explain how that is possible considering (d) there are large forces acting and (e) there is no motion beforehand and plenty of motion afterward.

5. Two blocks of masses m and $3m$
   QC are placed on a frictionless, hor-
   V izontal surface. A light spring is attached to the more massive block, and the blocks are pushed together with the spring between them (Fig. P9.5). A cord initially holding the blocks together is burned; after that happens, the block of mass $3m$ moves to the right with a speed of 2.00 m/s. (a) What is the velocity of the block of mass m? (b) Find the system's original elastic potential energy, taking

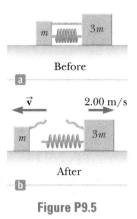

Before

$\vec{\mathbf{v}}$ 2.00 m/s

After

Figure P9.5

$m = 0.350$ kg. (c) Is the original energy in the spring or in the cord? (d) Explain your answer to part (c). (e) Is the momentum of the system conserved in the bursting-apart process? Explain how that is possible considering (f) there are large forces acting and (g) there is no motion beforehand and plenty of motion afterward?

6. When you jump straight up as high as you can, what is the order of magnitude of the maximum recoil speed that you give to the Earth? Model the Earth as a perfectly solid object. In your solution, state the physical quantities you take as data and the values you measure or estimate for them.

SECTION 9.3 Analysis Model: Nonisolated System (Momentum)

7. A glider of mass m is free to slide along a horizontal air
   QC track. It is pushed against a launcher at one end of the
   S track. Model the launcher as a light spring of force constant k compressed by a distance x. The glider is released from rest. (a) Show that the glider attains a speed of $v = x(k/m)^{1/2}$. (b) Show that the magnitude of the impulse imparted to the glider is given by the expression $I = x(km)^{1/2}$. (c) Is more work done on a cart with a large or a small mass?

8. You and your brother argue often about how to safely secure
   CR a toddler in a moving car. You insist that special toddler seats are critical in improving the chances of a toddler surviving a crash. Your brother claims that, as long as his wife is buckled in next to him with a seat belt while he drives, she can hold onto their toddler on her lap in a crash. You decide to perform a calculation to try to convince your brother. Consider a hypothetical collision in which the 12-kg toddler and his parents are riding in a car traveling at 60 mi/h relative to the ground. The car strikes a wall, tree, or another car, and is brought to rest in 0.10 s. You wish to demonstrate to your brother the magnitude of the force necessary for his wife to hold onto their child during the collision.

9. The front 1.20 m of a 1 400-kg car is designed as a "crumple
   T zone" that collapses to absorb the shock of a collision. If a car traveling 25.0 m/s stops uniformly in 1.20 m, (a) how long does the collision last, (b) what is the magnitude of the average force on the car, and (c) what is the magnitude of the acceleration of the car? Express the acceleration as a multiple of the acceleration due to gravity.

10. The magnitude of the net force exerted in the x direction on a 2.50-kg particle varies in time as shown in Figure P9.10 (page 244). Find (a) the impulse of the force over the 5.00-s time interval, (b) the final velocity the particle attains if it is

originally at rest, (c) its final velocity if its original velocity is $-2.00\hat{i}$ m/s, and (d) the average force exerted on the particle for the time interval between 0 and 5.00 s.

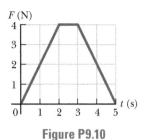

Figure P9.10

11. Water falls without splashing at a rate of 0.250 L/s from a height of 2.60 m into a bucket of mass 0.750 kg on a scale. If the bucket is originally empty, what does the scale read in newtons 3.00 s after water starts to accumulate in it?

SECTION 9.4 Collisions in One Dimension

12. A 1 200-kg car traveling initially at $v_{Ci} = 25.0$ m/s in an easterly direction crashes into the back of a 9 000-kg truck moving in the same direction at $v_{Ti} = 20.0$ m/s (Fig. P9.12). The velocity of the car immediately after the collision is $v_{Cf} = 18.0$ m/s to the east. (a) What is the velocity of the truck immediately after the collision? (b) What is the change in mechanical energy of the car–truck system in the collision? (c) Account for this change in mechanical energy.

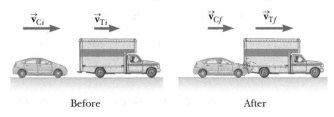

Before   After

Figure P9.12

13. A railroad car of mass 2.50×10^4 kg is moving with a speed of 4.00 m/s. It collides and couples with three other coupled railroad cars, each of the same mass as the single car and moving in the same direction with an initial speed of 2.00 m/s. (a) What is the speed of the four cars after the collision? (b) What is the decrease in mechanical energy in the collision?

14. Four railroad cars, each of mass 2.50×10^4 kg, are coupled together and coasting along horizontal tracks at speed v_i toward the south. A very strong but foolish movie actor, riding on the second car, uncouples the front car and gives it a big push, increasing its speed to 4.00 m/s southward. The remaining three cars continue moving south, now at 2.00 m/s. (a) Find the initial speed of the four cars. (b) By how much did the potential energy within the body of the actor change? (c) State the relationship between the process described here and the process in Problem 13.

15. A car of mass m moving at a speed v_1 collides and couples with the back of a truck of mass $2m$ moving initially in the same direction as the car at a lower speed v_2. (a) What is the speed v_f of the two vehicles immediately after the collision? (b) What is the change in kinetic energy of the car–truck system in the collision?

16. A 7.00-g bullet, when fired from a gun into a 1.00-kg block of wood held in a vise, penetrates the block to a depth of 8.00 cm. This block of wood is next placed on a frictionless horizontal surface, and a second 7.00-g bullet is fired from the gun into the block. To what depth will the bullet penetrate the block in this case?

17. A tennis ball of mass 57.0 g is held just above a basketball of mass 590 g as shown in Figure P9.17. With their centers

vertically aligned, both balls are released from rest at the same time, to fall through a distance of 1.20 m. (a) Find the magnitude of the downward velocity with which the basketball reaches the ground. (b) Assume that an elastic collision with the ground instantaneously reverses the velocity of the basketball while the tennis ball is still moving down. Next, the two balls meet in an elastic collision. To what height does the tennis ball rebound?

Figure P9.17

18. (a) Three carts of masses $m_1 = 4.00$ kg, $m_2 = 10.0$ kg, and $m_3 = 3.00$ kg move on a frictionless, horizontal track with speeds of $v_1 = 5.00$ m/s to the right, $v_2 = 3.00$ m/s to the right, and $v_3 = 4.00$ m/s to the left as shown in Figure P9.18. Velcro couplers make the carts stick together after colliding. Find the final velocity of the train of three carts. (b) **What If?** Does your answer in part (a) require that all the carts collide and stick together at the same moment? What if they collide in a different order?

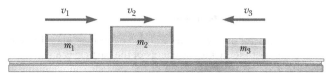

Figure P9.18

SECTION 9.5 Collisions in Two Dimensions

19. You have been hired as an expert witness by an attorney for a trial involving a traffic accident. The attorney's client, the plaintiff in this case, was traveling eastbound toward an intersection at 13.0 m/s as measured just before the accident by a roadside speed meter, and as seen by a trustworthy witness. As the plaintiff entered the intersection, his car was struck by a northbound driver, the defendant in this case, driving a car with identical mass to the plaintiff's. The vehicles stuck together after the collision and left parallel skid marks at an angle of $\theta = 55.0°$ north of east, as measured by accident investigators. The defendant is claiming that he was traveling within the 35-mi/h speed limit. What advice do you give to the attorney?

20. Two shuffleboard disks of equal mass, one orange and the other yellow, are involved in an elastic, glancing collision. The yellow disk is initially at rest and is struck by the orange disk moving with a speed of 5.00 m/s. After the collision, the orange disk moves along a direction that makes an angle of 37.0° with its initial direction of motion. The velocities of the two disks are perpendicular after the collision. Determine the final speed of each disk.

21. Two shuffleboard disks of equal mass, one orange and the other yellow, are involved in an elastic, glancing collision. The yellow disk is initially at rest and is struck by the orange disk moving with a speed v_i. After the collision, the orange disk moves along a direction that makes an angle θ with its initial direction of motion. The velocities of the two disks are perpendicular after the collision. Determine the final speed of each disk.

22. A 90.0-kg fullback running east with a speed of 5.00 m/s is tackled by a 95.0-kg opponent running north with a speed of 3.00 m/s. (a) Explain why the successful tackle constitutes a perfectly inelastic collision. (b) Calculate the velocity of the players immediately after the tackle. (c) Determine

the decrease in mechanical energy as a result of the collision. Account for this decrease.

23. A proton, moving with a velocity of $v_i\hat{\mathbf{i}}$, collides elastically
S with another proton that is initially at rest. Assuming that the two protons have equal speeds after the collision, find (a) the speed of each proton after the collision in terms of v_i and (b) the direction of the velocity vectors after the collision.

SECTION 9.6 The Center of Mass

24. A uniform piece of sheet
V metal is shaped as shown in Figure P9.24. Compute the x and y coordinates of the center of mass of the piece.

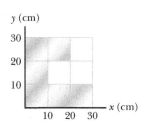

Figure P9.24

25. Explorers in the jungle find an ancient monument in the shape of a large isosceles triangle as shown in Figure P9.25. The monument is made from tens of thousands of small stone blocks of density 3 800 kg/m³. The monument is 15.7 m high and 64.8 m wide at its base and is everywhere 3.60 m thick from front to back. Before the monument was built many years ago, all the stone blocks lay on the ground. How much work did laborers do on the blocks to put them in position while building the entire monument? *Note:* The gravitational potential energy of an object–Earth system is given by $U_g = Mgy_{CM}$, where M is the total mass of the object and y_{CM} is the elevation of its center of mass above the chosen reference level.

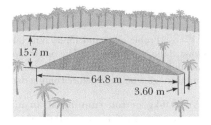

Figure P9.25

26. A rod of length 30.0 cm has linear density (mass per length) given by

$$\lambda = 50.0 + 20.0x$$

where x is the distance from one end, measured in meters, and λ is in grams/meter. (a) What is the mass of the rod? (b) How far from the $x = 0$ end is its center of mass?

SECTION 9.7 Systems of Many Particles

27. Consider a system of two particles in the xy plane: $m_1 = 2.00$ kg is at the location $\vec{\mathbf{r}}_1 = (1.00\hat{\mathbf{i}} + 2.00\hat{\mathbf{j}})$ m and has a velocity of $(3.00\hat{\mathbf{i}} + 0.500\hat{\mathbf{j}})$ m/s; $m_2 = 3.00$ kg is at $\vec{\mathbf{r}}_2 = (-4.00\hat{\mathbf{i}} - 3.00\hat{\mathbf{j}})$ m and has velocity $(3.00\hat{\mathbf{i}} - 2.00\hat{\mathbf{j}})$ m/s. (a) Plot these particles on a grid or graph paper. Draw their position vectors and show their velocities. (b) Find the position of the center of mass of the system and mark it on the grid. (c) Determine the velocity of the center of mass and also show it on the diagram. (d) What is the total linear momentum of the system?

28. The vector position of a 3.50-g particle moving in the xy plane varies in time according to $\vec{\mathbf{r}}_1 = (3\hat{\mathbf{i}} + 3\hat{\mathbf{j}})t + 2\hat{\mathbf{j}}t^2$, where t is in seconds and $\vec{\mathbf{r}}$ is in centimeters. At the same time, the vector position of a 5.50 g particle varies as $\vec{\mathbf{r}}_2 = 3\hat{\mathbf{i}} - 2\hat{\mathbf{i}}t^2 - 6\hat{\mathbf{j}}t$. At $t = 2.50$ s, determine (a) the vector position of the center of mass of the system, (b) the linear momentum of the system, (c) the velocity of the center of mass, (d) the acceleration of the center of mass, and (e) the net force exerted on the two-particle system.

29. You have been hired as an expert witness in an investiga-
CR tion of a quadcopter drone incident. The incident occurred during a very rare meteor shower during which several unusually massive chunks of meteoric material were passing through the atmosphere and striking the ground. The unmanned drone was hovering at rest over the center of a house on fire, having just dropped fire retardant, when it seemed to spontaneously explode into four large pieces. The locations of the four pieces on the ground were measured as follows, relative to the center of the house over which the drone was hovering:

Piece #	Mass (kg)	Distance from Center of House (m)	Direction from House
1	80.0	150	Due west
2	120	75.0	Due north
3	50.0	90.0	20.0° west of south
4	150	50.0	20.0° north of east

The fire department is suggesting that the drone was defective and exploded while in use. The drone manufacturer is suggesting that the drone was struck by a meteorite, causing the explosion. Perform a calculation that will show evidence suggesting agreement with one of these positions.

SECTION 9.8 Deformable Systems

30. For a technology project, a stu-
Q|C dent has built a vehicle, of total mass 6.00 kg, that moves itself. As shown in Figure P9.30, it runs on four light wheels. A reel is attached to one of the axles, and a cord originally wound on the reel goes up over a pulley attached to the vehicle to support an elevated load. After the vehicle is released from rest, the load descends very slowly, unwinding the cord to turn the axle and make the vehicle move forward (to the left in Fig. P9.30). Friction is negligible in the pulley and axle bearings. The wheels do not slip on the floor. The reel has been constructed with a conical shape so that the load descends at a constant low speed while the vehicle moves horizontally across the floor with constant acceleration, reaching a final velocity of $3.00\hat{\mathbf{i}}$ m/s. (a) Does the floor impart impulse to the vehicle? If so, how much? (b) Does the floor do work on the vehicle? If so, how much? (c) Does it make sense to say that the final momentum of the vehicle came from the floor? If not, where did it come from? (d) Does it make sense to say that the final kinetic energy of the vehicle came from the floor? If not, where did it come from? (e) Can we say that one particular force causes the forward acceleration of the vehicle? What does cause it?

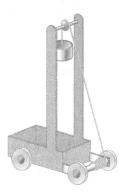

Figure P9.30

31. A 60.0-kg person bends his knees and then jumps straight up.
After his feet leave the floor, his motion is unaffected by air resistance and his center of mass rises by a maximum of 15.0 cm. Model the floor as completely solid and motionless. (a) Does the floor impart impulse to the person? (b) Does the floor do work on the person? (c) With what momentum does the person leave the floor? (d) Does it make sense to say that this momentum came from the floor? Explain. (e) With what kinetic energy does the person leave the floor? (f) Does it make sense to say that this energy came from the floor? Explain.

SECTION 9.9 Rocket Propulsion

32. A garden hose is held as shown in Figure P9.32. The hose is originally full of motionless water. What additional force is necessary to hold the nozzle stationary after the water flow is turned on if the discharge rate is 0.600 kg/s with a speed of 25.0 m/s?

Figure P9.32

33. A rocket for use in deep space is to be capable of boosting a total load (payload plus rocket frame and engine) of 3.00 metric tons to a speed of 10 000 m/s. (a) It has an engine and fuel designed to produce an exhaust speed of 2 000 m/s. How much fuel plus oxidizer is required? (b) If a different fuel and engine design could give an exhaust speed of 5 000 m/s, what amount of fuel and oxidizer would be required for the same task? (c) Noting that the exhaust speed in part (b) is 2.50 times higher than that in part (a), explain why the required fuel mass is not simply smaller by a factor of 2.50.

34. A rocket has total mass $M_i = 360$ kg, including $M_{fuel} = 330$ kg of fuel and oxidizer. In interstellar space, it starts from rest at the position $x = 0$, turns on its engine at time $t = 0$, and puts out exhaust with relative speed $v_e = 1\,500$ m/s at the constant rate $k = 2.50$ kg/s. The fuel will last for a burn time of $T_b = M_{fuel}/k = 330$ kg/(2.5 kg/s) = 132 s. (a) Show that during the burn the velocity of the rocket as a function of time is given by

$$v(t) = -v_e \ln\left(1 - \frac{kt}{M_i}\right)$$

(b) Make a graph of the velocity of the rocket as a function of time for times running from 0 to 132 s. (c) Show that the acceleration of the rocket is

$$a(t) = \frac{kv_e}{M_i - kt}$$

(d) Graph the acceleration as a function of time. (e) Show that the position of the rocket is

$$x(t) = v_e\left(\frac{M_i}{k} - t\right)\ln\left(1 - \frac{kt}{M_i}\right) + v_e t$$

(f) Graph the position during the burn as a function of time.

ADDITIONAL PROBLEMS

35. An amateur skater of mass M is trapped in the middle of an ice rink and is unable to return to the side where there is no ice. Every motion she makes causes her to slip on the ice and remain in the same spot. She decides to try to return to safety by throwing her gloves of mass m in the direction opposite the safe side. (a) She throws the gloves as hard as she can, and they leave her hand with a horizontal velocity $\vec{v}_{gloves}$. Explain whether or not she moves. (b) If she does move, calculate her velocity $\vec{v}_{girl}$ relative to the Earth after she throws the gloves. (c) Discuss her motion from the point of view of the forces acting on her.

36. (a) Figure P9.36 shows three points in the operation of the ballistic pendulum discussed in Example 9.6 (and shown in Fig. 9.10b). The projectile approaches the pendulum in Figure P9.36a. Figure P9.36b shows the situation just after the projectile is captured in the pendulum. In Figure P9.36c, the pendulum arm has swung upward and come to rest momentarily at a height h above its initial position. Prove that the ratio of the kinetic energy of the projectile–pendulum system immediately after the collision to the kinetic energy immediately before is $m_1/(m_1 + m_2)$. (b) What is the ratio of the momentum of the system immediately after the collision to the momentum immediately before? (c) A student believes that such a large decrease in mechanical energy must be accompanied by at least a small decrease in momentum. How would you convince this student of the truth?

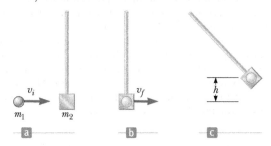

Figure P9.36 Problems 36 and 43. (a) A metal ball moves toward the pendulum. (b) The ball is captured by the pendulum. (c) The ball–pendulum combination swings up through a height h before coming to rest.

37. **Review.** A 60.0-kg person running at an initial speed of 4.00 m/s jumps onto a 120-kg cart initially at rest (Fig. P9.37). The person slides on the cart's top surface and finally comes to rest relative to the cart. The coefficient of kinetic friction between the person and the cart is 0.400. Friction between the cart and ground can be ignored. (a) Find the final velocity of the person and cart relative to the ground. (b) Find the friction force acting on the person while he is sliding across the top surface of the cart. (c) How long does the friction force act on the person? (d) Find the change in momentum of the person and the change in momentum of the cart. (e) Determine the displacement of the person relative to the ground while he is sliding on the cart. (f) Determine the displacement of the cart relative to the ground while the person is sliding. (g) Find the change in kinetic energy of

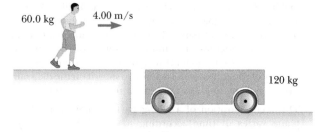

Figure P9.37

the person. (h) Find the change in kinetic energy of the cart. (i) Explain why the answers to (g) and (h) differ. (What kind of collision is this one, and what accounts for the loss of mechanical energy?)

38. A cannon is rigidly attached to a carriage, which can move along horizontal rails but is connected to a post by a large spring, initially unstretched and with force constant $k = 2.00 \times 10^4$ N/m, as shown in Figure P9.38. The cannon

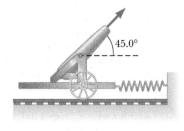

Figure P9.38

fires a 200-kg projectile at a velocity of 125 m/s directed 45.0° above the horizontal. (a) Assuming that the mass of the cannon and its carriage is 5 000 kg, find the recoil speed of the cannon. (b) Determine the maximum extension of the spring. (c) Find the maximum force the spring exerts on the carriage. (d) Consider the system consisting of the cannon, carriage, and projectile. Is the momentum of this system conserved during the firing? Why or why not?

39. A 1.25-kg wooden block rests on a table over a large hole as in Figure P9.39. A 5.00-g bullet with an initial velocity v_i is fired upward into the bottom of the block and remains in the block after the collision. The block and bullet rise to a maximum height of 22.0 cm. (a) Describe how you would find the initial velocity of the bullet using ideas you have learned in this chapter. (b) Calculate the initial velocity of the bullet from the information provided.

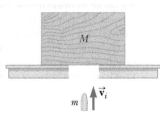

Figure P9.39
Problems 39 and 40.

40. A wooden block of mass M rests on a table over a large hole as in Figure P9.39. A bullet of mass m with an initial velocity of v_i is fired upward into the bottom of the block and remains in the block after the collision. The block and bullet rise to a maximum height of h. (a) Describe how you would find the initial velocity of the bullet using ideas you have learned in this chapter. (b) Find an expression for the initial velocity of the bullet.

41. Two gliders are set in motion on a horizontal air track. A light spring of force constant k is attached to the back end of the second glider. As shown in Figure P9.41, the first glider, of mass m_1, moves to the right with speed v_1, and the second glider, of mass m_2, moves more slowly to the right with speed v_2. When m_1 collides with the spring attached to m_2, the spring compresses by a distance x_{max}, and the gliders then move apart again. In terms of v_1, v_2, m_1, m_2, and k, find (a)

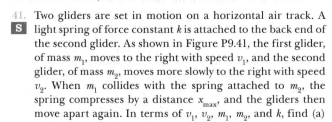

Figure P9.41

the speed v at maximum compression, (b) the maximum compression x_{max}, and (c) the velocity of each glider after m_1 has lost contact with the spring.

42. Pursued by ferocious wolves, you are in a sleigh with no horses, gliding without friction across an ice-covered lake. You take an action described by the equations

$$(270 \text{ kg})(7.50 \text{ m/s})\hat{\mathbf{i}} = (15.0 \text{ kg})(-v_{1f}\hat{\mathbf{i}}) + (255 \text{ kg})(v_{2f}\hat{\mathbf{i}})$$
$$v_{1f} + v_{2f} = 8.00 \text{ m/s}$$

(a) Complete the statement of the problem, giving the data and identifying the unknowns. (b) Find the values of v_{1f} and v_{2f}. (c) Find the amount of energy that has been transformed from potential energy stored in your body to kinetic energy of the system.

43. **Review.** A student performs a ballistic pendulum experiment using an apparatus similar to that discussed in Example 9.6 and shown in Figure P9.36. She obtains the following average data: $h = 8.68$ cm, projectile mass $m_1 = 68.8$ g, and pendulum mass $m_2 = 263$ g. (a) Determine the initial speed v_{1A} of the projectile. (b) The second part of her experiment is to obtain v_{1A} by firing the same projectile horizontally (with the pendulum removed from the path) and measuring its final horizontal position x and distance of fall y (Fig. P9.43). What numerical value does she obtain for v_{1A} based on her measured values of $x = 257$ cm and $y = 85.3$ cm? (c) What factors might account for the difference in this value compared with that obtained in part (a)?

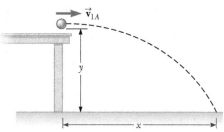

Figure P9.43

44. *Why is the following situation impossible?* An astronaut, together with the equipment he carries, has a mass of 150 kg. He is taking a space walk outside his spacecraft, which is drifting through space with a constant velocity. The astronaut accidentally pushes against the spacecraft and begins moving away at 20.0 m/s, relative to the spacecraft, without a tether. To return, he takes equipment off his space suit and throws it in the direction away from the spacecraft. Because of his bulky space suit, he can throw equipment at a maximum speed of 5.00 m/s relative to himself. After throwing enough equipment, he starts moving back to the spacecraft and is able to grab onto it and climb inside.

45. **Review.** A bullet of mass $m = 8.00$ g is fired into a block of mass $M = 250$ g that is initially at rest at the edge of a frictionless table of height $h = 1.00$ m (Fig. P9.45). The bullet

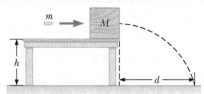

Figure P9.45 Problems 45 and 46.

remains in the block, and after the impact the block lands $d = 2.00$ m from the bottom of the table. Determine the initial speed of the bullet.

46. **Review.** A bullet of mass m is fired into a block of mass M
    S initially at rest at the edge of a frictionless table of height h (Fig. P9.45). The bullet remains in the block, and after impact the block lands a distance d from the bottom of the table. Determine the initial speed of the bullet.

47. A 0.500-kg sphere moving with a velocity expressed as $(2.00\hat{\mathbf{i}} - 3.00\hat{\mathbf{j}} + 1.00\hat{\mathbf{k}})$ m/s strikes a second, lighter sphere of mass 1.50 kg moving with an initial velocity of $(-1.00\hat{\mathbf{i}} + 2.00\hat{\mathbf{j}} - 3.00\hat{\mathbf{k}})$ m/s. (a) The velocity of the 0.500-kg sphere after the collision is $(-1.00\hat{\mathbf{i}} + 3.00\hat{\mathbf{j}} - 8.00\hat{\mathbf{k}})$ m/s. Find the final velocity of the 1.50-kg sphere and identify the kind of collision (elastic, inelastic, or perfectly inelastic). (b) Now assume the velocity of the 0.500-kg sphere after the collision is $(-0.250\hat{\mathbf{i}} + 0.750\hat{\mathbf{j}} - 2.00\hat{\mathbf{k}})$ m/s. Find the final velocity of the 1.50-kg sphere and identify the kind of collision. (c) **What If?** Take the velocity of the 0.500-kg sphere after the collision as $(-1.00\hat{\mathbf{i}} + 3.00\hat{\mathbf{j}} + a\hat{\mathbf{k}})$ m/s. Find the value of a and the velocity of the 1.50-kg sphere after an elastic collision.

48. **Review.** A metal cannonball of mass m rests next to a tree
    Q|C at the very edge of a cliff 36.0 m above the surface of the ocean. In an effort to knock the cannonball off the cliff, some children tie one end of a rope around a stone of mass 80.0 kg and the other end to a tree limb just above the cannonball. They tighten the rope so that the stone just clears the ground and hangs next to the cannonball. The children manage to swing the stone back until it is held at rest 1.80 m above the ground. The children release the stone, which then swings down and makes a head-on, elastic collision with the cannonball, projecting it horizontally off the cliff. The cannonball lands in the ocean a horizontal distance R away from its initial position. (a) Find the horizontal component R of the cannonball's displacement as it depends on m. (b) What is the maximum possible value for R, and (c) to what value of m does it correspond? (d) For the stone–cannonball–Earth system, is mechanical energy conserved throughout the process? Is this principle sufficient to solve the entire problem? Explain. (e) **What if?** Show that R does not depend on the value of the gravitational acceleration. Is this result remarkable? State how one might make sense of it.

49. **Review.** A light spring of force constant 3.85 N/m is compressed by 8.00 cm and held between a 0.250-kg block on the left and a 0.500-kg block on the right. Both blocks are at rest on a horizontal surface. The blocks are released simultaneously so that the spring tends to push them apart. Find the maximum velocity each block attains if the coefficient of kinetic friction between each block and the surface is (a) 0, (b) 0.100, and (c) 0.462. Assume the coefficient of static friction is greater than the coefficient of kinetic friction in every case.

50. Consider as a system the Sun with the Earth in a circular orbit around it. Find the magnitude of the change in the velocity of the Sun relative to the center of mass of the system over a six-month period. Ignore the influence of other celestial objects. You may obtain the necessary astronomical data from the endpapers of the book.

51. **Review.** There are (one can say) three coequal theories of
    GP motion for a single particle: Newton's second law, stating
    Q|C that the total force on the particle causes its acceleration; the work–kinetic energy theorem, stating that the total work on the particle causes its change in kinetic energy; and the impulse–momentum theorem, stating that the total impulse on the particle causes its change in momentum. In this problem, you compare predictions of the three theories in one particular case. A 3.00-kg object has velocity $7.00\hat{\mathbf{j}}$ m/s. Then, a constant net force $12.0\hat{\mathbf{i}}$ N acts on the object for 5.00 s. (a) Calculate the object's final velocity, using the impulse–momentum theorem. (b) Calculate its acceleration from $\vec{\mathbf{a}} = (\vec{\mathbf{v}}_f - \vec{\mathbf{v}}_i)/\Delta t$. (c) Calculate its acceleration from $\vec{\mathbf{a}} = \Sigma\vec{\mathbf{F}}/m$. (d) Find the object's vector displacement from $\Delta\vec{\mathbf{r}} = \vec{\mathbf{v}}_i t + \frac{1}{2}\vec{\mathbf{a}}t^2$. (e) Find the work done on the object from $W = \vec{\mathbf{F}} \cdot \Delta\vec{\mathbf{r}}$. (f) Find the final kinetic energy from $\frac{1}{2}mv_f^2 = \frac{1}{2}m\vec{\mathbf{v}}_f \cdot \vec{\mathbf{v}}_f$. (g) Find the final kinetic energy from $\frac{1}{2}mv_i^2 + W$. (h) State the result of comparing the answers to parts (b) and (c), and the answers to parts (f) and (g).

CHALLENGE PROBLEMS

52. Sand from a stationary hopper falls onto a moving con-
    Q|C veyor belt at the rate of 5.00 kg/s as shown in Figure P9.52. The conveyor belt is supported by frictionless rollers and moves at a constant speed of $v = 0.750$ m/s under the action of a constant horizontal external force $\vec{\mathbf{F}}_{ext}$ supplied by the motor that drives the belt. Find (a) the sand's rate of change of momentum in the horizontal direction, (b) the force of friction exerted by the belt on the sand, (c) the external force $\vec{\mathbf{F}}_{ext}$, (d) the work done by $\vec{\mathbf{F}}_{ext}$ in 1 s, and (e) the kinetic energy acquired by the falling sand each second due to the change in its horizontal motion. (f) Why are the answers to parts (d) and (e) different?

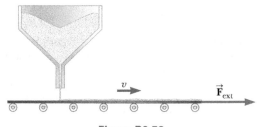

Figure P9.52

53. Two particles with masses m and $3m$ are moving toward each other along the x axis with the same initial speeds v_i. Particle m is traveling to the left, and particle $3m$ is traveling to the right. They undergo an elastic glancing collision such that particle m is moving in the negative y direction after the collision at a right angle from its initial direction. (a) Find the final speeds of the two particles in terms of v_i. (b) What is the angle θ at which the particle $3m$ is scattered?

54. On a horizontal air track, a glider of mass m carries a
    S Γ-shaped post. The post supports a small dense sphere, also of mass m, hanging just above the top of the glider on a cord of length L. The glider and sphere are initially at rest with the cord vertical. A constant horizontal force of magnitude F is applied to the glider, moving it through displacement x_1; then the force is removed. During the time interval when the force is applied, the sphere moves through a displacement with horizontal component x_2. (a) Find the horizontal component of the velocity of the center of mass of the glider–sphere system when the force is removed. (b) After the force is removed, the glider continues to move on the track and the sphere swings back and forth, both without friction. Find an expression for the largest angle the cord makes with the vertical.

A rusty bolt resists efforts to turn it with a wrench. How can you loosen the bolt? (*Scott Richardson/ Shutterstock*)

Rotation of a Rigid Object About a Fixed Axis

STORYLINE **You are back at home after your game of pool in the** previous chapter. You go back into your garage to work on another project. For this project, you need some pieces of metal from an older project. The pieces of metal have been joined together for years by nuts and bolts, which are now quite rusty. Using a wrench, you try to loosen a bolt. You are unable to do so because of the rust. You instinctively reach for a piece of hollow pipe that is longer than the handle of the wrench and slip it over the handle. Pushing on the far end of the pipe, you are now able to loosen the bolt. You say to yourself, "Wait a minute! How did I know to use a long piece of pipe? Why did the long pipe make it possible for me to loosen the rusted bolt?" Your project sits idle while you ponder this new development. Then your thoughts progress further. You applied a force on the pipe, like the forces studied in Chapter 5. But you didn't achieve an acceleration of something through space like the objects in Chapter 5. Something *rotated*: the bolt. This is new: force causes rotation. You have more thinking to do. Your project sits idle for the rest of the day.

CONNECTIONS We have focused our attention so far on particles in *translational* motion. When we analyzed the motion of objects with a size in previous chapters, we ignored any spinning motion of the object. It is now time to *not* ignore this spinning motion. In this chapter, we focus on the *rotational* motion of an object. We will be following the outline of earlier chapters for this new type of motion; we will find rotational analogs for position, speed, acceleration, mass, force, and energy. Many objects exhibit both translational and rotational motion at the same time. We will investigate how to reduce the apparently complicated motion of such an object to a combination of the two types of motion. In dealing with a rotating object, analysis is greatly simplified by assuming the object is rigid. A **rigid object** is one that is nondeformable; that is, the relative locations of all particles of which the object is composed remain constant. All real objects are deformable to some extent; our rigid-object model, however, is useful in many situations in which deformation is negligible. We have developed analysis models based on particles and systems. In this chapter, we introduce another class of analysis models based on the simplification model of a rigid object. In future chapters, we will see rotating objects for which we will need these models: the spinning Earth in Chapter 13, the armature of a motor in Chapter 30, and an HCl molecule in Chapter 42, for example.

10.1 Angular Position, Velocity, and Acceleration

As mentioned in the introduction, we will develop our understanding of rotational motion in a manner parallel to that used for translational motion in previous chapters. We began in Chapter 2 by defining kinematic variables for translational motion: position, velocity, and acceleration. We do the same here for rotational motion.

Figure 10.1 illustrates an overhead view of a rotating Blu-ray Disc. The disc rotates about a fixed axis perpendicular to the plane of the figure and passing through the center of the disc at O. A small element of the disc modeled as a particle at P is at a fixed distance r from the origin and rotates about it in a circle of radius r. (In fact, *every* element of the disc undergoes circular motion about O.) It is convenient to represent the position of P with its polar coordinates (r, θ), where r is the distance from the origin to P and θ is measured *counterclockwise* from some reference line fixed in space as shown in Figure 10.1a. In this representation, the angle θ changes in time while r remains constant. As the particle moves along the circle from the reference line, which is at angle $\theta = 0$, it moves through an arc of length s as in Figure 10.1b. We can define the angle θ as the ratio of the arc length s to the radius r:

$$\theta = \frac{s}{r} \tag{10.1a}$$

Because θ is the ratio of an arc length and the radius of the circle, it is a pure number. Usually, however, we give θ the artificial unit **radian** (rad), where one radian is the angle subtended by an arc length equal to the radius of the arc. Because the circumference of a circle is $2\pi r$, it follows from Equation 10.1a that $360°$ corresponds to an angle of $(2\pi r/r)$ rad $= 2\pi$ rad. Hence, 1 rad $= 360°/2\pi \approx 57.3°$. To convert an angle in degrees to an angle in radians, we use that π rad $= 180°$, so

$$\theta(\text{rad}) = \frac{\pi}{180°}\,\theta(\text{deg})$$

For example, $60°$ equals $\pi/3$ rad and $45°$ equals $\pi/4$ rad.

Based on the definition of the angle θ in Equation 10.1a, we can express the arc length s through which the particle at P moves in Figure 10.1b as

$$s = r\theta \tag{10.1b}$$

Because the disc in Figure 10.1 is a rigid object, as the particle moves through an angle θ from the reference line, every other particle on the object rotates through the same angle θ. Therefore, we can associate the angle θ with the entire rigid object as well as with an individual particle, which allows us to define the *angular position* of a rigid object in its rotational motion. We choose a reference line on the object, such as a line connecting O and a chosen particle on the object. The **angular position** of the rigid object is the angle θ between this reference line on the object and the fixed reference line in space, which is often chosen as the x axis. Such identification is similar to the way we define the position of an object in one-dimensional translational

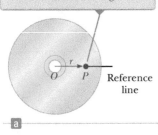

To define angular position for the disc, a reference line fixed in space is chosen. A particle at P is located at a distance r from the rotation axis through O.

a

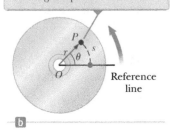

As the disc rotates, the particle at P moves through an arc length s on a circular path of radius r. The angular position of P is θ.

b

Figure 10.1 A Blu-ray Disc rotating about a fixed axis through O perpendicular to the plane of the figure.

PITFALL PREVENTION 10.1

Remember the Radian In rotational equations, you *must* use angles expressed in radians. Don't fall into the trap of using angles measured in degrees in rotational equations.

motion as the distance x between the object and the reference position, which is the origin, $x = 0$. Therefore, the angle θ plays the same role in rotational motion that the position x does in one-dimensional translational motion.

As the particle in question on our rigid object travels from position Ⓐ to position Ⓑ in a time interval Δt as in Figure 10.2, the reference line fixed to the object sweeps out an angle $\Delta\theta = \theta_f - \theta_i$. This quantity $\Delta\theta$ is defined as the **angular displacement** of the rigid object:

$$\Delta\theta \equiv \theta_f - \theta_i$$

◀ Angular displacement (Compare to Equation 2.1)

The rate at which this angular displacement occurs can vary. If the rigid object spins rapidly, this displacement can occur in a short time interval. If it rotates slowly, this displacement occurs in a longer time interval. These different rotation rates can be quantified by defining the **average angular speed** ω_{avg} (Greek letter omega) as the ratio of the angular displacement of a rigid object to the time interval Δt during which the displacement occurs:

$$\omega_{avg} \equiv \frac{\Delta\theta}{\Delta t} \qquad (10.2)$$

◀ Average angular speed (Compare to Equation 2.2)

In analogy to translational speed, the **instantaneous angular speed** ω is defined as the limit of the average angular speed as Δt approaches zero:

$$\omega \equiv \lim_{\Delta t \to 0} \frac{\Delta\theta}{\Delta t} = \frac{d\theta}{dt} \qquad (10.3)$$

◀ Instantaneous angular speed (Compare to Equation 2.5)

Angular speed has units of radians per second (rad/s), which can be written as s^{-1} because radians are not dimensional. We take ω to be positive when θ is increasing (counterclockwise motion in Fig. 10.2) and negative when θ is decreasing (clockwise motion in Fig. 10.2).

ⓆUICK QUIZ 10.1 A rigid object rotates in a counterclockwise sense around a fixed axis. Each of the following pairs of quantities represents an initial angular position and a final angular position of the rigid object. **(i)** Which of the sets can *only* occur if the rigid object rotates through more than 180°? (a) 3 rad, 6 rad (b) −1 rad, 1 rad (c) 1 rad, 5 rad **(ii)** Suppose the change in angular position for each of these pairs of values occurs in 1 s. Which choice represents the lowest average angular speed?

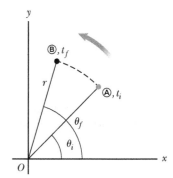

Figure 10.2 A particle on a rotating rigid object moves from Ⓐ to Ⓑ along the arc of a circle. In the time interval $\Delta t = t_f - t_i$, the radial line of length r moves through an angular displacement $\Delta\theta = \theta_f - \theta_i$.

If the instantaneous angular speed of an object changes from ω_i to ω_f in the time interval Δt, the object has an angular acceleration. The **average angular acceleration** α_{avg} (Greek letter alpha) of a rotating rigid object is defined as the ratio of the change in the angular speed to the time interval Δt during which the change in the angular speed occurs:

$$\alpha_{avg} \equiv \frac{\Delta\omega}{\Delta t} = \frac{\omega_f - \omega_i}{t_f - t_i} \qquad (10.4)$$

◀ Average angular acceleration (Compare to Equation 2.9)

In analogy to translational acceleration, the **instantaneous angular acceleration** is defined as the limit of the average angular acceleration as Δt approaches zero:

$$\alpha \equiv \lim_{\Delta t \to 0} \frac{\Delta\omega}{\Delta t} = \frac{d\omega}{dt} \qquad (10.5)$$

◀ Instantaneous angular acceleration (Compare to Equation 2.10)

Angular acceleration has units of radians per second squared (rad/s²), or simply s^{-2}. Notice that α is positive when a rigid object rotating counterclockwise is speeding up or when a rigid object rotating clockwise is slowing down during some time interval.

When a rigid object is rotating about a *fixed* axis, every particle on the object rotates through the same angle in a given time interval and has the same angular speed and the same angular acceleration. Therefore, like the angular position θ,

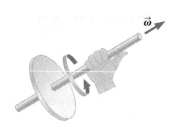

Figure 10.3 The right-hand rule for determining the direction of the angular velocity vector.

the quantities ω and α characterize the rotational motion of the entire rigid object as well as individual particles in the object.

Angular position (θ), angular speed (ω), and angular acceleration (α) are analogous to translational position (x), translational speed (v), and translational acceleration (a). The variables θ, ω, and α differ dimensionally from the variables x, v, and a only by a factor having the unit of length. (See Section 10.3.)

We have not specified any direction in space for angular speed and angular acceleration. Strictly speaking, ω and α are the magnitudes of the angular velocity and the angular acceleration vectors[1] $\vec{\omega}$ and $\vec{\alpha}$, respectively, and they should always be positive. Because we are considering rotation about a fixed axis, however, we can use nonvector notation and indicate the vectors' directions by assigning a positive or negative sign to ω and α as discussed earlier with regard to Equations 10.3 and 10.5. For rotation about a fixed axis, the only direction that uniquely specifies the rotational motion is the direction along the axis of rotation. Therefore, the directions of $\vec{\omega}$ and $\vec{\alpha}$ are along this axis. If a particle rotates in the xy plane as in Figure 10.2, the direction of $\vec{\omega}$ for the particle is out of the plane of the diagram when the rotation is counterclockwise and into the plane of the diagram when the rotation is clockwise. To illustrate this convention, it is convenient to use the *right-hand rule* demonstrated in Figure 10.3. When the four fingers of the right hand are wrapped in the direction of rotation, the extended right thumb points in the direction of $\vec{\omega}$. The direction of $\vec{\alpha}$ follows from its definition $\vec{\alpha} \equiv d\vec{\omega}/dt$. It is in the same direction as $\vec{\omega}$ if the angular speed is increasing in time, and it is antiparallel to $\vec{\omega}$ if the angular speed is decreasing in time.

10.2 Analysis Model: Rigid Object Under Constant Angular Acceleration

In our study of translational motion, after introducing the kinematic variables, we considered the special case of a particle under constant acceleration. We follow the same procedure here for a rigid object under constant angular acceleration.

Imagine a rigid object such as the disc in Figure 10.1 rotates about a fixed axis and has a constant angular acceleration. In parallel with our analysis model of the particle under constant acceleration, we generate a new analysis model for rotational motion called the **rigid object under constant angular acceleration.** We develop kinematic relationships for this model in this section. Writing Equation 10.5 in the form $d\omega = \alpha\, dt$ and integrating from $t_i = 0$ to $t_f = t$ gives

▶ Rotational kinematic equations

$$\omega_f = \omega_i + \alpha t \quad \text{(for constant } \alpha) \tag{10.6}$$

where ω_i is the angular speed of the rigid object at time $t = 0$. Equation 10.6 allows us to find the angular speed ω_f of the object at any later time t. Substituting Equation 10.6 into Equation 10.3 and integrating once more, we obtain

$$\theta_f = \theta_i + \omega_i t + \tfrac{1}{2}\alpha t^2 \quad \text{(for constant } \alpha) \tag{10.7}$$

where θ_i is the angular position of the rigid object at time $t = 0$. Equation 10.7 allows us to find the angular position θ_f of the object at any later time t. Eliminating t from Equations 10.6 and 10.7 gives

$$\omega_f^2 = \omega_i^2 + 2\alpha(\theta_f - \theta_i) \quad \text{(for constant } \alpha) \tag{10.8}$$

[1]Although we do not verify it here, the instantaneous angular velocity and instantaneous angular acceleration are vector quantities, but the corresponding average values are not because angular displacements do not add as vector quantities for finite rotations.

This equation allows us to find the angular speed ω_f of the rigid object for any value of its angular position θ_f. If we eliminate α between Equations 10.6 and 10.7, we obtain

$$\theta_f = \theta_i + \tfrac{1}{2}(\omega_i + \omega_f)t \quad \text{(for constant } \alpha) \tag{10.9}$$

Notice that these kinematic expressions for the rigid object under constant angular acceleration are of the same mathematical form as those for a particle under constant acceleration (see Table 10.1). They can be generated from the equations for translational motion by making the substitutions $x \rightarrow \theta$, $v \rightarrow \omega$, and $a \rightarrow \alpha$. Table 10.1 compares the kinematic equations for the rigid object under constant angular acceleration and particle under constant acceleration models.

QUICK QUIZ 10.2 Consider again the pairs of angular positions for the rigid object in Quick Quiz 10.1. If the object starts from rest at the initial angular position, moves counterclockwise with constant angular acceleration, and arrives at the final angular position with the same angular speed in all three cases, for which choice is the angular acceleration the highest?

TABLE 10.1 Kinematic Equations for Rotational and Translational Motion

Rigid Object Under Constant Angular Acceleration		Particle Under Constant Acceleration	
$\omega_f = \omega_i + \alpha t$	(10.6)	$v_f = v_i + at$	(2.13)
$\theta_f = \theta_i + \omega_i t + \tfrac{1}{2}\alpha t^2$	(10.7)	$x_f = x_i + v_i t + \tfrac{1}{2}at^2$	(2.16)
$\omega_f^2 = \omega_i^2 + 2\alpha(\theta_f - \theta_i)$	(10.8)	$v_f^2 = v_i^2 + 2a(x_f - x_i)$	(2.17)
$\theta_f = \theta_i + \tfrac{1}{2}(\omega_i + \omega_f)t$	(10.9)	$x_f = x_i + \tfrac{1}{2}(v_i + v_f)t$	(2.15)

ANALYSIS MODEL Rigid Object Under Constant Angular Acceleration

Imagine an object that undergoes a spinning motion such that its angular acceleration is constant. The equations describing its angular position and angular speed are analogous to those for the particle under constant acceleration model:

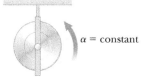
α = constant

$$\omega_f = \omega_i + \alpha t \tag{10.6}$$

$$\theta_f = \theta_i + \omega_i t + \tfrac{1}{2}\alpha t^2 \tag{10.7}$$

$$\omega_f^2 = \omega_i^2 + 2\alpha(\theta_f - \theta_i) \tag{10.8}$$

$$\theta_f = \theta_i + \tfrac{1}{2}(\omega_i + \omega_f)t \tag{10.9}$$

Examples:

- during its spin cycle, the tub of a clothes washer begins from rest and accelerates up to its final spin speed
- a workshop grinding wheel is turned off and comes to rest under the action of a constant friction force in the bearings of the wheel
- a gyroscope is powered up and approaches its operating speed (Chapter 11)
- the crankshaft of a diesel engine changes to a higher angular speed (Chapter 21)

Example 10.1 Rotating Wheel

A wheel rotates with a constant angular acceleration of 3.50 rad/s².

(A) If the angular speed of the wheel is 2.00 rad/s at $t_i = 0$, through what angular displacement does the wheel rotate in 2.00 s?

SOLUTION

Conceptualize Look again at Figure 10.1. Imagine that the disc rotates with its angular speed increasing at a constant rate. You start your stopwatch when the disc is rotating at 2.00 rad/s. This mental image is a model for the motion of the wheel in this example.

continued

10.1 continued

Categorize   The phrase "with a constant angular acceleration" tells us to apply the *rigid object under constant angular acceleration* model to the wheel.

Analyze   From the rigid object under constant angular acceleration model, choose Equation 10.7 and rearrange it so that it expresses the angular displacement of the wheel:

$$\Delta\theta = \theta_f - \theta_i = \omega_i t + \tfrac{1}{2}\alpha t^2$$

Substitute the known values to find the angular displacement at $t = 2.00$ s:

$$\Delta\theta = (2.00 \text{ rad/s})(2.00 \text{ s}) + \tfrac{1}{2}(3.50 \text{ rad/s}^2)(2.00 \text{ s})^2$$
$$= 11.0 \text{ rad} = (11.0 \text{ rad})(180°/\pi \text{ rad}) = 630°$$

(B) Through how many revolutions has the wheel turned during this time interval?

SOLUTION

Multiply the angular displacement found in part (A) by a conversion factor to find the number of revolutions:

$$\Delta\theta = 630°\left(\frac{1 \text{ rev}}{360°}\right) = 1.75 \text{ rev}$$

(C) What is the angular speed of the wheel at $t = 2.00$ s?

SOLUTION

Use Equation 10.6 from the rigid object under constant angular acceleration model to find the angular speed at $t = 2.00$ s:

$$\omega_f = \omega_i + \alpha t = 2.00 \text{ rad/s} + (3.50 \text{ rad/s}^2)(2.00 \text{ s})$$
$$= 9.00 \text{ rad/s}$$

Finalize   We could also obtain this result using Equation 10.8 and the results of part (A). (Try it!)

WHAT IF?   Suppose a particle moves along a straight line with a constant acceleration of 3.50 m/s². If the velocity of the particle is 2.00 m/s at $t_i = 0$, through what displacement does the particle move in 2.00 s? What is the velocity of the particle at $t = 2.00$ s?

Answer   Notice that these questions are translational analogs to parts (A) and (C) of the original problem. The mathematical solution follows exactly the same form. For the displacement, from the particle under constant acceleration model,

$$\Delta x = x_f - x_i = v_i t + \tfrac{1}{2}at^2$$
$$= (2.00 \text{ m/s})(2.00 \text{ s}) + \tfrac{1}{2}(3.50 \text{ m/s}^2)(2.00 \text{ s})^2 = 11.0 \text{ m}$$

and for the velocity,

$$v_f = v_i + at = 2.00 \text{ m/s} + (3.50 \text{ m/s}^2)(2.00 \text{ s}) = 9.00 \text{ m/s}$$

There is no translational analog to part (B) because translational motion under constant acceleration is not repetitive.

10.3   Angular and Translational Quantities

In this section, we derive some useful relationships between the angular speed and acceleration of a rotating rigid object and the translational speed and acceleration of a point in the object. To do so, we must keep in mind that when a rigid object rotates about a fixed axis, every particle of the object moves in a circle whose center is on the axis of rotation. We looked at a flat, circular object in Figure 10.1. Let us now generalize to an arbitrary, three-dimensional object, as in Figure 10.4. A reference axis fixed in space is chosen—the x axis in Figure 10.4—and we look at the motion of one point P contained within the object.

Because point P in Figure 10.4 moves in a circle, the translational velocity vector $\vec{v}$ is always tangent to the circular path and hence is called *tangential velocity.* The magnitude of the tangential velocity of the point P is by definition the tangential speed $v = ds/dt$, where s is the distance traveled by this point measured along the circular path. Recalling that $s = r\theta$ (Eq. 10.1b) and noting that, for a given point on the object, r is constant, we obtain

$$v = \frac{ds}{dt} = r\frac{d\theta}{dt}$$

Because $d\theta/dt = \omega$ (see Eq. 10.3), it follows that

$$v = r\omega \tag{10.10}$$

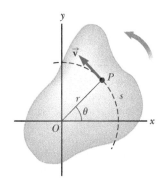

Figure 10.4 As a rigid object rotates about the fixed axis (the z axis) through O, the point P has a tangential velocity $\vec{v}$ that is always tangent to the circular path of radius r.

As we saw in Equation 4.24, the tangential speed of a particle moving in a circle equals the distance of the particle from the center of the circle multiplied by the angular speed. We find the same relationship for particles at every point on a rigid object. Although every point on the rigid object has the same *angular* speed, not every point has the same *tangential* speed because r is not the same for all points on the object. Equation 10.10 shows that the tangential speed of a point on the rotating object increases as one moves outward from the center of rotation, as we would intuitively expect. For example, the outer end of a swinging golf club moves much faster than a point near the handle.

We can relate the angular acceleration of the rotating rigid object to the tangential acceleration of the point P by taking the time derivative of v in Equation 10.10:

$$a_t = \frac{dv}{dt} = r\frac{d\omega}{dt}$$

$$a_t = r\alpha \tag{10.11}$$

◀ Relation between tangential acceleration and angular acceleration

That is, the tangential component of the translational acceleration of a point on a rotating rigid object equals the point's perpendicular distance from the axis of rotation multiplied by the angular acceleration.

In Section 4.4, we found that a point moving in a circular path undergoes a radial acceleration a_r directed toward the center of rotation and whose magnitude is that of the centripetal acceleration v^2/r (Fig. 10.5). Because $v = r\omega$ for a point P on a rotating object, we can express the centripetal acceleration at that point in terms of angular speed as we did for a particle moving in a circular path in Equation 4.25:

$$a_c = \frac{v^2}{r} = r\omega^2 \tag{10.12}$$

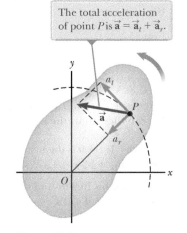

The total acceleration vector at the point is $\vec{a} = \vec{a}_t + \vec{a}_r$, where the magnitude of $\vec{a}_r$ is the centripetal acceleration a_c. Because $\vec{a}$ is a vector having a radial and a tangential component, the magnitude of $\vec{a}$ at the point P on the rotating rigid object is

$$a = \sqrt{a_t^2 + a_r^2} = \sqrt{r^2\alpha^2 + r^2\omega^4} = r\sqrt{\alpha^2 + \omega^4} \tag{10.13}$$

QUICK QUIZ 10.3 Ethan and Rebecca are riding on a merry-go-round. Ethan rides on a horse at the outer rim of the circular platform, twice as far from the center of the circular platform as Rebecca, who rides on an inner horse. **(i)** When the merry-go-round is rotating at a constant angular speed, what is Ethan's angular speed? (a) twice Rebecca's (b) the same as Rebecca's (c) half of Rebecca's (d) impossible to determine **(ii)** When the merry-go-round is rotating at a constant angular speed, describe Ethan's tangential speed from the same list of choices.

Figure 10.5 As a rigid object rotates about a fixed axis (the z axis) through O, the point P experiences a tangential component of translational acceleration a_t and a radial component of translational acceleration a_r.

Example 10.2 CD Player

Despite the availability of music in digital form, the compact disc, or CD, remains a popular format for music and data. On a CD (Fig. 10.6), audio information is stored digitally in a series of pits and flat areas on the surface of the disc. The alternations between pits and flat areas on the surface represent binary ones and zeros to be read by the CD player and converted back to sound waves. The pits and flat areas are detected by a system consisting of a laser and lenses. The length of a string of ones and zeros representing one piece of information is the same everywhere on the disc, whether the information is near the center of the disc or near its outer edge. So that this length of ones and zeros always passes by the laser–lens system in the same time interval, the tangential speed of the disc surface at the location of the lens must be constant. According to Equation 10.10, the angular speed must therefore vary as the laser–lens system moves radially along the disc. In a typical CD player, the constant speed of the surface at the point of the laser–lens system is 1.3 m/s.

(A) Find the angular speed of the disc in revolutions per minute when information is being read from the innermost first track ($r = 23$ mm) and the outermost final track ($r = 58$ mm).

SOLUTION

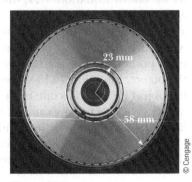

Conceptualize Figure 10.6 shows a photograph of a compact disc. Trace your finger around the circle marked "23 mm" and mentally estimate the time interval to go around the circle once. Now trace your finger around the circle marked "58 mm," moving your finger across the surface of the page at the same speed as you did when tracing the smaller circle. Notice how much longer in time it takes your finger to go around the larger circle. If your finger represents the laser reading the disc, you can see that the disc rotates once in a longer time interval when the laser reads the information in the outer circle. Therefore, the disc must rotate more slowly when the laser is reading information from this part of the disc.

Figure 10.6 (Example 10.2) A compact disc.

Categorize This part of the example is categorized as a simple substitution problem. In later parts, we will need to identify analysis models.

Use Equation 10.10 to find the angular speed that gives the required tangential speed at the position of the inner track:

$$\omega_i = \frac{v}{r_i} = \frac{1.3 \text{ m/s}}{2.3 \times 10^{-2} \text{ m}} = 57 \text{ rad/s}$$

$$= (57 \text{ rad/s})\left(\frac{1 \text{ rev}}{2\pi \text{ rad}}\right)\left(\frac{60 \text{ s}}{1 \text{ min}}\right) = 5.4 \times 10^2 \text{ rev/min}$$

Do the same for the outer track:

$$\omega_f = \frac{v}{r_f} = \frac{1.3 \text{ m/s}}{5.8 \times 10^{-2} \text{ m}} = 22 \text{ rad/s} = 2.1 \times 10^2 \text{ rev/min}$$

The CD player adjusts the angular speed ω of the disc within this range so that information moves past the objective lens at a constant rate.

(B) The maximum playing time of a standard music disc is 74 min and 33 s. How many revolutions does the disc make during that time?

SOLUTION

Categorize From part (A), the angular speed decreases as the disc plays. Let us assume it decreases steadily, with α constant. We can then apply the *rigid object under constant angular acceleration* model to the disc.

Analyze If $t = 0$ is the instant the disc begins rotating, with angular speed of 57 rad/s, the final value of the time t is (74 min) (60 s/min) + 33 s = 4 473 s. We are looking for the angular displacement $\Delta\theta$ during this time interval.

Use Equation 10.9 to find the angular displacement of the disc at $t = 4\,473$ s:

$$\Delta\theta = \theta_f - \theta_i = \tfrac{1}{2}(\omega_i + \omega_f)t$$

$$= \tfrac{1}{2}(57 \text{ rad/s} + 22 \text{ rad/s})(4\,473 \text{ s}) = 1.8 \times 10^5 \text{ rad}$$

Convert this angular displacement to revolutions:

$$\Delta\theta = (1.8 \times 10^5 \text{ rad})\left(\frac{1 \text{ rev}}{2\pi \text{ rad}}\right) = 2.8 \times 10^4 \text{ rev}$$

(c) What is the angular acceleration of the compact disc over the 4 473-s time interval?

SOLUTION

Categorize We again model the disc as a *rigid object under constant angular acceleration*. In this case, Equation 10.6 gives the value of the constant angular acceleration. Another approach is to use Equation 10.4 to find the average angular acceleration.

10.2 continued

In this case, we are not assuming the angular acceleration is constant. The answer is the same from both equations; only the interpretation of the result is different.

Analyze Use Equation 10.6 to find the angular acceleration:

$$\alpha = \frac{\omega_f - \omega_i}{t} = \frac{22 \text{ rad/s} - 57 \text{ rad/s}}{4\,473 \text{ s}} = -7.6 \times 10^{-3} \text{ rad/s}^2$$

Finalize The disc experiences a very gradual decrease in its rotation rate, as expected from the long time interval required for the angular speed to change from the initial value to the final value. In reality, the angular acceleration of the disc is not constant. Problem 46 allows you to explore the actual time behavior of the angular acceleration.

10.4 Torque

In our study of translational motion, after investigating the description of motion in Chapters 2–4, we studied the cause of changes in motion: force, in Chapters 5–6. We follow the same plan here: What is the cause of changes in rotational motion?

When a force is exerted on a rigid object pivoted about an axis, the object tends to rotate about that axis. Imagine trying to rotate a door by applying a force of magnitude F perpendicular to the door surface near the hinges and then at various distances from the hinges. You will achieve a more rapid rate of rotation for the door by applying the force near the doorknob than by applying it near the hinges. Because the *same* force was applied at different positions on the door, this experiment indicates that the cause of changes in rotational motion must also depend on the *location* at which the force is applied.

The cause of changes in the rotational motion of an object about some axis is measured by a quantity called **torque** $\vec{\tau}$ (Greek letter tau). Torque is a vector, but we will consider only its magnitude here; we will explore its vector nature in Chapter 11.

Consider the wrench and bolt from the opening storyline for this chapter. We show these objects with some geometry added in Figure 10.7. We wish to rotate the wrench around an axis that is perpendicular to the page and passes through the center of the bolt. The applied force $\vec{F}$ acts at an angle ϕ to the horizontal. We define the magnitude of the torque associated with the force $\vec{F}$ around the axis passing through O by the expression

$$\tau \equiv rF \sin \phi = Fd \tag{10.14}$$

where r is the distance between the rotation axis and the point of application of $\vec{F}$, and d is the perpendicular distance from the rotation axis to the line of action of $\vec{F}$. (The *line of action* of a force is an imaginary line extending out both ends of the vector representing the force. The dashed line extending from the tail of $\vec{F}$ in Fig. 10.7 is part of the line of action of $\vec{F}$.) From the right triangle in Figure 10.7 that has the wrench as its hypotenuse, we see that $d = r \sin \phi$. The quantity d is called the **moment arm** (or *lever arm*) of $\vec{F}$.

In Figure 10.7, the only component of $\vec{F}$ that tends to cause rotation of the wrench around an axis through O is $F \sin \phi$, the component perpendicular to a line drawn from the rotation axis to the point of application of the force. The horizontal component $F \cos \phi$, because its line of action passes through O, has no tendency to produce rotation about an axis passing through O. From the definition of torque in Equation 10.14, the cause of changes in rotational motion increases as F increases and as d increases, which explains why it is easier to rotate a door if we push at the doorknob rather than at a point close to the hinges. We also want to apply our push as closely perpendicular to the door as we can so that ϕ is close to 90°, which maximizes the moment arm. Pushing sideways on the doorknob ($\phi = 0$) will not cause the door to rotate.

Equation 10.14 allows us to understand the use of the pipe to turn the wrench in the opening storyline. The maximum force you can apply to the wrench is not

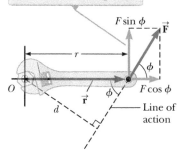

The component $F \sin \phi$ tends to rotate the wrench about an axis through O.

Figure 10.7 The force $\vec{F}$ has a greater rotating tendency about an axis through O as F increases and as the moment arm d increases.

PITFALL PREVENTION 10.4
Torque Depends on Your Choice of Axis There is no unique value of the torque on an object. Its value depends on your choice of rotation axis.

◀ Moment arm

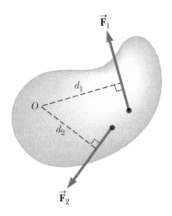

Figure 10.8 The force $\vec{F}_1$ tends to rotate the object counterclockwise about an axis through O, and $\vec{F}_2$ tends to rotate it clockwise.

enough to turn the bolt. You cannot apply more *force* $\vec{F}$, but you can increase the *torque* on the bolt by putting the pipe over the wrench handle. This allows you to apply the same force at a larger distance d from the axis of rotation. You have increased the moment arm of the force and, therefore, increased the torque applied by the *same* force.

If two or more forces act on a rigid object as in Figure 10.8, each tends to produce rotation about the axis through O. In this example, $\vec{F}_2$ tends to rotate the object clockwise and $\vec{F}_1$ tends to rotate it counterclockwise. We use the convention that the sign of the torque resulting from a force is positive if the turning tendency of the force is counterclockwise and negative if the turning tendency is clockwise. For Example, in Figure 10.8, the torque resulting from $\vec{F}_1$, which has a moment arm d_1, is positive and equal to $+F_1d_1$; the torque from $\vec{F}_2$ is negative and equal to $-F_2d_2$. Hence, the *net* torque about an axis through O is

$$\sum \tau = \tau_1 + \tau_2 = F_1d_1 - F_2d_2$$

Torque should not be confused with force. Forces can cause a change in translational motion as described by Newton's second law. Forces can also cause a change in rotational motion, but the effectiveness of the forces in causing this change depends on both the magnitudes of the forces and the moment arms of the forces, in the combination we call *torque*. Torque has units of force times length—newton meters (N · m) in SI units—and should be reported in these units. Do not confuse torque and work (Chapter 7), which have the same units but are very different concepts.

QUICK QUIZ 10.4 If you are trying to loosen a stubborn screw from a piece of wood with a screwdriver and fail, should you find a screwdriver for which the handle is (a) longer or (b) fatter?

Example 10.3 **The Net Torque on a Cylinder**

A one-piece cylinder is shaped as shown in Figure 10.9, with a core section protruding from the larger drum. The cylinder is free to rotate about the central z axis shown in the drawing. A rope wrapped around the drum, which has radius R_1, exerts a force $\vec{T}_1$ to the right on the cylinder. A rope wrapped around the core, which has radius R_2, exerts a force $\vec{T}_2$ downward on the cylinder.

(A) What is the net torque acting on the cylinder about the rotation axis (which is the z axis in Fig. 10.9)?

SOLUTION

Conceptualize Imagine that the cylinder in Figure 10.9 is a shaft in a machine. The force $\vec{T}_1$ could be applied by a drive belt wrapped around the drum. The force $\vec{T}_2$ could be applied by a friction brake at the surface of the core.

Categorize This example is a substitution problem in which we evaluate the net torque using Equation 10.14.

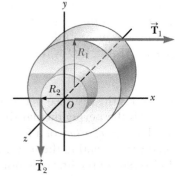

Figure 10.9 (Example 10.3) A solid cylinder pivoted about the z axis through O. The moment arm of $\vec{T}_1$ is R_1, and the moment arm of $\vec{T}_2$ is R_2.

The torque due to $\vec{T}_1$ about the rotation axis is $-R_1T_1$. (The sign is negative because the torque tends to produce clockwise rotation.) The torque due to $\vec{T}_2$ is $+R_2T_2$. (The sign is positive because the torque tends to produce counterclockwise rotation of the cylinder.)

Evaluate the net torque about the rotation axis: $\sum \tau = \tau_1 + \tau_2 = R_2T_2 - R_1T_1$

As a quick check, notice that if the two forces are of equal magnitude, the net torque is negative because $R_1 > R_2$. Starting from rest with both forces of equal magnitude acting on it, the cylinder would rotate clockwise because $\vec{T}_1$ would be more effective at turning it than would $\vec{T}_2$.

10.3 continued

(B) Suppose $T_1 = 5.0$ N, $R_1 = 1.0$ m, $T_2 = 15$ N, and $R_2 = 0.50$ m. What is the net torque about the rotation axis, and which way does the cylinder rotate starting from rest?

SOLUTION

Substitute the given values: $\qquad\sum \tau = (0.50 \text{ m})(15 \text{ N}) - (1.0 \text{ m})(5.0 \text{ N}) = 2.5 \text{ N} \cdot \text{m}$

Because this net torque is positive, the cylinder begins to rotate in the counterclockwise direction.

10.5 Analysis Model: Rigid Object Under a Net Torque

In Chapter 5, we learned that a net force on an object causes an acceleration of the object and that the acceleration is proportional to the net force. These facts are the basis of the particle under a net force model whose mathematical representation is Newton's second law. In this section, we show the rotational analog of Newton's second law: the angular acceleration of a rigid object rotating about a fixed axis is proportional to the net torque acting about that axis. Before discussing the more complex case of rigid-object rotation, however, it is instructive first to discuss the case of a particle moving in a circular path about some fixed point under the influence of an external force.

Consider a particle of mass m rotating in a circle of radius r under the influence of a tangential net force $\sum \vec{\mathbf{F}}_t$ and a radial net force $\sum \vec{\mathbf{F}}_r$ as shown in Figure 10.10. The radial net force causes the particle to move in the circular path with a centripetal acceleration. The tangential force provides a tangential acceleration $\vec{\mathbf{a}}_t$, and

$$\sum F_t = ma_t$$

The magnitude of the net torque due to $\sum \vec{\mathbf{F}}_t$ on the particle about an axis perpendicular to the page through the center of the circle is

$$\sum \tau = \sum F_t r = (ma_t)r$$

Because the tangential acceleration is related to the angular acceleration through the relationship $a_t = r\alpha$ (Eq. 10.11), the net torque can be expressed as

$$\sum \tau = (mr\alpha)r = (mr^2)\alpha \qquad (10.15)$$

Let us denote the quantity mr^2 with the symbol I for now. We will say more about this quantity below. Using this notation, Equation 10.15 can be written as

$$\sum \tau = I\alpha \qquad (10.16)$$

That is, the net torque acting on the particle is proportional to its angular acceleration. Notice that $\sum \tau = I\alpha$ has the same mathematical form as Newton's second law of motion, $\sum F = ma$ (Eq. 5.2).

Now let us extend this discussion to a rigid object of arbitrary shape rotating about a fixed axis passing through a point O as in Figure 10.11. The object can be regarded as a collection of particles of mass m_i. If we impose a Cartesian coordinate system on the object, each particle rotates in a circle about the origin and each has a tangential acceleration a_i produced by an external tangential force of magnitude F_i. For any given particle, we know from Newton's second law that

$$F_i = m_i a_i$$

The external torque $\vec{\boldsymbol{\tau}}_i$ associated with the force $\vec{\mathbf{F}}_i$ acts about the origin and its magnitude is given by

$$\tau_i = r_i F_i = r_i m_i a_i$$

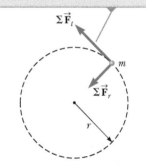

The tangential force on the particle results in a torque on the particle about an axis through the center of the circle.

Figure 10.10 A particle rotating in a circle under the influence of a tangential net force $\sum \vec{\mathbf{F}}_t$. A radial net force $\sum \vec{\mathbf{F}}_r$ also must be present to maintain the circular motion.

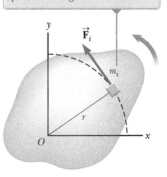

The particle of mass m_i of the rigid object experiences a torque in the same way that the particle in Figure 10.10 does.

Figure 10.11 A rigid object rotating about an axis through O. Each particle of mass m_i rotates about the axis with the same angular acceleration α.

Because $a_i = r_i \alpha$, the expression for τ_i becomes

$$\tau_i = m_i r_i^2 \alpha$$

Although different particles in the rigid object may have different translational accelerations a_i, they all have the *same* angular acceleration α. With that in mind, we can add the torques on all of the particles making up the rigid object to obtain the net torque on the object about an axis through O due to all external forces:

$$\sum \tau_{\text{ext}} = \sum_i \tau_i = \sum_i m_i r_i^2 \alpha = \left(\sum_i m_i r_i^2 \right) \alpha \qquad (10.17)$$

where α can be taken outside the summation because it is common to all particles. Calling the quantity in parentheses I, the expression for $\sum \tau_{\text{ext}}$ becomes

◀ Torque on a rigid object is proportional to angular acceleration

$$\sum \tau_{\text{ext}} = I\alpha \qquad (10.18)$$

This equation for a rigid object is the same as that found for a particle moving in a circular path (Eq. 10.16). The net torque about the rotation axis is proportional to the angular acceleration of the object, with the proportionality factor being I, a quantity that we have yet to describe fully. Equation 10.18 is the mathematical representation of the analysis model of a **rigid object under a net torque,** the rotational analog to the particle under a net force.

Let us now address the quantity I, defined as follows:

$$I = \sum_i m_i r_i^2 \qquad (10.19)$$

This quantity is called the **moment of inertia** of the object, and depends on the masses of the particles making up the object and their distances from the rotation axis. Notice that Equation 10.19 reduces to $I = mr^2$ for a single particle, consistent with our use of the notation I in going from Equation 10.15 to Equation 10.16. Note that moment of inertia has units of kg · m² in SI units.

Equation 10.18 has the same form as Newton's second law for a system of particles as expressed in Equation 9.39:

$$\sum \vec{F}_{\text{ext}} = M\vec{a}_{\text{CM}}$$

Consequently, the moment of inertia I must play the same role in rotational motion as the role that mass plays in translational motion: the moment of inertia is the resistance to changes in rotational motion. This resistance depends not only on the mass of the object, but also on how the mass is distributed around the rotation axis. Table 10.2 gives the moments of inertia[2] for a number of objects about specific axes. The moments of inertia of rigid objects with simple geometry (high symmetry) are relatively easy to calculate provided the rotation axis coincides with an axis of symmetry, as we show in the next section.

QUICK QUIZ 10.5 You turn off your electric drill and find that the time interval for the rotating bit to come to rest due to frictional torque in the drill is Δt. You replace the bit with a larger one that results in a doubling of the moment of inertia of the drill's entire rotating mechanism. When this larger bit is rotated at the same angular speed as the first and the drill is turned off, the frictional torque remains the same as that for the previous situation. What is the time interval for this second bit to come to rest? **(a)** $4\Delta t$ **(b)** $2\Delta t$ **(c)** Δt **(d)** $0.5\Delta t$ **(e)** $0.25\Delta t$ **(f)** impossible to determine

---

[2]Civil engineers use moment of inertia to characterize the elastic properties (rigidity) of such structures as loaded beams. Hence, it is often useful even in a nonrotational context.

TABLE 10.2 Moments of Inertia of Homogeneous Rigid Objects with Different Geometries

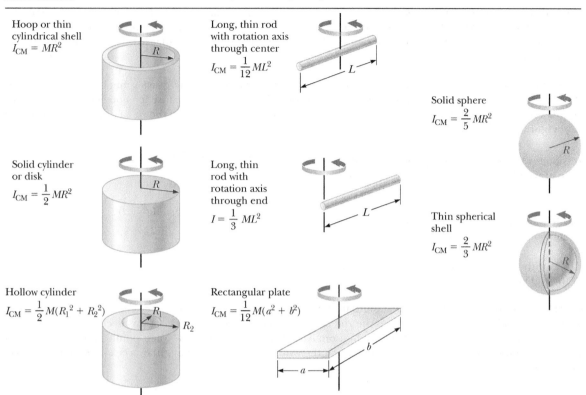

Hoop or thin cylindrical shell $I_{CM} = MR^2$

Long, thin rod with rotation axis through center $I_{CM} = \frac{1}{12}ML^2$

Solid sphere $I_{CM} = \frac{2}{5}MR^2$

Solid cylinder or disk $I_{CM} = \frac{1}{2}MR^2$

Long, thin rod with rotation axis through end $I = \frac{1}{3}ML^2$

Thin spherical shell $I_{CM} = \frac{2}{3}MR^2$

Hollow cylinder $I_{CM} = \frac{1}{2}M(R_1^2 + R_2^2)$

Rectangular plate $I_{CM} = \frac{1}{12}M(a^2 + b^2)$

ANALYSIS MODEL **Rigid Object Under a Net Torque**

Imagine you are analyzing the motion of an object that is free to rotate about a fixed axis. The cause of changes in rotational motion of this object is torque applied to the object and, in parallel to Newton's second law for translation motion, the torque is equal to the product of the moment of inertia of the object and the angular acceleration:

$$\sum \tau_{ext} = I\alpha \qquad \textbf{(10.18)}$$

The torque, the moment of inertia, and the angular acceleration must all be evaluated around the same rotation axis.

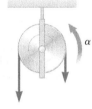

Examples:

- a bicycle chain around the sprocket of a bicycle causes the rear wheel of the bicycle to rotate
- an electric dipole moment in an electric field rotates due to the electric force from the field (Chapter 22)
- a magnetic dipole moment in a magnetic field rotates due to the magnetic force from the field (Chapter 28)
- the armature of a motor rotates due to the torque exerted by a surrounding magnetic field (Chapter 30)

Example 10.4 **Rotating Rod**

A uniform rod of length L and mass M is attached at one end to a frictionless pivot and is free to rotate about the pivot in the vertical plane as in Figure 10.12. The rod is released from rest in the horizontal position. What are the initial angular acceleration of the rod and the initial translational acceleration of its right end?

SOLUTION

Conceptualize Imagine what happens to the rod in Figure 10.12 when it is released. It rotates clockwise around the pivot at the left end. When an object is pivoted at a point other than its center of mass, the gravitational force, assumed to be acting through the center of mass, provides a torque about the pivot.

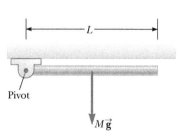

Figure 10.12 (Example 10.4) A rod is free to rotate around a pivot at the left end. The gravitational force on the rod acts at its center of mass.

continued

10.4 continued

Categorize The rod is categorized as a *rigid object under a net torque.* The torque is due only to the gravitational force on the rod if the rotation axis is chosen to pass through the pivot in Figure 10.12. We *cannot* categorize the rod as a rigid object under constant angular acceleration because the torque exerted on the rod and therefore the angular acceleration of the rod vary with its angular position.

Analyze The only force contributing to the torque about an axis through the pivot is the gravitational force $M\vec{g}$ exerted on the rod. (The force exerted by the pivot on the rod has zero torque about the pivot because its moment arm is zero.) To compute the torque on the rod, we assume the gravitational force acts at the center of mass of the rod a distance $L/2$ from the pivot as shown in Figure 10.12.

Write an expression for the magnitude of the net external torque due to the gravitational force about an axis through the pivot:

$$\sum \tau_{ext} = Mg\left(\frac{L}{2}\right)$$

Use Equation 10.18 to obtain the angular acceleration of the rod, using the moment of inertia for the rod from Table 10.2:

$$(1) \quad \alpha = \frac{\sum \tau_{ext}}{I} = \frac{Mg(L/2)}{\frac{1}{3}ML^2} = \frac{3g}{2L}$$

Use Equation 10.11 with $r = L$ to find the initial translational acceleration of the right end of the rod:

$$a_t = L\alpha = \tfrac{3}{2}g$$

Finalize These values are the *initial* values of the angular and translational accelerations. Once the rod begins to rotate, the gravitational force is no longer perpendicular to the rod and the values of the two accelerations decrease, going to zero at the moment the rod passes through the vertical orientation. Also, because the value of a_t at a point on the rod depends on the distance of that point from the pivot, every point along the rod will have the *same* angular acceleration but a *different* tangential acceleration.

Conceptual Example **10.5** Falling Smokestacks and Tumbling Blocks

When a tall smokestack falls over, it often breaks somewhere along its length before it hits the ground as shown in Figure 10.13. Why?

SOLUTION

As the smokestack rotates around its base, each higher portion of the smokestack falls with a larger tangential acceleration than the portion below it according to Equation 10.11. The angular acceleration increases as the smokestack tips farther. Eventually, higher portions of the smokestack experience a tangential acceleration greater than the acceleration that could result from gravity alone; this situation is similar to that for the end of the rod in Example 10.4. That can happen only if these portions are being pulled downward by a force in addition to the gravitational force. The force that causes that to occur is the shear force from lower portions of the smokestack. Eventually, the shear force that provides this acceleration is greater than the smokestack can withstand, and

Figure 10.13 (Conceptual Example 10.5) A falling smokestack breaks at some point along its length.

the smokestack breaks. The same thing happens with a tall tower of children's toy blocks. Borrow some blocks from a child and build such a tower. Push it over and watch it come apart at some point before it strikes the floor.

Example **10.6** Angular Acceleration of a Wheel

A wheel of radius R, mass M, and moment of inertia I is mounted on a frictionless, horizontal axle as in Figure 10.14. A light cord wrapped around the wheel supports an object of mass m. When the wheel is released, the object accelerates downward, the cord unwraps off the wheel, and the wheel rotates with an angular acceleration. Find expressions for the angular acceleration of the wheel, the translational acceleration of the object, and the tension in the cord.

SOLUTION

Conceptualize Imagine that the object is a bucket in an old-fashioned water well. It is tied to a cord that passes around a cylinder equipped with a crank for raising the bucket. After the bucket has been raised, the system is released and the bucket accelerates downward while the cord unwinds off the cylinder.

10.6 continued

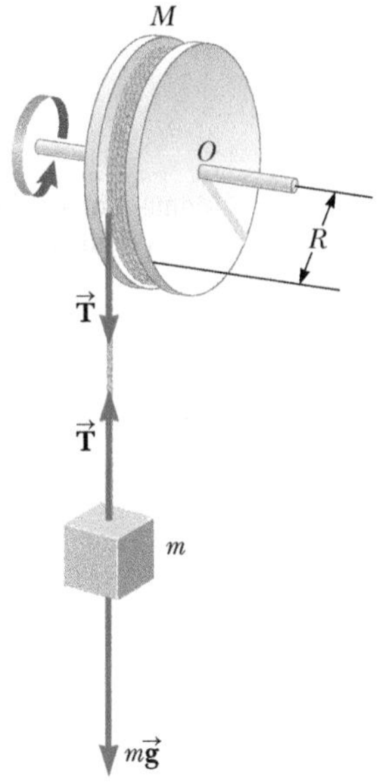

Categorize We apply two analysis models here. The object is modeled as a *particle under a net force*. The wheel is modeled as a *rigid object under a net torque*.

Analyze The magnitude of the torque acting on the wheel about its axis of rotation is $\tau = TR$, where T is the force exerted by the cord on the rim of the wheel. (The gravitational force exerted by the Earth on the wheel and the normal force exerted by the axle on the wheel both pass through the axis of rotation and therefore produce no torque about the axle.)

From the rigid object under a net torque model, write Equation 10.18:

$$\sum \tau_{ext} = I\alpha$$

Solve for α and substitute the net torque:

$$(1) \quad \alpha = \frac{\sum \tau_{ext}}{I} = \frac{TR}{I}$$

From the particle under a net force model, apply Newton's second law to the motion of the object, taking the downward direction to be positive:

$$\sum F_y = mg - T = ma$$

Solve for the acceleration a:

$$(2) \quad a = \frac{mg - T}{m}$$

Equations (1) and (2) have three unknowns: α, a, and T. Because the object and wheel are connected by a cord that does not slip, the translational acceleration of the suspended object is equal to the tangential acceleration of a point on the wheel's rim. Therefore, the angular acceleration α of the wheel and the translational acceleration of the object are related by $a = R\alpha$ (Eq. 10.11).

Use this fact together with Equations (1) and (2):

$$(3) \quad a = R\alpha = \frac{TR^2}{I} = \frac{mg - T}{m}$$

Solve for the tension T:

$$(4) \quad T = \frac{mg}{1 + (mR^2/I)}$$

Substitute Equation (4) into Equation (2) and solve for a:

$$(5) \quad a = \frac{g}{1 + (I/mR^2)}$$

Use $a = R\alpha$ and Equation (5) to solve for α:

$$\alpha = \frac{a}{R} = \frac{g}{R + (I/mR)}$$

Figure 10.14 (Example 10.6) An object hangs from a cord wrapped around a wheel.

Finalize We finalize this problem by imagining the behavior of the system in some extreme limits.

WHAT IF? What if the wheel were to become very massive so that I becomes very large? What happens to the acceleration a of the object and the tension T?

Answer If the wheel becomes infinitely massive, we can imagine that the object of mass m will simply hang from the cord without causing the wheel to rotate.

We can show that mathematically by taking the limit $I \rightarrow \infty$. Equation (5) then becomes

$$a = \lim_{I \to \infty} \frac{g}{1 + (I/mR^2)} = 0$$

which agrees with our conceptual conclusion that the object will hang at rest. Also, Equation (4) becomes

$$T = \lim_{I \to \infty} \frac{mg}{1 + (mR^2/I)} = mg$$

which is consistent because the object simply hangs at rest in equilibrium between the gravitational force and the tension in the string.

10.6 Calculation of Moments of Inertia

The moment of inertia of a system of discrete particles can be calculated in a straightforward way with Equation 10.19. On the other hand, suppose we consider a continuous rigid object. We can evaluate its moment of inertia by imagining the object to be divided into many small elements, each of which has mass Δm_i. We use the definition $I = \Sigma_i \, r_i^2 \, \Delta m_i$ and take the limit of this sum as $\Delta m_i \to 0$. In this limit, the sum becomes an integral over the volume of the object:

Moment of inertia ▶
of a rigid object

$$I = \lim_{\Delta m_i \to 0} \sum_i r_i^2 \, \Delta m_i = \int r^2 \, dm \qquad (10.20)$$

It is usually easier to calculate moments of inertia in terms of the volume of the elements rather than their mass, and we can easily make that change by using Equation 1.1, $\rho \equiv m/V$, where ρ is the density of the object and V is its volume. From this equation, the mass of a small element is $dm = \rho \, dV$. Substituting this result into Equation 10.20 gives

$$I = \int \rho r^2 \, dV \qquad (10.21)$$

If the object is homogeneous, ρ is constant and the integral can be evaluated for a known geometry. If ρ is not constant, its variation with position must be known to complete the integration.

The density given by $\rho = m/V$ sometimes is referred to as *volumetric mass density* because it represents mass per unit volume. Often we use other ways of expressing density. For instance, when dealing with a sheet of uniform thickness t, we can define a *surface mass density* $\sigma = m/A = \rho t$, which represents *mass per unit area*. Finally, when mass is distributed along a rod of uniform cross-sectional area A, we sometimes use *linear mass density* $\lambda = m/L = \rho A$, which is the *mass per unit length*.

Example 10.7 | Uniform Rigid Rod

Calculate the moment of inertia of a uniform thin rod of length L and mass M (Fig. 10.15) about an axis perpendicular to the rod (the y axis) and passing through its center of mass.

SOLUTION

Conceptualize Imagine twirling the rod in Figure 10.15 with your fingers around its midpoint. If you have a meterstick handy, use it to simulate the spinning of a thin rod and feel the resistance it offers to being spun.

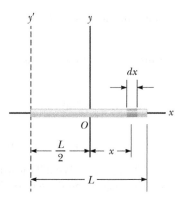

Figure 10.15 (Example 10.7) A uniform rigid rod of length L. The moment of inertia about the y axis is less than that about the y' axis. The latter axis is examined in Example 10.9.

Categorize This example is a substitution problem, using the definition of moment of inertia in Equation 10.20. As with any integration problem, the solution involves reducing the integrand to a single variable.

The shaded length element dx in Figure 10.15 has a mass dm equal to the mass per unit length λ multiplied by dx.

Express dm in terms of dx:

$$dm = \lambda \, dx = \frac{M}{L} \, dx$$

Substitute this expression into Equation 10.20, with $r^2 = x^2$:

$$I_y = \int r^2 \, dm = \int_{-L/2}^{L/2} x^2 \frac{M}{L} \, dx = \frac{M}{L} \int_{-L/2}^{L/2} x^2 \, dx$$

$$= \frac{M}{L} \left[\frac{x^3}{3} \right]_{-L/2}^{L/2} = \tfrac{1}{12} M L^2$$

Check this result in Table 10.2.

Example 10.8 | Uniform Solid Cylinder

A uniform solid cylinder has a radius R, mass M, and length L. Calculate its moment of inertia about its central axis (the z axis in Fig. 10.16).

10.8 c o n t i n u e d

SOLUTION

Conceptualize To simulate this situation, imagine twirling a can of frozen juice around its central axis. Don't twirl a nonfrozen can of vegetable soup; it is not a rigid object! The liquid is able to move relative to the metal can.

Categorize This example is a substitution problem, using the definition of moment of inertia. As with Example 10.7, we must reduce the integrand to a single variable.

It is convenient to divide the cylinder into many cylindrical shells, each having radius r, thickness dr, and length L as shown in Figure 10.16. The density of the cylinder is ρ. The volume dV of each shell is its cross-sectional area multiplied by its length: $dV = L\,dA = L(2\pi r)\,dr$.

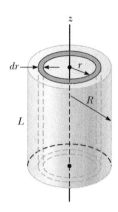

Figure 10.16 (Example 10.8) Calculating I about the z axis for a uniform solid cylinder.

Express dm in terms of dr:

$$dm = \rho\,dV = \rho L(2\pi r)\,dr$$

Substitute this expression into Equation 10.20:

$$I_z = \int r^2\,dm = \int r^2[\rho L(2\pi r)\,dr] = 2\pi\rho L \int_0^R r^3\,dr = \tfrac{1}{2}\pi\rho L R^4$$

Use the total volume $\pi R^2 L$ of the cylinder to express its density:

$$\rho = \frac{M}{V} = \frac{M}{\pi R^2 L}$$

Substitute this value into the expression for I_z:

$$I_z = \tfrac{1}{2}\pi\left(\frac{M}{\pi R^2 L}\right)LR^4 = \tfrac{1}{2}MR^2$$

Check this result in Table 10.2.

WHAT IF? What if the length of the cylinder in Figure 10.16 is increased to $2L$, while the mass M and radius R are held fixed? (The density becomes half as large.) How does that change the moment of inertia of the cylinder?

Answer Notice that the result for the moment of inertia of a cylinder does not depend on L, the length of the cylinder. It applies equally well to a long cylinder and a flat disk having the *same* mass M and radius R. Therefore, the moment of inertia of the cylinder around the central axis is not affected by how the mass is distributed along its length.

The calculation of moments of inertia of an object about an arbitrary axis can be cumbersome, even for a highly symmetric object. For example, imagine trying to find the moment of inertia of the cylinder in Figure 10.16 around an axis parallel to the z axis, but offset by the radius R of the cylinder, so that the axis just grazes along the outer surface of the cylinder. There is no symmetry around this axis! Fortunately, use of an important theorem, called the **parallel-axis theorem**, often simplifies the calculation.

To generate the parallel-axis theorem, suppose the object in Figure 10.17a (page 266) rotates about the z' axis. The moment of inertia does not depend on how the mass is distributed along the z' axis; as we found in Example 10.8, for example, the moment of inertia of a cylinder is independent of its length. Imagine collapsing the three-dimensional object in Figure 10.17a into a planar object of the same mass as in Figure 10.17b. In this imaginary process, all mass moves parallel to the z' axis until it lies in the $x'y'$ plane. The coordinates of the object's center of mass are now x'_{CM}, y'_{CM}, and $z'_{CM} = 0$. Let the mass element dm have coordinates $(x', y', 0)$ as shown in the view down the z' axis in Figure 10.17c. Because this element is a distance $r' = \sqrt{(x')^2 + (y')^2}$ from the z' axis, the moment of inertia of the entire object about the z' axis is

$$I = \int (r')^2\,dm = \int [(x')^2 + (y')^2]\,dm$$

We can relate the coordinates x', y' of the mass element dm to the coordinates of this same element located in a coordinate system having the object's center

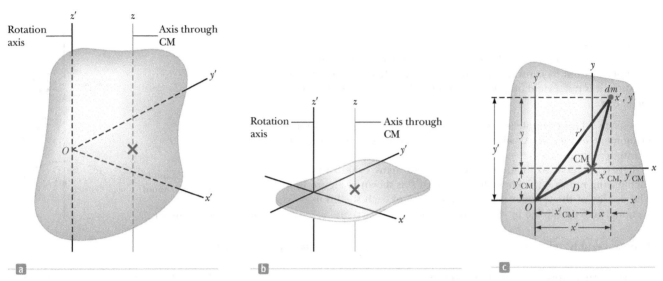

Figure 10.17 (a) An arbitrarily shaped rigid object. The origin of the coordinate system is not at the center of mass of the object. Imagine the object rotating about the z' axis. (b) All mass elements of the object are collapsed parallel to the z' axis to form a planar object. (c) An arbitrary mass element dm is indicated in blue in this view down the z' axis. The parallel axis theorem can be used with the geometry shown to determine the moment of inertia of the original object around the z' axis.

of mass as its origin. If the coordinates of the center of mass are x'_{CM}, and y'_{CM} in the original coordinate system centered on O, we see from Figure 10.17c that the relationships between the unprimed and primed coordinates are $x' = x + x'_{CM}$, $y' = y + y'_{CM}$, and $z' = z = 0$. Therefore,

$$I = \int \left[(x + x'_{CM})^2 + (y + y'_{CM})^2 \right] dm$$

$$= \int (x^2 + y)^2 \, dm + 2x'_{CM} \int x \, dm + 2y'_{CM} \int y \, dm + (x'^2_{CM} + y'^2_{CM}) \int dm$$

The first integral is, by definition, the moment of inertia I_{CM} about an axis that is parallel to the z' axis and passes through the center of mass. The second two integrals are zero because, by definition of the center of mass, $\int x \, dm = \int y \, dm = 0$. The last integral is simply MD^2 because $\int dm = M$ and $D^2 = x'^2_{CM} + y'^2_{CM}$. Therefore, we conclude that

Parallel-axis theorem ▶

$$I = I_{CM} + MD^2 \tag{10.22}$$

The parallel axis theorem allows us to evaluate the moment of inertia of an object of mass M about any axis that is parallel to its central axis as the moment of inertia around the central axis plus the term MD^2, where D is the perpendicular distance between the axes.

Example 10.9 | Applying the Parallel-Axis Theorem

Consider once again the uniform rigid rod of mass M and length L shown in Figure 10.15. Find the moment of inertia of the rod about an axis perpendicular to the rod through one end (the y' axis in Fig. 10.15).

SOLUTION

Conceptualize Imagine twirling the rod around an endpoint rather than the midpoint. If you have a meterstick handy, try it and notice the degree of difficulty in rotating it around the end compared with rotating it around the center.

10.9 continued

Categorize This example is a substitution problem, involving the parallel-axis theorem.

Intuitively, we expect the moment of inertia to be greater than the result $I_{CM} = \frac{1}{12}ML^2$ from Example 10.7 because there is mass up to a distance of L away from the rotation axis, whereas the farthest distance in Example 10.7 was only $L/2$. The distance between the center-of-mass axis and the y' axis is $D = L/2$.

Use the parallel-axis theorem:

$$I = I_{CM} + MD^2 = \tfrac{1}{12}ML^2 + M\left(\frac{L}{2}\right)^2 = \tfrac{1}{3}ML^2$$

Check this result in Table 10.2. As predicted in Example 10.7, it is more difficult to rotate the rod about one end than about the center of mass.

10.7 Rotational Kinetic Energy

After investigating the role of forces in our study of translational motion, we turned our attention to approaches involving energy in Chapters 7 and 8. We do the same thing in our current study of rotational motion.

In Chapter 7, we defined the kinetic energy of an object as the energy associated with its motion through space. An object rotating about a fixed axis remains stationary in space, so there is no kinetic energy associated with translational motion. The individual particles making up the rotating object, however, are moving through space; they follow circular paths. Consequently, there is kinetic energy associated with rotational motion.

Let us consider an object as a system of particles and assume it rotates about a fixed z axis with an angular speed ω. Figure 10.18 shows the rotating object and identifies one particle on the object located at a distance r_i from the rotation axis. If the mass of the ith particle is m_i and its tangential speed is v_i, its kinetic energy is

$$K_i = \tfrac{1}{2}m_i v_i^2$$

To proceed further, recall that although every particle in the rigid object has the same angular speed ω, the individual tangential speeds depend on the distance r_i from the axis of rotation according to Equation 10.10. The *total* kinetic energy of the rotating rigid object is the sum of the kinetic energies of the individual particles:

$$K_R = \sum_i K_i = \sum_i \tfrac{1}{2}m_i v_i^2 = \tfrac{1}{2}\sum_i m_i r_i^2 \omega^2$$

We can write this expression in the form

$$K_R = \tfrac{1}{2}\left(\sum_i m_i r_i^2\right)\omega^2 \tag{10.23}$$

where we have factored ω^2 from the sum because it is common to every particle. We recognize the quantity in parentheses as the moment of inertia of the object, introduced in Section 10.5.

Therefore, Equation 10.23 can be written

$$K_R = \tfrac{1}{2}I\omega^2 \tag{10.24}$$

◄ Rotational kinetic energy (Compare to Equation 7.16)

Compare Equation 10.24 to Equation 7.16 for the kinetic energy of an object in translational motion. Again, as in the discussion following Equation 10.19, we see that moment of inertia I plays the same role in rotational motion as mass m does in translational motion. Although we commonly refer to the quantity $\frac{1}{2}I\omega^2$ as **rotational kinetic energy,** it is not a new form of energy. It is ordinary kinetic energy because it is derived from a sum over individual kinetic energies of the particles contained in the rigid object. The mathematical form of the kinetic energy given by Equation 10.24 is convenient when we are dealing with rotational motion, provided we know how to calculate I.

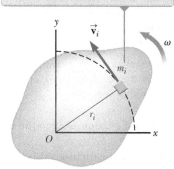

The particle of mass m_i of the rigid object has the same kinetic energy as if it were moving through space with the same speed.

Figure 10.18 A rigid object rotating about the z axis with angular speed ω. The kinetic energy of the particle of mass m_i is $\frac{1}{2}m_i v_i^2$. The total kinetic energy of the object is called its rotational kinetic energy.

QUICK QUIZ **10.6** A section of hollow pipe and a solid cylinder have the same radius, mass, and length. They both rotate about their long central axes with the same angular speed. Which object has the higher rotational kinetic energy? **(a)** The hollow pipe does. **(b)** The solid cylinder does. **(c)** They have the same rotational kinetic energy. **(d)** It is impossible to determine.

Example **10.10**   An Unusual Baton

Four tiny spheres are fastened to the ends of two rods of negligible mass lying in the *xy* plane to form an unusual baton (Fig. 10.19). We shall assume the radii of the spheres are small compared with the dimensions of the rods.

(A) If the system rotates about the *y* axis (Fig. 10.19a) with an angular speed ω, find the moment of inertia and the rotational kinetic energy of the system about this axis.

SOLUTION

Conceptualize Figure 10.19 is a pictorial representation that helps conceptualize the system of spheres and how it spins. Model the spheres as particles. Notice that only the blue spheres contribute to the moment of inertia around the *y* axis.

Categorize This example is a substitution problem because it is a straightforward application of the definitions discussed in this section.

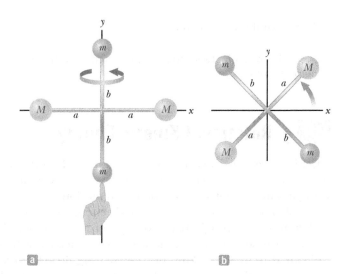

Figure 10.19 (Example 10.10) Four spheres form an unusual baton. (a) The baton is rotated about the *y* axis. (b) The baton is rotated about the *z* axis.

Apply Equation 10.19 to the system:

$$I_y = \sum_i m_i r_i^2 = Ma^2 + Ma^2 = 2Ma^2$$

Evaluate the rotational kinetic energy using Equation 10.24:

$$K_R = \tfrac{1}{2}I_y\omega^2 = \tfrac{1}{2}(2Ma^2)\omega^2 = Ma^2\omega^2$$

That the two spheres of mass *m* do not enter into this result makes sense because they have no motion about the axis of rotation; hence, they have no rotational kinetic energy. By similar logic, we expect the moment of inertia about the *x* axis to be $I_x = 2mb^2$ with a rotational kinetic energy about that axis of $K_R = mb^2\omega^2$.

(B) Suppose the system rotates in the *xy* plane about an axis (the *z* axis) through the center of the baton (Fig. 10.19b). Calculate the moment of inertia and rotational kinetic energy about this axis.

SOLUTION

Apply Equation 10.19 for this new rotation axis:

$$I_z = \sum_i m_i r_i^2 = Ma^2 + Ma^2 + mb^2 + mb^2 = 2Ma^2 + 2mb^2$$

Evaluate the rotational kinetic energy using Equation 10.24:

$$K_R = \tfrac{1}{2}I_z\omega^2 = \tfrac{1}{2}(2Ma^2 + 2mb^2)\omega^2 = (Ma^2 + mb^2)\omega^2$$

Comparing the results for parts (A) and (B), we conclude that the moment of inertia and therefore the rotational kinetic energy associated with a given angular speed depend on the axis of rotation. In part (B), we expect the result to include all four spheres and distances because all four spheres are rotating in the *xy* plane. Based on the work–kinetic energy theorem, the smaller rotational kinetic energy in part (A) than in part (B) indicates it would require less work to set the system into rotation about the *y* axis than about the *z* axis.

WHAT IF?   What if the mass *M* is much larger than *m*? How do the answers to parts (A) and (B) compare?

Answer   If $M \gg m$, then *m* can be neglected and the moment of inertia and the rotational kinetic energy in part (B) become

$$I_z = 2Ma^2 \quad \text{and} \quad K_R = Ma^2\omega^2$$

which are the same as the answers in part (A). If the masses *m* of the two tan spheres in Figure 10.19 are negligible, these spheres can be removed from the figure and rotations about the *y* and *z* axes are equivalent.

10.8 Energy Considerations in Rotational Motion

Having introduced rotational kinetic energy in Section 10.7, let us now see how an energy approach can be useful in solving rotational problems. We begin by considering the relationship between the torque acting on a rigid object and its resulting rotational motion so as to generate expressions for power and a rotational analog to the work–kinetic energy theorem. Consider the rigid object pivoted at O in Figure 10.20. Suppose a single external force $\vec{\mathbf{F}}$ is applied at P, where $\vec{\mathbf{F}}$ lies in the plane of the page. The work done on the object by $\vec{\mathbf{F}}$ as its point of application rotates through an infinitesimal distance $ds = r\,d\theta$ is

$$dW = \vec{\mathbf{F}} \cdot d\vec{\mathbf{s}} = (F\sin\phi)r\,d\theta$$

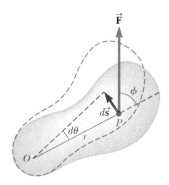

Figure 10.20 A rigid object rotates about an axis through O under the action of an external force $\vec{\mathbf{F}}$ applied at P.

where $F\sin\phi$ is the tangential component of $\vec{\mathbf{F}}$, or, in other words, the component of the force along the displacement. Notice that the radial component vector of $\vec{\mathbf{F}}$ does no work on the object because it is perpendicular to the displacement of the point of application of $\vec{\mathbf{F}}$.

Because the magnitude of the torque due to $\vec{\mathbf{F}}$ about an axis through O is defined as $rF\sin\phi$ by Equation 10.14, we can write the work done for the infinitesimal rotation as

$$dW = \tau\,d\theta \qquad (10.25)$$

The rate at which work is being done by $\vec{\mathbf{F}}$ as the object rotates about the fixed axis through the angle $d\theta$ in a time interval dt is

$$\frac{dW}{dt} = \tau\frac{d\theta}{dt}$$

Because dW/dt is the instantaneous power P (see Section 8.5) delivered by the force and $d\theta/dt = \omega$, this expression reduces to

$$P = \frac{dW}{dt} = \tau\omega \qquad (10.26)$$

◀ Power delivered to a rotating rigid object

This equation is analogous to $P = Fv$ (Eq. 8.18) in the case of translational motion, and Equation 10.25 is analogous to $dW = F_x\,dx$.

In studying translational motion, we have seen that models based on an energy approach can be extremely useful in describing a system's behavior. From what we learned of translational motion, we expect that when a symmetric object rotates frictionlessly about a fixed axis, the work done by external forces equals the change in the rotational energy of the object.

To prove that fact, let us begin with the rigid object under a net torque model, whose mathematical representation is $\sum \tau_{\text{ext}} = I\alpha$. Using the chain rule from calculus, we can express the net torque as

$$\sum \tau_{\text{ext}} = I\alpha = I\frac{d\omega}{dt} = I\frac{d\omega}{d\theta}\frac{d\theta}{dt} = I\frac{d\omega}{d\theta}\omega$$

Rearranging this expression and noting that $\sum \tau_{\text{ext}}\,d\theta = dW$ from Equation 10.25 gives

$$\sum \tau_{\text{ext}}\,d\theta = dW = I\omega\,d\omega$$

Integrating this expression, we obtain for the work W done by the net external force acting on a rotating system

$$W = \int_{\omega_i}^{\omega_f} I\omega\,d\omega = \tfrac{1}{2}I\omega_f^2 - \tfrac{1}{2}I\omega_i^2 \qquad (10.27)$$

◀ Work–kinetic energy theorem for rotational motion

where the angular speed changes from ω_i to ω_f. Equation 10.27 is the **work–kinetic energy theorem for rotational motion.** Similar to the work–kinetic energy theorem in translational motion (Section 7.5), Equation 10.27 states that the net work done by external forces in rotating a symmetric rigid object about a fixed friction-free axis equals the change in the object's rotational energy.

TABLE 10.3 Useful Equations in Rotational and Translational Motion

Rotational Motion About a Fixed Axis	Translational Motion
Angular speed $\omega = d\theta/dt$	Translational speed $v = dx/dt$
Angular acceleration $\alpha = d\omega/dt$	Translational acceleration $a = dv/dt$
Net torque $\Sigma\tau_{\text{ext}} = I\alpha$	Net force $\Sigma F = ma$
If $\alpha = \text{constant}$ $\begin{cases} \omega_f = \omega_i + \alpha t \\ \theta_f = \theta_i + \omega_i t + \frac{1}{2}\alpha t^2 \\ \omega_f^2 = \omega_i^2 + 2\alpha(\theta_f - \theta_i) \end{cases}$	If $a = \text{constant}$ $\begin{cases} v_f = v_i + at \\ x_f = x_i + v_i t + \frac{1}{2}at^2 \\ v_f^2 = v_i^2 + 2a(x_f - x_i) \end{cases}$
Work $W = \displaystyle\int_{\theta_i}^{\theta_f} \tau\, d\theta$	Work $W = \displaystyle\int_{x_i}^{x_f} F_x\, dx$
Rotational kinetic energy $K_R = \frac{1}{2}I\omega^2$	Kinetic energy $K = \frac{1}{2}mv^2$
Power $P = \tau\omega$	Power $P = Fv$
Angular momentum $L = I\omega$	Linear momentum $p = mv$
Net torque $\Sigma\tau = dL/dt$	Net force $\Sigma F = dp/dt$

Equation 10.27 is a form of the nonisolated system (energy) model discussed in Chapter 8. Work is done on the system of the rigid object, which represents a transfer of energy across the boundary of the system that appears as an increase in the object's rotational kinetic energy.

In general, we can combine Equation 10.27 with the translational form of the work–kinetic energy theorem from Chapter 7. Therefore, the net work done by external forces on an object is the change in its *total* kinetic energy, which is the sum of the translational and rotational kinetic energies. For example, when a pitcher throws a baseball, the work done by the pitcher's hands appears as kinetic energy associated with the ball moving through space as well as rotational kinetic energy associated with the spinning of the ball.

In addition to the work–kinetic energy theorem, other energy principles can also be applied to rotational situations. For example, if a system involving rotating objects is isolated and no nonconservative forces act within the system, the isolated system model and the principle of conservation of mechanical energy can be used to analyze the system as in Example 10.11 below. In general, Equation 8.2, the conservation of energy equation, applies to rotational situations, with the recognition that the change in kinetic energy ΔK will include changes in both translational and rotational kinetic energies.

Finally, in some situations an energy approach does not provide enough information to solve the problem and it must be combined with a momentum approach. Such a case is illustrated in Example 10.14 in Section 10.9.

Table 10.3 lists the various equations we have discussed pertaining to rotational motion together with the analogous expressions for translational motion. Notice the similar mathematical forms of the equations. The last two equations in the left-hand column of Table 10.3, involving angular momentum L, are discussed in Chapter 11 and are included here only for the sake of completeness.

Example **10.11** Rotating Rod Revisited

A uniform rod of length L and mass M is free to rotate on a frictionless pin passing through one end (Fig 10.21). The rod is released from rest in the horizontal position.

(A) What is its angular speed when the rod reaches its lowest position?

SOLUTION

Conceptualize Consider Figure 10.21 and imagine the rod rotating downward through a quarter turn about the pivot at the left end. Also look back at Example 10.4. This physical situation is the same.

10.11 continued

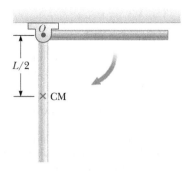

Figure 10.21 (Example 10.11)
A uniform rigid rod pivoted at O
rotates in a vertical plane under the
action of the gravitational force.

Categorize As mentioned in Example 10.4, the angular acceleration of the rod is not constant. Therefore, the kinematic equations for rotation (Section 10.2) cannot be used to solve this example. We categorize the system of the rod and the Earth as an *isolated system* in terms of *energy* with no nonconservative forces acting and use the principle of conservation of mechanical energy.

Analyze We choose the configuration in which the rod is hanging straight down as the reference configuration for gravitational potential energy and assign a value of zero for this configuration. When the rod is in the horizontal position, it has no rotational kinetic energy. The potential energy of the system in this configuration relative to the reference configuration is $MgL/2$ because the center of mass of the rod is at a height $L/2$ higher than its position in the reference configuration. When the rod reaches its lowest position, the energy of the system is entirely rotational energy $\frac{1}{2}I\omega^2$, where I is the moment of inertia of the rod about an axis passing through the pivot.

Using the isolated system (energy) model, write an appropriate reduction of Equation 8.2:

$$\Delta K + \Delta U = 0$$

Substitute for each of the final and initial energies:

$$(\tfrac{1}{2}I\omega^2 - 0) + (0 - \tfrac{1}{2}MgL) = 0$$

Solve for ω and use $I = \frac{1}{3}ML^2$ (see Table 10.2) for the rod:

$$\omega = \sqrt{\frac{MgL}{I}} = \sqrt{\frac{MgL}{\frac{1}{3}ML^2}} = \sqrt{\frac{3g}{L}}$$

(B) Determine the tangential speed of the center of mass and the tangential speed of the lowest point on the rod when it is in the vertical position.

SOLUTION

Use Equation 10.10 and the result from part (A):

$$v_{CM} = r\omega = \frac{L}{2}\omega = \tfrac{1}{2}\sqrt{3gL}$$

Because r for the lowest point on the rod is twice what it is for the center of mass, the lowest point has a tangential speed twice that of the center of mass:

$$v = 2v_{CM} = \sqrt{3gL}$$

Finalize The initial configuration in this example is the same as that in Example 10.4. In Example 10.4, however, we could only find the initial angular acceleration of the rod. Applying an energy approach in the current example allows us to find additional information, the angular speed of the rod at the lowest point. Convince yourself that you could find the angular speed of the rod at any angular position by knowing the location of the center of mass at this position.

WHAT IF? What if we want to find the angular speed of the rod when the angle it makes with the horizontal is 45.0°? Because this angle is half of 90.0°, for which we solved the problem above, is the angular speed at this configuration half the answer in the calculation above, that is, $\tfrac{1}{2}\sqrt{3g/L}$?

Answer Imagine the rod in Figure 10.21 at the 45.0° position. Use a pencil or a ruler to represent the rod at this position. Notice that the center of mass has dropped through more than half of the distance $L/2$ in this configuration. Therefore, more than half of the initial gravitational potential energy has been transformed to rotational kinetic energy. So, we should not expect the value of the angular speed to be as simple as proposed above.

Note that the center of mass of the rod drops through a distance of $0.500L$ as the rod reaches the vertical configuration. When the rod is at 45.0° to the horizontal, we can show that the center of mass of the rod drops through a distance of $0.354L$. Continuing the calculation, we find that the angular speed of the rod at this configuration is $0.841\sqrt{3g/L}$, (not $\tfrac{1}{2}\sqrt{3g/L}$).

Example 10.12 **Energy and the Atwood Machine**

Two blocks having different masses m_1 and m_2 are connected by a string passing over a pulley as shown in Figure 10.22 (page 272). The pulley has a radius R and moment of inertia I about its axis of rotation. The string does not slip on the pulley, and the system is released from rest. Find the translational speeds of the blocks after block 2 descends through a distance h and find the angular speed of the pulley at this time.

continued

10.12 continued

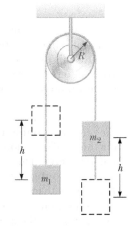

Conceptualize We have already seen the Atwood machine in Example 5.9, so the motion of the objects in Figure 10.22 should be easy to visualize.

Categorize Because the string does not slip, the pulley rotates about the axle. We can neglect friction in the axle because the axle's radius is small relative to that of the pulley. Hence, the frictional torque is much smaller than the net torque applied by the two blocks provided that their masses are significantly different. Consequently, the system consisting of the two blocks, the pulley, and the Earth is an *isolated system* in terms of *energy* with no nonconservative forces acting.

Figure 10.22 (Example 10.12) An Atwood machine with a massive pulley.

Analyze We define the zero configuration for gravitational potential energy as that which exists when the system is released. From Figure 10.22, we see that the descent of block 2 is associated with a decrease in system potential energy and that the rise of block 1 represents an increase in potential energy.

Using the isolated system (energy) model, write an appropriate reduction of the conservation of energy equation:

$$\Delta K + \Delta U = 0$$

Substitute for each of the energies:

$$[(\tfrac{1}{2}m_1 v_f^2 + \tfrac{1}{2}m_2 v_f^2 + \tfrac{1}{2}I\omega_f^2) - 0] + [(m_1 gh - m_2 gh) - 0] = 0$$

The two blocks, the string, and the outer rim of the pulley all move at the same speed. Therefore, use $v_f = R\omega_f$ to substitute for ω_f:

$$\tfrac{1}{2}m_1 v_f^2 + \tfrac{1}{2}m_2 v_f^2 + \tfrac{1}{2}I\frac{v_f^2}{R^2} = m_2 gh - m_1 gh$$

$$\tfrac{1}{2}\left(m_1 + m_2 + \frac{I}{R^2}\right)v_f^2 = (m_2 - m_1)gh$$

Solve for v_f:

$$(1) \quad v_f = \left[\frac{2(m_2 - m_1)gh}{m_1 + m_2 + I/R^2}\right]^{1/2}$$

Use $v_f = R\omega_f$ to solve for ω_f:

$$\omega_f = \frac{v_f}{R} = \frac{1}{R}\left[\frac{2(m_2 - m_1)gh}{m_1 + m_2 + I/R^2}\right]^{1/2}$$

Finalize Each block can be modeled as a *particle under constant acceleration* because it experiences a constant net force. Think about what you would need to do to use Equation (1) to find the acceleration of one of the blocks. Then imagine the pulley becoming massless and determine the acceleration of a block. How does this result compare with the result of Example 5.9?

10.9 Rolling Motion of a Rigid Object

In this section, we treat the motion of a rigid object rolling along a flat surface. In general, such motion is complex. For example, suppose a cylinder is rolling on a straight path such that the axis of rotation remains parallel to its initial orientation in space. As Figure 10.23 shows, a point on the rim of the cylinder moves in a complex path called a *cycloid*. We can simplify matters, however, by focusing on the center of mass rather than on a point on the rim of the rolling object. As shown in

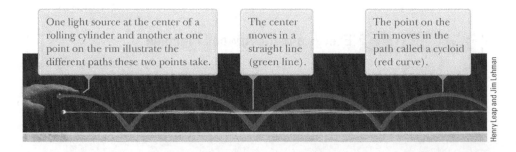

One light source at the center of a rolling cylinder and another at one point on the rim illustrate the different paths these two points take.

The center moves in a straight line (green line).

The point on the rim moves in the path called a cycloid (red curve).

Figure 10.23 Two points on a rolling object take different paths through space.

Figure 10.23, the center of mass moves in *translational* motion in a straight line. If an object such as a cylinder rolls without slipping on the surface (called *pure rolling motion*), a simple relationship exists between its rotational and translational motions.

Consider a uniform cylinder of radius R rolling without slipping on a horizontal surface (Fig. 10.24). As the cylinder rotates through an angle θ, its center of mass moves a linear distance $s = R\theta$ (see Eq. 10.1b). Therefore, the translational speed of the center of mass for pure rolling motion is given by

$$v_{CM} = \frac{ds}{dt} = R\frac{d\theta}{dt} = R\omega \qquad (10.28)$$

where ω is the angular speed of the cylinder. Equation 10.28 holds whenever a cylinder or sphere rolls without slipping and is the **condition for pure rolling motion.** The magnitude of the linear acceleration of the center of mass for pure rolling motion is

$$a_{CM} = \frac{dv_{CM}}{dt} = R\frac{d\omega}{dt} = R\alpha \qquad (10.29)$$

where α is the angular acceleration of the cylinder.

Imagine that you are moving along with a rolling object at speed v_{CM}, staying in a frame of reference at rest with respect to the center of mass of the object. As you observe the object, you will see the object in pure rotation around its center of mass. Figure 10.25a shows the velocities of points at the top, center, and bottom of the object as observed by you. In addition to these velocities, every point on the object moves in the same direction with speed v_{CM} relative to the surface on which it rolls. Figure 10.25b shows these velocities for a nonrotating object. In the reference frame at rest with respect to the surface, the velocity of a given point on the object is the sum of the velocities shown in Figures 10.25a and 10.25b. Figure 10.25c shows the results of adding these velocities.

Notice that the contact point between the surface and object in Figure 10.25c has a translational speed of zero. At this instant, the rolling object is moving in exactly the same way as if the surface were removed and the object were pivoted at point P and spun about an axis passing through P. We can express the total kinetic energy of this imagined spinning object as

$$K = \tfrac{1}{2}I_P\omega^2 \qquad (10.30)$$

where I_P is the moment of inertia about a rotation axis through P.

Because the motion of the imagined spinning object is the same at this instant as our actual rolling object, Equation 10.30 also gives the kinetic energy of the rolling object. Applying the parallel-axis theorem, we can substitute $I_P = I_{CM} + MR^2$ into Equation 10.30 to obtain

$$K = \tfrac{1}{2}I_{CM}\omega^2 + \tfrac{1}{2}MR^2\omega^2$$

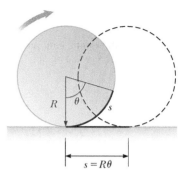

Figure 10.24 For pure rolling motion, as the cylinder rotates through an angle θ its center moves a linear distance $s = R\theta$.

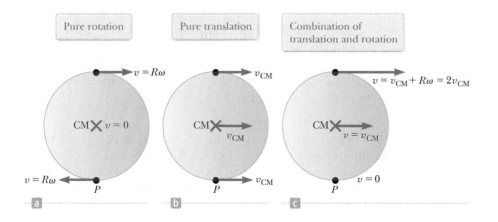

Figure 10.25 The motion of a rolling object can be modeled as a combination of pure translation and pure rotation. Equation 10.28 tells us that $v_{CM} = R\omega$.

Using $v_{CM} = R\omega$, this equation can be expressed as

$$K = \tfrac{1}{2}I_{CM}\omega^2 + \tfrac{1}{2}Mv_{CM}^2 \qquad (10.31)$$

The term $\tfrac{1}{2}I_{CM}\omega^2$ represents the rotational kinetic energy of the object about its center of mass, and the term $\tfrac{1}{2}Mv_{CM}^2$ represents the kinetic energy the object would have if it were just translating through space without rotating. Therefore, the total kinetic energy of a rolling object is the sum of the rotational kinetic energy *about* the center of mass and the translational kinetic energy *of* the center of mass. This statement is consistent with the situation illustrated in Figure 10.25, which shows that the velocity of a point on the object is the sum of the velocity of the center of mass and the tangential velocity around the center of mass.

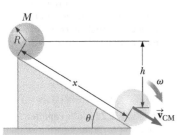

Energy methods can be used to treat a class of problems concerning the rolling motion of an object on a rough incline. For example, consider Figure 10.26, which shows a sphere rolling without slipping after being released from rest at the top of the incline. Accelerated rolling motion is possible only if a friction force is present between the sphere and the incline to produce a net torque about the center of mass. Despite the presence of friction, no loss of mechanical energy occurs because the contact point is at rest relative to the surface at any instant. (On the other hand, if the sphere were to slip, mechanical energy of the sphere–incline–Earth system would decrease due to the nonconservative force of kinetic friction.)

Figure 10.26 A sphere rolling down an incline. Mechanical energy of the sphere–Earth system is conserved if no slipping occurs.

In reality, *rolling friction* causes mechanical energy to transform to internal energy. Rolling friction is due to deformations of the surface and the rolling object. For example, automobile tires flex as they roll on a roadway, representing a transformation of mechanical energy to internal energy. The roadway also deforms a small amount, representing additional rolling friction. In our problem-solving models, we ignore rolling friction unless stated otherwise.

Using $v_{CM} = R\omega$ for pure rolling motion, we can express Equation 10.31 as

$$K = \tfrac{1}{2}I_{CM}\left(\frac{v_{CM}}{R}\right)^2 + \tfrac{1}{2}Mv_{CM}^2$$

$$K = \tfrac{1}{2}\left(\frac{I_{CM}}{R^2} + M\right)v_{CM}^2 \qquad (10.32)$$

For the sphere–Earth system in Figure 10.26, we define the zero configuration of gravitational potential energy to be when the sphere is at the bottom of the incline. Therefore, Equation 8.2 gives

$$\Delta K + \Delta U = 0$$

$$\left[\tfrac{1}{2}\left(\frac{I_{CM}}{R^2} + M\right)v_{CM}^2 - 0\right] + (0 - Mgh) = 0$$

$$v_{CM} = \left[\frac{2gh}{1 + (I_{CM}/MR^2)}\right]^{1/2} \qquad (10.33)$$

While this calculation was performed for the sphere in Figure 10.26, Equation 10.33 is general enough that it provides the speed of *any* object with a circular cross section that rolls from rest down an incline of height h.

ⓆUICK QUIZ 10.7 A ball rolls without slipping down incline A, starting from rest. At the same time, a box starts from rest and slides down incline B, which is identical to incline A except that it is frictionless. Which arrives at the bottom first? **(a)** The ball arrives first. **(b)** The box arrives first. **(c)** Both arrive at the same time. **(d)** It is impossible to determine.

Example **10.13** Sphere Rolling Down an Incline

Suppose the sphere shown in Figure 10.26 is solid and uniform. Calculate the translational speed of the center of mass at the bottom of the incline and the magnitude of the translational acceleration of the center of mass.

SOLUTION

Conceptualize Roll a golf ball or a marble down a ramp to visualize the motion of the sphere.

Categorize We model the sphere and the Earth as an *isolated system* in terms of *energy* with no nonconservative forces acting. This model is the one that led to Equation 10.33, so we can use that result. To find its acceleration, we will model the sphere as a *particle under constant acceleration.*

Analyze Evaluate the speed of the center of mass of the sphere from Equation 10.33, using the moment of inertia from Table 10.2:

$$(1) \quad v_{CM} = \left[\frac{2gh}{1 + (\frac{2}{5}MR^2/MR^2)} \right]^{1/2} = (\tfrac{10}{7}gh)^{1/2}$$

This result is less than $\sqrt{2gh}$, which is the speed an object would have if it simply slid down the incline without rotating. (Eliminate the rotation by setting $I_{CM} = 0$ in Eq. 10.33.)

To calculate the translational acceleration of the center of mass, notice that the vertical displacement of the sphere is related to the distance x it moves along the incline through the relationship $h = x \sin \theta$.

Use this relationship to rewrite Equation (1):

$$v_{CM}{}^2 = \tfrac{10}{7}gx \sin \theta$$

Write Equation 2.17 for an object starting from rest and moving through a distance x under constant acceleration:

$$v_{CM}{}^2 = 2a_{CM}x$$

Equate the preceding two expressions to find a_{CM}:

$$a_{CM} = \tfrac{5}{7}g \sin \theta$$

Finalize Both the speed and the acceleration of the center of mass are *independent* of the mass and the radius of the sphere. That is, all uniform solid spheres experience the same speed and acceleration on a given incline. Try to verify this statement experimentally with balls of different sizes, such as a marble and a croquet ball.

If we were to repeat the acceleration calculation for a hollow sphere, a solid cylinder, or a hoop, we would obtain similar results in which only the numerical factor in front of $g \sin \theta$ would differ. The numerical factors that appear in the expressions for v_{CM} and a_{CM} depend only on the moment of inertia about the center of mass for the specific object. In all cases, the acceleration of the center of mass is *less* than $g \sin \theta$, the value the acceleration would have if the incline were frictionless and no rolling occurred.

Example **10.14** Pulling on a Spool[3]

A cylindrically symmetric spool of mass m and radius R sits at rest on a horizontal table with friction (Fig. 10.27). With your hand on a light string wrapped around the axle of radius r, you pull on the spool with a constant horizontal force of magnitude T to the right. As a result, the spool rolls without slipping a distance L along the table with no rolling friction.

(A) Find the final translational speed of the center of mass of the spool.

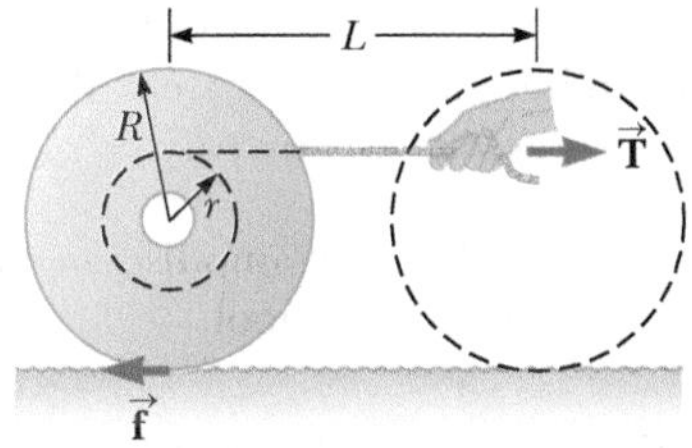

SOLUTION

Conceptualize Use Figure 10.27 to visualize the motion of the spool when you pull the string. For the spool to roll through a distance L, notice that your hand on the string must pull through a distance *different* from L.

Figure 10.27 (Example 10.14)
A spool rests on a horizontal table. A string is wrapped around the axle and is pulled to the right by a hand.

Categorize The spool is a *rigid object under a net torque,* but the net torque includes that due to the friction force at the bottom of the spool, about which we know nothing. Therefore, an approach based on the rigid object under a net torque model might be difficult. Work is done by your hand on the spool and string, which form a nonisolated system in terms of energy. Let's see if an approach based on the *nonisolated system (energy)* model is fruitful.

continued

[3]Example 10.14 was inspired in part by C. E. Mungan, "A primer on work–energy relationships for introductory physics," *The Physics Teacher,* **43**:10, 2005.

10.14 continued

Analyze The only type of energy that changes in the system is the kinetic energy of the spool. There is no rolling friction, so there is no change in internal energy. The only way that energy crosses the system's boundary is by the work done by your hand on the string. No work is done by the static force of friction on the bottom of the spool (to the left in Fig. 10.27) because the point of application of the force moves through no displacement.

Write the appropriate reduction of the conservation of energy equation, Equation 8.2:

$$(1) \quad W = \Delta K = \Delta K_{trans} + \Delta K_{rot}$$

In this expression, W is the work done on the string by your hand. To find this work, we need to find the displacement of your hand during the process. We first find the length of string that has unwound off the spool. If the spool rolls through a distance L, the total angle through which it rotates is $\theta = L/R$. The axle also rotates through this angle.

Use Equation 10.1b to find the total arc length through which the axle turns:

$$\ell = r\theta = \frac{r}{R}L$$

This result also gives the length of string pulled off the axle. Your hand will move through this distance *plus* the distance L through which the spool moves. Therefore, the magnitude of the displacement of the point of application of the force applied by your hand is $\ell + L = L(1 + r/R)$.

Evaluate the work done by your hand on the string:

$$(2) \quad W = TL\left(1 + \frac{r}{R}\right)$$

Substitute Equation (2) into Equation (1):

$$TL\left(1 + \frac{r}{R}\right) = \tfrac{1}{2}mv_{CM}^2 + \tfrac{1}{2}I\omega^2$$

where I is the moment of inertia of the spool about its center of mass and v_{CM} and ω are the final values after the wheel rolls through the distance L.

Apply the nonslip rolling condition $\omega = v_{CM}/R$:

$$TL\left(1 + \frac{r}{R}\right) = \tfrac{1}{2}mv_{CM}^2 + \tfrac{1}{2}I\frac{v_{CM}^2}{R^2}$$

Solve for v_{CM}:

$$(3) \quad v_{CM} = \sqrt{\frac{2TL(1 + r/R)}{m(1 + I/mR^2)}}$$

(B) Find the value of the friction force f.

SOLUTION

Categorize Because the friction force does no work, we cannot evaluate it from an energy approach. We model the spool as a *nonisolated system,* but this time in terms of *momentum.* The string applies a force across the boundary of the system, resulting in an impulse on the system. Because the forces on the spool are constant, we can model the spool's center of mass as a *particle under constant acceleration.*

Analyze Write the impulse–momentum theorem (Eq. 9.40) for the spool:

$$m(v_{CM} - 0) = (T - f)\Delta t$$
$$(4) \quad mv_{CM} = (T - f)\Delta t$$

For a particle under constant acceleration starting from rest, Equation 2.14 tells us that the average velocity of the center of mass is half the final velocity.

Use this fact and Equation 2.2 to find the time interval for the center of mass of the spool to move a distance L from rest to a final speed v_{CM}:

$$(5) \quad \Delta t = \frac{L}{v_{CM,avg}} = \frac{2L}{v_{CM}}$$

Substitute Equation (5) into Equation (4):

$$mv_{CM} = (T - f)\frac{2L}{v_{CM}}$$

Solve for the friction force f:

$$f = T - \frac{mv_{CM}^2}{2L}$$

10.14 continued

Substitute v_{CM} from Equation (3):

$$f = T - \frac{m}{2L}\left[\frac{2TL(1 + r/R)}{m(1 + I/mR^2)}\right]$$

$$= T - T\frac{(1 + r/R)}{(1 + I/mR^2)} = T\left[\frac{I - mrR}{I + mR^2}\right]$$

Finalize   Notice that we could use the impulse–momentum theorem for the translational motion of the spool while ignoring that the spool is rotating! This fact demonstrates the power of our growing list of approaches to solving problems. To challenge yourself, solve part (A) again, using the rigid object under a net torque model for the spool and the particle under constant acceleration model for the center of mass of the spool, to derive Equation (3). Calculate torque and moment of inertia around the base of the spool to eliminate the unknown friction force from the torque equation.

Summary

▶ Definitions

The **angular position** of a rigid object is defined as the angle θ between a reference line attached to the object and a reference line fixed in space. The **angular displacement** of a particle moving in a circular path or a rigid object rotating about a fixed axis is $\Delta\theta \equiv \theta_f - \theta_i$.

The **instantaneous angular speed** of a particle moving in a circular path or of a rigid object rotating about a fixed axis is

$$\omega \equiv \frac{d\theta}{dt} \qquad (10.3)$$

The **instantaneous angular acceleration** of a particle moving in a circular path or of a rigid object rotating about a fixed axis is

$$\alpha \equiv \frac{d\omega}{dt} \qquad (10.5)$$

When a rigid object rotates about a fixed axis, every part of the object has the same angular speed and the same angular acceleration.

The magnitude of the **torque** associated with a force $\vec{F}$ acting on an object at a distance r from the rotation axis is

$$\tau = rF\sin\phi = Fd \qquad (10.14)$$

where ϕ is the angle between the position vector of the point of application of the force and the force vector, and d is the moment arm of the force, which is the perpendicular distance from the rotation axis to the line of action of the force.

The **moment of inertia of a system of particles** is defined as

$$I \equiv \sum_i m_i r_i^2 \qquad (10.19)$$

where m_i is the mass of the ith particle and r_i is its distance from the rotation axis.

▶ Concepts and Principles

When a rigid object rotates about a fixed axis, the angular position, angular speed, and angular acceleration are related to the translational position, translational speed, and translational acceleration through the relationships

$$s = r\theta \qquad (10.1b)$$
$$v = r\omega \qquad (10.10)$$
$$a_t = r\alpha \qquad (10.11)$$

If a rigid object rotates about a fixed axis with angular speed ω, its **rotational kinetic energy** can be written

$$K_R = \tfrac{1}{2}I\omega^2 \qquad (10.24)$$

where I is the moment of inertia of the object about the axis of rotation.

The **moment of inertia of a rigid object** is

$$I = \int r^2\,dm \qquad (10.20)$$

where r is the distance from the mass element dm to the axis of rotation.

The rate at which work is done by an external force in rotating a rigid object about a fixed axis, or the **power** delivered, is

$$P = \tau\omega \qquad (10.26)$$

If work is done on a rigid object and the only result of the work is rotation about a fixed axis, the net work done by external forces in rotating the object equals the change in the rotational kinetic energy of the object:

$$W = \tfrac{1}{2}I\omega_f^2 - \tfrac{1}{2}I\omega_i^2 \qquad (10.27)$$

continued

The **total kinetic energy** of a rigid object rolling on a rough surface without slipping equals the rotational kinetic energy about its center of mass plus the translational kinetic energy of the center of mass:

$$K = \tfrac{1}{2}I_{CM}\omega^2 + \tfrac{1}{2}Mv_{CM}^{2}$$ (10.31)

▶ Analysis Models for Problem Solving

Rigid Object Under Constant Angular Acceleration. If a rigid object rotates about a fixed axis under constant angular acceleration, one can apply equations of kinematics that are analogous to those for translational motion of a particle under constant acceleration:

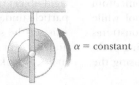

α = constant

$$\omega_f = \omega_i + \alpha t$$ (10.6)

$$\theta_f = \theta_i + \omega_i t + \tfrac{1}{2}\alpha t^2$$ (10.7)

$$\omega_f^2 = \omega_i^2 + 2\alpha(\theta_f - \theta_i)$$ (10.8)

$$\theta_f = \theta_i + \tfrac{1}{2}(\omega_i + \omega_f)t$$ (10.9)

Rigid Object Under a Net Torque. If a rigid object free to rotate about a fixed axis has a net external torque acting on it, the object undergoes an angular acceleration α, where

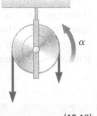

α

$$\sum \tau_{ext} = I\alpha$$ (10.18)

This equation is the rotational analog to Newton's second law in the particle under a net force model.

Think–Pair–Share

See the Preface for an explanation of the icons used in this problems set. For additional assessment items for this section, go to ⚡ **WEBASSIGN** From Cengage

1. You have a summer internship, working with other interns on an archeological dig. Your intern team has found a perfectly cylindrical object of an unknown material. Examination of the visible surface shows that the composition of the object seems to be uniform. The object has a mass of 15.7 kg and a radius of 5.00 cm. The lead archeologist wants to know if the artifact is hollow, but the x-ray machine and other scanning equipment have broken down, so there is no way to look inside. Your team comes up with the idea of building U-shaped supports from wood and laying the cylinder horizontally between the supports as shown in the end view in Figure TP10.1a. The wood can be sanded and oiled to almost eliminate friction. In this way, the cylindrical artifact is free to rotate around its long, horizontal axis. You wrap a long piece of twine several times around the cylinder

and attach a 2.00-kg pickax to the free hanging end of the twine as shown in the side view in Figure TP10.1b. When the pickax is released from rest, it descends and causes the cylinder to rotate. (a) You measure the falling of the pickax and find that it falls 1.50 m in 1.45 s. Is the cylinder hollow? (b) Suppose you measure the falling of the pickax through the same distance and find it to take 1.13 s. What can you conclude about the cylinder now?

2. In order to save money on construction costs, a circular race track has been built with a flat roadway rather than a banked roadway, like that discussed in Example 6.4. During testing of the track, several race cars start, one at a time, at the beginning of the track and at the same radial distance from the center of the track, and undergo constant translational acceleration of magnitude a. All cars have identical tires. Show that all of the cars skid outward off the track at the same angular position around the track, regardless of their mass. To solve this problem, the stubborn owner still does not want to spend the money on banked roadways, so he simply has a circular track built with the same road material but a larger radius. What happens?

3. **ACTIVITY** (a) Place ten pennies on a horizontal meterstick, with a penny at 10 cm, 20 cm, 30 cm, etc., out to 100 cm. Carefully pick up the meterstick, keeping it horizontal, and have a member of the group make a video recording of the following event, using a smartphone or other device. While the video recording is underway, release the 100-cm end of the meterstick while the 0-cm end rests on someone's finger or the edge of the desk. By stepping through the video images or watching the video in slow motion, determine which pennies first lose contact with the meterstick as it falls. (b) Make a theoretical determination of which pennies should first lose contact and compare to your experimental result.

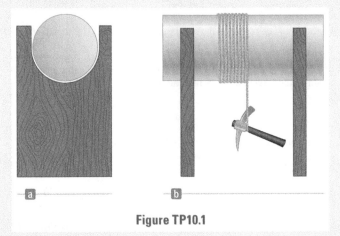

Figure TP10.1

Problems

See the Preface for an explanation of the icons used in this problems set. For additional assessment items for this section, go to **WEBASSIGN** From Cengage

SECTION 10.1 Angular Position, Velocity, and Acceleration

1. (a) Find the angular speed of the Earth's rotation about its axis. (b) How does this rotation affect the shape of the Earth?

2. A bar on a hinge starts from rest and rotates with an angular acceleration $\alpha = 10 + 6t$, where α is in rad/s² and t is in seconds. Determine the angle in radians through which the bar turns in the first 4.00 s.

SECTION 10.2 Analysis Model: Rigid Object Under Constant Angular Acceleration

3. A wheel starts from rest and rotates with constant angular acceleration to reach an angular speed of 12.0 rad/s in 3.00 s. Find (a) the magnitude of the angular acceleration of the wheel and (b) the angle in radians through which it rotates in this time interval.

4. A machine part rotates at an angular speed of 0.060 rad/s; its speed is then increased to 2.2 rad/s at an angular acceleration of 0.70 rad/s². (a) Find the angle through which the part rotates before reaching this final speed. (b) If both the initial and final angular speeds are doubled and the angular acceleration remains the same, by what factor is the angular displacement changed? Why?

5. A dentist's drill starts from rest. After 3.20 s of constant angular acceleration, it turns at a rate of 2.51×10^4 rev/min. (a) Find the drill's angular acceleration. (b) Determine the angle (in radians) through which the drill rotates during this period.

6. *Why is the following situation impossible?* Starting from rest, a disk rotates around a fixed axis through an angle of 50.0 rad in a time interval of 10.0 s. The angular acceleration of the disk is constant during the entire motion, and its final angular speed is 8.00 rad/s.

7. **Review.** Consider a tall building located on the Earth's equator. As the Earth rotates, a person on the top floor of the building moves faster than someone on the ground with respect to an inertial reference frame because the person on the ground is closer to the Earth's axis. Consequently, if an object is dropped from the top floor to the ground a distance h below, it lands east of the point vertically below where it was dropped. (a) How far to the east will the object land? Express your answer in terms of h, g, and the angular speed ω of the Earth. Ignore air resistance and assume the free-fall acceleration is constant over this range of heights. (b) Evaluate the eastward displacement for $h = 50.0$ m. (c) In your judgment, were we justified in ignoring this aspect of the *Coriolis effect* in our previous study of free fall? (d) Suppose the angular speed of the Earth were to decrease with constant angular acceleration due to tidal friction. Would the eastward displacement of the dropped object increase or decrease compared with that in part (b)?

SECTION 10.3 Angular and Translational Quantities

8. Make an order-of-magnitude estimate of the number of revolutions through which a typical automobile tire turns in one year. State the quantities you measure or estimate and their values.

9. A discus thrower (Fig. P10.9) accelerates a discus from rest to a speed of 25.0 m/s by whirling it through 1.25 rev. Assume the discus moves on the arc of a circle 1.00 m in radius. (a) Calculate the final angular speed of the discus. (b) Determine the magnitude of the angular acceleration of the discus, assuming it to be constant. (c) Calculate the time interval required for the discus to accelerate from rest to 25.0 m/s.

Figure P10.9

10. A straight ladder is leaning against the wall of a house. The ladder has rails 4.90 m long, joined by rungs 0.410 m long. Its bottom end is on solid but sloping ground so that the top of the ladder is 0.690 m to the left of where it should be, and the ladder is unsafe to climb. You want to put a flat rock under one foot of the ladder to compensate for the slope of the ground. (a) What should be the thickness of the rock? (b) Does using ideas from this chapter make it easier to explain the solution to part (a)? Explain your answer.

11. A car accelerates uniformly from rest and reaches a speed of 22.0 m/s in 9.00 s. Assuming the diameter of a tire is 58.0 cm, (a) find the number of revolutions the tire makes during this motion, assuming that no slipping occurs. (b) What is the final angular speed of a tire in revolutions per second?

12. **Review.** A small object with mass 4.00 kg moves counterclockwise with constant angular speed 1.50 rad/s in a circle of radius 3.00 m centered at the origin. It starts at the point with position vector $3.00\hat{i}$ m. It then undergoes an angular displacement of 9.00 rad. (a) What is its new position vector? Use unit-vector notation for all vector answers. (b) In what quadrant is the particle located, and what angle does its position vector make with the positive x axis? (c) What is its velocity? (d) In what direction is it moving? (e) What is its acceleration? (f) Make a sketch of its position, velocity, and acceleration vectors. (g) What total force is exerted on the object?

13. In a manufacturing process, a large, cylindrical roller is used to flatten material fed beneath it. The diameter of the roller is 1.00 m, and, while being driven into rotation around a fixed axis, its angular position is expressed as

$$\theta = 2.50t^2 - 0.600t^3$$

where θ is in radians and t is in seconds. (a) Find the maximum angular speed of the roller. (b) What is the maximum tangential speed of a point on the rim of the roller? (c) At what time t should the driving force be removed from the roller so that the roller does not reverse its direction of rotation? (d) Through how many rotations has the roller turned between $t = 0$ and the time found in part (c)?

SECTION 10.4 **Torque**

14. Find the net torque on the wheel in Figure P10.14 about the
T axle through O, taking $a = 10.0$ cm and $b = 25.0$ cm.

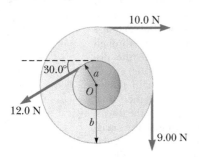

Figure P10.14

SECTION 10.5 **Analysis Model: Rigid Object Under a Net Torque**

15. A grinding wheel is in the form of a uniform solid disk of
AMT radius 7.00 cm and mass 2.00 kg. It starts from rest and
V accelerates uniformly under the action of the constant
torque of 0.600 N · m that the motor exerts on the wheel. (a)
How long does the wheel take to reach its final operating
speed of 1 200 rev/min? (b) Through how many revolutions
does it turn while accelerating?

16. **Review.** A block of mass $m_1 = 2.00$ kg and a block of mass
V $m_2 = 6.00$ kg are connected by a massless string over a pulley
in the shape of a solid disk having radius $R = 0.250$ m and
mass $M = 10.0$ kg. The fixed, wedge-shaped ramp makes an
angle of $\theta = 30.0°$ as shown in Figure P10.16. The coefficient
of kinetic friction is 0.360 for both blocks. (a) Draw force
diagrams of both blocks and of the pulley. Determine (b)
the acceleration of the two blocks and (c) the tensions in the
string on both sides of the pulley.

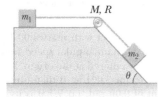

Figure P10.16

17. A model airplane with mass 0.750 kg is tethered to the
T ground by a wire so that it flies in a horizontal circle 30.0 m in
radius. The airplane engine provides a net thrust of 0.800 N
perpendicular to the tethering wire. (a) Find the torque the
net thrust produces about the center of the circle. (b) Find
the angular acceleration of the airplane. (c) Find the translational acceleration of the airplane tangent to its flight path.

18. A disk having moment of inertia 100 kg · m² is free to
QC rotate without friction, starting from rest, about a fixed axis
through its center. A tangential force whose magnitude can
range from $F = 0$ to $F = 50.0$ N can be applied at any distance ranging from $R = 0$ to $R = 3.00$ m from the axis of
rotation. (a) Find a pair of values of F and R that cause the
disk to complete 2.00 rev in 10.0 s. (b) Is your answer for part
(a) a unique answer? How many answers exist?

19. Your grandmother enjoys creating pottery as a hobby. She
CR uses a potter's wheel, which is a stone disk of radius $R =$
0.500 m and mass $M = 100$ kg. In operation, the wheel

rotates at 50.0 rev/min. While the wheel is spinning, your
grandmother works clay at the center of the wheel with
her hands into a pot-shaped object with circular symmetry. When the correct shape is reached, she wants to stop
the wheel in as short a time interval as possible, so that the
shape of the pot is not further distorted by the rotation. She
pushes continuously with a wet rag as hard as she can radially inward on the edge of the wheel and the wheel stops in
6.00 s. (a) You would like to build a brake to stop the wheel
in a shorter time interval, but you must determine the coefficient of friction between the rag and the wheel in order to
design a better system. You determine that the maximum
pressing force your grandmother can sustain for 6.00 s is
70.0 N. (b) **What If?** If your grandmother instead chooses to
press down on the upper surface of the wheel a distance $r =$
0.300 m from the axis of rotation, what is the force needed
to stop the wheel in 6.00 s? Assume that the coefficient of
kinetic friction between the wet rag and the wheel remains
the same as before.

20. At a local mine, a cave-in has trapped a number of miners.
CR You and some classmates rush to the scene to see how you
can help. The trapped miners have been able to reach a
point in the mine at the bottom of a tall vertical shaft to
the surface, allowing them access to fresh air. But they are
in desperate need of fresh water and bandages for injuries.
Some rescue workers ask you to help pack a light plastic
cylindrical container with bottles of water and bandages.
Simply dropping the container into the shaft risks damaging
the container and contents and injuring the miners. Tying
a rope to the container and lowering it on the end of the
rope takes a long time. A quick and relatively safe method
is to wrap a lightweight rope around the container. One
end of the rope will be secured and the container will be
released into the vertical shaft. The container will unroll off
the rope like a falling yo-yo. (a) If immediate access to the
lightweight bandages is needed due to injuries, so that you
want the container to reach the bottom of the shaft in the
shortest possible time interval, should you pack the heavy
water bottles at the center of the container or near the outer
edges? (b) If the medical necessity is not so urgent and, for
safety considerations, you want the container to arrive at the
bottom of the shaft with the lowest possible speed, should
you pack the heavy water bottles at the center of the container or near the outer edges? Assume that the center of
mass of the container is at its center.

21. You have just bought a new bicycle. On your first riding trip,
CR it seems that the bike comes to rest relatively quickly after
you stop pedaling and let the bicycle coast on flat ground.
You call the bicycle shop from which you purchased the
vehicle and describe the problem. The technician says that
they will replace the bearings in the wheels or do whatever
else is necessary if you can prove that the frictional torque
in the axle of the wheels is worse than −0.02 N · m. At first,
you are discouraged by the technical sound of what you have
been told and by the absence of any tool to measure torque
in your garage. But then you remember that you are taking
a physics class! You take your bike into the garage, turn it
upside down and start spinning the wheel while you think
about how to determine the frictional torque. The driveway
outside the garage had a small puddle, so you notice that
droplets of water are flying off the edge of one point on the
tire tangentially, including drops that are projected straight

upward, as shown in Figure P10.21. Ah-ha! Here is your torque-measuring method! The upward-projected drops leave the rim of the wheel at the same level as the axle. You measure the height to which a drop rises from the level of the axle: $h_1 = 54.0$ cm. The wet spot on the tire makes one revolution and another drop is projected upward. You measure its highest point: $h_2 = 51.0$ cm. You measure the radius of the wheel: $r = 0.381$ m. Finally, you take the wheel off the bike and find its mass: $m = 0.850$ kg. Because most of the mass of the wheel is at the tire, you model the wheel as a hoop. What do you tell the technician when you call back?

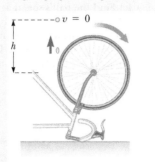

Figure P10.21

SECTION 10.6 Calculation of Moments of Inertia

22. Imagine that you stand tall and turn about a vertical axis through the top of your head and the point halfway between your ankles. Compute an order-of-magnitude estimate for the moment of inertia of your body for this rotation. In your solution, state the quantities you measure or estimate and their values.

23. **S** Following the procedure used in Example 10.7, prove that the moment of inertia about the y' axis of the rigid rod in Figure 10.15 is $\frac{1}{3}ML^2$.

24. **S** Two balls with masses M and m are connected by a rigid rod of length L and negligible mass as shown in Figure P10.24. For an axis perpendicular to the rod, (a) show that the system has the minimum moment of inertia when the axis passes through the center of mass. (b) Show that this moment of inertia is $I = \mu L^2$, where $\mu = mM/(m + M)$.

Figure P10.24

SECTION 10.7 Rotational Kinetic Energy

25. **QC** **V** Rigid rods of negligible mass lying along the y axis connect three particles (Fig. P10.25). The system rotates about the x axis with an angular speed of 2.00 rad/s. Find (a) the moment of inertia about the x axis, (b) the total rotational kinetic energy evaluated from $\frac{1}{2}I\omega^2$, (c) the tangential speed of each particle, and (d) the total kinetic energy evaluated from $\sum \frac{1}{2}m_iv_i^2$. (e) Compare the answers for kinetic energy in parts (a) and (b).

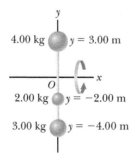

Figure P10.25

26. **QC** A *war-wolf* or *trebuchet* is a device used during the Middle Ages to throw rocks at castles and now sometimes used to fling large vegetables and pianos as a sport. A simple trebuchet is shown in Figure P10.26. Model it as a stiff rod of negligible mass, 3.00 m long, joining particles of mass $m_1 = 0.120$ kg and $m_2 = 60.0$ kg at its ends. It can turn on a frictionless, horizontal axle perpendicular to the rod and 14.0 cm from the large-mass particle. The operator releases the trebuchet from rest in a horizontal orientation. (a) Find the maximum speed that the small-mass object attains. (b) While the small-mass object is gaining speed, does it move with constant acceleration? (c) Does it move with constant tangential acceleration? (d) Does the trebuchet move with constant angular acceleration? (e) Does it have constant momentum? (f) Does the trebuchet–Earth system have constant mechanical energy?

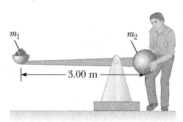

Figure P10.26

SECTION 10.8 Energy Considerations in Rotational Motion

27. Big Ben, the nickname for the clock in Elizabeth Tower (named after the Queen in 2012) in London, has an hour hand 2.70 m long with a mass of 60.0 kg and a minute hand 4.50 m long with a mass of 100 kg (Fig. P10.27). Calculate the total rotational kinetic energy of the two hands about the axis of rotation. (You may model the hands as long, thin rods rotated about one end. Assume the hour and minute hands are rotating at a constant rate of one revolution per 12 hours and 60 minutes, respectively.)

Figure P10.27 Problems 27 and 40.

28. Consider two objects with $m_1 > m_2$ connected by a light string that passes over a pulley having a moment of inertia of I about its axis of rotation as shown in Figure P10.28. The string does not slip on the pulley or stretch. The pulley turns without friction. The two objects are released from rest separated by a vertical distance $2h$. (a) Use the principle of conservation of energy to find the translational speeds of the objects as they pass each other. (b) Find the angular speed of the pulley at this time.

Figure P10.28

29. **Review.** **T** An object with a mass of $m = 5.10$ kg is attached to the free end of a light string wrapped around a reel of radius $R = 0.250$ m and mass $M = 3.00$ kg. The reel is a solid disk, free to rotate in a vertical plane about the horizontal axis passing through its center as shown in Figure P10.29. The suspended object is released from rest 6.00 m above the floor. Determine (a) the tension in the string, (b) the acceleration of the object, and (c) the speed with which the object hits the floor. (d) Verify your answer to part (c) by using the isolated system (energy) model.

Figure P10.29

30. *Why is the following situation impossible?* In a large city with an air-pollution problem, a bus has no combustion engine. It runs over its citywide route on energy drawn from a large, rapidly rotating flywheel under the floor of the bus. The flywheel is spun up to its maximum rotation rate of 3 000 rev/min by an electric motor at the bus terminal. Every time the bus speeds up, the flywheel slows down slightly. The bus is equipped with regenerative braking so that the flywheel can speed up when the bus slows down. The flywheel is a uniform solid cylinder with mass 1 200 kg and radius 0.500 m. The bus body does work against air resistance and rolling resistance at the average rate of 25.0 hp as it travels its route with an average speed of 35.0 km/h.

31. A uniform solid disk of radius R and mass M is free to rotate on a frictionless pivot through a point on its rim (Fig. P10.31). If the disk is released from rest in the position shown by the copper-colored circle, (a) what is the speed of its center of mass when the disk reaches the position indicated by the dashed circle? (b) What is the speed of the lowest point on the disk in the dashed position? (c) **What If?** Repeat part (a) using a uniform hoop.

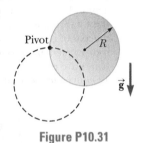

Figure P10.31

32. This problem describes one experimental method for determining the moment of inertia of an irregularly shaped object such as the payload for a satellite. Figure P10.32 shows a counterweight of mass m suspended by a cord wound around a spool of radius r, forming part of a turntable supporting the object. The turntable can rotate without friction. When the counterweight is released from rest, it descends through a distance h, acquiring a speed v. Show that the moment of inertia I of the rotating apparatus (including the turntable) is $mr^2(2gh/v^2 - 1)$.

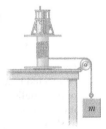

Figure P10.32

SECTION 10.9 Rolling Motion of a Rigid Object

33. A tennis ball is a hollow sphere with a thin wall. It is set rolling without slipping at 4.03 m/s on a horizontal section of a track as shown in Figure P10.33. It rolls around the inside of a vertical circular loop of radius r = 45.0 cm. As the ball nears the bottom of the loop, the shape of the track deviates from a perfect circle so that the ball leaves the track at a point h = 20.0 cm below the horizontal section. (a) Find the ball's speed at the top of the loop. (b) Demonstrate that the ball will not fall from the track at the top of the loop. (c) Find the ball's speed as it leaves the track at the bottom. (d) **What If?** Suppose that static friction between ball and track were negligible so that the ball slid instead of rolling. Describe the speed of the ball at the top of the loop in this situation. (e) Explain your answer to part (d).

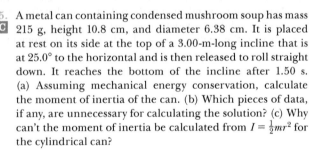

Figure P10.33

34. A smooth cube of mass m and edge length r slides with speed v on a horizontal surface with negligible friction. The cube then moves up a smooth incline that makes an angle θ with the horizontal. A cylinder of mass m and radius r rolls without slipping with its center of mass moving with speed v and encounters an incline of the same angle of inclination but with sufficient friction that the cylinder continues to roll without slipping. (a) Which object will go the greater distance up the incline? (b) Find the difference between the maximum distances the objects travel up the incline. (c) Explain what accounts for this difference in distances traveled.

35. A metal can containing condensed mushroom soup has mass 215 g, height 10.8 cm, and diameter 6.38 cm. It is placed at rest on its side at the top of a 3.00-m-long incline that is at 25.0° to the horizontal and is then released to roll straight down. It reaches the bottom of the incline after 1.50 s. (a) Assuming mechanical energy conservation, calculate the moment of inertia of the can. (b) Which pieces of data, if any, are unnecessary for calculating the solution? (c) Why can't the moment of inertia be calculated from $I = \frac{1}{2}mr^2$ for the cylindrical can?

ADDITIONAL PROBLEMS

36. You have been hired as an expert witness in the case of a factory owner suing a demolition company. The particular case involves a smokestack at a factory being demolished. In order to save money, the factory owner wanted to move the smokestack to a nearby factory that was being built. The demolition company guaranteed to deliver the undamaged smokestack to the new factory by toppling the smokestack freely onto a huge cushioned platform lying on the ground. The then-horizontal smokestack would have been loaded onto a long truck rig for transport to the new factory. However, as the smokestack toppled, it broke apart at a point along its length. The factory owner is blaming the demolition company for the destruction of his smokestack. The demolition company is claiming that there was a defect in the smokestack and that is the reason for its destruction. What advice do you give the attorney who is handling the case on the side of the factory owner?

37. A shaft is turning at 65.0 rad/s at time $t = 0$. Thereafter, its angular acceleration is given by

$$\alpha = -10.0 - 5.00t$$

where α is in rad/s² and t is in seconds. (a) Find the angular speed of the shaft at $t = 3.00$ s. (b) Through what angle does it turn between $t = 0$ and $t = 3.00$ s?

38. A shaft is turning at angular speed ω at time $t = 0$. Thereafter, its angular acceleration is given by

$$\alpha = A + Bt$$

(a) Find the angular speed of the shaft at time t. (b) Through what angle does it turn between $t = 0$ and t?

39. An elevator system in a tall building consists of a 800-kg car and a 950-kg counterweight joined by a light cable of constant length that passes over a pulley of mass 280 kg. The pulley, called a sheave, is a solid cylinder of radius 0.700 m turning on a horizontal axle. The cable does not slip on the sheave. A number n of people, each of mass 80.0 kg, are riding in the elevator car, moving upward at 3.00 m/s and approaching the floor where the car should stop. As an energy-conservation measure, a computer disconnects the elevator motor at just the right moment so that the sheave–car–counterweight system then coasts freely without friction and comes to rest at the floor desired. There it is caught by a simple latch rather than by a massive brake. (a) Determine the distance d the car coasts upward as a function of n. Evaluate the distance for (b) $n = 2$, (c) $n = 12$, and (d) $n = 0$. (e) For what integer values of n does the expression in part (a) apply? (f) Explain your answer to part (e). (g) If an infinite number of people could fit on the elevator, what is the value of d?

40. The hour hand and the minute hand of Big Ben, the Elizabeth Tower clock in London, are 2.70 m and 4.50 m long and have masses of 60.0 kg and 100 kg, respectively (see Fig. P10.27). (a) Determine the total torque due to the weight of these hands about the axis of rotation when the time reads (i) 3:00, (ii) 5:15, (iii) 6:00, (iv) 8:20, and (v) 9:45. (You may model the hands as long, thin, uniform rods.) (b) Determine all times when the total torque about the axis of rotation is zero. Determine the times to the nearest second, solving a transcendental equation numerically.

41. **Review.** A string is wound around a uniform disk of radius R and mass M. The disk is released from rest with the string vertical and its top end tied to a fixed bar (Fig. P10.41). Show that (a) the tension in the string is one third of the weight of the disk, (b) the magnitude of the acceleration of the center of mass is $2g/3$, and (c) the speed of the center of mass is $(4gh/3)^{1/2}$

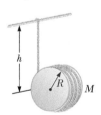

Figure P10.41

after the disk has descended through distance h. (d) Verify your answer to part (c) using the energy approach.

42. **Review.** A spool of wire of mass M and radius R is unwound under a constant force $\vec{F}$ (Fig. P10.42). Assuming the spool is a uniform, solid cylinder that doesn't slip, show that (a) the acceleration of the center of mass is $4\vec{F}/3M$ and (b) the force of friction is to the *right* and equal in magnitude to $F/3$. (c) If the cylinder starts from rest and rolls without slipping, what is the speed of its center of mass after it has rolled through a distance d?

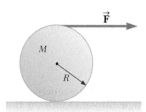

Figure P10.42

43. **Review.** A clown balances a small spherical grape at the top of his bald head, which also has the shape of a sphere. After drawing sufficient applause, the grape starts from rest and rolls down without slipping. It will leave contact with the clown's scalp when the radial line joining it to the center of curvature makes what angle with the vertical?

CHALLENGE PROBLEMS

44. As a gasoline engine operates, a flywheel turning with the crankshaft stores energy after each fuel explosion, providing the energy required to compress the next charge of fuel and air. For the engine of a certain lawn tractor, suppose a flywheel must be no more than 18.0 cm in diameter. Its thickness, measured along its axis of rotation, must be no larger than 8.00 cm. The flywheel must release energy 60.0 J when its angular speed drops from 800 rev/min to 600 rev/min. Design a sturdy steel (density 7.85×10^3 kg/m³) flywheel to meet these requirements with the smallest mass you can reasonably attain. Specify the shape and mass of the flywheel.

45. A spool of thread consists of a cylinder of radius R_1 with end caps of radius R_2 as depicted in the end view shown in Figure P10.45. The mass of the spool, including the thread, is m, and its moment of inertia about an axis through its center is I. The spool is placed on a rough, horizontal surface so that it rolls without slipping when a force $\vec{T}$ acting to the right is applied to the free end of the thread. (a) Show that the magnitude of the friction force exerted by the surface on the spool is given by

$$f = \left(\frac{I + mR_1 R_2}{I + mR_2^2}\right) T$$

(b) Determine the direction of the force of friction.

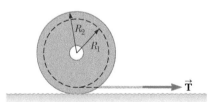

Figure P10.45

46. To find the total angular displacement during the playing time of the compact disc in part (B) of Example 10.2, the disc was modeled as a rigid object under constant angular acceleration. In reality, the angular acceleration of a disc is not constant. In this problem, let us explore the actual time dependence of the angular acceleration. (a) Assume the track on the disc is a spiral such that adjacent loops of the track are separated by a small distance h. Show that the radius r of a given portion of the track is given by

$$r = r_i + \frac{h\theta}{2\pi}$$

where r_i is the radius of the innermost portion of the track and θ is the angle through which the disc turns to arrive at the location of the track of radius r. (b) Show that the rate of change of the angle θ is given by

$$\frac{d\theta}{dt} = \frac{v}{r_i + (h\theta/2\pi)}$$

where v is the constant speed with which the disc surface passes the laser. (c) From the result in part (b), use integration to find an expression for the angle θ as a function of time. (d) From the result in part (c), use differentiation to find the angular acceleration of the disc as a function of time.

47. A uniform, hollow, cylindri-
S   cal spool has inside radius $R/2$, outside radius R, and mass M (Fig. P10.47). It is mounted so that it rotates on a fixed, horizontal axle. A counterweight of mass m is connected to the end of a string wound around the spool. The counterweight falls from rest at $t = 0$ to a position y at time t. Show that the torque due to the friction forces between spool and axle is

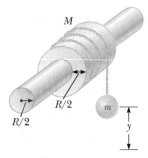

Figure P10.47

$$\tau_f = R\left[m\left(g - \frac{2y}{t^2} \right) - M\frac{5y}{4t^2} \right]$$

48. A cord is wrapped around a pulley that is shaped like a disk
S   of mass m and radius r. The cord's free end is connected to a block of mass M. The block starts from rest and then slides down an incline that makes an angle θ with the horizontal as shown in Figure P10.48. The coefficient of kinetic friction between block and incline is μ. (a) Use energy methods to show that the block's speed as a function of position d down the incline is

$$v = \sqrt{\frac{4Mgd(\sin\theta - \mu\cos\theta)}{m + 2M}}$$

(b) Find the magnitude of the acceleration of the block in terms of μ, m, M, g, and θ.

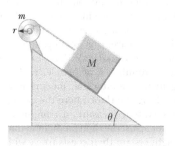

Figure P10.48

Angular Momentum

An Olympic diver does a fancy spinning dive. When she tucks into the position shown, she spins faster. Where does the extra rotational kinetic energy come from? (*Paolo Bona/Shutterstock.com*)

STORYLINE **While preparing to do your physics homework, you are** browsing among various YouTube videos and stumble on one of a spinning ice skater. You notice that he skates into a relatively slow spin, pulls his arms in, and then spins faster and faster. You wonder where the energy comes from to make him spin faster. Next, you find a slow-motion video of an Olympic high diver who is diving into a pool. You notice that after leaving the diving board, she spins slowly, but then tucks in and spins faster. Just like the ice skater, where did the additional rotational kinetic energy come from? In the suggested videos to the side of the web page, you see one about a falling cat. You watch that video and marvel about how a cat dropped upside down can always turn itself over and land on its feet. Just like the ice skater and the diver, there is rotational energy seemingly coming from nowhere. What's going on here? Spinning motion seems to have magical qualities associated with it!

CONNECTIONS The central topic of this chapter is *angular momentum*, a quantity that plays a key role in rotational dynamics. In analogy to the principle of conservation of linear momentum in Chapter 9, there is also a principle of conservation of angular momentum. The angular momentum of an isolated system is constant. For angular momentum, an isolated system is one for which no external torques act on the system. If a net external torque does act on a system, it is nonisolated, and the angular momentum of the system changes. Like the law of conservation of linear momentum, the law of conservation of angular momentum is a fundamental law of physics, equally valid for relativistic and quantum systems. This new fundamental principle allows us to understand more phenomena, such as the spinning skaters, divers, and cats in the opening

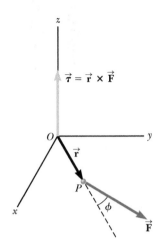

Figure 11.1 The torque vector $\vec{\tau}$ on a particle lies in a direction perpendicular to the plane formed by the position vector $\vec{r}$ of the particle and the applied force vector $\vec{F}$. In the situation shown, $\vec{r}$ and $\vec{F}$ lie in the xy plane, so the torque is along the z axis.

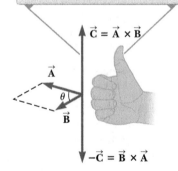

The direction of $\vec{C}$ is perpendicular to the plane formed by $\vec{A}$ and $\vec{B}$, and its direction is determined by the right-hand rule.

Figure 11.2 The vector product $\vec{A} \times \vec{B}$ is a third vector $\vec{C}$ having a magnitude $AB \sin \theta$ equal to the area of the parallelogram shown.

▶ Properties of the vector product

storyline. In addition, we will apply this new principle to the motion of planets in a solar system in Chapter 13, to atomic models in Chapter 41, and to molecular spectra in Chapter 42.

11.1 The Vector Product and Torque

An important consideration to address before defining angular momentum is the process of multiplying two vectors by means of the operation called the *vector product*. We will introduce the vector product by considering the vector nature of torque.

Consider a force $\vec{F}$ acting on a particle located at point P and described by the vector position $\vec{r}$ (Fig. 11.1). As we saw in Section 10.4, the *magnitude* of the torque on the particle due to this force about an axis through the origin is $rF \sin \phi$, where ϕ is the angle between $\vec{r}$ and $\vec{F}$. The axis about which $\vec{F}$ tends to produce rotation is perpendicular to the plane formed by $\vec{r}$ and $\vec{F}$.

The torque vector $\vec{\tau}$ is related to the two vectors $\vec{r}$ and $\vec{F}$. We can establish a mathematical relationship between $\vec{\tau}$, $\vec{r}$, and $\vec{F}$ using a mathematical operation called the **vector product:**

$$\vec{\tau} \equiv \vec{r} \times \vec{F} \qquad (11.1)$$

We now give a formal definition of the vector product. Given any two vectors $\vec{A}$ and $\vec{B}$, the vector product $\vec{A} \times \vec{B}$ is defined as a third vector $\vec{C}$, which has a magnitude of $AB \sin \theta$, where θ is the angle between $\vec{A}$ and $\vec{B}$. That is, if $\vec{C}$ is given by

$$\vec{C} = \vec{A} \times \vec{B} \qquad (11.2)$$

its magnitude is

$$C = AB \sin \theta \qquad (11.3)$$

The quantity $AB \sin \theta$ is equal to the area of the parallelogram formed by $\vec{A}$ and $\vec{B}$ as shown in Figure 11.2. The *direction* of $\vec{C}$ is perpendicular to the plane formed by $\vec{A}$ and $\vec{B}$, and the best way to determine this direction is to use the right-hand rule illustrated in Figure 11.2. The four fingers of the right hand are pointed along $\vec{A}$ and then "wrapped" in the direction that would rotate $\vec{A}$ into $\vec{B}$ through the angle θ. The direction of the upright thumb is the direction of $\vec{A} \times \vec{B} = \vec{C}$. Because of the notation, $\vec{A} \times \vec{B}$ is often read "$\vec{A}$ cross $\vec{B}$," so the vector product is also called the **cross product.**

Some properties of the vector product that follow from its definition are as follows:

1. Unlike the scalar product, the vector product is *not* commutative. Instead, the order in which the two vectors are multiplied in a vector product is important:

$$\vec{A} \times \vec{B} = -\vec{B} \times \vec{A} \qquad (11.4)$$

Therefore, if you change the order of the vectors in a vector product, you must change the sign. You can easily verify this relationship with the right-hand rule.

2. If $\vec{A}$ is parallel to $\vec{B}$ ($\theta = 0$ or $180°$), then $\vec{A} \times \vec{B} = 0$; therefore, it follows that $\vec{A} \times \vec{A} = 0$.

3. If $\vec{A}$ is perpendicular to $\vec{B}$, then $|\vec{A} \times \vec{B}| = AB$.

4. The vector product obeys the distributive law:

$$\vec{A} \times (\vec{B} + \vec{C}) = \vec{A} \times \vec{B} + \vec{A} \times \vec{C} \qquad (11.5)$$

5. The derivative of the vector product with respect to some variable such as t is

$$\frac{d}{dt}(\vec{\mathbf{A}} \times \vec{\mathbf{B}}) = \frac{d\vec{\mathbf{A}}}{dt} \times \vec{\mathbf{B}} + \vec{\mathbf{A}} \times \frac{d\vec{\mathbf{B}}}{dt} \qquad (11.6)$$

where it is important to preserve the multiplicative order of the terms on the right side in view of Equation 11.4.

It is left as an exercise (Problem 4) to show from Equations 11.3 and 11.4 and from the definition of unit vectors that the cross products of the unit vectors $\hat{\mathbf{i}}$, $\hat{\mathbf{j}}$, and $\hat{\mathbf{k}}$ obey the following rules:

$$\hat{\mathbf{i}} \times \hat{\mathbf{i}} = \hat{\mathbf{j}} \times \hat{\mathbf{j}} = \hat{\mathbf{k}} \times \hat{\mathbf{k}} = 0 \qquad (11.7a)$$

$$\hat{\mathbf{i}} \times \hat{\mathbf{j}} = -\hat{\mathbf{j}} \times \hat{\mathbf{i}} = \hat{\mathbf{k}} \qquad (11.7b)$$

$$\hat{\mathbf{j}} \times \hat{\mathbf{k}} = -\hat{\mathbf{k}} \times \hat{\mathbf{j}} = \hat{\mathbf{i}} \qquad (11.7c)$$

$$\hat{\mathbf{k}} \times \hat{\mathbf{i}} = -\hat{\mathbf{i}} \times \hat{\mathbf{k}} = \hat{\mathbf{j}} \qquad (11.7d)$$

◀ Cross products of unit vectors

Signs are interchangeable in cross products. For example, $\vec{\mathbf{A}} \times (-\vec{\mathbf{B}}) = -\vec{\mathbf{A}} \times \vec{\mathbf{B}}$ and $\hat{\mathbf{i}} \times (-\hat{\mathbf{j}}) = -\hat{\mathbf{i}} \times \hat{\mathbf{j}}$.

The cross product of any two vectors $\vec{\mathbf{A}}$ and $\vec{\mathbf{B}}$ can be expressed in unit-vector notation by using the following determinant form:

$$\vec{\mathbf{A}} \times \vec{\mathbf{B}} = \begin{vmatrix} \hat{\mathbf{i}} & \hat{\mathbf{j}} & \hat{\mathbf{k}} \\ A_x & A_y & A_z \\ B_x & B_y & B_z \end{vmatrix} = \begin{vmatrix} A_y & A_z \\ B_y & B_z \end{vmatrix} \hat{\mathbf{i}} + \begin{vmatrix} A_z & A_x \\ B_z & B_x \end{vmatrix} \hat{\mathbf{j}} + \begin{vmatrix} A_x & A_y \\ B_x & B_y \end{vmatrix} \hat{\mathbf{k}}$$

Expanding these determinants gives the result

$$\vec{\mathbf{A}} \times \vec{\mathbf{B}} = (A_y B_z - A_z B_y)\hat{\mathbf{i}} + (A_z B_x - A_x B_z)\hat{\mathbf{j}} + (A_x B_y - A_y B_x)\hat{\mathbf{k}} \qquad (11.8)$$

> **PITFALL PREVENTION 11.1**
> **The Vector Product Is a Vector**
> Remember that the result of taking a vector product between two vectors is *a third vector*. Equation 11.3 gives only the magnitude of this vector.

Given the definition of the cross product, we can now assign a direction to the torque vector. If the force lies in the xy plane as in Figure 11.1, the torque $\vec{\boldsymbol{\tau}}$ is represented by a vector parallel to the z axis. The force in Figure 11.1 creates a torque that tends to rotate the particle counterclockwise about the z axis; the direction of $\vec{\boldsymbol{\tau}}$ is toward increasing z, and $\vec{\boldsymbol{\tau}}$ is therefore in the positive z direction. If we reversed the direction of $\vec{\mathbf{F}}$ in Figure 11.1, $\vec{\boldsymbol{\tau}}$ would be in the negative z direction.

In Figure 11.1 and its discussion, we investigated the torque on a particle. Imagine that the particle is part of a rigid object free to rotate around the z axis. Then the torque that we found in Equation 11.1 is the torque applied to the entire rigid object due to the force $\vec{\mathbf{F}}$.

QUICK QUIZ 11.1 Which of the following statements about the relationship between the magnitude of the cross product of two vectors and the product of the magnitudes of the vectors is true? **(a)** $|\vec{\mathbf{A}} \times \vec{\mathbf{B}}|$ is larger than AB. **(b)** $|\vec{\mathbf{A}} \times \vec{\mathbf{B}}|$ is smaller than AB. **(c)** $|\vec{\mathbf{A}} \times \vec{\mathbf{B}}|$ could be larger or smaller than AB, depending on the angle between the vectors. **(d)** $|\vec{\mathbf{A}} \times \vec{\mathbf{B}}|$ could be equal to AB.

Example 11.1 The Vector Product

Two vectors lying in the xy plane are given by the equations $\vec{\mathbf{A}} = 2\hat{\mathbf{i}} + 3\hat{\mathbf{j}}$ and $\vec{\mathbf{B}} = -\hat{\mathbf{i}} + 2\hat{\mathbf{j}}$. Find $\vec{\mathbf{A}} \times \vec{\mathbf{B}}$ and verify that $\vec{\mathbf{A}} \times \vec{\mathbf{B}} = -\vec{\mathbf{B}} \times \vec{\mathbf{A}}$.

SOLUTION

Conceptualize Given the unit-vector notations of the vectors, think about the directions the vectors point in space. Draw them on graph paper and imagine the parallelogram shown in Figure 11.2 for these vectors.

continued

11.1 continued

Categorize Because we use the definition of the cross product discussed in this section, we categorize this example as a substitution problem.

Write the cross product of the two vectors:

$$\vec{A} \times \vec{B} = (2\,\hat{i} + 3\,\hat{j}) \times (-\hat{i} + 2\,\hat{j})$$

Perform the multiplication using the distributive law:

$$\vec{A} \times \vec{B} = 2\,\hat{i} \times (-\hat{i}) + 2\,\hat{i} \times 2\,\hat{j} + 3\,\hat{j} \times (-\hat{i}) + 3\,\hat{j} \times 2\,\hat{j}$$

Use Equations 11.7a through 11.7d to evaluate the various terms:

$$\vec{A} \times \vec{B} = 0 + 4\hat{k} + 3\hat{k} + 0 = 7\hat{k}$$

To verify that $\vec{A} \times \vec{B} = -\vec{B} \times \vec{A}$, evaluate $\vec{B} \times \vec{A}$:

$$\vec{B} \times \vec{A} = (-\hat{i} + 2\,\hat{j}) \times (2\,\hat{i} + 3\,\hat{j})$$

Perform the multiplication:

$$\vec{B} \times \vec{A} = (-\hat{i}) \times 2\,\hat{i} + (-\hat{i}) \times 3\,\hat{j} + 2\,\hat{j} \times 2\,\hat{i} + 2\,\hat{j} \times 3\,\hat{j}$$

Use Equations 11.7a through 11.7d to evaluate the various terms:

$$\vec{B} \times \vec{A} = 0 - 3\hat{k} - 4\hat{k} + 0 = -7\hat{k}$$

Therefore, $\vec{A} \times \vec{B} = -\vec{B} \times \vec{A}$. As an alternative method for finding $\vec{A} \times \vec{B}$, you could use Equation 11.8. Try it!

Example 11.2 **The Torque Vector**

A force of $\vec{F} = (2.00\,\hat{i} + 3.00\,\hat{j})$ N is applied to a rigid object that is pivoted about a fixed axis aligned along the z coordinate axis. The force is applied at a point located at $\vec{r} = (4.00\,\hat{i} + 5.00\,\hat{j})$ m relative to the axis. Find the torque $\vec{\tau}$ applied to the object.

SOLUTION

Conceptualize Given the unit-vector notations, think about the directions of the force and position vectors. If this force were applied at this position, in what direction would an object pivoted at the origin turn?

Categorize Because we use the definition of the cross product discussed in this section, we categorize this example as a substitution problem.

Set up the torque vector using Equation 11.1:

$$\vec{\tau} = \vec{r} \times \vec{F} = [(4.00\,\hat{i} + 5.00\,\hat{j})\text{ m}] \times [(2.00\,\hat{i} + 3.00\,\hat{j})\text{ N}]$$

Perform the multiplication using the distributive law:

$$\vec{\tau} = [(4.00)(2.00)\,\hat{i} \times \hat{i} + (4.00)(3.00)\,\hat{i} \times \hat{j} \\ + (5.00)(2.00)\hat{j} \times \hat{i} + (5.00)(3.00)\hat{j} \times \hat{j}]\text{ N} \cdot \text{m}$$

Use Equations 11.7a through 11.7d to evaluate the various terms:

$$\vec{\tau} = [0 + 12.0\hat{k} - 10.0\hat{k} + 0]\text{ N} \cdot \text{m} = 2.0\hat{k}\text{ N} \cdot \text{m}$$

Notice that both $\vec{r}$ and $\vec{F}$ are in the xy plane. As expected, the torque vector is perpendicular to this plane, having only a z component. We have followed the rules for significant figures discussed in Section 1.6, which lead to an answer with two significant figures. We have lost some precision because we ended up subtracting two numbers that are close.

11.2 Analysis Model: Nonisolated System (Angular Momentum)

Imagine a rigid pole sticking up through the ice on a frozen pond (Fig. 11.3). From the left in the figure, a skater glides rapidly along a straight line toward the pole, aiming a little to the side so that she does not hit it. As she passes the pole, she reaches out to her side and grabs it, an action that causes her to move in a circular path around the pole. Just as the idea of linear momentum helps us analyze translational motion, a rotational analog—*angular momentum*—helps us analyze the motion of this skater and other objects undergoing rotational motion.

In Chapter 9, we developed the mathematical form of linear momentum and then proceeded to show how this new quantity was valuable in problem solving. We will follow a similar procedure for angular momentum.

Consider a particle of mass m located at the vector position $\vec{r}$ and moving with linear momentum $\vec{p}$ as in Figure 11.4. In describing translational motion, we found that the net force on the particle equals the time rate of change of its linear momentum, $\sum \vec{F} = d\vec{p}/dt$ (see Eq. 9.3). Let us take the cross product of each side of Equation 9.3 with $\vec{r}$, which gives the net torque on the particle on the left side of the equation:

$$\vec{r} \times \sum \vec{F} = \sum \vec{\tau} = \vec{r} \times \frac{d\vec{p}}{dt} \tag{11.9}$$

Now, let's write Equation 11.6 with $\vec{A} = \vec{r}$ and $\vec{B} = \vec{p}$:

$$\frac{d}{dt}(\vec{r} \times \vec{p}) = \frac{d\vec{r}}{dt} \times \vec{p} + \vec{r} \times \frac{d\vec{p}}{dt} = \vec{r} \times \frac{d\vec{p}}{dt} \tag{11.10}$$

where we recognize that the cross product of $d\vec{r}/dt = \vec{v}$ with $\vec{p} = m\vec{v}$ is zero because $\vec{v}$ and $\vec{p}$ are parallel. Because the right sides of Equations 11.9 and 11.10 are the same, we equate the left sides:

$$\sum \vec{\tau} = \frac{d(\vec{r} \times \vec{p})}{dt} \tag{11.11}$$

which looks very similar in form to Equation 9.3, $\sum \vec{F} = d\vec{p}/dt$. Because torque plays the same role in rotational motion that force plays in translational motion, this result suggests that the combination $\vec{r} \times \vec{p}$ should play the same role in rotational motion that $\vec{p}$ plays in translational motion. We call this combination the *angular momentum* of the particle:

The instantaneous **angular momentum** $\vec{L}$ of a particle relative to an axis through a chosen origin O is defined by the cross product of the particle's instantaneous position vector $\vec{r}$ relative to that origin and its instantaneous linear momentum $\vec{p}$:

$$\vec{L} \equiv \vec{r} \times \vec{p} \tag{11.12}$$

We can now write Equation 11.11 as

$$\sum \vec{\tau} = \frac{d\vec{L}}{dt} \tag{11.13}$$

which is the rotational analog of Newton's second law, $\sum \vec{F} = d\vec{p}/dt$. Torque causes the angular momentum $\vec{L}$ to change just as force causes linear momentum $\vec{p}$ to change.

Notice that Equation 11.13 is valid only if $\sum \vec{\tau}$ and $\vec{L}$ are measured about the same axis. Furthermore, the expression is valid for any axis fixed in an inertial frame.

The SI unit of angular momentum is kg · m²/s. Notice also that both the magnitude and the direction of $\vec{L}$ depend on the choice of axis. Following the right-hand rule, we see that the direction of $\vec{L}$ is perpendicular to the plane formed by

Figure 11.3 As the skater passes the pole, she grabs hold of it, which causes her to swing around the pole rapidly in a circular path.

The angular momentum $\vec{L}$ of a particle about an axis is a vector perpendicular to both the particle's position $\vec{r}$ relative to the axis and its momentum $\vec{p}$.

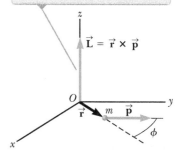

Figure 11.4 The angular momentum $\vec{L}$ of a particle is a vector given by $\vec{L} = \vec{r} \times \vec{p}$.

◀ Angular momentum of a particle

$\vec{\mathbf{r}}$ and $\vec{\mathbf{p}}$. In Figure 11.4, $\vec{\mathbf{r}}$ and $\vec{\mathbf{p}}$ are in the xy plane, so $\vec{\mathbf{L}}$ points in the z direction. Because $\vec{\mathbf{p}} = m\vec{\mathbf{v}}$, the magnitude of $\vec{\mathbf{L}}$ is

$$L = mvr \sin \phi \qquad (11.14)$$

where ϕ is the angle between $\vec{\mathbf{r}}$ and $\vec{\mathbf{p}}$. It follows that L is zero when $\vec{\mathbf{r}}$ is parallel to $\vec{\mathbf{p}}$ ($\phi = 0$ or $180°$). In other words, when the translational velocity of the particle is along a line that passes through the axis, the particle has zero angular momentum with respect to the axis. On the other hand, if $\vec{\mathbf{r}}$ is perpendicular to $\vec{\mathbf{p}}$ ($\phi = 90°$), then $L = mvr$. At that instant, the particle moves exactly as if it were on the rim of a wheel rotating about the axis in a plane defined by $\vec{\mathbf{r}}$ and $\vec{\mathbf{p}}$.

QUICK QUIZ 11.2 Recall the skater described at the beginning of this section. Let her mass be m. **(i)** What would be her angular momentum relative to the pole at the instant she is a distance d from the pole if she were skating directly toward it at speed v? (a) zero (b) mvd (c) impossible to determine **(ii)** What would be her angular momentum relative to the pole at the instant she is a distance d from the pole if she were skating at speed v along a straight path that is offset by a perpendicular distance a from the pole? (a) zero (b) mvd (c) mva (d) impossible to determine

---

Example 11.3 **Angular Momentum of a Particle in Uniform Circular Motion**

A particle moves at constant speed in the xy plane in a circular path of radius r as shown in Figure 11.5. Find the magnitude and direction of its angular momentum relative to an axis through O when its velocity is $\vec{\mathbf{v}}$.

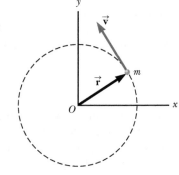

SOLUTION

Conceptualize The linear momentum of the particle is always changing in direction (but not in magnitude). You might therefore be tempted to conclude that the angular momentum of the particle is always changing. In this situation, however, that is not the case. Let's see why.

Figure 11.5 (Example 11.3) A particle moving in a circle of radius r has an angular momentum about an axis through O that has magnitude mvr. The vector $\vec{\mathbf{L}} = \vec{\mathbf{r}} \times \vec{\mathbf{p}}$ points *out* of the page.

Categorize We use the definition of the angular momentum of a particle discussed in this section, so we categorize this example as a substitution problem.

Use Equation 11.14 to evaluate the magnitude of $\vec{\mathbf{L}}$: $\qquad L = mvr \sin 90° = mvr$

This value of L is constant because all three factors on the right are constant. The direction of $\vec{\mathbf{L}}$ also is constant, even though the direction of $\vec{\mathbf{p}} = m\vec{\mathbf{v}}$ keeps changing. To verify this statement, apply the right-hand rule to find the direction of $\vec{\mathbf{L}} = \vec{\mathbf{r}} \times \vec{\mathbf{p}} = m\vec{\mathbf{r}} \times \vec{\mathbf{v}}$ in Figure 11.5. Your thumb points out of the page, so that is the direction of $\vec{\mathbf{L}}$. Hence, we can write the vector expression $\vec{\mathbf{L}} = (mvr)\hat{\mathbf{k}}$. If the particle were to move clockwise, $\vec{\mathbf{L}}$ would point downward and into the page and $\vec{\mathbf{L}} = -(mvr)\hat{\mathbf{k}}$. A particle in *uniform* circular motion has a *constant* angular momentum about an axis through the center of its path.

WHAT IF? The particle in Figure 11.4 moves in a straight line at constant speed along a path parallel to the linear momentum vector $\vec{\mathbf{p}}$. Is the angular momentum of the particle constant in this case?

Answer Yes. In Equation 11.14, m and v are constant while r and ϕ vary in time. However, the product $r \sin\phi$ represents the perpendicular distance between the y axis and the path of the particle. This distance is constant. Therefore, L in Equation 11.14 has a fixed value even though the distance between the particle and the origin changes.

Angular Momentum of a System of Particles

Using the techniques of Section 9.7, we can show that Newton's second law for a system of particles is

$$\sum \vec{\mathbf{F}}_{\text{ext}} = \frac{d\vec{\mathbf{P}}_{\text{tot}}}{dt}$$

This equation states that the net external force on a system of particles is equal to the time rate of change of the total linear momentum of the system. Let's see if a similar statement can be made for rotational motion. The total angular momentum of a system of particles about some axis is defined as the vector sum of the angular momenta of the individual particles:

$$\vec{\mathbf{L}}_{\text{tot}} = \vec{\mathbf{L}}_1 + \vec{\mathbf{L}}_2 + \cdots + \vec{\mathbf{L}}_n = \sum_i \vec{\mathbf{L}}_i$$

where the vector sum is over all n particles in the system.

Differentiating this equation with respect to time gives

$$\frac{d\vec{\mathbf{L}}_{\text{tot}}}{dt} = \sum_i \frac{d\vec{\mathbf{L}}_i}{dt} = \sum_i \vec{\boldsymbol{\tau}}_i$$

where we have used Equation 11.13 to replace the time rate of change of the angular momentum of each particle with the net torque on the particle.

The torques acting on the individual particles of the system are those associated with internal forces between particles and those associated with external forces. The net torque associated with all internal forces, however, is zero. Recall that Newton's third law tells us that internal forces between particles of the system occur in pairs that are equal in magnitude and opposite in direction. If we assume these forces lie along the line of separation of each pair of particles, the total torque around some axis passing through an origin O due to each action–reaction force pair is zero (that is, the moment arm d from O to the line of action of the forces is equal for both particles, and the forces are in opposite directions). In the summation, therefore, the net internal torque is zero. We conclude that the total angular momentum of a system can vary with time only if a net external torque is acting on the system:

$$\sum \vec{\boldsymbol{\tau}}_{\text{ext}} = \frac{d\vec{\mathbf{L}}_{\text{tot}}}{dt} \tag{11.15}$$

◀ The net external torque on a system equals the time rate of change of angular momentum of the system

This equation is indeed the rotational analog of $\sum \vec{\mathbf{F}}_{\text{ext}} = d\vec{\mathbf{p}}_{\text{tot}}/dt$ for a system of particles. Equation 11.15 is the mathematical representation of the **angular momentum version of the nonisolated system model.** If a system is nonisolated in the sense that there is a net external torque on it, the net external torque on the system is equal to the time rate of change of the angular momentum of the system.

Although we do not prove it here, this statement is true regardless of the motion of the center of mass. It applies even if the center of mass is accelerating, provided the torque and angular momentum are evaluated relative to an axis through the center of mass.

Equation 11.15 can be rearranged and integrated to give

$$\Delta\vec{\mathbf{L}}_{\text{tot}} = \int \left(\sum \vec{\boldsymbol{\tau}}_{\text{ext}} \right) dt$$

This equation represents the *angular impulse–angular momentum theorem.* Compare this equation to the translational version, Equation 9.40.

ANALYSIS MODEL **Nonisolated System (Angular Momentum)**

Imagine a system that rotates about an axis. If there is a net external torque acting on the system, the time rate of change of the angular momentum of the system is equal to the net external torque:

$$\sum \vec{\boldsymbol{\tau}}_{\text{ext}} = \frac{d\vec{\mathbf{L}}_{\text{tot}}}{dt} \tag{11.15}$$

continued

Examples:

- a flywheel in an automobile engine increases its angular momentum when the engine applies torque to it
- the tub of a washing machine decreases in angular momentum due to frictional torque after the machine is turned off
- the axis of the Earth undergoes a precessional motion due to the torque exerted on the Earth by the gravitational force from the Sun
- the armature of a motor increases its angular momentum due to the torque exerted by a surrounding magnetic field (Chapter 30)

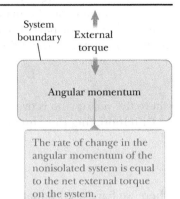

The rate of change in the angular momentum of the nonisolated system is equal to the net external torque on the system.

Example 11.4 **A System of Objects**

A sphere of mass m_1 and a block of mass m_2 are connected by a light cord that passes over a pulley as shown in Figure 11.6. The radius of the pulley is R, and the mass of the thin rim is M. The spokes of the pulley have negligible mass. The block slides on a frictionless, horizontal surface. Find an expression for the linear acceleration of the two objects, using the concepts of angular momentum and torque.

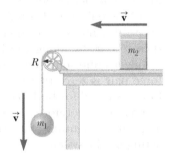

Figure 11.6 (Example 11.4) When the system is released, the sphere moves downward and the block moves to the left.

SOLUTION

Conceptualize When the system is released, the block slides to the left, the sphere drops downward, and the pulley rotates counterclockwise. This situation is similar to problems we have solved earlier except that now we want to use an angular momentum approach.

Categorize We identify the block, pulley, and sphere as a *nonisolated system* for *angular momentum*, subject to the external torque due to the gravitational force on the sphere. We shall calculate the angular momentum of the system about an axis that coincides with the axle of the pulley. The angular momentum of the system includes that of two objects moving translationally (the sphere and the block) and one object undergoing pure rotation (the pulley).

Analyze At any instant of time, the sphere and the block have a common speed v, so the angular momentum of the sphere about the pulley axle is m_1vR and that of the block is m_2vR. At the same instant, all points on the rim of the pulley also move with speed v, so the angular momentum of the pulley is MvR.

Now let's address the total external torque acting on the system about the pulley axle. Because it has a moment arm of zero, the force exerted by the axle on the pulley does not contribute to the torque. Furthermore, the normal force acting on the block is balanced by the gravitational force $m_2\vec{\mathbf{g}}$, so these forces do not contribute to the torque. The gravitational force $m_1\vec{\mathbf{g}}$ acting on the sphere produces a torque about the axle equal in magnitude to m_1gR, where R is the moment arm of the force about the axle. This result is the total external torque about the pulley axle; that is, $\sum \tau_{\text{ext}} = m_1gR$.

Write an expression for the total angular momentum of the system:

$$(1) \quad L = m_1vR + m_2vR + MvR = (m_1 + m_2 + M)vR$$

Substitute this expression and the total external torque into Equation 11.15, the mathematical representation of the nonisolated system model for angular momentum:

$$\sum \tau_{\text{ext}} = \frac{dL}{dt}$$

$$m_1gR = \frac{d}{dt}[(m_1 + m_2 + M)vR]$$

$$(2) \quad m_1gR = (m_1 + m_2 + M)R\frac{dv}{dt}$$

Recognizing that $dv/dt = a$, solve Equation (2) for a:

$$(3) \quad a = \frac{m_1g}{m_1 + m_2 + M}$$

11.4 continued

Finalize   When we evaluated the net torque about the axle, we did not include the forces that the cord exerts on the objects because these forces are internal to the system under consideration. Instead, we analyzed the system as a whole. Only *external* torques contribute to the change in the system's angular momentum. Let $M \rightarrow 0$ in Equation (3) and call the result Equation A. Now go back to Equation (5) in Example 5.10, let $\theta \rightarrow 0$, and call the result Equation B. Do Equations A and B match? Looking at Figures 5.16 and 11.6 in these limits, *should* the two equations match?

11.3 Angular Momentum of a Rotating Rigid Object

In Example 11.4, we considered the angular momentum of a deformable system of particles. Let us now restrict our attention to a nondeformable system, a rigid object. Consider a rigid object rotating about a fixed axis that coincides with the z axis of a coordinate system as shown in Figure 11.7. Let's determine the angular momentum of this object. Each *particle* of the object rotates in the xy plane about the z axis with an angular speed ω. The magnitude of the angular momentum of a particle of mass m_i about the z axis is $m_i v_i r_i$. Because $v_i = r_i \omega$ (Eq. 10.10), we can express the magnitude of the angular momentum of this particle as

$$L_i = m_i v_i r_i = m_i (r_i \omega) r_i = m_i r_i^2 \omega$$

The vector $\vec{L}_i$ for this particle is directed along the z axis, as is the vector $\vec{\omega}$.

We can now find the angular momentum (which in this situation has only a z component) of the whole object by taking the sum of L_i over all particles:

$$L_z = \sum_i L_i = \sum_i m_i r_i^2 \omega = \left(\sum_i m_i r_i^2 \right) \omega$$

$$L_z = I\omega \qquad (11.16)$$

where we have recognized $\sum_i m_i r_i^2$ as the moment of inertia I of the object about the z axis (Eq. 10.19). Notice that Equation 11.16 is mathematically similar in form to Equation 9.2 for linear momentum: $\vec{p} = m\vec{v}$.

Now let's differentiate Equation 11.16 with respect to time, noting that I is constant for a rigid object:

$$\frac{dL_z}{dt} = I \frac{d\omega}{dt} = I\alpha \qquad (11.17)$$

where α is the angular acceleration relative to the axis of rotation. Because dL_z/dt is equal to the net external torque (see Eq. 11.15), we can express Equation 11.17 as

$$\sum \tau_{\text{ext}} = I\alpha \qquad (11.18)$$

◀ Rotational form of Newton's second law

That is, the net external torque acting on a rigid object rotating about a fixed axis equals the moment of inertia about the rotation axis multiplied by the object's angular acceleration relative to that axis. This result is the same as Equation 10.18, which was derived using a force approach, but we derived Equation 11.18 using the concept of angular momentum. As we saw in Section 10.5, Equation 11.18 is the mathematical representation of the rigid object under a net torque analysis model. This equation is also valid for a rigid object rotating about a moving axis, provided the moving axis (1) passes through the center of mass and (2) is a symmetry axis.

If a symmetrical object rotates about a fixed axis passing through its center of mass, you can write Equation 11.16 in vector form as $\vec{L} = I\vec{\omega}$, where $\vec{L}$ is the total angular momentum of the object measured with respect to the axis of rotation.

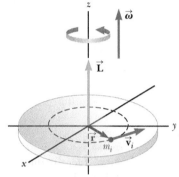

Figure 11.7 When a rigid object rotates about an axis, the angular momentum $\vec{L}$ is in the same direction as the angular velocity $\vec{\omega}$ according to the expression $\vec{L} = I\vec{\omega}$.

Furthermore, the expression is valid for any object, regardless of its symmetry, if $\vec{L}$ stands for the component of angular momentum along the axis of rotation.[1]

QUICK QUIZ 11.3 A solid sphere and a hollow sphere have the same mass and radius. They are rotating with the same angular speed. Which one has the higher angular momentum? **(a)** the solid sphere **(b)** the hollow sphere **(c)** both have the same angular momentum **(d)** impossible to determine

---

Example 11.5   The Seesaw

A father of mass m_f and his daughter of mass m_d sit on opposite ends of a seesaw at equal distances from the pivot at the center (Fig. 11.8). The seesaw is modeled as a rigid rod of mass M and length ℓ and is pivoted without friction. At a given moment, the combination rotates in a vertical plane with an angular speed ω.

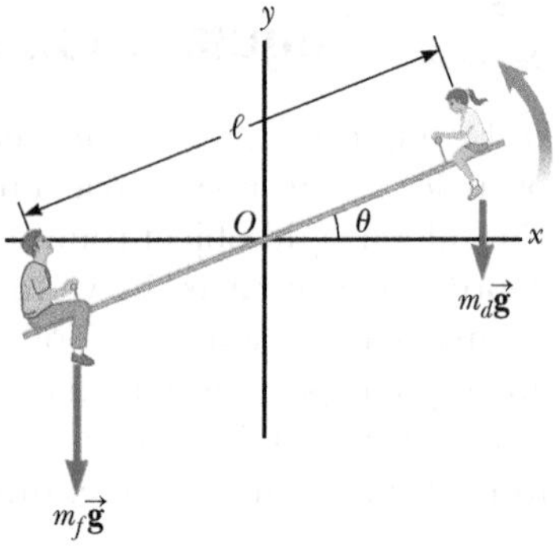

(A) Find an expression for the magnitude of the system's angular momentum.

Figure 11.8 (Example 11.5) A father and daughter demonstrate angular momentum on a seesaw.

SOLUTION

Conceptualize   Identify the z axis through O as the axis of rotation in Figure 11.8. The rotating system has angular momentum about that axis.

Categorize   Ignore any movement of arms or legs of the father and daughter and model them both as particles. The system is therefore modeled as a rigid object. This first part of the example is categorized as a substitution problem.

The moment of inertia of the system equals the sum of the moments of inertia of the three components: the seesaw and the two individuals. We can refer to Table 10.2 to obtain the expression for the moment of inertia of the rod and use the particle expression $I = mr^2$ for each person.

Find the total moment of inertia of the system about the z axis through O:

$$I = \tfrac{1}{12}M\ell^2 + m_f\left(\frac{\ell}{2}\right)^2 + m_d\left(\frac{\ell}{2}\right)^2 = \frac{\ell^2}{4}\left(\frac{M}{3} + m_f + m_d\right)$$

Find the magnitude of the angular momentum of the system:

$$L = I\omega = \frac{\ell^2}{4}\left(\frac{M}{3} + m_f + m_d\right)\omega$$

(B) Find an expression for the magnitude of the angular acceleration of the system when the seesaw makes an angle θ with the horizontal.

SOLUTION

Conceptualize   Generally, fathers are more massive than daughters, so the system is not in equilibrium and has an angular acceleration. We expect the angular acceleration to be positive in Figure 11.8.

Categorize   The combination of the board, father, and daughter is a *rigid object under a net torque* because of the external torque associated with the gravitational forces on the father and daughter. We again identify the axis of rotation as the z axis in Figure 11.8.

Analyze   To find the angular acceleration of the system at any angle θ, we first calculate the net torque on the system and then use $\Sigma \tau_{ext} = I\alpha$ from the rigid object under a net torque model to obtain an expression for α.

Evaluate the torque due to the gravitational force on the father:

$$\tau_f = m_f g \frac{\ell}{2}\cos\theta \quad (\vec{\tau}_f \text{ out of page})$$

Evaluate the torque due to the gravitational force on the daughter:

$$\tau_d = -m_d g \frac{\ell}{2}\cos\theta \quad (\vec{\tau}_d \text{ into page})$$

---

[1]In general, the expression $\vec{L} = I\vec{\omega}$ is not always valid. If a rigid object rotates about an *arbitrary* axis, then $\vec{L}$ and $\vec{\omega}$ may point in different directions. In this case, the moment of inertia cannot be treated as a scalar. Strictly speaking, $\vec{L} = I\vec{\omega}$ applies only to rigid objects of any shape that rotate about one of three mutually perpendicular axes (called *principal axes*) through the center of mass. This concept is discussed in more advanced texts on mechanics.

11.5 continued

Evaluate the net external torque exerted on the system:

$$\sum \tau_{\text{ext}} = \tau_f + \tau_d = \tfrac{1}{2}(m_f - m_d)g\ell \cos \theta$$

Use Equation 11.18 and I from part (A) to find α:

$$\alpha = \frac{\sum \tau_{\text{ext}}}{I} = \frac{2(m_f - m_d)g \cos \theta}{\ell \left[(M/3) + m_f + m_d \right]}$$

Finalize   For a father more massive than his daughter, the angular acceleration is positive as expected. If the seesaw begins in a horizontal orientation ($\theta = 0$) and is released, the rotation is counterclockwise in Figure 11.8 and the father's end of the seesaw drops, which is consistent with everyday experience.

WHAT IF?   Imagine the father moves inward on the seesaw to a distance d from the pivot to try to balance the two sides. What is the angular acceleration of the system in this case when it is released from an arbitrary angle θ?

Answer   The angular acceleration of the system should decrease if the system is more balanced.

Find the total moment of inertia about the z axis through O for the modified system:

$$I = \tfrac{1}{12}M\ell^2 + m_f d^2 + m_d \left(\frac{\ell}{2} \right)^2 = \frac{\ell^2}{4}\left(\frac{M}{3} + m_d \right) + m_f d^2$$

Find the net torque exerted on the modified system about an axis through O:

$$\sum \tau_{\text{ext}} = \tau_f + \tau_d = m_f g d \cos \theta - \tfrac{1}{2}m_d g\ell \cos \theta$$

Find the new angular acceleration of the system:

$$\alpha = \frac{\sum \tau_{\text{ext}}}{I} = \frac{(m_f d - \tfrac{1}{2}m_d \ell)g \cos \theta}{(\ell^2/4)\left[(M/3) + m_d \right] + m_f d^2}$$

WHAT IF?   Where must the father sit for the seesaw to be balanced?

Answer   The seesaw is balanced when the angular acceleration is zero. In this situation, both father and daughter can push off the ground and rise to the highest possible point.

For the seesaw to be balanced, the required position of the father is found by setting $\alpha = 0$:

$$\alpha = \frac{(m_f d - \tfrac{1}{2}m_d \ell)g \cos \theta}{(\ell^2/4)[(M/3) + m_d] + m_f d^2} = 0$$

$$m_f d - \tfrac{1}{2}m_d \ell = 0 \quad \rightarrow \quad d = \left(\frac{m_d}{m_f} \right)\frac{\ell}{2}$$

The heavier the father, the closer he must sit to the pivot to balance the seesaw. In the rare case that the father and daughter have the same mass, the father is located at the end of the seesaw, $d = \ell/2$.

11.4 Analysis Model: Isolated System (Angular Momentum)

In Chapter 9, we found that the total linear momentum of a system of particles remains constant if the system is isolated, that is, if the net external force acting on the system is zero. We have an analogous conservation law in rotational motion:

> The total angular momentum of a system is constant in both magnitude and direction if the net external torque acting on the system is zero, that is, if the system is isolated.

◀ Conservation of angular momentum

This statement is often called[2] the principle of **conservation of angular momentum** and is the basis of the **angular momentum version of the isolated system model**. This principle follows directly from Equation 11.15, which indicates that if

$$\sum \vec{\boldsymbol{\tau}}_{\text{ext}} = \frac{d\vec{\mathbf{L}}_{\text{tot}}}{dt} = 0$$

---

[2]The most general conservation of angular momentum equation is Equation 11.15, which describes how the system interacts with its environment.

then

$$\Delta \vec{L}_{tot} = 0 \qquad \text{(11.19)}$$

Equation 11.19 can be written as

$$\vec{L}_{tot} = \text{constant} \quad \text{or} \quad \vec{L}_i = \vec{L}_f \qquad \text{(11.20)}$$

For an isolated system consisting of a small number of particles, we write this conservation law as $\vec{L}_{tot} = \sum \vec{L}_n = \text{constant}$, where the index n denotes the nth particle in the system. If the system consists of a large number of particles, so that it is difficult to evaluate the individual L_n, then we can express the magnitude of the angular momentum of the system with Equation 11.16, $L = I\omega$.

If an isolated rotating system is deformable so that its mass undergoes redistribution in some way, the system's moment of inertia changes. Combining Equations 11.16 and 11.20, we see that conservation of angular momentum requires that the product of I and ω must remain constant. Therefore, a change in I for an isolated system requires a change in ω. In this case, we can express the principle of conservation of angular momentum as

$$I_i \omega_i = I_f \omega_f = \text{constant} \qquad \text{(11.21)}$$

This expression is valid both for rotation about a fixed axis and for rotation about an axis through the center of mass of a moving system as long as that axis remains fixed in direction. We require only that the net external torque be zero.

Many examples demonstrate conservation of angular momentum for a deformable system. You may have observed a figure skater spinning in the finale of a program (Fig. 11.9). The angular speed of the skater is large when his hands and feet are close to the trunk of his body. (Notice the skater's hair!) Ignoring friction between skater and ice, there are no external torques on the skater. The moment of inertia of his body increases as his hands and feet are moved away from his body, and therefore from the rotation axis, at the finish of the spin. According to the isolated system model for angular momentum, his angular speed must decrease, and he can perform his finishing flourish after coming to rest. In a similar way, when divers or acrobats wish to make several rotations, they pull their hands and feet close to their bodies to rotate at a higher rate, as shown in the photograph opening this chapter. In these cases, the external force due to gravity acts through the center of mass and hence exerts no torque about an axis through this point. Therefore, the angular momentum of the diver or acrobat about the center of mass must be conserved; that is, $I_i \omega_i = I_f \omega_f$. For example, when divers wish to double their angular speed, they must reduce their moment of inertia to half its initial value.

The introductory storyline to this chapter asked about the additional rotational kinetic energy possessed by spinning skaters and rotating divers when they pull their limbs inward. This energy comes from within the body. The muscles of the rotating athlete must do internal work to pull the limbs inward. This work is a transformation mechanism by which potential energy in the body from previous meals is transformed to rotational kinetic energy. The storyline also mentioned a falling cat, which is yet again an example of a deformable system, first introduced in Section 9.8. The cat is released with zero angular momentum, yet is able to rotate and right itself before landing. A number of theories have been proposed for this phenomenon, including a popular one modeling the cat as a pair of cylinders. Perform some online research to learn more about falling cats.

In Equation 11.20, we have a third version of the isolated system model. We can now state that the energy, linear momentum, and angular momentum of an

When his arms and legs are close to his body, the skater's moment of inertia is small and his angular speed is large.

Clive Rose/Getty Images

To slow down for the finish of his spin, the skater moves his arms and legs outward, increasing his moment of inertia.

Al Bello/Getty Images

Figure 11.9 Angular momentum is conserved as Russian gold medalist Evgeni Plushenko performs during the Turin 2006 Winter Olympic Games.

isolated system are all constant:

$$\Delta E_{\text{system}} = 0 \quad \text{(if there are no energy transfers across the system boundary)}$$

$$\Delta \vec{\mathbf{p}}_{\text{tot}} = 0 \quad \text{(if the net external force on the system is zero)}$$

$$\Delta \vec{\mathbf{L}}_{\text{tot}} = 0 \quad \text{(if the net external torque on the system is zero)}$$

Notice that the definition of an isolated system varies for the three conserved quantities. A system may be isolated in terms of one of these quantities but not in terms of another. If a system is nonisolated in terms of momentum or angular momentum, it will often be nonisolated also in terms of energy because the system has a net force or torque on it and the net force or torque will do work on the system. We can, however, identify systems that are nonisolated in terms of energy but isolated in terms of momentum. For example, imagine pushing inward on a balloon (the system) between your hands. Work is done in compressing the balloon, so the system is nonisolated in terms of energy, but there is zero net force on the system, so the system is isolated in terms of momentum. A similar statement could be made about twisting the ends of a long, flat, springy piece of metal with both hands. Work is done on the metal (the system), so energy is stored in the nonisolated system as elastic potential energy, but the net torque on the system is zero. Therefore, the system is isolated in terms of angular momentum. Other examples are collisions of macroscopic objects, which represent isolated systems in terms of momentum but nonisolated systems in terms of energy because of the output of energy from the system by mechanical waves (sound).

ⓆUICK QUIZ 11.4 A competitive diver leaves the diving board and falls toward the water with her body straight and rotating slowly. She pulls her arms and legs into a tight tuck position. What happens to her rotational kinetic energy? **(a)** It increases. **(b)** It decreases. **(c)** It stays the same. **(d)** It is impossible to determine.

ANALYSIS MODEL **Isolated System (Angular Momentum)**

Imagine a system rotates about an axis. If there is no net external torque on the system, there is no change in the angular momentum of the system:

$$\Delta \vec{\mathbf{L}}_{\text{tot}} = 0 \qquad \text{(11.19)}$$

Applying this law of conservation of angular momentum to a system whose moment of inertia changes gives

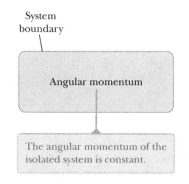

$$I_i \omega_i = I_f \omega_f = \text{constant} \qquad \text{(11.21)}$$

Examples:

- after a supernova explosion, the core of a star collapses to a small radius and spins at a much higher rate (Example 11.6, below).
- the square of the orbital period of a planet is proportional to the cube of its semimajor axis; Kepler's third law (Chapter 13)
- in atomic transitions, selection rules on the quantum numbers must be obeyed in order to conserve angular momentum (Chapter 41)
- in beta decay of a radioactive nucleus, a neutrino must be emitted in order to conserve angular momentum (Chapter 43)

Example **11.6** **Formation of a Neutron Star**

A star rotates with a period of 30 days about an axis through its center. The period is the time interval required for a point on the star's equator to make one complete revolution around the axis of rotation. After the star undergoes a supernova explosion, the stellar core, which had a radius of 1.0×10^4 km, collapses into a neutron star of radius 10.0 km. Determine the period of rotation of the neutron star.

SOLUTION

Conceptualize The change in the neutron star's motion is similar to that of the skater described earlier and illustrated in Figure 11.9, but in the reverse direction. As the mass of the star moves closer to the rotation axis, we expect the star to spin faster.

continued

11.6 c o n t i n u e d

Categorize   Let us assume that during the collapse of the stellar core, (1) no external torque acts on it, (2) it remains spherical with the same relative mass distribution, and (3) its mass remains constant. We categorize the star as an *isolated system* in terms of *angular momentum*. We do not know the mass distribution of the star, but we have assumed the distribution is symmetric, so the moment of inertia can be expressed as kMR^2, where k is some numerical constant. (From Table 10.2, for example, we see that $k = \frac{2}{5}$ for a solid sphere and $k = \frac{2}{3}$ for a spherical shell.)

Analyze   Let's use the symbol T for the period, with T_i being the initial period of the star and T_f being the period of the neutron star. The star's angular speed is given by $\omega = 2\pi/T$.

From the isolated system model for angular momentum, write Equation 11.21 for the star:

$$I_i \omega_i = I_f \omega_f$$

Use $\omega = 2\pi/T$ to rewrite this equation in terms of the initial and final periods:

$$I_i \left(\frac{2\pi}{T_i} \right) = I_f \left(\frac{2\pi}{T_f} \right)$$

Substitute the moments of inertia in the preceding equation:

$$kMR_i^2 \left(\frac{2\pi}{T_i} \right) = kMR_f^2 \left(\frac{2\pi}{T_f} \right)$$

Solve for the final period of the star:

$$T_f = \left(\frac{R_f}{R_i} \right)^2 T_i$$

Substitute numerical values:

$$T_f = \left(\frac{10.0 \text{ km}}{1.0 \times 10^4 \text{ km}} \right)^2 (30 \text{ days}) = 3.0 \times 10^{-5} \text{ days} = 2.6 \text{ s}$$

Finalize   The neutron star does indeed rotate faster after it collapses, as predicted. While it may seem difficult to believe that the core of a star could rotate as fast as once every 2.6 s, this is a relatively slow rotation rate. Some neutron stars rotate with a period of 1–2 milliseconds!

Example 11.7   **The Merry-Go-Round**

A horizontal platform in the shape of a circular disk rotates freely in a horizontal plane about a frictionless, vertical axle (Fig. 11.10). The platform has a mass $M = 100$ kg and a radius $R = 2.0$ m. A student whose mass is $m = 60$ kg walks slowly from the rim of the disk toward its center. If the angular speed of the system is 2.0 rad/s when the student is at the rim, what is the angular speed when she reaches a point $r = 0.50$ m from the center?

SOLUTION

Conceptualize   The speed change here is similar to those of the spinning skater and the neutron star in preceding discussions. This problem is different because part of the moment of inertia of the system changes (that of the student) while part remains fixed (that of the platform).

Categorize   Because the platform rotates on a frictionless axle, we identify the system of the student and the platform as an *isolated system* in terms of *angular momentum*.

Analyze   Let us denote the moment of inertia of the platform as I_p and that of the student as I_s. We model the student as a particle.

Write Equation 11.21 for the system:

$$I_i \omega_i = I_f \omega_f$$

Substitute the moments of inertia, using $r < R$ for the final position of the student:

$$(\tfrac{1}{2}MR^2 + mR^2)\omega_i = (\tfrac{1}{2}MR^2 + mr^2)\omega_f$$

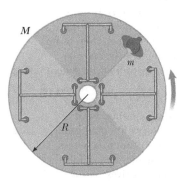

Figure 11.10 (Example 11.7) As the student walks toward the center of the rotating platform, the angular speed of the system increases because the angular momentum of the system remains constant.

11.7 continued

Solve for the final angular speed:
$$\omega_f = \left(\frac{\frac{1}{2}MR^2 + mR^2}{\frac{1}{2}MR^2 + mr^2} \right) \omega_i$$

Substitute numerical values:
$$\omega_f = \left[\frac{\frac{1}{2}(100 \text{ kg})(2.0 \text{ m})^2 + (60 \text{ kg})(2.0 \text{ m})^2}{\frac{1}{2}(100 \text{ kg})(2.0 \text{ m})^2 + (60 \text{ kg})(0.50 \text{ m})^2} \right](2.0 \text{ rad/s}) = \left[\frac{440 \text{ kg} \cdot \text{m}^2}{215 \text{ kg} \cdot \text{m}^2} \right](2.0 \text{ rad/s}) = 4.1 \text{ rad/s}$$

Finalize As expected, the angular speed increases. The fastest that this system could spin would be when the student moves to the center of the platform. Do this calculation to show that this maximum angular speed is 4.4 rad/s. Notice that the activity described in this problem is dangerous as discussed with regard to the Coriolis force in Section 6.3.

WHAT IF? What if you measured the kinetic energy of the system before and after the student walks inward? Are the initial kinetic energy and the final kinetic energy the same?

Answer You may be tempted to say yes because the system is isolated. Remember, however, that energy can be transformed among several forms, so we have to handle an energy question carefully.

Find the initial kinetic energy:
$$K_i = \tfrac{1}{2}I_i\omega_i^2 = \tfrac{1}{2}(440 \text{ kg} \cdot \text{m}^2)(2.0 \text{ rad/s})^2 = 880 \text{ J}$$

Find the final kinetic energy:
$$K_f = \tfrac{1}{2}I_f\omega_f^2 = \tfrac{1}{2}(215 \text{ kg} \cdot \text{m}^2)(4.1 \text{ rad/s})^2 = 1.80 \times 10^3 \text{ J}$$

Therefore, the kinetic energy of the system *increases* by more than a factor of 2. The student must perform muscular activity to move herself closer to the center of rotation, so this extra kinetic energy comes from potential energy stored in the student's body from previous meals. The system is isolated in terms of energy, but a transformation process within the system changes potential energy to kinetic energy.

Example 11.8 | Disk and Stick Collision

A 2.0-kg disk traveling at 3.0 m/s strikes a 1.0-kg stick of length 4.0 m that is lying flat on nearly frictionless ice as shown in the overhead view of Figure 11.11a. The disk strikes at the endpoint of the stick, at a distance $r = 2.0$ m from the stick's center. Assume the collision is elastic and the disk does not deviate from its original line of motion. Find the translational speed of the disk, the translational speed of the stick, and the angular speed of the stick after the collision. The moment of inertia of the stick about its center of mass is 1.33 kg · m².

SOLUTION

Conceptualize Examine Figure 11.11a and imagine what happens after the disk hits the stick. Figure 11.11b shows what you might expect: the disk continues to move at a slower speed, and the stick is in both translational and rotational motion. We assume the disk does not deviate from its original line of motion because the force exerted by the stick on the disk is parallel to the original path of the disk.

Categorize Because the ice is frictionless, the disk and stick form an *isolated system* in terms of *momentum* and *angular momentum*. Ignoring the sound made in the collision, we also model the system as an *isolated system* in terms of *energy*. In addition, because the collision is assumed to be elastic, the kinetic energy of the system is constant.

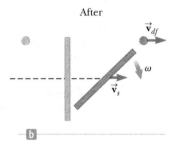

Figure 11.11 (Example 11.8) Overhead view of a disk striking a stick in an elastic collision. (a) Before the collision, the disk moves toward the stick. (b) The collision causes the stick to rotate and move to the right.

Analyze First notice that we have three unknowns, so we need three equations to solve simultaneously.

Apply the isolated system model for momentum to the system and then rearrange the result:
$$\Delta \vec{\mathbf{p}}_{tot} = 0 \rightarrow (m_d v_{df} + m_s v_s) - m_d v_{di} = 0$$
$$(1) \quad m_d(v_{di} - v_{df}) = m_s v_s$$

continued

11.8 continued

Apply the isolated system model for angular momentum to the system and rearrange the result. Use an axis passing through the center of the stick as the rotation axis so that the path of the disk is a distance r from the rotation axis:	$\Delta \vec{L}_{tot} = 0 \;\rightarrow\; (-rm_d v_{df} + I\omega) - (-rm_d v_{di}) = 0$ (2) $\;-rm_d(v_{di} - v_{df}) = I\omega$
Apply the isolated system model for energy to the system, rearrange the equation, and factor the combination of terms related to the disk:	$\Delta K = 0 \;\rightarrow\; (\tfrac{1}{2}m_d v_{df}^2 + \tfrac{1}{2}m_s v_s^2 + \tfrac{1}{2}I\omega^2) - \tfrac{1}{2}m_d v_{di}^2 = 0$ (3) $\; m_d(v_{di} - v_{df})(v_{di} + v_{df}) = m_s v_s^2 + I\omega^2$
Multiply Equation (1) by r and add to Equation (2):	$rm_d(v_{di} - v_{df}) = rm_s v_s$ $-rm_d(v_{di} - v_{df}) = I\omega$ $0 = rm_s v_s + I\omega$
Solve for ω:	(4) $\;\omega = -\dfrac{rm_s v_s}{I}$
Divide Equation (3) by Equation (1):	$\dfrac{m_d(v_{di} - v_{df})(v_{di} + v_{df})}{m_d(v_{di} - v_{df})} = \dfrac{m_s v_s^2 + I\omega^2}{m_s v_s}$ (5) $\; v_{di} + v_{df} = v_s + \dfrac{I\omega^2}{m_s v_s}$
Substitute Equation (4) into Equation (5):	(6) $\; v_{di} + v_{df} = v_s\left(1 + \dfrac{r^2 m_s}{I}\right)$
Substitute v_{df} from Equation (1) into Equation (6):	$v_{di} + \left(v_{di} - \dfrac{m_s}{m_d}v_s\right) = v_s\left(1 + \dfrac{r^2 m_s}{I}\right)$
Solve for v_s and substitute numerical values:	$v_s = \dfrac{2v_{di}}{1 + (m_s/m_d) + (r^2 m_s/I)}$ $= \dfrac{2(3.0 \text{ m/s})}{1 + (1.0 \text{ kg/2.0 kg}) + [(2.0 \text{ m})^2(1.0 \text{ kg})/1.33 \text{ kg} \cdot \text{m}^2]} = 1.3 \text{ m/s}$
Substitute numerical values into Equation (4):	$\omega = -\dfrac{(2.0 \text{ m})(1.0 \text{ kg})(1.3 \text{ m/s})}{1.33 \text{ kg} \cdot \text{m}^2} = -2.0 \text{ rad/s}$
Solve Equation (1) for v_{df} and substitute numerical values:	$v_{df} = v_{di} - \dfrac{m_s}{m_d}v_s = 3.0 \text{ m/s} - \dfrac{1.0 \text{ kg}}{2.0 \text{ kg}}(1.3 \text{ m/s}) = 2.3 \text{ m/s}$

Finalize These values seem reasonable. The disk is moving more slowly after the collision than it was before the collision. The stick has a small translational speed and is rotating clockwise. Table 11.1 summarizes the initial and final values of variables for the disk and the stick, and it verifies the conservation of linear momentum, angular momentum, and kinetic energy for the isolated system.

TABLE 11.1 Comparison of Values in Example 11.8 Before and After the Collision

	v (m/s)	ω (rad/s)	p (kg · m/s)	L (kg · m²/s)	K_{trans} (J)	K_{rot} (J)
Before						
Disk	3.0	—	6.0	−12	9.0	—
Stick	0	0	0	0	0	0
Total for system	—	—	6.0	−12	9.0	0
After						
Disk	2.3	—	4.7	−9.3	5.4	—
Stick	1.3	−2.0	1.3	−2.7	0.9	2.7
Total for system	—	—	6.0	−12	6.3	2.7

Note: Linear momentum, angular momentum, and total kinetic energy of the system are all conserved.

11.5 The Motion of Gyroscopes and Tops

An unusual and fascinating type of motion you have probably observed is that of a top spinning about its axis of symmetry as shown in Figure 11.12a. If the top spins rapidly, the symmetry axis rotates about the z axis, sweeping out a cone (see Fig. 11.12b). The motion of the symmetry axis about the vertical—known as **precessional motion**—is usually slow relative to the spinning motion of the top.

It is quite natural to wonder why the top does not fall over. Because the center of mass is not directly above the pivot point O, a net torque is acting on the top about an axis passing through O, a torque resulting from the gravitational force $M\vec{\mathbf{g}}$. The top would certainly fall over if it were not spinning. Because it is spinning, however, it has an angular momentum $\vec{\mathbf{L}}$ directed along its symmetry axis. We shall show that this symmetry axis moves about the z axis (precessional motion occurs) because the torque produces a change in the *direction* of the symmetry axis. This illustration is an excellent example of the importance of the vector nature of angular momentum.

The essential features of precessional motion can be illustrated by considering the top to act as a simple gyroscope. The two forces acting on the gyroscope are shown in Figure 11.12a: the downward gravitational force $M\vec{\mathbf{g}}$ and the normal force $\vec{\mathbf{n}}$ acting upward at the pivot point O. The normal force produces no torque about an axis passing through the pivot because its moment arm through that point is zero. The gravitational force, however, produces a torque $\vec{\boldsymbol{\tau}} = \vec{\mathbf{r}} \times M\vec{\mathbf{g}}$ about an axis passing through O, where the direction of $\vec{\boldsymbol{\tau}}$ is perpendicular to the plane formed by $\vec{\mathbf{r}}$ and $M\vec{\mathbf{g}}$. By necessity, the vector $\vec{\boldsymbol{\tau}}$ lies in a horizontal xy plane perpendicular to the angular momentum vector. The net torque and angular momentum of the gyroscope are related through Equation 11.15:

$$\sum \vec{\boldsymbol{\tau}}_{\text{ext}} = \frac{d\vec{\mathbf{L}}}{dt}$$

This expression shows that in the infinitesimal time interval dt, the nonzero torque produces a change in angular momentum $d\vec{\mathbf{L}}$, a change that is in the same direction as $\vec{\boldsymbol{\tau}}$. Therefore, like the torque vector, $d\vec{\mathbf{L}}$ must also be perpendicular to $\vec{\mathbf{L}}$. The overhead view in Figure 11.12c illustrates the resulting precessional motion of the symmetry axis of the gyroscope. In a time interval dt, the change in angular momentum is $d\vec{\mathbf{L}} = \vec{\mathbf{L}}_f - \vec{\mathbf{L}}_i = \vec{\boldsymbol{\tau}}\, dt$. Because $d\vec{\mathbf{L}}$ is perpendicular to $\vec{\mathbf{L}}$, the magnitude of $\vec{\mathbf{L}}$ does not change ($|\vec{\mathbf{L}}_i| = |\vec{\mathbf{L}}_f|$). Rather, what is changing is the *direction* of $\vec{\mathbf{L}}$. Because the change in angular momentum $d\vec{\mathbf{L}}$ is in the direction of $\vec{\boldsymbol{\tau}}$, which lies in the xy plane, the gyroscope undergoes precessional motion.

The vector diagram in Figure 11.12c shows that in the time interval dt, the angular momentum vector rotates through an angle $d\phi$, which is also the angle through which the gyroscope axle rotates. From the vector triangle formed by the vectors $\vec{\mathbf{L}}_i$, $\vec{\mathbf{L}}_f$, and $d\vec{\mathbf{L}}$, we see that

$$d\phi = \frac{dL}{L} = \frac{\sum \tau_{\text{ext}}\, dt}{L} = \frac{(Mgr_{\text{CM}})\, dt}{L}$$

Dividing through by dt and using the relationship $L = I\omega$, we find that the rate at which the axle rotates about the vertical axis is

$$\omega_p = \frac{d\phi}{dt} = \frac{Mgr_{\text{CM}}}{I\omega} \tag{11.22}$$

The angular speed ω_p is called the **precessional frequency**. This result is valid only when $\omega_p \ll \omega$. Otherwise, a much more complicated motion is involved. As you can see from Equation 11.22, the condition $\omega_p \ll \omega$ is met when ω is large, that is, when the wheel spins rapidly. Furthermore, notice that the precessional frequency decreases as ω increases, that is, as the wheel spins faster about its axis of symmetry.

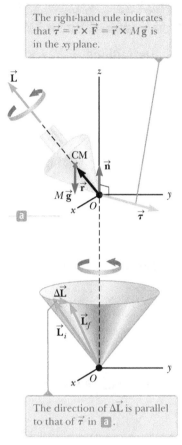

The right-hand rule indicates that $\vec{\boldsymbol{\tau}} = \vec{\mathbf{r}} \times \vec{\mathbf{F}} = \vec{\mathbf{r}} \times M\vec{\mathbf{g}}$ is in the xy plane.

a

The direction of $\Delta\vec{\mathbf{L}}$ is parallel to that of $\vec{\boldsymbol{\tau}}$ in **a**.

b

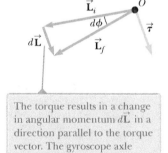

The torque results in a change in angular momentum $d\vec{\mathbf{L}}$ in a direction parallel to the torque vector. The gyroscope axle sweeps out an angle $d\phi$ in a time interval dt.

c

Figure 11.12 Precessional motion of a top spinning about its symmetry axis. (a) The only external forces acting on the top are the normal force $\vec{\mathbf{n}}$ and the gravitational force $M\vec{\mathbf{g}}$. The direction of the angular momentum $\vec{\mathbf{L}}$ is along the axis of symmetry. (b) Because $\vec{\mathbf{L}}_f = \Delta\vec{\mathbf{L}} + \vec{\mathbf{L}}_i$, the top precesses about the z axis. (c) Overhead view (looking down the z axis) of the gyroscope's initial and final angular momentum vectors for an infinitesimal time interval dt.

Summary

▶ Definitions

Given two vectors $\vec{A}$ and $\vec{B}$, the **vector product** $\vec{A} \times \vec{B}$ is a vector $\vec{C}$ having a magnitude

$$C = AB \sin \theta \qquad (11.3)$$

where θ is the angle between $\vec{A}$ and $\vec{B}$. The direction of the vector $\vec{C} = \vec{A} \times \vec{B}$ is perpendicular to the plane formed by $\vec{A}$ and $\vec{B}$, and this direction is determined by the right-hand rule.

The **torque** $\vec{\tau}$ on a particle due to a force $\vec{F}$ about an axis through the origin in an inertial frame is defined to be

$$\vec{\tau} \equiv \vec{r} \times \vec{F} \qquad (11.1)$$

The **angular momentum** $\vec{L}$ about an axis through the origin of a particle having linear momentum $\vec{p} = m\vec{v}$ is

$$\vec{L} \equiv \vec{r} \times \vec{p} \qquad (11.12)$$

where $\vec{r}$ is the vector position of the particle relative to the origin.

▶ Concepts and Principles

The z component of angular momentum of a rigid object rotating about a fixed z axis is

$$L_z = I\omega \qquad (11.16)$$

where I is the moment of inertia of the object about the axis of rotation and ω is its angular speed.

▶ Analysis Models for Problem Solving

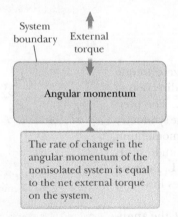

The rate of change in the angular momentum of the nonisolated system is equal to the net external torque on the system.

Nonisolated System (Angular Momentum). If a system interacts with its environment in the sense that there is an external torque on the system, the net external torque acting on a system is equal to the time rate of change of its angular momentum:

$$\sum \vec{\tau}_{\text{ext}} = \frac{d\vec{L}_{\text{tot}}}{dt} \qquad (11.15)$$

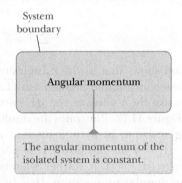

The angular momentum of the isolated system is constant.

Isolated System (Angular Momentum). If a system experiences no external torque from the environment, the total angular momentum of the system is conserved:

$$\Delta \vec{L}_{\text{tot}} = 0 \qquad (11.20)$$

Applying this law of conservation of angular momentum to a system whose moment of inertia changes gives

$$I_i\omega_i = I_f\omega_f = \text{constant} \qquad (11.21)$$

Think–Pair–Share

See the Preface for an explanation of the icons used in this problems set. For additional assessment items for this section, go to ⬧ **WEBASSIGN** From Cengage

1. **ACTIVITY** In your group, have someone donate temporarily a set of keys for the cause. Tie one end of a string about a meter long to the ring. To the other end attach a light weight, such as a paper clip or binder clip. If you hold the clip in your hand with the keys hanging on the string from your hand and drop it, everything of course falls to the ground. Now try something different. Suspend the keys from a horizontal pencil held in your left hand so that they hang down from the pencil on a short segment of the string as shown in Figure TP11.1. The rest of the string lies over the pencil and extends horizontally to the binder clip, which you hold with your right hand. Now release the binder clip. What happens? Explain this result.

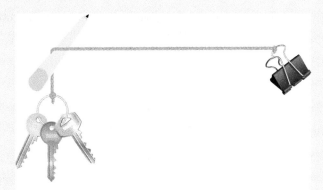

Figure TP11.1

2. A disk with moment of inertia I_1 rotates about a frictionless, vertical axle with angular speed ω_i. A second disk, this one having moment of inertia I_2 and initially not rotating, drops onto the first disk (Fig. TP11.2). Because of friction between the surfaces, the two eventually reach the same angular speed ω_f. Discuss in your group the following. (a) Calculate ω_f. (b) What fraction of the initial kinetic energy of the two-disk system remains after the disks rotate with the same angular speed? (c) Find the value of the answer in part (b) for the following limits: (i) $I_2 \rightarrow 0$, (ii) $I_1 = I_2$, (iii) $I_2 \rightarrow \infty$, and (iv) $I_1 \rightarrow \infty$. (d) Explain how each of the results in part (c) makes sense. (e) In the general case in which the kinetic energy of the system decreases in the process, where does that energy go? (f) **What If?** In Figure TP11.2, what is ω_f if the second disk is also rotating, but in the clockwise direction, opposite that of disk 1, with an angular speed of ω' before the collision?

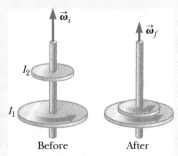

Before   After

Figure TP11.2

3. You are attending a county fair with your friend from your physics class. While walking around the fairgrounds, you discover a new game of skill. A thin rod of mass $M = 0.500$ kg and length $\ell = 2.00$ m hangs from a friction-free pivot at its upper end as shown in Figure TP11.3. The front surface of the rod is covered with Velcro. You are to throw a Velcro-covered ball of mass $m = 1.00$ kg at the rod in an attempt to make it swing backward and rotate all the way across the top. The ball must stick to the rod at all times after striking it. If you cause the rod to rotate over the top position, you win a stuffed animal. Your friend volunteers to try his luck. He feels that the most torque would be applied to the rod by striking it at its lowest end. After several tries, he fails to win the stuffed animal by throwing the ball so that it sticks at the end of the rod. He just couldn't throw the ball fast enough and accurately enough. (See Problem 43 to find out how fast he must throw the ball.) You analyze things differently from your friend. What if you were to throw the ball at a point *above* the end of the rod, a distance y below the pivot as shown in Figure TP11.3? This would reduce the torque on the rod, but torque is proportional to r, while moment of inertia is proportional to r^2. After the collision, the ball is part of the rotating system, so the moment of inertia of the system is reduced if the ball is stuck somewhere along the length of the rod, rather than at its end. (a) You pull out some sheets of paper and calculate an algebraic expression for the minimum required speed to spin the rod to the vertical position as a function of the point y along the rod at which the ball strikes and sticks to the rod. (b) Then, based on numerical values, you determine the point along the rod where you should strike it with the ball and make it go over the top by throwing the ball with the *lowest* speed. (c) Finally, you determine that lowest speed.

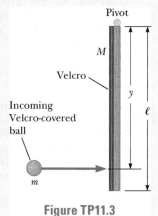

Figure TP11.3

Problems

See the Preface for an explanation of the icons used in this problems set. For additional assessment items for this section, go to ❖ **WEBASSIGN** From Cengage

SECTION 11.1 The Vector Product and Torque

1. Given $\vec{M} = 2\hat{i} - 3\hat{j} + \hat{k}$ and $\vec{N} = 4\hat{i} + 5\hat{j} - 2\hat{k}$, calculate the vector product $\vec{M} \times \vec{N}$.

2. The displacement vectors 42.0 cm at 15.0° and 23.0 cm at 65.0° both start from the origin and form two sides of a parallelogram. Both angles are measured counterclockwise from the x axis. (a) Find the area of the parallelogram. (b) Find the length of its longer diagonal.

3. If $|\vec{A} \times \vec{B}| = \vec{A} \cdot \vec{B}$, what is the angle between $\vec{A}$ and $\vec{B}$?

4. Use the definition of the vector product and the definitions of the unit vectors $\hat{i}$, $\hat{j}$, and $\hat{k}$ to prove Equations 11.7. You may assume the x axis points to the right, the y axis up, and the z axis horizontally toward you (not away from you). This choice is said to make the coordinate system a *right-handed system.*

5. Two forces $\vec{F}_1$ and $\vec{F}_2$ act along the two sides of an equilateral triangle as shown in Figure P11.5. Point O is the intersection of the altitudes of the triangle. (a) Find the magnitude of a third force $\vec{F}_3$ to be applied at B and along BC that will make the total torque zero about the point O. (b) **What If?** Will the total torque change if $\vec{F}_3$ is applied not at B but at any other point along BC?

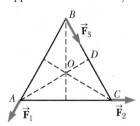

Figure P11.5

6. A student claims that he has found a vector $\vec{A}$ such that $(2\hat{i} - 3\hat{j} + 4\hat{k}) \times \vec{A} = (4\hat{i} + 3\hat{j} - \hat{k})$. (a) Do you believe this claim? (b) Explain why or why not.

7. A particle is located at a point described by the position vector $\vec{r} = (4.00\,\hat{i} + 6.00\,\hat{j})$ m, and a force exerted on it is given by $\vec{F} = (3.00\,\hat{i} + 2.00\,\hat{j})$ N. (a) What is the torque acting on the particle about the origin? (b) Can there be another point about which the torque caused by this force on this particle will be in the opposite direction and half as large in magnitude? (c) Can there be more than one such point? (d) Can such a point lie on the y axis? (e) Can more than one such point lie on the y axis? (f) Determine the position vector of one such point.

SECTION 11.2 Analysis Model: Nonisolated System (Angular Momentum)

8. A 1.50-kg particle moves in the xy plane with a velocity of $\vec{v} = (4.20\,\hat{i} - 3.60\,\hat{j})$ m/s. Determine the angular momentum of the particle about the origin when its position vector is $\vec{r} = (1.50\,\hat{i} + 2.20\,\hat{j})$ m.

9. A particle of mass m moves in the xy plane with a velocity of $\vec{v} = v_x\hat{i} + v_y\hat{j}$. Determine the angular momentum of the particle about the origin when its position vector is $\vec{r} = x\hat{i} + y\hat{j}$.

10. Heading straight toward the summit of Pike's Peak, an airplane of mass 12 000 kg flies over the plains of Kansas at nearly constant altitude 4.30 km with constant velocity 175 m/s west. (a) What is the airplane's vector angular momentum relative to a wheat farmer on the ground directly below the airplane? (b) Does this value change as the airplane continues its motion along a straight line? (c) **What If?** What is its angular momentum relative to the summit of Pike's Peak?

11. Review. A projectile of mass m is launched with an initial velocity $\vec{v}_i$ making an angle θ with the horizontal as shown in Figure P11.11. The projectile moves in the gravitational field of the Earth. Find the angular momentum of the projectile about the origin (a) when the projectile is at the origin, (b) when it is at the highest point of its trajectory, and (c) just before it hits the ground. (d) What torque causes its angular momentum to change?

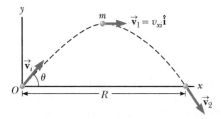

Figure P11.11

12. Review. A conical pendulum consists of a bob of mass m in motion in a circular path in a horizontal plane as shown in Figure P11.12. During the motion, the supporting wire of length ℓ maintains a constant angle θ with the vertical. Show that the magnitude of the angular momentum of the bob about the vertical dashed line is

$$L = \left(\frac{m^2 g \ell^3 \sin^4\theta}{\cos\theta}\right)^{1/2}$$

Figure P11.12

13. A particle of mass m moves in a circle of radius R at a constant speed v as shown in Figure P11.13. Time $t = 0$ is defined as

when the particle is at point Q. Determine the angular momentum of the particle about the axis perpendicular to the page through point P as a function of time.

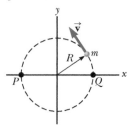

Figure P11.13
Problems 13 and 24.

14. A 5.00-kg particle starts from the origin at time zero. Its velocity as a function of time is given by

$$\vec{v} = 6t^2\hat{i} + 2t\hat{j}$$

where $\vec{v}$ is in meters per second and t is in seconds. (a) Find its position as a function of time. (b) Describe its motion qualitatively. Find (c) its acceleration as a function of time, (d) the net force exerted on the particle as a function of time, (e) the net torque about the origin exerted on the particle as a function of time, (f) the angular momentum of the particle as a function of time, (g) the kinetic energy of the particle as a function of time, and (h) the power injected into the system of the particle as a function of time.

15. A ball having mass m is fastened at the end of a flagpole that is connected to the side of a tall building at point P as shown in Figure P11.15. The length of the flagpole is ℓ, and it makes an angle θ with the x axis. The ball becomes loose and starts to fall with acceleration $-g\hat{j}$. (a) Determine the angular momentum of the ball about point P as a function of time. (b) For what physical reason does the angular momentum change? (c) What is the rate of change of the angular momentum of the ball about point P?

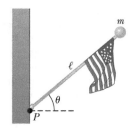

Figure P11.15

SECTION 11.3 Angular Momentum of a Rotating Rigid Object

16. A uniform solid sphere of radius $r = 0.500$ m and mass $m = 15.0$ kg turns counterclockwise about a vertical axis through its center. Find its vector angular momentum about this axis when its angular speed is 3.00 rad/s.

17. A uniform solid disk of mass $m = 3.00$ kg and radius $r = 0.200$ m rotates about a fixed axis perpendicular to its face with angular frequency 6.00 rad/s. Calculate the magnitude of the angular momentum of the disk when the axis of rotation (a) passes through its center of mass and (b) passes through a point midway between the center and the rim.

18. Show that the kinetic energy of an object rotating about a fixed axis with angular momentum $L = I\omega$ can be written as $K = L^2/2I$.

19. Big Ben (Fig. P10.27, page 281), the Parliament tower clock in London, has hour and minute hands with lengths of 2.70 m and 4.50 m and masses of 60.0 kg and 100 kg, respectively. Calculate the total angular momentum of these hands about the center point. (You may model the hands as long, thin rods rotating about one end. Assume the hour and minute hands are rotating at a constant rate of one revolution per 12 hours and 60 minutes, respectively.)

20. Model the Earth as a uniform sphere. (a) Calculate the angular momentum of the Earth due to its spinning motion about its axis. (b) Calculate the angular momentum of the Earth due to its orbital motion about the Sun. (c) Explain why the answer in part (b) is larger than that in part (a) even though it takes significantly longer for the Earth to go once around the Sun than to rotate once about its axis.

21. The distance between the centers of the wheels of a motorcycle is 155 cm. The center of mass of the motorcycle, including the rider, is 88.0 cm above the ground and halfway between the wheels. Assume the mass of each wheel is small compared with the body of the motorcycle. The engine drives the rear wheel only. What horizontal acceleration of the motorcycle will make the front wheel rise off the ground?

SECTION 11.4 Analysis Model: Isolated System (Angular Momentum)

22. You are working in an observatory, taking data on electromagnetic radiation from neutron stars. You happen to be analyzing results from the neutron star in Example 11.6, verifying that the period of the 10.0-km-radius neutron star is indeed 2.6 s. You go through weeks of data showing the same period. Suddenly, as you analyze the most recent data, you notice that the period has decreased to 2.3 s and remained at that level since that time. You ask your supervisor about this, who becomes excited and says that the neutron star must have undergone a *glitch,* which is a sudden shrinking of the radius of the star, resulting in a higher angular speed. As she runs to her computer to start writing a paper on the glitch, she calls back to you to calculate the new radius of the planet, assuming it has remained spherical. She is also talking about vortices and a superfluid core, but you don't understand those words.

23. A 60.0-kg woman stands at the western rim of a horizontal turntable having a moment of inertia of 500 kg · m² and a radius of 2.00 m. The turntable is initially at rest and is free to rotate about a frictionless, vertical axle through its center. The woman then starts walking around the rim clockwise (as viewed from above the system) at a constant speed of 1.50 m/s relative to the Earth. Consider the woman–turntable system as motion begins. (a) Is the mechanical energy of the system constant? (b) Is the momentum of the system constant? (c) Is the angular momentum of the system constant? (d) In what direction and with what angular speed does the turntable rotate? (e) How much potential energy in the woman's body is converted into mechanical energy of the woman–turntable system as the woman sets herself and the turntable into motion?

24. Figure P11.13 represents a small, flat puck with mass $m = 2.40$ kg sliding on a frictionless, horizontal surface. It is held in a circular orbit about a fixed axis by a rod with negligible mass and length $R = 1.50$ m, pivoted at one end. Initially, the puck has a speed of $v = 5.00$ m/s. A 1.30-kg ball of putty is dropped vertically onto the puck from a small distance above it and immediately sticks to the puck. (a) What is the new period of rotation? (b) Is the angular momentum of the puck–putty system about the axis of rotation constant in this process? (c) Is the momentum of the system constant in the process of the putty sticking to the

puck? (d) Is the mechanical energy of the system constant in the process?

25. A uniform cylindrical turntable of radius 1.90 m and mass 30.0 kg rotates counterclockwise in a horizontal plane with an initial angular speed of 4π rad/s. The fixed turntable bearing is frictionless. A lump of clay of mass 2.25 kg and negligible size is dropped onto the turntable from a small distance above it and immediately sticks to the turntable at a point 1.80 m to the east of the axis. (a) Find the final angular speed of the clay and turntable. (b) Is the mechanical energy of the turntable–clay system constant in this process? Explain and use numerical results to verify your answer. (c) Is the momentum of the system constant in this process? Explain your answer.

26. A puck of mass $m_1 = 80.0$ g and radius $r_1 = 4.00$ cm glides across an air table at a speed of $\vec{v} = 1.50$ m/s as shown in Figure P11.26a. It makes a glancing collision with a second puck of radius $r_2 = 6.00$ cm and mass $m_2 = 120$ g (initially at rest) such that their rims just touch. Because their rims are coated with instant-acting glue, the pucks stick together and rotate after the collision (Fig. P11.26b). (a) What is the angular momentum of the system relative to the center of mass? (b) What is the angular speed about the center of mass?

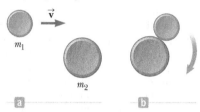

Figure P11.26

27. A wooden block of mass M resting on a frictionless, horizontal surface is attached to a rigid rod of length ℓ and of negligible mass (Fig. P11.27). The rod is pivoted at the other end. A bullet of mass m traveling parallel to the horizontal surface and perpendicular to the rod with speed v hits the block and becomes embedded in it. (a) What is the angular momentum of the bullet–block system about a vertical axis through the pivot? (b) What fraction of the original kinetic energy of the bullet is converted into internal energy in the system during the collision?

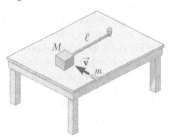

Figure P11.27

28. *Why is the following situation impossible?* A space station shaped like a giant wheel (Fig. P11.28, page 306) has a radius of $r = 100$ m and a moment of inertia of 5.00×10^8 kg · m². A crew of 150 people of average mass 65.0 kg is living on the rim, and the station's rotation causes the crew to experience an apparent free-fall acceleration of g. A research technician is assigned to perform an experiment in which a ball is dropped at the rim of the station every 15 minutes and the time interval for the ball to drop a given distance is measured as a test to make sure the apparent value of g is correctly maintained. One evening, 100 average people move to the center of the station for a union meeting. The research technician, who has already been performing his experiment for an hour before the meeting, is disappointed that he cannot attend the meeting, and his mood sours even further by his boring experiment in which every time interval for the dropped ball is identical for the entire evening.

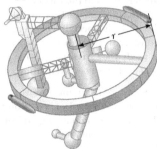

Figure P11.28

29. A wad of sticky clay with mass m and velocity $\vec{\mathbf{v}}_i$ is fired at a
 solid cylinder of mass M and radius R (Fig. P11.29). The cylinder is initially at rest and is mounted on a fixed horizontal axle that runs through its center of mass. The line of motion of the projectile is perpendicular to the axle and at a distance $d < R$ from the center. (a) Find the angular speed of the system just after the clay strikes and sticks to the surface of the cylinder. (b) Is the mechanical energy of the clay–cylinder system constant in this process? Explain your answer. (c) Is the momentum of the clay–cylinder system constant in this process? Explain your answer.

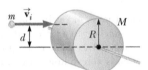

Figure P11.29

30. A 0.005 00-kg bullet traveling horizontally with a speed of
 1.00×10^3 m/s strikes an 18.0-kg door, embedding itself 10.0 cm from the side opposite the hinges as shown in Figure P11.30. The 1.00-m wide door is free to swing on its frictionless hinges. (a) Before it hits the door, does the bullet have angular momentum relative to the door's axis of rotation? (b) If so, evaluate this angular momentum. If not, explain why there is no angular momentum. (c) Is the mechanical energy of the bullet–door system constant during this collision? Answer without doing a calculation. (d) At what angular speed does the door swing open immediately after the collision? (e) Calculate the total energy of the bullet–door system and determine whether it is less than or equal to the kinetic energy of the bullet before the collision. (f) **What If?** Imagine now that the door is hanging vertically downward, hinged at the top, so that Figure P11.30 is a side view of the door and bullet during the collision. What is the maximum height that the bottom of the door will reach after the collision?

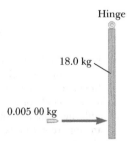

Figure P11.30 An overhead view of a bullet striking a door.

SECTION 11.5 The Motion of Gyroscopes and Tops

31. The angular momentum vector of a precessing gyroscope sweeps out a cone as shown in Figure P11.31. The angular speed of the tip of the angular momentum vector, called its precessional frequency, is given by $\omega_p = \tau/L$, where τ is the magnitude of the torque on the gyroscope and L is the magnitude of its angular momentum. In the motion called *precession of the equinoxes,* the Earth's axis of rotation precesses about the perpendicular to its orbital plane with a period of 2.58×10^4 yr. Model the Earth as a uniform sphere and calculate the torque on the Earth that is causing this precession.

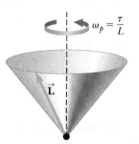

Figure P11.31 A precessing angular momentum vector sweeps out a cone in space.

ADDITIONAL PROBLEMS

32. A light rope passes over a light, fric-
 tionless pulley. One end is fastened to a bunch of bananas of mass M, and a monkey of mass M clings to the other end (Fig. P11.32). The monkey climbs the rope in an attempt to reach the bananas. (a) Treating the system as consisting of the monkey, bananas, rope, and pulley, find the net torque on the system about the pulley axis. (b) Using the result of part (a), determine the total angular momentum about the pulley axis and describe the motion of the system. (c) Will the monkey reach the bananas?

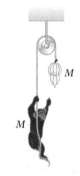

Figure P11.32

33. **Review.** A thin, uniform, rectangular signboard hangs vertically above the door of a shop. The sign is hinged to a stationary horizontal rod along its top edge. The mass of the sign is 2.40 kg, and its vertical dimension is 50.0 cm. The sign is swinging without friction, so it is a tempting target for children armed with snowballs. The maximum angular displacement of the sign is 25.0° on both sides of the vertical. At a moment when the sign is vertical and moving to the left, a snowball of mass 400 g, traveling horizontally with a velocity of 160 cm/s to the right, strikes perpendicularly at the lower edge of the sign and sticks there. (a) Calculate the angular speed of the sign immediately before the impact. (b) Calculate its angular speed immediately after the impact. (c) The spattered sign will swing up through what maximum angle?

34. You are advising a fellow student who wants to learn to
 perform multiple flips on the trampoline. You have him bounce vertically as high as he can, keeping his body perfectly straight and vertical. You determine that he can raise his center of mass by a distance of $h = 6.00$ m above its level when he initiates the jump. He can do a single flip by

bouncing gently, throwing his arms forward over his head, and tucking his body. You use your smartphone to make a video of him doing a single flip. Based on analysis of this video, you determine that his moment of inertia is $I_{\text{straight}} = 26.7 \text{ kg} \cdot \text{m}^2$ when his body is straight and $I_{\text{tuck}} = 5.62 \text{ kg} \cdot \text{m}^2$ in the tuck position. You suggest that he keep his body in the straight position for $\Delta t' = 0.400$ s after leaving the trampoline surface and then immediately go into a tuck position. As he lands, he should straighten his body out $\Delta t' = 0.400$ s before he lands. From analysis of the video recording, you determine that throwing his arms forward causes him to have an initial angular speed of $\omega_i = 2.88$ rad/s as he leaves the trampoline surface. If he tries to bounce as high as he can, do some flips, and land back on the same spot on the trampoline, predict how many flips he can safely do such that he lands on his feet on the trampoline.

35. We have all complained that there aren't enough hours in a day. In an attempt to fix that, suppose all the people in the world line up at the equator and all start running east at 2.50 m/s relative to the surface of the Earth. By how much does the length of a day increase? Assume the world population to be 7.00×10^9 people with an average mass of 55.0 kg each and the Earth to be a solid homogeneous sphere. In addition, depending on the details of your solution, you may need to use the approximation $1/(1 - x) \approx 1 + x$ for small x.

36. *Why is the following situation impossible?* A meteoroid strikes the Earth directly on the equator. At the time it lands, it is traveling exactly vertical and downward. Due to the impact, the time for the Earth to rotate once increases by 0.5 s, so the day is 0.5 s longer, undetectable to laypersons. After the impact, people on the Earth ignore the extra half-second each day and life goes on as normal. (Assume the density of the Earth is uniform.)

37. A rigid, massless rod has three particles with equal masses attached to it as shown in Figure P11.37. The rod is free to rotate in a vertical plane about a frictionless axle perpendicular to the rod through the point P and is released from rest in the horizontal position at $t = 0$. Assuming m and d are known, find (a) the moment of inertia of the system of three particles about the pivot, (b) the torque acting on the system at $t = 0$, (c) the angular acceleration of the system at $t = 0$, (d) the linear acceleration of the particle labeled 3 at $t = 0$, (e) the maximum kinetic energy of the system, (f) the maximum angular speed reached by the rod, (g) the maximum angular momentum of the system, and (h) the maximum speed reached by the particle labeled 2.

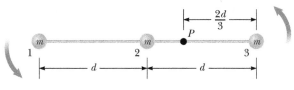

Figure P11.37

38. **Review.** Two boys are sliding toward each other on a frictionless, ice-covered parking lot. Jacob, mass 45.0 kg, is gliding to the right at 8.00 m/s, and Ethan, mass 31.0 kg, is gliding to the left at 11.0 m/s along the same line. When they meet, they grab each other and hang on. (a) What is their velocity immediately thereafter? (b) What fraction of their original kinetic energy is still mechanical energy after their collision? That was so much fun that the boys repeat the collision with the same original velocities, this time moving along

parallel lines 1.20 m apart. At closest approach, they lock arms and start rotating about their common center of mass. Model the boys as particles and their arms as a cord that does not stretch. (c) Find the velocity of their center of mass. (d) Find their angular speed. (e) What fraction of their original kinetic energy is still mechanical energy after they link arms? (f) Why are the answers to parts (b) and (e) so different?

39. Two astronauts (Fig. P11.39), each having a mass of 75.0 kg, are connected by a 10.0-m rope of negligible mass. They are isolated in space, orbiting their center of mass at speeds of 5.00 m/s. Treating the astronauts as particles, calculate (a) the magnitude of the angular momentum of the two-astronaut system and (b) the rotational energy of the system. By pulling on the rope, one astronaut shortens the distance between them to 5.00 m. (c) What is the new angular momentum of the system? (d) What are the astronauts' new speeds? (e) What is the new rotational energy of the system? (f) How much potential energy in the body of the astronaut was converted to mechanical energy in the system when he shortened the rope?

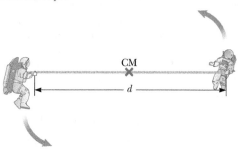

Figure P11.39 Problems 39 and 40.

40. Two astronauts (Fig. P11.39), each having a mass M, are connected by a rope of length d having negligible mass. They are isolated in space, orbiting their center of mass at speeds v. Treating the astronauts as particles, calculate (a) the magnitude of the angular momentum of the two-astronaut system and (b) the rotational energy of the system. By pulling on the rope, one of the astronauts shortens the distance between them to $d/2$. (c) What is the new angular momentum of the system? (d) What are the astronauts' new speeds? (e) What is the new rotational energy of the system? (f) How much potential energy in the body of the astronaut was converted to mechanical energy in the system when he shortened the rope?

41. Native people throughout North and South America used a *bola* to hunt for birds and animals. A bola can consist of three stones, each with mass m, at the ends of three light cords, each with length ℓ. The other ends of the cords are tied together to form a Y. The hunter holds one stone and swings the other two above his head (Figure P11.41a, page 308). Both these stones move together in a horizontal circle of radius 2ℓ with speed v_0. At a moment when the horizontal component of their velocity is directed toward the quarry, the hunter releases the stone in his hand. As the bola flies through the air, the cords quickly take a stable arrangement with constant 120-degree angles between them (Fig. P11.41b). In the vertical direction, the bola is in free fall. Gravitational forces exerted by the Earth make the junction of the cords move with the downward acceleration $\vec{g}$. You may ignore the vertical motion as you proceed to describe the horizontal motion of the bola. In terms of m, ℓ, and v_0, calculate (a) the magnitude of the momentum of

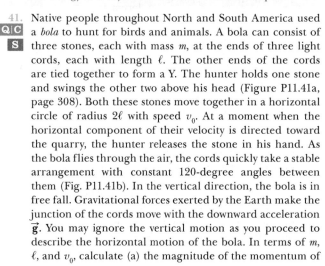

the bola at the moment of release and, after release, (b) the horizontal speed of the center of mass of the bola, and (c) the angular momentum of the bola about its center of mass. (d) Find the angular speed of the bola about its center of mass after it has settled into its Y shape. Calculate the kinetic energy of the bola (e) at the instant of release and (f) in its stable Y shape. (g) Explain how the conservation laws apply to the bola as its configuration changes. Robert Beichner suggested the idea for this problem.

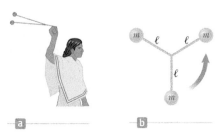

Figure P11.41

42. Two children are playing on stools at a restaurant counter.
Q|C Their feet do not reach the footrests, and the tops of the stools are free to rotate without friction on pedestals fixed to the floor. One of the children catches a tossed ball, in a process described by the equation

$$(0.730 \text{ kg} \cdot \text{m}^2)(2.40\hat{\mathbf{j}} \text{ rad/s})$$
$$+ (0.120 \text{ kg})(0.350\hat{\mathbf{i}} \text{ m}) \times (4.30\hat{\mathbf{k}} \text{ m/s})$$
$$= [0.730 \text{ kg} \cdot \text{m}^2 + (0.120 \text{ kg})(0.350 \text{ m})^2]\vec{\boldsymbol{\omega}}$$

(a) Solve the equation for the unknown $\vec{\boldsymbol{\omega}}$. (b) Complete the statement of the problem to which this equation applies. Your statement must include the given numerical information and specification of the unknown to be determined. (c) Could the equation equally well describe the other child throwing the ball? Explain your answer.

43. You are attending a county
CR fair with your friend from your physics class. While walking around the fairgrounds, you discover a new game of skill. A thin rod of mass $M = 0.500$ kg and length $\ell = 2.00$ m hangs from a friction-free pivot at its upper end as shown in Figure P11.43. The front surface of the rod is covered with Velcro. You are to throw a Velcro-covered ball of mass $m = 1.00$ kg at the rod in an attempt to make it swing backward and rotate all the way across the top. The ball must stick to the rod at all times after striking it. If you cause the rod to rotate over the top position, you win a stuffed animal. Your friend volunteers to try his luck. He feels that the most torque would be applied to the rod by striking it at its lowest end. While he prepares to aim at the lowest point on the rod, you calculate how fast he must throw the ball to win the stuffed animal with this technique.

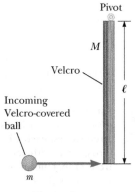

Figure P11.43

44. A uniform rod of mass 300 g and length 50.0 cm rotates in
Q|C a horizontal plane about a fixed, frictionless, vertical pin through its center. Two small, dense beads, each of mass m, are mounted on the rod so that they can slide without friction along its length. Initially, the beads are held by catches at positions 10.0 cm on each side of the center and the system is rotating at an angular speed of 36.0 rad/s. The catches are released simultaneously, and the beads slide outward along the rod. (a) Find an expression for the angular speed ω_f of the system at the instant the beads slide off the ends of the rod as it depends on m. (b) What are the maximum and the minimum possible values for ω_f and the values of m to which they correspond?

45. Global warming is a cause for concern because even small changes in the Earth's temperature can have significant consequences. For example, if the Earth's polar ice caps were to melt entirely, the resulting additional water in the oceans would flood many coastal areas. Model the polar ice as having mass 2.30×10^{19} kg and forming two flat disks of radius 6.00×10^5 m. Assume the water spreads into an unbroken thin, spherical shell after it melts. Calculate the resulting change in the duration of one day both in seconds and as a percentage.

46. The puck in Figure P11.46 has a mass of 0.120 kg. The distance of the puck from the center of rotation is originally 40.0 cm, and the puck is sliding with a speed of 80.0 cm/s. The string is pulled downward 15.0 cm through the hole in the frictionless table. Determine the work done on the puck. (*Suggestion:* Consider the change of kinetic energy.)

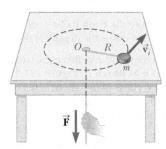

Figure P11.46

47. You operate a restaurant that has many large, circular
CR tables. At the center of each table is a Lazy Susan that can turn to deliver salt, pepper, jam, hot sauce, bread, and other items to diners on the other side of the table. A fancy flower arrangement is located at the center of each Lazy Susan, and the turning of the flower arrangement is beautiful to you. Because of your interest in model trains, you decide to replace each Lazy Susan with a circular track on the table around which a model train will run. You can load the various condiments in the cars of the train and press a button to operate the train, causing the train to begin moving around the circle and deliver the load to your fellow diners! The train is of mass 1.96 kg and moves at a speed of 0.18 m/s relative to the track. After a few days, you realize that you miss the beautiful turning flower arrangements. So you come up with a new scheme. You return the Lazy Susan to the table and mount the circular track on the platform of the Lazy Susan, which has a friction-free axle at its center. The radius of the circular track is 40.0 cm (measured halfway between the rails) and the platform of the Lazy Susan is a uniform

disk of mass 3.00 kg and radius 48.0 cm. You finally equip all of your tables with the new apparatus and open your restaurant. As a demonstration to the diners, you mount one salt shaker and one pepper shaker, having a mass of 0.100 kg each, onto a flatcar and push the button to deliver the condiments to the other side of the table! How long does it take to deliver the condiments to the exact opposite side of the table? Ignore the moment of inertia of the flower arrangement, since its mass is all close to the rotation axis.

CHALLENGE PROBLEMS

48. A solid cube of wood of side $2a$ and mass M is resting on a horizontal surface. The cube is constrained to rotate about a fixed axis AB (Fig. P11.48). A bullet of mass m and speed v is shot at the face opposite $ABCD$ at a height of $4a/3$. The bullet becomes embedded in the cube. Find the minimum value of v required to tip the cube so that it falls on face $ABCD$. Assume $m \ll M$.

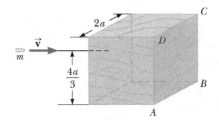

Figure P11.48

49. In Example 11.8, we investigated an elastic collision between a disk and a stick lying on a frictionless surface. Suppose everything is the same as in the example except that the collision is perfectly inelastic so that the disk adheres to the stick at the endpoint at which it strikes. Find (a) the speed of the center of mass of the system and (b) the angular speed of the system after the collision.

50. A solid cube of side $2a$ and mass M is sliding on a frictionless surface with uniform velocity $\vec{v}$ as shown in Figure P11.50a. It hits a small obstacle at the end of the table that causes the cube to tilt as shown in Figure P11.50b. Find the minimum value of the magnitude of $\vec{v}$ such that the cube tips over and falls off the table. *Note:* The cube undergoes an inelastic collision at the edge.

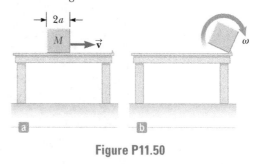

Figure P11.50

12

Support a meterstick near the ends on your fingers and move your hands toward each other. Your hands always meet at the 50-cm mark! (*Science Source*)

Static Equilibrium and Elasticity

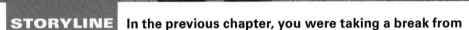

STORYLINE **In the previous chapter, you were taking a break from** your physics homework and browsing videos online about rotational motion. While thinking about the spinning phenomena in those videos, you pick up a meterstick and start twirling it and sliding it back and forth in your hands. At one point, something about the meterstick's behavior makes you forget about the rotational motion videos. You say, "Wait a minute! What just happened there?" You reproduce the behavior as follows. You support the meterstick horizontally on one finger of each hand, the fingers pointing out horizontally in a forward direction from your body. One finger supports the meterstick near the 0-cm end and the other near the 100-cm end. Now you slowly start moving your hands toward each other. The meterstick slides on one finger while sticking to the other finger. Then, it switches to slide on the other finger! And then back to the first finger! This alternation continues until your fingers meet. No matter what efforts you make to move only one finger at a time, this sticking and sliding behavior always occurs, the meterstick always stays supported on your fingers, and your fingers always meet at the 50-cm mark!

CONNECTIONS In Chapters 10 and 11, we studied the dynamics of rigid objects in motion. This chapter addresses the conditions under which a rigid object is in equilibrium. The term *equilibrium* implies that the object moves with both constant velocity and constant angular velocity relative to an observer in an inertial reference frame. In this chapter, we consider only the special case in which both of these velocities are equal to zero. In this case, the object is in what is called *static equilibrium*. In the meterstick phenomenon described in the storyline, anytime you momentarily stop your fingers so that the meterstick is at rest relative to the ground, it is in static equilibrium. Static equilibrium represents a common situation in engineering practice, and the principles it involves are of special interest to civil engineers, architects, and mechanical engineers. If you are an engineering student, you will undoubtedly take an advanced course in statics in the near future. The last section of this chapter deals with how objects deform under load conditions. An elastic object returns to its original shape when the deforming forces are removed. Several elastic constants are defined, each corresponding to a different type of deformation. In future chapters, we will see examples of rigid objects in static equilibrium: for example, polarized molecules in an electric field and a loop of wire carrying a current in a magnetic field.

12.1 Analysis Model: Rigid Object in Equilibrium

In Chapter 5, we discussed the particle in equilibrium model, in which a particle moves with constant velocity because the net force acting on it is zero. The situation with real (extended) objects is more complex because these objects often cannot be modeled as particles. For an extended object to be in equilibrium, a second condition must be satisfied. This second condition involves the rotational motion of the extended object.

Consider a single force $\vec{F}$ acting on a rigid object at point P as shown in Figure 12.1. Recall that the torque associated with the force $\vec{F}$ about an axis through O is given by Equation 11.1:

$$\vec{\tau} = \vec{r} \times \vec{F}$$

The magnitude of $\vec{\tau}$ is Fd (see Equation 10.14), where d is the moment arm shown in Figure 12.1. According to Equation 10.18, the net torque on a rigid object causes it to undergo an angular acceleration.

In this discussion, we investigate those rotational situations in which the angular acceleration of a rigid object is zero. Such an object is in **rotational equilibrium.** Because $\Sigma\,\tau_{\text{ext}} = I\alpha$ for rotation about a fixed axis, the necessary condition for rotational equilibrium is that the net torque about any axis must be zero. We now have two necessary conditions for equilibrium of a rigid object:

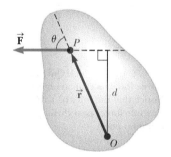

Figure 12.1 A single force $\vec{F}$ acts on a rigid object at the point P.

1. The net external force on the object must equal zero:

$$\sum \vec{F}_{\text{ext}} = 0 \qquad\qquad (12.1)$$

2. The net external torque on the object about *any* axis must be zero:

$$\sum \vec{\tau}_{\text{ext}} = 0 \qquad\qquad (12.2)$$

These conditions describe the **rigid object in equilibrium** analysis model. The first condition is a statement of translational equilibrium; it states that the translational acceleration of the object's center of mass must be zero when viewed from an inertial reference frame. The second condition is a statement of rotational equilibrium; it states that the angular acceleration about any axis must be zero. In the special case of **static equilibrium,** which is the main subject of this chapter, the object in equilibrium has a further requirement in addition to Equations 12.1 and 12.2: it is *at rest* relative to the observer and so has no translational or angular speed (that is, $v_{\text{CM}} = 0$ and $\omega = 0$).

PITFALL PREVENTION 12.1

Zero Torque Zero net torque does not mean an absence of rotational motion. An object that is rotating at a constant angular speed can be under the influence of a net torque of zero. This possibility is analogous to the translational situation: zero net force does not mean an absence of translational motion.

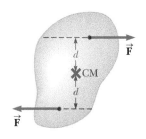

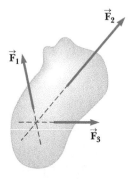

Figure 12.2 (Quick Quiz 12.1) Two forces of equal magnitude are applied at equal distances from the center of mass of a rigid object.

Figure 12.3 (Quick Quiz 12.2) Three forces act on an object. Notice that the lines of action of all three forces pass through a common point.

Ⓠ UICK QUIZ 12.1 Consider the object subject to the two forces of equal magnitude in Figure 12.2. Choose the correct statement with regard to this situation. **(a)** The object is in force equilibrium but not torque equilibrium. **(b)** The object is in torque equilibrium but not force equilibrium. **(c)** The object is in both force equilibrium and torque equilibrium. **(d)** The object is in neither force equilibrium nor torque equilibrium.

Ⓠ UICK QUIZ 12.2 Consider the object subject to the three forces in Figure 12.3. Choose the correct statement with regard to this situation from the choices **(a)**–**(d)** in Quick Quiz 12.1.

The two vector expressions given by Equations 12.1 and 12.2 are equivalent, in general, to six scalar equations: three from the first condition for equilibrium and three from the second (corresponding to x, y, and z components). Hence, in a complex system involving several forces acting in various directions, you could be faced with solving a set of equations with many unknowns. Here, we restrict our discussion to situations in which all the forces lie in the xy plane. (Forces whose vector representations are in the same plane are said to be *coplanar*.) With this restriction, we must deal with only three scalar equations. Two come from balancing the forces in the x and y directions. The third comes from the torque equation, namely that the net torque about a perpendicular axis through *any* point in the xy plane must be zero. This perpendicular axis will necessarily be parallel to the z axis, so the two conditions of the rigid object in equilibrium model provide the equations

$$\sum F_x = 0 \quad \sum F_y = 0 \quad \sum \tau_z = 0 \tag{12.3}$$

where the location of the axis of the torque equation is arbitrary.

ANALYSIS MODEL **Rigid Object in Equilibrium**

Imagine an object that can rotate, but is exhibiting no translational acceleration a and no rotational acceleration α. Such an object is in both translational *and* rotational equilibrium, so the net force *and* the net torque about any axis are both equal to zero:

$$\sum \vec{F}_{ext} = 0 \tag{12.1}$$

$$\sum \vec{\tau}_{ext} = 0 \tag{12.2}$$

Examples:

- a balcony juts out from a building and must support the weight of several humans without collapsing
- a gymnast performs the difficult *iron cross* maneuver in an Olympic event (Problem 37)
- a ship moves at constant speed through calm water and maintains a perfectly level orientation (Chapter 14)
- polarized molecules in a dielectric material in a constant electric field take on an average equilibrium orientation that remains fixed in time (Chapter 25)

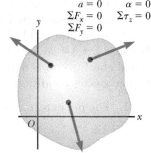

12.2 More on the Center of Gravity

Whenever we deal with a rigid object, one of the forces we must consider is the gravitational force acting on it, and we must know the point of application of this force. As we learned in Section 9.5, associated with every object is a special point called its center of gravity. The combination of the various gravitational forces acting on all the various mass elements of the object is equivalent to a single gravitational force acting through this point. Therefore, to compute the torque due to the gravitational force on an object of mass M, we need only consider the force $M\vec{g}$ acting at the object's center of gravity.

How do we find this special point? As mentioned in Section 9.5, if we assume $\vec{g}$ is uniform over the object, the center of gravity of the object coincides with its center of mass. To see why, consider an object of arbitrary shape lying in the xy plane as illustrated in Figure 12.4. Suppose the object is divided into a large number of particles of masses m_1, m_2, m_3, ... having coordinates (x_1, y_1), (x_2, y_2), (x_3, y_3), ... In Equation 9.29, we defined the x coordinate of the center of mass of such an object to be

$$x_{CM} = \frac{m_1 x_1 + m_2 x_2 + m_3 x_3 + \cdots}{m_1 + m_2 + m_3 + \cdots} = \frac{\sum_i m_i x_i}{\sum_i m_i} \qquad (9.29)$$

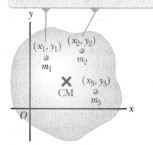

Figure 12.4 An object can be divided into many small particles. These particles can be used to locate the center of mass.

We use a similar equation to define the y coordinate of the center of mass, replacing each x with its y counterpart.

Let us now examine the situation from another point of view by considering the gravitational force exerted on each particle as shown in Figure 12.5. Each particle contributes a torque about an axis through the origin equal in magnitude to the particle's weight mg multiplied by its moment arm. For example, the magnitude of the torque due to the force $m_1 \vec{g}_1$ is $m_1 g_1 x_1$, where g_1 is the value of the gravitational acceleration at the position of the particle of mass m_1. We wish to locate the center of gravity, the point at which application of the single gravitational force $M\vec{g}_{CG}$ (where $M = m_1 + m_2 + m_3 + \cdots$ is the total mass of the object and $\vec{g}_{CG}$ is the acceleration due to gravity at the location of the center of gravity) has the same effect on rotation as does the combined effect of all the individual gravitational forces $m_i \vec{g}_i$. Equating the torque resulting from $M\vec{g}_{CG}$ acting at the center of gravity to the sum of the torques acting on the individual particles gives

$$(m_1 + m_2 + m_3 + \cdots)g_{CG}\, x_{CG} = m_1 g_1 x_1 + m_2 g_2 x_2 + m_3 g_3 x_3 + \cdots$$

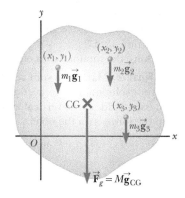

Figure 12.5 By dividing an object into many particles, we can find its center of gravity.

This expression accounts for the possibility that the value of g can in general vary over the object. If we assume uniform g over the object (as is usually the case), all the g factors are identical and cancel; we obtain

$$x_{CG} = \frac{m_1 x_1 + m_2 x_2 + m_3 x_3 + \cdots}{m_1 + m_2 + m_3 + \cdots} = \frac{\sum_i m_i x_i}{\sum_i m_i} \qquad (12.4)$$

Comparing this result with Equation 9.29 shows that the center of gravity is located at the center of mass as long as $\vec{g}$ is uniform over the entire object. Several examples in the next section deal with homogeneous, symmetric objects. The center of gravity for any such object coincides with its geometric center.

ⓆUICK QUIZ 12.3 A meterstick of uniform density is hung from a string tied at the 25-cm mark. A 0.50-kg object is hung from the zero end of the meterstick, and the meterstick is balanced horizontally. What is the mass of the meterstick? **(a)** 0.25 kg **(b)** 0.50 kg **(c)** 0.75 kg **(d)** 1.0 kg **(e)** 2.0 kg **(f)** impossible to determine

12.3 Examples of Rigid Objects in Static Equilibrium

The photograph of the one-bottle wine holder in Figure 12.6 shows one example of a balanced mechanical system that seems to defy gravity. For the system (wine holder plus bottle) to be in equilibrium, the net external force must be zero (see Eq. 12.1) and the net external torque around an axis passing through the support point must be zero (see Eq. 12.2). The second condition can be satisfied only when the center of gravity of the system in Figure 12.6 is directly over the support point.

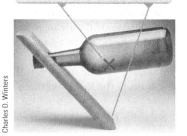

Charles D. Winters

Figure 12.6 This one-bottle wine holder is a surprising display of static equilibrium.

| PROBLEM-SOLVING STRATEGY | **Rigid Object in Equilibrium** |

When analyzing a rigid object in equilibrium under the action of several external forces, use the following procedure.

1. Conceptualize. Think about the object that is in equilibrium and identify all the forces on it. Imagine what effect each force would have on the rotation of the object if it were the only force acting.

2. Categorize. Confirm that the object under consideration is indeed a rigid object in equilibrium. The object must have zero translational acceleration and zero angular acceleration.

3. Analyze. Draw a diagram and label all external forces acting on the object. Try to guess the correct direction for any forces that are not specified. When using the particle under a net force model, the object on which forces act can be represented in a free-body diagram with a dot because it does not matter where on the object the forces are applied. When using the rigid object in equilibrium model, however, we cannot use a dot to represent the object because the location where forces act is important in the calculation. Therefore, in a diagram showing the forces on an object, we must show the actual object or a simplified version of it.

Resolve all forces into rectangular components, choosing a convenient coordinate system. Then apply the first condition for equilibrium, Equation 12.1. Remember to keep track of the signs of the various force components.

Choose a convenient axis for calculating the net torque on the rigid object. Remember that the choice of the axis for the torque equation is arbitrary; therefore, choose an axis that simplifies your calculation as much as possible. Usually, the most convenient axis for calculating torques is one through a point through which the lines of action of several forces pass, so their torques around this axis are zero. If you don't know a force or don't need to know a force, it is often beneficial to choose an axis through the point at which this force acts. Apply the second condition for equilibrium, Equation 12.2.

Solve the simultaneous equations for the unknowns in terms of the known quantities.

4. Finalize. Make sure your results are consistent with your diagram. If you selected a direction that leads to a negative sign in your solution for a force, do not be alarmed; it merely means that the direction of the force is the opposite of what you guessed. Add up the vertical and horizontal forces on the object and confirm that each set of components adds to zero. Add up the torques on the object and confirm that the sum equals zero.

| Example **12.1** | **The Seesaw Revisited** |

A seesaw consisting of a uniform board of mass M and length ℓ supports at rest a father and daughter with masses m_f and m_d, respectively, as shown in Figure 12.7. The support (called the *fulcrum*) is under the center of gravity of the board, the father is a distance d from the center, and the daughter is a distance $\ell/2$ from the center.

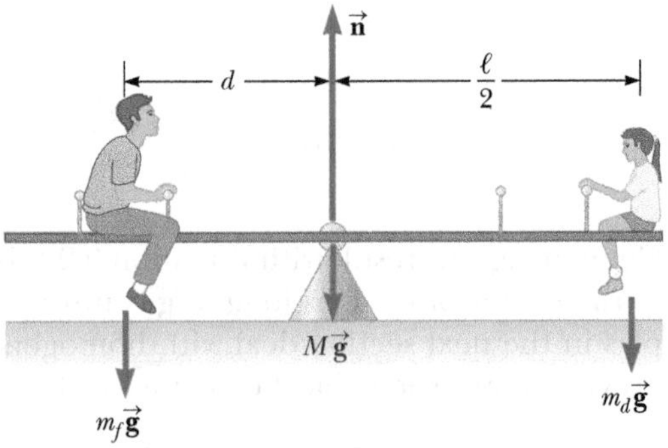

(A) Determine the magnitude of the upward force $\vec{\mathbf{n}}$ exerted by the support on the board.

Figure 12.7 (Example 12.1) A balanced system.

SOLUTION

Conceptualize Let us focus our attention on the board and consider the gravitational forces on the father and daughter as forces applied directly to the board. For force equilibrium, the location of the point of application of each force is not important.

Categorize Because the text of the problem states that the system is at rest, we model the board as a *rigid object in equilibrium*. Because we will only need the first condition of equilibrium to solve this part of the problem, however, we could also simply model the board as a *particle in equilibrium*.

Analyze Define upward as the positive y direction and substitute the forces on the board into Equation 12.1:

$$n - m_f g - m_d g - Mg = 0$$

Solve for the magnitude of the force $\vec{\mathbf{n}}$:

$$(1) \quad n = m_f g + m_d g + Mg = (m_f + m_d + M)g$$

(B) Determine where the father should sit to balance the system at rest.

12.1 continued

SOLUTION

Conceptualize For torque equilibrium, we need to pay attention to the location of the point of application of each force. The daughter would cause a clockwise rotation of the board around the support, whereas the father would cause a counterclockwise rotation.

Categorize This part of the problem requires the introduction of torque to find the position of the father, so we model the board as a *rigid object in equilibrium.*

Analyze The board's center of gravity is at its geometric center because we are told that the board is uniform. If we choose a rotation axis perpendicular to the page through the center of gravity of the board, the torques produced by $\vec{\mathbf{n}}$ and the gravitational force on the board about this axis are zero.

Substitute expressions for the torques on the board due to the father and daughter into Equation 12.2:

$$(m_f g)(d) - (m_d g)\frac{\ell}{2} = 0$$

Solve for d:

$$d = \left(\frac{m_d}{m_f}\right)\frac{\ell}{2}$$

Finalize This result is the same one we obtained in Example 11.5 by evaluating the angular acceleration of the system and setting the angular acceleration equal to zero.

WHAT IF? Suppose we had chosen another point through which the rotation axis were to pass. For example, suppose the axis is perpendicular to the page and passes through the location of the father. Does that change the results to parts (A) and (B)?

Answer Part (A) is unaffected because the calculation of the net force does not involve a rotation axis. In part (B), we would conceptually expect there to be no change if a different rotation axis is chosen because the second condition of equilibrium claims that the torque is zero about *any* rotation axis.

Let's verify this answer mathematically. Recall that the sign of the torque associated with a force is positive if that force tends to rotate the system counterclockwise, whereas the sign of the torque is negative if the force tends to rotate the system clockwise. Let's choose a rotation axis perpendicular to the page and passing through the location of the father.

Substitute expressions for the torques on the board around this axis into Equation 12.2:

$$n(d) - (Mg)(d) - (m_d g)\left(d + \frac{\ell}{2}\right) = 0$$

Substitute from Equation (1) in part (A) and solve for d:

$$(m_f + m_d + M)g(d) - (Mg)(d) - (m_d g)\left(d + \frac{\ell}{2}\right) = 0$$

$$(m_f g)(d) - (m_d g)\left(\frac{\ell}{2}\right) = 0 \rightarrow d = \left(\frac{m_d}{m_f}\right)\frac{\ell}{2}$$

This result is in agreement with the one obtained in part (B).

Example **12.2** **Standing on a Horizontal Beam**

A uniform horizontal beam with a length of $\ell = 8.00$ m and a weight of $W_b = 200$ N is attached to a wall by a pin connection. Its far end is supported by a cable that makes an angle of $\phi = 53.0°$ with the beam (Fig. 12.8a, page 316). A person of weight $W_p = 600$ N stands a distance $d = 2.00$ m from the wall. Find the tension in the cable as well as the magnitude and direction of the force exerted by the wall on the beam.

SOLUTION

Conceptualize Imagine the person in Figure 12.8a moving outward on the beam. It seems reasonable that the farther he moves outward, the larger the torque he applies about the pivot and the larger the tension in the cable must be to balance this torque.

Categorize Because the system is at rest, we categorize the beam as a *rigid object in equilibrium.*

continued

12.2 continued

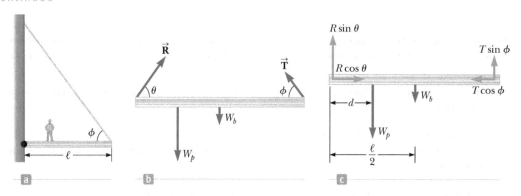

Figure 12.8 (Example 12.2) (a) A uniform beam supported by a cable. A person walks outward on the beam. (b) The force diagram for the beam. (c) The force diagram for the beam showing the components of $\vec{\mathbf{R}}$ and $\vec{\mathbf{T}}$.

Analyze We identify all the external forces acting on the beam: the 200-N gravitational force, the force $\vec{\mathbf{T}}$ exerted by the cable, the force $\vec{\mathbf{R}}$ exerted by the wall at the pivot, and the 600-N force that the person exerts on the beam. These forces are all indicated in the force diagram for the beam shown in Figure 12.8b. When we assign directions for forces, it is sometimes helpful to imagine what would happen if a force were suddenly removed. For example, if the wall were to vanish suddenly, the left end of the beam would move to the left as it begins to fall. This scenario tells us that the wall is not only holding the beam up but is also pressing outward against it. Therefore, we draw the vector $\vec{\mathbf{R}}$ in the direction shown in Figure 12.8b. Figure 12.8c shows the horizontal and vertical components of $\vec{\mathbf{T}}$ and $\vec{\mathbf{R}}$.

Applying the first condition of equilibrium, substitute expressions for the forces on the beam into component equations from Equation 12.1:

$$(1) \quad \sum F_x = R \cos \theta - T \cos \phi = 0$$

$$(2) \quad \sum F_y = R \sin \theta + T \sin \phi - W_p - W_b = 0$$

where we have chosen rightward and upward as our positive directions. Because R, T, and θ are all unknown, we cannot obtain a solution from these expressions alone. (To solve for the unknowns, the number of simultaneous equations must generally equal the number of unknowns.)

Now let's invoke the condition for rotational equilibrium. A convenient axis to choose for our torque equation is the one that passes through the pin connection. The feature that makes this axis so convenient is that the force $\vec{\mathbf{R}}$ and the horizontal component of $\vec{\mathbf{T}}$ both have a moment arm of zero; hence, these forces produce no torque about this axis.

Substitute expressions for the torques on the beam into Equation 12.2:

$$\sum \tau_z = (T \sin \phi)(\ell) - W_p d - W_b \left(\frac{\ell}{2}\right) = 0$$

This equation contains only T as an unknown because of our choice of rotation axis. Solve for T and substitute numerical values:

$$T = \frac{W_p d + W_b(\ell/2)}{\ell \sin \phi} = \frac{(600 \text{ N})(2.00 \text{ m}) + (200 \text{ N})(4.00 \text{ m})}{(8.00 \text{ m}) \sin 53.0°} = 313 \text{ N}$$

Rearrange Equations (1) and (2) and then divide:

$$\frac{R \sin \theta}{R \cos \theta} = \tan \theta = \frac{W_p + W_b - T \sin \phi}{T \cos \phi}$$

Solve for θ and substitute numerical values:

$$\theta = \tan^{-1}\left(\frac{W_p + W_b - T \sin \phi}{T \cos \phi}\right)$$

$$= \tan^{-1}\left[\frac{600 \text{ N} + 200 \text{ N} - (313 \text{ N}) \sin 53.0°}{(313 \text{ N}) \cos 53.0°}\right] = 71.1°$$

Solve Equation (1) for R and substitute numerical values:

$$R = \frac{T \cos \phi}{\cos \theta} = \frac{(313 \text{ N}) \cos 53.0°}{\cos 71.1°} = 581 \text{ N}$$

Finalize The positive value for the angle θ indicates that our estimate of the direction of $\vec{\mathbf{R}}$ was accurate.

Had we selected some other axis for the torque equation, the solution might differ in the details but the answers would be the same. For example, had we chosen an axis through the center of gravity of the beam, the torque equation would involve both T and R. This equation, coupled with Equations (1) and (2), however, could still be solved for the unknowns. Try it!

12.2 continued

▐ WHAT IF? ▌ What if the person walks farther out on the beam? Does T change? Does R change? Does θ change?

Answer T must increase because the gravitational force on the person exerts a larger torque about the pin connection, which must be countered by a larger torque in the opposite direction due to an increased value of T. If T increases, the vertical component of $\vec{R}$ decreases to maintain force equilibrium in the vertical direction. Force equilibrium in the horizontal direction, however, requires an increased horizontal component of $\vec{R}$ to balance the horizontal component of the increased $\vec{T}$. This fact suggests that θ becomes smaller, but it is hard to predict what happens to R. Problem 50 asks you to explore the behavior of R.

Example 12.3 The Leaning Ladder

A uniform ladder of length ℓ rests against a smooth, vertical wall (Fig. 12.9a). The mass of the ladder is m, and the coefficient of static friction between the ladder and the ground is $\mu_s = 0.40$. Find the minimum angle θ_{min} at which the ladder does not slip.

▐ SOLUTION ▌

Conceptualize Think about any ladders you have climbed. Do you want a large friction force between the bottom of the ladder and the surface or a small one? If the friction force is zero, will the ladder stay up? Simulate a ladder with a ruler leaning against a vertical surface. Does the ruler slip at some angles and stay up at others?

Categorize We do not wish the ladder to slip, so we model it as a *rigid object in equilibrium*.

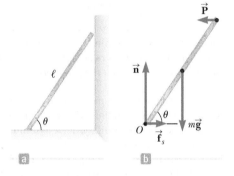

Figure 12.9 (Example 12.3) (a) A uniform ladder at rest, leaning against a smooth wall. The ground is rough. (b) The forces on the ladder.

Analyze A diagram showing all the external forces acting on the ladder is illustrated in Figure 12.9b. The force exerted by the ground on the ladder is the vector sum of a normal force $\vec{n}$ and the force of static friction $\vec{f}_s$. The wall exerts a normal force $\vec{P}$ on the top of the ladder, but there is no friction force here because the wall is smooth. So the net force on the top of the ladder is perpendicular to the wall and of magnitude P.

Apply the first condition for equilibrium to the ladder in both the x and the y directions:

(1) $\sum F_x = f_s - P = 0$

(2) $\sum F_y = n - mg = 0$

Solve Equation (1) for P:

(3) $P = f_s$

Solve Equation (2) for n:

(4) $n = mg$

When the ladder is on the verge of slipping, the force of static friction must have its maximum value, which is given by $f_{s,max} = \mu_s n$. Combine this equation with Equations (3) and (4):

(5) $P_{max} = f_{s,max} = \mu_s n = \mu_s mg$

Apply the second condition for equilibrium to the ladder, evaluating torques about an axis perpendicular to the page through O:

$$\sum \tau_O = P\ell \sin\theta - mg\frac{\ell}{2}\cos\theta = 0$$

Solve for $\tan\theta$:

$$\frac{\sin\theta}{\cos\theta} = \tan\theta = \frac{mg}{2P} \rightarrow \theta = \tan^{-1}\left(\frac{mg}{2P}\right)$$

Under the conditions that the ladder is just ready to slip, θ becomes θ_{min} and P_{max} is given by Equation (5). Substitute:

$$\theta_{min} = \tan^{-1}\left(\frac{mg}{2P_{max}}\right) = \tan^{-1}\left(\frac{1}{2\mu_s}\right) = \tan^{-1}\left[\frac{1}{2(0.40)}\right] = 51°$$

Finalize Notice that the angle depends only on the coefficient of friction, not on the mass or length of the ladder.

Example 12.4 **Negotiating a Curb**

(A) Estimate the magnitude of the force $\vec{\mathbf{F}}$ a person must apply to a wheelchair's main wheel to roll up over a sidewalk curb (Fig. 12.10a). This main wheel that comes in contact with the curb has a radius r, and the height of the curb is h.

SOLUTION

Conceptualize Think about wheelchair access to buildings. Generally, there are ramps built for individuals in wheelchairs or scooters. Steplike structures such as curbs are serious barriers to a wheelchair.

Categorize Imagine the person exerts enough force so that the bottom of the main wheel just loses contact with the lower surface and hovers at rest. We model the wheel in this situation as a *rigid object in equilibrium*.

Analyze Usually, the person's hands supply the required force to a slightly smaller wheel that is concentric with the main wheel. For simplicity, let's assume the radius of this second wheel is the same as the radius of the main wheel. Let's estimate a combined gravitational force of magnitude $mg = 1\,400$ N for the person and the wheelchair, acting along a line of action passing through the axle of the main wheel, and choose a wheel radius of $r = 30$ cm. We also pick a curb height of $h = 10$ cm. Let's also assume the wheelchair and occupant are symmetric and each wheel supports a weight of 700 N. We then proceed to analyze only one of the main wheels. Figure 12.10b shows the geometry for a single wheel.

When the wheel is just about to be raised from the street, the normal force exerted by the ground on the wheel at point B goes to zero. Hence, at this time only three forces act on the wheel as shown in the force diagram in Figure 12.10c. The force $\vec{\mathbf{R}}$, which is the force exerted by the curb on the wheel, acts at point A, so if we choose to have our axis of rotation be perpendicular to the page and pass through point A, we do not need to include $\vec{\mathbf{R}}$ in our torque equation. The moment arm of $\vec{\mathbf{F}}$ relative to an axis through A is given by $2r - h$ (see Fig. 12.10c).

Use the triangle OAC in Figure 12.10b to find the moment arm d of the gravitational force $m\vec{\mathbf{g}}$ acting on the wheel relative to an axis through point A:

Apply the second condition for equilibrium to the wheel, taking torques about an axis through A:

Substitute for d from Equation (1):

Solve for F:

Simplify:

Substitute numerical values:

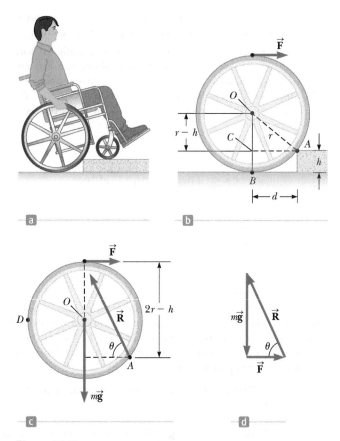

Figure 12.10 (Example 12.4) (a) A person in a wheelchair attempts to roll up over a curb. (b) Details of the wheel and curb. The person applies a force $\vec{\mathbf{F}}$ to the top of the wheel. (c) A force diagram for the wheel when it is just about to be raised. Three forces act on the wheel at this instant: $\vec{\mathbf{F}}$, which is exerted by the hand; $\vec{\mathbf{R}}$, which is exerted by the curb; and the gravitational force $m\vec{\mathbf{g}}$. (d) The vector sum of the three external forces acting on the wheel is zero.

(1) $d = \sqrt{r^2 - (r-h)^2} = \sqrt{2rh - h^2}$

(2) $\sum \tau_A = mgd - F(2r - h) = 0$

$mg\sqrt{2rh - h^2} - F(2r - h) = 0$

(3) $F = \dfrac{mg\sqrt{2rh - h^2}}{2r - h}$

$F = mg\dfrac{\sqrt{h}\sqrt{2r - h}}{2r - h} = mg\sqrt{\dfrac{h}{2r - h}}$

$F = (700\text{ N})\sqrt{\dfrac{0.1\text{ m}}{2(0.3\text{ m}) - 0.1\text{ m}}}$

$= 3 \times 10^2$ N

12.4 continued

(B) Determine the magnitude and direction of $\vec{\mathbf{R}}$.

SOLUTION

Apply the first condition for equilibrium to the x and y components of the forces on the wheel:

(4) $\sum F_x = F - R\cos\theta = 0$

(5) $\sum F_y = R\sin\theta - mg = 0$

Divide Equation (5) by Equation (4):

$$\frac{R\sin\theta}{R\cos\theta} = \tan\theta = \frac{mg}{F}$$

Solve for the angle θ:

$$\theta = \tan^{-1}\left(\frac{mg}{F}\right) = \tan^{-1}\left(\frac{700\text{ N}}{300\text{ N}}\right) = 70°$$

Solve Equation (5) for R and substitute numerical values:

$$R = \frac{mg}{\sin\theta} = \frac{700\text{ N}}{\sin 70°} = 8 \times 10^2\text{ N}$$

Finalize Notice that we have kept only one digit as significant because we have guessed at some numbers and made some assumptions. (We have written the angle as 70° because $7 \times 10^{1°}$ is awkward!) For example, we assumed that the center of mass of the wheelchair–person system was directly over the axle of the wheel. How likely do you think that is to be true? The results indicate that the force that must be applied to each wheel is substantial. You may want to estimate the force required to roll a wheelchair up a typical sidewalk accessibility ramp for comparison.

WHAT IF? Would it be easier to negotiate the curb if the person grabbed the wheel at point D in Figure 12.10c and pulled *upward*?

Answer If the force $\vec{\mathbf{F}}$ in Figure 12.10c is rotated counterclockwise by 90° and applied at D, its moment arm about an axis through A is $d + r$. Let's call the magnitude of this new force F'.

Modify Equation (2) for this situation:

$$\sum \tau_A = mgd - F'(d + r) = 0$$

Solve this equation for F' and substitute for d:

$$F' = \frac{mgd}{d + r} = \frac{mg\sqrt{2rh - h^2}}{\sqrt{2rh - h^2} + r}$$

Take the ratio of this force to the original force from Equation (3) and express the result in terms of h/r, the ratio of the curb height to the wheel radius:

$$\frac{F'}{F} = \frac{\dfrac{mg\sqrt{2rh - h^2}}{\sqrt{2rh - h^2} + r}}{\dfrac{mg\sqrt{2rh - h^2}}{2r - h}} = \frac{2r - h}{\sqrt{2rh - h^2} + r} = \frac{2 - \left(\dfrac{h}{r}\right)}{\sqrt{2\left(\dfrac{h}{r}\right) - \left(\dfrac{h}{r}\right)^2} + 1}$$

Substitute the ratio $h/r = 0.33$ from the given values:

$$\frac{F'}{F} = \frac{2 - 0.33}{\sqrt{2(0.33) - (0.33)^2} + 1} = 0.96$$

This result tells us that, *for these values,* it is slightly easier to pull upward at D than horizontally at the top of the wheel. For very high curbs, so that h/r is close to 1, the ratio F'/F drops to about 0.5 because point A is located near the right edge of the wheel in Figure 12.10b. The force at D is applied at a distance of about $2r$ from A, whereas the force at the top of the wheel has a moment arm of only about r. For high curbs, then, it is best to pull upward at D, although a large value of the force is required. For small curbs, it is best to apply the force at the top of the wheel. The ratio F'/F becomes larger than 1 at about $h/r = 0.3$ because point A is now close to the bottom of the wheel and the force applied at the top of the wheel has a larger moment arm than when applied at D.

Finally, let's comment on the validity of these mathematical results. Consider Figure 12.10d and imagine that the vector $\vec{\mathbf{F}}$ is upward instead of to the right. There is no way the three vectors can add to equal zero as required by the first equilibrium condition. Therefore, our results above may be qualitatively valid, but not exact quantitatively. To cancel the horizontal component of $\vec{\mathbf{R}}$, the force at D must be applied at an angle to the vertical rather than straight upward. This feature makes the calculation more complicated and requires both conditions of equilibrium.

12.4 Elastic Properties of Solids

In Section 9.8, we explored deformable systems consisting of masses and springs. We continue and generalize that discussion in this section. We have assumed objects remain rigid when external forces act on them. In reality, all objects are deformable to some extent. That is, it is possible to change the shape or the size (or both) of an object by applying external forces. As these changes take place, however, internal forces in the object resist the deformation.

We shall discuss the deformation of solids in terms of the concepts of *stress* and *strain*. **Stress** is a quantity that is proportional to the force causing a deformation; more specifically, stress is the external force acting on an object per unit cross-sectional area. The result of a stress is **strain,** which is a measure of the degree of deformation. It is found that, for sufficiently small stresses, stress is proportional to strain; the constant of proportionality depends on the material being deformed and on the nature of the deformation. We call this proportionality constant the **elastic modulus.** The elastic modulus is therefore defined as the ratio of the stress to the resulting strain:

$$\text{Elastic modulus} \equiv \frac{\text{stress}}{\text{strain}} \tag{12.5}$$

The elastic modulus in general relates what is done to a solid object (a force is applied) to how that object responds (it deforms to some extent). It is similar in nature to the spring constant k in Hooke's law (Eq. 7.9) that relates a force applied to a spring and the resultant deformation of the spring, measured by its extension or compression.

We consider three types of deformation and define an elastic modulus for each:

1. **Young's modulus** measures the resistance of a solid to a change in its length.
2. **Shear modulus** measures the resistance to motion of the planes within a solid parallel to each other.
3. **Bulk modulus** measures the resistance of solids or liquids to changes in their volume.

Young's Modulus: Elasticity in Length

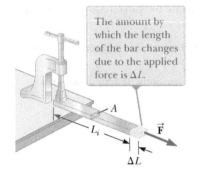

The amount by which the length of the bar changes due to the applied force is ΔL.

Figure 12.11 A force $\vec{\mathbf{F}}$ is applied to the free end of a bar clamped at the other end.

Consider a long bar of cross-sectional area A and initial length L_i that is clamped at one end as in Figure 12.11. When an external force is applied perpendicular to the cross section, internal molecular forces in the bar resist distortion ("stretching"), but the bar reaches an equilibrium situation in which its final length L_f is greater than L_i and in which the external force is exactly balanced by the internal forces. In such a situation, the bar is said to be stressed. We define the **tensile stress** as the ratio of the magnitude of the external force F to the cross-sectional area A, where the cross section is perpendicular to the force vector. The **tensile strain** in this case is defined as the ratio of the change in length ΔL to the original length L_i. We define **Young's modulus** by a combination of these two ratios:

Young's modulus ▶
$$Y \equiv \frac{\text{tensile stress}}{\text{tensile strain}} = \frac{F/A}{\Delta L/L_i} \tag{12.6}$$

Young's modulus is typically used to characterize a rod or wire stressed under either tension or compression. Because strain is a dimensionless quantity, Y has units of force per unit area. Typical values are given in Table 12.1.

TABLE 12.1 Typical Values for Elastic Moduli

Substance	Young's Modulus (N/m²)	Shear Modulus (N/m²)	Bulk Modulus (N/m²)
Tungsten	35×10^{10}	14×10^{10}	20×10^{10}
Steel	20×10^{10}	8.4×10^{10}	6×10^{10}
Copper	11×10^{10}	4.2×10^{10}	14×10^{10}
Brass	9.1×10^{10}	3.5×10^{10}	6.1×10^{10}
Aluminum	7.0×10^{10}	2.5×10^{10}	7.0×10^{10}
Glass	$6.5–7.8 \times 10^{10}$	$2.6–3.2 \times 10^{10}$	$5.0–5.5 \times 10^{10}$
Quartz	5.6×10^{10}	2.6×10^{10}	2.7×10^{10}
Water	—	—	0.21×10^{10}
Mercury	—	—	2.8×10^{10}

For relatively small stresses, the bar returns to its initial length when the force is removed. The **elastic limit** of a substance is defined as the maximum stress that can be applied to the substance before it becomes permanently deformed and does not return to its initial length. It is possible to exceed the elastic limit of a substance by applying a sufficiently large stress as seen in Figure 12.12. Initially, a stress-versus-strain curve is a straight line. As the stress increases, however, the curve is no longer a straight line. When the stress exceeds the elastic limit, the object is permanently distorted and does not return to its original shape after the stress is removed. As the stress is increased even further, the material ultimately breaks.

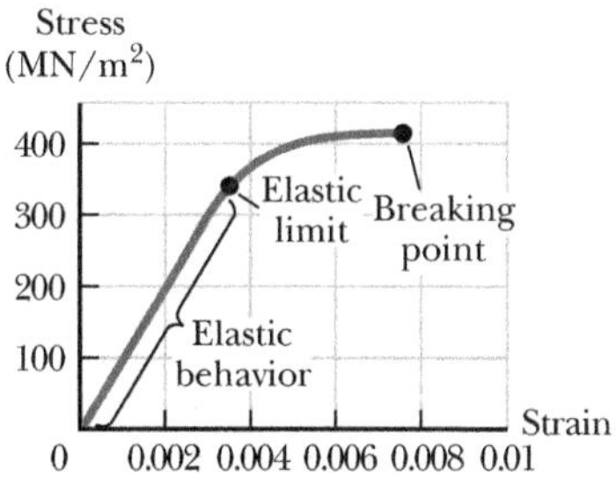

Stress (MN/m²)

Figure 12.12 Stress-versus-strain curve for an elastic solid.

Shear Modulus: Elasticity of Shape

Another type of deformation occurs when an object is subjected to a force parallel to one of its faces while the opposite face is held fixed by another force (Fig. 12.13a). The stress in this case is called a *shear stress*. If the object is originally a rectangular block, a shear stress results in a shape whose cross section is a parallelogram. A book pushed sideways as shown in Figure 12.13b is an example of an object subjected to a shear stress. To a first approximation (for small distortions), no change in volume occurs with this deformation.

We define the **shear stress** as F/A, the ratio of the tangential force to the area A of the face being sheared. The **shear strain** is defined as the ratio $\Delta x/h$, where Δx is the horizontal distance that the sheared face moves and h is the height of the object. In terms of these quantities, the **shear modulus** is

$$S \equiv \frac{\text{shear stress}}{\text{shear strain}} = \frac{F/A}{\Delta x/h} \qquad (12.7)$$

◀ Shear modulus

Values of the shear modulus for some representative materials are given in Table 12.1. Like Young's modulus, the unit of shear modulus is the ratio of that for force to that for area.

Bulk Modulus: Volume Elasticity

Bulk modulus characterizes the response of an object to changes in a force of uniform magnitude applied perpendicularly over the entire surface of the object as shown in Figure 12.14. (We assume here the object is made of a single substance.) As we shall see in Chapter 14, such a uniform distribution of forces occurs when an object is immersed in a fluid. An object subject to this type of stress undergoes a change in volume but no change in shape. The **volume stress** is defined as the ratio of the magnitude of the total force F exerted on a surface to the area A of the surface. The quantity $P = F/A$ is called **pressure,** which we shall study in more detail in Chapter 14. If the pressure on an object changes by an amount $\Delta P = \Delta F/A$, the object experiences a volume change ΔV. The **volume strain** is equal to the change in volume ΔV divided by the initial volume V_i. Therefore, from Equation 12.5,

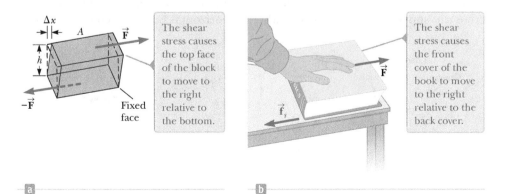

The shear stress causes the top face of the block to move to the right relative to the bottom.

The shear stress causes the front cover of the book to move to the right relative to the back cover.

Fixed face

Figure 12.13 (a) A shear deformation in which a rectangular block is distorted by two forces of equal magnitude but opposite directions applied to two parallel faces. (b) A book is under shear stress when a hand placed on the cover applies a horizontal force away from the spine.

we can characterize a volume ("bulk") compression in terms of the **bulk modulus,** which is defined as

Bulk modulus ▶

$$B \equiv \frac{\text{volume stress}}{\text{volume strain}} = -\frac{\Delta F/A}{\Delta V/V_i} = -\frac{\Delta P}{\Delta V/V_i} \qquad (12.8)$$

A negative sign is inserted in this defining equation so that B is a positive number. This maneuver is necessary because an increase in pressure (positive ΔP) causes a decrease in volume (negative ΔV) and vice versa.

Table 12.1 lists bulk moduli for some materials. If you look up such values in a different source, you may find the reciprocal of the bulk modulus listed. The reciprocal of the bulk modulus is called the **compressibility** of the material.

Notice from Table 12.1 that both solids and liquids have a bulk modulus. No shear modulus and no Young's modulus are given for liquids, however, because a liquid does not sustain a shearing stress or a tensile stress. If a shearing force or a tensile force is applied to a liquid, the liquid simply flows in response.

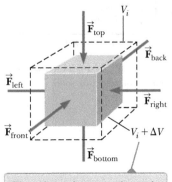

Figure 12.14 A cube is under uniform pressure and is therefore compressed on all sides by forces normal to its six faces. The arrowheads of force vectors on the sides of the cube that are not visible are hidden by the cube.

The cube undergoes a change in volume but no change in shape.

QUICK QUIZ 12.4 For the three parts of this Quick Quiz, choose from the following choices the correct answer for the elastic modulus that describes the relationship between stress and strain for the system of interest, which is in italics: (a) Young's modulus (b) shear modulus (c) bulk modulus (d) none of those choices **(i)** A *block of iron* is sliding across a horizontal floor. The friction force between the sliding block and the floor causes the block to deform. **(ii)** A trapeze artist swings through a circular arc. At the bottom of the swing, the *wires* supporting the trapeze are longer than when the trapeze artist simply hangs from the trapeze due to the increased tension in them. **(iii)** A spacecraft carries a *steel sphere* to a planet on which atmospheric pressure is much higher than on the Earth. The higher pressure causes the radius of the sphere to decrease.

Prestressed Concrete

If the stress on a solid object exceeds a certain value, the object fractures. The maximum stress that can be applied before fracture occurs—called the *tensile strength, compressive strength,* or *shear strength*—depends on the nature of the material and on the type of applied stress. For example, concrete has a tensile strength of about $2 \times 10^6 \text{ N/m}^2$, a compressive strength of $20 \times 10^6 \text{ N/m}^2$, and a shear strength of $2 \times 10^6 \text{ N/m}^2$. If the applied stress exceeds these values, the concrete fractures. It is common practice to use large safety factors to prevent failure in concrete structures.

Concrete is normally very brittle when it is cast in thin sections. Therefore, concrete slabs tend to sag and crack at unsupported areas as shown in Figure 12.15a. The slab can be strengthened by the use of steel rods to reinforce the concrete as illustrated in Figure 12.15b. Because concrete is much stronger under compression (squeezing) than under tension (stretching) or shear, vertical columns of concrete can support very heavy loads, whereas horizontal beams of concrete tend to

Figure 12.15 (a) A concrete slab with no reinforcement tends to crack under a heavy load. (b) The strength of the concrete is increased by using steel reinforcement rods. (c) The concrete is further strengthened by prestressing it with steel rods under tension.

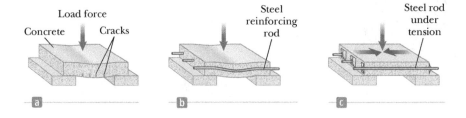

sag and crack. A significant increase in shear strength is achieved, however, if the reinforced concrete is prestressed as shown in Figure 12.15c. As the concrete is being poured, the steel rods are held under tension by external forces. The external forces are released after the concrete cures; the result is a permanent tension in the steel and hence a compressive stress on the concrete. The concrete slab can now support a much heavier load.

Example 12.5 | Stage Design

In Example 8.3, we analyzed a cable used to support an actor as he swings onto the stage. Now suppose the tension in the cable is 940 N as the actor reaches the lowest point. What diameter should a 10-m-long steel cable have if we do not want it to stretch more than 0.50 cm under these conditions?

SOLUTION

Conceptualize Look back at Example 8.3 to recall what is happening in this situation. We ignored any stretching of the cable there, but we wish to address this phenomenon in this example.

Categorize We perform a simple calculation involving Equation 12.6, so we categorize this example as a substitution problem.

Solve Equation 12.6 for the cross-sectional area of the cable:

$$A = \frac{FL_i}{Y\Delta L}$$

Assuming the cross section is circular, find the diameter of the cable from $d = 2r$ and $A = \pi r^2$:

$$d = 2r = 2\sqrt{\frac{A}{\pi}} = 2\sqrt{\frac{FL_i}{\pi Y \Delta L}}$$

Substitute numerical values:

$$d = 2\sqrt{\frac{(940 \text{ N})(10 \text{ m})}{\pi(20 \times 10^{10} \text{ N/m}^2)(0.005\,0 \text{ m})}} = 3.5 \times 10^{-3} \text{ m} = 3.5 \text{ mm}$$

To provide a large margin of safety, you would probably use a flexible cable made up of many smaller wires having a total cross-sectional area substantially greater than our calculated value.

Example 12.6 | Squeezing a Brass Sphere

A solid brass sphere is initially surrounded by air, and the air pressure exerted on it is 1.0×10^5 N/m^2 (normal atmospheric pressure). The sphere is lowered into the ocean to a depth where the pressure is 2.0×10^7 N/m^2. The volume of the sphere in air is 0.50 m^3. By how much does this volume change once the sphere is submerged?

SOLUTION

Conceptualize Think about movies or television shows you have seen in which divers go to great depths in the water in submersible vessels. These vessels must be very strong to withstand the large pressure under water. This pressure squeezes the vessel and reduces its volume.

Categorize We perform a simple calculation involving Equation 12.8, so we categorize this example as a substitution problem.

Solve Equation 12.8 for the volume change of the sphere:

$$\Delta V = -\frac{V_i \Delta P}{B}$$

Substitute numerical values:

$$\Delta V = -\frac{(0.50 \text{ m}^3)(2.0 \times 10^7 \text{ N/m}^2 - 1.0 \times 10^5 \text{ N/m}^2)}{6.1 \times 10^{10} \text{ N/m}^2}$$

$$= -1.6 \times 10^{-4} \text{ m}^3$$

The negative sign indicates that the volume of the sphere decreases.

Summary

▶ Definitions

The gravitational force exerted on an object can be considered as acting at a single point called the **center of gravity.** An object's center of gravity coincides with its center of mass if the object is in a uniform gravitational field.

We can describe the elastic properties of a substance using the concepts of stress and strain. **Stress** is a quantity proportional to the force producing a deformation; **strain** is a measure of the degree of deformation. Stress is proportional to strain, and the constant of proportionality is the **elastic modulus:**

$$\text{Elastic modulus} \equiv \frac{\text{stress}}{\text{strain}} \qquad (12.5)$$

▶ Concepts and Principles

Three common types of deformation are represented by (1) the resistance of a solid to elongation under a load, characterized by **Young's modulus** Y; (2) the resistance of a solid to the motion of internal planes sliding past each other, characterized by the **shear modulus** S; and (3) the resistance of a solid or fluid to a volume change, characterized by the **bulk modulus** B.

▶ Analysis Model for Problem Solving

Rigid Object in Equilibrium A rigid object in equilibrium exhibits no translational or angular acceleration. The net external force acting on it is zero, and the net external torque on it is zero about any axis:

$$\sum \vec{F}_{\text{ext}} = 0 \qquad (12.1)$$

$$\sum \vec{\tau}_{\text{ext}} = 0 \qquad (12.2)$$

The first condition is the condition for translational equilibrium, and the second is the condition for rotational equilibrium.

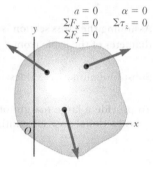

Think–Pair–Share

See the Preface for an explanation of the icons used in this problems set. For additional assessment items for this section, go to ⚙ **WEBASSIGN** From Cengage

1. A father and his son are painting a wall. To reach the higher portions of the wall, they place a 20.0-kg plank of wood, 3.50 m long and of uniform consistency, on two sawhorses. A sawhorse is placed 1.00 m from each end of the plank. The son has a mass 50.0 kg. He and the father, of mass 85.0 kg, climb up and stand on the plank. Discuss in your group and respond to the following. (a) The father stands at rest on the plank, directly over one of the sawhorses. Can the son move to *any* position he wishes on the plank without the plank tipping? (b) After learning about the tipping possibilities in part (a), the father and son decide to use a more massive plank. What must the mass of the plank be so that both father and son are free to move anywhere on the plank that they wish? (c) They find a plank of just the mass found in part (b) and test it by standing on the right-hand end together. Will they be safe from tipping if they both stand on the *left-hand* end together? (d) After using the plank of the required mass determined in part (b) and being exhausted from moving it to new positions along the

wall, the father and son decide to set it aside and continue to use the 20.0-kg plank, while being careful to stay far apart on the plank. But the use of the massive plank in part (b) has damaged one of the sawhorses so that it can only support a force of 1.75×10^3 N. When the father and son get back on the 20.0-kg plank and move around, will the damaged sawhorse collapse?

2. **ACTIVITY** If an object is set on a table such that part of it extends off the edge of the table, the center of mass of the object must be over part of the table surface to avoid the object falling. If the center of mass is beyond the edge of the table, the gravitational force will exert a torque on the object and tip it off the table. Gather four metersticks together. (a) Determine a way that you can stack all four metersticks so that (1) all four metersticks are parallel; (2) each higher meterstick is further out over the edge of the table than the one below it, and (3) the topmost meterstick has *no* part of its length above the table surface. *Hint:* Begin from the top; put the topmost meterstick on the second one so that the top one does not tip off. Then put the stack of two on the third meterstick so that the top

two do not tip, and so on. (b) After successfully building the appropriate stack, calculate the position of the center mass of the stack and show that it is over the table surface, not over the air beyond the edge of the table. (c) What if you were to rotate the top meterstick by 90° around a vertical axis so that it sits on the end of the third meterstick, perpendicular to the other three metersticks? Would the system still be in equilibrium? (d) Now stack the metersticks as follows with the zero ends of the metersticks all to the right in the diagram at the top of the next column:

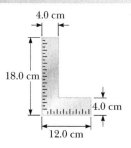

The right ends of the two metersticks at the upper left are above the 50-cm mark on the bottom meterstick. Place the system of metersticks on a table and move it off the edge to the right until a small additional outward movement would cause them to tip clockwise off the table. What reading on the bottom meterstick coincides with the edge of the table?

Problems

See the Preface for an explanation of the icons used in this problems set. For additional assessment items for this section, go to **WEBASSIGN** From Cengage

SECTION 12.1 **Analysis Model: Rigid Object in Equilibrium**

1. You are building additional storage space in your garage. You **CR** decide to suspend a 10.0-kg sheet of plywood of dimensions 0.600 m wide by 2.25 m long from the ceiling. The plywood will be held in a horizontal orientation by four light vertical chains attached to the plywood at its corners and mounted to the ceiling. After you complete the job of suspending the plywood from the ceiling, you choose three cubic boxes to place on the shelf. Each box is 0.750 m on a side. Box 1 has a mass of 50.0 kg, box 2 has a mass of 100 kg, and box 3 has a mass of 125 kg. The mass of each box is uniformly distributed within the box and each box is centered on the front-to-back width of the shelf. Unbeknownst to you, one of the chains on the right-hand end of your shelf is defective and will break if subjected to a force of more than 700 N. There are six possible arrangements of the three boxes on the shelf, for example, from left to right, Box 1, Box 2, Box 3, and Box 1, Box 3, Box, 2, and four more. Which arrangements are safe (that is, the defective chain will not break if the boxes are arranged in this way), and which arrangements are dangerous?

2. *Why is the following situation impossible?* A uniform beam of mass m_b = 3.00 kg and length ℓ = 1.00 m supports blocks with masses m_1 = 5.00 kg and m_2 = 15.0 kg at two positions as shown in Figure P12.2. The beam rests on two triangular blocks, with point P a distance d = 0.300 m to the right of the center of gravity of the beam. The position of the object of mass m_2 is adjusted along the length of the beam until the normal force on the beam at O is zero.

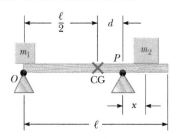

Figure P12.2

SECTION 12.2 **More on the Center of Gravity**

Problems 24 and 26 in Chapter 9 can also be assigned with this section.

3. A carpenter's square has the shape of an L as shown in **V** Figure P12.3. Locate its center of gravity.

Figure P12.3

4. A circular pizza of radius R has a circular piece of radius **S** $R/2$ removed from one side as shown in Figure P12.4. The center of gravity has moved from C to C' along the x axis. Show that the distance from C to C' is $R/6$. Assume the thickness and density of the pizza are uniform throughout.

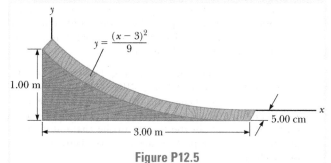

Figure P12.4

5. Your brother is opening a skateboard shop. He has created **CR** a sign for his shop made from a uniform material and in the shape shown in Figure P12.5. The shape of the sign represents one of the hills in the skateboard park he plans on building on land adjacent to the shop. The curve on the top of the sign is described by the function $y = (x-3)^2/9$. When the sign arrives in his shop, your brother wants to hang it from a single wire outside the shop. But he doesn't know where on the sign to attach the wire so that the bottom edge of the sign will hang in a horizontal orientation. He asks for your help.

$$y = \frac{(x-3)^2}{9}$$

Figure P12.5

SECTION 12.3 Examples of Rigid Objects in Static Equilibrium

Problems 14, 16, 18, 19, 34, 45, and 52 in Chapter 5 can also be assigned with this section.

6. A uniform beam of length 7.60 m and weight 4.50×10^2 N is carried by two workers, Sam and Joe, as shown in Figure P12.6. Determine the force that each person exerts on the beam.

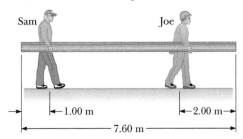

Figure P12.6

7. Find the mass m of the counterweight needed to balance a truck with mass $M = 1\,500$ kg on an incline of $\theta = 45°$ (Fig. P12.7). Assume both pulleys are frictionless and massless.

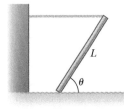

Figure P12.7

8. A uniform beam of length L and mass m shown in Figure P12.8 is inclined at an angle θ to the horizontal. Its upper end is connected to a wall by a rope, and its lower end rests on a rough, horizontal surface. The coefficient of static friction between the beam and surface is μ_s. Assume the angle θ is such that the static friction force is at its *maximum* value. (a) Draw a force diagram for the beam. (b) Using the condition of rotational equilibrium, find an expression for the tension T in the rope in terms of m, g, and θ. (c) Using the condition of translational equilibrium, find a second expression for T in terms of μ_s, m, and g. (d) Using the results from parts (a) through (c), obtain an expression for μ_s involving only the angle θ. (e) What happens if the ladder is lifted upward and its base is placed back on the ground slightly to the left of its position in Figure P12.8? Explain.

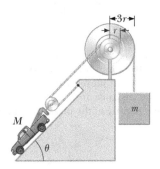

Figure P12.8

9. A flexible chain weighing 40.0 N hangs between two hooks located at the same height (Fig. P12.9). At each hook, the tangent to the chain makes an angle $\theta = 42.0°$ with the horizontal. Find (a) the magnitude of the force each hook exerts on the chain and (b) the tension in the chain at its midpoint. *Suggestion:* For part (b), make a force diagram for half of the chain.

Figure P12.9

10. A 20.0-kg floodlight in a park is supported at the end of a horizontal beam of negligible mass that is hinged to a pole as shown in Figure P12.10. A cable at an angle of $\theta = 30.0°$ with the beam helps support the light. (a) Draw a force diagram for the beam. By computing torques about an axis at the hinge at the left-hand end of the beam, find (b) the tension in the cable, (c) the horizontal component of the force exerted by the pole on the beam, and (d) the vertical component of this force. Now solve the same problem from the force diagram from part (a) by computing torques around the junction between the cable and the beam at the right-hand end of the beam. Find (e) the vertical component of the force exerted by the pole on the beam, (f) the tension in the cable, and (g) the horizontal component of the force exerted by the pole on the beam. (h) Compare the solution to parts (b) through (d) with the solution to parts (e) through (g). Is either solution more accurate?

Figure P12.10

11. Sir Lost-a-Lot dons his armor and sets out from the castle on his trusty steed (Fig. P12.11). Usually, the drawbridge is lowered to a horizontal position so that the end of the bridge rests on the stone ledge. Unfortunately, Lost-a-Lot's squire didn't lower the drawbridge far enough and stopped it at $\theta = 20.0°$ above the horizontal. The knight and his horse stop when their combined center of mass is $d = 1.00$ m from the end of the bridge. The uniform bridge is $\ell = 8.00$ m long and has mass 2 000 kg. The lift cable is attached to the bridge 5.00 m from the hinge at the castle end and to a point on the castle wall $h = 12.0$ m above the bridge. Lost-a-Lot's mass combined with his armor and steed is 1 000 kg. Determine (a) the tension in the cable and (b) the horizontal and (c) the vertical force components acting on the bridge at the hinge.

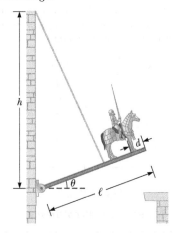

Figure P12.11 Problems 11 and 12.

12. **Review.** While Lost-a-Lot ponders his next move in the situation described in Problem 11 and illustrated in Figure P12.11, the enemy attacks! An incoming projectile breaks off the stone ledge so that the end of the drawbridge can be lowered past the wall where it usually rests. In addition, a fragment of the projectile bounces up and cuts the drawbridge cable! The hinge between the castle wall and the bridge is frictionless, and the bridge swings down freely until it is vertical and smacks into the vertical castle wall below the castle entrance. (a) How long does Lost-a-Lot stay in contact with the bridge while it swings downward? (b) Find the angular acceleration of the bridge just as it starts to move. (c) Find the angular speed of the bridge when it strikes the wall below the hinge. Find the force exerted by the hinge on the bridge (d) immediately after the cable breaks and (e) immediately before it strikes the castle wall.

13. Figure P12.13 shows a claw hammer being used to pull a nail
V out of a horizontal board. The mass of the hammer is 1.00 kg. A force of 150 N is exerted horizontally as shown, and the nail does not yet move relative to the board. Find (a) the force exerted by the hammer claws on the nail and (b) the force exerted by the surface on the point of contact with the hammer head. Assume the force the hammer exerts on the nail is parallel to the nail.

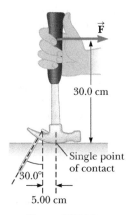

Figure P12.13

14. A 10.0-kg monkey climbs a uniform
QC ladder with weight 1.20×10^2 N and length $L = 3.00$ m as shown in Figure P12.14. The ladder rests against the wall and makes an angle of $\theta = 60.0°$ with the ground. The upper and lower ends of the ladder rest on frictionless surfaces. The lower end is connected to the wall by a horizontal rope that is frayed and can support a maximum tension of only 80.0 N. (a) Draw a force diagram for the ladder. (b) Find the normal force exerted on the bottom of the ladder. (c) Find the tension in the rope when the monkey is two-thirds of the way up the ladder. (d) Find the maximum distance d that the monkey can climb up the ladder before the rope breaks. (e) If the horizontal surface were rough and the rope were removed, how would your analysis of the problem change? What other information would you need to answer parts (c) and (d)?

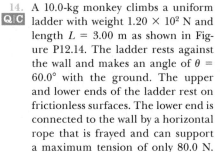
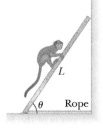

Figure P12.14

15. John is pushing his daughter Rachel in a wheelbarrow when it is stopped by a brick 8.00 cm high (Fig. P12.15). The handles make an angle of $\theta = 15.0°$ with the ground. Due to

the weight of Rachel and the wheelbarrow, a downward force of 400 N is exerted at the center of the wheel, which has a radius of 20.0 cm. (a) What force must John apply along the handles to just start the wheel over the brick? (b) What is the force (magnitude and direction) that the brick exerts on the wheel just as the wheel begins to lift over the brick? In both parts, assume the brick remains fixed and does not slide along the ground. Also assume the force applied by John is directed exactly toward the center of the wheel.

Figure P12.15 Problems 15 and 16.

16. John is pushing his daughter Rachel in a wheelbarrow when
S it is stopped by a brick of height h (Fig. P12.15). The handles make an angle of θ with the ground. Due to the weight of Rachel and the wheelbarrow, a downward force mg is exerted at the center of the wheel, which has a radius R. (a) What force F must John apply along the handles to just start the wheel over the brick? (b) What are the components of the force that the brick exerts on the wheel just as the wheel begins to lift over the brick? In both parts, assume the brick remains fixed and does not slide along the ground. Also assume the force applied by John is directed exactly toward the center of the wheel.

SECTION 12.4 Elastic Properties of Solids

17. The deepest point in the ocean is in the Mariana Trench,
QC about 11 km deep, in the Pacific. The pressure at this depth is huge, about 1.13×10^8 N/m². (a) Calculate the change in volume of 1.00 m³ of seawater carried from the surface to this deepest point. (b) The density of seawater at the surface is 1.03×10^3 kg/m³. Find its density at the bottom. (c) Explain whether or when it is a good approximation to think of water as incompressible.

18. A steel wire of diameter 1 mm can support a tension of 0.2 kN. A steel cable to support a tension of 20 kN should have diameter of what order of magnitude?

19. A child slides across a floor in a pair of rubber-soled shoes. The friction force acting on each foot is 20.0 N. The footprint area of each shoe sole is 14.0 cm², and the thickness of each sole is 5.00 mm. Find the horizontal distance by which the upper and lower surfaces of each sole are offset. The shear modulus of the rubber is 3.00 MN/m².

20. Evaluate Young's modulus for the material whose stress–strain curve is shown in Figure 12.12.

21. Assume if the shear stress in steel exceeds about $4.00 \times$
T 10^8 N/m², the steel ruptures. Determine the shearing force necessary to (a) shear a steel bolt 1.00 cm in diameter and (b) punch a 1.00-cm-diameter hole in a steel plate 0.500 cm thick.

22. When water freezes, it expands by about 9.00%. What pressure increase would occur inside your automobile engine

block if the water in it froze? (The bulk modulus of ice is 2.00×10^9 N/m².)

23. **Review.** A 30.0-kg hammer, moving with speed 20.0 m/s, strikes a steel spike 2.30 cm in diameter. The hammer rebounds with speed 10.0 m/s after 0.110 s. What is the average strain in the spike during the impact?

ADDITIONAL PROBLEMS

24. A uniform beam resting on two pivots has a length $L = 6.00$ m and mass $M = 90.0$ kg. The pivot under the left end exerts a normal force n_1 on the beam, and the second pivot located a distance $\ell = 4.00$ m from the left end exerts a normal force n_2. A woman of mass $m = 55.0$ kg steps onto the left end of the beam and begins walking to the right as in Figure P12.24. The goal is to find the woman's position when the beam begins to tip. (a) What is the appropriate analysis model for the beam before it begins to tip? (b) Sketch a force diagram for the beam, labeling the gravitational and normal forces acting on the beam and placing the woman a distance x to the right of the first pivot, which is the origin. (c) Where is the woman when the normal force n_1 is the greatest? (d) What is n_1 when the beam is about to tip? (e) Use Equation 12.1 to find the value of n_2 when the beam is about to tip. (f) Using the result of part (d) and Equation 12.2, with torques computed around the second pivot, find the woman's position x when the beam is about to tip. (g) Check the answer to part (e) by computing torques around the first pivot point.

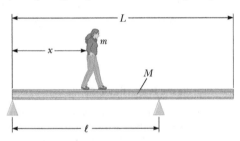

Figure P12.24

25. A bridge of length 50.0 m and mass 8.00×10^4 kg is supported on a smooth pier at each end as shown in Figure P12.25. A truck of mass 3.00×10^4 kg is located 15.0 m from one end. What are the forces on the bridge at the points of support?

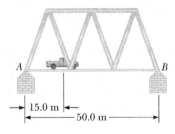

Figure P12.25

26. In exercise physiology studies, it is sometimes important to determine the location of a person's center of mass. This determination can be done with the arrangement shown in Figure P12.26. A light plank rests on two scales, which read $F_{g1} = 380$ N and $F_{g2} = 320$ N. A distance of 1.65 m separates the scales. How far from the woman's feet is her center of mass?

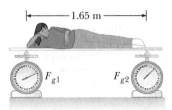

Figure P12.26

27. The lintel of prestressed reinforced concrete in Figure P12.27 is 1.50 m long. The concrete encloses one steel reinforcing rod with cross-sectional area 1.50 cm². The rod joins two strong end plates. The cross-sectional area of the concrete perpendicular to the rod is 50.0 cm². Young's modulus for the concrete is 30.0×10^9 N/m². After the concrete cures and the original tension T_1 in the rod is released, the concrete is to be under compressive stress 8.00×10^6 N/m². (a) By what distance will the rod compress the concrete when the original tension in the rod is released? (b) What is the new tension T_2 in the rod? (c) The rod will then be how much longer than its unstressed length? (d) When the concrete was poured, the rod should have been stretched by what extension distance from its unstressed length? (e) Find the required original tension T_1 in the rod.

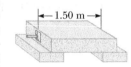

Figure P12.27

28. The following equations are obtained from a force diagram of a rectangular farm gate, supported by two hinges on the left-hand side. A bucket of grain is hanging from the latch.

$$-A + C = 0$$

$$+B - 392 \text{ N} - 50.0 \text{ N} = 0$$

$$A(0) + B(0) + C(1.80 \text{ m}) - 392 \text{ N}(1.50 \text{ m})$$

$$- 50.0 \text{ N}(3.00 \text{ m}) = 0$$

(a) Draw the force diagram and complete the statement of the problem, specifying the unknowns. (b) Determine the values of the unknowns and state the physical meaning of each.

29. A hungry bear weighing 700 N walks out on a beam in an attempt to retrieve a basket of goodies hanging at the end of the beam (Fig. P12.29). The beam is uniform, weighs 200 N, and is 6.00 m long, and it is supported by a wire at an angle of $\theta = 60.0°$. The basket weighs 80.0 N. (a) Draw a force diagram for the beam. (b) When the bear is at $x = 1.00$ m, find the tension in the wire supporting the beam and the components of the force exerted by the wall on the left end of the

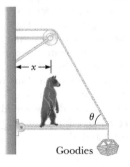

Figure P12.29

beam. (c) **What If?** If the wire can withstand a maximum tension of 900 N, what is the maximum distance the bear can walk before the wire breaks?

30. A 1 200-N uniform boom at $\phi = 65°$ to the vertical is supported by a cable at an angle $\theta = 25.0°$ to the horizontal as shown in Figure P12.30. The boom is pivoted at the bottom, and an object of weight $m = 2\,000$ N hangs from its top. Find (a) the tension in the support cable and (b) the components of the reaction force exerted by the floor on the boom.

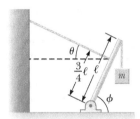

Figure P12.30

31. A uniform sign of weight F_g and width $2L$ hangs from a
S light, horizontal beam hinged at the wall and supported by a cable (Fig. P12.31). Determine (a) the tension in the cable and (b) the components of the reaction force exerted by the wall on the beam in terms of F_g, d, L, and θ.

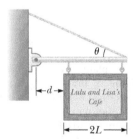

Figure P12.31

32. When a person stands on tiptoe on one foot (a strenuous
BIO position), the position of the foot is as shown in Figure P12.32a. The total gravitational force $\vec{F}_g$ on the body is supported by the normal force $\vec{n}$ exerted by the floor on the toes of one foot. A mechanical model of the situation is shown in Figure P12.32b, where $\vec{T}$ is the force exerted on the foot by the Achilles tendon and $\vec{R}$ is the force exerted on the foot by the tibia. Find the values of T, R, and θ when $F_g = 700$ N.

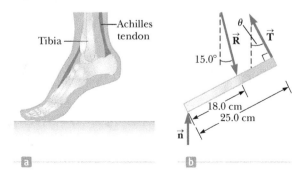

Figure P12.32

33. A 10 000-N shark is supported
T by a rope attached to a 4.00-m rod that can pivot at the base. (a) Calculate the tension in the cable between the rod and the wall, assuming the cable is holding the system in the position shown in Figure P12.33. Find (b) the horizontal force and (c) the vertical force exerted on the base of the rod. Ignore the weight of the rod.

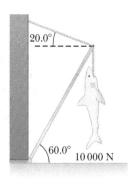

Figure P12.33

34. Assume a person bends for-
BIO ward to lift a load "with his
QC back" as shown in Figure P12.34a. The spine pivots mainly at the fifth lumbar vertebra, with the principal supporting force provided by the erector spinalis muscle in the back. To see the magnitude of the forces involved, consider the model shown in Figure P12.34b for a person bending forward to lift a 200-N object. The spine and upper body are represented as a uniform horizontal rod of weight 350 N, pivoted at the base of the spine. The erector spinalis muscle, attached at a point two-thirds of the way up the spine, maintains the position of the back. The angle between the spine and this muscle is $\theta = 12.0°$. Find (a) the tension T in the back muscle and (b) the compressional force in the spine. (c) Is this method a good way to lift a load? Explain your answer, using the results of parts (a) and (b). (d) Can you suggest a better method to lift a load?

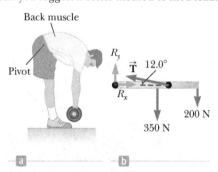

Figure P12.34

35. A uniform beam of mass m is inclined at an angle θ to the
S horizontal. Its upper end (point P) produces a 90° bend in a very rough rope tied to a wall, and its lower end rests on a rough floor (Fig. P12.35). Let μ_s represent the coefficient of static friction between beam and floor. Assume μ_s is less than the cotangent of θ. (a) Find an expression for the maximum mass M that can be suspended from the top before the beam slips. Determine (b) the magnitude of the reaction force at the floor and (c) the magnitude of the force exerted by the beam on the rope at P in terms of m, M, and μ_s.

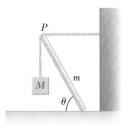

Figure P12.35

36. *Why is the following situation impossible?* A worker in a factory pulls a cabinet across the floor using a rope as shown in Figure P12.36a. The rope make an angle $\theta = 37.0°$ with the

floor and is tied $h_1 = 10.0$ cm from the bottom of the cabinet. The uniform rectangular cabinet has height $\ell = 100$ cm and width $w = 60.0$ cm, and it weighs 400 N. The cabinet slides with constant speed when a force $F = 300$ N is applied through the rope. The worker tires of walking backward. He fastens the rope to a point on the cabinet $h_2 = 65.0$ cm off the floor and lays the rope over his shoulder so that he can walk forward and pull as shown in Figure P12.36b. In this way, the rope again makes an angle of $\theta = 37.0°$ with the horizontal and again has a tension of 300 N. Using this technique, the worker is able to slide the cabinet over a long distance on the floor without tiring.

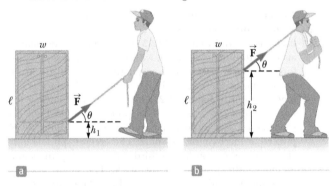

Figure P12.36 Problems 36 and 44.

37. When a circus performer performing on the rings executes the *iron cross,* he maintains the position at rest shown in Figure P12.37a. In this maneuver, the gymnast's feet (not shown) are off the floor. The primary muscles involved in supporting this position are the latissimus dorsi ("lats") and the pectoralis major ("pecs"). One of the rings exerts an upward force $\vec{\mathbf{F}}_h$ on a hand as shown in Figure P12.37b. The force $\vec{\mathbf{F}}_s$ is exerted by the shoulder joint on the arm. The latissimus dorsi and pectoralis major muscles exert a total force $\vec{\mathbf{F}}_m$ on the arm. (a) Using the information in the figure, find the

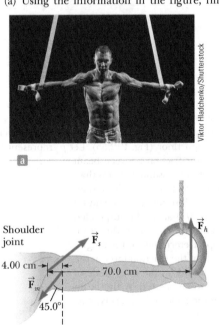

Figure P12.37

magnitude of the force $\vec{\mathbf{F}}_m$ for an athlete of weight 750 N. (b) Suppose a performer in training cannot perform the iron cross but can hold a position similar to the figure in which the arms make a 45° angle with the horizontal rather than being horizontal. Why is this position easier for the performer?

38. Figure P12.38 shows a light truss formed from three struts lying in a plane and joined by three smooth hinge pins at their ends. The truss supports a downward force of $\vec{\mathbf{F}} = 1\,000$ N applied at the point B. The truss has negligible weight. The piers at A and C are smooth. (a) Given $\theta_1 = 30.0°$ and $\theta_2 = 45.0°$, find n_A and n_C. (b) One can show that the force any strut exerts on a pin must be directed along the length of the strut as a force of tension or compression. Use that fact to identify the directions of the forces that the struts exert on the pins joining them. Find the force of tension or of compression in each of the three bars.

Figure P12.38

39. One side of a plant shelf is supported by a bracket mounted on a vertical wall by a single screw as shown in Figure P12.39. Ignore the weight of the bracket. (a) Find the horizontal component of the force that the screw exerts on the bracket when an 80.0 N vertical force is applied as shown. (b) As your grandfather waters his geraniums, the 80.0-N load force is increasing at the rate 0.150 N/s. At what rate is the force exerted by the screw changing? *Suggestion:* Imagine that the bracket is slightly loose.

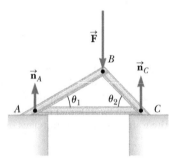

Figure P12.39

40. A stepladder of negligible weight is constructed as shown in Figure P12.40, with $AC = BC = \ell = 4.00$ m. A painter of mass $m = 70.0$ kg stands on the ladder $d = 3.00$ m from the bottom. Assuming the floor is frictionless, find (a) the tension in the horizontal bar DE connecting the two halves of the ladder, (b) the normal forces at A and B, and (c) the components of the reaction force at the single hinge C that the left half of the ladder exerts on the right half. *Suggestion:* Treat the

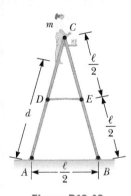

Figure P12.40
Problems 40 and 41.

ladder as a single object, but also treat each half of the ladder separately.

41. A stepladder of negligible weight is constructed as shown in Figure P12.40, with $AC = BC = \ell$. A painter of mass m stands on the ladder a distance d from the bottom. Assuming the floor is frictionless, find (a) the tension in the horizontal bar DE connecting the two halves of the ladder, (b) the normal forces at A and B, and (c) the components of the reaction force at the single hinge C that the left half of the ladder exerts on the right half. *Suggestion:* Treat the ladder as a single object, but also treat each half of the ladder separately.

42. **Review.** A wire of length L, Young's modulus Y, and cross-sectional area A is stretched elastically by an amount ΔL. By Hooke's law, the restoring force is $-k\,\Delta L$. (a) Show that $k = YA/L$. (b) Show that the work done in stretching the wire by an amount ΔL is $W = \frac{1}{2}YA(\Delta L)^2/L$.

43. Two racquetballs, each having a mass of 170 g, are placed in a glass jar as shown in Figure P12.43. Their centers lie on a straight line that makes a 45° angle with the horizontal. (a) Assume the walls are frictionless and determine P_1, P_2, and P_3. (b) Determine the magnitude of the force exerted by the left ball on the right ball.

Figure P12.43

44. Consider the rectangular cabinet of Problem 36 shown in Figure P12.36, but with a force $\vec{\mathbf{F}}$ applied horizontally at the upper edge. (a) What is the minimum force required to start to tip the cabinet? (b) What is the minimum coefficient of static friction required for the cabinet not to slide with the application of a force of this magnitude? (c) Find the magnitude and direction of the minimum force required to tip the cabinet if the point of application can be chosen *anywhere* on the cabinet.

45. **Review.** An aluminum wire is 0.850 m long and has a circular cross section of diameter 0.780 mm. Fixed at the top end, the wire supports a 1.20-kg object that swings in a horizontal circle. Determine the angular speed of the object required to produce a strain of 1.00×10^{-3}.

46. You have been hired as an expert witness in a case involving an injury in a factory. The attorney who hired you represents the injured worker. The worker was told to lift one end of a long, heavy crate that was lying horizontally on the floor and tilt it up so that it is standing on end. He began lifting the end of the crate, always applying a force that was perpendicular to the top of the crate. As the end of the crate got higher, at a certain angle, the bottom of the crate slipped on the floor, and the worker, in trying to recover, stepped forward and the crate landed on his foot, injuring it badly. As part of your investigation, you go to the factory and measure the coefficient of static friction between a crate and the smooth concrete floor. You find it to be 0.340. Prepare an argument for the attorney showing that it was impossible to lift the crate in the manner described without it slipping on the floor.

47. A 500-N uniform rectangular sign 4.00 m wide and 3.00 m high is suspended from a horizontal, 6.00-m-long, uniform, 100-N rod as indicated in Figure P12.47. The left end of the rod is supported by a hinge, and the right end is supported by a thin cable making a 30.0° angle with the vertical. (a) Find the tension T in the cable. (b) Find the horizontal and vertical components of force exerted on the left end of the rod by the hinge.

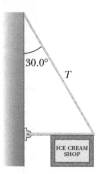

Figure P12.47

48. A steel cable 3.00 cm² in cross-sectional area has a mass of 2.40 kg per meter of length. If 500 m of the cable is hung over a vertical cliff, how much does the cable stretch under its own weight? Take $Y_{steel} = 2.00 \times 10^{11}$ N/m².

CHALLENGE PROBLEMS

49. A uniform rod of weight F_g and length L is supported at its ends by a frictionless trough as shown in Figure P12.49. (a) Show that the center of gravity of the rod must be vertically over point O when the rod is in equilibrium. (b) Determine the equilibrium value of the angle θ. (c) Is the equilibrium of the rod stable or unstable?

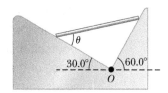

Figure P12.49

50. In the What If? section of Example 12.2, let d represent the distance in meters between the person and the hinge at the left end of the beam. (a) Show that the cable tension is given by $T = 93.9d + 125$, with T in newtons. (b) Show that the direction angle θ of the hinge force is described by

$$\tan\theta = \left(\frac{32}{3d+4} - 1\right)\tan 53.0°$$

(c) Show that the magnitude of the hinge force is given by

$$R = \sqrt{8.82 \times 10^3 d^2 - 9.65 \times 10^4 d + 4.96 \times 10^5}$$

(d) Describe how the changes in T, θ, and R as d increases differ from one another.

13

Universal Gravitation

Hubble Space Telescope image of a spiral galaxy, NGC 1566, taken in 2014. In the spiral arms of the galaxy, hydrogen gas is compressed to create new stars. It is theorized that our own galaxy, the Milky Way, has a similar structure with spiral arms. (*ESA/Hubble & NASA*)

STORYLINE **In Chapter 11, you were trying to do your physics** homework and were distracted by spinning skaters and divers. In Chapter 12, you were distracted by a surprising phenomenon with a meterstick. Now, you finally begin to work on your physics homework and you open your physics textbook. You look again at the tables in the front, before the title page. In the table of Solar System Data, you notice an entry for the mass of the Sun. After remarking to yourself that that's a lot of mass, you say, "Wait a minute! How did they find the mass of the Sun? In fact, how did they find the mass of any of the planets?" That leads you to think about the mass of the entire Milky Way galaxy. Looking online, you find different estimates of the mass of the galaxy, some in the range of hundreds of billions of solar masses, and others in the trillions of solar masses. Why can't we come up with a single number for the mass of the galaxy? Your physics homework goes undone as you ponder these new questions.

CONNECTIONS We first studied gravity in Section 2.8, where we talked about freely falling objects. There, and in Section 4.3 on projectile motion, we considered the effects of gravity on objects near the surface of the Earth. In Section 5.5, we related the gravitational force on such objects to their weight. In Chapter 7, we related the gravitational force on an object near the surface of the Earth to gravitational potential energy of the object–Earth system. In this chapter, we remove the assumption that objects are near the surface of the Earth. How does the gravitational force on an object vary as we move the object far from the surface of the Earth? The answer to that question will allow us to understand the motion of planets around the Sun and has allowed scientists to place many objects in orbit around the Earth, the Moon, and Mars. The principle that allows this understanding is the *law of universal gravitation*. We emphasize a description of planetary motion because astronomical data provide an important test of this law's validity. After introducing this law, we will show connections between it and the angular momentum of Chapter 11 and the energy techniques in Chapters 7 and 8. As preparation for the remainder of the book, we recognize gravitation as one of four "fundamental forces" of nature. The others are the electromagnetic force (Chapters 22–33), the nuclear strong force (Chapters 43–44), and the weak force (Chapter 44).

13.1 Newton's Law of Universal Gravitation

You may have heard the legend that, while napping under a tree, Newton was struck on the head by a falling apple. This alleged accident supposedly prompted him to imagine that perhaps all objects in the Universe were attracted to each other in the same way the apple was attracted to the Earth. Newton analyzed astronomical data on the motion of the Moon around the Earth. From that analysis, he made the bold assertion that the force law governing the motion of planets was the *same* as the force law that attracted a falling apple to the Earth. This assertion was contradictory to earlier thought that had lasted for centuries, which claimed that the laws of physics on the Earth did not apply to the heavens.

In 1687, Newton published his work on the law of gravity in his treatise *Mathematical Principles of Natural Philosophy*. **Newton's law of universal gravitation** states that

every particle in the Universe attracts every other particle with a force that is directly proportional to the product of their masses and inversely proportional to the square of the distance between them.

◀ The law of universal gravitation

If the particles have masses m_1 and m_2 and are separated by a distance r, the magnitude of this gravitational force is

$$F_g = G \frac{m_1 m_2}{r^2} \tag{13.1}$$

where G is a constant, called the *universal gravitational constant*. Its value in SI units is

$$G = 6.674 \times 10^{-11} \text{ N} \cdot \text{m}^2/\text{kg}^2 \tag{13.2}$$

The universal gravitational constant G was first evaluated in the late nineteenth century, based on results of an important experiment by Sir Henry Cavendish (1731–1810) in 1798. The law of universal gravitation was not expressed by Newton in the form of Equation 13.1, and Newton did not mention a constant such as G. In fact, even by the time of Cavendish, a unit of force had not yet been included in the existing system of units. Cavendish's goal was to measure the density of the Earth. His results were then used by other scientists 100 years later to generate a value for G.

Cavendish's apparatus consists of two small spheres, each of mass m, fixed to the ends of a light, horizontal rod suspended by a fine fiber or thin metal wire as illustrated in Figure 13.1. When two large spheres, each of mass M, are placed near the smaller ones, the attractive gravitational force between smaller and larger spheres causes the rod to rotate and twist the wire suspension to a new equilibrium orientation. The angle of rotation is measured by the deflection of a light beam reflected from a mirror attached to the vertical suspension.

The form of the force law given by Equation 13.1 is often referred to as an **inverse-square law** because the magnitude of the force varies as the inverse square of the separation of the particles.[1] We shall see other examples of this type of force

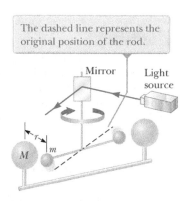

The dashed line represents the original position of the rod.

Figure 13.1 Cavendish apparatus for measuring gravitational forces.

---

[1] An *inverse* proportionality between two quantities x and y is one in which $y = k/x$, where k is a constant. A *direct* proportion between x and y exists when $y = kx$.

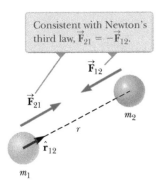

Figure 13.2 The gravitational force between two particles is attractive. The unit vector $\hat{\mathbf{r}}_{12}$ is directed from particle 1 toward particle 2.

PITFALL PREVENTION 13.1

Be Clear on g and G The symbol g represents the magnitude of the free-fall acceleration near a planet. At the surface of the Earth, g has an average value of 9.80 m/s². On the other hand, G is a universal constant that has the same value everywhere in the Universe.

---

law in subsequent chapters. We can express this force in vector form by defining a unit vector $\hat{\mathbf{r}}_{12}$ (Fig. 13.2). Because this unit vector is directed from particle 1 toward particle 2, the force exerted by particle 1 on particle 2 is

$$\vec{\mathbf{F}}_{12} = -G\frac{m_1 m_2}{r^2}\hat{\mathbf{r}}_{12} \qquad (13.3)$$

where the negative sign indicates that particle 2 is attracted to particle 1; hence, the force on particle 2 must be directed toward particle 1. By Newton's third law, the force exerted by particle 2 on particle 1, designated $\vec{\mathbf{F}}_{21}$, is equal in magnitude to $\vec{\mathbf{F}}_{12}$ and in the opposite direction. That is, these forces form an action–reaction pair, and $\vec{\mathbf{F}}_{21} = -\vec{\mathbf{F}}_{12}$.

Two features of Equation 13.3 deserve mention. First, the gravitational force is a field force that always exists between two particles, regardless of the medium that separates them. Second, because the force varies as the inverse square of the distance between the particles, it decreases rapidly with increasing separation.

Equation 13.3 can also be used to show that the gravitational force exerted by a finite-size, spherically symmetric mass distribution on a particle outside the distribution is the same as if the entire mass of the distribution were concentrated at the center. For example, the magnitude of the force exerted by the Earth on a particle of mass m near the Earth's surface is

$$F_g = G\frac{M_E m}{R_E{}^2} \qquad (13.4)$$

where M_E is the Earth's mass and R_E its radius. This force is directed toward the center of the Earth.

QUICK QUIZ 13.1 A planet has two moons of equal mass. Moon 1 is in a circular orbit of radius r. Moon 2 is in a circular orbit of radius $2r$. What is the magnitude of the gravitational force exerted by the planet on Moon 2? **(a)** four times as large as that on Moon 1 **(b)** twice as large as that on Moon 1 **(c)** equal to that on Moon 1 **(d)** half as large as that on Moon 1 **(e)** one-fourth as large as that on Moon 1

Example 13.1   **Billiards, Anyone?**

Three 0.300-kg billiard balls are placed on a table at the corners of a right triangle as shown in Figure 13.3. The sides of the triangle are of lengths $a = 0.400$ m, $b = 0.300$ m, and $c = 0.500$ m. Calculate the gravitational force vector on the cue ball (designated m_1) resulting from the other two balls as well as the magnitude and direction of this force.

SOLUTION

Conceptualize Notice in Figure 13.3 that the cue ball is attracted to both other balls by the gravitational force. We can see graphically that the net force should point upward and toward the right. We locate our coordinate axes as shown in Figure 13.3, placing our origin at the position of the cue ball.

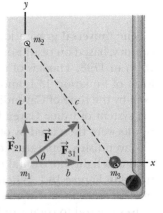

Figure 13.3 (Example 13.1) The resultant gravitational force acting on the cue ball is the vector sum $\vec{\mathbf{F}}_{21} + \vec{\mathbf{F}}_{31}$.

Categorize This problem involves evaluating the gravitational forces on the cue ball using Equation 13.3. Once these forces are evaluated, it becomes a vector addition problem to find the net force.

Analyze Find the force exerted by m_2 on the cue ball:

$$\vec{\mathbf{F}}_{21} = G\frac{m_2 m_1}{a^2}\hat{\mathbf{j}}$$

$$= (6.674 \times 10^{-11}\ \text{N} \cdot \text{m}^2/\text{kg}^2)\frac{(0.300\ \text{kg})(0.300\ \text{kg})}{(0.400\ \text{m})^2}\hat{\mathbf{j}}$$

$$= 3.75 \times 10^{-11}\hat{\mathbf{j}}\ \text{N}$$

13.1 continued

Find the force exerted by m_3 on the cue ball:

$$\vec{\mathbf{F}}_{31} = G\frac{m_3 m_1}{b^2}\,\hat{\mathbf{i}}$$

$$= (6.674 \times 10^{-11}\ \text{N} \cdot \text{m}^2/\text{kg}^2)\frac{(0.300\ \text{kg})(0.300\ \text{kg})}{(0.300\ \text{m})^2}\,\hat{\mathbf{i}}$$

$$= 6.67 \times 10^{-11}\,\hat{\mathbf{i}}\ \text{N}$$

Find the net gravitational force on the cue ball by adding these force vectors:

$$\vec{\mathbf{F}} = \vec{\mathbf{F}}_{31} + \vec{\mathbf{F}}_{21} = (6.67\,\hat{\mathbf{i}} + 3.75\,\hat{\mathbf{j}}) \times 10^{-11}\ \text{N}$$

Find the magnitude of this force:

$$F = \sqrt{F_{31}^2 + F_{21}^2} = \sqrt{(6.67)^2 + (3.75)^2} \times 10^{-11}\ \text{N}$$

$$= 7.66 \times 10^{-11}\ \text{N}$$

Find the tangent of the angle θ for the net force vector:

$$\tan\theta = \frac{F_y}{F_x} = \frac{F_{21}}{F_{31}} = \frac{3.75 \times 10^{-11}\ \text{N}}{6.67 \times 10^{-11}\ \text{N}} = 0.562$$

Evaluate the angle θ:

$$\theta = \tan^{-1}(0.562) = 29.4°$$

Finalize The result for F shows that the gravitational forces between everyday objects have extremely small magnitudes.

13.2 Free-Fall Acceleration and the Gravitational Force

We have called the magnitude of the gravitational force on an object near the Earth's surface the *weight* of the object, where the weight is given by Equation 5.6, $F = mg$. Equation 13.4 is another expression for this force. Therefore, we can set Equations 5.6 and 13.4 equal to each other to obtain

$$mg = G\frac{M_E m}{R_E^2}$$

$$g = G\frac{M_E}{R_E^2} \qquad (13.5)$$

Equation 13.5 relates the free-fall acceleration g to physical parameters of the Earth—its mass and radius—and explains the origin of the value of 9.80 m/s² that we have used in earlier chapters. Now consider an object of mass m located a distance h above the Earth's surface or a distance r from the Earth's center, where $r = R_E + h$. The magnitude of the gravitational force acting on this object is

$$F_g = G\frac{M_E m}{r^2} = G\frac{M_E m}{(R_E + h)^2}$$

The magnitude of the gravitational force acting on the object at this position is also $F_g = mg$, where g is the value of the free-fall acceleration at the altitude h. Substituting this expression for F_g into the last equation shows that g is given by

$$g = \frac{GM_E}{r^2} = \frac{GM_E}{(R_E + h)^2} \qquad (13.6)$$

TABLE 13.1 Free-Fall Acceleration g at Various Altitudes Above the Earth's Surface

Altitude h (km)	g (m/s²)
0	9.80
1 000	7.33
2 000	5.68
3 000	4.53
4 000	3.70
5 000	3.08
6 000	2.60
7 000	2.23
8 000	1.93
9 000	1.69
10 000	1.49
50 000	0.13
∞	0

◀ Variation of g with altitude

Therefore, it follows that g *decreases* with *increasing altitude*. Values of g for the Earth at various altitudes are listed in Table 13.1. Because an object's weight is mg, we see that as $r \rightarrow \infty$, the weight of the object approaches zero.

QUICK QUIZ 13.2 Superman stands on top of a very tall mountain and throws a baseball horizontally with a speed such that the baseball goes into a circular orbit around the Earth. While the baseball is in orbit, what is the magnitude of the acceleration of the ball? **(a)** It depends on how fast the baseball is thrown. **(b)** It is zero because the ball does not fall to the ground. **(c)** It is slightly less than 9.80 m/s². **(d)** It is equal to 9.80 m/s².

Example 13.2   The Density of the Earth

Using the known radius of the Earth and that $g = 9.80$ m/s² at the Earth's surface, find the average density of the Earth.

SOLUTION

Conceptualize   Assume the Earth is a perfect sphere. The density of material in the Earth varies, but let's adopt a simplified model in which we assume the density to be uniform throughout the Earth. The resulting density is the average density of the Earth.

Categorize   This example is a relatively simple substitution problem.

Using Equation 13.5, solve for the mass of the Earth:

$$M_E = \frac{gR_E^2}{G}$$

Substitute this mass and the volume of a sphere into the definition of density (Eq. 1.1):

$$\rho_E = \frac{M_E}{V_E} = \frac{gR_E^2/G}{\frac{4}{3}\pi R_E^3} = \frac{3}{4}\frac{g}{\pi G R_E}$$

$$= \frac{3}{4}\frac{9.80 \text{ m/s}^2}{\pi(6.674 \times 10^{-11} \text{ N} \cdot \text{m}^2/\text{kg}^2)(6.37 \times 10^6 \text{ m})} = 5.50 \times 10^3 \text{ kg/m}^3$$

WHAT IF?   What if you were told that a typical density of granite at the Earth's surface is 2.75×10^3 kg/m³? What would you conclude about the density of the material in the Earth's interior?

Answer   Because this value is about half the density we calculated as an average for the entire Earth, we would conclude that the inner core of the Earth has a density much higher than the average value. It is most amazing that the Cavendish experiment—which can be used to determine G and can be done today on a tabletop—combined with simple free-fall measurements of g provides information about the core of the Earth!

13.3   Analysis Model: Particle in a Field (Gravitational)

When Newton published his theory of universal gravitation, it was considered a success because it satisfactorily explained the motion of the planets. It represented strong evidence that the same laws that describe phenomena on the Earth can be used on large objects like planets and throughout the Universe. Since 1687, Newton's theory has been used to account for the motions of comets, the deflection of a Cavendish balance, the orbits of binary stars, and the rotation of galaxies. Nevertheless, both Newton's contemporaries and his successors found it difficult to accept the concept of a force that acts at a distance. They asked how it was possible for two objects such as the Sun and the Earth to interact when they were not in contact with each other. Newton himself could not answer that question.

An approach to describing interactions between objects that are not in contact came well after Newton's death. This approach enables us to look at the gravitational interaction in a different way, using the concept of a *gravitational field* that exists at every point in space. The concept of a *field* occurs often in physics. A **field** is a physical quantity that exists everywhere in space, is single-valued at all points, and is established by a source of some kind. For example, atmospheric pressure near the surface of the Earth is a field. At all points within the atmosphere, there is a value of the pressure. These values generally decrease with increasing altitude, and also change in time depending on current weather conditions. The source of atmospheric pressure is the air itself (see Chapter 14).

The source of a gravitational field is a *source particle* with mass M. Generally, this particle is planet- or star-sized, and can be modeled as a particle as long as we make observations outside of the planet or star. The source particle affects space about itself so that there is a quantity called the gravitational field everywhere in space.

This discussion of fields raises a couple of questions. First, how do we detect that a field exists at some point? And second, how do we define the value of the field at that point? To answer the first question, we must put a *test particle* at the point. A test particle is something that is sensitive to the altered space around the source. In the case of the atmosphere, imagine placing a helium-filled balloon at some point in the air. Because there is a pressure field at that point, and the pressure varies over the height of the balloon, the balloon will rise upward. (If a helium balloon were placed in empty space at a pressure of zero, it would remain stationary.) Therefore, the balloon detects the presence of the pressure field. In the case of the gravitational field, the test particle is a second particle, with mass m_0. If this particle is placed in the gravitational field, there is a gravitational force on the test particle. This force shows that a gravitational field exists at that point.

Now, how do we *define* the field so that we can assign a numerical value to it? For the balloon in the atmospheric pressure field, we could perhaps base the definition on the acceleration with which the balloon moves when released. In the case of gravity, we define the **gravitational field** $\vec{\mathbf{g}}$ as

$$\vec{\mathbf{g}} \equiv \frac{\vec{\mathbf{F}}_g}{m_0} \tag{13.7}$$

That is, the gravitational field at a point in space equals the gravitational force $\vec{\mathbf{F}}_g$ experienced by a test particle placed at that point divided by the mass m_0 of the test particle. Notice that the presence of the test particle is not necessary for the field to exist: the source particle creates the gravitational field. The gravitational field describes the effect that a source particle (for example, the Earth) has on the empty space around itself in terms of the force that *would* be present *if* a second object were somewhere in that space. It turns out to be useful to replace the direct gravitational force between two particles (Equation 13.1) with this "two-step" process: (1) one particle establishes a gravitational field, and (2) a second particle placed in the field experiences a force.[2]

The concept of a field is at the heart of the **particle in a field** analysis model. In Equation 13.7, the test particle of mass m_0 is placed in the field solely in order to determine the value of the gravitational field $\vec{\mathbf{g}}$. Once the value is determined, any arbitrary particle of mass m can be placed in the field and will experience a force $m\vec{\mathbf{g}}$. Therefore, the mathematical representation of the gravitational version of the particle in a field model is Equation 5.5:

$$\vec{\mathbf{F}}_g = m\vec{\mathbf{g}} \tag{5.5}$$

---

[2]We shall return to this idea of mass affecting the space around it when we discuss Einstein's theory of gravitation in Chapter 38.

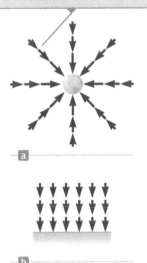

The field vectors point in the direction of the acceleration a particle would experience if it were placed in the field. The magnitude of the field vector at any location is the magnitude of the free-fall acceleration at that location.

Figure 13.4 (a) The gravitational field vectors in the vicinity of a uniform spherical mass such as the Earth vary in both direction and magnitude. (b) The gravitational field vectors in a small region near the Earth's surface are uniform in both direction and magnitude.

In the case of the pressure field in the atmosphere, we might recognize the force between the air and the balloon as a *contact force*, as discussed in Section 5.1. There is physical contact between the air and the balloon. What puzzled Newton and other scientists is that gravity is a *field force*: There is no physical contact between a star acting as a source particle and an orbiting planet placed in the resulting field.

In future chapters, we will see two other versions of the particle in a field model that turn out to be useful. In the electric version, the property of a source particle that results in an *electric field* is *electric charge*: when a second electrically-charged particle is placed in the electric field, it experiences a force. The magnitude of the force is the product of the electric charge and the field, in analogy with the gravitational force in Equation 5.5. In the magnetic version of the particle in a field model, a charged particle is placed in a *magnetic field*. One other property of this particle is required for the particle to experience a force: the particle must have a *velocity* at some nonzero angle to the magnetic field. The electric and magnetic versions of the particle in a field model are critical to the understanding of the principles of *electromagnetism*, which we will study in Chapters 22–33.

Because the gravitational force acting on a test particle of mass m_0 near the Earth has a magnitude $GM_E m_0/r^2$ (see Eq. 13.4), the gravitational field $\vec{\mathbf{g}}$ at a distance r from the center of the Earth is

$$\vec{\mathbf{g}} = \frac{\vec{\mathbf{F}}_g}{m_0} = -\frac{GM_E}{r^2}\hat{\mathbf{r}} \qquad (13.8)$$

where $\hat{\mathbf{r}}$ is a unit vector pointing radially outward from the Earth (see Fig. 3.15) and the negative sign indicates that the field points toward the center of the Earth as illustrated in Figure 13.4a. The field vectors at different points surrounding the Earth vary in both direction and magnitude. In a small region near the Earth's surface, the downward field $\vec{\mathbf{g}}$ is approximately constant and uniform as indicated in Figure 13.4b. Equation 13.8 is valid at all points *outside* the Earth's surface, assuming the Earth is spherical. At the Earth's surface, where $r = R_E$, $\vec{\mathbf{g}}$ has a magnitude of 9.80 N/kg. (The unit N/kg is the same as m/s².)

ANALYSIS MODEL   **Particle in a Field (Gravitational)**

Imagine an object with mass that we call a *source particle*. The source particle establishes a **gravitational field** $\vec{\mathbf{g}}$ throughout space. The gravitational field is evaluated by measuring the force on a test particle of mass m_0 and then using Equation 13.7. Now imagine a particle of mass m is placed in that field. The particle interacts with the gravitational field so that it experiences a gravitational force given by

$$\vec{\mathbf{F}}_g = m\vec{\mathbf{g}} \qquad (5.5)$$

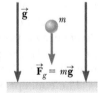

Examples:

- an object of mass m near the surface of the Earth has a *weight*, which is the result of the gravitational field established in space by the Earth
- a planet in the solar system is in orbit around the Sun, due to the gravitational force on the planet exerted by the gravitational field established by the Sun
- an object near a black hole is drawn into the black hole, never to escape, due to the tremendous gravitational field established by the black hole (Section 13.6)
- in the general theory of relativity, the gravitational field of a massive object is imagined to be described by a *curvature of spacetime* (Chapter 38)
- the gravitational field of a massive object is imagined to be mediated by particles called *gravitons*, which have never been detected (Chapter 44)

| Example **13.3** | The Weight of the Space Station |

The International Space Station operates at an altitude of 350 km. An online search for the station shows that a weight of 4.11×10^6 N, measured at the Earth's surface, has been lifted off the surface by various spacecraft during the construction process. What is the weight of the space station as it moves in its orbit?

SOLUTION

Conceptualize   The mass of the space station is fixed; it is independent of its location. Based on the discussions in this section and Section 13.2, we realize that the value of g is smaller at the height of the space station's orbit than at the surface of the Earth. Therefore, the weight of the Space Station is smaller than that at the surface of the Earth.

Categorize   We model the Space Station as a *particle in a gravitational field*.

Analyze   From the particle in a field model, find the mass of the space station from its weight at the surface of the Earth:

$$m = \frac{F_{g,\text{surface}}}{g_{\text{surface}}} = \frac{4.11 \times 10^6 \text{ N}}{9.80 \text{ m/s}^2} = 4.19 \times 10^5 \text{ kg}$$

Use Equation 13.6 with $h = 350$ km to find the magnitude of the gravitational field at the orbital location:

$$g_{\text{orbit}} = \frac{GM_E}{(R_E + h)^2}$$

$$= \frac{(6.674 \times 10^{-11} \text{ N} \cdot \text{m}^2/\text{kg}^2)(5.97 \times 10^{24} \text{ kg})}{(6.37 \times 10^6 \text{ m} + 0.350 \times 10^6 \text{ m})^2} = 8.82 \text{ m/s}^2$$

Use the particle in a field model again to find the space station's weight in orbit:

$$F_{g,\text{orbit}} = mg_{\text{orbit}} = (4.19 \times 10^5 \text{ kg})(8.82 \text{ m/s}^2) = 3.70 \times 10^6 \text{ N}$$

Finalize   Notice that the weight of the Space Station is less when it is in orbit, as we expected. It has about 10% less weight than it has when on the Earth's surface, representing a 10% decrease in the magnitude of the gravitational field.

13.4 Kepler's Laws and the Motion of Planets

Humans have observed the movements of the planets, stars, and other celestial objects for thousands of years. In early history, these observations led scientists to design a structural model in which the Earth was regarded as the center of the Universe. This *geocentric model* was elaborated and formalized by the Greek astronomer Claudius Ptolemy (c. 100–c. 170) in the second century and was accepted for the next 1 400 years. In 1543, Polish astronomer Nicolaus Copernicus (1473–1543) offered another structural model for the solar system that suggested that the Earth and the other planets revolved in circular orbits around the Sun (the *heliocentric model*).[3]

Danish astronomer Tycho Brahe (1546–1601) performed more observations to determine how the heavens were constructed and pursued a project to measure the positions of both stars and planets. Those observations of the planets and 777 stars visible to the naked eye were carried out with only a large sextant and a compass. (The telescope had not yet been invented.)

German astronomer Johannes Kepler was Brahe's assistant for a short while before Brahe's death, whereupon he acquired his mentor's astronomical data and spent 16 years trying to deduce a mathematical model for the motion of the planets. Such data are difficult to sort out because the moving planets are observed from a moving Earth. After many laborious calculations, Kepler found that Brahe's data on the revolution of Mars around the Sun led to a successful model.

---

[3]The heliocentric model was proposed by Aristarchus of Samos (c. 310 BC–c. 230 BC) several centuries before Copernicus, but the theory was not widely accepted.

Kepler's structural model of planetary motion is summarized in three statements known as **Kepler's laws:**

Kepler's laws ▶

1. All planets move in elliptical orbits with the Sun at one focus.
2. The radius vector drawn from the Sun to a planet sweeps out equal areas in equal time intervals.
3. The square of the orbital period of any planet is proportional to the cube of the semimajor axis of the elliptical orbit.

Kepler's First Law

Ptolemy's geocentric model and Copernicus's heliocentric model of the solar system both suggested circular orbits for heavenly bodies. Kepler's first law indicates that the circular orbit is a very special case and elliptical orbits are the general situation. This notion was difficult for scientists to accept because they believed, as had scientists for centuries before them, that perfect circular orbits of the planets reflected the perfection of heaven.

Figure 13.5 shows the geometry of an ellipse, which serves as our model for the elliptical orbit of a planet. An ellipse is mathematically defined by choosing two points F_1 and F_2, each of which is a called a **focus,** and then drawing a curve through points for which the sum of the distances r_1 and r_2 from F_1 and F_2, respectively, is a constant. The longest distance through the center between points on the ellipse (and passing through each focus) is called the **major axis,** and this distance is $2a$. In Figure 13.5, the major axis is drawn along the x direction. The distance a is called the **semimajor axis.** Similarly, the shortest distance through the center between points on the ellipse is called the **minor axis** of length $2b$, where the distance b is the **semiminor axis.** Either focus of the ellipse is located at a distance c from the center of the ellipse, where $a^2 = b^2 + c^2$. In the elliptical orbit of a planet around the Sun, the Sun is at one focus of the ellipse. There is nothing at the other focus.

The **eccentricity** of an ellipse is defined as $e = c/a$, and it describes the general shape of the ellipse. For a circle, $c = 0$, and the eccentricity is therefore zero. The smaller b is compared with a, the shorter the ellipse is along the y direction compared with its extent in the x direction in Figure 13.5. As b decreases, c increases and the eccentricity e increases. Therefore, higher values of eccentricity correspond to longer and thinner ellipses. The range of values of the eccentricity for an ellipse is $0 < e < 1$.

Eccentricities for planetary orbits vary widely in the solar system. The eccentricity of the Earth's orbit is 0.017, which makes it nearly circular. On the other hand, the eccentricity of Mercury's orbit is 0.21, the highest of the eight planets. Figure 13.6a shows an ellipse with an eccentricity equal to that of Mercury's orbit. Notice that even this highest-eccentricity orbit is difficult to distinguish from a circle, which is

Johannes Kepler
German astronomer (1571–1630)
Kepler is best known for developing the laws of planetary motion based on the careful observations of Tycho Brahe.

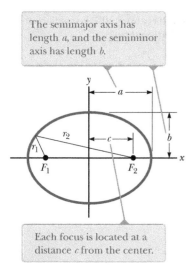

The semimajor axis has length a, and the semiminor axis has length b.

Each focus is located at a distance c from the center.

Figure 13.5 Plot of an ellipse.

Figure 13.6 (a) The shape of the orbit of Mercury, which has the highest eccentricity ($e = 0.21$) among the eight planets in the solar system. The broken line is *not* a circle. Measure the horizontal and vertical diameters. They differ by about 0.5 mm on the printed page. (Copy and enlarge to see the difference more easily!) (b) The shape of the orbit of Comet Halley. The shape of the orbit is correct; the comet and the Sun are shown larger than in reality for clarity.

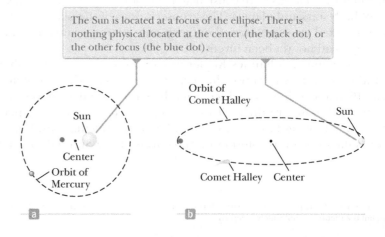

The Sun is located at a focus of the ellipse. There is nothing physical located at the center (the black dot) or the other focus (the blue dot).

one reason Kepler's first law is an admirable accomplishment. The eccentricity of the orbit of Comet Halley is 0.97, describing an orbit whose major axis is much longer than its minor axis, as shown in Figure 13.6b. As a result, Comet Halley spends much of its 76-year period far from the Sun and invisible from the Earth. It is only visible to the naked eye during a small part of its orbit when it is near the Sun.

Now imagine a planet in an elliptical orbit such as that shown in Figure 13.5, with the Sun at focus F_2. When the planet is at the far left in the diagram, the distance between the planet and the Sun is $a + c$. At this point, called the *aphelion,* the planet is at its maximum distance from the Sun. (For an object in orbit around the Earth, this point is called the *apogee.*) Conversely, when the planet is at the right end of the ellipse, the distance between the planet and the Sun is $a - c$. At this point, called the *perihelion* (for an Earth orbit, the *perigee*), the planet is at its minimum distance from the Sun.

Kepler's first law is a direct result of the inverse-square nature of the gravitational force. Circular and elliptical orbits correspond to objects that are *bound* to the gravitational force center. These objects include planets, asteroids, and comets that move repeatedly around the Sun as well as moons orbiting a planet. There are also *unbound* objects, such as a meteoroid from deep space that might pass by the Sun once and then never return. The gravitational force between the Sun and these objects also varies as the inverse square of the separation distance, and the allowed paths for these objects include parabolas ($e = 1$) and hyperbolas ($e > 1$).

Kepler's Second Law

Kepler's second law (page 340) can be shown to be a result of the isolated system model for angular momentum. Consider a planet of mass M_p moving about the Sun in an elliptical orbit (Fig. 13.7a). Let's consider the planet as a system. We model the Sun to be so much more massive than the planet that the Sun does not move. The gravitational force exerted by the Sun on the planet is a central force, always along the radius vector, directed toward the Sun (Fig. 13.7a). The torque on the planet due to this central force about an axis through the Sun is zero because $\vec{\mathbf{F}}_g$ is parallel to $\vec{\mathbf{r}}$.

Therefore, because the external torque on the planet is zero, it is modeled as an isolated system for angular momentum (Section 11.4), and the angular momentum $\vec{\mathbf{L}}$ of the planet is a constant of the motion:

$$\Delta \vec{\mathbf{L}} = 0 \;\rightarrow\; \vec{\mathbf{L}} = \text{constant}$$

Evaluating $\vec{\mathbf{L}}$ for the planet,

$$\vec{\mathbf{L}} = \vec{\mathbf{r}} \times \vec{\mathbf{p}} = M_p \vec{\mathbf{r}} \times \vec{\mathbf{v}} \;\rightarrow\; L = M_p |\vec{\mathbf{r}} \times \vec{\mathbf{v}}| \qquad (13.9)$$

We can relate this result to the following geometric consideration. In a time interval dt, the radius vector $\vec{\mathbf{r}}$ in Figure 13.7b sweeps out the area dA, which equals half the area $|\vec{\mathbf{r}} \times d\vec{\mathbf{r}}|$ of the parallelogram formed by the vectors $\vec{\mathbf{r}}$ and $d\vec{\mathbf{r}}$. Because the displacement of the planet in the time interval dt is given by $d\vec{\mathbf{r}} = \vec{\mathbf{v}}\, dt$,

$$dA = \tfrac{1}{2}|\vec{\mathbf{r}} \times d\vec{\mathbf{r}}| = \tfrac{1}{2}|\vec{\mathbf{r}} \times \vec{\mathbf{v}}\, dt| = \tfrac{1}{2}|\vec{\mathbf{r}} \times \vec{\mathbf{v}}|\, dt$$

Substitute for the absolute value of the cross product from Equation 13.9:

$$dA = \tfrac{1}{2}\left(\frac{L}{M_p}\right) dt$$

Divide both sides by dt to obtain

$$\frac{dA}{dt} = \frac{L}{2M_p} \qquad (13.10)$$

where L and M_p are both constants. This result shows that the derivative dA/dt is constant—the radius vector from the Sun to any planet sweeps out equal areas in equal time intervals as stated in Kepler's second law.

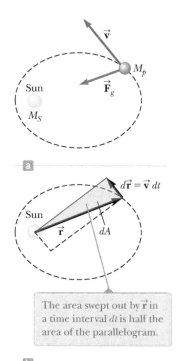

The area swept out by $\vec{\mathbf{r}}$ in a time interval dt is half the area of the parallelogram.

Figure 13.7 (a) The gravitational force acting on a planet is directed toward the Sun. (b) During a time interval dt, a parallelogram is formed by the vectors $\vec{\mathbf{r}}$ and $d\vec{\mathbf{r}} = \vec{\mathbf{v}}\, dt$.

This conclusion is a result of the gravitational force being a central force, which in turn implies that angular momentum of the planet is constant. Therefore, the law applies to *any* situation that involves a central force, whether inverse square or not.

Kepler's Third Law

Kepler's third law (page 340) can be predicted from the inverse-square law for circular orbits and our analysis models. Consider a planet of mass M_p that is assumed to be moving about the Sun (mass M_S) in a circular orbit as in Figure 13.8. Because the gravitational force provides the centripetal acceleration of the planet as it moves in a circle, we model the planet as a particle under a net force and as a particle in uniform circular motion and incorporate Newton's law of universal gravitation,

$$F_g = M_p a \;\; \rightarrow \;\; \frac{GM_S M_p}{r^2} = M_p \left(\frac{v^2}{r} \right)$$

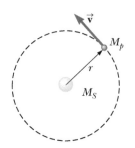

Figure 13.8 A planet of mass M_p moving in a circular orbit around the Sun. The orbits of all planets except Mercury are nearly circular.

The orbital speed of the planet is $2\pi r / T$, where T is the period; therefore, the preceding expression becomes

$$\frac{GM_S}{r^2} = \frac{(2\pi r / T)^2}{r}$$

$$T^2 = \left(\frac{4\pi^2}{GM_S} \right) r^3 = K_S r^3$$

where K_S is a constant given by

$$K_S = \frac{4\pi^2}{GM_S} = 2.97 \times 10^{-19} \text{ s}^2/\text{m}^3$$

This equation is also valid for elliptical orbits if we replace r with the length a of the semimajor axis (Fig. 13.5):

Kepler's third law ▶

$$T^2 = \left(\frac{4\pi^2}{GM_S} \right) a^3 = K_S a^3 \tag{13.11}$$

Equation 13.11 is Kepler's third law: the square of the period is proportional to the cube of the semimajor axis. Because the semimajor axis of a circular orbit is its radius, this equation is valid for both circular and elliptical orbits. Notice that the constant of proportionality K_S is independent of the mass of the planet.[4] Equation 13.11 is therefore valid for *any* planet. If we were to consider the orbit of a satellite such as the Moon about the Earth, the constant would have a different value, with the Sun's mass replaced by the Earth's mass; that is, $K_E = 4\pi^2/GM_E$.

Table 13.2 is a collection of useful data for planets and other objects in the solar system. The far-right column verifies that the ratio T^2/r^3 is constant for all objects orbiting the Sun. The small variations in the values in this column are the result of uncertainties in the data measured for the periods and semimajor axes of the objects.

Recent astronomical work has revealed the existence of a large number of solar system objects beyond the orbit of Neptune. In general, these objects lie in the *Kuiper belt,* a region that extends from about 30 AU (the orbital radius of Neptune) to 50 AU. (An AU is an *astronomical unit,* equal to the radius of the Earth's orbit.) Current estimates identify at least 100 000 objects in this region with diameters larger than 100 km. The first Kuiper belt object (KBO) is Pluto, discovered in 1930 and formerly classified as a planet. Starting in 1992, many more KBOs have been detected. Several have diameters in the 1 000-km range, such as Varuna (discovered in 2000), Ixion (2001), Quaoar (2002), Sedna (2003), Haumea (2004), Orcus (2004), and Makemake (2005). One KBO, Eris, discovered in 2005, is believed to be

---

[4]Equation 13.11 is indeed a proportion because the ratio of the two quantities T^2 and a^3 is a constant. The variables in a proportion are not required to be limited to the first power only.

TABLE 13.2 Useful Planetary Data

Body	Mass (kg)	Mean Radius (m)	Period of Revolution (s)	Mean Distance from the Sun (m)	$\dfrac{T^2}{r^3}$ (s²/m³)
Mercury	3.30×10^{23}	2.44×10^{6}	7.60×10^{6}	5.79×10^{10}	2.98×10^{-19}
Venus	4.87×10^{24}	6.05×10^{6}	1.94×10^{7}	1.08×10^{11}	2.99×10^{-19}
Earth	5.97×10^{24}	6.37×10^{6}	3.156×10^{7}	1.496×10^{11}	2.97×10^{-19}
Mars	6.42×10^{23}	3.39×10^{6}	5.94×10^{7}	2.28×10^{11}	2.98×10^{-19}
Jupiter	1.90×10^{27}	6.99×10^{7}	3.74×10^{8}	7.78×10^{11}	2.97×10^{-19}
Saturn	5.68×10^{26}	5.82×10^{7}	9.29×10^{8}	1.43×10^{12}	2.95×10^{-19}
Uranus	8.68×10^{25}	2.54×10^{7}	2.65×10^{9}	2.87×10^{12}	2.97×10^{-19}
Neptune	1.02×10^{26}	2.46×10^{7}	5.18×10^{9}	4.50×10^{12}	2.94×10^{-19}
Pluto[a]	1.25×10^{22}	1.20×10^{6}	7.82×10^{9}	5.91×10^{12}	2.96×10^{-19}
Moon	7.35×10^{22}	1.74×10^{6}	—	—	—
Sun	1.989×10^{30}	6.96×10^{8}	—	—	—

[a]In August 2006, the International Astronomical Union adopted a definition of a planet that separates Pluto from the other eight planets. Pluto is now defined as a "dwarf planet" like the asteroid Ceres.

similar in size to Pluto and about 27% more massive. Other KBOs do not yet have names, but are currently indicated by their year of discovery and a code, such as 2010 EK139 and 2015 FG345.

A subset of about 1 400 KBOs are called "Plutinos" because, like Pluto, they exhibit a resonance phenomenon, orbiting the Sun two times in the same time interval as Neptune revolves three times. The contemporary application of Kepler's laws and such exotic proposals as planetary angular momentum exchange and migrating planets suggest the excitement of this active area of current research.

QUICK QUIZ 13.3 An asteroid is in a highly eccentric elliptical orbit around the Sun. The period of the asteroid's orbit is 90 days. Which of the following statements is true about the possibility of a collision between this asteroid and the Earth? **(a)** There is no possible danger of a collision. **(b)** There is a possibility of a collision. **(c)** There is not enough information to determine whether there is danger of a collision.

Example **13.4** The Mass of the Sun

In the opening storyline, you were wondering how to determine the mass of the Sun. Now that we have discussed Kepler's third law, calculate the mass of the Sun.

SOLUTION

Conceptualize Based on the mathematical representation of Kepler's third law expressed in Equation 13.11, we realize that the mass of the central object in a gravitational system is related to the orbital size and period of objects in orbit around the central object.

Categorize This example is a relatively simple substitution problem.

Solve Equation 13.11 for the mass of the Sun:

$$M_S = \frac{4\pi^2 r^3}{GT^2}$$

Substitute numerical values, using data from Table 13.2:

$$M_S = \frac{4\pi^2(1.496 \times 10^{11} \text{ m})^3}{(6.674 \times 10^{-11} \text{ N} \cdot \text{m}^2/\text{kg}^2)(3.156 \times 10^7 \text{ s})^2} = 1.99 \times 10^{30} \text{ kg}$$

In Example 13.2, an understanding of gravitational forces enabled us to find out something about the density of the Earth's core, and now we have used this understanding to answer your question about the mass of the Sun! To answer your question about the masses of planets, we can perform the same calculation using the orbital size and period of a moon of a planet to find the planet mass. Neither Kepler's third law or Newton's law of universal gravitation can be used to determine the mass of the orbiting object, and the planets and KBOs for which we have precise mass data are the ones with moons or ones about which we have placed a spacecraft in orbit.

Example 13.5 A Geosynchronous Satellite

Consider a satellite of mass m moving in a circular orbit around the Earth at a constant speed v and at an altitude h above the Earth's surface as illustrated in Figure 13.9.

(A) Determine the speed of the satellite in terms of G, h, R_E (the radius of the Earth), and M_E (the mass of the Earth).

SOLUTION

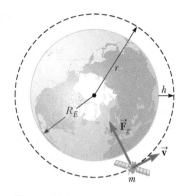

Figure 13.9 (Example 13.5) A satellite of mass m moving around the Earth in a circular orbit of radius r with constant speed v. The only force acting on the satellite is the gravitational force $\vec{F}_g$. (Not drawn to scale.)

Conceptualize Imagine the satellite moving around the Earth in a circular orbit under the influence of the gravitational force. Figure 13.9 is a polar view of this motion. This motion is similar to that of the International Space Station, the Hubble Space Telescope, and other objects in orbit around the Earth.

Categorize The satellite moves in a circular orbit at a constant speed. Therefore, we categorize the satellite as a *particle in uniform circular motion* as well as a *particle under a net force*.

Analyze The only external force acting on the satellite is the gravitational force from the Earth, which acts toward the center of the Earth and keeps the satellite in its circular orbit.

Apply the particle under a net force and particle in uniform circular motion models to the satellite:

$$F_g = ma \;\rightarrow\; G\frac{M_E m}{r^2} = m\left(\frac{v^2}{r}\right)$$

Solve for v, noting that the distance r from the center of the Earth to the satellite is $r = R_E + h$:

$$(1)\quad v = \sqrt{\frac{GM_E}{r}} = \sqrt{\frac{GM_E}{R_E + h}}$$

(B) If the satellite is to be *geosynchronous* (that is, appearing to remain over a fixed position on the Earth), how fast is it moving through space?

SOLUTION

To appear to remain over a fixed position on the Earth, the period of the satellite must be 24 h = 86 400 s and the satellite must be in orbit directly over the equator.

Solve Kepler's third law (Equation 13.11, with $a = r$ and $M_S \rightarrow M_E$) for r:

$$r = \left(\frac{GM_E T^2}{4\pi^2}\right)^{1/3}$$

Substitute numerical values:

$$r = \left[\frac{(6.674 \times 10^{-11}\,\text{N} \cdot \text{m}^2/\text{kg}^2)(5.97 \times 10^{24}\,\text{kg})(86\,400\,\text{s})^2}{4\pi^2}\right]^{1/3}$$

$$= 4.22 \times 10^7\,\text{m}$$

Use Equation (1) to find the speed of the satellite:

$$v = \sqrt{\frac{(6.674 \times 10^{-11}\,\text{N} \cdot \text{m}^2/\text{kg}^2)(5.97 \times 10^{24}\,\text{kg})}{4.22 \times 10^7\,\text{m}}}$$

$$= 3.07 \times 10^3\,\text{m/s}$$

Finalize The value of r calculated here translates to a height of the satellite above the surface of the Earth of almost 36 000 km. Therefore, geosynchronous satellites have the advantage of allowing an earthbound antenna to be aimed in a fixed direction, but there is a disadvantage in that the signals between the Earth and the satellite must travel a long distance. It is difficult to use geosynchronous satellites for optical observation of the Earth's surface because of their high altitude.

WHAT IF? What if the satellite motion in part (A) were taking place at height h above the surface of another planet more massive than the Earth but of the same radius? Would the satellite be moving at a higher speed or a lower speed than it does around the Earth?

Answer If the planet exerts a larger gravitational force on the satellite due to its larger mass, the satellite must move with a higher speed to avoid moving toward the surface. This conclusion is consistent with the predictions of Equation (1), which shows that because the speed v is proportional to the square root of the mass of the planet, the speed increases as the mass of the planet increases.

13.5 Gravitational Potential Energy

In Chapter 7, we introduced the concept of gravitational potential energy, which is the energy associated with the configuration of a system of objects interacting via the gravitational force. The gravitational potential energy function $U_g = mgy$ (Eq. 7.19) for a particle–Earth system is restricted to situations where a very massive object (such as the Earth) establishes a gravitational field of magnitude g and a particle of much smaller mass m resides in that field. It is also restricted to positions of the object near the surface of the Earth, where g is independent of y. In reality, however, because the gravitational field varies as $1/r^2$ as shown in Equation 13.8, we expect that a more general potential energy function—one that is valid without the restrictions mentioned above—will be different from $U_g = mgy$.

Recall from Equation 7.27 that the change in the potential energy of a system associated with a given displacement of a member of the system is defined as the negative of the internal work done by the force on that member during the displacement:

$$\Delta U = U_f - U_i = -\int_{r_i}^{r_f} F(r)\, dr \tag{13.12}$$

We can use this result to evaluate the general gravitational potential energy function. Consider a particle of mass m moving between two points Ⓐ and Ⓑ above the Earth's surface (Fig. 13.10). The particle is subject to the gravitational force given by Equation 13.1. We can express this force as

$$F(r) = -\frac{GM_E m}{r^2}$$

where the negative sign indicates that the force is attractive. Substituting this expression for $F(r)$ into Equation 13.12, we can compute the change in the gravitational potential energy function for the particle–Earth system as the separation distance r changes:

$$U_f - U_i = GM_E m \int_{r_i}^{r_f} \frac{dr}{r^2} = GM_E m \left[-\frac{1}{r} \right]_{r_i}^{r_f}$$

$$U_f - U_i = -GM_E m \left(\frac{1}{r_f} - \frac{1}{r_i} \right) \tag{13.13}$$

As always, the choice of a reference configuration for the potential energy is completely arbitrary. It is customary to choose the reference configuration for zero potential energy to be the same as that for which the force is zero. Taking $U_i = 0$ at $r_i = \infty$, we obtain the important result

$$U_g(r) = -\frac{GM_E m}{r} \tag{13.14}$$

This expression applies when the particle is separated from the center of the Earth by a distance r, provided that $r \geq R_E$. The result is not valid for particles inside the Earth, where $r < R_E$. Because of our choice of U_i, the function U_g is always negative (Fig. 13.11).

Although Equation 13.14 was derived for the particle–Earth system, a similar form of the equation can be applied to any two particles. That is, the gravitational potential energy associated with a system of two particles of masses m_1 and m_2 separated by a distance r is

$$U_g(r) = -\frac{Gm_1 m_2}{r} \tag{13.15}$$

This expression shows that the gravitational potential energy for any pair of particles varies as $1/r$, whereas the force between them varies as $1/r^2$. Furthermore,

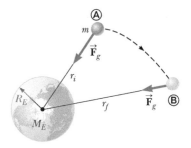

Figure 13.10 As a particle of mass m moves from Ⓐ to Ⓑ above the Earth's surface, the gravitational potential energy of the particle–Earth system changes according to Equation 13.12.

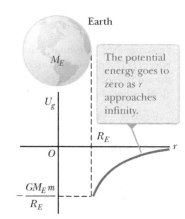

Figure 13.11 Graph of the gravitational potential energy U_g versus r for the system of an object above the Earth's surface.

◀ Gravitational potential energy of the Earth–particle system

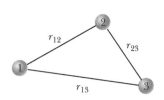

Figure 13.12 Three interacting particles.

the potential energy is negative because the force is attractive and we have chosen the potential energy as zero when the particle separation is infinite. Because the force between the particles is attractive, an external agent must do positive work to increase the separation between the particles. The work done by the external agent produces an increase in the potential energy as the two particles are separated. That is, U_g becomes less negative as r increases.

We can extend this concept to three or more particles. In this case, the total potential energy of the system is the sum over all pairs of particles. Each pair contributes a term of the form given by Equation 13.15. For example, if the system contains three particles as in Figure 13.12,

$$U_{total} = U_{12} + U_{13} + U_{23} = -G\left(\frac{m_1 m_2}{r_{12}} + \frac{m_1 m_3}{r_{13}} + \frac{m_2 m_3}{r_{23}}\right)$$

The absolute value of U_{total} represents the work needed to separate the three particles by an infinite distance.

Example 13.6 The Change in Potential Energy

A particle of mass m is displaced through a small vertical distance Δy near the Earth's surface. Show that in this situation the general expression for the change in gravitational potential energy given by Equation 13.13 reduces to the familiar relationship $\Delta U_g = mg\,\Delta y$.

SOLUTION

Conceptualize Compare the two different situations for which we have developed expressions for gravitational potential energy: (1) a planet and an object that are far apart for which the energy expression is Equation 13.14 and (2) a small object at the surface of a planet for which the energy expression is Equation 7.19. We wish to show that these two expressions are equivalent under the conditions described in the problem.

Categorize This example is a substitution problem.

Combine the fractions in Equation 13.13:

$$(1) \quad \Delta U_g = -GM_E m\left(\frac{1}{r_f} - \frac{1}{r_i}\right) = GM_E m\left(\frac{r_f - r_i}{r_i r_f}\right)$$

Evaluate $r_f - r_i$ and $r_i r_f$ if both the initial and final positions of the particle are close to the Earth's surface:

$$r_f - r_i = \Delta y \quad r_i r_f \approx R_E^2$$

Substitute these expressions into Equation (1):

$$\Delta U_g \approx GM_E m\left(\frac{\Delta y}{R_E^2}\right) = m\left(\frac{GM_E}{R_E^2}\right)\Delta y = mg\,\Delta y$$

where $g = GM_E/R_E^2$ from Equation 13.5.

WHAT IF? Suppose you are performing upper-atmosphere studies and are asked by your supervisor to find the height in the Earth's atmosphere at which the "surface equation" $\Delta U_g = mg\,\Delta y$ gives a 1.0% error in the change in the potential energy. What is this height?

Answer Because the surface equation assumes a constant value for g, it will give a ΔU_g value that is larger than the value given by the general equation, Equation 13.13.

Set up a ratio reflecting a 1.0% error:

$$\frac{\Delta U_{surface}}{\Delta U_{general}} = 1.010$$

Substitute the expressions for each of these changes ΔU_g:

$$\frac{mg\,\Delta y}{GM_E m(\Delta y/r_i r_f)} = \frac{gr_i r_f}{GM_E} = 1.010$$

Substitute for r_i, r_f, and g from Equation 13.5:

$$\frac{(GM_E/R_E^2)R_E(R_E + \Delta y)}{GM_E} = \frac{R_E + \Delta y}{R_E} = 1 + \frac{\Delta y}{R_E} = 1.010$$

Solve for Δy:

$$\Delta y = 0.010R_E = 0.010(6.37 \times 10^6\text{ m}) = 6.37 \times 10^4\text{ m} = 63.7\text{ km}$$

13.6 Energy Considerations in Planetary and Satellite Motion

Given the general expression for gravitational potential energy developed in Section 13.5, we can now apply the analysis models that we have developed for energy to gravitational systems. Consider an object of mass m moving with a speed v in the vicinity of a massive object of mass M, where $M \gg m$. The system might be a planet moving around the Sun, a satellite in orbit around the Earth, or a comet making a one-time flyby of the Sun. If we assume the object of mass M is at rest in an inertial reference frame, the total mechanical energy E of the two-object system when the objects are separated by a distance r is the sum of the kinetic energy of the object of mass m and the gravitational potential energy of the system, given by Equation 13.15:

$$E = K + U_g$$

$$E = \tfrac{1}{2}mv^2 - \frac{GMm}{r} \tag{13.16}$$

If the system of objects of mass m and M is isolated, and there are no nonconservative forces acting within the system, the mechanical energy of the system given by Equation 13.16 is the total energy of the system and this energy is conserved:

$$\Delta K + \Delta U_g = 0 \quad \rightarrow \quad E_i = E_f$$

Therefore, as the object of mass m moves from Ⓐ to Ⓑ in Figure 13.10, the total energy remains constant and Equation 13.16 gives

$$\tfrac{1}{2}mv_i^2 - \frac{GMm}{r_i} = \tfrac{1}{2}mv_f^2 - \frac{GMm}{r_f} \tag{13.17}$$

Combining this statement of energy conservation with our earlier discussion of conservation of angular momentum, we see that both the total energy and the total angular momentum of a gravitationally bound, two-object system are constants of the motion.

Equation 13.16 shows that E may be positive, negative, or zero, depending on the value of v. For a bound system such as the Earth–Sun system, however, E is necessarily *less than zero* because we have chosen the convention that $U_g \rightarrow 0$ as $r \rightarrow \infty$.

We can easily establish that $E < 0$ for the system consisting of an object of mass m moving in a circular orbit about an object of mass $M \gg m$ (Fig. 13.13). Modeling the object of mass m as a particle under a net force and a particle in uniform circular motion gives

$$F_g = ma \quad \rightarrow \quad \frac{GMm}{r^2} = \frac{mv^2}{r}$$

Multiplying both sides by r and dividing by 2 gives

$$\tfrac{1}{2}mv^2 = \frac{GMm}{2r} \tag{13.18}$$

Substituting this equation into Equation 13.16, we obtain

$$E = \frac{GMm}{2r} - \frac{GMm}{r}$$

$$E = -\frac{GMm}{2r} \quad \text{(circular orbits)} \tag{13.19}$$

This result shows that the total mechanical energy is negative in the case of circular orbits. Notice that the kinetic energy is positive and equal to half the absolute value of the potential energy.

The total mechanical energy is also negative in the case of elliptical orbits. The expression for E for elliptical orbits is the same as Equation 13.19 with r replaced by

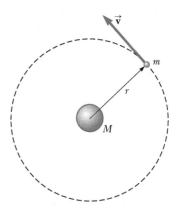

Figure 13.13 An object of mass m moving in a circular orbit about a much larger object of mass M.

◀ Total energy for circular orbits of an object of mass m around an object of mass $M \gg m$

the semimajor axis length a:

Total energy for elliptical ▶
orbits of an object of
mass m around an object of
mass $M \gg m$

$$E = -\frac{GMm}{2a} \quad \text{(elliptical orbits)} \tag{13.20}$$

QUICK QUIZ 13.4 A comet moves in an elliptical orbit around the Sun. Which point in its orbit (perihelion or aphelion) represents the highest value of **(a)** the speed of the comet, **(b)** the potential energy of the comet–Sun system, **(c)** the kinetic energy of the comet, and **(d)** the total energy of the comet–Sun system?

Example **13.7** **Changing the Orbit of a Satellite**

A space transportation vehicle releases a 470-kg communications satellite while in a circular orbit 280 km above the surface of the Earth. A rocket engine on the satellite boosts it into a geosynchronous orbit. How much energy does the engine have to provide for this boost?

SOLUTION

Conceptualize Notice that the height of 280 km is much lower than that for a geosynchronous satellite, 36 000 km, as mentioned in Example 13.5. Therefore, energy must be expended to raise the satellite to this much higher position.

Categorize This example is a substitution problem.

Find the initial radius of the satellite's orbit when it is still in the vehicle's cargo bay:

$$r_i = R_E + 280 \text{ km} = 6.65 \times 10^6 \text{ m}$$

Use Equation 13.19 to find the difference in energies for the satellite–Earth system with the satellite at the initial and final radii:

$$\Delta E = E_f - E_i = -\frac{GM_E m}{2r_f} - \left(-\frac{GM_E m}{2r_i}\right) = -\frac{GM_E m}{2}\left(\frac{1}{r_f} - \frac{1}{r_i}\right)$$

Substitute numerical values, using $r_f = 4.22 \times 10^7$ m from Example 13.5:

$$\Delta E = -\frac{(6.674 \times 10^{-11} \text{ N} \cdot \text{m}^2/\text{kg}^2)(5.97 \times 10^{24} \text{ kg})(470 \text{ kg})}{2}$$

$$\times \left(\frac{1}{4.22 \times 10^7 \text{ m}} - \frac{1}{6.65 \times 10^6 \text{ m}}\right) = 1.19 \times 10^{10} \text{ J}$$

which is the energy equivalent of 89 gal of gasoline. NASA engineers must account for the changing mass of the spacecraft as it ejects burned fuel, something we have not done here. Would you expect the calculation that includes the effect of this changing mass to yield a greater or a lesser amount of energy required from the engine?

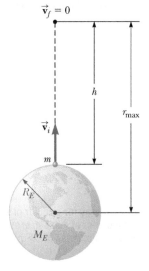

Figure 13.14 An object of mass m projected upward from the Earth's surface with an initial speed v_i reaches a maximum altitude h.

Escape Speed

Suppose an object of mass m is projected vertically upward from the Earth's surface with an initial speed v_i as illustrated in Figure 13.14. We can use energy considerations to find the value of the initial speed needed to allow the object to reach a certain distance away from the center of the Earth. Equation 13.16 gives the total energy of the system for any configuration. As the object is projected upward from the surface of the Earth, $v = v_i$ and $r = r_i = R_E$. When the object reaches its maximum altitude, $v = v_f = 0$ and $r = r_f = r_{max}$. Because the object–Earth system is isolated, we substitute these values into the isolated-system model expression given by Equation 13.17:

$$\tfrac{1}{2}mv_i^2 - \frac{GM_E m}{R_E} = -\frac{GM_E m}{r_{max}}$$

Solving for v_i^2 gives

$$v_i^2 = 2GM_E\left(\frac{1}{R_E} - \frac{1}{r_{max}}\right) \tag{13.21}$$

For a given maximum altitude $h = r_{max} - R_E$, we can use this equation to find the required initial speed.

We are now in a position to calculate the **escape speed,** which is the minimum speed the object must have at the Earth's surface to approach an infinite separation distance from the Earth. Traveling at this minimum speed, the object continues to move farther and farther away from the Earth as its speed asymptotically approaches zero. Letting $r_{max} \rightarrow \infty$ in Equation 13.21 and identifying v_i as v_{esc} gives

$$v_{esc} = \sqrt{\frac{2GM_E}{R_E}} \qquad (13.22)$$

◄ Escape speed from the Earth

This expression for v_{esc} is independent of the mass of the object. In other words, a spacecraft has the same escape speed as a molecule. Furthermore, the result is independent of the direction of the velocity and ignores air resistance. Keep in mind that an object already has some initial speed at the surface of the Earth due to its rotation.

If the object is given an initial speed equal to v_{esc}, the total energy of the system is equal to zero. Notice that when $r \rightarrow \infty$, the object's kinetic energy and the potential energy of the system are both zero. If v_i is greater than v_{esc}, however, the total energy of the system is greater than zero and the object has some residual kinetic energy as $r \rightarrow \infty$.

PITFALL PREVENTION 13.3
You Can't Really Escape Although Equation 13.22 provides the "escape speed" from the Earth, *complete* escape from the Earth's gravitational influence is impossible because the gravitational force is of infinite range. In addition, escape from the Earth to an infinite distance also requires escape from the Sun, requiring additional energy.

Example 13.8 Escape Speed of a Rock

Superman picks up a 20-kg rock and hurls it into space. What minimum speed must it have at the Earth's surface to move infinitely far away from the Earth?

SOLUTION

Conceptualize Imagine Superman throwing the rock from the Earth's surface so that it moves farther and farther away, traveling more and more slowly, with its speed approaching zero. Its speed will never reach zero, however, so the rock will never turn around and come back.

Categorize This example is a substitution problem.

Use Equation 13.22 to find the escape speed:

$$v_{esc} = \sqrt{\frac{2GM_E}{R_E}} = \sqrt{\frac{2(6.674 \times 10^{-11}\,\text{N}\cdot\text{m}^2/\text{kg}^2)(5.97 \times 10^{24}\,\text{kg})}{6.37 \times 10^6\,\text{m}}}$$

$$= 1.12 \times 10^4\,\text{m/s}$$

The calculated escape speed corresponds to about 25 000 mi/h. The mass of the rock does not appear in the calculation. Therefore, this is also the escape speed for Superman throwing a 5 000-kg spacecraft from the surface of the Earth. Furthermore, if a spacecraft is in an orbit around the Earth, its orbital radius r is close to that of the Earth, R_E, so the escape speed we have calculated is also valid for the non-superhero situation of a spacecraft in orbit firing its engines to escape that orbit.

Equations 13.21 and 13.22 can be applied to objects projected from any planet. That is, in general, the escape speed from the surface of any planet of mass M and radius R is

$$v_{esc} = \sqrt{\frac{2GM}{R}} \qquad (13.23)$$

◄ Escape speed from the surface of a planet of mass M and radius R

Escape speeds for the planets, the Moon, and the Sun are provided in Table 13.3 (page 350). The values vary from 2.3 km/s for the Moon to about 618 km/s for the Sun. These results, together with some ideas from the kinetic theory of gases (see Chapter 20), explain why some planets have atmospheres and others do not. As we shall see later, at a given temperature the average kinetic energy of a gas molecule depends only on the mass of the molecule. Lighter molecules, such as hydrogen and helium, have a higher average speed than heavier molecules at the same

TABLE 13.3 Escape Speeds from the Surfaces of the Planets, Moon, and Sun

Planet	v_{esc} (km/s)
Mercury	4.3
Venus	10.3
Earth	11.2
Mars	5.0
Jupiter	60
Saturn	36
Uranus	22
Neptune	24
Moon	2.3
Sun (from Earth orbit)	42
Sun (from Sun surface)	618

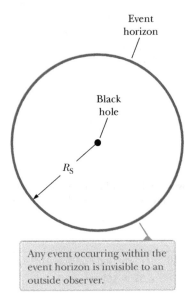

Any event occurring within the event horizon is invisible to an outside observer.

Figure 13.15 A black hole. The distance R_S equals the Schwarzschild radius.

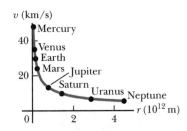

Figure 13.16 The orbital speed v as a function of distance r from the Sun for the eight planets of the solar system. The theoretical curve is in red-brown, and the data points for the planets are in black.

temperature. When the *average* speed of the lighter molecules is not much less than the escape speed of a planet, a significant fraction of them are moving faster than the average speed and have a chance to escape the planet.

This mechanism also explains why the Earth does not retain hydrogen molecules and helium atoms in its atmosphere but does retain heavier molecules, such as oxygen and nitrogen. On the other hand, the very large escape speed for Jupiter enables that planet to retain hydrogen, the primary constituent of its atmosphere.

Black Holes

In Example 11.6, we briefly described a rare event called a supernova, the catastrophic explosion at the end of the life of a very massive star. The material that remains in the central core of such an object continues to collapse, and the core's ultimate fate depends on its mass. As the star collapses inward, electrons and protons fuse to form neutrons, and an object made purely of neutrons, a neutron star, is born, as discussed in Example 11.6. The inward collapse of such a star is halted by the quantum mechanical repulsion of neutrons called *neutron degeneracy pressure*, and the mass of the star is compressed to a radius of about 10 km. (On the Earth, a teaspoon of this material would weigh about 5 billion tons!). The escape speed from a neutron star is typically $> 0.5c$, where c is the speed of light.

An even more unusual death occurs when the core has a mass greater than about 2–3 solar masses. Neutron degeneracy pressure is unable to halt the collapse of the star, and the star becomes a very small object in space, commonly referred to as a **black hole.** In effect, black holes are remains of stars that have collapsed under their own gravitational force. If an object such as a spacecraft comes close to a black hole, the object experiences an extremely strong gravitational force and is trapped forever.

The escape speed for a black hole is very high because of the concentration of the star's mass into a sphere of very small radius (see Eq. 13.23). If the escape speed exceeds the speed of light c, radiation from the object (such as visible light) cannot escape and the object appears to be black (hence the origin of the terminology "black hole"). The critical radius R_S at which the escape speed is c is called the **Schwarzschild radius** (Fig. 13.15). The imaginary surface of a sphere of this radius surrounding the black hole is called the **event horizon,** which is the limit of how close you can approach the black hole and hope to escape.

There is evidence that supermassive black holes exist at the centers of galaxies, with masses very much larger than the Sun. (There is strong evidence of a supermassive black hole of mass 4.0–4.3 million solar masses at the center of our galaxy.)

Dark Matter

Equation (1) in Example 13.5 shows that the speed of an object in orbit around the Earth decreases as the object is moved farther away from the Earth:

$$v = \sqrt{\frac{GM_E}{r}} \tag{13.24}$$

Using data in Table 13.2 to find the speeds of planets in their orbits around the Sun, we find the same behavior for the planets. Figure 13.16 shows this behavior for the eight planets of our solar system. The theoretical prediction of the planet speed as a function of distance from the Sun is shown by the red-brown curve, using Equation 13.24 with the mass of the Earth replaced by the mass of the Sun. Data points for the individual planets lie right on this curve. This behavior results from the vast majority (99.9%) of the mass of the solar system being concentrated in a small space, i.e., the Sun.

Extending this concept further, we might expect the same behavior in a galaxy. Much of the visible galactic mass, including that of a supermassive black hole, is near the central core of a galaxy. Measurements of the speeds of faraway objects in the galaxy can be used in Kepler's third law to estimate the mass of the entire galaxy, a topic raised in the opening storyline. These estimates range from 0.8×10^{12} to 4.5×10^{12} solar masses. Measurements made on these faraway objects are difficult and the results depend on the method used to make the observations.

The opening photograph for this chapter shows the central core of the galaxy NGC 1566 as a very bright area surrounded by the "arms" of the galaxy, which contain material in orbit around the central core. Based on this distribution of matter in the galaxy, the speed of an object in the outer part of the galaxy would be smaller than that for objects closer to the center, just like for the planets of the solar system.

That is *not* what is observed, however. Figure 13.17 shows the results of measurements of the speeds of objects in the Andromeda galaxy as a function of distance from the galaxy's center.[5] The red-brown curve shows the expected speeds for these objects if they were traveling in circular orbits around the mass concentrated in the central core. The data for the individual objects in the galaxy shown by the black dots are all well above the theoretical curve. These data, as well as an extensive amount of data taken over the past half century, show that for objects outside the central core of the galaxy, the curve of speed versus distance from the center of the galaxy is approximately flat rather than decreasing at larger distances. Therefore, these objects (including our own Solar System in the Milky Way) are rotating faster than can be accounted for by gravity due to the visible galaxy! This surprising result means that there must be additional mass in a more extended distribution, causing these objects to orbit so fast, and has led scientists to propose the existence of **dark matter.** This matter is proposed to exist in a large halo around each galaxy (with a radius up to 10 times as large as the visible galaxy's radius). Because it is not luminous (i.e., does not emit electromagnetic radiation) it must be either very cold or electrically neutral. Therefore, we cannot "see" dark matter, except through its gravitational effects.

The proposed existence of dark matter is also implied by earlier observations made on larger gravitationally bound structures known as galaxy clusters.[6] These observations show that the orbital speeds of galaxies in a cluster are, on average, too large to be explained by the luminous matter in the cluster alone. The speeds of the individual galaxies are so high, they suggest that there is 50 times as much dark matter in galaxy clusters as in the galaxies themselves!

Why doesn't dark matter affect the orbital speeds of planets like it does those of a galaxy? It seems that a solar system is too small a structure to contain enough dark matter to affect the behavior of orbital speeds. A galaxy or galaxy cluster, on the other hand, contains huge amounts of dark matter, resulting in the surprising behavior.

What, though, *is* dark matter? At this time, no one knows. One hypothesis claims that dark matter is based on a particle called a weakly interacting massive particle, or WIMP. If this theory is correct, calculations show that about 200 WIMPs pass through a human body at any given time. The new Large Hadron Collider in Europe (see Chapter 44) is the first particle accelerator with enough energy to possibly generate and detect the existence of WIMPs, which has generated much current interest in dark matter. Keeping an eye on this research in the future should be exciting, and the creativity of physicists in generating whimsical names for newly proposed objects should be entertaining.

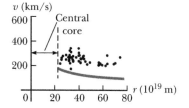

Figure 13.17 The orbital speed v of a galaxy object as a function of distance r from the center of the central core of the Andromeda galaxy. The theoretical curve is in red-brown, and the data points for the galaxy objects are in black. No data are provided on the left because the behavior inside the central core of the galaxy is more complicated.

[5]V. C. Rubin and W. K. Ford, "Rotation of the Andromeda Nebula from a Spectroscopic Survey of Emission Regions," *Astrophysical Journal* **159**: 379–403 (1970).

[6]F. Zwicky, "On the Masses of Nebulae and of Clusters of Nebulae," *Astrophysical Journal* **86**: 217–246 (1937).

Summary

▶ Definitions

The **gravitational field** at a point in space is defined as the gravitational force $\vec{\mathbf{F}}_g$ experienced by any test particle located at that point divided by the mass m_0 of the test particle:

$$\vec{\mathbf{g}} \equiv \frac{\vec{\mathbf{F}}_g}{m_0} \tag{13.7}$$

▶ Concepts and Principles

Newton's law of universal gravitation states that the gravitational force of attraction between any two particles of masses m_1 and m_2 separated by a distance r has the magnitude

$$F_g = G \frac{m_1 m_2}{r^2} \tag{13.1}$$

where $G = 6.674 \times 10^{-11}$ N · m^2/kg^2 is the **universal gravitational constant.** This equation enables us to calculate the force of attraction between masses under many circumstances.

An object at a distance h above the Earth's surface experiences a gravitational force of magnitude mg, where g is the free-fall acceleration at that elevation:

$$g = \frac{GM_E}{r^2} = \frac{GM_E}{(R_E + h)^2} \tag{13.6}$$

In this expression, M_E is the mass of the Earth and R_E is its radius. Therefore, the weight of an object decreases as the object moves away from the Earth's surface.

Kepler's laws of planetary motion state:

1. All planets move in elliptical orbits with the Sun at one focus.
2. The radius vector drawn from the Sun to a planet sweeps out equal areas in equal time intervals.
3. The square of the orbital period of any planet is proportional to the cube of the semimajor axis of the elliptical orbit.

Kepler's third law can be expressed as

$$T^2 = \left(\frac{4\pi^2}{GM_S} \right) a^3 \tag{13.11}$$

where M_S is the mass of the Sun and a is the semimajor axis. For a circular orbit, a can be replaced in Equation 13.11 by the radius r. Most planets have nearly circular orbits around the Sun.

The **gravitational potential energy** associated with a system of two particles of mass m_1 and m_2 separated by a distance r is

$$U_g(r) = -\frac{Gm_1 m_2}{r} \tag{13.15}$$

where U_g is taken to be zero as $r \rightarrow \infty$.

If an isolated system consists of an object of mass m moving with a speed v in the vicinity of a massive object of mass M, the total energy E of the system is the sum of the kinetic and potential energies:

$$E = \tfrac{1}{2}mv^2 - \frac{GMm}{r} \tag{13.16}$$

and the total energy of the system is a constant of the motion. If the object moves in an elliptical orbit of semimajor axis a around the massive object and $M \gg m$, the total energy of the system is

$$E = -\frac{GMm}{2a} \tag{13.20}$$

For a circular orbit, this same equation applies with $a = r$.

The **escape speed** for an object projected from the surface of a planet of mass M and radius R is

$$v_{\text{esc}} = \sqrt{\frac{2GM}{R}} \tag{13.23}$$

▶ Analysis Model for Problem Solving

Particle in a Field (Gravitational) A source particle with some mass establishes a **gravitational field** $\vec{\mathbf{g}}$ throughout space. When a particle of mass m is placed in that field, it experiences a gravitational force given by

$$\vec{\mathbf{F}}_g = m\vec{\mathbf{g}} \tag{5.5}$$

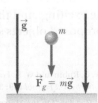

Think–Pair–Share

See the Preface for an explanation of the icons used in this problems set. For additional assessment items for this section, go to **WEBASSIGN** From Cengage

1. Kepler's third law should be obeyed for any group of objects orbiting a massive central object. Consider the five moons of Pluto: Charon, Styx, Nix, Kerberos, and Hydra. Data for these moons are given in the table below. If you evaluate the ratio T^2/a^3 for these moons, where T is the orbital period and a is the length of the semimajor axis, the results are all very close, *except* for Charon, which doesn't seem to follow Kepler's third law. Discuss with your group why the value of the ratio T^2/a^3 for Charon is different from those of the other moons.

Moon	Semimajor axis a (10^6 m)	Orbital period T (d)	Diameter (km)
Charon	17.54	6.387	1 208
Styx	42.66	20.16	~12
Nix	48.69	24.85	~40
Kerberos	57.78	32.17	~14
Hydra	64.74	38.20	~50

2. **ACTIVITY** Jupiter has over 60 moons, most of them discovered in the twenty-first century. The table shows astronomically measured data for the first fifteen moons, in order of semimajor axis of the orbit around Jupiter. (a) For these moons, show that Kepler's third law is satisfied within reasonable observational uncertainty for the data in the table. (b) Evaluate the ratio T^2/a^3, where T is the orbital period and a is the length of the semimajor axis, for these moons and compare the results to the value for the solar system (Table 13.2). Why is the value of this ratio larger for

the moons of Jupiter than for the solar system? (c) Before calculating the value of T^2/a^3 from the data, could you have *predicted* what it would be?

Moon	Semimajor axis a (10^9 m)	Orbital period T (d)	Eccentricity	Inclination Angle
Moons discovered by Galileo:				
Io	0.421 7	1.769 1	0.004 1	0.05
Europa	0.671 0	3.551 2	0.009 4	0.47
Ganymede	1.070 4	7.154 6	0.001 1	0.20
Callisto	1.882 7	16.689	0.007 4	0.20
Inner moons:				
Metis	0.127 7	0.294 8	0.000 02	0.06
Adrastea	0.128 7	0.298 3	0.001 5	0.03
Amalthea	0.181 4	0.498 2	0.003 2	0.37
Thebe	0.221 9	0.674 5	0.017 5	1.08
Outer moons:				
Themisto	7.393 2	129.87	0.215 5	45.8
Leda	11.187 8	240.82	0.167 3	27.6
Himalia	11.452 0	250.23	0.151 3	30.5
Lysithea	11.740 6	259.89	0.132 2	27.0
Elara	11.778 0	257.62	0.194 8	29.7
Dia	12.570 4	287.93	0.205 8	27.6
Carpo	17.144 9	458.62	0.273 5	56.0

Problems

See the Preface for an explanation of the icons used in this problems set. For additional assessment items for this section, go to **WEBASSIGN** From Cengage

SECTION 13.1 Newton's Law of Universal Gravitation

1. In introductory physics laboratories, a typical Cavendish balance for measuring the gravitational constant G uses lead spheres with masses of 1.50 kg and 15.0 g whose centers are separated by about 4.50 cm. Calculate the gravitational force between these spheres, treating each as a particle located at the sphere's center.

2. During a solar eclipse, the Moon, the Earth, and the Sun all lie on the same line, with the Moon between the Earth and the Sun. (a) What force is exerted by the Sun on the Moon? (b) What force is exerted by the Earth on the Moon? (c) What force is exerted by the Sun on the Earth? (d) Compare the answers to parts (a) and (b). Why doesn't the Sun capture the Moon away from the Earth?

3. Determine the order of magnitude of the gravitational force that you exert on another person 2 m away. In your solution, state the quantities you measure or estimate and their values.

4. *Why is the following situation impossible?* The centers of two homogeneous spheres are 1.00 m apart. The spheres are each made of the same element from the periodic table. The gravitational force between the spheres is 1.00 N.

SECTION 13.2 Free-Fall Acceleration and the Gravitational Force

5. **Review.** Miranda, a satellite of Uranus, is shown in Figure P13.5a. It can be modeled as a sphere of radius 242 km and mass 6.68×10^{19} kg. (a) Find the free-fall acceleration on its surface. (b) A cliff on Miranda is 5.00 km high. It appears on the limb at the 11 o'clock position in Figure P13.5a and is magnified in Figure P13.5b. If a devotee of extreme sports

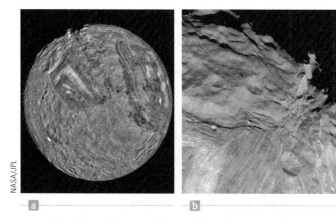

NASA/JPL

Figure P13.5

runs horizontally off the top of the cliff at 8.50 m/s, for what time interval is he in flight? (c) How far from the base of the vertical cliff does he strike the icy surface of Miranda? (d) What will be his vector impact velocity?

SECTION 13.3 Analysis Model: Particle in a Field (Gravitational)

6. (a) Compute the vector gravitational field at a point P on the perpendicular bisector of the line joining two objects of equal mass separated by a distance $2a$ as shown in Figure P13.6. (b) Explain physically why the field should approach zero as $r \to 0$. (c) Prove mathematically that the answer to part (a) behaves in this way. (d) Explain physically why the magnitude of the field should approach $2GM/r^2$ as $r \to \infty$. (e) Prove mathematically that the answer to part (a) behaves correctly in this limit.

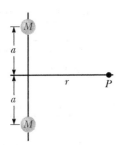

Figure P13.6

7. A spacecraft in the shape of a long cylinder has a length of 100 m, and its mass with occupants is 1 000 kg. It has strayed too close to a black hole having a mass 100 times that of the Sun (Fig. P13.7). The nose of the spacecraft points toward the black hole, and the distance between the nose and the center of the black hole is 10.0 km. (a) Determine the total force on the spacecraft. (b) What is the difference in the gravitational fields acting on the occupants in the nose of the ship and on those in the rear of the ship, farthest from the black hole? (This difference in accelerations grows rapidly as the ship approaches the black hole. It puts the body of the ship under extreme tension and eventually tears it apart.)

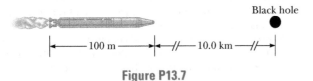

Black hole

|← 100 m →|←—//— 10.0 km —//→|

Figure P13.7

SECTION 13.4 Kepler's Laws and the Motion of Planets

8. An artificial satellite circles the Earth in a circular orbit at a location where the acceleration due to gravity is 9.00 m/s². Determine the orbital period of the satellite.

9. You are out on a date, eating dinner in a restaurant that has several television screens. Most of the screens are showing sports events, but one near you and your date is showing a discussion of an upcoming voyage to Mars. (a) Your date says, "I wonder how long it takes to get to Mars?" Wanting to impress your date, you grab a napkin and draw Figure P13.9 on it. Even more impressively, you tell your date that the minimum-energy transfer orbit from Earth to Mars is an elliptical trajectory with the departure planet corresponding to the perihelion of the ellipse and the arrival planet at

the aphelion. You pull out your smartphone, activate the calculator feature, and perform a calculation on another napkin to answer the question above that your date asked about the transfer time interval to Mars on this particular trajectory. (b) **What If?** Your date is impressed, but then asks you to determine the transit time to an *inner* planet, like Venus.

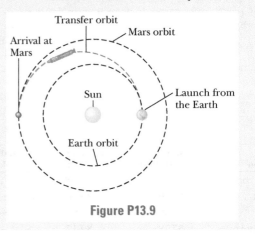

Figure P13.9

10. A particle of mass m moves along a straight line with constant velocity $\vec{v}_0$ in the x direction, a distance b from the x axis (Fig. P13.10). (a) Does the particle possess any angular momentum about the origin? (b) Explain why the amount of its angular momentum should change or should stay constant. (c) Show that Kepler's second law is satisfied by showing that the two shaded triangles in the figure have the same area when $t_{\text{Ⓓ}} - t_{\text{Ⓒ}} = t_{\text{Ⓑ}} - t_{\text{Ⓐ}}$.

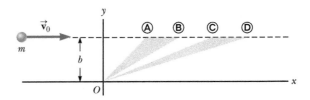

Figure P13.10

11. Use Kepler's third law to determine how many days it takes a spacecraft to travel in an elliptical orbit from a point 6 670 km from the Earth's center to the Moon, 385 000 km from the Earth's center.

12. The *Explorer VIII* satellite, placed into orbit November 3, 1960, to investigate the ionosphere, had the following orbit parameters: perigee, 459 km; apogee, 2 289 km (both distances above the Earth's surface); period, 112.7 min. Find the ratio v_p/v_a of the speed at perigee to that at apogee.

13. Suppose the Sun's gravity were switched off. The planets would leave their orbits and fly away in straight lines as described by Newton's first law. (a) Would Mercury ever be farther from the Sun than Pluto? (b) If so, find how long it would take Mercury to achieve this passage. If not, give a convincing argument that Pluto is always farther from the Sun than is Mercury.

14. (a) Given that the period of the Moon's orbit about the Earth is 27.32 days and the nearly constant distance between the center of the Earth and the center of the Moon is 3.84 × 10^8 m, use Equation 13.11 to calculate the mass of the Earth. (b) Why is the value you calculate a bit too large?

SECTION 13.5 Gravitational Potential Energy

Note: In Problems 15 through 23, assume $U_g = 0$ at $r = \infty$.

15. How much energy is required to move a 1 000-kg object from the Earth's surface to an altitude twice the Earth's radius?

16. An object is released from rest at an altitude h above the surface of the Earth. (a) Show that its speed at a distance r from the Earth's center, where $R_E \leq r \leq R_E + h$, is

$$v = \sqrt{2GM_E\left(\frac{1}{r} - \frac{1}{R_E + h}\right)}$$

(b) Assume the release altitude is 500 km. Perform the integral

$$\Delta t = \int_i^f dt = -\int_i^f \frac{dr}{v}$$

to find the time of fall as the object moves from the release point to the Earth's surface. The negative sign appears because the object is moving opposite to the radial direction, so its speed is $v = -dr/dt$. Perform the integral numerically.

17. A system consists of three particles, each of mass 5.00 g, located at the corners of an equilateral triangle with sides of 30.0 cm. (a) Calculate the gravitational potential energy of the system. (b) Assume the particles are released simultaneously. Describe the subsequent motion of each. Will any collisions take place? Explain.

SECTION 13.6 Energy Considerations in Planetary and Satellite Motion

18. A "treetop satellite" moves in a circular orbit just above the surface of a planet, assumed to offer no air resistance. Show that its orbital speed v and the escape speed from the planet are related by the expression $v_{\text{esc}} = \sqrt{2}\,v$.

19. A 500-kg satellite is in a circular orbit at an altitude of 500 km above the Earth's surface. Because of air friction, the satellite eventually falls to the Earth's surface, where it hits the ground with a speed of 2.00 km/s. How much energy was transformed into internal energy by means of air friction?

20. Derive an expression for the work required to move an Earth satellite of mass m from a circular orbit of radius $2R_E$ to one of radius $3R_E$.

21. An asteroid is on a collision course with Earth. An astronaut lands on the rock to bury explosive charges that will blow the asteroid apart. Most of the small fragments will miss the Earth, and those that fall into the atmosphere will produce only a beautiful meteor shower. The astronaut finds that the density of the spherical asteroid is equal to the average density of the Earth. To ensure its pulverization, she incorporates into the explosives the rocket fuel and oxidizer intended for her return journey. What maximum radius can the asteroid have for her to be able to leave it entirely simply by jumping straight up? On Earth she can jump to a height of 0.500 m.

22. (a) What is the minimum speed, relative to the Sun, necessary for a spacecraft to escape the solar system if it starts at the Earth's orbit? (b) *Voyager 1* achieved a maximum speed of 125 000 km/h on its way to photograph Jupiter. Beyond what distance from the Sun is this speed sufficient to escape the solar system?

23. Ganymede is the largest of Jupiter's moons. Consider a rocket on the surface of Ganymede, at the point farthest from the planet (Fig. P13.23). Model the rocket as a particle. (a) Does the presence of Ganymede make Jupiter exert a larger, smaller, or same size force on the rocket compared with the force it would exert if Ganymede were not interposed? (b) Determine the escape speed for the rocket from the planet–satellite system. The radius of Ganymede is 2.64×10^6 m, and its mass is 1.495×10^{23} kg. The distance between Jupiter and Ganymede is 1.071×10^9 m, and the mass of Jupiter is 1.90×10^{27} kg. Ignore the motion of Jupiter and Ganymede as they revolve about their center of mass.

Ganymede

Jupiter

Figure P13.23

ADDITIONAL PROBLEMS

24. A rocket is fired straight up through the atmosphere from the South Pole, burning out at an altitude of 250 km when traveling at 6.00 km/s. (a) What maximum distance from the Earth's surface does it travel before falling back to the Earth? (b) Would its maximum distance from the surface be larger if the same rocket were fired with the same fuel load from a launch site on the equator? Why or why not?

25. *Voyager 1* and *Voyager 2* surveyed the surface of Jupiter's moon Io and photographed active volcanoes spewing liquid sulfur to heights of 70 km above the surface of this moon. Find the speed with which the liquid sulfur left the volcano. Io's mass is 8.9×10^{22} kg, and its radius is 1 820 km.

26. After reading Sections 13.2 and 13.3, your classmate looks online to gather information about neutron stars. He finds that a typical radius of a neutron star is 10 km, a typical mass is two solar masses, and that a typical rotation period is as short as 1.4 ms. He suggests that if a spherical neutron star were spinning that fast, it seems that material at the equator of the sphere would be flung away because the gravity of the star could not supply the needed centripetal acceleration of the material. Prepare an argument that shows that the gravity at the surface of a neutron star is more than sufficient to provide the centripetal acceleration. (*Note:* Neutron stars typically have the same mass as our Sun.)

27. You are on a space station, in a circular orbit $h = 500$ km above the surface of the Earth. You complete your tasks several days early and must wait for the next mission from the surface to bring you home. After days of boredom, you decide to play some golf. Walking on the space station surface with magnetic shoes, you tee up a golf ball. You hit it with all of your might, sending it off with speed v_{rel}, relative to the space station, in a direction parallel to the velocity vector of the space station at the moment the ball is hit. You notice that you then orbit the Earth exactly $n = 2.00$ times and you reach up and catch the golf ball as it returns to the space station. With what speed v_{rel} was the golf ball hit? *Note:* Your result will be unrealistically high—much higher than it is possible for a human to hit a golf ball.

28. *Why is the following situation impossible?* A spacecraft is launched into a circular orbit around the Earth and circles the Earth once an hour.

29. Let Δg_M represent the difference in the gravitational fields produced by the Moon at the points on the Earth's surface nearest to and farthest from the Moon. Find the fraction $\Delta g_M/g$, where g is the Earth's gravitational field. (This difference is responsible for the occurrence of the *lunar tides* on the Earth.)

30. **QIC** A sleeping area for a long space voyage consists of two cabins each connected by a cable to a central hub as shown in Figure P13.30. The cabins are set spinning around the hub axis, which is connected to the rest of the spacecraft to generate artificial gravity in the cabins. A space traveler lies in a bed parallel to the outer wall as shown in Figure P13.30. (a) With $r = 10.0$ m, what would the angular speed of the 60.0-kg traveler need to be if he is to experience half his normal Earth weight? (b) If the astronaut stands up perpendicular to the bed, without holding on to anything with his hands, will his head be moving at a faster, a slower, or the same tangential speed as his feet? Why? (c) Why is the action in part (b) dangerous?

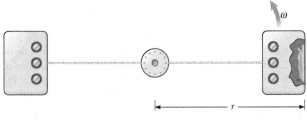

Figure P13.30

31. (a) A space vehicle is launched vertically upward from the Earth's surface with an initial speed of 8.76 km/s, which is less than the escape speed of 11.2 km/s. What maximum height does it attain? (b) A meteoroid falls toward the Earth. It is essentially at rest with respect to the Earth when it is at a height of 2.51×10^7 m above the Earth's surface. With what speed does the meteorite (a meteoroid that survives to impact the Earth's surface) strike the Earth?

32. **S** (a) A space vehicle is launched vertically upward from the Earth's surface with an initial speed of v_i that is comparable to but less than the escape speed v_{esc}. What maximum height does it attain? (b) A meteoroid falls toward the Earth. It is essentially at rest with respect to the Earth when it is at a height h above the Earth's surface. With what speed does the meteorite (a meteoroid that survives to impact the Earth's surface) strike the Earth? (c) **What If?** Assume a baseball is tossed up with an initial speed that is very small compared to the escape speed. Show that the result from part (a) is consistent with Equation 4.19.

33. **QIC** Assume you are agile enough to run across a horizontal surface at 8.50 m/s, independently of the value of the gravitational field. What would be (a) the radius and (b) the mass of an airless spherical asteroid of uniform density 1.10×10^3 kg/m^3 on which you could launch yourself into orbit by running? (c) What would be your period? (d) Would your running significantly affect the rotation of the asteroid? Explain.

34. **GP** **S** Two spheres having masses M and $2M$ and radii R and $3R$, respectively, are simultaneously released from rest when the distance between their centers is $12R$. Assume the two spheres interact only with each other and we wish to find the speeds with which they collide. (a) What *two* isolated system models are appropriate for this system? (b) Write an equation from one of the models and solve it for $\vec{v}_1$, the velocity of the sphere of mass M at any time after release in terms of $\vec{v}_2$, the velocity of $2M$. (c) Write an equation from the other model and solve it for speed v_1 in terms of speed v_2 when the spheres collide. (d) Combine the two equations to find the two speeds v_1 and v_2 when the spheres collide.

35. (a) Show that the rate of change of the free-fall acceleration with vertical position near the Earth's surface is

$$\frac{dg}{dr} = -\frac{2GM_E}{R_E^3}$$

This rate of change with position is called a *gradient*. (b) Assuming h is small in comparison to the radius of the Earth, show that the difference in free-fall acceleration between two points separated by vertical distance h is

$$|\Delta g| = \frac{2GM_E h}{R_E^3}$$

(c) Evaluate this difference for $h = 6.00$ m, a typical height for a two-story building.

36. **S** A certain quaternary star system consists of three stars, each of mass m, moving in the same circular orbit of radius r about a central star of mass M. The stars orbit in the same sense and are positioned one-third of a revolution apart from one another. Show that the period of each of the three stars is given by

$$T = 2\pi \sqrt{\frac{r^3}{G(M + m/\sqrt{3})}}$$

37. Studies of the relationship of the Sun to our galaxy—the Milky Way—have revealed that the Sun is located near the outer edge of the galactic disc, about 30 000 ly (1 ly $= 9.46 \times 10^{15}$ m) from the center. The Sun has an orbital speed of approximately 250 km/s around the galactic center. (a) What is the period of the Sun's galactic motion? (b) What is the order of magnitude of the mass of the Milky Way galaxy? (c) Suppose the galaxy is made mostly of stars of which the Sun is typical. What is the order of magnitude of the number of stars in the Milky Way?

38. **Review.** **S** Two identical hard spheres, each of mass m and radius r, are released from rest in otherwise empty space with their centers separated by the distance R. They are allowed to collide under the influence of their gravitational attraction. (a) Show that the magnitude of the impulse received by each sphere before they make contact is given by $[Gm^3(1/2r - 1/R)]^{1/2}$. (b) **What If?** Find the magnitude of the impulse each receives during their contact if they collide elastically.

39. The maximum distance from the Earth to the Sun (at aphelion) is 1.521×10^{11} m, and the distance of closest approach (at perihelion) is 1.471×10^{11} m. The Earth's orbital speed at perihelion is 3.027×10^4 m/s. Determine (a) the Earth's orbital speed at aphelion and the kinetic and potential energies of the Earth–Sun system (b) at

perihelion and (c) at aphelion. (d) Is the total energy of the system constant? Explain. Ignore the effect of the Moon and other planets.

40. Many people assume air resistance acting on a moving
Q|C object will always make the object slow down. It can, however, actually be responsible for making the object speed up. Consider a 100-kg Earth satellite in a circular orbit at an altitude of 200 km. A small force of air resistance makes the satellite drop into a circular orbit with an altitude of 100 km. (a) Calculate the satellite's initial speed. (b) Calculate its final speed in this process. (c) Calculate the initial energy of the satellite–Earth system. (d) Calculate the final energy of the system. (e) Show that the mechanical energy of the system has decreased and find the amount of the decrease due to friction. (f) What force makes the satellite's speed increase? *Hint:* You will find a free-body diagram useful in explaining your answer.

41. Consider an object of mass m, not necessarily small compared with the mass of the Earth, released at a distance of 1.20×10^7 m from the center of the Earth. Assume the Earth and the object behave as a pair of particles, isolated from the rest of the Universe. (a) Find the magnitude of the acceleration a_{rel} with which each starts to move relative to the other as a function of m. Evaluate the acceleration (b) for $m = 5.00$ kg, (c) for $m = 2\,000$ kg, and (d) for $m = 2.00 \times 10^{24}$ kg. (e) Describe the pattern of variation of a_{rel} with m.

42. Show that the minimum period for a satellite in orbit
S around a spherical planet of uniform density ρ is

$$T_{min} = \sqrt{\frac{3\pi}{G\rho}}$$

independent of the planet's radius.

43. As thermonuclear fusion proceeds in its core, the Sun loses mass at a rate of 3.64×10^9 kg/s. During the 5 000-yr period of recorded history, by how much has the length of the year changed due to the loss of mass from the Sun? *Suggestions:* Assume the Earth's orbit is circular. No external torque acts on the Earth–Sun system, so the angular momentum of the Earth is constant.

44. Two stars of masses M and
S m, separated by a distance d, revolve in circular orbits about their center of mass (Fig. P13.44). Show that each star has a period given by

$$T^2 = \frac{4\pi^2 d^3}{G(M + m)}$$

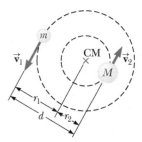

Figure P13.44

CHALLENGE PROBLEM

45. The Solar and Heliospheric Observatory (SOHO) spacecraft has a special orbit, located between the Earth and the Sun along the line joining them, and it is always close enough to the Earth to transmit data easily. Both objects exert gravitational forces on the observatory. It moves around the Sun in a near-circular orbit that is smaller than the Earth's circular orbit. Its period, however, is not less than 1 yr but just equal to 1 yr. Show that its distance from the Earth must be 1.48×10^9 m. In 1772, Joseph Louis Lagrange determined theoretically the special location allowing this orbit. *Suggestions:* Use data that are precise to four digits. The mass of the Earth is 5.974×10^{24} kg. You will not be able to easily solve the equation you generate; instead, use a computer to verify that 1.48×10^9 m is the correct value.

Fluid Mechanics

An airplane takes off from an airport runway. How long does the runway have to be?
(*F. JIMENEZ MECA/Shutterstock*)

STORYLINE **It is an academic holiday and you are spending some** time with your grandparents in Denver, Colorado. After visiting with them, your trip continues as you board an airplane to fly from Denver to see your other grandparents in Boston, Massachusetts. As the plane accelerates at the Denver airport, you notice that it is taking quite a while, compared to your previous flying experience, for the plane to leave the ground. You begin to worry that the plane will run out of runway length before it takes off. Finally, the plane lifts off and you breathe a sigh of relief. You think, "Why did it take so long for the plane to lift off? It didn't take that long when I took off for my flight from Los Angeles." Deciding that it would be worth it to pay for the Wi-Fi service on the plane, you look up runway lengths online. The longest runway at the airport at Los Angeles, at sea level, is 12 091 ft. The longest runway at Denver is 16 000 ft. The longest runway in the world is at Qamdo Bamda Airport in China: 18 045 ft. That airport is also at the second highest altitude in the world for an airport: 14 219 ft. (It was the highest until 2013.) Is there a relationship between airport altitude and runway length? Why?

CONNECTIONS In the previous chapters, we have considered the mechanics of particles, systems, and rigid objects. Forces on these particles and objects have been applied by hands, strings, inclined planes, gravity, etc. In this chapter, we consider the forces acting between an object and a *fluid*. A **fluid** is a collection of molecules that are randomly arranged and held together by weak cohesive forces between molecules and also by forces exerted by the container holding the fluid. Both liquids and gases are fluids. We discussed such a situation briefly in Section 6.4, when we considered the resistive forces on objects *moving* through fluids. Here we will discuss forces that fluids exert on objects that are *at rest* relative to the fluid. This discussion will lead to an important new quantity, *pressure,* and a force called the *buoyant force,* which is not a new type of force, but our familiar forces acting in a specific situation. We will also investigate the physics of moving fluids in the later sections of this chapter. Understanding the concepts of moving fluids is important for a wide range of applications, from plumbing systems to automobile aerodynamics to blood flow in veins and arteries.

14.1 Pressure

Fluids do not sustain shearing stresses or tensile stresses such as those discussed in Chapter 12; therefore, the only stress that can be exerted on an object submerged in a static fluid is one that tends to compress the object from all sides. In other words, the force exerted by a static fluid on an object is always perpendicular to the surfaces of the object as shown in Figure 14.1. We discussed this situation in Section 12.4.

The pressure in a fluid can be measured with the device pictured in Figure 14.2. The device consists of an evacuated cylinder that encloses a light piston connected to a spring. As the device is submerged in a fluid, the fluid presses on the piston and compresses the spring until the inward force exerted by the fluid is balanced by the outward force exerted by the spring. The fluid pressure can be measured directly if the spring is calibrated in advance. If F is the magnitude of the force exerted on the piston and A is the surface area of the piston, the **pressure** P of the fluid at the level to which the device has been submerged is defined as the ratio of the force exerted on the piston to its area:

$$P \equiv \frac{F}{A} \tag{14.1}$$

Pressure is a scalar quantity because it is proportional to the magnitude of the force on the piston.

If the pressure varies over an area, the infinitesimal force dF on an infinitesimal surface element of area dA is

$$dF = P \, dA \tag{14.2}$$

where P is the pressure at the location of the area dA. To calculate the total force exerted on a surface of a container, we must integrate Equation 14.2 over the surface.

The units of pressure are newtons per square meter (N/m^2) in the SI system. Another name for the SI unit of pressure is the **pascal** (Pa):

$$1 \text{ Pa} \equiv 1 \text{ N/m}^2 \tag{14.3}$$

For a tactile demonstration of the definition of pressure, hold a tack between your thumb and forefinger, with the point of the tack on your thumb and the head of the tack on your forefinger. Now *gently* press your thumb and forefinger together. Your thumb will begin to feel pain immediately while your forefinger will not. The tack is exerting the same force on both your thumb and forefinger, but the pressure on your thumb is much larger because of the small area over which the force is applied.

QUICK QUIZ 14.1 Suppose you are standing directly behind someone who steps back and accidentally stomps on your foot with the heel of one shoe. Would you be better off if that person were **(a)** a large, male professional basketball player wearing sneakers or **(b)** a petite woman wearing spike-heeled shoes?

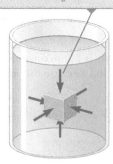

At any point on the surface of the object, the force exerted by the fluid is perpendicular to the surface of the object.

Figure 14.1 The forces exerted by a fluid on the surfaces of a submerged object.

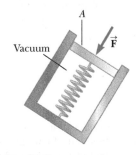

Figure 14.2 A simple device for measuring the pressure exerted by a fluid.

PITFALL PREVENTION 14.1
Force and Pressure Equations 14.1 and 14.2 make a clear distinction between force and pressure. Another important distinction is that *force is a vector* and *pressure is a scalar.* There is no direction associated with pressure, but the direction of the force associated with the pressure is perpendicular to the surface on which the pressure acts.

Example **14.1** The Water Bed

The mattress of a water bed is 2.00 m long by 2.00 m wide and 30.0 cm deep.

(A) Find the weight of the water in the mattress.

SOLUTION

Conceptualize Think about carrying a jug of water and how heavy it is. Now imagine a sample of water the size of a water bed. We expect the weight to be relatively large.

Categorize This example is a substitution problem.

Find the volume of the water filling the mattress:
$$V = \ell w h$$

Use Equation 1.1 and the density of fresh water (see Table 14.1) to find the weight of the water bed:
$$Mg = (\rho V)g = \rho g \ell w h$$

Substitute numerical values:
$$Mg = (1\,000 \text{ kg/m}^3)(9.80 \text{ m/s}^2)(2.00 \text{ m})(2.00 \text{ m})(0.300 \text{ m})$$
$$= 1.18 \times 10^4 \text{ N}$$

which is approximately 2 650 lb. (A regular bed, including mattress, box spring, and metal frame, weighs approximately 300 lb.) Because this load is so great, it is best to place a water bed in the basement or on a sturdy, well-supported floor.

(B) Find the pressure exerted by the water bed on the floor when the bed rests in its normal position. Assume the entire lower surface of the bed makes contact with the floor.

SOLUTION

When the water bed is in its normal position, the area in contact with the floor is $A = \ell w$. Use Equation 14.1 to find the pressure:
$$P = \frac{Mg}{\ell w} = \frac{1.18 \times 10^4 \text{ N}}{(2.00 \text{ m})(2.00 \text{ m})} = 2.94 \times 10^3 \text{ Pa}$$

WHAT IF? What if the water bed is replaced by a 300-lb regular bed that is supported by four legs? Each leg has a circular cross section of radius 2.00 cm. What pressure does this bed exert on the floor?

Answer The weight of the regular bed is distributed over four circular cross sections at the bottom of the legs. Therefore, the pressure is

$$P = \frac{F}{A} = \frac{mg}{4(\pi r^2)} = \frac{300 \text{ lb}}{4\pi(0.020\,0 \text{ m})^2}\left(\frac{1 \text{ N}}{0.225 \text{ lb}}\right)$$
$$= 2.65 \times 10^5 \text{ Pa}$$

This result is almost 100 times larger than the pressure due to the water bed! The weight of the regular bed, even though it is much less than the weight of the water bed, is applied over the very small area of the four legs. The high pressure on the floor at the feet of a regular bed could cause dents in wood floors or permanently crush carpet pile.

14.2 Variation of Pressure with Depth

As divers well know, water pressure increases with depth, resulting in a feeling of discomfort in the ears of the diver. Likewise, atmospheric pressure decreases with increasing altitude; for this reason, aircraft flying at high altitudes must have pressurized cabins for the comfort of the passengers.

We now show details of how the pressure in a liquid increases with depth. As Equation 1.1 describes, the *density* of a substance is defined as its mass per unit volume; Table 14.1 lists the densities of various substances. These values vary slightly with temperature because the volume of a substance is dependent on temperature (as shown in Chapter 18). Under standard conditions (at 0°C and at atmospheric pressure), the densities of gases are about $\frac{1}{1\,000}$ the densities of solids

TABLE 14.1 Densities of Some Common Substances at Standard Temperature (0°C) and Pressure (Atmospheric)

Substance	ρ (kg/m³)	Substance	ρ (kg/m³)
Air	1.29	Iron	7.86×10^3
Air (at 20°C and		Lead	11.3×10^3
atmospheric pressure)	1.20	Mercury	13.6×10^3
Aluminum	2.70×10^3	Nitrogen gas	1.25
Benzene	0.879×10^3	Oak	0.710×10^3
Brass	8.4×10^3	Osmium	22.6×10^3
Copper	8.92×10^3	Oxygen gas	1.43
Ethyl alcohol	0.806×10^3	Pine	0.373×10^3
Fresh water	1.00×10^3	Platinum	21.4×10^3
Glycerin	1.26×10^3	Seawater	1.03×10^3
Gold	19.3×10^3	Silver	10.5×10^3
Helium gas	1.79×10^{-1}	Tin	7.30×10^3
Hydrogen gas	8.99×10^{-2}	Uranium	19.1×10^3
Ice	0.917×10^3		

and liquids. This difference in densities implies that the average molecular spacing in a gas under these conditions is about ten times greater than that in a solid or liquid.

Now consider a liquid of density ρ at rest as shown in Figure 14.3. We assume ρ is uniform throughout the liquid, which means the liquid is *incompressible*. Let us select a parcel of the liquid contained within an imaginary block of cross-sectional area A extending from depth d to depth $d + h$. The liquid external to our parcel exerts forces at all points on the surface of the parcel, perpendicular to the surface. The pressure exerted by the liquid on the bottom face of the parcel is P, and the pressure on the top face is P_0. Therefore, the upward force exerted by the outside fluid on the bottom of the parcel has a magnitude PA, and the downward force exerted on the top has a magnitude $P_0 A$. The mass of liquid in the parcel is $M = \rho V = \rho A h$; therefore, the weight of the liquid in the parcel is $Mg = \rho A h g$. Because the parcel is at rest and remains at rest, it can be modeled as a particle in equilibrium, so that the net force acting on it must be zero. Choosing upward to be the positive y direction, we see that

$$\sum \vec{\mathbf{F}} = PA\hat{\mathbf{j}} - P_0 A\hat{\mathbf{j}} - Mg\hat{\mathbf{j}} = 0$$

or

$$PA - P_0 A - \rho A h g = 0$$

$$P = P_0 + \rho g h \qquad (14.4)$$

That is, the pressure P at a depth h below a point in the liquid at which the pressure is P_0 is greater by an amount $\rho g h$. If the liquid is open to the atmosphere and P_0 is the pressure at the surface of the liquid, then P_0 **is atmospheric pressure.** In our calculations and working of end-of-chapter problems, we usually take atmospheric pressure to be

$$P_0 = 1.00 \text{ atm} = 1.013 \times 10^5 \text{ Pa}$$

Equation 14.4 implies that the pressure is the same at all points having the same depth, independent of the shape of the container.

Because the pressure in a fluid depends on depth and on the value of P_0, any increase in pressure at the surface must be transmitted to every other point in the fluid. This concept was first recognized by French scientist Blaise Pascal (1623–1662) and is called **Pascal's law: a change in the pressure applied to a fluid is transmitted undiminished to every point of the fluid and to the walls of the container.**

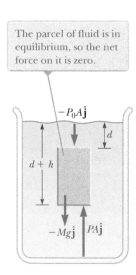

The parcel of fluid is in equilibrium, so the net force on it is zero.

Figure 14.3 A parcel of fluid in a larger volume of fluid is singled out.

◀ Variation of pressure with depth

◀ Pascal's law

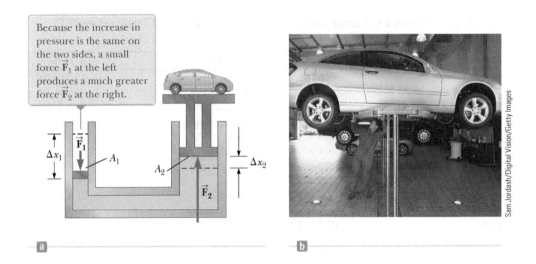

Figure 14.4 (a) Diagram of a hydraulic lift. (b) A vehicle undergoing repair is supported by a hydraulic lift in a garage.

Because the increase in pressure is the same on the two sides, a small force $\vec{\mathbf{F}}_1$ at the left produces a much greater force $\vec{\mathbf{F}}_2$ at the right.

a

b

Sam Jordash/Digital Vision/Getty Images

An important application of Pascal's law is the hydraulic lift illustrated in Figure 14.4a. A force of magnitude F_1 is applied to a small piston of surface area A_1. The pressure is transmitted through an incompressible liquid to a larger piston of surface area A_2. Because the pressure must be the same on both sides, $P = F_1/A_1 = F_2/A_2$. Therefore, the force F_2 is greater than the force F_1 by a factor of A_2/A_1. By designing a hydraulic lift with appropriate areas A_1 and A_2, a large output force can be applied by means of a small input force. Hydraulic brakes, car lifts, hydraulic jacks, and forklifts all make use of this principle (Fig. 14.4b).

Because liquid is neither added to nor removed from the system, the volume of liquid pushed down on the left in Figure 14.4a as the piston moves downward through a displacement Δx_1 equals the volume of liquid pushed up on the right as the right piston moves upward through a displacement Δx_2. That is, $A_1 \Delta x_1 = A_2 \Delta x_2$; therefore, $A_2/A_1 = \Delta x_1/\Delta x_2$. We have already shown that $A_2/A_1 = F_2/F_1$. Therefore, $F_2/F_1 = \Delta x_1/\Delta x_2$, so $F_1 \Delta x_1 = F_2 \Delta x_2$. Each side of this equation is the work done by the force on its respective piston. Therefore, the work done by $\vec{\mathbf{F}}_1$ on the input piston equals the work done by $\vec{\mathbf{F}}_2$ on the output piston, as it must to conserve energy. (The process can be modeled as a special case of the nonisolated system model: the *nonisolated system in steady state*. There is energy transfer into and out of the system, but these energy transfers balance, so that there is no net change in the energy of the system.) One could also consider the equation as indicating that you "trade" force for distance. Imagine jacking up a car. You can lift a heavy car with a relatively small force on the jack handle from your hand, but you have to move your hand through a very large total distance when you add up all the times you must move the end of the handle up and down!

QUICK QUIZ 14.2 The pressure at the bottom of a filled glass of water ($\rho = 1\,000$ kg/m^3) is P. The water is poured out, and the glass is filled with ethyl alcohol ($\rho = 806$ kg/m^3). What is the pressure at the bottom of the glass? **(a)** smaller than P **(b)** equal to P **(c)** larger than P **(d)** indeterminate

---

Example 14.2 **The Car Lift**

In a car lift used in a service station, compressed air exerts a force on a small piston that has a circular cross section of radius 5.00 cm. This pressure is transmitted by a liquid to a piston that has a radius of 15.0 cm.

(A) What force must the compressed air exert to lift a car weighing 13 300 N?

SOLUTION

Conceptualize Review the material just discussed about Pascal's law to understand the operation of a car lift.

14.2 continued

Categorize This example is a substitution problem.

Solve $F_1/A_1 = F_2/A_2$ for F_1:

$$F_1 = \left(\frac{A_1}{A_2}\right)F_2 = \frac{\pi(5.00 \times 10^{-2} \text{ m})^2}{\pi(15.0 \times 10^{-2} \text{ m})^2}(1.33 \times 10^4 \text{ N})$$

$$= 1.48 \times 10^3 \text{ N}$$

(B) What air pressure produces this force?

SOLUTION

Use Equation 14.1 to find the air pressure that produces this force:

$$P = \frac{F_1}{A_1} = \frac{1.48 \times 10^3 \text{ N}}{\pi(5.00 \times 10^{-2} \text{ m})^2}$$

$$= 1.88 \times 10^5 \text{ Pa}$$

This pressure is approximately twice atmospheric pressure.

Example **14.3** A Pain in Your Ear

Estimate the force exerted on your eardrum due to the water when you are swimming at the bottom of a pool that is 5.0 m deep.

SOLUTION

Conceptualize As you descend in the water, the pressure increases. You may have noticed this increased pressure in your ears while diving in a swimming pool, a lake, or the ocean. We can find the pressure difference exerted on the eardrum from the depth given in the problem; then, after estimating the ear drum's surface area, we can determine the net force the water exerts on it.

Categorize This example is a substitution problem.

The air inside the middle ear is normally at atmospheric pressure P_0. Therefore, to find the net force on the eardrum, we must consider the difference between the total pressure P_{bot} at the bottom of the pool and atmospheric pressure. Let's estimate the surface area of the eardrum to be approximately 1 cm² = 1 × 10⁻⁴ m².

Use Equation 14.4 to find this pressure difference:

$$P_{bot} - P_0 = \rho gh$$
$$= (1.00 \times 10^3 \text{ kg/m}^3)(9.80 \text{ m/s}^2)(5.0 \text{ m}) = 4.9 \times 10^4 \text{ Pa}$$

Use Equation 14.1 to find the magnitude of the net force on the ear:

$$F = (P_{bot} - P_0)A = (4.9 \times 10^4 \text{ Pa})(1 \times 10^{-4} \text{ m}^2) \approx 5 \text{ N}$$

Because a force of this magnitude on the eardrum is extremely uncomfortable, swimmers often "pop their ears" while under water, an action that pushes air from the lungs into the middle ear. Using this technique equalizes the pressure on the two sides of the eardrum and relieves the discomfort.

Example **14.4** The Force on a Dam

Water is filled to a height H behind a dam of width w (Fig. 14.5, page 364). Determine the resultant force exerted by the water on the dam.

SOLUTION

Conceptualize Because pressure varies with depth, we cannot calculate the force simply by multiplying the area by the pressure. As the pressure in the water increases with depth, the force on the adjacent portion of the dam also increases.

Categorize Because of the variation of pressure with depth, we must use integration to solve this example, so we categorize it as an analysis problem.

continued

14.4 continued

Analyze Let's imagine a vertical y axis, with $y = 0$ at the bottom of the dam. We divide the face of the dam into narrow horizontal strips at a distance y above the bottom, such as the red strip in Figure 14.5. The pressure on each such strip is due only to the water; atmospheric pressure acts on both sides of the dam.

Use Equation 14.4 to calculate the pressure due to the water at the depth h:

$$P = \rho g h = \rho g (H - y)$$

Use Equation 14.2 to find the force exerted on the shaded strip of area $dA = w\, dy$:

$$dF = P\, dA = \rho g (H - y) w\, dy$$

Integrate to find the total force on the dam:

$$F = \int P\, dA = \int_0^H \rho g (H - y) w\, dy = \tfrac{1}{2}\rho g w H^2$$

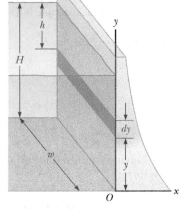

Figure 14.5 (Example 14.4) Water exerts a force on a dam.

Finalize Notice that the thickness of the dam shown in Figure 14.5 increases with depth. This design accounts for the greater force the water exerts on the dam at greater depths.

WHAT IF? What if you were asked to find this force without using calculus? How could you determine its value?

Answer We know from Equation 14.4 that pressure varies linearly with depth. Therefore, the average pressure due to the water over the face of the dam is the average of the pressure at the top and the pressure at the bottom:

$$P_{\text{avg}} = \frac{P_{\text{top}} + P_{\text{bottom}}}{2} = \frac{0 + \rho g H}{2} = \tfrac{1}{2}\rho g H$$

The total force on the dam is equal to the product of the average pressure and the area of the face of the dam:

$$F = P_{\text{avg}} A = \left(\tfrac{1}{2}\rho g H\right)(Hw) = \tfrac{1}{2}\rho g w H^2$$

which is the same result we obtained using calculus.

14.3 Pressure Measurements

During the weather report on a television news program, the *barometric pressure* is often provided. This reading is the current local pressure of the atmosphere, which varies over a small range from the standard value for P_0 provided earlier. How is this pressure measured?

One instrument used to measure atmospheric pressure is the common barometer, invented by Evangelista Torricelli (1608–1647). A long tube closed at one end is filled with mercury and then inverted into a container of mercury (Fig. 14.6a). The closed end of the tube is nearly a vacuum, so the pressure at the top of the mercury column can be taken as zero. In Figure 14.6a, the pressure at point A, due to the column of mercury, must equal the pressure at point B, due to the atmosphere. If that were not the case, there would be a net force that would move mercury from one point to the other until equilibrium is established. Therefore, $P_0 = \rho_{\text{Hg}} g h$, where ρ_{Hg} is the density of the mercury and h is the height of the mercury column. As atmospheric pressure varies, the height of the mercury column varies, so the height can be calibrated to measure atmospheric pressure. Let us determine the height of a mercury column for one atmosphere of pressure, $P_0 = 1\ \text{atm} = 1.013 \times 10^5\ \text{Pa}$:

$$P_0 = \rho_{\text{Hg}} g h \rightarrow h = \frac{P_0}{\rho_{\text{Hg}} g} = \frac{1.013 \times 10^5\ \text{Pa}}{(13.6 \times 10^3\ \text{kg/m}^3)(9.80\ \text{m/s}^2)} = 0.760\ \text{m}$$

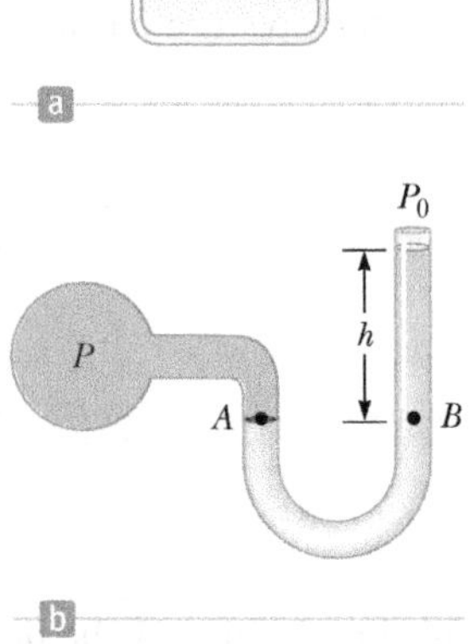

Figure 14.6 Two devices for measuring pressure: (a) a mercury barometer and (b) an open-tube manometer.

Based on such a calculation, one atmosphere of pressure is defined to be the pressure equivalent of a column of mercury that is exactly 0.760 0 m in height at 0°C.

A device for measuring the pressure of a gas contained in a vessel is the open-tube manometer illustrated in Figure 14.6b. One end of a U-shaped tube containing a liquid is open to the atmosphere, and the other end is connected to a container of gas at pressure P. In an equilibrium situation, the pressures at points A and B must be the same (otherwise, the curved portion of the liquid between points A and B would experience a net force and would accelerate), and the pressure at A is the unknown pressure of the gas. Therefore, equating the unknown pressure P to the pressure at point B, we see that $P = P_0 + \rho g h$. Again, we can calibrate the height h to the pressure P.

The difference in the pressures in each part of Figure 14.6 (that is, $P - P_0$) is equal to $\rho g h$. The pressure P is called the **absolute pressure,** and the difference $P - P_0$ is called the **gauge pressure.** For example, the pressure you measure in your bicycle tire is gauge pressure, the difference between the absolute pressure of the air inside the tire and the atmospheric pressure outside the tire.

QUICK QUIZ 14.3 Several common barometers are built, with a variety of fluids. For which of the following fluids will the column of fluid in the barometer be the highest? **(a)** mercury **(b)** water **(c)** ethyl alcohol **(d)** benzene

14.4 Buoyant Forces and Archimedes's Principle

Have you ever tried to push a beach ball down under water (Fig. 14.7a)? It is extremely difficult to do because of the large upward force exerted by the water on the ball. The upward force exerted by a fluid on any immersed object is called a **buoyant force.** The buoyant force is what allows huge ships made of steel to float on the surface of the ocean. We can determine the magnitude of a buoyant force by applying some logic. Imagine a beach ball–sized parcel of water beneath the water surface as in Figure 14.7b. Because this parcel can be modeled as a particle in equilibrium, there must be an upward force that balances the downward gravitational force on the parcel. This upward force is the buoyant force, and *its magnitude is equal to the weight of the water in the parcel.* The buoyant force is the resultant force on the parcel due to all forces applied on the parcel by the fluid surrounding the parcel.

Now imagine replacing the beach ball–sized parcel of water with an actual beach ball of the same size. The net force applied to the spherical volume indicated by the dashed line in Figure 14.7b is due to the surrounding fluid and is the same, regardless of whether it is applied to a beach ball or to a parcel of water. Consequently, **the magnitude of the buoyant force on an object always equals the weight of the fluid displaced by the object.** This statement is known as **Archimedes's principle.**

With the beach ball under water, the buoyant force, equal to the weight of a beach ball–sized parcel of water, is much larger than the weight of the beach ball.

Archimedes
Greek Mathematician, Physicist, and Engineer (c. 287–212 BC)
Archimedes was perhaps the greatest scientist of antiquity. He was the first to compute accurately the ratio of a circle's circumference to its diameter, and he also showed how to calculate the volume and surface area of spheres, cylinders, and other geometric shapes. He is well known for discovering the nature of the buoyant force and was also a gifted inventor. One of his practical inventions, still in use today, is Archimedes's screw, an inclined, rotating, coiled tube used originally to lift water from the holds of ships. He also invented the catapult and devised systems of levers, pulleys, and weights for raising heavy loads. Such inventions were successfully used to defend his native city, Syracuse, during a two-year siege by Romans.

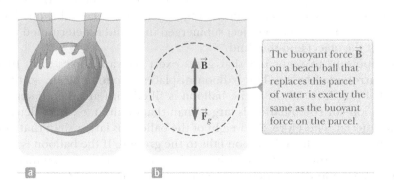

The buoyant force $\vec{\mathbf{B}}$ on a beach ball that replaces this parcel of water is exactly the same as the buoyant force on the parcel.

Figure 14.7 (a) A swimmer pushes a beach ball under water. (b) The forces on a beach ball–sized parcel of water.

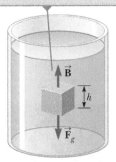

The buoyant force on the cube is the resultant of the forces exerted on its top and bottom faces by the liquid.

Figure 14.8 The external forces acting on an immersed cube are the gravitational force $\vec{F}_g$ and the buoyant force $\vec{B}$.

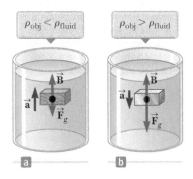

Figure 14.9 (a) A totally submerged object that is less dense than the fluid in which it is submerged experiences a net upward force and rises to the surface after it is released. (b) A totally submerged object that is denser than the fluid experiences a net downward force and sinks.

Therefore, there is a large net upward force on the ball, which explains why it is so hard to hold the beach ball under the water. Note that Archimedes's principle does not refer to the makeup of the object experiencing the buoyant force. The object's composition is not a factor in the buoyant force because the buoyant force is exerted by the surrounding fluid.

To better understand the origin of the buoyant force, consider a cube of solid material immersed in a liquid as in Figure 14.8. According to Equation 14.4, the pressure P_{bot} at the bottom of the cube is greater than the pressure P_{top} at the top by an amount $\rho_{\text{fluid}}gh$, where h is the height of the cube and ρ_{fluid} is the density of the fluid. The pressure at the bottom of the cube causes an *upward* force equal to $P_{\text{bot}}A$, where A is the area of the bottom face. The pressure at the top of the cube causes a *downward* force equal to $P_{\text{top}}A$. The resultant of these two forces is the buoyant force $\vec{B}$ with magnitude

$$B = (P_{\text{bot}} - P_{\text{top}})A = (\rho_{\text{fluid}}gh)A$$

$$B = \rho_{\text{fluid}}gV_{\text{disp}} \qquad (14.5)$$

where $V_{\text{disp}} = Ah$ is the volume of the fluid displaced by the cube. Because the product $\rho_{\text{fluid}}V_{\text{disp}}$ is equal to the mass of fluid displaced by the object,

$$B = M_{\text{fluid}}g$$

where $M_{\text{fluid}}g$ is the weight of the fluid displaced by the cube. This result is consistent with our initial statement about Archimedes's principle above, based on the discussion of the beach ball.

Buoyant forces are very important for the movement of fish through water. Under normal conditions, the weight of a fish is slightly greater than the buoyant force on the fish. Hence, the fish would sink if it did not have some mechanism for adjusting the buoyant force. The fish accomplishes that by internally regulating the size of its air-filled swim bladder to increase its volume and the magnitude of the buoyant force acting on it, according to Equation 14.5. In this manner, fish are able to swim to various depths.

Before we proceed with a few examples, it is instructive to discuss two common situations: a totally submerged object and a floating (partly submerged) object.

Case 1: Totally Submerged Object When an object is totally submerged in a fluid of density ρ_{fluid}, the volume V_{disp} of the displaced fluid is equal to the volume V_{obj} of the object; so, from Equation 14.5, the magnitude of the upward buoyant force is $B = \rho_{\text{fluid}}gV_{\text{obj}}$. If the object has a mass M and density ρ_{obj}, its weight is equal to $F_g = Mg = \rho_{\text{obj}}gV_{\text{obj}}$, and the net force on the object is $B - F_g = (\rho_{\text{fluid}} - \rho_{\text{obj}})gV_{\text{obj}}$. Hence, if the density of the object is *less* than the density of the fluid, the downward gravitational force is less than the buoyant force and the unsupported object accelerates upward (Fig. 14.9a). A block of wood held under water and released will rise to the surface. If the density of the object is *greater* than the density of the fluid, the upward buoyant force is less than the downward gravitational force and the unsupported object sinks (Fig. 14.9b). A rock will sink to the bottom when released in water. If the density of the submerged object *equals* the density of the fluid, the net force on the object is zero and the object remains in equilibrium. Therefore, the direction of motion of an object submerged in a fluid is determined *only* by the densities of the object and the fluid.

It is important to point out that gases exert buoyant forces also. Imagine a balloon surrounded by air. The balloon displaces a volume of air, so there is an upward buoyant force on it. If the balloon is filled with air, the effective density of the balloon–air combination is larger than that of air, due to the density of the balloon material. Therefore, the weight of the balloon is larger than that of the displaced air, and the released balloon falls to the ground. If the balloon is filled with helium, however, the effective density of the balloon-helium combination is less than that of air, and the balloon rises into the air when released.

Case 2: Floating Object Now consider the object of volume V_{obj} and density $\rho_{obj} <$ ρ_{fluid} in Figure 14.9a after it reaches the surface. After bobbing a bit, it will settle into static equilibrium on the surface of the fluid: it will *float*, and will be only *partially* submerged (Fig. 14.10). In this case, the object is modeled as a particle in equilibrium: the upward buoyant force is balanced by the downward gravitational force acting on the object. We no longer have $V_{disp} = V_{obj}$ as in Case 1, because only a portion of the object's volume is below the surface of the fluid. If V_{disp} is the volume of the fluid displaced by the object (this volume is the same as the volume of that part of the object beneath the surface of the fluid), the buoyant force has a magnitude $B = \rho_{fluid}gV_{disp}$. Because the weight of the object is $F_g = Mg = \rho_{obj}gV_{obj}$ and because $F_g = B$, we see that $\rho_{fluid}gV_{disp} = \rho_{obj}gV_{obj}$, or

$$\frac{V_{disp}}{V_{obj}} = \frac{\rho_{obj}}{\rho_{fluid}} \tag{14.6}$$

This equation shows that the fraction of the volume of a floating object that is below the fluid surface is equal to the ratio of the density of the object to that of the fluid. For example, the density of ice is less than that of liquid water. Therefore, when an ice cube floats in your water glass or an iceberg floats on the surface of the ocean, part of the ice is below the water surface and part is above. We explore this situation in Example 14.6.

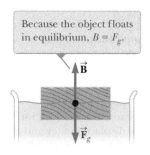

Figure 14.10 An object floating on the surface of a fluid experiences two forces, the gravitational force $\vec{\mathbf{F}}_g$ and the buoyant force $\vec{\mathbf{B}}$.

QUICK QUIZ 14.4 You are shipwrecked and floating in the middle of the ocean on a raft. Your cargo on the raft includes a treasure chest full of gold that you found before your ship sank, and the raft is just barely afloat. To keep you floating as high as possible in the water, should you **(a)** leave the treasure chest on top of the raft, **(b)** secure the treasure chest to the underside of the raft, or **(c)** hang the treasure chest in the water with a rope attached to the raft? (Assume throwing the treasure chest overboard is not an option you wish to consider.)

Example 14.5 Eureka!

Archimedes supposedly was asked to determine whether a crown made for the king consisted of pure gold. According to legend, he solved this problem by weighing the crown first in air and then in water as shown in Figure 14.11. Suppose the scale read 7.84 N when the crown was in air and 6.84 N when it was in water. What should Archimedes have told the king?

SOLUTION

Conceptualize Figure 14.11 helps us imagine what is happening in this example. Because of the upward buoyant force on the crown, the scale reading is smaller in Figure 14.11b than in Figure 14.11a.

Categorize This problem is an example of Case 1 discussed earlier because the crown is completely submerged. The scale reading is a measure of one of the forces on the crown, and the crown is stationary. Therefore, we can categorize the crown as a *particle in equilibrium*.

Analyze When the crown is suspended in air, the scale reads the true weight $T_1 = F_g$ (neglecting the small buoyant force due to the surrounding air). When the crown is immersed in water, the buoyant force $\vec{\mathbf{B}}$ due to the water reduces the scale reading to an *apparent* weight of $T_2 = F_g - B$.

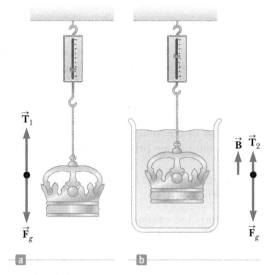

Figure 14.11 (Example 14.5) (a) When the crown is suspended in air, the scale reads its true weight because $T_1 = F_g$ (the buoyancy of air is negligible). (b) When the crown is immersed in water, the buoyant force $\vec{\mathbf{B}}$ changes the scale reading to a lower value $T_2 = F_g - B$.

continued

14.5 continued

Apply the particle in equilibrium model to the crown in water:

$$\sum F = B + T_2 - F_g = 0$$

Solve for B:

$$B = F_g - T_2 = m_c g - T_2$$

Because this buoyant force is equal in magnitude to the weight of the displaced water, $B = \rho_w g V_{disp}$, where V_{disp} is the volume of the displaced water and ρ_w is its density. Also, the volume of the crown V_c is equal to the volume of the displaced water because the crown is completely submerged, so $B = \rho_w g V_c$.

Find the density of the crown from Equation 1.1:

$$\rho_c = \frac{m_c}{V_c} = \frac{m_c g}{V_c g} = \frac{m_c g}{(B/\rho_w)} = \frac{m_c g \rho_w}{B} = \frac{m_c g \rho_w}{F_g - T_2} = \frac{m_c g \rho_w}{m_c g - T_2}$$

Substitute numerical values:

$$\rho_c = \frac{(7.84\ \text{N})(1\ 000\ \text{kg/m}^3)}{7.84\ \text{N} - 6.84\ \text{N}} = 7.84 \times 10^3\ \text{kg/m}^3$$

Finalize   From Table 14.1, we see that the density of gold is $19.3 \times 10^3\ \text{kg/m}^3$. Therefore, Archimedes should have reported that the king had been cheated. Either the crown was hollow, or it was not made of pure gold.

 WHAT IF?   Suppose the crown has the same weight but is indeed pure gold and not hollow. What would the scale reading be when the crown is immersed in water?

Answer   Find the buoyant force on the crown:

$$B = \rho_w g V_{disp} = \rho_w g V_c = \rho_w g \left(\frac{m_c}{\rho_c}\right) = \rho_w \left(\frac{m_c g}{\rho_c}\right)$$

Substitute numerical values:

$$B = (1.00 \times 10^3\ \text{kg/m}^3)\frac{7.84\ \text{N}}{19.3 \times 10^3\ \text{kg/m}^3} = 0.406\ \text{N}$$

Find the tension in the string hanging from the scale:

$$T_2 = m_c g - B = 7.84\ \text{N} - 0.406\ \text{N} = 7.43\ \text{N}$$

 Example **14.6** **A Titanic Surprise**

An iceberg floating in seawater as shown in Figure 14.12a is extremely dangerous because most of the ice is below the surface. This hidden ice can damage a ship that is still a considerable distance from the visible ice. What fraction of the iceberg lies below the water level?

 SOLUTION

Conceptualize   You are likely familiar with the phrase, "That's only the tip of the iceberg." The origin of this popular saying is that most of the volume of a floating iceberg is beneath the surface of the water (Fig. 14.12b).

Categorize   This example corresponds to Case 2 because only part of the iceberg is underneath the water. It is also a simple substitution problem involving Equation 14.6.

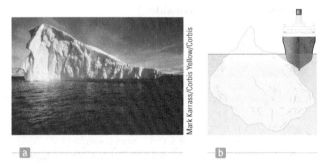

Figure 14.12   (Example 14.6) (a) Much of the volume of this iceberg is beneath the water. (b) A ship can be damaged even when it is not near the visible ice.

Evaluate Equation 14.6 using the densities of ice and seawater (Table 14.1):

$$f = \frac{V_{disp}}{V_{ice}} = \frac{\rho_{ice}}{\rho_{seawater}} = \frac{917\ \text{kg/m}^3}{1\ 030\ \text{kg/m}^3} = 0.890\ \text{or}\ 89.0\%$$

Therefore, the visible fraction of ice above the water's surface is about 11%. It is the unseen 89% below the water that represents the danger to a passing ship.

14.5 Fluid Dynamics

Thus far, our study of fluids has been restricted to fluids at rest. We now turn our attention to moving fluids. When fluid is in motion, its flow can be characterized as being one of two main types. The flow is said to be **steady,** or **laminar,** if each particle of the fluid follows a smooth path such that the paths of different particles never cross each other as shown in Figure 14.13. Laminar flow is predictable. If you determine the velocity vector of a fluid particle arriving at a certain position in space, every other particle arriving at that same position afterward will have the same velocity.

Above a certain critical speed, fluid flow becomes **turbulent.** Turbulent flow is irregular, unpredictable flow characterized by small whirlpool-like regions as shown in Figure 14.14.

The term *viscosity* is commonly used in the description of fluid flow to characterize the degree of internal friction in the fluid. This internal friction, or *viscous force,* is associated with the resistance that two adjacent layers of fluid have to moving relative to each other. Viscosity causes part of the fluid's kinetic energy to be transformed to internal energy. This mechanism is similar to the one by which the kinetic energy of an object sliding over a rough, horizontal surface decreases as discussed in Sections 8.3 and 8.4. We will address more details on viscosity in Section 14.7.

Because the motion of real fluids is very complex and not fully understood, we make some simplifying assumptions in our approach. In our simplification model of **ideal fluid flow,** we make the following four assumptions:

1. **The fluid is nonviscous.** In a nonviscous fluid, internal friction is neglected. An object moving through the fluid experiences no viscous force.
2. **The flow is laminar.** In laminar flow, all particles passing through a point have the same velocity and follow the same path.
3. **The fluid is incompressible.** The density of an incompressible fluid is the same throughout the fluid.
4. **The flow is irrotational.** In irrotational flow, the fluid has no angular momentum about any point. If a small paddle wheel placed anywhere in the fluid does not rotate about the wheel's center of mass, the flow is irrotational.

The path taken by a fluid particle under laminar flow is called a **streamline.** The velocity of the particle is always tangent to the streamline as shown in Figure 14.15. A set of streamlines like the ones shown in Figure 14.15 form a *tube of flow.* Fluid particles cannot flow into or out of the sides of this tube; if they could, the streamlines would cross one another.

Consider ideal fluid flow through a section of pipe of nonuniform size as illustrated in Figure 14.16. Let's focus our attention on a segment of fluid in the pipe. Figure 14.16a shows the segment at time $t = 0$ consisting of the gray portion between point 1 and point 2 and the short blue portion to the left of point 1. At this time, the fluid in the short blue portion is flowing through a cross section of area A_1 at speed v_1. During the time interval Δt, the small length Δx_1 of fluid in the blue portion moves into the section of pipe past point 1. During the same time interval, fluid at the right end of the segment moves out of the section of pipe past point 2.

Andy Sacks/Getty Images

Figure 14.13 Laminar flow of smoke over an automobile in a test wind tunnel.

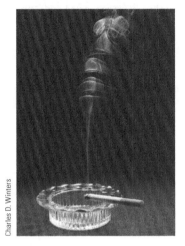

Charles D. Winters

Figure 14.14 Hot gases from a cigarette made visible by smoke particles. The smoke first moves in laminar flow at the bottom and then in turbulent flow above.

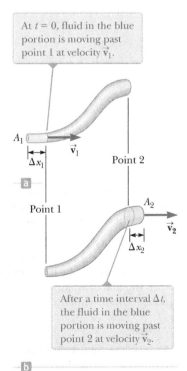

At $t = 0$, fluid in the blue portion is moving past point 1 at velocity $\vec{v}_1$.

A_1

Δx_1 $\vec{v}_1$

Point 2

a

Point 1 A_2

$\vec{v}_2$

Δx_2

After a time interval Δt, the fluid in the blue portion is moving past point 2 at velocity $\vec{v}_2$.

b

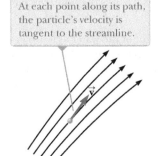

At each point along its path, the particle's velocity is tangent to the streamline.

$\vec{v}$

Figure 14.15 A particle in laminar flow follows a streamline.

Figure 14.16 A fluid moving with steady flow through a pipe of varying cross-sectional area. (a) At $t = 0$, the small blue-colored portion of the fluid at the left is moving into the section of pipe through area A_1. (b) After a time interval Δt, the blue-colored portion shown here is that fluid that has moved out of the section of pipe through area A_2.

The Language We Are Using Here To help understand this discussion, keep in mind three words that we are using: *section, segment,* and *portion. Section*: this word refers to the length of pipe between points 1 and 2 in Figure 14.16. *Segment*: this word refers to the total length of fluid that we are focusing on. In Figure 14.16a, this segment appears as the short blue portion to the left of point 1 *plus* the gray portion between points 1 and 2. In Figure 14.16b, the same segment of fluid has moved and appears as the gray portion plus the blue portion beyond point 2. *Portion*: this word refers to a piece of the segment of fluid. The portions appear either blue or gray in Figure 14.16.

Figure 14.17 The speed of water spraying from the end of a garden hose increases as the size of the opening is decreased with the thumb.

Figure 14.16b shows the situation at the end of the time interval Δt. The blue portion at the right end represents the fluid that was originally in the pipe and has moved past point 2 through an area A_2 at a speed v_2.

The mass of fluid contained in the blue portion in Figure 14.16a is given by $m_1 = \rho A_1 \,\Delta x_1 = \rho A_1 v_1 \,\Delta t$, where ρ is the (unchanging) density of the ideal fluid. Similarly, the fluid in the blue portion in Figure 14.16b has a mass $m_2 = \rho A_2 \,\Delta x_2 = \rho A_2 v_2 \,\Delta t$. Because the fluid is incompressible and the flow is laminar, however, the mass of fluid that passes point 1 in a time interval Δt must equal the mass that passes point 2 in the same time interval. That is, $m_1 = m_2$ or $\rho A_1 v_1 \,\Delta t = \rho A_2 v_2 \,\Delta t$, which means that

◄ Equation of Continuity for Fluids

$$A_1 v_1 = A_2 v_2 = \text{constant} \qquad (14.7)$$

This expression is called the **equation of continuity for fluids.** It states that the product of the area and the fluid speed at all points along a pipe is constant for an incompressible fluid. Equation 14.7 shows that the speed is high where the tube is constricted (small A) and low where the tube is wide (large A). The product Av, which has the dimensions of volume per unit time, is called either the *volume flux* or the *flow rate*. The condition $Av = $ constant is equivalent to the statement that the volume of fluid that enters one end of a tube in a given time interval equals the volume leaving the other end of the tube in the same time interval if no leaks are present.

You demonstrate the equation of continuity each time you water your garden with your thumb over the end of a garden hose as in Figure 14.17. By partially blocking the opening with your thumb, you reduce the cross-sectional area through which the water passes. As a result, the speed of the water increases as it exits the hose, and the water can be sprayed over a long distance.

Example 14.7 | **Watering a Garden**

A gardener uses a water hose to fill a 30.0-L bucket. The gardener notes that it takes 1.00 min to fill the bucket. A nozzle with an opening of cross-sectional area 0.500 cm² is then attached to the hose. The nozzle is held so that water is projected horizontally from a point 1.00 m above the ground. Over what horizontal distance can the water be projected?

SOLUTION

Conceptualize Imagine any past experience you have with projecting water from a horizontal hose or a pipe using either your thumb or a nozzle, which can be attached to the end of the hose. The faster the water is traveling as it leaves the hose, the farther it will land on the ground from the end of the hose.

Categorize Once the water leaves the hose, it is in free fall. Therefore, we categorize a given element of the water as a projectile. The element is modeled as a *particle under constant acceleration* (due to gravity) in the vertical direction and a *particle under constant velocity* in the horizontal direction. The horizontal distance over which the element is projected depends on the speed with which it is projected. This example involves a change in area for the pipe, so we also categorize it as one in which we use the continuity equation for fluids.

Analyze

Express the volume flow rate I_V in terms of area and speed of the water in the hose (we will discuss the origin for this notation in Section 14.7):

$$I_V = A_1 v_1$$

14.7 continued

Solve for the speed of the water in the hose:

$$v_1 = \frac{I_V}{A_1}$$

We have labeled this speed v_1 because we identify point 1 within the hose. We identify point 2 in the air just outside the nozzle. We must find the speed $v_2 = v_{xi}$ with which the water exits the nozzle (v_2) and begins its projectile motion (v_{xi}). The subscript i anticipates that it will be the *initial* velocity component of the water projected from the hose, and the subscript x indicates that the initial velocity vector of the projected water is horizontal.

Solve the continuity equation for fluids for v_2:

$$(1) \quad v_2 = v_{xi} = \frac{A_1}{A_2}v_1 = \frac{A_1}{A_2}\left(\frac{I_V}{A_1}\right) = \frac{I_V}{A_2}$$

We now shift our thinking away from fluids and to projectile motion. In the vertical direction, an element of the water starts from rest and falls through a vertical distance of 1.00 m.

Write Equation 2.16 for the vertical position of an element of water, modeled as a particle under constant acceleration:

$$y_f = y_i + v_{yi}t - \tfrac{1}{2}gt^2$$

Identify the origin as the initial position of the water as it leaves the hose, and recognize that the water begins with a vertical velocity component of zero. Solve for the time at which the water reaches the ground:

$$(2) \quad y_f = 0 + 0 - \tfrac{1}{2}gt^2 \;\rightarrow\; t = \sqrt{\frac{-2y_f}{g}}$$

Use Equation 2.7 to find the horizontal position of the element at this time, modeled as a particle under constant velocity:

$$x_f = x_i + v_{xi}t = 0 + v_2t = v_2t$$

Substitute from Equations (1) and (2):

$$x_f = \frac{I_V}{A_2}\sqrt{\frac{-2y_f}{g}}$$

Substitute numerical values:

$$x_f = \frac{30.0\ \text{L/min}}{0.500\ \text{cm}^2}\sqrt{\frac{-2(-1.00\ \text{m})}{9.80\ \text{m/s}^2}}\left(\frac{10^3\ \text{cm}^3}{1\ \text{L}}\right)\left(\frac{1\ \text{min}}{60\ \text{s}}\right) = 452\ \text{cm} = 4.52\ \text{m}$$

Finalize The time interval for the element of water to fall to the ground is unchanged if the projection speed is changed because the projection is horizontal. Increasing the projection speed results in the water hitting the ground farther from the end of the hose, but requires the same time interval to strike the ground.

14.6 Bernoulli's Equation

You have probably experienced driving on a highway and having a large truck pass you at high speed. In this situation, you may have had the frightening feeling that your car was being pulled in toward the truck as it passed. We will investigate the origin of this effect in this section.

As a fluid moves through a region where its speed or elevation above the Earth's surface changes, the pressure in the fluid varies with these changes. The relationship between fluid speed, pressure, and elevation was first derived in 1738 by Swiss physicist Daniel Bernoulli. Consider the flow of a segment of an ideal fluid through a non-uniform section of pipe in a time interval Δt as illustrated in Figure 14.18 (page 372). This figure is very similar to Figure 14.16, which we used to develop the continuity equation. We have added two features: the forces on the outer ends of the blue portions of fluid and the heights of these portions above the reference position $y = 0$.

The force exerted on the segment by the fluid to the left of the blue portion in Figure 14.18a has a magnitude P_1A_1. During a time interval Δt, the point of application of this force moves through a displacement of magnitude Δx_1, as the blue portion of fluid enters the section of pipe past point 1. The work done by this force on the segment in a time interval Δt is $W_1 = F_1\,\Delta x_1 = P_1A_1\,\Delta x_1 = P_1V$, where V is the volume of the blue portion of fluid passing point 1 in Figure 14.18a. In a similar manner, the work done on the segment by the fluid to the right of the segment in the same time interval Δt (Fig. 14.18b) is $W_2 = -P_2A_2\,\Delta x_2 = -P_2V$, where V is the

Daniel Bernoulli
Swiss physicist (1700–1782)
Bernoulli made important discoveries in fluid dynamics. Bernoulli's most famous work, *Hydrodynamica*, was published in 1738; it is both a theoretical and a practical study of equilibrium, pressure, and speed in fluids. He showed that as the speed of a fluid increases, its pressure decreases. Referred to as "Bernoulli's principle," Bernoulli's work is used to produce a partial vacuum in chemical laboratories by connecting a vessel to a tube through which water is running rapidly.

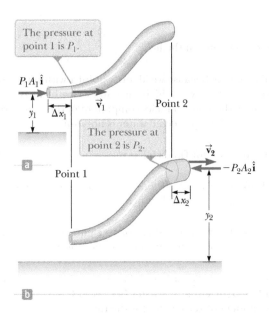

Figure 14.18 A fluid in laminar flow through a section of pipe. (a) A segment of the fluid at time $t = 0$. A small portion of the blue-colored fluid is at height y_1 above a reference position and is entering the section of pipe. (b) After a time interval Δt, the entire segment has moved to the right. The blue-colored portion of the fluid is that which has left the section of pipe at point 2 and is at height y_2.

volume of the blue portion of fluid passing point 2. (The volumes of the blue portions of fluid in Figures 14.18a and 14.18b are equal because the fluid is incompressible.) This work is negative because the force on the segment of fluid is to the left and the displacement of the point of application of the force is to the right. Therefore, the net work done on the segment by these forces in the time interval Δt is

$$W = (P_1 - P_2)V \tag{14.8}$$

Part of this work goes into changing the kinetic energy of the segment of fluid, and part goes into changing the gravitational potential energy of the segment–Earth system. The appropriate reduction of Equation 8.2 for the nonisolated system of the Earth and the segment of fluid is

$$\Delta K + \Delta U_g = W \tag{14.9}$$

Because we are assuming laminar flow, the kinetic energy K_{gray} of the gray portion of the segment is the same in both parts of Figure 14.18. Therefore, the change in the kinetic energy of the segment of fluid is

$$\Delta K = \left(\tfrac{1}{2}mv_2^2 + K_{gray}\right) - \left(\tfrac{1}{2}mv_1^2 + K_{gray}\right) = \tfrac{1}{2}mv_2^2 - \tfrac{1}{2}mv_1^2 \tag{14.10}$$

where m is the mass of the blue portions of fluid in both parts of Figure 14.18. (Because the volumes of both portions are the same, they also have the same mass.)

Considering the gravitational potential energy of the segment–Earth system, once again there is no change during the time interval for the gravitational potential energy U_{gray} associated with the gray portion of the fluid. Consequently, the change in gravitational potential energy of the system is

$$\Delta U_g = (mgy_2 + U_{gray}) - (mgy_1 + U_{gray}) = mgy_2 - mgy_1 \tag{14.11}$$

Substituting Equations 14.8, 14.10. and 14.11 into Equation 14.9 gives

$$\left(\tfrac{1}{2}mv_2^2 - \tfrac{1}{2}mv_1^2\right) - (mgy_2 - mgy_1) = (P_1 - P_2)V$$

If we divide each term by the blue portion volume V and recall that $\rho = m/V$, this expression reduces to

$$\tfrac{1}{2}\rho v_2^2 - \tfrac{1}{2}\rho v_1^2 + \rho gy_2 - \rho gy_1 = P_1 - P_2$$

Rearranging terms gives

$$P_1 + \tfrac{1}{2}\rho v_1^2 + \rho gy_1 = P_2 + \tfrac{1}{2}\rho v_2^2 + \rho gy_2 \tag{14.12}$$

which is **Bernoulli's equation** as applied to an ideal fluid. This equation is often expressed as

$$P + \tfrac{1}{2}\rho v^2 + \rho g y = \text{constant} \qquad (14.13)$$

◄ Bernoulli's equation

Bernoulli's equation shows that the pressure of a fluid decreases as the speed of the fluid increases. In addition, the pressure decreases as the elevation increases. This latter point explains why water pressure from faucets on the upper floors of a tall building is weak unless measures are taken to provide higher pressure for these upper floors.

When the fluid is at rest, $v_1 = v_2 = 0$ and Equation 14.12 becomes

$$P_1 - P_2 = \rho g(y_2 - y_1) = \rho g h$$

This result is in agreement with Equation 14.4.

Although Equation 14.13 was derived for an incompressible fluid, the general behavior of pressure with speed is true even for gases: as the speed increases, the pressure decreases. This *Bernoulli effect* explains the experience with the truck on the highway at the opening of this section. As air passes between you and the truck, it must pass through a relatively narrow channel. According to the continuity equation for fluids, the speed of the air in this channel is higher than that of the air on the other side of your car. According to the Bernoulli effect, this higher-speed air exerts less pressure on your car than the air on the other side. Therefore, there is a net force pushing you toward the truck!

QUICK QUIZ 14.5 You observe two helium balloons floating next to each other at the ends of strings secured to a table. The facing surfaces of the balloons are separated by 1–2 cm. You blow through the small space between the balloons. What happens to the balloons? **(a)** They move toward each other. **(b)** They move away from each other. **(c)** They are unaffected.

Example 14.8 The Venturi Tube

The horizontal constricted pipe illustrated in Figure 14.19, known as a *Venturi tube*, can be used to measure the flow speed of an incompressible fluid. Determine the flow speed at point 2 of Figure 14.19a if the pressure difference $P_1 - P_2$ is known.

SOLUTION

Conceptualize Bernoulli's equation shows how the pressure of an ideal fluid decreases as its speed increases. Therefore, we should be able to calibrate a device to give us the fluid speed if we can measure pressure.

Categorize Because the problem states that the fluid is incompressible, we can categorize it as one in which we can use the equation of continuity for fluids and Bernoulli's equation.

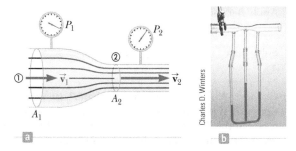

Charles D. Winters

Figure 14.19 (Example 14.8) (a) Pressure P_1 is greater than pressure P_2 because $v_1 < v_2$. This device can be used to measure the speed of fluid flow. (b) A Venturi tube, located at the top of the photograph. Air is blown through the tube from the left. The higher level of fluid in the middle column shows that the pressure of the moving air at the top of that column, which is in the constricted region of the Venturi tube, is lower.

Analyze Apply Equation 14.12 to points 1 and 2, noting that $y_1 = y_2$ because the pipe is horizontal:

$$(1) \quad P_1 + \tfrac{1}{2}\rho v_1^2 = P_2 + \tfrac{1}{2}\rho v_2^2$$

Solve the equation of continuity for v_1:

$$v_1 = \frac{A_2}{A_1} v_2$$

Substitute this expression into Equation (1):

$$P_1 + \tfrac{1}{2}\rho\left(\frac{A_2}{A_1}\right)^2 v_2^2 = P_2 + \tfrac{1}{2}\rho v_2^2$$

Solve for v_2:

$$v_2 = A_1 \sqrt{\frac{2(P_1 - P_2)}{\rho(A_1^2 - A_2^2)}}$$

continued

14.8 continued

Finalize From the design of the tube (areas A_1 and A_2) and measurements of the pressure difference $P_1 - P_2$, we can calculate the speed of the fluid with this equation. To see the relationship between fluid speed and pressure difference, place two empty soda cans on their sides about 2 cm apart on a table. Gently blow a stream of air horizontally between the cans and watch them roll together slowly due to a modest pressure difference between the stagnant air on their outside edges and the moving air between them. Now blow more strongly and watch the increased pressure difference move the cans together more rapidly.

Example 14.9 Torricelli's Law

An enclosed tank containing a liquid of density ρ has a hole in its side at a distance y_1 from the tank's bottom (Fig. 14.20). The hole is open to the atmosphere, and its diameter is much smaller than the diameter of the tank. The air above the liquid is maintained at a pressure P. Determine the speed of the liquid as it leaves the hole when the liquid's level is a distance h above the hole.

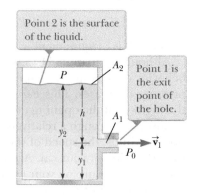

Figure 14.20 (Example 14.9) A liquid leaves a hole in a tank at speed v_1.

SOLUTION

Conceptualize Imagine that the tank is a fire extinguisher. When the hole is opened, liquid leaves the hole with a certain speed. If the pressure P at the top of the liquid is increased, the liquid leaves with a higher speed. If the pressure P falls too low, the liquid leaves with a low speed and the extinguisher must be replaced. Because $A_2 \gg A_1$, the liquid is approximately at rest at the top of the tank, where the pressure is P, so $v_2 = 0$. At the hole, the liquid is open to the external atmosphere, so P_1 is equal to atmospheric pressure P_0.

Categorize Looking at Figure 14.20, we know the pressure at two points and the velocity at point 2. We wish to find the velocity at point 1. Therefore, we can categorize this example as one in which we can apply Bernoulli's equation.

Analyze

Apply Bernoulli's equation between points 1 and 2:

$$P_0 + \tfrac{1}{2}\rho v_1^{\,2} + \rho g y_1 = P + \rho g y_2$$

Solve for v_1, noting that $y_2 - y_1 = h$:

$$v_1 = \sqrt{\frac{2(P - P_0)}{\rho} + 2gh}$$

Finalize When P is much greater than P_0 (so that the term $2gh$ can be neglected), the exit speed of the water is mainly a function of P. If the tank is open at the top to the atmosphere, then $P = P_0$ and $v_1 = \sqrt{2gh}$. In other words, for an open tank, the speed of the liquid leaving a hole a distance h below the surface is equal to that acquired by an object falling freely through a vertical distance h. This phenomenon is known as *Torricelli's law*.

WHAT IF? What if the position of the hole in Figure 14.20 could be adjusted vertically? If the top of the tank is open to the atmosphere and sitting on a table, what position of the hole would cause the water to land on the table at the farthest distance from the tank?

Answer Model a parcel of water exiting the hole as a projectile. From the *particle under constant acceleration* model, find the time at which the parcel strikes the table from a hole at an arbitrary position y_1:

$$y_f = y_i + v_{yi}t - \tfrac{1}{2}gt^2$$
$$0 = y_1 + 0 - \tfrac{1}{2}gt^2$$
$$t = \sqrt{\frac{2y_1}{g}}$$

From the *particle under constant velocity* model, find the horizontal position of the parcel at the time it strikes the table:

$$x_f = x_i + v_{xi}t = 0 + \sqrt{2g(y_2 - y_1)}\ \sqrt{\frac{2y_1}{g}}$$
$$= 2\sqrt{(y_2 y_1 - y_1^{\,2})}$$

Maximize the horizontal position by taking the derivative of x_f with respect to y_1 (because y_1, the height of the hole, is the variable that can be adjusted) and setting it equal to zero:

$$\frac{dx_f}{dy_1} = \tfrac{1}{2}(2)(y_2 y_1 - y_1^{\,2})^{-1/2}(y_2 - 2y_1) = 0$$

Solve for y_1:

$$y_1 = \tfrac{1}{2}y_2$$

14.9 c o n t i n u e d

Therefore, to maximize the horizontal distance, the hole should be halfway between the bottom of the tank and the upper surface of the water. Below this location, the water is projected at a higher speed but falls for a short time interval, reducing the horizontal range. Above this point, the water is in the air for a longer time interval but is projected with a smaller horizontal speed.

14.7 Flow of Viscous Fluids in Pipes

In Section 14.5, we discussed the flow of an ideal fluid. The results obtained there and in Section 14.6 are applicable to many situations. On the other hand, there are other situations in which we must investigate the flow of real, non-idealized fluids.

As an example, consider the flow of a fluid in a closed pipe, such as water in a plumbing system or blood in a human circulatory system. According to Bernoulli's principle, if the pipe were of uniform cross section, the pressure difference between two locations in a horizontal section of the pipe would be zero. Therefore, once set in motion, the fluid would flow without any external influence. This is not true in reality. If this were true, why would humans need hearts to continuously pump the blood?

In a real situation, the pressure differential to keep fluid moving at a fixed speed in a horizontal pipe is given by

$$\Delta P = I_V R \qquad (14.14)$$

In this equation, I_V represents the volume rate of fluid flow in m^3/s. This quantity is equal to the product Av in Equation 14.7, as mentioned in Example 14.7. The parameter R is a measure of the resistance of the system to the movement of fluid in the pipe.

The notation I_V may appear odd, but it is chosen in order to make a comparison with a similar equation in electricity that we will see in Chapter 26:

$$\Delta V = IR \qquad (14.15)$$

In this equation ΔV is an electric potential difference, which represents the external influence that attempts to move electrons through a wire. The quantity I is the current, representing the flow of electrons in the wire, and R is the resistance of the flow of those electrons through the wire. Compare this equation to Equation 14.14, in which ΔP represents a pressure difference that attempts to move fluid through a pipe. The quantity I_V represents the flow of fluid in the pipe, and R is the resistance of the flow of that fluid through the pipe. Equations 14.14 and 14.15 are both types of *transport equations,* in which an entity attempts to move something through space and encounters resistance to the effort. We will see a similar situation in Chapter 19, where a temperature difference attempts to move energy through a material by heat, and encounters resistance based on how good a thermal insulator the material is. We could even reverse the variables on the right side of Equation 5.2 to cast it in a form to compare with Equations 14.14 and 14.15:

$$\sum F = ma = am$$

Here, a net force on the left attempts to move an object through space, measured by its acceleration, and encounters resistance in the form of the mass of the object. Other transport situations also exist, such as a concentration difference driving a diffusion of molecules through another substance.

Now, what determines the resistance R for the fluid flow? One contribution to the physical origin of the resistance is the viscous resistive forces (Section 6.4) between the fluid and the inner wall of the pipe and between the layers of fluid that may be moving at different speeds relative to one another. To begin to evaluate the effect of these viscous forces, consider Figure 14.21 (page 376), which shows a layer of fluid of thickness h. Initially, the visible side of the layer forms a rectangle $ABCD$.

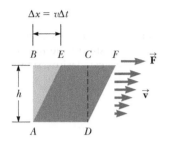

Figure 14.21 The lower surface of a layer of liquid is held fixed while a force is applied to the upper surface. As a result, the layer deforms in shape.

At $t = 0$, a force of magnitude F is applied to the right on the upper surface while the lower surface remains fixed. After a time interval Δt, the visible side of the layer now forms the parallelogram *AEFD*. The top surface of the liquid is moving with speed v, while lower portions of the layer move with progressively lower speeds.

Compare Figure 14.21 with Figure 12.13, in which we apply a shearing force to a solid material, and notice the similarities in the two situations. Based on this comparison, let us modify Equation 12.7 to fit the current circumstances. Solve Equation 12.7 for the force on the upper surface:

$$F = SA\frac{\Delta x}{h} \tag{14.16}$$

where S is the shear modulus of the material. This equation represents the deformation of a solid, but let us imagine it representing the deformation of the viscous liquid in Figure 14.21, with the top surface moving at speed v. The deformation behavior is the same, although we will ultimately replace the shear modulus with another parameter. Therefore, we can write Equation 14.16 as a proportionality:

$$F \propto A\frac{\Delta x}{h} \tag{14.17}$$

For a given time interval, Δx is proportional to the speed v of the upper surface, so we can write

$$F \propto A\frac{v}{h}$$

We can turn this inequality to an equality by introducing a proportionality constant η:

$$F = \eta A\frac{v}{h} \tag{14.18}$$

The constant η is called the **viscosity** of the fluid and has units of $\text{N} \cdot \text{s/m}^2 = \text{Pa} \cdot \text{s}$. Another common unit of viscosity is the *poise* (P), where $1 \text{ Pa} \cdot \text{s} = 10 \text{ P}$. Table 14.2 lists viscosities of some fluids. Notice that a "thick" fluid such as honey has a high viscosity, while fluids such as water and air have lower viscosity values.

Looking again at our representation of a viscous fluid in Figure 14.21, the speed of the bottom surface is zero and the speed of subsequent portions higher toward the top increases, with the highest surface having the highest speed. Applying this notion to the flow of fluid in a pipe, we find that, because of the viscous force between layers of fluid, the flow of a fluid in a pipe is not uniform across the area of the pipe. Figure 14.22 shows that the speed of the fluid is greatest at the center of the pipe and approaches zero at the pipe walls.

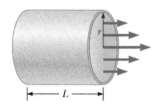

Figure 14.22 Flow of a viscous fluid in a pipe. The red velocity vectors show the variation in speed of the fluid across a diameter of the pipe. The fluid flows fastest at the center and slowly at the walls of the pipe.

TABLE 14.2   Viscosities of Various Fluids[a]

Fluid	Viscosity (mPa · s)
Air	0.018
Helium	0.020
Liquid nitrogen ($-196°$C)	0.158
Acetone	0.306
Water	0.894
Ethanol	1.07
Blood ($37.0°$C)	2.70
Olive oil	81
Motor oil (SAE 40, $20°$C)	319
Corn syrup	1 381
Glycerin	1 500
Honey[b]	2 000–10 000
Peanut butter	250 000

[a]All values at 25.0°C unless noted otherwise.
[b]Value depends on moisture content.

Let us return again to Equation 14.14. What is it that determines the resistance R to the flow of the fluid in the pipe? Clearly, the viscosity plays a role, but there are other factors. It can be shown that the resistance of the fluid in the segment of pipe in Figure 14.22 of length L and radius r is given by

$$R = \frac{8\eta L}{\pi r^4} \qquad (14.19)$$

Therefore, Equation 14.14 becomes

$$\Delta P = \frac{8\eta L}{\pi r^4} I_V \qquad (14.20)$$

◀ Poiseuille's Law (Hagen–Poiseuille equation)

This equation is known as **Poiseuille's law,** or the **Hagen–Poiseuille equation.**[1] Notice the important dependence of the pressure difference on r: the pressure difference is inversely proportional to r to the fourth power. Therefore, if the radius of the pipe drops by 50%, the pressure difference required to maintain the same flow through the pipe increases by a factor of 16.

This dependence is very important in blood flow in the human circulatory system. If a blood vessel becomes occluded by plaque so that the radius through which blood can flow is decreased, the pressure required to maintain the blood flow rises rapidly. Conversely, for a given pressure, the flow rate I_V of the blood is reduced.

14.8 Other Applications of Fluid Dynamics

Let's consider the opening storyline for this chapter. What forces are responsible for lifting an airplane into the air? Consider the streamlines that flow around an airplane wing as shown in Figure 14.23. Let's assume the airstream approaches the wing horizontally from the right with a velocity $\vec{v}_1$, which is equivalent to the airplane moving to the right through still air. The tilt of the wing causes the airstream to be deflected downward with a velocity $\vec{v}_2$. Because the airstream is deflected by the wing, the wing must exert a force on the airstream. According to Newton's third law, the airstream therefore exerts a force $\vec{F}$ on the wing that is equal in magnitude and opposite in direction. This force has a vertical component called **lift** (or aerodynamic lift) and a horizontal component called **drag.** The lift depends on several factors, such as the speed of the airplane, the area of the wing, the wing's curvature, and the angle between the wing and the horizontal. The curvature of the wing surfaces causes the pressure above the wing to be lower than that below the wing due to the Bernoulli effect. This pressure difference assists with the lift on the wing. As the angle between the wing and the horizontal increases, turbulent flow can set in above the wing to reduce the lift.

The lift force exerted by the air on the wing according to Newton's law and the pressure difference between the top and bottom of the wing caused by the Bernoulli effect will both depend on the density of the air surrounding the wings. What do we know about the location of Denver in our opening storyline? Denver is often called the "Mile-High City." That is because it is in the Rocky Mountains, at an altitude of 1 610 m above sea level. Because of that altitude, the air is of lower density than that at the airport at Los Angeles, by an average of about 15%. Consequently, aircraft must attain a higher speed in order for the forces and pressure differences to be sufficient to lift the aircraft. This leads to a longer distance on the runway for the aircraft to move before reaching this higher speed.

In general, an object moving through a fluid experiences lift as the result of any effect that causes the fluid to change its direction as it flows past the object. Some factors that influence lift are the shape of the object, its orientation with respect to the fluid flow, any spinning motion it might have, and the texture of its surface.

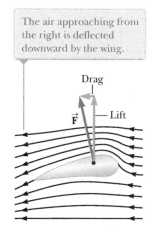

The air approaching from the right is deflected downward by the wing.

Figure 14.23 Streamline flow around a moving airplane wing. By Newton's third law, the air deflected by the wing results in an upward force on the wing from the air: *lift*. Because of air resistance, there is also a force opposite the velocity of the wing: *drag*.

[1]Named after Jean Leonard Marie Poiseuille (1797–1869), a French physicist, and Gotthilf Heinrich Ludwig Hagen (1797–1884), a German civil engineer. The unit *poise* is named after Poiseuille.

For example, a golf ball struck with a club is given a rapid backspin due to the slant of the club. The dimples on the ball increase the friction force between the ball and the air so that air adheres to the ball's surface and is deflected downward as a result. Because the ball pushes the air down, the air must push up on the ball. Without the dimples, the friction force is lower and the golf ball does not travel as far. It may seem counterintuitive to increase the range by increasing the friction force, but the lift gained by spinning the ball more than compensates for the loss of range due to the effect of friction on the translational motion of the ball. For the same reason, a baseball's cover helps the spinning ball "grab" the air rushing by and helps deflect it when a "curve ball" is thrown.

Summary

▶ Definitions

The **pressure** P in a fluid is the force per unit area exerted by the fluid on a surface:

$$P \equiv \frac{F}{A} \tag{14.1}$$

In the SI system, pressure has units of newtons per square meter (N/m^2), and 1 N/m^2 = 1 **pascal** (Pa).

▶ Concepts and Principles

The pressure in a fluid at rest varies with depth h in the fluid according to the expression

$$P = P_0 + \rho g h \tag{14.4}$$

where P_0 is the pressure at $h = 0$ and ρ is the density of the fluid, assumed uniform.

Pascal's law states that when pressure is applied to an enclosed fluid, the pressure is transmitted undiminished to every point in the fluid and to every point on the walls of the container.

When an object is partially or fully submerged in a fluid, the fluid exerts on the object an upward force called the **buoyant force.** According to **Archimedes's principle,** the magnitude of the buoyant force is equal to the weight of the fluid displaced by the object:

$$B = \rho_{\text{fluid}} g V_{\text{disp}} \tag{14.5}$$

The flow rate (volume flux) through a pipe that varies in cross-sectional area is constant; that is equivalent to stating that the product of the cross-sectional area A and the speed v at any point is a constant. This result is expressed in the **equation of continuity for fluids:**

$$A_1 v_1 = A_2 v_2 = \text{constant} \tag{14.7}$$

The sum of the pressure, kinetic energy per unit volume, and gravitational potential energy per unit volume has the same value at all points along a streamline for an ideal fluid. This result is summarized in **Bernoulli's equation:**

$$P + \tfrac{1}{2}\rho v^2 + \rho g y = \text{constant} \tag{14.13}$$

Think–Pair–Share

See the Preface for an explanation of the icons used in this problems set. For additional assessment items for this section, go to ⁙ **WEBASSIGN** From Cengage

1. You are a member of an expert witness group that provides scientific services to the legal community. Your group has been asked to defend an NFL football team, which has been caught in an embarrassing practice: they have been filling their teams' footballs with helium instead of air, believing that the helium would provide an increased buoyant force on the footballs, causing their passes and kicks to be longer. Despite your unhappiness with the efforts of the team to gain an unfair advantage, your legal group believes that

everyone deserves a defense, so you agree to take the case. (a) Develop an argument that filling the footballs with helium would *not* provide additional buoyant force on the footballs; therefore, the team was not trying to gain an advantage. (b) Develop a private argument to present to the team that filling the balls with helium will actually *decrease* the performance.

2. It is a warm day, and a student decides he would like to spend a few hours swimming in the pool. Using his snorkel equipment, he views the bottom of the pool. After a while, he notices a piece of PVC pipe leaning against the house and thinks about using it as a long snorkel. He takes the

mouthpiece off his snorkel and attaches it to the bottom of the PVC pipe. His plan is to have his mouth on the mouthpiece of the long snorkel and sink into the deep end of the pool, breathing all the way to the bottom! While he is preparing his long snorkel, his uncle, who is a pulmonologist, arrives for a visit. He asks the student what he is doing with the pipe and he explains. The uncle is horrified and says that it is very dangerous to attempt this activity. The student is disappointed, but then tells his uncle that he will hold his breath while he sinks to the bottom, keeping the mouthpiece closed with his thumb, put the mouthpiece on as he arrives at the bottom, and begin breathing. His uncle looks even more horrified, and tells him that that activity could be fatal! Discuss in your group: Why is deep snorkeling so *dangerous*?

3. **ACTIVITY** Perform the following activities and discuss the physics behind the results with your group members:

(a) Place a can of diet soda and regular soda of the same brand in a large container of water. Do they both float? Explain the results.

(b) Open a can of clear carbonated beverage and fill a transparent glass with the liquid. Now drop a few raisins into the beverage and record their behavior.

Problems

See the Preface for an explanation of the icons used in this problems set. For additional assessment items for this section, go to **WEBASSIGN** From Cengage

> *Note:* In all problems, assume the density of air is the 20°C value from Table 14.1, 1.20 kg/m³, unless noted otherwise.

SECTION 14.1 Pressure

1. A large man sits on a four-legged chair with his feet off the floor. The combined mass of the man and chair is 95.0 kg. If the chair legs are circular and have a radius of 0.500 cm at the bottom, what pressure does each leg exert on the floor?

2. **QC** The nucleus of an atom can be modeled as several protons and neutrons closely packed together. Each particle has a mass of 1.67×10^{-27} kg and radius on the order of 10^{-15} m. (a) Use this model and the data provided to estimate the density of the nucleus of an atom. (b) Compare your result with the density of a material such as iron. What do your result and comparison suggest concerning the structure of matter?

3. Estimate the total mass of the Earth's atmosphere. (The radius of the Earth is 6.37×10^{6} m, and atmospheric pressure at the surface is 1.013×10^{5} Pa.)

SECTION 14.2 Variation of Pressure with Depth

4. *Why is the following situation impossible?* Figure P14.4 shows Superman attempting to drink cold water through a straw of length $\ell = 12.0$ m. The walls of the tubular straw are very strong and do not collapse. With his great strength, he achieves maximum possible suction and enjoys drinking the cold water.

Figure P14.4

5. **AMT** **T** What must be the contact area between a suction cup (completely evacuated) and a ceiling if the cup is to support the weight of an 80.0-kg student?

6. For the cellar of a new house, a hole is dug in the ground, with vertical sides going down 2.40 m. A concrete foundation wall is built all the way across the 9.60-m width of the excavation. This foundation wall is 0.183 m away from the front of the cellar hole. During a rainstorm, drainage from the street fills up the space in front of the concrete wall, but not the cellar behind the wall. The water does not soak into the clay soil. Find the force the water causes on the foundation wall. For comparison, the weight of the water is given by 2.40 m $\times$ 9.60 m $\times$ 0.183 m $\times$ 1 000 kg/m³ $\times$ 9.80 m/s² = 41.3 kN.

7. **Review.** A solid sphere of brass (bulk modulus of 14.0×10^{10} N/m²) with a diameter of 3.00 m is thrown into the ocean. By how much does the diameter of the sphere decrease as it sinks to a depth of 1.00 km?

SECTION 14.3 Pressure Measurements

8. **BIO** **QC** The human brain and spinal cord are immersed in the cerebrospinal fluid. The fluid is normally continuous between the cranial and spinal cavities and exerts a pressure of 100 to 200 mm of H_2O above the prevailing atmospheric pressure. In medical work, pressures are often measured in units of millimeters of H_2O because body fluids, including the cerebrospinal fluid, typically have the same density as water. The pressure of the cerebrospinal fluid can be measured by means of a *spinal tap* as illustrated in Figure P14.8. A hollow tube is inserted into the spinal column, and the height to which the fluid rises is observed. If the fluid rises to a height of 160 mm, we write its gauge pressure as 160 mm H_2O. (a) Express this pressure in pascals, in atmospheres, and in millimeters of mercury. (b) Some conditions that block or inhibit the flow of cerebrospinal fluid can be investigated by means of *Queckenstedt's test*. In this procedure, the veins in the patient's neck are compressed to make the blood pressure rise in the brain, which in turn should be transmitted

Figure P14.8

to the cerebrospinal fluid. Explain how the level of fluid in the spinal tap can be used as a diagnostic tool for the condition of the patient's spine.

9. Blaise Pascal duplicated Torricelli's barometer using a
QC red Bordeaux wine, of density 984 kg/m³, as the working liquid (Fig. P14.9). (a) What was the height h of the wine column for normal atmospheric pressure? (b) Would you expect the vacuum above the column to be as good as for mercury?

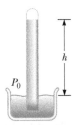

P_0

Figure P14.9

10. A tank with a flat bottom of area A and vertical sides is filled
S to a depth h with water. The pressure is P_0 at the top surface. (a) What is the absolute pressure at the bottom of the tank? (b) Suppose an object of mass M and density less than the density of water is placed into the tank and floats. No water overflows. What is the resulting increase in pressure at the bottom of the tank?

SECTION 14.4 Buoyant Forces and Archimedes's Principle

11. The gravitational force exerted on a solid object is 5.00 N. When the object is suspended from a spring scale and submerged in water, the scale reads 3.50 N (Fig. P14.11). Find the density of the object.

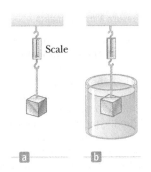

Scale

a   b

Figure P14.11 Problems 11 and 12.

12. A 10.0-kg block of metal measuring 12.0 cm by 10.0 cm by 10.0 cm is suspended from a scale and immersed in water as shown in Figure P14.11b. The 12.0-cm dimension is vertical, and the top of the block is 5.00 cm below the surface of the water. (a) What are the magnitudes of the forces acting on the top and on the bottom of the block due to the surrounding water? (b) What is the reading of the spring scale? (c) Show that the buoyant force equals the difference between the forces at the top and bottom of the block.

13. A plastic sphere floats in water with 50.0% of its volume sub-
T merged. This same sphere floats in glycerin with 40.0% of its volume submerged. Determine the densities of (a) the glycerin and (b) the sphere.

14. The weight of a rectangular block of low-density material
QC is 15.0 N. With a thin string, the center of the horizontal bottom face of the block is tied to the bottom of a beaker partly filled with water. When 25.0% of the block's volume is submerged, the tension in the string is 10.0 N. (a) Find the buoyant force on the block. (b) Oil of density 800 kg/m³ is now steadily added to the beaker, forming a layer above the water and surrounding the block. The oil exerts forces on each of the four sidewalls of the block that the oil touches. What are the directions of these forces? (c) What happens to the string tension as the oil is added? Explain how the oil has this effect on the string tension. (d) The string breaks when its tension reaches 60.0 N. At this moment, 25.0% of the block's volume is still below the water line. What additional fraction of the block's volume is below the top surface of the oil?

15. A wooden block of volume 5.24×10^{-4} m³ floats in water, and
QC a small steel object of mass m is placed on top of the block. When $m = 0.310$ kg, the system is in equilibrium and the top of the wooden block is at the level of the water. (a) What is the density of the wood? (b) What happens to the block when the steel object is replaced by an object whose mass is less than 0.310 kg? (c) What happens to the block when the steel object is replaced by an object whose mass is greater than 0.310 kg?

16. A *hydrometer* is an instrument used to determine liquid den-
S sity. A simple one is sketched in Figure P14.16. The bulb of a syringe is squeezed and released to let the atmosphere lift a sample of the liquid of interest into a tube containing a calibrated rod of known density. The rod, of length L and average density ρ_0, floats partially immersed in the liquid of density ρ. A length h of the rod protrudes above the surface of the liquid. Show that the density of the liquid is given by

$$\rho = \frac{\rho_0 L}{L - h}$$

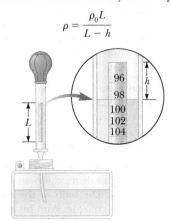

96
98
100
102
104

Figure P14.16 Problems 16 and 17

17. Refer to Problem 16 and Figure P14.16. A hydrometer is to
QC be constructed with a cylindrical floating rod. Nine fiduciary marks are to be placed along the rod to indicate densities of 0.98 g/cm³, 1.00 g/cm³, 1.02 g/cm³, 1.04 g/cm³, . . . , 1.14 g/cm³. The row of marks is to start 0.200 cm from the top end of the rod and end 1.80 cm from the top end. (a) What is the required length of the rod? (b) What must be its average density? (c) Should the marks be equally spaced? Explain your answer.

18. On October 21, 2001, Ian Ashpole of the United Kingdom
QC achieved a record altitude of 3.35 km (11 000 ft) powered by 600 toy balloons filled with helium. Each filled balloon had

a radius of about 0.50 m and an estimated mass of 0.30 kg. (a) Estimate the total buoyant force on the 600 balloons. (b) Estimate the net upward force on all 600 balloons. (c) Ashpole parachuted to the Earth after the balloons began to burst at the high altitude and the buoyant force decreased. Why did the balloons burst?

19. You have a job in a company that produces party supplies.
CR You are designing helium-filled balloons to be sold as gifts. To save money on production costs, part of this design is to choose from an available selection the least massive token necessary to be tied to the lower end of the string hanging from the balloon in order to keep the balloon from rising off a table, hospital food tray, bedroom dresser, etc. In this way, the token remains stationary on the flat surface and the balloon is buoyed above the token at a fixed height, with the string straight. You are working on a balloon whose envelope is very thin and has a mass of 0.150 kg. The envelope is filled to a volume of 0.2300 m³ with helium at atmospheric pressure. The string has a mass of 0.070 0 kg. Among the selection of tokens are those with masses 10.0 g, 20.0 g, 30.0 g, 40.0 g, and 50.0 g. Chose the appropriate token for the balloon you are working on.

SECTION 14.5 Fluid Dynamics

20. Water flowing through a garden hose of diameter 2.74 cm fills a 25-L bucket in 1.50 min. (a) What is the speed of the water leaving the end of the hose? (b) A nozzle is now attached to the end of the hose. If the nozzle diameter is one-third the diameter of the hose, what is the speed of the water leaving the nozzle?

21. Water falls over a dam of height h with a mass flow rate of I_V, in units of kilograms per second. (a) Show that the power available from the water is

$$P = I_V gh$$

where g is the free-fall acceleration. (b) Each hydroelectric unit at the Grand Coulee Dam takes in water at a rate of 8.50×10^5 kg/s from a height of 87.0 m. The power developed by the falling water is converted to electric power with an efficiency of 85.0%. How much electric power does each hydroelectric unit produce?

SECTION 14.6 Bernoulli's Equation

22. A legendary Dutch boy saved Holland by plugging a hole of diameter 1.20 cm in a dike with his finger. If the hole was 2.00 m below the surface of the North Sea (density 1 030 kg/m³), (a) what was the force on his finger? (b) If he pulled his finger out of the hole, during what time interval would the released water fill 1 acre of land to a depth of 1 ft? Assume the hole remained constant in size.

23. Water is pumped up from the Colorado River to supply Grand Canyon Village, located on the rim of the canyon. The river is at an elevation of 564 m, and the village is at an elevation of 2 096 m. Imagine that the water is pumped through a single long pipe 15.0 cm in diameter, driven by a single pump at the bottom end. (a) What is the minimum pressure at which the water must be pumped if it is to arrive at the village? (b) If 4 500 m³ of water is pumped per day, what is the speed of the water in the pipe? Note: Assume the free-fall acceleration and the

density of air are constant over this range of elevations. The pressures you calculate are too high for an ordinary pipe. The water is actually lifted in stages by several pumps through shorter pipes.

24. In ideal flow, a liquid of density 850 kg/m³ moves from a
Q|C horizontal tube of radius 1.00 cm into a second horizontal tube of radius 0.500 cm at the same elevation as the first tube. The pressure differs by ΔP between the liquid in one tube and the liquid in the second tube. (a) Find the volume flow rate as a function of ΔP. Evaluate the volume flow rate for (b) $\Delta P = 6.00$ kPa and (c) $\Delta P = 12.0$ kPa.

25. **Review.** Old Faithful Gey-
Q|C ser in Yellowstone National Park erupts at approximately one-hour intervals, and the height of the water column reaches 40.0 m (Fig. P14.25). (a) Model the rising stream as a series of separate droplets. Analyze the free-fall motion of one of the droplets to determine the speed at which the water leaves the ground. (b) **What If?** Model the rising stream as an ideal fluid in stream-line flow. Use Bernoulli's equation to determine the speed of the water as it leaves ground level. (c) How does

Figure P14.25

the answer from part (a) compare with the answer from part (b)? (d) What is the pressure (above atmospheric) in the heated underground chamber if its depth is 175 m? Assume the chamber is large compared with the geyser's vent.

26. You are working as an expert witness for the owner of a sky-
CR scraper complex in a downtown area. The owner is being sued by pedestrians on the streets below his buildings who were injured by falling glass when windows popped outward from the sides of the building. The Bernoulli effect can have important consequences for windows in such buildings. For example, wind can blow around a skyscraper at remarkably high speed, creating low pressure on the outside surface of the windows. The higher atmospheric pressure in the still air inside the buildings can cause windows to pop out. (a) In your research into the case, you find some overhead views of your client's project, as shown below. The project includes two tall skyscrapers and some park area on a square plot. Plan (i) (Fig. P14.26(i), page 382) was submitted by the original architects and planners. At the last minute, the owner decided he didn't want the park grounds to be divided into two areas and submitted Plan (ii) (Fig. P14.26(ii), which is the way the project was built. Explain to your client why Plan (ii) is a much more dangerous situation in terms of windows popping out than Plan (i). (b) Your client is not convinced by your conceptual argument in part (a), so you provide a numerical argument. Suppose a horizontal wind blows with a speed of 11.2 m/s outside a large pane of plate glass with dimensions 4.00 m × 1.50 m. Assume the density of the air to be constant at 1.20 kg/m³. The air inside the building is at atmospheric pressure. Calculate the total force exerted by air on the windowpane for your client. (c) **What If?** To further convince your client

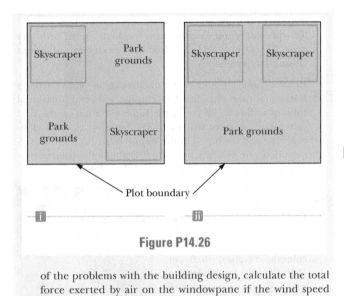

Figure P14.26

of the problems with the building design, calculate the total force exerted by air on the windowpane if the wind speed between the buildings is 22.4 m/s, twice as high as in part (b).

SECTION 14.7 Flow of Viscous Fluids in Pipes

27. A thin 1.50-mm coating of glycerin has been placed between two microscope slides of width 1.00 cm and length 4.00 cm. Find the force required to pull one of the microscope slides at a constant speed of 0.300 m/s relative to the other slide.

28. A hypodermic needle is 3.00 cm in length and 0.300 mm **BIO** in diameter. What pressure difference between the input and output of the needle is required so that the flow rate of water through it will be 1.00 g/s? (Use 1.00×10^{-3} Pa · s as the viscosity of water.)

29. What radius needle should be used to inject a volume of **BIO** 500 cm³ of a solution into a patient in 30.0 min? Assume the length of the needle is 2.50 cm and the solution is elevated 1.00 m above the point of injection. Further, assume the viscosity and density of the solution are those of pure water, and that the pressure inside the vein is atmospheric.

SECTION 14.8 Other Applications of Fluid Dynamics

30. An airplane has a mass of 1.60×10^4 kg, and each wing has **Q|C** an area of 40.0 m². During level flight, the pressure on the lower wing surface is 7.00×10^4 Pa. (a) Suppose the lift on the airplane were due to a pressure difference alone. Determine the pressure on the upper wing surface. (b) More realistically, a significant part of the lift is due to deflection of air downward by the wing. Does the inclusion of this force mean that the pressure in part (a) is higher or lower? Explain.

31. A siphon is used to drain water from a tank as illustrated in Figure P14.31. Assume steady flow without friction.

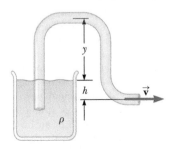

Figure P14.31

(a) If $h = 1.00$ m, find the speed of outflow at the end of the siphon. (b) **What If?** What is the limitation on the height of the top of the siphon above the end of the siphon? *Note:* For the flow of the liquid to be continuous, its pressure must not drop below its vapor pressure. Assume the water is at 20.0°C, at which the vapor pressure is 2.3 kPa.

ADDITIONAL PROBLEMS

32. Decades ago, it was thought that huge herbivorous dino-
BIO saurs such as *Apatosaurus* and *Brachiosaurus* habitually walked on the bottom of lakes, extending their long necks up to the surface to breathe. *Brachiosaurus* had its nostrils on the top of its head. In 1977, Knut Schmidt-Nielsen pointed out that breathing would be too much work for such a creature. For a simple model, consider a sample consisting of 10.0 L of air at absolute pressure 2.00 atm, with density 2.40 kg/m³, located at the surface of a freshwater lake. Find the work required to transport it to a depth of 10.3 m, with its temperature, volume, and pressure remaining constant. This energy investment is greater than the energy that can be obtained by metabolism of food with the oxygen in that quantity of air.

33. A helium-filled balloon (whose envelope **GP** has a mass of $m_b = 0.250$ kg) is tied to a uniform string of length $\ell = 2.00$ m and mass $m = 0.050\ 0$ kg. The balloon is spherical with a radius of $r = 0.400$ m. When released in air of temperature 20°C and density ρ_{air} = 1.20 kg/m³, it lifts a length h of string and then remains stationary as shown in Figure P14.33. We wish to find the length of string lifted by the balloon. (a) When the balloon remains stationary, what is the appropriate analysis model to describe it? (b) Write a force equation for the balloon from this model in terms of the buoyant force B, the weight F_b of the balloon, the weight F_{He} of the helium, and the weight F_s of the segment of string of length h. (c) Make an appropriate substitution for each of these forces and solve symbolically for the mass m_s of the segment of string of length h in terms of m_b, r, ρ_{air}, and the density of helium ρ_{He}. (d) Find the numerical value of the mass m_s. (e) Find the length h numerically.

Figure P14.33

34. The true weight of an object can be measured in a vacuum, **S** where buoyant forces are absent. A measurement in air, however, is disturbed by buoyant forces. An object of volume V is weighed in air on an equal-arm balance with the use of counterweights of density ρ. Representing the density of air as ρ_{air} and the balance reading as F'_g, show that the true weight F_g is

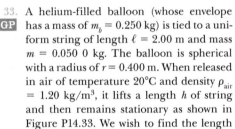

$$F_g = F'_g + \left(V - \frac{F'_g}{\rho g}\right)\rho_{air}g$$

35. To an order of magnitude, how many helium-filled toy balloons would be required to lift you? Because helium is an irreplaceable resource, develop a theoretical answer rather than an experimental answer. In your solution, state what physical quantities you take as data and the values you measure or estimate for them.

36. **Review.** Assume a certain liquid, with density 1 230 kg/m³,
Q|C exerts no friction force on spherical objects. A ball of mass 2.10 kg and radius 9.00 cm is dropped from rest into a deep tank of this liquid from a height of 3.30 m above the surface. (a) Find the speed at which the ball enters the

liquid. (b) Evaluate the magnitudes of the two forces that are exerted on the ball as it moves through the liquid. (c) Explain why the ball moves down only a limited distance into the liquid and calculate this distance. (d) With what speed will the ball pop up out of the liquid? (e) How does the time interval Δt_{down}, during which the ball moves from the surface down to its lowest point, compare with the time interval Δt_{up} for the return trip between the same two points? (f) **What If?** Now modify the model to suppose the liquid exerts a small friction force on the ball, opposite in direction to its motion. In this case, how do the time intervals Δt_{down} and Δt_{up} compare? Explain your answer with a conceptual argument rather than a numerical calculation.

37. Evangelista Torricelli was the first person to realize that we live at the bottom of an ocean of air. He correctly surmised that the pressure of our atmosphere is attributable to the weight of the air. The density of air at 0°C at the Earth's surface is 1.29 kg/m³. The density decreases with increasing altitude (as the atmosphere thins). On the other hand, if we assume the density is constant at 1.29 kg/m³ up to some altitude h and is zero above that altitude, then h would represent the depth of the ocean of air. (a) Use this model to determine the value of h that gives a pressure of 1.00 atm at the surface of the Earth. (b) Would the peak of Mount Everest rise above the surface of such an atmosphere?

38. A common parameter that can be used to predict turbu- **BIO** lence in fluid flow is called the *Reynolds number*. The Rey- **QC** nolds number for fluid flow in a pipe is a dimensionless quantity defined as

$$\text{Re} = \frac{\rho v d}{\eta}$$

where ρ is the density of the fluid, v is its speed, d is the inner diameter of the pipe, and η is the viscosity of the fluid. The criteria for the type of flow are as follows:

- If Re < 2 300, the flow is laminar.
- If 2 300 < Re < 4 000, the flow is in a transition region between laminar and turbulent.
- If Re > 4 000, the flow is turbulent.

(a) Let's model blood of density 1.06×10^3 kg/m³ and viscosity 3.00×10^{-3} Pa · s as a pure liquid, that is, ignore the fact that it contains red blood cells. Suppose it is flowing in a large artery of radius 1.50 cm with a speed of 0.067 0 m/s. Show that the flow is laminar. (b) Imagine that the artery ends in a *single* capillary so that the radius of the artery reduces to a much smaller value. What is the radius of the capillary that would cause the flow to become turbulent? (c) Actual capillaries have radii of about 5–10 micrometers, much smaller than the value in part (b). Why doesn't the flow in actual capillaries become turbulent?

39. In 1983, the United States began coining the one-cent piece out of copper-clad zinc rather than pure copper. The mass of the old copper penny is 3.083 g and that of the new cent is 2.517 g. The density of copper is 8.920 g/cm³ and that of zinc is 7.133 g/cm³. The new and old coins have the same volume. Calculate the percent of zinc (by volume) in the new cent.

40. **Review.** With reference to the dam studied in Example 14.4 **S** and shown in Figure 14.5, (a) show that the total torque exerted by the water behind the dam about a horizontal axis through O is $\frac{1}{6}\rho g w H^3$. (b) Show that the effective line of

action of the total force exerted by the water is at a distance $\frac{1}{3}H$ above O.

41. The *spirit-in-glass thermometer*, invented in Florence, Italy, around 1654, consists of a tube of liquid (the spirit) containing a number of submerged glass spheres with slightly different masses (Fig. P14.41). At sufficiently low temperatures, all the spheres float, but as the temperature rises, the spheres sink one after another. The device is a crude but interesting tool for measuring temperature. Suppose the tube is filled with ethyl alcohol, whose density is 0.789 45 g/cm³ at 20.0°C and decreases to 0.780 97 g/cm³ at 30.0°C. (a) Assuming that one of the spheres has a radius of 1.000 cm and is in equilibrium halfway up the tube at 20.0°C, determine its mass. (b) When the temperature increases to 30.0°C, what mass must a second sphere of the same radius have to be in equilibrium at the halfway point? (c) At 30.0°C, the first sphere has fallen to the bottom of the tube. What upward force does the bottom of the tube exert on this sphere?

Figure P14.41

42. A woman is draining her fish tank **S** by siphoning the water into an outdoor drain as shown in Figure P14.42. The rectangular tank has footprint area A and depth h. The drain is located a distance d below the surface of the water in the tank, where $d \gg h$. The cross-sectional area of the siphon tube is A'. Model the water as flowing without friction. Show that the time interval required to empty the tank is given by

$$\Delta t = \frac{Ah}{A'\sqrt{2gd}}$$

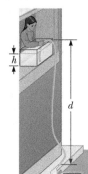
Figure P14.42

43. **Review.** You and your father are designing a waterfall for **CR** your backyard swimming pool. At the top of the waterfall is a tank containing water that is kept to a depth $d = 0.280$ m by a pump. As shown in Figure P14.43, there is a small hatch of height $h = 0.100$ m and width $w = 0.150$ m, hinged at the top, that can be used to turn on and turn off the supply of water to the waterfall. You want to attach a simple latch at the center of the bottom of the hatch and are trying to decide what type of latch to buy at your local hardware store. To make that decision, you need to determine the force that the latch must be able to withstand to keep the hatch closed.

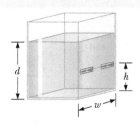
Figure P14.43 Problems 43 and 44.

44. Review. You and your father are designing a waterfall for your backyard swimming pool. At the top of the waterfall is a tank containing water that is kept to a depth d by a pump. As shown in Figure P14.43, there is a small hatch of height h and width w, hinged at the top, that can be used to turn on and turn off the supply of water to the waterfall. You want to attach a simple latch at the center of the bottom of the hatch and are trying to decide what type of latch to buy at your local hardware store. To make that decision, you need to determine the force that the latch must be able to withstand to keep the hatch closed.

45. Review. A uniform disk of mass 10.0 kg and radius 0.250 m spins at 300 rev/min on a low-friction axle. It must be brought to a stop in 1.00 min by a brake pad that makes contact with the disk at an average distance 0.220 m from the axis. The coefficient of friction between pad and disk is 0.500. A piston in a cylinder of diameter 5.00 cm presses the brake pad against the disk. Find the pressure required for the brake fluid in the cylinder.

46. Review. In a water pistol, a piston drives water through a large tube of area A_1 into a smaller tube of area A_2 as shown in Figure P14.46. The radius of the large tube is 1.00 cm and that of the small tube is 1.00 mm. The smaller tube is 3.00 cm above the larger tube. (a) If the pistol is fired horizontally at a height of 1.50 m, determine the time interval required for the water to travel from the nozzle to the ground. Neglect air resistance and assume atmospheric pressure is 1.00 atm. (b) If the desired range of the stream is 8.00 m, with what speed v_2 must the stream leave the nozzle? (c) At what speed v_1 must the plunger be moved to achieve the desired range? (d) What is the pressure at the nozzle? (e) Find the pressure needed in the larger tube. (f) Calculate the force that must be exerted on the trigger to achieve the desired range. (The force that must be exerted is due to pressure over and above atmospheric pressure.)

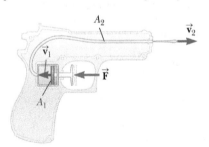

Figure P14.46

47. An incompressible, nonviscous fluid is initially at rest in the vertical portion of the pipe shown in Figure P14.47a, where

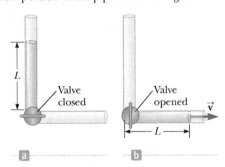

Figure P14.47

$L = 2.00$ m. When the valve is opened, the fluid flows into the horizontal section of the pipe. What is the fluid's speed when all the fluid is in the horizontal section as shown in Figure P14.47b? Assume the cross-sectional area of the entire pipe is constant.

48. The hull of an experimental boat is to be lifted above the water by a hydrofoil mounted below its keel as shown in Figure P14.48. The hydrofoil has a shape like that of an airplane wing. Its area projected onto a horizontal surface is A. When the boat is towed at sufficiently high speed, water of density ρ moves in streamline flow so that its average speed at the top of the hydrofoil is n times larger than its speed v_b below the hydrofoil. (a) Ignoring the buoyant force, show that the upward lift force exerted by the water on the hydrofoil has a magnitude

$$F \approx \tfrac{1}{2}(n^2 - 1)\rho v_b^2 A$$

(b) The boat has mass M. Show that the liftoff speed is given by

$$v \approx \sqrt{\dfrac{2Mg}{(n^2 - 1)A\rho}}$$

Figure P14.48

CHALLENGE PROBLEMS

49. Show that the variation of atmospheric pressure with altitude is given by $P = P_0 e^{-\alpha y}$, where $\alpha = \rho_0 g/P_0$, P_0 is atmospheric pressure at some reference level $y = 0$, and ρ_0 is the atmospheric density at this level. Assume the decrease in atmospheric pressure over an infinitesimal change in altitude (so that the density is approximately uniform over the infinitesimal change) can be expressed from Equation 14.4 as $dP = -\rho g\, dy$. Also assume the density of air is proportional to the pressure, which, as we will see in Chapter 18, is equivalent to assuming the temperature of the air is the same at all altitudes.

50. *Why is the following situation impossible?* A barge is carrying a load of small pieces of iron along a river. The iron pile is in the shape of a cone for which the radius r of the base of the cone is equal to the central height h of the cone. The barge is square in shape, with vertical sides of length $2r$, so that the pile of iron comes just up to the edges of the barge. The barge approaches a low bridge, and the captain realizes that the top of the pile of iron is not going to make it under the bridge. The captain orders the crew to shovel iron pieces from the pile into the water to reduce the height of the pile. As iron is shoveled from the pile, the pile always has the shape of a cone whose diameter is equal to the side length of the barge. After a certain volume of iron is removed from the barge, it makes it under the bridge without the top of the pile striking the bridge.

Oscillations and Mechanical Waves

In Part 1 of this text, we focused on one particular energy transfer term in Equation 8.2: work W. In Parts 2 through 5, we will focus our efforts in each part on a new term in Equation 8.2. Here in Part 2, we will investigate the term T_{MW}: transfer of energy by *mechanical waves*. We begin this new part of the text by studying a special type of motion called *periodic* motion, the repeating motion of an object in which it continues to return to a given position after a fixed time interval. The repetitive movements of such an object are called *oscillations*. We will focus our attention on a special case of periodic motion called *simple harmonic motion*. All periodic motions can be modeled as combinations of simple harmonic motions.

Simple harmonic motion also forms the basis for our understanding of mechanical waves. Sound waves, seismic waves, waves on stretched strings, and water waves are all produced by some source of oscillation. As a sound wave travels through the air, elements of the air oscillate back and forth; as a water wave travels across a pond, elements of the water oscillate up and down and backward and forward.

To explain many other phenomena in nature, we must understand the concepts of oscillations and waves. For instance, although skyscrapers and bridges appear to be rigid, they actually oscillate, something the architects and engineers who design and build them must take into account. To understand how radio and television work, we must understand the origin and nature of electromagnetic waves and how they propagate through space. Finally, much of what scientists have learned about atomic structure has come from information carried by waves. ■

Falling drops of water cause a water surface to oscillate. These oscillations are associated with circular waves moving away from the point at which the drops fall. In Part 2 of the text, we will explore the principles related to oscillations and waves. (*Ziga Camernik/Shutterstock*)

15

A grandfather clock keeps time in a room. The timing mechanism depends on the swinging of a pendulum. This repetitive swinging is an example of oscillatory motion. (*Antonio Gravante/Shutterstock*)

Oscillatory Motion

STORYLINE **In the previous chapter, you were taking off in an** airplane from Denver, Colorado, to Boston, Massachusetts. You are now visiting your grandparents in Boston. There is an antique grandfather clock keeping time in one of the rooms. The gentle clicking of the swinging pendulum is relaxing to you. You recall that your grandparents used to live in Denver, like your other set of grandparents, and then moved from Denver to Boston, bringing the clock with them. Your grandmother enters the room and you mention your childhood memories of the clock. She tells you that she had it calibrated professionally in Denver and it kept perfect time for years. After they moved it here to their Massachusetts house, it has not been accurate. It runs too fast and has to be reset to the correct time every few days. You ask your grandmother what the clock shop in Denver did to calibrate the clock, but she doesn't know. You wonder—could *you* do something to calibrate the clock?

CONNECTIONS This is a bridging chapter. For the most part so far, we have considered motion that occurs once and does not repeat—a thrown ball, an accelerating car, a pushed crate. In Section 4.4, we saw our first example of *repeating* motion: a particle moving in a circular path returns to the starting point and performs the same motion over and over. In this chapter, we will be applying the principles of mechanics to the special case of an *oscillating* object. From this point of view, this chapter is based on understanding a new and important situation based on material we have studied in previous chapters. On the other hand, oscillations are the basis for understanding all types of *waves*. We mentioned mechanical waves and electromagnetic waves briefly in Section 8.1 and will study mechanical waves in the next two chapters and electromagnetic waves in Chapter 33. Therefore, this chapter, while based on principles from the *past*, is preparing us for our *future* study of waves.

15.1 Motion of an Object Attached to a Spring

As a model for oscillatory motion, consider a block of mass m attached to the end of a spring, with the block free to move on a frictionless, horizontal surface (Fig. 15.1). When the spring is neither stretched nor compressed, the block is at rest at the position called the **equilibrium position** of the system, which we identify as $x = 0$ (Fig. 15.1b). We know from experience that such a system oscillates back and forth if disturbed from its equilibrium position.

We can understand the oscillating motion of the block in Figure 15.1 qualitatively by first recalling that when the block is displaced to a position x, the spring exerts on the block a force that is proportional to the position and given by **Hooke's law** (see Section 7.4):

$$F_s = -kx \tag{15.1}$$

◀ Hooke's law

We call F_s a **restoring force** because it is always directed toward the equilibrium position and therefore *opposite* the displacement of the block from equilibrium. That is, when the block is displaced to the right of $x = 0$ in Figure 15.1a, the position is positive and the restoring force is directed to the left. When the block is displaced to the left of $x = 0$ as in Figure 15.1c, the position is negative and the restoring force is directed to the right.

When the block is displaced from the equilibrium point and released, it is a particle under a net force and consequently undergoes an acceleration. Applying the particle under a net force model to the motion of the block, with Equation 15.1 providing the net force in the x direction, we obtain

$$\sum F_x = ma_x \rightarrow -kx = ma_x$$

$$a_x = -\frac{k}{m} x \tag{15.2}$$

That is, the acceleration of the block is proportional to its position, and the direction of the acceleration is opposite the direction of the displacement of the block from equilibrium. Systems that behave in this way are said to exhibit **simple harmonic motion.** An object moves with simple harmonic motion whenever its acceleration is proportional to its position and is oppositely directed to the displacement from equilibrium.

PITFALL PREVENTION 15.1
The Orientation of the Spring
Figure 15.1 shows a *horizontal* spring, with an attached block sliding on a frictionless surface. Another possibility is a block hanging from a *vertical* spring. All the results we discuss for the horizontal spring are the same for the vertical spring with one exception: when the block is placed on the vertical spring, its weight causes the spring to extend. If the resting position of the block on the extended spring is defined as $x = 0$, the results of this chapter also apply to this vertical system.

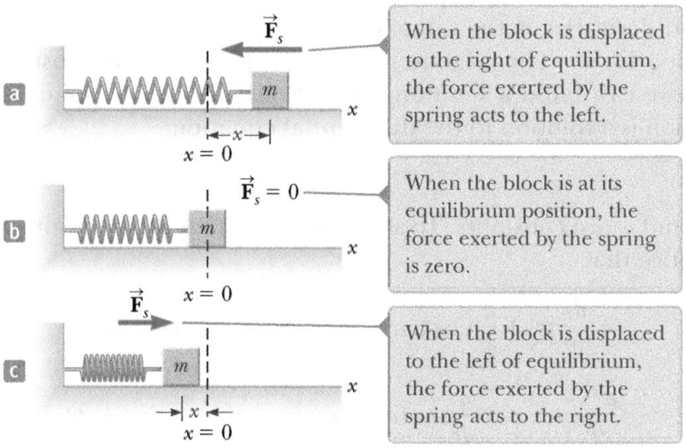

When the block is displaced to the right of equilibrium, the force exerted by the spring acts to the left.

When the block is at its equilibrium position, the force exerted by the spring is zero.

When the block is displaced to the left of equilibrium, the force exerted by the spring acts to the right.

Figure 15.1 A block attached to a spring and moving on a frictionless surface.

If the block in Figure 15.1 is displaced to a position $x = A$ and released from rest, its *initial* acceleration is $-kA/m$. When the block passes through the equilibrium position $x = 0$, its acceleration is zero. At this instant, its speed is a maximum because the acceleration changes sign. The block then continues to travel to the left of equilibrium with a positive acceleration and finally reaches $x = -A$, at which time its acceleration is $+kA/m$ and its speed is again zero as discussed in Sections 7.4 and 7.9. The block completes a full cycle of its motion by returning to the original position, again passing through $x = 0$ with maximum speed. Therefore, the block oscillates between the turning points $x = \pm A$. In the absence of friction, this idealized motion will continue forever because the force exerted by the spring is conservative. Real systems are generally subject to friction, so they do not oscillate forever. We shall explore the details of the situation with friction in Section 15.6.

ⓠUICK QUIZ 15.1 A block on the end of a spring is pulled to position $x = A$ and released from rest. In one full cycle of its motion, through what total distance does it travel? **(a)** $A/2$ **(b)** A **(c)** $2A$ **(d)** $4A$

15.2 Analysis Model: Particle in Simple Harmonic Motion

The idealized motion described in the preceding section is the basis for so many real motions of objects that we identify the **particle in simple harmonic motion** model to represent such situations. To develop a mathematical representation for this model, we will generally choose x as the axis along which the oscillation of an object occurs; hence, we will drop the subscript-x notation in this discussion. Recall that, by definition, $a = dv/dt = d^2x/dt^2$, so we can express Equation 15.2 as

$$\frac{d^2x}{dt^2} = -\frac{k}{m}x \tag{15.3}$$

> **PITFALL PREVENTION 15.2**
> **A Nonconstant Acceleration** The acceleration of a particle in simple harmonic motion is not constant. Equation 15.3 shows that its acceleration varies with position x. Therefore, we *cannot* apply the kinematic equations of Chapter 2 in this situation. Those equations describe a particle under *constant acceleration*.

If we denote the ratio k/m with the symbol ω^2 (we choose ω^2 rather than ω so as to make the solution we develop below simpler in form), then

$$\omega^2 = \frac{k}{m} \tag{15.4}$$

and Equation 15.3 can be written in the form

$$\frac{d^2x}{dt^2} = -\omega^2 x \tag{15.5}$$

Let's now find a mathematical solution to Equation 15.5, that is, a function $x(t)$ that satisfies this second-order differential equation and is a mathematical representation of the position of the particle as a function of time. We seek a function whose second derivative is the same as the original function with a negative sign and multiplied by ω^2. The trigonometric functions sine and cosine exhibit this behavior, so we can build a solution around one or both of them. The following cosine function is a solution to the differential equation:

▶ Position versus time for a particle in simple harmonic motion

$$x(t) = A \cos(\omega t + \phi) \tag{15.6}$$

where A, ω, and ϕ are constants. To show explicitly that this solution satisfies Equation 15.5, notice that

$$\frac{dx}{dt} = A\frac{d}{dt}\cos(\omega t + \phi) = -\omega A \sin(\omega t + \phi) \tag{15.7}$$

$$\frac{d^2x}{dt^2} = -\omega A\frac{d}{dt}\sin(\omega t + \phi) = -\omega^2 A \cos(\omega t + \phi) \tag{15.8}$$

Comparing Equations 15.6 and 15.8, we see that $d^2x/dt^2 = -\omega^2 x$ and Equation 15.5 is satisfied.

The parameters A, ω, and ϕ are constants of the motion. To give physical significance to these constants, it is convenient to form a graphical representation of the motion by plotting x as a function of t as in Figure 15.2a. First, A, called the **amplitude** of the motion, is simply the maximum value of the position of the particle in either the positive or negative x direction. The constant ω is called the **angular frequency,** and it has units[1] of radians per second. It is a measure of how rapidly the oscillations are occurring; the more oscillations per unit time, the higher the value of ω. From Equation 15.4, the angular frequency is

$$\omega = \sqrt{\frac{k}{m}} \qquad (15.9)$$

The quantity $(\omega t + \phi)$ in Equation 15.6 is called the **phase** of the motion. The constant angle ϕ is called the **phase constant** (or initial phase angle) and, along with the amplitude A, is determined uniquely by the position and velocity of the particle at $t = 0$. Therefore, A and ϕ are two parameters that define the *initial conditions* of the motion of an oscillating object, just as x_i and v_i describe the initial conditions of an object undergoing constant acceleration in Equation 2.16. If the particle is at its maximum position $x = A$ at $t = 0$, the phase constant is $\phi = 0$ and the graphical representation of the motion is as shown in Figure 15.2b. Notice that the function $x(t)$ is periodic and its value is the same each time ωt increases by 2π radians.

Equations 15.1, 15.5, and 15.6 form the basis of the mathematical representation of the particle in simple harmonic motion model. If you are analyzing a situation and find that the force on an object modeled as a particle is of the mathematical form of Equation 15.1, you know the motion is that of a simple harmonic oscillator and the position of the particle is described by Equation 15.6. If you analyze a system and find that it is described by a differential equation of the form of Equation 15.5, the motion is that of a simple harmonic oscillator. If you analyze a situation and find that the position of a particle is described by Equation 15.6, you know the particle undergoes simple harmonic motion.

QUICK QUIZ 15.2 Consider a graphical representation (Fig. 15.3) of simple harmonic motion as described mathematically in Equation 15.6. When the particle is at point Ⓐ on the graph, what can you say about its position and velocity? **(a)** The position and velocity are both positive. **(b)** The position and velocity are both negative. **(c)** The position is positive, and the velocity is zero. **(d)** The position is negative, and the velocity is zero. **(e)** The position is positive, and the velocity is negative. **(f)** The position is negative, and the velocity is positive.

QUICK QUIZ 15.3 Figure 15.4 shows two curves representing particles undergoing simple harmonic motion. The correct description of these two motions is that the simple harmonic motion of particle B is **(a)** of larger angular frequency and larger amplitude than that of particle A, **(b)** of larger angular frequency and smaller amplitude than that of particle A, **(c)** of smaller angular frequency and larger amplitude than that of particle A, or **(d)** of smaller angular frequency and smaller amplitude than that of particle A.

Let us investigate further the mathematical description of simple harmonic motion. The **period** T of the motion is the time interval required for the particle to go through one full cycle of its motion (Fig. 15.2a). That is, the values of x and v

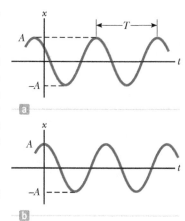

Figure 15.2 (a) An x–t graph for a particle undergoing simple harmonic motion. The amplitude of the motion is A, and the period is T. (b) The x–t graph for the special case in which $x = A$ at $t = 0$ and hence $\phi = 0$.

Figure 15.3 (Quick Quiz 15.2) An x–t graph for a particle undergoing simple harmonic motion. At a particular time, the particle's position is indicated by Ⓐ in the graph.

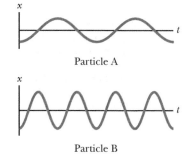

Particle A

Particle B

Figure 15.4 (Quick Quiz 15.3) Two x–t graphs for particles undergoing simple harmonic motion. The amplitudes and frequencies are different for the two particles.

---

[1]We have seen many examples in earlier chapters in which we evaluate a trigonometric function of an angle. The argument of a trigonometric function, such as the cosine function in Equation 15.6, *must* be a pure number with no units. The radian is a pure number because it is a ratio of lengths. Therefore, ω *must* be expressed in radians per second (and not, for example, in revolutions per second) if t is expressed in seconds.

for the particle at time *t* equal the values of *x* and *v* at time *t* + *T*. Because the phase increases by 2π radians in a time interval of *T*,

$$[\omega(t + T) + \phi] - (\omega t + \phi) = 2\pi$$

Simplifying this expression gives $\omega T = 2\pi$, or

$$T = \frac{2\pi}{\omega} \tag{15.10}$$

The inverse of the period is called the **frequency** *f* of the motion. Whereas the period is the time interval per oscillation, the frequency represents the number of oscillations the particle undergoes per unit time interval:

$$f = \frac{1}{T} = \frac{\omega}{2\pi} \tag{15.11}$$

The units of *f* are cycles per second, or **hertz** (Hz). Rearranging Equation 15.11 gives

$$\omega = 2\pi f = \frac{2\pi}{T} \tag{15.12}$$

Equations 15.9 through 15.11 can be used to express the period and frequency of the motion for the particle in simple harmonic motion in terms of the characteristics *m* and *k* of the system as

Period of a simple ▶
harmonic oscillator

$$T = \frac{2\pi}{\omega} = 2\pi\sqrt{\frac{m}{k}} \tag{15.13}$$

Frequency of a simple ▶
harmonic oscillator

$$f = \frac{1}{T} = \frac{1}{2\pi}\sqrt{\frac{k}{m}} \tag{15.14}$$

That is, the period and frequency depend *only* on the mass of the particle and the force constant of the spring and *not* on the parameters of the motion, such as *A* or ϕ. As we might expect, the frequency is larger for a stiffer spring (larger value of *k*) and decreases with increasing mass of the particle.

We can obtain the velocity and acceleration[2] of a particle undergoing simple harmonic motion from Equations 15.7 and 15.8:

Velocity as a function ▶
of time for a simple
harmonic oscillator

$$v = \frac{dx}{dt} = -\omega A \sin(\omega t + \phi) \tag{15.15}$$

Acceleration as a function ▶
of time for a simple
harmonic oscillator

$$a = \frac{d^2x}{dt^2} = -\omega^2 A \cos(\omega t + \phi) \tag{15.16}$$

From Equation 15.15, we see that because the sine and cosine functions oscillate between ±1, the extreme values of the velocity *v* are $\pm\omega A$. Likewise, Equation 15.16 shows that the extreme values of the acceleration *a* are $\pm\omega^2 A$. Therefore, the *maximum* values of the magnitudes of the velocity and acceleration are

Maximum magnitudes of ▶
velocity and acceleration in
simple harmonic motion

$$v_{\text{max}} = \omega A = \sqrt{\frac{k}{m}}\,A \tag{15.17}$$

$$a_{\text{max}} = \omega^2 A = \frac{k}{m}\,A \tag{15.18}$$

---

[2] Because the motion of a simple harmonic oscillator takes place in one dimension, we denote velocity as *v* and acceleration as *a*, with the direction indicated by a positive or negative sign as in Chapter 2.

Figure 15.5a plots position versus time for an arbitrary value of the phase constant. The associated velocity–time and acceleration–time curves are illustrated in Figures 15.5b and 15.5c, respectively. It is evident that all three curves have the same general shape. The phase of the velocity, however, differs from the phase of the position by $\pi/2$ rad, or 90°. That is, when x is a maximum or a minimum, the velocity is zero. Likewise, when x is zero, the speed is a maximum. Furthermore, notice that the phase of the acceleration differs from the phase of the position by π radians, or 180°. For example, when x is a maximum, a has a maximum magnitude in the opposite direction.

Ⓠ**UICK QUIZ 15.4** An object of mass m is hung from a spring and set into oscillation. The period of the oscillation is measured and recorded as T. The object of mass m is removed and replaced with an object of mass $2m$. When this object is set into oscillation, what is the period of the motion? **(a)** $2T$ **(b)** $\sqrt{2}\,T$ **(c)** T **(d)** $T/\sqrt{2}$ **(e)** $T/2$

Equation 15.6 describes simple harmonic motion of a particle in terms of three constants of the motion. Let's now see how to evaluate these constants. The angular frequency ω is evaluated using Equation 15.9. The constants A and ϕ are evaluated from the initial conditions, that is, the state of the oscillator at $t = 0$.

Suppose a block is set into motion by pulling it from equilibrium by a distance A and releasing it from rest at $t = 0$ as in Figure 15.6. We must then require our solutions for $x(t)$ and $v(t)$ (Eqs. 15.6 and 15.15) to obey the initial conditions that $x(0) = A$ and $v(0) = 0$:

$$x(0) = A \cos \phi = A$$

$$v(0) = -\omega A \sin \phi = 0$$

These conditions are met if $\phi = 0$, giving $x = A \cos \omega t$ as our solution. To check this solution, notice that it satisfies the condition that $x(0) = A$ because $\cos 0 = 1$.

The position, velocity, and acceleration of the block versus time are plotted in Figure 15.7a for this special case. The acceleration reaches extreme values of $\mp\omega^2 A$ when the position has extreme values of $\pm A$. Furthermore, the velocity has extreme values of $\pm\omega A$, which both occur at $x = 0$. Hence, the quantitative solution agrees with our qualitative description of this system.

Let's consider another possibility. Suppose the system is oscillating and we define $t = 0$ as the instant the block passes through the unstretched position of the spring while moving to the right (Fig. 15.8). In this case, our solutions for $x(t)$ and $v(t)$ must obey the initial conditions that $x(0) = 0$ and $v(0) = v_i$:

$$x(0) = A \cos \phi = 0$$

$$v(0) = -\omega A \sin \phi = v_i$$

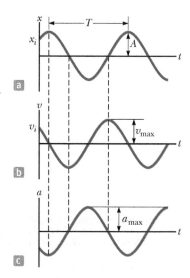

Figure 15.5 Graphical representation of simple harmonic motion. (a) Position versus time. (b) Velocity versus time. (c) Acceleration versus time. Notice that at any specified time the velocity is 90° out of phase with the position and the acceleration is 180° out of phase with the position.

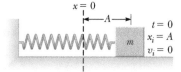

Figure 15.6 A block–spring system that begins its motion from rest with the block at $x = A$ at $t = 0$.

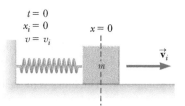

Figure 15.8 The block–spring system is undergoing oscillation, and $t = 0$ is defined at an instant when the block passes through the equilibrium position $x = 0$ and is moving to the right with speed v_i.

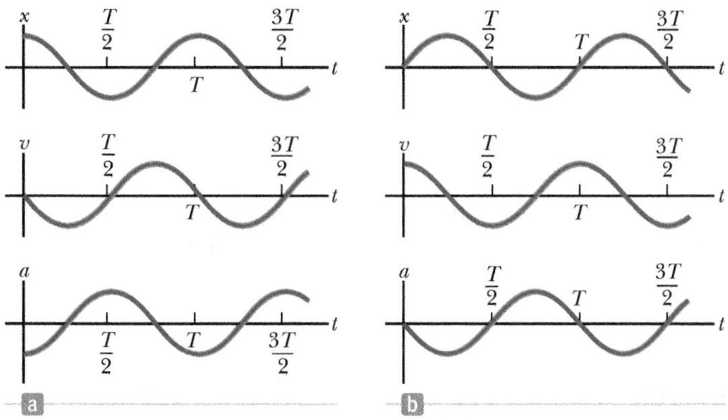

Figure 15.7 (a) Position, velocity, and acceleration versus time for the block in Figure 15.6 under the initial conditions that at $t = 0$, $x(0) = A$, and $v(0) = 0$. (b) Position, velocity, and acceleration versus time for the block in Figure 15.8 under the initial conditions that at $t = 0$, $x(0) = 0$, and $v(0) = v_i$.

The first of these conditions tells us that $\phi = \pm\pi/2$. With these choices for ϕ, the second condition tells us that $A = \mp v_i/\omega$. Because the initial velocity is positive and the amplitude must be positive, we must have $\phi = -\pi/2$. Hence, the solution is

$$x = \frac{v_i}{\omega} \cos \left(\omega t - \frac{\pi}{2} \right)$$

The graphs of position, velocity, and acceleration versus time for this choice of $t = 0$ are shown in Figure 15.7b. Notice that these curves are the same as those in Figure 15.7a, but shifted to the right by one-fourth of a cycle. This shift is described mathematically by the phase constant $\phi = -\pi/2$, which is one-fourth of a full cycle of 2π.

ANALYSIS MODEL   Particle in Simple Harmonic Motion

Imagine an object that is subject to a force that is proportional to the negative of the object's position, $F = -kx$ (Eq. 15.1). Such a force equation is known as Hooke's law, and it describes the force applied to an object attached to an ideal spring. The parameter k in Hooke's law is called the *spring constant* or the *force constant*. The position of an object acted on by a force described by Hooke's law is given by

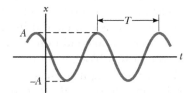

$$x(t) = A \cos (\omega t + \phi) \tag{15.6}$$

where A is the **amplitude** of the motion, ω is the **angular frequency**, and ϕ is the **phase constant**. The values of A and ϕ depend on the initial position and initial velocity of the particle.

The **period** of the oscillation of the particle is

$$T = \frac{2\pi}{\omega} = 2\pi \sqrt{\frac{m}{k}} \tag{15.13}$$

and the inverse of the period is the **frequency.**

Examples:

- a bungee jumper hangs from a bungee cord and oscillates up and down
- a guitar string vibrates back and forth in a standing wave, with each element of the string moving in simple harmonic motion (Chapter 17)
- a piston in a gasoline engine oscillates up and down within the cylinder of the engine (Chapter 21)
- an atom in a diatomic molecule vibrates back and forth as if it is connected by a spring to the other atom in the molecule (Chapter 42)

Example 15.1   A Block–Spring System

A 200-g block connected to a light spring for which the force constant is 5.00 N/m is free to oscillate on a frictionless, horizontal surface. The block is displaced 5.00 cm from equilibrium and released from rest as in Figure 15.6.

(A) Find the period of its motion.

SOLUTION

Conceptualize   Study Figure 15.6 and imagine the block moving back and forth in simple harmonic motion once it is released. Set up an experimental model in the vertical direction by hanging a heavy object such as a stapler from a strong rubber band.

Categorize   The block is modeled as a *particle in simple harmonic motion.*

Analyze

Use Equation 15.9 to find the angular frequency of the block–spring system:

$$\omega = \sqrt{\frac{k}{m}} = \sqrt{\frac{5.00 \text{ N/m}}{200 \times 10^{-3} \text{ kg}}} = 5.00 \text{ rad/s}$$

Use Equation 15.13 to find the period of the system:

$$T = \frac{2\pi}{\omega} = \frac{2\pi}{5.00 \text{ rad/s}} = 1.26 \text{ s}$$

15.1 continued

(B) Determine the maximum speed of the block.

SOLUTION

Use Equation 15.17 to find v_{max}:

$$v_{max} = \omega A = (5.00 \text{ rad/s})(5.00 \times 10^{-2} \text{ m}) = 0.250 \text{ m/s}$$

(C) What is the maximum acceleration of the block?

SOLUTION

Use Equation 15.18 to find a_{max}:

$$a_{max} = \omega^2 A = (5.00 \text{ rad/s})^2(5.00 \times 10^{-2} \text{ m}) = 1.25 \text{ m/s}^2$$

(D) Express the position, velocity, and acceleration as functions of time in SI units.

SOLUTION

Find the phase constant from the initial condition that $x = A$ at $t = 0$:

$$x(0) = A \cos \phi = A \rightarrow \phi = 0$$

Use Equation 15.6 to write an expression for $x(t)$:

$$x = A \cos (\omega t + \phi) = 0.050\,0 \cos 5.00t$$

Use Equation 15.15 to write an expression for $v(t)$:

$$v = -\omega A \sin (\omega t + \phi) = -0.250 \sin 5.00t$$

Use Equation 15.16 to write an expression for $a(t)$:

$$a = -\omega^2 A \cos (\omega t + \phi) = -1.25 \cos 5.00t$$

Finalize   Consider part (a) of Figure 15.7, which shows the graphical representations of the motion of the block in this problem. Make sure that the mathematical representations found above in part (D) are consistent with these graphical representations.

WHAT IF?   What if the block were released from the same initial position, $x_i = 5.00$ cm, but with an initial velocity of $v_i = -0.100$ m/s? Which parts of the solution change, and what are the new answers for those that do change?

Answers   Part (A) does not change because the period is independent of how the oscillator is set into motion. Parts (B), (C), and (D) will change.

Write position and velocity expressions for the initial conditions:

$$(1) \quad x(0) = A \cos \phi = x_i$$

$$(2) \quad v(0) = -\omega A \sin \phi = v_i$$

Divide Equation (2) by Equation (1) to find the phase constant:

$$\frac{-\omega A \sin \phi}{A \cos \phi} = \frac{v_i}{x_i}$$

$$\tan \phi = -\frac{v_i}{\omega x_i} = -\frac{-0.100 \text{ m/s}}{(5.00 \text{ rad/s})(0.050\,0 \text{ m})} = 0.400$$

$$\phi = \tan^{-1} (0.400) = 0.121\pi$$

Use Equation (1) to find A:

$$A = \frac{x_i}{\cos \phi} = \frac{0.050\,0 \text{ m}}{\cos (0.121\pi)} = 0.053\,9 \text{ m}$$

Find the new maximum speed:

$$v_{max} = \omega A = (5.00 \text{ rad/s})(5.39 \times 10^{-2} \text{ m}) = 0.269 \text{ m/s}$$

Find the new magnitude of the maximum acceleration:

$$a_{max} = \omega^2 A = (5.00 \text{ rad/s})^2(5.39 \times 10^{-2} \text{ m}) = 1.35 \text{ m/s}^2$$

Find new expressions for position, velocity, and acceleration in SI units:

$$x = 0.053\,9 \cos (5.00t + 0.121\pi)$$

$$v = -0.269 \sin (5.00t + 0.121\pi)$$

$$a = -1.35 \cos (5.00t + 0.121\pi)$$

As we saw in Chapters 7 and 8, many problems are easier to solve using an energy approach rather than one based on variables of motion. This particular What If? is easier to solve from an energy approach. Therefore, we shall investigate the energy of the simple harmonic oscillator in the next section.

| Example **15.2** | **More Details of the Block–Spring System** |

Consider again the block–spring system in Example 15.1, whose position, velocity, and acceleration are given in part (D) of the problem. Find a general expression for all times at which the block is located at $x = +\frac{1}{2}A$.

SOLUTION

Conceptualize An important factor to keep in mind is that the block will be located at the requested position *twice* during each cycle. Our general expression should reflect that fact.

Categorize As in Example 15.1, the block is modeled as a *particle in simple harmonic motion*.

Analyze Write an expression for the position of the block knowing that the phase constant is equal to zero:

$$x = A \cos\omega t$$

Enter the condition that the position be half the amplitude and solve for t:

$$\tfrac{1}{2}A = A \cos\omega t \quad \rightarrow \quad t = \frac{1}{\omega} \cos^{-1}\left(\tfrac{1}{2}\right)$$

Recognize that there are two angles in the first cycle at which the inverse cosine is one-half, plus additional angles can be found by adding integral multiples of 2π:

$$\begin{cases} \cos^{-1}\left(\tfrac{1}{2}\right) = \dfrac{\pi}{3} + 2\pi n = \dfrac{\pi}{3}(1 + 6n) & n = 0, 1, 2, \ldots \\[2mm] \cos^{-1}\left(\tfrac{1}{2}\right) = \dfrac{5\pi}{3} + 2\pi n = \dfrac{\pi}{3}(5 + 6n) & n = 0, 1, 2, \ldots \end{cases}$$

Substitute these angles into the expression for t:

$$t = \frac{1}{5.00 \text{ s}^{-1}} \cos^{-1}\left(\tfrac{1}{2}\right) = \frac{\pi}{15.0 \text{ s}^{-1}}(1 + 6n) \quad \text{or} \quad \frac{\pi}{15.0 \text{ s}^{-1}}(5 + 6n) \quad n = 0, 1, 2, \ldots$$

Finalize Use these expressions to show that the first two times at which the block is at this position are 0.209 s and 1.05 s. These instants are shortly after the block is released and shortly before one full cycle has been completed at 1.26 s.

WHAT IF? Suppose we measure the speed of the block at the instants found in the problem. At these instants, will the speed of the block be half the maximum speed?

Answer The velocity of the block depends on the sine function. The angles at which the cosine function is equal to one-half will not be the same as the angles at which the sine function is equal to one-half. Therefore, we expect the answer to be *no*. Notice that we asked about the *speed* of the block, not the *velocity*. Perform the calculation and show that there are four expressions for the times at which the speed is one-half the maximum speed:

$$t = \frac{\pi}{30.0 \text{ s}^{-1}}(1 + 12n) \quad \text{or} \quad \frac{\pi}{30.0 \text{ s}^{-1}}(5 + 12n) \quad \text{or} \quad \frac{\pi}{30.0 \text{ s}^{-1}}(7 + 12n) \quad \text{or} \quad \frac{\pi}{30.0 \text{ s}^{-1}}(11 + 12n) \quad n = 0, 1, 2, \ldots$$

15.3 Energy of the Simple Harmonic Oscillator

As we have done before, after studying the motion of an object modeled as a particle in a new situation (for example, as in Chapter 2) and investigating the forces involved in influencing that motion (for example, as in Chapter 5), we turn our attention to *energy* (for example, as in Chapter 7). Let us examine the mechanical energy of a system in which a particle undergoes simple harmonic motion, such as the block–spring system illustrated in Figure 15.1. Because the surface is frictionless and the normal and gravitational forces on the block cancel, the system can be modeled as isolated with no nonconservative forces acting, and we expect the total mechanical energy of the system to be constant. We assume a massless spring, so the kinetic energy of the system corresponds only to that of the block. We can use Equation 15.15 to express the kinetic energy of the block as

◀ Kinetic energy of a simple harmonic oscillator

$$K = \tfrac{1}{2}mv^2 = \tfrac{1}{2}m\omega^2 A^2 \sin^2(\omega t + \phi) \qquad (15.19)$$

The elastic potential energy stored in the spring for any elongation x is given by $\frac{1}{2}kx^2$ (see Eq. 7.22). Using Equation 15.6 gives

$$U_s = \tfrac{1}{2}kx^2 = \tfrac{1}{2}kA^2\cos^2(\omega t + \phi) \tag{15.20}$$

◄ Potential energy of a simple harmonic oscillator

We see that K and U_s are *always* positive quantities or zero. Because $\omega^2 = k/m$, we can express the total mechanical energy of the simple harmonic oscillator as

$$E = K + U_s = \tfrac{1}{2}kA^2[\sin^2(\omega t + \phi) + \cos^2(\omega t + \phi)]$$

From the identity $\sin^2\theta + \cos^2\theta = 1$, we see that the quantity in square brackets is unity. Therefore, this equation reduces to

$$E = \tfrac{1}{2}kA^2 \tag{15.21}$$

◄ Total energy of a simple harmonic oscillator

That is, the total mechanical energy of a simple harmonic oscillator is a constant of the motion and is proportional to the square of the amplitude. The total mechanical energy is equal to the maximum potential energy stored in the spring when $x = \pm A$ because $v = 0$ at these points and there is no kinetic energy. At the equilibrium position, where $U_s = 0$ because $x = 0$, the total energy, now all in the form of kinetic energy, still has the value $\frac{1}{2}kA^2$.

Plots of the kinetic and potential energies versus time appear in Figure 15.9a, where we have taken $\phi = 0$. At all times, the sum of the kinetic and potential energies is a constant equal to $\frac{1}{2}kA^2$, the total energy of the system.

The variations of K and U_s with the position x of the block are plotted in Figure 15.9b. Energy is continuously being transformed between potential energy stored in the spring and kinetic energy of the block.

Figure 15.10 (page 396) illustrates the position, velocity, acceleration, kinetic energy, and potential energy of the block–spring system for one full period of the motion. Most of the ideas discussed so far are incorporated in this important figure. Study it carefully.

Equation 15.15 gives the velocity of a particle in simple harmonic oscillation as function of time t. We can obtain the velocity of the block at an arbitrary *position* by expressing the total energy of the system at some arbitrary position x as

$$E = K + U_s = \tfrac{1}{2}mv^2 + \tfrac{1}{2}kx^2 = \tfrac{1}{2}kA^2$$

$$v = \pm\sqrt{\frac{k}{m}(A^2 - x^2)} = \pm\omega\sqrt{A^2 - x^2} \tag{15.22}$$

◄ Velocity as a function of position for a simple harmonic oscillator

When you check Equation 15.22 to see whether it agrees with known cases, you find that it verifies that the speed is a maximum at $x = 0$ and is zero at the turning points $x = \pm A$.

You may wonder why we are spending so much time studying simple harmonic oscillators. We do so because they are good models of a wide variety of physical phenomena.

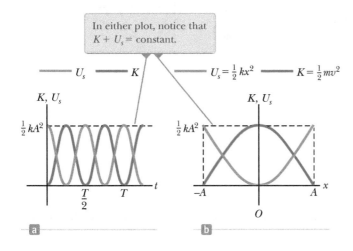

In either plot, notice that $K + U_s =$ constant.

—— U_s —— K —— $U_s = \frac{1}{2}kx^2$ —— $K = \frac{1}{2}mv^2$

Figure 15.9 (a) Kinetic energy and potential energy versus time for a simple harmonic oscillator with $\phi = 0$. (b) Kinetic energy and potential energy versus position for a simple harmonic oscillator.

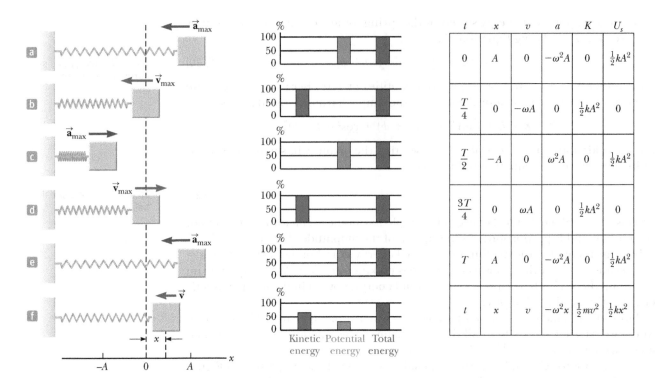

Figure 15.10 (a) through (e) Several instants in the simple harmonic motion for a block–spring system. Energy bar graphs show the distribution of the energy of the system at each instant. The parameters in the table at the right refer to the block–spring system, assuming at $t = 0$, $x = A$; hence, $x = A \cos \omega t$. For these five special instants, one of the types of energy is zero. (f) An arbitrary point in the motion of the oscillator. The system possesses both kinetic energy and potential energy at this instant as shown in the bar graph.

For example, recall the Lennard–Jones potential discussed in Example 7.9. This complicated function describes the forces holding atoms together. Figure 15.11a shows that for small displacements from the equilibrium position, the potential energy curve for this function approximates a parabola, which represents the potential energy function for a simple harmonic oscillator. Therefore, we can model the complex atomic binding forces as being due to tiny springs as depicted in Figure 15.11b.

The ideas presented in this chapter apply not only to block–spring systems and atoms, but also to a wide range of situations that include bungee jumping, playing a musical instrument, and viewing the light emitted by a laser. You will see more examples of simple harmonic oscillators as you work through this book.

Figure 15.11 (a) If the atoms in a molecule do not move too far from their equilibrium positions, a graph of potential energy versus separation distance between atoms is similar to the graph of potential energy versus position for a simple harmonic oscillator (dashed black curve). (b) The forces between atoms in a solid can be modeled by imagining springs between neighboring atoms.

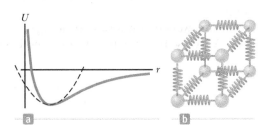

Example 15.3 Oscillations on a Horizontal Surface

A 0.500-kg cart connected to a light spring for which the force constant is 20.0 N/m oscillates on a frictionless, horizontal air track. Use an energy approach to respond to the questions below.

(A) Calculate the maximum speed of the cart if the amplitude of the motion is 3.00 cm.

SOLUTION

Conceptualize   The system oscillates in exactly the same way as the block in Figure 15.10, so use that figure in your mental image of the motion.

15.3 continued

Categorize The cart is modeled as a *particle in simple harmonic motion.*

Analyze Use Equation 15.21 to express the total energy of the oscillator system and equate it to the kinetic energy of the system when the cart is at $x = 0$:

$$E = \tfrac{1}{2}kA^2 = \tfrac{1}{2}mv_{max}^2$$

Solve for the maximum speed and substitute numerical values:

$$v_{max} = \sqrt{\frac{k}{m}}\, A = \sqrt{\frac{20.0\ \text{N/m}}{0.500\ \text{kg}}}(0.030\ 0\ \text{m}) = 0.190\ \text{m/s}$$

(B) What is the velocity of the cart when the position is 2.00 cm?

SOLUTION

Use Equation 15.22 to evaluate the velocity:

$$v = \pm\sqrt{\frac{k}{m}(A^2 - x^2)}$$

$$= \pm\sqrt{\frac{20.0\ \text{N/m}}{0.500\ \text{kg}}[(0.030\ 0\ \text{m})^2 - (0.020\ 0\ \text{m})^2]}$$

$$= \pm 0.141\ \text{m/s}$$

The positive and negative signs indicate that the cart could be moving to either the right or the left at this instant.

(C) Compute the kinetic and potential energies of the system when the position of the cart is 2.00 cm.

SOLUTION

Use the result of part (B) to evaluate the kinetic energy at $x = 0.020\ 0$ m:

$$K = \tfrac{1}{2}mv^2 = \tfrac{1}{2}(0.500\ \text{kg})(0.141\ \text{m/s})^2 = 5.00 \times 10^{-3}\ \text{J}$$

Evaluate the elastic potential energy at $x = 0.020\ 0$ m:

$$U_s = \tfrac{1}{2}kx^2 = \tfrac{1}{2}(20.0\ \text{N/m})(0.020\ 0\ \text{m})^2 = 4.00 \times 10^{-3}\ \text{J}$$

Finalize The sum of the kinetic and potential energies in part (C) is equal to the total energy, which can be found from Equation 15.21. That must be true for *any* position of the cart.

WHAT IF? The cart in this example could have been set into motion by releasing the cart from rest at $x = 3.00$ cm. What if the cart were released from the same position, but with an initial velocity of $v = -0.100$ m/s? What are the new amplitude and maximum speed of the cart?

Answer This question is of the same type we asked at the end of Example 15.1, but here we apply an energy approach.

First calculate the total energy of the system at $t = 0$:

$$E = \tfrac{1}{2}mv^2 + \tfrac{1}{2}kx^2$$

$$= \tfrac{1}{2}(0.500\ \text{kg})(-0.100\ \text{m/s})^2 + \tfrac{1}{2}(20.0\ \text{N/m})(0.030\ 0\ \text{m})^2$$

$$= 1.15 \times 10^{-2}\ \text{J}$$

Equate this total energy to the potential energy of the system when the cart is at the endpoint of the motion:

$$E = \tfrac{1}{2}kA^2$$

Solve for the amplitude A:

$$A = \sqrt{\frac{2E}{k}} = \sqrt{\frac{2(1.15 \times 10^{-2}\ \text{J})}{20.0\ \text{N/m}}} = 0.033\ 9\ \text{m}$$

Equate the total energy to the kinetic energy of the system when the cart is at the equilibrium position:

$$E = \tfrac{1}{2}mv_{max}^2$$

Solve for the maximum speed:

$$v_{max} = \sqrt{\frac{2E}{m}} = \sqrt{\frac{2(1.15 \times 10^{-2}\ \text{J})}{0.500\ \text{kg}}} = 0.214\ \text{m/s}$$

The amplitude and maximum velocity are larger than the previous values because the cart was given an initial velocity at $t = 0$.

The back edge of the treadle goes up and down as one's feet rock the treadle.

The oscillation of the treadle causes circular motion of the drive wheel, eventually resulting in additional up and down motion—of the sewing needle.

John W. Jewett, Jr

Figure 15.12 The bottom of a treadle-style sewing machine from the early twentieth century. The treadle is the wide, flat foot pedal with the metal grillwork.

15.4 Comparing Simple Harmonic Motion with Uniform Circular Motion

Some common devices in everyday life exhibit a relationship between oscillatory motion and circular motion. For example, consider the drive mechanism for a nonelectric sewing machine in Figure 15.12. The operator of the machine places her feet on the treadle and rocks them back and forth. This oscillatory motion causes the large wheel at the right to undergo circular motion. The red drive belt seen in the photograph transfers this circular motion to the sewing machine mechanism (above the photo) and eventually results in the oscillatory motion of the sewing needle. In this section, we explore this interesting relationship between these two types of motion.

Figure 15.13 is a view of an experimental arrangement that shows this relationship. A ball is attached to the rim of a turntable of radius A, which is illuminated from above by a lamp. The ball casts a shadow on a screen as the turntable rotates with constant angular speed. While the ball moves as a particle in uniform circular motion, the shadow of the ball moves back and forth on the screen as a particle in simple harmonic motion.

Consider a particle located at point P on the circumference of a circle of radius A as in Figure 15.14a, with the line OP making an angle ϕ with the x axis at $t = 0$. We call this circle a *reference circle* for comparing simple harmonic motion with uniform circular motion, and we choose the position of P at $t = 0$ as our reference position. If the particle moves counterclockwise along the circle with constant angular speed ω until OP makes an angle θ with the x axis as in Figure 15.14b, at some time $t > 0$ the angle between OP and the x axis is $\theta = \omega t + \phi$. As the particle moves along the circle, the projection of P on the x axis, labeled point Q, moves back and forth along the x axis between the limits $x = \pm A$.

Notice that points P and Q always have the same x coordinate. From the right triangle OPQ, we see that this x coordinate is

$$x(t) = A \cos (\omega t + \phi) \tag{15.23}$$

This expression is the same as Equation 15.6 and shows that the point Q moves with simple harmonic motion along the x axis. Therefore, the motion of an object described by the analysis model of a particle in simple harmonic motion along a straight line can be represented by the projection of an object that can be modeled as a particle in uniform circular motion along a diameter of a reference circle.

This geometric interpretation shows that the time interval for one complete revolution of the point P on the reference circle is equal to the period of motion T for

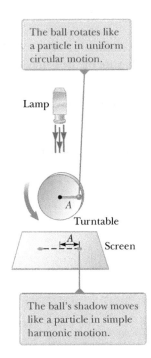

The ball rotates like a particle in uniform circular motion.

Lamp

A

Turntable

A Screen

The ball's shadow moves like a particle in simple harmonic motion.

Figure 15.13 An experimental setup for demonstrating the connection between a particle in simple harmonic motion and a corresponding particle in uniform circular motion.

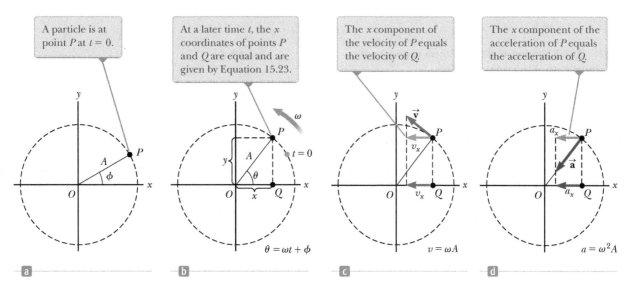

Figure 15.14 Relationship between the uniform circular motion of a point P and the simple harmonic motion of a point Q. A particle at P moves in a circle of radius A with constant angular speed ω.

simple harmonic motion between $x = \pm A$. Therefore, the angular speed ω of P is the same as the angular frequency ω of simple harmonic motion along the x axis (which is why we use the same symbol). The phase constant ϕ for simple harmonic motion corresponds to the initial angle OP makes with the x axis. The radius A of the reference circle equals the amplitude of the simple harmonic motion.

Because the relationship between linear and angular speed for circular motion is $v = r\omega$ (see Eq. 10.10), the particle moving on the reference circle of radius A has a velocity of magnitude ωA. From the geometry in Figure 15.14c, we see that the x component of this velocity is $-\omega A \sin(\omega t + \phi)$. By definition, point Q has a velocity given by dx/dt. Differentiating Equation 15.23 with respect to time, we find that the velocity of Q is the same as the x component of the velocity of P.

The acceleration of P on the reference circle is directed radially inward toward O and has a magnitude $v^2/A = \omega^2 A$. From the geometry in Figure 15.14d, we see that the x component of this acceleration is $-\omega^2 A \cos(\omega t + \phi)$. This value is also the acceleration of the projected point Q along the x axis, as you can verify by taking the second derivative of Equation 15.23.

ⓠUICK QUIZ 15.5 The ball in Figure 15.13 moves in a circle of radius 0.50 m. At $t = 0$, the ball is located on the left side of the turntable, exactly opposite its position in Figure 15.13. What are the correct values for the *amplitude* and *phase constant* (relative to an x axis to the right) of the simple harmonic motion of the shadow? **(a)** 0.50 m and 0 **(b)** 1.00 m and 0 **(c)** 0.50 m and π **(d)** 1.00 m and π

Example **15.4**	Circular Motion with Constant Angular Speed

The ball in Figure 15.13 rotates counterclockwise in a circle of radius 3.00 m with a constant angular speed of 8.00 rad/s. At $t = 0$, its shadow has an x coordinate of 2.00 m and is moving to the right.

(A) Determine the x coordinate of the shadow as a function of time in SI units.

SOLUTION

Conceptualize Be sure you understand the relationship between circular motion of the ball and simple harmonic motion of its shadow as described in Figure 15.13. Notice that the shadow is *not* at is maximum position at $t = 0$.

Categorize The ball on the turntable is a *particle in uniform circular motion*. The shadow is modeled as a *particle in simple harmonic motion*.

continued

15.4 continued

Analyze Use Equation 15.23 to write an expression for the x coordinate of the rotating ball:

$$x = A \cos(\omega t + \phi)$$

Solve for the phase constant:

$$\phi = \cos^{-1}\left(\frac{x}{A}\right) - \omega t$$

Substitute numerical values for the initial conditions:

$$\phi = \cos^{-1}\left(\frac{2.00 \text{ m}}{3.00 \text{ m}}\right) - 0 = \pm 48.2° = \pm 0.841 \text{ rad}$$

If we were to take $\phi = +0.841$ rad as our answer, the shadow would be moving to the left at $t = 0$. Because the shadow is moving to the right at $t = 0$, we must choose $\phi = -0.841$ rad.

Write the x coordinate as a function of time:

$$x = 3.00 \cos(8.00t - 0.841)$$

(B) Find the x components of the shadow's velocity and acceleration at any time t.

SOLUTION

Differentiate the x coordinate with respect to time to find the velocity at any time in m/s:

$$v_x = \frac{dx}{dt} = (-3.00 \text{ m})(8.00 \text{ rad/s}) \sin(8.00t - 0.841)$$
$$= -24.0 \sin(8.00t - 0.841)$$

Differentiate the velocity with respect to time to find the acceleration at any time in m/s²:

$$a_x = \frac{dv_x}{dt} = (-24.0 \text{ m/s})(8.00 \text{ rad/s}) \cos(8.00t - 0.841)$$
$$= -192 \cos(8.00t - 0.841)$$

Finalize Notice that the value of the phase constant puts the ball in the fourth quadrant of the xy coordinate system of Figure 15.14, which is consistent with the shadow having a positive value for x and moving toward the right.

15.5 The Pendulum

The **simple pendulum** is another mechanical system that exhibits periodic motion. It consists of a particle-like bob of mass m suspended by a light string of length L that is fixed at the upper end as shown in Figure 15.15. The motion occurs in the vertical plane and is driven by the gravitational force. We shall show that, provided the angle θ is small (less than about 10°), the motion is very close to that of a simple harmonic oscillator.

The forces acting on the bob are the force $\vec{T}$ exerted by the string and the gravitational force $m\vec{g}$. The tangential component $mg \sin \theta$ of the gravitational force always acts toward $\theta = 0$, opposite the displacement of the bob from the lowest position. Therefore, the tangential component is a restoring force, and we can apply Newton's second law for motion in the tangential direction:

$$F_t = ma_t \rightarrow -mg \sin \theta = m \frac{d^2 s}{dt^2}$$

where the negative sign indicates that the tangential force acts toward the equilibrium (vertical) position and s is the bob's position measured along the arc. We have expressed the tangential acceleration as the second derivative of the position s. Because $s = L\theta$ (Eq. 10.1b with $r = L$) and L is constant, this equation reduces to

$$\frac{d^2\theta}{dt^2} = -\frac{g}{L} \sin \theta$$

Considering θ as the position, let us compare this equation with Equation 15.3. Does it have the same mathematical form? No! The right side is proportional to $\sin \theta$ rather than to θ; hence, we would not expect simple harmonic motion because this expression is not of the same mathematical form as Equation 15.3. If we

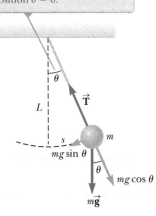

When θ is small, a simple pendulum's motion can be modeled as simple harmonic motion about the equilibrium position $\theta = 0$.

Figure 15.15 A simple pendulum.

assume θ is *small* (less than about 10° or 0.2 rad), however, we can use the **small angle approximation,** in which $\sin\theta \approx \theta$, where θ is measured in radians. Table 15.1 shows angles in degrees and radians and the sines of these angles. As long as θ is less than approximately 10°, the angle in radians and its sine are the same to within an accuracy of less than 1.0%. The table also shows the tangents of the angles, which we will use in the next chapter.

Therefore, for small angles, the equation of motion becomes

$$\frac{d^2\theta}{dt^2} = -\frac{g}{L}\theta \quad \text{(for small values of } \theta\text{)} \tag{15.24}$$

Equation 15.24 has the same mathematical form as Equation 15.3, so we conclude that the motion for small amplitudes of oscillation can be modeled as simple harmonic motion. Therefore, the solution of Equation 15.24 is modeled after Equation 15.6 and is given by $\theta = \theta_{max}\cos(\omega t + \phi)$, where θ_{max} is the *maximum angular position* and the angular frequency ω is

$$\omega = \sqrt{\frac{g}{L}} \tag{15.25}$$

◀ Angular frequency for a simple pendulum

The period of the motion is

$$T = \frac{2\pi}{\omega} = 2\pi\sqrt{\frac{L}{g}} \tag{15.26}$$

◀ Period of a simple pendulum

In other words, the period and frequency of a simple pendulum depend only on the length of the string and the acceleration due to gravity. Because the period is independent of the mass, we conclude that all simple pendula that are of equal length and are at the same location (so that g is the same) oscillate with the same period.

The simple pendulum can be used as a timekeeper because its period depends only on its length and the local value of g. It is also a convenient device for making precise measurements of the free-fall acceleration. Such measurements are important because variations in local values of g can provide information on the location of oil and other valuable underground resources.

QUICK QUIZ 15.6 The grandfather clock in the opening storyline depends on the period of a pendulum to keep correct time. **(i)** Suppose the clock is calibrated correctly and then a mischievous child slides the bob of the pendulum downward on the oscillating rod. Does the grandfather clock run (a) slow, (b) fast, or (c) correctly? **(ii)** Suppose a grandfather clock is calibrated correctly at sea level and is then taken to the top of a very tall mountain. Does the grandfather clock now run (a) slow, (b) fast, or (c) correctly?

Part (b) of Quick Quiz 15.6 relates to the grandfather clock at your grandparents' house in the opening storyline. The clock has been transferred from Denver,

PITFALL PREVENTION 15.4
Not True Simple Harmonic Motion
The pendulum *does not* exhibit true simple harmonic motion for *any* angle. If the angle is less than about 10°, the motion is close to and can be *modeled* as simple harmonic.

TABLE 15.1 Sines and Tangents of Angles

Angle in Degrees	Angle in Radians	Sine of Angle	Percent Difference	Tangent of Angle	Percent Difference
0°	0.000 0	0.000 0	0.0%	0.000 0	0.0%
1°	0.017 5	0.017 5	0.0%	0.017 5	0.0%
2°	0.034 9	0.034 9	0.0%	0.034 9	0.0%
3°	0.052 4	0.052 3	0.0%	0.052 4	0.1%
5°	0.087 3	0.087 2	0.1%	0.087 5	0.3%
10°	0.174 5	0.173 6	0.5%	0.176 3	1.0%
15°	0.261 8	0.258 8	1.2%	0.267 9	2.3%
20°	0.349 1	0.342 0	2.1%	0.364 0	4.3%
30°	0.523 6	0.500 0	4.7%	0.577 4	10.3%

at an altitude of one mile, to Boston, essentially at sea level. As a result, the value of *g*, the acceleration due to gravity, has increased. As we can see from Equation 15.26, this decreases the period of the clock so that it runs fast. What can you do to adjust the clock? You can look at part (a) of Quick Quiz 15.6! The bob of the pendulum should have an adjustment mechanism that allows you to move the bob downward to increase the effective length of the pendulum and therefore increase the period.

Example 15.5 | A Connection Between Length and Time

Christiaan Huygens (1629–1695), the greatest clockmaker in history, suggested that an international unit of length could be defined as the length of a simple pendulum having a period of exactly 1 s. How much shorter would our length unit be if his suggestion had been followed?

SOLUTION

Conceptualize   Imagine a pendulum that swings back and forth in exactly 1 second. Based on your experience in observing swinging objects, can you make an estimate of the required length? Hang a small object from a string and simulate the 1-s pendulum.

Categorize   This example involves a simple pendulum, so we categorize it as a substitution problem that applies the concepts introduced in this section.

Solve Equation 15.26 for the length and substitute numerical values:

$$L = \frac{T^2 g}{4\pi^2} = \frac{(1.00\ \text{s})^2(9.80\ \text{m/s}^2)}{4\pi^2} = 0.248\ \text{m}$$

The meter's length would be slightly less than one-fourth of its current length. Also, the number of significant digits depends only on how precisely we know *g* because the time has been defined to be exactly 1 s.

WHAT IF?   What if Huygens had been born on another planet? What would the value for *g* have to be on that planet such that the meter based on Huygens's pendulum would have the same value as our meter?

Answer   Solve Equation 15.26 for *g*:

$$g = \frac{4\pi^2 L}{T^2} = \frac{4\pi^2 (1.00\ \text{m})}{(1.00\ \text{s})^2} = 4\pi^2\ \text{m/s}^2 = 39.5\ \text{m/s}^2$$

No planet in our solar system has an acceleration due to gravity that large.

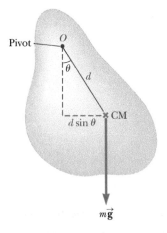

Figure 15.16 A physical pendulum pivoted at *O*.

Physical Pendulum

Suppose you balance a wire coat hanger so that the hook is supported by your extended index finger. When you give the hanger a small angular displacement with your other hand and then release it, it oscillates. If a hanging object oscillates about a fixed axis that does not pass through its center of mass and the object cannot be approximated as a point mass, we cannot treat the system as a simple pendulum. In this case, the system is called a **physical pendulum.**

Consider a rigid object pivoted at a point *O* that is a distance *d* from the center of mass (Fig. 15.16). The gravitational force provides a torque about an axis through *O*, and the magnitude of that torque is *mgd* sin θ, where θ is as shown in Figure 15.16. We apply the rigid object under a net torque analysis model to the object and use the rotational form of Newton's second law, $\Sigma\, \tau_{\text{ext}} = I\alpha$, where *I* is the moment of inertia of the object about the axis through *O*. The result is

$$-mgd \sin\theta = I\frac{d^2\theta}{dt^2}$$

The negative sign indicates that the torque about *O* tends to decrease θ. That is, the gravitational force produces a restoring torque. If we again assume θ is small, the approximation sin θ ≈ θ is valid and the equation of motion reduces to

$$\frac{d^2\theta}{dt^2} = -\left(\frac{mgd}{I}\right)\theta \tag{15.27}$$

Because this equation is of the same mathematical form as Equation 15.3, its solution is modeled after that of the simple harmonic oscillator. That is, the solution of Equation 15.27 is given by $\theta = \theta_{max}\cos(\omega t + \phi)$, where θ_{max} is the maximum angular position and

$$\omega = \sqrt{\frac{mgd}{I}}$$

The period is

$$T = \frac{2\pi}{\omega} = 2\pi\sqrt{\frac{I}{mgd}} \qquad \textbf{(15.28)} \qquad \blacktriangleleft \text{ Period of a physical pendulum}$$

This result can be used to measure the moment of inertia of a flat, rigid object. If the location of the center of mass—and hence the value of d—is known, the moment of inertia can be obtained by measuring the period. Finally, notice that Equation 15.28 reduces to the period of a simple pendulum (Eq. 15.26) when $I = md^2$, that is, when all the mass is concentrated at the center of mass.

Example **15.6** A Swinging Rod

A uniform rod of mass M and length L is pivoted about one end and oscillates in a vertical plane (Fig. 15.17).

(A) Find the period of oscillation if the amplitude of the motion is small.

SOLUTION

Conceptualize Imagine a rod swinging back and forth when pivoted at one end. Try it with a meterstick or a scrap piece of wood.

Categorize Because the rod is not a point particle, we categorize it as a physical pendulum.

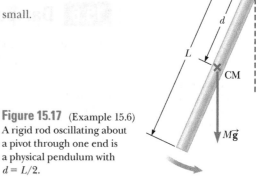

Figure 15.17 (Example 15.6) A rigid rod oscillating about a pivot through one end is a physical pendulum with $d = L/2$.

Analyze In Chapter 10, we found that the moment of inertia of a uniform rod about an axis through one end is $\frac{1}{3}ML^2$. The distance d from the pivot to the center of mass of the rod is $L/2$.

Substitute these quantities into Equation 15.28:

$$T = 2\pi\sqrt{\frac{\frac{1}{3}ML^2}{Mg(L/2)}} = 2\pi\sqrt{\frac{2L}{3g}}$$

(B) Suppose the pivot is moved to a small hole drilled in the rod at a distance $L/4$ from the upper end. What is the period of oscillation of the rod when it is hung from this pivot point and swings through small oscillations?

The moment of inertia in Equation 15.28 is now that about the new pivot point. Use the parallel axis theorem (Eq. 10.22):

$$I = I_{CM} + MD^2 = \frac{1}{12}ML^2 + M\left(\frac{1}{4}L\right)^2 = \frac{7}{48}ML^2$$

Substitute this moment of inertia and the new value of d into Equation 15.28:

$$T = 2\pi\sqrt{\frac{\frac{7}{48}ML^2}{Mg(L/4)}} = 2\pi\sqrt{\frac{7L}{12g}}$$

Finalize In one of the Moon landings, an astronaut walking on the Moon's surface had a belt hanging from his space suit, and the belt oscillated as a physical pendulum. A scientist on the Earth observed this motion on television and used it to estimate the free-fall acceleration on the Moon. How did the scientist make this calculation?

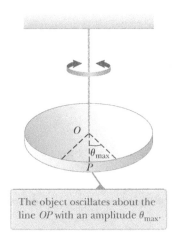

The object oscillates about the line *OP* with an amplitude θ_{max}.

Figure 15.18 A torsional pendulum.

Torsional Pendulum

Figure 15.18 shows a rigid object such as a disk suspended by a wire attached at the top to a fixed support. When the object is twisted through some angle θ, the twisted wire exerts on the object a restoring torque that is proportional to the angular position. That is,

$$\tau = -\kappa\theta$$

where κ (Greek letter kappa) is called the *torsion constant* of the support wire and is a rotational analog to the force constant k for a spring. The value of κ can be obtained by applying a known torque to twist the wire through a measurable angle θ. Applying Newton's second law for rotational motion, we find that

$$\sum \tau = I\alpha \quad \rightarrow \quad -\kappa\theta = I\frac{d^2\theta}{dt^2}$$

$$\frac{d^2\theta}{dt^2} = -\frac{\kappa}{I}\theta \tag{15.29}$$

Again, this result is the equation of motion for a simple harmonic oscillator, with $\omega = \sqrt{\kappa/I}$ and a period

$$T = 2\pi\sqrt{\frac{I}{\kappa}} \tag{15.30}$$

This system is called a *torsional pendulum*. There is no small-angle restriction in this situation as long as the elastic limit of the wire is not exceeded.

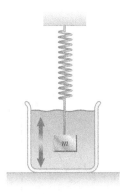

Figure 15.19 One example of a damped oscillator is an object attached to a spring and submersed in a viscous liquid.

15.6 Damped Oscillations

The oscillatory motions we have considered so far have been for ideal systems, that is, systems that oscillate indefinitely under the action of only one force, a linear restoring force. In many real systems, nonconservative forces such as friction or air resistance also act and retard the motion of the system. Consequently, the mechanical energy of the system diminishes in time, and the motion is said to be *damped*. The mechanical energy of the system is transformed into internal energy in the object and the retarding medium. Figure 15.19 depicts one such system: an object attached to a spring and submersed in a viscous liquid. Another example is a simple pendulum oscillating in air. After being set into motion, the pendulum eventually stops oscillating due to air resistance. Figure 15.20 depicts damped oscillations in practice. The spring-loaded devices mounted below the bridge are dampers that transform mechanical energy of the oscillating bridge into internal energy, reducing the swaying motion of the bridge.

One common type of retarding force is that discussed in Section 6.4, where the force is proportional to the speed of the moving object and acts in the direction opposite the velocity of the object with respect to the medium. This retarding force is often observed when an object moves through air, for instance. Because the retarding force can be expressed as $\vec{\mathbf{R}} = -b\vec{\mathbf{v}}$ (where b is a constant called the *damping coefficient*) and the restoring force of the system is $-kx$, we can write Newton's second law as

$$\sum F_x = -kx - bv_x = ma_x$$

which, by substituting derivatives for the velocity and acceleration, can be written as

$$m\frac{d^2x}{dt^2} + b\frac{dx}{dt} + kx = 0 \tag{15.31}$$

The solution to this equation requires mathematics that may be unfamiliar to you; we simply state it here without proof. When the retarding force is small compared

John W. Jewett, Jr.

Figure 15.20 The London Millennium Bridge over the River Thames in London. On opening day of the bridge, pedestrians noticed a swinging motion of the bridge, leading to its being named the "Wobbly Bridge." The bridge was closed after two days and remained closed for two years. Over 50 tuned mass dampers were added to the bridge: the pairs of spring-loaded structures on top of the cross members (arrow).

with the maximum restoring force—that is, when the damping coefficient b is small—the solution to Equation 15.31 is

$$x = Ae^{-(b/2m)t} \cos(\omega t + \phi) \tag{15.32}$$

where the angular frequency of oscillation is

$$\omega = \sqrt{\frac{k}{m} - \left(\frac{b}{2m}\right)^2} \tag{15.33}$$

This result can be verified by substituting Equation 15.32 into Equation 15.31. It is convenient to express the angular frequency of a damped oscillator in the form

$$\omega = \sqrt{\omega_0^2 - \left(\frac{b}{2m}\right)^2}$$

where $\omega_0 = \sqrt{k/m}$ represents the angular frequency in the absence of a retarding force (the undamped oscillator) and is called the **natural frequency** of the system.

Figure 15.21 shows the position as a function of time for an object oscillating in the presence of a retarding force. When the retarding force is small, the oscillatory character of the motion is preserved but the amplitude decreases exponentially in time, with the result that the motion ultimately becomes undetectable. Any system that behaves in this way is known as a **damped oscillator.** The dashed black lines in Figure 15.21, which define the *envelope* of the oscillatory curve, represent the exponential factor in Equation 15.32. This envelope shows that the amplitude decays exponentially with time. For motion with a given spring constant and object mass, the oscillations dampen more rapidly for larger values of the retarding force.

When the magnitude of the retarding force is small such that $b/2m < \omega_0$, the system is said to be **underdamped.** The resulting motion is represented by Figure 15.21 and the blue curve in Figure 15.22. As the value of b increases, the amplitude of the oscillations decreases more and more rapidly. When b reaches a critical value b_c such that $b_c/2m = \omega_0$, the system does not oscillate and is said to be **critically damped.** In this case, the system, once released from rest at some nonequilibrium position, approaches but does not pass through the equilibrium position. The graph of position versus time for this case is the red curve in Figure 15.22.

If the medium is so viscous that the retarding force is large compared with the restoring force—that is, if $b/2m > \omega_0$—the system is **overdamped.** Again, the displaced system, when free to move, does not oscillate but rather simply returns to its equilibrium position. As the damping increases, the time interval required for the system to approach equilibrium also increases as indicated by the black curve in Figure 15.22. For critically damped and overdamped systems, there is no angular frequency ω and the solution in Equation 15.32 is not valid.

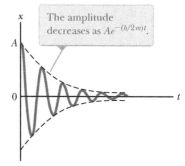

The amplitude decreases as $Ae^{-(b/2m)t}$.

Figure 15.21 Graph of position versus time for a damped oscillator.

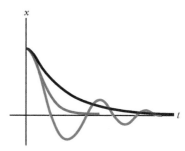

Figure 15.22 Graphs of position versus time for an underdamped oscillator (blue curve), a critically damped oscillator (red curve), and an overdamped oscillator (black curve).

15.7 Forced Oscillations

We have seen that the mechanical energy of a damped oscillator decreases in time as a result of the retarding force. It is possible to compensate for this energy decrease by applying a periodic external force that does positive work on the system. At any instant, energy can be transferred into the system by an applied force that acts in the direction of motion of the oscillator. For example, a child on a swing can be kept in motion by appropriately timed "pushes." The amplitude of motion remains constant if the energy input per cycle of motion exactly equals the decrease in mechanical energy in each cycle that results from retarding forces.

A common example of a forced oscillator is a damped oscillator driven by an external force that varies periodically, such as $F(t) = F_0 \sin \omega t$, where F_0 is a constant and ω is the angular frequency of the driving force. In general, the frequency ω of the driving force is variable, whereas the natural frequency ω_0 of the oscillator is fixed by the values of k and m. Modeling an oscillator with driving, retarding, and restoring forces as a particle under a net force, Newton's second law in this situation gives

$$\sum F_x = ma_x \quad \rightarrow \quad F_0 \sin \omega t - b\frac{dx}{dt} - kx = m\frac{d^2x}{dt^2} \qquad \text{(15.34)}$$

Again, the solution of this equation is rather lengthy and will not be presented. After the driving force on an initially stationary object begins to act, the amplitude of the oscillation will increase. The system of the oscillator and the surrounding medium is a nonisolated system: work is done by the driving force, such that the vibrational energy of the system (kinetic energy of the object, elastic potential energy in the spring) and internal energy of the object and the medium increase. After a sufficiently long period of time, when the energy input per cycle from the driving force equals the amount of mechanical energy transformed to internal energy for each cycle, a steady-state condition is reached in which the oscillations proceed with constant amplitude. In this situation, the solution of Equation 15.34 is

$$x = A \cos (\omega t + \phi) \qquad \text{(15.35)}$$

where

Amplitude of a ▶
driven oscillator

$$A = \frac{F_0/m}{\sqrt{(\omega^2 - \omega_0^2)^2 + \left(\dfrac{b\omega}{m}\right)^2}} \qquad \text{(15.36)}$$

and where $\omega_0 = \sqrt{k/m}$ is the natural frequency of the undamped oscillator ($b = 0$).

Equations 15.35 and 15.36 show that the forced oscillator vibrates at the frequency of the driving force and that the amplitude of the oscillator is constant for a given driving force because it is being driven in steady-state by an external force. For small damping, the amplitude is large when the frequency of the driving force is near the natural frequency of oscillation, or when $\omega \approx \omega_0$. The dramatic increase in amplitude near the natural frequency is called **resonance,** and the natural frequency ω_0 is also called the **resonance frequency** of the system.

The reason for large-amplitude oscillations at the resonance frequency is that energy is being transferred to the system under the most favorable conditions. We can better understand this concept by taking the first time derivative of x in Equation 15.35, which gives an expression for the velocity of the oscillator. We find that v is proportional to $\sin(\omega t + \phi)$, which is the same trigonometric function as that describing the driving force. Therefore, the applied force $\vec{\mathbf{F}}$ is in phase with the velocity. The rate at which work is done on the oscillator by $\vec{\mathbf{F}}$ equals the dot product $\vec{\mathbf{F}} \cdot \vec{\mathbf{v}}$; this rate is the power delivered to the oscillator. Because the product $\vec{\mathbf{F}} \cdot \vec{\mathbf{v}}$ is a maximum when $\vec{\mathbf{F}}$ and $\vec{\mathbf{v}}$ are in phase, we conclude that at resonance, the applied force is in phase with the velocity and the power transferred to the oscillator is a maximum.

Figure 15.23 is a graph of amplitude as a function of driving frequency for a forced oscillator with and without damping. Notice that the amplitude increases with decreasing damping ($b \rightarrow 0$) and that the resonance curve broadens as the damping increases. In the absence of a damping force ($b = 0$), we see from Equation 15.36 that the steady-state amplitude approaches infinity as ω approaches ω_0. In other words, if there are no losses in the system and we continue to drive an initially motionless oscillator with a periodic force that is in phase with the velocity, the amplitude of motion builds without limit (see the red-brown curve in Fig. 15.23). This limitless building does not occur in practice because some damping is always present in reality.

Later in this book we shall see that resonance appears in other areas of physics. For example, certain electric circuits have natural frequencies and can be set into strong resonance by a varying voltage applied at a given frequency. A bridge has

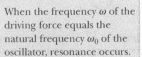

When the frequency ω of the driving force equals the natural frequency ω_0 of the oscillator, resonance occurs.

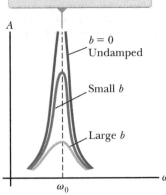

A

$b = 0$
Undamped

Small b

Large b

ω_0 ω

Figure 15.23 Graph of amplitude versus frequency for a damped oscillator when a periodic driving force is present. Notice that the shape of the resonance curve depends on the size of the damping coefficient b.

Figure 15.24 (a) In 1940, turbulent winds set up torsional vibrations in the Tacoma Narrows Bridge, causing it to oscillate at a frequency near one of the natural frequencies of the bridge structure. (b) Once established, this resonance condition led to the bridge's collapse. (Mathematicians and physicists are currently challenging some aspects of this interpretation.)

natural frequencies that can be set into resonance by an appropriate driving force. A dramatic example of such resonance occurred in 1940 when the Tacoma Narrows Bridge in the state of Washington was destroyed by resonant vibrations. Although the winds were not particularly strong on that occasion, the "flapping" of the wind across the roadway (think of the "flapping" of a flag in a strong wind) provided a periodic driving force whose frequency matched that of the bridge. The resulting oscillations of the bridge caused it to ultimately collapse (Fig. 15.24) because the bridge design had inadequate built-in safety features.

Many other examples of resonant vibrations can be cited. A resonant vibration you may have experienced is the "singing" of telephone wires in the wind. Machines often break if one vibrating part is in resonance with some other moving part. Soldiers marching in cadence across a bridge have been known to set up resonant vibrations in the structure and thereby cause it to collapse. Whenever any real physical system is driven near its resonance frequency, you can expect oscillations of very large amplitudes.

Summary

> ## Concepts and Principles

The kinetic energy and potential energy for an object of mass m oscillating at the end of a spring of force constant k vary with time and are given by

$$K = \tfrac{1}{2}mv^2 = \tfrac{1}{2}m\omega^2 A^2 \sin^2(\omega t + \phi) \qquad (15.19)$$

$$U_s = \tfrac{1}{2}kx^2 = \tfrac{1}{2}kA^2 \cos^2(\omega t + \phi) \qquad (15.20)$$

The total energy of a simple harmonic oscillator is a constant of the motion and is given by

$$E = \tfrac{1}{2}kA^2 \qquad (15.21)$$

A **simple pendulum** of length L can be modeled to move in simple harmonic motion for small angular displacements from the vertical. Its period is

$$T = 2\pi \sqrt{\frac{L}{g}} \qquad (15.26)$$

A **physical pendulum** is an extended object that, for small angular displacements, can be modeled to move in simple harmonic motion about a pivot that does not go through the center of mass. The period of this motion is

$$T = 2\pi \sqrt{\frac{I}{mgd}} \qquad (15.28)$$

where I is the moment of inertia of the object about an axis through the pivot and d is the distance from the pivot to the center of mass of the object.

If an oscillator experiences a damping force $\vec{\mathbf{R}} = -b\vec{\mathbf{v}}$, its position for small damping is described by

$$x = Ae^{-(b/2m)t} \cos(\omega t + \phi) \qquad (15.32)$$

where

$$\omega = \sqrt{\frac{k}{m} - \left(\frac{b}{2m}\right)^2} \qquad (15.33)$$

If an oscillator is subject to a sinusoidal driving force that is described by $F(t) = F_0 \sin \omega t$, it exhibits **resonance,** in which the amplitude is largest when the driving frequency ω matches the natural frequency $\omega_0 = \sqrt{k/m}$ of the oscillator.

continued

▶ Analysis Model for Problem Solving

Particle in Simple Harmonic Motion If a particle is subject to a force of the form of Hooke's law $F = -kx$, the particle exhibits **simple harmonic motion.** Its position is described by

$$x(t) = A \cos(\omega t + \phi) \qquad (15.6)$$

where A is the **amplitude** of the motion, ω is the **angular frequency,** and ϕ is the **phase constant.** The value of ϕ depends on the initial position and initial velocity of the particle.
    The **period** of the oscillation of the particle is

$$T = \frac{2\pi}{\omega} = 2\pi\sqrt{\frac{m}{k}} \qquad (15.13)$$

and the inverse of the period is the **frequency.**

Think–Pair–Share

See the Preface for an explanation of the icons used in this problems set. For additional assessment items for this section, go to **⚡WEBASSIGN** From Cengage

1. Two identical steel balls, each of mass $m = 67.4$ g and diameter $d = 25.4$ mm, are moving in opposite directions, each at $v = 5.00$ m/s. They collide head-on and bounce apart elastically. (a) Split your group in two and have each half find the total time interval that the balls are in contact, using different models. Group (i): Model a given ball as having kinetic energy that is then completely transformed to elastic potential energy at the instant that the balls have momentarily come to rest. Assume the acceleration of the ball during this time interval is constant and use the particle under constant acceleration model to find the total time interval that the balls are in contact. Group (ii): By squeezing one of the balls in a vise while precise measurements are made of the resulting amount of compression, you have found that Hooke's law is a good model of the ball's elastic behavior. A force of $F = 16.0$ kN exerted by each jaw of the vise reduces the diameter by a distance $s = 0.200$ mm. The diameter returns to the original value when the force from the vise is removed. Model the motion of each ball, while the balls are in contact, as one-half of a cycle of simple harmonic motion. Find the total time interval that the balls are in contact. (b) Which result is more accurate?

2. **ACTIVITY** Divide your group in half. Each subgroup should work on one of the situations below:

   (i) A hanging spring stretches by 35.0 cm when an object of mass 450 g is hung on it at rest. In this situation, we define its position as $x = 0$, with positive x upward. The object is pulled down an additional 18.0 cm and released from rest to oscillate without friction.

   (ii) Another hanging spring stretches by 35.5 cm when an object of mass 440 g is hung on it at rest. We define this new position as $x = 0$. This object is pulled down an additional 18.0 cm and released from rest to oscillate without friction.

   (a) For each of these situations, answer the following two questions: (1) What is the position x of the object at a moment 84.4 s later? (2) What total distance has the vibrating object traveled in the 84.4-s time interval? When the calculations are finished, compare the results for the two situations. (b) Why are the answers to question 1 so different when the initial data in situations (i) and (ii) are so similar and the answers to question 2 are relatively close? (c) Does this circumstance reveal a fundamental difficulty in calculating the future?

3. **ACTIVITY** Online, you read about a group of physics students doing a simple pendulum lab. They used a small object attached to the end of a string to form a simple pendulum. The students measured the total time intervals for 50 oscillations of its harmonic motion for small angular displacements and three lengths. They posted their data online:

Length L (m)	Time interval for 50 oscillations (s)
1.000	99.8
0.750	86.6
0.500	71.1

Split your group in two and have each half find a value for g, the acceleration due to gravity, using different approaches. Group (i): Determine the period of motion T for each length of the pendulum. From that length, use Equation 15.26 to find a value of g for each length. Determine the mean value of g obtained from these three independent measurements and compare it with the accepted value. Group (ii): Determine the period of motion T for each length of the pendulum. Plot T^2 versus length L and obtain a value for g from the slope of your best-fit straight-line graph, using Equation 15.26. How do the values of g for the two groups compare?

Problems

See the Preface for an explanation of the icons used in this problems set. For additional assessment items for this section, go to **WEBASSIGN** From Cengage

> *Note:* Ignore the mass of every spring.

SECTION 15.1 Motion of an Object Attached to a Spring

> Problems 11, 12, 41 in Chapter 7 can also be assigned with this section.

1. A 0.60-kg block attached to a spring with force constant 130 N/m is free to move on a frictionless, horizontal surface as in Figure 15.1. The block is released from rest when the spring is stretched 0.13 m. At the instant the block is released, find (a) the force on the block and (b) its acceleration.

SECTION 15.2 Analysis Model: Particle in Simple Harmonic Motion

2. A piston in a gasoline engine is in simple harmonic motion. The engine is running at the rate of 3 600 rev/min. Taking the extremes of its position relative to its center point as ± 5.00 cm, find the magnitudes of the (a) maximum velocity and (b) maximum acceleration of the piston.

3. **T** The position of a particle is given by the expression $x = 4.00 \cos (3.00\pi t + \pi)$, where x is in meters and t is in seconds. Determine (a) the frequency and (b) period of the motion, (c) the amplitude of the motion, (d) the phase constant, and (e) the position of the particle at $t = 0.250$ s.

4. **AMT** **V** A 7.00-kg object is hung from the bottom end of a vertical spring fastened to an overhead beam. The object is set into vertical oscillations having a period of 2.60 s. Find the force constant of the spring.

5. **Review.** A particle moves along the x axis. It is initially at the position 0.270 m, moving with velocity 0.140 m/s and acceleration -0.320 m/s². Suppose it moves as a particle under constant acceleration for 4.50 s. Find (a) its position and (b) its velocity at the end of this time interval. Next, assume it moves as a particle in simple harmonic motion for 4.50 s and $x = 0$ is its equilibrium position. Find (c) its position and (d) its velocity at the end of this time interval.

6. **Q|C** A ball dropped from a height of 4.00 m makes an elastic collision with the ground. Assuming no decrease in mechanical energy due to air resistance, (a) show that the ensuing motion is periodic and (b) determine the period of the motion. (c) Is the motion simple harmonic? Explain.

7. A particle moving along the x axis in simple harmonic motion starts from its equilibrium position, the origin, at $t = 0$ and moves to the right. The amplitude of its motion is 2.00 cm, and the frequency is 1.50 Hz. (a) Find an expression for the position of the particle as a function of time. Determine (b) the maximum speed of the particle and (c) the earliest time ($t > 0$) at which the particle has this speed. Find (d) the maximum positive acceleration of the particle and (e) the earliest time ($t > 0$) at which the particle has this acceleration. (f) Find the total distance traveled by the particle between $t = 0$ and $t = 1.00$ s.

8. **S** The initial position, velocity, and acceleration of an object moving in simple harmonic motion are x_i, v_i, and a_i; the angular frequency of oscillation is ω. (a) Show that the position and velocity of the object for all time can be written as

$$x(t) = x_i \cos \omega t + \left(\frac{v_i}{\omega}\right) \sin \omega t$$

$$v(t) = -x_i \omega \sin \omega t + v_i \cos \omega t$$

(b) Using A to represent the amplitude of the motion, show that

$$v^2 - ax = v_i^2 - a_i x_i = \omega^2 A^2$$

9. **Q|C** You attach an object to the bottom end of a hanging vertical spring. It hangs at rest after extending the spring 18.3 cm. You then set the object vibrating. (a) Do you have enough information to find its period? (b) Explain your answer and state whatever you can about its period.

SECTION 15.3 Energy of the Simple Harmonic Oscillator

10. **AMT** **T** To test the resiliency of its bumper during low-speed collisions, a 1 000-kg automobile is driven into a brick wall. The car's bumper behaves like a spring with a force constant 5.00×10^6 N/m and compresses 3.16 cm as the car is brought to rest. What was the speed of the car before impact, assuming no mechanical energy is transformed or transferred away during impact with the wall?

11. A particle executes simple harmonic motion with an amplitude of 3.00 cm. At what position does its speed equal half of its maximum speed?

12. The amplitude of a system moving in simple harmonic motion is doubled. Determine the change in (a) the total energy, (b) the maximum speed, (c) the maximum acceleration, and (d) the period.

13. **Q|C** **S** A simple harmonic oscillator of amplitude A has a total energy E. Determine (a) the kinetic energy and (b) the potential energy when the position is one-third the amplitude. (c) For what values of the position does the kinetic energy equal one-half the potential energy? (d) Are there any values of the position where the kinetic energy is greater than the maximum potential energy? Explain.

14. **Review.** **GP** A 65.0-kg bungee jumper steps off a bridge with a light bungee cord tied to her body and to the bridge. The unstretched length of the cord is 11.0 m. The jumper reaches the bottom of her motion 36.0 m below the bridge before bouncing back. We wish to find the time interval between her leaving the bridge and her arriving at the bottom of her motion. Her overall motion can be separated into an 11.0-m free fall and a 25.0-m section of simple harmonic oscillation. (a) For the free-fall part, what is the appropriate analysis model to describe her motion? (b) For what time interval is she in free fall? (c) For the simple harmonic oscillation part of the plunge, is the system of the bungee jumper, the spring, and the Earth isolated or nonisolated? (d) From your response in part (c) find the spring constant of the bungee cord. (e) What is the location of the equilibrium point where the spring force balances the gravitational force exerted on the jumper? (f) What is the angular frequency of the oscillation? (g) What time interval is required for the cord to stretch by 25.0 m? (h) What is the total time interval for the entire 36.0-m drop?

15. **Review.** A 0.250-kg block resting on a frictionless, horizontal surface is attached to a spring whose force constant is 83.8 N/m as in Figure P15.15. A horizontal force $\vec{F}$ causes the spring to stretch a distance of 5.46 cm from its equilibrium position. (a) Find the magnitude of $\vec{F}$. (b) What is the total energy stored in the system when the spring is stretched? (c) Find the magnitude of the acceleration of the block just after the applied force is removed. (d) Find the speed of the block when it first reaches the equilibrium position. (e) If the surface is not frictionless but the block still reaches the equilibrium position, would your answer to part (d) be larger or smaller? (f) What other information would you need to know to find the actual answer to part (d) in this case? (g) What is the largest value of the coefficient of friction that would allow the block to reach the equilibrium position?

Figure P15.15

SECTION 15.4 Comparing Simple Harmonic Motion with Uniform Circular Motion

16. While driving behind a car traveling at 3.00 m/s, you notice that one of the car's tires has a small hemispherical bump on its rim as shown in Figure P15.16. (a) Explain why the bump, from your viewpoint behind the car, executes simple harmonic motion. (b) If the radii of the car's tires are 0.300 m, what is the bump's period of oscillation? (c) **What If?** You hang a spring with spring constant $k = 100$ N/m from the rear view mirror of your car. What is the mass that needs to be hung from this spring to produce simple harmonic motion with the same period as the bump on the tire? (d) What would be the maximum speed of the hanging mass in your car if you initially pulled the mass down 8.00 cm beyond equilibrium before releasing it?

Bump

Figure P15.16

SECTION 15.5 The Pendulum

Problem 36 in Chapter 1 can also be assigned with this section.

17. A simple pendulum makes 120 complete oscillations in 3.00 min at a location where $g = 9.80$ m/s². Find (a) the period of the pendulum and (b) its length.

18. A particle of mass m slides without friction inside a hemispherical bowl of radius R. Show that if the particle starts from rest with a small displacement from equilibrium, it moves in simple harmonic motion with an angular frequency equal to that of a simple pendulum of length R. That is, $\omega = \sqrt{g/R}$.

19. A physical pendulum in the form of a planar object moves in simple harmonic motion with a frequency of 0.450 Hz. The pendulum has a mass of 2.20 kg, and the pivot is located 0.350 m from the center of mass. Determine the moment of inertia of the pendulum about the pivot point.

20. A physical pendulum in the form of a planar object moves in simple harmonic motion with a frequency f. The pendulum has a mass m, and the pivot is located a distance d from the center of mass. Determine the moment of inertia of the pendulum about the pivot point.

21. A simple pendulum has a mass of 0.250 kg and a length of 1.00 m. It is displaced through an angle of 15.0° and then released. Using the analysis model of a particle in simple harmonic motion, what are (a) the maximum speed of the bob, (b) its maximum angular acceleration, and (c) the maximum restoring force on the bob? (d) **What If?** Solve parts (a) through (c) again by using analysis models introduced in earlier chapters. (e) Compare the answers.

22. Consider the physical pendulum of Figure 15.16. (a) Represent its moment of inertia about an axis passing through its center of mass and parallel to the axis passing through its pivot point as I_{CM}. Show that its period is

$$T = 2\pi \sqrt{\frac{I_{CM} + md^2}{mgd}}$$

where d is the distance between the pivot point and the center of mass. (b) Show that the period has a minimum value when d satisfies $md^2 = I_{CM}$.

23. A watch balance wheel (Fig. P15.23) has a period of oscillation of 0.250 s. The wheel is constructed so that its mass of 20.0 g is concentrated around a rim of radius 0.500 cm. What are (a) the wheel's moment of inertia and (b) the torsion constant of the attached spring?

Balance wheel

Figure P15.23

SECTION 15.6 Damped Oscillations

24. Show that the time rate of change of mechanical energy for a damped, undriven oscillator is given by $dE/dt = -bv^2$ and hence is always negative. To do so, differentiate the expression for the mechanical energy of an oscillator, $E = \frac{1}{2}mv^2 + \frac{1}{2}kx^2$, and use Equation 15.31.

25. Show that Equation 15.32 is a solution of Equation 15.31 provided that $b^2 < 4mk$.

SECTION 15.7 Forced Oscillations

26. As you enter a fine restaurant, you realize that you have accidentally brought a small electronic timer from home instead of your cell phone. In frustration, you drop the timer into a side pocket of your suit coat, not realizing that the timer is operating. The arm of your chair presses the light cloth of your coat against your body at one spot. Fabric with a length L hangs freely below that spot, with the timer at the bottom. At one point during your dinner, the timer goes off and a buzzer and a vibrator turn on and off with a frequency of 1.50 Hz. It makes the hanging part of your coat swing back and forth with remarkably large amplitude, drawing everyone's attention. Find the value of L.

27. A 2.00-kg object attached to a spring moves without friction ($b = 0$) and is driven by an external force given by the expression $F = 3.00 \sin (2\pi t)$, where F is in newtons and t is in seconds. The force constant of the spring is 20.0 N/m. Find (a) the resonance angular frequency of the system,

(b) the angular frequency of the driven system, and (c) the amplitude of the motion.

28. Considering an undamped, forced oscillator ($b = 0$), show that Equation 15.35 is a solution of Equation 15.34, with an amplitude given by Equation 15.36.

29. You have scored a part-time job at a company that makes small probes to be released from satellites to study the very thin atmosphere at the location of satellite orbits. In order to keep the probes in a proper orientation in space, they will be spun about their axis before being released. It is important to know the moment of inertia of the odd-shaped probe. Your boss asks you to measure its moment of inertia. You

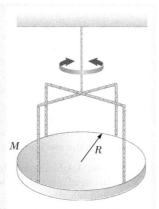

Figure P15.29

set up a system such as that in Figure 15.18, modifying it by adding a very light frame (Fig. P15.29) into which you can place objects, centering them on the disk. The frame is attached at the edges of the disk. The support wire is rigidly connected to the top of the frame so that it does not interfere with the objects you wish to place on the disk. The disk is of mass $M = 5.25$ kg and has a radius of $R = 25.8$ cm. You rotate the empty disk from its equilibrium position and let it operate as a torsional pendulum. You carefully measure its period of oscillation to be $T_{\text{empty}} = 10.8$ s. You then place the probe on the disk and adjust its position until the disk hangs exactly horizontal, so you know that the center of mass of the probe is directly over the center of the disk. You rotate the loaded disk from its equilibrium position and let it operate as a torsional pendulum. (a) You carefully measure its period of oscillation to be $T_{\text{loaded}} = 18.7$ s, and from this result you determine the moment of inertia of the probe about its center of mass. (b) When you present your results to your supervisor, she asks you about the moment of inertia of the frame you built. You go back to your desk and think about it. When you consider that the frame has some moment of inertia, is the value calculated in part (a) too high or too low?

30. You take on a research assistantship with a molecular physicist. She is studying the vibrations of diatomic molecules. In these vibrations, the two atoms in the molecule move back and forth along the line connecting them (see Figure 20.5c). As an introduction to her research, she asks you to familiarize yourself with the Lennard–Jones potential (see Example 7.9), which describes the potential energy function for a diatomic molecule. She asks you to determine the effective spring constant, in terms of the parameters σ and ϵ, for the bond holding the atoms together in the molecule for small vibrations around the equilibrium separation r_{eq}. After being stumped for a while, you ask her for a hint. She responds, "Example 7.9 provides the derivative of the potential energy function. Compare that to Equation 7.29 to find the force between the atoms. You want to show that F is of the form $-kx$, and find k. Let the separation distance $r = r_{\text{eq}} + x$, where x is

small and take advantage of the series approximations in Appendix Section B.5." Wow, that's several hints! You sit down and get to work.

ADDITIONAL PROBLEMS

31. An object of mass m moves in simple harmonic motion with amplitude 12.0 cm on a light spring. Its maximum acceleration is 108 cm/s². Regard m as a variable. (a) Find the period T of the object. (b) Find its frequency f. (c) Find the maximum speed v_{max} of the object. (d) Find the total energy E of the object–spring system. (e) Find the force constant k of the spring. (f) Describe the pattern of dependence of each of the quantities T, f, v_{max}, E, and k on m.

32. **Review.** This problem extends the reasoning of Problem 41 in Chapter 9. Two gliders are set in motion on an air track. Glider 1 has mass $m_1 = 0.240$ kg and moves to the right with speed 0.740 m/s. It will have a rear-end collision with glider 2, of mass $m_2 = 0.360$ kg, which initially moves to the right with speed 0.120 m/s. A light spring of force constant 45.0 N/m is attached to the back end of glider 2 as shown in Figure P9.41. When glider 1 touches the spring, superglue instantly and permanently makes it stick to its end of the spring. (a) Find the common speed the two gliders have when the spring is at maximum compression. (b) Find the maximum spring compression distance. The motion after the gliders become attached consists of a combination of (1) the constant-velocity motion of the center of mass of the two-glider system found in part (a) and (2) simple harmonic motion of the gliders relative to the center of mass. (c) Find the energy of the center-of-mass motion. (d) Find the energy of the oscillation.

33. An object attached to a spring vibrates with simple harmonic motion as described by Figure P15.33. For this motion, find (a) the amplitude, (b) the period, (c) the angular frequency, (d) the maximum speed, (e) the maximum acceleration, and (f) an equation for its position x as a function of time.

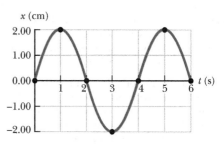

Figure P15.33

34. **Review.** A rock rests on a concrete sidewalk. An earthquake strikes, making the ground move vertically in simple harmonic motion with a constant frequency of 2.40 Hz and with gradually increasing amplitude. (a) With what amplitude does the ground vibrate when the rock begins to lose contact with the sidewalk? Another rock is sitting on the concrete bottom of a swimming pool full of water. The earthquake produces only vertical motion, so the water does not slosh from side to side. (b) Present a convincing argument that when the ground vibrates with the amplitude found in part (a), the submerged rock also barely loses contact with the floor of the swimming pool.

35. A pendulum of length L and mass
    [S] M has a spring of force constant
    k connected to it at a distance h
    below its point of suspension
    (Fig. P15.35). Find the frequency
    of vibration of the system for small
    values of the amplitude (small θ).
    Assume the vertical suspension rod
    of length L is rigid, but ignore its
    mass.

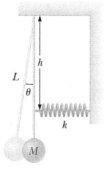

Figure P15.35

36. To account for the walking speed
    [BIO] of a bipedal or quadrupedal ani-
    mal, model a leg that is not con-
    tacting the ground as a uniform
    rod of length ℓ, swinging as a physical pendulum through
    one-half of a cycle, in resonance. Let θ_{max} represent its
    amplitude. (a) Show that the animal's speed is given by the
    expression

    $$v = \frac{\sqrt{6g\ell}\sin\theta_{max}}{\pi}$$

    if θ_{max} is sufficiently small that the motion is nearly simple
    harmonic. An empirical relationship that is based on the
    same model and applies over a wider range of angles is

    $$v = \frac{\sqrt{6g\ell}\cos(\theta_{max}/2)\sin\theta_{max}}{\pi}$$

    (b) Evaluate the walking speed of a human with leg length
    0.850 m and leg-swing amplitude 28.0°. (c) What leg
    length would give twice the speed for the same angular
    amplitude?

37. **Review.** A particle of mass 4.00 kg is attached to a spring
    [QC] with a force constant of 100 N/m. It is oscillating on a fric-
    tionless, horizontal surface with an amplitude of 2.00 m. A
    6.00-kg object is dropped vertically on top of the 4.00-kg
    object as it passes through its equilibrium point. The two
    objects stick together. (a) What is the new amplitude of the
    vibrating system after the collision? (b) By what factor has
    the period of the system changed? (c) By how much does
    the energy of the system change as a result of the collision?
    (d) Account for the change in energy.

38. People who ride motorcycles and bicycles learn to look out
    [QC] for bumps in the road and especially for *washboarding*, a con-
    dition in which many equally spaced ridges are worn into
    the road. What is so bad about washboarding? A motorcycle
    has several springs and shock absorbers in its suspension,
    but you can model it as a single spring supporting a block.
    You can estimate the force constant by thinking about
    how far the spring compresses when a heavy rider sits on
    the seat. A motorcyclist traveling at highway speed must be
    particularly careful of washboard bumps that are a certain
    distance apart. What is the order of magnitude of their sep-
    aration distance?

39. A ball of mass m is connected to two rubber bands of length
    [S] L, each under tension T as shown in Figure P15.39. The
    ball is displaced by a small distance y perpendicular to the
    length of the rubber bands. Assuming the tension does not
    change, show that (a) the restoring force is $-(2T/L)y$ and
    (b) the system exhibits simple harmonic motion with an
    angular frequency $\omega = \sqrt{2T/mL}$.

Figure P15.39

40. Consider the damped oscillator illustrated in Figure 15.19.
    The mass of the object is 375 g, the spring constant is
    100 N/m, and $b = 0.100$ N · s/m. (a) Over what time interval
    does the amplitude drop to half its initial value? (b) **What
    If?** Over what time interval does the mechanical energy
    drop to half its initial value? (c) Show that, in general,
    the fractional rate at which the amplitude decreases in a
    damped harmonic oscillator is one-half the fractional rate
    at which the mechanical energy decreases.

41. **Review.** A lobsterman's buoy is a solid wooden cylinder of
    [S] radius r and mass M. It is weighted at one end so that it floats
    upright in calm seawater, having density ρ. A passing shark
    tugs on the slack rope mooring the buoy to a lobster trap,
    pulling the buoy down a distance x from its equilibrium
    position and releasing it. (a) Show that the buoy will execute
    simple harmonic motion if the resistive effects of the water
    are ignored. (b) Determine the period of the oscillations.

42. Your thumb squeaks on a plate you have just washed. Your
    [S] sneakers squeak on the gym floor. Car tires squeal when
    you start or stop abruptly. You can make a goblet sing by
    wiping your moistened finger around its rim. When chalk
    squeaks on a blackboard, you can see that it makes a row of
    regularly spaced dashes. As these examples suggest, vibra-
    tion commonly results when friction acts on a moving elas-
    tic object. The oscillation is not simple harmonic motion,
    but is called *stick-and-slip*. This problem models stick-and-
    slip motion.

    A block of mass m is attached to a fixed support by a
    horizontal spring with force constant k and negligible mass
    (Fig. P15.42). Hooke's law describes the spring both in
    extension and in compression. The block sits on a long hor-
    izontal board, with which it has coefficient of static friction
    μ_s and a smaller coefficient of kinetic friction μ_k. The board
    moves to the right at constant speed v. Assume the block
    spends most of its time sticking to the board and moving to
    the right with it, so the speed v is small in comparison to the
    average speed the block has as it slips back toward the left.
    (a) Show that the maximum extension of the spring from its
    unstressed position is very nearly given by $\mu_s mg/k$. (b) Show
    that the block oscillates around an equilibrium position
    at which the spring is stretched by $\mu_k mg/k$. (c) Graph the
    block's position versus time. (d) Show that the amplitude of
    the block's motion is

    $$A = \frac{(\mu_s - \mu_k)mg}{k}$$

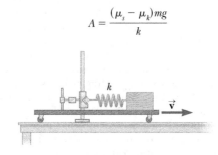

Figure P15.42

(e) Show that the period of the block's motion is

$$T = \frac{2(\mu_s - \mu_k)mg}{vk} + \pi\sqrt{\frac{m}{k}}$$

It is the excess of static over kinetic friction that is important for the vibration. "The squeaky wheel gets the grease" because even a viscous fluid cannot exert a force of static friction.

43. Your father is preparing the backyard for the installation
CR of new sod. He has finished cleaning the ground of roots and rocks, has raked it to the correct contours, and now must pull a heavy roller, shown in Figure P15.43a, over the ground several times to flatten and compact the dirt. He is tired after all of his work and asks you to do the rolling for him. He tells you that each section of the yard must be rolled over ten times with the roller. You are tired from your physics studying, but decide you can use your understanding of physics to make the job easier. You attach the roller to a spring as shown in Figure P15.43b, with the other end attached to a post pounded into the ground. You then just pull the roller out once and let it oscillate over each part of the yard for ten rolls while you sit back and relax. Before beginning, you wonder how much time you will have to relax at each location before you have to move the post and roller to a new location. The mass of the roller is $M =$ 400 kg, and the spring constant is $k = 3\,500$ N/m. The flat, smooth ground supplies enough friction that the roller rolls instead of sliding, but the rolling friction is negligible.

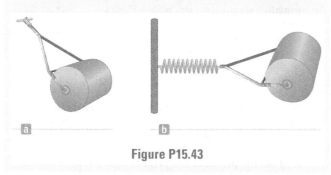

Figure P15.43

44. *Why is the following situation impossible?* Your job involves building very small damped oscillators. One of your designs involves a spring–object oscillator with a spring of force constant $k = 10.0$ N/m and an object of mass $m = 1.00$ g. Your design objective is that the oscillator undergo many oscillations as its amplitude falls to 25.0% of its initial value in a certain time interval. Measurements on your latest design show that the amplitude falls to the 25.0% value in 23.1 ms. This time interval is too long for what is needed in your project. To shorten the time interval, you double the damping constant b for the oscillator. This doubling allows you to reach your design objective.

45. A block of mass m is connected to two springs of force con-
S stants k_1 and k_2 in two ways as shown in Figure P15.45. In both cases, the block moves on a frictionless table after it is displaced from equilibrium and released. Show that in the two cases the block exhibits simple harmonic motion with periods

(a) $T = 2\pi\sqrt{\dfrac{m(k_1 + k_2)}{k_1 k_2}}$ and   (b) $T = 2\pi\sqrt{\dfrac{m}{k_1 + k_2}}$

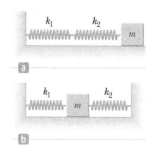

Figure P15.45

46. Review. A light balloon filled with helium of density 0.179 kg/m³ is tied to a light string of length $L = 3.00$ m. The string is tied to the ground forming an "inverted" simple pendulum (Fig. P15.46a). If the balloon is displaced slightly from equilibrium as in Figure P15.46b and released, (a) show that the motion is simple harmonic and (b) determine the period of the motion. Take the density of air to be 1.20 kg/m³. *Hint:* Use an analogy with the simple pendulum and see Chapter 14. Assume the air applies a buoyant force on the balloon but does not otherwise affect its motion.

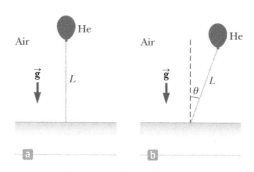

Figure P15.46

47. A particle with a mass of 0.500 kg is attached to a horizontal spring with a force constant of 50.0 N/m. At the moment $t = 0$, the particle has its maximum speed of 20.0 m/s and is moving to the left. (a) Determine the particle's equation of motion, specifying its position as a function of time. (b) Where in the motion is the potential energy three times the kinetic energy? (c) Find the minimum time interval required for the particle to move from $x = 0$ to $x =$ 1.00 m. (d) Find the length of a simple pendulum with the same period.

CHALLENGE PROBLEMS

48. A smaller disk of radius r and
S mass m is attached rigidly to the face of a second larger disk of radius R and mass M as shown in Figure P15.48. The center of the small disk is located at the edge of the large disk. The large disk is mounted at its center on a frictionless axle. The assembly is rotated through a small angle θ from its equilibrium position and

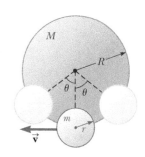

Figure P15.48

released. (a) Show that the speed of the center of the small disk as it passes through the equilibrium position is

$$v = 2\left[\frac{Rg(1 - \cos\theta)}{(M/m) + (r/R)^2 + 2}\right]^{1/2}$$

(b) Show that the period of the motion is

$$T = 2\pi\left[\frac{(M + 2m)R^2 + mr^2}{2mgR}\right]^{1/2}$$

49. **Review.** A system consists of a spring with force constant $k = 1\,250$ N/m, length $L = 1.50$ m, and an object of mass $m = 5.00$ kg attached to the end (Fig. P15.49). The object is placed at the level of the point of attachment with the spring unstretched, at position $y_i = L$, and then it is released so that it swings like a pendulum. (a) Find the y position of the object at the lowest point. (b) Will the pendulum's period be greater or less than the period of a simple pendulum with the same mass m and length L? Explain.

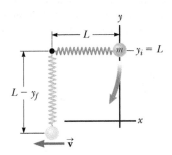

Figure P15.49

50. **Review.** *Why is the following situation impossible?* You are in the high-speed package delivery business. Your competitor in the next building gains the right-of-way to build an evacuated tunnel just above the ground all the way around the Earth. By firing packages into this tunnel at just the right speed, your competitor is able to send the packages into orbit around the Earth in this tunnel so that they arrive on the exact opposite side of the Earth in a very short time interval. You come up with a competing idea. Figuring that the distance *through* the Earth is shorter than the distance *around* the Earth, you obtain permits to build an evacuated tunnel through the center of the Earth (Fig. P15.50). By simply dropping packages into this tunnel, they fall downward and arrive at the other end of your tunnel, which is in a building right next to the other end of your competitor's tunnel. Because your packages arrive on the other side of the Earth in a shorter time interval, you win the competition and your business flourishes. *Note:* An object at a distance r from the center of the Earth is pulled toward the center of the Earth only by the mass within the sphere of radius r (the reddish region in Fig. P15.50). Assume the Earth has uniform density.

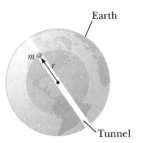

Figure P15.50

51. A light, cubical container of volume a^3 is initially filled with a liquid of mass density ρ as shown in Figure P15.51a. The cube is initially supported by a light string to form a simple pendulum of length L_i, measured from the center of mass of the filled container, where $L_i \gg a$. The liquid is allowed to flow from the bottom of the container at a constant rate (dM/dt). At any time t, the level of the liquid in the container is h and the length of the pendulum is L (measured relative to the instantaneous center of mass) as shown in Figure P15.51b. (a) Find the period of the pendulum as a function of time. (b) What is the period of the pendulum after the liquid completely runs out of the container?

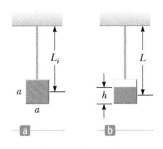

Figure P15.51

Wave Motion

16

At the Quincy Quarries Reservation in Quincy, Massachusetts, rainwater filled in an old granite quarry, so that the water was surrounded by rocks and cliffs. When Boston undertook the Big Dig, in which huge amounts of dirt were removed from beneath the city to make way for underground tunnels, the water at Quincy Quarry was filled in with that dirt. Consequently, there are now large flat areas between granite cliffs. (© Cengage)

STORYLINE **During your visit to your grandparents' house in Boston,** you and your grandfather take a day trip. You stop at Quincy Quarries Reservation, where there are large flat areas between the cliffs. While visiting, you notice that you can clap your hands and hear a distinct echo from a distant cliff. You say, "Watch this, Grandpa, let me show you something about cell phones." You pull out your smartphone and activate the digital recording app. You ask your grandfather to clap his hands once, just after you start recording. You stop the recording after the echo from the cliff arrives. Seeing the pulses on the app display representing both the clap and the echo, you determine the time interval for the sound of the clap to travel to the cliff and back. Then you use the GPS system on your phone to determine your latitude and longitude coordinates. At this point, you say, "Grandpa, let's go for a hike!" You hike across the former lake to the base of the cliff that provided the echo and determine your coordinates again. Based on the two sets of coordinates, you use a Web site to determine the distance between the cliff and your original position. From this distance and the time interval you measured for the echo to arrive, you make a reasonably accurate calculation of the speed of sound. Your grandfather is quite impressed with you.

CONNECTIONS In this chapter, we will continue crossing the bridge we mentioned at the beginning of Chapter 15. *Wave motion* represents phenomena in which a *disturbance* propagates through a medium. The disturbance carries energy from one point to another. But there is no matter that moves over that distance. For example, suppose you go bowling. You can knock the pins over by rolling the bowling ball at them. That is *not* wave motion. The energy is carried by the bowling ball—there is a transfer of matter. But suppose you could *shout* loud

415

enough to knock the pins over. (Do *not* try this!) *That* would be energy transfer by waves. The energy is carried by the sound wave of your voice—no matter transfers from your mouth to the pins. In our discussions in this chapter and the next, we discuss *mechanical* waves. These waves require a *medium.* For example, we will study one-dimensional waves traveling on a string. The string is the medium. We will also consider mechanical waves in three dimensions: the waves can travel in any direction through a bulk medium. When the medium is air, we call such mechanical waves *sound.* We will relate phenomena associated with sound waves to our sense of hearing. We will use our information from this chapter to study waves under boundary conditions in Chapter 17, which will lead to an understanding of musical instruments. Furthermore, the material in this chapter will form the foundation of our study of electromagnetic waves in Chapters 33–37 and quantum physics in Chapters 39–44.

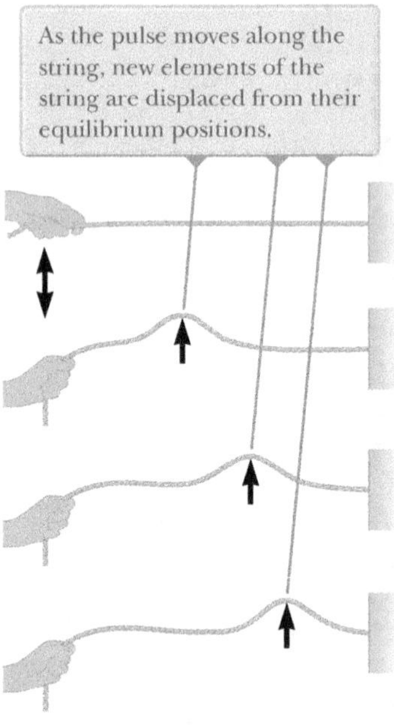

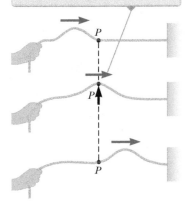

As the pulse moves along the string, new elements of the string are displaced from their equilibrium positions.

Figure 16.1 A hand moves the end of a stretched string up and down once (black, double-headed arrow), causing a pulse to travel along the string.

The direction of the displacement of any element at a point *P* on the string is perpendicular to the direction of propagation (red arrow).

Figure 16.2 The displacement of a particular string element for a transverse pulse traveling on a stretched string.

16.1 Propagation of a Disturbance

The introduction to this chapter alluded to the essence of wave motion: the transfer of energy through space without the accompanying transfer of matter. In the list of energy transfer mechanisms in Equation 8.2, two mechanisms—mechanical waves T_{MW} and electromagnetic radiation T_{ER}—depend on waves. By contrast, in another mechanism, matter transfer T_{MT}, the energy transfer is accompanied by a movement of matter through space with no wave character at all in the process.

All mechanical waves require (1) some source of disturbance, (2) a medium containing elements that can be disturbed, and (3) some physical mechanism through which elements of the medium can influence each other. One way to demonstrate wave motion is to flick one end of a long string that is under tension and has its opposite end fixed as shown in Figure 16.1. In this manner, a single bump (called a *pulse*) is formed and travels along the string with a constant speed. Figure 16.1 represents four consecutive "snapshots" of the creation and propagation of the traveling pulse. The hand is the source of the disturbance. The string is the medium through which the pulse travels—individual elements of the string are disturbed from their equilibrium position. Furthermore, the elements of the string are connected together so they influence each other: as one element goes up, it pulls the next one upward. The pulse has a definite height and a definite speed of propagation along the medium. The shape of the pulse changes very little as it travels along the string.[1]

As the pulse in Figure 16.1 travels to the right, each disturbed element of the string moves in the vertical direction, *perpendicular* to the direction of propagation. Figure 16.2 illustrates this point for one particular element, labeled *P.* Notice that no part of the string ever moves in the direction of the propagation. A traveling wave or pulse that causes the elements of the disturbed medium to move perpendicular to the direction of propagation is called **transverse.**

Compare the pulse in Figure 16.1 with another type of pulse, one moving down a long, stretched spring as shown in Figure 16.3. The left end of the spring is pushed briefly to the right and then pulled briefly to the left. This movement creates a sudden compression of a region of the coils. The compressed region travels along the spring (to the right in Fig. 16.3). Notice that the direction of the displacement of the coils is *parallel* to the direction of propagation of the compressed region. A traveling wave or pulse that causes the elements of the medium to move parallel to the direction of propagation is called **longitudinal.**

---

[1]In reality, the pulse changes shape and gradually spreads out during the motion. This effect, called *dispersion,* is common to many mechanical waves as well as to electromagnetic waves. We do not consider dispersion in this chapter.

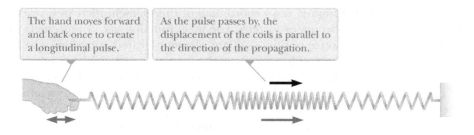

The hand moves forward and back once to create a longitudinal pulse.

As the pulse passes by, the displacement of the coils is parallel to the direction of the propagation.

Figure 16.3 A longitudinal pulse along a stretched spring.

If the end of the string in Figure 16.1 were moved up and down continuously, the hand would generate a series of pulses called a **transverse wave.** We will study the details of waves such as this in Section 16.2. Sound waves, which we shall discuss later in this chapter, are an example of **longitudinal waves.** The disturbance in a sound wave is a series of high-pressure and low-pressure regions that travel through air, as we shall see in Section 16.6.

Some waves in nature exhibit a combination of transverse and longitudinal displacements. Surface-water waves such as those in the ocean are a good example. When a water wave travels on the surface of deep water, elements of water at the surface move in nearly circular paths as shown in Figure 16.4. The disturbance has both transverse and longitudinal components. The transverse displacements seen in Figure 16.4 represent the variations in vertical position of the water elements. The longitudinal displacements represent elements of water moving back and forth in a horizontal direction. A point in Figure 16.4 at which the displacement of the element from its normal position is highest is called the **crest** of the wave. The lowest point is called the **trough.**

An earthquake represents a disturbance that results in *seismic waves.* Two types of three-dimensional seismic waves travel out from a point under the Earth's surface at which an earthquake occurs: transverse and longitudinal. The longitudinal waves are the faster of the two, traveling at speeds in the range of 7 to 8 km/s near the surface. They are called **P waves,** with "P" standing for *primary,* because they travel faster than the transverse waves and arrive first at a seismograph (a device used to detect waves due to earthquakes). The slower transverse waves, called **S waves,** with "S" standing for *secondary,* travel through the Earth at 4 to 5 km/s near the surface. By recording the time interval between the arrivals of these two types of waves at a seismograph, the distance from the seismograph to the point of origin of the waves can be determined. This distance is the radius of an imaginary sphere centered on the seismograph. The origin of the waves is located somewhere on that sphere. The imaginary spheres from three or more monitoring stations located far apart from one another intersect at one region of the Earth, and this region is where the earthquake occurred.

QUICK QUIZ 16.1 **(i)** In a long line of people waiting to buy tickets, the first person leaves and a pulse of motion occurs as people step forward to fill the gap. As each person steps forward, the gap moves through the line. Is the propagation of this gap (a) transverse or (b) longitudinal? **(ii)** Consider "the wave" at a baseball game: people stand up and raise their arms as the pulse arrives at their location, and the resultant pulse moves around the stadium. Is this pulse (a) transverse or (b) longitudinal?

Consider a pulse traveling to the right on a long string as shown in Figure 16.5. At any time, the pulse can be represented by some mathematical function that we will write as $y(x, t)$. At $t = 0$, as in Figure 16.5a, let's write this as $y(x, 0) = f(x)$, where $f(x)$ describes the shape of the pulse in space.

The function $y(x, t)$, sometimes called the **wave function,** depends on the two variables x and t. For this reason, it is described as "y as a function of x and t."

It is important to understand the meaning of y. Consider an element of the string at point P in Figure 16.5, identified by a particular value of its x coordinate.

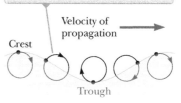

The elements at the surface move in nearly circular paths. Each element is displaced both horizontally and vertically from its equilibrium position.

Velocity of propagation

Crest

Trough

Figure 16.4 The motion of water elements on the surface of deep water in which a wave is propagating is a combination of transverse and longitudinal displacements.

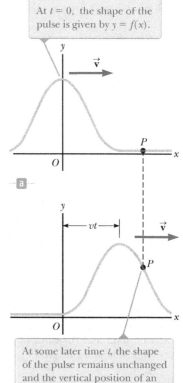

At $t = 0$, the shape of the pulse is given by $y = f(x)$.

$\vec{v}$

O

P

ⓐ

vt

$\vec{v}$

P

O

ⓑ

At some later time t, the shape of the pulse remains unchanged and the vertical position of an element of the medium at any point P is given by $y = f(x - vt)$.

Figure 16.5 A one-dimensional pulse traveling to the right on a string with a speed v.

As the pulse passes through P, the y coordinate of this element increases, reaches a maximum, and then decreases to zero. The wave function $y(x, t)$ represents the y coordinate—the transverse position—of any element located at position x at any time t. If we were to view the pulse at a particular instant of time, such as in the case of taking a snapshot of the pulse, we would see something like Figure 16.5a or 16.5b. The geometric shape $f(x)$ of the pulse at a particular instant is called the **waveform.**

Because the speed of the pulse is v, the crest of the pulse has traveled to the right a distance vt at the time t (Fig. 16.5b). We assume the shape of the pulse does not change with time. Therefore, at time t, the shape of the pulse is the same as it was at time $t = 0$ as in Figure 16.5a. Consequently, an element of the string at x at this time has the same y position as an element located at $x - vt$ had at time $t = 0$:

$$y(x, t) = y(x - vt, 0)$$

In general, then, we can represent the transverse position y for all positions and times, measured in a stationary frame with the origin at O, as

$$y(x, t) = f(x - vt) \qquad (16.1)$$

Similarly, if the pulse travels to the left, the transverse positions of elements of the string are described by

$$y(x, t) = f(x + vt) \qquad (16.2)$$

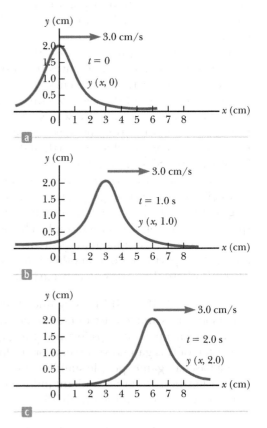

Figure 16.6
(Example 16.1) Graphs of the function $y(x, t) = 2/[(x - 3.0t)^2 + 1]$ at (a) $t = 0$, (b) $t = 1.0$ s, and (c) $t = 2.0$ s.

Example 16.1 A Pulse Moving to the Right

A pulse moving to the right along the x axis is represented by the wave function

$$y(x, t) = \frac{2}{(x - 3.0t)^2 + 1}$$

where x and y are measured in centimeters and t is measured in seconds. Find expressions for the wave function at $t = 0$, $t = 1.0$ s, and $t = 2.0$ s.

SOLUTION

Conceptualize Figure 16.6a shows the pulse represented by this wave function at $t = 0$. Imagine this pulse moving to the right and maintaining its shape as suggested by Figures 16.6b and 16.6c.

Categorize We categorize this example as a relatively simple analysis problem in which we interpret the mathematical representation of a pulse.

Analyze The wave function is of the form $y = f(x - vt)$. Inspection of the expression for $y(x, t)$ and comparison to Equation 16.1 reveal that the wave speed is $v = 3.0$ cm/s. Furthermore, we can maximize the value of y by letting $x - 3.0t = 0$, and find that $y_{max} = 2.0$ cm.

Write the wave function expression at $t = 0$:

$$y(x, 0) = \frac{2}{x^2 + 1}$$

Write the wave function expression at $t = 1.0$ s:

$$y(x, 1.0) = \frac{2}{(x - 3.0)^2 + 1}$$

Write the wave function expression at $t = 2.0$ s:

$$y(x, 2.0) = \frac{2}{(x - 6.0)^2 + 1}$$

16.1 continued

For each of these expressions, we can substitute various values of x and plot the wave function. This procedure yields the wave functions shown in the three parts of Figure 16.6.

Finalize These snapshots show that the pulse moves to the right without changing its shape and that it has a constant speed of 3.0 cm/s.

WHAT IF? What if the wave function were

$$y(x, t) = \frac{4}{(x + 3.0t)^2 + 1}$$

How would that change the situation?

Answer One new feature in this expression is the plus sign in the denominator rather than the minus sign. The new expression represents a pulse with a similar shape as that in Figure 16.6, but moving to the left as time progresses. Another new feature here is the numerator of 4 rather than 2. Therefore, the new expression represents a pulse with twice the height of that in Figure 16.6.

16.2 Analysis Model: Traveling Wave

To generate the pulse on the rope in Figure 16.1, we shook the end of the rope up and down *once*. In this section, we introduce an important wave function whose shape is shown in Figure 16.7 and is produced by shaking the end of the rope up and down *continuously* in simple harmonic motion. The wave represented by this curve is called a **sinusoidal wave** because the curve is the same as that of the function sin θ plotted against θ. Because shaking the end of the rope in simple harmonic motion leads to a sinusoidal wave, we see that there is a close relationship between simple harmonic motion and sinusoidal waves.

The sinusoidal wave is the simplest example of a periodic continuous wave and can be used to build more complex waves (see Section 17.8). The brown curve in Figure 16.7 represents a snapshot of a traveling sinusoidal wave at $t = 0$, and the blue curve represents a snapshot of the wave at some later time t. Imagine two types of motion that can occur. First, the entire waveform in Figure 16.7 moves to the right so that the brown curve moves toward the right and eventually reaches the position of the blue curve. This movement is the motion of the *wave*. If we focus on one element of the medium, such as the element at $x = 0$, we see that each element moves up and down along the y axis in simple harmonic motion. This movement is the motion of the *elements of the medium*. It is important to differentiate between the motion of the wave and the motion of the elements of the medium. An element of the medium is described by the particle in simple harmonic motion model. A point on the wave, such as the crest, can be described with the particle under constant velocity model.

In the early chapters of this book, we developed several analysis models based on three simplification models: the particle, the system, and the rigid object. With our introduction to waves, we can develop a new simplification model, the **wave,** that will allow us to explore more analysis models for solving problems. In what follows, we will develop the principal features and mathematical representations of the analysis model of a **traveling wave.** This model is used in situations in which a wave moves through space without interacting with other waves or particles.

Figure 16.8a (page 420) shows a snapshot of a traveling wave moving through a medium. Figure 16.8b shows a graph of the position of one element of the medium as a function of time. Recall from Section 16.1 that the highest point on a wave is called the crest of the wave, and the lowest point is the trough. The distance from one crest to the next is called the **wavelength** λ (Greek letter lambda). More generally, the wavelength is the minimum distance between any two identical points on adjacent waves as shown in Figure 16.8a.

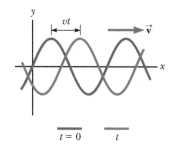

Figure 16.7 A one-dimensional sinusoidal wave traveling to the right with a speed v. The brown curve represents a snapshot of the wave at $t = 0$, and the blue curve represents a snapshot at some later time t.

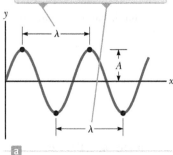

The wavelength λ of a wave is the distance between adjacent crests or adjacent troughs.

a

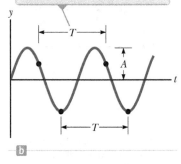

The period *T* of a wave is the time interval required for the element to complete one cycle of its oscillation and for the wave to travel one wavelength.

b

Figure 16.8 (a) A snapshot of a sinusoidal wave. (b) The position of one element of the medium as a function of time.

If you count the number of seconds between the arrivals of two adjacent crests at a given location in space, you measure the **period** *T* of the waves. In general, the period is the time interval required for an element of the medium to undergo a complete cycle and return to the same position as shown in Figure 16.8b. The period of the wave is the same as the period of the simple harmonic oscillation of one element of the medium.

The same information is more often given by the inverse of the period, which is called the **frequency** *f*. In general, the frequency of a periodic wave is the number of crests (or troughs, or any other point on the wave) that pass a given location in a unit time interval. The frequency of a sinusoidal wave is related to the period by the expression

$$f = \frac{1}{T} \tag{16.3}$$

The frequency of the wave is the same as the frequency of the simple harmonic oscillation of one element of the medium. The most common unit for frequency, as we learned in Chapter 15, is s^{-1}, or **hertz** (Hz). The corresponding unit for *T* is seconds.

An ideal particle has zero size. We can build physical objects with nonzero size as combinations of particles. Therefore, the particle can be considered a basic building block. An ideal wave has a single frequency and is infinitely long; that is, the wave exists throughout the Universe. (A wave of finite length must necessarily have a mixture of frequencies.) When this concept is explored in Section 17.8, we will find that ideal waves can be combined to build complex waves, just as we combined particles: the wave is a basic building block.

The maximum position of an element of the medium relative to its equilibrium position is called the **amplitude** *A* of the wave as indicated in Figure 16.8. Consider the sinusoidal wave in Figure 16.8a, which shows the position of the wave at *t* = 0. Because the wave is sinusoidal, we expect the wave function at this instant to be expressed as $y(x, 0) = A \sin ax$, where *A* is the amplitude and *a* is a constant to be determined. At *x* = 0, we see that $y(0, 0) = A \sin a(0) = 0$, consistent with Figure 16.8a. The next value of *x* for which *y* is zero is $x = \lambda/2$. Therefore,

$$y\left(\frac{\lambda}{2}, 0\right) = A \sin\left(a\frac{\lambda}{2}\right) = 0$$

For this equation to be true, we must have $a\lambda/2 = \pi$, or $a = 2\pi/\lambda$. Therefore, the function describing the positions of the elements of the medium through which the sinusoidal wave is traveling can be written

$$y(x, 0) = A \sin\left(\frac{2\pi}{\lambda} x\right) \tag{16.4}$$

where the constant *A* represents the wave amplitude and the constant λ is the wavelength. Notice that the vertical position of an element of the medium is the same whenever *x* is increased by an integral multiple of λ. Based on our discussion of Equation 16.1, if the wave moves to the right with a speed *v*, the wave function at some later time *t* is

$$y(x, t) = A \sin\left[\frac{2\pi}{\lambda}(x - vt)\right] \tag{16.5}$$

If the wave were traveling to the left, the quantity $x - vt$ would be replaced by $x + vt$ as we learned when we developed Equations 16.1 and 16.2.

By definition, the wave travels through a displacement Δ*x* equal to one wavelength λ in a time interval Δ*t* of one period *T*. Therefore, the wave speed, wavelength, and period are related by the expression

$$v = \frac{\Delta x}{\Delta t} = \frac{\lambda}{T} \tag{16.6}$$

Substituting this expression for v into Equation 16.5 gives

$$y(x, t) = A \sin\left[2\pi\left(\frac{x}{\lambda} - \frac{t}{T}\right)\right] \qquad (16.7)$$

This form of the wave function shows the *periodic* nature of $y(x, t)$. At any given time t, $y(x, t)$ has the *same* value at the positions x, $x + \lambda$, $x + 2\lambda$, and so on. Furthermore, at any given position x, the value of $y(x, t)$ is the same at times t, $t + T$, $t + 2T$, and so on.

We can express the wave function in a convenient form by defining two other quantities, the **angular wave number** k (usually called simply the **wave number**) and the **angular frequency** ω:

$$k \equiv \frac{2\pi}{\lambda} \qquad (16.8) \qquad \triangleleft \text{ Angular wave number}$$

$$\omega \equiv \frac{2\pi}{T} = 2\pi f \qquad (16.9) \qquad \triangleleft \text{ Angular frequency}$$

Using these definitions, Equation 16.7 can be written in the more compact form

$$y(x, t) = A \sin (kx - \omega t) \qquad (16.10) \qquad \triangleleft \begin{array}{l}\text{Wave function for a} \\ \text{sinusoidal wave}\end{array}$$

Using Equations 16.3, 16.8, and 16.9, the wave speed v originally given in Equation 16.6 can be expressed in the following alternative forms:

$$v = \frac{\omega}{k} \qquad (16.11)$$

$$v = \lambda f \qquad (16.12) \qquad \triangleleft \text{ Speed of a sinusoidal wave}$$

The wave function given by Equation 16.10 assumes the vertical position y of an element of the medium is zero at $x = 0$ and $t = 0$. That need not be the case. If it is not, we generally express the wave function in the form

$$y(x, t) = A \sin (kx - \omega t + \phi) \qquad (16.13) \qquad \triangleleft \begin{array}{l}\text{General expression for a} \\ \text{sinusoidal wave}\end{array}$$

where ϕ is the **phase constant,** just as we learned in our study of periodic motion in Chapter 15. This constant can be determined from the initial conditions. The primary equations in the mathematical representation of the traveling wave analysis model are Equations 16.3, 16.10, and 16.12.

Ⓠ**UICK QUIZ 16.2** A sinusoidal wave of frequency f is traveling along a stretched string. The string is brought to rest, and a second traveling wave of frequency $2f$ is established on the string. **(i)** What is the wave speed of the second wave? (a) twice that of the first wave (b) half that of the first wave (c) the same as that of the first wave (d) impossible to determine **(ii)** From the same choices, describe the wavelength of the second wave. **(iii)** From the same choices, describe the amplitude of the second wave.

Example 16.2   **A Traveling Sinusoidal Wave**

A sinusoidal wave traveling in the positive x direction has an amplitude of 15.0 cm, a wavelength of 40.0 cm, and a frequency of 8.00 Hz. The vertical position of an element of the medium at $t = 0$ and $x = 0$ is also 15.0 cm as shown in Figure 16.9 (page 422).

(A) Find the wave number k, period T, angular frequency ω, and speed v of the wave.

continued

16.2 continued

SOLUTION

Conceptualize Figure 16.9 shows the wave at $t = 0$. Imagine this wave moving to the right and maintaining its shape.

Categorize From the description in the problem statement, we see that we are analyzing a mechanical wave moving through a medium, so we categorize the problem with the *traveling wave* model.

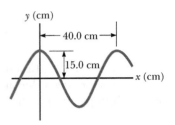

Figure 16.9 (Example 16.2) A sinusoidal wave of wavelength $\lambda = 40.0$ cm and amplitude $A = 15.0$ cm.

Analyze

Evaluate the wave number from Equation 16.8:

$$k = \frac{2\pi}{\lambda} = \frac{2\pi \text{ rad}}{40.0 \text{ cm}} = 15.7 \text{ rad/m}$$

Evaluate the period of the wave from Equation 16.3:

$$T = \frac{1}{f} = \frac{1}{8.00 \text{ s}^{-1}} = 0.125 \text{ s}$$

Evaluate the angular frequency of the wave from Equation 16.9:

$$\omega = 2\pi f = 2\pi(8.00 \text{ s}^{-1}) = 50.3 \text{ rad/s}$$

Evaluate the wave speed from Equation 16.12:

$$v = \lambda f = (40.0 \text{ cm})(8.00 \text{ s}^{-1}) = 3.20 \text{ m/s}$$

(B) Determine the phase constant ϕ and write a general expression for the wave function.

SOLUTION

Substitute $A = 15.0$ cm, $y = 15.0$ cm, $x = 0$, and $t = 0$ into Equation 16.13:

$$15.0 = (15.0) \sin \phi \rightarrow \sin \phi = 1 \rightarrow \phi = \frac{\pi}{2} \text{ rad}$$

Write the wave function:

$$y(x, t) = A \sin \left(kx - \omega t + \frac{\pi}{2} \right) = A \cos (kx - \omega t)$$

Substitute the values for A, k, and ω in SI units into this expression:

$$y(x, t) = 0.150 \cos (15.7x - 50.3t)$$

Finalize Review the results carefully and make sure you understand them. How would the graph in Figure 16.9 change if the phase angle were zero? How would the graph change if the amplitude were 30.0 cm? How would the graph change if the wavelength were 10.0 cm?

Sinusoidal Waves on Strings

In Figure 16.1, we demonstrated how to create a pulse by jerking a taut string up and down once. To create a series of such pulses—a wave—let's replace the hand with an oscillating blade whose end is vibrating in simple harmonic motion. Figure 16.10 represents snapshots of the wave created in this way at intervals of $T/4$. Because the end of the blade oscillates in simple harmonic motion, each element of the string, such as that at P, also oscillates vertically with simple harmonic motion. Therefore, every element of the string can be treated as a simple harmonic oscillator vibrating with a frequency equal to the frequency of oscillation of the blade.[2] Notice that while each element oscillates in the y direction, the wave travels to the right in the $+x$ direction with a speed v.

If we define $t = 0$ as the time for which the configuration of the string is as shown in Figure 16.10a, the wave function can be written from Equation 16.10 as

$$y = A \sin (kx - \omega t)$$

where we simplify $y(x, t)$ by writing it simply as y. We can use this expression to describe the motion of any element of the string. An element at point P (or any other element of the string) moves only vertically, and so its x coordinate remains

PITFALL PREVENTION 16.2

Two Kinds of Speed/Velocity
Do not confuse v, the speed of the wave as it propagates along the string, with v_y, the transverse velocity of a point on the string. The speed v is constant for a uniform medium, whereas v_y varies sinusoidally.

---

[2] In this arrangement, we are assuming that a string element always oscillates in a vertical line. The tension in the string would vary if an element were allowed to move sideways. Such motion would make the analysis very complex.

constant. Therefore, the **transverse speed** v_y (not to be confused with the wave speed v) and the **transverse acceleration** a_y of elements of the string are

$$v_y = \left.\frac{dy}{dt}\right]_{x=\text{constant}} = \frac{\partial y}{\partial t} = -\omega A \cos(kx - \omega t) \qquad (16.14)$$

$$a_y = \left.\frac{dv_y}{dt}\right]_{x=\text{constant}} = \frac{\partial v_y}{\partial t} = -\omega^2 A \sin(kx - \omega t) \qquad (16.15)$$

These expressions incorporate partial derivatives because y depends on both x and t. In the operation $\partial y/\partial t$, for example, we take a derivative with respect to t while holding x constant. The maximum magnitudes of the transverse speed and transverse acceleration are simply the absolute values of the coefficients of the cosine and sine functions:

$$v_{y,\text{max}} = \omega A \qquad (16.16)$$

$$a_{y,\text{max}} = \omega^2 A \qquad (16.17)$$

The transverse speed and transverse acceleration of elements of the string do not reach their maximum values simultaneously. The transverse speed reaches its maximum value (ωA) when $y = 0$, whereas the magnitude of the transverse acceleration reaches its maximum value ($\omega^2 A$) when $y = \pm A$. Finally, Equations 16.16 and 16.17 are identical in mathematical form to the corresponding equations for simple harmonic motion, Equations 15.17 and 15.18.

QUICK QUIZ 16.3 The amplitude of a wave is doubled, with no other changes made to the wave. As a result of this doubling, which of the following statements is correct? **(a)** The speed of the wave changes. **(b)** The frequency of the wave changes. **(c)** The maximum transverse speed of an element of the medium changes. **(d)** Statements **(a)** through **(c)** are all true. **(e)** None of statements **(a)** through **(c)** is true.

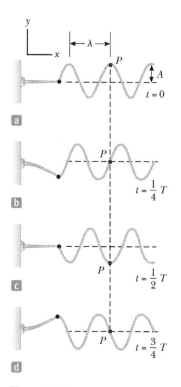

Figure 16.10 One method for producing a sinusoidal wave on a string. The left end of the string is connected to a blade that is set into oscillation. Every element of the string, such as that at point P, oscillates with simple harmonic motion in the vertical direction.

ANALYSIS MODEL **Traveling Wave**

Imagine a source vibrating such that it influences the medium that is in contact with the source. Such a source creates a disturbance that propagates through the medium. If the source vibrates in simple harmonic motion with period T, sinusoidal waves propagate through the medium at a speed given by

$$v = \frac{\lambda}{T} = \lambda f \qquad (16.6, 16.12)$$

where λ is the **wavelength** of the wave and f is its **frequency**. A sinusoidal wave can be expressed as

$$y(x, t) = A \sin(kx - \omega t) \qquad (16.10)$$

where A is the **amplitude** of the wave, k is its **wave number,** and ω is its **angular frequency.**

Examples:

- a vibrating blade sends a sinusoidal wave down a string attached to the blade
- a piston vibrates back and forth, emitting sound waves into a tube filled with gas (Section 16.6)
- a guitar body vibrates, emitting sound waves into the air (Chapter 17)
- a vibrating electric charge creates an electromagnetic wave that propagates into space at the speed of light (Chapter 33)

16.3 The Speed of Waves on Strings

One aspect of the behavior of *linear* mechanical waves is that the wave speed depends only on the properties of the medium through which the wave travels. Waves for which the amplitude A is small relative to the wavelength λ can be represented as linear waves. (See Section 16.5.) In this section, we determine the speed of a transverse wave traveling on a stretched string.

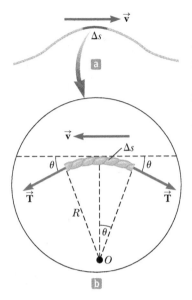

Figure 16.11 (a) In the reference frame of the Earth, a pulse moves to the right on a string with speed v. (b) In a frame of reference moving to the right with the pulse, the small element of length Δs moves to the left with speed v.

Let us use a mechanical analysis to derive the expression for the speed of a pulse traveling on a stretched string under tension T. Consider a pulse moving to the right with a uniform speed v, measured relative to a stationary (with respect to the Earth) inertial reference frame as shown in Figure 16.11a. Newton's laws are valid in any inertial reference frame. Therefore, let us view this pulse from a different inertial reference frame, one that moves along with the pulse at the same speed so that the pulse appears to be at rest in the frame as in Figure 16.11b. In this reference frame, the pulse remains fixed and each element of the string moves to the left through the pulse shape.

A short element of the string, of length Δs, forms an approximate arc of a circle of radius R as shown in the magnified view in Figure 16.11b. In our moving frame of reference, the element of the string moves to the left with speed v. As it travels through the arc, we can model the element as a particle in circular motion. This element has a centripetal (downward) acceleration of v^2/R, which is supplied by components of the force $\vec{T}$ whose magnitude is the tension in the string. The force $\vec{T}$ acts on each side of the element, tangent to the arc, as in Figure 16.11b. The horizontal components of $\vec{T}$ cancel, and each vertical component $T\sin\theta$ acts downward. Hence, the magnitude of the total radial force on the element is $2T\sin\theta$. Because the element is small, θ is small and we can use the small-angle approximation $\sin\theta \approx \theta$. Therefore, the magnitude of the total radial force is

$$F_r = 2T\sin\theta \approx 2T\theta$$

The element has mass $m = \mu\Delta s$, where μ is the mass per unit length of the string. Because the element forms part of a circle and subtends an angle of 2θ at the center, $\Delta s = R(2\theta)$, and

$$m = \mu\Delta s = 2\mu R\theta$$

The element of the string is modeled as a particle under a net force. Therefore, applying Newton's second law to this element in the radial direction gives

$$F_r = \frac{mv^2}{R} \quad\rightarrow\quad 2T\theta = \frac{2\mu R\theta v^2}{R} \quad\rightarrow\quad T = \mu v^2$$

Solving for v gives

◀ Speed of a wave on a stretched string

$$v = \sqrt{\frac{T}{\mu}} \tag{16.18}$$

PITFALL PREVENTION 16.3

Multiple T's Do not confuse the T in Equation 16.18 for the tension with the symbol T used in this chapter for the period of a wave. The context of the equation should help you identify which quantity is meant. There simply aren't enough letters in the alphabet to assign a unique letter to each variable!

Notice that this derivation is based on the assumption that the pulse height is small relative to the length of the pulse. Using this assumption, we were able to use the approximation $\sin\theta \approx \theta$. Furthermore, the model assumes that the tension T is not affected by the presence of the pulse, so T is the same at all points on the string. Finally, this proof does *not* assume any particular shape for the pulse. We therefore conclude that a pulse or a wave of *any shape* will travel on the string with speed $v = \sqrt{T/\mu}$, without any change in pulse shape.

QUICK QUIZ 16.4 Suppose you create a pulse by moving the free end of a taut string up and down once with your hand beginning at $t = 0$. The string is attached at its other end to a distant wall. The pulse reaches the wall at time t. Which of the following actions, taken by itself, decreases the time interval required for the pulse to reach the wall? More than one choice may be correct. **(a)** moving your hand more quickly, but still only up and down once by the same amount **(b)** moving your hand more slowly, but still only up and down once by the same amount **(c)** moving your hand a greater distance up and down in the same amount of time **(d)** moving your hand a lesser distance up and down in the same amount of time **(e)** using a heavier string of the same length and under the same tension **(f)** using a lighter string of the same length and under the same tension **(g)** using a string of the same linear mass density but under decreased tension **(h)** using a string of the same linear mass density but under increased tension

Example 16.3 The Speed of a Pulse on a Cord

A uniform string has a mass of 0.300 kg and a length of 6.00 m. The string passes over a pulley and supports a 2.00-kg object (Fig. 16.12). Find the speed of a pulse traveling along this string.

SOLUTION

Conceptualize In Figure 16.12, the hanging block establishes a tension in the horizontal string. This tension determines the speed with which waves move on the string.

Categorize To find the tension in the string, we model the hanging block as a *particle in equilibrium*. Then we use the tension to evaluate the wave speed on the string using Equation 16.18.

Figure 16.12 (Example 16.3) The tension T in the cord is maintained by the suspended object. The speed of any wave traveling along the cord is given by $v = \sqrt{T/\mu}$.

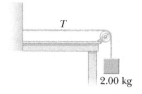

Analyze Apply the particle in equilibrium model to the block:

$$\sum F_y = T - m_{block}g = 0$$

Solve for the tension in the string:

$$T = m_{block}g$$

Use Equation 16.18 to find the wave speed, using $\mu = m_{string}/\ell$ for the linear mass density of the string:

$$v = \sqrt{\frac{T}{\mu}} = \sqrt{\frac{m_{block}g\ell}{m_{string}}}$$

Substitute numerical values:

$$v = \sqrt{\frac{(2.00 \text{ kg})(9.80 \text{ m/s}^2)(6.00 \text{ m})}{0.300 \text{ kg}}} = 19.8 \text{ m/s}$$

Finalize The calculation of the tension neglects the small mass of the string. Strictly speaking, the string can never be exactly straight due to its weight; therefore, the tension is not uniform.

WHAT IF? What if the block were swinging back and forth with respect to the vertical like a pendulum? How would that affect the wave speed on the string?

Answer The swinging block is categorized as a *particle under a net force*. The magnitude of one of the forces on the block is the tension in the string, which determines the wave speed. As the block swings, the tension changes, so the wave speed changes.

When the block is at the bottom of the swing, the string is vertical and the tension is larger than the weight of the block because the net force must be upward to provide the centripetal acceleration of the block. Therefore, the wave speed must be greater than 19.8 m/s.

When the block is at its highest point at the end of a swing, it is momentarily at rest, so there is no centripetal acceleration at that instant. The block is a particle in equilibrium in the radial direction. The tension is balanced by a component of the gravitational force on the block. Therefore, the tension is smaller than the weight and the wave speed is less than 19.8 m/s. With what frequency does the speed of the wave vary? Is it the same frequency as the pendulum?

Example 16.4 Rescuing the Hiker

An 80.0-kg hiker is trapped on a mountain ledge following a storm. A helicopter rescues the hiker by hovering above him and lowering a cable to him. The mass of the cable is 8.00 kg, and its length is 15.0 m. A sling of mass 70.0 kg is attached to the end of the cable. The hiker attaches himself to the sling, and the helicopter then accelerates upward. Terrified by hanging from the cable in midair, the hiker tries to signal the pilot by sending transverse pulses up the cable. A pulse takes 0.250 s to travel the length of the cable. What is the acceleration of the helicopter? Assume the tension in the cable is uniform.

SOLUTION

Conceptualize Imagine the effect of the acceleration of the helicopter on the cable. The greater the upward acceleration, the larger the tension in the cable. In turn, the larger the tension, the higher the speed of pulses on the cable.

Categorize This problem is a combination of one involving the speed of pulses on a string and one in which the hiker and sling are modeled as a *particle under a net force*.

continued

16.4 continued

Analyze Solve Equation 16.18 for the tension in the cable:

$$(1) \quad v = \sqrt{\frac{T}{\mu}} \quad \rightarrow \quad T = \mu v^2$$

Model the hiker and sling as a particle under a net force, noting that the acceleration of this particle of mass m is the same as the acceleration of the helicopter:

$$\sum F = ma \quad \rightarrow \quad T - mg = ma$$

Solve for the acceleration and substitute the tension from Equation (1):

$$a = \frac{T}{m} - g = \frac{\mu v^2}{m} - g = \frac{m_{\text{cable}}}{\ell_{\text{cable}}} \frac{v^2}{m} - g = \frac{m_{\text{cable}}}{\ell_{\text{cable}} m} \left(\frac{\Delta x}{\Delta t}\right)^2 - g$$

Substitute numerical values:

$$a = \frac{(8.00 \text{ kg})}{(15.0 \text{ m})(150.0 \text{ kg})}\left(\frac{15.0 \text{ m}}{0.250 \text{ s}}\right)^2 - 9.80 \text{ m/s}^2 = 3.00 \text{ m/s}^2$$

Finalize A real cable has stiffness in addition to tension. Stiffness tends to return a wire to its original straight-line shape even when it is not under tension. For example, a piano wire laid freely on a table straightens if released from a curved shape; package-wrapping string does not.

Stiffness represents a restoring force in addition to tension and increases the wave speed. Consequently, for a real cable, the acceleration of the helicopter is most likely smaller than what we calculated.

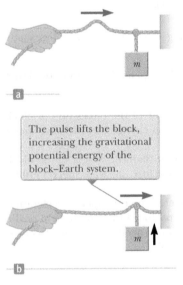

The pulse lifts the block, increasing the gravitational potential energy of the block–Earth system.

Figure 16.13 (a) A pulse travels to the right on a stretched string, carrying energy with it. (b) The energy of the pulse arrives at the hanging block.

Each element of the string is a simple harmonic oscillator and therefore has kinetic energy and potential energy associated with it.

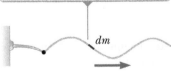

dm

Figure 16.14 A sinusoidal wave traveling along the x axis on a stretched string.

16.4 Rate of Energy Transfer by Sinusoidal Waves on Strings

Waves transport energy T_{MW} through a medium as they propagate. For example, suppose an object is hanging on a stretched string and a pulse is sent down the string as in Figure 16.13a. When the pulse meets the suspended object, the object is momentarily displaced upward as in Figure 16.13b. In the process, energy is transferred to the object and appears as an increase in the gravitational potential energy of the object–Earth system. This section examines the rate at which energy is transported along a string. We shall assume a one-dimensional sinusoidal wave in the calculation of the energy transferred.

Consider a sinusoidal wave traveling on a string (Fig. 16.14). The source of the energy is the vibrating blade at the left end of the string. We can consider the string to be a nonisolated system. As the blade performs work on the end of the string, moving it up and down, energy enters the system of the string and propagates along its length. Let's focus our attention on an infinitesimal element of the string of length dx and mass dm. We can model each element of the string as a particle in simple harmonic motion, with the oscillation in the y direction. All elements have the same angular frequency ω and the same amplitude A. The kinetic energy K associated with a moving particle is $K = \frac{1}{2}mv^2$. If we apply this equation to the infinitesimal element, the kinetic energy dK associated with the up and down motion of this element is

$$dK = \frac{1}{2}(dm)v_y^2$$

where v_y is the transverse speed of the element. If μ is the mass per unit length of the string, the mass dm of the element of length dx is equal to $\mu \, dx$. Hence, we can express the kinetic energy of an element of the string as

$$dK = \frac{1}{2}(\mu \, dx)v_y^2 \tag{16.19}$$

Substituting for the general transverse speed of an element of the medium using Equation 16.14 gives

$$dK = \frac{1}{2}\mu[-\omega A \cos (kx - \omega t)]^2 \, dx = \frac{1}{2}\mu\omega^2 A^2 \cos^2 (kx - \omega t) \, dx$$

If we take a snapshot of the wave at time $t = 0$, the kinetic energy of a given element is

$$dK = \tfrac{1}{2}\mu\omega^2 A^2 \cos^2 kx \, dx$$

Integrating this expression over all the string elements in a wavelength of the wave gives the total kinetic energy K_λ in one wavelength:

$$K_\lambda = \int dK = \int_0^\lambda \tfrac{1}{2}\mu\omega^2 A^2 \cos^2 kx \, dx = \tfrac{1}{2}\mu\omega^2 A^2 \int_0^\lambda \cos^2 kx \, dx$$

$$= \tfrac{1}{2}\mu\omega^2 A^2 \left[\tfrac{1}{2}x + \frac{1}{4k} \sin 2kx \right]_0^\lambda = \tfrac{1}{2}\mu\omega^2 A^2 \left[\tfrac{1}{2}\lambda \right] = \tfrac{1}{4}\mu\omega^2 A^2 \lambda$$

In addition to kinetic energy, there is potential energy associated with each element of the string due to its displacement from the equilibrium position and the restoring forces from neighboring elements. A similar analysis to that above for the total potential energy U_λ in one wavelength gives exactly the same result:

$$U_\lambda = \tfrac{1}{4}\mu\omega^2 A^2 \lambda$$

The total energy in one wavelength of the wave is the sum of the potential and kinetic energies:

$$E_\lambda = U_\lambda + K_\lambda = \tfrac{1}{2}\mu\omega^2 A^2 \lambda \tag{16.20}$$

As the wave moves along the string, this amount of energy passes by a given point on the string during a time interval of one period of the oscillation. Therefore, the power P, or rate of energy transfer T_{MW} associated with the mechanical wave, is

$$P = \frac{T_{MW}}{\Delta t} = \frac{E_\lambda}{T} = \frac{\tfrac{1}{2}\mu\omega^2 A^2 \lambda}{T} = \tfrac{1}{2}\mu\omega^2 A^2 \left(\frac{\lambda}{T} \right)$$

$$P = \tfrac{1}{2}\mu\omega^2 A^2 v \tag{16.21}$$

◀ Power of a wave

Equation 16.21 shows that the rate of energy transfer by a sinusoidal wave on a string is proportional to (a) the square of the frequency, (b) the square of the amplitude, and (c) the wave speed.

QUICK QUIZ 16.5 Which of the following, taken by itself, would be most effective in increasing the rate at which energy is transferred by a wave traveling along a string? **(a)** reducing the linear mass density of the string by one-half **(b)** doubling the wavelength of the wave **(c)** doubling the tension in the string **(d)** doubling the amplitude of the wave

---

Example 16.5 Power Supplied to a Vibrating String

A taut string for which $\mu = 5.00 \times 10^{-2}$ kg/m is under a tension of 80.0 N. How much power must be supplied to the string to generate sinusoidal waves at a frequency of 60.0 Hz and an amplitude of 6.00 cm?

SOLUTION

Conceptualize Consider Figure 16.14 again and notice that the vibrating blade supplies energy to the string at a certain rate. This energy then propagates to the right along the string.

Categorize We evaluate quantities from equations developed in the chapter, so we categorize this example as a substitution problem.

continued

16.5 continued

Use Equation 16.21 to evaluate the power: $P = \frac{1}{2}\mu\omega^2 A^2 v$

Use Equations 16.9 and 16.18 to substitute for ω and v:

$$P = \frac{1}{2}\mu(2\pi f)^2 A^2 \left(\sqrt{\frac{T}{\mu}}\right) = 2\pi^2 f^2 A^2 \sqrt{\mu T}$$

Substitute numerical values:

$$P = 2\pi^2(60.0 \text{ Hz})^2(0.060\,0 \text{ m})^2 \sqrt{(0.050\,0 \text{ kg/m})(80.0 \text{ N})} = 512 \text{ W}$$

WHAT IF?   What if the string is to transfer energy at a rate of 1 000 W? What must be the required amplitude if all other parameters remain the same?

Answer   Let us set up a ratio of the new and old power, reflecting only a change in the amplitude:

$$\frac{P_{new}}{P_{old}} = \frac{\frac{1}{2}\mu\omega^2 A_{new}^2 v}{\frac{1}{2}\mu\omega^2 A_{old}^2 v} = \frac{A_{new}^2}{A_{old}^2}$$

Solving for the new amplitude gives

$$A_{new} = A_{old}\sqrt{\frac{P_{new}}{P_{old}}} = (6.00 \text{ cm})\sqrt{\frac{1\,000 \text{ W}}{512 \text{ W}}} = 8.39 \text{ cm}$$

16.5 The Linear Wave Equation

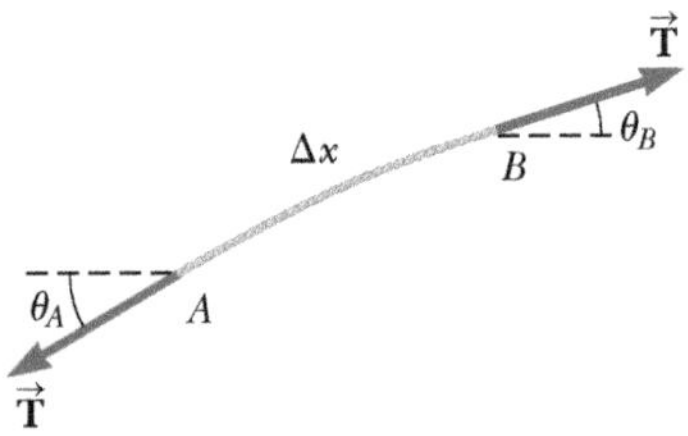

Figure 16.15 An element of a string under tension T.

In Section 16.1, we introduced the concept of the wave function to represent waves traveling on a string. All wave functions $y(x, t)$ represent solutions of an equation called the *linear wave equation*. This equation gives a complete description of the wave motion, and from it one can derive an expression for the wave speed. Furthermore, the linear wave equation is basic to many forms of wave motion. In this section, we derive this equation as applied to waves on strings.

Suppose a traveling wave is propagating along a string that is under a tension T. Let's consider one small string element of length Δx (Fig. 16.15). The ends of the element make small angles θ_A and θ_B with the x axis. Forces act on the string at its ends where it connects to neighboring elements. Therefore, the element is modeled as a particle under a net force. The net force acting on the element in the vertical direction is

$$\sum F_y = T\sin\theta_B - T\sin\theta_A = T(\sin\theta_B - \sin\theta_A)$$

Because the angles are small, we can use the approximation $\sin\theta \approx \tan\theta$ (see Table 15.1) to express the net force as

$$\sum F_y \approx T(\tan\theta_B - \tan\theta_A) \tag{16.22}$$

Imagine undergoing an infinitesimal displacement outward from the right end of the rope element in Figure 16.15 along the blue line representing the force $\vec{T}$. This displacement has infinitesimal x and y components and can be represented by the vector $dx\,\hat{\mathbf{i}} + dy\,\hat{\mathbf{j}}$. The tangent of the angle with respect to the x axis for this displacement is dy/dx. Because we evaluate this tangent at a particular instant of time, we must express it in partial derivative form as $\partial y/\partial x$. Substituting for the tangents in Equation 16.22 gives

$$\sum F_y \approx T\left[\left(\frac{\partial y}{\partial x}\right)_B - \left(\frac{\partial y}{\partial x}\right)_A\right] \tag{16.23}$$

Now, from the particle under a net force model, let's apply Newton's second law to the element, with the mass of the element given by $m = \mu\,\Delta x$:

$$\sum F_y = ma_y = \mu\,\Delta x\left(\frac{\partial^2 y}{\partial t^2}\right) \tag{16.24}$$

Combining Equation 16.23 with Equation 16.24 gives

$$\mu \, \Delta x \left(\frac{\partial^2 y}{\partial t^2} \right) = T \left[\left(\frac{\partial y}{\partial x} \right)_B - \left(\frac{\partial y}{\partial x} \right)_A \right]$$

$$\frac{\mu}{T} \frac{\partial^2 y}{\partial t^2} = \frac{(\partial y / \partial x)_B - (\partial y / dx)_A}{\Delta x} \qquad (16.25)$$

The right side of Equation 16.25 can be expressed in a different form if we note that the partial derivative of any function is defined as

$$\frac{\partial f}{\partial x} \equiv \lim_{\Delta x \to 0} \frac{f(x + \Delta x) - f(x)}{\Delta x}$$

Associating $f(x + \Delta x)$ with $(\partial y / \partial x)_B$ and $f(x)$ with $(\partial y / \partial x)_A$, we see that, in the limit $\Delta x \to 0$, Equation 16.25 becomes

$$\frac{\mu}{T} \frac{\partial^2 y}{\partial t^2} = \frac{\partial^2 y}{\partial x^2} \qquad (16.26)$$ ◀ Linear wave equation for a string

This expression is the linear wave equation as it applies to waves on a string.

The linear wave equation (Eq. 16.26) is often written in the form

$$\frac{\partial^2 y}{\partial x^2} = \frac{1}{v^2} \frac{\partial^2 y}{\partial t^2} \qquad (16.27)$$ ◀ Linear wave equation in general

Equation 16.27 applies in general to various types of traveling waves whose speed is v. For waves on strings, y represents the vertical position of elements of the string. For sound waves propagating through a gas, y corresponds to longitudinal position of elements of the gas from equilibrium or variations in either the pressure or the density of the gas. In the case of electromagnetic waves, y corresponds to electric or magnetic field components.

We have shown that the sinusoidal wave function (Eq. 16.10) is one solution of the linear wave equation (Eq. 16.27). Although we do not prove it here, the linear wave equation is satisfied by *any* wave function having the form $y = f(x \pm vt)$.

16.6 Sound Waves

We focus our attention now on **sound waves,** which travel through any material, but are most commonly experienced as the mechanical waves traveling through air that result in the human perception of hearing. As sound waves travel through air, elements of air are disturbed from their equilibrium positions. Accompanying these movements are changes in density and pressure of the air along the direction of wave motion. If the source of the sound waves vibrates sinusoidally, the density and pressure variations are also sinusoidal. The mathematical description of sinusoidal sound waves is very similar to that of sinusoidal waves on strings.

Sound waves are divided into three categories that cover different frequency ranges. (1) *Audible waves* lie within the range of sensitivity of the human ear. They can be generated in a variety of ways, such as by musical instruments, human voices, or loudspeakers. (2) *Infrasonic waves* have frequencies below the audible range. Elephants can use infrasonic waves to communicate with one another, even when separated by many kilometers. (3) *Ultrasonic waves* have frequencies above the audible range. You may have used a "silent" whistle to retrieve your dog. Dogs easily hear the ultrasonic sound this whistle emits, although humans cannot detect it at all. Ultrasonic waves are also used in medical imaging.

Earlier in the chapter, we began our investigation of waves by imagining the creation of a single pulse that traveled down a string (Figure 16.1) or a spring (Figure 16.3). Let's do something similar for sound. We describe pictorially the motion

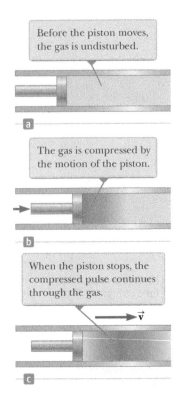

Before the piston moves, the gas is undisturbed.

a

The gas is compressed by the motion of the piston.

b

When the piston stops, the compressed pulse continues through the gas.

$\vec{v}$

c

Figure 16.16 Motion of a longitudinal pulse through a compressible gas. The compression (darker region) is produced by the moving piston.

of a one-dimensional longitudinal sound pulse moving through a long tube containing a compressible gas as shown in Figure 16.16. A piston at the left end can be quickly moved to the right to compress the gas and create the pulse. Before the piston is moved, the gas is undisturbed and of uniform density as represented by the uniformly shaded region in Figure 16.16a. When the piston is pushed to the right (Fig. 16.16b), the gas just in front of it is compressed (as represented by the more heavily shaded region); the pressure and density in this region are now higher than they were before the piston moved. When the piston comes to rest (Fig. 16.16c), the compressed region of the gas continues to move to the right, corresponding to a longitudinal pulse traveling through the tube with speed v.

One can produce a one-dimensional *periodic* sound wave in the tube of gas in Figure 16.16 by causing the piston to move in simple harmonic motion. The results are shown in Figure 16.17. The darker parts of the colored areas in this figure represent regions in which the gas is compressed and the density and pressure are above their equilibrium values. A compressed region is formed whenever the piston is pushed into the tube. This compressed region, called a **compression,** moves through the tube, continuously compressing the region just in front of itself. When the piston is pulled back, the gas in front of it expands and the pressure and density in this region fall below their equilibrium values (represented by the lighter parts of the colored areas in Fig. 16.17). These low-pressure regions, called **rarefactions,** also propagate along the tube, following the compressions. Both regions move at the speed of sound in the medium.

As the piston oscillates sinusoidally, regions of compression and rarefaction are continuously set up. The distance between two successive compressions (or two successive rarefactions) equals the wavelength λ of the sound wave. Because the sound wave is longitudinal, as the compressions and rarefactions travel through the tube, any small element of the gas moves with simple harmonic motion parallel to the direction of the wave. If $s(x, t)$ is the position of a small element relative to its equilibrium position,[3] we can express this harmonic position function as

$$s(x, t) = s_{max} \cos (kx - \omega t) \tag{16.28}$$

where s_{max} is the maximum position of the element relative to equilibrium. This parameter is often called the **displacement amplitude** of the wave. The parameter k is the wave number, and ω is the angular frequency of the wave as defined in Equations 16.8 and 16.9. Notice that the displacement of the element is along x, in the direction of propagation of the sound wave.

The variation in the gas pressure ΔP measured from the equilibrium value is also periodic with the same wave number and angular frequency as for the displacement in Equation 16.28. Therefore, we can write

$$\Delta P = \Delta P_{max} \sin (kx - \omega t) \tag{16.29}$$

where **the pressure amplitude ΔP_{max}** is the maximum change in pressure from the equilibrium value.

Notice that we have expressed the displacement by means of a cosine function and the pressure by means of a sine function. We will justify this choice in the procedure that follows and relate the pressure amplitude ΔP_{max} to the displacement amplitude s_{max}. Consider the piston–tube arrangement of Figure 16.16 once again. In Figure 16.18a, we focus our attention on a small cylindrical element of undisturbed gas of length Δx and area A. The volume of this element is $V_i = A \Delta x$.

Figure 16.18b shows this element of gas after a sound wave has moved it to a new position. The cylinder's two flat faces move through different distances s_1 and s_2. The change in volume ΔV of the element in the new position is equal to $A \Delta s$, where $\Delta s = s_1 - s_2$.

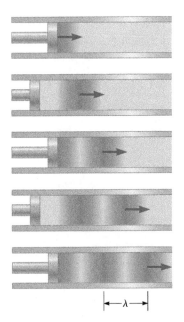

$\leftarrow\!\lambda\!\rightarrow$

Figure 16.17 A longitudinal wave propagating through a gas-filled tube. The source of the wave is an oscillating piston at the left.

---

[3] We use $s(x, t)$ here instead of $y(x, t)$ because the displacement of elements of the medium is not perpendicular to the x direction.

From the definition of bulk modulus (see Eq. 12.8), we express the pressure variation in the element of gas as a function of its change in volume:

$$\Delta P = -B \frac{\Delta V}{V_i}$$

Let's substitute for the initial volume and the change in volume of the element:

$$\Delta P = -B \frac{A \, \Delta s}{A \, \Delta x}$$

Let the length Δx of the cylinder approach zero so that the ratio $\Delta s / \Delta x$ becomes a partial derivative:

$$\Delta P = -B \frac{\partial s}{\partial x} \tag{16.30}$$

Substitute the position function given by Equation 16.28:

$$\Delta P = -B \frac{\partial}{\partial x} [s_{max} \cos(kx - \omega t)] = B s_{max} k \sin(kx - \omega t)$$

From this result, we see that a displacement described by a cosine function leads to a pressure described by a sine function. We also see that the displacement and pressure amplitudes are related by

$$\Delta P_{max} = B s_{max} k \tag{16.31}$$

This relationship depends on the bulk modulus of the gas, which is not as readily available as is the density of the gas. Once we determine the speed of sound in a gas in Section 16.7, we will be able to provide an expression that relates ΔP_{max} and s_{max} in terms of the density of the gas.

This discussion shows that a sound wave in a gas may be described equally well in terms of either pressure or displacement. A comparison of Equations 16.28 and 16.29 shows that the pressure wave is 90° out of phase with the displacement wave. Graphs of these functions are shown in Figure 16.19. The pressure variation is a maximum when the displacement from equilibrium is zero, and the displacement from equilibrium is a maximum when the pressure variation is zero.

QUICK QUIZ 16.6 If you blow across the top of an empty soft-drink bottle, a pulse of sound travels down through the air in the bottle. At the moment the pulse reaches the bottom of the bottle, what is the correct description of the displacement of elements of air from their equilibrium positions and the pressure of the air at this point? **(a)** The displacement and pressure are both at a maximum. **(b)** The displacement and pressure are both at a minimum. **(c)** The displacement is zero, and the pressure is a maximum. **(d)** The displacement is zero, and the pressure is a minimum.

16.7 Speed of Sound Waves

We now extend the discussion begun in Section 16.6 to evaluate the speed of sound in a gas. In Figure 16.20a (page 432), consider the cylindrical element of gas between the piston and the dashed line. This element of gas is in equilibrium under the influence of forces of equal magnitude, from the piston on the left and from the rest of the gas on the right. The magnitude of each of these forces is PA, where P is the pressure in the gas and A is the cross-sectional area of the tube. The length of the undisturbed element of gas is chosen to be $v \, \Delta t$, where v is the speed of sound in the gas and Δt is the time interval between the configurations in Figures 16.20a and 16.20b.

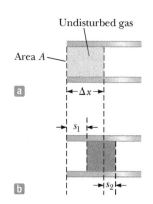

Figure 16.18 (a) An undisturbed cylindrical element of gas of length Δx in a tube of cross-sectional area A. (b) When a sound wave propagates through the gas, the element is moved to a new position and has a different length. The parameters s_1 and s_2 describe the displacements of the ends of the element from their equilibrium positions.

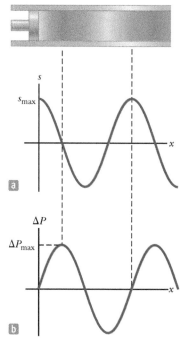

Figure 16.19 (a) Displacement amplitude and (b) pressure amplitude versus position for a sinusoidal longitudinal sound wave in a gas.

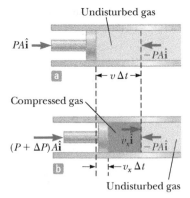

Figure 16.20 (a) An undisturbed element of gas of length $v\,\Delta t$ in a tube of cross-sectional area A. The element is in equilibrium between forces on either end. (b) When the piston moves inward at constant velocity v_x due to an increased force on the left, the element also moves with the same velocity.

Figure 16.20b shows the situation after this time interval Δt, during which the piston moves to the right at a constant speed v_x due to a force from the left on the piston that has increased in magnitude to $(P + \Delta P)A$. Because the speed of sound is v, the sound wave will just reach the right end of the cylindrical element of gas at the end of the time interval Δt. The gas to the right of the element is undisturbed because the sound wave has not reached it yet. At this moment every bit of gas in the element is moving with speed v_x. That will not be true in general for a macroscopic element of gas, but it will become true if we shrink the length of the element to an infinitesimal value.

The element of gas is modeled as a nonisolated system in terms of momentum. The force from the piston has provided an impulse to the element, which in turn exhibits a change in momentum. Therefore, we evaluate both sides of the impulse–momentum theorem, Equation 9.13:

$$\Delta \vec{\mathbf{p}} = \vec{\mathbf{I}} \tag{16.32}$$

On the right, the impulse is provided by the constant force due to the increased pressure on the piston:

$$\vec{\mathbf{I}} = \sum \vec{\mathbf{F}}\,\Delta t = (A\,\Delta P\,\Delta t)\hat{\mathbf{i}}$$

The pressure change ΔP can be related to the volume change and then to the speeds v and v_x through the bulk modulus:

$$\Delta P = -B\frac{\Delta V}{V_i} = -B\frac{(-v_x A\,\Delta t)}{vA\,\Delta t} = B\frac{v_x}{v}$$

Therefore, the impulse becomes

$$\vec{\mathbf{I}} = \left(AB\frac{v_x}{v}\,\Delta t\right)\hat{\mathbf{i}} \tag{16.33}$$

On the left-hand side of the impulse–momentum theorem, Equation 16.32, the change in momentum of the element of gas of mass m is as follows:

$$\Delta \vec{\mathbf{p}} = m\,\Delta \vec{\mathbf{v}} = (\rho V_i)(v_x \hat{\mathbf{i}} - 0) = (\rho v v_x A\,\Delta t)\hat{\mathbf{i}} \tag{16.34}$$

Substituting Equations 16.33 and 16.34 into Equation 16.32, we find

$$\rho v v_x A\,\Delta t = AB\frac{v_x}{v}\,\Delta t$$

which reduces to an expression for the speed of sound in a gas:

$$v = \sqrt{\frac{B}{\rho}} \tag{16.35}$$

It is interesting to compare this expression with Equation 16.18 for the speed of transverse waves on a string, $v = \sqrt{T/\mu}$. In both cases, the wave speed depends on an elastic property of the medium (bulk modulus B or string tension T) and on an inertial property of the medium (volume density ρ or linear density μ). In fact, the speed of all mechanical waves follows an expression of the general form

$$v = \sqrt{\frac{\text{elastic property}}{\text{inertial property}}}$$

For longitudinal sound waves in a solid rod of material, for example, the speed of sound depends on Young's modulus Y and the density ρ. Table 16.1 provides the speed of sound in several different materials.

TABLE 16.1 Speed of Sound in Various Media

Medium	v (m/s)	Medium	v (m/s)	Medium	v (m/s)
Gases		**Liquids at 25°C**		**Solids**[a]	
Hydrogen (0°C)	1 286	Glycerol	1 904	Pyrex glass	5 640
Helium (0°C)	972	Seawater	1 533	Iron	5 950
Air (20°C)	343	Water	1 493	Aluminum	6 420
Air (0°C)	331	Mercury	1 450	Brass	4 700
Oxygen (0°C)	317	Kerosene	1 324	Copper	5 010
		Methyl alcohol	1 143	Gold	3 240
		Carbon tetrachloride	926	Lucite	2 680
				Lead	1 960
				Rubber	1 600

[a]Values given are for propagation of longitudinal waves in bulk media. Speeds for longitudinal waves in thin rods are smaller, and speeds of transverse waves in bulk are smaller yet.

The speed of sound also depends on the temperature of the medium. For sound traveling through air, the relationship between wave speed and air temperature is

$$v = 331\sqrt{1 + \frac{T_C}{273}} \tag{16.36}$$

where v is in meters/second, 331 m/s is the speed of sound in air at 0°C, and T_C is the air temperature in degrees Celsius. Using this equation, one finds that at 20°C, the speed of sound in air is approximately 343 m/s.

Because the speed of sound is constant in a uniform medium, we can relate the speed to distance and time by modeling a pulse of sound as a particle under constant speed. For example, this model provides a convenient way to estimate the distance to a thunderstorm. First count the number of seconds between seeing the flash of lightning and hearing the thunder. Dividing this time interval by 3 gives the approximate distance to the lightning in kilometers because 343 m/s is approximately $\frac{1}{3}$ km/s. Dividing the time interval in seconds by 5 gives the approximate distance to the lightning in miles because the speed of sound is approximately $\frac{1}{5}$ mi/s.

Similarly, the particle under constant speed model allows the calculation described in the opening storyline. The GPS coordinates allow you to find the distance between you and the cliff. The sound of the clap echoes from the cliff and returns to you. So the distance traveled by the sound is twice the distance to the cliff. Dividing that distance by the time interval measured by the smartphone gives the speed of sound.

Having an expression (Eq. 16.35) for the speed of sound, we can now express the relationship between pressure amplitude and displacement amplitude for a sound wave (Eq. 16.31) as

$$\Delta P_{max} = Bs_{max}k = (\rho v^2)s_{max}\left(\frac{\omega}{v}\right) = \rho v \omega s_{max} \tag{16.37}$$

This expression is a bit more useful than Equation 16.31 because the density of a gas is more readily available than is the bulk modulus.

16.8 Intensity of Sound Waves

In Section 16.4, we showed that a wave traveling on a taut string transports energy, consistent with the notion of energy transfer T_{MW} by mechanical waves in Equation 8.2. Naturally, we would expect sound waves to also represent a transfer of energy. Consider the element of gas acted on by the piston in Figure 16.20. Imagine that the piston is moving back and forth in simple harmonic motion at angular frequency ω. Imagine also that the length of the element becomes very small so that the entire element moves with the same velocity as the piston. Then we can model the element as a particle on which the piston is doing work. The rate at which the piston is doing work on the element at any instant of time is given by Equation 8.18:

$$Power = \vec{\mathbf{F}} \cdot \vec{\mathbf{v}}_x$$

where we have used *Power* rather than *P* so that we don't confuse power *P* with pressure *P*! The force $\vec{\mathbf{F}}$ on the element of gas is related to the pressure and the velocity $\vec{\mathbf{v}}_x$ of the element is the derivative of the displacement function, so we find

$$Power = [\Delta P(x,\, t)A]\hat{\mathbf{i}} \cdot \frac{\partial}{\partial t}[s(x,\, t)\hat{\mathbf{i}}]$$

$$= [\rho v\omega As_{max} \sin(kx - \omega t)]\left\{\frac{\partial}{\partial t}[s_{max} \cos(kx - \omega t)]\right\}$$

$$= [\rho v\omega As_{max} \sin(kx - \omega t)][\omega s_{max} \sin(kx - \omega t)]$$

$$= \rho v\omega^2 As_{max}^2 \sin^2(kx - \omega t)$$

We now find the time average power over one period of the oscillation. For any given value of *x*, which we can choose to be $x = 0$, the average value of $\sin^2(kx - \omega t)$ over one period *T* is

$$\frac{1}{T}\int_0^T \sin^2(0 - \omega t)\, dt = \frac{1}{T}\int_0^T \sin^2 \omega t\, dt = \frac{1}{T}\left(\frac{t}{2} + \frac{\sin 2\omega t}{2\omega}\right)\Bigg|_0^T = \tfrac{1}{2}$$

Therefore,

$$(Power)_{avg} = \tfrac{1}{2}\rho A\omega^2 s_{max}^2 v$$

Compare this equation to that for power transmitted on a string, Equation 16.21. The two equations have the same form! Be careful, though: *A* in Equation 16.21 is the amplitude of the string wave, while *A* here is the area of the piston in Figure 16.20.

We define the **intensity** *I* of a wave, or the power per unit area, as the rate at which the energy transported by the wave transfers through a unit area *A* perpendicular to the direction of travel of the wave:

Intensity of a sound wave ▶

$$I \equiv \frac{(Power)_{avg}}{A} \tag{16.38}$$

In this case, the intensity is therefore

$$I = \tfrac{1}{2}\rho\omega^2 s_{max}^2 v$$

Hence, the intensity of a periodic sound wave is proportional to the square of the displacement amplitude and to the square of the angular frequency. This expression can also be written in terms of the pressure amplitude ΔP_{max}; in this case, we use Equation 16.37 to obtain

$$I = \frac{(\Delta P_{max})^2}{2\rho v} \tag{16.39}$$

The sound waves we have studied with regard to Figures 16.16 through 16.18 and 16.20 are constrained to move in one dimension along the length of the tube. Sound waves, however, can move through three-dimensional bulk media, so let's place a sound source in the open air and study the results with regard to intensity.

Consider the special case of a point source emitting sound waves equally in all directions. If the air around the source is perfectly uniform, the sound power radiated in all directions is the same, and the speed of sound in all directions is the same. The result in this situation is called a **spherical wave.** Figure 16.21 shows these spherical waves as a series of circular arcs concentric with the source. Each arc represents a surface over which the phase of the wave is constant. For example, the arcs may connect corresponding crests on all the waves. We call such a surface of constant phase a **wave front.** The radial distance between adjacent wave fronts that have the same phase is the wavelength λ of the wave. The radial lines pointing outward from the source, representing the direction of propagation of the waves, are called **rays.**

The average power emitted by the source must be distributed uniformly over each spherical wave front of area $4\pi r^2$, where r is the distance from the point source to the wave front. Hence, the wave intensity at a distance r from the source is

$$I = \frac{(Power)_{\text{avg}}}{A} = \frac{(Power)_{\text{avg}}}{4\pi r^2} \tag{16.40}$$

The intensity decreases as the square of the distance from the source. This inverse-square law is reminiscent of the behavior of gravity in Chapter 13.

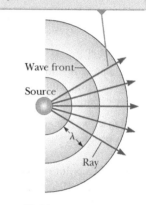

The rays are radial lines pointing outward from the source, perpendicular to the wave fronts.

Wave front

Source

λ

Ray

Figure 16.21 Spherical waves emitted by a point source. The circular arcs represent the spherical wave fronts that are concentric with the source.

○**UICK QUIZ 16.7** A vibrating guitar string makes very little sound if it is not mounted on the guitar body. Why does the sound have greater intensity if the string is attached to the guitar body? **(a)** The string vibrates with more energy. **(b)** The energy leaves the guitar more rapidly. **(c)** The sound power is spread over a larger area at the listener's position. **(d)** The sound power is concentrated over a smaller area at the listener's position. **(e)** The speed of sound is higher in the material of the guitar body. **(f)** None of these answers is correct.

Example 16.6   **Hearing Limits**

The faintest sounds the human ear can detect at a frequency of 1 000 Hz correspond to an intensity of about 1.00×10^{-12} W/m², which is called *threshold of hearing*. The loudest sounds the ear can tolerate at this frequency correspond to an intensity of about 1.00 W/m², the *threshold of pain*. Determine the pressure amplitude and displacement amplitude associated with these two limits.

SOLUTION

Conceptualize   Think about the quietest environment you have ever experienced. It is likely that the intensity of sound in even this quietest environment in your experience is higher than the threshold of hearing.

Categorize   Because we are given intensities and asked to calculate pressure and displacement amplitudes, this problem is a substitution problem requiring the concepts discussed in this section.

To find the amplitude of the pressure variation at the threshold of hearing, use Equation 16.39, taking the speed of sound waves in air to be $v = 343$ m/s and the density of air to be $\rho = 1.20$ kg/m³:

$$\Delta P_{\text{max}} = \sqrt{2\rho vI}$$
$$= \sqrt{2(1.20 \text{ kg/m}^3)(343 \text{ m/s})(1.00 \times 10^{-12} \text{ W/m}^2)}$$
$$= 2.87 \times 10^{-5} \text{ N/m}^2$$

Calculate the corresponding displacement amplitude using Equation 16.37, recalling that $\omega = 2\pi f$ (Eq. 16.9):

$$s_{\text{max}} = \frac{\Delta P_{\text{max}}}{\rho v \omega} = \frac{2.87 \times 10^{-5} \text{ N/m}^2}{(1.20 \text{ kg/m}^3)(343 \text{ m/s})(2\pi \times 1\,000 \text{ Hz})}$$
$$= 1.11 \times 10^{-11} \text{ m}$$

In a similar manner, one finds that the loudest sounds the human ear can tolerate (the threshold of pain) correspond to a pressure amplitude of 28.7 N/m² and a displacement amplitude equal to 1.11×10^{-5} m.

Because atmospheric pressure is about 10^5 N/m², the result for the pressure amplitude tells us that the ear is sensitive to pressure fluctuations as small as 3 parts in 10^{10}! The displacement amplitude is also a remarkably small number! If we compare this result for s_{max} to the size of an atom (about 10^{-10} m), we see that the ear is an extremely sensitive detector of sound waves.

Example **16.7**	**Intensity Variations of a Point Source**

A point source emits sound waves with an average power output of 80.0 W.

(A) Find the intensity 3.00 m from the source.

SOLUTION

Conceptualize Imagine a small loudspeaker sending sound out at an average rate of 80.0 W uniformly in all directions. You are standing 3.00 m away from the speakers. As the sound propagates, the energy of the sound waves is spread out over an ever-expanding sphere, so the intensity of the sound falls off with distance.

Categorize We evaluate the intensity from an equation generated in this section, so we categorize this example as a substitution problem.

Because a point source emits energy in the form of spherical waves, use Equation 16.40 to find the intensity:

$$I = \frac{(Power)_{avg}}{4\pi r^2} = \frac{80.0 \text{ W}}{4\pi (3.00 \text{ m})^2} = 0.707 \text{ W/m}^2$$

This intensity is close to the threshold of pain.

(B) Find the distance at which the intensity of the sound is 1.00×10^{-8} W/m².

SOLUTION

Solve for r in Equation 16.40 and use the given value for I:

$$r = \sqrt{\frac{(Power)_{avg}}{4\pi I}} = \sqrt{\frac{80.0 \text{ W}}{4\pi (1.00 \times 10^{-8} \text{ W/m}^2)}}$$

$$= 2.52 \times 10^4 \text{ m}$$

Sound Level in Decibels

Example 16.6 illustrates the wide range of intensities the human ear can detect. Because this range is so wide, it is convenient to use a logarithmic scale, where the **sound level** β (Greek letter beta) is defined by the equation

$$\beta \equiv 10 \log \left(\frac{I}{I_0} \right) \tag{16.41}$$

TABLE 16.2   Sound Levels

Source of Sound	β (dB)
Nearby jet airplane	150
Jackhammer; machine gun	130
Siren; rock concert	120
Subway; power lawn mower	100
Busy traffic	80
Vacuum cleaner	70
Normal conversation	60
Mosquito buzzing	40
Whisper	30
Rustling leaves	10
Threshold of hearing	0

This process compresses the range of hearing into a narrower scale of numbers. The constant I_0 is the *reference intensity*, taken to be at the threshold of hearing ($I_0 = 1.00 \times 10^{-12}$ W/m²), and I is the intensity in watts per square meter to which the sound level β corresponds, where β is measured[4] in **decibels** (dB). On this scale, the threshold of pain ($I = 1.00$ W/m²) corresponds to a sound level of $\beta = 10 \log [(1 \text{ W/m}^2)/(10^{-12} \text{ W/m}^2)] = 10 \log (10^{12}) = 120$ dB, and the threshold of hearing corresponds to $\beta = 10 \log [(10^{-12} \text{ W/m}^2)/(10^{-12} \text{ W/m}^2)] = 0$ dB.

Prolonged exposure to high sound levels may seriously damage the human ear. Ear plugs are recommended whenever sound levels exceed 90 dB. Recent evidence suggests that "noise pollution" may be a contributing factor to high blood pressure, anxiety, and nervousness. Table 16.2 gives some typical sound levels.

QUICK QUIZ 16.8 Increasing the intensity of a sound by a factor of 100 causes the sound level to increase by what amount? **(a)** 100 dB **(b)** 20 dB **(c)** 10 dB **(d)** 2 dB

---

[4]The unit *bel* is named after the inventor of the telephone, Alexander Graham Bell (1847–1922). The prefix *deci-* is the SI prefix that stands for 10^{-1}.

Example **16.8** Sound Levels

Two identical machines are positioned the same distance from a worker. The intensity of sound delivered by each operating machine at the worker's location is 2.0×10^{-7} W/m^2.

(A) Find the sound level heard by the worker when one machine is operating.

SOLUTION

Conceptualize Imagine a situation in which one source of sound is active and is then joined by a second identical source, such as one person speaking and then a second person speaking at the same time or one musical instrument playing and then being joined by a second instrument.

Categorize This example is a substitution problem requiring Equation 16.41.

Use Equation 16.41 to calculate the sound level at the worker's location with one machine operating:

$$\beta_1 = 10 \log \left(\frac{2.0 \times 10^{-7} \text{ W/m}^2}{1.00 \times 10^{-12} \text{ W/m}^2} \right) = 10 \log (2.0 \times 10^5) = 53 \text{ dB}$$

(B) Find the sound level heard by the worker when two machines are operating.

SOLUTION

Use Equation 16.41 to calculate the sound level at the worker's location with double the intensity:

$$\beta_2 = 10 \log \left(\frac{4.0 \times 10^{-7} \text{ W/m}^2}{1.00 \times 10^{-12} \text{ W/m}^2} \right) = 10 \log (4.0 \times 10^5) = 56 \text{ dB}$$

These results show that when the intensity is doubled, the sound level increases by only 3 dB. This 3-dB increase is independent of the original sound level. (Prove this to yourself!)

WHAT IF? *Loudness* is a psychological response to a sound. It depends on both the intensity and the frequency of the sound. As a rule of thumb, a doubling in loudness is approximately associated with an increase in sound level of 10 dB. (This rule of thumb is relatively inaccurate at very low or very high frequencies.) If the loudness of the machines in this example is to be doubled, how many machines at the same distance from the worker must be running?

Answer Using the rule of thumb, a doubling of loudness corresponds to a sound level increase of 10 dB. Therefore,

$$\beta_2 - \beta_1 = 10 \text{ dB} = 10 \log \left(\frac{I_2}{I_0} \right) - 10 \log \left(\frac{I_1}{I_0} \right) = 10 \log \left(\frac{I_2}{I_1} \right)$$

$$\log \left(\frac{I_2}{I_1} \right) = 1 \quad \rightarrow \quad I_2 = 10 I_1$$

Therefore, ten machines must be operating to double the loudness.

Loudness and Frequency

The discussion of sound level in decibels relates to a *physical* measurement of the strength of a sound. Let us now extend our discussion from the What If? section of Example 16.8 concerning the *psychological* "measurement" of the strength of a sound.

Of course, we don't have instruments in our bodies that can display numerical values of our reactions to stimuli. We have to "calibrate" our reactions somehow by comparing different sounds to a reference sound, but that is not easy to accomplish. For example, earlier we mentioned that the threshold intensity is 10^{-12} W/m^2, corresponding to an intensity level of 0 dB. In reality, this value is the threshold only for a sound of frequency 1 000 Hz, which is a standard reference frequency in acoustics. If we perform an experiment to measure the threshold intensity at other frequencies, we find a distinct variation of this threshold as a function of frequency. For example, at 100 Hz, a barely audible sound must have an intensity level of about

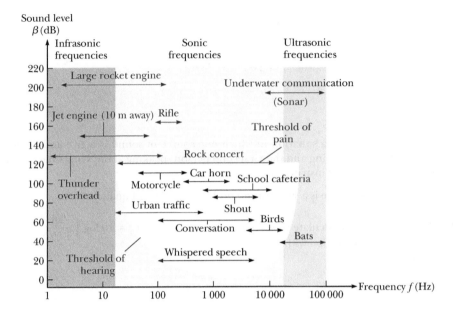

Figure 16.22 Approximate ranges of frequency and sound level of various sources and that of normal human hearing, shown by the white area. (From R. L. Reese, *University Physics*, Pacific Grove, Brooks/Cole, 2000.)

In all frames, the waves travel to the left, and their source is far to the right of the boat, out of the frame of the figure.

a

b

c

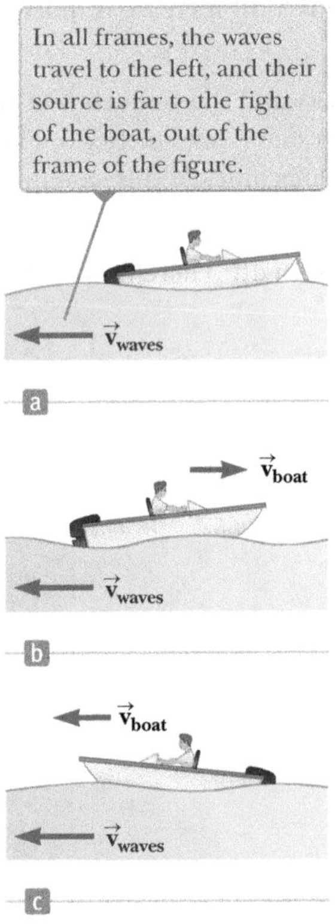

Figure 16.23 (a) Waves moving toward a stationary boat. (b) The boat moving toward the wave source. (c) The boat moving away from the wave source.

30 dB! Unfortunately, there is no simple relationship between physical measurements and psychological "measurements." The 100-Hz, 30-dB sound is psychologically "equal" in loudness to the 1 000-Hz, 0-dB sound (both are just barely audible), but they are not physically equal in sound level (30 dB ≠ 0 dB).

By using test subjects, the human response to sound has been studied, and the results are shown in the white area of Figure 16.22 along with the approximate frequency and sound-level ranges of other sound sources. The lower curve of the white area corresponds to the threshold of hearing. Its variation with frequency is clear from this diagram. Notice that humans are sensitive to frequencies ranging from about 20 Hz to about 20 000 Hz. The upper bound of the white area is the threshold of pain. Here the boundary of the white area appears straight because the psychological response is relatively independent of frequency at this high sound level.

16.9 The Doppler Effect

Perhaps you have noticed how the sound of a vehicle's horn changes as the vehicle moves past you. The frequency of the sound you hear as the vehicle approaches you is higher than the frequency you hear as it moves away from you. This experience is one example of the **Doppler effect**.[5]

To see what causes this apparent frequency change, imagine you are in a boat that is lying at anchor on a gentle sea where the waves have a period of $T = 3.0$ s. Hence, every 3.0 s a crest hits your boat. Figure 16.23a shows this situation, with the water waves moving toward the left. If you set your watch to $t = 0$ just as one crest hits, the watch reads 3.0 s when the next crest hits, 6.0 s when the third crest hits, and so on. From these observations, you conclude that the wave frequency is $f = 1/T = 1/(3.0 \text{ s}) = 0.33$ Hz. Now suppose you pull up the anchor, start your motor, and head directly into the oncoming waves as in Figure 16.23b. Again you set your watch to $t = 0$ as a crest hits the front (the bow) of your boat. Now, however, because you are moving toward the next wave crest as it moves toward you, it hits you less than 3.0 s after the first hit. In other words, the period you observe is shorter than the 3.0-s period you observed when you were stationary. Because $f = 1/T$, you observe a higher wave frequency than when you were at rest.

---

[5] Named after Austrian physicist Christian Johann Doppler (1803–1853), who in 1842 predicted the effect for both sound waves and light waves.

If you turn around and move in the same direction as the waves (Fig. 16.23c), you observe the opposite effect. You set your watch to $t = 0$ as a crest hits the back (the stern) of the boat. Because you are now moving away from the next crest, more than 3.0 s has elapsed on your watch by the time that crest catches you. Therefore, you observe a lower frequency than when you were at rest.

These effects occur because the *relative* speed between your boat and the waves depends on the direction of travel and on the speed of your boat. (See Section 4.6.) When you are moving toward the right in Figure 16.23b, this relative speed is higher than that of the wave speed, which leads to the observation of an increased frequency. When you turn around and move to the left, the relative speed is lower, as is the observed frequency of the water waves.

Let's now examine an analogous situation with sound waves in which the water waves become sound waves, the water becomes the air, and the person on the boat becomes an observer listening to the sound. In this case, an observer O is moving with speed v_O and a sound source S is stationary with respect to the medium, air (Fig. 16.24).

If a point source emits sound waves and the medium is uniform, the waves move at the same speed in all directions radially away from the source; the result is a spherical wave as mentioned in Section 16.8. The distance between adjacent wave fronts equals the wavelength λ. In Figure 16.24, the circles are the intersections of these three-dimensional wave fronts with the two-dimensional paper.

We take the frequency of the source in Figure 16.24 to be f, the wavelength to be λ, and the speed of sound to be v. When the observer moves toward the source, the speed of the waves relative to the observer is $v' = v + v_O$, as in the case of the boat in Figure 16.23, but the wavelength λ is unchanged. Hence, using Equation 16.12, $v = \lambda f$, we can say that the frequency f' heard by the observer is *increased* and is given by

$$f' = \frac{v'}{\lambda} = \frac{v + v_O}{\lambda}$$

Because $\lambda = v/f$, we can express f' as

$$f' = \left(\frac{v + v_O}{v}\right)f \quad \text{(observer moving toward source)} \qquad (16.42)$$

If the observer is moving away from the source, the speed of the wave relative to the observer is $v' = v - v_O$. The frequency heard by the observer in this case is *decreased* and is given by

$$f' = \left(\frac{v - v_O}{v}\right)f \quad \text{(observer moving away from source)} \qquad (16.43)$$

These last two equations can be reduced to a single equation by adopting a sign convention. Whenever an observer moves with a speed v_O relative to a stationary source, the frequency heard by the observer is given by Equation 16.42, with v_O interpreted as follows: a positive value is substituted for v_O when the observer moves toward the source, and a negative value is substituted when the observer moves away from the source.

Now suppose the *source* is in motion and the observer is at rest. If the source moves directly toward observer A in Figure 16.25a (page 440), each new wave is emitted from a position to the right of the origin of the previous wave. As a result, the wave fronts heard by the observer are closer together than they would be if the source were not moving. (Fig. 16.25b shows this effect for waves moving on the surface of water.) As a result, the wavelength λ' measured by observer A is shorter than the wavelength λ of the source. During each vibration, which lasts for a time interval T (the period), the source moves a distance $v_S T = v_S/f$ and the wavelength is *shortened* by this amount. Therefore, the observed wavelength λ' is

$$\lambda' = \lambda - \Delta\lambda = \lambda - \frac{v_S}{f}$$

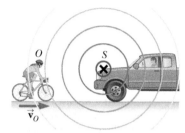

Figure 16.24 An observer O (the cyclist) moves with a speed v_O toward a stationary point source S, the horn of a parked truck. The observer hears a frequency f' that is greater than the source frequency f.

PITFALL PREVENTION 16.4
Doppler Effect Does Not Depend on Distance Some people think that the Doppler effect depends on the distance between the source and the observer. Although the *intensity* of a sound varies as the distance changes, the apparent *frequency* depends only on the relative speed of source and observer. As you listen to an approaching source, you will detect increasing intensity but constant frequency. As the source passes, you will hear the frequency suddenly drop to a new constant value and the intensity begin to decrease.

Figure 16.25 (a) A source *S* moving with a speed v_S toward a stationary observer A and away from a stationary observer B. Observer A hears an increased frequency, and observer B hears a decreased frequency. (b) The Doppler effect in water, observed in a ripple tank. Letters shown in the photo refer to Quick Quiz 16.9.

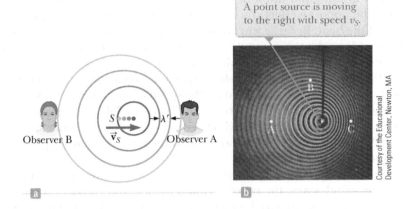

A point source is moving to the right with speed v_S.

Observer B   Observer A

S λ' $\vec{v}_S$

Courtesy of the Educational Development Center, Newton, MA

Because $\lambda = v/f$, the frequency f' heard by observer A is

$$f' = \frac{v}{\lambda'} = \frac{v}{\lambda - (v_S/f)} = \frac{v}{(v/f) - (v_S/f)}$$

$$f' = \left(\frac{v}{v - v_S}\right)f \quad \text{(source moving toward observer)} \tag{16.44}$$

That is, the observed frequency is *increased* whenever the source is moving toward the observer.

When the source moves away from a stationary observer, as is the case for observer B in Figure 16.25a, the observer measures a wavelength λ' that is *greater* than λ and hears a *decreased* frequency:

$$f' = \left(\frac{v}{v + v_S}\right)f \quad \text{(source moving away from observer)} \tag{16.45}$$

We can express the general relationship for the observed frequency when a source is moving and an observer is at rest as Equation 16.44, with the same sign convention applied to v_S as was applied to v_O: a positive value is substituted for v_S when the source moves toward the observer, and a negative value is substituted when the source moves away from the observer.

Finally, combining Equations 16.42 and 16.44 gives the following general relationship for the observed frequency that includes all four conditions described by Equations 16.42 through 16.45:

General Doppler-shift ▶
expression

$$f' = \left(\frac{v + v_O}{v - v_S}\right)f \tag{16.46}$$

In this expression, the signs for the values substituted for v_O and v_S depend on the direction of the velocity. A positive value is used for motion of the observer or the source *toward* the other (associated with an *increase* in observed frequency), and a negative value is used for motion of one *away from* the other (associated with a *decrease* in observed frequency).

Although the Doppler effect is most typically experienced with sound waves, it is a phenomenon common to all waves. For example, the relative motion of source and observer produces a frequency shift in light waves. The Doppler effect is used in police radar systems to measure the speeds of motor vehicles. Likewise, astronomers use the effect to determine the speeds of stars, galaxies, and other celestial objects relative to the Earth.

QUICK QUIZ 16.9 Consider detectors of water waves at three locations A, B, and C in Figure 16.25b. Which of the following statements is true? (a) The wave speed is highest at location A. (b) The wave speed is highest at location C. (c) The detected wavelength is largest at location B. (d) The detected wavelength is largest at location C. (e) The detected frequency is highest at location C. (f) The detected frequency is highest at location A.

QUICK QUIZ 16.10 You stand on a platform at a train station and listen to a train approaching the station at a constant velocity. While the train approaches, but before it arrives, what do you hear? (a) the intensity and the frequency of the sound both increasing (b) the intensity and the frequency of the sound both decreasing (c) the intensity increasing and the frequency decreasing (d) the intensity decreasing and the frequency increasing (e) the intensity increasing and the frequency remaining the same (f) the intensity decreasing and the frequency remaining the same

---

Example 16.9 | The Broken Clock Radio

Your clock radio awakens you with a steady and irritating sound of frequency 600 Hz. One morning, it malfunctions and cannot be turned off. In frustration, you drop the clock radio from rest out of your fourth-story dorm window, 15.0 m from the ground. Assume the speed of sound is 343 m/s. As you listen to the falling clock radio, what frequency do you hear just before you hear it striking the ground?

SOLUTION

Conceptualize The speed of the clock radio increases as it falls. Therefore, it is a source of sound moving away from you with an increasing speed so the frequency you hear should be less than 600 Hz.

Categorize We categorize this problem as one in which we combine the *particle under constant acceleration* model for the falling radio with our understanding of the frequency shift of sound due to the Doppler effect.

Analyze Because the clock radio is modeled as a particle under constant acceleration due to gravity, use Equation 2.13 to express the speed of the source of sound:

$$(1) \quad v_S = v_{yi} + a_y t = 0 - gt = -gt$$

From Equation 2.16, find the time at which the clock radio strikes the ground:

$$y_f = y_i + v_{yi}t - \tfrac{1}{2}gt^2 = 0 + 0 - \tfrac{1}{2}gt^2 \quad \rightarrow \quad t = \sqrt{-\frac{2y_f}{g}}$$

Substitute into Equation (1):

$$v_S = (-g)\sqrt{-\frac{2y_f}{g}} = -\sqrt{-2gy_f}$$

Use Equation 16.46 to determine the Doppler-shifted frequency heard from the falling clock radio:

$$f' = \left[\frac{v + 0}{v - (-\sqrt{-2gy_f})}\right]f = \left(\frac{v}{v + \sqrt{-2gy_f}}\right)f$$

Substitute numerical values:

$$f' = \left[\frac{343 \text{ m/s}}{343 \text{ m/s} + \sqrt{-2(9.80 \text{ m/s}^2)(-15.0 \text{ m})}}\right](600 \text{ Hz})$$

$$= 571 \text{ Hz}$$

Finalize The frequency is lower than the actual frequency of 600 Hz because the clock radio is moving away from you. If it were to fall from a higher floor so that it passes below $y = -15.0$ m, the clock radio would continue to accelerate and the frequency you hear would continue to drop.

Example 16.10 | Doppler Submarines

A submarine (sub A) travels through water at a speed of 8.00 m/s, emitting a sonar wave at a frequency of 1 400 Hz. The speed of sound in the water is 1 533 m/s. A second submarine (sub B) is located such that both submarines are traveling directly toward each other. The second submarine is moving at 9.00 m/s.

(A) What frequency is detected by an observer riding on sub B as the subs approach each other?

continued

16.10 continued

SOLUTION

Conceptualize   Even though the problem involves subs moving in water, there is a Doppler effect just like there is when you are in a moving car and listening to a sound moving through the air from another car.

Categorize   Because both subs are moving, we categorize this problem as one involving the Doppler effect for both a moving source and a moving observer.

Analyze   Use Equation 16.46 to find the Doppler-shifted frequency heard by the observer in sub B, being careful with the signs assigned to the source and observer speeds:

$$f' = \left(\frac{v + v_O}{v - v_S}\right)f$$

$$f' = \left[\frac{1\ 533\ \text{m/s} + (+9.00\ \text{m/s})}{1\ 533\ \text{m/s} - (+8.00\ \text{m/s})}\right](1\ 400\ \text{Hz}) = 1\ 416\ \text{Hz}$$

(B) The subs barely miss each other and pass. What frequency is detected by an observer riding on sub B as the subs recede from each other?

SOLUTION

Use Equation 16.46 to find the Doppler-shifted frequency heard by the observer in sub B, again being careful with the signs assigned to the source and observer speeds:

$$f' = \left(\frac{v + v_O}{v - v_S}\right)f$$

$$f' = \left[\frac{1\ 533\ \text{m/s} + (-9.00\ \text{m/s})}{1\ 533\ \text{m/s} - (-8.00\ \text{m/s})}\right](1\ 400\ \text{Hz}) = 1\ 385\ \text{Hz}$$

Notice that the frequency drops from 1 416 Hz to 1 385 Hz as the subs pass. This effect is similar to the drop in frequency you hear when a car passes by you while blowing its horn.

(C) While the subs are approaching each other, some of the sound from sub A reflects from sub B and returns to sub A. If this sound were to be detected by an observer on sub A, what is its frequency?

SOLUTION

The sound of apparent frequency 1 416 Hz found in part (A) is reflected from a moving source (sub B) and then detected by a moving observer (sub A). Find the frequency detected by sub A:

$$f'' = \left(\frac{v + v_O}{v - v_S}\right)f'$$

$$= \left[\frac{1\ 533\ \text{m/s} + (+8.00\ \text{m/s})}{1\ 533\ \text{m/s} - (+9.00\ \text{m/s})}\right](1\ 416\ \text{Hz}) = 1\ 432\ \text{Hz}$$

Finalize   This technique is used by police officers to measure the speed of a moving car. Microwaves are emitted from the police car and reflected by the moving car. By detecting the Doppler-shifted frequency of the reflected microwaves, the police officer can determine the speed of the moving car.

Shock Waves

Now consider what happens when the speed v_S of a source *exceeds* the wave speed v. This situation is depicted graphically in Figure 16.26a. The circles represent spherical wave fronts emitted by the source at various times during its motion. At $t = 0$,

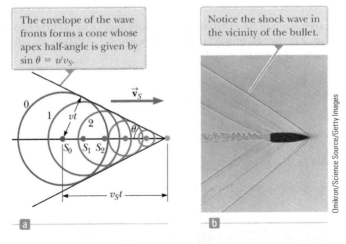

Figure 16.26 (a) A representation of a shock wave produced when a source moves from S_0 to the right with a speed v_S that is greater than the wave speed v in the medium. (b) A stroboscopic photograph of a bullet moving at supersonic speed through the hot air above a candle.

The envelope of the wave fronts forms a cone whose apex half-angle is given by $\sin \theta = v/v_S$.

Notice the shock wave in the vicinity of the bullet.

Omikron/Science Source/Getty Images

the source is at S_0 and moving toward the right. At later times, the source is at S_1, and then S_2, and so on. At the time t, the wave front centered at S_0 reaches a radius of vt. In this same time interval, the source travels a distance $v_S t$. Notice in Figure 16.26a that a straight line can be drawn tangent to all the wave fronts generated at various times. Therefore, the envelope of these wave fronts is a cone whose apex half-angle θ (the "Mach angle") is given by

$$\sin \theta = \frac{vt}{v_S t} = \frac{v}{v_S}$$

The inverse ratio v_S/v is referred to as the *Mach number,* and the conical wave front produced when $v_S > v$ (supersonic speeds) is known as a *shock wave.* An interesting analogy to shock waves is the V-shaped wave fronts produced by a boat when the boat's speed exceeds the speed of the surface-water waves (Fig. 16.27).

Jet airplanes traveling at supersonic speeds produce shock waves, which are responsible for the loud "sonic boom" one hears. The shock wave carries a great deal of energy concentrated on the surface of the cone, with correspondingly great pressure variations. Such shock waves are unpleasant to hear and can cause damage to buildings when aircraft fly supersonically at low altitudes. In fact, an airplane flying at supersonic speeds produces a double boom because two shock waves are formed, one from the nose of the plane and one from the tail.

QUICK QUIZ 16.11 An airplane flying with a constant velocity moves from a cold air mass into a warm air mass. Does the Mach number **(a)** increase, **(b)** decrease, or **(c)** stay the same?

© Robert Holland/Stone/Getty Images

Figure 16.27 The V-shaped bow wave of a boat is formed because the boat speed is greater than the speed of the water waves it generates. A bow wave is analogous to a shock wave formed by an airplane traveling faster than sound.

Summary

> ### Definitions

A **transverse wave** is one in which the elements of the medium move in a direction *perpendicular* to the direction of propagation.

A **one-dimensional sinusoidal wave** is one for which the positions of the elements of the medium vary sinusoidally. A sinusoidal wave traveling to the right can be expressed with a **wave function**

$$y(x, t) = A \sin\left[\frac{2\pi}{\lambda}(x - vt)\right] \quad (16.5)$$

where A is the **amplitude**, λ is the **wavelength**, and v is the **wave speed.**

The **angular wave number** k and **angular frequency** ω of a wave are defined as follows:

$$k \equiv \frac{2\pi}{\lambda} \quad (16.8)$$

$$\omega \equiv \frac{2\pi}{T} = 2\pi f \quad (16.9)$$

where T is the **period** of the wave and f is its **frequency.**

A **longitudinal wave** is one in which the elements of the medium move in a direction *parallel* to the direction of propagation.

The **intensity** of a periodic sound wave, which is the power per unit area, is

$$I \equiv \frac{(Power)_{avg}}{A} = \frac{(\Delta P_{max})^2}{2\rho v} \quad (16.38, 16.39)$$

The **sound level** of a sound wave in decibels is

$$\beta \equiv 10 \log\left(\frac{I}{I_0}\right) \quad (16.41)$$

The constant I_0 is a reference intensity, usually taken to be at the threshold of hearing ($1.00 \times 10^{-12}\,\text{W/m}^2$), and I is the intensity of the sound wave in watts per square meter.

continued

Concepts and Principles

Any one-dimensional wave traveling with a speed v in the x direction can be represented by a wave function of the form

$$y(x, t) = f(x \pm vt) \qquad (16.1, 16.2)$$

where the positive sign applies to a wave traveling in the negative x direction and the negative sign applies to a wave traveling in the positive x direction. The shape of the wave at any instant in time (a snapshot of the wave) is obtained by holding t constant.

The speed of a wave traveling on a taut string of mass per unit length μ and tension T is

$$v = \sqrt{\frac{T}{\mu}} \qquad (16.18)$$

The **power** transmitted by a sinusoidal wave on a stretched string is

$$P = \tfrac{1}{2}\mu\omega^2 A^2 v \qquad (16.21)$$

Wave functions are solutions to a differential equation called the **linear wave equation:**

$$\frac{\partial^2 y}{\partial x^2} = \frac{1}{v^2}\frac{\partial^2 y}{\partial t^2} \qquad (16.27)$$

Sound waves are longitudinal and travel through a compressible medium with a speed that depends on the elastic and inertial properties of that medium. The speed of sound in a gas having a bulk modulus B and density ρ is

$$v = \sqrt{\frac{B}{\rho}} \qquad (16.35)$$

For sinusoidal sound waves, the variation in the position of an element of the medium is

$$s(x, t) = s_{max}\cos(kx - \omega t) \qquad (16.28)$$

and the variation in pressure from the equilibrium value is

$$\Delta P = \Delta P_{max}\sin(kx - \omega t) \qquad (16.29)$$

where ΔP_{max} is the **pressure amplitude.** The pressure wave is 90° out of phase with the displacement wave. The relationship between s_{max} and ΔP_{max} is

$$\Delta P_{max} = \rho v \omega s_{max} \qquad (16.37)$$

The change in frequency heard by an observer whenever there is relative motion between a source of sound waves and the observer is called the **Doppler effect.** The observed frequency is

$$f' = \left(\frac{v + v_O}{v - v_S}\right)f \qquad (16.46)$$

In this expression, the signs for the values substituted for v_O and v_S depend on the direction of the velocity. A positive value for the speed of the observer or source is substituted if the velocity of one is toward the other, whereas a negative value represents a velocity of one in a direction away from the other.

Analysis Model for Problem Solving

Traveling Wave. The wave speed of a sinusoidal wave is

$$v = \frac{\lambda}{T} = \lambda f \qquad (16.6, 16.12)$$

A sinusoidal wave can be expressed as

$$y = A \sin(kx - \omega t) \qquad (16.10)$$

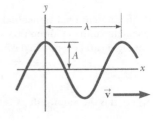

Think–Pair–Share

1. You and your friends are working for the National Oceanic and Atmospheric Administration (NOAA) and are learning about tsunamis to prepare you to help at the Pacific Tsunami Warning Center in Hawaii. Your instructor tells you that a

typical tsunami in the open ocean might have a speed of 800 km/h, a wavelength of 200 km, and an amplitude of 1.0 m. (a) If you were on a ship in the open ocean traveling to your position in Hawaii and a tsunami passed through the water around your ship, what effects would you experience? (b) If you were on the beach in Hawaii and saw the ocean water recede over a much larger distance than that due to normal wave activity, representing the trough of a tsunami, how much time would you have to warn everyone to get off the beach? Note that the frequency of a water wave remains the same as it passes through different depths of water. The amplitude and speed of the wave do change, however. (c) Assume that the energy of the tsunami is conserved. That is, ignore any reflection of energy as the water wave transmits into the shallow water, where its speed changes. In that case, from Equation 16.21, let us express the power of the wave as

$$P \sim \omega^2 A^2 v$$

Based on our assumption of no reflection of energy, what would the amplitude of the wave described above be when it is passing through a shallow region where the wave speed is 75 km/h? (d) If half the energy of the wave is reflected as it enters the shallow water, what would be the amplitude of the wave in the shallow water described in part (c)?

2. An entrepreneur is designing a new outdoor concert venue. He wants to procure loudspeakers that will provide a sound level of 83 dB at a location 100 m from the speakers. (a) What sound power output is required for the speakers? Assume that the sound radiates as a spherical wave and that there are no reflections from the ground. (b) After having someone perform the calculation in part (a) for him, the entrepreneur goes to the electronics store and purchases a speaker that the salesman tells him is rated at 150 W, which he thinks should clearly do what he needs. When he tests the speaker at the venue and turns the sound up to the maximum that

the speaker can handle, however, the sound level at 100 m is only 70.8 dB. The angry entrepreneur runs to his attorney, who hires your team as expert witnesses to see if litigation is appropriate against the salesman. What is the power output of the speaker, based on the sound level data? (c) After performing some research on loudspeakers, argue that litigation should *not* be initiated against the salesman.

3. **ACTIVITY** The epicenter of an earthquake can be located by looking at the difference in arrival times between P and S waves on a seismograph. A single seismograph station can determine how far away the earthquake occurs. With data from three seismograph stations, triangulation can be used to determine the exact location. For near earthquakes, seismic waves travel through the crust of the Earth at typical speeds of 4.00 km/s for S waves and 8.00 km/s for P waves. The following table shows times of day for the *first* arrival of P and S waves at three different California seismograph stations. (a) Print out a map of California from the Internet and use the data below to determine what large California city represents the epicenter of the earthquake. (b) At what time did the earthquake occur?

Seismic Station	Arrival Time of P Wave (h:min:s)	Arrival Time of S Wave (h:min:s)
Sacramento	3:21:34.4 PM	3:22:04.8 PM
San Francisco	3:21:35.9 PM	3:22:07.8 PM
Los Angeles	3:21:52.1 PM	3:22:40.2 PM

Problems

See the Preface for an explanation of the icons used in this problems set. For additional assessment items for this section, go to **WEBASSIGN** From Cengage

SECTION 16.1 **Propagation of a Disturbance**

1. A seismographic station receives S and P waves from an earthquake, separated in time by 17.3 s. Assume the waves have traveled over the same path at speeds of 4.50 km/s and 7.80 km/s. Find the distance from the seismograph to the focus of the quake.

2. Two points A and B on the surface of the Earth are at the same longitude and 60.0° apart in latitude as shown in Figure P16.2. Suppose an earthquake at point A creates a P wave that reaches point B by traveling straight through the body of the Earth at a constant speed of 7.80 km/s.

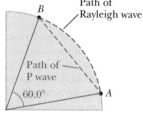

Figure P16.2

The earthquake also radiates a Rayleigh wave that travels at 4.50 km/s. In addition to P and S waves, Rayleigh waves are a third type of seismic wave that travels along the *surface* of the Earth rather than through the *bulk* of the Earth. (a) Which of these two seismic waves arrives at B first? (b) What is the time difference between the arrivals of these two waves at B?

3. You are working for a plumber who is laying very long sections of copper pipe for a large building project. He spends a lot of time measuring the lengths of the sections with a

measuring tape. You suggest a faster way to measure the length. You know that the speed of a one-dimensional compressional wave traveling along a copper pipe is 3.56 km/s. You suggest that a worker give a sharp hammer blow at one end of the pipe. Using an oscilloscope app on your smartphone, you will measure the time interval Δt between the arrival of the two sound waves due to the blow: one through the 20.0°C air and the other through the pipe. (a) To measure the length, you must derive an equation that relates the length L of the pipe numerically to the time interval Δt. (b) You measure a time interval of $\Delta t = 127$ ms between the arrivals of the pulses and, from this value, determine the length of the pipe. (c) Your smartphone app claims an accuracy of 1.0% in measuring time intervals. So you calculate by how many centimeters your calculation of the length might be in error.

4. You are working on a senior project and are analyzing a human "wave" at a sports stadium such as that shown in Figure P16.4 (page 446). You are trying to determine the effect of the wave on concession sales because people are standing up and sitting down while they participate in the wave, instead of buying food or drinks. You have made observations at a local stadium and have taken data on one particularly stable wave. This wave took 47.4 s to travel around a specific stadium row consisting of a circular ring of 974 seats. You also find that a typical time interval for spectators to stand and sit back down is 0.95 s. In this wave, how many people in the specific row were out of their seats at any given instant?

Figure P16.4 Problems 4 and 44.

JOE KLAMAR/AFP/Getty Images

SECTION 16.2 Analysis Model: Traveling Wave

5. When a particular wire is vibrating with a frequency of 4.00 Hz, a transverse wave of wavelength 60.0 cm is produced. Determine the speed of waves along the wire.

6. (a) Plot y versus t at $x = 0$ for a sinusoidal wave of the form $y = 0.150 \cos (15.7x - 50.3t)$, where x and y are in meters and t is in seconds. (b) Determine the period of vibration. (c) State how your result compares with the value found in Example 16.2.

7. Consider the sinusoidal wave of Example 16.2 with the wave function

$$y = 0.150 \cos (15.7x - 50.3t)$$

where x and y are in meters and t is in seconds. At a certain instant, let point A be at the origin and point B be the closest point to A along the x axis where the wave is 60.0° out of phase with A. What is the coordinate of B?

8. A sinusoidal wave traveling in the negative x direction (to the left) has an amplitude of 20.0 cm, a wavelength of 35.0 cm, and a frequency of 12.0 Hz. The transverse position of an element of the medium at $t = 0$, $x = 0$ is $y = -3.00$ cm, and the element has a positive velocity here. We wish to find an expression for the wave function describing this wave. (a) Sketch the wave at $t = 0$. (b) Find the angular wave number k from the wavelength. (c) Find the period T from the frequency. Find (d) the angular frequency ω and (e) the wave speed v. (f) From the information about $t = 0$, find the phase constant ϕ. (g) Write an expression for the wave function $y(x, t)$.

9. (a) Write the expression for y as a function of x and t in SI units for a sinusoidal wave traveling along a rope in the negative x direction with the following characteristics: $A = 8.00$ cm, $\lambda = 80.0$ cm, $f = 3.00$ Hz, and $y(0, t) = 0$ at $t = 0$. (b) **What If?** Write the expression for y as a function of x and t for the wave in part (a) assuming $y(x, 0) = 0$ at the point $x = 10.0$ cm.

SECTION 16.3 The Speed of Waves on Strings

10. Review. The elastic limit of a steel wire is 2.70×10^8 Pa. What is the maximum speed at which transverse wave pulses can propagate along this wire without exceeding this stress? (The density of steel is 7.86×10^3 kg/m³.)

11. Transverse waves travel with a speed of 20.0 m/s on a string under a tension of 6.00 N. What tension is required for a wave speed of 30.0 m/s on the same string?

12. *Why is the following situation impossible?* An astronaut on the Moon is studying wave motion using the apparatus discussed in Example 16.3 and shown in Figure 16.12. He measures the time interval for pulses to travel along the horizontal wire. Assume the horizontal wire has a mass of 4.00 g and a length of 1.60 m and assume a 3.00-kg object is suspended from its extension around the pulley. The astronaut finds that a pulse requires 26.1 ms to traverse the length of the wire.

13. Tension is maintained in a string as in Figure P16.13. The observed wave speed is $v = 24.0$ m/s when the suspended mass is $m = 3.00$ kg. (a) What is the mass per unit length of the string? (b) What is the wave speed when the suspended mass is $m = 2.00$ kg?

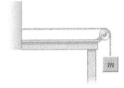

Figure P16.13
Problems 13 and 43.

14. Transverse pulses travel with a speed of 200 m/s along a taut copper wire whose diameter is 1.50 mm. What is the tension in the wire? (The density of copper is 8.92 g/cm³.)

SECTION 16.4 Rate of Energy Transfer by Sinusoidal Waves on Strings

15. Transverse waves are being generated on a rope under constant tension. By what factor is the required power increased or decreased if (a) the length of the rope is doubled and the angular frequency remains constant, (b) the amplitude is doubled and the angular frequency is halved, (c) both the wavelength and the amplitude are doubled, and (d) both the length of the rope and the wavelength are halved?

16. In a region far from the epicenter of an earthquake, a seismic wave can be modeled as transporting energy in a single direction without absorption, just as a string wave does. Suppose the seismic wave moves from granite into mudfill with similar density but with a much smaller bulk modulus. Assume the speed of the wave gradually drops by a factor of 25.0, with negligible reflection of the wave. (a) Explain whether the amplitude of the ground shaking will increase or decrease. (b) Does it change by a predictable factor? (This phenomenon led to the collapse of part of the Nimitz Freeway in Oakland, California, during the Loma Prieta earthquake of 1989.)

17. A long string carries a wave; a 6.00-m segment of the string contains four complete wavelengths and has a mass of 180 g. The string vibrates sinusoidally with a frequency of 50.0 Hz and a peak-to-valley displacement of 15.0 cm. (The "peak-to-valley" distance is the vertical distance from the farthest positive position to the farthest negative position.) (a) Write the function that describes this wave traveling in the positive x direction. (b) Determine the power being supplied to the string.

18. A two-dimensional water wave spreads in circular ripples. Show that the amplitude A at a distance r from the initial disturbance is proportional to $1/\sqrt{r}$. *Suggestion:* Consider the energy carried by one outward-moving ripple.

19. A horizontal string can transmit a maximum power P_0 (without breaking) if a wave with amplitude A and angular frequency ω is traveling along it. To increase this maximum power, a student folds the string and uses this "double string" as a medium. Assuming the tension in the two strands together is the same as the original tension in the single string and the angular frequency of the wave remains the same, determine the maximum power that can be transmitted along the "double string."

SECTION 16.5 The Linear Wave Equation

20. Show that the wave function $y = \ln[b(x - vt)]$ is a solution to
S Equation 16.27, where b is a constant.

21. Show that the wave function $y = e^{b(x-vt)}$ is a solution of the
S linear wave equation (Eq. 16.27), where b is a constant.

22. (a) Show that the function $y(x, t) = x^2 + v^2t^2$ is a solution to
S the wave equation. (b) Show that the function in part (a) can be written as $f(x + vt) + g(x - vt)$ and determine the functional forms for f and g. (c) **What If?** Repeat parts (a) and (b) for the function $y(x, t) = \sin(x)\cos(vt)$.

Note: In the rest of this chapter, for problems involving sound waves, pressure variations ΔP are measured relative to atmospheric pressure, 1.013×10^5 Pa.

SECTION 16.6 Sound Waves

23. A sinusoidal sound wave moves through a medium and is
V described by the displacement wave function

$$s(x, t) = 2.00\cos(15.7x - 858t)$$

where s is in micrometers, x is in meters, and t is in seconds. Find (a) the amplitude, (b) the wavelength, and (c) the speed of this wave. (d) Determine the instantaneous displacement from equilibrium of the elements of the medium at the position $x = 0.050\,0$ m at $t = 3.00$ ms. (e) Determine the maximum speed of the element's oscillatory motion.

SECTION 16.7 Speed of Sound Waves

Note: In the rest of this chapter, unless otherwise specified, the equilibrium density of air is $\rho = 1.20$ kg/m^3 and the speed of sound in air is $v = 343$ m/s. Use Table 16.1 to find speeds of sound in other media.

24. Earthquakes at fault lines in the Earth's crust create seismic waves, which are longitudinal (P waves) or transverse (S waves). The P waves have a speed of about 7 km/s. Estimate the average bulk modulus of the Earth's crust given that the density of rock is about 2 500 kg/m^3.

25. An experimenter wishes to generate in air a sound wave that
T has a displacement amplitude of 5.50×10^{-6} m. The pressure amplitude is to be limited to 0.840 Pa. What is the minimum wavelength the sound wave can have?

26. A sound wave propagates in air at 27°C with frequency
QC 4.00 kHz. It passes through a region where the temperature gradually changes and then moves through air at 0°C. Give numerical answers to the following questions to the extent possible and state your reasoning about what happens to the wave physically. (a) What happens to the speed of the wave? (b) What happens to its frequency? (c) What happens to its wavelength?

27. You are at Quincy Quarries Reservation with your grand-
CR father, performing the activity described in the opening storyline. The coordinates of your position when your grandfather claps his hands are N 42.244 34°, W 71.033 78°. Your smartphone stopwatch tells you that the time interval

between the clap and the echo is 0.47 s. When you walk to the cliff, your coordinates are N 42.244 06°, W 71.034 66°. What speed of sound do you report to your grandfather? (*Hint:* Use an online resource to calculate the distance between the coordinates.)

28. A rescue plane flies horizontally at a constant speed searching
V for a disabled boat. When the plane is directly above the boat, the boat's crew blows a loud horn. By the time the plane's sound detector receives the horn's sound, the plane has traveled a distance equal to half its altitude above the ocean. Assuming it takes the sound 2.00 s to reach the plane, determine (a) the speed of the plane and (b) its altitude.

29. The speed of sound in air (in meters per second) depends on temperature according to the approximate expression

$$v = 331.5 + 0.607T_C$$

where T_C is the Celsius temperature. In dry air, the temperature decreases about 1°C for every 150-m rise in altitude. (a) Assume this change is constant up to an altitude of 9 000 m. What time interval is required for the sound from an airplane flying at 9 000 m to reach the ground on a day when the ground temperature is 30°C? (b) **What If?** Compare your answer with the time interval required if the air were uniformly at 30°C. Which time interval is longer?

30. A sound wave moves down a cylinder as in Figure 16.17.
S Show that the pressure variation of the wave is described by $\Delta P = \pm\rho v\omega\sqrt{s_{max}^2 - s^2}$, where $s = s(x, t)$ is given by Equation 16.28.

SECTION 16.8 Intensity of Sound Waves

31. The intensity of a sound wave at a fixed distance from a speaker vibrating at 1.00 kHz is 0.600 W/m^2. (a) Determine the intensity that results if the frequency is increased to 2.50 kHz while a constant displacement amplitude is maintained. (b) Calculate the intensity if the frequency is reduced to 0.500 kHz and the displacement amplitude is doubled.

32. The intensity of a sound wave at a fixed distance from a
S speaker vibrating at a frequency f is I. (a) Determine the intensity that results if the frequency is increased to f' while a constant displacement amplitude is maintained. (b) Calculate the intensity if the frequency is reduced to $f/2$ and the displacement amplitude is doubled.

33. The power output of a certain public-address speaker is
V 6.00 W. Suppose it broadcasts equally in all directions. (a) Within what distance from the speaker would the sound be painful to the ear? (b) At what distance from the speaker would the sound be barely audible?

34. A fireworks rocket explodes at a height of 100 m above the ground. An observer on the ground directly under the explosion experiences an average sound intensity of 7.00×10^{-2} W/m^2 for 0.200 s. (a) What is the total amount of energy transferred away from the explosion by sound? (b) What is the sound level (in decibels) heard by the observer?

35. You are working at an open-air amphitheater, where rock
CR concerts occur regularly. The venue has powerful loudspeakers mounted on 10.6-m-tall columns at various locations surrounding the audience. The loudspeakers emit sound uniformly in all directions. There are ladder steps sticking out from the columns, to help workers service the loudspeakers. Many times, audience members break

through the protective fencing around the columns and climb upward on the columns to get a better view of the performers. The upcoming concert is by a group that states that several very-high-volume pulses of sound occur in their concerts, and these sounds are part of their artistic expression. The amphitheater owners are worried about people climbing the columns and being too close to the loudspeakers when these peak sounds are emitted. They do not want to be held responsible for injuries to audience members' ears. Based on past performances of the group, you determine that the peak sound level is 150 dB measured 20.0 cm from the speakers on the columns. The owners ask you to determine the heights on the columns at which to mount impassable barricades to keep people from getting too close to the speakers and hearing sound above the threshold of pain.

36. *Why is the following situation impossible?* It is early on a Saturday morning, and much to your displeasure your next-door neighbor starts mowing his lawn. As you try to get back to sleep, your next-door neighbor on the other side of your house also begins to mow the lawn with an identical mower the same distance away. This situation annoys you greatly because the total sound now has twice the loudness it had when only one neighbor was mowing.

37. Show that the difference between decibel levels β_1 and β_2 of a sound is related to the ratio of the distances r_1 and r_2 from the sound source by

$$\beta_2 - \beta_1 = 20 \log\left(\frac{r_1}{r_2}\right)$$

SECTION 16.9 The Doppler Effect

38. Submarine A travels horizontally at 11.0 m/s through ocean water. It emits a sonar signal of frequency $f = 5.27 \times 10^3$ Hz in the forward direction. Submarine B is in front of submarine A and traveling at 3.00 m/s relative to the water in the same direction as submarine A. A crewman in submarine B uses his equipment to detect the sound waves ("pings") from submarine A. We wish to determine what is heard by the crewman in submarine B. (a) An observer on which submarine detects a frequency f' as described by Equation 16.46? (b) In Equation 16.46, should the sign of v_S be positive or negative? (c) In Equation 16.46, should the sign of v_O be positive or negative? (d) In Equation 16.46, what speed of sound should be used? (e) Find the frequency of the sound detected by the crewman on submarine B.

39. When high-energy charged particles move through a transparent medium with a speed greater than the speed of light in that medium, a shock wave, or bow wave, of light is produced. This phenomenon is called the *Cerenkov effect*. When a nuclear reactor is shielded by a large pool of water, Cerenkov radiation can be seen as a blue glow in the vicinity of the reactor core due to high-speed electrons moving through the water (Fig. P16.39). In a particular

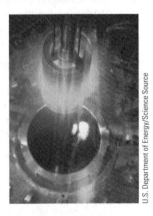

Figure P16.39

case, the Cerenkov radiation produces a wave front with an apex half-angle of 53.0°. Calculate the speed of the electrons in the water. The speed of light in water is 2.25×10^8 m/s.

40. *Why is the following situation impossible?* At the Summer Olympics, an athlete runs at a constant speed down a straight track while a spectator near the edge of the track blows a note on a horn with a fixed frequency. When the athlete passes the horn, she hears the frequency of the horn fall by the musical interval called a minor third. That is, the frequency she hears drops to five-sixths its original value.

41. **Review.** A block with a speaker bolted to it is connected to a spring having spring constant $k = 20.0$ N/m and oscillates as shown in Figure P16.41. The total mass of the block and speaker is 5.00 kg, and the amplitude of this unit's motion is 0.500 m. The speaker emits sound waves of frequency 440 Hz. Determine (a) the highest and (b) the lowest frequencies heard by the person to the right of the speaker. (c) If the maximum sound level heard by the person is 60.0 dB when the speaker is at its closest distance $d = 1.00$ m from him, what is the minimum sound level heard by the observer?

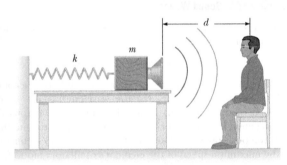

Figure P16.41 Problems 41 and 42.

42. **Review.** A block with a speaker bolted to it is connected to a spring having spring constant k and oscillates as shown in Figure P16.41. The total mass of the block and speaker is m, and the amplitude of this unit's motion is A. The speaker emits sound waves of frequency f. Determine (a) the highest and (b) the lowest frequencies heard by the person to the right of the speaker. (c) If the maximum sound level heard by the person is β when the speaker is at its closest distance d from him, what is the minimum sound level heard by the observer?

ADDITIONAL PROBLEMS

43. A sinusoidal wave in a rope is described by the wave function

$$y = 0.20 \sin(0.75\pi x + 18\pi t)$$

where x and y are in meters and t is in seconds. The rope has a linear mass density of 0.250 kg/m. The tension in the rope is provided by an arrangement like the one illustrated in Figure P16.13. What is the mass of the suspended object?

44. "The wave" is a particular type of pulse that can propagate through a large crowd gathered at a sports arena (Fig. P16.4). The elements of the medium are the spectators, with zero position corresponding to their being seated and maximum position corresponding to their standing and raising their arms. When a large fraction of the spectators

participates in the wave motion, a somewhat stable pulse shape can develop. The wave speed depends on people's reaction time, which is typically on the order of 0.1 s. Estimate the order of magnitude, in minutes, of the time interval required for such a pulse to make one circuit around a large sports stadium. State the quantities you measure or estimate and their values.

45. Some studies suggest that the upper frequency limit of hearing is determined by the diameter of the eardrum. The diameter of the eardrum is approximately equal to half the wavelength of the sound wave at this upper limit. If the relationship holds exactly, what is the diameter of the eardrum of a person capable of hearing 20 000 Hz? (Assume a body temperature of 37.0°C.)

46. An undersea earthquake or a landslide can produce an ocean wave of short duration carrying great energy, called a tsunami. When its wavelength is large compared to the ocean depth d, the speed of a water wave is given approximately by $v = \sqrt{gd}$. Assume an earthquake occurs all along a tectonic plate boundary running north to south and produces a straight tsunami wave crest moving everywhere to the west. (a) What physical quantity can you consider to be constant in the motion of any one wave crest? (b) Explain why the amplitude of the wave increases as the wave approaches shore. (c) If the wave has amplitude 1.80 m when its speed is 200 m/s, what will be its amplitude where the water is 9.00 m deep? (d) Explain why the amplitude at the shore should be expected to be still greater, but cannot be meaningfully predicted by your model.

47. A sinusoidal wave in a string is described by the wave function

$$y = 0.150 \sin (0.800x - 50.0t)$$

where x and y are in meters and t is in seconds. The mass per length of the string is 12.0 g/m. (a) Find the maximum transverse acceleration of an element of this string. (b) Determine the maximum transverse force on a 1.00-cm segment of the string. (c) State how the force found in part (b) compares with the tension in the string.

48. A rope of total mass m and length L is suspended vertically. Analysis shows that for short transverse pulses, the waves above a short distance from the free end of the rope can be represented to a good approximation by the linear wave equation discussed in Section 16.5. Show that a transverse pulse travels the length of the rope in a time interval that is given approximately by $\Delta t \approx 2\sqrt{L/g}$. *Suggestion:* First find an expression for the wave speed at any point a distance x from the lower end by considering the rope's tension as resulting from the weight of the segment below that point.

49. A wire of density ρ is tapered so that its cross-sectional area varies with x according to

$$A = 1.00 \times 10^{-5} x + 1.00 \times 10^{-6}$$

where A is in meters squared and x is in meters. The tension in the wire is T. (a) Derive a relationship for the speed of a wave as a function of position. (b) **What If?** Assume the wire is aluminum and is under a tension $T = 24.0$ N. Determine the wave speed at the origin and at $x = 10.0$ m.

50. *Why is the following situation impossible?* Tsunamis are ocean surface waves that have enormous wavelengths (100 to 200 km), and the propagation speed for these waves is $v \approx \sqrt{gd_{avg}}$, where d_{avg} is the average depth of the water. An earthquake on the ocean floor in the Gulf of Alaska produces a tsunami that reaches Hilo, Hawaii, 4 450 km away, in a time interval of 5.88 h. (This method was used in 1856 to estimate the average depth of the Pacific Ocean long before soundings were made to give a direct determination.)

51. A pulse traveling along a string of linear mass density μ is described by the wave function

$$y = [A_0 e^{-bx}] \sin (kx - \omega t)$$

where the factor in brackets is said to be the amplitude. (a) What is the power $P(x)$ carried by this wave at a point x? (b) What is the power $P(0)$ carried by this wave at the origin? (c) Compute the ratio $P(x)/P(0)$.

52. A train whistle ($f = 400$ Hz) sounds higher or lower in frequency depending on whether it approaches or recedes. (a) Prove that the difference in frequency between the approaching and receding train whistle is

$$\Delta f = \frac{2u/v}{1 - u^2/v^2} f$$

where u is the speed of the train and v is the speed of sound. (b) Calculate this difference for a train moving at a speed of 130 km/h. Take the speed of sound in air to be 340 m/s.

53. **Review.** A 150-g glider moves at $v_1 = 2.30$ m/s on an air track toward an originally stationary 200-g glider as shown in Figure P16.53. The gliders undergo a completely inelastic collision and latch together over a time interval of 7.00 ms. A student suggests roughly half the decrease in mechanical energy of the two-glider system is transferred to the environment by sound. Is this suggestion reasonable? To evaluate the idea, find the implied sound level at a position 0.800 m from the gliders. If the student's idea is unreasonable, suggest a better idea.

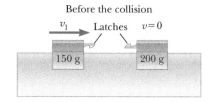

Before the collision

Figure P16.53

54. Consider the following wave function in SI units:

$$\Delta P(r, t) = \left(\frac{25.0}{r}\right) \sin (1.36r - 2\,030t)$$

Explain how this wave function can apply to a wave radiating from a small source, with r being the radial distance from the center of the source to any point outside the source. Give the most detailed description of the wave that you can. Include answers to such questions as the following and give representative values for any quantities that can be evaluated. (a) Does the wave move more toward the right or the left? (b) As it moves away from the source, what happens to its amplitude? (c) Its speed? (d) Its frequency? (e) Its wavelength? (f) Its power? (g) Its intensity?

55. With particular experimental methods, it is possible to produce and observe in a long, thin rod both a transverse wave whose speed depends primarily on tension in the rod and a longitudinal wave whose speed is determined by Young's modulus and the density of the material according to the expression $v = \sqrt{Y/\rho}$. The transverse wave can be modeled as a wave in a stretched string. A particular metal rod is 150 cm long and has a radius of 0.200 cm and a mass of 50.9 g. Young's modulus for the material is 6.80×10^{10} N/m². What must the tension in the rod be if the ratio of the speed of longitudinal waves to the speed of transverse waves is 8.00?

56. A large set of unoccupied football bleachers has solid seats and risers. You stand on the field in front of the bleachers and sharply clap two wooden boards together once. The sound pulse you produce has no definite frequency and no wavelength. The sound you hear reflected from the bleachers has an identifiable frequency and may remind you of a short toot on a trumpet, buzzer, or kazoo. (a) Explain what accounts for this sound. Compute order-of-magnitude estimates for (b) the frequency, (c) the wavelength, and (d) the duration of the sound on the basis of data you specify.

CHALLENGE PROBLEMS

57. A string on a musical instrument is held under tension T and extends from the point $x = 0$ to the point $x = L$. The string is overwound with wire in such a way that its mass per unit length $\mu(x)$ increases uniformly from μ_0 at $x = 0$ to μ_L at $x = L$. (a) Find an expression for $\mu(x)$ as a function of x over the range $0 \le x \le L$. (b) Find an expression for the time interval required for a transverse pulse to travel the length of the string.

58. Assume an object of mass M is suspended from the bottom of the rope of mass m and length L in Problem 48. (a) Show that the time interval for a transverse pulse to travel the length of the rope is

$$\Delta t = 2\sqrt{\frac{L}{mg}}(\sqrt{M + m} - \sqrt{M})$$

(b) **What If?** Show that the expression in part (a) reduces to the result of Problem 48 when $M = 0$. (c) Show that for $m \ll M$, the expression in part (a) reduces to

$$\Delta t = \sqrt{\frac{mL}{Mg}}$$

59. Equation 16.40 states that at distance r away from a point source with power $(Power)_{\text{avg}}$, the wave intensity is

$$I = \frac{(Power)_{\text{avg}}}{4\pi r^2}$$

Study Figure 16.25 and prove that at distance r straight in front of a point source with power $(Power)_{\text{avg}}$ moving with constant speed v_S the wave intensity is

$$I = \frac{(Power)_{\text{avg}}}{4\pi r^2}\left(\frac{v - v_S}{v}\right)$$

60. In Section 16.7, we derived the speed of sound in a gas using the impulse–momentum theorem applied to the cylinder of gas in Figure 16.20. Let us find the speed of sound in a gas using a different approach based on the element of gas in Figure 16.18. Proceed as follows. (a) Draw a force diagram for this element showing the forces exerted on the left and right surfaces due to the pressure of the gas on either side of the element. (b) By applying Newton's second law to the element, show that

$$-\frac{\partial(\Delta P)}{\partial x}A\,\Delta x = \rho A\,\Delta x\frac{\partial^2 s}{\partial t^2}$$

(c) By substituting $\Delta P = -(B\,\partial s/\partial x)$ (Eq. 16.30), derive the following wave equation for sound:

$$\frac{B}{\rho}\frac{\partial^2 s}{\partial x^2} = \frac{\partial^2 s}{\partial t^2}$$

(d) To a mathematical physicist, this equation demonstrates the existence of sound waves and determines their speed. As a physics student, you must take another step or two. Substitute into the wave equation the trial solution $s(x, t) = s_{\text{max}}\cos(kx - \omega t)$. Show that this function satisfies the wave equation, provided $\omega/k = v = \sqrt{B/\rho}$.

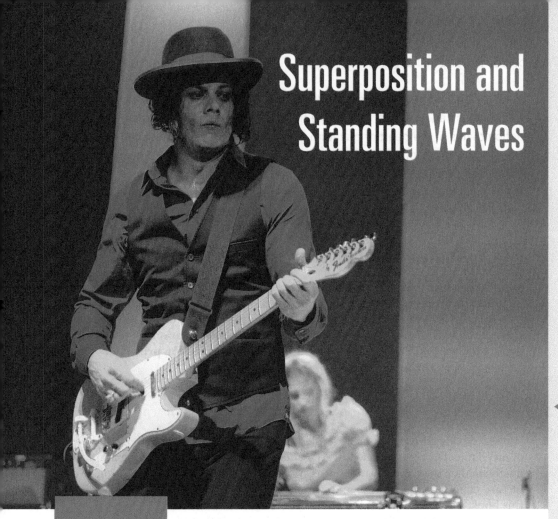

Superposition and Standing Waves

17

Guitarist Jack White takes advantage of standing waves on strings. He changes to higher notes on the guitar by pushing the strings against the frets on the fingerboard, shortening the lengths of the portions of the strings that vibrate. (*Mat Hayward/ Shutterstock.com*)

STORYLINE **Your previous roommate moved out to live in an** apartment. You are just getting to know your new roommate. One evening, your roommate shows you her guitar that she uses when she performs in a musical group. You have no idea how to play a guitar, but you do know about music from your choir experiences in high school. You absentmindedly start plucking the strings while your roommate starts looking through your physics textbook. You've seen guitar players pressing their fingers on the frets, as Jack White is doing above, so you do the same. During your explorations, you notice the following. You pluck an open string and then you place your finger *lightly* on the midpoint of the string. When you pluck the string now, the note is an octave above that of the open string. And there is something different about the nature of the sound, aside from it being an octave higher. You continue experimenting and find that you can generate higher notes that have a musical relationship to the open string by lightly touching at other points, such as one-third the length of the string and one-fourth the length. You ask your roommate about this phenomenon. She mentions something about "harmonics" and tells you to read Chapter 17 in your textbook.

CONNECTIONS This chapter continues our studies of waves begun in Chapter 16. We have seen that waves are very different from particles. A particle is of zero size, whereas a wave has a characteristic size, its wavelength. Another important difference between waves and particles is that we can explore the possibility of two or more waves combining at one point in the same medium. Particles can be combined to form extended objects, but the particles must be at *different* locations. In contrast, two waves can both be present at the *same* location. The ramifications of this possibility are explored in this chapter. When

451

waves are combined in systems with boundary conditions, only certain allowed frequencies can exist and we say the frequencies are *quantized*. Quantization is a notion that is at the heart of quantum mechanics, a subject introduced formally in Chapter 40. In Chapters 40–44, we show that analysis of waves under boundary conditions explains many of the quantum phenomena studied there. In this chapter, we use quantization to understand the behavior of the wide array of musical instruments that are based on strings and air columns.

17.1 Analysis Model: Waves in Interference

Many interesting wave phenomena in nature cannot be described by a single traveling wave. Instead, one must analyze these phenomena in terms of a combination of traveling waves. As noted in the introduction, waves have a remarkable difference from particles in that waves can be combined at the *same* location in space. To analyze such wave combinations, we make use of the **superposition principle:**

Superposition principle ▶

> If two or more traveling waves are moving through a medium, the resultant value of the wave function at any point where the waves both exist is the algebraic sum of the values of the wave functions of the individual waves at that point.

Waves that obey this principle are called *linear waves*. (See Section 16.5.) In the case of mechanical waves, linear waves are generally characterized by having amplitudes much smaller than their wavelengths. Waves that violate the superposition principle are called *nonlinear waves* and are often characterized by large amplitudes. In this book, we deal only with linear waves.

One consequence of the superposition principle is that two traveling waves can pass through each other without affecting one another. For instance, when two pebbles are thrown into a pond and hit the surface at different locations, the expanding circular surface waves from the two locations simply pass through each other with no permanent effect. The resulting complex pattern can be viewed as a combination of two independent sets of expanding circles.

Figure 17.1 is a pictorial representation of the superposition of two pulses moving on the same string. The wave function for the pulse moving to the right is y_1, and the wave function for the pulse moving to the left is y_2. The pulses have the same speed but different shapes, and the displacement of the elements of the medium is in the positive y direction for both pulses. When the waves overlap (Fig. 17.1b), the wave function for the resulting complex wave is given by $y_1 + y_2$. When the crests of the pulses coincide (Fig. 17.1c), the resulting wave given by $y_1 + y_2$ has a larger amplitude than that of the individual pulses. The two pulses finally separate and continue moving in their original directions (Fig. 17.1d). Notice that the pulse shapes remain unchanged after the interaction, as if the two pulses had never met!

The combination of separate waves in the same region of space to produce a resultant wave is called **interference.** When the displacements caused by the two pulses are in the same direction, as in Figure 17.1, we refer to their superposition as **constructive interference.**

Constructive interference ▶

Now consider two pulses traveling toward each other on a taut string where one pulse is inverted relative to the other as illustrated in Figure 17.2. When these pulses begin to overlap, the resultant pulse is given by $y_1 + y_2$, but the values of the function y_2 are negative. Therefore, at the instant shown in Figure 17.2c, the amplitude of the combined waves is *less* than that of the individual waves. Again, the two pulses pass through each other; because the displacements caused by the two pulses are in opposite directions, however, we refer to their superposition as **destructive interference.**

Destructive interference ▶

The superposition principle is the centerpiece of the analysis model called **waves in interference.** In many situations, both in acoustics and optics, waves

PITFALL PREVENTION 17.1
Do Waves Actually *Interfere*? In popular usage, the term *interfere* implies that an agent affects a situation in some way so as to preclude something from happening. For example, in American football, *pass interference* means that a defending player has affected the receiver so that the receiver is unable to catch the ball. This usage is very different from its use in physics, where waves pass through each other and interfere, but do not affect each other in any way. In physics, interference is similar to the notion of *combination* as described in this chapter.

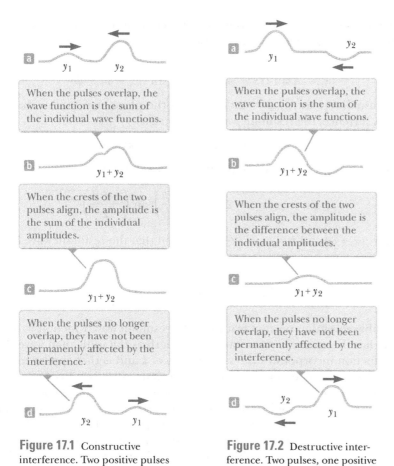

Figure 17.1 Constructive interference. Two positive pulses travel on a stretched string in opposite directions and overlap.

Figure 17.2 Destructive interference. Two pulses, one positive and one negative, travel on a stretched string in opposite directions and overlap.

combine according to this principle and exhibit interesting phenomena with practical applications.

QUICK QUIZ 17.1 Two pulses move in opposite directions on a string and are identical in shape and size except that one has positive displacements of the elements of the string and the other has negative displacements. At the moment the two pulses completely overlap on the string, what happens? **(a)** The energy associated with the pulses has disappeared. **(b)** The string is not moving. **(c)** The string forms a straight line. **(d)** The pulses have vanished and will not reappear.

Superposition of Sinusoidal Waves

Let us now apply the principle of superposition to two sinusoidal waves traveling in the *same* direction in a linear medium. Figure 17.3 shows a simple device that could create this situation for sound waves. Sound from a loudspeaker S is sent into a tube at point *P*, where there is a T-shaped junction. Half the sound energy travels in one direction, and half travels in the opposite direction. Therefore, the sound waves that reach the receiver R can travel along either of the two paths. The distance along any path from speaker to receiver is called the *path length r*. The lower path length r_1 is fixed, but the upper path length r_2 can be varied by sliding the U-shaped tube, which is similar to that on a slide trombone. This capability allows us to vary the phase difference between the waves arriving at *Q*. After the sound waves arrive at *Q*, they combine and travel together to the right from that point to the receiver R.

A sound wave from the speaker (S) propagates into the tube and splits into two parts at point *P*.

Path length r_2

Path length r_1

The two waves, which combine at the opposite side, are detected at the receiver (R).

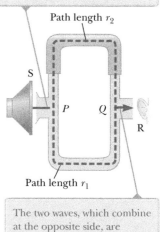

Figure 17.3 An acoustical system for demonstrating interference of sound waves. The upper path length r_2 can be varied by sliding the upper section.

If two waves travel together in the same direction and have the same frequency, wavelength, and amplitude but differ in phase, we can express their individual wave functions as

$$y_1 = A \sin (kx - \omega t) \quad y_2 = A \sin (kx - \omega t + \phi)$$

where, as usual, $k = 2\pi/\lambda$, $\omega = 2\pi f$, and ϕ is the phase constant as discussed in Section 16.2. Hence, the resultant wave function y is

$$y = y_1 + y_2 = A [\sin (kx - \omega t) + \sin (kx - \omega t + \phi)]$$

To simplify this expression, we use the trigonometric identity

$$\sin a + \sin b = 2 \cos \left(\frac{a - b}{2} \right) \sin \left(\frac{a + b}{2} \right)$$

Letting $a = kx - \omega t$ and $b = kx - \omega t + \phi$, we find that the resultant wave function y reduces to

Resultant of two traveling ▶
sinusoidal waves

$$y = 2A \cos \left(\frac{\phi}{2} \right) \sin \left(kx - \omega t + \frac{\phi}{2} \right)$$

This result has several important features. The resultant wave function y also is sinusoidal and has the same frequency and wavelength as the individual waves because the sine function incorporates the same values of k and ω that appear in the original wave functions. The amplitude of the resultant wave is $2A \cos (\phi/2)$, and its phase constant is $\phi/2$. Let's investigate the results for different values of ϕ. If the phase constant ϕ of the original wave equals 0, then $\cos (\phi/2) = \cos 0 = 1$ and the amplitude of the resultant wave is $2A$, twice the amplitude of either individual wave. This can occur in Figure 17.3 when the difference in the path lengths $\Delta r = |r_2 - r_1|$ is either zero or some integer multiple of the wavelength λ (that is, $\Delta r = n\lambda$, where $n = 0, 1, 2, 3, \ldots$). In this case, the crests of the two waves are at the same locations in space and the waves are said to be everywhere *in phase* and therefore interfere constructively. The individual waves y_1 and y_2 combine to form the red-brown curve y of amplitude $2A$ shown in Figure 17.4a. Because the individual waves are in phase, they are indistinguishable in Figure 17.4a, where they appear as a single blue curve. In general, constructive interference occurs when $\cos (\phi/2) = \pm 1$. That is true, for example, when $\phi = 0, 2\pi, 4\pi, \ldots$ rad, that is, when ϕ is an *even* multiple of π.

When ϕ is equal to π rad or to any *odd* multiple of π, then $\cos (\phi/2) = \cos (\pi/2) = 0$ and the crests of one wave occur at the same positions as the troughs of the second

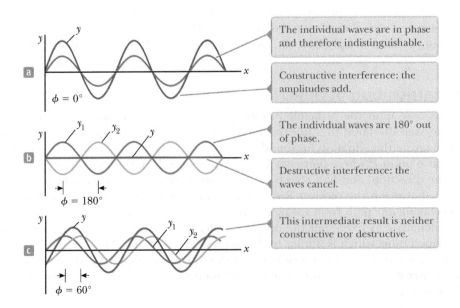

Figure 17.4 The superposition of two identical waves y_1 and y_2 (blue and green, respectively) to yield a resultant wave (red-brown).

wave, as shown by the blue and green curves in Figure 17.4b. This can be established in Figure 17.3 when the path length r_2 is adjusted so that the path difference $\Delta r = \lambda/2, 3\lambda/2, \dots, n\lambda/2$ (for n odd). In this case, as a consequence of destructive interference, the resultant wave has *zero* amplitude everywhere as shown by the straight red-brown line in Figure 17.4b. Finally, when the phase constant has an arbitrary value other than 0 or an integer multiple of π rad (Fig. 17.4c), the resultant wave has an amplitude whose value is somewhere between 0 and $2A$.

In the more general case in which the waves have the same wavelength but different amplitudes, the results are similar with the following exceptions. In the in-phase case, the amplitude of the resultant wave is not twice that of a single wave, but rather is the sum of the amplitudes of the two waves. (See the figure in the Analysis Model box below.) When the waves are π rad out of phase, they do not completely cancel as they do in Figure 17.4b. The result is a wave whose amplitude is the difference in the amplitudes of the individual waves.

ANALYSIS MODEL **Waves in Interference**

Imagine two waves traveling in the same location through a medium. The displacement of elements of the medium is affected by both waves. According to the **principle of super-position**, the displacement of an element is the sum of the individual displacements that would be caused by each wave. When the waves are in phase, **constructive interference** occurs and the resultant displacement is larger than the individual displacements. **Destructive interference** occurs when the waves are out of phase.

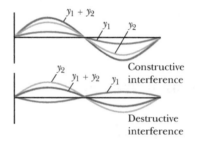

Examples:

- a piano tuner listens to a piano string and a tuning fork vibrating together and notices beats (Section 17.7)
- light waves from two coherent sources combine to form an interference pattern on a screen (Chapter 36)
- a thin film of oil on top of water shows swirls of color (Chapter 36)
- x-rays passing through a crystalline solid combine to form a Laue pattern (Chapter 37)

Example 17.1 **Two Speakers Driven by the Same Source**

Two identical loudspeakers placed 3.00 m apart are driven by the same oscillator (Fig. 17.5). A listener is originally at point O, located 8.00 m from the center of the line connecting the two speakers. The listener then moves to point P, which is a perpendicular distance 0.350 m from O, and she experiences the *first minimum* in sound intensity. What is the frequency of the oscillator?

SOLUTION

Conceptualize In Figure 17.3, a sound wave from one speaker enters a tube and is then *acoustically* split into two different paths before recombining at the other end. In this example, a signal representing the sound is *electrically* split and sent to two different loudspeakers. After leaving the speakers, the sound waves recombine at the position of the listener. Despite the difference in how the splitting occurs, the path difference discussion related to Figure 17.3 can be applied here.

Categorize Because the sound waves from two separate sources combine, we apply the *waves in interference* analysis model.

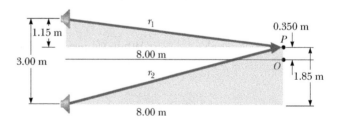

Figure 17.5 (Example 17.1) Two identical loudspeakers emit sound waves to a listener at P.

Analyze Figure 17.5 shows the physical arrangement of the speakers, along with two shaded right triangles that can be drawn on the basis of the lengths described in the problem. The first minimum occurs when the two waves reaching the listener at point P are 180° out of phase, in other words, when their path difference Δr equals $\lambda/2$.

continued

17.1 continued

From the shaded triangles, find the path lengths from the speakers to the listener:

$$r_1 = \sqrt{(8.00 \text{ m})^2 + (1.15 \text{ m})^2} = 8.08 \text{ m}$$

$$r_2 = \sqrt{(8.00 \text{ m})^2 + (1.85 \text{ m})^2} = 8.21 \text{ m}$$

Hence, the path difference is $r_2 - r_1 = 0.13$ m. Because this path difference must equal $\lambda/2$ for the first minimum, $\lambda = 0.26$ m.

To obtain the oscillator frequency, use Equation 16.12, $v = \lambda f$, where v is the speed of sound in air, 343 m/s:

$$f = \frac{v}{\lambda} = \frac{343 \text{ m/s}}{0.26 \text{ m}} = 1.3 \text{ kHz}$$

Finalize This example enables us to understand why the speaker wires in a stereo system should be connected properly. When connected the wrong way—that is, when the positive (or red) wire is connected to the negative (or black) terminal on one of the speakers and the other is correctly wired—the speakers are said to be "out of phase," with one speaker moving outward while the other moves inward. As a consequence, the sound wave coming from one speaker destructively interferes with the wave coming from the other at point O in Figure 17.5. A rarefaction region due to one speaker is superposed on a compression region from the other speaker. Although the two sounds probably do not completely cancel each other (because the left and right stereo signals are usually not identical), a substantial loss of sound quality occurs at point O.

WHAT IF? What if the speakers were connected out of phase? What happens at point P in Figure 17.5?

Answer In this situation, the path difference of $\lambda/2$ combines with a phase difference of $\lambda/2$ due to the incorrect wiring to give a full phase difference of λ. As a result, the waves are in phase and there is a *maximum* intensity at point P.

Figure 17.6 Two identical loudspeakers emit sound waves toward each other. When they overlap, identical waves traveling in opposite directions will combine to form standing waves.

17.2 Standing Waves

The sound waves from the pair of loudspeakers in Example 17.1 leave the speakers in the forward direction, and we considered interference at a point in front of the speakers. Suppose we turn each speaker by 90° so that they face each other as in Figure 17.6, and then have them emit sound of the same frequency and amplitude. In this situation, two identical waves travel in opposite directions in the same medium. These waves combine in accordance with the waves in interference model.

We can analyze such a situation by considering wave functions for two transverse sinusoidal waves having the same amplitude, frequency, and wavelength but traveling in opposite directions in the same medium:

$$y_1 = A \sin (kx - \omega t) \qquad y_2 = A \sin (kx + \omega t)$$

where y_1 represents a wave traveling in the positive x direction and y_2 represents one traveling in the negative x direction. Adding these two functions according to the superposition principle gives the resultant wave function y:

$$y = y_1 + y_2 = A \sin (kx - \omega t) + A \sin (kx + \omega t)$$

When we use the trigonometric identity $\sin (a \pm b) = \sin a \cos b \pm \cos a \sin b$, this expression reduces to

$$y = (2A \sin kx) \cos \omega t \tag{17.1}$$

Equation 17.1 represents the wave function of a **standing wave.** A standing wave such as the one on a string shown in Figure 17.7 is an oscillation pattern *with a stationary outline* that results from the superposition of two identical waves traveling in opposite directions.

Notice that Equation 17.1 does not contain a function of $kx - \omega t$. Therefore, it is not an expression for a traveling wave. When you observe a standing wave, there is no sense of motion in the direction of propagation of either original wave. If you were to observe the motion of the string in Figure 17.7, you would not see any motion to the left or right. You would only see up and down motion of the elements of the string. Comparing Equation 17.1 with Equation 15.6, we see that it describes

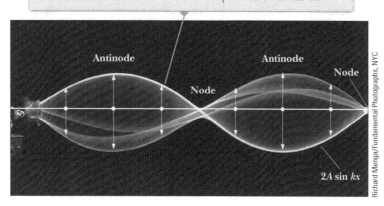

The amplitude of the vertical oscillation of any element of the string depends on the horizontal position of the element. Each element vibrates within the confines of the envelope function 2A sin kx.

Figure 17.7 Multiflash photograph of a standing wave on a string. The limits of motion of the string are seen as light blue and orange sine waves, while two intermediate positions of the string are seen as darker blue. The time behavior of the vertical displacement from equilibrium of an individual element of the string is given by cos ωt. That is, each element vibrates at an angular frequency ω.

a special kind of simple harmonic motion. Every element of the medium oscillates in simple harmonic motion with the same angular frequency ω (according to the cos ωt factor in the equation). The amplitude of the simple harmonic motion of a given element (given by the factor 2A sin kx, the coefficient of the cosine function) depends on the location x of the element in the medium, however.

If you can find a noncordless telephone with a coiled cord connecting the handset to the base unit, you can see the difference between a standing wave and a traveling wave. Stretch the coiled cord out and flick it with a finger. You will see a pulse traveling along the cord. Now shake the handset up and down and adjust your shaking frequency until every coil on the cord is moving up at the same time and then down. That is a standing wave, formed from the combination of waves moving away from your hand and reflected from the base unit toward your hand. Notice that there is no sense of traveling along the cord like there was for the pulse. You only see up-and-down motion of the elements of the cord.

Equation 17.1 shows that the amplitude of the simple harmonic motion of an element of the medium has a minimum value of zero when x satisfies the condition sin kx = 0, that is, when

$$kx = 0, \pi, 2\pi, 3\pi, \ldots$$

Because $k = 2\pi/\lambda$, these values for kx give

$$x = 0, \frac{\lambda}{2}, \lambda, \frac{3\lambda}{2}, \ldots = \frac{n\lambda}{2} \quad n = 0, 1, 2, \ldots \quad (17.2)$$

◀ Positions of nodes

These points of zero amplitude are called **nodes.** See if you can shake the coiled telephone cord at a higher frequency to generate a wave with a node in the middle, as shown in Figure 17.7.

The element of the medium with the *greatest* possible displacement from equilibrium has an amplitude of 2A, which we define as the amplitude of the standing wave. The positions in the medium at which this maximum displacement occurs are called **antinodes.** The antinodes are located at positions for which the coordinate x satisfies the condition sin kx = ±1, that is, when

$$kx = \frac{\pi}{2}, \frac{3\pi}{2}, \frac{5\pi}{2}, \ldots$$

Therefore, the positions of the antinodes are given by odd values of n:

$$x = \frac{\lambda}{4}, \frac{3\lambda}{4}, \frac{5\lambda}{4}, \ldots = \frac{n\lambda}{4} \quad n = 1, 3, 5, \ldots \quad (17.3)$$

◀ Positions of antinodes

PITFALL PREVENTION 17.2
Three Types of Amplitude We need to distinguish carefully here between the **amplitude of the individual waves,** which is A, and the **amplitude of the simple harmonic motion of the elements of the medium,** which is 2A sin kx. A given element in a standing wave vibrates within the constraints of the *envelope* function 2A sin kx, where x is that element's position in the medium. Such vibration is in contrast to traveling sinusoidal waves, in which elements at all positions oscillate with the same amplitude and the same frequency, and the amplitude A of the wave is the same as the amplitude A of the simple harmonic motion of the elements. Furthermore, we can identify the **amplitude of the standing wave** as 2A.

Figure 17.8 Standing-wave patterns produced at various times by two waves of equal amplitude traveling in opposite directions. For the resultant wave y, the nodes (N) are points of zero displacement and the antinodes (A) are points of maximum displacement. Two wavelengths are shown for each traveling wave, so the standing wave patterns show twice as many antinodes as that in Figure 17.7.

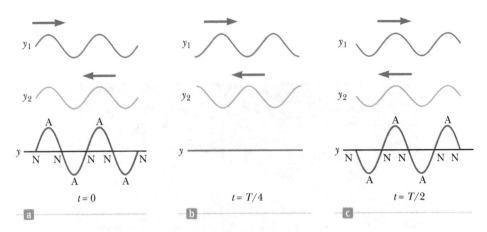

Two nodes and two antinodes are labeled in the standing wave in Figure 17.7. The light blue curve labeled $2A \sin kx$ in Figure 17.7 represents one wavelength of the traveling waves that combine to form the standing wave. Figure 17.7 and Equations 17.2 and 17.3 provide the following important features of the locations of nodes and antinodes:

The distance between adjacent antinodes is equal to $\lambda/2$.
The distance between adjacent nodes is equal to $\lambda/2$.
The distance between a node and an adjacent antinode is $\lambda/4$.

In the photograph in Figure 17.7, the frequency of the waves is so high that *several* oscillations of the elements of the string occur during the time interval during which the camera shutter is open. Let's slow things down a bit. Wave patterns of the elements of the medium produced at various times during half a cycle of oscillation for two transverse traveling waves moving in opposite directions are shown in Figures 17.8a–c. The blue and green curves are the wave patterns for the individual traveling waves, and the red-brown curves are the wave patterns for the resultant standing wave when they are combined. At $t = 0$ (Fig. 17.8a), the two traveling waves are in phase, giving a wave pattern in which each element of the medium is at rest and experiencing its maximum displacement from equilibrium. One-quarter of a period later, at $t = T/4$ (Fig. 17.8b), the traveling waves have moved one-fourth of a wavelength (one to the right and the other to the left). At this time, the traveling waves are out of phase, and each element of the medium is passing through the equilibrium position in its simple harmonic motion. The result is zero displacement for elements at all values of x; that is, the wave pattern is a straight line. At $t = T/2$ (Fig. 17.8c), the traveling waves are again in phase, producing a wave pattern that is inverted relative to the $t = 0$ pattern. In the standing wave, the elements of the medium alternate in time between the extremes shown in Figures 17.8a and 17.8c.

QUICK QUIZ 17.2 Consider the waves in Figure 17.8 to be waves on a stretched string. Define the velocity of elements of the string as positive if they are moving upward in the figure. **(i)** At the moment the string has the shape shown by the red-brown curve in Figure 17.8a, what is the instantaneous velocity of elements along the string? (a) zero for all elements (b) positive for all elements (c) negative for all elements (d) varies with the position of the element **(ii)** From the same choices, at the moment the string has the shape shown by the red-brown curve in Figure 17.8b, what is the instantaneous velocity of elements along the string?

Example **17.2** Formation of a Standing Wave

Two waves traveling in opposite directions produce a standing wave. The individual wave functions are

$$y_1 = 4.0 \sin (3.0x - 2.0t)$$
$$y_2 = 4.0 \sin (3.0x + 2.0t)$$

where x and y are measured in centimeters and t is in seconds.

(A) Find the amplitude of the simple harmonic motion of the element of the medium located at $x = 2.3$ cm.

SOLUTION

Conceptualize The waves described by the given equations are identical except for their directions of travel, so they indeed combine to form a standing wave as discussed in this section. We can represent the waves graphically by the blue and green curves in Figure 17.8.

Categorize We will substitute values into equations developed in this section, so we categorize this example as a substitution problem.

From the equations for the waves, we see that $A = 4.0$ cm, $k = 3.0$ rad/cm, and $\omega = 2.0$ rad/s. Use Equation 17.1 to write an expression for the standing wave:

$$y = (2A \sin kx) \cos \omega t = 8.0 \sin 3.0x \cos 2.0t$$

Find the amplitude of the simple harmonic motion of the element at the position $x = 2.3$ cm by evaluating the sine function at this position:

$$y_{max} = (8.0 \text{ cm}) \sin 3.0x \big|_{x = 2.3}$$
$$= (8.0 \text{ cm}) \sin (6.9 \text{ rad}) = 4.6 \text{ cm}$$

(B) Find the positions of the nodes and antinodes if one end of the string is at $x = 0$.

SOLUTION

Find the wavelength of the traveling waves:

$$k = \frac{2\pi}{\lambda} = 3.0 \text{ rad/cm} \rightarrow \lambda = \frac{2\pi}{3.0} \text{ cm}$$

Use Equation 17.2 to find the locations of the nodes:

$$x = n\frac{\lambda}{2} = n\left(\frac{\pi}{3.0}\right) \text{ cm} \quad n = 0, 1, 2, 3, \dots$$

Use Equation 17.3 to find the locations of the antinodes:

$$x = n\frac{\lambda}{4} = n\left(\frac{\pi}{6.0}\right) \text{ cm} \quad n = 1, 3, 5, 7, \dots$$

17.3 Boundary Effects: Reflection and Transmission

So far in our discussion of waves, we have primarily considered waves traveling through a medium without interacting with any boundaries of the medium. The only exceptions have been references to reflections of waves, such as the echoes from the cliffs in the opening storyline for Chapter 16 and the reflection of waves on the coiled telephone cord from the base unit in Section 17.2. We now address the details of the interactions of waves with boundaries. For example, consider a pulse traveling on a string that is rigidly attached to a support at one end as in Figure 17.9. When the pulse reaches the support, the string ends. As a result, the pulse undergoes **reflection**; that is, the pulse moves back along the string in the opposite direction.

Notice that the reflected pulse is *inverted*. This inversion can be explained as follows. When the pulse reaches the fixed end of the string, the string produces an upward force on the support. By Newton's third law, the support must exert an equal-magnitude and oppositely directed (downward) reaction force on the string. This downward force causes a downward-oriented reflected pulse.

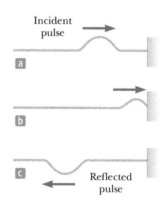

Figure 17.9 The reflection of a traveling pulse at the fixed end of a stretched string. The reflected pulse is inverted, but its shape is otherwise unchanged.

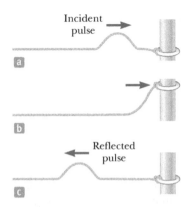

Figure 17.10 The reflection of a traveling pulse at the free end of a stretched string. The reflected pulse is not inverted.

Now consider another case. This time, the pulse arrives at the end of a string that is free to move vertically as in Figure 17.10. The tension at the free end is maintained because the string is tied to a ring of negligible mass that is free to slide vertically on a smooth post without friction. Again, the pulse is reflected, but this time it is not inverted. When it reaches the post, the pulse exerts a force on the free end of the string, causing the ring to accelerate upward. The ring rises as high as the incoming pulse, and then the downward component of the tension force pulls the ring back down. This movement of the ring produces a reflected pulse that is upward-oriented and that has the same amplitude as the incoming pulse.

Finally, consider a situation in which the boundary is intermediate between these two extremes. In this case, the medium does not end, but rather it changes in some way and continues. When there is a change in the medium, part of the energy in the incident pulse is reflected and part undergoes **transmission;** that is, some of the energy passes through the boundary. For instance, suppose a light string is attached to a heavier string as in Figure 17.11. When a pulse traveling on the light string reaches the boundary between the two strings, part of the pulse is reflected and inverted and part is transmitted to the heavier string. The reflected pulse is inverted for the same reasons described earlier in the case of the string rigidly attached to a support.

The reflected pulse has a smaller amplitude than the incident pulse. In Section 16.4, we showed that the energy carried by a wave is related to its amplitude. According to the principle of conservation of energy, when the pulse breaks up into a reflected pulse and a transmitted pulse at the boundary, the sum of the energies of these two pulses must equal the energy of the incident pulse. Because the reflected pulse contains only part of the energy of the incident pulse, its amplitude must be smaller.

When a pulse traveling on a heavy string strikes the boundary between the heavy string and a lighter one as in Figure 17.12, again part is reflected and part is transmitted. In this case, the reflected pulse is not inverted.

According to Equation 16.18, the speed of a wave on a string increases as the mass per unit length of the string decreases. In other words, a wave travels more rapidly on a light string than on a heavy string if both are under the same tension. The following general rules apply to reflected waves: When a wave or pulse travels from medium A to medium B and $v_A > v_B$ (that is, when B is denser than A), it is inverted upon reflection. When a wave or pulse travels from medium A to medium B and $v_A < v_B$ (that is, when A is denser than B), it is not inverted upon reflection.

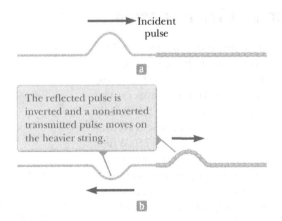

Figure 17.11 (a) A pulse traveling to the right on a light string approaches the junction with a heavier string. (b) The situation after the pulse reaches the junction.

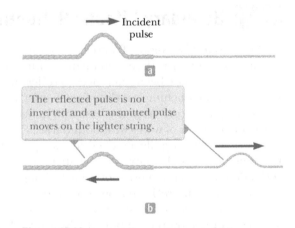

Figure 17.12 (a) A pulse traveling to the right on a heavy string approaches the junction with a lighter string. (b) The situation after the pulse reaches the junction.

17.4 Analysis Model: Waves Under Boundary Conditions

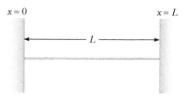

Figure 17.13 A string of length L fixed at both ends.

In Section 17.2, we studied standing waves in a medium with no boundaries. In Section 17.3, we investigated the effect of a rigid boundary on waves in a medium: the waves reflect from the boundary. In this section, let us combine these ideas to see how the existence of boundaries affects the standing wave.

Consider a string of length L fixed at both ends as shown in Figure 17.13. We will use this system as a model for a guitar string or piano string. Waves can travel in both directions on the string due to reflections from the ends. Therefore, standing waves can be set up in the string by a continuous superposition of incident and reflected waves. Notice that there is a *boundary condition* for the waves on the string: because the ends of the string are fixed, they must necessarily have zero displacement and are therefore nodes by definition. The condition that both ends of the string must be nodes fixes the wavelength of the standing wave on the string: at the right end of the string, where $x = L$, Equation 17.2 gives us

$$L = \frac{n\lambda_n}{2} \tag{17.4}$$

where the subscript on λ indicates that different values of n will result in different values of the wavelength. The wavelength, in turn, determines the frequency of the wave according to Equation 16.12. The boundary condition results in the string having a number of discrete natural patterns of oscillation, called **normal modes,** each of which has a characteristic frequency that is easily calculated. This situation in which only certain frequencies of oscillation are allowed is called **quantization.** Quantization is a common occurrence when waves are subject to boundary conditions and is a central feature in our discussions of quantum physics in the extended version of this text. Notice in Figure 17.8 that there are no boundary conditions, so standing waves of *any* frequency can be established; there is no quantization without boundary conditions. Because boundary conditions occur so often for waves, we identify an analysis model called **waves under boundary conditions** for the discussion that follows.

The normal modes of oscillation for the string in Figure 17.13 can be described by imposing the boundary conditions that the ends be nodes and that the nodes be separated by one-half of a wavelength with antinodes halfway between the nodes. The first normal mode that is consistent with these requirements, shown in Figure 17.14a, has nodes at its ends and one antinode in the middle. This normal mode is the longest-wavelength mode that is consistent with our boundary conditions. The first normal mode occurs when the wavelength λ_1 is equal to twice the length of the string, or, from Equation 17.4, $\lambda_1 = 2L$. The section of a standing wave from one node to the next node is called a *loop*. In the first normal mode, the string is vibrating in one loop. In the second normal mode (see Fig. 17.14b), the string vibrates in two loops. When the left half of the string is moving upward, the right

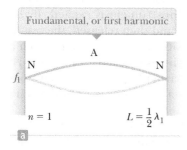

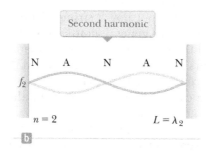

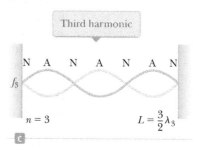

Figure 17.14 The normal modes of vibration of the string in Figure 17.13 form a harmonic series. The string vibrates between the extremes shown.

half is moving downward. In this case, from Equation 17.4 with $n = 2$, the wavelength λ_2 is equal to the length of the string: $\lambda_2 = L$. The third normal mode (see Fig. 17.14c) corresponds to the case in which $\lambda_3 = 2L/3$, and the string vibrates in three loops. In general, the wavelengths of the various normal modes for a string of length L fixed at both ends are found by rearranging Equation 17.4:

Wavelengths of ▶
normal modes

$$\lambda_n = \frac{2L}{n} \quad n = 1, 2, 3, \ldots \tag{17.5}$$

where the index n refers to the nth normal mode of oscillation. These modes are *possible*. The *actual* modes that are excited on a string are discussed shortly.

The natural frequencies associated with the modes of oscillation are obtained from Equation 16.12, $f = v/\lambda$, where the wave speed v is the same for all frequencies. Using Equation 17.5, we find that the natural frequencies f_n of the normal modes are

Natural frequencies of ▶
normal modes as functions
of wave speed and length
of string

$$f_n = \frac{v}{\lambda_n} = n\frac{v}{2L} \quad n = 1, 2, 3, \ldots \tag{17.6}$$

These natural frequencies are also called the *quantized frequencies* associated with the vibrating string fixed at both ends.

Because $v = \sqrt{T/\mu}$ (see Eq. 16.18) for waves on a string, where T is the tension in the string and μ is its linear mass density, we can also express the natural frequencies of a taut string as

Natural frequencies of ▶
normal modes as functions
of string tension and
linear mass density

$$f_n = \frac{n}{2L}\sqrt{\frac{T}{\mu}} \quad n = 1, 2, 3, \ldots \tag{17.7}$$

The lowest frequency f_1, which corresponds to $n = 1$, is called either the **fundamental** or the **fundamental frequency** and is given by

Fundamental frequency ▶
of a taut string

$$f_1 = \frac{1}{2L}\sqrt{\frac{T}{\mu}} \tag{17.8}$$

The frequencies of the remaining normal modes are integer multiples of the fundamental frequency (Eq. 17.6). Frequencies of normal modes that exhibit such an integer-multiple relationship form a **harmonic series,** and the normal modes are called **harmonics.** The fundamental frequency f_1 is the frequency of the first harmonic, the frequency $f_2 = 2f_1$ is that of the second harmonic, and the frequency $f_n = nf_1$ is that of the nth harmonic. Other oscillating systems, such as a drumhead, exhibit normal modes, but the frequencies are not related as integer multiples of a fundamental. Therefore, we do not use the term *harmonic* in association with those types of systems.

Let us examine now how the various harmonics are actually excited in a string. To excite only a single harmonic, the string would have to be distorted into a shape that corresponds to that of the desired harmonic. After being released, the string would vibrate at the frequency of that harmonic. This maneuver is difficult to perform, however, and is not how a string of a musical instrument is excited. If the string is distorted into a general, nonsinusoidal shape and then released, the resulting vibration of the string includes a combination of its various harmonics. Such a distortion occurs in musical instruments when the string is plucked (as in a guitar), bowed (as in a cello), or struck (as in a piano). The particular mixture of harmonics in the string can be changed by plucking the guitar string or bowing the cello string at different locations.

The frequency of a string that defines the musical note that it plays is that of the fundamental, even though other harmonics are present. The additional harmonics determine the *quality*, or the *timbre*, of the sound without altering its frequency, as discussed further in Section 17.8. The quality of the sound is part of what allows you to identify instruments playing the same note. For example, you can differentiate between a guitar, banjo, or a sitar playing the same note.

The string's frequency can be varied by changing the string's tension or its length. For example, the tension in guitar and violin strings is varied by a screw adjustment mechanism or by tuning pegs located on the neck of the instrument. As the tension is increased, the frequency of the normal modes increases in accordance with Equation 17.7. Once the instrument is "tuned," players vary the frequency by moving their fingers along the neck, thereby changing the length L of the oscillating portion of the string. As the length is shortened, the frequency increases because, as Equation 17.7 specifies, the normal-mode frequencies are inversely proportional to string length.

In the opening storyline, when you pluck an open string on your roommate's guitar, the fundamental mode is that shown in Figure 17.4a. Then, you place your finger lightly at the midpoint of the string. Because your finger is pressing only lightly on the string, the entire string can still vibrate when you pluck it. But your finger imposes a node at the center of the string. Therefore, the fundamental mode of vibration now looks like Figure 17.4b. This is the $n = 2$ harmonic of the open string, so the frequency is twice as high: an octave.

QUICK QUIZ 17.3 When a standing wave is set up on a string fixed at both ends, which of the following statements is true? **(a)** The number of nodes is equal to the number of antinodes. **(b)** The wavelength is equal to the length of the string divided by an integer. **(c)** The frequency is equal to the number of nodes times the fundamental frequency. **(d)** The shape of the string at any instant shows a symmetry about the midpoint of the string.

ANALYSIS MODEL | **Waves Under Boundary Conditions**

Imagine a wave that is not free to travel throughout all space as in the traveling wave model. If the wave is subject to boundary conditions, such that certain requirements must be met at specific locations in space, the wave is limited to a set of **normal modes** with quantized wavelengths and quantized natural frequencies.

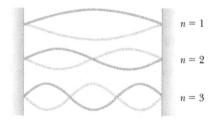

For waves on a string fixed at both ends, the natural frequencies are

$$f_n = \frac{n}{2L}\sqrt{\frac{T}{\mu}} \quad n = 1, 2, 3, \ldots \qquad \textbf{(17.7)}$$

where T is the tension in the string and μ is its linear mass density.

Examples:

- waves traveling back and forth on a guitar string combine to form a standing wave
- sound waves traveling back and forth in a clarinet combine to form standing waves (Section 17.6)
- a microscopic particle confined to a small region of space is modeled as a wave and exhibits quantized energies (Chapter 40)
- the Fermi energy of a metal is determined by modeling electrons as wave-like particles in a box (Chapter 42)

Example **17.3** | **Give Me a C Note!**

The middle C string on a piano has a fundamental frequency of 262 Hz, and the string for the first A above middle C has a fundamental frequency of 440 Hz.

(A) Calculate the frequencies of the next two harmonics of the C string.

SOLUTION

Conceptualize Remember that the harmonics of a vibrating string have frequencies that are related by integer multiples of the fundamental.

continued

17.3 continued

Categorize This first part of the example is a simple substitution problem.

Knowing that the fundamental frequency is $f_1 = 262$ Hz, find the frequencies of the next harmonics by multiplying by integers:

$$f_2 = 2f_1 = 524 \text{ Hz}$$

$$f_3 = 3f_1 = 786 \text{ Hz}$$

(B) If the A and C strings have the same linear mass density μ and length L, determine the ratio of tensions in the two strings.

SOLUTION

Categorize This part of the example is more of an analysis problem than is part (A) and uses the *waves under boundary conditions* model.

Analyze Use Equation 17.8 to write expressions for the fundamental frequencies of the two strings:

$$f_{1A} = \frac{1}{2L}\sqrt{\frac{T_A}{\mu}} \quad \text{and} \quad f_{1C} = \frac{1}{2L}\sqrt{\frac{T_C}{\mu}}$$

Divide the first equation by the second and solve for the ratio of tensions:

$$\frac{f_{1A}}{f_{1C}} = \sqrt{\frac{T_A}{T_C}} \rightarrow \frac{T_A}{T_C} = \left(\frac{f_{1A}}{f_{1C}}\right)^2 = \left(\frac{440 \text{ Hz}}{262 \text{ Hz}}\right)^2 = 2.82$$

Finalize If the frequencies of piano strings were determined solely by tension, this result suggests that the ratio of tensions from the lowest string to the highest string on the piano would be enormous. Such large tensions would make it difficult to design a frame to support the strings. In reality, the frequencies of piano strings vary due to additional parameters, including the mass per unit length and the length of the string. The What If? below explores a variation in length.

WHAT IF? If you look inside a real piano, you'll see that the assumption made in part (B) is only partially true. The strings are not likely to have the same length. The string densities for the given notes might be equal, but suppose the length of the A string is only 64% of the length of the C string. What is the ratio of their tensions?

Answer Using Equation 17.8 again, we set up the ratio of frequencies:

$$\frac{f_{1A}}{f_{1C}} = \frac{L_C}{L_A}\sqrt{\frac{T_A}{T_C}} \rightarrow \frac{T_A}{T_C} = \left(\frac{L_A}{L_C}\right)^2\left(\frac{f_{1A}}{f_{1C}}\right)^2$$

$$\frac{T_A}{T_C} = (0.64)^2\left(\frac{440 \text{ Hz}}{262 \text{ Hz}}\right)^2 = 1.16$$

Notice that this result represents only a 16% increase in tension, compared with the 182% increase in part (B).

Example 17.4 Changing String Vibration with Water

One end of a horizontal string is attached to a vibrating blade, and the other end passes over a pulley as in Figure 17.15a. A sphere of mass 2.00 kg hangs on the end of the string. The string is vibrating in its second harmonic. A container of water is raised under the sphere so that the sphere is completely submerged. In this configuration, the string vibrates in its fifth harmonic as shown in Figure 17.15b. What is the radius of the sphere?

SOLUTION

Conceptualize Imagine what happens when the sphere is immersed in the water. The buoyant force acts upward on the sphere, reducing the tension in the string. The change in tension causes a change in the speed of waves on the string, which in turn causes a change in the wavelength. This altered wavelength results in the string vibrating in its fifth normal mode rather than the second.

Categorize The hanging sphere is modeled as a *particle in equilibrium*. One of the forces acting on it is the buoyant force from the water. We also apply the *waves under boundary conditions* model to the string.

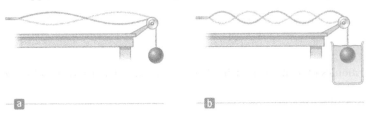

a   b

Figure 17.15 (Example 17.4) (a) When the sphere hangs in air, the string vibrates in its second harmonic. (b) When the sphere is immersed in water, the string vibrates in its fifth harmonic.

17.4 continued

Analyze Apply the particle in equilibrium model to the sphere in Figure 17.15a, identifying T_1 as the tension in the string as the sphere hangs in air:

$$\sum F = T_1 - mg = 0$$

$$T_1 = mg$$

Apply the particle in equilibrium model to the sphere in Figure 17.15b, where T_2 is the tension in the string as the sphere is immersed in water:

$$T_2 + B - mg = 0$$

$$(1) \quad B = mg - T_2$$

The desired quantity, the radius of the sphere, will appear in the expression for the buoyant force B. Before proceeding in this direction, however, we must evaluate T_2 from the information about the standing wave.

Write the equation for the frequency of a standing wave on a string (Eq. 17.7) twice, once before the sphere is immersed and once after. Notice that the frequency f is the same in both cases because it is determined by the vibrating blade. In addition, the linear mass density μ and the length L of the vibrating portion of the string are the same in both cases. Divide the equations:

$$f = \frac{n_1}{2L}\sqrt{\frac{T_1}{\mu}} \quad \rightarrow \quad 1 = \frac{n_1}{n_2}\sqrt{\frac{T_1}{T_2}}$$
$$f = \frac{n_2}{2L}\sqrt{\frac{T_2}{\mu}}$$

Solve for T_2:

$$T_2 = \left(\frac{n_1}{n_2}\right)^2 T_1 = \left(\frac{n_1}{n_2}\right)^2 mg$$

Substitute this result into Equation (1):

$$(2) \quad B = mg - \left(\frac{n_1}{n_2}\right)^2 mg = mg\left[1 - \left(\frac{n_1}{n_2}\right)^2\right]$$

Using Equation 14.5, express the buoyant force in terms of the radius of the sphere:

$$B = \rho_{water} g V_{sphere} = \rho_{water} g(\tfrac{4}{3}\pi r^3)$$

Solve for the radius of the sphere and substitute from Equation (2):

$$r = \left(\frac{3B}{4\pi\rho_{water}g}\right)^{1/3} = \left\{\frac{3m}{4\pi\rho_{water}}\left[1 - \left(\frac{n_1}{n_2}\right)^2\right]\right\}^{1/3}$$

Substitute numerical values:

$$r = \left\{\frac{3(2.00\text{ kg})}{4\pi(1\,000\text{ kg/m}^3)}\left[1 - \left(\frac{2}{5}\right)^2\right]\right\}^{1/3}$$

$$= 0.073\,7\text{ m} = 7.37\text{ cm}$$

Finalize Notice that only certain radii of the sphere will result in the string vibrating in a normal mode; the speed of waves on the string must be changed to a value such that the length of the string is an integer multiple of half wavelengths. This limitation is a feature of the *quantization* that was introduced earlier in this chapter: the sphere radii that cause the string to vibrate in a normal mode are *quantized*.

17.5 Resonance

We have seen that a system such as a taut string is capable of oscillating in one or more normal modes of oscillation. We find that if a periodic force is applied to such a system, the amplitude of the resulting motion of the string is greatest when the frequency of the applied force is equal to one of the natural frequencies of the system. This phenomenon, known as *resonance,* was discussed in Section 15.7 with regard to a simple harmonic oscillator. Although a block–spring system or a simple pendulum has only one natural frequency, standing-wave systems have a whole set of natural frequencies, such as that given by Equation 17.7 for a string. Because an oscillating system exhibits a large amplitude when driven at any of its natural frequencies, these frequencies are often referred to as **resonance frequencies.**

Consider Figure 17.16, which shows a string being driven by a vibrating blade. When the frequency of the blade equals one of the natural frequencies of the string,

When the blade vibrates at one of the natural frequencies of the string, large-amplitude standing waves are created.

Vibrating blade

Figure 17.16 Standing waves are set up in a string when one end is connected to a vibrating blade.

standing waves are produced and the string oscillates with a large amplitude. In this resonance case, the wave generated by the oscillating blade is in phase with the reflected wave and the string absorbs energy from the blade. If the string is driven at a frequency that is not one of its natural frequencies, the oscillations are of low amplitude and exhibit no stable pattern.

Resonance is very important in the excitation of musical instruments based on air columns. We shall discuss this application of resonance in Section 17.6.

17.6 Standing Waves in Air Columns

The waves under boundary conditions model can also be applied to sound waves in a column of air such as that inside an organ pipe or a clarinet. Standing waves in this case are the result of interference between longitudinal sound waves traveling in opposite directions.

In a pipe closed at one end, the closed end is a **displacement node** because the rigid barrier at this end does not allow longitudinal motion of the air. Because the pressure wave is 90° out of phase with the displacement wave (see Section 16.6), the closed end of an air column corresponds to a **pressure antinode** (that is, a point of maximum pressure variation).

The open end of an air column is approximately a **displacement antinode**[1] and a pressure node. We can understand why no pressure variation occurs at an open end by noting that the end of the air column is open to the atmosphere; therefore, the pressure at this end must remain constant at atmospheric pressure.

You may wonder how a sound wave can reflect from an open end because there may not appear to be a change in the medium at this point: the medium through which the sound wave moves is *air*, both inside and outside the pipe. Sound can be represented as a pressure wave, however, and a compression region of the sound wave is constrained by the sides of the pipe as long as the region is inside the pipe. As the compression region exits at the open end of the pipe, the constraint of the pipe is removed and the compressed air is free to expand into the atmosphere. Therefore, there is a change in the *character* of the medium between the inside of the pipe and the outside even though there is no change in the *material* of the medium. This change in character is sufficient to allow some reflection.

With the boundary conditions of nodes or antinodes at the ends of the air column, we have a set of normal modes of oscillation as is the case for the string fixed at both ends. Therefore, the air column has quantized frequencies.

The first three normal modes of oscillation of a pipe open at both ends are shown in Figure 17.17a. The diagrams in the left column show *graphical* representations of the *displacement* of elements of air from their equilibrium positions. The second column shows *pictorial* representations of the *pressure* in the air at various locations in the pipe, following the technique used in Figure 16.17. There is a lot of information in Figure 17.17. Study it carefully.

Notice that both ends of the pipe in Figure 17.17a are displacement antinodes (approximately) or pressure nodes. In the first normal mode, the standing wave extends between two adjacent displacement antinodes or two adjacent pressure nodes, which is a distance of half a wavelength. Therefore, the wavelength is twice the length of the pipe, and the fundamental frequency is $f_1 = v/2L$. As Figure 17.17a

---

[1] Strictly speaking, the open end of an air column is not exactly a displacement antinode. A compression reaching an open end does not reflect until it passes beyond the end. For a tube of circular cross section, an end correction equal to approximately $0.6R$, where R is the tube's radius, must be added to the length of the air column. Hence, the effective length of the air column is longer than the true length L. We ignore this end correction in this discussion.

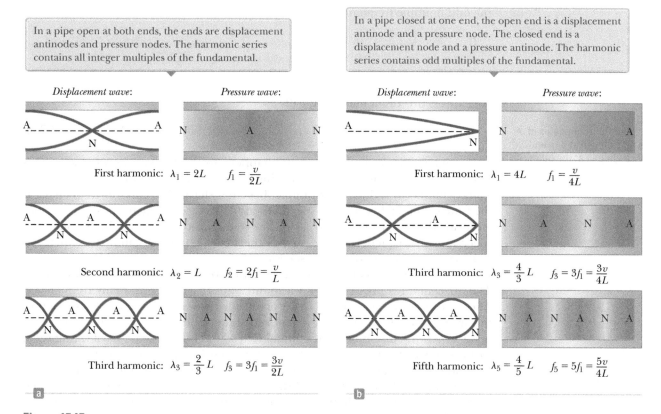

In a pipe open at both ends, the ends are displacement antinodes and pressure nodes. The harmonic series contains all integer multiples of the fundamental.

In a pipe closed at one end, the open end is a displacement antinode and a pressure node. The closed end is a displacement node and a pressure antinode. The harmonic series contains odd multiples of the fundamental.

Displacement wave: *Pressure wave:*

First harmonic: $\lambda_1 = 2L$ $f_1 = \dfrac{v}{2L}$

Second harmonic: $\lambda_2 = L$ $f_2 = 2f_1 = \dfrac{v}{L}$

Third harmonic: $\lambda_3 = \dfrac{2}{3}L$ $f_3 = 3f_1 = \dfrac{3v}{2L}$

Displacement wave: *Pressure wave:*

First harmonic: $\lambda_1 = 4L$ $f_1 = \dfrac{v}{4L}$

Third harmonic: $\lambda_3 = \dfrac{4}{3}L$ $f_3 = 3f_1 = \dfrac{3v}{4L}$

Fifth harmonic: $\lambda_5 = \dfrac{4}{5}L$ $f_5 = 5f_1 = \dfrac{5v}{4L}$

a   **b**

Figure 17.17 Standing longitudinal sound waves in air columns, showing the wave patterns for the three lowest frequencies. (a) In an open column, the standing waves are symmetric around the mid-point of the column. On the left are graphical representations of the displacement of elements of the air. On the right are pictorial representations of the pressure at various points in the wave. (b) In a column closed at one end, the standing waves are not symmetric. Again, the left-hand diagrams show graphical representations of the displacement of elements of the air, while the right-hand diagrams are pictorial representation of the pressure.

shows, the frequencies of the higher harmonics are $2f_1, 3f_1, \ldots.$

In a pipe open at both ends, the natural frequencies of oscillation form a harmonic series that includes all integral multiples of the fundamental frequency.

Because the fundamental frequency is given by the same expression as that for a string (see Eq. 17.6), we can express the natural frequencies of oscillation as

$$f_n = n\frac{v}{2L} \quad n = 1, 2, 3, \ldots \tag{17.9}$$

◀ Natural frequencies of a pipe open at both ends

Despite the similarity between Equations 17.6 and 17.9, you must remember that v in Equation 17.6 is the speed of waves on the string, whereas v in Equation 17.9 is the speed of sound in air.

If a pipe is closed at one end and open at the other, the closed end is a displacement node or a pressure antinode (see Fig. 17.17b). In this case, the standing wave for the fundamental mode extends from an antinode to the adjacent node, which is one-fourth of a wavelength. Hence, the wavelength for the first normal mode is $4L$, and the fundamental frequency is $f_1 = v/4L$. As Figure 17.17b shows, the higher-frequency waves that satisfy our conditions are those that have a node at the closed end and an antinode at the open end; hence, the higher harmonics do *not* include *all* integer multiples of the fundamental frequency, but rather have only the odd-multiple frequencies $3f_1, 5f_1, \ldots$.

> In a pipe closed at one end, the natural frequencies of oscillation form a harmonic series that includes only odd integral multiples of the fundamental frequency.

We express this result mathematically as

Natural frequencies of ▷
a pipe closed at one end
and open at the other

$$f_m = m\frac{v}{4L} \quad m = 1, 3, 5, \ldots \quad \text{or} \quad f_n = (2n - 1)\frac{v}{4L} \quad n = 1, 2, 3, \ldots \quad \textbf{(17.10)}$$

It is interesting to investigate what happens to the frequencies of instruments based on air columns and strings during a concert as the temperature rises. The sound emitted by a flute, for example, becomes sharp (increases in frequency) as the flute warms up because the speed of sound increases in the increasingly warmer air inside the flute (consider Eq. 17.9). The sound produced by a violin becomes flat (decreases in frequency) as the strings thermally expand because the expansion causes their tension to decrease (see Eq. 17.7).

Musical instruments based on air columns are generally excited by resonance. The air column is presented with a sound wave that is rich in many frequencies. The air column then responds by resonance with a large-amplitude oscillation to the frequencies that match the quantized frequencies in its set of harmonics. In many woodwind instruments, the initial rich sound is provided by a vibrating reed. In brass instruments, this excitation is provided by the sound coming from the vibration of the player's lips. In a flute, the initial excitation comes from blowing over an edge at the mouthpiece of the instrument in a manner similar to blowing across the opening of a bottle with a narrow neck. The sound of the air rushing across the bottle opening has many frequencies, including one that sets the air cavity in the bottle into resonance.

QUICK QUIZ 17.4 A pipe open at both ends resonates at a fundamental frequency f_{open}. When one end is covered and the pipe is again made to resonate, the fundamental frequency is f_{closed}. Which of the following expressions describes how these two resonant frequencies compare? (a) $f_{\text{closed}} = f_{\text{open}}$ (b) $f_{\text{closed}} = \frac{1}{2}f_{\text{open}}$ (c) $f_{\text{closed}} = 2f_{\text{open}}$ (d) $f_{\text{closed}} = \frac{3}{2}f_{\text{open}}$

QUICK QUIZ 17.5 Balboa Park in San Diego has an outdoor organ. When the air temperature increases, the fundamental frequency of one of the organ pipes (a) stays the same, (b) goes down, (c) goes up, or (d) is impossible to determine.

Example 17.5 **Wind in a Culvert**

A section of drainage culvert 1.23 m in length makes a howling noise when the wind blows across its open ends.

Determine the frequencies of the first three harmonics of the culvert if it is cylindrical in shape and open at both ends. Take $v = 343$ m/s as the speed of sound in air.

SOLUTION

Conceptualize The sound of the wind blowing across the end of the pipe contains many frequencies, and the culvert responds to the sound by resonance, vibrating at the natural frequencies of the air column.

Categorize This example is a relatively simple substitution problem.

Find the frequency of the first harmonic of the culvert, modeling it as an air column open at both ends:

$$f_1 = \frac{v}{2L} = \frac{343 \text{ m/s}}{2(1.23 \text{ m})} = 139 \text{ Hz}$$

Find the next harmonics by multiplying by integers:

$$f_2 = 2f_1 = 279 \text{ Hz}$$

$$f_3 = 3f_1 = 418 \text{ Hz}$$

Example 17.6 Measuring the Frequency of a Tuning Fork

A simple apparatus for demonstrating resonance in an air column is depicted in Figure 17.18. A vertical pipe open at both ends is partially submerged in water, and a tuning fork vibrating at an unknown frequency is placed near the top of the pipe. The length L of the air column can be adjusted by moving the pipe vertically. The sound waves generated by the fork are reinforced when L corresponds to one of the resonance frequencies of the pipe. For a certain pipe, the smallest value of L for which a peak occurs in the sound intensity is 9.00 cm.

(A) What is the frequency of the tuning fork?

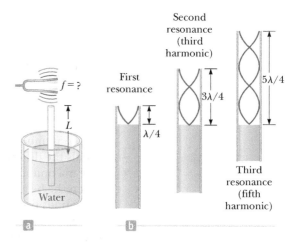

SOLUTION

Conceptualize Sound waves from the tuning fork enter the pipe at its upper end. Although the pipe is open at its lower end to allow the water to enter, the water's surface acts like a barrier, as if the end of the part of the pipe that is above the water were closed. The waves reflect from the water surface and combine with those moving downward to form a standing wave.

Figure 17.18 (Example 17.6) (a) Apparatus for demonstrating the resonance of sound waves in a pipe closed at one end. The length L of the air column is varied by moving the pipe vertically while it is partially submerged in water. (b) The first three normal modes of the system shown in (a).

Categorize Because of the reflection of the sound waves from the water surface, we can model the part of the pipe that is above the water as open at the upper end and closed at the lower end. Therefore, we can apply the *waves under boundary conditions* model to this situation.

Analyze

Use Equation 17.10 to find the fundamental frequency for $L = 0.090\ 0$ m:

$$f_1 = \frac{v}{4L} = \frac{343 \text{ m/s}}{4(0.090\ 0 \text{ m})} = 953 \text{ Hz}$$

Because the tuning fork causes the air column to resonate at this frequency, this frequency must also be that of the tuning fork.

(B) What are the values of L for the next two resonance conditions?

SOLUTION

Use Equation 16.12 to find the wavelength of the sound wave from the tuning fork:

$$\lambda = \frac{v}{f} = \frac{343 \text{ m/s}}{953 \text{ Hz}} = 0.360 \text{ m}$$

Notice from Figure 17.18b that the length of the air column above the water for the second resonance is $3\lambda/4$:

$$L = 3\lambda/4 = 0.270 \text{ m}$$

Notice from Figure 17.18b that the length of the air column above the water for the third resonance is $5\lambda/4$:

$$L = 5\lambda/4 = 0.450 \text{ m}$$

Finalize Consider how this problem differs from the preceding example. In the culvert, the length was fixed and the air column was presented with a mixture of many frequencies. The pipe in this example is presented with one single frequency from the tuning fork, and the length of the pipe above the water is varied until resonance is achieved.

17.7 Beats: Interference in Time

The interference phenomena we have studied so far involve the superposition of two or more waves having the same frequency. Because the amplitude of the oscillation of elements of the medium varies with the position in space of the element in such a wave, we refer to the phenomenon as *spatial interference*. Standing waves in strings and pipes are common examples of spatial interference.

Now let's consider another type of interference, one that results from the superposition of two waves having slightly *different* frequencies. In this case, when the two waves are observed at a point in space, they are periodically in and out of phase. That is, there is a *temporal* (time) alternation between constructive and destructive interference. As a consequence, we refer to this phenomenon as *interference in time* or *temporal interference*. For example, if two tuning forks of slightly different frequencies are struck, one hears a sound of periodically varying amplitude. This phenomenon is called **beating.**

Definition of beating ▶   Beating is the periodic variation in amplitude at a given point due to the superposition of two waves having slightly different frequencies.

The number of amplitude maxima one hears per second, or the *beat frequency,* equals the difference in frequency between the two sources as we shall show below. The maximum beat frequency that the human ear can detect is about 20 beats/s. When the beat frequency exceeds this value, the beats blend indistinguishably with the sounds producing them.

Consider two sound waves of equal amplitude and slightly different frequencies f_1 and f_2 traveling through a medium. We use equations similar to Equation 16.13 to represent the wave functions for these two waves at a point that we identify as $x = 0$. We also choose the phase angle in Equation 16.13 as $\phi = \pi/2$:

$$y_1 = A \sin\left(\frac{\pi}{2} - \omega_1 t\right) = A \cos\left(2\pi f_1 t\right)$$

$$y_2 = A \sin\left(\frac{\pi}{2} - \omega_2 t\right) = A \cos\left(2\pi f_2 t\right)$$

Using the superposition principle, we find that the resultant wave function at this point is

$$y = y_1 + y_2 = A\left(\cos 2\pi f_1 t + \cos 2\pi f_2 t\right)$$

The trigonometric identity

$$\cos a + \cos b = 2\cos\left(\frac{a-b}{2}\right)\cos\left(\frac{a+b}{2}\right)$$

allows us to write the expression for y as

Resultant of two waves of ▶
different frequencies but
equal amplitude

$$y = \left[2A\cos 2\pi\left(\frac{f_1 - f_2}{2}\right)t\right]\cos 2\pi\left(\frac{f_1 + f_2}{2}\right)t \qquad (17.11)$$

Graphs of the individual waves and the resultant wave are shown in Figure 17.19. From the factors in Equation 17.11, we see that the resultant wave has an effective frequency equal to the average frequency $(f_1 + f_2)/2$. This wave is multiplied by an envelope wave given by the expression in the square brackets:

$$y_{\text{envelope}} = 2A\cos 2\pi\left(\frac{f_1 - f_2}{2}\right)t \qquad (17.12)$$

That is, the amplitude and therefore the intensity of the resultant sound vary in time. The dashed black line in Figure 17.19b is a graphical representation of the envelope wave in Equation 17.12 and is a sine wave varying with frequency $(f_1 - f_2)/2$.

A maximum in the amplitude of the resultant sound wave is detected whenever

$$\cos 2\pi\left(\frac{f_1 - f_2}{2}\right)t = \pm 1$$

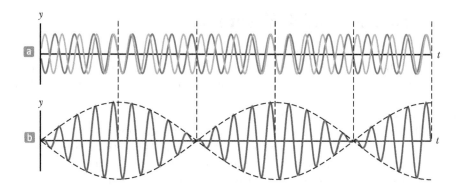

Figure 17.19 Beats are formed by the combination of two waves of slightly different frequencies. (a) The individual waves, shown in blue and green. (b) The combined wave. The envelope wave (dashed line) represents the beating of the combined sounds.

Hence, there are *two* maxima in each period of the envelope wave. Because the amplitude varies with frequency as $(f_1 - f_2)/2$, the number of beats per second, or the **beat frequency** f_{beat}, is twice this value. That is,

$$f_{\text{beat}} = |f_1 - f_2| \qquad (17.13)$$ ◀ Beat frequency

For instance, if one tuning fork vibrates at 438 Hz and a second one vibrates at 442 Hz, the resultant sound wave of the combination has a frequency of 440 Hz (the musical note A) and a beat frequency of 4 Hz. A listener would hear a 440-Hz sound wave go through an intensity maximum four times every second.

Example 17.7 **The Mistuned Piano Strings**

Two identical piano strings of length 0.750 m are each tuned exactly to 440 Hz. The tension in one of the strings is then increased by 1.0%. If they are now struck, what is the beat frequency between the fundamentals of the two strings?

SOLUTION

Conceptualize As the tension in one of the strings is changed, its fundamental frequency changes. Therefore, when both strings are played, they will have different frequencies and beats will be heard.

Categorize We must combine our understanding of the *waves under boundary conditions* model for strings with our new knowledge of beats.

Analyze Set up a ratio of the fundamental frequencies of the two strings using Equation 17.6 with $n = 1$:	$\dfrac{f_2}{f_1} = \dfrac{(v_2/2L)}{(v_1/2L)} = \dfrac{v_2}{v_1}$
Use Equation 16.18 to substitute for the wave speeds on the strings:	$\dfrac{f_2}{f_1} = \dfrac{\sqrt{T_2/\mu}}{\sqrt{T_1/\mu}} = \sqrt{\dfrac{T_2}{T_1}}$
Incorporate that the tension in one string is 1.0% larger than the other; that is, $T_2 = 1.010 T_1$:	$\dfrac{f_2}{f_1} = \sqrt{\dfrac{1.010 T_1}{T_1}} = 1.005$
Solve for the frequency of the tightened string:	$f_2 = 1.005 f_1 = 1.005(440 \text{ Hz}) = 442 \text{ Hz}$
Find the beat frequency using Equation 17.13:	$f_{\text{beat}} = 442 \text{ Hz} - 440 \text{ Hz} = 2 \text{ Hz}$

Finalize Notice that a 1.0% mistuning in tension leads to an easily audible beat frequency of 2 Hz. A piano tuner can use beats to tune a stringed instrument by "beating" a note against a reference tone of known frequency. The tuner can then adjust the string tension until the frequency of the sound it emits equals the frequency of the reference tone. The tuner does so by tightening or loosening the string until the beats produced by it and the reference source become too infrequent to notice.

17.8 Nonsinusoidal Waveforms

It is relatively easy to distinguish the sounds coming from a violin and a saxophone even when they are both playing the same note. On the other hand, a person untrained in music may have difficulty distinguishing a note played on a clarinet from the same note played on an oboe. We can use the pattern of the sound waves from various sources to explain these effects.

Recall that when a system under boundary conditions vibrates, it does so with a combination of frequencies occurring simultaneously. When those frequencies are integer multiples of a fundamental frequency, such as from a string or an air column, the result is a *musical* sound. A listener can assign a pitch to the sound based on the fundamental frequency. Pitch is a psychological reaction to a sound that allows the listener to place the sound on a scale from low to high (bass to treble). Combinations of frequencies that are not integer multiples of a fundamental, such as from a drumhead, result in a *noise* rather than a musical sound. It is much harder for a listener to assign a pitch to a noise than to a musical sound.

The wave patterns produced by a musical instrument are the result of the superposition of frequencies that are integer multiples of a fundamental. This superposition results in the corresponding richness of musical tones. The human perceptive response associated with various mixtures of harmonics is the *quality* or *timbre* of the sound. For instance, the sound of the trumpet is perceived to have a "brassy" quality (that is, we have learned to associate the adjective *brassy* with that sound); this quality enables us to distinguish the sound of the trumpet from that of the saxophone, whose quality is perceived as "reedy." The clarinet and oboe, however, both contain air columns excited by reeds; because of this similarity, they have similar mixtures of frequencies and it is more difficult for the human ear to distinguish them on the basis of their sound quality.

The sound waveforms produced by the majority of musical instruments are nonsinusoidal. Characteristic waveforms produced by a tuning fork, a flute, and a clarinet, each playing the same note, are shown in Figure 17.20. Each instrument has its own characteristic waveform. Notice, however, that despite the differences in the waveforms, each one is periodic. This point is important for our analysis of these waves. Notice that the frequencies at which the waveforms repeat are the same; the addition of higher harmonics does *not* affect the fundamental frequency of the sound.

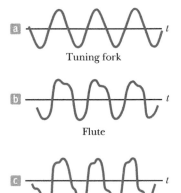

Figure 17.20 Sound waveforms produced by (a) a tuning fork, (b) a flute, and (c) a clarinet, each at approximately the same frequency.

The problem of analyzing nonsinusoidal waveforms appears at first sight to be a formidable task. If the waveform is periodic, however, it can be represented as closely as desired by the combination of a sufficiently large number of sinusoidal waves that form a harmonic series. In fact, we can represent any periodic function as a series of sine and cosine terms by using a mathematical technique based on **Fourier's theorem.**[2] The corresponding sum of terms that represents the periodic waveform is called a **Fourier series.** Let $y(t)$ be any function that is periodic in time with period T such that $y(t + T) = y(t)$. Fourier's theorem states that this function can be written as

Fourier's theorem ▶

$$y(t) = \sum (A_n \sin 2\pi f_n t + B_n \cos 2\pi f_n t) \qquad (17.14)$$

where the lowest frequency is $f_1 = 1/T$. The higher frequencies are integer multiples of the fundamental, $f_n = nf_1$, and the coefficients A_n and B_n represent the amplitudes of the various harmonics. Figure 17.21 represents a harmonic analysis of the waveforms shown in Figure 17.20. Each bar in the graph represents one of the terms in the series in Equation 17.14 up to $n = 9$. Notice that a struck tuning fork produces only one harmonic (the first), so that all coefficients except for A_1 are zero in Equation 17.14, and the waveform is a pure sine wave. On the other hand, the flute and clarinet produce the first harmonic and many higher ones.

---

[2]Developed by Jean Baptiste Joseph Fourier (1768–1830), a French physicist and mathematician.

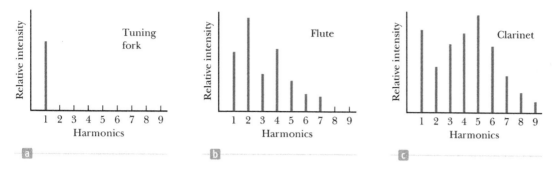

Figure 17.21 Harmonics of the waveforms shown in Figure 17.20. Notice the variations in intensity of the various harmonics. Parts (a), (b), and (c) correspond to those in Figure 17.20.

Notice the variation in relative intensity of the various harmonics for the flute and the clarinet. In general, any musical sound consists of a fundamental frequency f plus other frequencies that are integer multiples of f, all having different intensities.

We have discussed the *analysis* of a waveform using Fourier's theorem. The analysis involves determining the coefficients of the harmonics in Equation 17.14 from a knowledge of the waveform. The reverse process, called *Fourier synthesis,* can also be performed. In this process, various harmonics are added together to form a resultant waveform. As an example of Fourier synthesis, consider the building of a square wave as shown in Figure 17.22. The symmetry of the square wave results in only odd multiples of the fundamental frequency combining in its synthesis. In Figure 17.22a, the blue curve shows the combination of f and $3f$, shown as black curves. In Figure 17.22b, we have added $5f$ to the combination and obtained the green curve. Notice how the general shape of the square wave is approximated, even though the upper and lower portions are not flat as they should be.

Figure 17.22c shows the result of adding odd frequencies up to $9f$. This approximation (red-brown curve) to the square wave is better than the approximations in Figures 17.22a and 17.22b. To approximate the square wave as closely as possible, we must add all odd multiples of the fundamental frequency, up to infinite frequency.

Using modern technology, musical sounds can be generated electronically by mixing different amplitudes of any number of harmonics. These widely used electronic music synthesizers are capable of producing an infinite variety of musical tones.

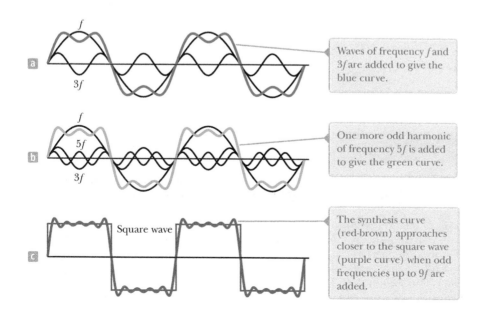

Waves of frequency f and $3f$ are added to give the blue curve.

One more odd harmonic of frequency $5f$ is added to give the green curve.

The synthesis curve (red-brown) approaches closer to the square wave (purple curve) when odd frequencies up to $9f$ are added.

Figure 17.22 Fourier synthesis of a square wave, represented by the sum of odd multiples of the first harmonic, which has frequency f.

Summary

▶ Concepts and Principles

The **superposition principle** specifies that when two or more waves move through a medium, the value of the resultant wave function equals the algebraic sum of the values of the individual wave functions.

The phenomenon of **beating** is the periodic variation in intensity at a given point due to the superposition of two waves having slightly different frequencies. The **beat frequency** is

$$f_{\text{beat}} = |f_1 - f_2| \tag{17.13}$$

where f_1 and f_2 are the frequencies of the individual waves.

Standing waves are formed from the combination of two sinusoidal waves having the same frequency, amplitude, and wavelength but traveling in opposite directions. The resultant standing wave is described by the wave function

$$y = (2A \sin kx) \cos \omega t \tag{17.1}$$

Hence, the amplitude of the standing wave is $2A$, and the amplitude of the simple harmonic motion of any element of the medium varies according to its position as $2A \sin kx$. The points of zero amplitude (called **nodes**) occur at $x = n\lambda/2$ ($n = 0, 1, 2, 3, \ldots$). The maximum amplitude points (called **antinodes**) occur at $x = n\lambda/4$ ($n = 1, 3, 5, \ldots$). Adjacent antinodes are separated by a distance $\lambda/2$. Adjacent nodes also are separated by a distance $\lambda/2$.

▶ Analysis Models for Problem Solving

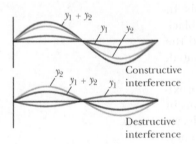

$y_1 + y_2$

y_1 y_2

Constructive interference

y_2 $y_1 + y_2$ y_1

Destructive interference

Waves in Interference. When two traveling waves having equal frequencies superimpose, the resultant wave is described by the **principle of superposition** and has an amplitude that depends on the phase angle ϕ between the two waves. **Constructive interference** occurs when the two waves are in phase, corresponding to $\phi = 0, 2\pi, 4\pi, \ldots$ rad. **Destructive interference** occurs when the two waves are 180° out of phase, corresponding to $\phi = \pi, 3\pi, 5\pi, \ldots$ rad.

Waves Under Boundary Conditions. When a wave is subject to boundary conditions, only certain natural frequencies are allowed; we say that the frequencies are *quantized*.

For waves on a string fixed at both ends, the natural frequencies are

$n = 1$

$n = 2$

$n = 3$

$$f_n = \frac{n}{2L} \sqrt{\frac{T}{\mu}} \quad n = 1, 2, 3, \ldots \tag{17.7}$$

where T is the tension in the string and μ is its linear mass density.

For sound waves with speed v in an air column of length L open at both ends, the natural frequencies are

$$f_n = n \frac{v}{2L} \quad n = 1, 2, 3, \ldots \tag{17.9}$$

If an air column is open at one end and closed at the other, only odd harmonics are present and the natural frequencies are

$$f_m = m \frac{v}{4L} \quad m = 1, 3, 5, \ldots \quad \text{or} \quad f_n = (2n - 1) \frac{v}{4L} \quad n = 1, 2, 3, \ldots \tag{17.10}$$

Think–Pair–Share

See the Preface for an explanation of the icons used in this problems set. For additional assessment items for this section, go to ⚡ **WEBASSIGN** From Cengage

1. **ACTIVITY** You and your friends decide to found a small start-up business in your garage. Your business will design and build acoustic guitars with steel strings. You perform research online and decide on the particular strings you will use. The table shows data from your research for the six strings that will be used on your guitars.

Open String Note	Fundamental Frequency (Hz)	String weight/unit length (10^{-5} lb/in)
e'	329.6	2.000
b	246.9	2.930
g	196.0	5.870
d	146.8	9.180
A	110.0	14.70
E	82.41	32.20

The system of naming notes used in the first column of the table is such that middle C (as played on a piano) is notated with lowercase c′. The notes in the octave above middle C use this lower case/primed notation. The C below middle C (and the notes in the octave starting with this C) is notated with a simple lower case c. The next C down (and the notes in the octave starting with this C) is notated with a capital letter C. Therefore, the E-note of the lowest string on the guitar is two octaves below the e′-note of the highest string.

You design your guitars so that the scale length of the string (the length of the vibrating portion of the string) is 25.50 in for all six strings. (a) After selecting the strings that will be used, your team needs to choose the particular wood that will be used in the guitar, and then design the thickness of the wood on the front face of the guitar. This choice and this design will depend on the total tension exerted by the strings on the front face. What is the total tension in all six strings? (b) Another part of your design relates to the wave speed in the strings. For your design, the wave speed in the e′-string should be about four times that in the E-string. Does the data above satisfy this design criterion? (c) Would any guitar designed like this one *not* satisfy the design criteria in part (b)?

2. ACTIVITY Set up four identical glass bottles filled with increasing levels of water from left to right. Have part of your group strike the bottles with a spoon from left to right and listen to how the frequencies change. Now have the other part blow into the top of each bottle from left to right and listen to how the frequencies change. Why do the frequencies change in opposite directions for these two experiments? What's vibrating in each case?

3. ACTIVITY Cost-free signal generator or function generator apps are available for download to your smartphone. Have two members of your group download an app that has the capability of performing a frequency sweep. Set one of the phones to play a continuous sine wave of frequency 4 000 Hz. Set the other to begin a downward sine wave sweep from 3 800 Hz to 3 000 Hz. Start both phones playing at the same time. Have each member of the group listen carefully to the combined sound. While you will clearly hear the frequency of the second phone going *down*, some members of the group may also hear a sound going *up* in frequency. It will be different in nature; it will sound as if it is coming from inside your ear rather than from the smartphones. Rotate your head back and forth, which may help you hear it. What is causing this sound?

Problems

See the Preface for an explanation of the icons used in this problems set. For additional assessment items for this section, go to ✲ WEBASSIGN From Cengage

Note: Unless otherwise specified, assume the speed of sound in air is 343 m/s, its value at an air temperature of 20.0°C. At any other Celsius temperature T_C, the speed of sound in air is described by

$$v = 331 \sqrt{1 + \frac{T_C}{273}}$$

where v is in m/s and T is in °C.

SECTION 17.1 Analysis Model: Waves in Interference

1. Two waves on one string are described by the wave functions

$$y_1 = 3.0 \cos (4.0x - 1.6t) \qquad y_2 = 4.0 \sin (5.0x - 2.0t)$$

where x and y are in centimeters and t is in seconds. Find the values of $y_1 + y_2$ at the points (a) $x = 1.00$, $t = 1.00$; (b) $x = 1.00$, $t = 0.500$; and (c) $x = 0.500$, $t = 0$. *Note:* Remember that the arguments of the trigonometric functions are in radians.

2. Two pulses of different amplitudes approach each other, each having a speed of $v = 1.00$ m/s. Figure P17.2 shows the

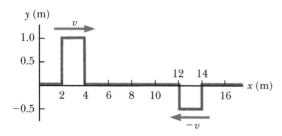

Figure P17.2

positions of the pulses at time $t = 0$. (a) Sketch the resultant wave at $t = 2.00$ s, 4.00 s, 5.00 s, and 6.00 s. (b) **What If?** If the pulse on the right is inverted so that it is upright, how would your sketches of the resultant wave change?

3. Two wave pulses A and B are moving in opposite directions, each with a speed $v = 2.00$ cm/s. The amplitude of A is twice the amplitude of B. The pulses are shown in Figure P17.3 at $t = 0$. Sketch the resultant wave at $t = 1.00$ s, 1.50 s, 2.00 s, 2.50 s, and 3.00 s.

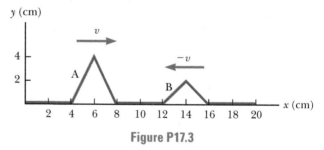

Figure P17.3

4. *Why is the following situation impossible?* Two identical loudspeakers are driven by the same oscillator at frequency 200 Hz. They are located on the ground a distance $d = 4.00$ m from each other. Starting far from the speakers, a man walks straight toward the right-hand speaker as shown in Figure P17.4. After passing through three minima in sound intensity, he walks to the next maximum and stops. Ignore any sound reflection from the ground.

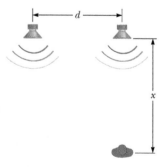

Figure P17.4

5. Two pulses traveling on the same string are described by

$$y_1 = \frac{5}{(3x - 4t)^2 + 2} \qquad y_2 = \frac{-5}{(3x + 4t - 6)^2 + 2}$$

(a) In which direction does each pulse travel? (b) At what instant do the two pulses cancel for all x? (c) At what point do the two pulses cancel at all times t?

6. Two identical loudspeakers 10.0 m apart are driven by the same oscillator with a frequency of $f = 21.5$ Hz (Fig. P17.6) in an area where the speed of sound is 344 m/s. (a) Show that a receiver at point A records a minimum in sound intensity from the two speakers. (b) If the receiver is moved in the plane of the speakers, show that the path it should take so that the intensity remains at a minimum is along the hyperbola $9x^2 - 16y^2 = 144$ (shown in red-brown in Fig. P17.6). (c) Can the receiver remain at a minimum and move very far away from the two sources? If so, determine the limiting form of the path it must take. If not, explain how far it can go.

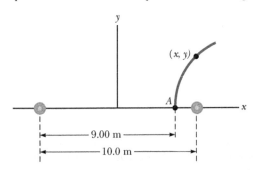

Figure P17.6

7. Two sinusoidal waves on a string are defined by the wave functions

$$y_1 = 2.00 \sin (20.0x - 32.0t) \qquad y_2 = 2.00 \sin (25.0x - 40.0t)$$

where x, y_1, and y_2 are in centimeters and t is in seconds. (a) What is the phase difference between these two waves at the point $x = 5.00$ cm at $t = 2.00$ s? (b) What is the positive x value closest to the origin for which the two phases differ by $\pm\pi$ at $t = 2.00$ s? (At that location, the two waves add to zero.)

SECTION 17.2 **Standing Waves**

8. Verify by direct substitution that the wave function for a standing wave given in Equation 17.1,

$$y = (2A \sin kx) \cos \omega t$$

is a solution of the general linear wave equation, Equation 16.27:

$$\frac{\partial^2 y}{\partial x^2} = \frac{1}{v^2} \frac{\partial^2 y}{\partial t^2}$$

9. Two waves simultaneously present on a long string have a phase difference ϕ between them so that a standing wave formed from their combination is described by

$$y(x, t) = 2A \sin\left(kx + \frac{\phi}{2}\right) \cos\left(\omega t - \frac{\phi}{2}\right)$$

(a) Despite the presence of the phase angle ϕ, is it still true that the nodes are one-half wavelength apart? Explain. (b) Are the nodes different in any way from the way they would be if ϕ were zero? Explain.

10. A standing wave is described by the wave function

$$y = 6 \sin\left(\frac{\pi}{2} x\right) \cos (100\pi t)$$

where x and y are in meters and t is in seconds. (a) Prepare graphs showing y as a function of x for five instants: $t = 0$, 5 ms, 10 ms, 15 ms, and 20 ms. (b) From the graph, identify the wavelength of the wave and explain how to do so. (c) From the graph, identify the frequency of the wave and explain how to do so. (d) From the equation, directly identify the wavelength of the wave and explain how to do so. (e) From the equation, directly identify the frequency and explain how to do so.

SECTION 17.4 **Analysis Model: Waves Under Boundary Conditions**

11. A standing wave is established in a 120-cm-long string fixed at both ends. The string vibrates in four segments when driven at 120 Hz. (a) Determine the wavelength. (b) What is the fundamental frequency of the string?

12. A taut string has a length of 2.60 m and is fixed at both ends. (a) Find the wavelength of the fundamental mode of vibration of the string. (b) Can you find the frequency of this mode? Explain why or why not.

13. A string that is 30.0 cm long and has a mass per unit length of 9.00×10^{-3} kg/m is stretched to a tension of 20.0 N. Find (a) the fundamental frequency and (b) the next three frequencies of possible standing-wave patterns on the string.

14. In the arrangement shown in Figure P17.14, an object of mass $m = 5.00$ kg hangs from a cord around a light pulley. The length of the cord between point P and the pulley is $L = 2.00$ m. (a) When the vibrator is set to a frequency of 150 Hz, a standing wave with six loops is formed. What must be the linear mass density of the cord? (b) How many loops (if any) will result if m is changed to 45.0 kg? (c) How many loops (if any) will result if m is changed to 10.0 kg?

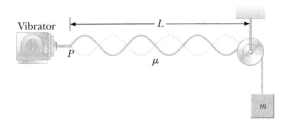

Figure P17.14

15. **Review.** A sphere of mass $M = 1.00$ kg is supported by a string that passes over a pulley at the end of a horizontal rod of length $L = 0.300$ m (Fig. P17.15). The string makes an angle $\theta = 35.0°$ with the rod. The fundamental frequency of standing waves in the portion of the string above the rod is $f = 60.0$ Hz. Find the mass of the portion of the string above the rod.

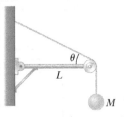

Figure P17.15
Problems 15 and 16.

16. **Review.** A sphere of mass M is supported by a string that passes over a pulley at the end of a horizontal rod of length L (Fig. P17.15). The string makes an angle θ with the rod. The fundamental frequency of standing waves in the portion of the string above the rod is f. Find the mass of the portion of the string above the rod.

17. A violin string has a length of 0.350 m and is tuned to concert G, with $f_G = 392$ Hz. (a) How far from the end of the string must the violinist place her finger to play concert A, with $f_A = 440$ Hz? (b) If this position is to remain correct to one-half the width of a finger (that is, to within 0.600 cm), what is the maximum allowable percentage change in the string tension?

18. **Review.** A solid copper object hangs at the bottom of a steel wire of negligible mass. The top end of the wire is fixed. When the wire is struck, it emits sound with a fundamental frequency of 300 Hz. The copper object is then submerged in water so that half its volume is below the water line. Determine the new fundamental frequency.

SECTION 17.5 Resonance

19. The Bay of Fundy, Nova Scotia, has the highest tides in the world. Assume in midocean and at the mouth of the bay the Moon's gravity gradient and the Earth's rotation make the water surface oscillate with an amplitude of a few centimeters and a period of 12 h 24 min. At the head of the bay, the amplitude is several meters. Assume the bay has a length of 210 km and a uniform depth of 36.1 m. The speed of long-wavelength water waves is given by $v = \sqrt{gd}$, where d is the water's depth. Argue for or against the proposition that the tide is magnified by standing-wave resonance.

SECTION 17.6 Standing Waves in Air Columns

20. The windpipe of one typical whooping crane is 5.00 feet long. What is the fundamental resonant frequency of the bird's trachea, modeled as a narrow pipe closed at one end? Assume a temperature of 37°C.

21. The fundamental frequency of an open organ pipe corresponds to middle C (261.6 Hz on the chromatic musical scale). The third resonance of a closed organ pipe has the same frequency. What is the length of (a) the open pipe and (b) the closed pipe?

22. Ever since seeing Figure 16.22 in the previous chapter, you have been fascinated with the hearing response in humans. You have set up an apparatus that allows you to determine your own threshold of hearing as a function of frequency. After performing the experiment and recording the results, you graph the results, which look like Figure P17.22. You are

intrigued by the two dips in the curve at the right-hand side of the graph. You measure carefully and find that the minimum values of these dips occur at 3 800 Hz and 11 500 Hz. Performing some online research, you discover that the outer canal of the human ear can be modeled as an air column open at the outer end and closed at the inner end by the eardrum. You use this information to determine the length of the outer canal in your ear.

23. An air column in a glass tube is open at one end and closed at the other by a movable piston. The air in the tube is warmed above room temperature, and a 384-Hz tuning fork is held at the open end. Resonance is heard when the piston is at a distance $d_1 = 22.8$ cm from the open end and again when it is at a distance $d_2 = 68.3$ cm from the open end. (a) What speed of sound is implied by these data? (b) How far from the open end will the piston be when the next resonance is heard?

24. A shower stall has dimensions 86.0 cm × 86.0 cm × 210 cm. Assume the stall acts as a pipe closed at both ends, with nodes at opposite sides. Assume singing voices range from 130 Hz to 2 000 Hz and let the speed of sound in the hot air be 355 m/s. For someone singing in this shower, which frequencies would sound the richest (because of resonance)?

25. A glass tube (open at both ends) of length L is positioned near an audio speaker of frequency $f = 680$ Hz. For what values of L will the tube resonate with the speaker?

26. A tunnel under a river is 2.00 km long. (a) At what frequencies can the air in the tunnel resonate? (b) Explain whether it would be good to make a rule against blowing your car horn when you are in the tunnel.

27. As shown in Figure P17.27, water is pumped into a tall, vertical cylinder at a volume flow rate $R = 1.00$ L/min. The radius of the cylinder is $r = 5.00$ cm, and at the open top of the cylinder a tuning fork is vibrating with a frequency $f = 512$ Hz. As the water rises, what time interval elapses between successive resonances?

28. As shown in Figure P17.27, water is pumped into a tall, vertical cylinder at a volume flow rate R. The radius of the cylinder is r, and at the open top of the cylinder a tuning fork is vibrating with a frequency f. As the water rises, what time interval elapses between successive resonances?

Figure P17.27
Problems 27 and 28.

29. You are a flutist in a local orchestra. On a cold winter day, you are late to a performance. Arriving at the orchestra hall, you know that you have missed the group tune-up before the performance, so you tune your instrument in the cold air outside the stage door. After tuning, you run inside the auditorium, where the temperature is 22.2°C, take your seat, and begin playing the first song with the rest of the orchestra. You are quite embarrassed to notice that you are playing the song a half-step higher than your colleagues in the orchestra. Your excitement about physics overcomes your musical embarrassment as you realize that you can use this information to calculate the temperature outside. (Assume that the length of the instrument does not change with temperature. A half-step represents a frequency ratio of $2^{1/12}$.)

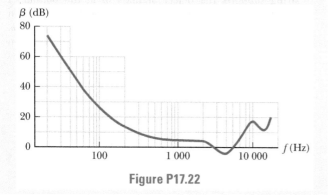

β (dB)

Figure P17.22

30. *Why is the following situation impossible?* A student is listening to the sounds from an air column that is 0.730 m long. He doesn't know if the column is open at both ends or open at only one end. He hears resonance from the air column at frequencies 235 Hz and 587 Hz.

SECTION 17.7 Beats: Interference in Time

31. **Review.** A student holds a tuning fork oscillating at 256 Hz. He walks toward a wall at a constant speed of 1.33 m/s. (a) What beat frequency does he observe between the tuning fork and its echo? (b) How fast must he walk away from the wall to observe a beat frequency of 5.00 Hz?

32. While attempting to tune the note C at 523 Hz, a piano tuner hears 2.00 beats/s between a reference oscillator and the string. (a) What are the possible frequencies of the string? (b) When she tightens the string slightly, she hears 3.00 beats/s. What is the frequency of the string now? (c) By what percentage should the piano tuner now change the tension in the string to bring it into tune?

SECTION 17.8 Nonsinusoidal Waveforms

33. Suppose a flutist plays a 523-Hz C note with first harmonic displacement amplitude A_1 = 100 nm. From Figure 17.21b read, by proportion, the displacement amplitudes of harmonics 2 through 7. Take these as the values A_2 through A_7 in the Fourier analysis of the sound and assume $B_1 = B_2 = \cdots = B_7 = 0$. Construct a graph of the waveform of the sound. Your waveform will not look exactly like the flute waveform in Figure 17.20b because you simplify by ignoring cosine terms; nevertheless, it produces the same sensation to human hearing.

ADDITIONAL PROBLEMS

34. Two strings are vibrating at the same frequency of 150 Hz. After the tension in one of the strings is decreased, an observer hears four beats each second when the strings vibrate together. Find the new frequency in the adjusted string.

35. The ship in Figure P17.35 travels along a straight line parallel to the shore and a distance d = 600 m from it. The ship's radio receives simultaneous signals of the same frequency from antennas A and B, separated by a distance L = 800 m. The signals interfere constructively at point C, which is equidistant from A and B. The signal goes through the first minimum at point D, which is directly outward from the shore from point B. Determine the wavelength of the radio waves.

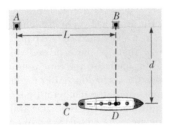

Figure P17.35

36. A 2.00-m-long wire having a mass of 0.100 kg is fixed at both ends. The tension in the wire is maintained at 20.0 N. (a) What are the frequencies of the first three allowed modes of vibration? (b) If a node is observed at a point 0.400 m from one end, in what mode and with what frequency is it vibrating?

37. A string fixed at both ends and having a mass of 4.80 g, a length of 2.00 m, and a tension of 48.0 N vibrates in its second (n = 2) normal mode. (a) Is the wavelength in air of the sound emitted by this vibrating string larger or smaller than the wavelength of the wave on the string? (b) What is the ratio of the wavelength in air of the sound emitted by this vibrating string and the wavelength of the wave on the string?

38. You are working as an assistant to a landscape architect, who is designing the landscaping around a new commercial building. The architect plans to have a large rectangular water basin as part of his design. When you see this design, you mention to the architect that the project is located in an area prone to earthquakes. You point out that an earthquake could create a seiche in the basin by resonance, causing the water in the basin to spill out and enter nearby underground electrical transformers. A *seiche* is a standing wave in a body of water, in which the water sloshes back and forth with antinodes at the ends of the basin. (You may have created a seiche in a bathtub as a child by sliding your body back and forth along the length of the tub, leaving water on the floor for your parents to wipe up.) The architect dismisses your comments as unrealistic. While visiting your cousin the previous week in a non-earthquake-prone area, you had seen a water basin similar to the one planned by the architect. You call your cousin and find out that the water basin in his town has the same depth of water as that planned by the architect. You ask your cousin to create a pulse in the water by dropping a pebble, and determine how long the pulse takes to cross the basin. Based on this time interval and the length of your cousin's basin, you determine that a pulse will take 2.50 s to cross the basin planned by the architect. Show the architect that there will be several possible seiche resonances in the water basin for typical low frequencies of earthquakes in the range of 0–4 Hz.

39. **Review.** Consider the apparatus shown in Figure 17.15 and described in Example 17.4. Suppose the number of antinodes in Figure 17.15b is an arbitrary value n. (a) Find an expression for the radius of the sphere in the water as a function of only n. (b) What is the minimum allowed value of n for a sphere of nonzero size? (c) What is the radius of the largest sphere that will produce a standing wave on the string? (d) What happens if a larger sphere is used?

40. **Review.** For the arrangement shown in Figure P17.40, the inclined plane and the small pulley are frictionless; the string supports the object of mass M at the bottom of the plane; and the entire string has mass m. The system is in equilibrium, and the vertical part of the string has a length h. We wish to study standing waves set up in the vertical section

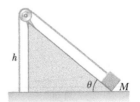

Figure P17.40

of the string. (a) What analysis model describes the object of mass M? (b) What analysis model describes the waves on the vertical part of the string? (c) Find the tension in the string. (d) Model the shape of the string as one leg and the hypotenuse of a right triangle. Find the whole length of the string. (e) Find the mass per unit length of the string. (f) Find the speed of waves on the string. (g) Find the lowest frequency for a standing wave on the vertical section of the string. (h) Evaluate this result for $M = 1.50$ kg, $m = 0.750$ g, $h = 0.500$ m, and $\theta = 30.0°$. (i) Find the numerical value for the lowest frequency for a standing wave on the sloped section of the string.

41. **Review.** A loudspeaker at the front of a room and an identical loudspeaker at the rear of the room are being driven by the same oscillator at 456 Hz. A student walks at a uniform rate of 1.50 m/s along the length of the room. She hears a single tone repeatedly becoming louder and softer. (a) Model these variations as beats between the Doppler-shifted sounds the student receives. Calculate the number of beats the student hears each second. (b) Model the two speakers as producing a standing wave in the room and the student as walking between antinodes. Calculate the number of intensity maxima the student hears each second.

42. Two speakers are driven by the same oscillator of frequency f. They are located a distance d from each other on a vertical pole. A man walks straight toward the lower speaker in a direction perpendicular to the pole as shown in Figure P17.42. (a) How many times will he hear a minimum in sound intensity? (b) How far is he from the pole at these moments? Let v represent the speed of sound and assume that the ground does not reflect sound. The man's ears are at the same level as the lower speaker.

Figure P17.42

43. A standing wave is set up in a string of variable length and tension by a vibrator of variable frequency. Both ends of the string are fixed. When the vibrator has a frequency f, in a string of length L and under tension T, n antinodes are set up in the string. (a) If the length of the string is doubled, by what factor should the frequency be changed so that the same number of antinodes is produced? (b) If the frequency and length are held constant, what tension will produce $n + 1$ antinodes? (c) If the frequency is tripled and the length of the string is halved, by what factor should the tension be changed so that twice as many antinodes are produced?

44. **Review.** The top end of a yo-yo string is held stationary. The yo-yo itself is much more massive than the string. It starts from rest and moves down with constant acceleration 0.800 m/s² as it unwinds from the string. The rubbing of the string against the edge of the yo-yo excites transverse standing-wave vibrations in the string. Both ends of the string are nodes even as the length of the string increases. Consider the instant 1.20 s after the motion begins from rest. (a) Show that the rate of change with time of the wavelength of the fundamental mode of oscillation is 1.92 m/s. (b) **What if?** Is the rate of change of the wavelength of the second harmonic also 1.92 m/s at this moment? Explain your answer. (c) **What if?** The experiment is repeated after more mass has been added to the yo-yo body. The mass distribution is kept the same so that the yo-yo still moves with downward acceleration 0.800 m/s². At the 1.20-s point in this case, is the rate of change of the fundamental wavelength of the string vibration still equal to 1.92 m/s? Explain. (d) Is the rate of change of the second harmonic wavelength the same as in part (b)? Explain.

45. **Review.** Consider the copper object hanging from the steel wire in Problem 18. The top end of the wire is fixed. When the wire is struck, it emits sound with a fundamental frequency of 300 Hz. The copper object is then submerged in water. If the object can be positioned with any desired fraction of its volume submerged, what is the lowest possible new fundamental frequency?

46. A string of linear density 1.60 g/m is stretched between clamps 48.0 cm apart. The string does not stretch appreciably as the tension in it is steadily raised from 15.0 N at $t = 0$ to 25.0 N at $t = 3.50$ s. Therefore, the tension as a function of time is given by the expression $T = 15.0 + 10.0t/3.50$, where T is in newtons and t is in seconds. The string is vibrating in its fundamental mode throughout this process. Find the number of oscillations it completes during the 3.50-s interval.

47. **Review.** A 12.0-kg object hangs in equilibrium from a string with a total length of $L = 5.00$ m and a linear mass density of $\mu = 0.001\ 00$ kg/m. The string is wrapped around two light, frictionless pulleys that are separated by a distance of $d = 2.00$ m (Fig. P17.47a). (a) Determine the tension in the string. (b) At what frequency must the string between the pulleys vibrate to form the standing-wave pattern shown in Figure P17.47b?

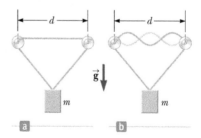

Figure P17.47 Problems 47 and 48.

48. **Review.** An object of mass m hangs in equilibrium from a string with a total length L and a linear mass density μ. The string is wrapped around two light, frictionless pulleys that are separated by a distance d (Fig. P17.47a). (a) Determine the tension in the string. (b) At what frequency must the string between the pulleys vibrate to form the standing-wave pattern shown in Figure P17.47b?

49. Two waves are described by the wave functions

$$y_1(x, t) = 5.00 \sin (2.00x - 10.0t)$$

$$y_2(x, t) = 10.0 \cos (2.00x - 10.0t)$$

where x, y_1, and y_2 are in meters and t is in seconds. (a) Show that the wave resulting from their superposition can be expressed as a single sine function. (b) Determine the amplitude and phase angle for this sinusoidal wave.

CHALLENGE PROBLEM

50. In Figures 17.22a and 17.22b, notice that the amplitude
[S] of the component wave for frequency f is large, that for $3f$ is smaller, and that for $5f$ smaller still. How do we know exactly how much amplitude to assign to each frequency component to build a square wave? This problem helps us find the answer to that question. Let the square wave in Figure 17.22c have an amplitude A and let $t = 0$ be at the extreme left of the figure. So, one period T of the square wave is described by

$$y(t) = \begin{cases} A & 0 < t < \dfrac{T}{2} \\ -A & \dfrac{T}{2} < t < T \end{cases}$$

Express Equation 17.14 with angular frequencies:

$$y(t) = \sum_n (A_n \sin n\omega t + B_n \cos n\omega t)$$

Now proceed as follows. (a) Multiply both sides of Equation 17.14 by $\sin m\omega t$ and integrate both sides over one period T. Show that the left-hand side of the resulting equation is equal to 0 if m is even and is equal to $4A/m\omega$ if m is odd. (b) Using trigonometric identities, show that all terms on the right-hand side involving B_n are equal to zero. (c) Using trigonometric identities, show that all terms on the right-hand side involving A_n are equal to zero *except* for the one case of $m = n$. (d) Show that the entire right-hand side of the equation reduces to $\frac{1}{2}A_m T$. (e) Show that the Fourier series expansion for a square wave is

$$y(t) = \sum_n \frac{4A}{n\pi} \sin n\omega t$$

PART 3

Thermodynamics

We now direct our attention to the study of thermodynamics, which involves situations in which the temperature or state (solid, liquid, gas) of a system changes due to energy transfers. In this part of the book, we will focus on the heat Q in Equation 8.2 and its effects on the thermal conditions of a system. We will also look at work W performed on deformable systems, such as an enclosed gas, as well as electromagnetic radiation T_{ER} across a system boundary. Each of these energy transfers can cause a change in the internal energy E_{int} of the system, which we can relate to *temperature*.

Historically, the development of thermodynamics paralleled the development of the atomic theory of matter. By the 1820s, chemical experiments had provided solid evidence for the existence of atoms. At that time, scientists recognized that a connection between thermodynamics and the structure of matter must exist. In 1827, botanist Robert Brown reported that grains of pollen suspended in a liquid move erratically from one place to another as if under constant agitation. In 1905, Albert Einstein used kinetic theory to explain the cause of this erratic motion, known today as *Brownian motion*. Einstein explained this phenomenon by assuming the grains are under constant bombardment by "invisible" molecules in the liquid, which themselves move erratically. The motion of the molecules is related to the temperature of the liquid. A connection was thus forged between the everyday world and the tiny, invisible building blocks that make up this world.

Thermodynamics also addresses more practical questions. Have you ever wondered how a refrigerator is able to cool its contents, or what types of transformations occur in a power plant or in the engine of your automobile, or what happens to the kinetic energy of a moving object when the object comes to rest? The laws of thermodynamics can be used to provide explanations for these and other phenomena. ∎

A bubble in one of the many mud pots in Yellowstone National Park is caught just at the moment of popping. A mud pot is a pool of bubbling hot mud that demonstrates the existence of thermodynamic processes below the Earth's surface. (*Adambooth/Dreamstime.com*)

481

18

Temperature

A brick sidewalk exhibits buckling. In some cases, this is caused by a *mechanical* phenomenon: the growth of tree roots under the sidewalk. But we see no trees here that are close enough to the sidewalk to cause this effect. This buckling is caused by a *thermal* process, related to a high temperature. (*John W. Jewett, Jr.*)

STORYLINE **You have discovered that you are out of potato chips.** While driving to the store, you look at the high-voltage electric power transmission lines crossing the road ahead of you. You have seen these lines almost every day, but there is something different about them today. The power lines between the towers on either side of the road seem to be sagging lower today than they have in the past. Then you notice that a brick sidewalk on the side of the road has buckled, as shown above. What's causing these effects? Your hometown is experiencing severely high temperatures that have lasted for several days; it has been hotter than you can ever remember. Could this be related to these effects? Will the power lines rise back up when the weather cools? Will the sidewalk "unbuckle" when the temperature drops? What does the lizard checking out the situation from the curb think? Wait a minute! In all these questions, what exactly *is* temperature, what do *hot* and *cool* actually mean? This question haunts your visit to the store and you drive home without purchasing your potato chips.

CONNECTIONS Up to this point in the text, we have focused on *mechanical* situations, which generally involve macroscopic objects. For example, we looked at kinetic energies of cars, billiard balls, planets, and rolling wheels. We performed calculations using potential energies in systems of springs, a ball and the Earth, a planet and the Sun. In this chapter, we begin to investigate *thermal* phenomena. We introduced internal energy in Chapter 7, where we talked about something becoming warmer due to friction. That was our first hint of a thermal process. The hallmark of thermal processes is that they involve energy on a microscopic scale. As we shall see in Chapter 20, we can relate the temperature of an object to the kinetic energy of the molecules of the object. We introduced the energy transfer process of heat in Chapter 8 as a means of transferring energy into or out of a system; in Chapter 19, we will discuss this process in terms of microscopic collisions between molecules at the boundary of the system. To establish the basis for these discussions in the next chapters, we will first embark in this chapter on a macroscopic understanding of the concept of temperature and its effects. The chapter concludes with a study of ideal gases on the macroscopic scale. In this study, we connect the quantity of pressure from Chapter 14 to that of temperature from this chapter. Once we understand thermal phenomena, we will see, for example, thermal effects on electrical resistance in Chapter 26, on magnetic properties of materials in Chapter 29, on the radiation from a hot surface in Chapter 39, and so on.

18.1 Temperature and the Zeroth Law of Thermodynamics

We often associate the concept of temperature with how hot or cold an object feels when we touch it. In this way, our senses provide us with a qualitative indication of temperature. Our senses, however, are unreliable and often mislead us. For example, if you stand in bare feet with one foot on carpet and the other on an adjacent tile floor, the tile feels colder than the carpet *even though both are at the same temperature*. The two objects feel different because tile transfers energy by heat at a higher rate than carpet does. Your skin "measures" the rate of energy transfer by heat rather than the actual temperature. What we need is a reliable and reproducible method for measuring the relative hotness or coldness of objects rather than the rate of energy transfer. Scientists have developed a variety of thermometers for making such quantitative measurements.

Two objects at different initial temperatures eventually reach some intermediate temperature when placed in contact with each other. For example, when hot water and cold water are mixed in a bathtub, energy is transferred from the hot water to the cold water and the final temperature of the mixture is somewhere between the initial hot and cold temperatures.

The energy-transfer mechanisms from Chapter 8 that we will focus on in this current discussion are heat, Q in Eq. 8.2, and electromagnetic radiation, T_{ER}. For purposes of this discussion, let's assume two objects are in **thermal contact** with each other if energy can be exchanged between them by these processes due to a temperature difference. **Thermal equilibrium** is a situation in which two objects would not exchange energy by heat or electromagnetic radiation if they were placed in thermal contact.

Let's consider two objects A and B, which are not in thermal contact, and a third object C, which is our thermometer. We wish to determine whether A and B are in thermal equilibrium with each other. The thermometer (object C) is first placed in thermal contact with object A until thermal equilibrium is reached[1] as shown in Figure 18.1a (page 484). From that moment on, the thermometer's reading remains constant and we record this reading. The thermometer is then removed from object A and placed in thermal contact with object B as shown in Figure 18.1b. The reading is again recorded after thermal equilibrium is reached. If the two readings are the same, we can conclude that object A and object B are in thermal equilibrium with each other. If they are placed in contact with each other as in Figure 18.1c, there will be no exchange of energy between them.

---

[1]We assume a negligible amount of energy transfers between the thermometer and object A in the time interval during which they are in thermal contact. Without this assumption, which is also made for the thermometer and object B, the measurement of the temperature of an object disturbs the system so that the measured temperature is different from the initial temperature of the object. In practice, whenever you measure a temperature with a thermometer, you measure the disturbed system, not the original system.

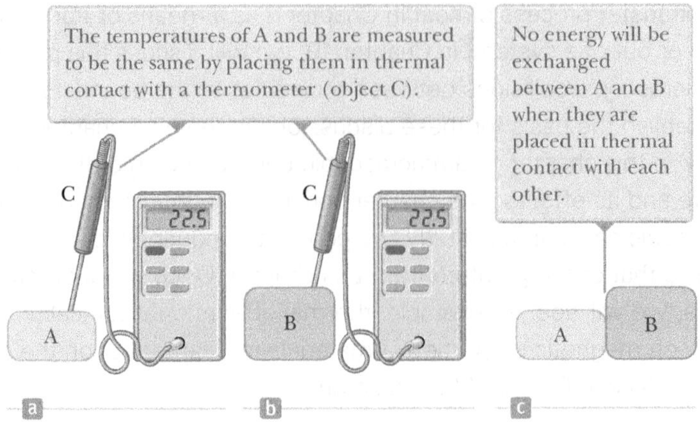

The temperatures of A and B are measured to be the same by placing them in thermal contact with a thermometer (object C).

No energy will be exchanged between A and B when they are placed in thermal contact with each other.

Figure 18.1 The zeroth law of thermodynamics. In general, objects A and B can be of different sizes, different masses, and different materials. The zeroth law allows us to identify something that is the *same* for both objects: temperature.

We can summarize these results in a statement known as the **zeroth law of thermodynamics** (the law of equilibrium):

Zeroth law ▶
of thermodynamics

If objects A and B are separately in thermal equilibrium with a third object C, then A and B are in thermal equilibrium with each other.

This statement can easily be proved experimentally and is very important because *it enables us to define temperature.* We can think of **temperature** as the property that determines whether or not energy will transfer between two objects when they are in thermal contact. Two objects in thermal equilibrium with each other are at the same temperature. Conversely, if two objects have different temperatures, they are not in thermal equilibrium and energy will transfer between them when they are placed in thermal contact. In Figure 18.1, it is *only* the temperatures of A and B that determine whether energy will transfer from one to the other when they are placed in thermal contact—not *size, mass, material, density,* or anything else. For now, temperature is only defined for us in terms of the zeroth law. We will relate temperature to molecular motion in Chapter 20.

QUICK QUIZ 18.1 Two objects, with different sizes, masses, and temperatures, are placed in thermal contact. In which direction does the energy travel? **(a)** Energy travels from the larger object to the smaller object. **(b)** Energy travels from the object with more mass to the one with less mass. **(c)** Energy travels from the object at higher temperature to the object at lower temperature.

18.2 Thermometers and the Celsius Temperature Scale

In Figure 18.1, we used a *thermometer* to measure the temperatures of A and B. All thermometers are based on the principle that some physical property of a system changes as the system's temperature changes. Some physical properties that change with temperature are (1) the volume of a liquid, (2) the dimensions of a solid, (3) the pressure of a gas at constant volume, (4) the volume of a gas at constant pressure, (5) the electric resistance of a conductor, and (6) the color of an object.

A common thermometer in everyday use consists of a mass of liquid—usually mercury or alcohol—that expands into a glass capillary tube when heated (Fig. 18.2). In this case, the physical property that changes is the volume of a liquid. Any temperature change in the range of the thermometer can be defined as being proportional to the change in length of the liquid column. The thermometer can be calibrated by placing it in thermal contact with a natural system that remains

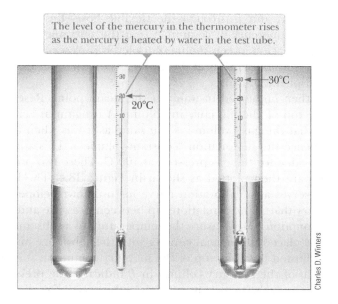

The level of the mercury in the thermometer rises as the mercury is heated by water in the test tube.

20°C

30°C

Figure 18.2 A mercury thermometer before and after increasing its temperature.

at constant temperature. One such system is a mixture of water and ice in thermal equilibrium at atmospheric pressure. On the **Celsius temperature scale,** this mixture is defined to have a temperature of zero degrees Celsius, which is written as 0°C; this temperature is called the *ice point* of water. Another commonly used system is a mixture of water and steam in thermal equilibrium at atmospheric pressure; its temperature is defined as 100°C, which is the *steam point* of water. Once the liquid levels in the thermometer have been established at these two points, the length of the liquid column between the two points is divided into 100 equal segments to create the Celsius scale. Therefore, each segment denotes a change in temperature of one Celsius degree.

Thermometers calibrated in this way present problems when extremely accurate readings are needed. For instance, the readings given by an alcohol thermometer calibrated at the ice and steam points of water might agree with those given by a mercury thermometer only at the calibration points. Because mercury and alcohol have different thermal expansion properties, when one thermometer reads a temperature of, for example, 50°C, the other may indicate a slightly different value. The discrepancies between thermometers are especially large when the temperatures to be measured are far from the calibration points.[2]

An additional practical problem of any thermometer is the limited range of temperatures over which it can be used. A mercury thermometer, for example, cannot be used below the freezing point of mercury, which is −39°C, and an alcohol thermometer is not useful for measuring temperatures above 85°C, the boiling point of alcohol. To surmount this problem, we need a universal thermometer whose readings are independent of the substance used in it. The gas thermometer, discussed in the next section, approaches this requirement.

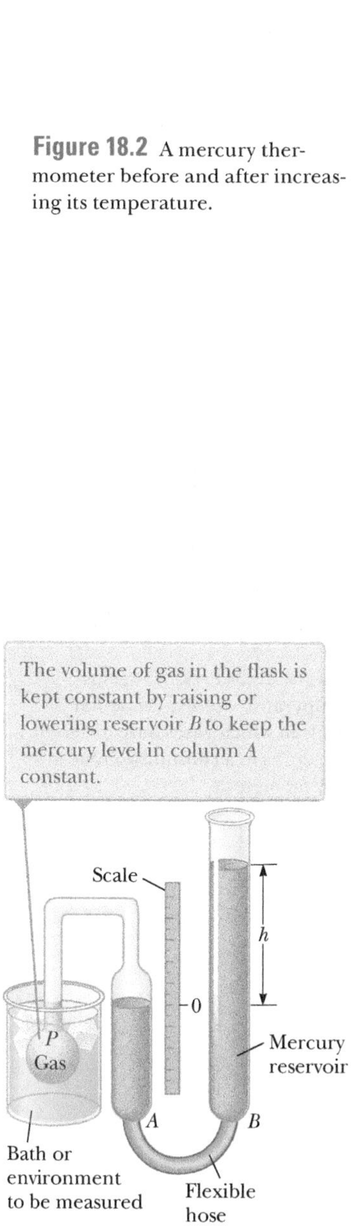

18.3 The Constant-Volume Gas Thermometer and the Absolute Temperature Scale

One version of a gas thermometer is the constant-volume apparatus shown in Figure 18.3. The physical change exploited in this device is the variation of pressure of a fixed volume of gas with temperature. The flask is immersed in an ice-water bath, and mercury reservoir *B* is raised or lowered. This will cause mercury to transfer

The volume of gas in the flask is kept constant by raising or lowering reservoir *B* to keep the mercury level in column *A* constant.

Scale

h

0

P
Gas

A

B

Mercury reservoir

Bath or environment to be measured

Flexible hose

Figure 18.3 A constant-volume gas thermometer measures the pressure of the gas contained in the flask immersed in the bath.

---

[2] Two thermometers that use the same liquid may also give different readings, due in part to difficulties in constructing uniform-bore glass capillary tubes.

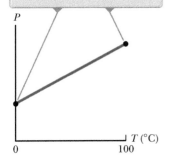

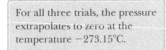

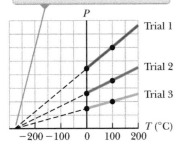

Figure 18.4 A typical graph of pressure versus temperature taken with a constant-volume gas thermometer.

Figure 18.5 Pressure versus temperature for experimental trials in which gases have different pressures in a constant-volume gas thermometer.

between reservoirs A and B through the flexible hose. Reservoir B is adjusted until the top of the mercury in column A is at the zero point on the scale. The height h, the difference between the mercury levels in reservoir B and column A, indicates the pressure in the flask at 0°C by means of Equation 14.4, $P = P_0 + \rho g h$, where P_0 is atmospheric pressure.

The flask is then immersed in water at the steam point. Reservoir B is readjusted until the top of the mercury in column A is again at zero on the scale, which ensures that the gas's volume is the same as it was when the flask was in the ice bath (hence the designation "constant-volume"). This adjustment of reservoir B gives a value for the gas pressure at 100°C. These two pressure and temperature values are then plotted as shown in Figure 18.4. The line connecting the two points serves as a calibration curve for unknown temperatures. (Other experiments show that a linear relationship between pressure and temperature is a very good assumption.) To measure the temperature of a substance, the gas flask of Figure 18.3 is placed in thermal contact with the substance and the height of reservoir B is adjusted until the top of the mercury column in A is at zero on the scale. The height of the mercury column in B indicates the pressure of the gas; knowing the pressure, the temperature of the substance is found using the graph in Figure 18.4.

Now suppose temperatures of different gases at different initial pressures are measured with gas thermometers. Experiments show that the thermometer readings are nearly independent of the type of gas used as long as the gas pressure is low and the temperature is well above the point at which the gas liquefies (Fig. 18.5). The agreement among thermometers using various gases improves as the pressure is reduced.

If we extend the solid-color straight lines in Figure 18.5 toward negative temperatures, we find a remarkable result: **in every case, the pressure is zero when the temperature is −273.15°C!** This finding suggests some special role that this particular temperature must play. It is used as the basis for the **absolute temperature scale**, which sets −273.15°C as its zero point. This temperature is often referred to as **absolute zero.** It is indicated as a zero because at a lower temperature, the pressure of the gas would become negative, which is meaningless. Therefore, absolute zero is a true, naturally defined zero of temperature. The size of one degree on the absolute temperature scale is chosen to be identical to the size of one degree on the Celsius scale. Therefore, the conversion between these temperatures is

$$T_C = T - 273.15 \qquad (18.1)$$

where T_C is the Celsius temperature and T is the absolute temperature.

Because the ice and steam points are experimentally difficult to duplicate and depend on atmospheric pressure, an absolute temperature scale based on two new fixed points was adopted in 1954 by the International Committee on Weights and Measures. The first point is absolute zero, which does not depend on atmospheric pressure or on any particular material. The second reference temperature for this new scale was chosen as the **triple point of water,** which is the single combination of temperature and pressure at which liquid water, gaseous water, and ice (solid water) coexist in equilibrium. This triple point occurs at a temperature of 0.01°C and a pressure of 4.58 mm of mercury. The triple point of water is the same everywhere in the Universe. On the new scale, which uses the unit *kelvin,* the temperature of water at the triple point was set at 273.16 kelvins, abbreviated 273.16 K. This choice was made so that the old absolute temperature scale based on the ice and steam points would agree closely with the new scale based on the triple point. This new **absolute temperature scale** (also called the **Kelvin scale**) employs the SI unit of absolute temperature, the **kelvin,** which is defined to be 1/273.16 of the difference between absolute zero and the temperature of the triple point of water.

Figure 18.6 gives the absolute temperature for various physical processes and structures. The temperature of absolute zero (0 K) cannot be achieved, although laboratory experiments have come very close, reaching temperatures of less than one nanokelvin.

The Celsius, Fahrenheit, and Kelvin Temperature Scales[3]

Equation 18.1 shows that the Celsius temperature T_C is shifted from the absolute (Kelvin) temperature T by 273.15°. Because the size of one degree is the same on the two scales, a temperature difference of 5°C is equal to a temperature difference of 5 K. The two scales differ only in the choice of the zero point. Therefore, the ice-point temperature on the Kelvin scale, 273.15 K, corresponds to 0.00°C, and the Kelvin-scale steam point, 373.15 K, is equivalent to 100.00°C.

A common temperature scale in everyday use in the United States is the **Fahrenheit scale.** This scale sets the temperature of the ice point at 32°F and the temperature of the steam point at 212°F. The relationship between the Celsius and Fahrenheit temperature scales is

$$T_F = \tfrac{9}{5}T_C + 32°F \tag{18.2}$$

We can use Equations 18.1 and 18.2 to find a relationship between changes in temperature on the Celsius, Kelvin, and Fahrenheit scales:

$$\Delta T_C = \Delta T = \tfrac{5}{9}\Delta T_F \tag{18.3}$$

Of these three temperature scales, only the Kelvin scale is based on a true zero value of temperature. The Celsius and Fahrenheit scales are based on an arbitrary zero associated with one particular substance, water, on one particular planet, the Earth. Therefore, if you encounter an equation that calls for a temperature T or that involves a ratio of temperatures, you *must* convert all temperatures to kelvins. If the equation contains a change in temperature ΔT, using Celsius temperatures will give you the correct answer, in light of Equation 18.3, but it is always *safest* to convert temperatures to the Kelvin scale.

QUICK QUIZ 18.2 Consider the following pairs of materials. Which pair represents two materials, one of which is twice as hot as the other? **(a)** boiling water at 100°C, a glass of water at 50°C **(b)** boiling water at 100°C, frozen methane at −50°C **(c)** an ice cube at −20°C, flames from a circus fire-eater at 233°C **(d)** none of those pairs

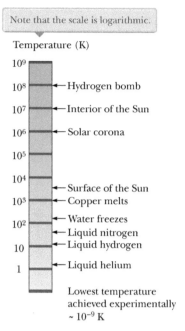

Note that the scale is logarithmic.

Temperature (K)

- 10^9
- 10^8 ← Hydrogen bomb
- 10^7 ← Interior of the Sun
- 10^6 ← Solar corona
- 10^5
- 10^4
- 10^3 ← Surface of the Sun / Copper melts
- 10^2 ← Water freezes / Liquid nitrogen / Liquid hydrogen
- 10 ← Liquid helium
- 1

Lowest temperature achieved experimentally ∼ 10^{-9} K

Figure 18.6 Absolute temperatures at which various physical processes occur.

Example 18.1 Converting Temperatures

On a day when the temperature reaches 50°F, what is the temperature in degrees Celsius and in kelvins?

SOLUTION

Conceptualize In the United States, a temperature of 50°F is well understood. In many other parts of the world, however, this temperature might be meaningless because people are familiar with the Celsius temperature scale.

Categorize This example is a simple substitution problem.

Solve Equation 18.2 for the Celsius temperature and substitute numerical values:

$$T_C = \tfrac{5}{9}(T_F - 32) = \tfrac{5}{9}(50 - 32) = 10°C$$

Use Equation 18.1 to find the Kelvin temperature:

$$T = T_C + 273.15 = 10°C + 273.15 = 283\ K$$

A convenient set of weather-related temperature equivalents to keep in mind is that 0°C is (literally) freezing at 32°F, 10°C is cool at 50°F, 20°C is room temperature, 30°C is warm at 86°F, and 40°C is a hot day at 104°F.

---

[3]Named after Anders Celsius (1701–1744), Daniel Gabriel Fahrenheit (1686–1736), and William Thomson, Lord Kelvin (1824–1907), respectively.

18.4 Thermal Expansion of Solids and Liquids

Our discussion of the liquid thermometer makes use of one of the best-known thermal changes in a substance: as its temperature increases, its volume increases. This phenomenon, known as **thermal expansion,** plays an important role in numerous engineering applications. For example, thermal-expansion joints such as those shown in Figure 18.7 must be included in buildings, concrete highways, railroad tracks, brick walls, and bridges to compensate for dimensional changes that occur as the temperature changes.

Thermal expansion is responsible for the effects you saw in the opening storyline. On a hot day the power lines expand. The distance between the ends of the lines is fixed at the positions of the poles. Therefore, when the power line lengthens, it sags downward from its fixed ends. The brick sidewalk in the chapter opening photograph was likely installed with no expansion joints. As the temperature rises, the expansion of the sections of sidewalk causes them to buckle upward.

Thermal expansion is a consequence of the change in the *average* separation between the atoms in an object. To understand this concept, let's model the atoms as being connected by stiff springs as discussed in Section 15.3 and shown in Figure 15.11b. At ordinary temperatures, the atoms in a solid oscillate about their equilibrium positions with an amplitude of approximately 10^{-11} m and a frequency of approximately 10^{13} Hz. The average spacing between the atoms is about 10^{-10} m. As the temperature of the solid increases, the atoms oscillate with greater amplitudes; as a result, the average separation between them increases.[4] Consequently, the object expands.

If thermal expansion is sufficiently small relative to an object's initial dimensions, the change in any dimension is, to a good approximation, proportional to the first power of the temperature change. Suppose an object has an initial length L_i along some direction at some temperature and the length changes by an amount ΔL for a change in temperature ΔT. Because it is convenient to consider the fractional change in length per degree of temperature change, we define the **average coefficient of linear expansion** as

$$\alpha \equiv \frac{\Delta L / L_i}{\Delta T} \tag{18.4}$$

Without these joints to separate sections of roadway on bridges, the surface would buckle due to thermal expansion on very hot days or crack due to contraction on very cold days.

The long, vertical joint is filled with a soft material that allows the wall to expand and contract as the temperature of the bricks changes.

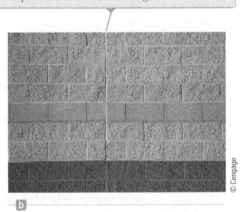

© Cengage

© Cengage

Figure 18.7 Thermal-expansion joints in (a) bridges and (b) walls.

---

[4]More precisely, thermal expansion arises from the *asymmetrical* nature of the potential energy curve for the atoms in a solid as shown in Figure 15.11a. If the oscillators were truly harmonic, the average atomic separations would not change regardless of the amplitude of vibration.

Experiments show that α is constant for small changes in temperature. For purposes of calculation, this equation is usually rewritten as

$$\Delta L = \alpha L_i \Delta T \qquad (18.5)$$

or as

$$L_f - L_i = \alpha L_i (T_f - T_i) \qquad (18.6)$$

◀ Thermal expansion in one dimension

where L_f is the final length, T_i and T_f are the initial and final temperatures, respectively, and the proportionality constant α is the average coefficient of linear expansion for a given material and has units of $(°C)^{-1}$. Equation 18.5 can be used for both thermal expansion, when the temperature of the material increases, and thermal contraction, when its temperature decreases.

It may be helpful to think of thermal expansion as an effective magnification or as a photographic enlargement of an object. For example, as a metal washer is heated (Fig. 18.8), all dimensions, including the radius of the hole, increase according to Equation 18.5. A cavity in a piece of material expands in the same way as if the cavity were filled with the material.

Table 18.1 lists the average coefficients of linear expansion for various materials. For these materials, α is positive, indicating an increase in length with increasing temperature. That is not always the case, however. Some substances—calcite $(CaCO_3)$ is one example—expand along one dimension (positive α) and contract along another (negative α) as their temperatures are increased.

Because the linear dimensions of an object change with temperature, it follows that surface area and volume change as well. The change in volume is proportional to the initial volume V_i and to the change in temperature according to the relationship

$$\Delta V = \beta V_i \Delta T \qquad (18.7)$$

◀ Thermal expansion in three dimensions

where β is the **average coefficient of volume expansion**. To find the relationship between β and α, assume the average coefficient of linear expansion of the solid is the same in all directions; that is, assume the material is *isotropic*. Consider a solid box of dimensions ℓ, w, and h. Its volume at some temperature T_i is $V_i = \ell w h$. If the temperature changes to $T_i + \Delta T$, its volume changes to $V_i + \Delta V$, where each dimension changes according to Equation 18.5. Therefore,

$$V_i + \Delta V = (\ell + \Delta \ell)(w + \Delta w)(h + \Delta h)$$

$$= (\ell + \alpha \ell \, \Delta T)(w + \alpha w \, \Delta T)(h + \alpha h \, \Delta T)$$

$$= \ell w h (1 + \alpha \, \Delta T)^3$$

$$= V_i [1 + 3\alpha \, \Delta T + 3(\alpha \, \Delta T)^2 + (\alpha \, \Delta T)^3]$$

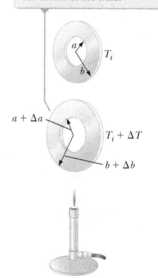

As the washer is heated, all dimensions increase, including the radius of the hole.

Figure 18.8 Thermal expansion of a homogeneous metal washer. (The expansion is exaggerated in this figure.)

TABLE 18.1 Average Expansion Coefficients for Some Materials Near Room Temperature

Material (Solids)	Average Linear Expansion Coefficient $(\alpha)(°C)^{-1}$	Material (Liquids and Gases)	Average Volume Expansion Coefficient $(\beta)(°C)^{-1}$
Aluminum	24×10^{-6}	Acetone	1.5×10^{-4}
Brass and bronze	19×10^{-6}	Alcohol, ethyl	1.12×10^{-4}
Concrete	12×10^{-6}	Benzene	1.24×10^{-4}
Copper	17×10^{-6}	Gasoline	9.6×10^{-4}
Glass (ordinary)	9×10^{-6}	Glycerin	4.85×10^{-4}
Glass (Pyrex)	3.2×10^{-6}	Mercury	1.82×10^{-4}
Invar (Ni–Fe alloy)	0.9×10^{-6}	Turpentine	9.0×10^{-4}
Lead	29×10^{-6}	Air[a] at 0°C	3.67×10^{-3}
Steel	11×10^{-6}	Helium[a]	3.665×10^{-3}

[a]Gases do not have a specific value for the volume expansion coefficient because the amount of expansion depends on the type of process through which the gas is taken. The values given here assume the gas undergoes an expansion at constant pressure.

Dividing both sides by V_i and isolating the term $\Delta V/V_i$, we obtain the fractional change in volume:

$$\frac{\Delta V}{V_i} = 3\alpha \, \Delta T + 3(\alpha \, \Delta T)^2 + (\alpha \, \Delta T)^3$$

Because $\alpha \, \Delta T \ll 1$ for typical values of ΔT ($< \sim 100°C$), we can neglect the terms $3(\alpha \, \Delta T)^2$ and $(\alpha \, \Delta T)^3$. Upon making this approximation, we see that

$$\frac{\Delta V}{V_i} = 3\alpha \, \Delta T \quad \rightarrow \quad \Delta V = (3\alpha) V_i \, \Delta T$$

Comparing this expression to Equation 18.7 shows that

$$\beta = 3\alpha$$

In a similar way, you can show that the change in area of a rectangular plate is given by $\Delta A = 2\alpha A_i \, \Delta T$ (see Problem 37).

QUICK QUIZ 18.3 If you are asked to make a very sensitive glass thermometer, which of the following working liquids would you choose? **(a)** mercury **(b)** alcohol **(c)** gasoline **(d)** glycerin

QUICK QUIZ 18.4 Two spheres are made of the same metal and have the same radius, but one is hollow and the other is solid. The spheres are taken through the same temperature increase. Which sphere expands more? **(a)** The solid sphere expands more. **(b)** The hollow sphere expands more. **(c)** They expand by the same amount. **(d)** There is not enough information to say.

The Unusual Behavior of Water

Liquids generally increase in volume with increasing temperature and have average coefficients of volume expansion about ten times greater than those of solids. Water follows this general behavior *except* near 0°C, as you can see from its density-versus-temperature curve shown in Figure 18.9. As the temperature increases from 0°C to 4°C, water contracts and its density therefore increases. Above 4°C, water expands normally with increasing temperature and so its density decreases. Therefore, the density of water reaches a maximum value of 1.000 g/cm^3 at 4°C.

We can use this unusual thermal-expansion behavior of water to explain why a pond begins freezing at the surface rather than at the bottom. When the air temperature drops from, for example, 7°C to 6°C, the surface water also cools and consequently decreases in volume. The surface water is denser than the water below it, which has not cooled and decreased in volume. As a result, the surface water sinks, and warmer water from below moves to the surface. When the air temperature is between 4°C and 0°C, however, the surface water expands as it cools, becoming less dense than the water below it. The mixing process stops, and eventually the surface water freezes. As the water freezes, the ice remains on the surface because ice is less dense than water. The ice continues to build up at the surface, while water near the bottom remains at 4°C. If that were not the case, fish and other forms of marine life would not survive.

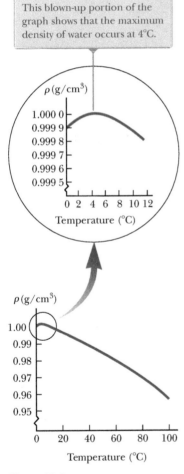

This blown-up portion of the graph shows that the maximum density of water occurs at 4°C.

Figure 18.9 The variation in the density of water at atmospheric pressure with temperature.

Example 18.2 **Expansion of a Railroad Track**

A segment of steel railroad track has a length of 30.000 m when the temperature is 0.0°C. What is its length when the temperature is 40.0°C?

18.2 continued

SOLUTION

Conceptualize Because the rail is relatively long, we expect to obtain a measurable increase in length for a 40°C temperature increase.

Categorize We will evaluate a length increase using the discussion of this section, so this example is a substitution problem.

Use Equation 18.5 and the value of the coefficient of linear expansion from Table 18.1:

$$\Delta L = \alpha L_i \Delta T = [11 \times 10^{-6} \ (°C)^{-1}](30.000 \ \text{m})(40.0°C) = 0.013 \ \text{m}$$

Find the new length of the track:

$$L_f = 30.000 \ \text{m} + 0.013 \ \text{m} = 30.013 \ \text{m}$$

The expansion of 1.3 cm is indeed measurable as predicted in the Conceptualize step. If the section of rail is butted right against another rail, this expansion cannot occur and a *thermal stress* is developed in the rail. The thermal stress could bend the rail. This stress can be avoided by leaving small expansion gaps between the rails.

WHAT IF? What if the temperature drops to −40.0°C? What is the length of the unclamped segment?

Answer The expression for the change in length in Equation 18.5 is the same whether the temperature increases or decreases. Therefore, if there is an increase in length of 0.013 m when the temperature increases by 40°C, there is a decrease in length of 0.013 m when the temperature decreases by 40°C. (We assume α is constant over the entire range of temperatures.) The new length at the colder temperature is 30.000 m − 0.013 m = 29.987 m.

Example 18.3 **The Thermal Electrical Short**

A poorly designed electronic device has two bolts attached to different parts of the device that almost touch each other in its interior as in Figure 18.10. The steel and brass bolts are at different electric potentials, and if they touch, a short circuit will develop, damaging the device. (We will study electric potential in Chapter 24.) The initial gap between the ends of the bolts is $d = 5.0 \ \mu\text{m}$ at 27°C. At what temperature will the bolts touch? Assume the distance between the walls of the device is not affected by the temperature change.

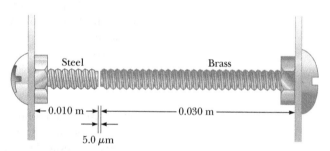

Figure 18.10 (Example 18.3) Two bolts attached to different parts of an electrical device are almost touching when the temperature is 27°C. As the temperature increases, the ends of the bolts move toward each other.

SOLUTION

Conceptualize Imagine the ends of both bolts expanding into the gap between them as the temperature rises.

Categorize We categorize this example as a thermal expansion problem in which the *sum* of the changes in length of the two bolts must equal the length d of the initial gap between the ends.

Analyze Set the sum of the length changes equal to the width of the gap:

$$\Delta L_{br} + \Delta L_{st} = \alpha_{br} L_{i,br} \ \Delta T + \alpha_{st} L_{i,st} \ \Delta T = d$$

Solve for ΔT:

$$\Delta T = \frac{d}{\alpha_{br} L_{i,br} + \alpha_{st} L_{i,st}}$$

Substitute numerical values:

$$\Delta T = \frac{5.0 \times 10^{-6} \ \text{m}}{[19 \times 10^{-6} \ (°C)^{-1}](0.030 \ \text{m}) + [11 \times 10^{-6} \ (°C)^{-1}](0.010 \ \text{m})} = 7.4°C$$

Find the temperature at which the bolts touch:

$$T = 27°C + 7.4°C = 34°C$$

Finalize This temperature is possible if the air conditioning in the building housing the device fails for a long period on a very hot summer day.

18.5 Macroscopic Description of an Ideal Gas

The volume expansion equation $\Delta V = \beta V_i \Delta T$ (Eq. 18.7) is based on the assumption that the material has an initial volume V_i before the temperature change occurs. Such is the case for solids and liquids because a sample of solid or liquid has a fixed volume at a given temperature.

The case for gases is completely different. The interatomic forces within gases are very weak, and, in many cases, we can imagine these forces to be nonexistent and still make very good approximations. Therefore, *there is no equilibrium separation* for the atoms and no "standard" volume at a given temperature; the volume depends on the size of the container. As a result, we cannot express changes in volume ΔV in a process on a gas with Equation 18.7 because we have no defined volume V_i at the beginning of the process. Equations involving gases contain the volume V, rather than a *change* in the volume from an initial value, as a variable.

For a gas, it is useful to know how the quantities volume V, pressure P, and temperature T are related for a sample of gas of mass m. In general, the equation that interrelates these quantities, called the *equation of state,* is very complicated. If the gas is maintained at a very low pressure (or low density), however, the equation of state is quite simple and can be determined from experimental results. Such a low-density gas is commonly referred to as an **ideal gas.**[5] We can use a simplification model called the **ideal gas model** to make predictions that are adequate to describe the behavior of real gases at low pressures.

It is convenient to express the amount of gas in a given volume in terms of the number of moles n. One **mole** of any substance is that amount of the substance that contains **Avogadro's number** $N_A = 6.022 \times 10^{23}$ of constituent particles (atoms or molecules). The number of moles n of a substance is related to its mass m through the expression

$$n = \frac{m}{M} \tag{18.8}$$

Figure 18.11 An ideal gas confined to a cylinder whose volume can be varied by means of a movable piston.

where M is the molar mass of the substance. The molar mass of each chemical element is the atomic mass (from the periodic table; see Appendix C) expressed in grams per mole. For example, the mass of one helium (He) atom is 4.00 u (atomic mass units), so the molar mass of He is 4.00 g/mol.

Now suppose an ideal gas is confined to a cylindrical container whose volume can be varied by means of a movable piston as in Figure 18.11. If we assume the cylinder does not leak, the mass (or the number of moles) of the gas remains constant. For such a system, experiments measuring pressure, volume, and temperature provide the following information:

- When the gas is kept at a constant temperature, its pressure is inversely proportional to the volume. (This behavior is described historically as Boyle's law.)
- When the pressure of the gas is kept constant, the volume is directly proportional to the temperature. (This behavior is described historically as Charles's law.)
- When the volume of the gas is kept constant, the pressure is directly proportional to the temperature. (This behavior is described historically as Gay–Lussac's law, and justifies the straight lines we drew through the data points in the graph in Figure 18.5.)

These observations are summarized by the **equation of state for an ideal gas:**

Equation of state for ▶
an ideal gas

$$PV = nRT \tag{18.9}$$

---

[5] To be more specific, the assumptions here are that the temperature of the gas must not be too low (the gas must not condense into a liquid) or too high and that the pressure must be low. The concept of an ideal gas implies that the gas molecules do not interact except upon collision and that the molecular volume is negligible compared with the volume of the container. In reality, an ideal gas does not exist. The concept of an ideal gas is nonetheless very useful because real gases at low pressures are well-modeled as ideal gases.

In this expression, also known as the **ideal gas law,** n is the number of moles of gas in the sample and R is a constant. Experiments on numerous gases show that as the pressure approaches zero, the quantity PV/nT approaches the same value R for all gases. For this reason, R is called the **universal gas constant.** In SI units, in which pressure is expressed in pascals ($1 \text{ Pa} = 1 \text{ N/m}^2$) and volume in cubic meters, the product PV has units of newton · meters, or joules, and R has the value

$$R = 8.314 \text{ J/mol} \cdot \text{K} \qquad (18.10)$$

If the pressure is expressed in atmospheres and the volume in liters ($1 \text{ L} = 10^3 \text{ cm}^3 = 10^{-3} \text{ m}^3$), then R has the value

$$R = 0.082\ 06 \text{ L} \cdot \text{atm/mol} \cdot \text{K}$$

Figure 18.12 A bottle of champagne is shaken and opened. Liquid spews out of the opening. A common misconception is that the pressure inside the bottle is increased by the shaking.

Using this value of R and Equation 18.9 shows that the volume occupied by 1 mol of *any* gas at atmospheric pressure and at 0°C (273 K) is 22.4 L.

The ideal gas law states that if the volume and temperature of a fixed amount of gas do not change, the pressure also remains constant. Consider a bottle of champagne that is shaken and then spews liquid when opened as shown in Figure 18.12. A common misconception is that the pressure inside the bottle is increased when the bottle is shaken. On the contrary, because the temperature of the bottle and its contents remains constant as long as the bottle is sealed, so does the pressure, as can be shown by replacing the cork with a pressure gauge. The correct explanation is as follows. Carbon dioxide gas resides in the volume between the liquid surface and the cork. The pressure of the gas in this volume is set higher than atmospheric pressure in the bottling process. Shaking the bottle displaces some of the carbon dioxide gas into the liquid, where it forms bubbles, and these bubbles become attached to the inside of the bottle. (No new gas is generated by shaking.) When the bottle is opened, the pressure is reduced to atmospheric pressure, which causes the volume of the bubbles to increase suddenly. If the bubbles are attached to the bottle (beneath the liquid surface), their rapid expansion expels liquid from the bottle. If the sides and bottom of the bottle are first tapped until no bubbles remain beneath the surface, however, the drop in pressure does not force liquid from the bottle when the champagne is opened.

The ideal gas law is often expressed in terms of the total number of molecules N. Because the number of moles n equals the ratio of the total number of molecules and Avogadro's number N_A, we can write Equation 18.9 as

$$PV = nRT = \frac{N}{N_A} RT$$

$$PV = Nk_B T \qquad (18.11)$$

where k_B is **Boltzmann's constant,** which has the value

$$k_B = \frac{R}{N_A} = 1.38 \times 10^{-23} \text{ J/K} \qquad (18.12)$$

◀ Boltzmann's constant

It is common to call quantities such as P, V, and T the **thermodynamic variables** of an ideal gas. If the equation of state is known, one of the variables can always be expressed as some function of the other two.

QUICK QUIZ 18.5 A common material for cushioning objects in packages is made by trapping bubbles of air between sheets of plastic. Is this material more effective at keeping the contents of the package from moving around inside the package on **(a)** a hot day, **(b)** a cold day, or **(c)** either hot or cold days?

QUICK QUIZ 18.6 On a winter day, you turn on your furnace and the temperature of the air inside your home increases. Assume your home has the normal amount of leakage between inside air and outside air. Is the number of moles of air in your room at the higher temperature **(a)** larger than before, **(b)** smaller than before, or **(c)** the same as before?

Example 18.4 Heating a Spray Can

A spray can containing a propellant gas at twice atmospheric pressure (202 kPa) and having a volume of 125.00 cm³ is at 22°C. It is then tossed into an open fire. (*Warning:* Do not do this experiment; it is very dangerous.) When the temperature of the gas in the can reaches 195°C, what is the pressure inside the can? Assume any change in the volume of the can is negligible.

SOLUTION

Conceptualize Intuitively, you should expect that the pressure of the gas in the container increases because of the increasing temperature.

Categorize We model the gas in the can as ideal and use the ideal gas law to calculate the new pressure.

Analyze Rearrange Equation 18.9:

$$(1) \quad \frac{PV}{T} = nR$$

No air escapes during the compression, so n, and therefore nR, remains constant. Hence, set the initial value of the left side of Equation (1) equal to the final value:

$$(2) \quad \frac{P_i V_i}{T_i} = \frac{P_f V_f}{T_f}$$

Because the initial and final volumes of the gas are assumed to be equal, cancel the volumes:

$$(3) \quad \frac{P_i}{T_i} = \frac{P_f}{T_f}$$

Solve for P_f:

$$P_f = \left(\frac{T_f}{T_i}\right)P_i = \left(\frac{468 \text{ K}}{295 \text{ K}}\right)(202 \text{ kPa}) = 320 \text{ kPa}$$

Finalize The higher the temperature, the higher the pressure exerted by the trapped gas as expected. If the pressure increases sufficiently, the can may explode. Because of this possibility, you should never dispose of spray cans in a fire.

WHAT IF? Suppose we include a volume change due to thermal expansion of the steel can as the temperature increases. Does that alter our answer for the final pressure significantly?

Answer Because the thermal expansion coefficient of steel is very small, we do not expect much of an effect on our final answer.

Find the change in the volume of the can using Equation 18.7 and the value for α for steel from Table 18.1:

$$\Delta V = \beta V_i \Delta T = 3\alpha V_i \Delta T$$
$$= 3[11 \times 10^{-6} \text{ (°C)}^{-1}](125.00 \text{ cm}^3)(173°C) = 0.71 \text{ cm}^3$$

Start from Equation (2) again and find an equation for the final pressure:

$$P_f = \left(\frac{T_f}{T_i}\right)\left(\frac{V_i}{V_f}\right)P_i$$

This result differs from Equation (3) only in the factor V_i/V_f. Evaluate this factor:

$$\frac{V_i}{V_f} = \frac{125.00 \text{ cm}^3}{(125.00 \text{ cm}^3 + 0.71 \text{ cm}^3)} = 0.994 = 99.4\%$$

Therefore, the final pressure will differ by only 0.6% from the value calculated without considering the thermal expansion of the can. Taking 99.4% of the previous final pressure, the final pressure including thermal expansion is 318 kPa.

Summary

> ## Definitions

Two objects are in **thermal equilibrium** with each other if they do not exchange energy when in thermal contact.

Temperature is the property that determines whether an object is in thermal equilibrium with other objects. Two objects in thermal equilibrium with each other are at the same temperature. The SI unit of absolute temperature is the **kelvin,** which is defined to be 1/273.16 of the difference between absolute zero and the temperature of the triple point of water.

❯ Concepts and Principles

The **zeroth law of thermodynamics** states that if objects A and B are separately in thermal equilibrium with a third object C, then objects A and B are in thermal equilibrium with each other.

When the temperature of an object is changed by an amount ΔT, its length changes by an amount ΔL that is proportional to ΔT and to its initial length L_i:

$$\Delta L = \alpha L_i \Delta T \tag{18.5}$$

where the constant α is the **average coefficient of linear expansion.** The **average coefficient of volume expansion** β for a solid is approximately equal to 3α.

An **ideal gas** is one for which PV/nT is constant. An ideal gas is described by the **equation of state,**

$$PV = nRT \tag{18.9}$$

where n equals the number of moles of the gas, P is its pressure, V is its volume, R is the universal gas constant ($8.314 \, \text{J/mol} \cdot \text{K}$), and T is the absolute temperature of the gas. A real gas behaves approximately as an ideal gas if it has a low density.

Think–Pair–Share

See the Preface for an explanation of the icons used in this problems set. For additional assessment items for this section, go to ❖ **WEBASSIGN** From Cengage

1. **Review.** In the storyline opening Chapter 15, we discussed the change in timing for a grandfather clock due to a change in the value of the acceleration g due to gravity. Have your group think about a change in the timing of the pendulum clock due to a change in temperature. Suppose the pendulum is made of brass and has a period of 1.000 s when the temperature is 20.0°C. During a warm summer week, the temperature remains in a very small range with an average of 30.0°C. (a) Does the clock lose time or gain time during that week? (b) By how much is the clock in error by the end of the week?

2. Your expert witness team has been hired by the city council in a lawsuit filed by a truck driver. While driving down a city street on a hot day, the top of the truck driven by the driver struck a power line hanging across the street, causing damage to his truck and injury to himself. The driver claims that the power line was sagging so much because of the temperature. As a result, it was hanging below the height clearance limit of 14 feet, 0 inches posted on a sign on the street. You go to the site of the accident and take the following measurements. The poles supporting the ends of the power line are 40.0 ft apart. Both ends of the line are supported at a height of 16.0 ft above the roadway surface. On the winter day you visit the site, the temperature is −25.0°C and the copper power line shows essentially no sag. On the day of the accident, the temperature was 38.0°C. Prepare an argument that the power line did not sag below the height clearance, and that the truck driver must have loaded his truck to a point higher than the posted clearance. (*Suggestion:* the shape of the sagging power line will be a curve, but assume an unrealistic shape for the line that will allow a simple calculation for the lowest possible point on the line.)

3. **ACTIVITY** For this activity, you'll need some containers, some straws, and some water. (a) Fill various containers to different levels with water and measure the depth h of the water in each cup carefully. Now immerse a straw vertically in each cup with the bottom of the straw resting on the bottom of the cup. Place your finger over the top end of the straw to seal it and lift the straw vertically out of the cup. Measure the length h' of the column of water trapped in the straw. You may need to take a photo with your smartphone of the straw and a ruler and carefully analyze an enlargement of the photo to make this measurement. Does the length of the column of water trapped in the straw agree with the depth of water in the cup? Should it agree? (b) In order for the column of water to be suspended in the straw, the pressure above the column (within the straw) must be less than atmospheric pressure. For that to happen, the column of water must move downward a bit when the straw is raised from the water, so that the air above the column expands in volume and the pressure decreases. Therefore, the two measured lengths should not be the same. But is the difference detectable? Show that the length h' of water suspended in the straw of length ℓ should be related to the depth h of the water in the cup by

$$h' = \tfrac{1}{2}\left(\ell + \frac{P_0}{\rho g}\right) - \tfrac{1}{2}\sqrt{\ell^2 + 2\left(\frac{P_0}{\rho g}\right)(\ell - 2h) + \left(\frac{P_0}{\rho g}\right)^2}$$

where P_0 is atmospheric pressure and ρ is the density of the water. (c) For a 30-cm straw, use the equation to determine the difference $h - h'$ for various levels of h from zero to 30 cm in increments of 1 cm. (d) For what values of h is the *percentage* drop in the water column the greatest when the straw is drawn out of the cup? (e) For what value of h is the value of $h - h'$ a maximum?

Problems

See the Preface for an explanation of the icons used in this problems set. For additional assessment items for this section, go to :'• **WEBASSIGN** From Cengage

SECTION 18.2 Thermometers and the Celsius Temperature Scale

1. You are working as a research assistant for a professor CR whose research area is thermodynamics. He points out to you that Daniel Fahrenheit used the best estimate of normal human body temperature as one of the points in defining the original Fahrenheit temperature scale. On the revised scale we now use, normal human body temperature is 98.6°F. Your professor proposes a new scale on which normal human body temperature would be exactly 100°N, where the unit °N is a degree on the *New* scale. The temperature of freezing water would be 0°N, as on the Celsius scale. Your professor asks you to determine the following temperatures on his new scale: (a) absolute zero, (b) the melting point of mercury (−37.9°F), (c) the boiling point of water, and, for publicity at his expected future press conference, (d) the highest recorded air temperature on the Earth's surface, 134.1°F on July 10, 1913, in Death Valley, California.

SECTION 18.3 The Constant-Volume Gas Thermometer and the Absolute Temperature Scale

2. A nurse measures the temperature of a patient to be 41.5°C. BIO (a) What is this temperature on the Fahrenheit scale? (b) Do QIC you think the patient is seriously ill? Explain.

3. Convert the following temperatures to their values on the Fahrenheit and Kelvin scales: (a) the sublimation point of dry ice, −78.5°C; (b) human body temperature, 37.0°C.

4. Liquid nitrogen has a boiling point of −195.81°C at atmospheric pressure. Express this temperature (a) in degrees Fahrenheit and (b) in kelvins.

5. Death Valley holds the record for the highest recorded temperature in the United States. On July 10, 1913, at a place called Furnace Creek Ranch, the temperature rose to 134°F. The lowest U.S. temperature ever recorded occurred at Prospect Creek Camp in Alaska on January 23, 1971, when the temperature plummeted to −79.8°F. (a) Convert these temperatures to the Celsius scale. (b) Convert the Celsius temperatures to Kelvin.

SECTION 18.4 Thermal Expansion of Solids and Liquids

Note: Table 18.1 is available for use in solving problems in this section.

6. **Review.** Inside the wall of a house, an L-shaped section of hot-water pipe consists of three parts: a straight, horizontal piece $h = 28.0$ cm long; an elbow; and a straight, vertical piece $\ell = 134$ cm long (Fig. P18.6). A stud and a second-story floorboard hold the ends of this section of copper pipe stationary. Find the magnitude and direction of the displacement of the pipe elbow when the water flow is turned on, raising the temperature of the pipe from 18.0°C to 46.5°C.

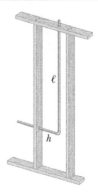

Figure P18.6

7. A copper telephone wire has essentially no sag between T poles 35.0 m apart on a winter day when the temperature is −20.0°C. How much longer is the wire on a summer day when the temperature is 35.0°C?

8. A pair of eyeglass frames is made of epoxy plastic. At room BIO temperature (20.0°C), the frames have circular lens holes 2.20 cm in radius. To what temperature must the frames be heated if lenses 2.21 cm in radius are to be inserted in them? The average coefficient of linear expansion for epoxy is 1.30×10^{-4} (°C)$^{-1}$.

9. The Trans-Alaska pipeline is 1 300 km long, reaching from Prudhoe Bay to the port of Valdez. It experiences temperatures from −73°C to +35°C. How much does the steel pipeline expand because of the difference in temperature? How can this expansion be compensated for?

10. A square hole 8.00 cm along each side is cut in a sheet of copper. (a) Calculate the change in the area of this hole resulting when the temperature of the sheet is increased by 50.0 K. (b) Does this change represent an increase or a decrease in the area enclosed by the hole?

11. You are watching a new CR bridge being built near your house. You notice during the construction that two concrete spans of the bridge of total length $L_i = 250$ m are placed end to end so that no room is allowed for expansion (Fig. P18.11a). In the opening storyline for this chapter, we talked about buckling sidewalks. The same thing will happen with spans on bridges if allowance is not made for expansion (Fig. P18.11b). You want to warn the construction crew about this dangerous situation, so you calculate the height y to which the spans will rise when they buckle in response to a temperature increase of $\Delta T = 20.0$°C.

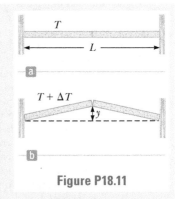

Figure P18.11

12. You are watching a new bridge being built near your house. CR You notice during the construction that two concrete spans S are placed end to end to form a span of length L_i. However, they are placed end to end so that no room is allowed for expansion (Fig. P18.11a). In the opening storyline for this chapter, we talked about buckling sidewalks. The same thing will happen with spans on bridges if allowance is not made for expansion (Fig. P18.11b). You want to warn the construction crew about this dangerous situation, so you calculate the height y to which the spans will rise when they buckle in response to a temperature increase of ΔT.

13. At 20.0°C, an aluminum ring has an inner diameter of 5.000 0 cm and a brass rod has a diameter of 5.050 0 cm. (a) If only the ring is warmed, what temperature must it reach so that it will just slip over the rod? (b) **What If?** If both the ring and the rod are warmed together, what temperature must they both reach so that the ring barely slips over the rod? (c) Would this latter process work? Explain. *Hint:* Consult Table 19.2 in the next chapter.

14. *Why is the following situation impossible?* A thin brass ring has an inner diameter 10.00 cm at 20.0°C. A solid aluminum cylinder has diameter 10.02 cm at 20.0°C. Assume the average coefficients of linear expansion of the two metals are constant. Both metals are cooled together to a temperature at which the ring can be slipped over the end of the cylinder.

15. A volumetric flask made of Pyrex is calibrated at 20.0°C. It is filled to the 100-mL mark with 35.0°C acetone. After the flask is filled, the acetone cools and the flask warms so that the combination of acetone and flask reaches a uniform temperature of 32.0°C. The combination is then cooled back to 20.0°C. (a) What is the volume of the acetone when it cools to 20.0°C? (b) At the temperature of 32.0°C, does the level of acetone lie above or below the 100-mL mark on the flask? Explain.

16. **Review.** On a day that the temperature is 20.0°C, a concrete walk is poured in such a way that the ends of the walk are unable to move. Take Young's modulus for concrete to be 7.00×10^9 N/m² and the compressive strength to be 2.00×10^9 N/m². (a) What is the stress in the cement on a hot day of 50.0°C? (b) Does the concrete fracture?

17. **Review.** The Golden Gate Bridge in San Francisco has a main span of length 1.28 km, one of the longest in the world. Imagine that a steel wire with this length and a cross-sectional area of 4.00×10^{-6} m² is laid in a straight line on the bridge deck with its ends attached to the towers of the bridge. On a summer day the temperature of the wire is 35.0°C. (a) When winter arrives, the towers stay the same distance apart and the bridge deck keeps the same shape as its expansion joints open. When the temperature drops to -10.0°C, what is the tension in the wire? Take Young's modulus for steel to be 20.0×10^{10} N/m². (b) Permanent deformation occurs if the stress in the steel exceeds its elastic limit of 3.00×10^8 N/m². At what temperature would the wire reach its elastic limit? (c) **What If?** Explain how your answers to parts (a) and (b) would change if the Golden Gate Bridge were twice as long.

SECTION 18.5 Macroscopic Description of an Ideal Gas

18. Your father and your younger brother are confronted with the same puzzle. Your father's garden sprayer and your brother's water cannon both have tanks with a capacity of 5.00 L (Fig. P18.18). Your father puts a negligible amount of concentrated fertilizer into his tank. They both pour in 4.00 L of water and seal up their tanks, so the tanks also contain air at atmospheric pressure. Next, each uses a hand-operated pump to inject more air until the absolute pressure in the tank reaches 2.40 atm. Now each uses his device to spray out water—not air—until the stream becomes feeble, which it does when the pressure in the tank reaches 1.20 atm. To accomplish spraying out all the water, each finds he must pump up the tank three times. Here is the puzzle: most of the water sprays out after the second pumping. The first and the third pumping-up processes seem just as difficult as the second but result in a much smaller amount of water coming out. Account for this phenomenon.

Figure P18.18

19. An auditorium has dimensions 10.0 m × 20.0 m × 30.0 m. How many molecules of air fill the auditorium at 20.0°C and a pressure of 101 kPa (1.00 atm)?

20. A container in the shape of a cube 10.0 cm on each edge contains air (with equivalent molar mass 28.9 g/mol) at atmospheric pressure and temperature 300 K. Find (a) the mass of the gas, (b) the gravitational force exerted on it, and (c) the force it exerts on each face of the cube. (d) Why does such a small sample exert such a great force?

21. (a) Find the number of moles in one cubic meter of an ideal gas at 20.0°C and atmospheric pressure. (b) For air, Avogadro's number of molecules has mass 28.9 g. Calculate the mass of one cubic meter of air. (c) State how this result compares with the tabulated density of air at 20.0°C.

22. Use the definition of Avogadro's number to find the mass of a helium atom.

23. In state-of-the-art vacuum systems, pressures as low as 1.00×10^{-9} Pa are being attained. Calculate the number of molecules in a 1.00-m³ vessel at this pressure and a temperature of 27.0°C.

24. You have scored a great internship with NASA, working on planning for an upcoming mission to Mars. The transfer orbit to Mars will last for several months and will require reclamation of the oxygen in the carbon dioxide exhaled by the crew. In one method of reclamation, 1.00 mol of carbon dioxide produces 1.00 mol of oxygen and 1.00 mol of methane as a byproduct. The methane is stored in a tank under pressure and is available to control the orientation of the spacecraft by controlled venting. A single astronaut exhales 1.09 kg of carbon dioxide each day. If the methane generated in the respiration recycling of three astronauts during one week of flight is stored in an originally empty 150-L tank at -45.0°C, what is the final pressure in the tank?

25. **Review.** The mass of a hot-air balloon and its cargo (not including the air inside) is 200 kg. The air outside is at 10.0°C and 101 kPa. The volume of the balloon is 400 m³. To what temperature must the air in the balloon be warmed before the balloon will lift off? (Air density at 10.0°C is 1.244 kg/m³.)
   AMT
   T

26. A room of volume V contains air having equivalent molar mass M (in g/mol). If the temperature of the room is raised from T_1 to T_2, what mass of air will leave the room? Assume that the air pressure in the room is maintained at P_0.
   S

27. Estimate the mass of the air in your bedroom. State the quantities you take as data and the value you measure or estimate for each.

28. You are applying for a position with a sea rescue unit and are taking the qualifying exam. One question on the exam is about the use of a diving bell. The diving bell is in the shape of a cylinder with a vertical length of $L = 2.50$ m. It is closed at the upper circular end and open at the lower circular end. The bell is lowered from air into seawater ($\rho = 1.025$ g/cm³) and kept in its upright orientation as it is lowered. The air in the bell is initially at temperature $T_i = 20.0$°C. The bell, with two humans inside, is lowered to a depth (measured to the bottom of the bell) of 27.0 fathoms, or $h = 49.4$ m. At this depth the water temperature is $T_f = 4.0$°C, and the bell is in thermal equilibrium with the water. The exam question asks you to compare two situations: (i) No additional gas is added to the interior of the bell as it is submerged. Therefore, water enters the open bottom of the bell and the volume of the enclosed air decreases. (ii) The bell is fitted with pressurized air tanks, which deliver high-pressure air into the interior of the bell to keep the level of water at the bottom edge of the bell. This choice requires money and effort to attach the tanks. The exam question asks: Which scenario is better?
   CR

29. The pressure gauge on a cylinder of gas registers the gauge pressure, which is the difference between the interior pressure and the exterior pressure P_0. Let's call the gauge pressure P_g. When the cylinder is full, the mass of the gas in it is m_i at a gauge pressure of P_{gi}. Assuming the temperature of the cylinder remains constant, show that the mass of the gas *remaining* in the cylinder when the pressure reading is P_{gf} is given by
   S

$$m_f = m_i \left(\frac{P_{gf} + P_0}{P_{gi} + P_0} \right)$$

ADDITIONAL PROBLEMS

30. A steel beam being used in the construction of a skyscraper has a length of 35.000 m when delivered on a cold day at a temperature of 15.000°F. What is the length of the beam when it is being installed later on a warm day when the temperature is 90.000°F?

31. Two metal bars are made of invar and a third bar is made of aluminum. At 0°C, each of the three bars is drilled with two holes 40.0 cm apart. Pins are put through the holes to assemble the bars into an equilateral triangle as in Figure P18.31. (a) First ignore the expansion of the invar. Find the angle between the invar bars as a function of Celsius temperature. (b) Is your answer accurate for negative as well as positive temperatures? (c) Is it accurate for 0°C? (d) Solve
   QC

the problem again, including the expansion of the invar. Aluminum melts at 660°C and invar at 1 427°C. Assume the tabulated expansion coefficients are constant. What are (e) the greatest and (f) the smallest attainable angles between the invar bars?

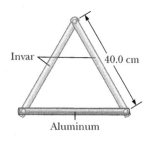

Figure P18.31

32. *Why is the following situation impossible?* An apparatus is designed so that steam initially at $T = 150$°C, $P = 1.00$ atm, and $V = 0.500$ m³ in a piston–cylinder apparatus undergoes a process in which (1) the volume remains constant and the pressure drops to 0.870 atm, followed by (2) an expansion in which the pressure remains constant and the volume increases to 1.00 m³, followed by (3) a return to the initial conditions. It is important that the pressure of the gas never fall below 0.850 atm so that the piston will support a delicate and very expensive part of the apparatus. Without such support, the delicate apparatus can be severely damaged and rendered useless. When the design is turned into a working prototype, it operates perfectly.

33. A student measures the length of a brass rod with a steel tape at 20.0°C. The reading is 95.00 cm. What will the tape indicate for the length of the rod when the rod and the tape are at (a) −15.0°C and (b) 55.0°C?

34. The density of gasoline is 730 kg/m³ at 0°C. Its average coefficient of volume expansion is 9.60×10^{-4} (°C)$^{-1}$. Assume 1.00 gal of gasoline occupies 0.003 80 m³. How many extra kilograms of gasoline would you receive if you bought 10.0 gal of gasoline at 0°C rather than at 20.0°C from a pump that is not temperature compensated?

35. A liquid has a density ρ. (a) Show that the fractional change in density for a change in temperature ΔT is $\Delta\rho/\rho = -\beta \Delta T$. (b) What does the negative sign signify? (c) Fresh water has a maximum density of 1.000 0 g/cm³ at 4.0°C. At 10.0°C, its density is 0.999 7 g/cm³. What is β for water over this temperature interval? (d) At 0°C, the density of water is 0.999 9 g/cm³. What is the value for β over the temperature range 0°C to 4.00°C?
   QC

36. (a) Take the definition of the coefficient of volume expansion to be
   QC

$$\beta = \frac{1}{V} \frac{dV}{dT}\bigg|_{P=\text{constant}} = \frac{1}{V} \frac{\partial V}{\partial T}$$

Use the equation of state for an ideal gas to show that the coefficient of volume expansion for an ideal gas at constant pressure is given by $\beta = 1/T$, where T is the absolute temperature. (b) What value does this expression predict for β at 0°C? State how this result compares with the experimental values for (c) helium and (d) air in Table 18.1. *Note:* These values are much larger than the coefficients of volume expansion for most liquids and solids.

37. The rectangular plate shown in Figure P18.37 has an area A_i equal to ℓw. If the temperature increases by ΔT, each dimension increases according to Equation 18.5, where α is the average coefficient of linear expansion. (a) Show that
   QC

the increase in area is $\Delta A = 2\alpha A_i \Delta T$. (b) What approximation does this expression assume?

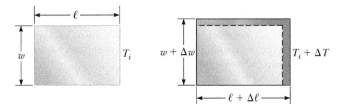

Figure P18.37

38. A bimetallic strip of length L is made of two ribbons of different metals bonded together. (a) First assume the strip is originally straight. As the strip is warmed, the metal with the greater average coefficient of expansion expands more than the other, forcing the strip into an arc with the outer radius having a greater circumference (Fig. P18.38). Derive an expression for the angle of bending θ as a function of the initial length of the strips, their average coefficients of linear expansion, the change in temperature, and the separation of the centers of the strips ($\Delta r = r_2 - r_1$). (b) Show that the angle of bending decreases to zero when ΔT decreases to zero and also when the two average coefficients of expansion become equal. (c) **What If?** What happens if the strip is cooled?

Figure P18.38

39. A copper rod and a steel rod are different in length by 5.00 cm at 0°C. The rods are warmed and cooled together. (a) Is it possible that the length difference remains constant at all temperatures? Explain. (b) If so, describe the lengths at 0°C as precisely as you can. Can you tell which rod is longer? Can you tell the lengths of the rods?

40. A vertical cylinder of cross-sectional area A is fitted with a tight-fitting, frictionless piston of mass m (Fig. P18.40). The piston is not restricted in its motion in any way and is supported by the gas at pressure P below it. Atmospheric pressure is P_0. We wish to find the height h in Figure P18.40. (a) What analysis model is appropriate to describe the piston? (b) Write an appropriate force equation for the piston from this analysis model in terms of P, P_0, m, A, and g. (c) Suppose n moles of an ideal gas are in the cylinder at a temperature of T. Substitute for P in your answer to part (b) to find the height h of the piston above the bottom of the cylinder.

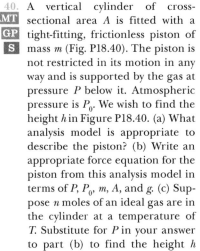

Figure P18.40

41. **Review.** Consider an object with any one of the shapes displayed in Table 10.2. What is the percentage increase in the moment of inertia of the object when it is warmed from 0°C to 100°C if it is composed of (a) copper or (b) aluminum? Assume the average linear expansion coefficients shown in Table 18.1 do not vary between 0°C and 100°C. (c) Why are the answers for parts (a) and (b) the same for all the shapes?

42. **Review.** Following a collision in outer space, a copper disk at 850°C is rotating about its axis with an angular speed of 25.0 rad/s. As the disk radiates infrared light, its temperature falls to 20.0°C. No external torque acts on the disk. (a) Does the angular speed change as the disk cools? Explain how it changes or why it does not. (b) What is its angular speed at the lower temperature?

43. Starting with Equation 18.11, show that the total pressure P in a container filled with a mixture of several ideal gases is $P = P_1 + P_2 + P_3 + \cdots$, where $P_1, P_2, \cdots$ are the pressures that each gas would exert if it alone filled the container. (These individual pressures are called the *partial pressures* of the respective gases.) This result is known as *Dalton's law of partial pressures.*

CHALLENGE PROBLEMS

44. **Review.** A house roof is a perfectly flat plane that makes an angle θ with the horizontal. When its temperature changes, between T_c before dawn each day and T_h in the middle of each afternoon, the roof expands and contracts uniformly with a coefficient of thermal expansion α_1. Resting on the roof is a flat, rectangular metal plate with expansion coefficient α_2, greater than α_1. The length of the plate is L, measured along the slope of the roof. The component of the plate's weight perpendicular to the roof is supported by a normal force uniformly distributed over the area of the plate. The coefficient of kinetic friction between the plate and the roof is μ_k. The plate is always at the same temperature as the roof, so we assume its temperature is continuously changing. Because of the difference in expansion coefficients, each bit of the plate is moving relative to the roof below it, except for points along a certain horizontal line running across the plate called the stationary line. If the temperature is rising, parts of the plate below the stationary line are moving down relative to the roof and feel a force of kinetic friction acting up the roof. Elements of area above the stationary line are sliding up the roof, and on them kinetic friction acts downward parallel to the roof. The stationary line occupies no area, so we assume no force of static friction acts on the plate while the temperature is changing. The plate as a whole is very nearly in equilibrium, so the net friction force on it must be equal to the component of its weight acting down the incline. (a) Prove that the stationary line is at a distance of

$$\frac{L}{2}\left(1 - \frac{\tan\theta}{\mu_k}\right)$$

below the top edge of the plate. (b) Analyze the forces that act on the plate when the temperature is falling and prove that the stationary line is at that same distance above the bottom edge of the plate. (c) Show that the plate steps down the roof like an inchworm, moving each day by the distance

$$\frac{L}{\mu_k}(\alpha_2 - \alpha_1)(T_h - T_c)\tan\theta$$

(d) Evaluate the distance an aluminum plate moves each day if its length is 1.20 m, the temperature cycles between

4.00°C and 36.0°C, and if the roof has slope 18.5°, coefficient of linear expansion 1.50×10^{-5} (°C)$^{-1}$, and coefficient of friction 0.420 with the plate. (e) **What If?** What if the expansion coefficient of the plate is less than that of the roof? Will the plate creep up the roof?

45. A 1.00-km steel railroad rail is fastened securely at both ends when the temperature is 20.0°C. As the temperature increases, the rail buckles, taking the shape of an arc of a vertical circle. Find the height h of the center of the rail when the temperature is 25.0°C. (You will need to solve a transcendental equation.)

46. Helium gas is sold in steel tanks that will rupture if subjected to tensile stress greater than its yield strength of $5 \times 10^8 \, \text{N/m}^2$. If the helium is used to inflate a balloon, could the balloon lift the spherical tank the helium came in? Justify your answer. *Suggestion:* You may consider a spherical steel shell of radius r and thickness t having the density of iron and on the verge of breaking apart into two hemispheres because it contains helium at high pressure.

The First Law of Thermodynamics

A cake is pulled from the oven and it has *fallen*. What causes a cake to fall and why is this question being asked in a chapter on thermodynamics? (*bonchan/Shutterstock*)

STORYLINE **It's a three-day weekend, and you decide to go RV** camping with other members of the Physics Club at Whitney Portal, California, the gateway to Mount Whitney, the tallest peak in the contiguous United States. This community is at an altitude of 2 393 m above sea level, so it should be a good place to do some astronomy observations. As your Club advisor's RV progresses toward Whitney Portal, gaining altitude with each minute, you notice a sign that says, "Caution: Bridge Freezes Before Road Surface." You wonder why that would happen. As you reach your destination, you climb out of your car and marvel at how cold it is. But then you think, "Wait a minute! I'm closer to the Sun than I was at sea level. Why isn't it warmer at the top of a mountain?" You set up camp and offer to make dinner for the group. You boil some eggs for three minutes, fry some hamburgers, and place a sheet of cookies in the oven. After the cookies are done, you put a homemade cake in the oven to bake. At the end of the meal, the results are mixed. The hamburgers are great, the eggs were not quite cooked enough, the cookies were too well done near the edges of the cookie sheet, and the cake fell. Why was your dinner so unsuccessful? You are unhappy with your cooking performance and go to bed. The next morning, you arise for a brisk walk and notice that there is frost on the cars, mailboxes, and the like, but only on the top surfaces of those items, not the sides. Why is the frost only on the top surfaces? There have been so many mysteries associated with this mountain trip and it's only the first morning! You hope there is cell phone service when you return to your RV from your walk so that you can spend some time investigating these mysteries online.

CONNECTIONS Equation 8.2, the conservation of energy equation, shows how the energy of a system can change due to mechanical transfers of energy, like work, and thermal transfers, like heat. It also shows that the energy of a system is divided between mechanical types (kinetic and potential energy) and a thermal type (internal energy). But this is our modern-day understanding of energy. Until about 1850, the fields of thermodynamics and mechanics were considered to be two distinct branches of science. The principle of conservation of energy seemed to describe only certain kinds of mechanical systems. Mid-19th-century experiments performed by Englishman James Joule and others, however, showed a strong connection between the transfer of energy by heat in thermal processes and the transfer of energy by work in mechanical processes. This connection led to what we know as Equation 8.2. This current chapter focuses on a reduced form of Equation 8.2, known as the *first law of thermodynamics*. The first law of thermodynamics describes systems in which the only energy change is that of internal energy and the transfers of energy are by heat and work. A major difference in our discussion of work in this chapter from that in most of the chapters on mechanics is that we will consider work done on *deformable* systems. We will see energy transfers associated with temperature and internal energy in a number of cases in the future, including, among others, the warming of electrical resistors in Chapter 26, cooking a potato in a microwave oven in Chapter 33, and thermal radiation from a black body in Chapter 39.

19.1 Heat and Internal Energy

In Chapter 7, we introduced *internal energy* E_{int}, which exhibits changes on the left side of Equation 8.2, and in Chapter 8, we introduced *heat Q*, which is a mechanism for energy transfer on the right hand side of the equation. These terms are often incorrectly used interchangeably in popular language. Therefore, let us define them carefully:

> **Internal energy** is all the energy of a system that is associated with its microscopic components—atoms and molecules—when viewed from a reference frame at rest with respect to the center of mass of the system.

The last part of this sentence ensures that any bulk kinetic energy of the system due to its motion through space is not included in internal energy. Internal energy includes kinetic energy of random translational, rotational, and vibrational motion of molecules; vibrational potential energy associated with forces between atoms in molecules; and electric potential energy associated with forces between molecules. In Chapter 7, we related internal energy to the temperature of an object, but this relationship is limited. We show in Section 19.3 that internal energy changes can also occur in the absence of temperature changes. In that discussion, we will investigate the internal energy of the system when there is a *physical change*, most often related to a phase change, such as melting or boiling.

We assign energy associated with *chemical changes*, related to chemical reactions, to the potential energy term in Equation 8.2, not to internal energy. Therefore, we discuss the *chemical potential energy* in, for example, a human body (due to previous meals), the gas tank of a car (due to an earlier transfer of fuel), and a battery of an electric circuit (stored in the battery during its construction in the manufacturing process).

Compare this description of internal energy with the following for heat:

> **Heat** is defined as a process of transferring energy across the boundary of a system because of a temperature difference between the system and its surroundings. It is also the amount of energy Q transferred by this process.

PITFALL PREVENTION 19.1

Internal Energy, Thermal Energy, and Bond Energy When reading other physics books, you may see terms such as *thermal energy* and *bond energy*. Thermal energy can be interpreted as that part of the internal energy associated with random motion of molecules and therefore related to temperature. Bond energy is the intermolecular potential energy. Therefore,

Internal energy =
  thermal energy + bond energy

Although this breakdown is presented here for clarification with regard to other books, we will not use these terms because there is no need for them.

PITFALL PREVENTION 19.2

Heat, Temperature, and Internal Energy Are Different As you read the newspaper or explore online, be alert for incorrectly used phrases including the word *heat* and think about the proper word to be used in place of *heat*. Incorrect examples include "As the truck braked to a stop, a large amount of heat was generated by friction" and "The heat of a hot summer day . . . ."

When you *heat* a substance, you are transferring energy into it by placing it in contact with surroundings that have a higher temperature. Such is the case, for example, when you place a pan of cold water on a stove burner. The burner is at a higher temperature than the water, and so the water gains energy by heat. For your cake in the oven in the opening storyline, energy transfers by heat from the hot air in the oven to the cake mixture.

Read this definition of heat (Q in Eq. 8.2) very carefully. In particular, notice what heat is *not* in the following common quotes. (1) Heat is *not* energy in a hot substance. For example, "The boiling water has a lot of heat" is incorrect; the boiling water has *internal energy* E_{int}. (2) Heat is *not* radiation. For example, "It was so hot during the bicycle race because the black roadway was radiating heat" is incorrect; energy is leaving the roadway by *electromagnetic radiation*, T_{ER} in Equation 8.2. (3) Heat is *not* warmth of an environment. For example, "The heat in the air was so oppressive" is incorrect; on a hot day, the air has a high *temperature T.*

As an analogy to the distinction between heat and internal energy, consider the distinction between work and mechanical energy discussed in Chapter 7. The work done on a system is a measure of the amount of energy transferred to the system from its surroundings, whereas the mechanical energy (kinetic energy plus potential energy) of a system is a consequence of the motion and configuration of the system. Therefore, when a person does work on a system, energy is transferred from the person to the system. It makes no sense to talk about the work *of* a system; one can refer only to the work done *on* or *by* a system when some process has occurred in which energy has been transferred to or from the system. Likewise, it makes no sense to talk about the heat *of* a system; one can refer to heat only when energy has been transferred to or from the system as a result of a temperature difference. Both heat and work are ways of transferring energy between a system and its surroundings, which is why they both appear on the right-hand side of Equation 8.2.

Units of Heat

Early studies of heat focused on the resultant increase in temperature of a substance, which was often water. Initial notions of heat were based on a fluid called *caloric* that flowed from one substance to another and caused changes in temperature. From the name of this mythical fluid came an energy unit related to thermal processes, the **calorie (cal),** which is defined as the amount of energy transfer necessary to raise the temperature of 1 g of water from 14.5°C to 15.5°C.[1] (The "Calorie," written with a capital "C" and used in describing the energy content of foods, is actually a kilocalorie.) The unit of energy in the U.S. customary system is the **British thermal unit (Btu),** which is defined as the amount of energy transfer required to raise the temperature of 1 lb of water from 63°F to 64°F.

Once the relationship between energy in thermal and mechanical processes became clear, there was no need for a separate unit related to thermal processes. We have already defined the *joule* as an energy unit based on mechanical processes. Scientists are increasingly turning away from the calorie and the Btu and are using the joule when describing thermal processes. In this textbook, heat, work, and internal energy are usually measured in joules.

The Mechanical Equivalent of Heat

In Chapters 7 and 8, we found that whenever friction is present in a mechanical system, the mechanical energy in the system decreases; in other words, mechanical energy is not conserved in the presence of nonconservative forces. Various experiments show that this mechanical energy does not simply disappear but is transformed into internal energy. You can perform such an experiment at home

James Prescott Joule
British physicist (1818–1889)
Joule received some formal education in mathematics, philosophy, and chemistry from John Dalton but was in large part self-educated. Joule's research led to the establishment of the principle of conservation of energy. His study of the quantitative relationship among electrical, mechanical, and chemical effects of heat culminated in his announcement in 1843 of the amount of work required to produce a unit of energy, called the mechanical equivalent of heat.

---

[1] Originally, the calorie was defined as the energy transfer necessary to raise the temperature of 1 g of water by 1°C. Careful measurements, however, showed that the amount of energy required to produce a 1°C change depends somewhat on the initial temperature; hence, a more precise definition evolved.

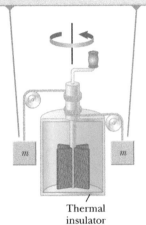

The falling blocks rotate the paddles, causing the temperature of the water to increase.

Thermal insulator

Figure 19.1 Joule's experiment for determining the mechanical equivalent of heat.

by hammering a nail into a scrap piece of wood. What happens to all the kinetic energy of the hammer once you have finished? For the nail and board as a nonisolated system, Equation 8.2 becomes $\Delta E_{int} = W + T_{MW}$, where W is the work done by the hammer on the nail, T_{MW} is the energy leaving the system by sound waves when the nail is struck, and ΔE_{int} represents the warmer nail and wood. Notice that there is *no* transfer of energy by heat in this process. Although this connection between mechanical and internal energy was first suggested by Benjamin Thompson, it was James Prescott Joule who established the equivalence of the decrease in mechanical energy and the increase in internal energy.

A schematic diagram of Joule's most famous experiment is shown in Figure 19.1. The system of interest is the Earth, the two blocks, and the water in a thermally insulated container. Work is done within the system on the water by a rotating paddle wheel, which is driven by heavy blocks falling at a constant speed. The energy transformed in the bearings and the energy passing through the walls by heat are neglected. After the blocks and paddle stop moving, the decrease in gravitational potential energy during the fall of the blocks equals the internal work done by the paddle wheel on the water and, in turn, the increase in internal energy of the water. If the two blocks fall through a distance h, the decrease in potential energy of the system is $2mgh$, where m is the mass of one block; this energy transforms to internal energy E_{int} of the water. By varying the conditions of the experiment, Joule found that the decrease in mechanical energy is proportional to the product of the mass of the water and the increase in water temperature. The proportionality constant was found to be approximately 4.18 J/g · °C. Hence, 4.18 J of mechanical energy raises the temperature of 1 g of water by 1°C. More precise measurements taken later demonstrated the proportionality to be 4.186 J/g · °C when the temperature of the water was raised from 14.5°C to 15.5°C. We adopt this "15-degree calorie" value:

$$1 \text{ cal} = 4.186 \text{ J} \qquad (19.1)$$

This equality is known, for purely historical reasons, as the **mechanical equivalent of heat.** A more proper name would be the *conversion factor between calories and joules,* but the historical name is well entrenched in our language, despite the incorrect use of the word *heat.*

Example 19.1 Losing Weight the Hard Way

A student eats a dinner rated at 2 000 Calories. He wishes to do an equivalent amount of work in the gymnasium by lifting a 50.0-kg barbell. How many times must he raise the barbell to expend this much energy? Assume he raises the barbell 2.00 m each time he lifts it and he transfers no energy when he lowers the barbell.

SOLUTION

Conceptualize Imagine the student raising the barbell. He is doing work on the system of the barbell and the Earth, so energy is leaving his body. The total amount of work that the student must do is 2 000 Calories.

Categorize We model the system of the barbell and the Earth as a *nonisolated system* for *energy.*

Analyze Reduce the conservation of energy equation, Equation 8.2, to the appropriate expression for the system of the barbell and the Earth:

(1) $\Delta U_{total} = W_{total}$

Express the change in gravitational potential energy of the system after the barbell is raised once:

$\Delta U = mgh$

Express the total amount of energy that must be transferred into the system by work for lifting the barbell n times, assuming energy is not regained when the barbell is lowered:

(2) $\Delta U_{total} = nmgh$

19.1 continued

Substitute Equation (2) into Equation (1):	$nmgh = W_{total}$
Solve for n:	$n = \dfrac{W_{total}}{mgh}$
Substitute numerical values:	$n = \dfrac{(2\,000\ \text{Cal})}{(50.0\ \text{kg})(9.80\ \text{m/s}^2)(2.00\ \text{m})}\left(\dfrac{1.00 \times 10^3\ \text{cal}}{\text{Calorie}}\right)\left(\dfrac{4.186\ \text{J}}{1\ \text{cal}}\right)$
	$= 8.54 \times 10^3\ \text{times}$

Finalize If the student is in good shape and lifts the barbell once every 5 s, it will take him about 12 h to perform this feat. Clearly, it is much easier for this student to lose weight by dieting.

In reality, the human body is not 100% efficient. Therefore, not all the energy transformed within the body from the dinner transfers out of the body by work done on the barbell. Some of this energy is used to pump blood and perform other functions within the body. Therefore, the 2 000 Calories can be worked off in less time than 12 h when these other energy processes are included.

19.2 Specific Heat and Calorimetry

When energy is added to a system and there is no change in the kinetic or potential energy of the system, the temperature of the system usually rises. (An exception to this statement is the case in which a system undergoes a change of state—also called a *phase transition*—as discussed in the next section.) If the system consists of a sample of a substance, we find that the quantity of energy required to raise the temperature of a given mass of the substance by some amount varies from one substance to another. For example, the quantity of energy required to raise the temperature of 1 kg of water by 1°C is 4 186 J, but the quantity of energy required to raise the temperature of 1 kg of copper by 1°C is only 387 J. In the discussion that follows, we shall use heat as our example of energy transfer, but keep in mind that the temperature of the system could be changed by means of any method of energy transfer.

The **heat capacity** C of a particular sample is defined as the amount of energy needed to raise the temperature of that sample by 1°C. From this definition, we see that if energy Q produces a change ΔT in the temperature of a sample, then

$$Q = C\,\Delta T \tag{19.2}$$

The **specific heat** c of a substance is the heat capacity per unit mass. Therefore, if energy Q transfers to a sample of a substance with mass m and the temperature of the sample changes by ΔT, the specific heat of the substance is

$$c \equiv \frac{Q}{m\,\Delta T} \tag{19.3}$$

◀ Specific heat

Specific heat is essentially a measure of how thermally insensitive a substance is to the addition of energy. The greater a material's specific heat, the more energy must be added to a given mass of the material to cause a particular temperature change. Table 19.1 (page 506) lists representative specific heats.

From this definition, we can relate the energy Q transferred between a sample of mass m of a material and its surroundings to a temperature change ΔT as

$$Q = mc\,\Delta T \tag{19.4}$$

For example, the energy required to raise the temperature of 0.500 kg of water by 3.00°C is $Q = (0.500\ \text{kg})(4\,186\ \text{J/kg}\cdot°\text{C})(3.00°\text{C}) = 6.28 \times 10^3\ \text{J}$. Notice that when the temperature increases, Q and ΔT are taken to be positive and energy transfers into the system. When the temperature decreases, Q and ΔT are negative and energy transfers out of the system.

PITFALL PREVENTION 19.3
An Unfortunate Choice of Terminology The name *specific heat* is an unfortunate holdover from the days when thermodynamics and mechanics developed separately. A better name would be *specific energy transfer*, but the existing term is too entrenched to be replaced.

PITFALL PREVENTION 19.4
Energy Can Be Transferred by Any Method The symbol Q represents the amount of energy transferred, but keep in mind that the energy transfer in Equation 19.4 could be by *any* of the methods introduced in Chapter 8; it does not have to be heat. For example, repeatedly bending a wire coat hanger raises the temperature at the bending point by *work*.

TABLE 19.1 Specific Heats of Some Substances at 25°C and Atmospheric Pressure

Substance	Specific Heat (J/kg · °C)	Substance	Specific Heat (J/kg · °C)
Elemental solids		*Other solids*	
Aluminum	900	Brass	380
Beryllium	1 830	Glass	837
Cadmium	230	Ice (−5°C)	2 090
Copper	387	Marble	860
Germanium	322	Wood	1 700
Gold	129	*Liquids*	
Iron	448	Alcohol (ethyl)	2 400
Lead	128	Mercury	140
Silicon	703	Water (15°C)	4 186
Silver	234		
		Gas	
		Steam (100°C)	2 010

Note: To convert values to units of cal/g · °C, divide by 4 186.

We can identify $mc\,\Delta T$ as the change in internal energy of the system if we ignore any thermal expansion or contraction of the system, and if there are no phase changes. (Thermal expansion or contraction would result in a very small amount of work being done on the system by the surrounding air.) Then, Equation 19.4 is a reduced form of Equation 8.2: $\Delta E_{int} = Q$. The internal energy of the system can be changed by transferring energy into the system by any mechanism. For example, if the system is a baked potato in a microwave oven, Equation 8.2 reduces to the following analog to Equation 19.4: $\Delta E_{int} = T_{ER} = mc\,\Delta T$, where T_{ER} is the energy transferred to the potato from the microwave oven by electromagnetic radiation. If the system is the air in a bicycle pump, which becomes hot when the pump is operated, Equation 8.2 reduces to the following analog to Equation 19.4: $\Delta E_{int} = W = mc\,\Delta T$, where W is the work done on the pump by the operator. By identifying $mc\,\Delta T$ as ΔE_{int}, we have taken a step toward a better understanding of temperature: temperature is related to the energy of the molecules of a system. We will learn more details of this relationship in Chapter 20.

Specific heat varies with temperature. If, however, temperature intervals are not too great, the temperature variation can be ignored and c can be treated as a constant.[2] For example, the specific heat of water varies by only about 1% from 0°C to 100°C at atmospheric pressure. Unless stated otherwise, we shall neglect such variations.

QUICK QUIZ 19.1 Imagine you have 1 kg each of iron, glass, and water, and all three samples are at 10°C. **(a)** Rank the samples from highest to lowest temperature after 100 J of energy is added to each sample. **(b)** Rank the samples from greatest to least amount of energy transferred by heat if each sample increases in temperature by 20°C.

Notice from Table 19.1 that water has the highest specific heat of common materials. This high specific heat is in part responsible for the moderate climates found near large bodies of water. As the temperature of a body of water decreases during the winter, energy is transferred from the cooling water to the air by heat, increasing the internal energy of the air. Because of the high specific heat of water, a relatively large amount of energy is transferred to the air for even modest temperature changes of the water. The prevailing winds on the West Coast of

---

[2]The definition given by Equation 19.4 assumes the specific heat does not vary with temperature over the interval $\Delta T = T_f - T_i$. In general, if c varies with temperature over the interval, the correct expression for Q is $Q = m \int_{T_i}^{T_f} c(T)\, dT$.

the United States are toward the land (eastward). Hence, the energy liberated by the Pacific Ocean as it cools keeps coastal areas much warmer than they would otherwise be. As a result, West Coast states generally have more favorable winter weather than East Coast states, where the prevailing winds carry the energy away from land.

Calorimetry

One technique for measuring specific heat involves heating a sample to some known temperature T_x, placing it in a vessel containing water of known mass and temperature $T_w < T_x$, and measuring the temperature of the water after equilibrium has been reached. This technique is called **calorimetry,** and devices in which this energy transfer occurs are called **calorimeters.** Figure 19.2 shows the hot sample in the cold water and the resulting energy transfer by heat from the high-temperature part of the system to the low-temperature part. If the system of the sample and the water is isolated, the principle of conservation of energy requires that the amount of energy Q_{hot} that leaves the sample (of unknown specific heat) equal the amount of energy Q_{cold} that enters the water.[3] Conservation of energy allows us to write the mathematical representation of this energy statement as

$$Q_{cold} = -Q_{hot} \tag{19.5}$$

Suppose m_x is the mass of a sample of some substance whose specific heat we wish to determine. Let's call its specific heat c_x and its initial temperature T_x as shown in Figure 19.2. Likewise, let m_w, c_w, and T_w represent corresponding values for the water. If T_f is the final temperature after the system comes to equilibrium, Equation 19.4 shows that the energy transfer for the water is $m_w c_w (T_f - T_w)$, which is positive because $T_f > T_w$, and that the energy transfer for the sample of unknown specific heat is $m_x c_x (T_f - T_x)$, which is negative. Substituting these expressions into Equation 19.5 gives

$$m_w c_w (T_f - T_w) = -m_x c_x (T_f - T_x) \tag{19.6}$$

This equation can be solved for the unknown specific heat c_x.

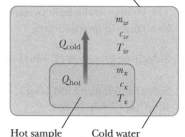

Isolated system boundary

Hot sample   Cold water

Figure 19.2 In a calorimetry experiment, a hot sample whose specific heat is unknown is placed in cold water in a container that isolates the system from the environment.

PITFALL PREVENTION 19.5
Remember the Negative Sign It is *critical* to include the negative sign in Equation 19.5. The negative sign in the equation is necessary for consistency with our sign convention for energy transfer. The energy transfer Q_{hot} has a negative value because energy is leaving the hot substance. The negative sign in the equation ensures that the right side is a positive number, consistent with the left side, which is positive because energy is entering the cold water.

| Example **19.2** | **Fun Time for a Cowboy** |

A cowboy fires a silver bullet with a muzzle speed of 200 m/s into the pine wall of a saloon. Assume all the internal energy generated by the impact remains with the bullet. What is the temperature change of the bullet?

S O L U T I O N

Conceptualize Imagine similar experiences you may have had in which mechanical energy is transformed to internal energy when a moving object is stopped. For example, as mentioned in Section 19.1, a nail becomes warm after it is hit a few times with a hammer.

Categorize The bullet is modeled as an *isolated system*. No work is done on the system because the force from the wall moves through no displacement. This example is similar to the skateboarder pushing off a wall in Section 9.8. There, no work is done on the skateboarder by the wall, and potential energy stored in the body from previous meals is transformed to kinetic energy. Here, no work is done by the wall on the bullet, and kinetic energy of the bullet is transformed to internal energy of the silver comprising the bullet.

Analyze Reduce the conservation of energy equation, Equation 8.2, to the appropriate expression for the system of the bullet:

$$(1) \quad \Delta K + \Delta E_{int} = 0$$

continued

---

[3]For precise measurements, the water container should be included in our calculations because it also exchanges energy with the sample. Doing so would require that we know the container's mass and composition, however. If the mass of the water is much greater than that of the container, we can neglect the effects of the container.

19.2 continued

The change in the bullet's internal energy is related to its change in temperature:

$$(2) \quad \Delta E_{int} = mc\,\Delta T$$

Substitute Equation (2) into Equation (1):

$$(0 - \tfrac{1}{2}mv^2) + mc\,\Delta T = 0$$

Solve for ΔT, using 234 J/kg · °C as the specific heat of silver (see Table 19.1):

$$(3) \quad \Delta T = \frac{\tfrac{1}{2}mv^2}{mc} = \frac{v^2}{2c} = \frac{(200 \text{ m/s})^2}{2(234 \text{ J/kg} \cdot °\text{C})} = 85.5°\text{C}$$

Finalize   Notice that the result does not depend on the mass of the bullet. (In reality, the wall also becomes warmer, so our analysis is simplified.)

WHAT IF?   Suppose the cowboy runs out of silver bullets and fires a lead bullet at the same speed into the wall. Will the temperature change of the bullet be larger or smaller?

Answer   Table 19.1 shows that the specific heat of lead is 128 J/kg · °C, which is smaller than that for silver. Therefore, a given amount of energy input or transformation raises lead to a higher temperature than silver and the final temperature of the lead bullet will be larger. In Equation (3), let's substitute the new value for the specific heat:

$$\Delta T = \frac{v^2}{2c} = \frac{(200 \text{ m/s})^2}{2(128 \text{ J/kg} \cdot °\text{C})} = 156°\text{C}$$

There is no requirement that the silver and lead bullets have the same mass to determine this change in temperature. The only requirement is that they have the same speed.

Example **19.3**   Cooling a Hot Ingot

A 0.050 0-kg ingot of metal is heated to 200.0°C and then dropped into a calorimeter containing 0.400 kg of water initially at 20.0°C. The final equilibrium temperature of the mixed system is 22.4°C. Find the specific heat of the metal.

SOLUTION

Conceptualize   Imagine the process occurring in the isolated system of Figure 19.2. Energy leaves the hot ingot and goes into the cold water, so the ingot cools off and the water warms up. Once both are at the same temperature, the energy transfer stops.

Categorize   We use an equation developed in this section, so we categorize this example as a substitution problem.

Solve Equation 19.6 for c_x:

$$c_x = \frac{m_w c_w (T_f - T_w)}{m_x (T_x - T_f)}$$

Substitute numerical values:

$$c_x = \frac{(0.400 \text{ kg})(4\,186 \text{ J/kg} \cdot °\text{C})(22.4°\text{C} - 20.0°\text{C})}{(0.050\,0 \text{ kg})(200.0°\text{C} - 22.4°\text{C})}$$

$$= 453 \text{ J/kg} \cdot °\text{C}$$

The ingot is most likely iron as you can see by comparing this result with the data given in Table 19.1. The temperature of the ingot is initially above the steam point. Therefore, some of the water may vaporize when the ingot is dropped into the water. We assume the system is sealed and this steam cannot escape. Because the final equilibrium temperature is lower than the steam point, any steam that does result recondenses back into water.

WHAT IF?   Suppose you are performing an experiment in the laboratory that uses this technique to determine the specific heat of a sample and you wish to decrease the overall uncertainty in your final result for c_x. Of the data given in this example, changing which value would be most effective in decreasing the uncertainty?

Answer   The largest experimental uncertainty is associated with the small difference in temperature of 2.4°C for the water. For example, using the rules for propagation of uncertainty in Appendix Section B.8, an uncertainty of 0.1°C in each of T_f and T_w leads to an 8% uncertainty in their difference. For this temperature difference to be larger experimentally, the most effective change is to *decrease the amount of water*.

19.3 Latent Heat

As we have seen in the preceding section, a substance can undergo a change in temperature when energy is transferred between it and its surroundings. In some situations, however, the transfer of energy does not result in a change in temperature. That is the case whenever the physical characteristics of the substance change from one form to another; such a change is commonly referred to as a **phase change.** Two common phase changes are from solid to liquid (melting) and from liquid to gas (boiling); another is a change in the crystalline structure of a solid. All such phase changes involve a change in the system's internal energy but no change in its temperature. The increase in internal energy in boiling, for example, is represented by the breaking of bonds between molecules in the liquid state; this bond breaking allows the molecules to move farther apart in the gaseous state, with a corresponding increase in intermolecular potential energy.

As you might expect, different substances respond differently to the addition or removal of energy as they change phase because their internal molecular arrangements vary. Also, the amount of energy transferred during a phase change depends on the amount of substance involved. (It takes less energy to melt an ice cube than it does a frozen lake.) When discussing two phases of a material, we will use the term *higher-phase material* to mean the material existing at the higher temperature. So, for example, if we discuss water and ice, water is the higher-phase material, whereas steam is the higher-phase material in a discussion of steam and water. Consider a system containing a substance in two phases in equilibrium such as water and ice. The initial amount of the higher-phase material, water, in the system is m_i. Now imagine that energy Q enters the system. As a result, the final amount of water is m_f due to the melting of some of the ice. Therefore, the amount of ice that melted, equal to the amount of *new* water, is $\Delta m = m_f - m_i$. We define the **latent heat** for this phase change as

$$L \equiv \frac{Q}{\Delta m} \qquad (19.7)$$

This parameter is called latent heat (literally, the "hidden" heat) because this added or removed energy does not result in a temperature change. The value of L for a substance depends on the nature of the phase change as well as on the properties of the substance. If the entire amount of the lower-phase material undergoes a phase change, the change in mass Δm of the higher-phase material is equal to the initial mass of the lower-phase material. For example, if an ice cube of mass m on a plate melts completely, the change in mass of the water is $\Delta m = m_f - 0 = m$, which is the mass of new water and is also equal to the initial mass of the ice cube.

From the definition of latent heat, and again choosing heat as our energy transfer mechanism, the energy required to change the phase of a pure substance is

$$Q = L\,\Delta m \qquad (19.8)$$

where Δm is the change in mass of the higher-phase material.

Latent heat of fusion L_f is the term used when the phase change is from solid to liquid (*to fuse* means "to combine by melting"), and **latent heat of vaporization** L_v is the term used when the phase change is from liquid to gas (the liquid "vaporizes").[4] When energy enters a system, causing melting or vaporization, the amount of the higher-phase material increases, so Δm is positive and Q is positive, consistent with our sign convention. When energy is extracted from a system, causing freezing or condensation, the amount of the higher-phase material decreases, so Δm is negative and Q is negative, again consistent with our sign convention. Keep

PITFALL PREVENTION 19.6

Signs Are Critical Sign errors occur very often when students apply calorimetry equations. For phase changes, remember that Δm in Equation 19.8 is always the change in mass of the higher-phase material. In Equation 19.4, be sure your ΔT is *always* the final temperature minus the initial temperature. In addition, you must *always* include the negative sign on the right side of Equation 19.5.

◄ Energy transferred to a substance during a phase change

---

[4]When a gas cools, it eventually *condenses;* that is, it returns to the liquid phase. The energy given up per unit mass is called the *latent heat of condensation* and is numerically equal to the latent heat of vaporization. Likewise, when a liquid cools, it eventually solidifies, and the *latent heat of solidification* is numerically equal to the latent heat of fusion.

TABLE 19.2   Latent Heats of Fusion and Vaporization

Substance	Melting Point (°C)	Latent Heat of Fusion (J/kg)	Boiling Point (°C)	Latent Heat of Vaporization (J/kg)
Helium[a]	−272.2	5.23×10^3	−268.93	2.09×10^4
Oxygen	−218.79	1.38×10^4	−182.97	2.13×10^5
Nitrogen	−209.97	2.55×10^4	−195.81	2.01×10^5
Ethyl alcohol	−114	1.04×10^5	78	8.54×10^5
Water	0.00	3.33×10^5	100.00	2.26×10^6
Sulfur	119	3.81×10^4	444.60	3.26×10^5
Lead	327.3	2.45×10^4	1 750	8.70×10^5
Aluminum	660	3.97×10^5	2 450	1.14×10^7
Silver	960.80	8.82×10^4	2 193	2.33×10^6
Gold	1 063.00	6.44×10^4	2 660	1.58×10^6
Copper	1 083	1.34×10^5	1 187	5.06×10^6

[a]Helium does not solidify at atmospheric pressure. The melting point given here corresponds to a pressure of 2.5 MPa.

in mind that Δm in Equation 19.8 always refers to the higher-phase material. The latent heats of various substances vary considerably as data in Table 19.2 show.

To understand the role of latent heat in phase changes, consider the energy required to convert a system consisting of a 1.00-g cube of ice at −30.0°C to steam at 120.0°C. Figure 19.3 indicates the experimental results obtained when energy is gradually added to the ice. The results are presented as a graph of temperature of the system versus energy added to the system. Let's examine each portion of the red-brown curve, which is divided into parts A through E.

Part A. On this portion of the curve, the temperature of the system changes from −30.0°C to 0.0°C. Equation 19.4 indicates that the temperature varies linearly with the energy added, so the experimental result is a straight line on the graph. Because the specific heat of ice is 2 090 J/kg · °C, we can calculate the amount of energy added by using Equation 19.4:

$$Q = m_{ice} c_{ice} \, \Delta T = (1.00 \times 10^{-3} \, \text{kg})(2\,090 \, \text{J/kg} \cdot °C)(30.0°C) = 62.7 \, \text{J}$$

Part B. When the temperature of the system reaches 0.0°C, the ice–water mixture remains at this temperature—even though energy is being added—until all the ice melts. The energy required to melt 1.00 g of ice at 0.0°C is, from Equation 19.8,

$$Q = L_f \Delta m_w = L_f m_{ice} = (3.33 \times 10^5 \, \text{J/kg})(1.00 \times 10^{-3} \, \text{kg}) = 333 \, \text{J}$$

At this point, we have moved to the 396 J (= 62.7 J + 333 J) mark on the energy axis in Figure 19.3.

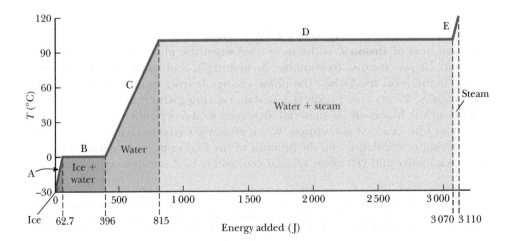

Figure 19.3 A plot of temperature versus energy added when a system initially consisting of 1.00 g of ice at −30.0°C is converted to steam at 120.0°C.

Part C. Between 0.0°C and 100.0°C, nothing surprising happens. No phase change occurs, and so all energy added to the system, which is now water, is used to increase its temperature. The amount of energy necessary to increase the temperature from 0.0°C to 100.0°C is

$$Q = m_w c_w \Delta T = (1.00 \times 10^{-3}\ \text{kg})(4.19 \times 10^3\ \text{J/kg} \cdot °\text{C})(100.0°\text{C}) = 419\ \text{J}$$

where m_w is the mass of the water in the system, which is the same as the mass m_{ice} of the original ice.

Part D. At 100.0°C, another phase change occurs as the system changes from water at 100.0°C to steam at 100.0°C. Similar to the ice–water mixture in part B, the water–steam mixture remains at a fixed temperature, this time 100.0°C—even though energy is being added—until all the liquid has been converted to steam. The energy required to convert 1.00 g of water to steam at 100.0°C is

$$Q = L_v \Delta m_s = L_v m_w = (2.26 \times 10^6\ \text{J/kg})(1.00 \times 10^{-3}\ \text{kg}) = 2.26 \times 10^3\ \text{J}$$

Part E. On this portion of the curve, as in parts A and C, no phase change occurs; therefore, all energy added is used to increase the temperature of the system, which is now steam. The energy that must be added to raise the temperature of the steam from 100.0°C to 120.0°C is

$$Q = m_s c_s \Delta T = (1.00 \times 10^{-3}\ \text{kg})(2.01 \times 10^3\ \text{J/kg} \cdot °\text{C})(20.0°\text{C}) = 40.2\ \text{J}$$

The total amount of energy that must be added to the system to change 1 g of ice at −30.0°C to steam at 120.0°C is the sum of the results from all five parts of the curve, which is 3.11×10^3 J. Conversely, to cool 1 g of steam at 120.0°C to ice at −30.0°C, we must remove 3.11×10^3 J of energy.

Notice in Figure 19.3 the relatively large amount of energy that is transferred into the water to vaporize it to steam. Imagine reversing this process, with a large amount of energy transferred out of steam to condense it into water. That is why a burn to your skin from steam at 100°C is much more damaging than exposure of your skin to water at 100°C. A very large amount of energy enters your skin from the steam, and the steam remains at 100°C for a long time while it condenses. Conversely, when your skin makes contact with water at 100°C, the water immediately begins to drop in temperature as energy transfers from the water to your skin.

QUICK QUIZ 19.2 Suppose the same process of adding energy to the ice cube is performed as discussed above with regard to Figure 19.3, but instead we graph the internal energy of the system as a function of energy input. What would this graph look like?

If liquid water is held perfectly still in a very clean container, it is possible for the water to drop below 0°C without freezing into ice. This phenomenon, called **supercooling,** arises because the water requires a disturbance of some sort for the molecules to move apart and start forming the large, open ice structure that makes the density of ice lower than that of water as discussed in Section 18.4. If supercooled water is disturbed, it suddenly freezes. The system drops into the lower-energy configuration of bound molecules of the ice structure, and the energy released raises the temperature back to 0°C.

Commercial hand warmers consist of liquid sodium acetate in a sealed plastic pouch. The solution in the pouch is in a stable supercooled state. When a disk in the pouch is clicked by your fingers, the liquid solidifies and the temperature increases, just like the supercooled water just mentioned. In this case, however, the freezing point of the liquid is higher than body temperature, so the pouch feels warm to the touch. To reuse the hand warmer, the pouch must be boiled until the solid liquefies. Then, as it cools, it passes below its freezing point into the supercooled state.

It is also possible to create **superheating.** For example, clean water in a very clean cup placed in a microwave oven can sometimes rise in temperature beyond 100°C without boiling because the formation of a bubble of steam in the water requires scratches in the cup or some type of impurity in the water to serve as a nucleation site. When the cup is removed from the microwave oven, the superheated water can become explosive as bubbles form immediately and the hot water is forced upward out of the cup.

Example 19.4   Cooling the Steam

What mass of steam initially at 130°C is needed to warm 200 g of water in a 100-g glass container from 20.0°C to 50.0°C?

SOLUTION

Conceptualize   Imagine placing water and steam together in a closed insulated container. The steam cools and condenses into liquid water, and the system eventually reaches a uniform state of water with a final temperature of 50.0°C in equilibrium with the glass at the same temperature.

Categorize   Based on our conceptualization of this situation, we categorize this example as one involving calorimetry in which a phase change occurs. The calorimeter is an *isolated system* for *energy:* energy transfers between the components of the system but does not cross the boundary between the system and the environment.

Analyze   Write Equation 19.5 to describe the calorimetry process:

$$(1)\quad Q_{\text{cold}} = -Q_{\text{hot}}$$

The steam undergoes three processes: first a decrease in temperature to 100°C, then condensation into liquid water, and finally a decrease in temperature of the water to 50.0°C. Find the energy transfer in the first process using the unknown mass m_s of the steam:

$$Q_1 = m_s c_s \Delta T_s$$

Find the energy transfer in the second process:

$$Q_2 = L_v \Delta m_s = L_v(0 - m_s) = -m_s L_v$$

Find the energy transfer in the third process:

$$Q_3 = m_s c_w \Delta T_{\text{hot water}}$$

Add the energy transfers in these three stages:

$$(2)\quad Q_{\text{hot}} = Q_1 + Q_2 + Q_3 = m_s(c_s \Delta T_s - L_v + c_w \Delta T_{\text{hot water}})$$

The 20.0°C water and the glass undergo only one process, an increase in temperature to 50.0°C. Find the energy transfer in this process:

$$(3)\quad Q_{\text{cold}} = m_w c_w \Delta T_{\text{cold water}} + m_g c_g \Delta T_{\text{glass}}$$

Substitute Equations (2) and (3) into Equation (1):

$$m_w c_w \Delta T_{\text{cold water}} + m_g c_g \Delta T_{\text{glass}} = -m_s(c_s \Delta T_s - L_v + c_w \Delta T_{\text{hot water}})$$

Solve for m_s:

$$m_s = -\frac{m_w c_w \Delta T_{\text{cold water}} + m_g c_g \Delta T_{\text{glass}}}{c_s \Delta T_s - L_v + c_w \Delta T_{\text{hot water}}}$$

Substitute numerical values:

$$m_s = -\frac{(0.200\text{ kg})(4\,186\text{ J/kg} \cdot {}^\circ\text{C})(50.0{}^\circ\text{C} - 20.0{}^\circ\text{C}) + (0.100\text{ kg})(837\text{ J/kg} \cdot {}^\circ\text{C})(50.0{}^\circ\text{C} - 20.0{}^\circ\text{C})}{(2\,010\text{ J/kg} \cdot {}^\circ\text{C})(100{}^\circ\text{C} - 130{}^\circ\text{C}) - (2.26 \times 10^6\text{ J/kg}) + (4\,186\text{ J/kg} \cdot {}^\circ\text{C})(50.0{}^\circ\text{C} - 100{}^\circ\text{C})}$$

$$= 1.09 \times 10^{-2}\text{ kg} = 10.9\text{ g}$$

WHAT IF?   What if the final state of the system is water at 100°C? Would we need more steam or less steam? How would the analysis above change?

Answer   More steam would be needed to raise the temperature of the water and glass to 100°C instead of 50.0°C. There would be two major changes in the analysis. First, we would not have a term Q_3 for the steam because the water that condenses from the steam does not cool below 100°C. Second, in Q_{cold}, the temperature change would be 80.0°C instead of 30.0°C. For practice, show that the result is a required mass of steam of 31.8 g.

19.4 Work in Thermodynamic Processes

In thermodynamics, we describe the *state* of a system using such variables as pressure, volume, temperature, and internal energy. As a result, these quantities belong to a category called **state variables.** For any given configuration of the system, we can identify values of the state variables. (For mechanical systems, the state variables include kinetic energy K and potential energy U. For a single particle as a system, we could identify more state variables: its position x, its velocity v, and its acceleration a.) A state of a system can be specified only if the system is in thermal equilibrium internally. In the case of a gas in a container, internal thermal equilibrium requires that every part of the gas be at the same pressure and temperature.

A second category of variables in situations involving energy is **transfer variables.** These variables are those that appear on the right side of the conservation of energy equation, Equation 8.2. Such a variable has a nonzero value if a process occurs in which energy is transferred across the system's boundary. The transfer variable is positive or negative, depending on whether energy is entering or leaving the system. Because a transfer of energy across the boundary represents a change in the system, transfer variables are not associated with a given state of the system, but rather with a *change* in the state of the system.

In the previous sections, we discussed heat as a transfer variable. In this section, we study another important transfer variable for thermodynamic systems, work. Work performed on particles and nondeformable objects was studied extensively in Chapter 7, and here we investigate the work done on a deformable system, a gas. Consider a gas contained in a cylinder fitted with a movable piston (Fig. 19.4a). At equilibrium, the gas occupies a volume V and exerts a uniform pressure P on the cylinder's walls and on the piston. If the piston has a cross-sectional area A, the magnitude of the force exerted by the gas on the piston is $F = PA$. By Newton's third law, the magnitude of the force exerted by the piston on the gas is also PA. Now let's assume we push the piston inward and compress the gas **quasi-statically,** that is, slowly enough to allow the system to remain essentially in internal thermal equilibrium at all times. The point of application of the force on the gas is the bottom face of the piston. As the piston is pushed downward by an external force $\vec{\mathbf{F}} = -F\hat{\mathbf{j}}$ through a displacement of $d\vec{\mathbf{r}} = dy\,\hat{\mathbf{j}}$ (Fig. 19.4b), the work done on the gas is, according to our definition of work in Chapter 7,

$$dW = \vec{\mathbf{F}} \cdot d\vec{\mathbf{r}} = -F\hat{\mathbf{j}} \cdot dy\,\hat{\mathbf{j}} = -F\,dy = -PA\,dy$$

Because $A\,dy$ is the change in volume of the gas dV, we can express the work done on the gas as

$$dW = -P\,dV \tag{19.9}$$

If the gas is compressed, dV is negative and the work done on the gas is positive. If the gas expands, dV is positive and the work done on the gas is negative. If the volume remains constant, the work done on the gas is zero. The total work done on the gas as its volume changes from V_i to V_f is given by the integral of Equation 19.9:

$$W = -\int_{V_i}^{V_f} P\,dV \tag{19.10}$$

To evaluate this integral, you must know how the pressure varies with volume during the process.

In general, the pressure is not constant during a process followed by a gas, but depends on the volume and temperature. If the pressure and volume are known at each step of the process, the state of the gas at each step can be plotted on an important graphical representation called a ***PV* diagram** as in Figure 19.5. This type of diagram allows us to visualize a process through which a gas is progressing. The curve on a *PV* diagram is called the *path* taken between the initial and final states.

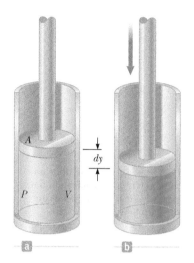

Figure 19.4 Work is done on a gas contained in a cylinder at a pressure P as the piston is pushed downward so that the gas is compressed.

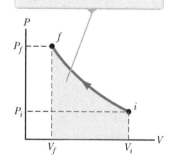

The work done on a gas equals the negative of the area under the *PV* curve. The area is negative here because the volume is decreasing, resulting in positive work.

Figure 19.5 A gas is compressed quasi-statically (slowly) from state i to state f. An outside agent must do positive work on the gas to compress it.

◀ Work done on a gas

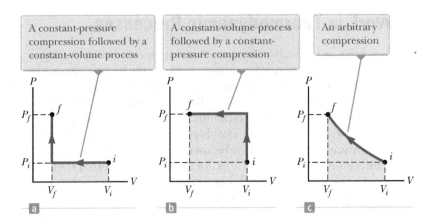

Figure 19.6 The work done on a gas as it is taken from an initial state to a final state depends on the path between these states.

Notice that the integral in Equation 19.10 is equal to the area under a curve on a *PV* diagram. Therefore, we can identify an important use for *PV* diagrams:

> The work done on a gas in a quasi-static process that takes the gas from an initial state to a final state is the negative of the area under the curve on a *PV* diagram, evaluated between the initial and final states.

For the process of compressing a gas in a cylinder, the work done depends on the particular path taken between the initial and final states as Figure 19.5 suggests. To illustrate this important point, consider several different paths connecting *i* and *f* (Fig. 19.6). In the process depicted in Figure 19.6a, the volume of the gas is first reduced from V_i to V_f at constant pressure P_i and the pressure of the gas then increases from P_i to P_f by heating at constant volume V_f. The work done on the gas along this path is $-P_i(V_f - V_i)$. In Figure 19.6b, the pressure of the gas is increased from P_i to P_f at constant volume V_i and then the volume of the gas is reduced from V_i to V_f at constant pressure P_f. The work done on the gas is $-P_f(V_f - V_i)$. This value is greater than that for the process described in Figure 19.6a because the piston is moved through the same displacement by a larger force. Finally, for the process described in Figure 19.6c, where both *P* and *V* change continuously, the work done on the gas has some value between the values obtained in the first two processes. To evaluate the work in this case, the function $P(V)$ must be known so that we can evaluate the integral in Equation 19.10.

The energy transfer *Q* into or out of a system by heat also depends on the process. For example, in Chapter 20, we will show that a constant-volume process between two temperatures requires a different amount of heat than a constant-pressure process between the same temperatures.

19.5 The First Law of Thermodynamics

When we introduced the law of conservation of energy in Chapter 8, we stated that the change in the energy of a system is equal to the sum of all transfers of energy across the system's boundary (Eq. 8.2). The **first law of thermodynamics** is a special case of the law of conservation of energy that describes processes in which only the internal energy[5] changes and the only energy transfers are by heat and work:

First law of thermodynamics ▶

$$\Delta E_{\text{int}} = Q + W \tag{19.11}$$

Look back at Equation 8.2 to see that the first law of thermodynamics is contained within that more general equation.

---

[5] It is an unfortunate accident of history that the traditional symbol for internal energy is *U*, which is also the traditional symbol for potential energy as introduced in Chapter 7. To avoid confusion between potential energy and internal energy, we use the symbol E_{int} for internal energy in this book. If you take an advanced course in thermodynamics, however, be prepared to see *U* used as the symbol for internal energy in the first law.

Let's discuss each of the three terms in the first law for various processes through which a gas is taken. As a model, let's consider the sample of gas contained in the piston–cylinder apparatus in Figure 19.7. This figure shows work being done on the gas and energy transferring in by heat, so the internal energy of the gas is rising. In the following discussion of various processes, refer back to this figure and mentally alter the directions of the transfer of energy to reflect what is happening in the process.

First, consider an *isolated system,* that is, one that does not interact with its surroundings, as we have seen before. In this case, no energy transfer by heat takes place and the work done on the system is zero; hence, the internal energy remains constant. That is, because $Q = W = 0$, it follows that $\Delta E_{int} = 0$; therefore, $E_{int,i} = E_{int,f}$. We conclude that the internal energy E_{int} of an isolated system remains constant.

Next, consider the case of a system that can exchange energy with its surroundings and is taken through a **cyclic process,** that is, a process that starts and ends at the same state. On a *PV* diagram, a cyclic process appears as a closed curve, as shown in Figure 19.8. In this case, the change in the internal energy must again be zero because E_{int} is a state variable; therefore, the energy Q added to the system must equal the negative of the work W done on the system during the cycle. That is, in a cyclic process,

$$\Delta E_{int} = 0 \quad \text{and} \quad Q = -W \quad \text{(cyclic process)}$$

It can be shown that in a cyclic process for a gas, the net work done on the system per cycle equals the area enclosed by the path representing the process on a *PV* diagram.

A process that occurs at constant temperature is called an **isothermal process.** This process can be established by immersing the cylinder in Figure 19.7 in an ice–water bath or by putting the cylinder in contact with some other constant-temperature reservoir. A plot of *P* versus *V* at constant temperature for an ideal gas yields a hyperbolic curve called an *isotherm,* as shown in Figure 19.9. The ideal gas law (Eq. 18.9) with *T* constant indicates that the equation of this curve is $PV = nRT = $ constant. We show in Chapter 20 that the internal energy of an ideal gas is a function of temperature only. Hence, because the temperature does not change in an isothermal process involving an ideal gas, we must have $\Delta E_{int} = 0$. For an isothermal process, we conclude from the first law that the energy transfer Q must be equal to the negative of the work done on the gas; that is, $Q = -W$. Any energy that enters the system by heat is transferred out of the system by work.

Let's calculate the work done on the gas in the isothermal expansion from state *i* to state *f* in Figure 19.9. The work done on the gas is given by Equation 19.10.

Figure 19.7 The first law of thermodynamics equates the change in internal energy E_{int} in a system to the net energy transfer to the system by heat Q and work W. In the situation shown here, the internal energy of the gas increases.

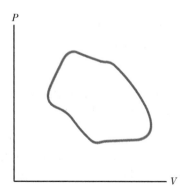

Figure 19.8 A cyclic process on a gas forms a closed curve on a *PV* diagram.

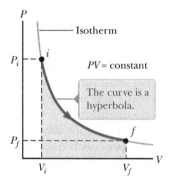

Figure 19.9 The *PV* diagram for an isothermal expansion of an ideal gas from an initial state to a final state.

Because the gas is ideal and the process is quasi-static, the ideal gas law is valid for each point on the path. Therefore,

$$W = -\int_{V_i}^{V_f} P \, dV = -\int_{V_i}^{V_f} \frac{nRT}{V} \, dV$$

Because T is constant in this case, it can be removed from the integral along with n and R:

$$W = -nRT \int_{V_i}^{V_f} \frac{dV}{V} = -nRT \ln V \Big|_{V_i}^{V_f}$$

To evaluate the integral, we used $\int(dx/x) = \ln x$. (See Appendix B.) Evaluating the result at the initial and final volumes gives

$$W = nRT \ln \left(\frac{V_i}{V_f}\right) \quad \text{(isothermal process)} \tag{19.12}$$

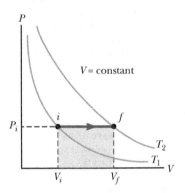

Figure 19.10 An isobaric process takes a gas between temperatures T_1 and T_2.

Numerically, this work W equals the negative of the shaded area under the PV curve shown in Figure 19.9. Because the gas expands, $V_f > V_i$ and the value for the work done on the gas is negative as we expect. If the gas is compressed, then $V_f < V_i$ and the work done on the gas is positive.

A process that occurs at constant pressure is called an **isobaric process.** In Figure 19.7, an isobaric process could be established by allowing the piston to move freely so that it is always in equilibrium between the net force from the gas pushing upward and the weight of the piston plus the force due to atmospheric pressure pushing downward. An isobaric process appears as a horizontal line on a PV diagram as shown in Figure 19.10. Find the isobaric processes in Figure 19.6.

In such a process, the values of the heat and the work are both usually nonzero. The work done on the gas in an isobaric process is simply

$$W = -P(V_f - V_i) \quad \text{(isobaric process)} \tag{19.13}$$

where P is the constant pressure of the gas during the process.

A process that takes place at constant volume is called an **isovolumetric process.** Another name for this type of process is *isochoric*. In Figure 19.7, clamping the piston at a fixed position would ensure an isovolumetric process. An isovolumetric process appears as a vertical line on a PV diagram as shown in Figure 19.11. Find the isovolumetric processes in Figure 19.6.

Because the volume of the gas does not change in such a process, the work given by Equation 19.10 is zero. Hence, from the first law we see that in an isovolumetric process, because $W = 0$,

$$\Delta E_{\text{int}} = Q \quad \text{(isovolumetric process)} \tag{19.14}$$

This expression specifies that if energy is added by heat to a system kept at constant volume, all the transferred energy remains in the system as an increase in its internal energy. For example, when a spray can is thrown in a fire, as in Example 18.4, energy enters the system (the gas in the can) by heat through the metal walls of the can. Consequently, the temperature, and therefore the pressure, in the can increases until the can possibly explodes.

An **adiabatic process** is one during which no energy enters or leaves the system by heat; that is, $Q = 0$. An adiabatic process can be achieved either by thermally insulating the walls of the system or by performing the process rapidly so that there is negligible time for energy to transfer by heat. Applying the first law of thermodynamics to an adiabatic process gives

$$\Delta E_{\text{int}} = W \quad \text{(adiabatic process)} \tag{19.15}$$

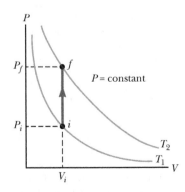

Figure 19.11 An isovolumetric process takes a gas between temperatures T_1 and T_2.

This result shows that if a gas is compressed adiabatically such that W is positive, then ΔE_{int} is positive and the temperature of the gas increases. Conversely, the temperature of a gas decreases when the gas expands adiabatically.

Adiabatic processes are very important in engineering practice. Some common examples are the expansion of hot gases in an internal combustion engine, the liquefaction of gases in a cooling system, and the compression stroke in a diesel engine. We will see PV diagrams for adiabatic processes and study them in more detail in Chapter 20.

QUICK QUIZ 19.3 In the last three columns of the following table, fill in the boxes with the correct signs ($-$, $+$, or 0) for Q, W, and ΔE_{int}. For each situation, the system to be considered is identified.

Situation	System	Q	W	ΔE_{int}
(a) Rapidly pumping up a bicycle tire	Air in the pump			
(b) Pan of room-temperature water sitting on a hot stove	Water in the pan			
(c) Air quickly leaking out of a balloon	Air originally in the balloon			

QUICK QUIZ 19.4 Characterize the paths in Figure 19.12 as isobaric, isovolumetric, isothermal, or adiabatic. For path B, $Q = 0$. The blue curves are isotherms.

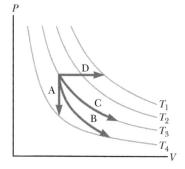

Figure 19.12 (Quick Quiz 19.4) Identify the nature of paths A, B, C, and D.

Example 19.5 **An Isothermal Expansion**

A 1.0-mol sample of an ideal gas is kept at 0.0°C during an expansion from 3.0 L to 10.0 L.

(A) How much work is done on the gas during the expansion?

SOLUTION

Conceptualize Run the process in your mind: the cylinder in Figure 19.7 is immersed in an ice-water bath, and the piston moves outward so that the volume of the gas increases. You can also use the graphical representation in Figure 19.9 to conceptualize the process.

Categorize We will evaluate parameters using equations developed in the preceding sections, so we categorize this example as a substitution problem. Because the temperature of the gas is fixed, the process is isothermal.

Substitute the given values into Equation 19.12:

$$W = nRT \ln\left(\frac{V_i}{V_f}\right)$$

$$= (1.0 \text{ mol})(8.31 \text{ J/mol} \cdot \text{K})(273 \text{ K}) \ln\left(\frac{3.0 \text{ L}}{10.0 \text{ L}}\right)$$

$$= -2.7 \times 10^3 \text{ J}$$

(B) How much energy transfer by heat occurs between the gas and its surroundings in this process?

SOLUTION

Find the heat from the first law:

$$\Delta E_{int} = Q + W$$

$$0 = Q + W$$

$$Q = -W = 2.7 \times 10^3 \text{ J}$$

continued

19.5 continued

(C) If the gas is returned to the original volume by means of an isobaric process, how much work is done on the gas?

SOLUTION

Use Equation 19.13. The pressure is not given, so incorporate the ideal gas law:

$$W = -P(V_f - V_i) = -\frac{nRT_i}{V_i}(V_f - V_i)$$

$$= -\frac{(1.0 \text{ mol})(8.31 \text{ J/mol} \cdot \text{K})(273 \text{ K})}{10.0 \times 10^{-3} \text{ m}^3}(3.0 \times 10^{-3} \text{ m}^3 - 10.0 \times 10^{-3} \text{ m}^3)$$

$$= 1.6 \times 10^3 \text{ J}$$

We used the initial temperature and volume to calculate the work done because the final temperature was unknown. The work done on the gas is positive because the gas is being compressed.

Example 19.6   Boiling Water

Suppose 1.00 g of water vaporizes isobarically at atmospheric pressure (1.013×10^5 Pa). Its volume in the liquid state is $V_i = V_{liquid} = 1.00 \text{ cm}^3$, and its volume in the vapor state is $V_f = V_{vapor} = 1\,671 \text{ cm}^3$. Find the work done in the expansion and the change in internal energy of the system. Ignore any mixing of the steam and the surrounding air; imagine that the steam simply pushes the surrounding air out of the way.

SOLUTION

Conceptualize   Notice that the temperature of the system does not change. There is a phase change occurring as the water evaporates to steam.

Categorize   Because the expansion takes place at constant pressure, we categorize the process as isobaric. We will use equations developed in the preceding sections, so we categorize this example as a substitution problem.

Use Equation 19.13 to find the work done on the system as the air is pushed out of the way:

$$W = -P(V_f - V_i)$$

$$= -(1.013 \times 10^5 \text{ Pa})(1\,671 \times 10^{-6} \text{ m}^3 - 1.00 \times 10^{-6} \text{ m}^3)$$

$$= -169 \text{ J}$$

Use Equation 19.8 and the latent heat of vaporization for water to find the energy transferred into the system by heat:

$$Q = L_v \Delta m_s = m_s L_v = (1.00 \times 10^{-3} \text{ kg})(2.26 \times 10^6 \text{ J/kg})$$

$$= 2\,260 \text{ J}$$

Use the first law to find the change in internal energy of the system:

$$\Delta E_{int} = Q + W = 2\,260 \text{ J} + (-169 \text{ J}) = 2.09 \text{ kJ}$$

The positive value for ΔE_{int} indicates that the internal energy of the system increases. The largest fraction of the energy ($2\,090 \text{ J}/ 2\,260 \text{ J} = 93\%$) transferred to the liquid goes into increasing the internal energy of the system. The remaining 7% of the energy transferred leaves the system by work done by the steam on the surrounding atmosphere.

19.6 Energy Transfer Mechanisms in Thermal Processes

In Chapter 8, we introduced a global approach to the energy analysis of physical processes through Equation 8.2, where the energy transfer on the right hand side of the equation can occur by several mechanisms. Earlier in this chapter, we discussed two of the terms on the right side of this equation, work W and heat Q. In this section, we explore more details about heat as a means of energy transfer and two other energy transfer methods often related to temperature changes: convection (a form of matter transfer T_{MT}) and electromagnetic radiation T_{ER}.

Thermal Conduction

The process of energy transfer by heat (Q in Eq. 8.2) can also be called **conduction** or **thermal conduction.** In this process, the transfer can be represented on an atomic scale as an exchange of kinetic energy between microscopic particles—molecules, atoms, and free electrons—in which less-energetic particles gain energy in collisions with more-energetic particles. For example, if you hold one end of a long metal bar and insert the other end into a flame, you will find that the temperature of the metal in your hand soon increases. The energy reaches your hand by means of conduction. Initially, before the rod is inserted into the flame, the microscopic particles in the metal are vibrating about their equilibrium positions. As the flame raises the temperature of the rod, the particles near the flame begin to vibrate with greater and greater amplitudes. These particles, in turn, collide with their neighbors and transfer some of their energy in the collisions. Slowly, the amplitudes of vibration of metal atoms and electrons farther and farther from the flame increase until eventually those in the metal near your hand are affected. This increased vibration is detected by an increase in the temperature of the metal and of your potentially burned hand.

The rate of thermal conduction through a material depends on the properties of the material. For example, it is possible to hold a piece of asbestos in a flame indefinitely, which implies that very little energy is conducted through the asbestos. In general, metals are good thermal conductors and materials such as asbestos, cork, paper, and fiberglass are poor conductors. Gases also are poor conductors because the separation distance between the particles is so great. Metals are good thermal conductors because they contain large numbers of electrons that are relatively free to move through the metal and so can transport energy over large distances. Therefore, in a good conductor such as copper, conduction takes place by means of both the vibration of atoms and the motion of free electrons. The presence of free electrons in metals is also the reason that metals are good *electrical* conductors. We will study electrical conduction in metals in Chapter 26.

Conduction occurs only if there is a difference in temperature between two parts of the conducting medium. Consider a slab of material of thickness L and cross-sectional area A. One face of the slab is at a temperature T_c, and the other face is at a temperature $T_h > T_c$ (Fig. 19.13). Experimentally, it is found that energy Q transfers in a time interval Δt from the hotter face to the colder one. The energy Q that transfers is found to be proportional to the cross-sectional area, the temperature difference $\Delta T = T_h - T_c$, and the time interval, and inversely proportional to the thickness:

$$Q = kA\frac{\Delta T}{L}\Delta t \tag{19.16}$$

where the proportionality constant k is the **thermal conductivity** of the material.

For a slab of infinitesimal thickness dx and temperature difference dT, we can write the **law of thermal conduction** as

$$P = kA\left|\frac{dT}{dx}\right| \tag{19.17}$$

where $|dT/dx|$ is the **temperature gradient** (the rate at which temperature varies with position). Notice that P has units of watts when Q is in joules and Δt is in seconds. That is not surprising because P is power, the rate of energy transfer by heat.

Substances that are good thermal conductors have large thermal conductivity values, whereas good thermal insulators have low thermal conductivity values. Table 19.3 lists thermal conductivities for various substances. Notice that metals are generally better thermal conductors than nonmetals.

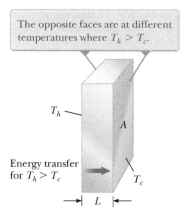

The opposite faces are at different temperatures where $T_h > T_c$.

T_h

A

Energy transfer for $T_h > T_c$

T_c

L

Figure 19.13 Energy transfer through a conducting slab with a cross-sectional area A and a thickness L.

TABLE 19.3 Thermal Conductivities

Substance	Thermal Conductivity (W/m · °C)
Metals (at 25°C)	
Aluminum	238
Copper	397
Gold	314
Iron	79.5
Lead	34.7
Silver	427
Nonmetals (approximate values)	
Asbestos	0.08
Concrete	0.8
Diamond	2 300
Glass	0.8
Ice	2
Rubber	0.2
Water	0.6
Wood	0.08
Gases (at 20°C)	
Air	0.023 4
Helium	0.138
Hydrogen	0.172
Nitrogen	0.023 4
Oxygen	0.023 8

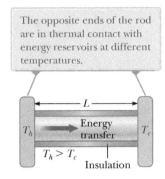

The opposite ends of the rod are in thermal contact with energy reservoirs at different temperatures.

Figure 19.14 Conduction of energy through a uniform, insulated rod of length L.

Suppose a long, uniform rod of length L is thermally insulated so that energy cannot escape by heat from its surface except at the ends as shown in Figure 19.14. One end is in thermal contact with an energy reservoir at temperature T_c, and the other end is in thermal contact with a reservoir at temperature $T_h > T_c$. When a steady state has been reached, the temperature at each point along the rod is constant in time. In this case, if we assume k is not a function of temperature, the temperature gradient is the same everywhere along the rod and is

$$\left|\frac{dT}{dx}\right| = \frac{T_h - T_c}{L}$$

Therefore, the rate of energy transfer by conduction through the rod is

$$P = kA\left(\frac{T_h - T_c}{L}\right) \tag{19.18}$$

For a compound slab containing several materials of thicknesses L_1, L_2, ... and thermal conductivities k_1, k_2, ..., the rate of energy transfer through the slab at steady state is

$$P = \frac{A(T_h - T_c)}{\sum_i (L_i/k_i)} \tag{19.19}$$

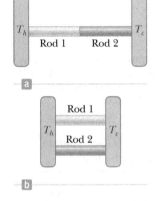

Figure 19.15 (Quick Quiz 19.5) In which case is the rate of energy transfer larger?

where T_h and T_c are the temperatures of the outer surfaces (which are held constant) and the summation is over all slabs. Example 19.7 shows how Equation 19.19 results from a consideration of two thicknesses of materials.

QUICK QUIZ 19.5 You have two rods of the same length and diameter, but they are formed from different materials. The rods are used to connect two regions at different temperatures so that energy transfers through the rods by heat. They can be connected in series as in Figure 19.15a or in parallel as in Figure 19.15b. In which case is the rate of energy transfer by heat larger? **(a)** The rate is larger when the rods are in series. **(b)** The rate is larger when the rods are in parallel. **(c)** The rate is the same in both cases.

Example 19.7   **Energy Transfer Through Two Slabs**

Two slabs of thickness L_1 and L_2 and thermal conductivities k_1 and k_2 are in thermal contact with each other as shown in Figure 19.16. The temperatures of their outer surfaces are T_c and T_h, respectively, and $T_h > T_c$. Determine the temperature at the interface and the rate of energy transfer by conduction through an area A of the slabs in the steady-state condition.

SOLUTION

Conceptualize Notice the phrase "in the steady-state condition." We interpret this phrase to mean that energy transfers through the compound slab at the same rate at all points. Otherwise, energy would be building up or disappearing at some point. Furthermore, the temperature varies with position in the two slabs, most likely at different rates in each part of the compound slab. When the system is in steady state, the interface is at some fixed temperature T.

Categorize We categorize this example as a thermal conduction problem and impose the condition that the power is the same in both slabs of material.

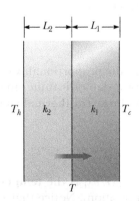

Figure 19.16 (Example 19.7) Energy transfer by conduction through two slabs in thermal contact with each other. At steady state, the rate of energy transfer through slab 1 equals the rate of energy transfer through slab 2.

19.7 continued

Analyze Use Equation 19.18 to express the rate at which energy is transferred through an area A of slab 1:

$$(1) \quad P_1 = k_1 A \left(\frac{T - T_c}{L_1} \right)$$

Express the rate at which energy is transferred through the same area of slab 2:

$$(2) \quad P_2 = k_2 A \left(\frac{T_h - T}{L_2} \right)$$

Set these two rates equal to represent the steady-state situation:

$$k_1 A \left(\frac{T - T_c}{L_1} \right) = k_2 A \left(\frac{T_h - T}{L_2} \right)$$

Solve for T:

$$(3) \quad T = \frac{k_1 L_2 T_c + k_2 L_1 T_h}{k_1 L_2 + k_2 L_1}$$

Substitute Equation (3) into either Equation (1) or Equation (2):

$$(4) \quad P = \frac{A(T_h - T_c)}{(L_1/k_1) + (L_2/k_2)}$$

Finalize Extension of this procedure to several slabs of materials leads to Equation 19.19. Equation (4) is Equation 19.19 with i ranging from 1 to 2.

WHAT IF? Suppose you are building an insulated container with two layers of insulation and the rate of energy transfer determined by Equation (4) turns out to be too high. You have enough room to increase the thickness of one of the two layers by 20%. How would you decide which layer to choose?

Answer To decrease the power as much as possible, you must increase the denominator in Equation (4) as much as possible. Whichever thickness you choose to increase, L_1 or L_2, you increase the corresponding term L/k in the denominator by 20%. For this percentage change to represent the largest absolute change, you want to take 20% of the larger term. Therefore, you should increase the thickness of the layer that has the larger value of L/k.

Home Insulation

In engineering practice, the term L/k for a particular substance is referred to as the **R-value** of the material. Therefore, Equation 19.19 reduces to

$$P = \frac{A(T_h - T_c)}{\sum_i R_i} \qquad (19.20)$$

where $R_i = L_i/k_i$. The R-values for a few common building materials are given in Table 19.4. In the United States, the insulating properties of materials used in buildings are usually expressed in U.S. customary units, not SI units. Therefore,

TABLE 19.4 R-Values for Some Common Building Materials

Material	R-value (ft$^2 \cdot$ °F $\cdot$ h/Btu)
Hardwood siding (1 in. thick)	0.91
Wood shingles (lapped)	0.87
Brick (4 in. thick)	4.00
Concrete block (filled cores)	1.93
Fiberglass insulation (3.5 in. thick)	10.90
Fiberglass insulation (6 in. thick)	18.80
Fiberglass board (1 in. thick)	4.35
Cellulose fiber (1 in. thick)	3.70
Flat glass (0.125 in. thick)	0.89
Insulating glass (0.25-in. space)	1.54
Air space (3.5 in. thick)	1.01
Stagnant air layer	0.17
Drywall (0.5 in. thick)	0.45
Sheathing (0.5 in. thick)	1.32

in Table 19.4, *R*-values are given as a combination of British thermal units, feet, hours, and degrees Fahrenheit.

At any vertical surface open to the air, a very thin stagnant layer of air adheres to the surface. One must consider this layer when determining the *R*-value for a wall. The thickness of this stagnant layer on an outside wall depends on the speed of the wind. Energy transfer through the walls of a house on a windy day is greater than that on a day when the air is calm. A representative *R*-value for this stagnant layer of air is given in Table 19.4.

Example **19.8**   The *R*-Value of a Typical Wall

Calculate the total *R*-value for a wall constructed as shown in Figure 19.17a. Starting outside the house (toward the front in the figure) and moving inward, the wall consists of 4 in. of brick, 0.5 in. of sheathing, an air space 3.5 in. thick, and 0.5 in. of drywall.

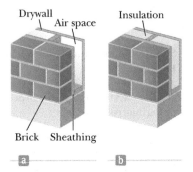

Figure 19.17 (Example 19.8) An exterior house wall containing (a) an air space and (b) insulation.

SOLUTION

Conceptualize Use Figure 19.17 to help conceptualize the structure of the wall. Do not forget the stagnant air layers inside and outside the house.

Categorize We will use specific equations developed in this section on home insulation, so we categorize this example as a substitution problem.

Use Table 19.4 to find the *R*-value of each layer:

$$R_1 \text{ (outside stagnant air layer)} = 0.17 \text{ ft}^2 \cdot {}^\circ\text{F} \cdot \text{h/Btu}$$

$$R_2 \text{ (brick)} = 4.00 \text{ ft}^2 \cdot {}^\circ\text{F} \cdot \text{h/Btu}$$

$$R_3 \text{ (sheathing)} = 1.32 \text{ ft}^2 \cdot {}^\circ\text{F} \cdot \text{h/Btu}$$

$$R_4 \text{ (air space)} = 1.01 \text{ ft}^2 \cdot {}^\circ\text{F} \cdot \text{h/Btu}$$

$$R_5 \text{ (drywall)} = 0.45 \text{ ft}^2 \cdot {}^\circ\text{F} \cdot \text{h/Btu}$$

$$R_6 \text{ (inside stagnant air layer)} = 0.17 \text{ ft}^2 \cdot {}^\circ\text{F} \cdot \text{h/Btu}$$

Add the *R*-values to obtain the total *R*-value for the wall:

$$R_{\text{total}} = R_1 + R_2 + R_3 + R_4 + R_5 + R_6 = 7.12 \text{ ft}^2 \cdot {}^\circ\text{F} \cdot \text{h/Btu}$$

WHAT IF? Suppose you are not happy with this total *R*-value for the wall. You cannot change the overall structure, but you can fill the air space as in Figure 19.17b. To *maximize* the total *R*-value, what material should you choose to fill the air space?

Answer Looking at Table 19.4, we see that 3.5 in. of fiberglass insulation is more than ten times as effective as 3.5 in. of air. Therefore, we should fill the air space with fiberglass insulation. The result is that we add 10.90 ft$^2 \cdot {}^\circ$F $\cdot$ h/Btu of *R*-value, and we lose 1.01 ft$^2 \cdot {}^\circ$F $\cdot$ h/Btu due to the air space we have replaced. The new total *R*-value is equal to 7.12 ft$^2 \cdot {}^\circ$F $\cdot$ h/Btu + 9.89 ft$^2 \cdot {}^\circ$F $\cdot$ h/Btu = 17.01 ft$^2 \cdot {}^\circ$F $\cdot$ h/Btu.

Convection

At one time or another, you may have warmed your hands on a cold day by holding them over a toaster while it is operating. In this situation, the air in the toaster is warmed and expands. As a result, the density of this air decreases and the air rises. This hot air warms your hands as it flows by. Energy transferred by the movement of a warm substance is said to have been transferred by **convection,** which is a form of matter transfer, T_{MT} in Equation 8.2. When resulting from differences in density, as with air in the toaster, the process is referred to as *natural convection*. Airflow at an ocean coast (Section 19.2) is an example of natural convection, as is the mixing that occurs as surface water in a lake cools and sinks (see Section 18.4). When the heated substance is forced to move by a fan or pump, as in some hot-air and hot-water heating systems, the process is called *forced convection.*

If it were not for convection currents, it would be very difficult to boil water. As water is heated in a teakettle, the lower layers are warmed first. This water expands and rises to the top because its density is lowered. At the same time, the denser, cool water at the surface sinks to the bottom of the kettle and is heated.

Radiation

The third means of energy transfer we shall discuss is **electromagnetic radiation, T_{ER}** in Equation 8.2. All objects radiate energy continuously in the form of electromagnetic waves (see Chapter 33) produced by thermal vibrations of the molecules. You are likely familiar with electromagnetic radiation in the form of the orange glow from an electric stove burner or an electric space heater. In the toaster mentioned in the section on convection, energy reaches the bread by electromagnetic radiation from the glowing coils, which you can see if you look downward into the toaster.

The rate at which the surface of an object radiates energy is proportional to the fourth power of the absolute temperature of the surface. Known as **Stefan's law,** this behavior is expressed in equation form as

$$P = \sigma A e T^4 \qquad (19.21) \qquad \blacktriangleleft \text{ Stefan's law}$$

where P is the power in watts of electromagnetic waves radiated from the surface of the object, σ is a constant equal to $5.669\,6 \times 10^{-8}$ W/m² · K⁴, A is the surface area of the object in square meters, e is the **emissivity,** and T is the surface temperature in kelvins. The value of e can vary between zero and unity depending on the properties of the surface of the object. The emissivity is equal to the **absorptivity,** which is the fraction of the incoming radiation that the surface absorbs. A mirror has very low absorptivity because it reflects almost all incident light. Therefore, a mirror surface also has a very low emissivity. At the other extreme, a black surface has high absorptivity and high emissivity. An **ideal absorber** is defined as an object that absorbs all the energy incident on it, and for such an object, $e = 1$. An object for which $e = 1$ is often referred to as a **black body.** We shall investigate experimental and theoretical approaches to radiation from a black body in Chapter 39.

Every second, approximately 1 370 J of electromagnetic radiation from the Sun passes perpendicularly through each 1 m² at the top of the Earth's atmosphere. This radiation is primarily visible, infrared, and ultraviolet. We shall study these types of radiation in detail in Chapter 33. Enough energy arrives at the surface of the Earth each day to supply all our energy needs on this planet hundreds of times over, if only it could be captured and used efficiently. The growth in the number of solar energy–powered houses and solar energy "farms" in the world reflects the increasing efforts being made to use this abundant energy.

What happens to the atmospheric temperature at night is another example of the effects of energy transfer by radiation. If there is a cloud cover above the Earth, the water vapor in the clouds absorbs part of the infrared radiation emitted by the Earth and re-emits it back to the surface. Consequently, temperature levels at the surface remain moderate. In the absence of this cloud cover, there is less in the way to prevent this radiation from escaping into space; therefore, the temperature decreases more on a clear night than on a cloudy one.

As an object radiates energy at a rate given by Equation 19.21, it also absorbs electromagnetic radiation from the surroundings, which consist of other objects that radiate energy. If the latter process did not occur, an object would eventually radiate all its energy and its temperature would reach absolute zero. If an object is at a temperature T and its surroundings are at an average temperature T_0, the net rate of energy gained or lost by the object as a result of radiation is

$$P_{net} = \sigma A e (T^4 - T_0{}^4) \qquad \text{(19.22)}$$

When an object is in equilibrium with its surroundings, it radiates and absorbs energy at the same rate and its temperature remains constant. When an object is hotter than its surroundings, it radiates more energy than it absorbs and its temperature decreases.

Let's revisit your trip to the mountain in the opening storyline. Your first experience was seeing a sign, "Caution: Bridge Freezes Before Road Surface." A major contribution to this effect is that a roadway on the ground has energy transferring to it by heat Q from the warm ground underneath the roadway. A bridge has cold air underneath it, so it does not have this source of energy. Another factor is that the bridge roadway can radiate energy T_{ER} into the air from both upper and lower surfaces, losing internal energy E_{int} more rapidly than a roadway on the ground. Therefore, the bridge cools faster than the roadway, and water freezes on the bridge first. Your next thought was why mountain air is cold even though you are closer to the Sun. The change in distance to the Sun is miniscule compared to the distance to the Sun; that has no effect. Imagine air moving up to the mountain from sea level by convection T_{MT}. Because air is a poor thermal conductor, as a parcel of air moves from high pressure surroundings at sea level to lower-pressure surroundings on the mountain, it undergoes an adiabatic expansion, $Q = 0$. As mentioned in Section 19.5, an adiabatic expansion causes the temperature of the air to decrease: cold air at high altitudes.

Now, why was your meal on the mountain so unsuccessful? The eggs were undercooked. At lower atmospheric pressure on the mountain, the phase transition from water to steam takes place at a lower temperature. Therefore, you cooked your eggs at a temperature lower than 100°C, and they did not cook completely in the three-minute time interval. At higher altitudes, you need to boil food longer. The cookies near the edge of the baking sheet were too well done. The baking sheet is an object with a high temperature, so it radiates energy T_{ER} perpendicularly away from its surface. If the baking sheet has turned-up edges, this perpendicular direction from the edges is toward the cookies near the edge. Therefore, these cookies receive more energy by radiation than those near the center of the baking sheet, and they bake faster, even possibly burning while the ones near the center are perfect. Why did your cake fall? High-altitude baking is an art and requires careful adjustment of ingredients for successful cakes. One consideration in baking cakes is again the reduced boiling point of water. As a result, when the cake batter is placed in the oven, the water evaporates more rapidly than at sea level. As the too-dry batter rises, it cannot form "bubbles" of steam that build the regular cellular structure that supports the weight of the upper part of the cake.

After waking up the next morning and going for a walk, you noticed frost on cars and mailboxes, but only on the top surfaces. This effect is a demonstration of Equation 19.22. The side surfaces of cars and mailboxes are emitting energy T_{ER} horizontally. These surfaces are also absorbing radiation T_{ER} from other surrounding objects: houses, trees, other cars, and so on. As a result, the temperature of the side surfaces is relatively high, and the frost melts. On the other hand, upward-facing surfaces on top of the cars and mailboxes are radiating energy T_{ER} upward, but above them is open sky. There are no objects radiating energy downward into the top surfaces. As a result, the top surfaces are colder, and the frost doesn't melt as soon as that on the sides.

Notice that all of these effects involve transfers of energy like those discussed in this chapter and especially in this section. The only effect that does not depend on altitude is that of the well-done cookies near the edge of the baking sheet. There are many such thermal effects all around you: look for others!

The Dewar Flask

The *Dewar flask*[6] is a container designed to minimize energy transfers by conduction, convection, and radiation. Such a container is used to store cold or hot liquids for long periods of time. (An insulated bottle, such as a Thermos, is a common household equivalent of a Dewar flask.) The standard construction (Fig. 19.18) consists of a double-walled Pyrex glass vessel with silvered walls. The space between the walls is evacuated to minimize energy transfer by conduction and convection. The silvered surfaces minimize energy transfer by radiation because silver is a very good reflector and has very low emissivity. A further reduction in energy loss is obtained by reducing the size of the neck. Dewar flasks are commonly used to store liquid nitrogen (boiling point 77 K) and liquid oxygen (boiling point 90 K).

To confine liquid helium (boiling point 4.2 K), which has a very low heat of vaporization, it is often necessary to use a double Dewar system in which the Dewar flask containing the liquid is surrounded by a second Dewar flask. The space between the two flasks is filled with liquid nitrogen.

Newer designs of storage containers use "superinsulation" that consists of many layers of reflecting material separated by fiberglass. All this material is in a vacuum, and no liquid nitrogen is needed with this design.

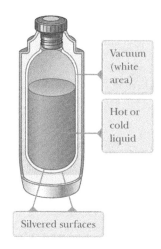

Figure 19.18 A cross-sectional view of a Dewar flask, which is used to store hot or cold substances.

---

[6] Invented by Sir James Dewar (1842–1923).

Summary

▶ Definitions

Internal energy is a system's energy associated with its temperature and its physical state (solid, liquid, gas). Internal energy includes kinetic energy of random translation, rotation, and vibration of molecules; vibrational potential energy within molecules; and potential energy between molecules.

Heat is the process of energy transfer across the boundary of a system resulting from a temperature difference between the system and its surroundings. The symbol Q represents the amount of energy transferred by this process.

A **calorie** is the amount of energy necessary to raise the temperature of 1 g of water from 14.5°C to 15.5°C.

The **heat capacity** C of any sample is the amount of energy needed to raise the temperature of the sample by 1°C.

The **specific heat** c of a substance is the heat capacity per unit mass:

$$c \equiv \frac{Q}{m \, \Delta T} \qquad (19.3)$$

The **latent heat** of a substance is defined as the ratio of the energy input to a substance to the change in mass of the higher-phase material:

$$L \equiv \frac{Q}{\Delta m} \qquad (19.7)$$

▶ Concepts and Principles

The energy Q required to change the temperature of a mass m of a substance by an amount ΔT is

$$Q = mc \, \Delta T \qquad (19.4)$$

where c is the specific heat of the substance.

The energy required to change the phase of a pure substance is

$$Q = L \, \Delta m \qquad (19.8)$$

where L is the latent heat of the substance, which depends on the nature of the phase change and the substance, and Δm is the change in mass of the higher-phase material.

The **work** done on a gas as its volume changes from some initial value V_i to some final value V_f is

$$W = -\int_{V_i}^{V_f} P \, dV \qquad (19.10)$$

where P is the pressure of the gas, which may vary during the process. To evaluate W, the process must be fully specified; that is, P and V must be known during each step. The work done depends on the path taken between the initial and final states.

continued

The **first law of thermodynamics** is a specific reduction of the conservation of energy equation (Eq. 8.2) and states that when a system undergoes a change from one state to another, the change in its internal energy is

$$\Delta E_{int} = Q + W \tag{19.11}$$

where Q is the energy transferred into the system by heat and W is the work done on the system. Although Q and W both depend on the path taken from the initial state to the final state, the quantity ΔE_{int} does not depend on the path.

In a **cyclic process** (one that originates and terminates at the same state), $\Delta E_{int} = 0$ and therefore $Q = -W$. That is, the energy transferred into the system by heat equals the negative of the work done on the system during the process.

In an **adiabatic process,** no energy is transferred by heat between the system and its surroundings ($Q = 0$). In this case, the first law gives $\Delta E_{int} = W$.

An **isothermal process** is one that occurs at constant temperature. The work done on an ideal gas during an isothermal process is

$$W = nRT \ln\left(\frac{V_i}{V_f}\right) \tag{19.12}$$

An **isobaric process** is one that occurs at constant pressure. The work done on a gas in such a process is $W = -P(V_f - V_i)$.

An **isovolumetric process** is one that occurs at constant volume. No work is done in such a process, so $\Delta E_{int} = Q$.

Conduction can be viewed as an exchange of kinetic energy between colliding molecules or electrons. The rate of energy transfer by conduction through a slab of area A is

$$P = kA\left|\frac{dT}{dx}\right| \tag{19.17}$$

where k is the **thermal conductivity** of the material from which the slab is made and $|dT/dx|$ is the **temperature gradient.**

In **convection,** a warm substance transfers energy from one location to another.

All objects emit **electromagnetic radiation** in the form of electromagnetic waves at the rate

$$P = \sigma AeT^4 \tag{19.21}$$

Think–Pair–Share

See the Preface for an explanation of the icons used in this problems set. For additional assessment items for this section, go to **WEBASSIGN** From Cengage

1. Your team has been hired by a major builder who is designing simple homes for a new housing tract. He asks you to estimate the amount of natural gas that will be required to heat each house during the winter months. Figure TP19.1 shows the house you are currently working on. The average thermal conductivity of the walls (including the windows) and roof of the house depicted in the figure is 0.480 W/m · °C, and their average thickness is 21.0 cm. The heat of combustion (that is, the energy provided per cubic meter) of natural gas is 3.89×10^7 J/m³. (a) How many cubic meters of gas must be burned each day to maintain an inside temperature of 25.0°C in this house if the outside temperature is 0.0°C? Disregard radiation and the energy transferred by heat through the ground. (b) How will the answer to part (a) be affected (*increase* or *decrease* the gas requirements?) by the inclusion of (i) thermal conduction through the floor; (ii) radiation incident on the roof, walls, and windows during the daytime;

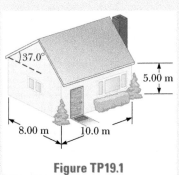

Figure TP19.1

(iii) operation of appliances, computers, entertainment systems; and (iv) leakage of air through cracks around doors and windows.

2. **ACTIVITY** Consider a spherical object of radius r with no atmosphere at a distance d from the Sun. Assume its emissivity is $e = 1$ for all kinds of electromagnetic waves and its temperature is uniform over its surface. At Earth's distance R from the Sun, the intensity of solar radiation is $I_S = 1\ 370$ W/m². This intensity varies as $1/d^2$ for distances other than R. A typical spherical object will *absorb* 70.0% of the solar radiation over its circular cross section πr^2. (The object will reflect about 30.0% of the incident radiation; the object appears circular when viewed from the Sun.) It will *emit* primarily infrared radiation from its entire surface area $4\pi r^2$. (a) Show that the equilibrium surface temperature of an object at a distance d from the Sun is

$$T = \left[\frac{(0.700)I_S}{4\sigma}\left(\frac{R}{d}\right)^2\right]^{1/4} = (255\ \text{K})\sqrt{\frac{R}{d}}$$

(b) Use the equation in part (a) to determine a theoretical surface temperature for the eight planets plus the dwarf planet Pluto, using the mean distance from the Sun given in Table 13.2. Also include the dwarf planet Ceres, at a distance of $d = 4.14 \times 10^{11}$ m from the Sun, for a total of ten objects in our solar system. (c) Make a bar graph of the temperatures found in part (c). (d) Add to your bar graph the measured and estimated surface temperatures, in kelvins, as provided by the Lunar and Planetary Institute, which are shown in the accompanying table. (e) Look first at our own planet, Earth. Is there a significance, in terms of life on this planet, to the fact that the theoretical temperature is below the freezing point of water, while the measured temperature is above it? (f) The actual temperature of Earth is raised by the atmospheric absorption of infrared radiation emitted from the surface. This effect is sometimes called the *greenhouse effect*. Consider the objects with the thinnest atmospheres: Mercury, Ceres, and Pluto. What do you notice about the comparison of theoretical and measured temperatures for these planets? (g) Consider the gas giants: Jupiter, Saturn, Uranus, and Neptune. These planets have no solid surface; the temperature data is provided for a point in the atmosphere where the pressure is the same as that at sea level on Earth. What do you notice about the comparison of theoretical and measured temperatures for these planets? (h) The clearest discrepancy between theoretical and measured temperatures in your graph is for Venus. Why is the measured temperature

so much higher than the theoretical temperature? (i) What can you conclude about the atmosphere of Mars from your graph?

Object	Surface Temperature (K) (from the Lunar and Planetary Institute)
Mercury	440
Venus	741
Earth	288
Mars	244
Ceres	173
Jupiter	165
Saturn	134
Uranus	77
Neptune	70
Pluto	40

Problems

See the Preface for an explanation of the icons used in this problems set. For additional assessment items for this section, go to ⚡ **WEBASSIGN**
From Cengage

SECTION 19.1 Heat and Internal Energy

1. A 55.0-kg woman eats a 540 Calorie (540 kcal) jelly doughnut for breakfast. (a) How many joules of energy are the equivalent of one jelly doughnut? (b) How many steps must the woman climb on a very tall stairway to change the gravitational potential energy of the woman–Earth system by a value equivalent to the food energy in one jelly doughnut? Assume the height of a single stair is 15.0 cm. (c) If the human body is only 25.0% efficient in converting chemical potential energy to mechanical energy, how many steps must the woman climb to work off her breakfast?
BIO

SECTION 19.2 Specific Heat and Calorimetry

2. The highest waterfall in the world is the Salto Angel in Venezuela. Its longest single falls has a height of 807 m. If water at the top of the falls is at 15.0°C, what is the maximum temperature of the water at the bottom of the falls? Assume all the kinetic energy of the water as it reaches the bottom goes into raising its temperature.

3. A combination of 0.250 kg of water at 20.0°C, 0.400 kg of aluminum at 26.0°C, and 0.100 kg of copper at 100°C is mixed in an insulated container and allowed to come to thermal equilibrium. Ignore any energy transfer to or from the container. What is the final temperature of the mixture?

4. The temperature of a silver bar rises by 10.0°C when it absorbs 1.23 kJ of energy by heat. The mass of the bar is 525 g. Determine the specific heat of silver from these data.
T

5. You are working in your kitchen preparing lunch for your family. You have decided to make egg salad sandwiches and are boiling six eggs, each of mass 55.5 g, in 0.750 L of water at 100°C. You wish to take all the eggs out of the boiling water and immediately place them in 23.0°C water to cool them down to a comfortable temperature to hold them
CR

and peel them. You decide that you wish the mixture of the water and the eggs to reach an equilibrium temperature of 40.0°C. Explaining this to a family member, she challenges you to determine *exactly* how much water at 23.0°C you need to achieve your desired equilibrium temperature. Take the average specific heat of an egg over the expected temperature range to be 3.27×10^3 J/kg · °C.

6. If water with a mass m_h at temperature T_h is poured into an aluminum cup of mass m_{Al} containing mass m_c of water at T_c, where $T_h > T_c$, what is the equilibrium temperature of the system?
S

7. An aluminum calorimeter with a mass of 100 g contains 250 g of water. The calorimeter and water are in thermal equilibrium at 10.0°C. Two metallic blocks are placed into the water. One is a 50.0-g piece of copper at 80.0°C. The other has a mass of 70.0 g and is originally at a temperature of 100°C. The entire system stabilizes at a final temperature of 20.0°C. (a) Determine the specific heat of the unknown sample. (b) Using the data in Table 19.1, can you make a positive identification of the unknown material? Can you identify a possible material? (c) Explain your answers for part (b).
Q|C V

8. An electric drill with a steel drill bit of mass $m = 27.0$ g and diameter 0.635 cm is used to drill into a cubical steel block of mass $M = 240$ g. Assume steel has the same properties as iron. The cutting process can be modeled as happening at one point on the circumference of the bit. This point moves in a helix at constant tangential speed 40.0 m/s and exerts a force of constant magnitude 3.20 N on the block. As shown in Figure P19.8 (page 528), a groove in the bit carries the chips up to the top of the block, where they form a pile around the hole. The drill is turned on and drills into the block for a time interval of 15.0 s. Let's assume this time interval is long enough for conduction within the steel to bring it all to a uniform temperature. Furthermore, assume the steel objects lose a negligible amount of energy by conduction, convection, and radiation into their environment. (a) Suppose the
Q|C

drill bit cuts three-quarters of the way through the block during 15.0 s. Find the temperature change of the whole quantity of steel. (b) **What If?** Now suppose the drill bit is dull and cuts only one-eighth of the way through the block in 15.0 s. Identify the temperature change of the whole quantity of steel in this case. (c) What pieces of data, if any, are unnecessary for the solution? Explain.

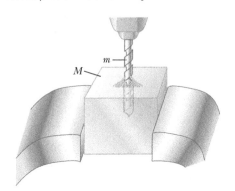

Figure P19.8

9. A 3.00-g copper coin at 25.0°C drops 50.0 m to the ground. **Q|C** (a) Assuming 60.0% of the change in gravitational potential energy of the coin–Earth system goes into increasing the internal energy of the coin, determine the coin's final temperature. (b) **What If?** Does the result depend on the mass of the coin? Explain.

SECTION 19.3 Latent Heat

10. How much energy is required to change a 40.0-g ice cube **V** from ice at −10.0°C to steam at 110°C?

11. A 75.0-kg cross-country skier **T** glides over snow as in Figure P19.11. The coefficient of friction between skis and snow is 0.200. Assume all the snow beneath his skis is at 0°C and that all the internal energy generated by friction is added to snow, which sticks to his skis until it melts. How far would he have to ski to melt 1.00 kg of snow?

Figure P19.11

12. A 3.00-g lead bullet at **AMT** 30.0°C is fired at a speed of **T** 240 m/s into a large block of ice at 0°C, in which it becomes embedded. What quantity of ice melts?

13. In an insulated vessel, 250 g of ice at 0°C is added to 600 g of water at 18.0°C. (a) What is the final temperature of the system? (b) How much ice remains when the system reaches equilibrium?

14. An automobile has a mass of 1 500 kg, and its aluminum **Q|C** brakes have an overall mass of 6.00 kg. (a) Assume all the mechanical energy that transforms into internal energy when the car stops is deposited in the brakes and no energy is transferred out of the brakes by heat. The brakes are originally at 20.0°C. How many times can the car be stopped

from 25.0 m/s before the brakes start to melt? (b) Identify some effects ignored in part (a) that are important in a more realistic assessment of the warming of the brakes.

SECTION 19.4 Work in Thermodynamic Processes

15. One mole of an ideal gas is warmed slowly so that it goes **Q|C** from the PV state (P_i, V_i) to $(3P_i, 3V_i)$ in such a way that **S** the pressure of the gas is directly proportional to the volume. (a) How much work is done on the gas in the process? (b) How is the temperature of the gas related to its volume during this process?

16. (a) Determine the work done on a gas that expands from i **V** to f as indicated in Figure P19.16. (b) **What If?** How much work is done on the gas if it is compressed from f to i along the same path?

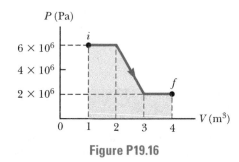

Figure P19.16

SECTION 19.5 The First Law of Thermodynamics

17. A thermodynamic system undergoes a process in which its internal energy decreases by 500 J. Over the same time interval, 220 J of work is done on the system. Find the energy transferred from it by heat.

18. *Why is the following situation impossible?* An ideal gas undergoes a process with the following parameters: $Q = 10.0$ J, $W = 12.0$ J, and $\Delta T = -2.00$°C.

19. A 2.00-mol sample of helium gas initially at 300 K, and **T** 0.400 atm is compressed isothermally to 1.20 atm. Noting that the helium behaves as an ideal gas, find (a) the final volume of the gas, (b) the work done on the gas, and (c) the energy transferred by heat.

20. (a) How much work is done on the steam when 1.00 mol of water at 100°C boils and becomes 1.00 mol of steam at 100°C at 1.00 atm pressure? Assume the steam to behave as an ideal gas. (b) Determine the change in internal energy of the system of the water and steam as the water vaporizes.

21. A 1.00-kg block of aluminum is warmed at atmospheric pressure so that its temperature increases from 22.0°C to 40.0°C. Find (a) the work done on the aluminum, (b) the energy added to it by heat, and (c) the change in its internal energy.

22. In Figure P19.22, the change in internal energy of a gas that is taken from A to C along the blue path is +800 J. The work done on the gas along the red path ABC is −500 J. (a) How much energy must be added to the system by heat as it goes from A through B to C? (b) If the pressure at point A is five times that of point C, what is the work done on the system in going from C to D?

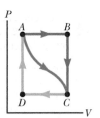

Figure P19.22

(c) What is the energy exchanged with the surroundings by heat as the gas goes from *C* to *A* along the green path? (d) If the change in internal energy in going from point *D* to point *A* is +500 J, how much energy must be added to the system by heat as it goes from point *C* to point *D*?

SECTION 19.6 Energy Transfer Mechanisms in Thermal Processes

23. A student is trying to decide what to wear. His bedroom is
BIO at 20.0°C. His skin temperature is 35.0°C. The area of his exposed skin is 1.50 m². People all over the world have skin that is dark in the infrared, with emissivity about 0.900. Find the net energy transfer from his body by radiation in 10.0 min.

24. A concrete slab is 12.0 cm thick and has an area of 5.00 m². Electric heating coils are installed under the slab to melt the ice on the surface in the winter months. What minimum power must be supplied to the coils to maintain a temperature difference of 20.0°C between the bottom of the slab and its surface? Assume all the energy transferred is through the slab.

25. Two lightbulbs have cylindrical filaments much greater in length than in diameter. The evacuated bulbs are identical except that one operates at a filament temperature of 2 100°C and the other operates at 2 000°C. (a) Find the ratio of the power emitted by the hotter lightbulb to that emitted by the cooler lightbulb. (b) With the bulbs operating at the same respective temperatures, the cooler lightbulb is to be altered by making its filament thicker so that it emits the same power as the hotter one. By what factor should the radius of this filament be increased?

26. The human body must maintain its core temperature inside
BIO a rather narrow range around 37°C. Metabolic processes, notably muscular exertion, convert potential energy into internal energy deep in the interior. From the interior, energy must flow out to the skin or lungs to be expelled to the environment. During moderate exercise, an 80-kg man can metabolize food energy at the rate 300 kcal/h, do 60 kcal/h of mechanical work, and put out the remaining 240 kcal/h of energy by heat. Most of the energy is carried from the body interior out to the skin by forced convection, whereby blood is warmed in the interior and then cooled at the skin, which is a few degrees cooler than the body core. Without blood flow, living tissue is a good thermal insulator, with thermal conductivity about 0.210 W/m · °C. Show that blood flow is essential to cool the man's body by calculating the rate of energy conduction in kcal/h through the tissue layer under his skin. Assume that its area is 1.40 m², its thickness is 2.50 cm, and it is maintained at 37.0°C on one side and at 34.0°C on the other side.

27. (a) Calculate the *R*-value of a thermal window made of two single panes of glass each 0.125 in. thick and separated by a 0.250-in. air space. (b) By what factor is the transfer of energy by heat through the window reduced by using the thermal window instead of the single-pane window? Include the contributions of inside and outside stagnant air layers.

28. For bacteriological testing of water supplies and in medi-
BIO cal clinics, samples must routinely be incubated for 24 h at
QC 37°C. Peace Corps volunteer and MIT engineer Amy Smith invented a low-cost, low-maintenance incubator. The incubator consists of a foam-insulated box containing a waxy material that melts at 37.0°C interspersed among tubes, dishes, or bottles containing the test samples and growth medium (bacteria food). Outside the box, the waxy material is first melted by a stove or solar energy collector. Then the waxy material is put into the box to keep the test samples warm as the material solidifies. The heat of fusion of the phase-change material is 205 kJ/kg. Model the insulation as a panel with surface area 0.490 m², thickness 4.50 cm, and conductivity 0.012 0 W/m · °C. Assume the exterior temperature is 23.0°C for 12.0 h and 16.0°C for 12.0 h. (a) What mass of the waxy material is required to conduct the bacteriological test? (b) Explain why your calculation can be done without knowing the mass of the test samples or of the insulation.

ADDITIONAL PROBLEMS

29. Gas in a container is at a pressure of 1.50 atm and a vol-
T ume of 4.00 m³. What is the work done on the gas (a) if it expands at constant pressure to twice its initial volume, and (b) if it is compressed at constant pressure to one-quarter its initial volume?

30. You are reading your textbook on Greek mythology. You
CR find a story about Daedalus and Icarus. Daedalus built two sets of wings out of feathers and wax, one set for him and one for his son Icarus. The father and son planned to use the wings to escape from their imprisonment on the island of Crete. The father warned Icarus not to fly too high because the proximity to the Sun might melt the wax in his wings. Of course, Icarus was overtaken by the thrill of flying and flew too close to the Sun. His wings melted and he fell into the sea. While reading this information, you think about your physics class, where your instructor has just discussed the equilibrium temperature of an object with no atmosphere at a given distance from the Sun. You look in your notes and find the following equation for this equilibrium temperature:

$$T = (255 \text{ K})\sqrt{\frac{R}{r}}$$

where *R* is the distance from the Sun to the Earth, *r* is the distance from the Sun to the object, and *T* is in kelvins. This raises a conundrum in your mind: If Icarus flew so close to the Sun that the wax in his wings melted, would there still be air at that location to allow him to fly to that location? Take the melting point of wax to be 65°C.

31. You have a particular interest in automobile engines, so you
CR have secured a co-op position at an automobile company while you attend school. Your supervisor is helping you to learn about the operation of an internal combustion engine. She gives you the following assignment, related to a simulation of a new engine she is designing. A gas, beginning at P_A = 1.00 atm, V_A = 0.500 L, and T_A = 27.0°C, is compressed from point *A* on the *PV* diagram in Figure P19.31 (page 530) to point *B*. This represents the compression stroke in a four-cycle gasoline engine. At that point, 132 J of energy is delivered to the gas at constant volume, taking the gas to point *C*. This represents the transformation of potential energy in the gasoline to internal energy when the spark plug fires. Your supervisor tells you that the internal energy of a gas is proportional to temperature (as we shall find in Chapter 20), the internal energy of the gas at point *A* is 200 J, and she wants to know what the temperature of the gas is at point *C*.

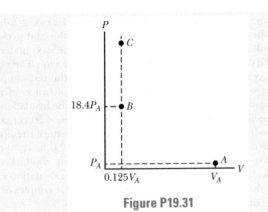

Figure P19.31

32. You are working in a condensed-matter laboratory for your senior project. Several of the ongoing projects use liquid helium, which is contained in a thermally insulated vessel that can hold up to a maximum of V_{max} = 240 L of the liquid at T_c = 4.20 K. Because some of the liquid helium has already been used, someone asks you to check to see if there is enough for the next day, on which four different experimental groups will need liquid helium. You are not sure how to measure the amount of liquid remaining, so you insert an aluminum rod of length L = 2.00 m and with a cross-sectional area A = 2.50 cm^2 into the vessel. By seeing how much of the lower end of the rod is frosted when you pull it out, you can estimate the depth of the liquid helium. After inserting the rod, however, one of the experimenters calls you over to perform a task and you forget about the rod, leaving it in the liquid helium until the next morning. How much liquid helium is available for the next day's experiments? (Aluminum has thermal conductivity of 3 100 W/m · K at 4.20 K; ignore its temperature variation. The density of liquid helium is 125 kg/m^3.) Assume that gaseous helium can escape from the top of the vessel.

33. A *flow calorimeter* is an apparatus used to measure the specific heat of a liquid. The technique of flow calorimetry involves measuring the temperature difference between the input and output points of a flowing stream of the liquid while energy is added by heat at a known rate. A liquid of density 900 kg/m^3 flows through the calorimeter with volume flow rate of 2.00 L/min. At steady state, a temperature difference 3.50°C is established between the input and output points when energy is supplied at the rate of 200 W. What is the specific heat of the liquid?

34. A *flow calorimeter* is an apparatus used to measure the specific heat of a liquid. The technique of flow calorimetry involves measuring the temperature difference between the input and output points of a flowing stream of the liquid while energy is added by heat at a known rate. A liquid of density ρ flows through the calorimeter with volume flow rate R. At steady state, a temperature difference ΔT is established between the input and output points when energy is supplied at the rate P. What is the specific heat of the liquid?

35. **Review.** Following a collision between a large spacecraft and an asteroid, a copper disk of radius 28.0 m and thickness 1.20 m at a temperature of 850°C is floating in space, rotating about its symmetry axis with an angular speed of 25.0 rad/s. As the disk radiates infrared light, its temperature falls to 20.0°C. No external torque acts on the disk.

(a) Find the change in kinetic energy of the disk. (b) Find the change in internal energy of the disk. (c) Find the amount of energy it radiates.

36. **Review.** Two speeding lead bullets, one of mass 12.0 g moving to the right at 300 m/s and one of mass 8.00 g moving to the left at 400 m/s, collide head-on, and all the material sticks together. Both bullets are originally at temperature 30.0°C. Assume the change in kinetic energy of the system appears entirely as increased internal energy. We would like to determine the temperature and phase of the bullets after the collision. (a) What two analysis models are appropriate for the system of two bullets for the time interval from before to after the collision? (b) From one of these models, what is the speed of the combined bullets after the collision? (c) How much of the initial kinetic energy has transformed to internal energy in the system after the collision? (d) Does all the lead melt due to the collision? (e) What is the temperature of the combined bullets after the collision? (f) What is the phase of the combined bullets after the collision?

37. An ice-cube tray is filled with 75.0 g of water. After the filled tray reaches an equilibrium temperature of 20.0°C, it is placed in a freezer set at −8.00°C to make ice cubes. (a) Describe the processes that occur as energy is being removed from the water to make ice. (b) Calculate the energy that must be removed from the water to make ice cubes at −8.00°C.

38. The rate at which a resting person converts food energy is called one's *basal metabolic rate* (BMR). Assume that the resulting internal energy leaves a person's body by radiation and convection of dry air. When you jog, most of the food energy you burn above your BMR becomes internal energy that would raise your body temperature if it were not eliminated. Assume that evaporation of perspiration is the mechanism for eliminating this energy. Suppose a person is jogging for "maximum fat burning," converting food energy at the rate 400 kcal/h above his BMR, and putting out energy by work at the rate 60.0 W. Assume that the heat of evaporation of water at body temperature is equal to its heat of vaporization at 100°C. (a) Determine the hourly rate at which water must evaporate from his skin. (b) When you metabolize fat, the hydrogen atoms in the fat molecule are transferred to oxygen to form water. Assume that metabolism of 1.00 g of fat generates 9.00 kcal of energy and produces 1.00 g of water. What fraction of the water the jogger needs is provided by fat metabolism?

39. An iron plate is held against an iron wheel so that a kinetic friction force of 50.0 N acts between the two pieces of metal. The relative speed at which the two surfaces slide over each other is 40.0 m/s. (a) Calculate the rate at which mechanical energy is converted to internal energy. (b) The plate and the wheel each have a mass of 5.00 kg, and each receives 50.0% of the internal energy. If the system is run as described for 10.0 s and each object is then allowed to reach a uniform internal temperature, what is the resultant temperature increase?

40. One mole of an ideal gas is contained in a cylinder with a movable piston. The initial pressure, volume, and temperature are P_i, V_i, and T_i, respectively. Find the work done on the gas in the following processes. In operational terms, describe how to carry out each process and show each process on a *PV* diagram. (a) an isobaric compression in which the final volume is one-half the initial volume (b) an isothermal compression in which the final pressure is four times the initial pressure (c) an isovolumetric process in which the final pressure is three times the initial pressure

41. During periods of high activity, the Sun has more sunspots
    Q|C than usual. Sunspots are cooler than the rest of the luminous layer of the Sun's atmosphere (the photosphere). Paradoxically, the total power output of the active Sun is not lower than average but is the same or slightly higher than average. Work out the details of the following crude model of this phenomenon. Consider a patch of the photosphere with an area of 5.10×10^{14} m². Its emissivity is 0.965. (a) Find the power it radiates if its temperature is uniformly 5 800 K, corresponding to the quiet Sun. (b) To represent a sunspot, assume 10.0% of the patch area is at 4 800 K and the other 90.0% is at 5 890 K. Find the power output of the patch. (c) State how the answer to part (b) compares with the answer to part (a). (d) Find the average temperature of the patch. Note that this cooler temperature results in a higher power output.

42. *Why is the following situation impossible?* A group of campers arises at 8:30 a.m. and uses a solar cooker, which consists of a curved, reflecting surface that concentrates sunlight onto the object to be warmed (Fig. P19.42). During the day, the maximum solar intensity reaching the Earth's surface at the cooker's location is $I = 600$ W/m². The cooker faces the Sun and has a face diameter of $d = 0.600$ m. Assume a fraction f of 40.0% of the incident energy is transferred to 1.50 L of water in an open container, initially at 20.0°C. The water comes to a boil, and the campers enjoy hot coffee for breakfast before hiking ten miles and returning by noon for lunch.

Figure P19.42

43. A cooking vessel on a slow burner contains 10.0 kg of water and an unknown mass of ice in equilibrium at 0°C at time $t = 0$. The temperature of the mixture is measured at various times, and the result is plotted in Figure P19.43.

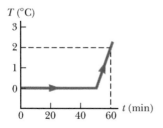

Figure P19.43

During the first 50.0 min, the mixture remains at 0°C. From 50.0 min to 60.0 min, the temperature increases to 2.00°C. Ignoring the heat capacity of the vessel, determine the initial mass of the ice.

44. A student measures the following data in a calorimetry
    Q|C experiment designed to determine the specific heat of aluminum:

Initial temperature of water and calorimeter:	70.0°C
Mass of water:	0.400 kg
Mass of calorimeter:	0.040 kg
Specific heat of calorimeter:	0.63 kJ/kg · °C
Initial temperature of aluminum:	27.0°C
Mass of aluminum:	0.200 kg
Final temperature of mixture:	66.3°C

(a) Use these data to determine the specific heat of aluminum. (b) Explain whether your result is within 15% of the value listed in Table 19.1.

CHALLENGE PROBLEMS

45. (a) The inside of a hollow cylinder is maintained at a temperature T_a, and the outside is at a lower temperature, T_b (Fig. P19.45). The wall of the cylinder has a thermal conductivity k. Ignoring end effects, show that the rate of energy conduction from the inner surface to the outer surface in the radial direction is

$$\frac{dQ}{dt} = 2\pi L k \left[\frac{T_a - T_b}{\ln(b/a)} \right]$$

Suggestions: The temperature gradient is dT/dr. A radial energy current passes through a concentric cylinder of area $2\pi rL$. (b) The passenger section of a jet airliner is in the shape of a cylindrical tube with a length of 35.0 m and an inner radius of 2.50 m. Its walls are lined with an insulating material 6.00 cm in thickness and having a thermal conductivity of 4.00×10^{-5} cal/s · cm · °C. A heater must maintain the interior temperature at 25.0°C while the outside temperature is −35.0°C. What power must be supplied to the heater?

Figure P19.45

46. A spherical shell has inner radius 3.00 cm and outer radius
    Q|C 7.00 cm. It is made of material with thermal conductivity $k = 0.800$ W/m · °C. The interior is maintained at temperature 5°C and the exterior at 40°C. After an interval of time, the shell reaches a steady state with the temperature at each point within it remaining constant in time. (a) Explain why

the rate of energy transfer P must be the same through each spherical surface, of radius r, within the shell and must satisfy

$$\frac{dT}{dr} = \frac{P}{4\pi kr^2}$$

(b) Next, prove that

$$\int_5^{40} dT = \frac{P}{4\pi k} \int_{0.03}^{0.07} r^{-2}\, dr$$

where T is in degrees Celsius and r is in meters. (c) Find the rate of energy transfer through the shell. (d) Prove that

$$\int_5^T dT = 1.84 \int_{0.03}^r r^{-2}\, dr$$

where T is in degrees Celsius and r is in meters. (e) Find the temperature within the shell as a function of radius. (f) Find the temperature at $r = 5.00$ cm, halfway through the shell.

47. A pond of water at 0°C is covered with a layer of ice 4.00 cm thick. If the air temperature stays constant at -10.0°C, what time interval is required for the ice thickness to increase to 8.00 cm? *Suggestion:* Use Equation 19.18 in the form

$$\frac{dQ}{dt} = kA\frac{\Delta T}{x}$$

and note that the incremental energy dQ extracted from the water through the thickness x of ice is the amount required to freeze a thickness dx of ice. That is, $dQ = L_f \rho A\, dx$, where ρ is the density of the ice, A is the area, and L_f is the latent heat of fusion.

The Kinetic Theory of Gases

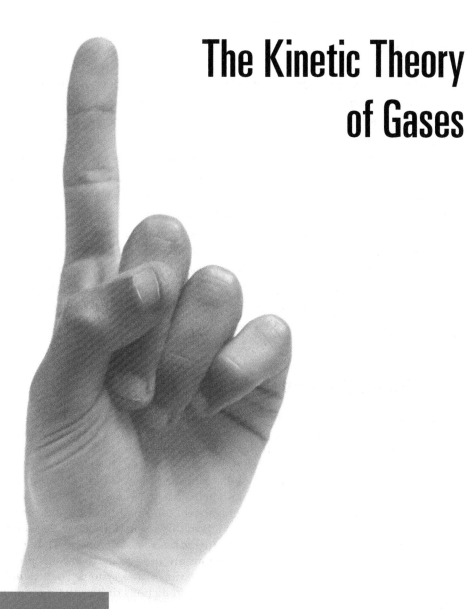

20

A wet finger is held upward to test the direction of the wind. Why is the finger cold on the side from which the wind is blowing? (*Joel Calheiros/Shutterstock*)

STORYLINE **You are still on the Physics Club camping trip to Whitney** Portal described at the beginning of Chapter 19. The evening plan is to build a campfire and huddle around it revisiting Physics Department stories. You are in charge of gathering the wood and setting up the fire. In order to locate the fire in the best place possible, you need to know the wind direction. As someone taught you years ago, you put your index finger in your mouth and then hold it vertically, knowing that the coldest side of your finger will be the direction from which the wind is coming. Your physics course kicks in again and you say, "Wait a minute! Why is that side of the finger cold in the wind?" While thinking about an answer, you reach down and grab a piece of wood. You scrape your finger painfully on it and go into the RV to receive medical treatment from your Club advisor. He puts some alcohol on the wound. Even though you know the alcohol is at the same temperature as the rest of the interior of the RV, the alcohol feels cold on your finger. Could the cold feeling of the alcohol be related to the cold feeling of your finger in the wind?

CONNECTIONS In Chapter 18, we discussed the properties of an ideal gas by using such *macroscopic* variables as pressure, volume, and temperature. Such large-scale properties can be related to a description of the gas on a *microscopic*

533

scale, where matter is treated as a collection of a huge number of molecules rather than as a single macroscopic sample. Applying Newton's laws of motion in a statistical manner to a collection of particles provides a reasonable description of thermodynamic processes. To keep the mathematics relatively simple, we shall consider primarily the behavior of gases because in gases the interactions between molecules are much weaker than they are in liquids or solids. We shall begin by relating pressure and temperature directly to the details of molecular motion in a sample of gas. Based on these results, we will make predictions of molar specific heats of gases. Some of these predictions will be correct and some will not. We will extend our model to explain those values that are not predicted correctly by the simpler model. Finally, we discuss the distribution of molecular speeds in a gas, and apply the results to a liquid. We shall find the concepts discussed in this chapter useful in the future when we analyze a situation on a microscopic scale, such as, for example, the analysis of the electrical characteristics of an *electron gas* in a conducting wire.

20.1 Molecular Model of an Ideal Gas

In Section 1.2, we introduced a number of types of models, one of which is the *structural model*. A structural model is a theoretical construct designed to represent a system that cannot be observed directly because it is too large or too small. For example, here on Earth we can only observe the solar system from the inside; we cannot travel outside the solar system and look back to see how it works. This restricted vantage point has led to the geocentric and heliocentric models of the solar system discussed in Section 13.4. An example of a system too small to observe directly is the hydrogen atom. Various structural models of this system have been developed, including the *Bohr model* (Section 41.3) and the *quantum model* (Section 41.4). Once a structural model is developed, its assumptions are used to make various predictions for experimental observations of the behavior of the system. For example, the geocentric model of the solar system makes predictions of how the movement of Mars should appear from the Earth. It turns out that those predictions do not match the actual observations. When this mismatch occurs with a structural model, the model must be modified or replaced with another model.

In this chapter, we will consider a structural model for an ideal gas, with the goal of relating *macroscopic* measurements of the gas (pressure, volume, temperature, etc.) to the behavior of its *microscopic* components—molecules. The structural model that we will develop is called **kinetic theory**. This model treats an ideal gas as a collection of molecules with the following assumptions:

1. *Physical components:*
   The gas consists of a number of identical molecules within a cubic container of side length d (Fig. 20.1). The number of molecules in the gas is large, and the average separation between them is large compared with their dimensions. Therefore, the molecules occupy a negligible volume in the container. This assumption is consistent with the ideal gas model, in which we imagine the molecules to be point-like.

2. *Behavior of the components:*
   (a) The molecules obey Newton's laws of motion, but as a whole their motion is isotropic: any molecule can move in any direction with any speed.
   (b) The molecules interact only by short-range forces during elastic collisions. This assumption is consistent with the ideal gas model, in which the molecules exert no long-range forces on one another.
   (c) The molecules make elastic collisions with the walls of the container.

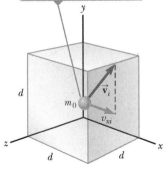

One molecule of the gas moves with velocity $\vec{\mathbf{v}}$ on its way toward a collision with the wall.

Figure 20.1 A cubical box with sides of length d containing an ideal gas.

Although we often picture an ideal gas as consisting of single atoms modeled as particles, the behavior of molecular gases approximates that of ideal gases rather well at low pressures. Usually, the internal structure of the molecule has no effect on the motions considered here.

For our first application of kinetic theory, let us relate the macroscope variable of pressure P to microscopic quantities. This will be a relatively long process, but each step will be based on a simple mathematical calculation, or on a principle or analysis model that we have studied in a previous chapter. Consider a collection of N molecules of an ideal gas in a container of volume V. As indicated in assumption 1, the container is a cube with edges of length d. We shall first focus our attention on one of these molecules of mass m_0 and assume it is moving so that its component of velocity in the x direction is v_{xi} as in Figure 20.1. (The subscript i here refers to the ith molecule in the collection, not to an initial value. We will combine the effects of all the molecules shortly.) Figure 20.2 shows the molecule making a collision with the wall of the container. As the molecule collides elastically with the wall (assumption 2(c) above), its velocity component perpendicular to the wall is reversed because the mass of the wall is far greater than the mass of the molecule. The molecule is modeled as a nonisolated system for which the impulse from the wall causes a change in the molecule's momentum. Because the momentum component p_{xi} of the molecule is $m_0 v_{xi}$ before the collision and $-m_0 v_{xi}$ after the collision, the change in the x component of the momentum of the molecule is

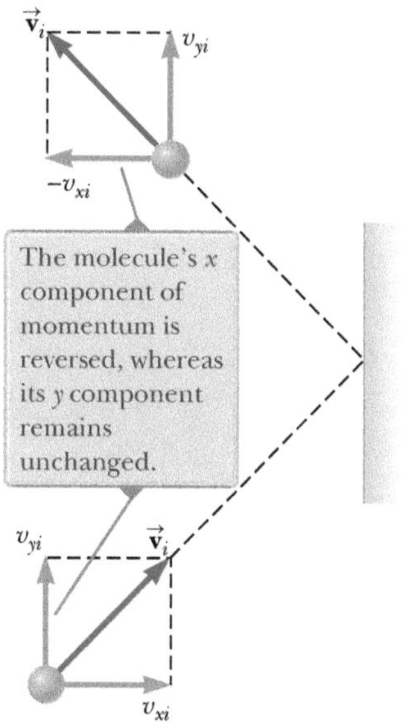

$$\Delta p_{xi} = -m_0 v_{xi} - (m_0 v_{xi}) = -2m_0 v_{xi} \tag{20.1}$$

From the nonisolated system model for momentum, we can apply the impulse-momentum theorem (Eqs. 9.11 and 9.13) to the molecule to give

$$\overline{F}_{i,\text{on molecule}} \, \Delta t_{\text{collision}} = \Delta p_{xi} = -2m_0 v_{xi} \tag{20.2}$$

where $\overline{F}_{i,\text{on molecule}}$ is the x component of the average force[1] the wall exerts on the molecule during the collision and $\Delta t_{\text{collision}}$ is the duration of the collision. For the molecule to make another collision with the same wall after this first collision, it must travel a distance of $2d$ in the x direction (across the cube and back). Therefore, from the particle under constant velocity model, the time interval between two collisions with the same wall is

$$\Delta t = \frac{2d}{v_{xi}} \tag{20.3}$$

The force that causes the change in momentum of the molecule in the collision with the wall occurs only during the collision. We can, however, find the long-term average force for many back-and-forth trips across the cube by averaging the force in Equation 20.2 over the time interval for the molecule to move across the cube and back once, Equation 20.3. The average change in momentum per trip for the time interval for many trips is the same as that for the short duration of the collision. Therefore, we can rewrite Equation 20.2 as

$$\overline{F}_i \, \Delta t = -2m_0 v_{xi} \tag{20.4}$$

where $\overline{F}_i$ is the average force component over the time interval Δt for the molecule to move across the cube and back. Because exactly one collision occurs with the given wall for each such time interval, this result is also the long-term average force on the molecule over long time intervals containing any number of multiples of Δt.

The molecule's x component of momentum is reversed, whereas its y component remains unchanged.

Figure 20.2 A molecule makes an elastic collision with the wall of the container. In this construction, we assume the molecule moves in the xy plane.

---

[1]For this discussion, we use a bar over a variable to represent the average value of the variable, such as $\overline{F}$ for the average force, rather than the subscript "avg" that we have used before. This notation is to save confusion because we already have a number of subscripts on variables.

Equations 20.3 and 20.4 enable us to express the x component of the long-term average force exerted by the wall on the molecule as

$$\bar{F}_i = -\frac{2m_0 v_{xi}}{\Delta t} = -\frac{2m_0 v_{xi}^2}{2d} = -\frac{m_0 v_{xi}^2}{d} \tag{20.5}$$

Now, by Newton's third law, the x component of the long-term average force exerted by the *molecule* on the *wall* is equal in magnitude and opposite in direction:

$$\bar{F}_{i,\text{on wall}} = -\bar{F}_i = -\left(-\frac{m_0 v_{xi}^2}{d}\right) = \frac{m_0 v_{xi}^2}{d} \tag{20.6}$$

The total average force $\bar{F}$ exerted by the gas on the wall is found by adding the average forces exerted by *all* the individual molecules striking the wall. Adding terms such as those in Equation 20.6 for all molecules gives

$$\bar{F} = \sum_{i=1}^{N} \frac{m_0 v_{xi}^2}{d} = \frac{m_0}{d} \sum_{i=1}^{N} v_{xi}^2 \tag{20.7}$$

where we have factored out the length of the box and the mass m_0 because assumption 1 tells us that all the molecules are the same. We now impose an additional feature from assumption 1, that the number of molecules is large. For a small number of molecules, the actual force on the wall would vary with time. It would be nonzero during the short interval of a collision of a molecule with the wall and zero when no molecule happens to be hitting the wall. For a very large number of molecules such as Avogadro's number, however, these variations in force are smoothed out so that the average force given above is the same over *any* time interval. Therefore, the *constant* force F on the wall due to the molecular collisions is

$$F = \frac{m_0}{d} \sum_{i=1}^{N} v_{xi}^2 \tag{20.8}$$

To proceed further, let's consider part of the right-hand side of Equation 20.8: how do we express the average value of the square of the x component of the velocity for N molecules? The traditional average of a set of values is the sum of the values over the number of values:

$$\overline{v_x^2} = \frac{\sum_{i=1}^{N} v_{xi}^2}{N} \quad \rightarrow \quad \sum_{i=1}^{N} v_{xi}^2 = N\overline{v_x^2} \tag{20.9}$$

Using Equation 20.9 to substitute for the sum in Equation 20.8 gives

$$F = \frac{m_0}{d} N\overline{v_x^2} \tag{20.10}$$

Now let's focus again on one molecule with velocity components v_{xi}, v_{yi}, and v_{zi}. The Pythagorean theorem relates the square of the speed of the molecule to the squares of the velocity components:

$$v_i^2 = v_{xi}^2 + v_{yi}^2 + v_{zi}^2 \tag{20.11}$$

Hence, the average value of v^2 for all the molecules in the container is related to the average values of v_x^2, v_y^2, and v_z^2 according to the expression

$$\overline{v^2} = \overline{v_x^2} + \overline{v_y^2} + \overline{v_z^2} \tag{20.12}$$

Because the motion is isotropic (assumption 2(a) above), the average values $\overline{v_x^2}$, $\overline{v_y^2}$, and $\overline{v_z^2}$ are equal to one another. Using this fact and Equation 20.12, we find that

$$\overline{v^2} = 3\overline{v_x^2} \tag{20.13}$$

Therefore, from Equation 20.10, the total force exerted on the wall is

$$F = \tfrac{1}{3}N\,\frac{m_0\overline{v^2}}{d}$$

(20.14)

Using this expression, we can find the total pressure exerted on the wall:

$$P = \frac{F}{A} = \frac{F}{d^2} = \tfrac{1}{3}N\,\frac{m_0\overline{v^2}}{d^3} = \tfrac{1}{3}\left(\frac{N}{V}\right)m_0\overline{v^2}$$

$$P = \tfrac{2}{3}\left(\frac{N}{V}\right)(\tfrac{1}{2}m_0\overline{v^2})$$

(20.15)

◀ Relationship between pressure and molecular kinetic energy

where we have recognized the volume V of the cube as d^3.

We have finished the long process initiated at the beginning of this section. The reward for our patience and diligence is something profound: Equation 20.15 indicates that the pressure of a gas is proportional to (1) the number of molecules per unit volume and (2) the average translational kinetic energy of the molecules, $\tfrac{1}{2}m_0\overline{v^2}$. In analyzing this structural model of an ideal gas, we obtain an important result that relates the macroscopic quantity of pressure to a microscopic quantity, the average value of the square of the molecular speed. Therefore, a key link between the molecular world and the large-scale world has been established.

Notice that Equation 20.15 verifies some features of pressure with which you are probably familiar. One way to increase the pressure inside a container is to increase the number of molecules per unit volume N/V in the container. That is what you do when you add air to a tire. We will return to discuss the second set of parentheses in Equation 20.15 very shortly, after we discuss the macroscopic quantity of temperature.

We can gain some insight into the meaning of temperature by first writing Equation 20.15 in the form

$$PV = \tfrac{2}{3}N(\tfrac{1}{2}m_0\overline{v^2})$$

(20.16)

Let's now compare this expression with the equation of state for an ideal gas (Eq. 18.11):

$$PV = Nk_BT$$

(20.17)

Equating the right sides of Equations 20.16 and 20.17 and solving for T gives

$$T = \frac{2}{3k_B}(\tfrac{1}{2}m_0\overline{v^2})$$

(20.18)

◀ Relationship between temperature and molecular kinetic energy

This result tells us that *temperature is a direct measure of average molecular kinetic energy.* In Chapter 18, we could only define temperature macroscopically, in terms of the transfer of energy between two objects. In Equation 20.18, we have a deeper definition of temperature in terms of the microscopic motion of the molecules of a substance.

By rearranging Equation 20.18, we can relate the translational molecular kinetic energy to the temperature:

$$\tfrac{1}{2}m_0\overline{v^2} = \tfrac{3}{2}k_BT$$

(20.19)

◀ Average kinetic energy per molecule

Now look back at Equation 20.15. The quantity in the second set of parentheses in that equation is the same as the left-hand side of Equation 20.19. Therefore, we see that the pressure in Equation 20.15 depends on the temperature of the gas. With regard to the discussion of the air pressure in an automobile tire, the pressure can be raised by increasing the temperature of that air, which is why the pressure inside a tire increases as the tire warms up during long road trips. The continuous flexing of the tire as it moves along the road surface results in work done on the rubber as parts of the tire distort, causing an increase in internal energy of the rubber. The increased temperature of the rubber results in the transfer of energy by heat into the air inside the tire. This transfer increases the air's temperature, and this increase in temperature in turn produces an increase in pressure.

Equation 20.19 shows us that the average translational kinetic energy per molecule is $\frac{3}{2}k_B T$. Because $\overline{v_x^2} = \frac{1}{3}\overline{v^2}$ (Eq. 20.13), it follows that

$$\tfrac{1}{2}m_0\overline{v_x^2} = \tfrac{1}{2}k_B T \tag{20.20}$$

In a similar manner, for the y and z directions,

$$\tfrac{1}{2}m_0\overline{v_y^2} = \tfrac{1}{2}k_B T \quad \text{and} \quad \tfrac{1}{2}m_0\overline{v_z^2} = \tfrac{1}{2}k_B T$$

Therefore, each translational degree of freedom contributes an equal amount of energy, $\frac{1}{2}k_B T$, to the gas. (In general, a "degree of freedom" refers to an independent means by which a molecule can possess energy.) A generalization of this result, known as the **theorem of equipartition of energy,** is as follows:

◀ Theorem of equipartition of energy

Each degree of freedom contributes $\frac{1}{2}k_B T$ to the energy of a system, where possible degrees of freedom are those associated with translation, rotation, and vibration of molecules.

The total translational kinetic energy of N molecules of gas is simply N times the average energy per molecule, which is given by Equation 20.19:

◀ Total translational kinetic energy of N molecules

$$K_{\text{tot trans}} = N(\tfrac{1}{2}m_0\overline{v^2}) = \tfrac{3}{2}Nk_B T = \tfrac{3}{2}nRT \tag{20.21}$$

where we have used $k_B = R/N_A$ for Boltzmann's constant and $N = nN_A$ for the number of molecules of gas. If the gas molecules possess only translational kinetic energy, *Equation 20.21 represents the internal energy of the gas.* This result implies that the internal energy of an ideal gas depends *only* on the temperature. We will follow up on this point in Section 20.2.

The square root of $\overline{v^2}$ is called the **root-mean-square (rms) speed** of the molecules. From Equation 20.19, we find that the rms speed is

◀ Root-mean-square speed

$$v_{\text{rms}} = \sqrt{\overline{v^2}} = \sqrt{\frac{3k_B T}{m_0}} = \sqrt{\frac{3RT}{M}} \tag{20.22}$$

where M is the molar mass in kilograms per mole and is equal to $m_0 N_A$. This expression shows that, at a given temperature, lighter molecules move faster, on the average, than do heavier molecules. For example, at a given temperature, hydrogen molecules, whose molar mass is 2.02×10^{-3} kg/mol, have an average speed approximately four times that of oxygen molecules, whose molar mass is 32.0×10^{-3} kg/mol. Table 20.1 lists the rms speeds for various molecules at 20°C.

PITFALL PREVENTION 20.1
The Square Root of the Square?
Taking the square root of $\overline{v^2}$ in Equation 20.22 does not "undo" the square because we have taken an average *between* squaring and taking the square root. Although the square root of $(\overline{v})^2$ is $\overline{v} = v_{\text{avg}}$ because the squaring is done after the averaging, the square root of $\overline{v^2}$ is *not* v_{avg}, but rather v_{rms}.

QUICK QUIZ 20.1 Two containers hold an ideal gas at the same temperature and pressure. Both containers hold the same type of gas, but container B has twice the volume of container A. **(i)** What is the average translational kinetic energy per molecule in container B? (a) twice that of container A (b) the same as that of container A (c) half that of container A (d) impossible to determine **(ii)** From the same choices, describe the internal energy of the gas in container B.

TABLE 20.1 Some Root-Mean-Square (rms) Speeds

Gas	Molar Mass (g/mol)	v_{rms} at 20°C (m/s)	Gas	Molar Mass (g/mol)	v_{rms} at 20°C (m/s)
H_2	2.02	1902	NO	30.0	494
He	4.00	1352	O_2	32.0	478
H_2O	18.0	637	CO_2	44.0	408
Ne	20.2	602	SO_2	64.1	338
N_2 or CO	28.0	511			

Example **20.1** A Tank of Helium

A tank used for filling helium balloons has a volume of 0.300 m³ and contains 2.00 mol of helium gas at 20.0°C. Assume the helium behaves like an ideal gas.

(A) What is the total translational kinetic energy of the gas molecules?

SOLUTION

Conceptualize Imagine a microscopic model of a gas in which you can watch the molecules move about the container more rapidly as the temperature increases. Because the gas is monatomic, the only type of motion the particles of the gas can exhibit is translation, and the total translational kinetic energy of the molecules is the internal energy of the gas.

Categorize We evaluate parameters with equations developed in the preceding discussion, so this example is a substitution problem.

Use Equation 20.21 with $n = 2.00$ mol and $T = 293$ K:

$$E_{\text{int}} = K_{\text{tot trans}} = \tfrac{3}{2}nRT = \tfrac{3}{2}(2.00 \text{ mol})(8.31 \text{ J/mol} \cdot \text{K})(293 \text{ K})$$

$$= 7.30 \times 10^3 \text{ J}$$

(B) What is the average kinetic energy per molecule?

SOLUTION

Use Equation 20.19:

$$K_{\text{avg}} = \tfrac{1}{2}m_0\overline{v^2} = \tfrac{3}{2}k_{\text{B}}T = \tfrac{3}{2}(1.38 \times 10^{-23} \text{ J/K})(293 \text{ K})$$

$$= 6.07 \times 10^{-21} \text{ J}$$

WHAT IF? What if the temperature is raised from 20.0°C to 40.0°C? Because 40.0 is twice as large as 20.0, is the total translational energy of the molecules of the gas twice as large at the higher temperature?

Answer The expression for the total translational energy depends on the temperature, and the value for the temperature must be expressed in kelvins, not in degrees Celsius. Therefore, the ratio of 40.0 to 20.0 is *not* the appropriate ratio. Converting the Celsius temperatures to kelvins, 20.0°C is 293 K and 40.0°C is 313 K. Therefore, the total translational energy increases by a factor of only 313 K/293 K = 1.07.

20.2 Molar Specific Heat of an Ideal Gas

Let's use the results of Section 20.1 to investigate a macroscopic quantity associated with a gas: its specific heat. Consider an ideal gas undergoing several processes such that the change in temperature is $\Delta T = T_f - T_i$ for all processes. The temperature change can be achieved by taking a variety of paths from one isotherm to another as shown in Figure 20.3. Because ΔT is the same for all paths, the change in internal energy ΔE_{int} is the same for all paths. The work W done on the gas (the negative of the area under the curves), however, is different for each path, as we found in Section 19.4. Therefore, from the first law of thermodynamics, we can argue that the heat $Q = \Delta E_{\text{int}} - W$ associated with a given change in temperature does *not* have a unique value: the heat Q for a process taking place between two temperatures depends on the process. Therefore, in $Q = mc\Delta T$, the specific heat c does not have a unique value for a gas!

We can address this difficulty by defining specific heats for two special processes that we have studied: isovolumetric ($i \to f$ in Figure 20.4 on page 540) and isobaric ($i \to f'$ in Figure 20.4). Because the number of moles n is a convenient measure of the amount of gas, we define the **molar specific heats** associated with these processes as follows:

$$Q = nC_V \Delta T \quad \text{(constant volume)} \tag{20.23}$$

$$Q = nC_P \Delta T \quad \text{(constant pressure)} \tag{20.24}$$

where C_V is the **molar specific heat at constant volume** and C_P is the **molar specific heat at constant pressure.** At constant volume, no work is done on the gas; the

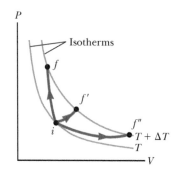

Figure 20.3 An ideal gas is taken from one isotherm at temperature T to another at temperature $T + \Delta T$ along three different paths.

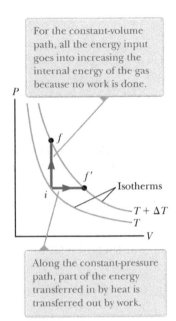

For the constant-volume path, all the energy input goes into increasing the internal energy of the gas because no work is done.

Along the constant-pressure path, part of the energy transferred in by heat is transferred out by work.

Figure 20.4 Energy is transferred by heat to an ideal gas in two ways.

only energy change in the system is the change of internal energy. When energy is added to a gas by heat at constant pressure, however, not only does the internal energy of the gas increase, but (negative) work is done on the gas: the volume of the gas must increase to keep the pressure constant. Therefore, the heat Q in Equation 20.24 must account for both the increase in internal energy and the transfer of energy out of the system by work. For this reason, Q in Equation 20.24 is greater than that in Equation 20.23 for given values of n and ΔT. Therefore, C_p is greater than C_V.

Because no work is done on the gas in the constant-volume process in Figure 20.4, the first law of thermodynamics gives us

$$Q = \Delta E_{\text{int}} \quad (20.25)$$

Substituting the expression for Q given by Equation 20.23 into Equation 20.25, we obtain

$$\Delta E_{\text{int}} = nC_V \Delta T \quad (20.26)$$

where ΔT is the temperature difference between the two isotherms. This equation applies to all ideal gases, those gases having more than one atom per molecule as well as monatomic ideal gases. It also applies to *all processes* taking place between the same temperatures. All five processes in Figures 20.3 and 20.4 have the same change in internal energy because they all take place through the same temperature difference ΔT. This statement might seem surprising, because Equation 20.26 contains a molar specific heat for a specific process at constant volume. It is true, however, due to the fact that internal energy is a state variable, and its change depends only on the temperature change, so it is independent of the particular process. Equation 20.26 gives us the change in internal energy for all processes, and allows us to evaluate this change using the constant-volume specific heat.

In the limit of infinitesimal changes, we can use Equation 20.26 to express the molar specific heat at constant volume as

$$C_V = \frac{1}{n} \frac{dE_{\text{int}}}{dT} \quad (20.27)$$

Let's now consider the simplest case of an ideal monatomic gas, that is, a gas containing one atom per molecule such as helium, neon, or argon. When energy is added to a monatomic gas in a container of fixed volume, all the added energy goes into increasing the translational kinetic energy of the atoms. There is no other way to store the energy in a monatomic gas. Therefore, from Equation 20.21, we see that the internal energy E_{int} of N molecules (or n mol) of an ideal monatomic gas is

Internal energy of an ideal ▶ monatomic gas

$$E_{\text{int}} = K_{\text{tot trans}} = \tfrac{3}{2} N k_B T = \tfrac{3}{2} n R T \quad (20.28)$$

For a monatomic ideal gas, E_{int} is a function of T only and the functional relationship is given by Equation 20.28. In general, the internal energy of any ideal gas is a function of T only and the exact relationship depends on the type of gas.

Substituting the internal energy from Equation 20.28 into Equation 20.27 gives

$$C_V = \tfrac{3}{2} R = 12.5 \text{ J/mol} \cdot \text{K} \quad (20.29)$$

This expression predicts a value of $C_V = \tfrac{3}{2} R$ for *all* monatomic gases. This prediction is in excellent agreement with measured values of molar specific heats for such gases as helium, neon, argon, and xenon over a wide range of temperatures (see the C_V column in Table 20.2). Small variations in Table 20.2 from the predicted values are because real gases are not ideal gases. In real gases, weak intermolecular interactions occur, which are not addressed in our ideal gas model.

Now suppose the gas is taken along the constant-pressure path $i \rightarrow f'$ shown in Figure 20.4. Along this path, the temperature again increases by ΔT. The energy that must be transferred by heat to the gas in this process is $Q = nC_p \Delta T$. Because the volume changes in this process, the work done on the gas is $W = -P \Delta V$, where

TABLE 20.2   Molar Specific Heats of Various Gases

Gas	Molar Specific Heat (J/mol · K)[a]			
	C_P	C_V	$C_P - C_V$	$\gamma = C_P/C_V$
Monatomic gases				
He	20.8	12.5	8.33	1.67
Ar	20.8	12.5	8.33	1.67
Ne	20.8	12.7	8.12	1.64
Kr	20.8	12.3	8.49	1.69
Diatomic gases				
H_2	28.8	20.4	8.33	1.41
N_2	29.1	20.8	8.33	1.40
O_2	29.4	21.1	8.33	1.40
CO	29.3	21.0	8.33	1.40
Cl_2	34.7	25.7	8.96	1.35
Polyatomic gases				
CO_2	37.0	28.5	8.50	1.30
SO_2	40.4	31.4	9.00	1.29
H_2O	35.4	27.0	8.37	1.30
CH_4	35.5	27.1	8.41	1.31

[a] All values except that for water were obtained at 300 K.

P is the constant pressure at which the process occurs. Applying the first law of thermodynamics to this process, we have

$$\Delta E_{\text{int}} = Q + W = nC_P \Delta T + (-P\Delta V) \qquad (20.30)$$

As discussed above, the change in internal energy for the process $i \rightarrow f'$ in Figure 20.4 is equal to that for process $i \rightarrow f$ because ΔT is the same for both processes. In addition, because $PV = nRT$, note that for a constant-pressure process, $P\Delta V = nR\,\Delta T$. Substituting this value for $P\Delta V$ into Equation 20.30 with $\Delta E_{\text{int}} = nC_V \Delta T$ (Eq. 20.26) gives

$$nC_V \Delta T = nC_P \Delta T - nR\,\Delta T$$

$$C_P - C_V = R \qquad (20.31)$$

This expression applies to *any* ideal gas. It predicts that the molar specific heat of an ideal gas at constant pressure is greater than the molar specific heat at constant volume by an amount R, the universal gas constant (which has the value 8.31 J/mol · K). This expression is applicable to real gases as the data in the $C_P - C_V$ column in Table 20.2 show.

Because $C_V = \frac{3}{2}R$ for a monatomic ideal gas, Equation 20.31 predicts a value $C_P = \frac{5}{2}R = 20.8$ J/mol · K for the molar specific heat of a monatomic gas at constant pressure. The ratio of these molar specific heats is a dimensionless quantity γ (Greek letter gamma):

$$\gamma = \frac{C_P}{C_V} = \frac{5R/2}{3R/2} = \frac{5}{3} = 1.67 \qquad (20.32)$$

◀ Ratio of molar specific heats for a monatomic ideal gas

Theoretical values of C_V, C_P, and γ are in excellent agreement with experimental values obtained for monatomic gases, but they are in serious disagreement with the values for the more complex gases (see Table 20.2). That is not surprising; the value $C_V = \frac{3}{2}R$ was derived for a monatomic ideal gas, and we expect some additional contribution to the molar specific heat from the internal structure of the more complex molecules. In Section 20.3, we describe the effect of molecular structure on the molar specific heat of a gas. The internal energy—and hence the molar specific heat—of a complex gas must include contributions from the rotational and the vibrational motions of the molecule.

In the case of solids and liquids heated at constant pressure, very little work is done during such a process because the thermal expansion is small. Consequently, C_P and C_V are approximately equal for solids and liquids.

QUICK QUIZ 20.2 **(i)** How does the internal energy of an ideal gas change as it follows path $i \rightarrow f$ in Figure 20.4? (a) E_{int} increases. (b) E_{int} decreases. (c) E_{int} stays the same. (d) There is not enough information to determine how E_{int} changes. **(ii)** From the same choices, how does the internal energy of an ideal gas change as it follows path $f \rightarrow f'$ along the isotherm labeled $T + \Delta T$ in Figure 20.4?

Example 20.2 **Heating a Cylinder of Helium**

A cylinder contains 3.00 mol of helium gas at a temperature of 300 K.

(A) If the gas is heated at constant volume, how much energy must be transferred by heat to the gas for its temperature to increase to 500 K?

SOLUTION

Conceptualize Run the process in your mind with the help of the piston–cylinder arrangement in Figure 19.7. Imagine that the piston is clamped in position to maintain the constant volume of the gas.

Categorize We evaluate parameters with equations developed in the preceding discussion, so this example is a substitution problem.

Use Equation 20.23 to find the energy transfer: $Q_1 = nC_V \Delta T$

Substitute the given values: $Q_1 = (3.00 \text{ mol})(12.5 \text{ J/mol} \cdot \text{K})(500 \text{ K} - 300 \text{ K})$
$= 7.50 \times 10^3 \text{ J}$

(B) How much energy must be transferred by heat to the gas at constant pressure to raise the temperature to 500 K?

SOLUTION

Use Equation 20.24 to find the energy transfer: $Q_2 = nC_P \Delta T$

Substitute the given values: $Q_2 = (3.00 \text{ mol})(20.8 \text{ J/mol} \cdot \text{K})(500 \text{ K} - 300 \text{ K})$
$= 12.5 \times 10^3 \text{ J}$

This value is larger than Q_1 because of the transfer of energy out of the gas by work to raise the piston in the constant pressure process.

20.3 The Equipartition of Energy

In Section 20.1, we found that the temperature of a gas is a measure of the average translational kinetic energy of the gas molecules. This kinetic energy is associated with the motion of the center of mass of each molecule. It does not include the energy associated with the internal motion of the molecule, namely, vibrations and rotations about the center of mass. In this section, we introduce the contributions from rotation and vibration of the molecule to the specific heats of the gas.

Predictions based on our model for molar specific heat agree quite well with the behavior of monatomic gases, but not with the behavior of complex gases (see Table 20.2). The value predicted by the model for the quantity $C_P - C_V = R$, however, is the same for all gases. This similarity is not surprising because this difference is the result of the work done on the gas, which is independent of its molecular structure.

To clarify the variations in C_V and C_P in gases more complex than monatomic gases, let's explore further the origin of molar specific heat. So far, we have assumed the sole contribution to the internal energy of a gas is the translational kinetic energy of the molecules. The internal energy of a gas, however, includes contributions from the translational, vibrational, and rotational motion of the molecules. The rotational and vibrational motions of molecules can be activated by collisions and therefore are "coupled" to the translational motion of the molecules. The branch of physics known as *statistical mechanics* has shown that, for a large number of particles obeying the laws of Newtonian mechanics, the available energy is, on average, shared equally by each independent degree of freedom. Recall from Section 20.1 that the equipartition theorem states that, at equilibrium, each degree of freedom contributes $\frac{1}{2}k_B T$ of energy per molecule.

Let's consider a diatomic gas whose molecules have the shape of a dumbbell (Fig. 20.5). In this model, the center of mass of the molecule can translate in the x, y, and z directions. The gray arrow in Figure 20.5a shows a translation in the x direction. In addition, the molecule can rotate about three mutually perpendicular axes (Fig. 20.5b). The rotation about the y axis can be neglected because the molecule's moment of inertia I_y and its rotational energy $\frac{1}{2}I_y\omega^2$ about this axis are negligible compared with those associated with the x and z axes. (If the two atoms are modeled as particles, then I_y is identically zero.) Therefore, there are five degrees of freedom for translation and rotation: three associated with the translational motion (x, y, and z) and two associated with the rotational motion (x and z). Because each degree of freedom contributes, on average, $\frac{1}{2}k_B T$ of energy per molecule, the internal energy for a system of N molecules, ignoring vibration for now, is

$$E_{\text{int}} = 3N(\tfrac{1}{2}k_B T) + 2N(\tfrac{1}{2}k_B T) = \tfrac{5}{2}Nk_B T = \tfrac{5}{2}nRT \qquad (20.33)$$

We can use this result and Equation 20.27 to find the molar specific heat at constant volume:

$$C_V = \frac{1}{n}\frac{dE_{\text{int}}}{dT} = \frac{1}{n}\frac{d}{dT}(\tfrac{5}{2}nRT) = \tfrac{5}{2}R = 20.8 \text{ J/mol} \cdot \text{K} \qquad (20.34)$$

From Equations 20.31 and 20.32, we find that

$$C_P = C_V + R = \tfrac{7}{2}R = 29.1 \text{ J/mol} \cdot \text{K}$$

$$\gamma = \frac{C_P}{C_V} = \frac{\tfrac{7}{2}R}{\tfrac{5}{2}R} = \frac{7}{5} = 1.40$$

These results agree quite well with most of the data for diatomic molecules given in Table 20.2. That is rather surprising, however, because we have not yet accounted for the possible vibrations of the molecule.

In the model for vibration, the two atoms are joined by an imaginary spring (see Fig. 20.5c). The vibrational motion adds two more degrees of freedom, which correspond to the kinetic energy and the potential energy associated with vibrations along the length of the molecule. Hence, a model that includes all three types of motion predicts a total internal energy of

$$E_{\text{int}} = 3N(\tfrac{1}{2}k_B T) + 2N(\tfrac{1}{2}k_B T) + 2N(\tfrac{1}{2}k_B T) = \tfrac{7}{2}Nk_B T = \tfrac{7}{2}nRT$$

and a molar specific heat at constant volume of

$$C_V = \frac{1}{n}\frac{dE_{\text{int}}}{dT} = \frac{1}{n}\frac{d}{dT}(\tfrac{7}{2}nRT) = \tfrac{7}{2}R = 29.1 \text{ J/mol} \cdot \text{K} \qquad (20.35)$$

This value is inconsistent with experimental data for molecules such as H_2 and N_2 (see Table 20.2) and suggests a breakdown of our model based on classical physics.

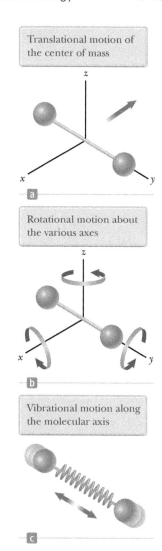

Translational motion of the center of mass

Rotational motion about the various axes

Vibrational motion along the molecular axis

Figure 20.5 Possible motions of a diatomic molecule.

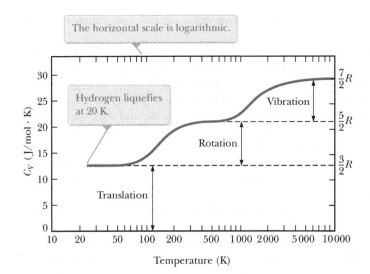

Figure 20.6 The molar specific heat of hydrogen as a function of temperature.

It might seem that our model is a failure for predicting molar specific heats for diatomic gases. We can claim some success for our model, however, if measurements of molar specific heat are made over a wide temperature range rather than at the single temperature that gives us the values in Table 20.2. Figure 20.6 shows the molar specific heat of hydrogen as a function of temperature. The remarkable feature about the three plateaus in the graph's curve is that they are at the values of the molar specific heat predicted by Equations 20.29, 20.34, and 20.35! For low temperatures, the diatomic hydrogen gas behaves like a monatomic gas. As the temperature rises to room temperature, its molar specific heat rises to a value for a diatomic gas, consistent with the inclusion of rotation but not vibration. For high temperatures, the molar specific heat is consistent with a model including all types of motion.

Before addressing the reason for this mysterious behavior, let's make some brief remarks about polyatomic gases. For nonlinear molecules with more than two atoms, three axes of rotation are available. The vibrations are more complex than for diatomic molecules. Therefore, the number of degrees of freedom is even larger. The result is an even higher predicted molar specific heat, which is in qualitative agreement with experiment. The molar specific heats for the polyatomic gases in Table 20.2 are higher than those for diatomic gases. The more degrees of freedom available to a molecule, the more "ways" there are to store energy, resulting in a higher molar specific heat.

A Hint of Energy Quantization

Our model for molar specific heats has been based so far on purely classical notions. It predicts a value of the specific heat for a diatomic gas that, according to Figure 20.6, only agrees with experimental measurements made at high temperatures. To explain why this value is only true at high temperatures and why the plateaus in Figure 20.6 exist, we must go beyond classical physics and introduce some quantum physics into the model. In Chapter 17, we discussed quantization of frequency for vibrating strings and air columns; only certain frequencies of standing waves can exist. That is a natural result whenever waves are subject to boundary conditions.

Quantum physics (Chapters 39 through 42) shows that atoms and molecules have a wavelike nature and can be analyzed with the waves under boundary conditions analysis model. Consequently, these waves have quantized frequencies. Furthermore, in quantum physics, the energy of a system is proportional to the frequency of the wave representing the system. Hence, **the energies of atoms and molecules are quantized:** only certain energies are allowed.

For a molecule, quantum physics tells us that the rotational and vibrational energies are quantized. Figure 20.7 shows an **energy-level diagram** for the rotational and vibrational quantum states of a diatomic molecule. The lowest allowed state is called the **ground state.** The three longer black lines represent allowed vibrational energies. These states are widely spaced in energy. Associated with each allowed vibrational energy is a set of more narrowly spaced rotational energies, represented by the shorter black lines.

If a molecule is in the ground state for rotation or vibration, then rotation or vibration does not contribute to the molar specific heat. These types of motion only contribute when there is a *transition* to an excited state. Section 20.1 tells us that molecular energies are proportional to temperature. Therefore, we can add labels on the right of Figure 20.7 that correspond roughly to temperatures at which the energy levels will be excited.

At low temperatures, the energy a molecule gains in collisions with its neighbors is generally not large enough to raise it to the first excited state of either rotation or vibration. Therefore, even though rotation and vibration are allowed according to classical physics, they do not occur in reality at low temperatures. All molecules are in the ground state for rotation and vibration. The only contribution to the molecules' average energy is from translation, and the specific heat is that predicted by Equation 20.29. The temperature T_A in Figure 20.7 might be 50 K for hydrogen: only the ground states for vibration or rotation are occupied; we are on the lowest plateau in Figure 20.6.

As the temperature is raised, the average energy of the molecules increases. In some collisions, a molecule may have enough energy transferred to it from another molecule to excite the first rotational state. As the temperature is raised further, more molecules can be excited to this state. The result is that rotation begins to contribute to the internal energy, and the molar specific heat rises. For hydrogen, the temperature T_B in Figure 20.7 might be 500 K: excited rotational levels are occupied, but for vibration, only the ground state is occupied; we are on the second plateau in Figure 20.6. The molar specific heat is now equal to the value predicted by Equation 20.34.

There is no contribution at room temperature from vibration because the molecules are still in the ground vibrational state. The temperature must be raised even further to excite the first vibrational state. For hydrogen, the temperature T_C in Figure 20.7 might be 5 000 K: excited rotational and vibrational levels are occupied; we are on the highest plateau in Figure 20.6 and the molar specific heat has the value predicted by Equation 20.35.

The predictions of this model are supportive of the theorem of equipartition of energy. In addition, the inclusion in the model of energy quantization from quantum physics allows a full understanding of Figure 20.6.

QUICK QUIZ 20.3 The molar specific heat of a diatomic gas is measured at constant volume and found to be 29.1 J/mol · K. What are the types of energy that are contributing to the molar specific heat? **(a)** translation only **(b)** translation and rotation only **(c)** translation and vibration only **(d)** translation, rotation, and vibration

QUICK QUIZ 20.4 The molar specific heat of a gas is measured at constant volume and found to be $11R/2$. Is the gas most likely to be **(a)** monatomic, **(b)** diatomic, or **(c)** polyatomic?

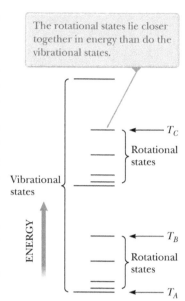

The rotational states lie closer together in energy than do the vibrational states.

Figure 20.7 An energy-level diagram for vibrational and rotational states of a diatomic molecule.

20.4 Adiabatic Processes for an Ideal Gas

Our study of molar specific heats allows us to complete our discussion of adiabatic processes begun in Section 19.5. As noted there, an **adiabatic process** is one in which no energy is transferred by heat between a system and its surroundings.

For example, if a gas is compressed (or expanded) rapidly, very little energy is transferred out of (or into) the system by heat, so the process is nearly adiabatic. Another example of an adiabatic process is the slow expansion of a gas that is thermally insulated from its surroundings. All three variables in the ideal gas law—P, V, and T—change during an adiabatic process.

Let's imagine an adiabatic gas process involving an infinitesimal change in volume dV and an accompanying infinitesimal change in temperature dT. The work done on the gas is $-P\,dV$. Because the internal energy of an ideal gas depends only on temperature, the change in the internal energy in an adiabatic process is the same as that for an isovolumetric process between the same temperatures, $dE_{int} = nC_V\,dT$ (Eq. 20.26). Hence, the first law of thermodynamics, $\Delta E_{int} = Q + W$, with $Q = 0$, becomes the infinitesimal form

$$dE_{int} = nC_V\,dT = -P\,dV \qquad (20.36)$$

Taking the total differential of the equation of state of an ideal gas, $PV = nRT$, gives

$$P\,dV + V\,dP = nR\,dT \qquad (20.37)$$

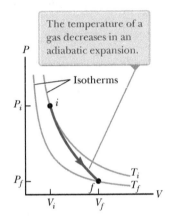

The temperature of a gas decreases in an adiabatic expansion.

Figure 20.8 The PV diagram for an adiabatic expansion of an ideal gas.

Eliminating dT between Equations 20.36 and 20.37, we find that

$$P\,dV + V\,dP = -\frac{R}{C_V}P\,dV$$

Substituting $R = C_P - C_V$ and dividing by PV gives

$$\frac{dV}{V} + \frac{dP}{P} = -\left(\frac{C_P - C_V}{C_V}\right)\frac{dV}{V} = (1 - \gamma)\frac{dV}{V}$$

$$\frac{dP}{P} + \gamma\frac{dV}{V} = 0$$

Integrating this expression, we have

$$\ln P + \gamma \ln V = \text{constant}$$

which is equivalent to

$$PV^\gamma = \text{constant} \qquad (20.38)$$

The PV diagram for an adiabatic expansion is shown in Figure 20.8. Because $\gamma > 1$, the PV curve is steeper than it would be for an isothermal expansion, for which $PV = \text{constant}$. By the definition of an adiabatic process, no energy is transferred by heat into or out of the system. Hence, from the first law, we see that ΔE_{int} is negative (work is done *by* the gas, so its internal energy decreases) and so ΔT also is negative. Therefore, the temperature of the gas decreases $(T_f < T_i)$ during an adiabatic expansion. Conversely, the temperature increases if the gas is compressed adiabatically. Applying Equation 20.38 to the initial and final states, we see that

$$P_i V_i^\gamma = P_f V_f^\gamma \qquad (20.39)$$

Using the ideal gas law, we can express Equation 20.38 as

$$TV^{\gamma-1} = \text{constant} \qquad (20.40)$$

◀ Relationship between P and V for an adiabatic process involving an ideal gas

◀ Relationship between T and V for an adiabatic process involving an ideal gas

Example 20.3 A Diesel Engine Cylinder

Air at 20.0°C in the cylinder of a diesel engine is compressed from an initial pressure of 1.00 atm and volume of 800.0 cm³ to a volume of 60.0 cm³. Assume air behaves as an ideal gas with $\gamma = 1.40$ and the compression is adiabatic. Find the final pressure and temperature of the air.

20.3 continued

SOLUTION

Conceptualize Imagine what happens if a gas is compressed into a smaller volume. Our discussion above and Figure 20.8 tell us that the pressure and temperature both increase.

Categorize We categorize this example as a problem involving an adiabatic process.

Analyze Use Equation 20.39 to find the final pressure:

$$P_f = P_i \left(\frac{V_i}{V_f}\right)^{\gamma} = (1.00 \text{ atm})\left(\frac{800.0 \text{ cm}^3}{60.0 \text{ cm}^3}\right)^{1.40}$$

$$= 37.6 \text{ atm}$$

Use the ideal gas law to find the final temperature:

$$\frac{P_i V_i}{T_i} = \frac{P_f V_f}{T_f}$$

$$T_f = \frac{P_f V_f}{P_i V_i} T_i = \frac{(37.6 \text{ atm})(60.0 \text{ cm}^3)}{(1.00 \text{ atm})(800.0 \text{ cm}^3)}(293 \text{ K})$$

$$= 826 \text{ K} = 553°C$$

Finalize The temperature of the gas increases by a factor of 826 K/293 K = 2.82. The high compression in a diesel engine raises the temperature of the gas enough to cause the combustion of fuel without the use of spark plugs.

20.5 Distribution of Molecular Speeds

Thus far, we have considered only average values of the energies of all the molecules in a gas and have not addressed the distribution of energies among individual molecules. The motion of the molecules is extremely chaotic. Any individual molecule collides with others at an enormous rate, typically a billion times per second. Each collision results in a change in the speed and direction of motion of each of the participant molecules. Equation 20.22 shows that rms molecular speeds increase with increasing temperature. At a given time, what is the relative number of molecules that possess some characteristic such as energy within a certain range?

We shall address this question by considering the **number density** $n_V(E)$. This quantity, called a *distribution function*, is defined so that $n_V(E)\, dE$ is the number of molecules per unit volume with energy between E and $E + dE$. In general, the number density is found from statistical mechanics to be

$$n_V(E) = n_0 e^{-E/k_B T} \tag{20.41}$$

where n_0 is defined such that $n_0\, dE$ is the number of molecules per unit volume having energy between $E = 0$ and $E = dE$. This equation, known as the **Boltzmann distribution law,** is important in describing the statistical mechanics of a large number of molecules. It states that the probability of finding the molecules in a particular energy state varies exponentially as the negative of the energy divided by $k_B T$. All the molecules would fall into the ground state if the thermal agitation at a temperature T did not excite the molecules to higher energy levels.

> **PITFALL PREVENTION 20.2**
> **The Distribution Function**
> The distribution function $n_V(E)$ is defined in terms of the number of molecules with energy in the range E to $E + dE$ rather than in terms of the number of molecules with a specific energy E. Because the number of molecules is finite and the number of possible values of the energy is infinite, the number of molecules with an *exact* energy E may be zero.

◀ Boltzmann distribution law

Example 20.4 | Thermal Excitation of Atomic Energy Levels

As discussed in Section 20.3, atoms can occupy only certain discrete energy levels. Consider a gas at a temperature of 2 500 K whose atoms can occupy only two energy levels separated by 1.50 eV, where 1 eV (electron volt) is an energy unit equal to 1.60×10^{-19} J. Determine the ratio of the number of atoms in the higher energy level to the number in the lower energy level.

continued

20.4 continued

SOLUTION

Conceptualize In your mental representation of this example, remember that only two possible states are allowed for the system of the atom. Figure 20.9 helps you visualize the two states on an energy-level diagram. In this case, the atom has two possible energies, E_1 and E_2, where $E_1 < E_2$.

Categorize We categorize this example as one in which we focus on particles in a two-state quantized system. We will apply the Boltzmann distribution law to this system.

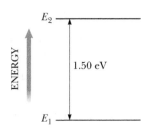

Figure 20.9 (Example 20.4) Energy-level diagram for a gas whose atoms can occupy two energy states.

Analyze Set up the ratio of the number density of atoms in the higher energy level to the number density in the lower energy level and use Equation 20.41 to express each number:

$$(1) \quad \frac{n_V(E_2)}{n_V(E_1)} = \frac{n_0 e^{-E_2/k_B T}}{n_0 e^{-E_1/k_B T}} = e^{-(E_2 - E_1)/k_B T}$$

Evaluate $k_B T$ in the exponent:

$$k_B T = (1.38 \times 10^{-23} \, \text{J/K})(2\,500 \, \text{K})\left(\frac{1 \, \text{eV}}{1.60 \times 10^{-19} \, \text{J}}\right) = 0.216 \, \text{eV}$$

Substitute this value into Equation (1):

$$\frac{n_V(E_2)}{n_V(E_1)} = e^{-1.50 \, \text{eV}/0.216 \, \text{eV}} = e^{-6.96} = 9.52 \times 10^{-4}$$

Finalize This result indicates that at $T = 2\,500$ K, only a small fraction of the atoms are in the higher energy level. In fact, for every atom in the higher energy level, there are about 1 000 atoms in the lower level. The number of atoms in the higher level increases at even higher temperatures, but the distribution law specifies that at equilibrium there are always more atoms in the lower level than in the higher level.

Ludwig Boltzmann
Austrian physicist (1844–1906)
Boltzmann made many important contributions to the development of the kinetic theory of gases, electromagnetism, and thermodynamics. His pioneering work in the field of kinetic theory led to the branch of physics known as statistical mechanics.

INTERFOTO/Alamy

Now that we have discussed the distribution of energies among molecules in a gas, let's think about the distribution of molecular speeds. In 1860, James Clerk Maxwell (1831–1879) derived an expression that describes the distribution of molecular speeds in a very definite manner. His work and subsequent developments by other scientists were highly controversial because direct detection of molecules could not be achieved experimentally at that time. About 60 years later, however, experiments were devised that confirmed Maxwell's predictions.

Let's consider a container of gas whose molecules have some distribution of speeds. Suppose we want to determine how many gas molecules have a speed in the range from, for example, 400 to 401 m/s. Intuitively, we expect the speed distribution to depend on temperature. Furthermore, we expect the distribution to peak in the vicinity of v_{rms}. That is, few molecules are expected to have speeds much less than or much greater than v_{rms} because these extreme speeds result only from an unlikely chain of collisions.

The observed speed distribution of gas molecules in thermal equilibrium is shown in Figure 20.10. The quantity N_v, called the **Maxwell–Boltzmann speed distribution function,** is defined as follows. If N is the total number of molecules, the number of molecules with speeds between v and $v + dv$ is $dN = N_v \, dv$, where the quantity N_v is dependent on temperature. This number is also equal to the area of the shaded rectangle in Figure 20.10. Furthermore, the fraction of molecules with speeds between v and $v + dv$ is $(N_v \, dv)/N$. This fraction is also equal to the probability that a molecule has a speed in the range v to $v + dv$.

The quantity N_v that describes the distribution of speeds of N gas molecules is

$$N_v = 4\pi N \left(\frac{m_0}{2\pi k_B T}\right)^{3/2} v^2 e^{-m_0 v^2/2k_B T} \tag{20.42}$$

where m_0 is the mass of a gas molecule, k_B is Boltzmann's constant, and T is the absolute temperature.[2] Observe the appearance of the Boltzmann factor $e^{-E/k_B T}$ with $E = \frac{1}{2}m_0 v^2$.

---

[2] For the derivation of this expression, see an advanced textbook on thermodynamics.

As indicated in Figure 20.10, the average speed is somewhat lower than the rms speed. The *most probable speed* v_{mp} is the speed at which the distribution curve reaches a peak. Using Equation 20.42, we find that

$$v_{rms} = \sqrt{\overline{v^2}} = \sqrt{\frac{3k_B T}{m_0}} = 1.73\sqrt{\frac{k_B T}{m_0}} \tag{20.43}$$

$$v_{avg} = \sqrt{\frac{8k_B T}{\pi m_0}} = 1.60\sqrt{\frac{k_B T}{m_0}} \tag{20.44}$$

$$v_{mp} = \sqrt{\frac{2k_B T}{m_0}} = 1.41\sqrt{\frac{k_B T}{m_0}} \tag{20.45}$$

Equation 20.43 has previously appeared as Equation 20.22. The details of the derivations of these equations from Equation 20.42 are left for the end-of-chapter problems (see Problems 24 and 41). From these equations, we see that

$$v_{rms} > v_{avg} > v_{mp}$$

Figure 20.11 represents speed distribution curves for nitrogen, N_2, at two temperatures. Notice that the peak in the curve shifts to the right as T increases, indicating that the average speed increases with increasing temperature, as do the rms speed and the most probable speed, as expected from Equations 20.43–20.45. Because the lowest speed possible is zero and the upper classical limit of the speed is infinity, the curves are asymmetrical. (In Chapter 38, we show that the actual upper limit is the speed of light.)

Equation 20.42 shows that the distribution of molecular speeds in a gas depends on temperature as well as on the mass m_0 of the molecule. At a given temperature, the fraction of molecules with speeds exceeding a fixed value increases as the mass decreases. Hence, lighter molecules such as H_2 and He escape into space more readily from the Earth's atmosphere than do heavier molecules such as N_2 and O_2. (See the discussion of escape speed in Chapter 13. Gas molecules escape even more readily from the Moon's surface than from the Earth's because the escape speed on the Moon is lower than that on the Earth, leaving essentially no atmosphere.)

The speed distribution curves for molecules in a liquid are similar to those shown in Figure 20.11. We can understand the phenomenon of *evaporation* of a liquid from this distribution in speeds, given that some molecules in the liquid

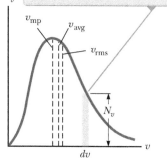

Figure 20.10 The speed distribution of gas molecules at some temperature. The function N_v approaches zero as v approaches infinity.

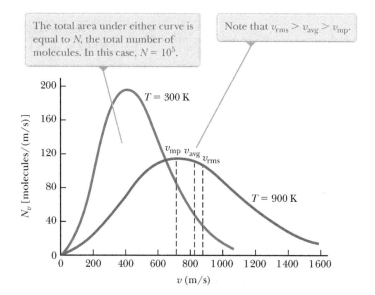

Figure 20.11 The speed distribution function for 10^5 nitrogen molecules at 300 K and 900 K.

are more energetic than others. Some of the faster-moving molecules in the liquid penetrate the surface and even leave the liquid at temperatures well below the boiling point. The molecules that escape the liquid by evaporation are those that have sufficient energy to overcome the attractive forces of the molecules in the liquid phase. Consequently, the molecules left behind in the liquid phase have a lower average kinetic energy; as a result, the temperature of the liquid decreases. Hence, evaporation is a cooling process.

That evaporation is a cooling process explains the effects you noticed in the opening storyline. When the wind blows on one side of your upturned finger, the evaporation process is accelerated. Molecules of water vapor are blown away from the surface of your finger, reducing the water vapor pressure at the surface and making it easier for more molecules to leave the surface of the water. As a result, the cooling process is augmented, and that side of your finger feels cool. The opposite side of your finger is shielded from the wind, so that the evaporation is not as rapid. When your Club advisor put alcohol on your wound, it felt cold. Alcohol evaporates at a higher rate than water, so the evaporative cooling process makes your skin feel colder than the surrounding skin that is dry.

Example **20.5** Molecular Speeds in a Hydrogen Gas

A 0.500-mol sample of hydrogen gas is at 300 K.

(A) Find the average speed, the rms speed, and the most probable speed of the H_2 molecules.

SOLUTION

Conceptualize   Imagine the huge number of particles in a macroscopic sample of gas, all moving in random directions with different speeds.

Categorize   We are dealing with a very large number of particles, so we can use the Maxwell–Boltzmann speed distribution function.

Analyze   Use Equation 20.44 to find the average speed:

$$v_{avg} = 1.60\sqrt{\frac{k_B T}{m_0}} = 1.60\sqrt{\frac{(1.38\times10^{-23}\,\text{J/K})(300\,\text{K})}{2(1.67\times10^{-27}\,\text{kg})}}$$

$$= 1.78\times10^3\,\text{m/s}$$

Use Equation 20.43 to find the rms speed:

$$v_{rms} = 1.73\sqrt{\frac{k_B T}{m_0}} = 1.73\sqrt{\frac{(1.38\times10^{-23}\,\text{J/K})(300\,\text{K})}{2(1.67\times10^{-27}\,\text{kg})}}$$

$$= 1.93\times10^3\,\text{m/s}$$

Use Equation 20.45 to find the most probable speed:

$$v_{mp} = 1.41\sqrt{\frac{k_B T}{m_0}} = 1.41\sqrt{\frac{(1.38\times10^{-23}\,\text{J/K})(300\,\text{K})}{2(1.67\times10^{-27}\,\text{kg})}}$$

$$= 1.57\times10^3\,\text{m/s}$$

(B) Find the number of molecules with speeds between 400 m/s and 401 m/s.

SOLUTION

Use Equation 20.42 to evaluate the number of molecules in a narrow speed range between v and $v + dv$:

$$(1)\quad N_v\,dv = 4\pi N\left(\frac{m_0}{2\pi k_B T}\right)^{3/2} v^2 e^{-m_0 v^2/2k_B T}\,dv$$

Evaluate the constant in front of v^2:

$$4\pi N\left(\frac{m_0}{2\pi k_B T}\right)^{3/2} = 4\pi n N_A\left(\frac{m_0}{2\pi k_B T}\right)^{3/2}$$

$$= 4\pi(0.500\,\text{mol})(6.02\times10^{23}\,\text{mol}^{-1})\left[\frac{2(1.67\times10^{-27}\,\text{kg})}{2\pi(1.38\times10^{-23}\,\text{J/K})(300\,\text{K})}\right]^{3/2}$$

$$= 1.74\times10^{14}\,\text{s}^3/\text{m}^3$$

20.5 continued

Evaluate the exponent of e that appears in Equation (1):

$$-\frac{m_0 v^2}{2k_B T} = -\frac{2(1.67 \times 10^{-27} \text{ kg})(400 \text{ m/s})^2}{2(1.38 \times 10^{-23} \text{ J/K})(300 \text{ K})} = -0.064\,5$$

Evaluate $N_v \, dv$ using these values in Equation (1) and evaluating v and dv:

$$N_v \, dv = (1.74 \times 10^{14} \text{ s}^3/\text{m}^3)(400 \text{ m/s})^2 e^{-0.064\,5}(1 \text{ m/s})$$

$$= 2.61 \times 10^{19} \text{ molecules}$$

Finalize In this evaluation, we could calculate the result without integration because $dv = 1$ m/s is much smaller than $v = 400$ m/s. Had we sought the number of particles between, say, 400 m/s and 500 m/s, we would need to integrate Equation (1) between these speed limits.

Summary

▶ Concepts and Principles

The pressure of N molecules of an ideal gas contained in a volume V is

$$P = \frac{2}{3}\left(\frac{N}{V}\right)\left(\tfrac{1}{2}m_0 \overline{v^2}\right) \qquad (20.15)$$

The average translational kinetic energy per molecule of a gas, $\frac{1}{2}m_0 \overline{v^2}$, is related to the temperature T of the gas through the expression

$$\tfrac{1}{2}m_0 \overline{v^2} = \tfrac{3}{2}k_B T \qquad (20.19)$$

where k_B is Boltzmann's constant. Each translational degree of freedom (x, y, or z) has $\frac{1}{2}k_B T$ of energy associated with it.

The molar specific heat of an ideal monatomic gas at constant volume is $C_V = \frac{3}{2}R$; the molar specific heat at constant pressure is $C_P = \frac{5}{2}R$. The ratio of specific heats is given by $\gamma = C_P/C_V = \frac{5}{3}$.

The **Boltzmann distribution law** describes the distribution of particles among available energy states. The relative number of particles having energy between E and $E + dE$ is $n_V(E)\, dE$, where

$$n_V(E) = n_0 e^{-E/k_B T} \qquad (20.41)$$

The **Maxwell–Boltzmann speed distribution function** describes the distribution of speeds of molecules in a gas:

$$N_v = 4\pi N\left(\frac{m_0}{2\pi k_B T}\right)^{3/2} v^2 e^{-m_0 v^2/2k_B T} \qquad (20.42)$$

The change in internal energy for n mol of any ideal gas that undergoes a change in temperature ΔT is

$$\Delta E_{int} = nC_V \Delta T \qquad (20.26)$$

where C_V is the **molar specific heat at constant volume.**

The internal energy of N molecules (or n mol) of an ideal monatomic gas is

$$E_{int} = \tfrac{3}{2}Nk_B T = \tfrac{3}{2}nRT \qquad (20.28)$$

If an ideal gas undergoes an adiabatic expansion or compression, the first law of thermodynamics, together with the equation of state, shows that

$$PV^\gamma = \text{constant} \qquad (20.38)$$

Equation 20.42 enables us to calculate the **root-mean-square speed**, the **average speed**, and the **most probable speed** of molecules in a gas:

$$v_{rms} = \sqrt{\overline{v^2}} = \sqrt{\frac{3k_B T}{m_0}} = 1.73\sqrt{\frac{k_B T}{m_0}} \qquad (20.43)$$

$$v_{avg} = \sqrt{\frac{8k_B T}{\pi m_0}} = 1.60\sqrt{\frac{k_B T}{m_0}} \qquad (20.44)$$

$$v_{mp} = \sqrt{\frac{2k_B T}{m_0}} = 1.41\sqrt{\frac{k_B T}{m_0}} \qquad (20.45)$$

Think–Pair–Share

See the Preface for an explanation of the icons used in this problems set. For additional assessment items for this section, go to **⋮ WEBASSIGN** From Cengage

1. Your group has been hired to do some consulting for a new baseball stadium that is being built in your city. It is going to be a totally enclosed stadium, and the architect is concerned about temperature regulation in the interior of the stadium. An eccentric architect on the development team has asked you to consider the following: every time a

baseball is hit or thrown, it eventually comes to rest because it is caught, it hits the fence, it lands in the stands, or it rolls on the ground to a stop. Whatever stops the ball, its initial kinetic energy is eventually transformed to internal energy in the interior of the stadium. The fear has been raised by this oddball character that an exciting game with lots of stopped baseballs will warm up the interior of the stadium too much for the air conditioning system to handle it. Perform a calculation to show the architect that even an exciting game will not challenge the air conditioning system for this reason.

2. **ACTIVITY** Consider the ten objects in our solar system listed below. Using the surface temperatures listed for these objects, determine the average speed of helium atoms located in the atmospheres of these objects. Using data in Table 13.2 and Equation 13.22, determine the escape speed for each of the objects. Finally, take the ratio of the escape speed to the average speed for helium atoms for each object. (a) What is a typical value for this ratio for an object with little to no atmosphere? (b) What is a typical value for this ratio for an object to have a robust atmosphere, but with little to no helium present? (c) What is a typical value for this ratio for an object to have a robust atmosphere with a significant amount of helium?

Object	Surface Temperature (K) (from the Lunar and Planetary Institute)	Atmosphere
Mercury	440	Little to none
Venus	741	Robust, little He
Earth	288	Robust, little He
Mars	244	Robust, little He
Ceres	173	Little to none
Jupiter	165	Robust, much He
Saturn	134	Robust, much He
Uranus	77	Robust, much He
Neptune	70	Robust, much He
Pluto	40	Little to none

3. **ACTIVITY** You are working as a teaching assistant to a physics professor. He has provided you with the following data on exam scores for two different sections of his physics class:

Section 1 Exam Scores	Section 2 Exam Scores
65	99
65	95
65	90
65	85
65	77
65	75
65	75
65	73
65	70
65	67
64	66
64	65
64	63
64	58
64	52
64	49
64	48
64	35
64	28
64	20

(a) Calculate the average score for each section and the rms average score for each section. (b) How the do the average scores for the two sections compare? (c) Compare the average and rms average scores for Section 1. Why do you think these two quantities have this relationship? (d) Compare the average and rms average scores for Section 2. (e) Will the rms average of a set of numbers always be larger than the straight average? Why or why not?

Problems

See the Preface for an explanation of the icons used in this problems set. For additional assessment items for this section, go to ✷ **WEBASSIGN** *From Cengage*

SECTION 20.1 Molecular Model of an Ideal Gas

Problem 20 in Chapter 18 can be assigned with this section.

1. **T** A spherical balloon of volume 4.00×10^3 cm^3 contains helium at a pressure of 1.20×10^5 Pa. How many moles of helium are in the balloon if the average kinetic energy of the helium atoms is 3.60×10^{-22} J?

2. **S** A spherical balloon of volume V contains helium at a pressure P. How many moles of helium are in the balloon if the average kinetic energy of the helium atoms is K_{avg}?

3. **V** A 2.00-mol sample of oxygen gas is confined to a 5.00-L vessel at a pressure of 8.00 atm. Find the average translational kinetic energy of the oxygen molecules under these conditions.

4. Oxygen, modeled as an ideal gas, is in a container and has a temperature of 77.0°C. What is the rms-average magnitude of the momentum of the gas molecules in the container?

5. A 5.00-L vessel contains nitrogen gas at 27.0°C and 3.00 atm. Find (a) the total translational kinetic energy of the gas molecules and (b) the average kinetic energy per molecule.

6. Calculate the mass of an atom of (a) helium, (b) iron, and (c) lead. Give your answers in kilograms. The atomic masses of these atoms are 4.00 u, 55.9 u, and 207 u, respectively.

7. **T** **V** In a period of 1.00 s, 5.00×10^{23} nitrogen molecules strike a wall with an area of 8.00 cm^2. Assume the molecules move with a speed of 300 m/s and strike the wall head-on in elastic collisions. What is the pressure exerted on the wall? *Note:* The mass of one N$_2$ molecule is 4.65×10^{-26} kg.

8. **Q|C** A 7.00-L vessel contains 3.50 moles of gas at a pressure of 1.60×10^6 Pa. Find (a) the temperature of the gas and (b) the average kinetic energy of the gas molecules in the vessel. (c) What additional information would you need if you were asked to find the average speed of the gas molecules?

SECTION 20.2 Molar Specific Heat of an Ideal Gas

Note: You may use data in Table 20.2 about particular gases. Here we define a "monatomic ideal gas" to have molar specific heats $C_V = \frac{3}{2}R$ and $C_P = \frac{5}{2}R$, and a "diatomic ideal gas" to have $C_V = \frac{5}{2}R$ and $C_P = \frac{7}{2}R$.

9. Calculate the change in internal energy of 3.00 mol of helium gas when its temperature is increased by 2.00 K.

10. Your sister, who is a realtor, is quite interested in your studies
CR in physics. Part of her job in selling properties involves being
aware of the details of heating systems for the houses. She
comes to you one day and says that a client told her the following: "If you measure the internal energy in the air in a house
and then turn up the thermostat to a higher temperature, the
internal energy of the air in the house is exactly the same as it
was at the lower temperature." She finds this hard to believe,
since you have added energy to the air by running the furnace.
Help her to figure out if this statement is true or not.

11. In a constant-volume process, 209 J of energy is transferred
V by heat to 1.00 mol of an ideal monatomic gas initially at
300 K. Find (a) the work done on the gas, (b) the increase
in internal energy of the gas, and (c) its final temperature.

12. A vertical cylinder with a heavy piston contains air at 300 K.
The initial pressure is 2.00×10^5 Pa, and the initial volume
is 0.350 m^3. Take the molar mass of air as 28.9 g/mol and
assume $C_V = \frac{5}{2}R$. (a) Find the specific heat of air at constant
volume in units of J/kg $\cdot$ °C. (b) Calculate the mass of the air
in the cylinder. (c) Suppose the piston is held fixed. Find the
energy input required to raise the temperature of the air to
700 K. (d) **What If?** Assume again the conditions of the initial
state and assume the heavy piston is free to move. Find the
energy input required to raise the temperature to 700 K.

13. A 1.00-L insulated bottle is full of tea at 90.0°C. You pour
out one cup of tea and immediately screw the stopper back
on the bottle. Make an order-of-magnitude estimate of the
change in temperature of the tea remaining in the bottle
that results from the admission of air at room temperature.
State the quantities you take as data and the values you measure or estimate for them.

SECTION 20.3 The Equipartition of Energy

14. A certain molecule has f degrees of freedom. Show that an
S ideal gas consisting of such molecules has the following
properties: (a) its total internal energy is $fnRT/2$, (b) its
molar specific heat at constant volume is $fR/2$, (c) its molar
specific heat at constant pressure is $(f + 2)R/2$, and (d) its
specific heat ratio is $\gamma = C_P/C_V = (f + 2)/f$.

15. You are working for an automobile tire company. Your super-
CR visor is studying the effects of molecules striking the inner
surface of the tire due to their thermal motion. He gives you
the following data from a recent experiment. The air in a tire
on a parked car was measured to have a gauge pressure of
$P_i = 1.65$ atm on a day when the temperature was $T = 6.5$°C.
The car was then driven for a while and then measurements
were taken again. The gauge pressure in the tire was then
$P_f = 1.95$ atm and the interior volume of the tire had increased
by 5.00%. (a) Your supervisor asks you to determine by what
factor the rms speed of the air molecules had increased from
the first measurement to the second. (b) He also hints at a
proposal he is going to make to replace air in tires with argon.
Will this change the factor by which the average speed of the
molecules changes in the conditions described?

16. *Why is the following situation impossible?* A team of researchers
discovers a new gas, which has a value of $\gamma = C_P/C_V$ of 1.75.

17. You and your younger brother are designing an air rifle
CR that will shoot a lead pellet with mass $m = 1.10$ g and cross-
sectional area $A = 0.030\ 0$ cm^2. The rifle works by allowing
high-pressure air to expand, propelling the pellet down the
rifle barrel. Because this process happens very quickly, no
appreciable thermal conduction occurs and the expansion

is essentially adiabatic. Your design is such that, once the
pressure begins pushing on the pellet, it moves a distance of
$L = 50.0$ cm before leaving the open end of the rifle at your
desired speed of $v = 120$ m/s.

Your design also includes a chamber of volume $V = 12.0$ cm^3
in which the high-pressure air is stored until it is released.
Your brother reminds you that you need to purchase a
pump to pressurize the chamber. To determine what kind
of pump to buy, you need to find what the pressure of the
air must be in the chamber to achieve your desired muzzle
speed. Ignore the effects of the air in front of the bullet and
friction with the inside walls of the barrel.

SECTION 20.4 Adiabatic Processes for an Ideal Gas

18. During the compression stroke of a certain gasoline engine,
V the pressure increases from 1.00 atm to 20.0 atm. If the process is adiabatic and the air–fuel mixture behaves as a diatomic
ideal gas, (a) by what factor does the volume change and (b) by
what factor does the temperature change? Assuming the compression starts with 0.016 0 mol of gas at 27.0°C, find the values
of (c) Q, (d) ΔE_{int}, and (e) W that characterize the process.

19. Air in a thundercloud expands as it rises. If its initial temper-
T ature is 300 K and no energy is lost by thermal conduction
on expansion, what is its temperature when the initial volume has doubled?

20. *Why is the following situation impossible?* A new diesel engine
that increases fuel economy over previous models is designed.
Automobiles fitted with this design become incredible best
sellers. Two design features are responsible for the increased
fuel economy: (1) the engine is made entirely of aluminum to
reduce the weight of the automobile, and (2) the exhaust of the
engine is used to prewarm the air to 50°C before it enters the
cylinder to increase the final temperature of the compressed
gas. The engine has a *compression ratio*—that is, the ratio of
the initial volume of the air to its final volume after compression—of 14.5. The compression process is adiabatic, and the
air behaves as a diatomic ideal gas with $\gamma = 1.40$.

21. Air (a diatomic ideal gas) at 27.0°C and atmospheric pressure
GP is drawn into a bicycle pump that has a cylinder with an inner
diameter of 2.50 cm and length 50.0 cm. The downstroke adiabatically compresses the air, which reaches a gauge pressure
of 8.00×10^5 Pa before entering the tire. We wish to investigate the temperature increase of the pump. (a) What is the
initial volume of the air in the pump? (b) What is the number
of moles of air in the pump? (c) What is the absolute pressure
of the compressed air? (d) What is the volume of the compressed air? (e) What is the temperature of the compressed
air? (f) What is the increase in internal energy of the gas during the compression? **What If?** The pump is made of steel that
is 2.00 mm thick. Assume 4.00 cm of the cylinder's length is
allowed to come to thermal equilibrium with the air. (g) What
is the volume of steel in this 4.00-cm length? (h) What is the
mass of steel in this 4.00-cm length? (i) Assume the pump is
compressed once. After the adiabatic expansion, conduction
results in the energy increase in part (f) being shared between
the gas and the 4.00-cm length of steel. What will be the
increase in temperature of the steel after one compression?

SECTION 20.5 Distribution of Molecular Speeds

22. Two gases in a mixture diffuse through a filter at rates proportional to their rms speeds. (a) Find the ratio of speeds
for the two isotopes of chlorine, ^{35}Cl and ^{37}Cl, as they diffuse
through the air. (b) Which isotope moves faster?

23. **Review.** At what temperature would the average speed of helium atoms equal (a) the escape speed from the Earth, 1.12 $\times 10^4$ m/s, and (b) the escape speed from the Moon, 2.37 $\times 10^3$ m/s? *Note:* The mass of a helium atom is 6.64×10^{-27} kg.

24. From the Maxwell–Boltzmann speed distribution, show that **S** the most probable speed of a gas molecule is given by Equation 20.45. *Note:* The most probable speed corresponds to the point at which the slope of the speed distribution curve dN_v/dv is zero.

25. Assume the Earth's atmosphere has a uniform temperature of 20.0°C and uniform composition, with an effective molar mass of 28.9 g/mol. (a) Show that the number density of molecules depends on height y above sea level according to

$$n_V(y) = n_0 e^{-m_0 g y / k_B T}$$

where n_0 is the number density at sea level (where $y = 0$). This result is called the *law of atmospheres.* (b) Commercial jetliners typically cruise at an altitude of 11.0 km. Find the ratio of the atmospheric density there to the density at sea level.

26. The law of atmospheres states that the number density of molecules in the atmosphere depends on height y above sea level according to

$$n_V(y) = n_0 e^{-m_0 g y / k_B T}$$

where n_0 is the number density at sea level (where $y = 0$). The average height of a molecule in the Earth's atmosphere is given by

$$y_{avg} = \frac{\int_0^\infty y n_V(y)\, dy}{\int_0^\infty n_V(y)\, dy} = \frac{\int_0^\infty y e^{-m_0 g y / k_B T}\, dy}{\int_0^\infty e^{-m_0 g y / k_B T}\, dy}$$

(a) Prove that this average height is equal to $k_B T / m_0 g$. (b) Evaluate the average height, assuming the temperature is 10.0°C and the molecular mass is 28.9 u, both uniform throughout the atmosphere.

ADDITIONAL PROBLEMS

27. Eight molecules have speeds of 3.00 km/s, 4.00 km/s, 5.80 km/s, 2.50 km/s, 3.60 km/s, 1.90 km/s, 3.80 km/s, and 6.60 km/s. Find (a) the average speed of the molecules and (b) the rms speed of the molecules.

28. In a sample of a solid metal, each atom is free to vibrate **Q|C** about some equilibrium position. The atom's energy consists of kinetic energy for motion in the x, y, and z directions plus elastic potential energy associated with the Hooke's law forces exerted by neighboring atoms on it in the x, y, and z directions. According to the theorem of equipartition of energy, assume the average energy of each atom is $\frac{1}{2} k_B T$ for each degree of freedom. (a) Prove that the molar specific heat of the solid is $3R$. The *Dulong–Petit law* states that this result generally describes pure solids at sufficiently high temperatures. (You may ignore the difference between the specific heat at constant pressure and the specific heat at constant volume.) (b) Evaluate the specific heat c of iron. Explain how it compares with the value listed in Table 19.1. (c) Repeat the evaluation and comparison for gold.

29. The dimensions of a classroom are 4.20 m $\times$ 3.00 m $\times$ 2.50 m. **Q|C** (a) Find the number of molecules of air in the classroom at atmospheric pressure and 20.0°C. (b) Find the mass of this air, assuming the air consists of diatomic molecules with molar mass 28.9 g/mol. (c) Find the average kinetic energy of the molecules. (d) Find the rms molecular speed. (e) **What If?** Assume the molar specific heat of the air is independent of temperature. Find the change in internal energy of the air in the room as the temperature is raised to 25.0°C. (f) Explain how you could convince a fellow student that your answer to part (e) is correct, even though it sounds surprising.

30. The compressibility κ of a substance is defined as the fractional change in volume of that substance for a given change in pressure:

$$\kappa = -\frac{1}{V} \frac{dV}{dP}$$

(a) Explain why the negative sign in this expression ensures κ is always positive. (b) Show that if an ideal gas is compressed isothermally, its compressibility is given by $\kappa_1 = 1/P$. (c) **What If?** Show that if an ideal gas is compressed adiabatically, its compressibility is given by $\kappa_2 = 1/(\gamma P)$. Determine values for (d) κ_1 and (e) κ_2 for a monatomic ideal gas at a pressure of 2.00 atm.

31. The Earth's atmosphere consists primarily of oxygen (21%) **Q|C** and nitrogen (78%). The rms speed of oxygen molecules (O_2) in the atmosphere at a certain location is 535 m/s. (a) What is the temperature of the atmosphere at this location? (b) Would the rms speed of nitrogen molecules (N_2) at this location be higher, equal to, or lower than 535 m/s? Explain. (c) Determine the rms speed of N_2 at his location.

32. **Review.** As a sound wave passes through a gas, the compres- **Q|C** sions are either so rapid or so far apart that thermal conduction is prevented by a negligible time interval or by effective thickness of insulation. The compressions and rarefactions are adiabatic. (a) Show that the speed of sound in an ideal gas is

$$v = \sqrt{\frac{\gamma R T}{M}}$$

where M is the molar mass. The speed of sound in a gas is given by Equation 16.35; use that equation and the definition of the bulk modulus from Section 12.4. (b) Compute the theoretical speed of sound in air at 20.0°C and state how it compares with the value in Table 16.1. Take $M = 28.9$ g/mol. (c) Show that the speed of sound in an ideal gas is

$$v = \sqrt{\frac{\gamma k_B T}{m_0}}$$

where m_0 is the mass of one molecule. (d) State how the result in part (c) compares with the most probable, average, and rms molecular speeds.

33. Examine the data for polyatomic gases in Table 20.2 and **Q|C** give a reason why sulfur dioxide has a higher specific heat at constant volume than the other polyatomic gases at 300 K.

34. In a cylinder, a sample of an ideal gas with number of moles **Q|C** n undergoes an adiabatic process. (a) Starting with the **S** expression $W = -\int P\, dV$ and using the condition $PV^\gamma = $ constant, show that the work done on the gas is

$$W = \left(\frac{1}{\gamma - 1}\right)(P_f V_f - P_i V_i)$$

(b) Starting with the first law of thermodynamics, show that the work done on the gas is equal to $nC_V(T_f - T_i)$. (c) Are these two results consistent with each other? Explain.

35. As a 1.00-mol sample of a monatomic ideal gas expands adiabatically, the work done on it is -2.50×10^3 J. The initial temperature and pressure of the gas are 500 K and 3.60 atm. Calculate (a) the final temperature and (b) the final pressure.

36. A sample consists of an amount n in moles of a monatomic ideal gas. The gas expands adiabatically, with work W done on it. (Work W is a negative number.) The initial temperature and pressure of the gas are T_i and P_i. Calculate (a) the final temperature and (b) the final pressure.

37. The latent heat of vaporization for water at room temperature is 2 430 J/g. Consider one particular molecule at the surface of a glass of liquid water, moving upward with sufficiently high speed that it will be the next molecule to join the vapor. (a) Find its translational kinetic energy. (b) Find its speed. Now consider a thin gas made only of molecules like that one. (c) What is its temperature? (d) Why are you not burned by water evaporating from a vessel at room temperature?

38. A vessel contains 1.00×10^4 oxygen molecules at 500 K. (a) Make an accurate graph of the Maxwell speed distribution function versus speed with points at speed intervals of 100 m/s. (b) Determine the most probable speed from this graph. (c) Calculate the average and rms speeds for the molecules and label these points on your graph. (d) From the graph, estimate the fraction of molecules with speeds in the range 300 m/s to 600 m/s.

39. For a Maxwellian gas, use a computer or programmable calculator to find the numerical value of the ratio $N_v(v)/N_v(v_{mp})$ for the following values of v: (a) $v = (v_{mp}/50.0)$, (b) $(v_{mp}/10.0)$, (c) $(v_{mp}/2.00)$, (d) v_{mp}, (e) $2.00v_{mp}$, (f) $10.0v_{mp}$, and (g) $50.0v_{mp}$. Give your results to three significant figures.

40. A triatomic molecule can have a linear configuration, as does CO_2 (Fig. P20.40a), or it can be nonlinear, like H_2O (Fig. P20.40b). Suppose the temperature of a gas of triatomic molecules is sufficiently low that vibrational motion is negligible. What is the molar specific heat at constant volume, expressed as a multiple of the universal gas constant, (a) if the molecules are linear and (b) if the molecules are nonlinear? At high temperatures, a triatomic molecule has two modes of vibration, and each contributes $\frac{1}{2}R$ to the molar specific heat for its kinetic energy and another $\frac{1}{2}R$ for its potential energy. Identify the high-temperature molar specific heat at constant volume for a triatomic ideal gas of (c) linear molecules and (d) nonlinear molecules. (e) Explain how specific heat data can be used to determine whether a triatomic molecule is linear or nonlinear. Are the data in Table 20.2 sufficient to make this determination?

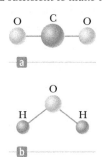

Figure P20.40

41. Using the Maxwell–Boltzmann speed distribution function, verify Equations 20.43 and 20.44 for (a) the rms speed and (b) the average speed of the molecules of a gas at a temperature T. The average value of v^n is

$$\overline{v^n} = \frac{1}{N} \int_0^\infty v^n N_v \, dv$$

Use the table of integrals B.6 in Appendix B.

42. On the PV diagram for an ideal gas, one isothermal curve and one adiabatic curve pass through each point as shown in Figure P20.42. Prove that the slope of the adiabatic curve is steeper than the slope of the isotherm at that point by the factor γ.

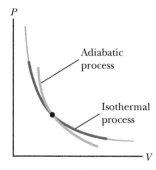

Figure P20.42

43. Using multiple laser beams, physicists have been able to cool and trap sodium atoms in a small region. In one experiment, the temperature of the atoms was reduced to 0.240 mK. (a) Determine the rms speed of the sodium atoms at this temperature. The atoms can be trapped for about 1.00 s. The trap has a linear dimension of roughly 1.00 cm. (b) Over what approximate time interval would an atom wander out of the trap region if there were no trapping action?

44. Consider the particles in a gas centrifuge, a device used to separate particles of different mass by whirling them in a circular path of radius r at angular speed ω. The force acting on a gas molecule toward the center of the centrifuge is $m_0\omega^2 r$. (a) Discuss how a gas centrifuge can be used to separate particles of different mass. (b) Suppose the centrifuge contains a gas of particles of identical mass. Show that the density of the particles as a function of r is

$$n(r) = n_0 e^{m_0 r^2 \omega^2 / 2k_B T}$$

CHALLENGE PROBLEM

45. Equations 20.43 and 20.44 show that $v_{rms} > v_{avg}$ for a collection of gas particles, which turns out to be true whenever the particles have a distribution of speeds. Let us explore this inequality for a two-particle gas. Let the speed of one particle be $v_1 = av_{avg}$ and the other particle have speed $v_2 = (2 - a)v_{avg}$. (a) Show that the average of these two speeds is v_{avg}. (b) Show that

$$v_{rms}^2 = v_{avg}^2 (2 - 2a + a^2)$$

(c) Argue that the equation in part (b) proves that, in general, $v_{rms} > v_{avg}$. (d) Under what special condition will $v_{rms} = v_{avg}$ for the two-particle gas?

Heat Engines, Entropy, and the Second Law of Thermodynamics

▲
A refrigerator in a recreational vehicle. How does a refrigerator work?
(*Adam Bronkhorst/Alamy*)

STORYLINE **You are still on your Physics Club camping trip. Uh-oh.** The electric refrigerator in your Club advisor's RV has suddenly stopped working. You help your advisor take the refrigerator out from its mounting and inspect the workings. He looks around a bit, tries a few things, and then says that he suspects the compressor has gone bad. You ask him what the compressor does and he says that it compresses the gaseous refrigerant to a high temperature and pressure. This starts you thinking. If you are trying to keep food *cold* in a refrigerator, why would you want to make the refrigerant *hot*? Your advisor goes on to explain that air conditioners work the same way. Then he says that he thinks the refrigerator cannot be repaired and they should go to the camping supply store and buy an inexpensive propane-powered refrigerator to use for the rest of your trip. You say, "What!? You can *cool* food by *burning* propane? How can that possibly work?" You spend the next couple of hours online investigating refrigeration cycles.

CONNECTIONS Although the first law of thermodynamics, which we studied in Chapter 19, is very important, it makes no distinction between processes that occur spontaneously and those that do not. Only certain types of energy transformation and transfer processes actually take place in nature. The second law of thermodynamics, the major topic in this chapter, establishes which processes do and do not occur. In general, for example, it is common to see processes in which mechanical energy is transformed into internal energy. As a book sliding across a surface comes to rest, its kinetic energy has transformed to internal energy, which spreads out in the book and the surface. One would never expect this internal energy to somehow gather itself back into the book, so that the book begins to move again. Books at rest and in static equilibrium *always* remain at rest. It is also common to see energy transfer by heat from a hot object to a cold object with which it is in contact. One would never expect to add ice to room-temperature water and see the water become warmer and the ice colder. Energy *always* transfers from the warm water to the cold ice. The expected processes described here are *irreversible*; that is, they are processes that occur naturally in one direction only. No irreversible process has ever been observed to run backward. If it were to do so, it would violate the second law of thermodynamics.[1] The second law of thermodynamics is at work in all natural processes that we will study in future chapters. In this chapter, we study this law and a closely related quantity, *entropy*. We begin our quest to understand both the second law and entropy by investigating the thermodynamics of *heat engines*.

Lord Kelvin
British physicist and mathematician
(1824–1907)
Born William Thomson in Belfast, Kelvin was the first to propose the use of an absolute scale of temperature. The Kelvin temperature scale is named in his honor. Kelvin's work in thermodynamics led to the idea that energy cannot pass spontaneously from a colder object to a hotter object.

21.1 Heat Engines and the Second Law of Thermodynamics

A **heat engine** is a device that takes in energy by heat[2] and, operating in a cyclic process, expels a fraction of that energy by means of work. For instance, in a typical process by which a power plant produces electricity, a fuel such as natural gas is burned and the high-temperature gases produced are used to convert liquid water to steam. This steam is directed at the blades of a turbine. The steam does work on the blades of the turbine, setting it into rotation. The mechanical energy associated with this rotation is used to drive an electric generator. Another device that can be modeled as a heat engine is the internal combustion engine in an automobile. This device uses energy from a burning fuel to perform work on pistons that results in the motion of the automobile.

Let us consider the fundamental operation of a heat engine in more detail. A heat engine carries some working substance through a cyclic process during which (1) the working substance absorbs energy by heat from a high-temperature energy reservoir, (2) work is done by the engine, and (3) energy is expelled by heat to a lower-temperature reservoir. As an example, consider the operation of a steam engine (Fig. 21.1), which uses water as the working substance. The water in a boiler absorbs energy from burning fuel and evaporates to steam, which then does work by expanding against a piston. After the steam cools and condenses, the liquid water produced returns to the boiler and the cycle repeats.

It is useful to represent a heat engine schematically as in Figure 21.2 (page 558). The engine absorbs a quantity of energy $|Q_h|$ from the hot reservoir. For the mathematical discussion of heat engines, we use absolute values to make all energy transfers by heat positive, and the direction of transfer is indicated with an explicit positive or negative sign. The engine does work W_{eng} (so that *negative* work $W = -W_{eng}$ is done *on* the engine) and then gives up a quantity of energy $|Q_c|$ to the cold reservoir. Because the working substance in the engine goes through a cycle, its initial and final internal energies are equal: $\Delta E_{int} = 0$. Hence, from the first law of thermodynamics, for each cycle of the engine, $\Delta E_{int} = Q + W = Q_{net} - W_{eng} = 0$, and

Figure 21.1 A steam-driven locomotive obtains its energy by burning wood or coal. The generated energy vaporizes water into steam, which powers the locomotive. Modern locomotives use diesel fuel instead of wood or coal. Whether old-fashioned or modern, such locomotives can be modeled as heat engines, which extract energy from a burning fuel and convert a fraction of it to mechanical energy.

---

[1]Although a process occurring in the time-reversed sense has never been *observed*, it is *possible* for it to occur. As we shall see later in this chapter, however, the probability of such a process occurring is infinitesimally small. From this viewpoint, processes occur with a vastly greater probability in one direction than in the opposite direction.

[2]We use heat as our model for energy transfer into a heat engine. Other methods of energy transfer are possible in the model of a heat engine, however. For example, the Earth's atmosphere can be modeled as a heat engine in which the input energy transfer is by means of electromagnetic radiation from the Sun. The output of the atmospheric heat engine causes the wind structure in the atmosphere.

the net work W_{eng} done by a heat engine is equal to the net energy Q_{net} transferred to it. As you can see from Figure 21.2, $Q_{net} = |Q_h| - |Q_c|$; therefore,

$$W_{eng} = |Q_h| - |Q_c| \qquad (21.1)$$

The **thermal efficiency** e of a heat engine is defined as the ratio of the net work done by the engine during one cycle to the energy input at the higher temperature during the cycle:

Thermal efficiency of ▶
a heat engine

$$e \equiv \frac{W_{eng}}{|Q_h|} = \frac{|Q_h| - |Q_c|}{|Q_h|} = 1 - \frac{|Q_c|}{|Q_h|} \qquad (21.2)$$

You can think of the efficiency as the ratio of what you gain (work) to what you give (energy transfer at the higher temperature). In practice, all heat engines expel only a fraction of the input energy Q_h by mechanical work; consequently, their efficiency is always less than 100%. For example, a good automobile engine has an efficiency of about 20%, and diesel engines have efficiencies ranging from 35% to 40%.

Equation 21.2 shows that a heat engine has 100% efficiency ($e = 1$) only if $|Q_c| = 0$, that is, if no energy is expelled to the cold reservoir. In other words, a heat engine with perfect efficiency would have to expel all the input energy by work. Figure 21.3 is a schematic diagram of the "perfect" heat engine. The efficiencies of real engines are well below 100%, which is related to one form of the second law of thermodynamics. The **Kelvin–Planck form of the second law of thermodynamics** states the impossibility of an engine with 100% efficiency:

It is impossible to construct a heat engine that, operating in a cycle, produces no effect other than the input of energy by heat from a reservoir and the performance of an equal amount of work.

This statement of the second law means that during the operation of a heat engine, W_{eng} can never be equal to $|Q_h|$ or, alternatively, that some energy $|Q_c|$ *must* be rejected to the environment.

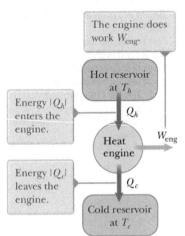

Figure 21.2 Schematic representation of a heat engine.

The engine does work W_{eng}.

Hot reservoir at T_h

Energy $|Q_h|$ enters the engine.

Q_h

Heat engine

W_{eng}

Energy $|Q_c|$ leaves the engine.

Q_c

Cold reservoir at T_c

PITFALL PREVENTION 21.1

The First and Second Laws Notice the distinction between the first and second laws of thermodynamics. If a gas undergoes a *one-time isothermal process*, then $\Delta E_{int} = Q + W = 0$ and $W = -Q$. Therefore, the first law allows *all* energy input by heat to be expelled by work. In a heat engine, however, in which a substance undergoes a *cyclic* process, only a *portion* of the energy input by heat can be expelled by work according to the second law.

⊙UICK QUIZ 21.1 The energy input to an engine is 4.00 times greater than the work it performs. **(i)** What is its thermal efficiency? (a) 4.00 (b) 1.00 (c) 0.250 (d) impossible to determine **(ii)** What fraction of the energy input is expelled to the cold reservoir? (a) 0.250 (b) 0.750 (c) 1.00 (d) impossible to determine

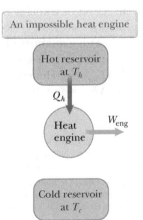

An impossible heat engine

Hot reservoir at T_h

Q_h

Heat engine

W_{eng}

Cold reservoir at T_c

Figure 21.3 Schematic diagram of a heat engine that takes in energy from a hot reservoir and does an equivalent amount of work. It is impossible to construct such a perfect engine.

| Example **21.1** | The Efficiency of an Engine |

An engine transfers 2.00×10^3 J of energy from a hot reservoir during a cycle and transfers 1.50×10^3 J as exhaust to a cold reservoir.

(A) Find the efficiency of the engine.

SOLUTION

Conceptualize Review Figure 21.2; think about energy going into the engine from the hot reservoir and splitting, with part coming out by work and part by heat into the cold reservoir.

Categorize This example involves evaluation of quantities from the equations introduced in this section, so we categorize it as a substitution problem.

Find the efficiency of the engine from Equation 21.2:

$$e = 1 - \frac{|Q_c|}{|Q_h|} = 1 - \frac{1.50 \times 10^3 \text{ J}}{2.00 \times 10^3 \text{ J}} = 0.250, \text{ or } 25.0\%$$

(B) How much work does this engine do in one cycle?

SOLUTION

Find the work done by the engine by taking the difference between the input and output energies:

$$W_{eng} = |Q_h| - |Q_c| = 2.00 \times 10^3 \text{ J} - 1.50 \times 10^3 \text{ J}$$

$$= 5.0 \times 10^2 \text{ J}$$

WHAT IF? Suppose you were asked for the power output of this engine. Do you have sufficient information to answer this question?

Answer No, you do not have enough information. The power of an engine is the *rate* at which work is done by the engine. You know how much work is done per cycle, but you have no information about the time interval associated with one cycle. If you were told that the engine operates at 2 000 rpm (revolutions per minute), however, you could relate this rate to the period of rotation T of the mechanism of the engine. Assuming there is one thermodynamic cycle per revolution, the power is

$$P = \frac{W_{eng}}{T} = \frac{5.0 \times 10^2 \text{ J}}{\left(\frac{1}{2\,000} \text{ min}\right)} \left(\frac{1 \text{ min}}{60 \text{ s}}\right) = 1.7 \times 10^4 \text{ W}$$

21.2 Heat Pumps and Refrigerators

In a heat engine, the direction of energy transfer is from the hot reservoir to the cold reservoir, which is the natural direction. The role of the heat engine is to process the energy from the hot reservoir so as to expel part of it by useful work. What if we wanted to transfer energy from the cold reservoir to the hot reservoir? Because that is not the natural direction of energy transfer, we must put some energy into a device to perform this task. Devices that transfer energy from a cold reservoir to a warm reservoir are called **heat pumps** and **refrigerators.** For example, homes in summer are cooled using heat pumps called *air conditioners.* The air conditioner transfers energy from the cool room in the home to the warm air outside.

In a refrigerator or a heat pump, the engine takes in energy $|Q_c|$ from a cold reservoir and expels energy $|Q_h|$ to a hot reservoir (Fig. 21.4), which can be accomplished only if work is done *on* the engine. From the first law, we know that the energy given up to the hot reservoir must equal the sum of the work done and the energy taken in from the cold reservoir. Therefore, the refrigerator or heat pump transfers energy from a colder body (for example, the contents of a kitchen refrigerator or the winter air outside a building) to a hotter body (the air in the kitchen or a room in the building). In practice, it is desirable to carry out this process with a minimum of work. If the process could be accomplished without doing

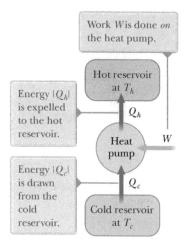

Figure 21.4 Schematic representation of a heat pump.

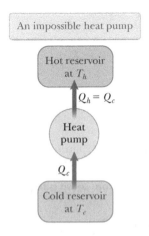

Figure 21.5 Schematic diagram of an impossible heat pump or refrigerator, that is, one that takes in energy from a cold reservoir and expels an equivalent amount of energy to a hot reservoir without the input of energy by work.

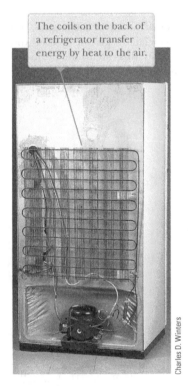

The coils on the back of a refrigerator transfer energy by heat to the air.

Charles D. Winters

Figure 21.6 The back of a household refrigerator. The air surrounding the coils is the hot reservoir.

any work, the refrigerator or heat pump would be "perfect" (Fig. 21.5). Again, the existence of such a device is impossible and is related to another form of the second law of thermodynamics. The **Clausius statement**[3] **of the second law of thermodynamics** states:

It is impossible to construct a cyclical machine whose sole effect is to transfer energy continuously by heat from one object to another object at a higher temperature without the input of energy by work.

In simpler terms, energy does not transfer spontaneously by heat from a cold object to a hot object. Work input is required to run a refrigerator.

The Clausius and Kelvin–Planck statements of the second law of thermodynamics appear at first sight to be unrelated, but in fact they are equivalent in all respects. Although we do not prove so here, if either statement is false, so is the other.[4]

In practice, a heat pump includes a circulating fluid that passes through two sets of metal coils that can exchange energy with the surroundings. The fluid is cold and at low pressure when it is in the coils located in a cool environment, where it absorbs energy by heat. The resulting warm fluid is then compressed and enters the other coils as a hot, high-pressure fluid. There it releases its stored energy to the warm surroundings. In an air conditioner, energy is absorbed into the fluid in coils located in a building's interior; after the fluid is compressed, energy leaves the fluid through coils located outdoors. In a refrigerator, the external coils are behind the unit (Fig. 21.6) or underneath the unit. The internal coils are in the walls of the refrigerator and absorb energy from the food.

In the preceding paragraph, we see why we want the refrigerant to be hot, a question we asked in the opening storyline with regard to the refrigerator in your RV. But what about the propane refrigerator? This type of refrigerator also takes a substance through a cycle. In this case, the substance is ammonia, which combines at various parts of the cycle with water and hydrogen. The propane burner warms up the ammonia–water combination, which then travels to external coils to release energy into the air. The ammonia is separated from the water and then combines with hydrogen. It then evaporates. As we discussed in Chapter 20, evaporation is a cooling process. The cool ammonia is passed through the cooling coils, where it absorbs energy from the interior of the refrigerator, is mixed with water again, and then proceeds back to the beginning of the cycle.

The effectiveness of a heat pump is described in terms of a number called the **coefficient of performance** (COP). The COP is similar to the thermal efficiency for a heat engine in that it is a ratio of what you gain (energy transferred to or from a reservoir) to what you give (work input). For a heat pump operating in the cooling mode, "what you gain" is energy removed from the cold reservoir. The most effective refrigerator or air conditioner is one that removes the greatest amount of energy from the cold reservoir in exchange for the least amount of work. Therefore, for these devices operating in the cooling mode, we define the COP in terms of $|Q_c|$:

$$\text{COP (cooling mode)} = \frac{\text{energy transferred at low temperature}}{\text{work done on heat pump}} = \frac{|Q_c|}{W} \quad (21.3)$$

A good refrigerator should have a high COP, typically 5 or 6.

In addition to cooling applications, heat pumps are becoming increasingly popular for heating purposes. In the heating mode, energy is absorbed from the cool air outside a building and warm air is released inside the building. The COP of a

[3]First expressed by Rudolf Clausius (1822–1888), a German physicist and mathematician.

[4]See an advanced textbook on thermodynamics for this proof.

heat pump is defined as the ratio of the energy transferred to the hot reservoir to the work required to transfer that energy:

$$\text{COP (heating mode)} = \frac{\text{energy transferred at high temperature}}{\text{work done on heat pump}} = \frac{|Q_h|}{W} \quad \text{(21.4)}$$

If the outside temperature is 25°F (−4°C) or higher, a typical value of the COP for a heat pump is about 4. That is, the amount of energy transferred to the building is about four times greater than the work done by the motor in the heat pump. As the outside temperature decreases, however, it becomes more difficult for the heat pump to extract sufficient energy from the air and so the COP decreases. Therefore, the use of heat pumps that extract energy from the air, although satisfactory in moderate climates, is not appropriate in areas where winter temperatures are very low. It is possible to use heat pumps in colder areas by burying the external coils deep in the ground. In that case, the energy is extracted from the ground, which tends to be warmer than the air in the winter.

ⓠ**UICK QUIZ 21.2** The energy entering an electric heater by electrical transmission can be converted to internal energy with an efficiency of 100%. By what factor does the cost of heating your home change when you replace your electric heating system with an electric heat pump that has a COP of 4.00? Assume the motor running the heat pump is 100% efficient. **(a)** 4.00 **(b)** 2.00 **(c)** 0.500 **(d)** 0.250

Example **21.2** **Freezing Water**

A certain refrigerator has a COP of 5.00. When the refrigerator is running, its power input is 500 W. A sample of water of mass 500 g and temperature 20.0°C is placed in the freezer compartment. How long does it take to freeze the water to ice at 0°C? Assume all other parts of the refrigerator stay at the same temperature and there is no leakage of energy from the exterior, so the operation of the refrigerator results only in energy being extracted from the water.

SOLUTION

Conceptualize Energy leaves the water, reducing its temperature and then freezing it into ice. The time interval required for this entire process is related to the rate at which energy is withdrawn from the water, which, in turn, is related to the power input of the refrigerator.

Categorize We categorize this example as one that combines our understanding of temperature changes and phase changes from Chapter 19 and our understanding of heat pumps from this chapter.

Analyze Use the power rating of the refrigerator to find the time interval Δt required for the freezing process to occur:

$$P = \frac{W}{\Delta t} \;\rightarrow\; \Delta t = \frac{W}{P}$$

Use Equation 21.3 to relate the work W done on the heat pump to the energy $|Q_c|$ extracted from the water:

$$\Delta t = \frac{|Q_c|}{P(\text{COP})}$$

Use Equations 19.4 and 19.8 to substitute the amount of energy $|Q_c|$ that must be extracted from the water of mass m:

$$\Delta t = \frac{|mc\,\Delta T + L_f\,\Delta m|}{P(\text{COP})}$$

Recognize that the amount of water that freezes is $\Delta m = -m$ because all the water freezes:

$$\Delta t = \frac{|m(c\,\Delta T - L_f)|}{P(\text{COP})}$$

continued

21.2 continued

Substitute numerical values:

$$\Delta t = \frac{|(0.500 \text{ kg})[(4\ 186 \text{ J/kg} \cdot °C)(-20.0°C) - 3.33 \times 10^5 \text{ J/kg}]|}{(500 \text{ W})(5.00)}$$

$$= 83.3 \text{ s}$$

Finalize In reality, the time interval for the water to freeze in a refrigerator is much longer than 83.3 s, which suggests that the assumptions of our model are not valid. Only a small part of the energy extracted from the refrigerator interior in a given time interval comes from the water. Energy must also be extracted from the container in which the water is placed, and energy that continuously leaks into the interior from the exterior must be extracted.

21.3 Reversible and Irreversible Processes

PITFALL PREVENTION 21.2
All Real Processes Are Irreversible
The reversible process is an idealization; all real processes on the Earth are irreversible.

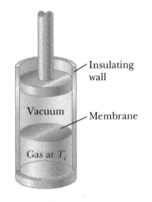

Figure 21.7 Adiabatic free expansion of a gas.

In the next section, we will discuss a theoretical heat engine that is the most efficient possible. To understand its nature, we must first examine the meaning of reversible and irreversible processes. In a **reversible** process, the system undergoing the process can be returned to its initial conditions along the same path on a *PV* diagram, and every point along this path is an equilibrium state. A process that does not satisfy these requirements is **irreversible.**

All natural processes are known to be irreversible. Let's examine a unique process called the **adiabatic free expansion** of a gas, and show that it cannot be reversible. Consider a gas in a thermally insulated container as shown in Figure 21.7. A membrane separates the gas from a vacuum. When the membrane is punctured, the gas expands freely into the vacuum. As a result of the puncture, the system has changed because it occupies a greater volume after the expansion. Because the gas does not exert a force through a displacement, it does no work on the surroundings as it expands: $W = 0$. In addition, no energy is transferred to or from the gas by heat because the container is insulated from its surroundings: $Q = 0$. Therefore, from the first law of thermodynamics, the internal energy E_{int} of the gas does not change and, as a result, its temperature is the same after the expansion. In this process, the system has changed, but the surroundings have not.

For this process to be reversible, we must return the gas to its original volume and temperature without changing the surroundings. Imagine trying to reverse the process by compressing the gas to its original volume. To do so, we use an engine to force the piston shown in Figure 21.7 inward. During this process, the surroundings change because work is being done by an outside agent on the system. In addition, the system changes because the compression increases the temperature of the gas. The temperature of the gas can be lowered by allowing it to come into contact with an external energy reservoir. Although this step returns the gas to its original conditions, the surroundings are again affected because energy is being added to the surroundings from the gas. If this energy could be used to drive the engine that compressed the gas, the net energy transfer to the surroundings would be zero. In this way, the system and its surroundings could be returned to their initial conditions and we could identify the process as reversible. The Kelvin–Planck statement of the second law, however, specifies that the energy removed from the gas to return the temperature to its original value cannot be completely converted to mechanical energy by the process of work done by the engine in compressing the gas. Therefore, we must conclude that the process is irreversible.

We could also argue that the adiabatic free expansion is irreversible by relying on the portion of the definition of a reversible process that refers to equilibrium states. For example, during the sudden expansion, significant variations in pressure occur throughout the gas. Therefore, there is no single, well-defined value of the pressure for the entire system at any time between the initial and final states. In fact, the process cannot even be represented as a path on a *PV* diagram. The *PV* diagram for an adiabatic free expansion would show the initial and final conditions as points, but these points would not be connected by a path. Therefore, because the intermediate conditions between the initial and final states are not equilibrium states, the process is irreversible.

Although all real processes are irreversible, some are almost reversible. If a real process occurs very slowly such that the system is always very nearly in an equilibrium state, the process can be approximated as being reversible.

A general characteristic of a reversible process is that no nonconservative effects (such as turbulence or friction) that transform mechanical energy to internal energy can be present. Such effects can be impossible to eliminate completely. Hence, it is not surprising that real processes in nature are irreversible.

21.4 The Carnot Engine

In 1824, a French engineer named Sadi Carnot described a theoretical engine, now called a **Carnot engine,** that is of great importance from both practical and theoretical viewpoints. He showed that a heat engine operating in an ideal, reversible cycle—called a **Carnot cycle**—between two energy reservoirs is the most efficient engine possible. Such an ideal engine establishes an upper limit on the efficiencies of all other engines. That is, the net work done by a working substance taken through the Carnot cycle is the greatest amount of work possible for a given amount of energy supplied to the substance at the higher temperature. **Carnot's theorem** can be stated as follows:

No real heat engine operating between two energy reservoirs can be more efficient than a Carnot engine operating between the same two reservoirs.

In this section, we will show that the efficiency of a Carnot engine depends only on the temperatures of the reservoirs. In turn, that efficiency represents the maximum possible efficiency for real engines. Let us confirm that the Carnot engine is the most efficient. We imagine a hypothetical engine with an efficiency greater than that of the Carnot engine. Consider Figure 21.8, which shows the hypothetical engine with $e > e_C$ on the left connected between hot and cold reservoirs. In addition, let us attach a Carnot engine between the same reservoirs. Because the Carnot cycle is reversible, the Carnot engine can be run in reverse as a Carnot heat pump as shown on the right in Figure 21.8. We match the output work of the engine to the input work of the heat pump, $W = W_C$, so there is no exchange of energy by work between the surroundings and the engine–heat pump combination.

Because of the proposed relation between the efficiencies of the heat engine and heat pump when both are operated as engines, we must have

$$e > e_C \;\rightarrow\; \frac{|W|}{|Q_h|} > \frac{|W_C|}{|Q_{hC}|}$$

The numerators of these two fractions cancel because the works have been matched in the configuration in Figure 21.8. Therefore, this expression becomes

$$|Q_{hC}| > |Q_h| \qquad (21.5)$$

From Equation 21.1, the equality of the works gives us

$$|W| = |W_C| \;\rightarrow\; |Q_h| - |Q_c| = |Q_{hC}| - |Q_{cC}|$$

Sadi Carnot
French engineer (1796–1832)
Carnot was the first to show the quantitative relationship between work and heat. In 1824, he published his only work, *Reflections on the Motive Power of Heat*, which reviewed the industrial, political, and economic importance of the steam engine. In it, he defined work as "weight lifted through a height."

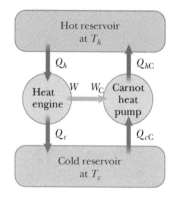

Figure 21.8 Two engines operate between two energy reservoirs: a Carnot engine operating as a heat pump and another engine with an efficiency that is proposed to be higher than that of the Carnot engine. The work output and input are matched.

which can be rewritten to put the energies exchanged with the cold reservoir on the left and those with the hot reservoir on the right:

$$|Q_{hC}| - |Q_h| = |Q_{cC}| - |Q_c| \tag{21.6}$$

Note that, in light of Equation 21.5, the left side of Equation 21.6 is positive, so the right side must be positive also. We see that the net energy exchange with the hot reservoir is equal to the net energy exchange with the cold reservoir. As a result, for the combination of the heat engine and the heat pump, energy is transferring from the cold reservoir to the hot reservoir by heat with no input of energy by work from the surroundings.

This result is in violation of the Clausius statement of the second law. Therefore, our original assumption that $e > e_C$ must be incorrect, and we must conclude that the Carnot engine represents the highest possible efficiency for an engine. The key feature of the Carnot engine that makes it the most efficient is its *reversibility;* it can be run in reverse as a heat pump. All real engines are less efficient than the Carnot engine because they do not operate through a reversible cycle. The efficiency of a real engine is further reduced by such practical difficulties as friction and energy losses by conduction.

Let's now look at the details of the Carnot cycle for an engine operating between temperatures T_c and T_h. Assume the working substance is an ideal gas contained in a cylinder fitted with a movable piston at one end. The cylinder's walls and the piston are thermally nonconducting. Four stages of the Carnot cycle are shown in Figure 21.9,

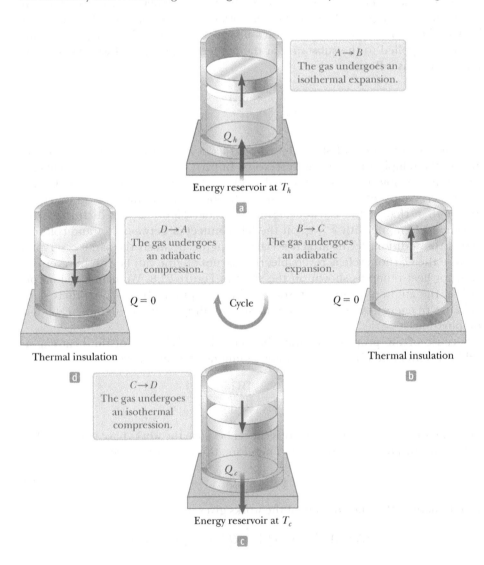

Figure 21.9 A pictorial representation of the Carnot cycle. The letters A, B, C, and D refer to the states of the gas shown in Figure 21.10. The arrows on the piston indicate the direction of its motion during each process. Compare to the graphical representation in Figure 21.10.

and the *PV* diagram for the cycle is shown in Figure 21.10. The Carnot cycle consists of two adiabatic processes and two isothermal processes, all reversible:

1. Process $A \rightarrow B$ (Fig. 21.9a) is an isothermal expansion at temperature T_h. The gas is placed in thermal contact with an energy reservoir at temperature T_h. During the expansion, the gas absorbs energy $|Q_h|$ from the reservoir through the base of the cylinder and does work W_{AB} in raising the piston.

2. In process $B \rightarrow C$ (Fig. 21.9b), the base of the cylinder is replaced by a thermally nonconducting wall and the gas expands adiabatically; that is, no energy enters or leaves the system by heat. During the expansion, the temperature of the gas decreases from T_h to T_c and the gas does work W_{BC} in raising the piston.

3. In process $C \rightarrow D$ (Fig. 21.9c), the gas is placed in thermal contact with an energy reservoir at temperature T_c and is compressed isothermally at temperature T_c. During this time, the gas expels energy $|Q_c|$ to the reservoir and the work done by the piston on the gas is W_{CD}.

4. In the final process $D \rightarrow A$ (Fig. 21.9d), the base of the cylinder is replaced by a nonconducting wall and the gas is compressed adiabatically. The temperature of the gas increases to T_h, and the work done by the piston on the gas is W_{DA}.

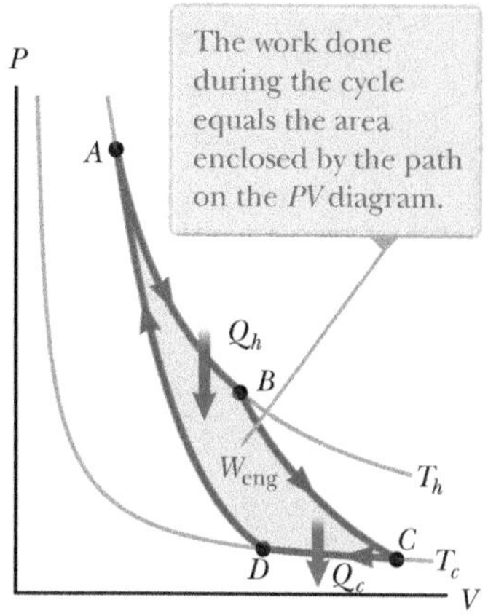

The work done during the cycle equals the area enclosed by the path on the *PV* diagram.

Figure 21.10 *PV* diagram for the Carnot cycle represented pictorially in Figure 21.9. This is a graphical representation of the cycle. The net work done W_{eng} equals the net energy transferred into the Carnot engine in one cycle, $|Q_h| - |Q_c|$.

The thermal efficiency of the engine is given by Equation 21.2:

$$e = 1 - \frac{|Q_c|}{|Q_h|}$$

In Example 21.3, we show that for a Carnot cycle,

$$\frac{|Q_c|}{|Q_h|} = \frac{T_c}{T_h} \tag{21.7}$$

Hence, the thermal efficiency of a Carnot engine is

$$e_C = 1 - \frac{T_c}{T_h} \tag{21.8}$$

◀ Efficiency of a Carnot engine

This result indicates that all Carnot engines operating between the same two temperatures have the same efficiency.[5]

Equation 21.8 can be applied to any working substance operating in a Carnot cycle between two energy reservoirs. According to this equation, the efficiency is zero if $T_c = T_h$, as one would expect. The efficiency increases as T_c is lowered and T_h is raised. The efficiency can be unity (100%), however, only if $T_c = 0$ K. A reservoir at absolute zero is not available; therefore, the maximum efficiency is always less than 100%. In most practical cases, T_c is near room temperature, which is about 300 K. Therefore, one usually strives to increase the efficiency by raising T_h.

Theoretically, a Carnot-cycle heat engine run in reverse constitutes the most effective heat pump possible, and it determines the maximum COP for a given

---

[5]For the processes in the Carnot cycle to be reversible, they must be carried out infinitesimally slowly. Therefore, although the Carnot engine is the most efficient engine possible, it has zero power output because it takes an infinite time interval to complete one cycle! For a real engine, the short time interval for each cycle results in the working substance reaching a high temperature lower than that of the hot reservoir and a low temperature higher than that of the cold reservoir. An engine undergoing a Carnot cycle between this narrower temperature range was analyzed by F. L. Curzon and B. Ahlborn ("Efficiency of a Carnot engine at maximum power output," *Am. J. Phys.* **43**(1), 22, 1975), who found that the efficiency at maximum power output depends only on the reservoir temperatures T_c and T_h and is given by $e_{C-A} = 1 - (T_c/T_h)^{1/2}$. The Curzon–Ahlborn efficiency e_{C-A} provides a closer approximation to the efficiencies of real engines than does the Carnot efficiency.

combination of hot and cold reservoir temperatures. Using Equations 21.1 and 21.4, we see that the maximum COP for a heat pump in its heating mode is

$$\text{COP}_C \text{ (heating mode)} = \frac{|Q_h|}{W}$$

$$= \frac{|Q_h|}{|Q_h| - |Q_c|} = \frac{1}{1 - \dfrac{|Q_c|}{|Q_h|}} = \frac{1}{1 - \dfrac{T_c}{T_h}} = \frac{T_h}{T_h - T_c}$$

The Carnot COP for a heat pump in the cooling mode is

$$\text{COP}_C \text{ (cooling mode)} = \frac{T_c}{T_h - T_c}$$

As the difference between the temperatures of the two reservoirs approaches zero in this expression, the theoretical COP approaches infinity. In practice, the low temperature of the cooling coils and the high temperature at the compressor limit the COP to values below 10.

> **QUICK QUIZ 21.3** Three engines operate between reservoirs separated in temperature by 300 K. The reservoir temperatures are as follows: Engine A: $T_h = 1\,000$ K, $T_c = 700$ K; Engine B: $T_h = 800$ K, $T_c = 500$ K; Engine C: $T_h = 600$ K, $T_c = 300$ K. Rank the engines in order of theoretically possible efficiency from highest to lowest.

Example 21.3   Efficiency of the Carnot Engine

Show that the ratio of energy transfers by heat in a Carnot engine is equal to the ratio of reservoir temperatures, as given by Equation 21.7.

SOLUTION

Conceptualize   Make use of Figures 21.9 and 21.10 to help you visualize the processes in the Carnot cycle.

Categorize   Because of our understanding of the Carnot cycle, we can categorize the processes in the cycle as isothermal and adiabatic.

Analyze   For the isothermal expansion (process $A \rightarrow B$ in Fig. 21.9), find the energy transfer by heat from the hot reservoir using Equation 19.12 and the first law of thermodynamics:

$$|Q_h| = |\Delta E_{\text{int}} - W_{AB}| = |0 - W_{AB}| = nRT_h \ln \frac{V_B}{V_A}$$

In a similar manner, find the energy transfer to the cold reservoir during the isothermal compression $C \rightarrow D$:

$$|Q_c| = |\Delta E_{\text{int}} - W_{CD}| = |0 - W_{CD}| = nRT_c \ln \frac{V_C}{V_D}$$

Divide the second expression by the first:

$$(1) \quad \frac{|Q_c|}{|Q_h|} = \frac{T_c \ln (V_C/V_D)}{T_h \ln (V_B/V_A)}$$

Apply Equation 20.40 to the adiabatic processes $B \rightarrow C$ and $D \rightarrow A$:

$$T_h V_B^{\gamma-1} = T_c V_C^{\gamma-1}$$
$$T_h V_A^{\gamma-1} = T_c V_D^{\gamma-1}$$

Divide the first equation by the second:

$$\left(\frac{V_B}{V_A}\right)^{\gamma-1} = \left(\frac{V_C}{V_D}\right)^{\gamma-1}$$

$$(2) \quad \frac{V_B}{V_A} = \frac{V_C}{V_D}$$

Substitute Equation (2) into Equation (1):

$$\frac{|Q_c|}{|Q_h|} = \frac{T_c \ln (V_C/V_D)}{T_h \ln (V_B/V_A)} = \frac{T_c \ln (V_C/V_D)}{T_h \ln (V_C/V_D)} = \frac{T_c}{T_h}$$

Finalize   This last equation is Equation 21.7, the one we set out to prove.

Example 21.4 The Steam Engine

A steam engine has a boiler that operates at 500 K. The energy from the burning fuel changes water to steam, and this steam then drives a piston. The cold reservoir's temperature is that of the outside air, approximately 300 K. What is the maximum thermal efficiency of this steam engine?

SOLUTION

Conceptualize In a steam engine, the gas pushing on the piston in Figure 21.9 is steam. A real steam engine does not operate in a Carnot cycle, but, to find the maximum possible efficiency, imagine a Carnot steam engine.

Categorize We calculate an efficiency using Equation 21.8, so we categorize this example as a substitution problem.

Substitute the reservoir temperatures into Equation 21.8:
$$e_C = 1 - \frac{T_c}{T_h} = 1 - \frac{300 \text{ K}}{500 \text{ K}} = 0.400 \quad \text{or} \quad 40.0\%$$

This result is the highest *theoretical* efficiency of the engine. In practice, the efficiency is considerably lower.

WHAT IF? Suppose we wished to increase the theoretical efficiency of this engine. This increase can be achieved by raising T_h by ΔT or by decreasing T_c by the same ΔT. Which would be more effective?

Answer A given ΔT would have a larger fractional effect on a smaller temperature, so you would expect a larger change in efficiency if you alter T_c by ΔT. Let's test that numerically. Raising T_h by 50 K, corresponding to T_h = 550 K, would give a maximum efficiency of

$$e_C = 1 - \frac{T_c}{T_h} = 1 - \frac{300 \text{ K}}{550 \text{ K}} = 0.455$$

Decreasing T_c by 50 K, corresponding to T_c = 250 K, would give a maximum efficiency of

$$e_C = 1 - \frac{T_c}{T_h} = 1 - \frac{250 \text{ K}}{500 \text{ K}} = 0.500$$

Although changing T_c is *mathematically* more effective, often changing T_h is *practically* more feasible.

21.5 Gasoline and Diesel Engines

In a gasoline engine, four *strokes* occur in each cycle; in addition, two *events* have a significant effect on the state of the gas in the cylinder. These strokes and events are illustrated with a pictorial representation in Figure 21.11 (page 568). In this discussion, let's consider the interior of the cylinder above the piston to be the system that is taken through repeated cycles in the engine's operation. For a given cycle, the piston moves up and down twice, which represents a four-stroke cycle consisting of two upstrokes and two downstrokes. The processes in the cycle can be approximated by the **Otto cycle** shown in the *PV* diagram in Figure 21.12 (page 568), which is a graphical representation of the cycle. In the following discussion, note that the letter designations next to the piston in Figure 21.11 correspond to the states on the *PV* diagram in Figure 21.12.

1. During the *intake stroke* ($O \rightarrow A$ in Figures 21.11a and 21.12), the piston moves downward and a gaseous mixture of air and fuel is drawn into the cylinder at atmospheric pressure. That is the energy input part of the cycle: energy enters the system (the interior of the cylinder) by matter transfer as potential energy stored in the fuel. In this process, the volume increases from V_2 to V_1. This apparent backward numbering is based on the compression stroke (see 2 below), in which the air–fuel mixture is compressed from V_1 to V_2.
2. During the *compression stroke* ($A \rightarrow B$ in Figures 21.11b and 21.12), the piston moves upward, the air–fuel mixture is compressed adiabatically from volume V_1 to volume V_2, and the temperature increases from T_A to T_B. The work done on the gas is positive, and its value is equal to the negative of the area under the curve AB in Figure 21.12.

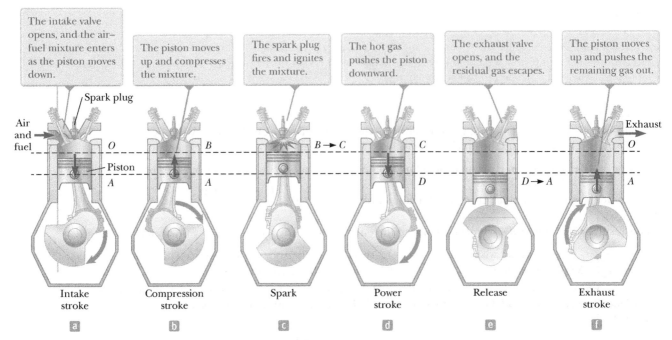

Figure 21.11 The processes occurring during one cycle of a conventional gasoline engine. The broken lines show the extreme positions of the top of the piston and, therefore, represent the largest and smallest volumes of the gas in the cylinder. Parts a, b, d, and f represent *strokes* in the cycle, justifying the name of the device as a four-stroke engine. In a stroke, the piston moves up or down between its extreme positions. The red arrows show the direction of travel of the piston, and the letters next to the piston correspond to the states on the *PV* diagram in Figure 21.12. Parts c and e in the figure represent *events*, during which the piston does not move. In part c, the spark plug fires and the pressure and temperature of the gas shoot upward. In part e, the exhaust valve opens and the pressure and temperature of the gas plummet. The events in this figure correspond to the constant-volume processes in Figure 21.12. By comparing that figure with this one, convince yourself that the volumes at *O*, *B*, and *C* are all the same, as indicated by their positions on the upper broken line. Similarly, the volumes at *A* and *D* are the same.

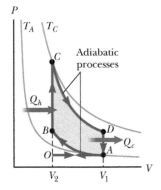

Figure 21.12 *PV* diagram for the Otto cycle, which approximately represents the processes occurring in an internal combustion engine.

3. Combustion occurs when the spark plug fires ($B \rightarrow C$ in Figures 21.11c and 21.12). This is the *combustion event* in the cycle; it occurs in a very short time interval while the piston is at its highest position. The combustion represents a rapid energy transformation from potential energy stored in chemical bonds in the fuel to internal energy associated with molecular motion, which is related to temperature. During this time interval, the mixture's pressure and temperature increase rapidly, with the temperature rising from T_B to T_C. The volume, however, remains approximately constant because of the short time interval. As a result, approximately no work is done on or by the gas. We can model this process in the *PV* diagram (Fig. 21.12) as that process in which the energy $|Q_h|$ enters the system. (In reality, however, this process is a *transformation* of energy already in the cylinder from process $O \rightarrow A$.)

4. In the *power stroke* ($C \rightarrow D$ in Figures 21.11d and 21.12), the gas expands adiabatically from V_2 to V_1. This expansion causes the temperature to drop from T_C to T_D. Work is done by the gas in pushing the piston downward, and the value of this work is equal to the area under the curve *CD*.

5. The *release event* in the cycle occurs when an exhaust valve is opened ($D \rightarrow A$ in Figures 21.11e and 21.12). The pressure suddenly drops for a short time interval. During this time interval, the piston is almost stationary and the volume is approximately constant. Energy is expelled from the interior of the cylinder and continues to be expelled during the next process.

6. In the final process, the *exhaust stroke* ($A \rightarrow O$ in Figures 21.11f and 21.12), the piston moves upward while the exhaust valve remains open. Residual gases are exhausted at atmospheric pressure, and the volume decreases from V_1 to V_2. The cycle then repeats.

If the air–fuel mixture is assumed to be an ideal gas, the efficiency of the Otto cycle is

$$e = 1 - \frac{1}{(V_1/V_2)^{\gamma-1}} \quad \text{(Otto cycle)} \tag{21.9}$$

where V_1/V_2 is the **compression ratio** and γ is the ratio of the molar specific heats C_P/C_V for the air–fuel mixture. Equation 21.9, which is derived in Example 21.5, shows that the efficiency increases as the compression ratio increases. For a typical compression ratio of 8 and with $\gamma = 1.4$, Equation 21.9 predicts a theoretical efficiency of 56% for an engine operating in the idealized Otto cycle. This value is much greater than that achieved in real engines (15% to 20%) because of such effects as friction, energy transfer by conduction through the cylinder walls, and incomplete combustion of the air–fuel mixture.

Diesel engines operate on a cycle similar to the Otto cycle, but they do not employ a spark plug. The compression ratio for a diesel engine is much greater than that for a gasoline engine. Air in the cylinder is compressed to a very small volume, and, as a consequence, the cylinder temperature at the end of the compression stroke is very high. At this point, fuel is injected into the cylinder. The temperature is high enough for the air–fuel mixture to ignite without the assistance of a spark plug. Diesel engines are more efficient than gasoline engines because of their greater compression ratios and resulting higher combustion temperatures.

Example **21.5** Efficiency of the Otto Cycle

Show that the thermal efficiency of an engine operating in an idealized Otto cycle (see Figs. 21.11 and 21.12) is given by Equation 21.9. Treat the working substance as an ideal gas.

SOLUTION

Conceptualize Study Figures 21.11 and 21.12 to make sure you understand the working of the Otto cycle.

Categorize As seen in Figure 21.12, we categorize the processes in the Otto cycle as isovolumetric and adiabatic.

Analyze Model the energy input and output as occurring by heat in processes $B \rightarrow C$ and $D \rightarrow A$. (In reality, most of the energy enters and leaves by matter transfer as the air–fuel mixture enters and leaves the cylinder.) Use Equation 20.23 to find the energy transfers by heat for these processes, which take place at constant volume:

$$B \rightarrow C \quad |Q_h| = nC_V(T_C - T_B)$$
$$D \rightarrow A \quad |Q_c| = nC_V(T_D - T_A)$$

Substitute these expressions into Equation 21.2:

$$(1) \quad e = 1 - \frac{|Q_c|}{|Q_h|} = 1 - \frac{T_D - T_A}{T_C - T_B}$$

Apply Equation 20.40 to the adiabatic processes $A \rightarrow B$ and $C \rightarrow D$:

$$A \rightarrow B \quad T_A V_A^{\gamma-1} = T_B V_B^{\gamma-1}$$
$$C \rightarrow D \quad T_C V_C^{\gamma-1} = T_D V_D^{\gamma-1}$$

Solve these equations for the temperatures T_A and T_D, noting that $V_A = V_D = V_1$ and $V_B = V_C = V_2$:

$$(2) \quad T_A = T_B \left(\frac{V_B}{V_A}\right)^{\gamma-1} = T_B \left(\frac{V_2}{V_1}\right)^{\gamma-1}$$

$$(3) \quad T_D = T_C \left(\frac{V_C}{V_D}\right)^{\gamma-1} = T_C \left(\frac{V_2}{V_1}\right)^{\gamma-1}$$

Subtract Equation (2) from Equation (3) and rearrange:

$$(4) \quad \frac{T_D - T_A}{T_C - T_B} = \left(\frac{V_2}{V_1}\right)^{\gamma-1}$$

Substitute Equation (4) into Equation (1):

$$e = 1 - \frac{1}{(V_1/V_2)^{\gamma-1}}$$

Finalize This final expression is Equation 21.9.

21.6 Entropy

The zeroth law of thermodynamics involves the concept of temperature, and the first law involves the concept of internal energy. Temperature and internal energy are both state variables; that is, the value of each depends only on the thermodynamic state of a system, not on the process that brought it to that state. Another state variable—this one related to the second law of thermodynamics—is *entropy*.

Entropy was originally formulated as a useful concept in thermodynamics. Its importance grew, however, as the field of statistical mechanics developed. In statistical mechanics, the behavior of a substance is described in terms of the statistical behavior of its atoms and molecules, as we did with our study of kinetic theory in Chapter 20. The analytical techniques of statistical mechanics provide an alternative means of interpreting entropy and a more global significance to the concept.

We will develop our understanding of entropy by first considering some nonthermodynamic systems, such as a pair of dice and poker hands. We will then expand on these ideas and use them to understand the concept of entropy as applied to thermodynamic systems.

We begin this process by distinguishing between *microstates* and *macrostates* of a system. A **microstate** is a particular configuration of the individual constituents of the system. A **macrostate** is a description of the system's conditions from a macroscopic point of view.

For any given macrostate of the system, a number of microstates are possible. For example, the macrostate of a 4 when a pair of six-sided dice are rolled can be formed from the possible microstates 1–3, 2–2, and 3–1. The macrostate of 2 has only one microstate, 1–1. It is assumed all microstates are equally probable. We can compare the two macrostates just mentioned in three ways. (1) *Uncertainty:* If we know that a macrostate of 4 exists, there is some uncertainty as to the microstate that exists, because there are multiple microstates that will result in a 4. In comparison, there is lower uncertainty (in fact, *zero* uncertainty) for a macrostate of 2 because only one microstate is possible. (2) *Choice:* There are more choices of microstates for a 4 than for a 2. (3) *Probability:* The macrostate of 4 has a higher probability than a macrostate of 2 because there are more ways (microstates) of achieving a 4. The notions of uncertainty, choice, and probability are central to the concept of entropy, as we discuss below.

Let's look at another example related to a five-card poker hand. There is only one microstate associated with the macrostate of a "royal flush" of five spades, laid out in order from ten to ace (Fig. 21.13a). Figure 21.13b shows another poker hand. The macrostate here is "worthless hand." The *particular* hand (the microstate) in Figure 21.13b and the hand in Figure 21.13a are equally probable. There are, however,

Figure 21.13 (a) A royal flush has low probability of occurring. (b) A worthless poker hand, one of many.

many other hands similar in value to that in Figure 21.13b; that is, there are many microstates that also qualify as worthless hands. If you, as a poker player, are told your opponent holds a macrostate of a royal flush in spades, there is *zero uncertainty* as to what five cards are in the hand, only *one choice* of what those cards are, and *low probability* that the hand actually occurred. In contrast, if you are told that your opponent has the macrostate of "worthless hand," there is *high uncertainty* as to what the five cards are, *many choices* of what they could be, and a *high probability* that a worthless hand occurred. Another variable in poker, of course, is the value of the hand, related to the probability: the higher the probability, the lower the value. The important point to take away from this discussion is that uncertainty, choice, and probability are related in these situations: if one is high, the others are high, and vice versa.[6]

For thermodynamic systems, the variable **entropy** S is used to represent the level of uncertainty, choice, and probability in the system. Consider Configuration 1 (a macrostate) in which all the oxygen molecules in the air in your room are located in the west half of the room and the nitrogen molecules in the east half. Compare that macrostate to the more common Configuration 2, in which the oxygen and nitrogen molecules are distributed uniformly throughout the room. Configuration 2 has the higher uncertainty as to where the molecules are located because they could be anywhere, not just in one-half of the room according to the type of molecule. Configuration 2 also represents more choices as to where to locate molecules. It also has a much higher probability of occurring; have you ever noticed your half of the room suddenly being empty of oxygen? Therefore, Configuration 2 represents a higher entropy.

For systems of dice and poker hands, the comparisons between probabilities for various macrostates involve relatively small numbers. For example, a macrostate of a 4 on a pair of dice is only three times as probable as a macrostate of 2. When we are talking about a macroscopic thermodynamic system containing on the order of Avogadro's number of molecules, however, the ratios of probabilities can be astronomical.

Let's explore this concept by considering 100 molecules in a container. Half of the molecules are oxygen and the other half are nitrogen. At any given moment, the probability of one molecule being in the left part of the container shown in Figure 21.14a as a result of random motion is $\frac{1}{2}$. If there are two molecules as shown in Figure 21.14b, the probability of both being in the left part is $\left(\frac{1}{2}\right)^2$, or 1 in 4. If there are three molecules (Fig. 21.14c), the probability of them all being in the left portion at the same moment is $\left(\frac{1}{2}\right)^3$, or 1 in 8. For 100 independently moving molecules, the probability that the 50 oxygen molecules will be found in the left part at any moment is $\left(\frac{1}{2}\right)^{50}$. Likewise, the probability that the remaining

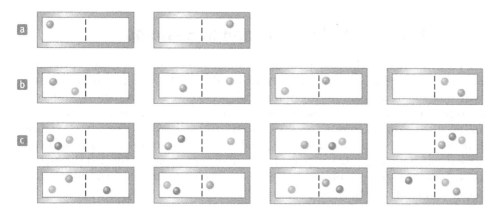

Figure 21.14 Possible distributions of identical molecules in a container. The colors used here exist only to allow us to distinguish among the molecules.
(a) One molecule in a container has a 1-in-2 chance of being on the left side. (b) Two molecules have a 1-in-4 chance of being on the left side at the same time.
(c) Three molecules have a 1-in-8 chance of being on the left side at the same time.

---

[6]Another way of describing macrostates is by means of "missing information." For high-probability macrostates with many microstates, there is a large amount of missing information, meaning we have very little information about what microstate actually exists.

50 nitrogen molecules will be found in the right part at any moment is $\left(\frac{1}{2}\right)^{50}$. Therefore, the probability of finding this oxygen–nitrogen separation as a result of random motion is the product $\left(\frac{1}{2}\right)^{50}\left(\frac{1}{2}\right)^{50} = \left(\frac{1}{2}\right)^{100}$, which corresponds to about 1 in 10^{30}. When this calculation is extrapolated from 100 molecules to the number in 1 mol of gas (6.02×10^{23}), the separated arrangement is found to be *extremely* improbable!

QUICK QUIZ 21.4 **(a)** Suppose you select four cards at random from a standard deck of playing cards and end up with a macrostate of four deuces. How many microstates are associated with this macrostate? **(b)** Suppose you pick up two cards and end up with a macrostate of two aces. How many microstates are associated with this macrostate?

Conceptual Example 21.6 **Let's Play Marbles!**

Suppose you have a bag of 100 marbles of which 50 are red and 50 are green. You are allowed to draw four marbles from the bag according to the following rules. Draw one marble, record its color, and return it to the bag. Shake the bag and then draw another marble. Continue this process until you have drawn and returned four marbles. What are the possible macrostates for this set of events? What is the most likely macrostate? What is the least likely macrostate?

SOLUTION

Because each marble is returned to the bag before the next one is drawn and the bag is then shaken, the probability of drawing a red marble is always the same as the probability of drawing a green one. All the possible microstates and macrostates are shown in Table 21.1. As this table indicates, there is only one way to draw a macrostate of four red marbles, so there is only one microstate for that macrostate. There are, however, four possible microstates that correspond to the macrostate of one green marble and three red marbles, six microstates that correspond to two green mar-

TABLE 21.1 Possible Results of Drawing Four Marbles from a Bag

Macrostate	Possible Microstates	Total Number of Microstates
All R	RRRR	1
1G, 3R	RRRG, RRGR, RGRR, GRRR	4
2G, 2R	RRGG, RGRG, GRRG, RGGR, GRGR, GGRR	6
3G, 1R	GGGR, GGRG, GRGG, RGGG	4
All G	GGGG	1

bles and two red marbles, four microstates that correspond to three green marbles and one red marble, and one microstate that corresponds to four green marbles. The most likely macrostate—two red marbles and two green marbles—corresponds to the largest number of choices of microstates, and, therefore, the most uncertainty as to what the exact microstate is. The least likely macrostates—four red marbles or four green marbles—correspond to only one choice of microstate and, therefore, zero uncertainty. There is no uncertainty for the least likely states: we know the colors of all four marbles.

21.7 Entropy in Thermodynamic Systems

We have investigated the notions of uncertainty, number of choices, and probability for some non-thermodynamic systems such as dice and cards, as well as for a small system of 100 oxygen and nitrogen molecules. We have argued that the concept of entropy can be related to these notions for macroscopic thermodynamic systems. In our discussion of entropy, there are two things we have *not* done yet: (1) indicate how to evaluate entropy numerically, and (2) discuss entropy for a macroscopic system with a huge number of particles. Both of these were performed through statistical means by Boltzmann in the 1870s and the numerical evaluation of entropy appears in its currently accepted form as

$$S = k_B \ln W \qquad (21.10)$$

where k_B is Boltzmann's constant. Boltzmann intended W, standing for *Wahrscheinlichkeit*, the German word for probability, to be proportional to the probability that a given macrostate exists. It is equivalent to let W be the number of micro-states associated with the macrostate, so we can interpret W as representing the number of "ways" of achieving the macrostate. Therefore, macrostates with larger numbers of microstates have higher probability and, equivalently, higher entropy. Notice that the units of entropy are those of Boltzmann's constant, J/K.

In the kinetic theory of gases, gas molecules are represented as particles moving randomly. Suppose the gas is confined to a volume V. For a uniform distribution of gas in the volume, there are a large number of equivalent microstates, and the entropy of the gas can be related to the number of microstates corresponding to a given macrostate. Let us count the number of microstates by considering the variety of molecular locations available to the molecules. Let us assume each molecule occupies some microscopic volume V_m. The total number of possible locations of a single molecule in a macroscopic volume V is the ratio $w = V/V_m$, which is a huge number. We use lowercase w here to represent the number of ways a single molecule can be placed in the volume or the number of microstates for a single molecule, which is equivalent to the number of available locations. We assume the probabilities of a molecule occupying any of these locations are equal. As more molecules are added to the system, the number of possible ways the molecules can be positioned in the volume multiplies, as we saw in Figure 21.14. For example, if you consider two molecules, for every possible placement of the first, all possible placements of the second are available. Therefore, there are w ways of locating the first molecule, and for each way, there are w ways of locating the second molecule. The total number of ways of locating the two molecules is $W = w \times w = w^2 = (V/V_m)^2$. (Uppercase W represents the number of ways of putting multiple molecules into the volume and is not to be confused with work.)

Now consider placing N molecules of gas in the volume V. Neglecting the very small probability of having two molecules occupy the same location, each molecule may go into any of the V/V_m locations, and so the number of ways of locating N molecules in the volume becomes $W = w^N = (V/V_m)^N$. Therefore, the spatial part of the entropy of the gas, from Equation 21.10, is

$$S = k_B \ln W = k_B \ln \left(\frac{V}{V_m} \right)^N = N k_B \ln \left(\frac{V}{V_m} \right) = nR \ln \left(\frac{V}{V_m} \right) \qquad (21.11)$$

We will use this expression in the next section as we investigate changes in entropy for processes occurring in thermodynamic systems.

Notice that we have indicated Equation 21.11 as representing only the *spatial* portion of the entropy of the gas. There is also a temperature-dependent portion of the entropy that the discussion above does not address. For example, imagine an isovolumetric process in which the temperature of the gas increases. Equation 21.11 above shows no change in the spatial portion of the entropy for this situation. There *is* a change in entropy, however, associated with the increase in temperature. We can understand this by appealing again to a bit of quantum physics. Recall from Section 20.3 that the energies of the gas molecules are quantized. When the temperature of a gas changes, the distribution of energies of the gas molecules changes according to the Boltzmann distribution law, discussed in Section 20.5. Therefore, as the temperature of the gas increases, there is more uncertainty about the particular microstate that exists as gas molecules distribute themselves into higher available quantum states.

Thermodynamic systems are constantly in flux, changing continuously from one microstate to another. If the system is in equilibrium, a given macrostate exists, described by variables such as P, V, T, and E_{int} and the system fluctuates from one microstate associated with that macrostate to another. This change is unobservable because we are only able to detect the macrostate. Equilibrium states have tremendously higher probability than nonequilibrium states, so it is highly unlikely that an equilibrium state will spontaneously change to a nonequilibrium state. For example, we do not observe a spontaneous split into the oxygen–nitrogen separation discussed in Section 21.6.

What happens, however, if the system begins in a low-probability macrostate? What if the room *begins* with an oxygen–nitrogen separation? In this case, the system will progress from this low-probability macrostate to the much-higher probability state: the gases will disperse and mix throughout the room. Because entropy is related to probability, a spontaneous *increase* in entropy, such as in the latter situation, is natural. If the oxygen and nitrogen molecules were initially spread evenly throughout the room, the entropy of the mixture would *decrease* if the spontaneous splitting of molecules occurred.

One way of conceptualizing a change in entropy is to relate it to *energy spreading*. A natural tendency is for energy to undergo spatial spreading in time, representing an increase in entropy. If a basketball is dropped onto a floor, it bounces several times and eventually comes to rest. The initial gravitational potential energy in the basketball–Earth system has been transformed to internal energy in the ball and the floor. That energy is spreading outward by heat into the air and into regions of the floor farther from the drop point. In addition, some of the energy has spread throughout the room by sound. It would be unnatural for energy in the room and floor to reverse this motion and concentrate into the stationary ball so that it spontaneously begins to bounce again.

In the adiabatic free expansion discussed in Section 21.3, the spreading of energy accompanies the spreading of the molecules as the gas rushes into the evacuated half of the container. If a warm object is placed in thermal contact with a cool object, energy transfers from the warm object to the cool one by heat, representing a spread of internal energy until it is distributed more evenly between the two objects.

Now consider a mathematical representation of this spreading of energy or, equivalently, the change in entropy. The original formulation of entropy in thermodynamics involves the transfer of energy by heat during a reversible process. Consider any infinitesimal process in which a system changes from one equilibrium state to another. If dQ_r is the amount of energy transferred by heat when the system follows a reversible path between the states, the change in entropy dS can be shown to be equal to this amount of energy divided by the absolute temperature of the system:

◀ Change in entropy for an infinitesimal process

$$dS = \frac{dQ_r}{T}$$

(21.12)

We have assumed the temperature is constant because the process is infinitesimal. Because entropy is a state variable, the change in entropy during a process depends only on the endpoints and therefore is independent of the actual path followed. Consequently, the entropy change for an irreversible process can be determined by calculating the entropy change for a *reversible* process that connects the same initial and final states.

Equation 21.10 defines entropy statistically. Evaluating W, however, is extremely difficult for a macroscopic system with a huge number of particles, on the order of Avogadro's number. On the other hand, Equation 21.12 defines changes in entropy in terms of macroscopic quantities, Q_r and T. Therefore, this equation is more practical than Equation 21.10.

The subscript r on the quantity dQ_r is a reminder that the transferred energy is to be measured along a reversible path even though the system may actually have followed some irreversible path. When energy is absorbed by the system, dQ_r is positive and the entropy of the system increases. When energy is expelled by the system, dQ_r is negative and the entropy of the system decreases. Notice that Equation 21.12 does not define entropy but rather the *change* in entropy. Hence, the meaningful quantity in describing a process is the *change* in entropy.

To calculate the change in entropy for a *finite* process, first recognize that T is generally not constant during the process. Therefore, we must integrate Equation 21.12:

$$\Delta S = \int_i^f dS = \int_i^f \frac{dQ_r}{T} \qquad (21.13)$$

◀ Change in entropy for a finite process

As with an infinitesimal process, the change in entropy ΔS of a system going from one state to another has the same value for *all* paths connecting the two states. That is, the finite change in entropy ΔS of a system depends only on the properties of the initial and final equilibrium states. Therefore, we are free to choose any convenient reversible path over which to evaluate the entropy in place of the actual path as long as the initial and final states are the same for both paths. This point is explored further on in this section.

From Equation 21.10, we see that a change in entropy is represented in the Boltzmann formulation as

$$\Delta S = k_B \ln\left(\frac{W_f}{W_i}\right) \qquad (21.14)$$

where W_i and W_f represent the initial and final numbers of microstates, respectively, for the initial and final configurations of the system. If $W_f > W_i$, the final state is more probable than the initial state (there are more choices of microstates), and the entropy increases. As mentioned above, however, evaluating W is extremely difficult for macroscopic systems.

QUICK QUIZ 21.5 An ideal gas is taken from an initial temperature T_i to a higher final temperature T_f along two different reversible paths as shown in Figure 21.15. Path A is at constant pressure, and path B is at constant volume. What is the relation between the entropy changes of the gas for these paths? (a) $\Delta S_A > \Delta S_B$ (b) $\Delta S_A = \Delta S_B$ (c) $\Delta S_A < \Delta S_B$

QUICK QUIZ 21.6 True or False: The entropy change in an adiabatic process must be zero because $Q = 0$.

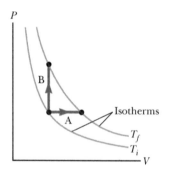

Figure 21.15 (Quick Quiz 21.5) An ideal gas is taken from temperature T_i to T_f via two different paths.

Example 21.7 **Change in Entropy: Melting**

A solid that has a latent heat of fusion L_f melts at a temperature T_m. Calculate the change in entropy of this substance when a mass m of the substance melts.

SOLUTION

Conceptualize We can choose any convenient reversible path to follow that connects the initial and final states. It is not necessary to identify the process or the path because, whatever it is, the effect is the same: energy enters the substance by heat and the substance melts. The mass m of the substance that melts is equal to Δm, the change in mass of the higher-phase (liquid) substance.

Categorize Because the melting takes place at a fixed temperature, we categorize the process as isothermal.

Analyze Use Equation 19.8 in Equation 21.13, noting that the temperature remains fixed:

$$\Delta S = \int \frac{dQ_r}{T} = \frac{1}{T_m}\int dQ_r = \frac{Q_r}{T_m} = \frac{L_f \Delta m}{T_m} = \frac{L_f m}{T_m}$$

Finalize Notice that Δm is positive so that ΔS is positive, representing that energy is added to the substance.

Entropy Change in a Carnot Cycle

Now that we have some understanding of entropy, let's consider the changes in entropy that occur in a Carnot heat engine that operates between the temperatures

T_c and T_h. In one cycle, the engine takes in energy $|Q_h|$ from the hot reservoir and expels energy $|Q_c|$ to the cold reservoir. These energy transfers occur only during the reversible, isothermal portions of the Carnot cycle; therefore, the constant temperature can be brought out in front of the integral sign in Equation 21.13. The integral then simply has the value of the total amount of energy transferred by heat. During the two adiabatic processes, for which $Q = 0$, the entropy changes are zero because these processes are reversible. Therefore, the total change in entropy for one cycle is

$$\Delta S = \frac{|Q_h|}{T_h} - \frac{|Q_c|}{T_c} \tag{21.15}$$

where the minus sign represents that energy is leaving the engine at temperature T_c. In Example 21.3, we showed that for a Carnot engine,

$$\frac{|Q_c|}{|Q_h|} = \frac{T_c}{T_h}$$

Using this result in Equation 21.15, we find that the total change in entropy for a Carnot engine operating in a cycle is *zero:*

$$\Delta S = 0$$

Now consider a system taken through an arbitrary (non-Carnot) reversible cycle. Because entropy is a state variable—and hence depends only on the properties of a given equilibrium state—we conclude that $\Delta S = 0$ for *any* reversible cycle. In general, we can write this condition as

$$\oint \frac{dQ_r}{T} = 0 \quad \text{(reversible cycle)} \tag{21.16}$$

where the symbol $\oint$ indicates that the integration is over a closed path.

Entropy Change in a Free Expansion

Let's again consider the adiabatic free expansion of a gas occupying an initial volume V_i (Fig. 21.16). In this situation, a membrane separating the gas from an evacuated region is broken and the gas expands to a volume V_f. This process is irreversible; the gas would not spontaneously crowd into half the volume after filling the entire volume. What is the change in entropy of the gas during this process? The process is neither reversible nor quasi-static. As argued in Section 21.3, the initial and final temperatures of the gas are the same.

To apply Equation 21.13, we cannot take $Q = 0$, the value for the irreversible process, but must instead find Q_r; that is, we must find an equivalent reversible path that shares the same initial and final states. A simple choice is an isothermal, reversible expansion in which the gas pushes slowly against a piston while energy enters the gas by heat from a reservoir to hold the temperature constant. Because T is constant in this process, Equation 21.13 gives

$$\Delta S = \int_i^f \frac{dQ_r}{T} = \frac{1}{T} \int_i^f dQ_r$$

For an isothermal process, the first law of thermodynamics specifies that $\int_i^f dQ_r$ is equal to the negative of the work done on the gas during the expansion from V_i to V_f, which is given by Equation 19.12. Using this result, we find that the entropy change for the gas is

$$\Delta S = nR \ln \left(\frac{V_f}{V_i} \right) \tag{21.17}$$

Because $V_f > V_i$, we conclude that ΔS is positive. This positive result indicates that the entropy of the gas *increases* as a result of the irreversible, adiabatic expansion.

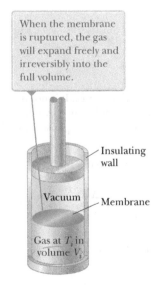

When the membrane is ruptured, the gas will expand freely and irreversibly into the full volume.

Insulating wall

Vacuum

Membrane

Gas at T_i in volume V_i

Figure 21.16 Adiabatic free expansion of a gas. The container is thermally insulated from its surroundings; therefore, $Q = 0$.

It is easy to see that the energy has spread after the expansion. Instead of being concentrated in a relatively small space, the molecules and the energy associated with them are scattered over a larger region. In addition, there are more choices of the locations of the molecules, higher uncertainty as to their locations, and a higher probability for the molecules to be spread throughout the volume. The probability is indeed low for the molecules, in the absence of the membrane, to concentrate spontaneously in the lower half of the container.

Entropy Change in Thermal Conduction

Let us now consider a system consisting of a hot reservoir and a cold reservoir that are in thermal contact with each other and isolated from the rest of the Universe. A process occurs during which energy Q is transferred by heat from the hot reservoir at temperature T_h to the cold reservoir at temperature T_c. The process as described is irreversible (energy would not spontaneously flow from cold to hot), so we must find an equivalent reversible process. The overall process is a combination of two processes: energy leaving the hot reservoir and energy entering the cold reservoir. We will calculate the entropy change for the reservoir in each process and add to obtain the overall entropy change.

Consider first the process of energy entering the cold reservoir. Although the reservoir has absorbed some energy, the temperature of the reservoir has not changed. The energy that has entered the reservoir is the same as that which would enter by means of a reversible, isothermal process. The same is true for energy leaving the hot reservoir.

Because the cold reservoir absorbs energy Q, its entropy increases by Q/T_c. At the same time, the hot reservoir loses energy Q, so its entropy change is $-Q/T_h$. Therefore, the change in entropy of the system is

$$\Delta S = \frac{Q}{T_c} + \frac{-Q}{T_h} = Q\left(\frac{1}{T_c} - \frac{1}{T_h}\right) > 0 \qquad \text{(21.18)}$$

This increase is consistent with our interpretation of entropy changes as representing the spreading of energy. In the initial configuration, the hot reservoir has excess internal energy relative to the cold reservoir. The process that occurs spreads the energy into a more equitable distribution between the two reservoirs.

Example 21.8 **Adiabatic Free Expansion: Revisited**

Let's verify that the macroscopic and microscopic approaches to the calculation of entropy lead to the same conclusion for the adiabatic free expansion of an ideal gas. Suppose the ideal gas in Figure 21.16 expands to four times its initial volume. As we have seen for this process, the initial and final temperatures are the same.

(A) Using a macroscopic approach, calculate the entropy change for the gas.

SOLUTION

Conceptualize Look back at Figure 21.16, which is a diagram of the system before the adiabatic free expansion. Imagine breaking the membrane so that the gas moves into the evacuated area. The expansion is irreversible.

Categorize We can replace the irreversible process with a reversible isothermal process between the same initial and final states. This approach is macroscopic, so we use a thermodynamic variable, in particular, the volume V.

Analyze Use Equation 21.17 to evaluate the entropy change:

$$\Delta S = nR \ln\left(\frac{V_f}{V_i}\right) = nR \ln\left(\frac{4V_i}{V_i}\right) = nR \ln 4$$

(B) Using statistical considerations, calculate the change in entropy for the gas and show that it agrees with the answer you obtained in part (A).

continued

21.8 continued

SOLUTION

Categorize This approach is microscopic, so we use variables related to the individual molecules.

Analyze As in the discussion leading to Equation 21.11, the number of microstates available to a single molecule in the initial volume V_i is $w_i = V_i/V_m$, where V_i is the initial volume of the gas and V_m is the microscopic volume occupied by the molecule. Use this number to find the number of available microstates for N molecules:

$$W_i = w_i^N = \left(\frac{V_i}{V_m}\right)^N$$

Find the number of available microstates for N molecules in the final volume $V_f = 4V_i$:

$$W_f = \left(\frac{V_f}{V_m}\right)^N = \left(\frac{4V_i}{V_m}\right)^N$$

Use Equation 21.14 to find the entropy change:

$$\Delta S = k_B \ln \left(\frac{W_f}{W_i}\right)$$

$$= k_B \ln \left(\frac{4V_i}{V_i}\right)^N = k_B \ln(4^N) = Nk_B \ln 4 = nR \ln 4$$

Finalize The answer is the same as that for part (A), which dealt with macroscopic parameters.

WHAT IF?   In part (A), we used Equation 21.17, which was based on a reversible isothermal process connecting the initial and final states. Would you arrive at the same result if you chose a different reversible process?

Answer You *must* arrive at the same result because entropy is a state variable. For example, consider the two-step process in Figure 21.17: a reversible adiabatic expansion from V_i to $4V_i$ ($A \rightarrow B$) during which the temperature drops from T_1 to T_2 and a reversible isovolumetric process ($B \rightarrow C$) that takes the gas back to the initial temperature T_1. During the reversible adiabatic process, $\Delta S = 0$ because $Q_r = 0$.

For the reversible isovolumetric process ($B \rightarrow C$), use Equation 21.13:

Find the ratio of temperature T_1 to T_2 from Equation 20.40 for the adiabatic process:

Substitute to find ΔS:

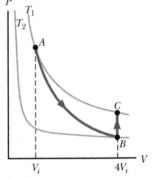

Figure 21.17 (Example 21.8) A gas expands to four times its initial volume and back to the initial temperature by means of a two-step process.

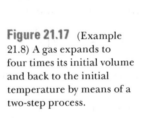

$$\Delta S = \int_i^f \frac{dQ_r}{T} = \int_{T_2}^{T_1} \frac{nC_V dT}{T} = nC_V \ln\left(\frac{T_1}{T_2}\right)$$

$$\frac{T_1}{T_2} = \left(\frac{4V_i}{V_i}\right)^{\gamma-1} = (4)^{\gamma-1}$$

$$\Delta S = nC_V \ln(4)^{\gamma-1} = nC_V(\gamma - 1)\ln 4$$

$$= nC_V\left(\frac{C_P}{C_V} - 1\right)\ln 4 = n(C_P - C_V)\ln 4 = nR \ln 4$$

We do indeed obtain the exact same result for the entropy change.

21.8 Entropy and the Second Law

If we consider a system and its surroundings to include the entire Universe, the Universe is always moving toward a higher-probability macrostate, corresponding to the continuous spreading of energy. An alternative way of stating this behavior is yet another wording of the second law of thermodynamics:

◀ Entropy statement of
the second law of
thermodynamics

The entropy of the Universe increases in all real processes.

This statement can be shown to be equivalent to the Kelvin-Planck and Clausius statements.

Let us show this equivalence first for the Clausius statement. Looking at Figure 21.5, we see that, if the heat pump operates in the manner shown in the figure, energy is spontaneously flowing from the cold reservoir to the hot reservoir without an input of energy by work. As a result, the energy in the system is not spreading evenly between the two reservoirs, but is *concentrating* in the hot reservoir. Consequently, if the Clausius statement of the second law is not true, then the entropy statement is also not true, demonstrating their equivalence.

For the equivalence of the Kelvin–Planck statement, consider Figure 21.18, which shows the impossible engine of Figure 21.3 connected to a heat pump operating between the same reservoirs. The output work of the engine is used to drive the heat pump. The net effect of this combination is that energy leaves the cold reservoir and is delivered to the hot reservoir without the input of work. (The work done by the engine on the heat pump is *internal* to the system of both devices.) This is forbidden by the Clausius statement of the second law, which we have shown to be equivalent to the entropy statement. Therefore, the Kelvin–Planck statement of the second law is also equivalent to the entropy statement.

When dealing with a system that is not isolated from its surroundings, remember that the increase in entropy described in the second law is that of the system *and* its surroundings. When a system and its surroundings interact in an irreversible process, the increase in entropy of one is greater than the decrease in entropy of the other. Hence, the change in entropy of the Universe must be greater than zero for an irreversible process and equal to zero for a reversible process.

We can check this statement of the second law for the calculations of entropy change that we made in Section 21.7. Consider first the entropy change in a free expansion, described by Equation 21.17. Because the free expansion takes place in an insulated container, no energy is transferred by heat from the surroundings. Therefore, Equation 21.17 represents the entropy change of the entire Universe. Because $V_f > V_i$, the entropy change of the Universe is positive, consistent with the second law.

Now consider the entropy change in thermal conduction, described by Equation 21.18. Let each reservoir be half the Universe. (The larger the reservoir, the better is the assumption that its temperature remains constant!) Then the entropy change of the Universe is represented by Equation 21.18. Because $T_h > T_c$, this entropy change is positive, again consistent with the second law. The positive entropy change is also consistent with the notion of energy spreading. The warm portion of the Universe has excess internal energy relative to the cool portion. Thermal conduction represents a spreading of the energy more equitably throughout the Universe.

Finally, let us look at the entropy change in a Carnot cycle, given by Equation 21.15. The entropy change of the engine itself is zero. The entropy change of the reservoirs is

$$\Delta S = \frac{|Q_c|}{T_c} - \frac{|Q_h|}{T_h}$$

In light of Equation 21.7, this entropy change is also zero. Therefore, the entropy change of the Universe is only that associated with the work done by the engine. A portion of that work will be used to change the mechanical energy of a system external to the engine: speed up the shaft of a machine, raise a weight, and so on. There is no change in internal energy of the external system due to this portion of the work, or, equivalently, no energy spreading, so the entropy change is again zero. The other portion of the work will be used to overcome various friction forces or other nonconservative forces in the external system. This process will cause an increase in internal energy of that system. That same increase in internal energy could have happened via a reversible thermodynamic process in which energy Q_r is

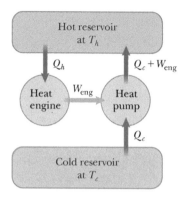

Figure 21.18 The impossible engine of Figure 21.3 transfers energy by work to a heat pump operating between two energy reservoirs. This situation is forbidden by the Clausius statement of the second law of thermodynamics.

transferred by heat, so the entropy change associated with that part of the work is positive. As a result, the overall entropy change of the Universe for the operation of the Carnot engine is positive, again consistent with the second law.

Ultimately, because real processes are irreversible, the entropy of the Universe should increase steadily and eventually reach a maximum value. At this value, assuming that the second law of thermodynamics, as formulated here on Earth, applies to the entire expanding Universe, the Universe will be in a state of uniform temperature and density. The total energy of the Universe will have spread more evenly throughout the Universe. All physical, chemical, and biological processes will have ceased at this time. This gloomy state of affairs is sometimes referred to as the *heat death* of the Universe.

Summary

▶ Definitions

The **thermal efficiency** e of a heat engine is

$$e \equiv \frac{W_{eng}}{|Q_h|} = \frac{|Q_h| - |Q_c|}{|Q_h|} = 1 - \frac{|Q_c|}{|Q_h|} \qquad (21.2)$$

From a microscopic viewpoint, the **entropy** of a given macrostate is defined as

$$S \equiv k_B \ln W \qquad (21.10)$$

where k_B is Boltzmann's constant and W is the number of microstates of the system corresponding to the macrostate.

The **microstate** of a system is the description of its individual components. The **macrostate** is a description of the system from a macroscopic point of view. A given macrostate can have many microstates.

In a **reversible** process, the system can be returned to its initial conditions along the same path on a PV diagram, and every point along this path is an equilibrium state. A process that does not satisfy these requirements is **irreversible**.

▶ Concepts and Principles

A **heat engine** is a device that takes in energy by heat and, operating in a cyclic process, expels a fraction of that energy by means of work. The net work done by a heat engine in carrying a working substance through a cyclic process ($\Delta E_{int} = 0$) is

$$W_{eng} = |Q_h| - |Q_c| \qquad (21.1)$$

where $|Q_h|$ is the energy taken in from a hot reservoir and $|Q_c|$ is the energy expelled to a cold reservoir.

Two ways the **second law of thermodynamics** can be stated are as follows:

- It is impossible to construct a heat engine that, operating in a cycle, produces no effect other than the input of energy by heat from a reservoir and the performance of an equal amount of work (the Kelvin–Planck statement).
- It is impossible to construct a cyclical machine whose sole effect is to transfer energy continuously by heat from one object to another object at a higher temperature without the input of energy by work (the Clausius statement).

Carnot's theorem states that no real heat engine operating (irreversibly) between the temperatures T_c and T_h can be more efficient than an engine operating reversibly in a Carnot cycle between the same two temperatures.

The thermal efficiency of a heat engine operating in the Carnot cycle is

$$e_C = 1 - \frac{T_c}{T_h} \qquad (21.8)$$

The macroscopic state of a system that has a large number of microstates has three qualities that are all related: (1) *uncertainty:* because of the large number of microstates, there is a large uncertainty as to which one actually exists; (2) *choice:* again because of the large number of microstates, there is a large number of choices from which to select as to which one exists; (3) *probability:* a macrostate with a large number of microstates is more likely to exist than one with a small number of microstates. For a thermodynamic system, all three of these can be related to the state variable of **entropy.**

The second law of thermodynamics states that when real (irreversible) processes occur, there is a spatial spreading of energy. This spreading of energy is related to a thermodynamic state variable called **entropy** S. Therefore, yet another way the second law can be stated is as follows:

- The entropy of the Universe increases in all real processes.

The **change in entropy** dS of a system during a process between two infinitesimally separated equilibrium states is

$$dS = \frac{dQ_r}{T} \qquad (21.12)$$

where dQ_r is the energy transfer by heat for the system for a reversible process that connects the initial and final states.

The change in entropy of a system during an arbitrary finite process between an initial state and a final state is

$$\Delta S = \int_i^f \frac{dQ_r}{T} \qquad (21.13)$$

The value of ΔS for the system is the same for all paths connecting the initial and final states. The change in entropy for a system undergoing any reversible, cyclic process is zero.

Think–Pair–Share

See the Preface for an explanation of the icons used in this problems set. For additional assessment items for this section, go to **WEBASSIGN** From Cengage

1. An engine operates in a Carnot cycle as follows. At point A in the cycle, 2.34 mol of a monatomic ideal gas has a pressure of 1 400 kPa, a volume of 10.0 L, and a temperature of 720 K. The gas expands isothermally to point B and then expands adiabatically to point C, where its volume is 24.0 L. An isothermal compression brings it to point D, where its volume is 15.0 L. An adiabatic process returns the gas to point A. (a) Fill in the following table with the pressures, volumes, and temperatures at each of the four points in the cycle:

Point in Cycle	P(kPa)	V(L)	T(K)
A			
B			
C			
D			

(b) Fill in the following table with the energy transfer by heat, work done on the gas, and the change in internal energy of the gas for each of the four processes in the cycle:

Process in Cycle	Q(kJ)	W(kJ)	ΔE_{int} (kJ)
$A \rightarrow B$			
$B \rightarrow C$			
$C \rightarrow D$			
$D \rightarrow A$			

(c) Find the efficiency of the engine from the data in the table in part (b). (d) Find the efficiency of the engine from the data in the table in part (a).

2. **ACTIVITY** If two six-sided dice are rolled, possible results of adding up the number of dots on the upper faces range from 2 to 12. The probabilities of these results vary due to the number of ways a particular result can be achieved. For example, there is only one way that a 2 can be achieved: 1–1. But there are six ways to achieve a result of 7: 1–6, 2–5, 3–4, 4–3, 5–2, and 6–1. Therefore, a result of 7 is six times more probable than a 2. The bar graph in Figure TP21.2 shows the theoretical probability of all possible results for two dice. Experimentally, if the two dice were thrown 100 times and a histogram of the possible results were drawn, it would have a shape very similar to this probability graph. (a) What if we throw *three* dice? What will the histogram look like? Do this in your group. Throw three dice 100 times and make a histogram of the results. How does the shape of the resulting histogram differ from the probability curve shown in Figure TP21.2 for two dice? Make a theoretical probability graph for three dice like that

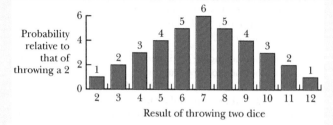

Figure TP21.2

in Figure TP21.2 and compare to your histogram. (b) What if you rolled Avogadro's number of dice? (Don't try this!) What would the histogram look like? (Make a prediction based on how the graph for three dice varies from that for two dice.) (c) Suppose you laboriously set up Avogadro's number of dice on a table, with all dice having a 1 on their upper face. Then you shook the table for a few seconds. When you added up all the numbers on the upper faces, what is the most likely result? (d) Now imagine shaking the table again. How likely is it that the dice could all return to having a 1 on their upper faces? (e) What does all this have to do with entropy?

3. **ACTIVITY** Let's consider the various liquids in the table below at their boiling points. The table provides the latent heat of vaporization of each liquid in kJ/mol (note the units), and the boiling point in °C. For each of the liquids, evaluate the entropy change of the liquid per mole when it vaporizes at the boiling point. What do you notice about the results?

	L_v (kJ/mol)	Boiling Point (°C)
Polar compounds		
HF	25.2	19.7
HCl	16.2	−84.8
HI	19.8	−35.6
H_2O	40.7	100
Nonpolar compounds		
C_3H_8	19.0	−42.1
C_4H_{10}	22.4	−0.50
Elements		
Hg	54.7	357
Pb	178	1 749
Cl_2	20.4	−34.0
Br_2	30.0	58.8

Problems

See the Preface for an explanation of the icons used in this problems set. For additional assessment items for this section, go to ⚡ **WEBASSIGN** From Cengage

SECTION 21.1 Heat Engines and the Second Law of Thermodynamics

1. A particular heat engine has a mechanical power output of 5.00 kW and an efficiency of 25.0%. The engine expels 8.00×10^3 J of exhaust energy in each cycle. Find (a) the energy taken in during each cycle and (b) the time interval for each cycle.

2. The work done by an engine equals one-fourth the energy it absorbs from a reservoir. (a) What is its thermal efficiency? (b) What fraction of the energy absorbed is expelled to the cold reservoir?

3. Suppose a heat engine is connected to two energy reservoirs, one a pool of molten aluminum (660°C) and the other a block of solid mercury (−38.9°C). The engine runs by freezing 1.00 g of aluminum and melting 15.0 g of mercury during each cycle. The heat of fusion of aluminum is 3.97×10^5 J/kg; the heat of fusion of mercury is 1.18×10^4 J/kg. What is the efficiency of this engine?

SECTION 21.2 Heat Pumps and Refrigerators

4. During each cycle, a refrigerator ejects 625 kJ of energy to a high-temperature reservoir and takes in 550 kJ of energy from a low-temperature reservoir. Determine (a) the work done on the refrigerant in each cycle and (b) the coefficient of performance of the refrigerator.

5. A freezer has a coefficient of performance of 6.30. It is advertised as using electricity at a rate of 457 kWh/yr. (a) On average, how much energy does it use in a single day? (b) On average, how much energy does it remove from the refrigerator in a single day? (c) What maximum mass of water at 20.0°C could the freezer freeze in a single day? *Note:* One kilowatt-hour (kWh) is an amount of energy equal to running a 1-kW appliance for one hour.

6. A heat pump has a coefficient of performance equal to 4.20 and requires a power of 1.75 kW to operate. (a) How much energy does the heat pump add to a home in one hour? (b) If the heat pump is reversed so that it acts as an air conditioner in the summer, what would be its coefficient of performance?

SECTION 21.4 The Carnot Engine

7. One of the most efficient heat engines ever built is a coal-fired steam turbine in the Ohio River valley, operating between 1 870°C and 430°C. (a) What is its maximum theoretical efficiency? (b) The actual efficiency of the engine is 42.0%. How much mechanical power does the engine deliver if it absorbs 1.40×10^5 J of energy each second from its hot reservoir?

8. *Why is the following situation impossible?* An inventor comes to a patent office with the claim that her heat engine, which employs water as a working substance, has a thermodynamic efficiency of 0.110. Although this efficiency is low compared with typical automobile engines, she explains that her engine operates between an energy reservoir at room temperature and a water–ice mixture at atmospheric pressure and therefore requires no fuel other than that to make the ice. The patent is approved, and working prototypes of the engine prove the inventor's efficiency claim.

9. If a 35.0%-efficient Carnot heat engine (Fig. 21.2) is run in reverse so as to form a refrigerator (Fig. 21.4), what would be this refrigerator's coefficient of performance?

10. An ideal refrigerator or ideal heat pump is equivalent to a Carnot engine running in reverse. That is, energy $|Q_c|$ is taken in from a cold reservoir and energy $|Q_h|$ is rejected to a hot reservoir. (a) Show that the work that must be supplied to run the refrigerator or heat pump is

$$W = \frac{T_h - T_c}{T_c} |Q_c|$$

(b) Show that the coefficient of performance (COP) of the ideal refrigerator is

$$\text{COP} = \frac{T_c}{T_h - T_c}$$

11. A heat engine is being designed to have a Carnot efficiency of 65.0% when operating between two energy reservoirs. (a) If the temperature of the cold reservoir is 20.0°C, what must be the temperature of the hot reservoir? (b) Can the actual efficiency of the engine be equal to 65.0%? Explain.

12. A power plant operates at a 32.0% efficiency during the summer when the seawater used for cooling is at 20.0°C. The plant uses 350°C steam to drive turbines. If the plant's efficiency changes in the same proportion as the ideal efficiency, what would be the plant's efficiency in the winter, when the seawater is at 10.0°C?

13. You are working on a summer job at a company that designs non-traditional energy systems. The company is working on a proposed electric power plant that would make use of the temperature gradient in the ocean. The system includes a heat engine that would operate between 20.0°C (surface-water temperature) and 5.00°C (water temperature at a depth of about 1 km). (a) Your supervisor asks you to determine the maximum efficiency of such a system. (b) In addition, if the electric power output of the plant is 75.0 MW and it operates at the maximum theoretically possible efficiency, you must determine the rate at which energy is taken in from the warm reservoir. (c) From this information, if an electric bill for a typical home shows a use of 950 kWh per month, your supervisor wants to know how many homes can be provided with power from this energy system operating at its maximum efficiency. (d) As energy is drawn from the warm surface water to operate the engine, it is replaced by energy absorbed from sunlight on the surface. If the average intensity absorbed from sunlight is 650 W/m² for 12 daylight hours on a clear day, you need to find the area of the ocean surface that is necessary for sunlight to replace the energy absorbed into the engine. (e) From this information, you need to determine if there is enough ocean surface on the Earth to use such engines to supply the electrical needs for all the homes associated with the Earth's population. Assume the energy use for a home in part (c) is an average over the entire planet. (f) In view of your results in this problem, your supervisor has asked for your conclusion as to whether such a system is worthwhile to pursue. Note that the "fuel" (sunlight) is free.

14. A Carnot heat engine operates between temperatures T_h and T_c. (a) If $T_h = 500$ K and $T_c = 350$ K, what is the efficiency of the engine? (b) What is the change in its efficiency for each degree of increase in T_h above 500 K? (c) What is the change in its efficiency for each degree of change in T_c? (d) Does the answer to part (c) depend on T_c? Explain.

15. An electric generating station is designed to have an electric output power of 1.40 MW using a turbine with two-thirds the efficiency of a Carnot engine. The exhaust energy is transferred by heat into a cooling tower at 110°C. (a) Find the rate at which the station exhausts energy by heat as a function of the fuel combustion temperature T_h. (b) If the firebox is modified to run hotter by using more advanced combustion technology, how does the amount of energy exhaust change? (c) Find the exhaust power for $T_h = 800$°C. (d) Find the value of T_h for which the exhaust power would be only half as large as in part (c). (e) Find the value of T_h for which the exhaust power would be one-fourth as large as in part (c).

16. Suppose you build a two-engine device with the exhaust energy output from one heat engine supplying the input energy for a second heat engine. We say that the two engines are running *in series*. Let e_1 and e_2 represent the efficiencies of the two engines. (a) The overall efficiency of the two-engine device is defined as the total work output divided by the energy put into the first engine by heat. Show that the overall efficiency e is given by

$$e = e_1 + e_2 - e_1 e_2$$

What If? For parts (b) through (e) that follow, assume the two engines are Carnot engines. Engine 1 operates between temperatures T_h and T_i. The gas in engine 2 varies in temperature between T_i and T_c. In terms of the temperatures, (b) what is the efficiency of the combination engine? (c) Does an improvement in net efficiency result from the use of two engines instead of one? (d) What value of the intermediate temperature T_i results in equal work being done by each of the two engines in series? (e) What value of T_i results in each of the two engines in series having the same efficiency?

17. A heat pump used for heating shown in Figure P21.17 is essentially an air conditioner installed backward. It extracts energy from colder air outside and deposits it in a warmer room. Suppose the ratio of the actual energy entering the room to the work done by the device's motor is 10.0% of the theoretical maximum ratio. Determine the energy entering the room per joule of work done by the motor given that the inside temperature is 20.0°C and the outside temperature is −5.00°C.

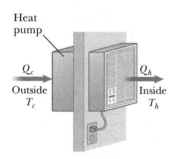

Figure P21.17

SECTION 21.5 Gasoline and Diesel Engines

Note: For problems in this section, assume the gas in the engine is diatomic with $\gamma = 1.40$.

18. A gasoline engine has a compression ratio of 6.00. (a) What is the efficiency of the engine if it operates in an idealized Otto cycle? (b) **What If?** If the actual efficiency is 15.0%, what fraction of the fuel is wasted as a result of friction and energy transfers by heat that could be avoided in a reversible engine? Assume complete combustion of the air–fuel mixture.

19. An idealized diesel engine operates in a cycle known as the *air-standard diesel cycle* shown in Figure P21.19. Fuel is sprayed into the cylinder at the point of maximum compression, *B*. Combustion occurs during the expansion $B \rightarrow C$, which is modeled as an isobaric process. Show that the efficiency of an engine operating in this idealized diesel cycle is

$$e = 1 - \frac{1}{\gamma}\left(\frac{T_D - T_A}{T_C - T_B}\right)$$

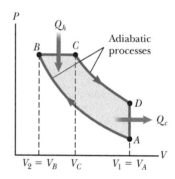

Figure P21.19

SECTION 21.6 Entropy

20. (a) Prepare a table like Table 21.1 for the following occurrence. You toss four coins into the air simultaneously and then record the results of your tosses in terms of the numbers of heads (H) and tails (T) that result. For example, HHTH and HTHH are two possible ways in which three heads and one tail can be achieved. (b) On the basis of your table, what is the most probable result recorded for a toss?

21. Prepare a table like Table 21.1 by using the same procedure (a) for the case in which you draw three marbles from your bag rather than four and (b) for the case in which you draw five marbles rather than four.

SECTION 21.7 Entropy in Thermodynamic Systems

22. A Styrofoam cup holding 125 g of hot water at 100°C cools to room temperature, 20.0°C. What is the change in entropy of the room? Neglect the specific heat of the cup and any change in temperature of the room.

23. A 1 500-kg car is moving at 20.0 m/s. The driver brakes to a stop. The brakes cool off to the temperature of the surrounding air, which is nearly constant at 20.0°C. What is the total entropy change?

24. A 2.00-L container has a center partition that divides it into two equal parts as shown in Figure P21.24. The left side

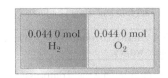

Figure P21.24

contains 0.044 0 mol of H_2 gas, and the right side contains 0.044 0 mol of O_2 gas. Both gases are at room temperature and at atmospheric pressure. The partition is removed, and the gases are allowed to mix. What is the entropy increase of the system?

25. Calculate the change in entropy of 250 g of water warmed slowly from 20.0°C to 80.0°C.

26. What change in entropy occurs when a 27.9-g ice cube at −12°C is transformed into steam at 115°C?

SECTION 21.8   Entropy and the Second Law

27. When an aluminum bar is connected between a hot reservoir at 725 K and a cold reservoir at 310 K, 2.50 kJ of energy is transferred by heat from the hot reservoir to the cold reservoir. In this irreversible process, calculate the change in entropy of (a) the hot reservoir, (b) the cold reservoir, and (c) the Universe, neglecting any change in entropy of the aluminum rod.

28. When a metal bar is connected between a hot reservoir at T_h and a cold reservoir at T_c, the energy transferred by heat from the hot reservoir to the cold reservoir is Q. In this irreversible process, find expressions for the change in entropy of (a) the hot reservoir, (b) the cold reservoir, and (c) the Universe, neglecting any change in entropy of the metal rod.

29. How fast are you personally making the entropy of the Universe increase right now? Compute an order-of-magnitude estimate, stating what quantities you take as data and the values you measure or estimate for them.

ADDITIONAL PROBLEMS

30. Every second at Niagara Falls, some 5.00×10^3 m³ of water
    AMT falls a distance of 50.0 m. What is the increase in entropy of the Universe per second due to the falling water? Assume the mass of the surroundings is so great that its temperature and that of the water stay nearly constant at 20.0°C. Also assume a negligible amount of water evaporates.

31. The energy absorbed by an engine is three times greater than the work it performs. (a) What is its thermal efficiency? (b) What fraction of the energy absorbed is expelled to the cold reservoir?

32. In 1993, the U.S. government instituted a requirement that
    QIC all room air conditioners sold in the United States must have an energy efficiency ratio (EER) of 10 or higher. The EER is defined as the ratio of the cooling capacity of the air conditioner, measured in British thermal units per hour, or Btu/h, to its electrical power requirement in watts. (a) Convert the EER of 10.0 to dimensionless form, using the conversion 1 Btu = 1 055 J. (b) What is the appropriate name for this dimensionless quantity? (c) In the 1970s, it was common to find room air conditioners with EERs of 5 or lower. State how the operating costs compare for 10 000-Btu/h air conditioners with EERs of 5.00 and 10.0. Assume each air

conditioner operates for 1 500 h during the summer in a city where electricity costs 17.0¢ per kWh.

33. In 1816, Robert Stirling, a Scottish clergyman, patented the
    GP *Stirling engine,* which has found a wide variety of applications
    S ever since, including current use in solar energy collectors to transform sunlight into electricity. Fuel is burned externally to warm one of the engine's two cylinders. A fixed quantity of inert gas moves cyclically between the cylinders, expanding in the hot one and contracting in the cold one. Figure P21.33 represents a model for its thermodynamic cycle. Consider n moles of an ideal monatomic gas being taken once through the cycle, consisting of two isothermal processes at temperatures $3T_i$ and T_i and two constant-volume processes. Let us find the efficiency of this engine. (a) Find the energy transferred by heat into the gas during the isovolumetric process AB. (b) Find the energy transferred by heat into the gas during the isothermal process BC. (c) Find the energy transferred by heat into the gas during the isovolumetric process CD. (d) Find the energy transferred by heat into the gas during the isothermal process DA. (e) Identify which of the results from parts (a) through (d) are positive and evaluate the energy input to the engine by heat. (f) From the first law of thermodynamics, find the work done by the engine. (g) From the results of parts (e) and (f), evaluate the efficiency of the engine. A Stirling engine is easier to manufacture than an internal combustion engine or a turbine. It can run on burning garbage. It can run on the energy transferred by sunlight and produce no material exhaust. Stirling engines are not currently used in automobiles due to long startup times and poor acceleration response.

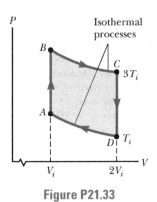

Figure P21.33

34. Suppose an ideal (Carnot) heat pump could be constructed
    QIC for use as an air conditioner. (a) Obtain an expression for the coefficient of performance (COP) for such an air conditioner in terms of T_h and T_c. (b) Would such an air conditioner operate on a smaller energy input if the difference in the operating temperatures were greater or smaller? (c) Compute the COP for such an air conditioner if the indoor temperature is 20.0°C and the outdoor temperature is 40.0°C.

35. **Review.** This problem complements Problem 44 in Chapter 10. In the operation of a single-cylinder internal combustion piston engine, one charge of fuel explodes to drive the piston outward in the *power stroke.* Part of its energy output is stored in a turning flywheel. This energy is then used to push the piston inward to compress the next charge of fuel and air. In this compression process, assume an original volume of 0.120 L of a diatomic ideal gas at atmospheric pressure

is compressed adiabatically to one-eighth of its original volume. (a) Find the work input required to compress the gas. (b) Assume the flywheel is a solid disk of mass 5.10 kg and radius 8.50 cm, turning freely without friction between the power stroke and the compression stroke. How fast must the flywheel turn immediately after the power stroke? This situation represents the minimum angular speed at which the engine can operate without stalling. (c) When the engine's operation is well above the point of stalling, assume the flywheel puts 5.00% of its maximum energy into compressing the next charge of fuel and air. Find its maximum angular speed in this case.

36. A firebox is at 750 K, and the ambient temperature is 300 K. **QC** The efficiency of a Carnot engine doing 150 J of work as it transports energy between these constant-temperature baths is 60.0%. The Carnot engine must take in energy 150 J/0.600 = 250 J from the hot reservoir and must put out 100 J of energy by heat into the environment. To follow Carnot's reasoning, suppose some other heat engine S could have an efficiency of 70.0%. (a) Find the energy input and exhaust energy output of engine S as it does 150 J of work. (b) Let engine S operate as in part (a) and run the Carnot engine in reverse between the same reservoirs. The output work of engine S is the input work for the Carnot refrigerator. Find the total energy transferred to or from the firebox and the total energy transferred to or from the environment as both engines operate together. (c) Explain how the results of parts (a) and (b) show that the Clausius statement of the second law of thermodynamics is violated. (d) Find the energy input and work output of engine S as it puts out exhaust energy of 100 J. Let engine S operate as in part (c) and contribute 150 J of its work output to running the Carnot engine in reverse. Find (e) the total energy the firebox puts out as both engines operate together, (f) the total work output, and (g) the total energy transferred to the environment. (h) Explain how the results show that the Kelvin–Planck statement of the second law is violated. Therefore, our assumption about the efficiency of engine S must be false. (i) Let the engines operate together through one cycle as in part (d). Find the change in entropy of the Universe. (j) Explain how the result of part (i) shows that the entropy statement of the second law is violated.

37. A 1.00-mol sample of an ideal monatomic gas is taken **QC** through the cycle shown in Figure P21.37. The process $A \rightarrow B$ is a reversible isothermal expansion. Calculate (a) the net work done by the gas, (b) the energy added to the gas by heat, (c) the energy exhausted from the gas by heat, and (d) the efficiency of the cycle. (e) Explain how the efficiency compares with that of a Carnot engine operating between the same temperature extremes.

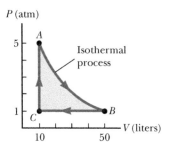

Figure P21.37

38. A system consisting of n moles of an ideal gas with molar **QC** specific heat at constant pressure C_p undergoes two revers-**S** ible processes. It starts with pressure P_i and volume V_i, expands isothermally, and then contracts adiabatically to reach a final state with pressure P_i and volume $3V_i$. (a) Find its change in entropy in the isothermal process. (The entropy does not change in the adiabatic process.) (b) **What If?** Explain why the answer to part (a) must be the same as the answer to Problem 46. (You do not need to solve Problem 46 to answer this question.)

39. A heat engine operates between two reservoirs at $T_2 = 600$ K and $T_1 = 350$ K. It takes in 1.00×10^3 J of energy from the higher-temperature reservoir and performs 250 J of work. Find (a) the entropy change of the Universe ΔS_U for this process and (b) the work W that could have been done by an ideal Carnot engine operating between these two reservoirs. (c) Show that the difference between the amounts of work done in parts (a) and (b) is $T_1 \Delta S_U$.

40. You are working as an assistant to a physics professor. She **CR** has seen some presentations you have made to your classes and is aware of your expertise in preparing presentation slides. Her laptop has crashed and she cannot access the presentation slides she needs for her lecture coming up in one hour. Her lecture is on entropy in engine cycles. She asks you to quickly generate two slides on your laptop, both showing *TS* diagrams, (a) one for the Carnot cycle and (b) one for the Otto cycle. As she leaves, you think, "Uh-oh. What's a *TS diagram*?" Quick, you have no time to waste! Get to work!

41. You are working as an expert witness for an environmen-**CR** tal agency. A utility in a neighboring town has proposed a new power plant that produces 1.00 GW of electrical power from turbines. The utility claims that the plant will take in steam at 500 K and reject water at 300 K into a flowing cold-water river. The flow rate of the river is 6.00×10^4 kg/s. The agency supervisor is concerned about the effect of dumping warm water on the fish in the river. (a) The utility claims that the power plant operates with Carnot efficiency. With that assumption, you need to determine for a trial presentation by how much the temperature of the water downstream from the power plant will rise due to the rejected energy from the power plant. (b) If you abandon the utility's claim that the power plant operates at Carnot efficiency and assume a more realistic efficiency, you need to testify whether the increase in water temperature will be higher or lower than that found in part (a). (c) Finally, determine the increase in water temperature in the stream if the actual efficiency of the power plant were estimated by you and the agency physicist to be 15.0%.

42. You are working as an expert witness for an environmental **CR** agency. A utility in a neighboring town has proposed a new **S** power plant that produces electrical power P from turbines. The utility claims that the plant will take in steam at temperature T_h and reject water at temperature T_c into a flowing cold-water river. The flow rate of the river is $\Delta m / \Delta t$. The agency supervisor is concerned about the effect of dumping warm water on the fish in the river. (a) The utility claims that the power plant operates with Carnot efficiency. With that assumption, you need to determine for a trial presentation by how much the temperature of the water downstream from the power plant will rise due to the rejected energy from the power plant. (b) If you abandon the utility's claim that the power plant operates at Carnot efficiency and assume a more

realistic efficiency e, you need to determine the increase in water temperature in the stream. (c) Finally, you need to testify whether the increase in water temperature in part (b) will be higher or lower than that found in part (a).

43. An athlete whose mass is 70.0 kg drinks 16.0 ounces (454 g) of refrigerated water. The water is at a temperature of 35.0°F. (a) Ignoring the temperature change of the body that results from the water intake (so that the body is regarded as a reservoir always at 98.6°F), find the entropy increase of the entire system. (b) **What If?** Assume the entire body is cooled by the drink and the average specific heat of a person is equal to the specific heat of liquid water. Ignoring any other energy transfers by heat and any metabolic energy release, find the athlete's temperature after she drinks the cold water given an initial body temperature of 98.6°F. (c) Under these assumptions, what is the entropy increase of the entire system? (d) State how this result compares with the one you obtained in part (a).

44. *Why is the following situation impossible?* Two samples of water are mixed at constant pressure inside an insulated container: 1.00 kg of water at 10.0°C and 1.00 kg of water at 30.0°C. Because the container is insulated, there is no exchange of energy by heat between the water and the environment. Furthermore, the amount of energy that leaves the warm water by heat is equal to the amount that enters the cool water by heat. Therefore, the entropy change of the Universe is zero for this process.

45. A sample of an ideal gas expands isothermally, doubling in volume. (a) Show that the work done on the gas in expanding is $W = -nRT \ln 2$. (b) Because the internal energy E_{int} of an ideal gas depends solely on its temperature, the change in internal energy is zero during the expansion. It follows from the first law that the energy input to the gas by heat during the expansion is equal to the energy output by work. Does this process have 100% efficiency in converting energy input by heat into work output? (c) Does this conversion violate the second law? Explain.

46. A sample consisting of n moles of an ideal gas undergoes a reversible isobaric expansion from volume V_i to volume $3V_i$. Find the change in entropy of the gas by calculating $\int_i^f dQ/T$, where $dQ = nC_p \, dT$.

CHALLENGE PROBLEM

47. The compression ratio of an Otto cycle as shown in Figure 21.12 is $V_A/V_B = 8.00$. At the beginning A of the compression process, 500 cm³ of gas is at 100 kPa and 20.0°C. At the beginning of the adiabatic expansion, the temperature is $T_C = 750°C$. Model the working fluid as an ideal gas with $\gamma = 1.40$. (a) Fill in this table to follow the states of the gas:

	T (K)	P (kPa)	V (cm³)
A	293	100	500
B			
C	1 023		
D			

(b) Fill in this table to follow the processes:

	Q	W	ΔE_{int}
$A \to B$			
$B \to C$			
$C \to D$			
$D \to A$			
$ABCDA$			

(c) Identify the energy input $|Q_h|$, (d) the energy exhaust $|Q_c|$, and (e) the net output work W_{eng}. (f) Calculate the thermal efficiency. (g) Find the number of crankshaft revolutions per minute required for a one-cylinder engine to have an output power of 1.00 kW = 1.34 hp. *Note:* The thermodynamic cycle involves four piston strokes.

Appendix A Tables

TABLE A.1 Conversion Factors

Length

	m	cm	km	in.	ft	mi
1 meter	1	10^2	10^{-3}	39.37	3.281	6.214×10^{-4}
1 centimeter	10^{-2}	1	10^{-5}	0.393 7	3.281×10^{-2}	6.214×10^{-6}
1 kilometer	10^3	10^5	1	3.937×10^4	3.281×10^3	0.621 4
1 inch	2.540×10^{-2}	2.540	2.540×10^{-5}	1	8.333×10^{-2}	1.578×10^{-5}
1 foot	0.304 8	30.48	3.048×10^{-4}	12	1	1.894×10^{-4}
1 mile	1 609	1.609×10^5	1.609	6.336×10^4	5 280	1

Mass

	kg	g	slug	u
1 kilogram	1	10^3	6.852×10^{-2}	6.024×10^{26}
1 gram	10^{-3}	1	6.852×10^{-5}	6.024×10^{23}
1 slug	14.59	1.459×10^4	1	8.789×10^{27}
1 atomic mass unit	1.660×10^{-27}	1.660×10^{-24}	1.137×10^{-28}	1

Note: 1 metric ton = 1 000 kg.

Time

	s	min	h	day	yr
1 second	1	1.667×10^{-2}	2.778×10^{-4}	1.157×10^{-5}	3.169×10^{-8}
1 minute	60	1	1.667×10^{-2}	6.994×10^{-4}	1.901×10^{-6}
1 hour	3 600	60	1	4.167×10^{-2}	1.141×10^{-4}
1 day	8.640×10^4	1 440	24	1	2.738×10^{-5}
1 year	3.156×10^7	5.259×10^5	8.766×10^3	365.2	1

Speed

	m/s	cm/s	ft/s	mi/h
1 meter per second	1	10^2	3.281	2.237
1 centimeter per second	10^{-2}	1	3.281×10^{-2}	2.237×10^{-2}
1 foot per second	0.304 8	30.48	1	0.681 8
1 mile per hour	0.447 0	44.70	1.467	1

Note: 1 mi/min = 60 mi/h = 88 ft/s.

Force

	N	lb
1 newton	1	0.224 8
1 pound	4.448	1

(Continued)

TABLE A.1   Conversion Factors (*continued*)

Energy, Energy Transfer

	J	ft · lb	eV
1 joule	1	0.737 6	6.242×10^{18}
1 foot-pound	1.356	1	8.464×10^{18}
1 electron volt	1.602×10^{-19}	1.182×10^{-19}	1
1 calorie	4.186	3.087	2.613×10^{19}
1 British thermal unit	1.055×10^{3}	7.779×10^{2}	6.585×10^{21}
1 kilowatt-hour	3.600×10^{6}	2.655×10^{6}	2.247×10^{25}

	cal	Btu	kWh
1 joule	0.238 9	9.481×10^{-4}	2.778×10^{-7}
1 foot-pound	0.323 9	1.285×10^{-3}	3.766×10^{-7}
1 electron volt	3.827×10^{-20}	1.519×10^{-22}	4.450×10^{-26}
1 calorie	1	3.968×10^{-3}	1.163×10^{-6}
1 British thermal unit	2.520×10^{2}	1	2.930×10^{-4}
1 kilowatt-hour	8.601×10^{5}	3.413×10^{2}	1

Pressure

	Pa	atm
1 pascal	1	9.869×10^{-6}
1 atmosphere	1.013×10^{5}	1
1 centimeter mercury[a]	1.333×10^{3}	1.316×10^{-2}
1 pound per square inch	6.895×10^{3}	6.805×10^{-2}
1 pound per square foot	47.88	4.725×10^{-4}

	cm Hg	lb/in.^{2}	lb/ft^2
1 pascal	7.501×10^{-4}	1.450×10^{-4}	2.089×10^{-2}
1 atmosphere	76	14.70	2.116×10^{3}
1 centimeter mercury[a]	1	0.194 3	27.85
1 pound per square inch	5.171	1	144
1 pound per square foot	3.591×10^{-2}	6.944×10^{-3}	1

[a]At 0°C and at a location where the free-fall acceleration has its "standard" value, 9.806 65 m/s^2.

TABLE A.2   Symbols, Dimensions, and Units of Physical Quantities

Quantity	Common Symbol	Unit[a]	Dimensions[b]	Unit in Terms of Base SI Units
Acceleration	$\vec{a}$	m/s^2	L/T^2	m/s^2
Amount of substance	n	MOLE		mol
Angle	θ, ϕ	radian (rad)		
Angular acceleration	$\vec{\alpha}$	rad/s^2	T^{-2}	s^{-2}
Angular frequency	ω	rad/s	T^{-1}	s^{-1}
Angular momentum	$\vec{L}$	kg · m^2/s	ML^2/T	kg · m^2/s
Angular velocity	$\vec{\omega}$	rad/s	T^{-1}	s^{-1}
Area	A	m^2	L^2	m^2
Atomic number	Z			
Capacitance	C	farad (F)	Q^{2}T^2/ML^2	A^2 · s^4/kg · m^2
Charge	q, Q, e	coulomb (C)	Q	A · s

TABLE A.2 Symbols, Dimensions, and Units of Physical Quantities (*continued*)

Quantity	Common Symbol	Unit[a]	Dimensions[b]	Unit in Terms of Base SI Units
Charge density				
Line	λ	C/m	Q/L	A · s/m
Surface	σ	C/m^2	Q/L^2	A · s/m^2
Volume	ρ	C/m^3	Q/L^3	A · s/m^3
Conductivity	σ	1/Ω · m	Q^{2}T/ML^3	A^2 · s^3/kg · m^3
Current	I	AMPERE	Q/T	A
Current density	J	A/m^2	Q/TL^2	A/m^2
Density	ρ	kg/m^3	M/L^3	kg/m^3
Dielectric constant	κ			
Electric dipole moment	$\vec{p}$	C · m	QL	A · s · m
Electric field	$\vec{E}$	V/m	ML/QT^2	kg · m/A · s^3
Electric flux	Φ_E	V · m	ML^3/QT^2	kg · m^3/A · s^3
Electromotive force	ε	volt (V)	ML^2/QT^2	kg · m^2/A · s^3
Energy, energy transfer	E, U, K, T	joule (J)	ML^2/T^2	kg · m^2/s^2
Entropy	S	J/K	ML^2/T^{2}K	kg · m^2/s^2 · K
Force	$\vec{F}$	newton (N)	ML/T^2	kg · m/s^2
Frequency	f	hertz (Hz)	T^{-1}	s^{-1}
Heat	Q	joule (J)	ML^2/T^2	kg · m^2/s^2
Inductance	L	henry (H)	ML^2/Q^2	kg · m^2/A^2 · s^2
Length	ℓ, L	METER	L	m
Displacement	$\Delta x, \Delta \vec{r}$			
Distance	d, h			
Position	$x, y, z, \vec{r}$			
Width, height, radius	w, h, r, R, a, b			
Magnetic dipole moment	$\vec{\mu}$	N · m/T	QL^2/T	A · m^2
Magnetic field	$\vec{B}$	tesla (T) (= Wb/m^2)	M/QT	kg/A · s^2
Magnetic flux	Φ_B	weber (Wb)	ML^2/QT	kg · m^2/A · s^2
Mass	m, M	KILOGRAM	M	kg
Moment of inertia	I	kg · m^2	ML^2	kg · m^2
Momentum	$\vec{p}$	kg · m/s	ML/T	kg · m/s
Period	T	s	T	s
Permeability of free space	μ_0	N/A^2 (= H/m)	ML/Q^2	kg · m/A^2 · s^2
Permittivity of free space	ϵ_0	C^2/N · m^2 (= F/m)	Q^{2}T^2/ML^3	A^2 · s^4/kg · m^3
Potential	V	volt (V)(= J/C)	ML^2/QT^2	kg · m^2/A · s^3
Power	P	watt (W)(= J/s)	ML^2/T^3	kg · m^2/s^3
Pressure	P	pascal (Pa)(= N/m^2)	M/LT^2	kg/m · s^2
Resistance	R	ohm (Ω)(= V/A)	ML^2/Q^{2}T	kg · m^2/A^2 · s^3
Specific heat	c	J/kg · K	L^2/T^{2}K	m^2/s^2 · K
Speed	v	m/s	L/T	m/s
Temperature	T	KELVIN	K	K
Time	t	SECOND	T	s
Torque	$\vec{\tau}$	N · m	ML^2/T^2	kg · m^2/s^2
Velocity	$\vec{v}$	m/s	L/T	m/s
Volume	V	m^3	L^3	m^3
Wavelength	λ	m	L	m
Work	W	joule (J)(= N · m)	ML^2/T^2	kg · m^2/s^2

[a]The base SI units are given in uppercase letters.
[b]The symbols M, L, T, K, and Q denote mass, length, time, temperature, and charge, respectively.

Appendix B Mathematics Review

This appendix in mathematics is intended as a brief review of operations and methods. Early in this course, you should be totally familiar with basic algebraic techniques, analytic geometry, and trigonometry. The sections on differential and integral calculus are more detailed and are intended for students who have difficulty applying calculus concepts to physical situations.

B.1 Scientific Notation

Many quantities used by scientists often have very large or very small values. The speed of light, for example, is about 300 000 000 m/s, and the ink required to make the dot over an i in this textbook has a mass of about 0.000 000 001 kg. Obviously, it is very cumbersome to read, write, and keep track of the numbers of zeros in such quantities. We avoid this problem by using a method incorporating powers of the number 10:

$$10^0 = 1$$
$$10^1 = 10$$
$$10^2 = 10 \times 10 = 100$$
$$10^3 = 10 \times 10 \times 10 = 1\,000$$
$$10^4 = 10 \times 10 \times 10 \times 10 = 10\,000$$
$$10^5 = 10 \times 10 \times 10 \times 10 \times 10 = 100\,000$$

and so on. The number of zeros corresponds to the power to which ten is raised, called the **exponent** of ten. For example, the speed of light, 300 000 000 m/s, can be expressed as 3.00×10^8 m/s.

In this method, some representative numbers smaller than unity are the following:

$$10^{-1} = \frac{1}{10} = 0.1$$

$$10^{-2} = \frac{1}{10 \times 10} = 0.01$$

$$10^{-3} = \frac{1}{10 \times 10 \times 10} = 0.001$$

$$10^{-4} = \frac{1}{10 \times 10 \times 10 \times 10} = 0.000\,1$$

$$10^{-5} = \frac{1}{10 \times 10 \times 10 \times 10 \times 10} = 0.000\,01$$

In these cases, the number of places the decimal point is to the left of the digit 1 equals the value of the (negative) exponent. Numbers expressed as some power of ten multiplied by another number between one and ten are said to be in **scientific notation**. For example, the scientific notation for 5 943 000 000 is 5.943×10^9 and that for 0.000 083 2 is 8.32×10^{-5}.

When numbers expressed in scientific notation are being multiplied, the following general rule is very useful:

$$10^n \times 10^m = 10^{n+m} \qquad \textbf{(B.1)}$$

where n and m can be *any* numbers (not necessarily integers). For example, $10^2 \times 10^5 = 10^7$. The rule also applies if one of the exponents is negative: $10^3 \times 10^{-8} = 10^{-5}$.

When dividing numbers expressed in scientific notation, note that

$$\frac{10^n}{10^m} = 10^n \times 10^{-m} = 10^{n-m} \qquad \textbf{(B.2)}$$

Exercises

With help from the preceding rules, verify the answers to the following equations:

1. $86\ 400 = 8.64 \times 10^4$
2. $9\ 816\ 762.5 = 9.816\ 762\ 5 \times 10^6$
3. $0.000\ 000\ 039\ 8 = 3.98 \times 10^{-8}$
4. $(4.0 \times 10^8)(9.0 \times 10^9) = 3.6 \times 10^{18}$
5. $(3.0 \times 10^7)(6.0 \times 10^{-12}) = 1.8 \times 10^{-4}$
6. $\dfrac{75 \times 10^{-11}}{5.0 \times 10^{-3}} = 1.5 \times 10^{-7}$
7. $\dfrac{(3 \times 10^6)(8 \times 10^{-2})}{(2 \times 10^{17})(6 \times 10^5)} = 2 \times 10^{-18}$

B.2   Algebra

Some Basic Rules

When algebraic operations are performed, the laws of arithmetic apply. Symbols such as x, y, and z are usually used to represent unspecified quantities, called the **unknowns.**

First, consider the equation

$$8x = 32$$

If we wish to solve for x, we can divide (or multiply) each side of the equation by the same factor without destroying the equality. In this case, if we divide both sides by 8, we have

$$\frac{8x}{8} = \frac{32}{8}$$

$$x = 4$$

Next, consider the equation

$$x + 2 = 8$$

In this type of expression, we can add or subtract the same quantity from each side. If we subtract 2 from each side, we have

$$x + 2 - 2 = 8 - 2$$

$$x = 6$$

In general, if $x + a = b$, then $x = b - a$.

Now consider the equation

$$\frac{x}{5} = 9$$

If we multiply each side by 5, we are left with x on the left by itself and 45 on the right:

$$\left(\frac{x}{5}\right)(5) = 9 \times 5$$

$$x = 45$$

In all cases, *whatever operation is performed on the left side of the equality must also be performed on the right side.*

The following rules for multiplying, dividing, adding, and subtracting fractions should be recalled, where a, b, c, and d are four numbers:

	Rule	Example
Multiplying	$\left(\dfrac{a}{b}\right)\left(\dfrac{c}{d}\right) = \dfrac{ac}{bd}$	$\left(\dfrac{2}{3}\right)\left(\dfrac{4}{5}\right) = \dfrac{8}{15}$
Dividing	$\dfrac{(a/b)}{(c/d)} = \dfrac{ad}{bc}$	$\dfrac{2/3}{4/5} = \dfrac{(2)(5)}{(4)(3)} = \dfrac{10}{12}$
Adding	$\dfrac{a}{b} \pm \dfrac{c}{d} = \dfrac{ad \pm bc}{bd}$	$\dfrac{2}{3} - \dfrac{4}{5} = \dfrac{(2)(5) - (4)(3)}{(3)(5)} = -\dfrac{2}{15}$

Exercises

In the following exercises, solve for x:

Answers

1. $a = \dfrac{1}{1 + x}$ $x = \dfrac{1 - a}{a}$

2. $3x - 5 = 13$ $x = 6$

3. $ax - 5 = bx + 2$ $x = \dfrac{7}{a - b}$

4. $\dfrac{5}{2x + 6} = \dfrac{3}{4x + 8}$ $x = -\dfrac{11}{7}$

Powers

When powers of a given quantity x are multiplied, the following rule applies:

$$x^n \, x^m = x^{n+m} \tag{B.3}$$

For example, $x^2 x^4 = x^{2+4} = x^6$.

When dividing the powers of a given quantity, the rule is

$$\frac{x^n}{x^m} = x^{n-m} \tag{B.4}$$

For example, $x^8/x^2 = x^{8-2} = x^6$.

A power that is a fraction, such as $\frac{1}{3}$, corresponds to a root as follows:

$$x^{1/n} = \sqrt[n]{x} \tag{B.5}$$

For example, $4^{1/3} = \sqrt[3]{4} = 1.587\,4$. (A scientific calculator is useful for such calculations.)

Finally, any quantity x^n raised to the mth power is

$$(x^n)^m = x^{nm} \tag{B.6}$$

Table B.1 summarizes the rules of exponents.

TABLE B.1   Rules of Exponents

$$x^0 = 1$$
$$x^1 = x$$
$$x^n \, x^m = x^{n+m}$$
$$x^n/x^m = x^{n-m}$$
$$x^{1/n} = \sqrt[n]{x}$$
$$(x^n)^m = x^{nm}$$

Exercises

Verify the following equations:

1. $3^2 \times 3^3 = 243$
2. $x^5 x^{-8} = x^{-3}$

3. $x^{10}/x^{-5} = x^{15}$
4. $5^{1/3} = 1.709\ 976$ (Use your calculator.)
5. $60^{1/4} = 2.783\ 158$ (Use your calculator.)
6. $(x^4)^3 = x^{12}$

Factoring

Some useful formulas for factoring an equation are the following:

$ax + ay + az = a(x + y + z)$	common factor
$a^2 + 2ab + b^2 = (a + b)^2$	perfect square
$a^2 - b^2 = (a + b)(a - b)$	differences of squares

Quadratic Equations

The general form of a quadratic equation is

$$ax^2 + bx + c = 0 \qquad \text{(B.7)}$$

where x is the unknown quantity and a, b, and c are numerical factors referred to as **coefficients** of the equation. This equation has two roots, given by

$$x = \frac{-b \pm \sqrt{b^2 - 4ac}}{2a} \qquad \text{(B.8)}$$

If $b^2 \geq 4ac$, the roots are real.

Example **B.1**

Find the roots of the equation $x^2 + 5x + 4 = 0$.

SOLUTION

Use Equation B.8 to find the roots:

$$x = \frac{-5 \pm \sqrt{5^2 - (4)(1)(4)}}{2(1)} = \frac{-5 \pm \sqrt{9}}{2} = \frac{-5 \pm 3}{2}$$

Evaluate the root for each of the two possibilities of the sign:

$$x_+ = \frac{-5 + 3}{2} = -1 \qquad x_- = \frac{-5 - 3}{2} = -4$$

where x_+ refers to the root corresponding to the positive sign and x_- refers to the root corresponding to the negative sign.

Exercises

Solve the following quadratic equations:

Answers

1. $x^2 + 2x - 3 = 0$	$x_+ = 1$	$x_- = -3$
2. $2x^2 - 5x + 2 = 0$	$x_+ = 2$	$x_- = \frac{1}{2}$
3. $2x^2 - 4x - 9 = 0$	$x_+ = 1 + \sqrt{22}/2$	$x_- = 1 - \sqrt{22}/2$

Linear Equations

A linear equation has the general form

$$y = mx + b \qquad \text{(B.9)}$$

where m and b are constants. This equation is referred to as linear because the graph of y versus x is a straight line as shown in Figure B.1. The constant b, called the **y-intercept,** represents the value of y at which the straight line intersects

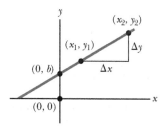

Figure B.1 A straight line graphed on an xy coordinate system. The slope of the line is the ratio of Δy to Δx.

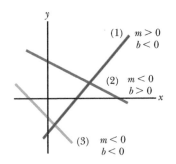

Figure B.2 The brown line has a positive slope and a negative y-intercept. The blue line has a negative slope and a positive y-intercept. The green line has a negative slope and a negative y-intercept.

the y axis. The constant m is equal to the **slope** of the straight line. If any two points on the straight line are specified by the coordinates (x_1, y_1) and (x_2, y_2) as in Figure B.1, the slope of the straight line can be expressed as

$$\text{Slope} = \frac{y_2 - y_1}{x_2 - x_1} = \frac{\Delta y}{\Delta x} \qquad \textbf{(B.10)}$$

Note that m and b can have either positive or negative values. If $m > 0$, the straight line has a *positive* slope as in Figure B.1. If $m < 0$, the straight line has a *negative* slope. In Figure B.1, both m and b are positive. Three other possible situations are shown in Figure B.2.

Exercises

1. Draw graphs of the following straight lines: (a) $y = 5x + 3$ (b) $y = -2x + 4$ (c) $y = -3x - 6$
2. Find the slopes of the straight lines described in Exercise 1.

Answers   (a) 5   (b) -2 (c) -3

3. Find the slopes of the straight lines that pass through the following sets of points: (a) $(0, -4)$ and $(4, 2)$ (b) $(0, 0)$ and $(2, -5)$ (c) $(-5, 2)$ and $(4, -2)$

Answers   (a) $\frac{3}{2}$ (b) $-\frac{5}{2}$ (c) $-\frac{4}{9}$

Solving Simultaneous Linear Equations

Consider the equation $3x + 5y = 15$, which has two unknowns, x and y. Such an equation does not have a unique solution. For example, $(x = 0, y = 3)$, $(x = 5, y = 0)$, and $(x = 2, y = \frac{9}{5})$ are all solutions to this equation.

If a problem has two unknowns, a unique solution is possible only if we have *two* pieces of information. In most common cases, those two pieces of information are equations. In general, if a problem has n unknowns, its solution requires n equations. To solve two simultaneous equations involving two unknowns, x and y, we solve one of the equations for x in terms of y and substitute this expression into the other equation.

In some cases, the two pieces of information may be (1) one equation and (2) a condition on the solutions. For example, suppose we have (1) the equation $m = 3n$ and (2) the condition that m and n must be the smallest positive nonzero integers possible. Then, the single equation does not allow a unique solution, but the addition of the condition gives us that $n = 1$ and $m = 3$.

Example **B.2**

Solve the two simultaneous equations

$$(1)\quad 5x + y = -8$$

$$(2)\quad 2x - 2y = 4$$

SOLUTION

Solve Equation (2) for x:

$$(3)\qquad x = y + 2$$

Substitute Equation (3) into Equation (1):

$$5(y + 2) + y = -8$$

$$6y = -18$$

$$y = -3$$

Use Equation (3) to find x:

$$x = y + 2 = -1$$

B.2 continued

Alternative Solution

Multiply each term in Equation (1) by 2: $10x + 2y = -16$

Add Equation (2): $2x - 2y = 4$

$$12x = -12$$

Solve for x: $x = -1$

Use Equation (3) to find y: $y = x - 2 = -3$

Two linear equations containing two unknowns can also be solved by a graphical method. If the straight lines corresponding to the two equations are plotted in a conventional coordinate system, the intersection of the two lines represents the solution. For example, consider the two equations

$$x - y = 2$$
$$x - 2y = -1$$

These equations are plotted in Figure B.3. The intersection of the two lines has the coordinates $x = 5$ and $y = 3$, which represents the solution to the equations. You should check this solution by the analytical technique discussed earlier.

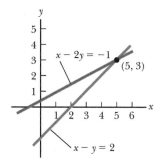

Figure B.3 A graphical solution for two linear equations.

Exercises

Solve the following pairs of simultaneous equations involving two unknowns:

Answers

1. $x + y = 8$ $x = 5, y = 3$
   $x - y = 2$

2. $98 - T = 10a$ $T = 65, a = 3.27$
   $T - 49 = 5a$

3. $6x + 2y = 6$ $x = 2, y = -3$
   $8x - 4y = 28$

Logarithms

Suppose a quantity x is expressed as a power of some quantity a:

$$x = a^y \tag{B.11}$$

The number a is called the **base** number. The **logarithm** of x with respect to the base a is equal to the exponent to which the base must be raised to satisfy the expression $x = a^y$:

$$y = \log_a x \tag{B.12}$$

Conversely, the **antilogarithm** of y is the number x:

$$x = \text{antilog}_a y \tag{B.13}$$

In practice, the two bases most often used are base 10, called the *common* logarithm base, and base $e = 2.718\,282$, called Euler's constant or the *natural* logarithm base. When common logarithms are used,

$$y = \log_{10} x \quad (\text{or } x = 10^y) \tag{B.14}$$

When natural logarithms are used,

$$y = \ln x \quad (\text{or } x = e^y) \tag{B.15}$$

For example, $\log_{10} 52 = 1.716$, so antilog$_{10}$ $1.716 = 10^{1.716} = 52$. Likewise, $\ln 52 = 3.951$, so antiln $3.951 = e^{3.951} = 52$.

In general, note you can convert between base 10 and base e with the equality

$$\ln x = (2.302\ 585) \log_{10} x \qquad \textbf{(B.16)}$$

Finally, some useful properties of logarithms are the following:

$$\left.\begin{array}{l} \log(ab) = \log a + \log b \\ \log(a/b) = \log a - \log b \\ \log(a^n) = n \log a \end{array}\right\} \text{any base}$$

$$\ln e = 1$$

$$\ln e^a = a$$

$$\ln\left(\frac{1}{a}\right) = -\ln a$$

B.3   Geometry

The **distance** d between two points having coordinates (x_1, y_1) and (x_2, y_2) is

$$d = \sqrt{(x_2 - x_1)^2 + (y_2 - y_1)^2} \qquad \textbf{(B.17)}$$

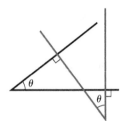

Figure B.4 The angles are equal because their sides are perpendicular.

Two angles are equal if their sides are perpendicular, right side to right side and left side to left side. For example, the two angles marked θ in Figure B.4 are the same because of the perpendicularity of the sides of the angles. To distinguish the left and right sides of an angle, imagine standing at the angle's apex and facing into the angle.

Radian measure: The arc length s of a circular arc (Fig. B.5) is proportional to the radius r for a fixed value of θ (in radians):

$$s = r\theta$$
$$\theta = \frac{s}{r} \qquad \textbf{(B.18)}$$

Figure B.5 The angle θ in radians is the ratio of the arc length s to the radius r of the circle.

Table B.2 gives the **areas** and **volumes** for several geometric shapes used throughout this text.

TABLE B.2   Useful Information for Geometry

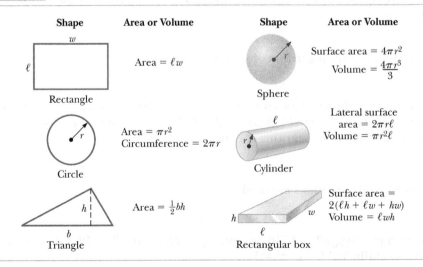

Shape	Area or Volume	Shape	Area or Volume
Rectangle	Area $= \ell w$	Sphere	Surface area $= 4\pi r^2$ Volume $= \frac{4\pi r^3}{3}$
Circle	Area $= \pi r^2$ Circumference $= 2\pi r$	Cylinder	Lateral surface area $= 2\pi r\ell$ Volume $= \pi r^2 \ell$
Triangle	Area $= \frac{1}{2}bh$	Rectangular box	Surface area $= 2(\ell h + \ell w + hw)$ Volume $= \ell wh$

The equation of a **straight line** (Fig. B.6) is

$$y = mx + b \qquad \text{(B.19)}$$

where b is the y-intercept and m is the slope of the line.

The equation of a **circle** of radius R centered at the origin is

$$x^2 + y^2 = R^2 \qquad \text{(B.20)}$$

The equation of an **ellipse** having the origin at its center (Fig. B.7) is

$$\frac{x^2}{a^2} + \frac{y^2}{b^2} = 1 \qquad \text{(B.21)}$$

where a is the length of the semimajor axis (the longer one) and b is the length of the semiminor axis (the shorter one).

The equation of a **parabola** the vertex of which is at $y = b$ (Fig. B.8) is

$$y = ax^2 + b \qquad \text{(B.22)}$$

The equation of a **rectangular hyperbola** (Fig. B.9) is

$$xy = \text{constant} \qquad \text{(B.23)}$$

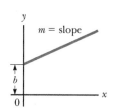

Figure B.6 A straight line with a slope of m and a y-intercept of b.

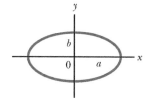

Figure B.7 An ellipse with semimajor axis a and semiminor axis b.

B.4 Trigonometry

That portion of mathematics based on the special properties of the right triangle is called trigonometry. By definition, a right triangle is a triangle containing a 90° angle. Consider the right triangle shown in Figure B.10, where side a is opposite the angle θ, side b is adjacent to the angle θ, and side c is the hypotenuse of the triangle. The three basic trigonometric functions defined by such a triangle are the sine (sin), cosine (cos), and tangent (tan). In terms of the angle θ, these functions are defined as follows:

$$\sin\theta = \frac{\text{side opposite }\theta}{\text{hypotenuse}} = \frac{a}{c} \qquad \text{(B.24)}$$

$$\cos\theta = \frac{\text{side adjacent to }\theta}{\text{hypotenuse}} = \frac{b}{c} \qquad \text{(B.25)}$$

$$\tan\theta = \frac{\text{side opposite }\theta}{\text{side adjacent to }\theta} = \frac{a}{b} \qquad \text{(B.26)}$$

The Pythagorean theorem provides the following relationship among the sides of a right triangle:

$$c^2 = a^2 + b^2 \qquad \text{(B.27)}$$

From the preceding definitions and the Pythagorean theorem, it follows that

$$\sin^2\theta + \cos^2\theta = 1$$

$$\tan\theta = \frac{\sin\theta}{\cos\theta}$$

The cosecant, secant, and cotangent functions are defined by

$$\csc\theta = \frac{1}{\sin\theta} \quad \sec\theta = \frac{1}{\cos\theta} \quad \cot\theta = \frac{1}{\tan\theta}$$

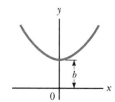

Figure B.8 A parabola with its vertex at $y = b$.

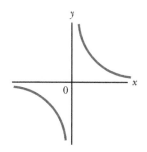

Figure B.9 A hyperbola.

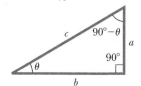

a = opposite side
b = adjacent side
c = hypotenuse

Figure B.10 A right triangle, used to define the basic functions of trigonometry.

TABLE B.3   Some Trigonometric Identities

$\sin^2 \theta + \cos^2 \theta = 1$	$\csc^2 \theta = 1 + \cot^2 \theta$
$\sec^2 \theta = 1 + \tan^2 \theta$	$\sin^2 \dfrac{\theta}{2} = \frac{1}{2}(1 - \cos \theta)$
$\sin 2\theta = 2 \sin \theta \cos \theta$	$\cos^2 \dfrac{\theta}{2} = \frac{1}{2}(1 + \cos \theta)$
$\cos 2\theta = \cos^2 \theta - \sin^2 \theta$	$1 - \cos \theta = 2 \sin^2 \dfrac{\theta}{2}$
$\tan 2\theta = \dfrac{2 \tan \theta}{1 - \tan^2 \theta}$	$\tan \dfrac{\theta}{2} = \sqrt{\dfrac{1 - \cos \theta}{1 + \cos \theta}}$

$\sin (A \pm B) = \sin A \cos B \pm \cos A \sin B$

$\cos (A \pm B) = \cos A \cos B \mp \sin A \sin B$

$\sin A \pm \sin B = 2 \sin \left[\frac{1}{2}(A \pm B)\right] \cos \left[\frac{1}{2}(A \mp B)\right]$

$\cos A + \cos B = 2 \cos \left[\frac{1}{2}(A + B)\right] \cos \left[\frac{1}{2}(A - B)\right]$

$\cos A - \cos B = 2 \sin \left[\frac{1}{2}(A + B)\right] \sin \left[\frac{1}{2}(B - A)\right]$

The following relationships are derived directly from the right triangle shown in Figure B.10:

$$\sin \theta = \cos (90° - \theta)$$
$$\cos \theta = \sin (90° - \theta)$$
$$\cot \theta = \tan (90° - \theta)$$

Some properties of trigonometric functions are the following:

$$\sin (-\theta) = -\sin \theta$$
$$\cos (-\theta) = \cos \theta$$
$$\tan (-\theta) = -\tan \theta$$

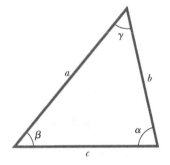

Figure B.11 An arbitrary, non-right triangle.

The following relationships apply to *any* triangle as shown in Figure B.11:

$$\alpha + \beta + \gamma = 180°$$

$$\text{Law of cosines} \begin{cases} a^2 = b^2 + c^2 - 2bc \cos \alpha \\ b^2 = a^2 + c^2 - 2ac \cos \beta \\ c^2 = a^2 + b^2 - 2ab \cos \gamma \end{cases}$$

$$\text{Law of sines} \qquad \frac{a}{\sin \alpha} = \frac{b}{\sin \beta} = \frac{c}{\sin \gamma}$$

Table B.3 lists a number of useful trigonometric identities.

Example B.3

Consider the right triangle in Figure B.12 in which $a = 2.00$, $b = 5.00$, and c is unknown. **(A)** Find c.

SOLUTION

Use the Pythagorean theorem: $c^2 = a^2 + b^2 = 2.00^2 + 5.00^2 = 4.00 + 25.0 = 29.0$

$c = \sqrt{29.0} = 5.39$

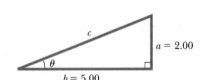

Figure B.12 (Example B.3)

(B) Find the angle θ.

Use the tangent function: $\tan \theta = \dfrac{a}{b} = \dfrac{2.00}{5.00} = 0.400$

B.3 continued

Use your calculator to find the angle: $\theta = \tan^{-1}(0.400) = 21.8°$

where $\tan^{-1}(0.400)$ is the notation for "angle whose tangent is 0.400," sometimes written as arctan (0.400).

Exercises

1. In Figure B.13, identify (a) the side opposite θ (b) the side adjacent to ϕ and then find (c) $\cos \theta$, (d) $\sin \phi$, and (e) $\tan \phi$.

Answers   (a) 3   (b) 3   (c) $\frac{4}{5}$ (d) $\frac{4}{5}$ (e) $\frac{4}{3}$

2. In a certain right triangle, the two sides that are perpendicular to each other are 5.00 m and 7.00 m long. What is the length of the third side?

Answer   8.60 m

3. A right triangle has a hypotenuse of length 3.0 m, and one of its angles is 30°. (a) What is the length of the side opposite the 30° angle? (b) What is the side adjacent to the 30° angle?

Answers   (a) 1.5 m   (b) 2.6 m

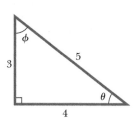

Figure B.13 (Exercise 1)

B.5   Series Expansions

$$(a + b)^n = a^n + \frac{n}{1!} a^{n-1} b + \frac{n(n-1)}{2!} a^{n-2} b^2 + \cdots$$

$$(1 + x)^n = 1 + nx + \frac{n(n-1)}{2!} x^2 + \cdots$$

$$e^x = 1 + x + \frac{x^2}{2!} + \frac{x^3}{3!} + \cdots$$

$$\ln (1 \pm x) = \pm x - \tfrac{1}{2} x^2 \pm \tfrac{1}{3} x^3 - \cdots$$

$$\left.\begin{array}{l} \sin x = x - \dfrac{x^3}{3!} + \dfrac{x^5}{5!} - \cdots \\[2mm] \cos x = 1 - \dfrac{x^2}{2!} + \dfrac{x^4}{4!} - \cdots \\[2mm] \tan x = x + \dfrac{x^3}{3} + \dfrac{2x^5}{15} + \cdots \quad |x| < \dfrac{\pi}{2} \end{array}\right\} \; x \text{ in radians}$$

The following approximations can be used:

For $x \ll 1$: $(1 + x)^n \approx 1 + nx$ For $x \le 0.1$ rad: $\sin x \approx x$

$\qquad\qquad\quad e^x \approx 1 + x$ $\qquad\qquad\qquad\qquad\quad \cos x \approx 1$

$\qquad\qquad\quad \ln (1 \pm x) \approx \pm x$ $\qquad\qquad\qquad\quad \tan x \approx x$

B.6   Differential Calculus

In various branches of science, it is sometimes necessary to use the basic tools of calculus, invented by Newton, to describe physical phenomena. The use of calculus is fundamental in the treatment of various problems in Newtonian mechanics, electricity, and magnetism. In this section, we simply state some basic properties and "rules of thumb" that should be a useful review to the student.

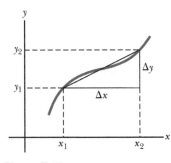

Figure B.14 The lengths Δx and Δy are used to define the derivative of this function at a point.

First, a **function** must be specified that relates one variable to another (e.g., position as a function of time). Suppose one of the variables is called y (the dependent variable), and the other x (the independent variable). We might have a function relationship such as

$$y(x) = ax^3 + bx^2 + cx + d$$

If a, b, c, and d are specified constants, y can be calculated for any value of x. We usually deal with continuous functions, that is, those for which y varies "smoothly" with x.

The **derivative** of y with respect to x is defined as the limit as Δx approaches zero of the slopes of chords drawn between two points on the y versus x curve. Mathematically, we write this definition as

$$\frac{dy}{dx} = \lim_{\Delta x \to 0} \frac{\Delta y}{\Delta x} = \lim_{\Delta x \to 0} \frac{y(x + \Delta x) - y(x)}{\Delta x} \tag{B.28}$$

where Δy and Δx are defined as $\Delta x = x_2 - x_1$ and $\Delta y = y_2 - y_1$ (Fig. B.14). Note that dy/dx does *not* mean dy divided by dx, but rather is simply a notation of the limiting process of the derivative as defined by Equation B.28.

A useful expression to remember when $y(x) = ax^n$, where a is a *constant* and n is *any* positive or negative number (integer or fraction), is

$$\frac{dy}{dx} = nax^{n-1} \tag{B.29}$$

If $y(x)$ is a polynomial or algebraic function of x, we apply Equation B.29 to *each* term in the polynomial and take $d[\text{constant}]/dx = 0$. In Examples B.4 through B.7, we evaluate the derivatives of several functions.

Special Properties of the Derivative

A. Derivative of the product of two functions If a function $f(x)$ is given by the product of two functions—say, $g(x)$ and $h(x)$—the derivative of $f(x)$ is defined as

$$\frac{d}{dx} f(x) = \frac{d}{dx} [g(x)h(x)] = g\frac{dh}{dx} + h\frac{dg}{dx} \tag{B.30}$$

B. Derivative of the sum of two functions If a function $f(x)$ is equal to the sum of two functions, the derivative of the sum is equal to the sum of the derivatives:

$$\frac{d}{dx} f(x) = \frac{d}{dx} [g(x) + h(x)] = \frac{dg}{dx} + \frac{dh}{dx} \tag{B.31}$$

C. Chain rule of differential calculus If $y = f(x)$ and $x = g(z)$, then dy/dz can be written as the product of two derivatives:

$$\frac{dy}{dz} = \frac{dy}{dx} \frac{dx}{dz} \tag{B.32}$$

D. The second derivative The second derivative of y with respect to x is defined as the derivative of the function dy/dx (the derivative of the derivative). It is usually written as

$$\frac{d^2y}{dx^2} = \frac{d}{dx}\left(\frac{dy}{dx}\right) \tag{B.33}$$

TABLE B.4 Derivative for Several Functions

$\dfrac{d}{dx}(a) = 0$

$\dfrac{d}{dx}(ax^n) = nax^{n-1}$

$\dfrac{d}{dx}(e^{ax}) = ae^{ax}$

$\dfrac{d}{dx}(\sin ax) = a\cos ax$

$\dfrac{d}{dx}(\cos ax) = -a\sin ax$

$\dfrac{d}{dx}(\tan ax) = a\sec^2 ax$

$\dfrac{d}{dx}(\cot ax) = -a\csc^2 ax$

$\dfrac{d}{dx}(\sec x) = \tan x \sec x$

$\dfrac{d}{dx}(\csc x) = -\cot x \csc x$

$\dfrac{d}{dx}(\ln ax) = \dfrac{1}{x}$

$\dfrac{d}{dx}(\sin^{-1} ax) = \dfrac{a}{\sqrt{1 - a^2x^2}}$

$\dfrac{d}{dx}(\cos^{-1} ax) = \dfrac{-a}{\sqrt{1 - a^2x^2}}$

$\dfrac{d}{dx}(\tan^{-1} ax) = \dfrac{a}{1 + a^2x^2}$

Note: The symbols a and n represent constants.

Some of the more commonly used derivatives of functions are listed in Table B.4.

Example **B.4**

Use Equation B.28 to find the derivative of the following function: $y(x) = ax^3 + bx + c$, where a, b, and c are constants.

SOLUTION

Evaluate the function at $x + \Delta x$:

$$y(x + \Delta x) = a(x + \Delta x)^3 + b(x + \Delta x) + c$$

$$= a(x^3 + 3x^2\,\Delta x + 3x\,\Delta x^2 + \Delta x^3) + b(x + \Delta x) + c$$

Evaluate the numerator of Equation B.28:

$$\Delta y = y(x + \Delta x) - y(x) = a(3x^2\,\Delta x + 3x\,\Delta x^2 + \Delta x^3) + b\,\Delta x$$

Substitute into Equation B.28 and take the limit:

$$\frac{dy}{dx} = \lim_{\Delta x \to 0} \frac{\Delta y}{\Delta x} = \lim_{\Delta x \to 0}\left[a(3x^2 + 3x\,\Delta x + \Delta x^2)\right] + b$$

$$\frac{dy}{dx} = 3ax^2 + b$$

Example **B.5**

Find the derivative of

$$y(x) = 8x^5 + 4x^3 + 2x + 7$$

SOLUTION

Apply Equation B.29 to each term separately and remember that the derivative of a constant is zero:

$$\frac{dy}{dx} = 8(5)x^4 + 4(3)x^2 + 2(1)x^0 + 0$$

$$\frac{dy}{dx} = 40x^4 + 12x^2 + 2$$

Example **B.6**

Find the derivative of $y(x) = x^3/(x + 1)^2$ with respect to x.

SOLUTION

Rewrite the function as a product:

$$y(x) = x^3(x + 1)^{-2}$$

Use Equation B.30 to find the derivative:

$$\frac{dy}{dx} = (x + 1)^{-2}\frac{d}{dx}(x^3) + x^3\frac{d}{dx}(x + 1)^{-2}$$

$$= (x + 1)^{-2}\,3x^2 + x^3\,(-2)(x + 1)^{-3}$$

$$\frac{dy}{dx} = \frac{3x^2}{(x + 1)^2} - \frac{2x^3}{(x + 1)^3} = \frac{x^2(x + 3)}{(x + 1)^3}$$

Example **B.7**

A useful formula that follows from Equation B.30 is the derivative of the quotient of two functions. Show that

$$\frac{d}{dx}\left[\frac{g(x)}{h(x)}\right] = \frac{h\dfrac{dg}{dx} - g\dfrac{dh}{dx}}{h^2}$$

continued

B.6 continued

SOLUTION

Write the quotient as gh^{-1} and use Equations B.29 and B.30:

$$\frac{d}{dx}\left(\frac{g}{h}\right) = \frac{d}{dx}(gh^{-1}) = g\frac{d}{dx}(h^{-1}) + h^{-1}\frac{d}{dx}(g)$$

$$= -gh^{-2}\frac{dh}{dx} + h^{-1}\frac{dg}{dx}$$

$$= \frac{h\frac{dg}{dx} - g\frac{dh}{dx}}{h^2}$$

B.7 Integral Calculus

We think of integration as the inverse of differentiation. As an example, consider the expression

$$f(x) = \frac{dy}{dx} = 3ax^2 + b \tag{B.34}$$

which was the result of differentiating the function

$$y(x) = ax^3 + bx + c$$

in Example B.4. We can write Equation B.34 as $dy = f(x)\,dx = (3ax^2 + b)\,dx$ and obtain $y(x)$ by "summing" over all values of x. Mathematically, we write this inverse operation as

$$y(x) = \int f(x)\,dx$$

For the function $f(x)$ given by Equation B.34, we have

$$y(x) = \int (3ax^2 + b)\,dx = ax^3 + bx + c$$

where c is a constant of the integration. This type of integral is called an *indefinite integral* because its value depends on the choice of c.

A general **indefinite integral** $I(x)$ is defined as

$$I(x) = \int f(x)\,dx \tag{B.35}$$

where $f(x)$ is called the *integrand* and $f(x) = dI(x)/dx$.

For a *general continuous* function $f(x)$, the integral can be interpreted geometrically as the area under the curve bounded by $f(x)$ and the x axis, between two specified values of x, say, x_1 and x_2, as in Figure B.15.

The area of the blue element in Figure B.15 is approximately $f(x_i)\,\Delta x_i$. If we sum all these area elements between x_1 and x_2 and take the limit of this sum as $\Delta x_i \to 0$, we obtain the *true* area under the curve bounded by $f(x)$ and the x axis, between the limits x_1 and x_2:

$$\text{Area} = \lim_{\Delta x_i \to 0}\sum_i f(x_i)\Delta x_i = \int_{x_1}^{x_2} f(x)\,dx \tag{B.36}$$

Integrals of the type defined by Equation B.36 are called **definite integrals.**

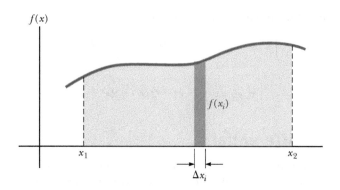

Figure B.15 The definite integral of a function is the area under the curve of the function between the limits x_1 and x_2.

One common integral that arises in practical situations has the form

$$\int x^n \, dx = \frac{x^{n+1}}{n+1} + c \quad (n \neq -1) \tag{B.37}$$

This result is obvious, being that differentiation of the right-hand side with respect to x gives $f(x) = x^n$ directly. If the limits of the integration are known, this integral becomes a *definite integral* and is written

$$\int_{x_1}^{x_2} x^n \, dx = \frac{x^{n+1}}{n+1} \Big|_{x_1}^{x_2} = \frac{x_2^{\,n+1} - x_1^{\,n+1}}{n+1} \quad (n \neq -1) \tag{B.38}$$

Exercises

In the following exercises, evaluate the integral:

	Answer			**Answer**
1. $\displaystyle\int_0^a x^2 \, dx$	$\dfrac{a^3}{3}$		3. $\displaystyle\int_3^5 x \, dx$	8
2. $\displaystyle\int_0^b x^{3/2} \, dx$	$\dfrac{2}{5} b^{5/2}$			

Partial Integration

Sometimes it is useful to apply the method of *partial integration* (also called "integrating by parts") to evaluate certain integrals. This method uses the property

$$\int u \, dv = uv - \int v \, du \tag{B.39}$$

where u and v are *carefully* chosen so as to reduce a complex integral to a simpler one. In many cases, several reductions have to be made. Consider the function

$$I(x) = \int x^2 \, e^x \, dx$$

which can be evaluated by integrating by parts twice. First, if we choose $u = x^2$, $v = e^x$, we obtain

$$\int x^2 \, e^x \, dx = \int x^2 \, d(e^x) = x^2 \, e^x - 2 \int e^x \, x \, dx + c_1$$

Now, in the second term, choose $u = x$, $v = e^x$, which gives

$$\int x^2 e^x \, dx = x^2 e^x - 2x \, e^x + 2\int e^x \, dx + c_1$$

or

$$\int x^2 e^x \, dx = x^2 e^x - 2xe^x + 2e^x + c_2$$

The Perfect Differential

Another useful method to remember is that of the *perfect differential*, in which we look for a change of variable such that the differential of the function is the differential of the independent variable appearing in the integrand. For example, consider the integral

$$I(x) = \int \cos^2 x \sin x \, dx$$

This integral becomes easy to evaluate if we rewrite the differential as $d(\cos x) = -\sin x \, dx$. The integral then becomes

$$\int \cos^2 x \sin x \, dx = -\int \cos^2 x \, d(\cos x)$$

If we now change variables, letting $y = \cos x$, we obtain

$$\int \cos^2 x \sin x \, dx = -\int y^2 \, dy = -\frac{y^3}{3} + c = -\frac{\cos^3 x}{3} + c$$

Table B.5 lists some useful indefinite integrals. Table B.6 gives Gauss's probability integral and other definite integrals. A more complete list can be found in various handbooks, such as *The Handbook of Chemistry and Physics* (Boca Raton, FL: CRC Press, published annually).

TABLE B.5   Some Indefinite Integrals (An arbitrary constant should be added to each of these integrals.)

$\displaystyle\int x^n \, dx = \frac{x^{n+1}}{n+1}$ (provided $n \neq 1$)	$\displaystyle\int \ln ax \, dx = (x \ln ax) - x$
$\displaystyle\int \frac{dx}{x} = \int x^{-1} \, dx = \ln x$	$\displaystyle\int xe^{ax} \, dx = \frac{e^{ax}}{a^2}(ax - 1)$
$\displaystyle\int \frac{dx}{a + bx} = \frac{1}{b} \ln (a + bx)$	$\displaystyle\int \frac{dx}{a + be^{cx}} = \frac{x}{a} - \frac{1}{ac} \ln (a + be^{cx})$
$\displaystyle\int \frac{x \, dx}{a + bx} = \frac{x}{b} - \frac{a}{b^2} \ln (a + bx)$	$\displaystyle\int \sin ax \, dx = -\frac{1}{a} \cos ax$
$\displaystyle\int \frac{dx}{x(x + a)} = -\frac{1}{a} \ln \frac{x + a}{x}$	$\displaystyle\int \cos ax \, dx = \frac{1}{a} \sin ax$
$\displaystyle\int \frac{dx}{(a + bx)^2} = -\frac{1}{b(a + bx)}$	$\displaystyle\int \tan ax \, dx = -\frac{1}{a} \ln (\cos ax) = \frac{1}{a} \ln (\sec ax)$
$\displaystyle\int \frac{dx}{a^2 + x^2} = \frac{1}{a} \tan^{-1} \frac{x}{a}$	$\displaystyle\int \cot ax \, dx = \frac{1}{a} \ln (\sin ax)$
$\displaystyle\int \frac{dx}{a^2 - x^2} = \frac{1}{2a} \ln \frac{a + x}{a - x} \, (a^2 - x^2 > 0)$	$\displaystyle\int \sec ax \, dx = \frac{1}{a} \ln (\sec ax + \tan ax) = \frac{1}{a} \ln \left[\tan \left(\frac{ax}{2} + \frac{\pi}{4} \right) \right]$
$\displaystyle\int \frac{dx}{x^2 - a^2} = \frac{1}{2a} \ln \frac{x - a}{x + a} \, (x^2 - a^2 > 0)$	$\displaystyle\int \csc ax \, dx = \frac{1}{a} \ln (\csc ax - \cot ax) = \frac{1}{a} \ln \left(\tan \frac{ax}{2} \right)$

continued

TABLE B.5 Some Indefinite Integrals *(continued)*

$$\int \frac{x \, dx}{a^2 \pm x^2} = \pm\tfrac{1}{2} \ln (a^2 \pm x^2)$$

$$\int \sin^2 ax \, dx = \frac{x}{2} - \frac{\sin 2ax}{4a}$$

$$\int \frac{dx}{\sqrt{a^2 - x^2}} = \sin^{-1} \frac{x}{a} = -\cos^{-1} \frac{x}{a} \; (a^2 - x^2 > 0)$$

$$\int \cos^2 ax \, dx = \frac{x}{2} + \frac{\sin 2ax}{4a}$$

$$\int \frac{dx}{\sqrt{x^2 \pm a^2}} = \ln (x + \sqrt{x^2 \pm a^2})$$

$$\int \frac{dx}{\sin^2 ax} = -\frac{1}{a} \cot ax$$

$$\int \frac{x \, dx}{\sqrt{a^2 - x^2}} = -\sqrt{a^2 - x^2}$$

$$\int \frac{dx}{\cos^2 ax} = \frac{1}{a} \tan ax$$

$$\int \frac{x \, dx}{\sqrt{x^2 \pm a^2}} = \sqrt{x^2 \pm a^2}$$

$$\int \tan^2 ax \, dx = \frac{1}{a} (\tan ax) - x$$

$$\int \sqrt{a^2 - x^2} \, dx = \tfrac{1}{2} \left(x\sqrt{a^2 - x^2} + a^2 \sin^{-1} \frac{x}{|a|} \right)$$

$$\int \cot^2 ax \, dx = -\frac{1}{a} (\cot ax) - x$$

$$\int x\sqrt{a^2 - x^2} \, dx = -\tfrac{1}{3} (a^2 - x^2)^{3/2}$$

$$\int \sin^{-1} ax \, dx = x(\sin^{-1} ax) + \frac{\sqrt{1 - a^2 x^2}}{a}$$

$$\int \sqrt{x^2 \pm a^2} \, dx = \tfrac{1}{2} x\sqrt{x^2 \pm a^2} \pm a^2 \ln (x + \sqrt{x^2 \pm a^2})$$

$$\int \cos^{-1} ax \, dx = x(\cos^{-1} ax) - \frac{\sqrt{1 - a^2 x^2}}{a}$$

$$\int x(\sqrt{x^2 \pm a^2}) \, dx = \tfrac{1}{3} (x^2 \pm a^2)^{3/2}$$

$$\int \frac{dx}{(x^2 + a^2)^{3/2}} = \frac{x}{a^2 \sqrt{x^2 + a^2}}$$

$$\int e^{ax} \, dx = \frac{1}{a} e^{ax}$$

$$\int \frac{x \, dx}{(x^2 + a^2)^{3/2}} = -\frac{1}{\sqrt{x^2 + a^2}}$$

TABLE B.6 Gauss's Probability Integral and Other Definite Integrals

$$\int_0^\infty x^n e^{-ax} \, dx = \frac{n!}{a^{n+1}}$$

$$I_0 = \int_0^\infty e^{-ax^2} \, dx = \frac{1}{2} \sqrt{\frac{\pi}{a}} \quad \text{(Gauss's probability integral)}$$

$$I_1 = \int_0^\infty x e^{-ax^2} \, dx = \frac{1}{2a}$$

$$I_2 = \int_0^\infty x^2 e^{-ax^2} \, dx = -\frac{dI_0}{da} = \frac{1}{4} \sqrt{\frac{\pi}{a^3}}$$

$$I_3 = \int_0^\infty x^3 e^{-ax^2} \, dx = -\frac{dI_1}{da} = \frac{1}{2a^2}$$

$$I_4 = \int_0^\infty x^4 e^{-ax^2} \, dx = \frac{d^2 I_0}{da^2} = \frac{3}{8} \sqrt{\frac{\pi}{a^5}}$$

$$I_5 = \int_0^\infty x^5 e^{-ax^2} \, dx = \frac{d^2 I_1}{da^2} = \frac{1}{a^3}$$

$$\vdots$$

$$I_{2n} = (-1)^n \frac{d^n}{da^n} I_0$$

$$I_{2n+1} = (-1)^n \frac{d^n}{da^n} I_1$$

B.8   Propagation of Uncertainty

In laboratory experiments, a common activity is to take measurements that act as raw data. These measurements are of several types—length, time interval, temperature, voltage, and so on—and are taken by a variety of instruments. Regardless of the measurement and the quality of the instrumentation, **there is always uncertainty associated with a physical measurement.** This uncertainty is a combination of that associated with the instrument and that related to the system being measured. An example of the former is the inability to exactly determine the position of a length measurement between the lines on a meterstick. An example of uncertainty related to the system being measured is the variation of temperature within a sample of water so that a single temperature for the sample is difficult to determine.

Uncertainties can be expressed in two ways. **Absolute uncertainty** refers to an uncertainty expressed in the same units as the measurement. Therefore, the length of a computer disk label might be expressed as (5.5 ± 0.1) cm. The uncertainty of ± 0.1 cm by itself is not descriptive enough for some purposes, however. This uncertainty is large if the measurement is 1.0 cm, but it is small if the measurement is 100 m. To give a more descriptive account of the uncertainty, **fractional uncertainty** or **percent uncertainty** is used. In this type of description, the uncertainty is divided by the actual measurement. Therefore, the length of the computer disk label could be expressed as

$$\ell = 5.5 \text{ cm} \pm \frac{0.1 \text{ cm}}{5.5 \text{ cm}} = 5.5 \text{ cm} \pm 0.018 \quad \text{(fractional uncertainty)}$$

or as

$$\ell = 5.5 \text{ cm} \pm 1.8\% \quad \text{(percent uncertainty)}$$

When combining measurements in a calculation, the percent uncertainty in the final result is generally larger than the uncertainty in the individual measurements. This is called **propagation of uncertainty** and is one of the challenges of experimental physics.

Some simple rules can provide a reasonable estimate of the uncertainty in a calculated result:

Multiplication and division: When measurements with uncertainties are multiplied or divided, add the *percent uncertainties* to obtain the percent uncertainty in the result.

Example B.8

Find the area, with associated uncertainty, of a rectangular plate of dimensions 5.5 cm $\pm$ 1.8% by 6.4 cm $\pm$ 1.6%.

SOLUTION

Because the result is a multiplication, add the percent uncertainties:

$$A = \ell w = (5.5 \text{ cm} \pm 1.8\%)(6.4 \text{ cm} \pm 1.6\%)$$
$$= 35 \text{ cm}^2 \pm 3.4\% = (35 \pm 1) \text{ cm}^2$$

Addition and subtraction: When measurements with uncertainties are added or subtracted, add the *absolute uncertainties* to obtain the absolut uncertainty in the result.

Example B.9

Find the change in temperature, with associated uncertainty, when the temperature increases from $(27.6 \pm 1.5)°C$ to $(99.2 \pm 1.5)°C$

SOLUTION

Because the result is a subtraction, add the absolute uncertainties:

$$\Delta T = T_2 - T_1 = (99.2 \pm 1.5)°C - (27.6 \pm 1.5)°C$$
$$= (71.6 \pm 3.0)°C = 1.6°C \pm 4.2\%$$

Powers: If a measurement is taken to a power, the percent uncertainty is multiplied by that power to obtain the percent uncertainty in the result.

Example B.10

Find the volume of a sphere of radius 6.20 cm $\pm 2.0\%$.

SOLUTION

Because the result is determined by raising a quantity to a power, multiply the power by the percent uncertainty:

$$V = \tfrac{4}{3}\pi r^3 = \tfrac{4}{3}\pi(6.20 \text{ cm} \pm 2.0\%)^3$$
$$= 998 \text{ cm}^3 \pm 6.0\% = (998 \pm 60) \text{ cm}^3$$

For complicated calculations, many uncertainties are added together, which can cause the uncertainty in the final result to be undesirably large. Experiments should be designed such that calculations are as simple as possible.

Notice that uncertainties in a calculation always add. As a result, an experiment involving a subtraction should be avoided if possible, especially if the measurements being subtracted are close together. The result of such a calculation is a small difference in the measurements and uncertainties that add together. It is possible that the uncertainty in the result could be larger than the result itself!

Appendix C Periodic Table of the Elements

| Group I | Group II | | | | | Transition elements | | | |

Legend box:
Symbol — **Ca** 20 — Atomic number
Atomic mass† — 40.078
$4s^2$ — Electron configuration

Group I	Group II
H 1 1.007 9 $1s$	
Li 3 6.941 $2s^1$	**Be** 4 9.0122 $2s^2$
Na 11 22.990 $3s^1$	**Mg** 12 24.305 $3s^2$

K 19 39.098 $4s^1$	**Ca** 20 40.078 $4s^2$	**Sc** 21 44.956 $3d^14s^2$	**Ti** 22 47.867 $3d^24s^2$	**V** 23 50.942 $3d^34s^2$	**Cr** 24 51.996 $3d^54s^1$	**Mn** 25 54.938 $3d^54s^2$	**Fe** 26 55.845 $3d^64s^2$	**Co** 27 58.933 $3d^74s^2$
Rb 37 85.468 $5s^1$	**Sr** 38 87.62 $5s^2$	**Y** 39 88.906 $4d^15s^2$	**Zr** 40 91.224 $4d^25s^2$	**Nb** 41 92.906 $4d^45s^1$	**Mo** 42 95.96 $4d^55s^1$	**Tc** 43 (98) $4d^55s^2$	**Ru** 44 101.07 $4d^75s^1$	**Rh** 45 102.91 $4d^85s^1$
Cs 55 132.91 $6s^1$	**Ba** 56 137.33 $6s^2$	57–71*	**Hf** 72 178.49 $5d^26s^2$	**Ta** 73 180.95 $5d^36s^2$	**W** 74 183.84 $5d^46s^2$	**Re** 75 186.21 $5d^56s^2$	**Os** 76 190.23 $5d^66s^2$	**Ir** 77 192.2 $5d^76s^2$
Fr 87 (223) $7s^1$	**Ra** 88 (226) $7s^2$	89–103**	**Rf** 104 (267) $6d^27s^2$	**Db** 105 (268) $6d^37s^2$	**Sg** 106 (269) $6d^47s^2$	**Bh** 107 (270) $6d^57s^2$	**Hs** 108 (277) $6d^67s^2$	**Mt**†† 109 (278) $6d^77s^2$

*Lanthanide series

La 57 138.91 $5d^16s^2$	**Ce** 58 140.12 $5d^14f^16s^2$	**Pr** 59 140.91 $4f^36s^2$	**Nd** 60 144.24 $4f^46s^2$	**Pm** 61 (145) $4f^56s^2$	**Sm** 62 150.36 $4f^66s^2$

**Actinide series

Ac 89 (227) $6d^17s^2$	**Th** 90 232.04 $6d^27s^2$	**Pa** 91 231.04 $5f^26d^17s^2$	**U** 92 238.03 $5f^36d^17s^2$	**Np** 93 (237) $5f^46d^17s^2$	**Pu** 94 (244) $5f^67s^2$

Note: Atomic mass values given are averaged over isotopes in the percentages in which they exist in nature.
† For an unstable element, mass number of the most stable known isotope is given in parentheses.
†† For elements 109 and higher, electron configurations are theoretically predicted.

		Group III	Group IV	Group V	Group VI	Group VII	Group 0
						H 1 1.007 9 $1s^1$	**He** 2 4.002 6 $1s^2$
		B 5 10.811 $2p^1$	**C** 6 12.011 $2p^2$	**N** 7 14.007 $2p^3$	**O** 8 15.999 $2p^4$	**F** 9 18.998 $2p^5$	**Ne** 10 20.180 $2p^6$
		Al 13 26.982 $3p^1$	**Si** 14 28.086 $3p^2$	**P** 15 30.974 $3p^3$	**S** 16 32.066 $3p^4$	**Cl** 17 35.453 $3p^5$	**Ar** 18 39.948 $3p^6$
Ni 28 58.693 $3d^84s^2$	**Cu** 29 63.546 $3d^{10}4s^1$	**Zn** 30 65.39 $3d^{10}4s^2$	**Ga** 31 69.723 $4p^1$	**Ge** 32 72.64 $4p^2$	**As** 33 74.922 $4p^3$	**Se** 34 78.96 $4p^4$	**Br** 35 79.904 $4p^5$	**Kr** 36 83.80 $4p^6$
Pd 46 106.42 $4d^{10}$	**Ag** 47 107.87 $4d^{10}5s^1$	**Cd** 48 112.41 $4d^{10}5s^2$	**In** 49 114.82 $5p^1$	**Sn** 50 118.71 $5p^2$	**Sb** 51 121.76 $5p^3$	**Te** 52 127.60 $5p^4$	**I** 53 126.90 $5p^5$	**Xe** 54 131.29 $5p^6$
Pt 78 195.08 $5d^96s^1$	**Au** 79 196.97 $5d^{10}6s^1$	**Hg** 80 200.59 $5d^{10}6s^2$	**Tl** 81 204.38 $6p^1$	**Pb** 82 207.2 $6p^2$	**Bi** 83 208.98 $6p^3$	**Po** 84 (209) $6p^4$	**At** 85 (210) $6p^5$	**Rn** 86 (222) $6p^6$
Ds 110 (281) $6d^87s^2$	**Rg** 111 (282) $6d^97s^2$	**Cn** 112 (285) $6d^{10}7s^2$	**Nh** 113 (286) $7p^1$	**Fl** 114 (289) $7p^2$	**Mc** 115 (289) $7p^3$	**Lv** 116 (293) $7p^4$	**Ts** 117 (294) $7p^5$	**Og** 118 (294) $7p^6$

Eu 63 151.96 $4f^76s^2$	**Gd** 64 157.25 $4f^75d^16s^2$	**Tb** 65 158.93 $4f^85d^16s^2$	**Dy** 66 162.50 $4f^{10}6s^2$	**Ho** 67 164.93 $4f^{11}6s^2$	**Er** 68 167.26 $4f^{12}6s^2$	**Tm** 69 168.93 $4f^{13}6s^2$	**Yb** 70 173.04 $4f^{14}6s^2$	**Lu** 71 174.97 $4f^{14}5d^16s^2$
Am 95 (243) $5f^77s^2$	**Cm** 96 (247) $5f^76d^17s^2$	**Bk** 97 (247) $5f^86d^17s^2$	**Cf** 98 (251) $5f^{10}7s^2$	**Es** 99 (252) $5f^{11}7s^2$	**Fm** 100 (257) $5f^{12}7s^2$	**Md** 101 (258) $5f^{13}7s^2$	**No** 102 (259) $5f^{14}7s^2$	**Lr** 103 (262) $5f^{14}6d^17s^2$

Note: For a description of the atomic data, visit *physics.nist.gov/PhysRefData/Elements/per_text.html.*

Appendix D SI Units

TABLE D.1 SI Units

| Base Quantity | SI Base Unit | |
	Name	Symbol
Length	meter	m
Mass	kilogram	kg
Time	second	s
Electric current	ampere	A
Temperature	kelvin	K
Amount of substance	mole	mol
Luminous intensity	candela	cd

TABLE D.2 Some Derived SI Units

Other Quantity	Name	Symbol	Expression in Terms of Base Units	Expression in Terms of SI Units
Plane angle	radian	rad	m/m	
Frequency	hertz	Hz	s^{-1}	
Force	newton	N	$kg \cdot m/s^2$	J/m
Pressure	pascal	Pa	$kg/m \cdot s^2$	N/m^2
Energy	joule	J	$kg \cdot m^2/s^2$	$N \cdot m$
Power	watt	W	$kg \cdot m^2/s^3$	J/s
Electric charge	coulomb	C	$A \cdot s$	
Electric potential	volt	V	$kg \cdot m^2/A \cdot s^3$	W/A
Capacitance	farad	F	$A^2 \cdot s^4/kg \cdot m^2$	C/V
Electric resistance	ohm	Ω	$kg \cdot m^2/A^2 \cdot s^3$	V/A
Magnetic flux	weber	Wb	$kg \cdot m^2/A \cdot s^2$	$V \cdot s$
Magnetic field	tesla	T	$kg/A \cdot s^2$	
Inductance	henry	H	$kg \cdot m^2/A^2 \cdot s^2$	$T \cdot m^2/A$

Answers to Quick Quizzes and Odd-Numbered Problems

Chapter 1

Answers to Quick Quizzes

1. (a)
2. False
3. (b)

Answers to Odd-Numbered Problems

1. (a) $5.52 \times 10^3 \text{ kg/m}^3$ (b) It is between the density of aluminum and that of iron and is greater than the densities of typical surface rocks.
3. 7.69 cm
5. The angle subtended by the Great Wall is less than the visual acuity of the eye.
7. 0.141 nm
9. (b) only
11. $11.4 \times 10^3 \text{ kg/m}^3$
13. 2.86 cm
15. 151 μm
17. (a) $\sim 10^2 \text{ kg}$ (b) $\sim 10^3 \text{ kg}$
19. The average distance between asteroids in the asteroid belt is about 400 000 km.
21. 31 556 926.0 s
23. 19
25. 63
27. ± 3.46
29. 316 m
31. 10^{11} stars
33. Answers may vary.   (a) $\sim 10^{29}$ prokaryotes   (b) $\sim 10^{14}$ kg
35. (a) 478 cm^3/s (b) 0.225 cm/s   (c) When the balloon radius is twice as large, its surface area is four times larger. The new volume added in one second in the inflation process is equal to this larger area times an extra radial thickness that is one-fourth as large as it was when the balloon was smaller.
37. $V = 0.579t + (1.19 \times 10^{-9})t^2$, where V is in cubic feet and t is in seconds
39. $\dfrac{d \tan\phi \tan\theta}{\tan\phi - \tan\theta}$

Chapter 2

Answers to Quick Quizzes

1. (b)
2. (c)
3. (b)
4. False. Your graph should look something like the one shown in the next column. This v_x–t graph shows that the maximum speed is about 5.0 m/s, which is 18 km/h (= 11 mi/h), so the driver was not speeding.

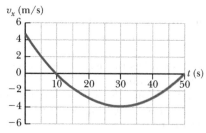

v_x (m/s)

5. (b)
6. (c)
7. (a)–(e), (b)–(d), (c)–(f)
8. (i) (e)   (ii) (d)

Answers to Odd-Numbered Problems

1. 0.02 s
3. (a) 2.30 m/s   (b) 16.1 m/s   (c) 11.5 m/s
5. (a) -2.4 m/s   (b) -3.8 m/s   (c) 4.0 s
7. (a) 2.80 h   (b) 218 km
9. (a) 1.3 m/s^2 (b) $t = 3$ s, $a = 2 \text{ m/s}^2$ (c) $t = 6$ s, $t > 10$ s
   (d) $a = -1.5 \text{ m/s}^2$, $t = 8$ s
11. (a) 20 m/s, 5 m/s   (b) 263 m
13.

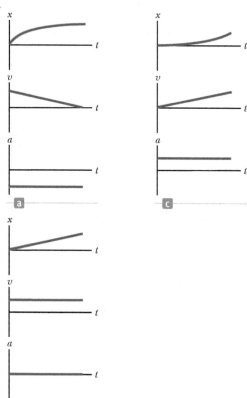

15. (a) 9.00 m/s   (b) -3.00 m/s   (c) 17.0 m/s   (d) The graph of velocity versus time is a straight line passing through 13 m/s at 10:05 a.m. and sloping downward, decreasing by 4 m/s for each second thereafter.   (e) If and only if we know the object's velocity at one instant of time, knowing its acceleration tells us its velocity at every other moment as long as the acceleration is constant.

17. -16.0 cm/s^2

19. (a) The idea is false unless the acceleration is zero. We define constant acceleration to mean that the velocity is changing steadily in time. So, the velocity cannot be changing steadily in space.   (b) This idea is true. Because the velocity is changing steadily in time, the velocity halfway through an interval is equal to the average of its initial and final values.

21. (a) 19.7 cm/s   (b) 4.70 cm/s^2 (c) The length of the glider is used to find the average velocity during a known time interval.

23. (a) 3.75 s   (b) 5.50 cm/s   (c) 0.604 s   (d) 13.3 cm, 47.9 cm   (e) The cars are initially moving toward each other, so they soon arrive at the same position x when their speeds are quite different, giving one answer to (c) that is not an answer to (a). The first car slows down in its motion to the left, turns around, and starts to move toward the right, slowly at first and gaining speed steadily. At a particular moment its speed will be equal to the constant rightward speed of the second car, but at this time the accelerating car is far behind the steadily moving car; thus, the answer to (a) is not an answer to (c). Eventually the accelerating car will catch up to the steadily coasting car, but passing it at higher speed, and giving another answer to (c) that is not an answer to (a).

25. David will be unsuccessful. The average human reaction time is about 0.2 s (research on the Internet) and a dollar bill is about 15.5 cm long, so David's fingers are about 8 cm from the end of the bill before it is dropped. The bill will fall about 20 cm before he can close his fingers.

27. 7.96 s

29. (a) 10.0 m/s up   (b) 4.68 m/s down

31. (a) The box could reach the window according to the data provided.   (b) Answers will vary.

33. (a) $a_x(t) = a_{xi} + Jt$; $v_x(t) = v_{xi} + a_{xi}t + \frac{1}{2}Jt^2$; $x(t) = x_i + v_{xi}t + \frac{1}{2}a_{xi}t^2 + \frac{1}{6}Jt^3$

35. (a) 4.00 m/s   (b) 1.00 ms   (c) 0.816 m

37. (a) Here, v_1 must be greater than v_2 and the distance between the leading athlete and the finish line must be great enough so that the trailing athlete has time to catch up.

   (b) $t = \dfrac{d_1}{v_1 - v_2}$ (c) $d_2 = \dfrac{v_2 d_1}{v_1 - v_2}$

39. (a) 5.46 s   (b) 73.0 m   (c) $v_{Stan} = 22.6$ m/s, $v_{Kathy} = 26.7$ m/s

41. 1.60 m/s^2

43. (a) 5.32 m/s^2 for Laura and 3.75 m/s^2 for Healan   (b) 10.6 m/s for Laura and 11.2 m/s for Healan   (c) Laura, by 2.63 m   (d) 4.47 m at $t = 2.84$ s

Chapter 3

Answers to Quick Quizzes

1. vectors: (b), (c); scalars: (a), (d), (e)
2. (c)
3. (b) and (c)
4. (b)
5. (c)

Answers to Odd-Numbered Problems

1. (a) 8.60 m   (b) 4.47 m, $-63.4°$; 4.24 m, 135°

3. (a) $(-3.56$ cm, -2.40 cm)   (b) $(r = 4.30$ cm, $\theta = 326°)$ (c) $(r = 8.60$ cm, $\theta = 34.0°)$ (d) $(r = 12.9$ cm, $\theta = 146°)$

5. This situation can *never* be true because the distance is the length of an arc of a circle between two points, whereas the magnitude of the displacement vector is a straight-line chord of the circle between the same points.

7. 9.5 N, 57° above the x axis

9. (a) 5.2 m at 60°   (b) 3.0 m at 330°   (c) 3.0 m at 150°   (d) 5.2 m at 300°

11. (a) yes   (b) The speed of the camper should be 28.3 m/s or more to satisfy this requirement.

13. 9.48 m at 166°

15. (a) 185 N at 77.8° from the positive x axis   (b) $(-39.3\hat{\mathbf{i}} - 181\hat{\mathbf{j}})$ N

17. (a) 2.83 m at $\theta = 315°$ (b) 13.4 m at $\theta = 117°$

19. (a) $8.00\hat{\mathbf{i}} + 12.0\hat{\mathbf{j}} - 4.00\hat{\mathbf{k}}$ (b) $2.00\hat{\mathbf{i}} + 3.00\hat{\mathbf{j}} - 1.00\hat{\mathbf{k}}$ (c) $-24.0\hat{\mathbf{i}} - 36.0\hat{\mathbf{j}} + 12.0\hat{\mathbf{k}}$

21. (a) $-3.00\hat{\mathbf{i}} + 2.00\hat{\mathbf{j}}$ (b) 3.61 at 146°   (c) $3.00\hat{\mathbf{i}} - 6.00\hat{\mathbf{j}}$

23. (a) $a = 5.00$ and $b = 7.00$ (b) For vectors to be equal, all their components must be equal. A vector equation contains more information than a scalar equation.

25. $(2.60\hat{\mathbf{i}} + 4.50\hat{\mathbf{j}})$m

27. 196 cm at 345°

29. (a) $(-20.5\hat{\mathbf{i}} + 35.5\hat{\mathbf{j}})$ m/s   (b) $25.0\hat{\mathbf{j}}$ m/s   (c) $(-61.5\hat{\mathbf{i}} + 107\hat{\mathbf{j}})$ m   (d) $37.5\hat{\mathbf{j}}$ m   (e) 157 km

31. 1.43×10^4 m at 32.2° above the horizontal

33. (a) $(5 + 11f)\hat{\mathbf{i}} + (3 + 9f)\hat{\mathbf{j}}$ meters   (b) $(5 + 0)\hat{\mathbf{i}} + (3 + 0)\hat{\mathbf{j}}$ meters   (c) This is reasonable because it is the location of the starting point, $5\hat{\mathbf{i}} + 3\hat{\mathbf{j}}$ meters.   (d) $16\hat{\mathbf{i}} + 12\hat{\mathbf{j}}$ meters   (e) This is reasonable because we have completed the trip, and this is the position vector of the endpoint.

35. 240 m at 237°

37. 1.15°

39. (a) 25.4 s   (b) 15.0 km/h

41. (a) The x, y, and z components are, respectively, 2.00, 1.00, and 3.00.   (b) 3.74   (c) $\theta_x = 57.7°$, $\theta_y = 74.5°$, $\theta_z = 36.7°$

43. (a) $-2.00\hat{\mathbf{k}}$ m/s   (b) its velocity vector

45. (a) $\vec{\mathbf{R}}_1 = a\hat{\mathbf{i}} + b\hat{\mathbf{j}}$ (b) $R_1 = (a^2 + b^2)^{1/2}$ (c) $\vec{\mathbf{R}}_2 = a\hat{\mathbf{i}} + b\hat{\mathbf{j}} + c\hat{\mathbf{k}}$

Chapter 4

Answers to Quick Quizzes

1. (a)
2. (i) (b)   (ii) (a)
3. 15°, 30°, 45°, 60°, 75°

4. (i) (d) (ii) (b)
5. (i) (b) (ii) (d)

Answers to Odd-Numbered Problems

1. (a) $(1.00\hat{\mathbf{i}} + 0.750\hat{\mathbf{j}})$ m/s (b) $(1.00\hat{\mathbf{i}} + 0.500\hat{\mathbf{j}})$ m/s, 1.12 m/s
3. (a) $\vec{\mathbf{v}} = -12.0t\,\hat{\mathbf{j}}$, where $\vec{\mathbf{v}}$ is in meters per second and t is in seconds (b) $\vec{\mathbf{a}} = -12.0\hat{\mathbf{j}}$ m/s^2 (c) $\vec{\mathbf{r}} = (3.00\hat{\mathbf{i}} - 6.00\hat{\mathbf{j}})$ m; $\vec{\mathbf{v}} = -12.0\hat{\mathbf{j}}$ m/s
5. (a) $\vec{\mathbf{v}}_f = (3.45 - 1.79t)\hat{\mathbf{i}} + (2.89 - 0.650t)\hat{\mathbf{j}}$
   (b) $\vec{\mathbf{r}}_f = (-25.3 + 3.45t - 0.893t^2)\hat{\mathbf{i}} + (28.9 + 2.89t - 0.325t^2)\hat{\mathbf{j}}$
7. 12.0 m/s
9. 67.8°
11. $d\tan\theta_i - \dfrac{gd^2}{2v_i^2\cos^2\theta_i}$
13. (a) (0, 50.0 m) (b) $v_{xi} = 18.0$ m/s; $v_{yi} = 0$ (c) Particle under constant acceleration (d) Particle under constant velocity (e) $v_{xf} = v_{xi}$; $v_{yf} = -gt$ (f) $x_f = v_{xi}t$; $y_f = y_i - \frac{1}{2}gt^2$ (g) 3.19 s (h) 36.1 m/s, −60.1°
15. (a) 41.7 m/s (b) 3.81 s (c) $v_x = 34.1$ m/s, $v_y = -13.4$ m/s, $v = 36.7$ m/s
17. 1.92 s
19. 7.58×10^3 m/s, 5.80×10^3 s
21. 377 m/s^2
23. (a) Yes. The particle can be either speeding up or slowing down, with a tangential component of acceleration of magnitude $\sqrt{6^2 - 4.5^2} = 3.97$ m/s^2. (b) No. The magnitude of the acceleration cannot be less than $v^2/r = 4.5$ m/s^2.
25. (a) 9.80 m/s^2 down and 2.50 m/s^2 south (b) 9.80 m/s^2 down (c) The bolt moves on a parabola with its axis downward and tilting to the south. It lands south of the point directly below its starting point. (d) The bolt moves on a parabola with a vertical axis.
27. 18.2°
29. 15.3 m
31. (a) $\dfrac{2d/c}{1 - v^2/c^2}$ (b) $\dfrac{2d}{c}$
   (c) The trip in flowing water takes a longer time interval. The swimmer travels at the low upstream speed for a longer time interval, so his average speed is reduced below c. Mathematically, $1/(1 - v^2/c^2)$ is always greater than 1. In the extreme, as $v \to c$, the time interval becomes infinite. In that case, the student can never return to the starting point because he cannot swim fast enough to overcome the river current.
33. (a) straight up, at 0° to the vertical (b) 8.25 m/s (c) a straight up and down line (d) a symmetric parabola opening downward (e) 12.6 m/s north at $\tan^{-1}(8.25/9.5) = 41.0°$ above the horizontal
35. The relationship between the height h and the walking speed is $h = (4.16 \times 10^{-3})v_x^2$, where h is in meters and v_x is in meters per second. At a typical walking speed of 4 to 5 km/h, the ball would have to be dropped from a height of about 1 cm, clearly much too low for a person's hand. Even at Olympic-record speed for the 100-m run (confirm on the Internet), this situation would only occur if

the ball is dropped from about 0.4 m, which is also below the hand of a normally proportioned person.
37. (a) 26.9 m/s (b) 67.3 m (c) $(2.00\hat{\mathbf{i}} - 5.00\hat{\mathbf{j}})$ m/s^2
39. The initial height of the ball when struck is 3.94 m, which is too high for the batter to hit the ball.
41. (a) 1.69 km/s (b) 1.80 h
43. (a) $x = v_i(0.164\,3 + 0.002\,299v_i^2)^{1/2} + 0.047\,94v_i^2$, where x is in meters and v_i is in meters per second (b) 0.041 0 m (c) 961 m (d) $x \approx 0.405v_i$ (e) $x \approx 0.095\,9v_i^2$ (f) The graph of x versus v_i starts from the origin as a straight line with slope 0.405 s. Then it curves upward above this tangent line, becoming closer and closer to the parabola $x = 0.095\,9v_i^2$, where x is in meters and v_i is in meters per second.
45. (a) 4.00 km/h (b) 4.00 km/h
47. $\sim 10^2$ m/s^2
49. (a) 43.2 m (b) $(9.66\hat{\mathbf{i}} - 25.6\,\hat{\mathbf{j}})$ m/s (c) Air resistance would ordinarily make the jump distance smaller and the final horizontal and vertical velocity components both somewhat smaller. If a skilled jumper shapes her body into an airfoil, however, she can deflect downward the air through which she passes so that it deflects her upward, giving her more time in the air and a longer jump.
51. (a) $\Delta t_1 = \dfrac{L}{c + v} + \dfrac{L}{c - v} = \dfrac{2L/c}{1 - v^2/c^2}$
   (b) $\Delta t_2 = \dfrac{2L}{\sqrt{c^2 - v^2}} = \dfrac{2L/c}{\sqrt{1 - v^2/c^2}}$
   (c) Sarah, who swims cross-stream, returns first.
53. $\tan^{-1}\left(\dfrac{\sqrt{2gh}}{v}\right)$

Chapter 5

Answers to Quick Quizzes

1. (d)
2. (a)
3. (d)
4. (b)
5. (i) (c) (ii) (a)
6. (b)
7. (b) Pulling up on the rope decreases the normal force, which, in turn, decreases the force of kinetic friction.

Answers to Odd-Numbered Problems

1. 8.71 N
3. (a) $(6.00\hat{\mathbf{i}} + 15.0\hat{\mathbf{j}})$ N (b) 16.2 N
5. (a) $(-45.0\hat{\mathbf{i}} + 15.0\,\hat{\mathbf{j}})$ m/s (b) 162° from the $+x$ axis (c) $(-225\hat{\mathbf{i}} + 75.0\,\hat{\mathbf{j}})$ m (d) $(-227\hat{\mathbf{i}} + 79.0\,\hat{\mathbf{j}})$ m
7. (a) $\hat{\mathbf{a}}$ is at 181° (b) 11.2 kg (c) 37.5 m/s (d) $(-37.5\hat{\mathbf{i}} - 0.893\hat{\mathbf{j}})$ m/s
9. (a) 1.53 m (b) 24.0 N forward and upward at 5.29° with the horizontal
11. (a) 3.64×10^{-18} N (b) 8.93×10^{-30} N is 408 billion times smaller
13. (a) $\sim 10^{-22}$ m/s^2 (b) $d \sim 10^{-23}$ m
15. (a) 3.43 kN (b) 0.967 m/s horizontally forward

17. (a)

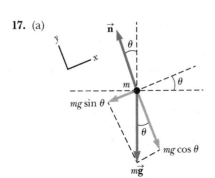

(b) -2.54 m/s^2 (c) 3.19 m/s

19. (a)

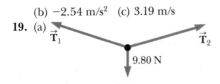

9.80 N

(b) 613 N

21. (a) $a = g \tan \theta$ (b) 4.16 m/s^2

23. (a) $F_x > 19.6$ N   (b) $F_x \leq -78.4$ N

(c)

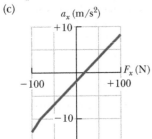

25. (a) $a_2 = 2a_1$ (b) $T_2 = \dfrac{m_1 m_2}{2m_2 + \frac{1}{2}m_1} g$ and $T_2 = \dfrac{m_1 m_2}{m_2 + \frac{1}{4}m_1} g$

(c) $\dfrac{m_1 g}{2m_2 + \frac{1}{2}m_1}$ and $\dfrac{m_1 g}{4m_2 + m_1}$

27. (a) 14.7 m (b) neither mass is necessary

29. 37.8 N

31. (a)

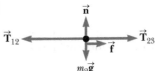

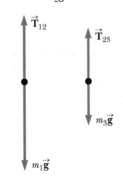

(b) 2.31 m/s^2, down for m_1, left for m_2, and up for m_3

(c) $T_{12} = 30.0$ N and $T_{23} = 24.2$ N

(d) T_{12} decreases and T_{23} increases

33. Driver was traveling at 67.1 mi/h

35. 834 N

37. (a) 3.43 m/s^2 toward the scrap iron   (b) 3.43 m/s^2 toward the scrap iron; -6.86 m/s^2 toward the magnet

39. (a)

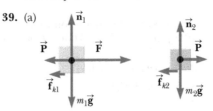

(b) F (c) $F - P$ (d) P (e) m_1: $F - P = m_1 a$; m_2: $P = m_2 a$

(f) $a = \dfrac{F - \mu_1 m_1 g - \mu_2 m_2 g}{m_1 + m_2}$

(g) $P = \dfrac{m_2}{m_1 + m_2}[F + m_1(\mu_2 - \mu_1)g]$

41. (a)

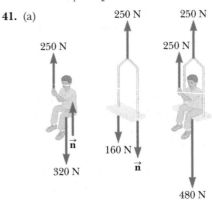

(b) 0.408 m/s^2 (c) 83.3 N

43. (a)

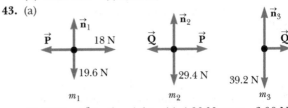

(b) 2.00 m/s^2 to the right   (c) 4.00 N on m_1, 6.00 N right on m_2, 8.00 N right on m_3 (d) 14.0 N between m_1 and m_2, 8.00 N between m_2 and m_3 (e) The m_2 block models the heavy block of wood. The contact force on your back is modeled by the force between the m_2 and the m_3 blocks, which is much less than the force F. The difference between F and this contact force is the net force causing the acceleration of the 5-kg pair of objects. The acceleration is real and nonzero, but it lasts for so short a time that it is never associated with a large velocity. The frame of the building and your legs exert forces, small in magnitude relative to the hammer blow, to bring the partition, block, and you to rest again over a time interval large relative to the hammer blow.

45. (b) If θ is greater than $\tan^{-1}(1/\mu_s)$, motion is impossible.

47. Ship requires 1.5 km to come to rest.

49. $(M + m_1 + m_2)(m_1 g/m_2)$

51. (a) 0.931 m/s^2 (b) From a value of 0.625 m/s^2 for large x, the acceleration gradually increases, passes through a

Answers to Quick Quizzes and Odd-Numbered Problems

maximum, and then drops more rapidly, becoming negative and reaching -2.10 m/s² at $x = 0$.   (c) 0.976 m/s² at $x = 25.0$ cm   (d) 6.10 cm

53. (a) $m_2g\left[\dfrac{m_1M}{m_2M + m_1(m_2 + M)}\right]$ (b) $\left[\dfrac{gm_1(m_2 + M)}{m_2M + m_1(m_2 + M)}\right]$

   (c) $\left[\dfrac{m_1m_2g}{m_2M + m_1(m_2 + M)}\right]$ (d) $\left[\dfrac{m_1Mg}{m_2M + m_1(m_2 + M)}\right]$

55. $\vec{\mathbf{R}} = [m\cos\theta \sin\theta\hat{\mathbf{i}} + (M + m\cos^2\theta)\hat{\mathbf{j}}]g$, where the x axis is horizontal and the y axis is vertical in Figure P5.55.

Chapter 6

Answers to Quick Quizzes

1. (i) (a)   **(ii)** (b)
2. (i) Because the speed is constant, the only direction the force can have is that of the centripetal acceleration. The force is larger at Ⓒ than at Ⓐ because the radius at Ⓒ is smaller. There is no force at Ⓑ because the wire is straight.   **(ii)** In addition to the forces in the centripetal direction in part (a), there are now tangential forces to provide the tangential acceleration. The tangential force is the same at all three points because the tangential acceleration is constant.

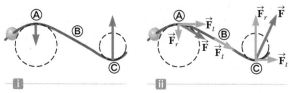

3. (c)
4. (a)

Answers to Odd-Numbered Problems

1. (a) 8.33×10^{-8} N toward the nucleus
   (b) 9.15×10^{22} m/s² inward
3. (a) $(-0.233\hat{\mathbf{i}} + 0.163\hat{\mathbf{j}})$ m/s²
   (b) 6.53 m/s, $(-0.181\hat{\mathbf{i}} + 0.181\hat{\mathbf{j}})$ m/s²
5. 6.22×10^{-12} N
7. (a) no   (b) yes
9. (a) 1.33 m/s²   (b) 1.79 m/s² at 48.0° inward from the direction of the velocity
11. (a) $v = \sqrt{R\left(\dfrac{2T}{m} - g\right)}$ (b) $2T$ up
13. (a) 8.62 m   (b) Mg, downward   (c) 8.45 m/s²   (d) Calculation of the normal force shows it to be negative, which is impossible. We interpret it to mean that the normal force goes to zero at some point and the passengers will fall out of their seats near the top of the ride if they are not restrained in some way. We could arrive at this same result without calculating the normal force by noting that the acceleration in part (c) is smaller than that due to gravity. The teardrop shape has the advantage of a larger acceleration of the riders at the top of the arc for a path having the same height as the circular path, so the passengers stay in the cars.

15. (a) 491 N   (b) 50.1 kg   (c) 2.00 m/s²
17. 0.527°
19. (a) 2.03 N down   (b) 3.18 m/s² down   (c) 0.205 m/s down
21. (a) 1.47 N · s/m   (b) 2.04×10^{-3} s   (c) 2.94×10^{-2} N
23. 10^1 N
25. 781 N
27. (a) $mg - \dfrac{mv^2}{R}$ (b) $\sqrt{gR}$
29. (a) $v = v_i e^{-bt/m}$ (b)

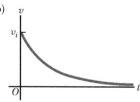

   (c) In this model, the object keeps moving forever.
   (d) It travels a finite distance in an infinite time interval.
31. (a) the downward gravitational force and the tension force in the string, always directed toward the center of the path

   (b)

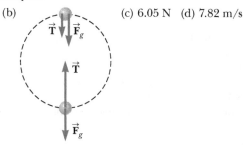

   (c) 6.05 N   (d) 7.82 m/s

33. (a) 1 975 lb, directed upward   (b) 647 lb, directed downward   (c) When $F'_g = 0$, then $mg = \dfrac{mv^2}{R}$.
35. (a) The only horizontal force on the car is the force of friction, with a maximum value determined by the surface roughness (described by the coefficient of static friction) and the normal force (here equal to the gravitational force on the car).   (b) 34.3 m   (c) 68.6 m   (d) Braking is better. You should not turn the wheel. If you used any of the available friction force to change the direction of the car, it would be unavailable to slow the car and the stopping distance would be greater.   (e) The conclusion is true in general. The radius of the curve you can barely make is twice your minimum stopping distance.
37. (a) 735 N   (b) 732 N   (c) The gravitational force is larger. The normal force is smaller, just like it is when going over the top of a Ferris wheel.
39. (a) $\sum\vec{\mathbf{F}} = km\vec{\mathbf{v}}$ (b) In general, the possibility of k positive is unrealistic in nature. You might be able to imagine some device with a feedback mechanism that could be used to apply a force to cause the velocity to increase in magnitude. In this case the speed would increase exponentially, so such a situation could only exist temporarily.   (c) Think of a duck landing on a lake, where the water exerts a resistive force on the duck proportional to its speed.
41. (a) $v_{\min} = \sqrt{\dfrac{Rg(\tan\theta - \mu_s)}{1 + \mu_s\tan\theta}}$, $v_{\max} = \sqrt{\dfrac{Rg(\tan\theta + \mu_s)}{1 - \mu_s\tan\theta}}$
   (b) $\mu_s = \tan\theta$

43. (a) Particle under constant acceleration   (b) $\Delta t = \dfrac{2v}{g}$

(c) Particle in uniform circular motion   (d) $T = \dfrac{2\pi R}{v}$
(e) $v = \sqrt{\pi Rg}$ (f) $F = \pi mg$

45. (a) 78.3 m/s   (b) 11.1 s   (c) 121 m

47. (a) 8.04 s   (b) 379 m/s   (c) 1.19×10^{-2} m/s   (d) 9.55 cm

49. 0.092 8°

Chapter 7

Answers to Quick Quizzes

1. (a)
2. (c), (a), (d), (b)
3. (d)
4. (a)
5. (b)
6. (c)
7. (i) (c)   (ii) (a)
8. (d)

Answers to Odd-Numbered Problems

1. (a) 1.59×10^3 J   (b) smaller   (c) the same
3. (a) 472 J   (b) 2.76 kN
7. 5.33 J
9. (a) 7.50 J   (b) 15.0 J   (c) 7.50 J   (d) 30.0 J
11. (a) 0.938 cm   (b) 1.25 J
13. Each spring should have a spring constant of 316 N/m.
15. (b) mgR
17. (a)

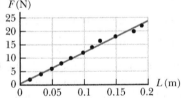

(b) The slope of the line is 116 N/m.   (c) We use all the points listed and also the origin. There is no visible evidence for a bend in the graph or nonlinearity near either end.   (d) 116 N/m   (e) 12.7 N
19. (a) 50.0 J   (b) 87.5 J; path independent.
21. (a) 1.20 J   (b) 5.00 m/s   (c) 6.30 J
23. 878 kN up
25. (a) 4.56 kJ   (b) 4.56 kJ   (c) 6.34 kN
(d) 422 km/s²   (e) 6.34 kN   (f) The two theories agree.
27. (a) 97.8 J   (b) $(-4.31\hat{\mathbf{i}} + 31.6\,\hat{\mathbf{j}})$ N   (c) 8.73 m/s
29. (a) 2.5 J   (b) -9.8 J   (c) -12 J
31. (a) -196 J   (b) -196 J   (c) -196 J   (d) The gravitational force is conservative.
33. (a) 125 J   (b) 50.0 J   (c) 66.7 J   (d) nonconservative
(e) The work done on the particle depends on the path followed by the particle.
35. (a) 40.0 J   (b) -40.0 J   (c) 62.5 J
37. A/r^2 away from the other particle
39.

Stable   Unstable   Neutral

41. 0.559 m/s
43. 0.799 N · m
45. (a) $\vec{\mathbf{F}}_1 = (20.5\hat{\mathbf{i}} + 14.3\hat{\mathbf{j}})$ N, $\vec{\mathbf{F}}_2 = (-36.4\hat{\mathbf{i}} + 21.0\hat{\mathbf{j}})$ N
(b) $\Sigma\vec{\mathbf{F}} = (-15.9\hat{\mathbf{i}} + 35.3\hat{\mathbf{j}})$ N
(c) $\vec{\mathbf{a}} = (-3.18\hat{\mathbf{i}} + 7.07\hat{\mathbf{j}})$ m/s²
(d) $\vec{\mathbf{v}} = (-5.54\hat{\mathbf{i}} + 23.7\hat{\mathbf{j}})$ m/s
(e) $\vec{\mathbf{r}} = (-2.30\hat{\mathbf{i}} + 39.3\hat{\mathbf{j}})$ m   (f) 1.48 kJ   (g) 1.48 kJ
(h) The work–kinetic energy theorem is consistent with Newton's second law.
47. 0.131 m
49. (a) 19.3°   (b) 1.39×10^4 J

Chapter 8

Answers to Quick Quizzes

1. (i) (b)   (ii) (b)   (iii) (a)
2. (a)
3. $v_1 = v_2 = v_3$
4. (c)

Answers to Odd-Numbered Problems

1. (a) $\Delta K + \Delta U = 0$, $v = \sqrt{2gh}$ (b) $\Delta K = W$, $v = \sqrt{2gh}$
3. (a) 5.94 m/s, 7.67 m/s   (b) 147 J
5. 5.49 m/s
7. (a) -168 J   (b) 184 J   (c) 500 J   (d) 148 J   (e) 5.65 m/s
9. (a) 5.60 J   (b) 2.29 rev
11. (a) 22.0 J, 40.0 J   (b) Yes   (c) The total mechanical energy has decreased, so a nonconservative force must have acted.
13. (a) Isolated. The only external influence on the system is the normal force from the slide, but this force is always perpendicular to its displacement so it performs no work on the system.   (b) No, the slide is frictionless.
(c) $E_{\text{system}} = mgh$ (d) $E_{\text{system}} = \frac{1}{5}mgh + \frac{1}{2}mv_i^2$
(e) $E_{\text{system}} = mgy_{\text{max}} + \frac{1}{2}mv_{xi}^2$
(f) $v_i = \sqrt{\dfrac{8gh}{5}}$ (g) $y_{\text{max}} = h(1 - \frac{4}{5}\cos^2\theta)$ (h) If friction is present, mechanical energy of the system would *not* be conserved, so the child's kinetic energy at all points after leaving the top of the waterslide would be reduced when compared with the frictionless case. Consequently, her launch speed and maximum height would be reduced as well.
15. Both trails result in the same speed.
17. $145
19. $\sim 10^4$ W
21. (a) 423 mi/gal   (b) 776 mi/gal
23. (a) 0.225 J   (b) -0.363 J   (c) no   (d) It is possible to find an effective coefficient of friction but not the actual value of μ since n and f vary with position.
25. (a) 1.29×10^4 N   (b) 45.4 m/s   (c) 3.72×10^4 N ; 46.1 m/s
(d) 45 m   (e) No
27. (a) $x = -4.0$ mm   (b) -1.0 cm
29. (a) -6.08×10^3 J   (b) -4.59×10^3 J   (c) 4.59×10^3 J
31. (a) 1.38×10^4 J   (b) 5.51×10^3 W
(c) The value in part (b) represents only energy that leaves the engine and is transformed to kinetic energy of the car. Additional energy leaves the engine by sound and heat. More energy from the engine is transformed to internal energy by friction forces and air resistance.

33. (a) 0.403 m or −0.357 m   (b) From a perch at a height of 2.80 m above the top of a pile of mattresses, a 46.0-kg child jumps upward at 2.40 m/s. The mattresses behave as a linear spring with force constant 19.4 kN/m. Find the maximum amount by which they are compressed when the child lands on them.   (c) 0.023 2 m   (d) This result is the distance by which the mattresses compress if the child just stands on them. It is the location of the equilibrium position of the oscillator.

35. (a) 1.53 J at $x = 6.00$ cm, 0 J at $x = 0$ (b) 1.75 m/s   (c) 1.51 m/s   (d) The answer to part (c) is not half the answer to part (b), because the equation for the speed of an oscillator is not linear in position

37. 48.2°

39. (a) No, mechanical energy is not conserved in this case.
(b) 77.0 m/s

43. (b) 0.342

45. (a) $-\mu_k gx/L$ (b) $(\mu_k gL)^{1/2}$

47. Less dangerous

Chapter 9

Answers to Quick Quizzes

1. (d)

2. (b), (c), (a)

3. (i) (c), (e)   (ii) (b), (d)

4. (a) All three are the same. (b) dashboard, seat belt, air bag

5. (a)

6. (b)

7. (b)

8. (i) (a)   (ii) (b)

Answers to Odd-Numbered Problems

1. (b) $p = \sqrt{2mK}$

3. $\vec{\mathbf{F}}_{\text{on bat}} = (3.26\hat{\mathbf{i}} - 3.99\hat{\mathbf{j}})$ kN, where positive x is from the pitcher toward home plate and positive y is upward.

5. (a) $-6.00\hat{\mathbf{i}}$ m/s   (b) 8.40 J   (c) The original energy is in the spring.   (d) A force had to be exerted over a displacement to compress the spring, transferring energy into it by work. The cord exerts force, but over no displacement.   (e) System momentum is conserved with the value zero.   (f) The forces on the two blocks are internal forces, which cannot change the momentum of the system; the system is isolated.   (g) Even though there is motion afterward, the final momenta are of equal magnitude in opposite directions, so the final momentum of the system is still zero.

7. (c) no difference

9. (a) 9.60×10^{-2} s   (b) 3.65×10^5 N   (c) $26.6g$

11. 16.5 N

13. (a) 2.50 m/s   (b) 37.5 kJ

15. (a) $v_f = \dfrac{1}{3}(v_1 + 2v_2)$ (b) $\Delta K = -\dfrac{m}{3}(v_1^2 + v_2^2 - 2v_1 v_2)$

17. (a) 4.85 m/s   (b) 8.41 m

19. The defendant was traveling at 41.5 mi/h.

21. $v_O = v_i \cos\theta,\ v_Y = v_i \sin\theta$

23. $v = \dfrac{v_i}{\sqrt{2}},\ 45.0°,\ -45.0°$

25. 3.57×10^8 J

27. (a)

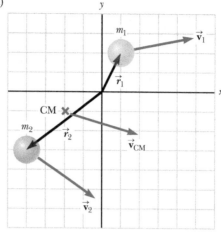

(b) $(-2.00\hat{\mathbf{i}} - 1.00\hat{\mathbf{j}})$ m   (c) $(3.00\hat{\mathbf{i}} - 1.00\hat{\mathbf{j}})$ m/s
(d) $(15.0\hat{\mathbf{i}} - 5.00\hat{\mathbf{j}})$ kg · m/s

29. The drone was struck by a meteorite.

31. (a) yes   (b) no   (c) 103 kg · m/s, up   (d) yes   (e) 88.2 J
(f) No, the energy came from potential energy stored in the person from previous meals.

33. (a) 442 metric tons   (b) 19.2 metric tons   (c) It is much less than the suggested value of 442/2.50. Mathematically, the logarithm in the rocket propulsion equation is not a linear function. Physically, a higher exhaust speed has an extra-large cumulative effect on the rocket body's final speed by counting again and again in the speed the body attains second after second during its burn.

35. (a) She moves, just like the archer in Example 9.1.
(b) $-\left(\dfrac{m}{M-m}\right)\vec{\mathbf{v}}_{\text{gloves}}$ (c) As she throws the gloves and exerts a force on them, the gloves exert an equal and opposite force on her that causes her to accelerate from rest to reach the velocity $\vec{\mathbf{v}}_{\text{girl}}$.

37. (a) $1.33\hat{\mathbf{i}}$ m/s   (b) $-235\hat{\mathbf{i}}$ N   (c) 0.680 s   (d) $-160\hat{\mathbf{i}}$ N · s and $+160\hat{\mathbf{i}}$ N · s   (e) 1.81 m   (f) 0.454 m   (g) -427 J   (h) $+107$ J   (i) Let's imagine an ideal situation in which the person and the cart have a perfect thermal insulator between them, so that no energy can transfer by heat Q between the person and the cart. Then, the change in kinetic energy of one member of the system, according to Equation 8.2, will be equal to the negative of the change in internal energy for that member: $\Delta K = -\Delta E_{\text{int}}$. The change in internal energy, in turn, is the product of the friction force and the distance through which the member moves while experiencing that force. Equal-magnitude friction forces act on the person and the cart, but the person and the cart move through different distances, as we see in parts (e) and (f). Therefore, there are different changes in internal energy for the person and the cart and, in turn, different changes in kinetic energy. The person and the cart will experience different changes in internal energy and, therefore, in temperature, which, in the real situation without the thermal insulator, will equalize after the event by means of the transfer of energy by heat Q between the person and the cart. The total change in kinetic energy of the system, -320 J, becomes $+320$ J of extra internal energy in the entire system in this perfectly inelastic collision.

39. (a) Momentum of the bullet–block system is conserved in the collision, so you can relate the speed of the block and bullet immediately after the collision to the initial speed of the bullet. Then, you can use conservation of mechanical energy for the bullet–block–Earth system to relate the speed after the collision to the maximum height. (b) 521 m/s upward

41. (a) $\dfrac{m_1 v_1 + m_2 v_2}{m_1 + m_2}$ (b) $(v_1 - v_2)\sqrt{\dfrac{m_1 m_2}{k(m_1 + m_2)}}$

(c) $v_{1f} = \dfrac{(m_1 - m_2)v_1 + 2m_2 v_2}{m_1 + m_2}$,

$v_{2f} = \dfrac{2m_1 v_1 + (m_2 - m_1)v_2}{m_1 + m_2}$

43. (a) 6.29 m/s (b) 6.16 m/s (c) Most of the 2% difference between the values for speed could be accounted for by air resistance.

45. 143 m/s

47. (a) 0; inelastic (b) $(-0.250\hat{\mathbf{i}} + 0.75\hat{\mathbf{j}} - 2.00\hat{\mathbf{k}})$ m/s; perfectly inelastic (c) either $a = -6.74$ with $\vec{\mathbf{v}} = -0.419\hat{\mathbf{k}}$ m/s or $a = 2.74$ with $\vec{\mathbf{v}} = -3.58\hat{\mathbf{k}}$ m/s

49. (a) $-0.256\hat{\mathbf{i}}$ m/s and $0.128\hat{\mathbf{i}}$ m/s
(b) $-0.064\,2\hat{\mathbf{i}}$ m/s and 0 (c) 0 and 0

51. (a) $(20.0\hat{\mathbf{i}} + 7.00\hat{\mathbf{j}})$ m/s (b) $4.00\hat{\mathbf{i}}$ m/s² (c) $4.00\hat{\mathbf{i}}$ m/s²
(d) $(50.0\hat{\mathbf{i}} + 35.0\hat{\mathbf{j}})$ m (e) 600 J (f) 674 J (g) 674 J
(h) The accelerations computed in different ways agree. The kinetic energies computed in different ways agree. The three theories are consistent.

53. (a) particle of mass m: $\sqrt{2}v_i$; particle of mass $3m$: $\sqrt{\frac{2}{3}}v_i$
(b) 35.3°

Chapter 10

Answers to Quick Quizzes

1. (i) (c) (ii) (b)
2. (b)
3. (i) (b) (ii) (a)
4. (b)
5. (b)
6. (a)
7. (b)

Answers to Odd-Numbered Problems

1. (a) 7.27×10^{-5} rad/s (b) Because of its angular speed, the Earth bulges at the equator.
3. (a) 4.00 rad/s² (b) 18.0 rad
5. (a) 8.21×10^2 rad/s² (b) 4.21×10^3 rad

7. (a) $\omega h^{3/2}\left(\dfrac{2}{g}\right)^{1/2}$ (b) 1.16 cm (c) The deflection is only 0.02% of the original height, so it is negligible in many practical cases. (d) decrease

9. (a) 25.0 rad/s (b) 39.8 rad/s² (c) 0.628 s
11. (a) 54.3 rev (b) 12.1 rev/s
13. (a) 3.47 rad/s (b) 1.74 m/s (c) 2.78 s (d) 1.02 rotations
15. (a) 1.03 s (b) 10.3 rev
17. (a) 24.0 N · m (b) 0.035 6 rad/s² (c) 1.07 m/s²
19. (a) 0.312 (b) 117 N

21. $\tau_f = -0.039\,8$ N · m

23. $I_{y'} = \displaystyle\int_{\text{all mass}} r^2\, dm = \int_0^L x^2\dfrac{M}{L}\, dx = \dfrac{M}{L}\dfrac{x^3}{3}\Big|_0^L = \tfrac{1}{3}ML^2$

25. (a) 92.0 kg · m² (b) 184 J (c) 6.00 m/s, 4.00 m/s, 8.00 m/s (d) 184 J (e) The kinetic energies computed in parts (b) and (d) are the same.

27. 1.03×10^{-3} J

29. (a) 11.4 N (b) 7.57 m/s² (c) 9.53 m/s (d) 9.53 m/s

31. (a) $2(Rg/3)^{1/2}$ (b) $4(Rg/3)^{1/2}$ (c) $(Rg)^{1/2}$

33. (a) 2.38 m/s (b) The centripetal acceleration at the top is $\dfrac{v_2^2}{r} = \dfrac{(2.38 \text{ m/s})^2}{0.450 \text{ m}} = 12.6 \text{ m/s}^2 > g$. Therefore, the ball must be in contact with the track, with the track pushing downward on it. (c) 4.31 m/s (d) The speed of the ball turns out to be imaginary. (e) When the ball is projected with the same speed as before, but with only translational kinetic energy, there is insufficient kinetic energy for the ball to arrive at the top of the track.

35. (a) 1.21×10^{-4} kg · m² (b) Knowing the height of the can is unnecessary. (c) The mass is not uniformly distributed; the density of the metal can is larger than that of the soup.

37. (a) 12.5 rad/s (b) 128 rad

39. (a) $d = (1\,890 + 80n)\left(\dfrac{0.459 \text{ m}}{80n - 150}\right)$ (b) 94.1 m (c) 1.62 m
(d) -5.79 m (e) The rising car will coast to a stop only for $n \geq 2$. (f) For $n = 0$ or $n = 1$, the mass of the elevator is less than the counterweight, so the car would accelerate upward if released. (g) 0.459 m

43. 54.0°
45. (b) to the left

Chapter 11

Answers to Quick Quizzes

1. (d)
2. (i) (a) (ii) (c)
3. (b)
4. (a)

Answers to Odd-Numbered Problems

1. $\hat{\mathbf{i}} + 8.00\hat{\mathbf{j}} + 22.0\hat{\mathbf{k}}$
3. 45.0°
5. (a) $F_3 = F_1 + F_2$ (b) no
7. (a) $(-10.0 \text{ N} \cdot \text{m})\hat{\mathbf{k}}$ (b) yes (c) yes (d) yes (e) no
(f) $5.00\hat{\mathbf{j}}$ m
9. $m(xv_y - yv_x)\hat{\mathbf{k}}$
11. (a) zero (b) $(-mv_i^3 \sin^2 \theta \cos \theta/2g)\hat{\mathbf{k}}$
(c) $(-2mv_i^3 \sin^2 \theta \cos \theta/g)\hat{\mathbf{k}}$
(d) The downward gravitational force exerts a torque on the projectile in the negative z direction.
13. $mvR[\cos(vt/R) + 1]\hat{\mathbf{k}}$
15. (a) $-m\ell gt\cos\theta\,\hat{\mathbf{k}}$ (b) The Earth exerts a gravitational torque on the ball. (c) $-mg\ell\cos\theta\,\hat{\mathbf{k}}$
17. (a) 0.360 kg · m²/s (b) 0.540 kg · m²/s
19. 1.20 kg · m²/s
21. 8.63 m/s²
23. (a) The mechanical energy of the system is not constant. Some potential energy in the woman's body from

previous meals is converted into mechanical energy. (b) The momentum of the system is not constant. The turntable bearing exerts an external northward force on the axle. (c) The angular momentum of the system is constant. (d) 0.360 rad/s counterclockwise (e) 99.9 J

25. (a) 11.1 rad/s counterclockwise (b) No; 507 J is transformed into internal energy in the system. (c) No; the turntable bearing promptly imparts impulse 44.9 kg · m/s north into the turntable–clay system and thereafter keeps changing the system momentum as the velocity vector of the clay continuously changes direction.

27. (a) $mv\ell$ down (b) $M/(M + m)$

29. (a) $\omega = 2mv_id/[M + 2m]R^2$ (b) No; some mechanical energy of the system changes into internal energy. (c) The momentum of the system is not constant. The axle exerts a backward force on the cylinder when the clay strikes.

31. 5.46×10^{22} N · m

33. (a) 2.35 rad/s (b) 0.498 rad/s (c) 5.58°

35. 7.50×10^{-11} s

37. (a) $7md^2/3$ (b) $mgd\hat{\mathbf{k}}$ (c) $3g/7d$ counterclockwise (d) $2g/7$ upward (e) mgd (f) $\sqrt{6g/7d}$ (g) $m\sqrt{14gd^3/3}$ (h) $\sqrt{2gd/21}$

39. (a) 3 750 kg · m²/s (b) 1.88 kJ (c) 3 750 kg · m²/s (d) 10.0 m/s (e) 7.50 kJ (f) 5.62 kJ

41. (a) $2mv_0$ (b) $2v_0/3$ (c) $4m\ell v_0/3$ (d) $4v_0/9\ell$ (e) mv_0^2 (f) $26mv_0^2/27$ (g) No horizontal forces act on the bola from outside after release, so the horizontal momentum stays constant. Its center of mass moves steadily with the horizontal velocity it had at release. No torques about its axis of rotation act on the bola, so the angular momentum stays constant. Internal forces cannot affect momentum conservation and angular momentum conservation, but they can affect mechanical energy.

43. 10.7 m/s

45. an increase of 6.368×10^{-4} % or 0.550 s, which is not significant

47. 14.0 s

49. (a) 2.0 m/s (b) 1.0 rad/s

Chapter 12

1. (a)
2. (b)
3. (b)
4. (i) (b) (ii) (a) (iii) (c)

1. Safe arrangements: 2-3-1, 3-1-2, 3-2-1; dangerous arrangements: 1-2-3, 1-3-2, 2-1-3

3. (3.85 cm, 6.85 cm)

5. $x = 0.750$ m

7. 177 kg

9. (a) 29.9 N (b) 22.2 N

11. (a) 27.7 kN (b) 11.5 kN (c) 4.19 kN

13. (a) 1.04 kN at 60.0° upward and to the right (b) $(370\hat{\mathbf{i}} + 910\hat{\mathbf{j}})$ N

15. (a) 859 N (b) 1.04 kN at 36.9° to the left and upward

17. (a) $-0.053\,8$ m³ (b) 1.09×10^3 kg/m³ (c) With only a 5% change in volume in this extreme case, liquid water can be modeled as incompressible in biological and student laboratory situations.

19. 23.8 μm

21. (a) 3.14×10^4 N (b) 6.28×10^4 N

23. 9.85×10^{-5}

25. $n_A = 5.98 \times 10^5$ N, $n_B = 4.80 \times 10^5$ N

27. (a) 0.400 mm (b) 40.0 kN (c) 2.00 mm (d) 2.40 mm (e) 48.0 kN

29. (a)

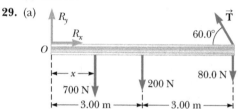

(b) $T = 343$ N, $R_x = 171$ N to the right, $R_y = 683$ N up (c) 5.14 m

31. (a) $T = F_g(L + d)/[\sin\theta\,(2L + d)]$ (b) $R_x = F_g(L + d)\cot\theta/(2L + d)$; $R_y = F_gL/(2L + d)$

33. (a) 5.08 kN (b) 4.77 kN (c) 8.26 kN

35. (a) $\frac{1}{2}m\left(\dfrac{2\mu_s\sin\theta - \cos\theta}{\cos\theta - \mu_s\sin\theta}\right)$ (b) $(m + M)g\sqrt{1 + \mu_s^2}$ (c) $g\sqrt{M^2 + \mu_s^2(m + M)^2}$

37. (a) 9.28 kN (b) The moment arm of the force $\vec{\mathbf{F}}_h$ is no longer 70 cm from the shoulder joint but only 49.5 cm, therefore reducing $\vec{\mathbf{F}}_m$ to 6.56 kN.

39. (a) 66.7 N (b) increasing at 0.125 N/s

41. (a) $\dfrac{1}{\sqrt{15}}\dfrac{mgd}{\ell}$ (b) $n_A = mg\left(\dfrac{2\ell - d}{2\ell}\right)$, $n_B = \dfrac{mgd}{2\ell}$ (c) $R_x = \dfrac{1}{\sqrt{15}}\dfrac{mgd}{\ell}$ to the right, $R_y = \dfrac{mgd}{2\ell}$ downward

43. (a) $P_1 = P_3 = 1.67$ N, $P_2 = 3.33$ N (b) 2.36 N

45. 5.73 rad/s

47. (a) 443 N (b) 221 N (to the right), 217 N (upward)

49. (b) 60.0° (c) unstable

Chapter 13

1. (e)
2. (c)
3. (a)
4. (a) perihelion (b) aphelion (c) perihelion (d) all points

1. 7.41×10^{-10} N

3. $\sim 10^{-7}$ N

5. (a) 7.61 cm/s² (b) 363 s (c) 3.08 km (d) 28.9 m/s at 72.9° below the horizontal

7. (a) 1.31×10^{17} N (b) 2.62×10^{12} N/kg

9. (a) 0.708 yr (b) 0.399 yr

11. 4.99 days

13. (a) yes (b) 3.93 yr

15. 4.17×10^{10} J

17. (a) -1.67×10^{-14} J (b) The particles collide at the center of the triangle.

19. 1.58×10^{10} J

21. 1.78×10^3 m

23. (a) same size force (b) 15.6 km/s

25. 492 m/s

27. 1.30×10^3 m/s

29. 2.25×10^{-7}

31. (a) 1.00×10^7 m (b) 1.00×10^4 m/s

33. (a) 15.3 km (b) 1.66×10^{16} kg (c) 1.13×10^4 s (d) No; its mass is so large compared with yours that you would have a negligible effect on its rotation.

35. (c) 1.85×10^{-5} m/s²

37. (a) 2×10^8 yr (b) $\sim10^{41}$ kg (c) 10^{11}

39. (a) 2.93×10^4 m/s (b) $K = 2.74 \times 10^{33}$ J, $U = -5.39 \times 10^{33}$ J (c) $K = 2.56 \times 10^{33}$ J, $U = -5.21 \times 10^{33}$ J (d) Yes; $E = -2.65 \times 10^{33}$ J at both aphelion and perihelion.

41. (a) $(2.77 \text{ m/s}^2)\left(1 + \dfrac{m}{5.98 \times 10^{24} \text{ kg}}\right)$ (b and c) 2.77 m/s² (d) 3.70 m/s² (e) Any object with mass small compared to the Earth starts to fall with acceleration 2.77 m/s². As m increases to become comparable to the mass of the Earth, the acceleration increases and can become arbitrarily large. It approaches a direct proportionality to m.

43. 18.2 ms

Chapter 14

Answers to Quick Quizzes

1. (a)

2. (a)

3. (c)

4. (b) or (c)

5. (a)

Answers to Odd-Numbered Problems

1. 2.96×10^6 Pa

3. 5.27×10^{18} kg

5. 7.74×10^{-3} m²

7. 0.072 1 mm

9. (a) 10.5 m (b) No. The vacuum is not as good because some alcohol and water in the wine will evaporate. The equilibrium vapor pressures of alcohol and water are higher than the vapor pressure of mercury.

11. 3.33×10^3 kg/m³

13. (a) 1 250 kg/m³ (b) 500 kg/m³

15. (a) 408 kg/m³ (b) When m is less than 0.310 kg, the wooden block will be only partially submerged in the water. (c) When m is greater than 0.310 kg, the wooden block and steel object will sink.

17. (a) 11.6 cm (b) 0.963 g/cm³ (c) No; the density ρ is not linear in h.

19. 20.0 g

21. (b) 616 MW

23. (a) 15.1 MPa (b) 2.95 m/s

25. (a) 28.0 m/s (b) 28.0 m/s (c) The answers agree precisely. The models are consistent with each other. (d) 2.11 MPa

27. 0.120 N

29. 0.200 mm

31. (a) 4.43 m/s (b) 10.1 m

33. (a) particle in equilibrium

(b) $\sum F_y = B - F_b - F_{He} - F_s = 0$

(c) $m_s = \frac{4}{3}(\rho_{air} - \rho_{He})\pi r^3 - m_b$

(d) 0.023 7 kg (e) 0.948 m

35. $\sim10^4$

37. (a) 8.01 km (b) yes

39. 91.64%

41. (a) 3.307 g (b) 3.271 g (c) 3.48×10^{-4} N

43. 18.1 N

45. 758 Pa

47. 4.43 m/s

Chapter 15

Answers to Quick Quizzes

1. (d)

2. (f)

3. (a)

4. (b)

5. (c)

6. (i) (a) (ii) (a)

Answers to Odd-Numbered Problems

1. (a) 17 N to the left (b) 28 m/s² to the left

3. (a) 1.50 Hz (b) 0.667 s (c) 4.00 m (d) π rad (e) 2.83 m

5. (a) -2.34 m (b) -1.30 m/s (c) -0.076 3 m (d) 0.315 m/s

7. (a) $x = 2.00 \cos(3.00\pi t - 90°)$ or $x = 2.00 \sin(3.00\pi t)$ where x is in centimeters and t is in seconds (b) 18.8 cm/s (c) 0.333 s (d) 178 cm/s² (e) 0.500 s (f) 12.0 cm

9. (a) yes (b) The value of k in Equation 15.13 is proportional to the mass m, so the mass cancels in the equation, leaving only the extension of the spring and the acceleration due to gravity in the equation: $T = 0.859$ s.

11. 2.60 cm or -2.60 cm

13. (a) $\frac{8}{9}E$ (b) $\frac{1}{9}E$ (c) $x = \pm\sqrt{\frac{2}{3}}A$ (d) No; the maximum potential energy is equal to the total energy of the system. Because the total energy must remain constant, the kinetic energy can never be greater than the maximum potential energy.

15. (a) 4.58 N (b) 0.125 J (c) 18.3 m/s² (d) 1.00 m/s (e) smaller (f) the coefficient of kinetic friction between the block and surface (g) 0.934

17. (a) 1.50 s (b) 0.559 m

19. 0.944 kg · m²

21. (a) 0.820 m/s (b) 2.57 rad/s² (c) 0.641 N (d) $v_{max} = 0.817$ m/s, $\alpha_{max} = 2.54$ rad/s², $F_{max} = 0.634$ N (e) The answers are close but not exactly the same. The angular amplitude of 15° is not a small angle, so the simple harmonic oscillation model is not accurate. The answers computed from conservation of energy and from Newton's second law are more accurate.

23. (a) 5.00×10^{-7} kg · m² (b) 3.16×10^{-4} N · m/rad

27. (a) 3.16 s^{-1} (b) 6.28 s^{-1} (c) 5.09 cm

29. (a) 0.349 kg · m² (b) too low

31. (a) 2.09 s (b) 0.477 Hz (c) 36.0 cm/s (d) $E = 0.064$ 8m, where E is in joules and m is in kilograms (e) $k = 9.00m$, where k is in newtons/meter and m is in kilograms (f) Period, frequency, and maximum speed are all independent of mass in this situation. The energy and the force constant are directly proportional to mass.

33. (a) 2.00 cm (b) 4.00 s (c) $\dfrac{\pi}{2}$ rad/s (d) π cm/s (e) 4.93 cm/s² (f) $x = 2.00 \sin\left(\dfrac{\pi}{2}t\right)$, where x is in centimeters and t is in seconds

35. $\dfrac{1}{2\pi L}\sqrt{gL + \dfrac{kh^2}{M}}$

37. (a) 1.26 m (b) 1.58 (c) The energy decreases by 120 J.
(d) Mechanical energy is transformed into internal energy in the perfectly inelastic collision.

41. (b) $T = \dfrac{2}{r}\sqrt{\dfrac{\pi M}{\rho g}}$

43. 13.0 s

47. (a) $x = 2\cos\left(10t + \dfrac{\pi}{2}\right)$ (b) ± 1.73 m (c) 0.105 s = 105 ms
(d) 0.098 0 m

49. (a) $y_f = -0.110$ m (b) greater

51. (a) $\dfrac{2\pi}{\sqrt{g}}\sqrt{L_i + \dfrac{1}{2\rho a^2}\left(\dfrac{dM}{dt}\right)t}$ (b) $2\pi\sqrt{\dfrac{L_i}{g}}$

Chapter 16

Answers to Quick Quizzes

1. (i) (b) (ii) (a)
2. (i) (c) (ii) (b) (iii) (d)
3. (c)
4. (f) and (h)
5. (d)
6. (c)
7. (b)
8. (b)
9. (e)
10. (e)
11. (b)

Answers to Odd-Numbered Problems

1. 184 km
3. (a) $L = (380$ m/s$)\Delta t$ (b) 48.2 m (c) 48 cm
5. 2.40 m/s
7. ± 6.67 cm
9. (a) $y = 0.080\,0\sin(2.5\pi x + 6\pi t)$
(b) $y = 0.080\,0\sin(2.5\pi x + 6\pi t - 0.25\pi)$
11. 13.5 N
13. (a) 0.051 0 kg/m (b) 19.6 m/s
15. (a) 1 (b) 1 (c) 1 (d) increased by a factor of 4
17. (a) $y = 0.075\sin(4.19x - 314t)$, where x and y are in meters and t is in seconds (b) 625 W
19. $\sqrt{2}P_0$
23. (a) 2.00 μm (b) 40.0 cm (c) 54.6 m/s (d) -0.433 μm
(e) 1.72 mm/s
25. 5.81 m
27. 335 m/s
29. (a) 27.2 s (b) 25.7 s; the time interval in part (a) is longer.
31. (a) 3.75 W/m² (b) 0.600 W/m²
33. (a) 0.691 m (b) 691 km
35. 4.28 m
39. 2.82×10^8 m/s
41. (a) 441 Hz (b) 439 Hz (c) 54.0 dB
43. 14.7 kg
45. 0.883 cm
47. (a) 375 m/s² (b) 0.045 0 N (c) The maximum transverse force is very small compared to the tension of 46.9 N in the string, more than a thousand times smaller.

49. (a) $v = \sqrt{\dfrac{T}{\rho(1.00 \times 10^{-5}\,x + 1.00 \times 10^{-6})}}$, where v is in meters per second, T is in newtons, ρ is in kilograms per meter cubed, and x is in meters (b) $v(0) = 94.3$ m/s, $v(10.0$ m$) = 9.38$ m/s

51. (a) $\dfrac{\mu\omega^3}{2k}A_0^2 e^{-2bx}$ (b) $\dfrac{\mu\omega^3}{2k}A_0^2$ (c) e^{-2bx}

53. It is unreasonable, implying a sound level of 123 dB. Nearly all the decrease in mechanical energy becomes internal energy in the latch.

55. 1.34×10^4 N

57. (a) $\mu(x) = \dfrac{(\mu_L - \mu_0)x}{L} + \mu_0$

(b) $\Delta t = \dfrac{2L}{3\sqrt{T}\,(\mu_L - \mu_0)}(\mu_L^{3/2} - \mu_0^{3/2})$

Chapter 17

Answers to Quick Quizzes

1. (c)
2. (i) (a) (ii) (d)
3. (d)
4. (b)
5. (c)

Answers to Odd-Numbered Problems

1. (a) -1.65 cm (b) -6.02 cm (c) 1.15 cm
3.

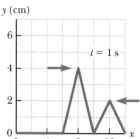

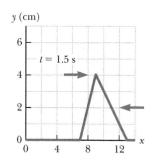

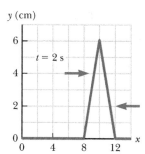

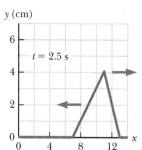

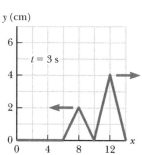

5. (a) y_1: positive x direction; y_2: negative x direction
   (b) 0.750 s (c) 1.00 m
7. (a) 2.72 rad = 156° (b) 0.058 4 cm
9. (a) The separation of adjacent nodes is $\Delta x = \dfrac{\pi}{k} = \dfrac{\lambda}{2}$. The
   nodes are still separated by half a wavelength (b) Yes.
   The nodes are located at $kx + \dfrac{\phi}{2} = n\pi$, so that
   $x = \dfrac{n\pi}{k} - \dfrac{\phi}{2k}$, which means that each node is shifted $\dfrac{\phi}{2k}$
   to the left by the phase difference between the traveling
   waves in comparison to the case in which $\phi = 0$.
11. (a) 0.600 m (b) 30.0 Hz
13. (a) 78.6 Hz (b) 157 Hz, 236 Hz, 314 Hz
15. 1.86 g
17. (a) 3.8 cm (b) 3.85%
19. The resonance frequency of the bay calculated from the
    data provided is 12 h, 24 min. The natural frequency of
    the water sloshing in the bay agrees precisely with that
    of lunar excitation, so we identify the extra-high tides as
    amplified by resonance.
21. (a) 0.656 m (b) 1.64 m
23. (a) 349 m/s (b) 1.14 m
25. $n(0.252 \text{ m})$ with $n = 1, 2, 3, \ldots$
27. 158 s
29. −10.0°C
31. (a) 1.99 beats/s (b) 3.38 m/s
33. The coefficients beyond $n = 1$ are approximate: $A_1 = 100$,
    $A_2 = 156$, $A_3 = 62$, $A_4 = 104$, $A_5 = 52$, $A_6 = 29$, $A_7 = 25$.

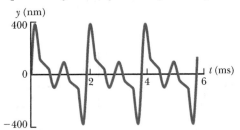

35. 800 m
37. (a) larger (b) 2.43
39. (a) $r = 0.078\,2\left(1 - \dfrac{4}{n^2}\right)^{1/3}$ (b) 3 (c) 0.078 2 m
    (d) The sphere floats on the water.
41. (a) 3.99 beats/s (b) 3.99 beats/s
43. (a) Frequency should be halved. (b) $\left[\dfrac{n}{n+1}\right]^2 T$
    (c) $\dfrac{T'}{T} = \dfrac{9}{16}$
45. 283 Hz
47. (a) 78.9 N (b) 211 Hz
49. (b) $A = 11.2$ m, $\phi = 1.11$ rad $= 63.4°$

Chapter 18

Answers to Quick Quizzes

1. (c)
2. (c)
3. (c)
4. (c)
5. (a)
6. (b)

Answers to Odd-Numbered Problems

1. (a) −738°N (b) −105°N (c) 270°N (d) 153°N
3. (a) −109°F, 195 K (b) 98.6°F, 310 K
5. (a) 56.7°C and −62.1°C (b) 330 K and 211 K
7. 3.27 cm
9. 1.54 km. The pipeline can be supported on rollers. In
   addition, Ω-shaped loops can be built between straight
   sections; these loops bend as the steel changes length.
11. 2.74 m
13. (a) 437°C (b) 2.1×10^3 °C (c) No; aluminum melts at
    660°C (Table 19.2). Also, although it is not in Table 19.2,
    Internet research shows that brass (an alloy of copper
    and zinc) melts at about 900°C.
15. (a) 99.8 mL (b) It lies below the mark. The acetone
    has reduced in volume, and the flask has increased in
    volume.
17. (a) 396 N (b) −101°C (c) The original length divides
    out of the equations in the calculation, so the answers
    would not change.
19. 1.50×10^{29} molecules
21. (a) 41.6 mol (b) 1.20 kg (c) This value is in agreement
    with the tabulated density.
23. 2.42×10^{11} molecules
25. 473 K
27. $\sim 10^2$ kg

31. (a) $\theta = 2\sin^{-1}\left(\dfrac{1 + \alpha_{Al} T_C}{2}\right)$ (b) yes (c) yes
    (d) $\theta = 2\sin^{-1}\left(\dfrac{1 + \alpha_{Al} T_C}{2(1 + \alpha_{invar} T_C)}\right)$ (e) 61.0° (f) 59.6°

33. (a) 94.97 cm (b) 95.03 cm
35. (b) As the temperature increases, the density decreases
    (assuming β is positive). (c) 5×10^{-5} (°C)$^{-1}$
    (d) -2.5×10^{-5} (°C)$^{-1}$
37. (b) It assumes $\alpha \Delta T$ is much less than 1.
39. (a) yes, as long as the coefficients of expansion remain
    constant (b) The lengths L_{Cu} and L_{St} at 0°C need to
    satisfy $17L_{Cu} = 11L_{St}$. Then the steel rod must be longer.
    With $L_{St} - L_{Cu} = 5.00$ cm, the only possibility is $L_{St} =$
    14.2 cm and $L_{Cu} = 9.17$ cm.
41. (a) 0.34% (b) 0.48% (c) All the moments of inertia
    have the same mathematical form: the product of a con-
    stant, the mass, and a length squared.
45. 4.54 m

Chapter 19

Answers to Quick Quizzes

1. (i) iron, glass, water (ii) water, glass, iron
2. The figure on the next page shows a graphical representation of the internal energy of the system as a function
   of energy added. Notice that this graph looks quite different from Figure 19.3 in that it doesn't have the flat
   portions during the phase changes. Regardless of how
   the temperature is varying in Figure 19.3, the internal
   energy of the system simply increases linearly with energy
   input; the line in the graph below has a slope of 1.

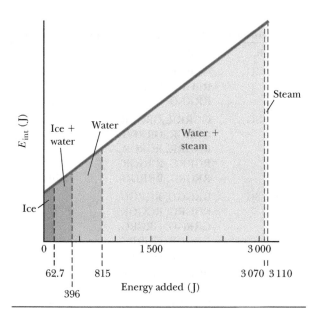

3.

Situation	System	Q	W	ΔE_{int}
(a) Rapidly pumping up a bicycle tire	Air in the pump	0	+	+
(b) Pan of room-temperature water sitting on a hot stove	Water in the pan	+	0	+
(c) Air quickly leaking out of a balloon	Air originally in the balloon	0	−	−

4. Path A is isovolumetric, path B is adiabatic, path C is isothermal, and path D is isobaric.

5. (b)

Answers to Odd-Numbered Problems

1. (a) 2.26×10^6 J (b) 2.80×10^4 steps (c) 6.99×10^3 steps
3. 23.6°C
5. 0.918 kg
7. (a) 1 822 J/kg · °C (b) We cannot make a definite identification. It might be beryllium. (c) The material might be an unknown alloy or a material not listed in the table.
9. (a) 25.8°C (b) The symbolic result from part (a) shows no dependence on mass. Both the change in gravitational potential energy and the change in internal energy of the system depend on the mass, so the mass cancels.
11. 2.27 km
13. (a) 0°C (b) 114 g
15. (a) $-4P_iV_i$ (b) According to $T = (P_i/nRV_i)V^2$, it is proportional to the square of the volume.
17. 720 J
19. (a) 0.041 0 m³ (b) +5.48 kJ (c) −5.48 kJ
21. (a) −0.048 6 J (b) 16.2 kJ (c) 16.2 kJ
23. 74.8 kJ
25. (a) 1.19 (b) 1.19
27. (a) 1.85 ft² · °F · h/Btu (b) 1.78
29. (a) -6.08×10^5 J (b) 4.56×10^5 J
31. 888 K
33. 1.90×10^3 J/kg · °C
35. (a) 9.31×10^{10} J (b) -8.47×10^{12} J (c) 8.38×10^{12} J
37. (a) First, energy must be removed from the liquid water to cool it to 0° C. Next, energy must be removed from the water at 0° C to freeze it, which corresponds to a liquid-to-solid phase transition. Finally, once all the water has frozen, additional energy must be removed from the ice to cool it from 0° to −8.00°C (b) 32.5 kJ
39. (a) 2 000 W (b) 4.46°C
41. (a) 3.16×10^{22} W (b) 3.17×10^{22} W (c) It is 0.408% larger. (d) 5.78×10^3 K
43. 1.44 kg
45. (b) 9.32 kW
47. 3.66×10^4 s = 10.2 h

Chapter 20

Answers to Quick Quizzes

1. (i) (b) (ii) (a)
2. (i) (a) (ii) (c)
3. (d)
4. (c)

Answers to Odd-Numbered Problems

1. 3.32 mol
3. 5.05×10^{-21} J
5. (a) 2.28 kJ (b) 6.21×10^{-21} J
7. 17.4 kPa
9. 74.8 J
11. (a) $W = 0$ (b) $\Delta E_{int} = 209$ J (c) 317 K
13. between 10^{-3} °C and 10^{-2} °C
15. (a) 1.08 (b) no
17. 5.74×10^6 Pa
19. 227 K
21. (a) 2.45×10^{-4} m³ (b) 9.97×10^{-3} mol (c) 9.01×10^5 Pa (d) 5.15×10^{-5} m³ (e) 560 K (f) 53.9 J (g) 6.79×10^{-6} m³ (h) 53.3 g (i) 2.24 K
23. (a) 2.37×10^4 K (b) 1.06×10^3 K
25. (b) 0.278
27. (a) 3.90 km/s (b) 4.18 km/s
29. (a) 7.89×10^{26} molecules (b) 37.9 kg (c) 6.07×10^{-21} J (d) 503 m/s (e) 0 (f) When the furnace operates, air expands and some of it leaves the room. The smaller mass of warmer air left in the room contains the same internal energy as the cooler air initially in the room.
31. (a) 367 K (b) The rms speed of nitrogen would be higher because the molar mass of nitrogen is less than that of oxygen. (c) 572 m/s
33. Sulfur dioxide is the gas with the greatest molecular mass of those listed. If the effective spring constants for various chemical bonds are comparable, SO_2 can then be expected to have low frequencies of atomic vibration. Vibration can be excited at lower temperature than for other gases. Some vibration may be going on at 300 K. With more degrees of freedom for molecular motion, the material has higher specific heat.
35. (a) 300 K (b) 1.00 atm
37. (a) 7.27×10^{-20} J (b) 2.20 km/s (c) 3.51×10^3 K (d) The evaporating particles emerge with much less kinetic energy, as negative work is performed on them by restraining forces as they leave the liquid. Much of the initial kinetic energy is used up in overcoming the latent heat of vaporization. There are also very few of these escaping at any moment in time.

39. (a) 1.09×10^{-3} (b) 2.69×10^{-2} (c) 0.529 (d) 1.00 (e) 0.199 (f) 1.01×10^{-41} (g) $1.25 \times 10^{-1\,082}$

43. (a) 0.510 m/s (b) 20 ms

45. (c) $2 - 2a + a^2$

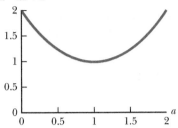

The graph above shows the behavior of the factor in parentheses in part (b) between the possible limits of $a = 0$ and $a = 2$. Except at the value of $a = 1$, the factor is always greater than 1. Therefore, the equation shows that, in general (except for the special case of $a = 1$), $v_{\text{rms}} > v_{\text{avg}}$. (d) $a = 1$

Chapter 21

Answers to Quick Quizzes

1. (i) (c) (ii) (b)
2. (d)
3. C, B, A
4. (a) one (b) six
5. (a)
6. false (The adiabatic process must be *reversible* for the entropy change to be equal to zero.)

Answers to Odd-Numbered Problems

1. (a) 10.7 kJ (b) 0.533 s
3. 55.4%
5. (a) 4.51×10^6 J (b) 2.84×10^7 J (c) 68.1 kg
7. (a) 67.2% (b) 58.8 kW
9. 1.86
11. (a) $564°C$ (b) No; a real engine will always have an efficiency *less* than the Carnot efficiency because it operates in an irreversible manner.
13. (a) 5.12% (b) 5.27×10^{12} J/h (c) 5.68×10^4 (d) 4.50×10^6 m² (e) yes (f) numerically, yes; feasibly, probably not
15. (a) $\dfrac{Q_c}{\Delta t} = 1.40 \left(\dfrac{0.5 T_h + 383}{T_h - 383} \right)$, where $Q_c/\Delta t$ is in megawatts and T_h is in kelvins (b) The exhaust power decreases as the firebox temperature increases. (c) 1.87 MW (d) 3.84×10^3 K (e) No answer exists. The energy exhaust cannot be that small.
17. 1.17
21. (a)

Macrostate	Microstates	Number of ways to draw
All R	RRR	1
2 R, 1 G	GRR, RGR, RRG	3
1 R, 2 G	GGR, GRG, RGG	3
All G	GGG	1

(b)

Macrostate	Microstates	Number of ways to draw
All R	RRRR	1
4R, 1G	GRRRR, RGRRR, RRGRR, RRRGR, RRRRG	5
3R, 2G	GGRRR, GRGRR, GRRGR, GRRRG, RGGRR, RGRGR, RGRRG, RRGGR, RRGRG, RRRGG	10
2R, 3G	RRGGG, RGRGG, RGGRG, RGGGR, GRRGG, GRGRG, GRGGR, GGRRG, GGRGR, GGGRR	10
1R, 4G	RGGGG, GRGGG, GGRGG, GGGRG, GGGGR	5
All G	GGGGG	1

23. 1.02 kJ/K
25. 195 J/K
27. (a) -3.45 J/K (b) $+8.06$ J/K (c) $+4.62$ J/K
29. 1 W/K
31. (a) $\frac{1}{3}$ (b) $\frac{2}{3}$
33. (a) $3nRT_i$ (b) $3nRT_i \ln 2$ (c) $-3nRT_i$ (d) $-nRT_i \ln 2$ (e) $3nRT_i(1 + \ln 2)$ (f) $2nRT_i \ln 2$ (g) 0.273
35. (a) 39.4 J (b) 65.4 rad/s $= 625$ rev/min (c) 293 rad/s $= 2.79 \times 10^3$ rev/min
37. (a) 4.10×10^3 J (b) 1.42×10^4 J (c) 1.01×10^4 J (d) 28.8% (e) Because $e_C = 80.0\%$, the efficiency of the cycle is much lower than that of a Carnot engine operating between the same temperature extremes.
39. (a) 0.476 J/K (b) 417 J
41. (a) 5.97 K (b) higher (c) 22.6 K
43. (a) 13.4 J/K (b) 310 K (c) 13.3 J/K (d) smaller by less than 1%
45. (b) yes (c) No; the second law refers to an engine operating in a cycle, whereas this problem involves only a single process.
47. (a)

	T (K)	P (kPa)	V (cm³)
A	293	100	500
B	673	1.84×10^3	62.5
C	1 023	2.79×10^3	62.5
D	445	152	500

(b)

	Q	W_{eng}	ΔE_{int}
$A{\rightarrow}B$	0	−162	162
$B{\rightarrow}C$	149	0	149
$C{\rightarrow}D$	0	246	−246
$D{\rightarrow}A$	−65.0	0	−65.0
$ABCD$	84.3	84.3	0

(c) 149 J (d) 65.0 J (e) 84.3 J (f) 0.565 (g) 1.42×10^3 rev/min

Index

Locator note: **boldface** indicates a definition; *italics* indicates a figure; *t* indicates a table; *n* indicates a footnote

Some Physical Constants

Quantity	Symbol	Value[a]
Atomic mass unit	u	$1.660\ 539\ 040\ (20) \times 10^{-27}$ kg
		$931.494\ 095\ 4\ (57)$ MeV/c^2
Avogadro's number	N_A	$6.022\ 140\ 857\ (74) \times 10^{23}$ particles/mol
Bohr magneton	$\mu_B = \dfrac{e\hbar}{2m_e}$	$9.274\ 009\ 994\ (57) \times 10^{-24}$ J/T
Bohr radius	$a_0 = \dfrac{\hbar^2}{m_e e^2 k_e}$	$5.291\ 772\ 106\ 7\ (12) \times 10^{-11}$ m
Boltzmann's constant	$k_B = \dfrac{R}{N_A}$	$1.380\ 648\ 52\ (79) \times 10^{-23}$ J/K
Compton wavelength	$\lambda_C = \dfrac{h}{m_e c}$	$2.426\ 310\ 236\ 7\ (11) \times 10^{-12}$ m
Coulomb constant	$k_e = \dfrac{1}{4\pi\epsilon_0}$	$8.987\ 551\ 788\ldots \times 10^9$ N·m^2/C^2 (exact)
Deuteron mass	m_d	$3.343\ 583\ 719\ (41) \times 10^{-27}$ kg
		$2.013\ 553\ 212\ 745\ (40)$ u
Electron mass	m_e	$9.109\ 383\ 56\ (11) \times 10^{-31}$ kg
		$5.485\ 799\ 090\ 70\ (16) \times 10^{-4}$ u
		$0.510\ 998\ 946\ 1\ (31)$ MeV/c^2
Electron volt	eV	$1.602\ 176\ 620\ 8\ (98) \times 10^{-19}$ J
Elementary charge	e	$1.602\ 176\ 620\ 8\ (98) \times 10^{-19}$ C
Gas constant	R	$8.314\ 459\ 8\ (48)$ J/mol·K
Gravitational constant	G	$6.674\ 08\ (31) \times 10^{-11}$ N·m^2/kg^2
Neutron mass	m_n	$1.674\ 927\ 471\ (21) \times 10^{-27}$ kg
		$1.008\ 664\ 915\ 88\ (49)$ u
		$939.565\ 413\ 3\ (58)$ MeV/c^2
Nuclear magneton	$\mu_n = \dfrac{e\hbar}{2m_p}$	$5.050\ 783\ 699\ (31) \times 10^{-27}$ J/T
Permeability of free space	μ_0	$4\pi \times 10^{-7}$ T·m/A (exact)
Permittivity of free space	$\epsilon_0 = \dfrac{1}{\mu_0 c^2}$	$8.854\ 187\ 817\ldots \times 10^{-12}$ C^2/N·m^2 (exact)
Planck's constant	h	$6.626\ 070\ 040\ (81) \times 10^{-34}$ J·s
	$\hbar = \dfrac{h}{2\pi}$	$1.054\ 571\ 800\ (13) \times 10^{-34}$ J·s
Proton mass	m_p	$1.672\ 621\ 898\ (21) \times 10^{-27}$ kg
		$1.007\ 276\ 466\ 879\ (91)$ u
		$938.272\ 081\ 3\ (58)$ MeV/c^2
Rydberg constant	R_H	$1.097\ 373\ 156\ 850\ 8\ (65) \times 10^7$ m^{-1}
Speed of light in vacuum	c	$2.997\ 924\ 58 \times 10^8$ m/s (exact)

Note: These constants are the values recommended in 2014 by CODATA, based on a least-squares adjustment of data from different measurements. For a more complete list, see P. J. Mohr, B. N. Taylor, and D. B. Newell, "CODATA Recommended Values of the Fundamental Physical Constants: 2014." *Rev. Mod. Phys.* **88**:3, 035009, 2016.

[a]The numbers in parentheses for the values represent the uncertainties of the last two digits.

Solar System Data

Body	Mass (kg)	Mean Radius (m)	Period (s)	Mean Distance from the Sun (m)
Mercury	3.30×10^{23}	2.44×10^6	7.60×10^6	5.79×10^{10}
Venus	4.87×10^{24}	6.05×10^6	1.94×10^7	1.08×10^{11}
Earth	5.97×10^{24}	6.37×10^6	3.156×10^7	1.496×10^{11}
Mars	6.42×10^{23}	3.39×10^6	5.94×10^7	2.28×10^{11}
Jupiter	1.90×10^{27}	6.99×10^7	3.74×10^8	7.78×10^{11}
Saturn	5.68×10^{26}	5.82×10^7	9.29×10^8	1.43×10^{12}
Uranus	8.68×10^{25}	2.54×10^7	2.65×10^9	2.87×10^{12}
Neptune	1.02×10^{26}	2.46×10^7	5.18×10^9	4.50×10^{12}
Pluto[a]	1.25×10^{22}	1.20×10^6	7.82×10^9	5.91×10^{12}
Moon	7.35×10^{22}	1.74×10^6	—	—
Sun	1.989×10^{30}	6.96×10^8	—	—

[a]In August 2006, the International Astronomical Union adopted a definition of a planet that separates Pluto from the other eight planets. Pluto is now defined as a "dwarf planet" (like the asteroid Ceres).

Physical Data Often Used

Average Earth–Moon distance	3.84×10^8 m
Average Earth–Sun distance	1.496×10^{11} m
Average radius of the Earth	6.37×10^6 m
Density of air (20°C and 1 atm)	1.20 kg/m^3
Density of air (0°C and 1 atm)	1.29 kg/m^3
Density of water (20°C and 1 atm)	1.00×10^3 kg/m^3
Free-fall acceleration on the Earth	9.80 m/s^2
Mass of the Earth	5.97×10^{24} kg
Mass of the Moon	7.35×10^{22} kg
Mass of the Sun	1.99×10^{30} kg
Standard atmospheric pressure on the Earth	1.013×10^5 Pa

Note: These values are the ones used in the text.

Some Prefixes for Powers of Ten

Power	Prefix	Abbreviation	Power	Prefix	Abbreviation
10^{-24}	yocto	y	10^1	deka	da
10^{-21}	zepto	z	10^2	hecto	h
10^{-18}	atto	a	10^3	kilo	k
10^{-15}	femto	f	10^6	mega	M
10^{-12}	pico	p	10^9	giga	G
10^{-9}	nano	n	10^{12}	tera	T
10^{-6}	micro	μ	10^{15}	peta	P
10^{-3}	milli	m	10^{18}	exa	E
10^{-2}	centi	c	10^{21}	zetta	Z
10^{-1}	deci	d	10^{24}	yotta	Y

Standard Abbreviations and Symbols for Units

Symbol	Unit	Symbol	Unit
A	ampere	K	kelvin
u	atomic mass unit	kg	kilogram
atm	atmosphere	kmol	kilomole
Btu	British thermal unit	L	liter
C	coulomb	lb	pound
°C	degree Celsius	ly	light-year
cal	calorie	m	meter
d	day	min	minute
eV	electron volt	mol	mole
°F	degree Fahrenheit	N	newton
F	farad	Pa	pascal
ft	foot	rad	radian
G	gauss	rev	revolution
g	gram	s	second
H	henry	T	tesla
h	hour	V	volt
hp	horsepower	W	watt
Hz	hertz	Wb	weber
in.	inch	yr	year
J	joule	Ω	ohm

Mathematical Symbols Used in the Text and Their Meaning

Symbol	Meaning
$=$	is equal to
$\equiv$	is defined as
$\neq$	is not equal to
$\propto$	is proportional to
$\sim$	is on the order of
$>$	is greater than
$<$	is less than
$\gg (\ll)$	is much greater (less) than
$\approx$	is approximately equal to
Δx	the change in x
$\displaystyle\sum_{i=1}^{N} x_i$	the sum of all quantities x_i from $i = 1$ to $i = N$
$|x|$	the absolute value of x (always a nonnegative quantity)
$\Delta x \to 0$	Δx approaches zero
$\dfrac{dx}{dt}$	the derivative of x with respect to t
$\dfrac{\partial x}{\partial t}$	the partial derivative of x with respect to t
$\displaystyle\int$	integral